Subaqueous Mass Movements and their Consequences: Advances in Process Understanding, Monitoring and Hazard Assessments

Geological Society books refereeing procedures

This volume is published under an agreement between the International Union of Geological Sciences and the Geological Society of London and arises from IGCP-640, Significance of Modern and Ancient Subaqueous Slope Landslides.

GSL is the publisher of choice for books related to IUGS activities, and the IUGS receives a fee for all books published under this agreement.

Books published under this agreement are subject to the Society's standard rigorous proposal and manuscript review procedures.

Submitting a book proposal
More information about submitting a proposal and producing a book for the Society can be found at https://www.geolsoc.org.uk/proposals

It is recommended that reference to all or part of this book should be made in one of the following ways:

Georgiopoulou, A., Amy, L. A., Benetti, S., Chaytor, J. D., Clare, M. A., Gamboa, D., Haughton, P. D. W., Moernaut, J. and Mountjoy, J. J. (eds) 2020. *Subaqueous Mass Movements and their Consequences: Advances in Process Understanding, Monitoring and Hazard Assessments*. Geological Society, London, Special Publications, **500**, https://doi.org/10.1144/SP500.0

Geersen, J., Festa, A. and Remitti, F. 2020. Structural constraints on the subduction of mass-transport deposits in convergent margins. *Geological Society, London, Special Publications*, **500**, 115–128, https://doi.org/10.1144/SP500-2019-174

Geological Society Special Publication No. 500

Subaqueous Mass Movements and their Consequences: Advances in Process Understanding, Monitoring and Hazard Assessments

Edited by

A. Georgiopoulou
University of Brighton, UK

L. A. Amy
University College Dublin, Ireland

S. Benetti
University of Ulster, UK

J. D. Chaytor
United States Geological Survey, USA

M. A. Clare
National Oceanography Centre, UK

D. Gamboa
Instituto Português do Mar e da Atmosfera, I.P., Portugal

P. D. W. Haughton
University College Dublin, Ireland

J. Moernaut
University of Innsbruck, Austria

and

J. J. Mountjoy
National Institute of Water and Atmospheric Research, New Zealand

2020
Published by
The Geological Society
London

The Geological Society of London

The Geological Society of London is a not-for-profit organisation, and a registered charity (no. 210161). Our aims are to improve knowledge and understanding of the Earth, to promote Earth science education and awareness, and to promote professional excellence and ethical standards in the work of Earth scientists, for the public good. Founded in 1807, we are the oldest geological society in the world. Today, we are a world-leading communicator of Earth science – through scholarly publishing, library and information services, cutting-edge scientific conferences, education activities and outreach to the general public. We also provide impartial scientific information and evidence to support policy-making and public debate about the challenges facing humanity. For more about the Society, please go to https://www.geolsoc.org.uk/

The Geological Society Publishing House (Bath, UK) produces the Society's international journals and books, and acts as European distributor for selected publications of the American Association of Petroleum Geologists (AAPG), the Geological Society of America (GSA), the Society for Sedimentary Geology (SEPM) and the Geologists' Association (GA). GSL Fellows may purchase these societies' publications at a discount. The Society's online bookshop is at https://www.geolsoc.org.uk/bookshop

To find out about joining the Society and benefiting from substantial discounts on publications of GSL and other Societies go to https://www.geolsoc.org.uk/membership or contact the Fellowship Department at: The Geological Society, Burlington House, Piccadilly, London W1J 0BG: Tel. +44 (0)20 7434 9944; Fax +44 (0)20 7439 8975; E-mail: enquiries@geolsoc.org.uk

For information about the Society's meetings, go to https://www.geolsoc.org.uk/events. To find out more about the Society's Corporate Patrons Scheme visit https://www.geolsoc.org.uk/patrons

Proposing a book
If you are interested in proposing a book then please visit: https://www.geolsoc.org.uk/proposals

Published by The Geological Society from:
The Geological Society Publishing House, Unit 7, Brassmill Enterprise Centre, Brassmill Lane, Bath BA1 3JN, UK

The Lyell Collection: www.lyellcollection.org
Online bookshop: www.geolsoc.org.uk/bookshop
Orders: Tel. +44 (0)1225 445046, Fax +44 (0)1225 442836

British Library Cataloguing in Publication Data

A catalogue record for this book is available from the British Library.
ISBN 978-1-78620-477-6
ISSN 0305-8719

Distributors
For details of international agents and distributors see:
www.geolsoc.org.uk/agentsdistributors

Typeset by Nova Techset Private Limited, Bengaluru & Chennai, India
Printed and bound by CPI Group (UK) Ltd, Croydon CR0 4YY, UK

Contents

Subaqueous mass movements in the context of observations of contemporary slope failure

Joshu J. Mountjoy[1]*, Aggeliki Georgiopoulou[2], Jason Chaytor[3], Michael A. Clare[4], Davide Gamboa[5] and Jasper Moernaut[6]

[1]National Institute of Water and Atmospheric Research, Wellington, New Zealand

[2]School of Environment and Technology, University of Brighton, UK

[3]United States Geological Survey, Woods Hole, USA

[4]National Oceanography Centre, Southampton, UK

[5]Instituto Português do Mar e da Atmosfera, Lisbon, Portugal

[6]Institute of Geology, University of Innsbruck, Austria

JJM, 0000-0002-3703-0006; AG, 0000-0003-4298-5090; DG, 0000-0003-3323-4172
*Correspondence: joshu.mountjoy@niwa.co.nz

Abstract: The consequences of subaqueous landslides have been at the forefront of societal conscience more than ever in the last few years, with devastating and fatal events in the Indonesian Archipelago making global news. The new research presented in this volume demonstrates the breadth of ongoing investigation into subaqueous landslides, and shows that while events like the recent ones can be devastating, they are smaller in scale than those Earth has experienced in the past. Understanding the spectrum of subaqueous landslide processes, and therefore the potential societal impact, requires research across all spatial and temporal scales. This volume delivers a compilation of state-of-the-art papers covering regional landslide databases, advanced techniques for *in situ* measurements, numerical modelling of processes and hazards.

This Geological Society Special Publication volume comes at a key time in the discipline of subaqueous mass movement research. Catastrophic recent events (e.g. at Anak Krakatau and during the Sulawesi Earthquake in 2018) have demonstrated the substantial hazard that can be associated with subaqueous landslides, with multiple fatalities directly attributed to landslide tsunami. Public awareness is greater than ever given the easy proliferation of information through new media formats (e.g. social media, live streaming, blogging). While a significant amount can be learned from these contemporary events, in the geological past there is a much larger range of mass movement scales and variability in processes that is preserved in seafloor and lakefloor geomorphology and in the sedimentary record. Untangling the controls on past slope failures remains arguably the main challenge in subaqueous mass movement research, but one that can benefit greatly from considering the context of direct observations of contemporary subaqueous slope failure.

There have been many instances over the past 50 years where subaqueous or partially subaqueous slope failure have been directly observed and the hazard has been significant (Dan *et al.* 2007; Parsons *et al.* 2014). The events that have happened in the past decade may not be as large or as devastating, but they occurred in a time of rapidly advancing technology in the field of Earth observation.

In this introductory paper, we review recent subaqueous mass failure events, some of which have received widespread media coverage. We look at these events in terms of how they can help us better study subaqueous mass movements, particularly from a hazard and risk perspective. We follow this review with a summary of the papers in this Special Publication volume and emphasize why, in light of contemporary events, they offer important and timely contributions to the discipline.

Volcanic flank collapse at Anak Krakatau

On 22 December 2018 a volcanic flank collapse of the Anak Krakatau volcanic island in the Sunda Straits, Indonesia, created a tsunami that generated waves up to 1.4 m high at the coast, killing over 400 people (Williams *et al.* 2019). This event is defined by unprecedented remote sensing observations when compared with previous historical volcanic sector collapses, most notably the 1888 Ritter Island event (Day *et al.* 2015). Anak Krakatau first

From: Georgiopoulou, A., Amy, L. A., Benetti, S., Chaytor, J. D., Clare, M. A., Gamboa, D., Haughton, P. D. W., Moernaut, J. and Mountjoy, J. J. (eds) 2020. *Subaqueous Mass Movements and their Consequences: Advances in Process Understanding, Monitoring and Hazard Assessments.* Geological Society, London, Special Publications, **500**, 1–12.
First published online May 26, 2020, https://doi.org/10.1144/SP500-2019-237

grew above sea-level in 1929 and had risen to 300 m elevation by December 2018, with changes documented by repeat satellite observations. Sentinel satellite footage from just 8 h after the tsunami shows the western flank failure and the collapse of the summit (Williams *et al.* 2019). Using these observations, combined with geomorphic interpretation, a subaqueous component of the failure is calculated to be a modest 0.1 km^3 and the subaerial component 0.004 km^3. Previous work had identified the tsunami hazard associated with flank collapse on Anak Krakatau, modelling a 0.28 km^3 failure in approximately the same location, which generated the same tsunami height in some locations but with significantly slower arrival times (Giachetti *et al.* 2012). For post-event modelling of the tsunami, Grilli *et al.* (2019) used a similar volume (0.27 km^3 failure). In both of these studies, a large component of the subaerial cone is assumed to have failed concurrently with the subaqueous flank; however, Williams *et al.* (2019) demonstrate that the cone did not fail and the subaerial component of the flank collapse was minor. New unpublished bathymetric data show that blocks up to 90 m high lie on the seafloor and the total slide mass volume is calculated to be 0.19 km^3 (http://www.bbc.com/news/science-environment-50798253). This event most probably will, and should, become the touchstone for tsunami model validation and volcanic island flank collapse hazard assessment. It was a catastrophic event in terms of loss of life, injury and displacement, despite the tsunami waves being comparatively small. The fact that this hazard had been well modelled in advance provides affirmation of the value of undertaking research into subaqueous mass movement hazards. A key outstanding question in our understanding of the Anak Krakatau landslide tsunami is the discrepancy in the calculated volumes of the failure, which is something to which studies of the failure history on volcanic cones can make a significant contribution. Understanding the mechanism behind contemporary events such as this one provides crucial context for landslide tsunami hazard studies when specific single-event failure mechanisms can be demonstrated. Conversely, the study of historical cone collapse can reveal long-term behaviour, as **Barrett *et al.* (2020)** show for past flank collapses in the Cape Verde Islands, where the deposition of volcanic debris on the slope may have triggered a chain of slope failures.

Earthquake triggering of subaqueous landslides

Shortly preceding the Anak Krakatau tsunami, in the same region, the 2018 M_w 7.5 Sulawesi earthquake on the 28 September 2018 was accompanied by a tsunami with run-up of 6–11 m (Takagi *et al.* 2019). The combined events resulted in >4000 deaths, but the number associated with the tsunami alone remains unclear. Slope failures have been inferred to have a role in tsunami generation, supported by video observations of localized wave generation and evidence of coastal failures (Arikawa *et al.* 2018; Carvajal *et al.* 2019). While it is clear from the extraordinary video footage collected at the time that delta collapse and coastal landslides caused tsunamis (Fig. 1), the role that subaqueous slope failures played in the main wave generation remains speculative, and no more conclusive than for comparative previous events where earthquake mechanisms could not clearly be reconciled with tsunami generation (e.g. Tappin *et al.* 2001). In essence, we are still in a situation where there is no direct evidence for large, wholly subaqueous landslide-generated tsunami. To make this link in the future, with no doubt as to the mechanism, will most probably require an event to occur in a location where high-resolution bathymetry has previously been collected. This becomes ever more likely as regional bathymetric datasets covering active continental margins are being collected and analysed. In this volume, **Watson *et al.* (2020)** analyse regional bathymetric coverage across the entire offshore component of the Hikurangi Subduction margin, New Zealand; **Hill *et al.* (2020)** present a study of full-coverage multibeam data along almost 300 km of the Cascadia Subduction Zone; **Stacey *et al.* (2020)** and **Lintern *et al.* (2020)** present analyses of full-coverage bathymetric datasets in two active margin fjord systems; and **León *et al.* (2020)** present a landslide database for offshore Spain. These comprehensive studies, amongst others, will set a valuable baseline for analysing the impact of large earthquakes in the future.

Although no tsunami has been directly linked to subaqueous landslides, the November 2016 Kaikoura Earthquake, New Zealand, is one of the best documented earthquake-triggered, wholly subaqueous seafloor-failure events to date (Mountjoy *et al.* 2018). The difference between 2 m-resolution pre- and post-earthquake bathymetry shows widespread shallow-seated slope failures around the rim of Kaikoura Canyon in close proximity to the documented seafloor rupture of the Hundalee Fault (Mountjoy *et al.* 2018). Landslide failure depth is shallow (2–10 m) so it is unsurprising that no clear landslide tsunami signal was detected. The significant information that this event provides is that the extent of landslide occurrence around the fault rupture could be mapped, partly using multibeam differencing and partly from the geomorphic signature of fresh slope failures. Identifying the distribution of slope failures enabled the ground motion threshold for landslide triggering in this area to be defined,

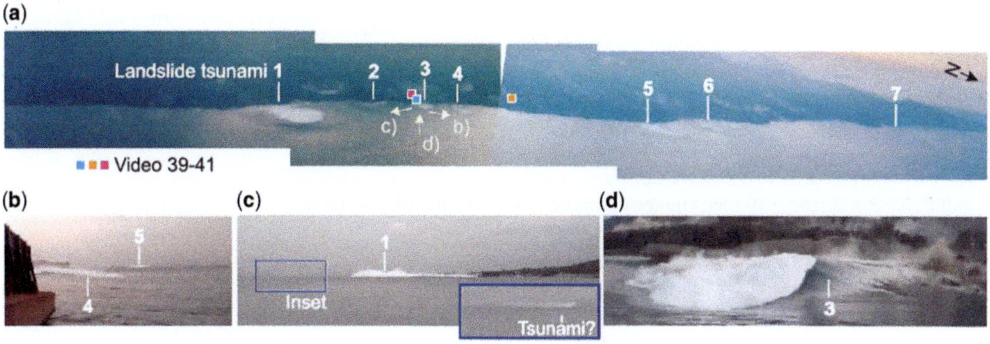

Fig. 1. Some of the only images ever captured of natural landslide tsunami generation, taken at the time of the Sulawesi earthquake in 2018 (Carvajal *et al.* 2019). Frames taken from (**a**) video footage and (**b–d**) near shoreline photos of the resulting waves. The numbers 1–7 refer to seven separate locations of wave generation.

by a peak ground acceleration of 0.32g, and a return period to be calculated (Mountjoy *et al.* 2018). The definition of a slope failure threshold has significant implications for hazard assessments where estimating the likelihood and distribution of subaqueous slope failure can be approached based on the proximity to fault rupture. An outstanding question, and one that observations following the 2016 Kaikoura Earthquake do little to address, is 'what causes very large subaqueous landslides?'. Despite the widespread occurrence of deep-seated landslides onland (Massey *et al.* 2018), multiple offshore fault ruptures (Litchfield *et al.* 2018), repeat mapping of over 4000 km^2 of seafloor – in an area with evidence for multiple deep-seated landslides – no landslides greater than 0.1 km^3 were documented for composite failure of the canyon floor, rather than for canyon wall areas where past landslide evidence is preserved. This follows from several repeat observations of the seafloor in the area of large earthquakes where no large landslides have been detected (Tappin *et al.* 2007; Völker *et al.* 2011). Comparisons of passive v. active margin open slopes show strengthening of the slope material on active margins due to earthquake shaking, demonstrating counter-intuitively that when earthquakes do not cause slope instability, they in fact have the opposite, strengthening effect (Sawyer and DeVore 2015; DeVore and Sawyer 2016; ten Brink *et al.* 2016; Sawyer *et al.* 2017; Molenaar *et al.* 2019). Whether this is the case for highly incised, canyon-dissected margins remains to be explored. This really highlights the value of advancing analytical, laboratory and case study research on the preconditioning and triggering of deep-seated landslides to advance our understanding of what makes large parts of the seafloor fail. Studying initiation, triggering and preconditioning of subaqueous mass movements forms a main focus in this volume.

Active landslides in dynamic sedimentary environments

Repeated failure (or contemporary 'active' landslide regions) in highly dynamic environments can significantly affect seafloor cables and oil field developments. This has been shown to be the case in submarine canyons (Clare *et al.* 2017; Pope *et al.* 2017) and on large deltas (Coleman *et al.* 1980). Significant efforts are being made to monitor and measure slope failure processes in these dynamic environments that include landslides, sediment bedform change and turbidity current processes. **Clare *et al.* (2020)** provide an overview of lessons learned from recent field campaigns, including those at submarine deltas (e.g. the Squamish Delta, British Columbia, and the Var Delta, Mediterranean). Repeat seafloor surveys and sediment flow monitoring at the Squamish submarine delta have revealed that short periods of rapid sediment accumulation can precondition the delta to fail and that landslides of up to 150 000 m^3 are ultimately triggered by tidally controlled pore pressure fluctuations (Hughes Clarke *et al.* 2012; Clare *et al.* 2016; Hizzett *et al.* 2018). **Chaytor *et al.* (2020)** document pipeline breaks on the Mississippi Delta, and use repeat seafloor mapping to track object displacement (shipwrecks, blocks). Their study shows that repeated seafloor failure results in downslope movement rates of 85 m a^{-1} for blocks measured in the 2016–17 period and potentially 150 m a^{-1} for the wreck of the SS *Virginia*. These repeat observations of seafloor change demonstrate that the Mississippi Delta is subject to highly dynamic conditions, and that slope failure can take place without major cyclonic events. It is now being recognized that repeated failures of larger deep-seated landslides can occur in the same location, but such events differ as they involve reactivation of the same material. Seafloor instrumentation has enabled measurements of downslope movement on the submerged flank of Mt Etna (Urlaub *et al.* 2018), and repeated movement is hypothesized for the Tuaheni Landslides offshore New Zealand (Mountjoy *et al.* 2009). **Couvin *et al.* (2020)** use new IODP drilling data in combination with P-Cable 3D seismic to propose a new model for the failure history of the Tuaheni Landslides. The new data show that only the top 40 m of the landslide is likely to be repeatedly reactivated, and that this portion is dominated by decimetre-scale sand units.

Glacial alpine environments: locus of climate sensitivity

The largest potential tsunami hazard related to subaqueous mass movements comes from large rockfalls entering small water bodies (e.g. fjords and lakes) where an extraordinarily large amount of water can be displaced. This is exemplified by the 1958 Lituya Bay, Alaska, tsunami where a 1 km^3 rockfall generated a wave over 500 m high (Miller 1960; Weiss *et al.* 2009). These extreme events are rare and have mostly happened in remote areas; however, it is definitely a significant concern that something like this might affect populated areas (Harbitz *et al.* 2014) and improved understanding of the preconditioning factors, frequency and dynamics of these events is therefore very important. In 2015 in Taan Fjord, Alaska, a 0.147 km^3 rockfall generated a large tsunami, fortunately with no casualties (Haeussler *et al.* 2018). The tsunami run-up from this event was an astounding 193 m, making it among the largest ever recorded (Higman *et al.* 2018). Post-event mapping has enabled a detailed calculation of the scale of the deposit, and the characteristics of the source area (Dufresne *et al.* 2018; Haeussler *et al.* 2018). It is likely that landslides in glaciated areas will become more frequent as the climate changes and glaciers retreat (Grämiger *et al.* 2017). Furthermore, increasingly dynamic landscape processes in response to climate change will more rapidly obscure the onshore evidence for such events (Dufresne *et al.* 2018), further highlighting the value of the subaqueous record.

Given these points, the depositional characteristics of large landslides in alpine environments, as well as their extent, age and frequency, become increasingly important. Several papers in this volume directly address this topic. **Stacey *et al.* (2020)** derive a magnitude frequency relationship for slope stability in the 140 km-long Douglas Channel Fjord system in Canada, and show that potentially damaging landslides occur throughout the fjord. Several studies make basin-wide assessments for lakes, illustrating the widespread occurrence of lacustrine landslides, which are commonly little considered but potentially catastrophic, even when occurring on entirely submerged slopes (**Daxer *et al.* 2020; Moernaut *et al.* 2020; Strasser *et al.* 2020; Strupler *et al.* 2020**).

State of the art

While the recent events reviewed here highlight the consequences of subaqueous mass movements, they capture just a small part of the broader spectrum of scales, processes and event hazards. The revealing information about event timing and consequences (e.g. tsunami or infrastructure damage) is of great value to practitioners, but leaves many gaps that need to be filled in by the study of past events, through numerical modelling and experimental research in the laboratory. The power of this volume lies in the breadth of studies from across the world's ocean basins, fjords and lake environments (Fig. 2), going back through the depositional record to build up a picture of the controls and triggers of subaqueous mass movements, their flow behaviour and their impact on society. To develop a logical and easily readable structure to the volume, the papers are presented in four thematic areas.

Section A: consequences and implications

Understanding and quantifying tsunami hazards related to subaqueous landslides remains a significant challenge. Apart from fjord-wall instabilities, the most significant landslide-generated tsunamis are often attributed to the collapse of volcanic island flanks, typically involving both subaerial and subaqueous slope failure. In this issue, a case study at the base of Fogo Volcano (Cape Verde Islands; **Barrett *et al.* 2020**) highlights the need for high-resolution mapping of complex landslide deposits to accurately constrain the number of failures, the geometry of their deposits and the extent of possible loading-induced deformation of pre-existing seafloor sediments. This combined information is crucial for evaluating the tsunamigenic potential associated with volcanic islands. In addition to landslide geometry, the characterization of landslide material forms a basic input parameter for landslide modelling, determining how sediment rheology affects landslide processes. Such sedimentological analysis is presented for the Byron landslide, located on the east Australian continental margin, and serves to inform hydrodynamic models constraining the associated tsunami hazard (**Mollison *et al.* 2020**). Tsunami models are often based on very simplified concepts considering the landslide source, such as the sliding block model. A more sophisticated approach is presented in this volume, in which

Fig. 2. Distribution of studies in this volume. Red stars indicate studies at a specific site, whereas blue polygons show regional mapping studies. Some studies are represented by multiple stars. The papers by **Kaminski *et al.* (2020)** and **Silver and Dugan (2020)** are not related to a specific geographical area so are not shown on the map.

slump motion is modelled using a viscoplastic flow, allowing the influence of soil parameters and failure plane geometry on frontal tsunami wave height to be evaluated by simulating the 1929 Grand Banks tsunami using updated geological source information (**Zengaffinen *et al.* 2020**).

The main motivation for reconstructing past landslide and tsunami events is the prognostic forecasting of future events on different spatiotemporal scales. This can involve scenario-based tsunami inundation analysis, often adopting a worst-case credible scenario in terms of landslide size and location. Here, such analysis is carried out for the geodynamically very active north Sicily Continental Margin, including both subaqueous landslides and earthquake ruptures as potential tsunami sources (**Dignan *et al.* 2020**), improving the tsunami hazard assessment for this vulnerable, low-lying coast with numerous coastal villages and important tourist sites. High population density also characterizes the shoreline of large perialpine lakes in Europe, and although lake tsunamis do not occur often, a robust hazard evaluation is important. **Strupler *et al.* (2020)** present a workflow for rapid screening of landslide-generated tsunami hazard related to poorly investigated lakes, building on knowledge gained from previous lacustrine slope stability studies and mass transport deposit (MTD) mapping in a few well-investigated lakes. Another step towards developing adequate mitigation, prevention and adaptation strategies is taken by the Geological Survey of Canada, which developed a national subaqueous landslide database as part of the national tsunami strategy (**Lintern *et al.* 2020**). This major effort from a federal government geological survey incorporates the morphometrics and processes of dozens of major landslides and hundreds of smaller events.

Apart from the significant tsunami hazard, subaqueous landslides can have other consequences that may even be indirectly beneficial to society. For example, it has been suggested that the subduction of large, chaotic MTDs can influence megathrust seismogenic behaviour at convergent margins by forming a rough boundary that may impede the propagation of earthquake ruptures. On the basis of a data comparison from different convergent margins, a study in this volume presents an evaluation of different controlling factors that define whether trench MTDs are subducted with the downgoing plate or are accreted to the upper plate (**Geersen *et al.* 2020**). Moreover, deeply buried MTDs can be important (economically relevant) elements in hydrocarbon systems, acting as potential reservoirs or seals for migrating fluids. A specific case study analyses the (*in situ*) physical properties of very young and shallowly buried MTDs in a lake to better understand the role of MTDs in fluid flow during their earliest stage of burial (**Moernaut *et al.* 2020**), concluding that shallow MTDs can form a relatively rapid seal for fluid migration, but this seal can be locally degraded by rafted blocks.

Finally, **Bull *et al.* (2020)** show that MTD topography can have a controlling influence on sediment routing and deposition, based on the 3D seismic analysis of the relative timing of the emplacement of exceptionally large MTDs and high sediment flux slope channels offshore Taranaki, New Zealand.

Section B: initiation, triggers, and preconditioning

Subaqueous landslides occur when a large number of physical conditions for sediment failure are met or exceeded. Understanding which physical conditions precondition the sediment, trigger and initiate movement, and control the dynamics of the resulting landslide, continues to be of fundamental importance in evaluating past events and forecasting future instability. Papers in this volume continue the effort to illuminate the roles that different geological, oceanographic, climatic and seismic conditions play in subaqueous landslide occurrence.

The contribution of oceanic bottom current erosion and deposition to controlling the mechanical properties of submarine slopes and localizing failure is addressed for seismically passive (**Gatter *et al.* 2020**) and active (**Brackenridge *et al.* 2020**) regions. **Gatter *et al.* (2020)** demonstrate that failure on sheeted contourite drifts localizes at contrasting lithological interfaces. **Brackenridge *et al.* (2020)** reveal the landslide hazard posed by the combination of high sediment supply from a delta and reworking and redistribution of that sediment by bottom currents to a specific location. **Locat *et al.* (2020)** look in detail at the effect of channel erosion and knickpoint formation in influencing the overall stability of coarse-grained deltas and find that breaching and liquefaction may occur when the knickpoint slope direction is close to the bedding plane dip direction. The response of seafloor environments to changes in global climatic conditions remains an area of intense debate and active study. The impacts of the indirect (**Daxer *et al.* 2020**) and direct influence of ice sheets on sediment pore pressure (**Urlaub *et al.* 2020**) and sea-level variation on sedimentation patterns (**Micallef *et al.* 2020**) continue to inform the discussion on the mechanisms by which these climate-influenced processes impact instability and the magnitude of their effects.

Defining the roles of gases and fluids in preconditioning sediment for failure, especially in terms of capturing the nature of *in situ* conditions, is fundamental in current subaqueous landslide research. **Kaminski *et al.* (2020)** investigate the role of free gas in influencing sediment shear strength via alteration of pore-space conditions and propose that the effect of trapped gas in sediment pores on the sediment's shear resistance is insufficient to trigger large-scale instability, but high gas pressures could lead to liquefaction. Modelling of sediment stability under changing fluid, pore pressure and sedimentological conditions based on field (**Mencaroni *et al.* 2020**) and laboratory observations (**Silver and Dugan 2020**) furthers our understanding of the way in which each of these physical parameters controls preconditioning and initiation.

Investigating submarine landslides along subduction margins, and relating them to characteristics of the incoming plate and the strength of the sediments on the overriding plate, continues to be a challenging task due to the large scale and heterogeneity of subduction margins. **Vargas *et al.* (2020)** explore the relationships between subduction of the extinct Sandra Ridge under the South American Plate and the location and timing of landslides on the trench margins.

Section C: characterization and regional controls

The triggering and characteristics of MTDs are primarily influenced by the region in which they occur. Regional tectonics is a key aspect to consider as seismicity is widely implicated as a major trigger for subaqueous landslides (Masson *et al.* 2006), although higher seismicity may not result in more landslides (ten Brink *et al.* 2016). Climatic controls play an important role, as well as hinterland precipitation, and the presence of ice sheets influences the volume and rate of sediment input to the basin. Locally, slope instability is controlled by sediment input, seismicity, slope steepness, geometry, geology and geotechnical properties (McAdoo *et al.* 2000; ten Brink *et al.* 2016). All of these parameters will influence the final character of the slide mass and associated kinematics, which can be analysed at multiple scales. The acquisition of data is key for the understanding of subaqueous landslide events in relatively unknown areas. New bathymetry and sub-bottom profiler data acquired during recent ice-breaker expeditions to the Alpha Ridge in the Arctic Ocean revealed a highly disturbed seabed with numerous MTDs with varying degrees of lateral transport, characterized by blocks, ridges and scarps (**Boggild *et al.* 2020**). The new information also sheds light on the likely origin of the deposits, ruling out a bolide impact, as previously hypothesized, and instead supporting the presence of a seismic trigger associated with tectonism in or around the Alpha Ridge.

New high-resolution data bring new insights into MTDs occurring even in well-studied regions, such as the Mediterranean Basin. **Cattaneo *et al.* (2020)** revisit a turbidite megabed in the Balearic Basin deposited during the Last Glacial Maximum. They used high-resolution geophysical data in conjunction with geotechnical data to remap the deposit and obtain a new age constraint. Within the same area, **Badhani *et al.* (2020)** investigate recurrent mass-wasting in the Gulf of Lions and revisit the Rhône western and eastern MTDs, triggered during the peak of the Last Glacial Maximum. They reveal previously unidentified internal structures in the MTD and, through integration with *in situ* measurements, they demonstrate that the recurrence of landslides in the area is influenced by the presence of clay-rich sediments. At the other end of the Mediterranean, **Katz *et al.* (2020)** use the taphonomy of foraminiferal assemblages to evaluate the deformation in debrites offshore Israel that are estimated to be contemporaneous with, or slightly predating, the transition to the Holocene. A case study from NE Sicily by **Casalbore *et al.* (2020)** focuses on the morphometrics of numerous landslide scars at submarine canyons and shows a prominent role of slope gradient in the size and character of landslide scars, particularly on steep slopes.

Through a wealth of high-resolution data, including historical records, **Strasser *et al.* (2020)** present a detailed study of MTDs in Lake Hallstatt (Austria), where earthquake-induced MTDs can be tied to historical events. These MTDs are not only larger than ones derived from flood events, rock falls or debris flows, but also present sedimentological and geochemical differences with evidence of mixing with intralake sediments. The larger dimensions of such MTDs present a significant geohazard that could impact structures around the lake.

Landslides in fjords are comparable to lacustrine MTDs; mass failures derived from steep slopes entering shallow-water environments can cause very large waves (e.g. Miller 1960). **Stacey *et al.* (2020)** summarize a five-year-long investigation of the Douglas Channel in Canada, where steep slopes, high precipitation and seismicity are responsible for the widespread occurrence of MTDs in this fjord environment. The larger and more frequent slope failures were active during deglaciation, but smaller and less frequent ones have also been occurring during the Holocene over the whole Douglas Channel system, all with tsunamigenic potential.

The relationship between tectonism, sediment bypass and slope morphology along the tectonically active Cascadia Margin is presented by **Hill *et al.* (2020)**. The results suggest that despite the presence of several canyons on the margin, few act as efficient sediment conduits to the deep sea. Instead, MTDs are likely to be the main process of sediment accumulation on the abyssal plains, with implications for interpretation of palaeoseismicity records on the margin.

On the other side of the Pacific Ocean, on the tectonic margins of New Zealand, **Watson *et al.* (2020)** investigated over 2000 MTDs following the subaqueous landslide characterization methodology in Clare *et al.* (2018). Based on morphological parameters, an assessment of triggers and areas of high landslide density, the authors demonstrate that landslides are most likely to occur in submarine canyons and, contrary to other studies, show that twice as many landslides occur on the active margin portion of the study area (the Hikurangi Margin) compared with the passive margin setting to the south. Another large-scale approach following Clare *et al.* (2018) is presented by **León *et al.* (2020)** for the entire Spanish offshore territory. Three types of source have been identified (deep-ocean ridges, volcanic islands and sedimentary continental margins) and, using a statistical analysis of MTD morphometrics, they distinguish between tectonic and non-tectonic triggers for the subaqueous mass failure.

Section D: mobility and kinematics

MTDs have shown a great variety of shapes and sizes that relates to their transport dynamics and internal heterogeneity. A series of intraMTD features can be used as kinematic markers to understand their movement (Bull *et al.* 2009) and assess whether MTDs derive from unidirectional flows or higher complexity is involved in a single episode or within complexes.

Internal MTD features such as blocks, folds, internal thrusts and extensional faults are key elements for understanding mass-movement kinematics. Blocks exhibit different shapes and sizes, as they are often associated with glide tracks that provide quantitative clues to constrain the remobilization distance often on the order of several kilometres (**Nwoko *et al.* 2020**). However, a full assessment of total MTD remobilization distances in three dimensions is an ongoing challenge. **Bull and Cartwright (2020)** show how simple structural restoration techniques of length-balancing help address this, and demonstrate the degree of underestimation in volume removal associated with the Storegga Slide. The availability of core and log data intersecting MTDs is generally limited. Thanks to a recent IODP expedition focusing on MTDs (Expedition 372, Pecher *et al.* 2019), invaluable additional data have been made available that have proved crucial to understanding landslide dynamics. This dataset, in combination with a P-Cable 3D seismic cube, allowed **Couvin *et al.* (2020)** to reinterpret the dynamics of the Tuaheni Landslide Complex offshore New Zealand and propose a new depositional model for it.

The occurrence of multiple stacked MTDs allows a more comprehensive understanding of not only their emplacement, but also wider controls on slope failure and sediment transport. **Roy *et al.* (2020)**, using internal deformation structures as kinematic indicators, show how five megascale MTDs on the Rockall Trough associated with different episodes of expansion of the British–Irish Ice Sheet interacted in a flow convergence zone of the palaeoslope.

Most of the published studies used to understand the dynamics, run-out velocity and timing of MTDs available in scientific literature have the major limitation of relying on static ‘snapshots’ in time of what is effectively a dynamic system. More than a challenge, there is a necessity to improve the current technology to allow a 4D study of seafloor dynamics through repeated surveys to assess the mobility and associated geohazards, as exemplified by **Chaytor *et al.* (2020)**. The authors present high-resolution data from repeat surveys to evaluate and quantify the complex mobility of the Mississippi River delta front on a decadal to annual timescale. By quantifying the displacement on the order of tens or hundreds of metres from seabed infrastructures, shipwrecks and MTD blocks, a novel monitoring methodology is introduced that can increase our understanding of seasonal, annual or progressive mass-flow triggers. Monitoring technologies will be key in the future of MTD research, yet substantial barriers still lie ahead to mitigate equipment damage or loss on the harsh and challenging subaqueous environment. **Clare *et al.* (2020)** present insights on lessons learned and propose future directions regarding the design, preservation and safety of mooring devices for gravity flow monitoring and highlight how these can be applied to realize future monitoring techniques capable of high-quality data acquisition and applicable to a wider range of subaqueous settings.

Perspectives for the future

In the first section of this paper, we reviewed cases of subaqueous mass movements in the past few years. It is clear that the technology available to document subaqueous landslides is significantly better than it has been in the past, from terrestrial remote sensing to seafloor imaging to real time monitoring. What we have still not experienced in recent history is devastating loss of life from tsunami that can unequivocally be shown to be generated by large mass-failure events. Given what we know about the scale of ancient events, e.g. the tsunami associated with the subaqueous Storegga landslide in Norway (Dawson *et al.* 1988) and the potential of historical events that have happened in remote areas, e.g. the Lituya Bay tsunami (Miller 1960), it is probable that some populated part of the world will be catastrophically affected by this hazard in the foreseeable future.

The papers in this GSL volume cover a very broad range of approaches into studying subaqueous mass movement that are necessary to grapple with the processes, triggers, preconditioning factors, hazards, mobility and kinematics, and the temporal and spatial occurrence that will ultimately assist in building resilient societies through better hazard and risk assessments.

As we move forward into an increasingly advanced technological age, direct measurement and monitoring of subaqueous slopes is going to become routine. We are already making significant progress in this area and the research of the IGCP Project 640 – S4SLIDE (Significance of Modern and Ancient Submarine Slope LandSLIDEs) community is playing a leading role. Developments in monitoring now include subsurface sensors that measure pore pressure fluctuations (Sultan *et al.* 2004; Lintern and Hill 2010; **Clare *et al.* 2020**), small-scale geodetic measurements of seafloor displacement (e.g. Urlaub *et al.* 2018) and novel uses of fibre-optic cables to monitor ground accelerations (e.g. Hartog *et al.* 2018). It is hoped that these proven new technologies, and other emerging systems, will provide early warning systems to reduce the risk posed to coastal communities and increase the resilience of critical seafloor infrastructure. These highly detailed and advanced measurements will always need to be considered in the context of the broad understanding of subaqueous failure that the geological record provides.

Acknowledgements We thank reviewers Lawrence Amy, Uri ten Brink and Jim Griffiths who provided feedback to improve this paper. We thank the authors of all the papers in this volume, as well as all the reviewers and editors that contributed to the process.

Funding Joshu Mountjoy's contribution to this paper and the editorial work for the volume was supported by New Zealand's Ministry for Business Innovation and Employment via SSIF provided to NIWA.

Author contributions **JJM**: conceptualization (lead), formal analysis (lead), investigation (lead), writing – original draft (lead), writing – review & editing (lead); **AG**: conceptualization (equal), formal analysis (equal), writing – original draft (equal), writing – review & editing (equal); **JC**: writing – original draft (supporting), writing – review & editing (supporting); **MAC**: writing – original draft (supporting), writing – review & editing (supporting); **DG**: writing – original draft (supporting), writing – review & editing (supporting); **JM**: writing – original draft (supporting), writing – review & editing (supporting).

References

Arikawa, T., Muhari, A., Okumura, Y., Dohi, Y., Afriyanto, B., Sujatmiko, K.A. and Imamura, F. 2018. Coastal subsidence induced several tsunamis during the 2018 Sulawesi earthquake. *Journal of Disaster Research*, **13**, 1–3, https://doi.org/10.20965/jdr.2018.sc20181204

Badhani, S., Cattaneo, A. *et al.* 2020. Integrated geophysical, sedimentological and geotechnical investigation of submarine landslides in the Gulf of Lions (Western Mediterranean). *Geological Society, London, Special Publications*, **500**, https://doi.org/10.1144/SP500-2019-175

Barrett, R., Lebas, E. *et al.* 2020. Revisiting the tsunamigenic volcanic flank collapse of Fogo Island in the Cape Verdes, offshore West Africa. *Geological Society, London, Special Publications*, **500**, https://doi.org/10.1144/SP500-2019-187

Boggild, K., Mosher, D.C., Travaglini, P., Gebhardt, C. and Mayer, L. 2020. Mass wasting on Alpha Ridge in the Arctic Ocean: new insights from multibeam bathymetry and sub-bottom profiler data. *Geological Society, London, Special Publications*, **500**, https://doi.org/10.1144/SP500-2019-196

Brackenridge, R.E., Nicholson, U., Sapiie, B., Stow, D. and Tappin, D.R. 2020. Indonesian throughflow as a preconditioning mechanism for submarine landslides in the Makassar Strait. *Geological Society, London, Special Publications*, **500**, https://doi.org/10.1144/SP500-2019-171

Bull, S. and Cartwright, J.A. 2020. Line length balancing to evaluate multi-phase submarine landslide development: an example from the Storegga Slide, Norway. *Geological Society, London, Special Publications*, **500**, https://doi.org/10.1144/SP500-2019-168

Bull, S., Cartwright, J. and Huuse, M. 2009. A review of kinematic indicators from mass-transport complexes using 3D seismic data. *Marine and Petroleum Geology*, **26**, 1132–1151, https://doi.org/10.1016/j.marpetgeo.2008.09.011

Bull, S., Browne, G.H., Arnot, M.J. and Strachan, L.J. 2020. Influence of mass transport deposit (MTD) surface topography on deep-water deposition: an example from a predominantly fine-grained continental margin, New Zealand. *Geological Society, London, Special Publications*, **500**, https://doi.org/10.1144/SP500-2019-192

Carvajal, M., Araya-Cornejo, C., Sepúlveda, I., Melnick, D. and Haase, J.S. 2019. Nearly instantaneous tsunamis following the Mw 7.5 2018 Palu earthquake. *Geophysical Research Letters*, **46**, 5117–5126, https://doi.org/10.1029/2019GL082578

Casalbore, D., Clementucci, R., Bosman, A., Chiocci, F.L., Martorelli, E. and Ridente, D. 2020. Widespread mass-wasting processes off NE Sicily (Italy): insights from morpho-bathymetric analysis. *Geological Society, London, Special Publications*, **500**, https://doi.org/10.1144/SP500-2019-195

Cattaneo, A., Badhani, S. *et al.* 2020. The Balearic Abyssal Plain megabed of the Last Glacial Maximum revisited. *Geological Society, London, Special Publications*, **500**, https://doi.org/10.1144/SP500-2019-188

Chaytor, J.D., Baldwin, W.E. *et al.* 2020. Short- and long-term movement of mudflows of the Mississippi River Delta Front and their known and potential impacts on oil and gas infrastructure. *Geological Society, London, Special Publications*, **500**, https://doi.org/10.1144/SP500-2019-183

Clare, M.A., Talling, P.J., Challenor, P.G. and Hunt, J.E. 2016. Tempo and triggering of large submarine landslides: statistical analysis for hazard assessment. *In*: Lamarche, G. *et al.* (eds) *Submarine Mass Movements and Their Consequences*. Advances in Natural and Technological Hazards Research, **41**. Springer, Cham, 509–517, https://doi.org/10.1007/978-3-319-20979-1_51

Clare, M.A., Vardy, M.E. *et al.* 2017. Direct monitoring of active geohazards: Emerging geophysical tools for deep-water assessments. *Near Surface Geophysics*, **15**, 427–444, https://doi.org/10.3997/1873-0604.2017033

Clare, M., Chaytor, J. *et al.* 2018. A consistent global approach for the morphometric characterization of subaqueous landslides. *Geological Society, London, Special Publications*, **477**, 455–477, https://doi.org/10.1144/SP477.15

Clare, M., Lintern, D.G. *et al.* 2020. Lessons learned from monitoring of turbidity currents and guidance for future platform designs. *Geological Society, London, Special Publications*, **500**, https://doi.org/10.1144/SP500-2019-173

Coleman, J.M., Prior, D.B. and Garrison, L.E. 1980. *Subaqueous Sediment Instabilities in the Offshore Mississippi River Delta*. United States Department of Interior, Bureau of Land Management, New Orleans Outer Continental Shelf Office Open-File Report **80-01**.

Couvin, B., Georgiopoulou, A. *et al.* 2020. A new depositional model for the Tuaheni Landslide Complex, Hikurangi Margin, New Zealand. *Geological Society, London, Special Publications*, **500**, https://doi.org/10.1144/SP500-2019-180

Dan, G., Sultan, N. and Savoye, B. 2007. The 1979 Nice harbour catastrophe revisited: trigger mechanism inferred from geotechnical measurements and numerical modelling. *Marine Geology*, **245**, 40–64, https://doi.org/10.1016/j.margeo.2007.06.011

Dawson, A.G., Long, D. and Smith, D.E. 1988. The Storegga slides: evidence from eastern Scotland for a possible tsunami. *Marine Geology*, **82**, 271–276, https://doi.org/10.1016/0025-3227(88)90146-6

Daxer, C., Sammartini, M., Molenaar, A., Piechl, T., Strasser, M. and Moernaut, J. 2020. Morphology and spatio-temporal distribution of lacustrine mass-transport deposits in Wörthersee, Eastern Alps, Austria. *Geological Society, London, Special Publications*, **500**, https://doi.org/10.1144/SP500-2019-179

Day, S., Llanes, P., Silver, E., Hoffmann, G., Ward, S. and Driscoll, N. 2015. Submarine landslide deposits of the historical lateral collapse of Ritter Island, Papua New Guinea. *Marine and Petroleum Geology*, **67**, 419–438, https://doi.org/10.1016/j.marpetgeo.2015.05.017

DeVore, J.R. and Sawyer, D.E. 2016. Shear strength of siliciclastic sediments from passive and active margins (0–100 m below seafloor): insights into seismic strengthening. *In*: Lamarche, G. *et al.* (eds) *Submarine Mass Movements and Their Consequences*. Advances in Natural and Technological Hazards Research, **41**. Springer, Cham, 173–180, https://doi.org/10.1007/978-3-319-20979-1_17

Dignan, J., Micallef, A., Mueller, C., Sulli, A., Zizzo, E. and Spatola, D. 2020. A scenario-based assessment of the tsunami hazard in Palermo, northern Sicily, and the southern Tyrrhenian Sea. *Geological Society, London, Special Publications*, **500**, https://doi.org/10.1144/SP500-2019-181

Dufresne, A., Geertsema, M. *et al.* 2018. Sedimentology and geomorphology of a large tsunamigenic landslide, Taan Fiord, Alaska. *Sedimentary Geology*, **364**, 302–318, https://doi.org/10.1016/j.sedgeo.2017.10.004

Gatter, R., Clare, M.A., Hunt, J.E., Watts, M., Madhusudhan, B.N., Talling, P.J. and Huhn, K. 2020. A multi-disciplinary investigation of the AFEN Slide: the relationship between contourites and submarine landslides. *Geological Society, London, Special Publications*, **500**, https://doi.org/10.1144/SP500-2019-184

Geersen, J., Festa, A. and Remitti, F. 2020. Structural constraints on the subduction of mass-transport deposits in convergent margins. *Geological Society, London, Special Publications*, **500**, https://doi.org/10.1144/SP500-2019-174

Giachetti, T., Paris, R., Kelfoun, K. and Ontowirjo, B. 2012. Tsunami hazard related to a flank collapse of Anak Krakatau volcano, Sunda Strait, Indonesia. *Geological Society, London, Special Publications*, **361**, 79–90, https://doi.org/10.1144/SP361.7

Grämiger, L.M., Moore, J.R., Gischig, V.S., Ivy-Ochs, S. and Loew, S. 2017. Beyond debuttressing: mechanics of paraglacial rock slope damage during repeat glacial cycles. *Journal of Geophysical Research: Earth Surface*, **122**, 1004–1036, https://doi.org/10.1002/2016JF003967

Grilli, S.T., Tappin, D.R. *et al.* 2019. Modelling of the tsunami from the December 22, 2018 lateral collapse of Anak Krakatau volcano in the Sunda Straits, Indonesia. *Scientific Reports*, **9**, 1–13, https://doi.org/10.1038/s41598-019-48327-6

Haeussler, P., Gulick, S. *et al.* 2018, Submarine deposition of a subaerial landslide in Taan Fiord, Alaska. *Journal of Geophysical Research: Earth Surface*, **123**, 2443–2463, https://doi.org/10.1029/2018JF004608

Harbitz, C., Glimsdal, S., Løvholt, F., Kveldsvik, V., Pedersen, G. and Jensen, A. 2014. Rockslide tsunamis in complex fjords: from an unstable rock slope at Åkerneset to tsunami risk in western Norway. *Coastal Engineering*, **88**, 101–122, https://doi.org/10.1016/j.coastaleng.2014.02.003

Hartog, A.H., Belal, M. and Clare, M.A. 2018. Advances in distributed fiber-optic sensing for monitoring marine infrastructure, measuring the deep ocean, and quantifying the risks posed by seafloor hazards. *Marine Technology Society Journal*, **52**, 58–73, https://doi.org/10.4031/MTSJ.52.5.7

Higman, B., Shugar, D.H. *et al.* 2018. The 2015 landslide and tsunami in Taan Fiord, Alaska. *Scientific Reports*, **8**, 12993, https://doi.org/10.1038/s41598-018-30475-w

Hill, J.C., Watt, J.T., Brothers, D.S. and Kluesner, J.W. 2020. Submarine canyons, slope failures and mass

transport processes in southern Cascadia. *Geological Society, London, Special Publications*, **500**, https://doi.org/10.1144/SP500-2019-169
Hizzett, J.L., Hughes Clarke, J.E., Sumner, E.J., Cartigny, M.J.B., Talling, P.J. and Clare, M.A. 2018. Which triggers produce the most erosive, frequent, and longest runout turbidity currents on deltas? *Geophysical Research Letters*, **45**, 855–863, https://doi.org/10.1002/2017GL075751
Hughes Clarke, J.E., Brucker, S. *et al.* 2012. The Squamish ProDelta: monitoring active landslides and turbidity currents. *Proceedings of the Canadian Hydrographic Conference 2012*, 'The Arctic, Old Challenges, New Approaches', Niagara Falls, Canada, 15–17 May 2012.
Kaminski, P., Urlaub, M., Grabe, J. and Berndt, C. 2020. Geomechanical behaviour of gassy soils and implications for submarine slope stability: a literature analysis. *Geological Society, London, Special Publications*, **500**, https://doi.org/10.1144/SP500-2019-149
Katz, O., Ashkenazi, L., Sultan-Levi, S., Abramovich, S., Almogi-Labin, A. and Hyams-Kaphzan, O. 2020. Characterization of recent deep-sea debrites in the eastern Mediterranean based on foraminiferal taphonomy. *Geological Society, London, Special Publications*, **500**, https://doi.org/10.1144/SP500-2019-170
León, R., Urgeles, R., Pérez-López, R., Payo, E., Vázquez-Izquierdo, A., Giménez-Moreno, C.J. and Casas, D. 2020. Geological and tectonic controls on morphometrics of submarine landslides of the Spanish margins. *Geological Society, London, Special Publications*, **500**, https://doi.org/10.1144/SP500-2019-153
Lintern, D.G. and Hill, P.R. 2010. An underwater laboratory at the Fraser River delta. Eos. *Transactions of the American Geophysical Union*, **91**, 333–334, https://doi.org/10.1029/2010EO380001
Lintern, D.G., Rutherford, J., Hill, P.R., Campbell, C. and Normandeau, A. 2020. Towards a national-scale assessment of the subaqueous mass movement hazard in Canada. *Geological Society, London, Special Publications*, **500**, https://doi.org/10.1144/SP500-2019-206
Litchfield, N.J., Villamor, P. *et al.* 2018. Surface rupture of multiple crustal faults in the 2016 M_w 7.8 Kaikōura, New Zealand, earthquake. *Bulletin of the Seismological Society of America*, **108**, 1496–1520, https://doi.org/10.1785/0120170300
Locat, J., Azizian, A., Stronach, J., Hospital, A., Young, C., Turmel, D. and Bevan, A. 2020. Morphological signature of gully development by rapid slide retrogression in a layered coarse-grained delta foreslope. *Geological Society, London, Special Publications*, **500**, https://doi.org/10.1144/SP500-2019-159
Massey, C., Townsend, D. *et al.* 2018. Landslides Triggered by the 14 November 2016 Mw 7.8 Kaikōura Earthquake, New Zealand. *Bulletin of the Seismological Society of America*, **108**, 1630–1648, https://doi.org/10.1785/0120170305
Masson, D.G., Harbitz, C.B., Wynn, R.B., Pedersen, G. and Løvholt, F. 2006. Submarine landslides: processes, triggers and hazard prediction. *Philosophical Transactions of the Royal Society A*, **364**, 2009–2039, https://doi.org/10.1098/rsta.2006.1810
McAdoo, B.G., Pratson, L.F. and Orange, D.L. 2000. Submarine landslide geomorphology, US continental slope. *Marine Geology*, **169**, 103–136, https://doi.org/10.1016/S0025-3227(00)00050-5
Mencaroni, D., Llopart, J., Urgeles, R., Lafuerza, S., Gràcia, E., Le Friant, A. and Urlaub, M. 2020. From gravity cores to overpressure history: the importance of measured sediment physical properties in hydrogeological models. *Geological Society, London, Special Publications*, **500**, https://doi.org/10.1144/SP500-2019-176
Micallef, A., Georgiopoulou, A., Green, A. and Maselli, V. 2020. Impact of sea-level fluctuations on the sedimentation patterns of the SE African margin: implications for slope instability. *Geological Society, London, Special Publications*, **500**, https://doi.org/10.1144/SP500-2019-172
Miller, D.J. 1960. The Alaska earthquake of July 10, 1958: giant wave in Lituya Bay. *Bulletin of the Seismological Society of America*, **50**, 253–266.
Moernaut, J., Wiemer, G., Kopf, A. and Strasser, M. 2020. Evaluating the sealing potential of young and thin mass-transport deposits: Lake Villarrica, Chile. *Geological Society, London, Special Publications*, **500**, https://doi.org/10.1144/SP500-2019-155
Molenaar, A., Moernaut, J., Wiemer, G., Dubois, N. and Strasser, M. 2019. Earthquake impact on active margins: tracing surficial remobilization and seismic strengthening in a slope sedimentary sequence. *Geophysical Research Letters*, **41**, 1195, https://doi.org/10.1029/2019GL082350
Mollison, K.C., Power, H.E., Clarke, S.L., Baxter, A.T., Lane, E.M. and Hubble, T.C.T. 2020. The sedimentology and tsunamigenic potential of the Byron submarine landslide off New South Wales, Australia. *Geological Society, London, Special Publications*, **500**, https://doi.org/10.1144/SP500-2019-160
Mountjoy, J.J., McKean, J., Barnes, P.M. and Pettinga, J.R. 2009. Terrestrial-style slow-moving earthflow kinematics in a submarine landslide complex. *Marine Geology*, **267**, 114–127, https://doi.org/10.1016/j.margeo.2009.09.007
Mountjoy, J.J., Howarth, J.D. *et al.* 2018. Earthquakes drive large-scale submarine canyon development and sediment supply to deep-ocean basins. *Science Advances*, **4**, 3748, https://doi.org/10.1126/sciadv.aar3748
Nwoko, J., Kane, I. and Huuse, M. 2020. Megaclasts within mass-transport deposits: their origin, characteristics and effect on substrates and succeeding flows. *Geological Society, London, Special Publications*, **500**, https://doi.org/10.1144/SP500-2019-146
Parsons, T., Geist, E.L. *et al.* 2014. Source and progression of a submarine landslide and tsunami: the 1964 Great Alaska earthquake at Valdez. *Journal of Geophysical Research: Solid Earth*, **119**, 8502–8516, https://doi.org/10.1002/2014JB011514
Pecher, I.A., Barnes, P.M., LeVay, L.J. and the Expedition 372A Scientists 2019. Creeping gas hydrate slides. *Proceedings of the International Ocean Discovery Program*, **372A**, College Station, TX (International Ocean Discovery Program), https://doi.org/10.14379/iodp.proc.372A.2019
Pope, E.L., Talling, P.J., Carter, L., Clare, M.A. and Hunt, J.E. 2017. Damaging sediment density flows triggered by tropical cyclones. *Earth and Planetary Science*

Letters, **458**, 161–169, https://doi.org/10.1016/j.epsl.2016.10.046

Roy, S., Georgiopoulou, A., Benetti, S. and Sacchetti, F. 2020. Mass transport deposits in the Donegal Barra Fan and their association with British–Irish Ice Sheet dynamics. *Geological Society, London, Special Publications*, **500**, https://doi.org/10.1144/SP500-2019-177

Sawyer, D.E. and DeVore, J.R. 2015. Elevated shear strength of sediments on active margins: Evidence for seismic strengthening. *Geophysical Research Letters*, **42**, 10 216–10 221, https://doi.org/10.1002/2015GL066603

Sawyer, D.E., Reece, R.S., Gulick, S.P.S. and Lenz, B.L. 2017. Submarine landslide and tsunami hazards offshore southern Alaska: Seismic strengthening versus rapid sedimentation. *Geophysical Research Letters*, **44**, 8435–8442, https://doi.org/10.1002/2017GL074537

Silver, M.M.W. and Dugan, B. 2020. The influence of clay content on submarine slope failure: insights from laboratory experiments and numerical models. *Geological Society, London, Special Publications*, **500**, https://doi.org/10.1144/SP500-2019-186

Stacey, C.D., Lintern, D.G., Shaw, J. and Conway, K.W. 2020. Slope stability hazard in a fjord environment: Douglas Channel, Canada. *Geological Society, London, Special Publications*, **500**, https://doi.org/10.1144/SP500-2019-191

Strasser, M., Berberich, T. *et al.* 2020. Geomorphology and event-stratigraphy of recent mass-movement processes in Lake Hallstatt (UNESCO World Heritage Cultural Landscape, Austria). *Geological Society, London, Special Publications*, **500**, https://doi.org/10.1144/SP500-2019-178

Strupler, M., Anselmetti, F.S., Hilbe, M., Kremer, K. and Wiemer, S. 2020. A workflow for the rapid assessment of the landslide-tsunami hazard in peri-alpine lakes. *Geological Society, London, Special Publications*, **500**, https://doi.org/10.1144/SP500-2019-166

Sultan, N., Cochonat, P. *et al.* 2004. Triggering mechanisms of slope instability processes and sediment failures on continental margins: a geotechnical approach. *Marine Geology*, **213**, 291–321, https://doi.org/10.1016/j.margeo.2004.10.011

Takagi, H., Pratama, M.B., Kurobe, S., Esteban, M., Aránguiz, R. and Ke, B. 2019. Analysis of generation and arrival time of landslide tsunami to Palu City due to the 2018 Sulawesi Earthquake. *Landslides*, **16**, 983–991, https://doi.org/10.1007/s10346-019-01166-y

Tappin, D.R., Watts, P., McMurtry, G.M., Lafoy, Y. and Matsumoto, T. 2001. The Sissano, Papua New Guinea tsunami of July 1998; offshore evidence on the source mechanism. *Marine Geology*, **175**, 1–23, https://doi.org/10.1016/S0025-3227(01)00131-1

Tappin, D., McNeil, L., Henstock, T. and Mosher, D. 2007. Mass Wasting Processes-offshore Sumatra. *In*: Lykousis, V., Sakellariou, D. and Locat, J. (eds) *Submarine Mass Movements and Their Consequences*. Advances in Natural and Technological Hazards Research, **27**. Springer, Dordrecht, 327–336, https://doi.org/10.1007/978-1-4020-6512-5_34

ten Brink, U.S., Andrews, B.D. and Miller, N.C. 2016. Seismicity and sedimentation rate effects on submarine slope stability. *Geology*, **44**, 563–566, https://doi.org/10.1130/G37866.1

Urlaub, M., Petersen, F. *et al.* 2018. Gravitational collapse of Mount Etna's southeastern flank. *Science Advances*, **4**, eaat9700, https://doi.org/10.1126/sciadv.aat9700

Urlaub, M., Kratzke, I. and Hjelstuen, B.O. 2020. A numerical investigation of excess pore pressures and continental slope stability in response to ice-sheet dynamics. *Geological Society, London, Special Publications*, **500**, https://doi.org/10.1144/SP500-2019-185

Vargas, C.A., Gutiérrez, G.A. and Sarmiento, G.A. 2020. Subduction of an extinct rift and its role in the formation of submarine landslides in NW South America. *Geological Society, London, Special Publications*, **500**, https://doi.org/10.1144/SP500-2019-189

Völker, D., Scholz, F. and Geersen, J. 2011. Analysis of submarine landsliding in the rupture area of the 27 February 2010 Maule earthquake, Central Chile. *Marine Geology*, **288**, 79–89, https://doi.org/10.1016/j.margeo.2011.08.003

Watson, S.J., Mountjoy, J.J. and Crutchley, G.J. 2020. Tectonic and geomorphic controls on the distribution of submarine landslides across active and passive margins, eastern New Zealand. *Geological Society, London, Special Publications*, **500**, https://doi.org/10.1144/SP500-2019-165

Weiss, R., Fritz, H.M. and Wünnemann, K. 2009, Hybrid modeling of the mega-tsunami runup in Lituya Bay after half a century. *Geophysical Research Letters*, **36**, L09602, https://doi.org/10.1029/2009GL037814

Williams, R., Rowley, P. and Garthwaite, M.C. 2019. Reconstructing the Anak Krakatau flank collapse that caused the December 2018 Indonesian tsunami. *Geology*, **47**, 973–976, https://doi.org/10.1130/G46517.1

Zengaffinen, T., Løvholt, F., Pedersen, G. and Harbitz, C.B. 2020. Effects of rotational submarine slump dynamics on tsunami genesis: new insight from idealized models and the 1929 Grand Banks event. *Geological Society, London, Special Publications*, **500**, https://doi.org/10.1144/SP500-2019-201

Revisiting the tsunamigenic volcanic flank collapse of Fogo Island in the Cape Verdes, offshore West Africa

Rachel Barrett[1]*, Elodie Lebas[1], Ricardo Ramalho[2,3,4,5], Ingo Klaucke[6], Steffen Kutterolf[6], Andreas Klügel[7], Katja Lindhorst[1], Felix Gross[1,8] and Sebastian Krastel[1]

[1]Institute of Geosciences, Christian-Albrechts-University, 24118 Kiel, Germany

[2]Instituto Dom Luiz, Faculdade de Ciências da Universidade de Lisboa, 1749-016 Lisbon, Portugal

[3]Departamento de Geologia, Faculdade de Ciências da Universidade de Lisboa, 1749-016 Lisbon, Portugal

[4]School of Earth Sciences, University of Bristol, Bristol BS8 1RL, UK

[5]Lamont-Doherty Earth Observatory of Columbia University, New York, NY 10964, USA

[6]GEOMAR Helmholtz Centre for Ocean Research Kiel, 24148 Kiel, Germany

[7]Department of Geosciences, University of Bremen, 28359 Bremen, Germany

[8]Center for Ocean and Society, Kiel Marine Science, Christian-Albrechts-University, 24118 Kiel, Germany

RB, 0000-0001-6463-4473; EL, 0000-0003-1617-143X; RR, 0000-0002-2338-5535; SKu, 0000-0002-0645-3399; AK, 0000-0002-5136-4774; FG, 0000-0002-0749-829X; SKr, 0000-0002-5899-9748

*Correspondence: rachel.barrett@ifg.uni-kiel.de

Abstract: Volcanic archipelagos are a source of numerous on- and offshore geohazards, including explosive eruptions and potentially tsunamigenic large-scale flank collapses. Fogo Island in the southern Cape Verdes is one of the most active volcanoes in the world, making it both prone to collapse (as evidenced by the *c.* 73 ka Monte Amarelo volcanic flank collapse), and a source of widely distributed tephra and volcanic material. The offshore distribution of the Monte Amarelo debris avalanche deposits and the surrounding volcaniclastic apron were previously mapped using only medium-resolution bathymetric data. Here, using recently acquired, higher-resolution acoustic data, we revisit Fogo's flank collapse and find evidence suggesting that the deposition of hummocky volcanic debris originating from the failed eastern flank most likely triggered the contemporaneous, multi-phase failure of pre-existing seafloor sediments. Additionally, we identify, for the first time, multiple mass-transport deposits in the southern part of the volcaniclastic apron of Fogo and Santiago based on the presence of acoustically chaotic deposits in parametric echo sounder data and volcaniclastic turbiditic sands in recovered cores. These preliminary findings indicate a long and complex history of instability on the southern slopes of Fogo and suggest that Fogo may have experienced multiple flank collapses.

Volcanic archipelagos are home to numerous on- and offshore hazards such as changes in sea-level, storms, volcanic eruptions, slope instabilities, large flank collapses and tsunamis (Casalbore 2018). These events can have devastating consequences for people, nature and infrastructure, both onshore and offshore. A variety of internal and external factors can precondition and trigger flank instabilities. These include: dyke and sill intrusions; volcanic eruptions and tremor; earthquakes; flank over-steepening; the weight of new volcanic material on the island flanks; weakening of the volcanic edifice by weathering and hydrothermal activity and, for smaller landslides, by the effects of wave, wind and storm activity (e.g. Siebert 1984; Begét and Kienle 1992; McGuire 1996, 2003; Murray and Voight 1996; Tibaldi 2001; Cervelli *et al.* 2002; Casalbore *et al.* 2011, 2015; Gross *et al.* 2014; Clare *et al.* 2018). Such factors occur over timescales ranging from seconds to thousands of years and are globally widespread (McGuire 1996; Blahůt *et al.* 2019).

The volume of volcanic flank collapses varies significantly, but can be as large as thousands of cubic kilometres, for example, in the Hawaiian archipelago (Moore *et al.* 1989). Many such events have volumes in the order of tens to hundreds of cubic kilometres, for example at Nisyros Volcano in the Aegean Sea (Tibaldi *et al.* 2008; Livanos *et al.*

From: Georgiopoulou, A., Amy, L. A., Benetti, S., Chaytor, J. D., Clare, M. A., Gamboa, D., Haughton, P. D. W., Moernaut, J. and Mountjoy, J. J. (eds) 2020. *Subaqueous Mass Movements and their Consequences: Advances in Process Understanding, Monitoring and Hazard Assessments*. Geological Society, London, Special Publications, **500**, 13–26.
First published online March 13, 2020, https://doi.org/10.1144/SP500-2019-187

2013), in the Lesser Antilles Arc (Lebas *et al.* 2011; Le Friant *et al.* 2015; Brunet *et al.* 2016), in the Canary Islands (Krastel *et al.* 2001; Masson *et al.* 2002; León *et al.* 2017) and in the Cape Verde Islands (Masson *et al.* 2008). However, even comparatively small-volume volcanic flank collapses, such as the 0.22–0.3 km^3 Anak Krakatau flank collapse in December 2018 (Grilli *et al.* 2019), may result in catastrophic tsunamis. The hazard potential of such flank collapses is widely recognized, but the magnitude, and therefore hazard potential, of the tsunamis that can be triggered by flank collapses is heavily debated (e.g. Moore and Moore 1984; McMurtry *et al.* 2004; Watt *et al.* 2012*a*; Goff *et al.* 2014; Ramalho *et al.* 2015; Paris *et al.* 2018).

Other factors that can contribute to the instability of slopes offshore are the deposition of centimetre- to decimetre-thick discrete layers of volcanic ash across a wide region and the presence of buried turbidites in the volcanic apron. Studies in lacustrine settings (e.g. Wiemer *et al.* 2015; Moernaut *et al.* 2019) and on active offshore margins (e.g. Harders *et al.* 2010; Lafuerza *et al.* 2014; Hornbach *et al.* 2015; Kuhlmann *et al.* 2016; Sammartini *et al.* 2018) have indicated a relationship between tephra layers or turbidites and slide failure planes. Although the exact nature of this relationship is disputed (Wiemer and Kopf 2016), it is thought that tephra could behave as 'weak layers' – layers of inherently lower strength than adjacent layers, which are thereby prone to failure (Locat *et al.* 2014). Using core logging, sedimentological and geotechnical data from the IODP Expedition 340, Lafuerza *et al.* (2014) showed that low hydraulic conductivity of hemipelagic sediments offshore Martinique (Lesser Antilles) could cause low rates of dewatering in turbidites and tephra layers, allowing excess pore fluid pressures to persist at depth. Moreover, Hornbach *et al.* (2015) suggested that even small changes in the stress regime of these layers, such as that resulting from regional strain and grain reorganization during the compaction of sediments, might trigger

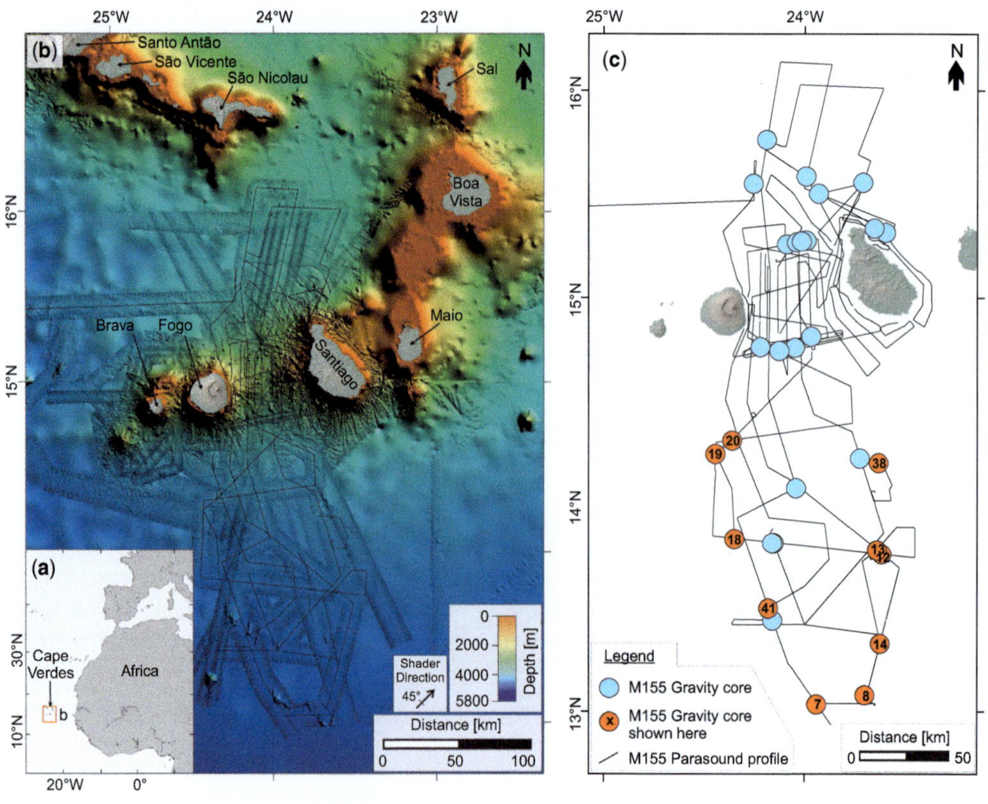

Fig. 1. (**a**) Location of the Cape Verdes offshore West Africa. Box outlines the region presented in (b). (**b**) Multibeam bathymetric data collected during R/V *Meteor* cruises M155 and M80/3 overlain on GEBCO2019 (British Oceanographic Data Centre, http://www.gebco.net) bathymetry of the southern Cape Verdean archipelago. Black lines show the locations of Parasound profiles collected during cruise M155. (**c**) Location of the newly acquired Parasound data and of the gravity cores collected during cruise M155.

motion. Effectively, results from IODP 340 showed that sand layers (i.e. tephra ash and turbidites), as well as boundaries between sand and mud layers, may act as multiple decollement surfaces that promote and enhance the mobility of landslide deposits (Le Friant *et al.* 2015).

Geological setting

The Cape Verdes archipelago, offshore West Africa (Fig. 1), is the surface expression of a mantle hotspot (Crough 1978; Holm *et al.* 2008; Ramalho *et al.* 2010). The Island of Fogo, in the southern part of the archipelago, is one of the most active oceanic intraplate volcanoes in the world, having erupted 28 times in the last 520 years (Ribeiro 1960; Torres *et al.* 1997; González *et al.* 2015). Such volcanically active areas are well known to be associated with seismicity and the southern Cape Verdes are no different, recording frequent volcano-tectonic earthquakes (Grevemeyer *et al.* 2010; Faria and Fonseca 2014; Vales *et al.* 2014). Sediment cores collected in the region provide evidence of at least 43 large, explosive eruptions in the area in the last 150 kyr (Eisele *et al.* 2015), attesting to Fogo's vigorous volcanic activity. Such frequent volcanic activity, along with the accumulation of volcanic deposits on the submarine flanks, means that Fogo's flanks are potentially unstable and prone to collapsing.

A semi-circular depression (Borderia), up to 1 km high and open to the east on central Fogo, was interpreted by Day *et al.* (1999) as the scar of a large flank collapse, referred to as 'Monte Amarelo' (Fig. 2). Other authors, however, interpret the same morphology as two partially overlapping volcanic calderas that were later cut by a flank collapse that affected the eastern portion of the edifice (Brum da Silveira *et al.* 1997; Torres *et al.* 1997; Madeira *et al.* 2008;

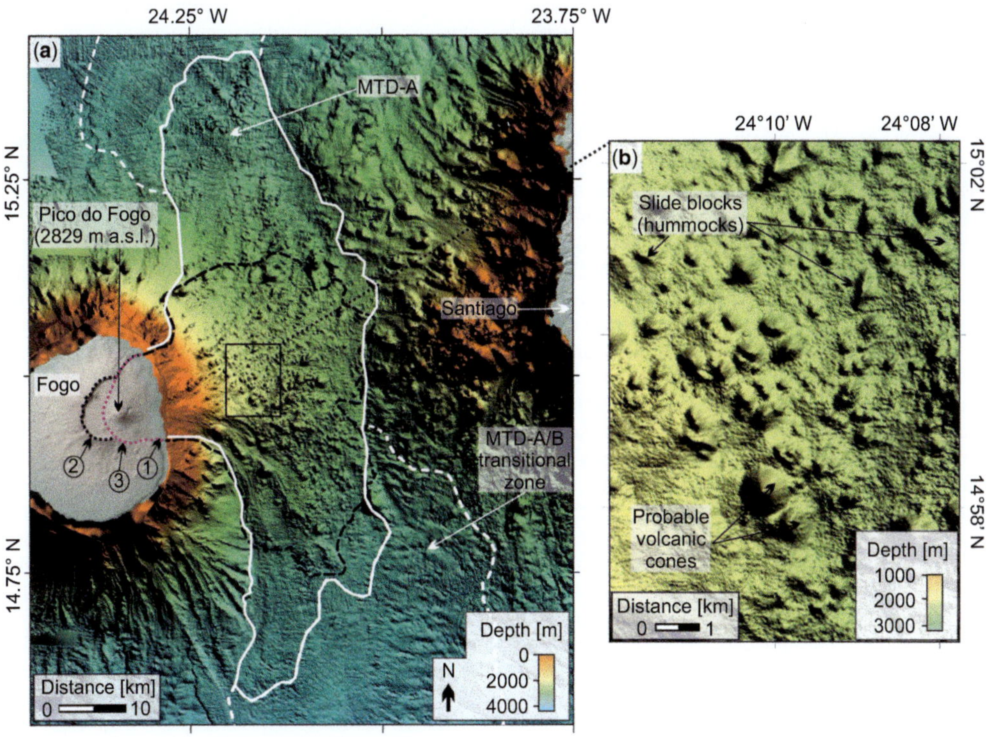

Fig. 2. High-resolution, multibeam bathymetry data highlighting the hummocky nature of the Monte Amarelo debris avalanche deposits (MTD-A) proximal to Fogo. (**a**) Compilation of the M155 and M80/3 multibeam bathymetry data overlain on GEBCO2019 bathymetry and DEM topography of Fogo Island. (1) Espigão escarpment; (2) Bordeira escarpment (dotted black line) is regarded as either the Monte Amarelo headwall scarp (Day *et al.* 1999), or the edge of a caldera (Madeira *et al.* 2008; Martínez-Moreno *et al.* 2018); (3) Monte Amarelo headwall scarp (dotted pink line; Madeira *et al.* 2008; Martínez-Moreno *et al.* 2018); dashed black line: previously estimated extent of the Monte Amarelo debris avalanche deposits (Masson *et al.* 2008); solid white line: revised extent of the Monte Amarelo debris avalanche deposits (MTD-A); dashed white line: transitional zone where the size of the hummocks decreases with distance from the source. (**b**) Enlarged view of the hummocky topography of the Monte Amarelo debris avalanche deposits. Note the angular shape and variable size of the hummocks.

Martínez-Moreno *et al.* 2018). Fogo's Monte Amarelo flank collapse, however, is strongly supported by the presence of a preserved lateral ramp at Espigão (Brum da Silveira *et al.* 1997) and a debris avalanche deposit located between the islands of Fogo and Santiago (Le Bas *et al.* 2007; Masson *et al.* 2008) (Figs 2 & 3). Using a combined magnetotelluric- and multibeam-based approach, Martínez-Moreno *et al.* (2018) estimated a volume of 110 km^3 for the Monte Amarelo debris avalanche deposit. This estimate corresponds well with previous bathymetric-based estimates, which ranged between 80 and 160 km^3 (Le Bas *et al.* 2007; Madeira *et al.* 2008; Masson *et al.* 2008). Tsunami deposits found on the nearby islands of Santiago and Maio indicate that the Monte Amarelo flank collapse was tsunamigenic, with the resulting tsunami achieving a run-up in excess of 270 m above coeval sea-level on Santiago (Ramalho *et al.* 2015; Madeira *et al.* 2019). The exact age of the collapse, however, is the topic of ongoing debate. On the basis of ^{3}He geochronology of lava flows from Fogo, Foeken *et al.* (2009) first estimated the collapse to have occurred between 62 and 123 ka. Paris *et al.* (2011) suggested a narrower window of 86–124 ka, based, respectively, on Ar/Ar ages of lava flows from Fogo thought to be post-collapse, and U–Th dating of corals from tsunami deposits on Santiago. More recently, based on the results of cosmogenic ^{3}He dating of tsunami megaclasts from Santiago Island, Ramalho *et al.* (2015) proposed that the collapse and ensuing tsunami took place between 65 and 84 ka, with a most probable age of 73 $\pm$ 7 ka. This agrees, within uncertainty, with the age recently reported by Maderia *et al.* (2019; 78 $\pm$ 0.9 ka) for a set of tsunami deposits found on the coast of Maio Island. Dating of turbidite material attributed to tsunami-triggered sediment transport along the flanks of Fogo and Brava islands, however, led Eisele *et al.* (2015) to favour an older age of 86–117 ka. Finally, and most recently, Marques *et al.* (2019) suggested a much younger age of 43–59 ka, based on K/Ar dating of Fogo lava flows that they considered to be pre- and post-collapse.

Whether Fogo has collapsed only once or multiple times is also the subject of discussion (Day *et al.* 1999; Ramalho *et al.* 2015; Martínez-Moreno *et al.* 2018; Marques *et al.* 2019). Subsequent eruptions on Fogo have largely been constrained to the central and eastern parts of the island, and a prominent stratovolcano – Pico do Fogo – presently rises to 2829 m above sea-level within the landslide scar (Fig. 2; Torres *et al.* 1997).

Objectives

A detailed analysis of the distribution of the landslide deposits and failure mechanism are critical for constraining the hazard linked to the collapse of a volcanic flank. Moreover, this characterization is crucial for tsunami hazard modelling, given that the mode, volume and run-out of a collapse will have a profound effect on the resulting tsunami waves (Grilli *et al.* 1997; Abadie *et al.* 2012; Watt *et al.* 2012*a*). The offshore distribution of the Monte Amarelo flank-collapse debris (i.e. debris avalanche deposits) and the surrounding volcaniclastic apron were previously mapped using only medium-resolution (100 m grid cell size) multibeam bathymetric data (Masson *et al.* 2008). Consequently, the distribution of the debris avalanche deposits and of any additional slope instabilities in the area are still poorly constrained. As Fogo's volcanic flank collapse likely triggered a mega-tsunami with a recognizable impact on the adjacent islands' coastlines (Paris *et al.* 2011, 2018; Ramalho *et al.* 2015; Madeira *et al.* 2019), a full characterization of this landslide and its related volume is crucial for improving numerical models of tsunami generation, propagation and inundation, and for constraining the hazard potential associated with large, tsunamigenic volcanic flank collapses.

Here, using recently acquired multibeam bathymetric data (50 m grid cell size) in conjunction with parametric sediment echo-sounder data and sediment gravity cores, we revisit the Monte Amarelo volcanic flank collapse and consider general slope stability in the southern distal region of the volcanic apron. We aim to (1) map out the lateral extent and characterize the acoustic nature of the Monte Amarelo debris avalanche deposits; (2) identify possible additional landslides on the slopes south of the islands of Fogo and Santiago; and (3) consider the related preconditioning and triggering processes.

Data and methodology

The data presented in this paper were collected during R/V *Meteor* cruise M155 (May–June 2019; Krastel *et al.* 2019) and are supplemented by multibeam bathymetric data collected during R/V *Meteor* cruise M80/3 (Hansteen *et al.* 2014). The bathymetric data were acquired using hull-mounted Kongsberg EM120 (M80/3), EM122 (M155) and EM710 (M155) multibeam echosounders. The EM120/EM122 system has a swath coverage of up to 150° and a nominal sonar frequency of 12 kHz, and is designed to perform seabed mapping to full ocean depth. During cruise M155, the swath width was reduced to 120° in order to increase the quality and resolution of the data. The EM710 system has a swath coverage of up to 140°, and a nominal sonar frequency of 70–100 kHz. As such, the EM710 was used only in water depths less than 700 m close to the islands of Fogo and Santiago. All bathymetric data were filtered for outliers and manually edited. The data

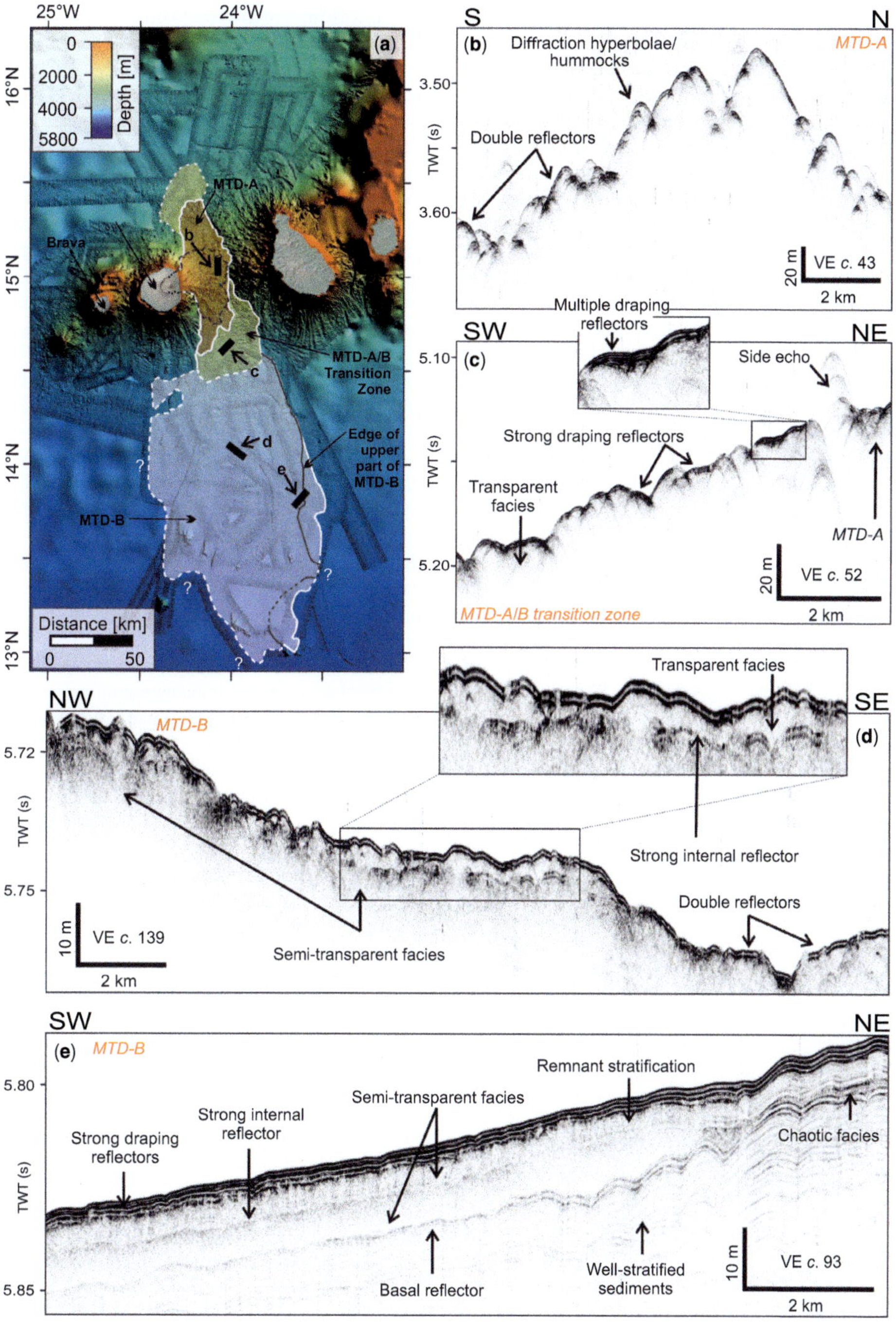

Fig. 3. (**a**) Map showing the distribution of the debris avalanche deposits (MTD-A), the acoustically transparent/ seismically chaotic MTD-B associated with the Monte Amarelo flank collapse, and a transitional zone where the size of the hummocks decreases with distance from the source. The lateral extent of MTD-B may extend further south and west of the surveyed area. Dashed black line: previous mapped extent of the Monte Amarelo deposits (Masson *et al.* 2008); grey line: eastern margin of the upper part of MTD-B; solid black lines: location of the Parasound lines shown in (b)–(e). (**b–e**) Parasound profiles highlighting the acoustic character of (**b**) MTD-A, (**c**) the transitional zone, and (d–e) MTD-B. See the text for discussion on the nature of these deposits. VE, vertical exaggeration.

from all cruises and multibeam systems are combined and gridded at 50 m.

Parametric sediment echo-sounder data were collected using a parametric ATLAS DS-3/P70 system (Parasound). This system has an opening angle of 4° and operates at primary high frequencies of 18.5 and 22.5 kHz, resulting in a parametric low frequency of 4 kHz. The vertical resolution of this system is in the decimetre range. All depth scales on Parasound images presented in this study were calculated using a constant velocity of 1500 m s^{-1}.

Sediment cores were collected during cruise M155 using a gravity corer with tube lengths of 3 to 15 m. These cores were visually described onboard and smear slides were analysed to obtain further microscopic information about the core mineralogy, texture and composition. Dating and further geochemical and sedimentological analysis of these cores is ongoing and will form the basis of future work on the subject.

Results

The Monte Amarelo debris avalanche deposits (MTD-A)

The Monte Amarelo deposits proximal to Fogo (MTD (mass-transport deposit)-A; distribution shown in Fig. 3a) are characterized by overlapping diffraction hyperbolae (Fig. 3b). This hummocky character is typical of debris avalanche deposits from volcanic flank collapses (Siebert 1984). The sedimentary drape covering MTD-A is relatively thin (<1.5 m thick) or not imaged, and is often characterized by two strong, positive reflections. Irregularly shaped blocks that are up to 100 m higher than the surrounding seafloor are present within the landslide debris (Fig. 2). Further away from Fogo, the hummocky topography is less prevalent in the bathymetric data, despite the presence of diffraction hyperbolae in the Parasound data. This highlights a decrease in the size of the hummocks with distance from the island. We note a progressive transition in the acoustic signature of the deposits from a hyperbolic facies with metre-scale acoustic penetration (Fig. 3b), to a mounded facies draped by *c.* 1.5 m of sediment characterized by the aforementioned prominent double reflections (Fig. 3d). In a few places, these double reflections are replaced by a succession of finely stratified layers (inset of Fig. 3c).

Distal deposits related to the Monte Amarelo volcanic flank collapse (MTD-B)

Southwards, with increasing distance from Fogo, the acoustic character of the Monte Amarelo deposits changes from being hyperbolae-dominated to being characterized by an acoustically transparent/semi-transparent facies (Fig. 3d; hereafter referred to as MTD-B; distribution shown in Fig. 3a). In the northern reaches of MTD-B, the upper surface of the deposits is undulating, with metre-scale variations in depth over wavelengths of tens to hundreds of metres (Fig. 3d). The overlying sedimentary drape mantles the surface of the deposits and a strong internal reflector is imaged in places within the deposits (Fig. 3d). This internal reflector clearly contrasts with the overlying transparent facies, separating it from the lower semi-transparent facies. In the northern and western reaches of MTD-B, this internal reflector is either clearly imaged or not present at all, and the base of the lower part of MTD-B is not resolved (Fig. 3d).

In the eastern and southern reaches of MTD-B, the aforementioned internal reflector is continuous over a relatively long distance (Fig. 3e). In these areas, a thicker sedimentary drape (up to *c.* 4 m thick), containing multiple strong reflections alternating with relatively thin, transparent layers, overlies MTD-B. The upper surface of MTD-B is diffuse and undulating in parts. The base of MTD-B is sometimes marked by a prominent reflection that has a similar acoustic signature, although of lower amplitude, as the double reflector that overlies the deposits (e.g. Fig. 3e). A series of well-stratified sediments is imaged below the base of MTD-B (Fig. 3e). At the eastern lateral margin of MTD-B, the internal reflector within the slide deposits is more diffuse (Fig. 3e). The lowermost part of MTD-B remains transparent in nature, but remnant stratifications are imaged in some areas of the uppermost part. The eastern boundary is gradational over 2–3 km, and is marked by a progressive thinning of the transparent facies, and an increase in stratification within the deposits (Fig. 3e).

Constraining the thickness and lateral extent of MTD-B is challenging. In the northern reaches, where the strong internal reflector is imaged, the base of the deposits is not resolved by the acoustic system (Fig. 3c). This means that only the thickness of the upper transparent facies related to MTD-B (up to *c.* 7 m thick, but variable) can be estimated. In its eastern and southern extents, however, the bases of the upper and lower units of MTD-B are imaged, leading to thickness estimates of 4–6 m and 5–15 m for the upper and lower units, respectively (Fig. 3e). The large spacings between the Parasound profiles in the central and southern parts of the working area, south of Fogo and Santiago, mean that we cannot constrain the thickness of the deposits in these areas with certainty. In addition, MTD-B deposits might extend beyond the limits of the surveyed area, especially to the west and south (Fig. 3a).

Additional mass wasting events on the shallow slopes south of Fogo and Santiago

Visual analysis of gravity cores taken in the southern part of the volcaniclastic apron of Fogo and Santiago reveals mud- and nanofossil-rich facies interbedded with multiple sand units that are defined by parallel and cross-laminated, seldom normally graded beds (Fig. 4). These sandy intervals generally range from fine- to coarse-sand deposits, and are typically of centimetre to decimetre scale (Fig. 4b). Microscopic analyses show that these shallow sandy turbidites are predominantly of volcaniclastic (volcanic lithics that are mostly lava fragments, tachylitic to brown glass, crystal fragments), and/or mixed volcaniclastic–bioclastic composition, rather than solely bioclastic.

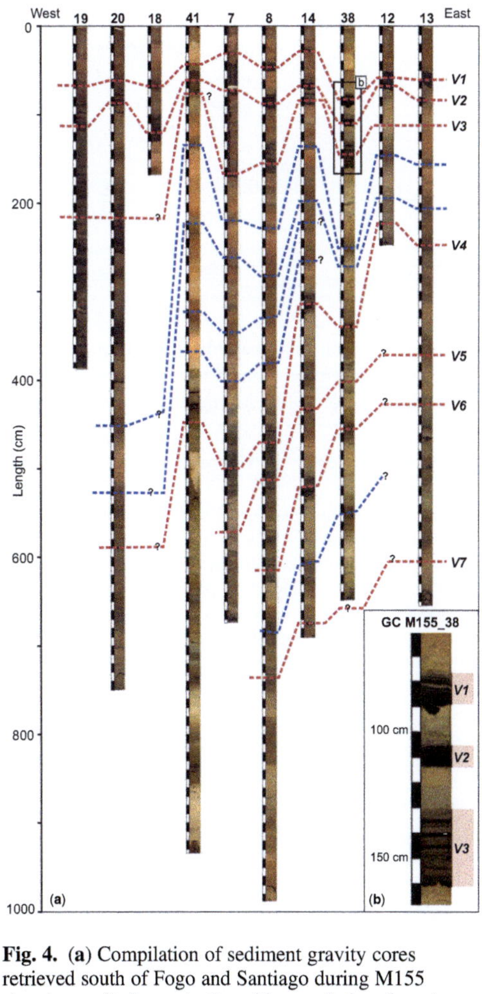

Fig. 4. (**a**) Compilation of sediment gravity cores retrieved south of Fogo and Santiago during M155 showing preliminary stratigraphic correlations based on visual analysis. Dashed red lines: correlations between different volcaniclastic sand layers (V1–V7); dashed blue lines: correlation of background sediment layers (reddish-brown horizons) that helped to correlate the volcaniclastic sand layers. The locations of these cores are shown in Figure 1c. (**b**) Zoom of gravity core M155_38 (62–168 cm) showing the visually identified volcaniclastic layers V1–V3.

Multibeam-bathymetry and sediment echo-sounder data on the shallow slopes south of Fogo and Santiago reveal multiple morphological steps and several MTDs (Fig. 5). These additional MTDs are characterized by acoustically transparent and/or hyperbolic facies and occur both above and below MTD-B. Several of these MTDs are exposed at the surface and have remobilized previously failed sediments, including those of MTD-B (e.g. Figs 5 & 6). However, as for MTD-B, their full lateral extent is generally poorly constrained due to large profile spacings and their possible continuation outside the surveyed area. The morphological steps vary in length, height, strike and orientation (Fig. 5a), but are typically near vertical (Fig. 5c, d). In the following section, we describe the character of, and relation between, several of these features in more detail.

Tectonic escarpments along the western and eastern margins. A series of escarpments is imaged to the west, along the western margin of MTD-B, and stretching between 40 and 110 km south of Fogo (Fig. 5a, b). The most prominent escarpment is orientated roughly north–south and extends across *c.* 55 km. It is *c.* 50 m high and near vertical (Fig. 5c). Monte Amarelo MTD-B deposits are imaged on the eastern side (footwall) of this step. On the western (hanging-wall) side, the sediments are characterized by a semi-transparent facies with a hummocky upper surface (hereafter referred to as MTD-X; Fig. 5c). In places, a pair of strong reflectors is identified within the sedimentary drape that overlies MTD-X. Proximal to the escarpment, the base of MTD-X cannot be resolved. Further west of the scarp, however, stratified sediment is imaged below the base of MTD-X (inset 1 of Fig. 5c). The lateral extent of MTD-X is constrained to both the north and east by the morphological steps (Fig. 5b, c), resulting in a minimum area of 900 km^2 for MTD-X.

Southern Scour Complex. A prominent, elongated scour-shaped feature is evident in the central part of the southern distal region (hereafter referred to as the Southern Scour Complex (SSC), Fig. 6a). This feature is *c.* 60 km long and covers an area of *c.* 340 km^2. The western and eastern sides of the SSC are constrained by up to 40 and 55 m high escarpments, respectively (Fig. 6). In Parasound data crossing the complex, acoustically transparent facies with a strong internal reflector characterizing the Monte Amarelo MTD-B are evident on both the eastern and western sides of the SSC (Fig. 6b & c). In the central part of the SSC (Fig. 6c), acoustically transparent

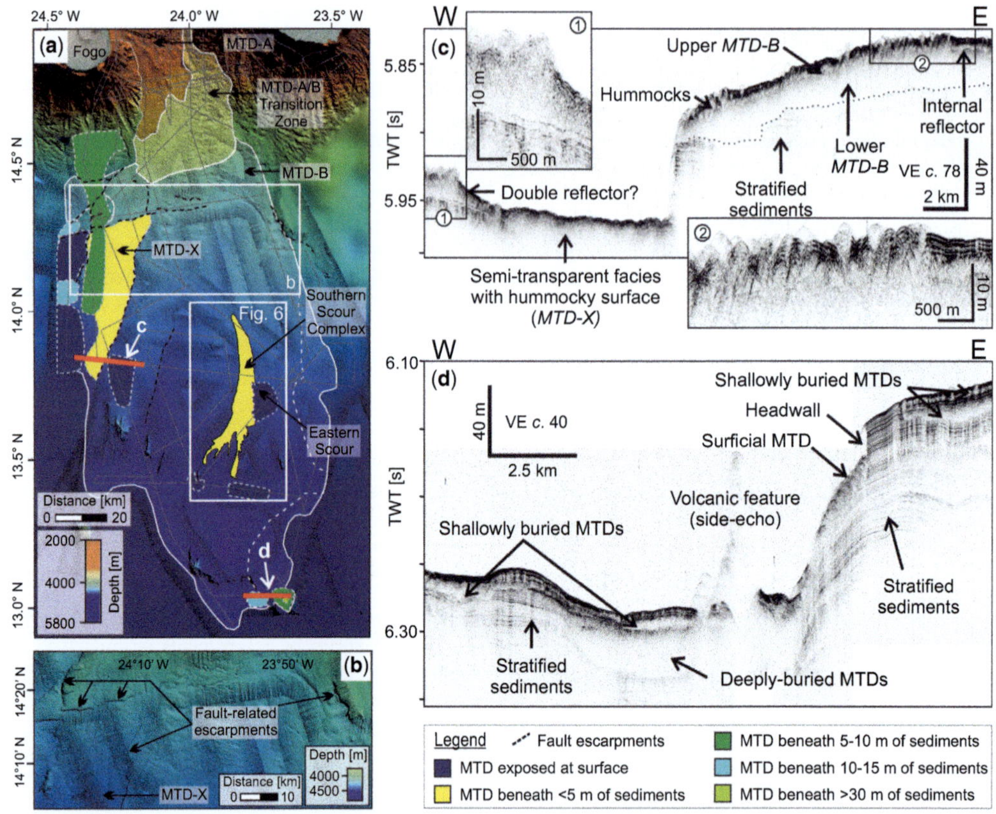

Fig. 5. (**a**) Multibeam data overlain on GEBCO2019 bathymetry highlighting the presence of multiple fault-related escarpments and volcanic features on the shallow slopes south of Fogo. Additional MTDs are also identified in the Parasound data and coloured according to their burial depth (see legend). Solid red lines: location of the Parasound profiles shown in (c) and (d). (**b**) Multibeam data showing the nature of the fault-related escarpments on the slopes south of Fogo and Santiago in more detail. (**c**) Parasound profile crossing one of the most striking escarpments visible on the western side (see (b)), highlights the relation between tectonic processes and landsliding in the area. Monte Amarelo *MTD-B* deposits (eastern side) are remobilized across the escarpment. (**d**) Parasound profile crossing an escarpment in the southern reaches of the working area highlights acoustically transparent MTDs at varying depths.

MTDs with some internal structure are covered by *c.* 3 m of stratified sediment. In the northern part, the material within the SSC is characterized by a semi-transparent facies with a hummocky upper surface, and the deposits are largely exposed at the surface (Fig. 6b). An additional instability is visible on the eastern side of the SSC (hereafter referred to as the Eastern Scour (ES); Fig. 6d). The ES reaches a thickness of up to 35 m thick in its centre and covers an area of *c.* 120 km^2. Parasound data across the ES show that this failure primarily remobilized the Monte Amarelo MTD-B deposits in this region along the same basal glide plane as MTD-B (Fig. 6d). To the south, the SCC is fan-shaped, with fingers of unfailed stratified sediments standing up to *c.* 20 m above failed sediment (Fig. 6d). Semi-transparent deposits characteristic of MTD-B are imaged on the tops of these stratified fingers.

Discussion

The two-fold nature of Fogo's Monte Amarelo flank collapse

Based on the difference in their acoustic character, we distinguish between two main types of deposits related to the Monte Amarelo volcanic flank collapse: hummocky, debris avalanche deposits proximal to Fogo (MTD-A), and finer-grained, acoustically transparent landslide deposits (MTD-B) at greater distances from Fogo that were previously not recognized (Fig. 3). The blocky debris avalanche deposits of MTD-A cover a surface area of 3180 km^2. If we also include the region in which the size of the hummocks decreases with distance from the source (marked by the yellow shaded region in Fig. 3a), this increases the total area of MTD-A to 6820 km^2;

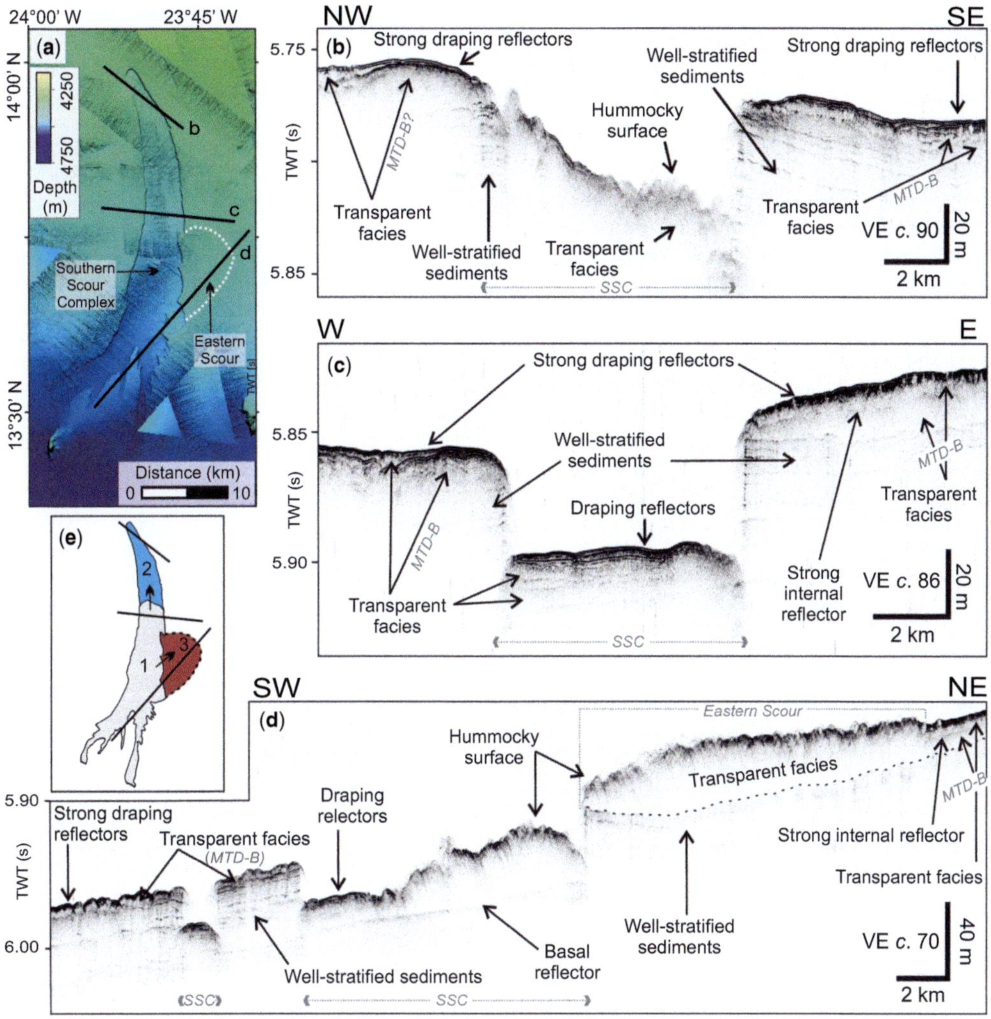

Fig. 6. Southern Scour Complex (SSC). (**a**) Multibeam data highlighting the surface morphology of the SCC and the extension of the Eastern Scour (dashed white line). The location of Figure 6a with respect to the other features discussed in this article is shown in Figure 5. (**b–d**) Parasound profiles crossing the SSC showing a complex relationship between former and more recent landsliding processes, along with the variable acoustic character of the deposits and of the adjacent unfailed material. (**e**) Conceptual model showing the proposed retrogressive development of the SSC towards the north and east. Numbers 1–3 highlight the sequence of events.

more than four times the previous estimate (*c.* 1470 km^2; Masson *et al.* 2008). As the base of the volcanic debris avalanche deposits is not resolved by the Parasound data, we cannot, at this stage, revisit estimates of the volume involved.

Monte Amarelo MTD-B is characterized by an acoustically transparent to semi-transparent facies that is overlain by a series of two to four strong reflectors (Figs 3, 5c & 6b–d). A prominent, internal reflection is clearly observed in places within the deposits; separating MTD-B into two main parts (Fig. 3d, e). The uppermost section is of relatively constant thickness (4–6 m). Contrastingly the thickness of the lower part of MTD-B is highly variable (up to 15 m thick). Interestingly, the upper and lower parts of MTD-B have a similar areal distribution within the volcaniclastic apron; covering a minimum of 18 400 km^2 (upper) and 19 500 km^2 (lower) (Fig. 3a). This implies minimum volumes of 92 km^3 (upper) and 195 km^3 (lower), assuming average thicknesses of 5 m (upper) and 10 m (lower). Their similar distribution implies that the upper and lower parts of MTD-B might share the same source or, at least, be genetically linked. We therefore infer that the upper and lower part of MTD-B might have been emplaced (almost) contemporaneously. The absolute timing of these two phases, however, remains unclear. Of particular interest is whether these two failures occurred as two phases of the same event, or as two distinct events separated by some time. Further analytical work on the sediment cores will help to reveal more information about the nature of the prominent internal reflection, along with the relative timing of these two depositional episodes.

A similar two-fold nature of volcanic flank-collapse deposits – hummocky debris avalanche deposits accompanied by acoustically transparent to seismically chaotic deposits with a comparatively smooth upper surface – has also been reported for flank collapses at other locations, including in the Lesser Antilles (Watt *et al.* 2012*a*, *b*; Le Friant *et al.* 2015; Brunet *et al.* 2016), at La Réunion (Indian Ocean; Lebas *et al.* 2018) and at Ritter Island (Papua New Guinea; Karstens *et al.* 2019; Watt *et al.* 2019). Drilled cores retrieved during IODP Expedition 340 in the Lesser Antilles indicated that widespread, seismically chaotic deposits (interpreted as equivalent to our MTD-B) primarily consist of hemipelagic mud interbedded with a combination of tephra, volcaniclastic layers, or bioclastic turbiditic deposits, which have undergone varying degrees of deformation (Le Friant *et al.* 2015; Brunet *et al.* 2016). To explain these findings, Le Friant *et al.* (2015) proposed a failure model where the loading of seafloor sediment by volcanic debris avalanche deposits triggered sediment destabilization and progressive downslope-propagating failure along a decollement. In this way, the deformation can propagate great distances away from the flank collapse, affecting seafloor sediments that were otherwise stable (Le Friant *et al.* 2015). The acoustic character of our MTD-B bears a strong resemblance to the seismically chaotic deposits in the Lesser Antilles. We, therefore, interpret MTD-B to be the result of the failure of pre-existing seafloor sediments following the loading of the Monte Amarelo debris avalanche deposits (MTD-A).

A history of mass wasting and remobilization in the southern Cape Verdes

Repeated mass-wasting events at Fogo? Preliminary stratigraphic correlations based on visual similarities of cores in the southern distal part of the working area indicate at least seven volcaniclastic or mixed volcaniclastic–bioclastic sandy turbidite layers above and below the Monte Amarelo flank-collapse deposits (Fig. 4). These turbidite layers are dominated by mafic glass, crystals and lava fragments, which, based on their petrography, suggests that they originated from Fogo volcano. This indicates that smaller mass-wasting events may have occurred at Fogo in addition to the Monte Amarelo volcanic flank collapse, as also recently suggested by Marques *et al.* (2019). Correlation and origin of these volcaniclastic deposits will, however, be verified by future analytical work on the sediment cores.

Regional tectonic and volcanic influences south of Fogo? The multiple fault-related escarpments and MTDs identified in the acoustic data south of Fogo further highlight a long and complex history of instability and slope failure in the region (Fig. 5). Ramalho *et al.* (2010) found that the island of Santiago has uplifted at a rate of *c.* 100 m Ma^{-1} over the past 4 Ma, and attributed this uplift to a combination of magmatic intrusions under or within the island edifice, together with episodic swell-wide uplift that affected the whole archipelago. Although geologically recent uplift has not been reported at Fogo, the neighbouring island of Brava has experienced one of the most dramatic intrusion-related uplift trends of any ocean island in the world, with up to 400 m of uplift in the last 1.8 Ma (Madeira *et al.* 2010). The presence of such widespread uplift across the archipelago – and particularly around the southern Cape Verdes – thus suggests that the series of escarpments and landslides observed on the slopes south of Fogo may be the surface expression of a combination of regional tectonics, associated with the growth of the Cape Verdes hotspot swell, and crustal intrusions in the vicinity of the islands. As several of the exposed faults have landslide deposits on their hanging wall side, we interpret that these MTDs are most likely the results of movement along the faults and, as such, the result of neotectonic activity.

The SSC, which is exposed at the seafloor, postdates the Monte Amarelo flank collapse (Fig. 6). Monte Amarelo MTD-B deposits are imaged on the footwall east of the SSC (Fig. 6b–d), and on the tops of the unfailed, stratified fingers at the southern extent of the SSC (Fig. 6d). This indicates that the SSC formed after the Monte Amarelo flank collapse, remobilizing MTD-B. In the northern and eastern parts of the SSC, acoustically transparent mass-wasting deposits are exposed at the seafloor, highlighting more recent mass wasting and remobilization of MTD-B (Fig. 6b). We interpret the more recent mass-wasting deposits in the northern and eastern parts of the SSC as evidence of retrogressive development of the SSC in these directions (Fig. 6e).

The presence of fault-related escarpments on the slopes south of Fogo, frequent volcanic episodes at Fogo (Ribeiro 1960; Torres *et al.* 1997), and ongoing uplift at the neighbouring islands of Santiago and Brava (Madeira *et al.* 2010; Ramalho *et al.* 2010), together imply the occurrence of frequent earthquakes in the past. The volcanic activity at Fogo has resulted in the widespread deposition of numerous tephra layers (Eisele *et al.* 2015). Subsequent earthquake shaking may have resulted in liquefaction of overlying layers, which, together with sedimentary overpressure following further sediment deposition, can help to precondition the shallow slopes for failure (Moernaut *et al.* 2019). Even small changes in the stress regime of tephra layers have been found to be able to trigger failure (Hornbach *et al.* 2015). Consequently, it is possible that some of the MTDs observed on the shallowly dipping slopes south of Fogo, and unrelated to escarpments, may result from such a process. Future work, particularly on the sediment cores collected during cruise M155, will help to shed light on the sources and processes responsible for the MTDs mapped on the shallow slopes south of Fogo and Santiago.

Implications and conclusions

In this study, we revisited the Monte Amarelo volcanic flank collapse of Fogo Island and found that the deposition of the debris avalanche material may have triggered subsequent failures of pre-existing seafloor sediments in (at least) two phases. This is similar to what has been observed in the Lesser Antilles (Le Friant *et al.* 2015; Brunet *et al.* 2016), at La Réunion (Lebas *et al.* 2018) and at Ritter Island (Karstens *et al.* 2019; Watt *et al.* 2019). It is not yet clear whether these two phases of seafloor sediment failure were synchronous as a result of a single flank collapse, or if they reflect multiple, distinct events. The question of whether the main flank collapse occurred as a single or as multiple events is of utmost importance for tsunami modelling because the volume and timing of individual failures are the main factors controlling the tsunamigenic potential (Løvholt *et al.* 2015). Water depth and landslide kinematics play a key role in controlling the tsunamigenic potential of a subaqueous landslide (Watts *et al.* 2000; Ward 2001; Watt *et al.* 2012*a*; Harbitz *et al.* 2014). As MTD-B occurred in depths exceeding 3000 m, we consider that its influence on the resulting tsunami magnitude was negligible, and that only the hummocky debris avalanche deposits (MTD-A) contributed to the mega-tsunami that inundated nearby islands.

Multibeam bathymetric, sediment echo-sounder and sediment gravity core data acquired during cruise M155 of R/V *Meteor* provide seafloor evidence that shows, for the first time, the presence of multiple additional mass-wasting events on the shallow slopes south of Fogo. Turbidite sands recovered in the cores have a volcaniclastic or mixed volcaniclastic–bioclastic composition, which suggests that prevailing mass-transport processes in the region may be dominated by volcanic eruptions or volcanic mass-wasting events. However, it is also possible that some of the additional mass-wasting events are unrelated to flank collapse events or submarine failures, and could instead have been generated by sedimentary overpressure and/or by the presence of a mechanically weak layer such as tephra or turbidites. Further analytical work is required to verify the sources of these events on the shallow slopes south of the Cape Verdean archipelago. As these layers are thin, have limited lateral extent and occurred on low-gradient slopes in deep water, we consider their associated tsunamigenic potential to be low.

Acknowledgements We are grateful to the captain, crew and scientific teams of R/V *Meteor* (M155) for their enthusiasm and assistance onboard. We are also thankful for the thoughtful reviews of the editor, Lawrence Amy, as well as Paraskevi Nomikou and two anonymous reviewers, whose comments improved this manuscript.

Funding Data acquisition during R/V *Meteor*'s cruise M155 was funded by the DFG and BMBF. R. Barrett is funded by the European Union's Horizon 2020 research and innovation programme under the Marie Skłodowska-Curie grant agreement No. 721403 (SLATE – 'Submarine LAndslides and Their impact on European continental margins'). R. Ramalho acknowledges his IF/01641/2015 contract funded by Fundação para a Ciência e Tecnologia (FCT), as well as the following projects funded by FCT: IF/01641/2015, LISBOA-01-0145-FEDER-028588 and PTDC/CTA-GEO/28588/2017 UNTIeD, PTDC/GEO-GEO/1123/2014 FIRE, and UID/GEO/50019/2019 – Instituto Dom Luiz.

Author contributions **RB**: conceptualization (lead), formal analysis (lead), investigation (supporting), methodology (lead), validation (lead), visualization (lead), writing – original draft (lead), writing – review & editing (lead); **EL**: conceptualization (supporting), formal analysis (supporting), investigation (supporting), methodology (supporting), project administration (supporting), supervision (supporting), validation (supporting), visualization (supporting), writing – review & editing (supporting); **RR**: conceptualization (supporting), formal analysis (supporting), funding acquisition (supporting), investigation (supporting), project administration (supporting), writing – review & editing (supporting); **IK**: conceptualization (supporting), formal analysis (supporting), investigation (supporting), project administration (supporting), writing – review & editing (supporting); **SKu**: conceptualization (supporting), data curation (equal), formal analysis (supporting), funding acquisition (supporting), investigation (supporting), project administration (supporting), writing – review & editing (supporting); **AK**: conceptualization (supporting), formal analysis (supporting), funding acquisition (supporting), investigation (supporting), project administration (supporting), writing – review & editing (supporting); **KL**: conceptualization (supporting), formal analysis (supporting), investigation (supporting), writing – review & editing (supporting); **FG**: conceptualization (supporting), formal analysis (supporting), methodology (supporting), supervision (supporting), writing – review & editing (supporting); **SKr**: conceptualization (supporting), data curation (lead), formal analysis (supporting), funding acquisition (lead), investigation (lead), project administration (lead), supervision (lead), validation (supporting), writing – review & editing (supporting).

References

Abadie, S.M., Harris, J.C., Grilli, S.T. and Fabre, R. 2012. Numerical modeling of tsunami waves generated by the flank collapse of the Cumbre Vieja Volcano (La Palma, Canary Islands): tsunami source and near field effects. *Journal of Geophysical Research: Oceans*, **117**, 1–26, https://doi.org/10.1029/2011JC007646

Begét, J.E. and Kienle, J. 1992. Cyclic formation of debris avalanches at Mount St Augustine volcano. *Nature*, **356**, 701–704, https://doi.org/10.1038/356701a0

Blahůt, J., Balek, J., Klimeš, J., Rowberry, M., Kusák, M. and Kalina, J. 2019. A comprehensive global database of giant landslides on volcanic islands. *Landslides*, **16**, 2045–2052, https://doi.org/10.1007/s10346-019-01275-8

Brum da Silveira, A., Madeira, J., Serralheiro, A., Torres, P.C., da Silva, L.C. and Mendes, M.H. 1997. O controlo estrutural da erupção de Abril de 1995 na Ilha do Fogo, Cabo Verde. *In*: *A Erupção Vulcânica de 1995 na Ilha do Fogo, Cabo Verde*. Publ. IICT, Lisbon, 51–61.

Brunet, M., Le Friant, A. *et al.* and IODP Expedition 340 Science Party. 2016. Composition, geometry, and emplacement dynamics of a large volcanic island landslide offshore Martinique: from volcano flank- collapse to seafloor sediment failure? *Geochememistry, Geophysics, Geosystems*, **17**, 699–724, https://doi.org/10.1002/2015GC006034

Casalbore, D. 2018. Volcanic islands and seamounts. *In*: Micallef, A., Krastel, S. and Savini, A. (eds) *Submarine Geomorphology*. Springer Geology. Springer, Cham, 333–347, https://doi.org/10.1007/978-3-319-57852-1_17

Casalbore, D., Romagnoli, C., Bosman, A. and Chiocci, F.L. 2011. Potential tsunamigenic landslides at Stromboli Volcano (Italy): insight from marine DEM analysis. *Geomorphology*, **126**, 42–50, https://doi.org/10.1016/j.geomorph.2010.10.026

Casalbore, D., Romagnoli, C. *et al.* 2015. Volcanic, tectonic and mass-wasting processes offshore Terceira Island (Azores) revealed by high-resolution seafloor mapping. *Bulletin of Volcanology*, **77**, 1–19, https://doi.org/10.1007/s00445-015-0905-3

Cervelli, P., Segall, P., Johnson, K., Lisowski, M. and Miklius, A. 2002. Sudden aseismic fault slip on the south flank of Kilauea volcano. *Nature*, **415**, 1014–1018, https://doi.org/10.1038/4151014a

Clare, M.A., Le Bas, T. *et al.* 2018. Complex and cascading triggering of submarine landslides and turbidity currents at volcanic islands revealed from integration of high-resolution onshore and offshore surveys. *Frontiers in Earth Science*, **6**, https://doi.org/10.3389/feart.2018.00223

Crough, S.T. 1978. Thermal origin of mid-plate hot-spot swells. *Geophysical Journal of the Royal Astronomical Society*, **55**, 451–469, https://doi.org/10.1111/j.1365-246X.1978.tb04282

Day, S.J., Heleno da Silva, S.I.N. and Fonseca, J.F.B.D. 1999. A past giant lateral collapse and present-day flank instability of Fogo, Cape Verde Islands. *Journal of Volcanology and Geothermal Research*, **94**, 191–218, https://doi.org/10.1016/S0377-0273(99)00103-1

Eisele, S., Reißig, S., Freundt, A., Kutterolf, S., Nürnberg, D., Wang, K.L. and Kwasnitschka, T. 2015. Pleistocene to Holocene offshore tephrostratigraphy of highly explosive eruptions from the southwestern Cape Verde Archipelago. *Marine Geology*, **369**, 233–250, https://doi.org/10.1016/j.margeo.2015.09.006

Faria, B. and Fonseca, J.F.B.D. 2014. Investigating volcanic hazard in Cape Verde Islands through geophysical monitoring: network description and first results. *Natural Hazards Earth System Sciences*, **14**, 485–499, https://doi.org/10.5194/nhess-14-485-2014

Foeken, J.P.T., Day, S. and Stuart, F.M. 2009. Cosmogenic ^{3}He exposure dating of the Quaternary basalts from Fogo, Cape Verdes: implications for rift zone and magmatic reorganization. *Quaternary Geochronology*, **4**, 37–49, https://doi.org/10.1016/j.quageo.2008.07.002

Goff, J., Terry, J.P., Chagué-Goff, C. and Goto, K. 2014. What is a mega-tsunami? *Marine Geology*, **358**, 12–17, https://doi.org/10.1016/j.margeo.2014.03.013

González, P.J., Bagnardi, M., Hooper, A.J., Larsen, Y., Marinkovic, P., Samsonov, S.V. and Wright, T.J. 2015. The 2014–2015 eruption of Fogo volcano: geodetic modeling of Sentinel-1 TOPS interferometry. *Geophysical Research Letters*, **42**, 9239–9246, https://doi.org/10.1002/2015GL066003

Grevemeyer, I., Helffrich, G., Faria, B., Booth-Rea, G., Schnabel, M. and Weinrebe, W. 2010. Seismic activity at Cadamosto seamount near Fogo Island, Cape Verdes: formation of a new ocean island? *Geophysical Journal International*, **180**, 552–558, https://doi.org/10.1111/j.1365-246X.2009.04440.x

Grilli, S.T., Svendsen, I.A. and Subramanya, R. 1997. Breaking criterion and characteristics for solitary waves on slopes. *Journal of Waterway, Port, Coastal, and Ocean Engineering*, **123**, 102–112, https://doi.org/10.1061/(ASCE)0733-950X(1997)123:3(102)

Grilli, S.T., Tappin, D.R. *et al.* 2019. Modelling of the tsunami from the December 22, 2018 lateral collapse of Anak Krakatau volcano in the Sunda Straits, Indonesia. *Science Reports*, **9**, 11946, https://doi.org/10.1038/s41598-019-48327-6

Gross, F., Krastel, S. *et al.* 2014. Evidence for submarine landslides offshore Mt. Etna, Italy. *In*: Krastel, S. *et al.* (eds) *Submarine Mass Movements and Their Consequences*. Advances in Natural and Technological Hazards Research, **37**. Springer, Cham, 307–316, https://doi.org/10.1007/978-3-319-00972-8_27

Hansteen, T.H., Kwasnitschka, T. and Klügel, A. 2014. *Cape Verde Seamounts – Cruise No. M80/3 – December 29 2009 – February 1 2010 – Dakar (Senegal) – Las Palmas de Gran Canaria (Spain)*. METEOR-Berichte, **M80/3**, DFG-Senatskommission für Ozeanographie, Bremen, https://doi.org/10.2312/cr_m80_3

Harbitz, C.B., Løvholt, F. and Bungum, H. 2014. Submarine landslide tsunamis: how extreme and how likely?. *Natural Hazards*, **72**, 1341–1374, https://doi.org/10.1007/s11069-013-0681-3

Harders, R., Kutterolf, S., Hensen, C., Moerz, T. and Brueckmann, W. 2010. Tephra layers: a controlling factor on submarine translational sliding?. *Geochemistry, Geophysics, Geosystems*, **11**, 1–18, https://doi.org/10.1029/2009GC002844

Holm, P.M., Grandvuinet, T., Friis, J., Wilson, J.R., Barker, A.K. and Plesner, S. 2008. An ^{40}Ar-^{39}Ar study of the Cape Verde hot spot: temporal evolution in a semistationary plate environment. *Journal of Geophysical Research*, **113**, 1–22, https://doi.org/10.1029/2007JB005339

Hornbach, M.J., Manga, M. *et al.* 2015. Permeability and pressure measurements in Lesser Antilles submarine slides: evidence for pressure-driven slow-slip failure. *Journal of Geophysical Research: Solid Earth*, **120**, 7986–8011, https://doi.org/10.1002/2015JB012061

Krastel, S., Schmincke, H.-U., Jacobs, C.L., Rihm, R., Le Bas, T.P. and Alibes, B. 2001. Submarine landslides around the Canary Islands. *Journal of Geophysical Research*, **106**, 3977–3998, https://doi.org/10.1029/2000JB900413

Krastel, S., Andrade, M. *et al.* 2019. *The tsunamigenic gravitational flank-collapse of Fogo volcano, Cape Verde Islands, Seismic pre-site survey for an IODP site on the Cape Verde Plateau, Cruise No. M155, 26.05.19–30.06.19, Pointe-à-Pitre (Guadeloupe) – Mindelo (Cape Verde)*. METEOR-Berichte, Gutachterpanel Forschungsschiffe, Bonn, **M155**, 1–44, https://doi.org/10.2312/cr_m155

Karstens, J., Berndt, C. *et al.* 2019. From gradual spreading to catastrophic collapse: Reconstruction of the 1888 Ritter Island volcanic sector collapse from high-resolution 3D seismic data. *Earth and Planetary Science Letters*, **517**, 1–13, https://doi.org/10.1016/j.epsl.2019.04.009

Kuhlmann, J., Huhn, K. and Ikari, M.J. 2016. Do embedded volcanoclastic layers serve as potential glide planes? An integrated analysis from the Gela Basin offshore southern Sicily. *In*: Lamarche, G. *et al.* (eds) *Submarine Mass Movements and Their Consequences*. Advances in Natural and Technological Hazards Research, **41**. Springer, Cham, 273–280, https://doi.org/10.1007/978-3-319-20979-1_27

Lafuerza, S., Le Friant, A. and the Expedition 340 Scientific Party 2014. Geomechanical characterizations of submarine volcano flank sediments, Martinique, Lesser Antilles Arc. *In*: Krastel, S. *et al.* (eds) *Submarine Mass Movements and Their Consequences*. Advances in Natural and Technological Hazards Research, **37**. Springer, Cham, https://doi.org/10.1007/978-3-319-00972-8_7

Lebas, E., Le Friant, A. *et al.* 2011. Multiple widespread landslides during the long-term evolution of a volcanic island: insights from high-resolution seismic data. Montserrat, Lesser Antilles. *Geochemistry, Geophysics, Geosystems*, **12**, https://doi.org/10.1029/2010GC003451

Lebas, E., Le Friant, A., Deplus, C. and de Voogd, B. 2018. Understanding the evolution of an oceanic intraplate volcano from seismic reflection data: a new model for La Réunion, Indian Ocean. *Journal of Geophysical Research*, **123**, 1035–1059, https://doi.org/10.1002/2017JB014959

Le Bas, T.P., Masson, D.G., Holtom, R.T. and Grevemeyer, I. 2007. Slope failures of the flanks of the southern Cape Verde Islands. *In*: Lykousis, V., Sakellariou, D. and Locat, J. (eds) *Submarine Mass Movements and Their Consequences*. Advances in Natural and Technological Hazards Research, **27**. Springer, Dordrecht, 337–345, https://doi.org/10.1007/978-1-4020-6512-5_35

Le Friant, A., Ishizuka, O., Boudon, G. *et al.* 2015. Submarine record of volcanic island construction and collapse in the Lesser Antilles arc: first scientific drilling of submarine volcanic island landslides by IODP Expedition 340. *Geochemistry, Geophysics, Geosystems*, **16**, 420–442, https://doi.org/10.1002/2014GC005652

León, R., Somoza, L. *et al.* 2017. Multi-event oceanic island landslides: new onshore-offshore insights from El Hierro Island, Canary Archipelago. *Marine Geology*, **393**, 156–175, https://doi.org/10.1016/j.margeo.2016.07.001

Livanos, I., Nomikou, P., Papanikolaou, D. and Rousakis, G. 2013. The volcanic debris avalanche on the SE submarine slope of Nisyros volcano, Greece: geophysical exploration and implications for subaerial eruption history. *Geo-Marine Letters*, **33**, 419, https://doi.org/10.1007/s00367-013-0338-y

Locat, J., Leroueil, S., Locat, A. and Lee, H.J. 2014. Weak layers: their definition and classification from a geotechnical perspective. *In*: Krastel, S. *et al.* (eds) *Submarine Mass Movements and Their Consequences*. Advances in Natural and Technological Hazards Research, **37**. Springer, Cham, 3–12, https://doi.org/10.1007/978-3-319-00972-8_1

Løvholt, F., Pedersen, G., Harbitz, C.B., Glimsdal, S. and Kim, J. 2015. On the characteristics of landslide tsunamis. *Philosophical Transactions of the Royal Society A*, **373**, https://doi.org/10.1098/rsta.2014.0376

Madeira, J., Brum da Silveira, A., Mata, J., Mourão, C. and Martins, S. 2008. The role of mass movements on the geomorphologic evolution of island volcanoes: examples from Fogo and Brava in the Cape Verde archipelago. *Comunicações Geológicas*, **95**, 93–106.

Madeira, J., Mata, J., Mourão, C., Brum dal Silveira, A., Martins, S., Ramalho, R. and Hoffman, D.L. 2010. Volcano-stratigraphic and structural evolution of Brava Island (Cape Verde) based on ^{40}Ar/^{39}Ar, U–Th and field constraints. *Journal of Volcanology and Geothermal Research*, **196**, 219–235, https://doi.org/10.1016/j.jvolgeores.2010.07.010

Madeira, J., Ramalho, R.S., Hoffmann, D.L., Mata, J. and Moreira, M. 2019. A geological record of multiple Pleistocene tsunami inundations in an oceanic island: the case of Maio, Cape Verde. *Sedimentology*, https://doi.org/10.1111/sed.12612

Marques, F.O., Hildenbrand, A., Victória, S.S., Cunha, C. and Dias, P. 2019. Caldera or flank collapse in the Fogo volcano? What age? Consequences for risk assessment in volcanic islands. *Journal of Volcanology and Geothermal Research*, **388**, 106686. https://doi.org/10.1016/j.jvolgeores.2019.106686

Martínez-Moreno, F.J., Monteiro Santos, F.A. *et al.* 2018. Investigating collapse structures in oceanic islands using magnetotelluric surveys: the case of Fogo Island in Cape Verde. *Journal of Volcanology and*

Geothermal Research, **357**, 152–162, https://doi.org/10.1016/j.jvolgeores.2018.04.028
Masson, D.G., Watts, A.B., Gee, M.J.R., Urgeles, R., Mitchell, N.C., Le Bas, T.P. and Canals, M. 2002. Slope failures on the flanks of the western Canary Islands. *Earth-Science Reviews*, **57**, 1–35, https://doi.org/10.1016/S0012-8252(01)00069-1
Masson, D.G., Le Bas, T.P., Grevemeyer, I. and Weinrebe, W. 2008. Flank collapse and large-scale landsliding in the Cape Verde Islands, off West Africa. *Geochemistry, Geophysics, Geosystems*, **9**, 16, https://doi.org/10.1029/2008GC001983
McGuire, W.J. 1996. Volcano instability: a review of contemporary themes. *Geological Society, London, Special Publications*, **110**, 1–23, https://doi.org/10.1144/GSL.SP.1996.110.01.01
McGuire, W.J. 2003. Volcano instability and lateral collapse. *Revista*, **1**, 33–45.
McMurtry, G.M., Fryer, G.J. *et al.* 2004. Megatsunami deposits on Kohala volcano, Hawaii, from flank collapse of Mauna Loa. *Geology*, **32**, 741–744, https://doi.org/10.1130/G20642.1
Moernaut, J., Van Daele, M. *et al.* 2019. The subaqueous landslide cycle in south-central Chilean lakes: the role of tephra, slope gradient and repeated seismic shaking. *Sedimentary Geology*, **381**, 84–105, https://doi.org/10.1016/j.sedgeo.2019.01.002
Moore, J.G. and Moore, G.W. 1984. Deposit from a giant wave on the island of Lanai, Hawaii. *Science*, **226**, 1312–1315, https://doi.org/10.1126/science.226.4680.1312
Moore, J.G., Clague, D.A., Holcomb, R.T., Lipman, P.W., Normark, W.R. and Torresan, M.E. 1989. Prodigious submarine landslides on the Hawaiian Ridge. *Journal of Volcanology and Geothermal Research*, **94**, 17465–17485, https://doi.org/10.1029/JB094iB12p17465
Murray, J.B. and Voight, B. 1996. Slope stability and eruption prediction on the eastern flank of Mount Etna. *Geological Society, London, Special Publications*, **110**, 111–114, https://doi.org/10.1144/GSL.SP.1996.110.01.08
Paris, R., Geichetti, T., Chevalier, J., Guillou, H. and Frank, N. 2011. Tsunami deposits in Santiago Island (Cape Verde archipelago) as possible evidence of a massive flank failure of Fogos volcano. *Sedimentary Geology*, **239**, 129–145, https://doi.org/10.1016/j.sedgeo.2011.06.006
Paris, R., Ramalho, R.S. *et al.* 2018. Mega-tsunami conglomerates and flank collapses of ocean island volcanoes. *Marine Geology*, **395**, 168–187, https://doi.org/10.1016/j.margeo.2017.10.004
Ramalho, R., Helffrich, G., Cosca, M., Vance, D., Hoffmann, D. and Schmidt, D.N. 2010. Episodic swell growth inferred from variable uplift of the Cape Verde hotspot islands. *Nature Geoscience*, **3**, 774–777, https://doi.org/10.1038/NGEO982
Ramalho, R.S., Winckler, G. *et al.* 2015. Hazard potential of volcanic flank collapses raised by new megatsunami evidence. *Scientific Advances*, **1**, e1500456, https://doi.org/10.1126/sciadv.1500456
Ribeiro, O. 1960. *A ilha do Fogo e as suas erupiones*. 2nd edn. Memorias, serie geografica I. Publ. Junta de Investigacoes do Ultramar, Ministerio do Ultramar, Lisbon.
Sammartini, M., Camerlenghi, A. *et al.* 2018. Open-slope, translational submarine landslide in a tectonically active volcanic continental margin (Licosa submarine landslide, southern Tyrrhenian Sea). *Geological Society, London, Special Publications*, **477**, 133–150, https://doi.org/10.1144/SP477.34
Siebert, L. 1984. Large volcanic debris avalanches: characteristics of source areas, deposits, and associated eruptions. *Journal of Volcanology and Geothermal Research*, **22**, 163–197, https://doi.org/10.1016/0377-0273(84)90002-7
Tibaldi, A. 2001. Multiple sector collapses at Stromboli volcano, Italy: how they work. *Bulletin of Volcanology*, **63**, 112–125, https://doi.org/10.1016/S1464-1909(99)00022-2
Tibaldi, A., Pasquarè, F.A., Papanikolaou, D. and Nomikou, P. 2008. Discovery of a huge sector collapse at the Nisyros volcano, Greece, by on-land and offshore geological-structural data. *Journal of Volcanology and Geothermal Research*, **177**, 485, https://doi.org/10.1016/j.jvolgeores.2008.06.014
Torres, P., Madeira, J., Silva, L., Brum da Silveira, A., Serralheiro, A. and Mota Gomes, A. 1997. *Carta geológica das erupções históricas da Ilha do Fogo: revisão e actualização: A erupcao vulcanica de 1995 na Ilha do Fogo, Cabo Verde*. Publ IICT, Lisbon, 119–132.
Vales, D., Dias, N.A. *et al.* 2014. Intraplate seismicity across the Cape Verde swell: a contribution from a temporary seismic network. *Tectonophysics*, **636**, 325–337, https://doi.org/10.1016/j.tecto.2014.09.014
Ward, S.N. 2001. Landslide tsunami. *Journal of Geophysical Research*, **106**, 11201–11215, https://doi.org/10.1029/2000JB900450
Watt, S.F.L., Talling, P.J. *et al.* 2012*a*. Combinations of volcanic-flank and seafloor-sediment failure offshore Montserrat, and their implications for tsunami generation. *Earth and Planetary Science Letters*, **319**, 228–240, https://doi.org/10.1016/j.epsl.2011.11.032
Watt, S.F.L., Talling, P.J. *et al.* 2012*b*. Widespread and progressive seafloor-sediment failure following volcanic debris avalanche emplacement: landslide dynamics and timing offshore Montserrat, Lesser Antilles. *Marine Geology*, **323–325**, 69–94, https://doi.org/10.1016/j.margeo.2012.08.002
Watt, S.F.L., Karstens, J. *et al.* 2019. From catastrophic collapse to multi-phase deposition: flow transformation, seafloor interaction and triggered eruption following a volcanic-island landslide. *Earth and Planetary Science Letters*, **517**, 135–147, https://doi.org/10.1016/j.epsl.2019.04.024
Watts, P., Imamura, F. and Grilli, S.T. 2000. Comparing model simulations of three benchmark tsunami generation cases. *Science of Tsunami Hazards*, **18**, 107–124.
Wiemer, G. and Kopf, A. 2016. On the role of volcanic ash deposits as preferential submarine slope failure planes. *Landslides*, **14**, https://doi.org/10.1007/s10346-016-0706-6
Wiemer, G., Moernaut, J. *et al.* 2015. The role of sediment composition and behaviour under dynamic loading conditions on slope failure initiation: a study of a subaqueous landslide in earthquake-prone South-Central Chile. *International Journal of Earth Sciences*, **104**, 1439–1457, https://doi.org/10.1007/s00531-015-1144-8

The sedimentology and tsunamigenic potential of the Byron submarine landslide off New South Wales, Australia

Kendall C. Mollison[1]*, Hannah E. Power[1], Samantha L. Clarke[2], Alan T. Baxter[3], Emily M. Lane[4] and Thomas C. T. Hubble[2]

[1]School of Environmental and Life Sciences, University of Newcastle, Newcastle, Australia

[2]School of Geosciences, Faculty of Science, University of Sydney, Sydney, Australia

[3]School of Environmental and Rural Science, University of New England, Armidale, Australia

[4]National Institute of Water and Atmospheric Research, Christchurch, New Zealand

KCM, 0000-0003-0545-1460; HEP, 0000-0003-4332-8607;
SLC, 0000-0003-4997-1580; ATB, 0000-0002-1278-6536; EML, 0000-0002-9261-6640
*Correspondence: Kendall.Mollison@uon.edu.au

Abstract: Extensive evidence for submarine landslide failure is found along the east Australian continental margin. This paper assesses the sedimentological properties and models the failure event that created the Byron landslide scar, located on the SE Australian continental margin, *c.* 34 km off the coast of Byron Bay, New South Wales. Sedimentological analyses and dating (radiocarbon and biostratigraphic) were conducted on three gravity cores collected from within the Byron landslide scar. A paraconformity, identified in one of the three cores by a distinct colour change, was found to represent a distinct radiocarbon age gap of at least 25 ka and probably represents the detachment surface of the slide plane. The core-derived sediment properties for the Byron landslide scar were used to inform hydrodynamic modelling using the freely available numerical modelling software, Basilisk. Model results highlight the important role of sediment rheology on the tsunamigenic potential of the slide and on the resulting inundation along the east Australian coastline, therefore providing a greater understanding of the modern hazard posed by comparable future submarine landslide events for the east Australian coastline.

Submarine landslides are a type of underwater gravity-driven mass movement. They are often identified through the scars they leave upslope and/or the deposits transferred to deeper water on both active and passive continental margins and on the flanks of ocean islands (McAdoo *et al.* 2000; Masson *et al.* 2006). Submarine landslides range in size from small slides <0.1 km^3 to gigantic slides of >2400 km^3, such as the Storegga slide (Haflidason *et al.* 2005). Evidence for submarine landslides along the SE Australian continental margin (SEACM) was first identified in GLORIA sidescan sonar data (Jenkins and Keene 1992). More recent studies on this margin have demonstrated that submarine landslide scars are common between Ulladulla, on the mid-south coast of New South Wales, and Fraser Island, Queensland (Glenn 2008; Boyd *et al.* 2010; Hubble *et al.* 2016), with submarine landslides now considered a common and ongoing characteristic of the SEACM (Keene *et al.* 2008; Boyd *et al.* 2010; Clarke *et al.* 2012, 2016; Hubble *et al.* 2012). The identified slides range in volume from <0.1 to 20 km^3. The extensive submarine landsliding is unexpected owing to the relatively low sediment accumulation rates along the SEACM, the absence of Pliocene or Pleistocene ice sheets, the inherently stable continental slope sediments and the relatively aseismic passive margin tectonic setting (Boyd *et al.* 2004; Hubble *et al.* 2012; Clarke 2014; Clarke *et al.* 2016).

Understanding submarine landslides is important as they are the second most frequent source of tsunami globally (Harbitz *et al.* 2013) and they have been shown to have potentially devastating impacts (Assier-Rzadkieaicz *et al.* 2000; Tappin *et al.* 2001; Masson *et al.* 2006; Harbitz *et al.* 2013). While there is consensus on the hazard posed by submarine landslide events and the tsunami they may generate, their triggering mechanisms and the factors contributing to their tsunamigenic potential are less well understood. Of the submarine landslide scars along the SEACM, approximately 36 upper and mid-slope slides have been identified as having been tsunamigenic during the Pliocene and Quaternary (Clarke *et al.* 2019). The largest of these tsunamigenic slides is the Bulli slide, which is 10 km wide,

From: Georgiopoulou, A., Amy, L. A., Benetti, S., Chaytor, J. D., Clare, M. A., Gamboa, D., Haughton, P. D. W., Moernaut, J. and Mountjoy, J. J. (eds) 2020. *Subaqueous Mass Movements and their Consequences: Advances in Process Understanding, Monitoring and Hazard Assessments.* Geological Society, London, Special Publications, **500**, 27–40.
First published online April 7, 2020, https://doi.org/10.1144/SP500-2019-160

13 km long, approximately 20 km^3 in volume, and with a centre of mass located in 1700 m water depth (Clarke 2014; Clarke *et al.* 2019).

Tsunami generation from a submarine landslide source is dependent on the landslide kinematics, which depend on the material properties of the landslide (Kim *et al.* 2019). These material properties are determined by factors such as the depositional environment and history, and the sediment composition, which in turn determine the landslide rheology during failure (Jeong 2010; Kelfoun *et al.* 2010; Grilli *et al.* 2017; Locat 2019). Continental slope sediments examined from the SEACM are composed primarily of calcareous sandy-silts and, while they are inherently stable, they have been shown to be highly deformable (Clarke 2014; Fletcher 2015; Yu 2017). Because submarine landslides are always deformable in real cases, it has been recommended that landslide deformation be incorporated when modelling submarine landslide-generated tsunami by taking into account properties of the slide (Grilli and Watts 2005; Locat 2019). Traditionally, submarine landslide-generated tsunami have been modelled with a simplified block modelling approach (Fine *et al.* 2005; Grilli and Watts 2005; Løvholt *et al.* 2005). While this approach has been successful in replicating some submarine landslide-generated tsunami (Tappin *et al.* 2008), it does not account for the complexities of landslide deformation and flow determined by the landslide properties. Modelling a submarine landslide as a fluid, with a specified density, allows the deformation and flow of the slide to be represented during modelling (Smith *et al.* 2016).

The east Australian coastline is thought to have a low to moderate submarine landslide-generated tsunami hazard (Hubble *et al.* 2019), but the high population concentration and critical infrastructure within 50 km of the coastline (Short and Woodroffe 2009) make it particularly vulnerable, which results in a relatively high risk (Clarke *et al.* 2019). In this study, we investigate the sediment properties and tsunami hazard posed by the failure of one of the SEACM submarine landslides, the Byron slide, to understand the tsunami hazard it poses to the adjacent coastline.

Study area and geological setting

The SEACM is a narrow, deep and sediment-deficient passive margin when compared with continental margins globally (Boyd *et al.* 2010), with sedimentary deposits generally less than 1 km thick and often less than 500 m thick (Keene *et al.* 2008; Boyd *et al.* 2010). The continental slope here is particularly steep (5–10°) and narrow, extending from the shelf break at a water depth of 110 m to the Tasman Abyssal Plain and contrasting with typical Atlantic-style passive margins where slopes are typically <5° (McAdoo *et al.* 2000). The shelf in the Byron Bay region is only 30 km wide such that the slope and site of failures are relatively close to the modern coastline.

The study area is offshore the coastal town of Byron Bay in northern New South Wales (Fig. 1). The Byron slide is 34 km offshore and is located on the upper continental slope, centred at a water depth of 800 m. It is thought to be a translational slab failure, based on its bathymetric expression and comparison with other similar and better

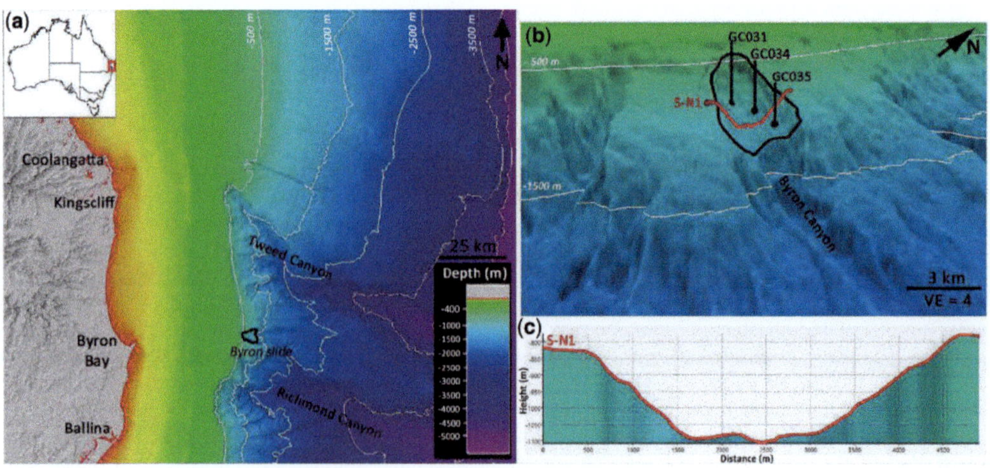

Fig. 1. Maps of (**a**) the Byron Bay region of study, including its location within Australia, and (**b**) the continental slope in which the Byron slide scar is located, highlighting the position from which gravity cores GC031, GC034 and GC035 were collected and the profile line S–N1. (**c**) A cross-section of depths along the line S–N1 as shown in (**b**). VE is the vertical exaggeration of the bathymetry.

Table 1. *Summary of water depth (m), core recovery length (m), and location/target feature of the sediment cores retrieved from the Byron Bay region of study*

Core name	Latitude (°S)	Longitude (°E)	Water depth (m)	Length (m)	Location and target feature
GC031	28°37.468′	153°56.817	893.2	1.8	Within Byron slide scar on the upper slope
GC034	28°36.894′	153°57.206	977.2	3.4	Within Byron slide scar on the upper slope
GC035	28°36.686′	153°57.887	1072.6	1.77	Within Byron slide scar on the upper slope

characterized slides identified along the SEACM (Clarke *et al.* 2016; Hubble *et al.* 2019). The slide scar is 3.6 km wide and 4 km long with a distinct U-shaped cross-section (Fig. 1c). The slide has a maximum thickness of 222 m and an average thickness of 76 m. The Byron slide was identified by Clarke *et al.* (2019) as being of a size and depth sufficient to generate a tsunami with a flow depth ≥5 m at the coastline. The slide is located on the upper slope of the Byron canyon, which has an average canyon width of 3 km and a canyon length of 20 km. The Byron canyon is a blind canyon and presents no dendritic canyon head to indicate top-down incision; rather the morphology of the canyon suggests retrogressive failure driven by toe erosion (Yu 2017). Despite this erosion, continental slope sediments from the Byron region have been shown to be inherently stable (Clarke 2014; Yu 2017). Three gravity cores (GC) have been collected from within the Byron slide scar (GC031, GC034 and GC035) (Fig. 1b; Table 1).

Methods

Bathymetric and sedimentological data were used in this study to define the Byron slide, its pre-slide bathymetry, its sediment characteristics and its likely age. This information, together with topographic data, were utilized in the hydrodynamic modelling of the Byron slide in order to assess the tsunami hazard posed for the adjacent coastline.

Bathymetry

All available multibeam bathymetry data for the Byron region were collated into a digital bathymetric model for analysis in this study using Fledermaus V7.8.6b and ArcGIS 10.5.1 (Fig. 2). Higher-resolution datasets covered areas of the continental slope (50 m resolution) and the nearshore region of Byron Bay (5 m resolution; Fig. 2). Approximately 35 000 km^2 of bathymetric data were collected by the RV *Southern Surveyor* on three voyages in 2006 (Glenn 2008), 2012 (Boyd 2009) and 2013 (Hubble 2013), which were processed into a 50 m gridded digital bathymetric model and utilized in this study. The 5 m nearshore dataset, sourced from the New South Wales Department of Planning, Industry, and Environment, was collected in 2012. Submarine landslide morphology is described according to the definitions and nomenclature in Clare *et al.* (2019), updated from McAdoo *et al.* (2000). The compiled bathymetry for the Byron region, in addition to the 5 m resolution topographic data also sourced from the New South Wales Department of Planning, Industry, and Environment, was used for hydrodynamic modelling of the Byron slide and resulting tsunami (see 'Basilisk model and setup' section). A background 100 m resolution dataset was used in regions where no higher-resolution data were available (Fig. 2). These 100 m resolution background data were sourced from Beaman (2010).

The pre-failure bathymetry for the Byron slide was estimated by clipping out the slide area from the 50 m elevation dataset in ArcGIS 10.5.1 and then interpolating a new hydraulically correct surface over the slide area. The difference in elevation between the pre-slide bathymetry and current post-failure bathymetry was then calculated to create a thickness file (Fig. 2). This file was used to calculate the maximum and average thickness of the Byron slide as well as the total slide volume. The thickness file was input into Basilisk to allow for accurate modelling of the Byron slide failure, from the estimated pre-slide bathymetry to the current post-failure bathymetry (see 'Basilisk model and setup' section for more information).

Sedimentology

The sediment cores examined in this study were collected using a gravity corer. The gravity cores (GC031, GC034 and GC035) were cut into 1 m lengths, split longitudinally, photographed, logged and subsampled at various intervals down core. Additional samples were taken at points of interest above and below unconformities. Grain size samples were collected at 20 cm intervals and were determined using a Malvern Mastersizer 2000 laser particle analyser and statistically analysed using

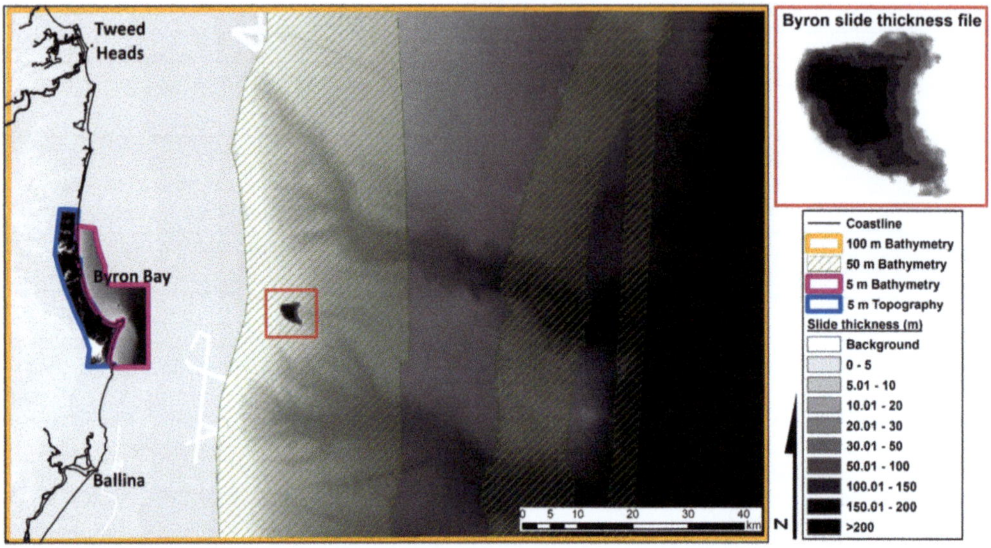

Fig. 2. The model domain showing the location of the Byron slide scar. The elevation data for the Byron model domain were derived from four datasets: a 5 m nearshore; a 5 m onshore; a 50 m continental slope; and a 100 m background resolution dataset. See 'Methods' section for further detail.

GRADISTAT 8.0 (Blott and Pye 2001). Prior to geotechnical testing, brass ring cutters (*c.* 13 cm^3) were numbered, weighed and measured. These brass ring cutters were pushed into the centre of the cores at 30 cm intervals to extract a sediment sample. The sediment sample and brass ring were then weighed to calculate wet weight and then reweighed after 24 h of drying in a 60°C oven to calculated dry weight. From this, bulk density, unit weight and moisture content were calculated. The relative density (RD) of the sediment, sometimes called specific gravity, is the ratio of sediment density to the density of ocean water, and was calculated by dividing the bulk density of sediment samples by an ocean water density of 1028 kg m^{-3}. Total organic carbon (TOC) and total inorganic carbon (TIC) were calculated using the sequential loss on ignition method.

Sediment dating

Radiocarbon dates were obtained by CHRONO Centre (Queen's University, Belfast, UK) from picked foraminifera assemblages for six samples. The *Globigerinoides ruber* foraminifera species was chosen for dating as it is common along the SEACM and has been used for ^{14}C dating in previous sedimentological studies (Troedson and Davies 2001; Clarke *et al.* 2016; Mollison *et al.* 2020). Conventional ^{14}C years BP were converted into calibrated calendar ages following Stuiver and Reimer (1993). Median calibrated ages (cal BP) were calculated with CALIB V7.0.4 (Stuiver and Reimer 1993) using the marine calibration curve Marine13 dataset (Reimer *et al.* 2013) with a reservoir correction (ΔR) value of 25 $\pm$ 8 years (the average for eastern Australian data offshore Stradbroke Island; Ulm *et al.* 2009), reported with 2σ errors (Table 2).

Biostratigraphic dating was completed using calcareous nanofossils for six sediment samples (Table 3). Smear slides were made using standard techniques and samples were examined using a light microscope with an oil immersion lens at a magnification of 1250× under brightfield and cross-polarized light. Nanofossils were identified to species level where possible. Nanofossil age data were evaluated using the standard nanofossil zonations of Martini (1971) and checked against Nannotax, an online database of nanofossil species age ranges (Young *et al.* 2017; http://ina.tmsoc.org/Nannotax3). The zonal scheme of Martini (1971) was selected for the biozones and correlated to the geological timescale of Gradstein (2012).

Basilisk model and setup

Hydrodynamic modelling of the Byron slide failure and resulting tsunami was conducted using the Basilisk model for the submarine landslide tsunami source, tsunami wave propagation and onshore inundation. Basilisk builds upon the Gerris flow solver detailed by Popinet (2011) and has been used extensively for tsunami modelling (Popinet 2015; Orpin *et al.* 2016). This model utilizes a fully coupled two-layer scheme, with the submarine landslide modelled

Table 2. *Radiocarbon ages of six core samples from cores GC031, GC034 and GC035*

Core ID	Sample ID	Depth (cm)	Conventional ^{14}C Age (years BP)	^{14}C error (1σ) (years)	2σ calibrated age range (years BP)	Median calibrated ^{14}C age (years BP)	^{14}C error (2σ) (years)	Additional information
GC031	GC031/2A/ 93–94.5 cm	94	9244	±42	9890–10 182	10 052	±49	No feature
GC034	GC034/4A/ 79–80 cm	79.5	14 730	±62	17 193–17 649	17 445	±67	No feature
GC035	GC035/2A/ 73–74 cm	73.5	7710	±42	8030–8204	8167	±49	No feature
	GC035/1B/ 7–8 cm	86.5	21 400	±120	25 007–25 646	25 339	±123	Above paraconformity
	GC035/1B/ 10.5–11.5 cm	90	>51 560	–	–	>50 000	–	Below paraconformity
	GC035/1B/ 96–97 cm	175.5	49 851	±3951	–	>50 000	–	No feature

as a dense Newtonian fluid of a specified density that fails under gravity, whose motion is coupled with that of the water in the upper layer. The RD values incorporated into each model scenario were based on the range of RDs recorded for continental slope sediments along the SEACM in this study as well as previous studies (Fletcher 2015; Clarke *et al.* 2016; Hubble *et al.* 2019; Mollison *et al.* 2020; Table 4). The Byron slide failure was modelled from the pre-slide bathymetry to the current post-failure bathymetry using the slide thickness file (see 'Bathymetry' section). From this, a total volume of 1.3 km^3 of sediment was modelled to fail during each simulation. Friction is applied to the model between the slide and the substrate, between the overlying water and substrate, and between the slide and the overlying water. The Saint-Venant (or non-linear shallow water) equations used for numerical tsunami simulations have been integrated into Basilisk and utilized in the inundation modelling to accurately model the wetting and drying of the coast during tsunami inundation.

Basilisk generates an adaptive grid for time-dependent tsunami simulations, which refines and coarsens to select the appropriate grid resolution to capture the tsunami wave evolution. This dynamically adaptive mesh refinement can lead to orders of magnitude gains in computational efficiency when compared with a non-adaptive grid (Popinet 2011). Despite having an elevation dataset that includes up to 5 m resolution in coastal regions, simulations were run with a maximum resolution of 30 m for the model scenarios in this study owing to time constraints on computation.

Results

Core sedimentology and ages

The sediments recovered in GC031, GC034 and GC035 primarily comprise clay-bearing sandy-silt, with a dominant grain size distribution of 3.9–62.5 µm (Fig. 3). The GCs all present uniform, weakly layered, hemipelagic muds with sediment compositions and structures consistent with other GCs collected on the SEACM (Troedson and Davies 2001; Clarke *et al.* 2012; Fletcher 2015; Hubble *et al.* 2019), including another GC from within the Byron slide scar (core GC012 in Clarke *et al.* 2016). Images of the full cores, annotated with radiocarbon (see Table 2) and biostratigraphic (see Table 3) dates, are presented in Figure 4.

Table 3. *Nanofossil biostratigraphy ages of six core samples taken from GC031, GC034 and GC035*

Core	Section	Depth (cm)	Age (Ma BP)	Biomarkers used
GC031	2A	93–94.5	0.44–1.93	Presence of *Pseudoemiliania lacunosa*. Absence of *Discoaster* spp.
	1B	187–189	0.44–1.93	Presence of *P. lacunosa*. Absence of *Discoaster* spp.
GC034	4A	79–80	0.44–1.93	Presence of *P. lacunosa*. Absence of *Discoaster* spp.
	2C	205.5–206.5	0.44–1.93	Presence of *P. lacunosa*. Absence of *Discoaster* spp.
GC035	2A	73–74	>0.44	Presence of *P. lacunosa*
	1B	175–177	0.44–1.93	Presence of *P. lacunosa*. Absence of *Discoaster* spp.

Table 4. *Summary of tsunami modelling results for five different relative density (RD) model scenarios conducted for the Byron slide.*

Model scenario	RD1.4	RD1.6	RD1.8	RD2	RD2.2
Maximum wave amplitude at Cape Byron (m)	7.7	8.6	9.3	*11.3*	10.8
Maximum water velocity at coast (m)	19	19	27.1	27.9	*28*
Maximum wave height 500 m landward of headscarp (m)	9.5	11.7	12.7	13.7	*14.5*
Arrival time (min)	17.6	17.4	17.4	17.2	*17*
Leading negative wave speed (m s^{-1})	28.4	28.7	28.7	29.1	*29.4*
Average velocity of slide centre of mass (m s^{-1})	8.54	9.15	9.83	10.25	*11.16*
Average velocity of slide frontal mass (m s^{-1})	15.97	16.97	17.62	18.20	*18.38*
Runout distance of slide centre of mass (km)	17.3	18.6	18.5	18.9	*18.9*
Runout distance of slide frontal mass (km)	28.8	30.5	31.7	32.8	*33.1*
Total onshore inundation area (km^2)	8.0	13.1	12.2	12.9	*14.9*

Italicized values show the most severe of the model scenarios for each measure of tsuanmi impact.

GC031 is 1.8 m in length (Table 1) and was collected from within the Byron slide scar, closer to the headscarp than GC034 and GC035. GC031 comprises a uniform sandy-silt matrix from 0 to 159 cm depth that grades in colour from greyish-olive (5Y 5/3) to dark olive (5Y 4/3). A 2 cm-thick loose, medium-grained silty-sand lens is present from 159 to 161 cm depth and is reflected by an increase in grain size and change in grain size distribution to 72.5% sand volume (Fig. 3). The sediment matrix below the sand lens is greyish-olive (5Y 4/2) throughout, with some mottling of greyish-olive (5Y 6/2) from 165 cm depth to the base of the core. An overall increase in both bulk density and unit weight is seen down core, from 790 to 1106 kg m^{-3} and from 7.8 to 10.8 kN m^{-3}, respectively. In contrast, a gradual decrease in moisture content is seen down core from 99 to 53%. The TOC and TIC values for these sediments average 10.7 and 18.7% respectively. The RD of the sampled sediments ranges from 1.4 to 1.6 (Fig. 3). A sediment sample taken from 94 cm depth returned an age of 10.1 $\pm$ 42 cal ka BP (Fig. 4).

GC034 is 3.4 m in length (Table 1) and is the longest of the three cores analysed; however, only two of the four core sections (4A and 2C) were available for analysis in this study (Fig. 2) as sections 3B and 1D were claimed for geotechnical testing which will be reported elsewhere. The top 2 cm of section 4A of GC034 is a yellowish-brown (2.5Y 5/4) clay-bearing silt, which then grades into a greyish-olive (5Y 6/2) clay-bearing sandy-silt from 2 to 5 cm depth, which then grades in colour to greyish-olive (5Y 4/2) from 5 to 82 cm depth. Mottling of greyish-olive (5Y 6/2) is observed from 18 to 30 cm depth and speckles of light grey (5Y 7/1) are observed from 50 to 82 cm depth. Mean grain size increases with depth in section 4A, with an increase of sand volume from 5.5 to 33.5%. A sediment sample taken from 79.5 cm depth returned an age of 17.4 $\pm$ 67 cal ka BP (Fig. 4). Section 2C of GC034 is characterized by a greyish-olive (5Y 5/2) clay-bearing sandy-silt sediment matrix that is speckled with olive black (5Y 3/1) and dark olive (5Y 4/3) and grades from silt (121–141 cm depth) to a sandy-silt for the remainder of section 2C An overall increase in both bulk density and unit weight is seen down core, from 658 to 1028 kg m^{-3} and from 6.45 to 10.08 kN m^{-3}, respectively, from section 4A to 2C. In contrast, a gradual decrease in moisture content is seen down core from 107 to 79% from section 4A to 2C. The TOC and TIC values for these sediments average 9.5 and 19.7% respectively. These overall trends are inferred for the missing core sections based on the trends seen in available core sections and the general consistency of slope sediment previously reported for the margin (Fletcher 2015; Clarke *et al.* 2016; Hubble *et al.* 2019). However, this cannot be claimed with certainty. The RD of the sampled sediments range from 1.4 to 1.8 (Fig. 3).

GC035 is 1.77 m in length (Table 1) and has evidence of a paraconformity at *c.* 88 cm depth (Fig. 4). The feature is visually identified by a distinct colour change across the surface from dark olive (5Y 4/3) to greyish-olive (5Y 5/2). There is a step increase in bulk density and unit weight and a decrease in moisture content across the surface (Fig. 3). Radiocarbon dates across the feature show that sediments directly overlying the feature return an age of 25.3 $\pm$ 120 cal ka BP and sediments directly underlying the feature are radiocarbon dead (Fig. 4), indicating a depositional hiatus of at least 25 ka. The sediment matrix overlying this paraconformity is primarily dark olive (5Y 4/3) in colour, with some mottling of light grey (5Y 7/2) from 70 to 77 cm depth. Some dark olive (5Y 4/3) lamination is present within the greyish-olive (5Y 5/3) matrix from 77 to 87 cm followed by the distinct horizontal paraconformity which delineates a stiffer, primarily

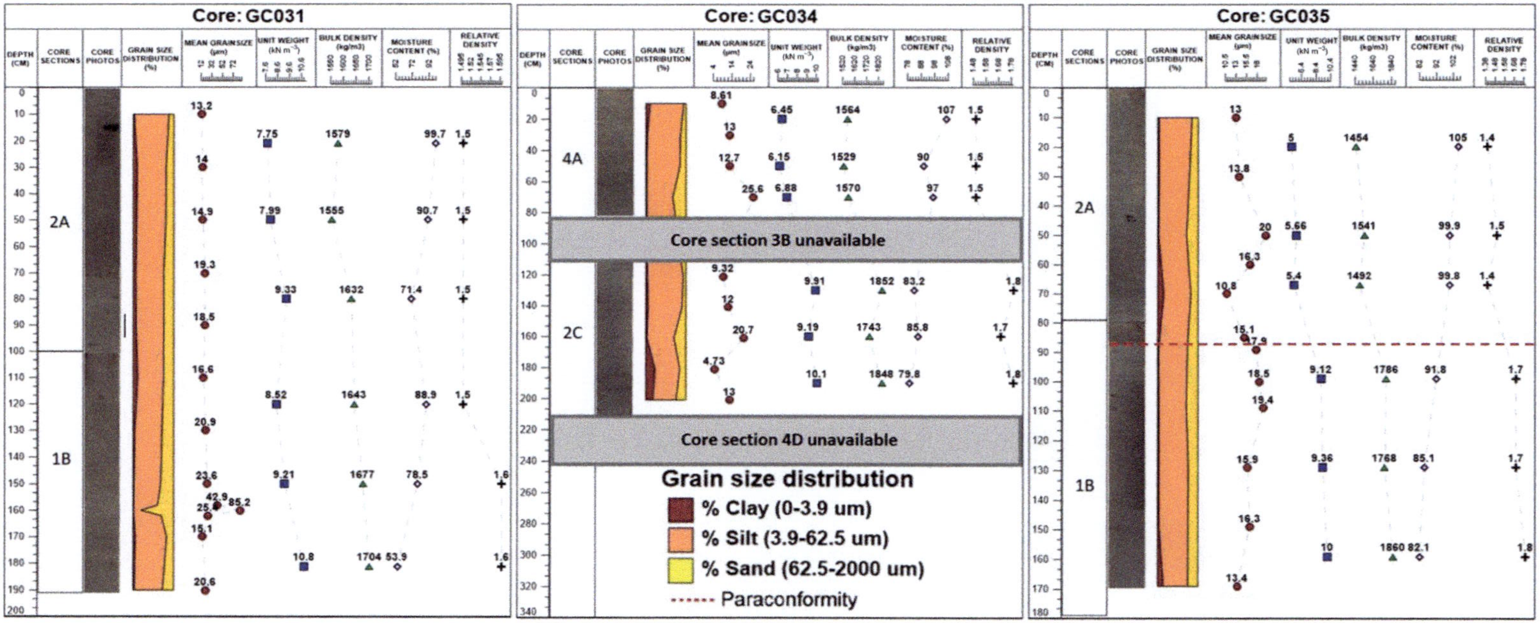

Fig. 3. Core logs of GC031, GC034 and GC035 collected from within the Byron slide scar. The core logs show penetration depth (cm), core sections, full core photos, grain size distribution (%), mean grain size (µm), unit weight (kN m^{-3}), bulk density (kg m^{-3}), moisture content (%) and relative density. The location of the paraconformity in GC035 is also indicated by the red dashed line.

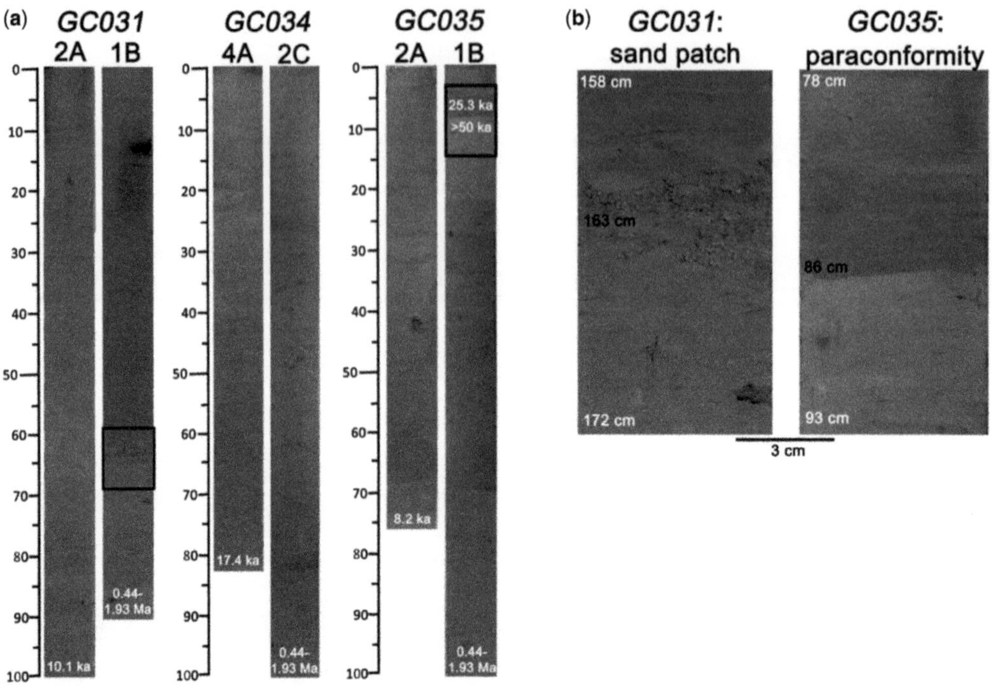

Fig. 4. (**a**) Full core images of the three within-Byron slide scar gravity cores GC031, GC034 and GC035, with radiocarbon and biostratigraphic dates shown (white text). (**b**) High-resolution images of the sand patch at *c.* 163 cm depth in GC031 and the paraconformity at *c.* 86 cm depth in GC035.

greyish-olive (5Y 4/2) matrix below. The grain size distribution of the sandy-silt matrix is uniform down core and no significant change in grain size or grain size distribution is observed across the paraconformity (Fig. 3). Similar to GC031 and GC034, an overall increase in both bulk density and unit weight is seen down core, from 510 to 1021 kg m^{-3} and from 5 to 10.1 kN m^{-3}, respectively, as well as a decrease in moisture content from 105 to 82%. The TOC and TIC values for these sediments average 10.4 and 19.2% respectively. The RD of the sampled sediments ranges from 1.4 to 1.8 (Fig. 3).

The nanofossil contents of the six samples from cores GC031, GC034 and GC035 imply they are between 0.44 and 1.93 myr old as they contain *Pseudoemiliania lacunosa,* while *Discoaster* spp. are absent (Table 3). The nanofossil dates (Table 3) thus imply older ages than the radiocarbon dates (Table 2) for the same sampled intervals. This suggests that sediments may include a reworked component having been remobilized from elsewhere and deposited in the current location within the Byron slide scar. The limited radiocarbon dates obtained for GC031, GC034 and GC035 did not enable reliable rates of sediment accumulation to be determined.

Hydrodynamic modelling of the Byron slide

Numerical modelling of the Byron slide was conducted to understand how changes to the sedimentology of the slide affect its tsunamigenic potential and to assess the hazard that a slide of this size would pose for the modern east Australian coastline. The parameter incorporated into the models to assess the impact of sedimentological change was RD. RD is used within numerical submarine landslide modelling as density of the landslide is a key parameter in controlling landslide motion (Grilli and Watts 1999; Kelfoun *et al.* 2010; Smith *et al.* 2016).

Five different scenarios modelling Byron slide failure were conducted for the region shown in Figure 2, incorporating RD values ranging from 1.4 to 2.2 (see ‘Methods’ section for more detail on RD). The model domain includes the continental slope on which the Byron slide is located, part of the Tasman Basin offshore of the slide and *c.* 100 km of the adjacent coastline north and south of the slide.

The results clearly demonstrate that the RD of slide sediments directly impacts tsunamigenic potential, with models with larger slide RDs generating more extreme tsunami events (Table 4). Changes in RD also affected the runout of the slide, increasing the average speed of the frontal mass of the slide from 15.97 to 18.38 m s^{-1} with increasing RD (Table 4). The average speed of the centre of mass of the slide also increased from 8.54 to 11.16 m s^{-1} as the RD increased from 1.4 to 2.2, with a subsequent increase in the runout distance of the centre of mass from 17.7 to 19 km after 1 h simulation time (Table 4). Initial acceleration of the slide centre of mass also increased from 0.22 to 0.3 m s^{-2} as the RD increased from 1.4 to 2.2. As a result, a 53% increase in the initial height of the leading negative tsunami wave, from 9.5 to 14.5 m, was observed 500 m landwards of the headscarp with an increase in RD from 1.4 to 2.2 (Table 4). Similarly, increases of 21, 31 and 4% were observed for the maximum wave amplitude along the coastline (m), maximum water velocity at the coastline (m s^{-1}) and average speed of the negative-leading wave (m s^{-1}), respectively, with increases in RD from 1.4 to 2.2. A slight decrease in arrival time was observed at Cape Byron for RD1.4 when compared with RD2.2, from 17.6 to 17 min (Table 4). An 85% increase in inundation area (km^2) for the adjacent coastline was observed with an increase in RD from 1.4 to 2.2 (Table 4).

The most extreme onshore inundation was observed in model scenario RD2.2 (Fig. 5). This model scenario resulted in a 14.5 m wave 500 m landward of the slide headscarp which attenuated at a rate of 1.7 m for every 10 km propagated towards the coastline, such that a 9.3 m wave height was modelled at Cape Byron 17 min after failure initialization. Average velocities of 11.16 and 18.38 m s^{-1} were recorded for the slide centre of mass and the slide frontal mass of the slide, respectively (Table 4). The slide frontal mass had a runout distance 14.1 km longer than the slide centre of mass.

Maximum water velocities of 28 m s^{-1} were recorded at the coastline for this model scenario. These velocities are an order of magnitude greater than those which have been demonstrated to cause significant damage to maritime and coastal infrastructure (Uslu *et al.* 2010). A total onshore inundation area of 14.9 km^2 was recorded for this scenario, with estuaries identified as the coastal areas most vulnerable to inland tsunami penetration. Of the total onshore inundation, 50% occurred within 200 m of a coastal estuary, despite estuaries and their immediate surrounds representing only 31% of the coastal plain. The maximum observed inundation distance of 2.4 km was seen at Mooball Creek (Fig. 5).

Submarine canyons have been generally shown within the literature to reduce both tsunami wave heights and arrival times at the coastline adjacent to the canyon head, while slightly increasing wave heights at the coastline adjacent to either side of the canyon head (Aranguiz and Shibayama 2013; Iglesias *et al.* 2014; Seelam *et al.* 2016). Maximum wave amplitude at the coastline 5 km south of the Tweed Canyon head is 5.01–7.2 m compared with 1.01–3 m directly adjacent to the canyon head (Fig. 5). Although no clear patterns in maximum wave amplitude are seen along the coastline adjacent to the Richmond Canyon, arrival times were slightly affected, with the arrival time either side of the canyon head up to 2 s faster than that at the coastline adjacent to the Richmond canyon head (Fig. 5). These observations appear to be consistent with current understanding of the role submarine canyons play in tsunami impacts for coastal regions.

Discussion

The results of the sedimentological analyses presented here show that the sediments collected from the continental slope within the Byron slide scar primarily comprise weakly layered hemipelagic muds of Pleistocene age, similar to continental slope sediments analysed in previous studies of the SEACM (Boyd *et al.* 2004; Fletcher 2015; Clarke *et al.* 2016; Yang 2018; Hubble *et al.* 2019; Mollison *et al.* 2020). The average sediment matrix of each core is a clay-bearing sandy-silt, with the 2 cm-thick medium-grained silty-sand lens present in GC031 a notable exception. This sand lens is interpreted to be a grain flow deposit originating from continental shelf sands, similar deposits of which are commonly seen in cores collected from the Fraser Island region of the SEACM, offshore of Queensland (Boyd *et al.* 2008; Fletcher 2015; Mollison *et al.* 2020), but are less common in cores analysed from further south on the SEACM. This may be due to a reduced sediment supply from the continental shelf in this part of the SEACM.

Significance of the Byron slide paraconformity

Abundant foraminifera and nanofossils place constraints on the age of sediment draping the slide scar and timing of the paraconformity observed in GC035. This paraconformity resembles other surfaces observed in sediment cores collected from within submarine landslide features along the SEACM (Boyd *et al.* 2010; Clarke *et al.* 2016; Hubble *et al.* 2019; Mollison *et al.* 2020) and it is interpreted to represent a slide plane. The new ages indicate that a gap of at least 25 kyr is present across the slide plane surface. This is consistent with the age gap identified across another slide plane identified within the Byron slide by Clarke *et al.* (2016); however, the new data imply that the minimum age of the slide is 9.5 cal ka BP older than that reported in the earlier study. This may partly reflect the closer

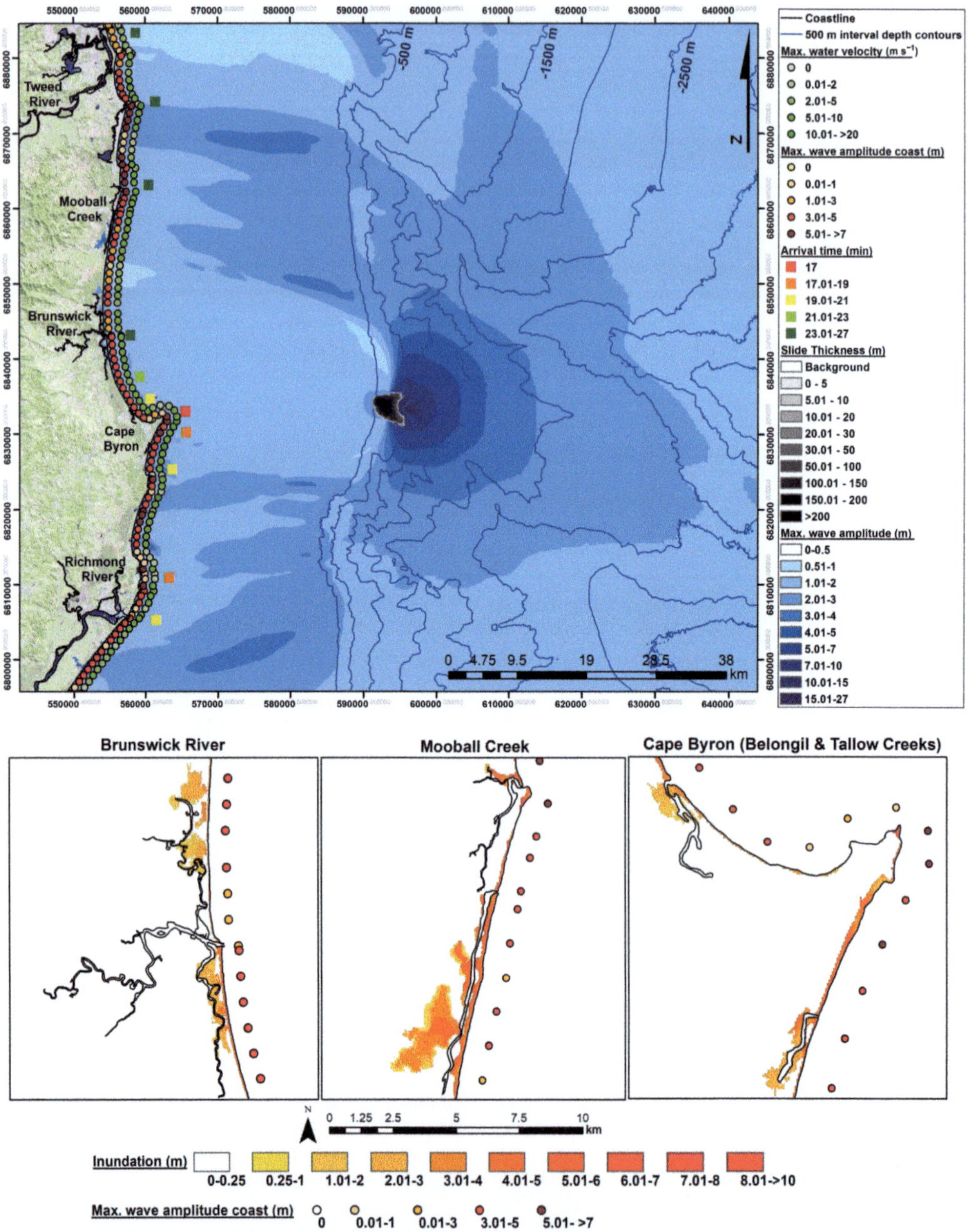

Fig. 5. Model scenario RD2.2 for the Byron slide failure. The top panel shows the maximum water velocity (m s^{-1}) along the coastline, maximum wave amplitude (m) along the coastline, arrival time of the tsunami wave at the coastline (min), the location and thickness (m) of the Byron slide, and the overall maximum wave amplitude (m) across the Byron model domain after 30 min simulation time. The bottom panels show onshore inundation depths (m) and maximum wave amplitudes (m) at the coastline at the four most exposed estuarine systems in the Byron region: Brunswick River, Mooball Creek, Belongil Creek and Tallow Creek.

proximity of the sample taken for dating in GC035 to the slide plane surface (2 cm closer) compared with that discussed in Clarke *et al.* (2016). The minimum age for the slide plane surface in this study therefore provides a conservative estimate of timing of the slide, which was a minimum of 25 ka ago.

Tsunamigenic potential of the Byron slide and the tsunami hazard for the east Australian coastline

The tsunami record for Australia identifies 47 tsunami since the historical period began in AD 1858 (Dominey-Howes 2007; Goff and Chagué-Goff 2014). Of these 47 events, 28.1% were generated by unknown causes. Submarine landslides provide a likely source for these recorded tsunami as Clarke *et al.* (2019) identified 260 distinct submarine landslide scars along the SEACM, 36 of which were shown, through empirical modelling, to have potential to generate a tsunami with a flow depth ≥5 m at the coastline. This is further confirmed by Talukder and Völker (2014), whose empirical modelling indicates the potential for submarine landslides to be a source for local tsunami generation along the New South Wales coastline. Previous studies on submarine landslide-generated tsunami have indicated that the tsunamigenic potential of a submarine landslide event is directly linked to the sediment properties of the landslide (Grilli *et al.* 2017). This study is the first to investigate the tsunami hazard posed by the Byron slide for the adjacent coastline by incorporating a range of known RD values for SEACM continental slope sediments into the properties of the slide.

The results of the modelling conducted here show that the overall tsunami hazard generated by submarine landslide failures increases with increasing RD values if all else is equal, with model scenario RD2.2 resulting in the greatest onshore inundation distance and area of inundation (Fig. 5). Clarke *et al.* (2019) empirically modelled the failure of the Byron slide for four different maximum slide velocities. Their preferred maximum slide velocity of 20 m s^{-1}, on which their hazard analysis was based, most closely represents the slide velocities obtained in this study and generated a 4.3 m flow depth at the coastline. This is significantly less than the average coastal flow depth of 9.5 m modelled in this study. Furthermore, the flow depth estimated by Clarke *et al.* (2019) is significantly less than the maximum wave amplitude at the coastline of 11.3 m as modelled in this study for the same event (see Table 4; model scenario RD2). Wave amplitude values modelled along the coastline in this study range from 0.01 to 11.3 m, highlighting the spatial variability of impact from a submarine landslide-generated tsunami along a coastline, owing to the interaction of complex factors such as wave attenuation, changes in local bathymetry and variations in onshore topography. This demonstrates the importance of high-resolution numerical modelling, the need for high-resolution bathymetry and topography, and the importance of accurately replicating the submarine landslide-generated tsunami hazard for site specific locations along the coastline.

Average RD values for SEACM continental slope sediments range from 1.4 to 2.2, as modelled in this study, but the range of RD values within each gravity core was shown to vary (Fig. 3). The average RD value used when modelling a submarine landslide was found to be subject to an error margin of ± 0.2. Overall, an increase in RD was shown to increase both wave height 500 m landward of the slide headscarp as well as maximum wave amplitude at the coastline; however, this trend is not linear. A 20% increase in RD leads to a 12% increase in maximum wave height 500 m landward of the headscarp. In contrast, a 20% increase in RD leads to a 14% increase in maximum wave amplitude at the coastline. The sensitivity of tsunami parameters to changing RD values will be investigated in detail in a subsequent publication.

The absence of obvious slide deposits downslope from the evacuated Byron slide scar places limits on the submarine landslide modelling, as runout deposits are useful in the calibration of sedimentary parameters (Kim *et al.* 2019). Future modelling of submarine landslide generated tsunami along the SEACM would benefit from calibrating the modelled runout distances to down-slope slide deposits to verify the most appropriate sediment characteristics for model inputs. Greater multibeam coverage of the SEACM is required for slide deposits to be identified in order to achieve a better calibration of the models. Another consideration would be to further refine the maximum resolution of the model grid in order to provide greater detail of the coastal inundation.

Triggering of landslide events

The inherently stable nature of continental slope sediments along the SEACM (Clarke 2014; Yu 2017) suggests that external triggering, either in the short term or long term, is responsible for the extensive landsliding on this margin. The ongoing erosion and destabilization of the SEACM, through erosion of the toe of the continental slope through canyon incision, preconditions the lower and middle slopes to fail owing to a large external triggering event such as an earthquake (Boyd *et al.* 2010; Hubble *et al.* 2012). The recurrence interval of tsunamigenic submarine landslide events along the SEACM is estimated to be between 1 in 9000 years and 1 in 60 000 years (Clarke *et al.* 2019) while the recurrence interval of magnitude ≥6 earthquakes for the Australian continent is *c.* 1 in 8 years, the majority of which are focused around the coastline (Geoscience Australia 2019). It is therefore plausible that an earthquake triggered the Byron slide event. Sand layers, such as the 2 cm grain flow observed in GC031, may also contribute to the destabilization of overlying sediment through the creation of weak layers; however, Mollison *et al.* (2020) suggest that because they are rarely observed in sediment cores along the SEACM, despite the abundance of submarine landslide scars, they are unlikely to play an important role in the extensive submarine landslide events on this margin.

Conclusions

We have investigated a submarine landslide off the coast of Byron Bay, New South Wales, Australia, to understand its sedimentology, tsunamigenic potential and the resultant tsunami hazard for the adjacent east Australian coastline. With a volume of 1.3 km^3, the Byron slide size is relatively large when compared with other tsunamigenic upper- and middle-slope submarine landslides identified along the SEACM. Continental slope sediments from within the Byron slide scar were found to be consistent with other studies of SEACM sedimentology. This slide is therefore representative of the tsunami hazard for the SEACM posed by submarine landslides. Sediments from the Byron slide contain weakly layered, hemipelagic Pleistocene muds, with a distinct slide plane surface identified in core GC035. Radiocarbon dating of this slide plane indicates that the Byron slide failure occurred a minimum of 25.3 cal ka BP ago. Numerical modelling of the slide failure and the resulting tsunami generation showed the direct effect of the relative density of slide sediments on the slide's tsunamigenic potential. As relative density increased, so did negative impacts of tsunami at the coastline. The most extreme inundation event modelled suggests the Byron slide had the potential to generate a 14.5 m wave directly shoreward of the slide headscarp, resulting in a 9.8 m amplitude wave reaching the adjacent coastline in 17 min. The failure of this submarine landslide had the potential to inundate 14.9 km^2 of land, with coastal estuaries highlighted as being particularly vulnerable to tsunami inundation. Submarine landslides are recognized as an ongoing process along the SEACM and with the possibility of earthquakes acting as a primary trigger, it can be concluded that submarine landslides, such as the Byron slide, pose a moderate hazard to the east Australian coastline.

Acknowledgements The authors would like to acknowledge the P&O crew and scientific crew of the SS2013-V01 voyage, which was funded by a grant from the Marine National Facility, CSIRO, Australia. The authors are also grateful to CHRONO Centre, Queen's University, Belfast, UK for conducting the radiocarbon dating for this study. The authors are also grateful to Anna M. Helfensdorfer for her assistance in the laboratory during the sediment analysis for this study and to Aaron Scott for his technical support in running multiple model simulations on the University of Newcastle High Performance Computer Grid. This manuscript was greatly improved thanks to the feedback provided by Peter Haughton, Morelia Urlaub and one other, anonymous, reviewer.

Funding This work was funded by a grant from the Commonwealth Scientific and Industrial Research Organization (SS2013-VO1) to Thomas C.T Hubble; and from the Australian Government (Australian Postgraduate Award) to Kendall C. Mollison. Funding for radiocarbon dating was provided by the University of Newcastle Higher Graduate Research (HDR) fund.

Author contributions **KCM**: conceptualization (equal), formal analysis (lead), funding acquisition (supporting), investigation (lead), methodology (lead), writing – original draft (lead), writing – review & editing (lead); **HEP**: conceptualization (equal), supervision (lead), writing – review & editing (supporting); **SLC**: conceptualization (supporting), methodology (supporting), supervision (supporting), writing – review & editing (supporting); **ATB**: formal analysis (supporting), methodology (supporting), writing – review & editing (supporting); **EML**: methodology (equal), software (equal), writing – review & editing (supporting); **TCTH**: conceptualization (lead), funding acquisition (lead), supervision (supporting), writing – review & editing (supporting).

References

Aranguiz, R. and Shibayama, T. 2013. Effect of submarine canyons on tsunami propagation: a case study of the Biobio canyon, Chile. *Coastal Engineering Journal*, **55**, https://doi.org/10.1142/s0578563413500162

Assier-Rzadkieaicz, S., Heinrich, P., Sabatier, P., Savoye, B. and Bourillet, J. 2000. Numerical modelling of a landslide-generated tsunami: the 1979 Nice event. *Pure and Applied Geophysics*, **157**, 1707–1727, https://doi.org/10.1007/PL00001057

Beaman, R. 2010. *3DGBR: A high-resolution depth model for the Great Barrier Reef and Coral Sea*, http://www.deepreef.org/images/stories/publications/reports/Project3DGBRFinal_RRRC2010.pdf

Blott, S.J. and Pye, K. 2001. GRADISTAT: a grain size distribution and statistics package for the analysis of unconsolidated sediments. *Earth Surface Processes and Landforms*, **26**, 1237–1248, https://doi.org/10.1002/esp.261

Boyd, R. 2009. *SS12/2008 Voyage Summary: Marine Geology and Geohazard Survey of the SE Australian Margin off Northern NSW and Southern Queensland.* CSIRO Division of Marine and Atmospheric Research, Hobart.

Boyd, R., Ruming, K. and Roberts, J. 2004. Geomorphology and surficial sediments of the southeast Australian continental margin. *Australian Journal of Earth Sciences*, **51**, 743–764.

Boyd, R., Ruming, K., Goodwin, I., Sandstrom, M. and Schröder-Adams, C. 2008. Highstand transport of coastal sand to the deep ocean: a case study from Fraser Island, southeast Australia. *Geology*, **36**, 15–18, https://doi.org/10.1130/G24211A.1

Boyd, R., Keene, J., Hubble, T., Gardner, J., Glenn, K., Ruming, K. and Exon, N. 2010. Southeast Australia: a Cenozoic continental margin dominated by mass transport. *In*: Mosher, D.C. *et al.* (eds) *Submarine Mass Movements and their Consequences*. Springer, Dordrecht, 491–502.

Clare, M., Chaytor, J. *et al.* 2019. A consistent global approach for the morphometric characterization of subaqueous landslides. *Geological Society, London, Special Publications*, **477**, 455–477, https://doi.org/10.1144/sp477.15

Clarke, S. 2014. *Submarine landslides of the eastern Australian upper continental margin*. PhD thesis, University of Sydney.

Clarke, S., Hubble, T. *et al.* 2012. Submarine landslides on the upper southeast Australian passive continental margin–preliminary findings. *In*: Yamada, Y. *et al.* (eds) *Submarine Mass Movements and Their Consequences*. Springer, Dordrecht, 55–66.

Clarke, S., Hubble, T. *et al.* 2016. Sedimentology, structure and age estimate of five continental slope submarine landslides, eastern Australia. *Australian Journal of Earth Sciences*, **63**, 631–652.

Clarke, S.L., Hubble, T.C.T., Miao, G., Airey, D.W. and Ward, S.N. 2019. Eastern Australia's submarine landslides: implications for tsunami hazard between Jervis Bay and Fraser Island. *Landslides*, **16**, 2059–2085, https://doi.org/10.1007/s10346-019-01223-6.

Dominey-Howes, D. 2007. Geological and historical records of tsunami in Australia. *Marine Geology*, **239**, 99–123, https://doi.org/10.1016/j.margeo.2007.01.010

Fine, I.V., Rabinovich, A.B., Bornhold, B.D., Thomson, R.E. and Kulikov, E.A. 2005. The Grand Banks landslide-generated tsunami of November 18, 1929: preliminary analysis and numerical modeling. *Marine Geology*, **215**, 45–57, https://doi.org/10.1016/j.margeo.2004.11.007

Fletcher, M. 2015. *An analysis of the morphology and submarine landslide potential of the upper and middle continental slope offshore Fraser Island, Queensland, Australia*. MSc thesis, University of Sydney.

Geoscience Australia 2019. *Where do earthquakes occur?* https://www.ga.gov.au/scientific-topics/community-safety/earthquake

Glenn, K. 2008. *NSW Continental Slope Survey: Post-Cruise Report. RV Southern Surveyor* **10/2006**. Geoscience Australia.

Goff, J. and Chagué-Goff, C. 2014. The Australian tsunami database: a review. *Progress in Physical Geography*, **38**, 218–240, https://doi.org/10.1177/0309133314522282

Gradstein, F.M. 2012. Introduction. *In*: *The Geologic Time Scale*. Elsevier, 1–29.

Grilli, S.T. and Watts, P. 1999. Modeling of waves generated by a moving submerged body. Applications to underwater landslides. *Engineering Analysis with Boundary Elements*, **23**, 645–656, https://doi.org/10.1016/S0955-7997(99)00021-1

Grilli, S.T. and Watts, P. 2005. Tsunami generation by submarine mass failure. I: modeling, experimental validation, and sensitivity analyses. *Journal of Waterway, Port, Coastal, and Ocean Engineering*, **131**, 283–297, https://doi.org/10.1061/(ASCE)0733-950X(2005)131:6(283)

Grilli, S.T., Shelby, M. *et al.* 2017. Modeling coastal tsunami hazard from submarine mass failures: effect of slide rheology, experimental validation, and case studies off the US East Coast. *Natural Hazards*, **86**, 353–391, https://doi.org/10.1007/s11069-016-2692-3

Haflidason, H., Lien, R., Sejrup, H.P., Forsberg, C.F. and Bryn, P. 2005. The dating and morphometry of the storegga slide. *Marine and Petroleum Geology*, **22**, 123–136, https://doi.org/10.1016/j.marpetgeo.2004.10.008

Harbitz, C.B., Løvholt, F. and Bungum, H. 2013. Submarine landslide tsunamis: how extreme and how likely? *Natural Hazards*, **72**, 1341–1374, https://doi.org/10.1007/s11069-013-0681-3.

Hubble, T. 2013. *Voyage Summary SS2013-V01: Marine Geology and Geohazard Survey of the SE Australian Margin off Northern NSW and Southern Queensland*. CSIRO, Hobart.

Hubble, T., Yu, P. *et al.* 2012. Physical properties and age of continental slope sediments dredged from the eastern Australian continental margin – implications for timing of slope failure. *In*: Yamada, Y. *et al.* (eds) *Submarine Mass Movements and their Consequences*. Springer, Dordrecht, 43–54.

Hubble, T., Webster, J. *et al.* 2016. Submarine landslides and incised canyons of the southeast Queensland continental margin. *In*: Lamarche, G. *et al.* (eds) *Submarine Mass Movements and their Consequences*. Springer, Cham, 125–134.

Hubble, T., Yeung, S., Clarke, S., Baxter, A. and De Blasio, F. 2019. Submarine landslides offshore Yamba, NSW, Australia: an analysis of their timing, downslope motion and possible causes. *Geological Society, London, Special Publications*, **477**, 207–222, https://doi.org/10.1144/sp477.11

Iglesias, O., Lastras, G., Souto, C., Costa, S. and Canals, M. 2014. Effects of coastal submarine canyons on tsunami propagation and impact. *Marine Geology*, **350**, 39–51, https://doi.org/10.1016/j.margeo.2014.01.013

Jenkins, C.J. and Keene, J.B. 1992. Submarine slope failures of the southeast Australian continental slope: a thinly sedimented margin. *Deep Sea Research Part A. Oceanographic Research Papers*, **39**, 121–136, https://doi.org/10.1016/0198-0149(92)90100-8

Jeong, S.W. 2010. Grain size dependent rheology on the mobility of debris flows. *Geosciences Journal*, **14**, 359–369, https://doi.org/10.1007/s12303-010-0036-y

Keene, J., Baker, C., Tran, M. and Potter, A. 2008. Sedimentology and geomorphology of the East Marine region of Australia. *Geoscience Australia, Record*, **10**.

Kelfoun, K., Giachetti, T. and Labazuy, P. 2010. Landslide-generated tsunamis at Réunion Island. *Journal of Geophysical Research: Earth Surface*, **115**, https://doi.org/10.1029/2009jf001381

Kim, J., Løvholt, F., Issler, D. and Forsberg, C.F. 2019. Landslide material control on tsunami genesis – the Storegga slide and Tsunami (8,100 years BP). *Journal of Geophysical Research: Oceans*, **124**, 3607–3627, https://doi.org/10.1029/2018JC014893

Locat, J. 2019. Failure and post-failure analysis of submarine mass movements using geomorphology and geomechanical concepts. *Geological Society, London, Special Publications*, **477**, 333–351, https://doi.org/10.1144/SP477.27

Løvholt, F., Harbitz, C.B. and Haugen, K.B. 2005. A parametric study of tsunamis generated by submarine slides in the Ormen Lange/Storegga area off western Norway. *Marine and Petroleum Geology*, **22**, 219–231, https://doi.org/10.1016/j.marpetgeo.2004.10.017

Martini, E. 1971. Standard Tertiary and Quaternary calcareous nannoplankton zonation. *Proceedings of the II Planktonic Conference, Roma 1970*, Roma, Tecnoscienza, 739–785.

Masson, D.G., Harbitz, C.B., Wynn, R.B., Pedersen, G. and Løvholt, F. 2006. Submarine landslides: processes, triggers and hazard prediction. *Philosophical Transactions of the Royal Society A: Mathematical, Physical and Engineering Sciences*, **364**, 2009–2039, https://doi.org/10.1098/rsta.2006.1810

McAdoo, B.G., Pratson, L.F. and Orange, D.L. 2000. Submarine landslide geomorphology, US continental slope. *Marine Geology*, **169**, 103–136, https://doi.org/10.1016/S0025-3227(00)00050-5

Mollison, K.C., Power, H.E., Clarke, S.L., Baxter, A.T. and Hubble, T.C. 2020. Sedimentology, structure, and age of the Wide Bay Canyon submarine landslide on the southeast Australian continental slope. *Marine Geology*, **419**, https://doi.org/10.1016/j.margeo.2019.106063

Orpin, A.R., Rickard, G.J. and Gerring, P.K. 2016. Tsunami hazard potential for the equatorial southwestern Pacific atolls of Tokelau from scenario-based simulations. *Natural Hazards and Earth System Sciences*, **16**, 1239–1257, https://doi.org/10.5194/nhess-16-1239-2016

Popinet, S. 2011. Quadtree-adaptive tsunami modelling. *Ocean Dynamics*, **61**, 1261–1285.

Popinet, S. 2015. A quadtree-adaptive multigrid solver for the Serre–Green–Naghdi equations. *Journal of Computational Physics*, **302**, 336–358, https://doi.org/10.1016/j.jcp.2015.09.009

Reimer, P.J., Bard, E. *et al.* 2013. IntCal13 and Marine13 radiocarbon age calibration curves 0–50,000 years cal BP. *Radiocarbon*, **55**, 1869–1887, https://doi.org/10.2458/azu_js_rc.55.16947

Seelam, J.K., Murali, R.M. and Baldock, T. 2016. Effect of submarine canyons on tsunami heights, currents and run-up off the southeast coast of India. *Current Science*, **111**, 1990–2004, https://www.jstor.org/stable/24911583

Short, A.D. and Woodroffe, C.D. 2009. *The Coast of Australia*. Cambridge University Press, Cambridge.

Smith, R.C., Hill, J., Collins, G.S., Piggott, M.D., Kramer, S.C., Parkinson, S.D. and Wilson, C. 2016. Comparing approaches for numerical modelling of tsunami generation by deformable submarine slides. *Ocean Modelling*, **100**, 125–140, https://doi.org/10.1016/j.ocemod.2016.02.007

Stuiver, M. and Reimer, P. 1993. Extended ^{14}C data base and revised CALIB 3.0 14C age calibration program. *Radiocarbon*, **35**, 215–230, https://doi.org/10.1017/S0033822200013904

Talukder, A.R. and Völker, D. 2014. The tsunami generation potential of shovel and bulli slides in the continental margin SE Australia. *In*: Krastel, S. *et al.* (eds) *Submarine Mass Movements and their Consequences*. Springer, Cham, 539–548.

Tappin, D., Watts, P., McMurtry, G., Lafoy, Y. and Matsumoto, T. 2001. The Sissano, Papua New Guinea tsunami of July 1998 – offshore evidence on the source mechanism. *Marine Geology*, **175**, 1–23.

Tappin, D.R., Watts, P. and Grilli, S.T. 2008. The Papua New Guinea tsunami of 17 July 1998: anatomy of a catastrophic event. *Natural Hazards and Earth System Science*, **8**, 243–266, https://doi.org/10.5194/nhess-8-243-2008

Troedson, A. and Davies, P. 2001. Contrasting facies patterns in subtropical and temperate continental slope sediments: inferences from east Australian late Quaternary records. *Marine Geology*, **172**, 265–285, https://doi.org/10.1016/S0025-3227(00)00132-8

Ulm, S., Petchey, F. and Ross, A. 2009. Marine reservoir corrections for Moreton Bay, Australia. *Archaeology in Oceania*, **44**, 160–166, https://doi.org/10.1002/j.1834-4453.2009.tb00060.x

Uslu, B., Eble, M., Titov, V.V. and Bernard, E.N. 2010. *Distant tsunami threats to the ports of Los Angeles and Long Beach*, California. NOAA Special Report.

Yang, S. 2018. *Submarine landslides at the Broom' s Head Area: An analysis of their morphological, sedimentological, and geotechnical characteristics*. Honours thesis, University of Sydney.

Young, J.R., Bown, P.R. and Lees, J.A. 2017. Cylindralithus' Nannotax3 website. http://www.mikrotax.org/Nannotax3/index.php?taxon=Cylindralithus&module=Mesozoic

Yu, P.W.C. 2017. *Submarine Landslides, Canyons, and Morphological Evolution of the East Australian Continental Margin*. PhD thesis, University of Sydney.

Effects of rotational submarine slump dynamics on tsunami genesis: new insight from idealized models and the 1929 Grand Banks event

T. Zengaffinen[1,2]*, F. Løvholt[1], G. K. Pedersen[2] and C. B. Harbitz[1]

[1]Norwegian Geotechnical Institute, Sognsveien 72, 0806 Oslo, Norway

[2]Department of Mathematics, University of Oslo, Moltke Moes vei 35, 0851 Oslo, Norway

TZ, 0000-0003-0696-7776; FL, 0000-0003-1019-7321; CBH, 0000-0003-4280-7497
*Correspondence: thomas.zengaffinen@ngi.no

Abstract: Sediment slumps are known to have generated important tsunamis such as the 1998 Papua New Guinea (PNG) and the 1929 Grand Banks events. Tsunami modellers commonly use solid blocks with short run-out distances to simulate these slumps. While such methods have the obvious advantage of being simple to use, they offer little or no insight into physical processes that drive the events. The importance of rotational slump motion to tsunamigenic potential is demonstrated in this study by employing a viscoplastic landslide model with Herschel–Bulkley rheology. A large number of simulations for different material properties and landslide configurations are carried out to link the slump's deformation, rheology, its translational and rotational kinematics, to its tsunami genesis. The yield strength of the slump is shown to be the primary material property that determines the tsunami genesis. This viscoplastic model is further employed to simulate the 1929 Grand Banks tsunami using updated geological source information. The results of this case study suggest that the viscoplastic model can be used to simulate complex slump-induced tsunami. The simulations of the 1929 Grand Banks event also indicate that a pure slump mechanism is more tsunamigenic than a corresponding translational landslide mechanism.

Landslides constitute the second-most important tsunami source worldwide after earthquakes (Tappin 2010; Harbitz *et al.* 2014; Yavari-Ramshe and Ataie-Ashtiani 2016). Most recently, the 2018 Anak Krakatoa event caused several hundred fatalities (Grilli *et al.* 2019). Between 2007 and 2017 a string of at least five additional large subaerial landslides impacted water and generated run-up heights in the range of 30 to 150 m (Sepúlveda and Serey 2009; Wang *et al.* 2015; George *et al.* 2017; Gylfadóttir *et al.* 2017; Paris *et al.* 2019). Submarine landslide tsunamis are less frequent than these subaerial landslide tsunamis, but the largest recognized events worldwide indisputably illustrate their destructive potential and importance for society. Fatal examples of such submarine landslides are the 1998 Papua New Guinea (PNG) (Synolakis *et al.* 2002), 1992 Flores Island (Yeh *et al.* 1993), 1979 Lembata Island (Yudhicara *et al.* 2015) and 1929 Grand Banks landslides (Løvholt *et al.* 2019).

Slumps constitute a subset of landslides that are typically characterized by a rotational impulsive slope failure, a relatively coherent mass displacement, and a short landslide run-out distance. At least two of the above-mentioned events, the 1998 PNG and the 1929 Grand Banks events, were caused by rotational slumps. The study of the PNG event also led to acknowledgement in the scientific community that submarine slumps can cause large tsunamis (Bardet *et al.* 2003; Tappin *et al.* 2008). This tsunami has been successfully modelled using an approach where the landslide motion is a rigid block that follows a prescribed motion (Synolakis *et al.* 2002; Tappin *et al.* 2008), by tuning the block motion to comply with wave observations. A similar approach was adopted for modelling the slump part of the 1929 Grand Banks event (Løvholt *et al.* 2019). The rigid block approach was successful in these studies, because the block could mimic the rotational motion of the slump causing the tsunami genesis in an idealized and simple way, but did not include the updated geological source information from Schulten *et al.* (2019*b*), which envisaged a slump that partly evacuated the source area. Although this block modelling approach can help to shed light on the slump motion of past events, it has several obvious shortcomings. Firstly, this method does not include landslide deformation effects that are evident from geophysical data. Secondly, these models cannot be used to take into account the landslide material properties such as the yield strength, and its effect on the landslide dynamics.

From: Georgiopoulou, A., Amy, L. A., Benetti, S., Chaytor, J. D., Clare, M. A., Gamboa, D., Haughton, P. D. W., Moernaut, J. and Mountjoy, J. J. (eds) 2020. *Subaqueous Mass Movements and their Consequences: Advances in Process Understanding, Monitoring and Hazard Assessments*. Geological Society, London, Special Publications, **500**, 41–61.
First published online May 11, 2020, https://doi.org/10.1144/SP500-2019-201

Recent modelling efforts show that the landslide rheology and deformation is important for quantifying and understanding landslide tsunami genesis (Løvholt *et al.* 2017; Yavari-Ramshe and Ataie-Ashtiani 2019). Traditionally, such landslide tsunami studies are based on translational landslide models. However, translational landslides are believed to give rise to a different generation mechanism than slumps, as they do not exhibit a rotational motion as slumps do (Løvholt *et al.* 2015). Until recently, slump models that include a more sophisticated deformation and rheology had not been applied to slump-induced tsunamis. Schambach *et al.* (2018) provided back-to-back analysis with a viscous landslide model and a rigid block model simulating slumps, with both models showing similar results. Ren *et al.* (2019) used a viscoplastic landslide model to generate the slump tsunami due to the 1998 PNG failure, with simulation results that compare favourably with tsunami inundation observations. These studies (Schambach *et al.* 2018; Ren *et al.* 2019) show that a slump tsunami can be effectively modelled using a landslide dynamics model. This method allows for a more flexible, general modelling treatment of the slump tsunami genesis, including material properties, deformation and complex topography, which will be utilized herein.

In this paper, we will use the viscoplastic model BingClaw (Løvholt *et al.* 2017; Kim *et al.* 2019), coupled to the dispersive long-wave solver GloBouss (Løvholt *et al.* 2008), to study slump-induced tsunamis. We will first study landslide dynamics and tsunami genesis in an idealized geometry in one-horizontal dimension (1HD). The main aims of this idealized study are, for the first time, to:

(1) quantify relationships between landslide material yield strength, the resulting slump kinematics and dynamics, and slump tsunamigenic potential; and
(2) identify the extent to which slump tsunamigenic potential can be attributed to translational and rotational slump kinematics, such as the angular momentum.

We will apply the same model setup in two horizontal dimensions (2HD) to study a real case, namely the 1929 Grand Banks landslide and tsunami. The main emphasis of the real case example is to ensure that the landslide parameters and settings in the idealized study can yield a realistic range of analysis. However, a detailed study of the event is left for future investigations.

The 1929 Grand Banks landslide and tsunami

On 8 November 1929 a M_w 7.2 earthquake caused a massive landslide on the Grand Banks south of Newfoundland (Heezen and Ewing 1952; Piper *et al.* 1999) (see Fig. 1). This submarine mass failure comprises by far the largest landslide volume (*c.* 500 km^3) in historical time, worldwide. Deposits far from the landslide failure area and cable breaks (Heezen *et al.* 1954) suggest that the landslide evolved into a turbidity current. The landslide caused a tsunami several metres high at the Burin Peninsula on the south coast of Newfoundland, and waves were also recorded along the entire US East Coast, Bermuda and the Azores (Fine *et al.* 2005). Initial field evidence of the landslide deposits suggested that only turbidity current masses were available in the far field (Schulten *et al.* 2019*a*). Piper *et al.* (1999) noted that the Grand Banks landslide was a widely distributed surficial sediment failure, and Mosher and Piper (2007) noted from newly acquired multibeam bathymetric data that there was no evidence of a massive slump failure on the St Pierre Slope. As the turbidity current itself is likely not the cause of the tsunami, it has been difficult to link the tsunami genesis directly to landslide field evidence. Based on new field investigations of the slope failure, however, Schulten *et al.* (2019*a*) and Løvholt *et al.* (2019) suggested that the near-field tsunami was caused by a massive slump. Løvholt *et al.* (2019) further hypothesized that the more widespread near-surface landslide failure as mapped by Piper *et al.* (1999) and Schulten *et al.* (2019*a*) caused the far-field tsunamis, and that the landslide possibly disintegrated into the turbidity current. Løvholt *et al.* (2019) used a simplified block source and a slump volume of 17 km^3 to model the slump. However, the analysis of newly identified faults and horizons in the St Pierre Slope by Schulten *et al.* (2019*b*) suggest a much larger slump volume of *c.* 390 km^3 for the primary southward slump motion. This new interpretation for the 1929 Grand Banks slump is crucial for testing whether or not our viscoplastic flow model is suited to simulate slumps. Moreover, Schulten *et al.* (2019*b*) suggest that the slump was not confined only between the structural faults containing the slump mass, but also that parts of the landslide transgressed the downslope end of the slump source area through the channel systems, which is different from the assumption of Schulten *et al.* (2019*a*) and Løvholt *et al.* (2019).

Methods

Landslide model

In this paper, the viscoplastic landslide model BingClaw (Løvholt *et al.* 2017; Kim *et al.* 2019; Vanneste *et al.* 2019) is used to simulate the slump dynamics. The model implements the Herschel–Bulkley rheology in a two-layer depth-averaged formulation. Under simple shear conditions, the shear

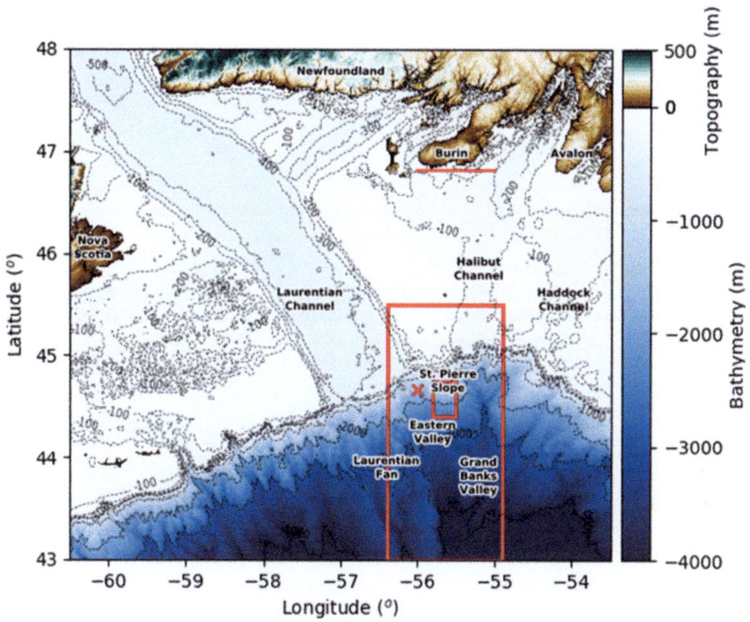

Fig. 1. Bathymetric map of the computational domain for the 2HD tsunami simulations. The bathymetry inside the large red rectangle is used for the simulation of the 2HD landslide dynamics and the small red rectangle is the slump source area. The red line just south of the Burin Peninsula represents the transect used to extract simulation results shown in Figure 15. The red cross shows the epicentre of the M_w 7.2 earthquake on 8 November 1929.

strain in the Herschel–Bulkley fluid is described as:

$$\left|\frac{\dot{\gamma}}{\dot{\gamma}_r}\right|^n = \begin{cases} 0, & \text{if} \quad |\tau| \le \tau_y \\ \dfrac{\tau}{\tau_y \, \mathrm{sgn}(\dot{\gamma})} - 1, & \text{if} \quad |\tau| > \tau_y \end{cases} \qquad (1)$$

where $\dot{\gamma}$ is strain rate, $\dot{\gamma}_r$ a reference strain rate defined as

$$\dot{\gamma}_r = (\tau_y/\mu)^{1/n} \qquad (2)$$

with dynamic consistency μ. τ and τ_y are shear stress and yield strength, respectively, and n the flow exponent. For a detailed description and derivation of the model, see Kim *et al.* (2019).

BingClaw solves the mass conservation equation integrated over the landslide depth (equation 3), the momentum conservation equation integrated separately over the plug layer depth (equation 4), and shear layer depth (equation 5), in 2HD. The unknown variables are bed-normal plug layer thickness d_p, bed-normal shear layer thickness d_s, plug layer volume flux per unit length $d_p\vec{v}_p$ with slope-parallel plug layer velocity $\vec{v}_p$, and shear layer volume flux per unit length $d_s\vec{v}_s$ with slope-parallel shear layer velocity $\vec{v}_s$. $d = d_p + d_s$ is the total thickness of the layers. Indices 'p' and 's' indicate plug and shear layer, respectively (see Fig. 2).

$$\frac{\partial}{\partial t}(d_p + d_s) + \nabla \cdot (d_p\vec{v}_p + d_s\vec{v}_s) = 0 \qquad (3)$$

$$\left(1 + C_m\frac{\rho_w}{\rho_d}\right)\left(\frac{\partial(d_p\vec{v}_p)}{\partial t} + \nabla \cdot (d_p\vec{v}_p\vec{v}_p)\right) + \vec{v}_p\left(\frac{\partial d_s}{\partial t} + \nabla \cdot (d_s\vec{v}_s)\right) = -g' d_p \nabla(d_p + d_s + b) - \frac{\tau_y + \tau_d}{\rho_d}\frac{\vec{v}_p}{||\vec{v}_p||} \qquad (4)$$

$$\left(1 + C_m\frac{\rho_w}{\rho_d}\right)\left(\frac{\partial(d_s\vec{v}_s)}{\partial t} + \nabla \cdot (\alpha d_s\vec{v}_s\vec{v}_s)\right) - \vec{v}_p\left(\frac{\partial d_s}{\partial t} + \nabla \cdot (d_s\vec{v}_s)\right) = -g' d_s \nabla(d_p + d_s + b) - \frac{\tau_y f_s}{\rho_d}\frac{\vec{v}_p}{||\vec{v}_p||} \qquad (5)$$

where C_m is the added-mass coefficient, ρ_w the density of ambient water, ρ_d the density of the slump material, α the velocity form factor, and t the time coordinate. The reduced gravitational acceleration

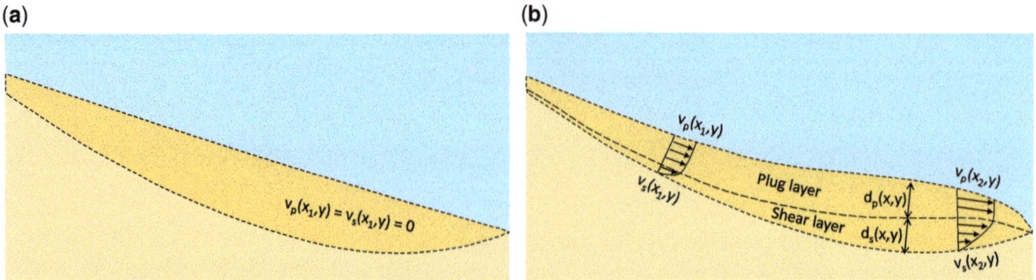

Fig. 2. Simplified schematic plot of the velocity profile (**a**) before and (**b**) during the slump motion simulated with the depth-averaged BingClaw model (modified after Kim *et al.* (2019) for our slump model). The velocity profile is uniform in the plug layer but follows a power law with exponent $n+1$ in the bottom shear layer. Velocities v_p and v_s, and thicknesses d_p and d_s vary spatially and temporally.

is given by $g' = g(1 - \rho_w/\rho_d)$ where g is the gravitational acceleration, b is the bathymetric depth, τ_d is the viscous drag at the free surface, split into a skin friction term τ_f given by

$$\tau_f = \frac{1}{2} C_F \rho_w \vec{v}_p ||\vec{v}_p|| \quad (6)$$

and a pressure drag term τ_p given by

$$\tau_p = \frac{1}{2} C_P \rho_w \max(0, \ -\vec{v}_p \cdot \nabla d)\vec{v}_p \quad (7)$$

where C_F and C_P are skin friction and pressure drag coefficients, respectively, and the viscous contribution of the net shear stress at the bed is given by $\tau_y f_s$ where

$$f_s = \beta \cdot \left(\frac{||\vec{v}_p||}{\dot{\gamma}_r d_s} \right)^n .$$

β is a shape factor depending on the rheological flow exponent n (Huang and Garcia 1998; Imran *et al.* 2001; Kim *et al.* 2019).

BingClaw combines a finite volume method for the leading-order terms with a finite difference model for the source terms. The model is implemented employing the conservation law package ClawPack (Mandli *et al.* 2016) using the GeoClaw module (Berger *et al.* 2011). If the earth pressure $p = \rho_d g' d \nabla(d + b)$ does not exceed the material's shear strength in a given computational cell, no motion is imposed in that cell. Otherwise a Godunov fractional step method is used for the dynamic equations. First the equations without friction terms are solved using the finite volume method in ClawPack, then the frictional terms are accounted for the next fractional step.

Tsunami model

We use the dispersive long wave model GloBouss (Løvholt *et al.* 2008, 2010; Pedersen and Løvholt 2008) to propagate the tsunami over varying bathymetry. In this study, we only use the model in linearized mode as we mainly study the tsunami in deep water, where non-linearities are unimportant.

When terms and factors that are not used herein are omitted (non-linear terms, Coriolis terms, spherical coordinate map-factors and dispersion enhancement terms), the hydrodynamic equations used in this paper read

$$\frac{\partial \eta}{\partial t} + \nabla \cdot (h\vec{u}) = q \quad (8)$$

$$\frac{\partial \vec{u}}{\partial t} = -\nabla\eta + \frac{1}{2} h \nabla\nabla \cdot (h \frac{\partial \vec{u}}{\partial t}) - \frac{1}{6} h^2 \nabla^2 \frac{\partial \vec{u}}{\partial t} \quad (9)$$

where q is a source flux term, which relates the landslide model to the tsunami model through the landslide volumetric displacement (explained below); h is the water depth relative to the mean sea-surface elevation, η the sea-surface elevation and $\vec{u}$ the wave speed.

In GloBouss the equations are discretized on a staggered C-grid (Mesinger and Arakawa 1976) in space and time to give an implicit finite difference method. An alternating direction implicit method (ADI) is used for solving the implicit algebraic equation systems for each time step. The model does not incorporate features like drying or wetting, so we cannot use this model to simulate dry-land inundation.

The slump causes a temporal volumetric change of the bathymetry, which is the primary source for the tsunami genesis. These source fields are then run through a low pass filter that conveys seabed displacements to sea-surface displacements based on full potential wave theory (Kajiura 1963; Løvholt *et al.* 2015) that transfers $\partial d/\partial t$ into $q(x, y, t)$.

Model setup

The geometrical setup is based on the most recent 1929 Grand Banks landslide information provided by Schulten *et al.* (2019*b*). Our first objective is to link a rotational slump motion to tsunami genesis in a systematic fashion, where the slump is confined between an upslope and a downslope fault. To force the slump to stay between these structures, we choose to excavate the slump mass from the seabed, replace it with our viscoplastic material for the initial setup, and elevate the face of the downslope fault (see Fig. 3a, b). While we acknowledge that this geometrical description would likely differ from more complex field observations, this was a necessary simplification to force the viscoplastic material not to evacuate the structure. To this end, we first simulate the slump tsunami in 1HD. The aim is to study idealized effects of kinematics and landslide parameters on tsunami genesis. Secondly, we study a 2HD scenario for the 1929 Grand Banks event, for which the purpose is to provide a realistic parameter range for the 1HD study.

1HD study. The 1HD geometries applied here are simplified from slope transects taken from the general Laurentian Fan bathymetry. As shown in Figure 3a and b, different bathymetries are investigated to study the sensitivity to the slope configuration of the slump source. The bathymetry outside the slump towards the shore is gentler with a constant inclination of 0.05° in all cases.

The computational domain for the landslide model has a total length of 50 km in the x-direction with a spatial resolution of $\Delta x = 80$ m. However, due to the computational stencil of BingClaw, several cells in the azimuthal y-direction are required. Non-reflecting outflow conditions are applied at the boundaries. The Kajiura-type full potential filter is run over the same length as the landslide model. Grid resolution for the Kajiura filtered output is

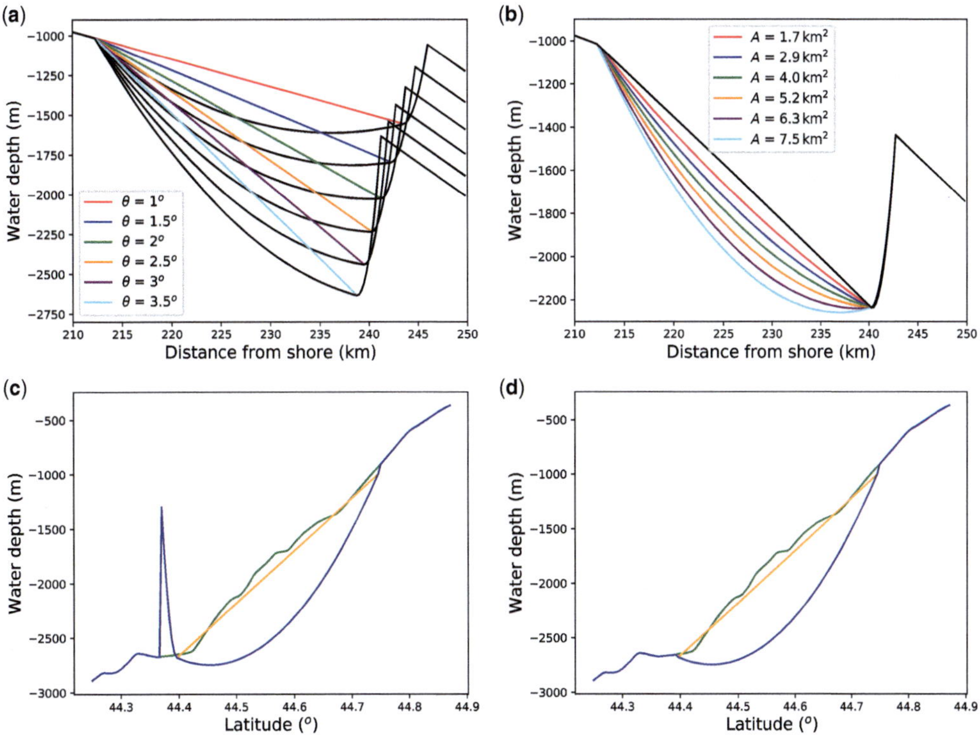

Fig. 3. Initial 1HD bathymetry with slump masses for (**a**) set S1 and (**b**) set S2. Bathymetry with orange lines indicates the same geometrical setup. Transects through longitude −55.77° over the initial 2HD bathymetry for (**c**) the pure slump that includes side walls and a bottom wall, and (**d**) the over-topping slump. The orange line indicates the initial slump surface and the blue line the seabed for the simulations. The green line represents the seabed surface prior to excavation.

Table 1. *Default parameters used for the 1HD and 2HD simulations*

Parameter	Symbol	Value	Units
Seawater density	ρ_w	1000	kg m^{-3}
Landslide density	ρ_d	2000	kg m^{-3}
Gravitational acceleration	g	9.81	m s^{-2}
Added-mass coefficient	C_m	0.1	
Skin friction coefficient	C_F	0.001	
Pressure drag coefficient	C_P	0.25	

also 80 m. For GloBouss we cover a computational domain extending 450 km horizontally, and with a resolution of 220 m. We apply a sponge layer at the right boundary, from 250 to 450 km, that relaxes the offshore-going waves (Pedersen and Løvholt 2008). No-flux conditions are applied at the other boundaries. In GloBouss a 1HD computation involves a single wet row of cells between two dry rows of ghost cells. Spatial and temporal grid refinement tests on the landslide model BingClaw, the full potential Kajiura-type filter, and the tsunami model GloBouss are described in Appendix A.

Default model input parameters (i.e. density and hydrodynamic resistance) are listed in Table 1, and geometrical and geotechnical model input parameters used for the sensitivity analysis are listed in Table 2. In order to establish the list of landslide input parameters, we ran several simulations to achieve a full parameter range that spans the relevant sensitivity range for tsunami genesis. By combining all relevant geotechnical and hydrodynamic resistance parameters for each geometrical setting, we ended up running 2640 simulations for the sensitivity analysis. We refer to 1440 model runs with constant slump volumes per unit width and variable initial slump surface slope angle as set S1, and 1440 model runs with a constant initial slump surface slope angle and variable volumes per unit width as set S2; 240 simulations overlap in set S1 and set S2.

A simplified basic geometry is defined by an initial slope angle $\theta = 2.5°$, as retrieved from the Laurentian fan, and a volume per unit width $A = 5.2$ km^2, which multiplied with a slump width of $W = 33$ km yields a total volume $V = 175$ km^3 as suggested by Schulten *et al.* (2019*b*) for the upper part of the 1929 Grand Banks slump. Then, in simulation set S1, θ is varied between 1 and 3.5°, while keeping A constant. Likewise, in set S2, A is varied between 1.7 km^2 and 7.5 km^2, while keeping θ constant. In each case the parabolic shape of the rigid seabed is adjusted accordingly.

For a very soft slump material (e.g. low values of τ_y in Table 2), the mass can be so mobile that it artificially reflects from the lower fault face and propagates back upslope and may even continue to slosh back and forth. This spurious sloshing occurs partly due to simplifications in the applied slump model, partly due to the geometrical setup and partly due to too small values employed for the landslide strength. Time-series of two examples of the centre-of-mass motions, which is used to filter events, are shown in Figure 4. The centre-of-mass velocities have a smoother time evolution than the maximum velocities. If an event gets a negative centre-of-mass velocity, it is removed from the analysis to avoid the artificial sloshing. This criterion was based on analyses of the wave generation for the sloshing events, where it transpired that events with negative centre-of-mass velocities influenced the wave generation significantly. An example of the artificial scaling behaviour that can be expected is discussed in one of our analyses below. The number of non-sloshing events as well as events where the yield strength is too large for the mass to mobilize the landslide (i.e. stable sediments), are shown for both set S1 and set S2 in Figure 5.

2HD study. The slump configuration with the new information provided by Schulten *et al.* (2019*b*) is used to simulate the slump dynamics. We distinguish between two different scenarios, an over-topping (where a part of the material escapes in the lower extremity) and a pure slump. For the pure slump the mass is confined to a source area limited by walls at the downslope extremity and at the two

Table 2. *Geometrical and geotechnical parameters used for the 1HD simulations*

Parameter	Symbol	Values								Units
Slump volume per unit width	A	1.7	2.9	4.0	5.2	6.3	7.5			km^2
Slump surface slope angle	θ	1	1.5	2	**2.5**	3	3.5			°
Yield strength	τ_y	10	**25**	40	**55**	70	**85**	100	115	kPa
Herschel–Bulkley flow exponent	n	0.1	**0.25**	0.5	0.75	1.0				
Dynamic landslide consistency	μ	1	4	7	**10**	13	16			kPa s^n

The bold values are also used for the 2HD simulations.
Set S1 with a constant volume per unit width $A = 5.2$ km^2 and set S2 with a constant slump surface slope angle $\theta = 2.5°$ each combine to 1440 scenarios.

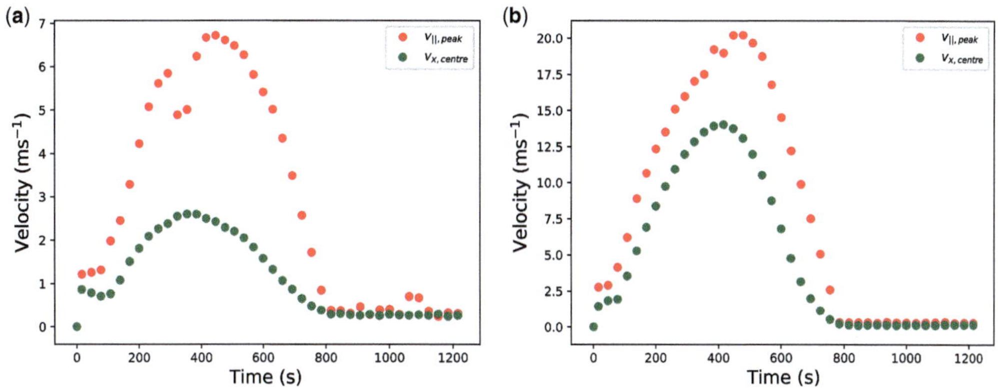

Fig. 4. Time-series of centre-of-mass velocity $v_{x,\mathrm{centre}}$ and peak bed-parallel velocity over the entire slump body $v_{\parallel,\mathrm{peak}}$. Maximum velocities are used for further analyses. Employed parameters are $\mu = 10$ kPa s^n, $n = 0.25$, the slump surface slope angle is $\theta = 2.5°$, the slump volume per unit width is $A = 5.2$ km^2, and (**a**) $\tau_y = 70$ kPa, (**b**) $\tau_y = 40$ kPa.

sides. It generates a rotational slump motion in a similar way as in the 1HD study (see Fig. 3c). We note that an over-topping scenario is considered as most likely by Schulten *et al.* (2019*b*) (see Fig. 3d). In the case of over-topping, the model geometry is set up to allow the slide material to continue as a translational landslide outside the region of mass failure. The further disintegration into the turbidity current observed in the field is, however, not included in the model. We note that the main orientation of this slump geometry is southward, which was also assumed by Løvholt *et al.* (2019). Yet, the revised slump volume used in the 2HD analysis here (390 km^3) is considerably larger than what was assumed by Løvholt *et al.* (2019).

Model bathymetries are based on the online geographical GEBCO 2014 Grid with 463 m cell size in longitude and latitude. The depth matrix for the landslide and source computations covers a rectangle with lengths of 114 and 255 km in the longitudinal and latitudinal directions, respectively. For the landslide model, a grid resolution of 185 m is used, while a resolution of 463 m suffices for the surface response. As in the 1HD slump model, there is

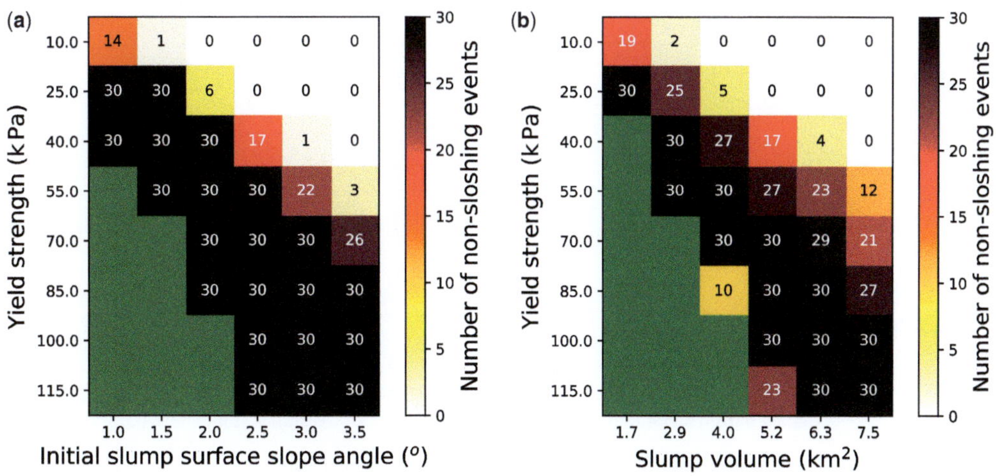

Fig. 5. Number of non-sloshing events and events with velocities greater than zero as a function of yield strength τ_y, (**a**) initial slump surface slope angle θ and (**b**) slump volume per unit width A. All combinations of flow exponent n and dynamic viscosity μ combine to a total of 30 events. Low τ_y, large A, and large θ indicate sloshing events. Large τ_y, low A, and low θ indicate stable sediments and are coloured in green.

non-reflecting outflow at the four boundaries. The grid for the tsunami computations is larger and covers a rectangle of 616 km (longitude) by 555 km (latitude). It has a resolution of 463 m and includes the source area, the southern coast of Newfoundland and the eastern coast of Nova Scotia (see Fig. 1). At all four boundaries we apply a sponge layer of 22 km width where the waves are relaxed and apply a minimum computation depth of 10 m in order to avoid spurious oscillations in shallow waters. Spatial and temporal grid refinement tests on the landslide model BingClaw, the full potential Kajiura-type filter and the tsunami model GloBouss are discussed in Appendix A. Default model parameters are presented in Table 1, geotechnical and geometrical parameters are given in bold in Table 2.

Results

1HD parametric sensitivity study

Example of tsunami-genesis mechanism. We first analyse, through one single simulation, the slump tsunami-genesis mechanism. We use the following BingClaw parameters, namely $\tau_y = 70$ kPa, $\mu = 10$ kPa s^n, and $n = 0.25$. The slump surface slope angle is $\theta = 2.5°$, the slump volume per unit width is $A = 5.2$ km^2 and the water depth of the initial centre of mass is *c.* 1750 m. At *c.* 236 km from shore, a maximum vertical landslide displacement of *c.* 100 m is obtained, which is similar to what was suggested by Schulten *et al.* (2019*b*). Figure 6a shows the slump motion at different times. The corresponding generated waves at different times are displayed in Figure 6b. While the slump mass rotates around its mass centre, the downslope part of the rotational slump pushes water upwards creating a positive wave at the surface, whereas the upslope part of the slump pulls down water and causes a trough at the surface.

Next, we ran two separate simulations for the same example, one using only the positive flux part of the slump source term, and another only using the negative flux part. Figure 7 shows the generated total wave (in solid lines), as well as the wave component only due to the slump uplift (in long dashed lines), and the wave component due to the slump depression (in short dashed lines). Both the generated wave elevations and wave troughs continuously split into landward and offshore travelling waves as long as the slump motion continues and add to the already propagated waves. Because the slump's upward and downward motions are spatially shifted, the landward wave-elevation travels slightly behind the landward wave-trough. Only a partial overlap of this wave-trough and wave-elevation occurs, which results in a landward trough followed by an elevation. The positive and negative amplitudes of this total wave, when travelled out of the source area, are roughly half of the maximum/minimum elevations from pure positive and negative source components.

This mechanism was discussed by Løvholt *et al.* (2005, 2015) and Haugen *et al.* (2005), but mainly for translational landslides. Based on analyses of the 1998 PNG event, Løvholt *et al.* (2015) suggested that the interaction between rear and frontal waves was limited for slumps, and that their wave generation was more efficient than for translational landslides. However, the present analysis shows that

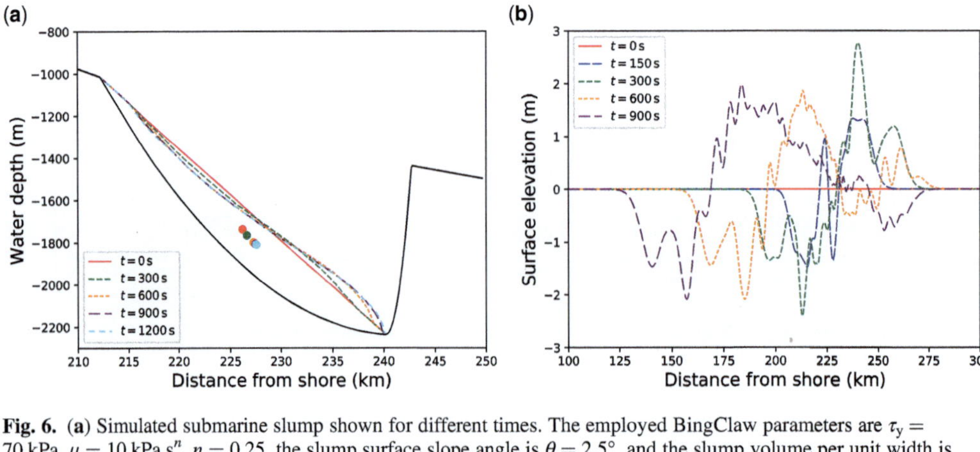

Fig. 6. (**a**) Simulated submarine slump shown for different times. The employed BingClaw parameters are $\tau_y =$ 70 kPa, $\mu = 10$ kPa s^n, $n = 0.25$, the slump surface slope angle is $\theta = 2.5°$, and the slump volume per unit width is $A = 5.2$ km^2. We show the slump from its initial state until it stops moving, 1200 s after failure. The dots indicate the centre of mass of the slump as a function of time. (**b**) Tsunami genesis and propagation until 900 s, which is the time we evaluate the maximum and absolute minimum landward sea-surface elevation. The offshore-going wave has been relaxed by the sponge layer at the right boundary starting at 250 km from the shore.

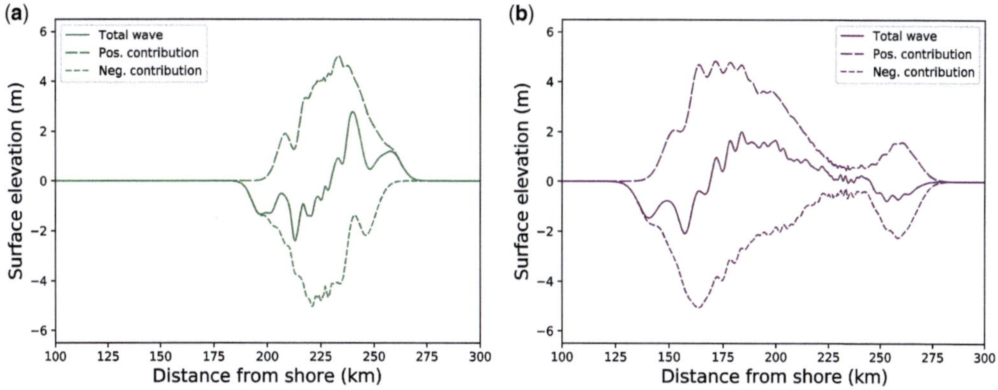

Fig. 7. Tsunami split into the total wave (in solid lines), due to slump uplifts (in long dashed lines) and slump depressions (in short dashed lines). Elapsed times are (**a**) 300 s and (**b**) 900 s. The latter time is when we evaluate the maximum and absolute minimum landward sea-surface elevation. The offshore-going wave has been relaxed by the sponge layer at the right boundary starting at 250 km from the shore.

the interaction between the frontal and rear wave clearly reduces the maximum elevation of the total wave for the 1929 Grand Banks slump. We stress that for other slump configurations and material parameters the picture could be different.

Relationship between geotechnical parameters and tsunami genesis. Figure 8 shows the sensitivity of the maximum landward sea-surface elevation η_{max} to various input parameters. η_{max} is evaluated 900 s after the slump mass release such that the wave

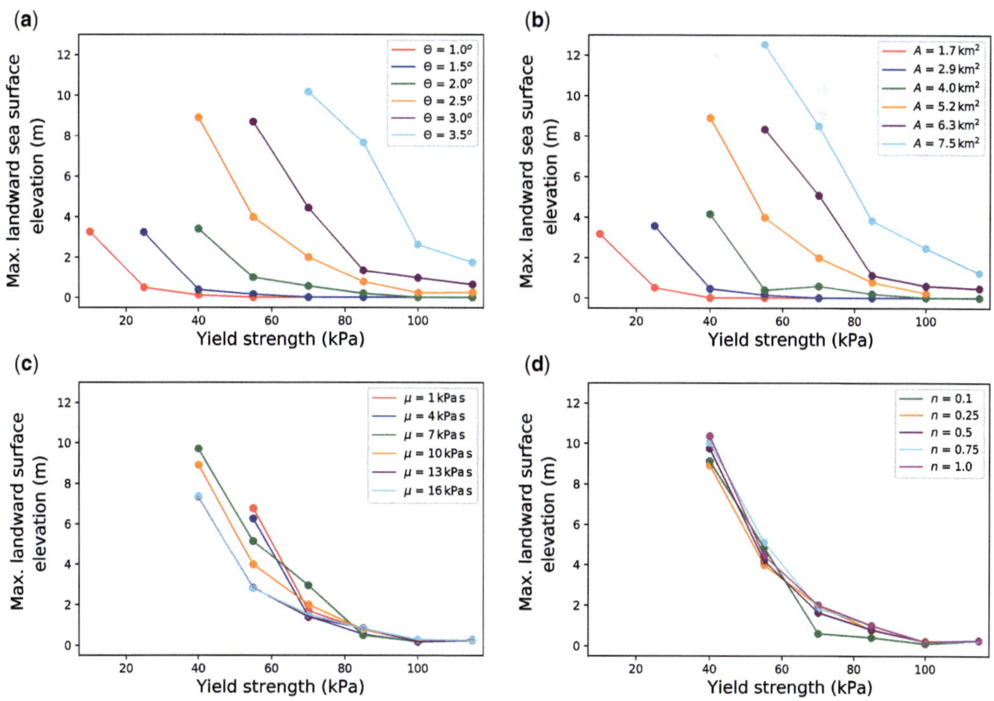

Fig. 8. Maximum landward sea-surface elevations η_{max} as a function of yield strength τ_y for a selection from (**a**) set S1, (**b**) set S2 and (**c**, **d**) common scenarios from both sets. Orange lines in all subplots refer to the same scenarios. Fixed parameters (except where parameters are subject to variation) are $\mu = 10$ kPa s^n, $n = 0.25$, $\theta = 2.5°$ and $A = 5.2$ km^2.

with the highest crest has propagated out of the source area. The various input parameters include the slump material's yield strength τ_y, the volume per unit width A, the initial slump surface slope angle θ, the dynamic consistency μ, and the flow exponent n. In all cases, $\eta_{\max}$ is plotted as a function of τ_y, and increases consistently with decreasing τ_y. As expected, $\eta_{\max}$ also increases with θ and A. Furthermore, we see that $\eta_{\max}$ is only moderately dependent on μ. The flow exponent n has a negligible influence on tsunami genesis, except when very small.

Relationship between landslide translational kinematics and tsunami genesis. Figure 9 shows relationships between maximum bed-parallel and vertical slump kinematics, and maximum and absolute minimum landward sea-surface elevations $\eta_{\max}$ and $\eta_{\min}$ for set S1. We recall that for S1, the initial slump surface slope angle θ is variable and the volume per unit width is constant at $A = 5.2\ \mathrm{km}^2$. The maximum kinematic quantities are calculated over the full computational domain for all times, whereas $\eta_{\max}$ and $\eta_{\min}$ are evaluated at a time of 900 s. Figure 9a shows scaled $\eta_{\max}$ and $\eta_{\min}$ as a function of the scaled maximum bed-parallel velocity $v_{||_{\max}}$ and a least-square power-law fit is included in some panels. $\eta_{\max}$ increases with $v_{||_{\max}}$ following fairly well a power-law behaviour with exponent of 0.9. There is more scattering for lower $v_{||_{\max}}$ values. Noticing that the quantity $v_{||\max}/\sqrt{(gH)}$ is closely related to the Froude number (see below), we point out that the growth rate of that quantity is less than the linear Froude scaling proposed by Løvholt *et al.* (2015) for slumps with small Froude numbers. The linear scaling relation should exist when there is no interaction between the frontal-wave elevation and rear-wave trough. However, in this case, there is clearly a destructive interference (see Fig. 7), which leads to a less effective wave generation. Figure 9c shows the relationship between the scaled maximum vertical velocity $v_{z_{\max}}$, scaled $\eta_{\max}$ and $\eta_{\min}$. Unlike in Figure 9a, we do not observe a simple power-law relationship. There is also clearly more scatter in the vertical velocity plot. Further, processing of the

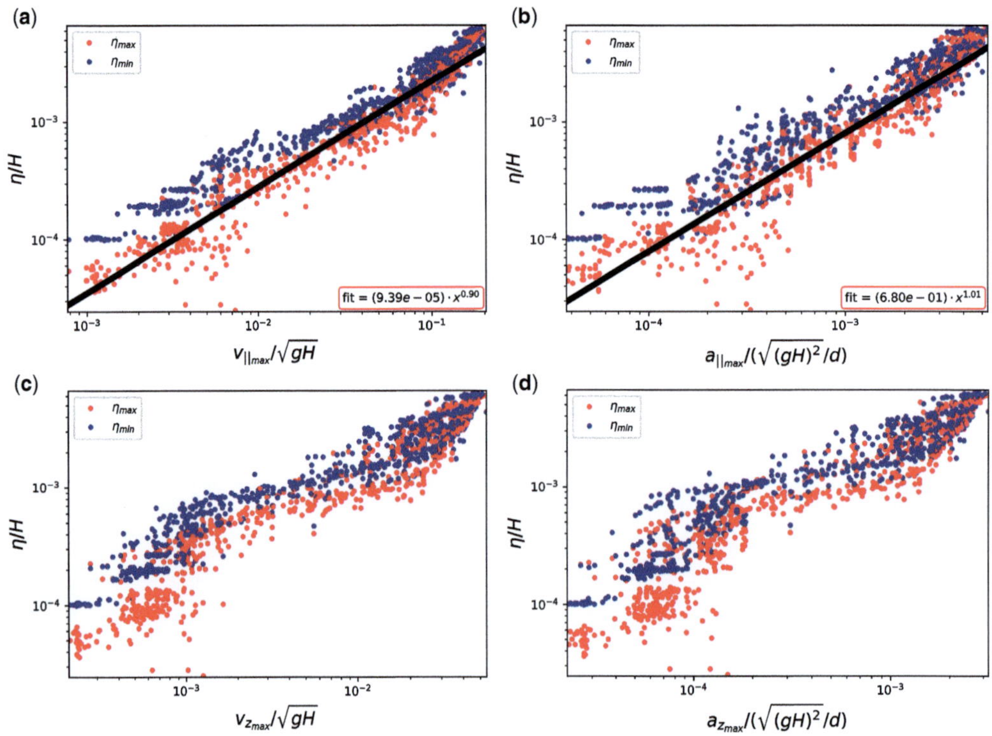

Fig. 9. Scaled maximum and absolute minimum landward sea-surface elevation $\eta_{\max}$ and $\eta_{\min}$ against (**a**) scaled maximum bed-parallel slump velocity $v_{||_{\max}}$, (**b**) scaled maximum bed-parallel slump acceleration $a_{||_{\max}}$, (**c**) scaled maximum vertical slump velocity $v_{z_{\max}}$ and (**d**) scaled maximum vertical slump acceleration $a_{z_{\max}}$ for set S1. The scale for the sea-surface elevation is the typical water depth $H = 2000$ m, the velocity scale is the linear wave speed $\sqrt{gH}$, and the acceleration scale is the square linear wave speed $\sqrt{gH}$ divided by the typical slump thickness $d = 250$ m. The power-law fits apply to $\eta_{\max}$ with x representing the x-axes.

kinematic output also verifies that maximum velocities, $v_{||_{max}}$ and $v_{z_{max}}$, and maximum accelerations, $a_{||_{max}}$ and $a_{z_{max}}$, depend strictly on each other (results not shown). Consequently, η_{max} shows a similar power-law dependency on $a_{||_{max}}$ as on $v_{||_{max}}$, with an exponent of 1.01, but with a lack of a simple power-law dependency on $a_{z_{max}}$ (see Fig. 9b, d). The almost linear relationship with the acceleration agrees with previous investigations that heavily relied on landslide block motion (Hammack 1973; Watts 2000; Løvholt *et al.* 2005, 2015). These studies concluded that the horizontal acceleration strongly influences tsunami genesis and, in particular, Løvholt *et al.* (2005, 2015) suggest the same linear relationship between η_{max} and $a_{||_{max}}$ as we find here.

We recall that set S2 has a constant initial slump surface slope angle $\theta = 2.5°$, but has different values for the volume per unit width A. The velocity is multiplied by the slump's total mass per width to quantify the momentum and to analyse how the momentum correlates with η_{max} and η_{min}. Figure 10a shows that η_{max} and η_{min} as functions of $m\, v_{||_{max}}$ follow a power-law fit, however, with a gentler growth rate and more scattering for small $m\, v_{||_{max}}$ than for high $m\, v_{||_{max}}$. The exponent for η_{max} is 0.9. Figure 10c shows that η_{max} and η_{min} have a similar relationship with the vertical maximum momentum $m\, v_{z_{max}}$, but that the relationship does not follow a simple power-law behaviour and with more scatter for the smallest values of the maximum vertical momentum. Figure 10b and d shows that the relationships between the rate of $m\, v_{||_{max}}$, the rate of $m\, v_{z_{max}}$, η_{max} and η_{min} follow similar relationships as the ones derived for $m\, v_{||_{max}}$ and $m\, v_{z_{max}}$, respectively. The fitted exponent between η_{max} and the rate of $m\, v_{||_{max}}$ is 1.01. For the mass times acceleration terms, we find a similar conclusion as for set S1 with a constant volume per unit width. We even remark that the power-law exponents for the mass dependent terms $m\, v_{||_{max}}$ and its rate for set S2 are almost identical to the fitted power-law exponents for $v_{||_{max}}$ and $a_{||_{max}}$ for set S1. However, the plots showing η_{max} and η_{min} against vertical momentum and momentum rates show less variability than the

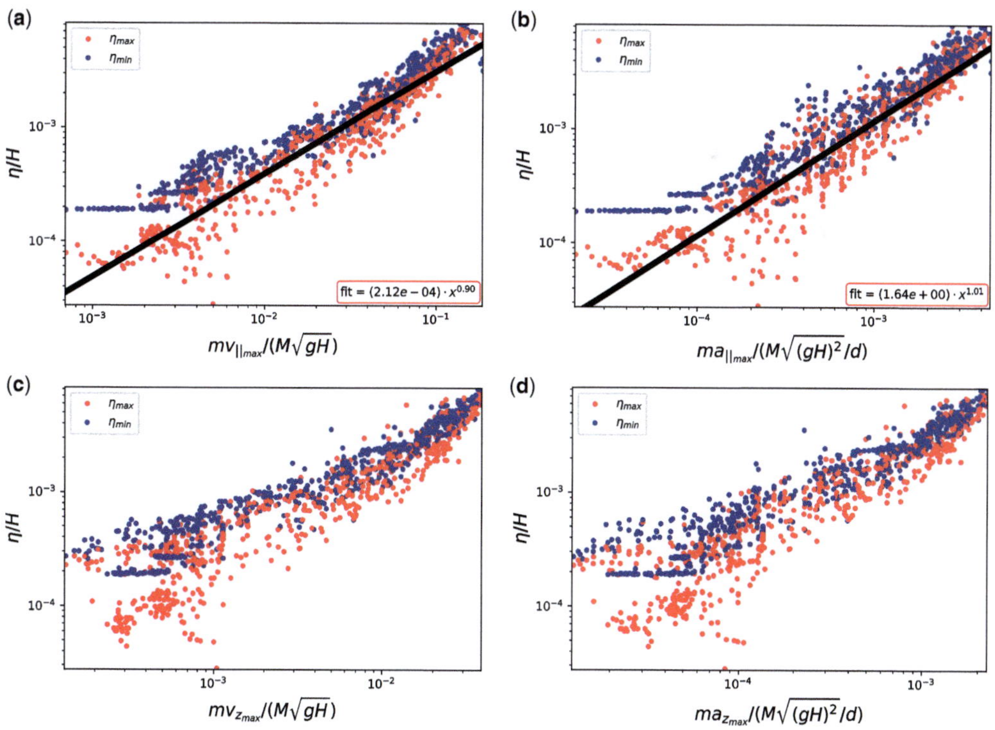

Fig. 10. Scaled maximum and absolute minimum landward sea-surface elevation η_{max} and η_{min} against (**a**) scaled maximum bed-parallel slump momentum $m\, v_{||_{max}}$, (**b**) scaled maximum bed-parallel slump momentum rate $m\, a_{||_{max}}$, (**c**) scaled maximum vertical slump momentum $m\, v_{z_{max}}$ and (**d**) scaled maximum vertical slump momentum rate $m\, a_{z_{max}}$ for set S2. The scale for the sea-surface elevation is the typical water depth $H = 2000$ m, the momentum scale is the largest mass M (from the $A = 7.5$ km^2 scenarios) multiplied by the linear wave speed $\sqrt{gH}$, and the scale for the momentum rate is the largest mass M multiplied by the square linear wave speed $\sqrt{gH}$ divided by the typical slump thickness $d = 250$ m. The power-law fits apply to η_{max} with x representing the x-axes.

corresponding plots for η_{max} against vertical velocities and accelerations for set S1.

A Froude number, *Fr*, is defined as the maximum horizontal central mass velocity divided by the linear wave speed $\sqrt{gH}$ at a typical water depth $H = 2000$ m. A nearly unitary *Fr* means the slump's horizontal central mass speed and the tsunami speed are the same, which represents the most efficient tsunami-genesis mechanism (Løvholt *et al.* 2015). In our study, *Fr* is invariably much smaller than unity. Figure 11 shows scaled η_{max} and η_{min} as a function of *Fr* for set S1, which represents the left side of the height–velocity curve peak in Ward (2001, fig. 3). We see that the growth rate of η_{max} as a function of *Fr* is slower than when we use the maximum landslide velocity (i.e. in Fig. 9a). On the other hand, we visually observe a slight misfit for the largest values of *Fr*, which may suggest that the exponent is not linear, possibly increasing with larger *Fr*. We note that Figure 11b also shows the results for the unfiltered simulations (i.e. including spurious sloshing events). Investigating Figure 11a and b, we see that the filter removes scenarios above $Fr \approx 0.13$. For larger Froude numbers, the scaling of the unfiltered maximum landward sea-surface elevation η_{max} separates from the scaling of the absolute minimum sea-surface elevation η_{min}, and the separation occurs above $Fr \approx 0.15$, say. The more rapid increase in the η_{max} with *Fr* is interpreted as a spurious result of the model (and hence filtered). On the other hand, we see that the scaling relationship for η_{min} is virtually unchanged for high Froude numbers (filtered events). The leading landward troughs are unaffected by the sloshing, which hints that a linear Froude scaling should also be expected for somewhat larger Froude numbers than those analysed elsewhere in this paper.

Relationship between landslide rotational kinematics and tsunami genesis. Slumps are mainly rotational and display different kinematics compared to translational landslides with long run-out. Here, we analyse to which extent the slump's scaled maximum angular momentum L_{max} is attributed to the slump's tsunamigenic potential. The technical derivation of this quantity is given in Appendix A. Figure 12a shows a power-law relationship between L_{max}, η_{max} and η_{min} for set S1. The exponent for η_{max} is 0.76. Figure 12b shows that the dependency between L_{max}, η_{max} and η_{min} for set S2 has significantly more scatter, and a less clear correlation. The fitted exponent is 0.66 for η_{max}. In both cases, the data exhibit little scatter for large L_{max}.

2HD study related to the 1929 Grand Banks event

Slump scenarios with over-topping. Figure 13 shows the simulated motion of the slump with over-topping for a volume of $V = 390$ km^3 and a yield strength of $\tau_y = 85$ kPa. At 300 s, the slump is still confined in the fault structure. Around $t = 600$ s the slump has its maximum vertical uplift of *c.* 400 m at its downslope extremity while parts of the slump mass escape the faulted pit and continue downslope as a translational landslide. This over-topping results in a 100 m high frontal landslide height. The output at 1380 s shows the landslide flowing into the Laurentian Fan region.

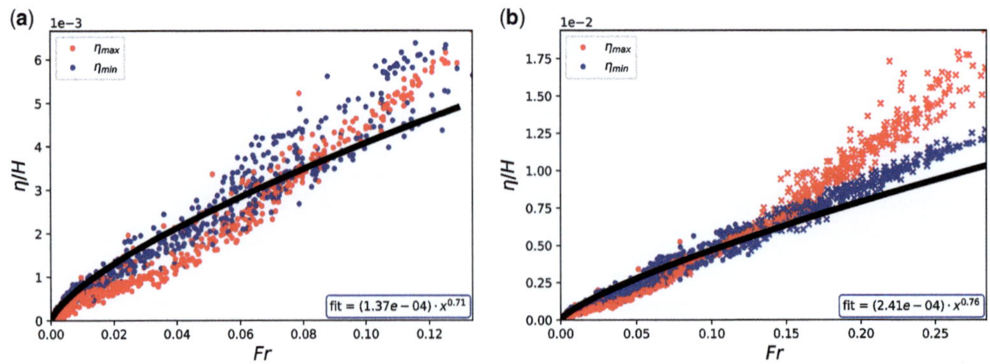

Fig. 11. Scaled maximum and absolute minimum landward sea-surface elevation η_{max} and η_{min} as a function of the Froude number *Fr* for set S1, with (**a**) filtered events only and (**b**) unfiltered events with dots representing events with no negative centre-of-mass velocities and crosses representing events with negative centre-of-mass velocities. The scale for the sea-surface elevation is the typical water depth $H = 2000$ m, and *Fr* is the maximum horizontal velocity of the centre of mass scaled with the linear wave speed $\sqrt{gH}$. The power-law fits apply to η_{min} with x representing the x-axes.

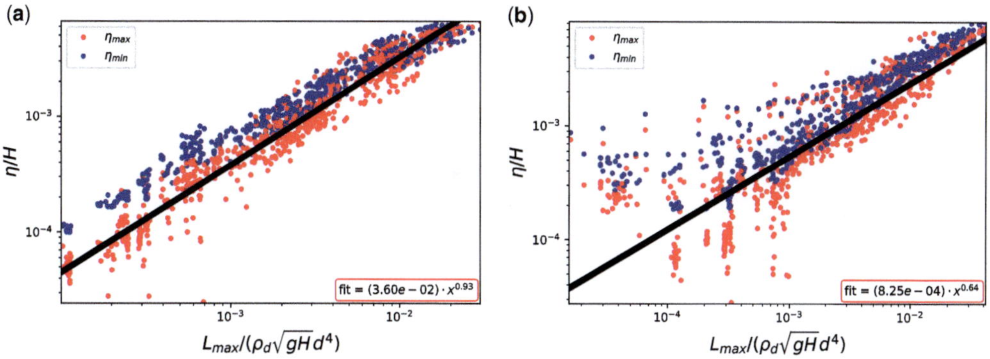

Fig. 12. Scaled maximum and absolute minimum landward sea-surface elevation $\eta_{\max}$ and $\eta_{\min}$ against scaled maximum angular momentum L for (**a**) set S1 and (**b**) set S2. The scale for the sea-surface elevation is the typical water depth $H = 2000$ m, and the scale for the angular momentum is the slump's density $\rho_{\rm d}$ multiplied by the square root of the linear wave speed $\sqrt{gH}$ multiplied by the fourth power of the typical slump thickness $d = 250$ m. The power-law fits apply to $\eta_{\max}$ with x representing the x-axes.

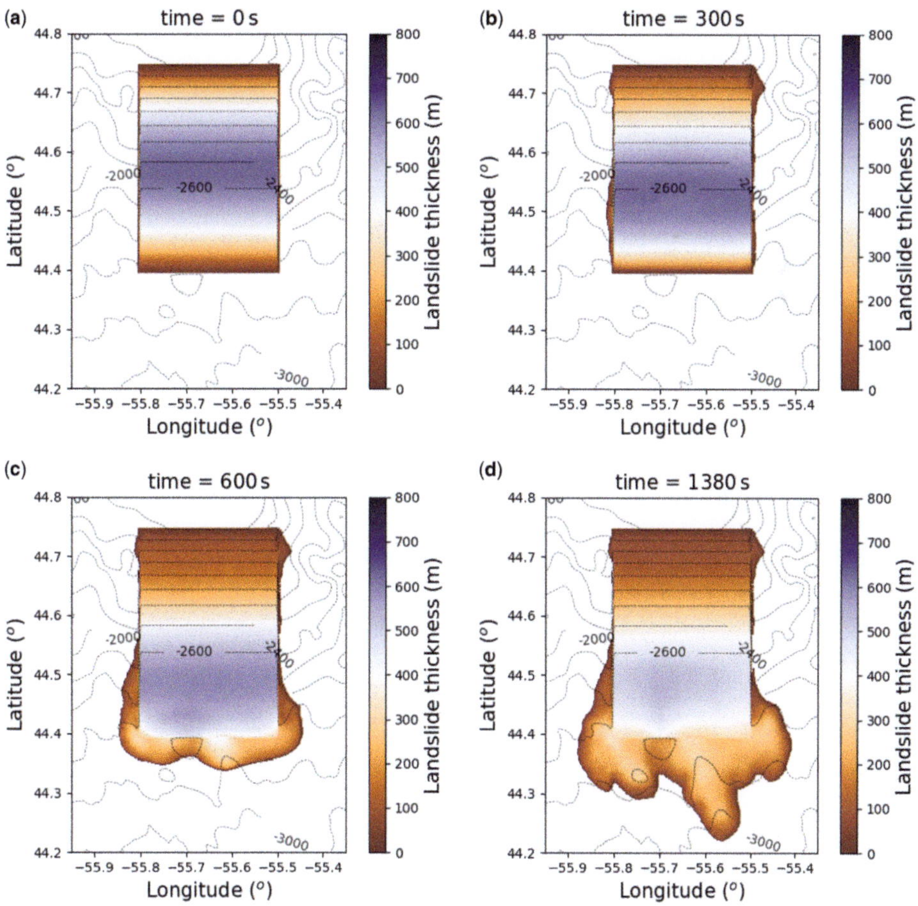

Fig. 13. Snapshots of the landslide thickness for the 1929 Grand Banks over-topping slump scenario at different times. The slump mass over-tops its bounding faults and transforms into a translational landslide as Schulten *et al.* (2019*b*) propose. The employed BingClaw parameters are $\tau_{\rm y} = 85$ kPa, $\mu = 10$ kPa s^n and $n = 0.25$.

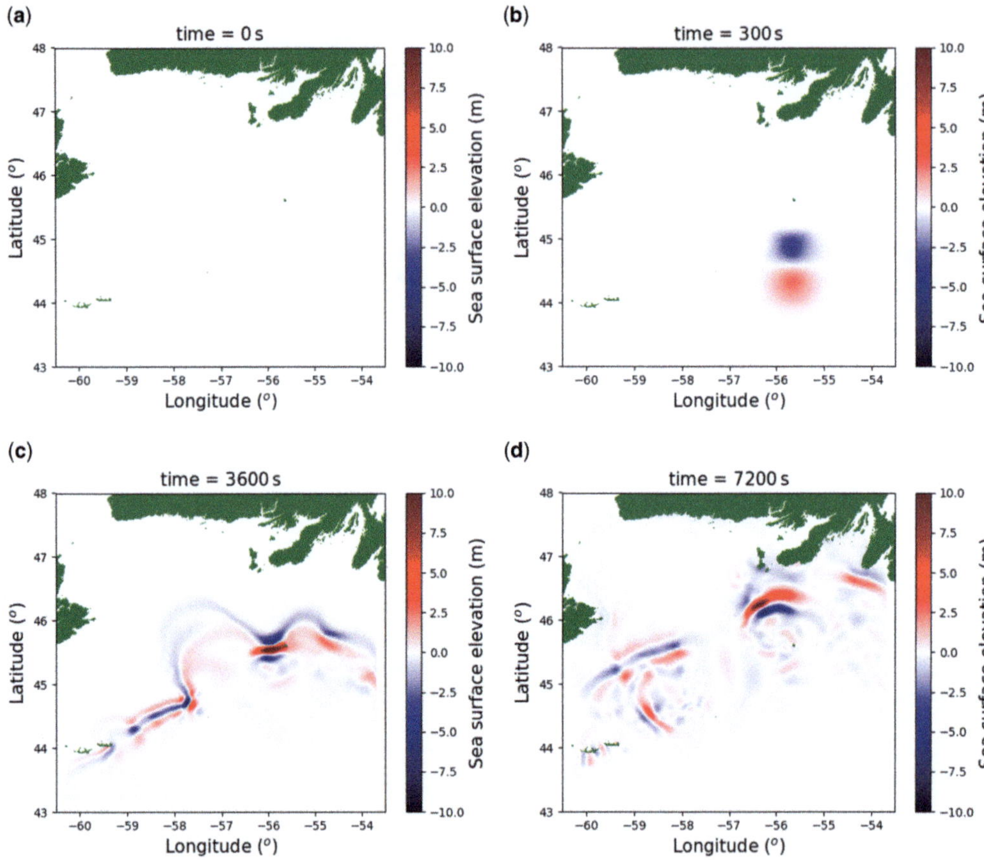

Fig. 14. Snapshots of the spreading waves for the 1929 Grand Banks over-topping slump source shown in Figure 13. Land is represented in green.

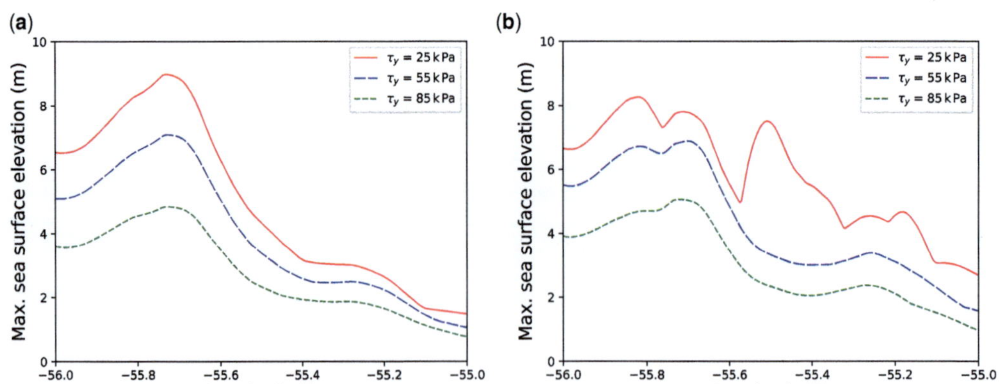

Fig. 15. Maximum sea-surface elevation until 8 h 20 min in a transect (see Fig. 1) near the Burin Peninsula for three different sediment yield strengths τ_y, and for both (**a**) the 1929 Grand Banks over-topping slump and (**b**) the 1929 Grand Banks pure slump. Other employed BingClaw parameters for both slump events are $\mu = 10$ kPa s^n and $n = 0.25$. The water depth along this transect is between 20 and 50 m.

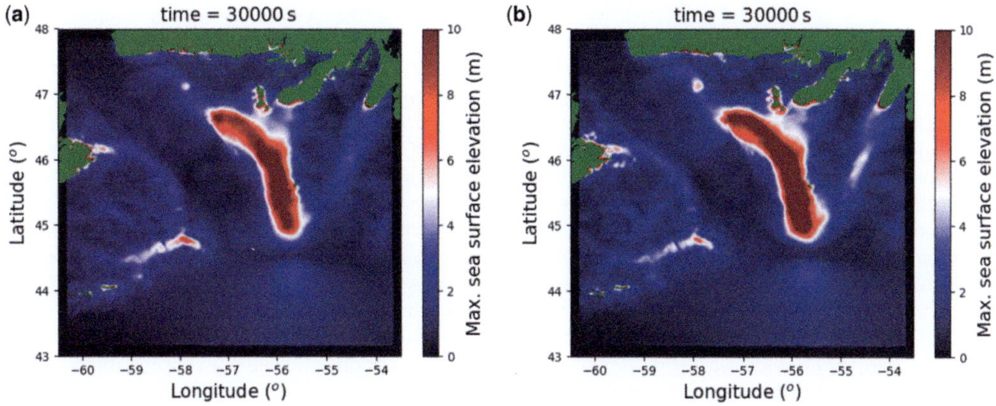

Fig. 16. Maximum sea-surface elevation until 8 h 20 min for the total wave field for (**a**) the 1929 Grand Banks over-topping slump and (**b**) the 1929 Grand Banks pure slump. The employed BingClaw parameters are $\tau_y = 85$ kPa, $\mu = 10$ kPa s^n and $n = 0.25$. Land is represented in green.

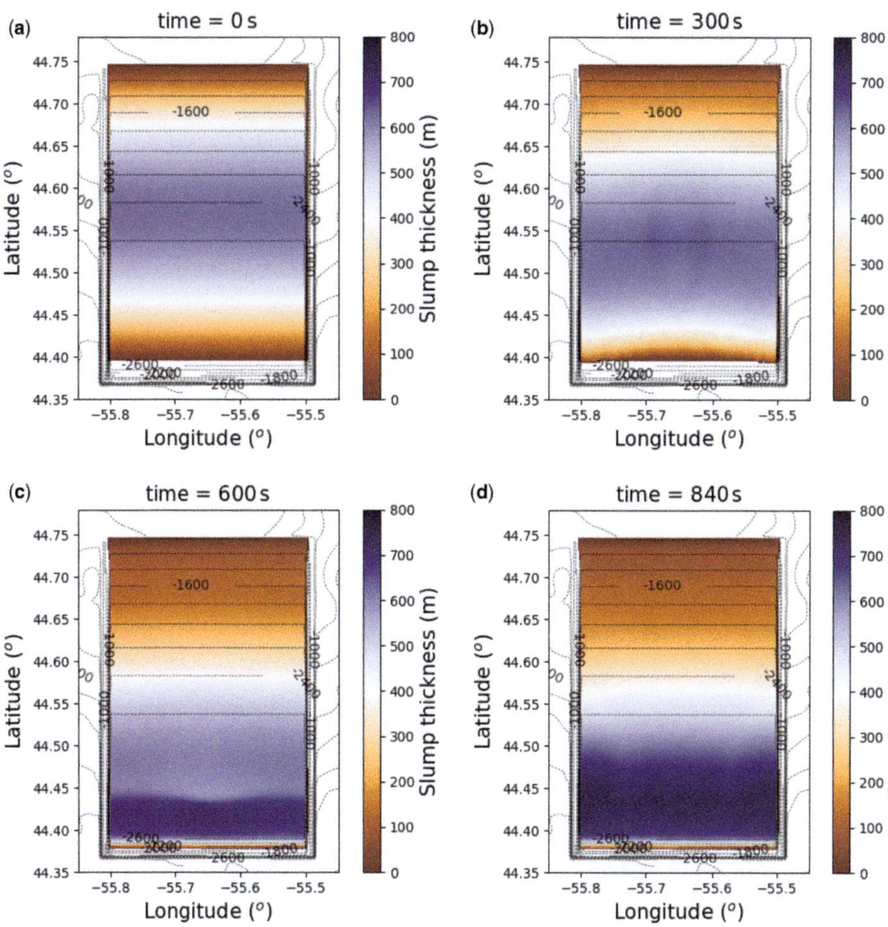

Fig. 17. Snapshots of the slump thickness for the 1929 Grand Banks pure slump scenario at different times. The slump mass stays inside the source area with employed BingClaw parameters $\tau_y = 85$ kPa, $\mu = 10$ kPa s^n and $n = 0.25$.

In the early phase, the generated wave (see Fig. 14) has a positive sea-surface elevation at the southern end of the slump area and a negative elevation at the northern end of the slump area. It is aligned north–south along the failure surface slope orientation. One hour after the slump mass release, the wave has started to turn gradually northwards and reaches the latitude 46° N after two hours. The main wave direction is towards the Burin Peninsula, whereas there is also a focus towards the Avalon Peninsula further east. Results extracted over the transect just south of Burin further show that maximum offshore sea-surface elevations range from 4 to 9 m for different landslide yield strengths (see Fig. 15a), which are in the same range or, for the lowest yield strengths, somewhat higher than those found by Løvholt *et al.* (2019). Figure 16a shows the maximum sea-surface elevations over the full simulation time, which coincides with the large waves observed near the Burin Peninsula (see, for example, Fine *et al.* 2005). Field observations of run-up elsewhere were mostly below 2 m; however, our simulations show waves just as large near Nova Scotia and the Avalon Peninsula as found near Burin.

More tuning would be necessary to provide a closer agreement with the data. For instance, Schulten *et al.* (2019*b*) found a vertical uplift of the slump mass of 100 m at its downslope extremity, although our example with $\tau_y = 85$ kPa produces a much larger vertical uplift of *c.* 400 m. Our simulations merely provide a first attempt. However, the simulations clearly show that the viscoplastic model is capable of producing sufficiently strong slump-induced waves to produce a tsunami at least of the size of the 1929 Grand Banks event. We re-emphasize that our objective here was primarily to investigate whether the material parameter ranges for the 1HD case were representative of a real example, and this analysis shows that they are.

Slump scenarios without over-topping. We turn our attention to the pure slump, which is confined by the outreaching fault at the lower extremity, and to its tsunami using the same volume and material parameters as for the over-topping slump. Figure 17

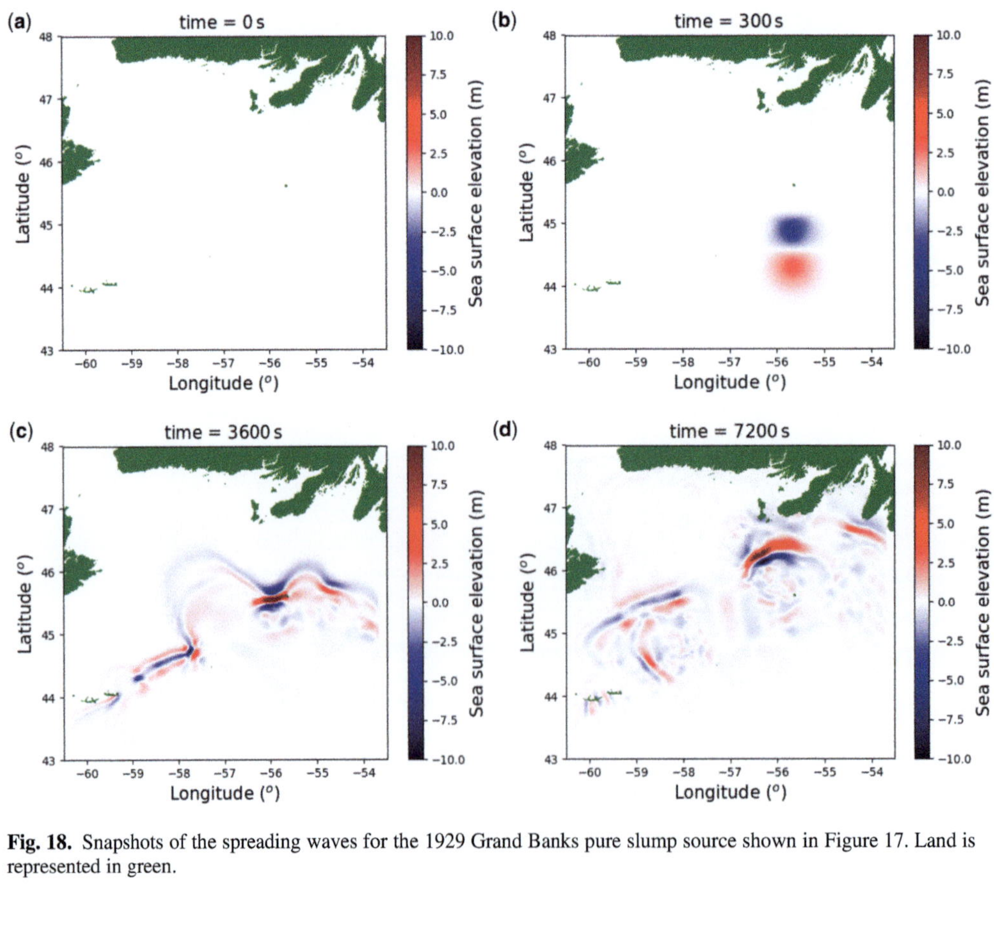

Fig. 18. Snapshots of the spreading waves for the 1929 Grand Banks pure slump source shown in Figure 17. Land is represented in green.

shows the slump thickness 0, 300, 600 and 840 s after the mass release. At the last time the slump motion has stopped. The maximum vertical uplift is *c.* 800 m, which is twice as much as for the over-topping slump due to confinement. The spreading waves (see Fig. 18) and the total wave field (see Fig. 16b) have a similar radiation pattern as the over-topping slump tsunami; however, the positive generated waves are significantly larger along the ridges between the Laurentian Channel, the Halibut Channel and the Haddock Channel than the waves for the over-topping slump source (see Fig. 16). Figure 15b shows a transect just south of Burin with maximum offshore sea-surface elevations ranging from 1 to 5 m, which are, however, in the same range as the sea-surface elevations for the over-topping case. Near longitude 55.7°, we see that the over-topping scenario produces slightly larger waves than the pure slump. Still, on an overall basis, we suggest that the pure slump event seems to be a slightly more efficient tsunami generator than the over-topping event. This was confirmed by our own preliminary work on simulating Grand Banks (results not shown) with other slump configurations, where the difference was even clearer.

Concluding remarks

In this paper, we have conducted a study of slump-induced tsunamis using a depth-averaged viscoplastic landslide model as the tsunami source, and a linear dispersive long-wave model for the tsunami propagation. Our main emphasis has been to study the sensitivity to slump material properties in 1HD on idealized geometries and the resulting slump kinematics on tsunami genesis. Contrary to most previous studies, our use of a viscoplastic landslide model allows us to link the tsunami directly to slump material properties, and avoid *ad-hoc* assumptions commonly made using a block model approach where the slump motion is prescribed. This refined model allows a more generalized treatment of slump sources, and hence is not limited to models that retrofit block source properties to simulate past events.

This study has shown that the material parameter that influences tsunami genesis the most is the initial yield strength of the sediment. Similar conclusions were reached for translational landslides in studies of the tsunami genesis of the Storegga landslide, for example (Kim *et al.* 2019). Moreover, our range of the dynamic landslide consistency (related to the viscosity) shows a more moderate influence on tsunami genesis. Naturally, geometrical factors, such as the slope angle and volume of the slump, were found to have a strong influence on the tsunami genesis too. Several kinematic properties were found to correlate well with the maximum landward sea-surface elevation. For the case of constant slide volume, the maximum landward sea-surface elevation increases monotonically with both scaled bed-parallel maximum velocity and acceleration mimicking a power-law relationship. The maximum landward sea-surface elevation also increases monotonically with vertical acceleration and velocity, but a less systematic relationship was found in this case. For the more general cases where variable volumes were investigated, the maximum bed-parallel momentum and momentum rates correlate well with the maximum landward sea-surface elevation, while the maximum landward sea-surface elevation had a somewhat less systematic relationship with corresponding vertical momentum and momentum rates.

Some of the findings of this study have been identified already in past studies (Tinti *et al.* 2001; Ward 2001; Løvholt *et al.* 2005), but only for translational landslides with a simplified block source representation. Here, we show that similar relationships between landslide velocities, accelerations and momentum apply for slumps. In particular, we find the scaling between the maximum height of the generated wave and the maximum bed-parallel landslide speed divided by the wave celerity, $\eta_{max} \propto Fr^{0.9}$. We note that the exponent of 0.9 is less than the linear relationship (i.e. exponent 1) expected for small Froude numbers for frontal wave elevations and rear wave troughs without any interference (Løvholt *et al.* 2015). In our study, we clearly have destructive interference between the waves caused at the front and rear part of the slump which reduced the tsunamigenic potential. However, we find, similar to Løvholt *et al.* (2005, 2015), an almost linear scaling with the horizontal landslide acceleration, which is hence clearly a good proxy for tsunamigenic potential. An additional finding from our study is that the angular momentum shows a particularly good correlation with the maximum landward sea-surface elevation. This suggests that the tsunamigenic potential can be directly linked to rotational kinematic properties of the slump. We are unaware of previous studies that identify such a relationship.

A second part of the study is devoted to studying the 1929 Grand Banks slump and tsunami in a real topographical setting. This was primarily done to investigate whether the parameter ranges used in the viscoplastic slump model in 1HD were realistic. A detailed analysis of the 1929 Grand Banks event with an emphasis on obtaining a close match with field observations of the tsunami was not attempted. Nevertheless, our model was set up with new field observations by Schulten *et al.* (2019*b*) to illustrate how the geological interpretation provided a significantly revised explanation for the slump event. Schulten *et al.* (2019*b*) concluded that the 1929

Grand Banks slump failed mainly southwards, and that the main slump volume was much larger than previously thought (390 km^3). Our tsunami modelling suggests that a viscoplastic model indeed should be capable of producing sufficiently large waves. The 1929 Grand Bank event also served the purpose of testing how a complex event with slump failure and over-topping compares with a pure slump event with respect to tsunami genesis. We found that the pure slump produced larger overall waves compared to the over-topping scenario. All in all, the 1929 Grand Banks model including new field observations for the slump event and an idealized study in 1HD could revise our understanding of tsunami genesis.

Acknowledgements Computations were done on a computer located at the Norwegian Geotechnical Institute in Oslo, Norway. The authors are very grateful for positive criticism from the editor Aggeliki Georgiopoulou, and the reviewers David Mosher and Alberto Armigliato.

Funding This project has received funding from the European Union's Horizon 2020 research and innovation programme under the Marie Sklodowska-Curie grant agreement No.721403.

Author contributions **TZ**: data curation (lead), formal analysis (lead), investigation (lead), methodology (equal), software (equal), validation (equal), visualization (lead), writing – original draft (lead), writing – review & editing (equal); **FL**: conceptualization (lead), data curation (supporting), formal analysis (supporting), investigation (supporting), methodology (equal), project administration (lead), software (supporting), supervision (lead), validation (equal), writing – original draft (supporting), writing – review & editing (lead); **GKP**: formal analysis (supporting), investigation (supporting), methodology (equal), software (equal), supervision (equal), validation (equal), writing – original draft (supporting); **CBH**: conceptualization (supporting), data curation (supporting), formal analysis (supporting), investigation (supporting), methodology (equal), project administration (equal), supervision (equal), validation (equal), writing – review & editing (equal).

Appendix A

Grid refinement tests

For the 1HD simulations, we conducted grid refinement tests on the spatial grid for the slump model, tsunami model, and the Kajiura-type filter (resolutions and parameters in Table A1). For the slump model, we tested soft slump materials, low τ_y and low μ. The slump thickness depended strongly on the grid resolution for $\Delta x > 80$ m. After 240 s, for instance, the slump thickness at the lower extremity is 8% thinner for $\Delta x = 160$ m than for $\Delta x = 26$ m, and for resolutions $\Delta x \leqslant 80$ m, the slump thickness varies maximum by 4%. The slump thickness at the upslope part coincides for resolutions $\Delta x \leqslant 80$ m, but gives twice the corresponding slump thickness for $\Delta x = 160$ m. Thus, a spatial grid resolution of 80 m is chosen for further use. The time step, Δt, is adapted during the simulation to keep the Courant–Friedrichs–Lewy (CFL) number (Courant *et al.* 1967),

Table A1. *Numerical parameters for the 1HD grid refinement tests*

Physical process	Numerical parameter	Value	Units
Landslide	cell size	26.7, 40, **80**, 160	m
Landslide	CFL number	**0.45**	
Kajiura-type filter	time interval	5, 10, 15, 20, **30**, 40, 50, 60	s
Kajiura-type filter	cell size	**80**	m
Wave propagation	cell size	55, 110, **220**	m
Wave propagation	CFL number	**0.5**	

Applied models are BingClaw, Kajiura filter and GloBouss in 1HD. The values in bold were used in our study.

$$\mathrm{CFL} = \frac{U_o\,\Delta t}{\Delta x}, \tag{A1}$$

constant. Here U_0 is the maximum particle speed in the slide body. In all our 1HD model runs, we use a CFL = 0.45, which yields stable behaviour (greatest landslide velocities are *c.* 70 m s^{-1}). When the source input is fed into the tsunami model each 30 s, we have a deviation of less than 2% from the smallest interval tested (5 s) at $t = 480$ s. Hence, we stay with 30 s. Since the surface response is smoother than the slide surface, application of the same spatial grid resolution for the Kajiura-type filter as for the landslide model, 80 m, is more than adequate. We tested spatial grid resolutions for the tsunami model 980 s after slump mass release. The maximum landward sea-surface elevation of a resolution of 220 m only deviated by 0.6% from the elevation of a finer resolution of 55 m. Therefore, we further used the 220 m resolution. The CFL number used is 0.5.

In 2HD, we executed spatial and temporal grid refinement tests of the landslide model BingClaw, the Kajiura-type filter, and the tsunami model GloBouss. All numerical parameters can be found in Table A2. For BingClaw we evaluated the grid dependency on the slump thickness after 600 s in a transect striking north–south. At the location of the thickest slump mass, the thickness obtained with $\Delta x = 185$ m deviated only by 1.7% from that of $\Delta x =$

Table A2. *Numerical parameters for the 2HD grid refinement tests*

Physical process	Numerical parameter	Value	Units
Landslide	cell size (m)	93, **185**, 370, 741, 148	m
Landslide	CFL number	0.45, **0.65**, 0.85	
Kajiura type filter	time interval	20, **30**, 50, 80, 300	s
Kajiura type filter	cell size	232, **463**, 926	m
Wave propagation	cell size	**463**, 926, 1852	m
Wave propagation	CFL number	0.4, 0.6, **0.8**, 1.0	

Applied models are BingClaw, Kajiura filter, and GloBouss in 2HD. The values in bold were used in our study.

93 m. The double the resolution of 370 m caused a corresponding deviation of 6.8%. Thus, we used a spatial resolution of 185 m. The slump was stable with a CFL number of 0.65.

We evaluated the spatial resolution of the input fluxes into GloBouss 120 s after failure by analysing the first wave amplitudes of the propagated waves along the same transect striking north–south. The wave height of the 926 m resolution differed only by 3.1% from that of 232 m. However, since it was feasible to use even 463 m in the modelling, we chose that. These sources were fed into GloBouss at various time intervals (see Table A2), whereas the resulting wave field 3000 s after failure was analysed. The amplitude of a resolution of 50 s deviated by 8% from a 30 s resolution. The wave amplitude of a finer resolution of 20 s deviated by 2% from the 30 s resolution. Thus, the flux sources were fed into GloBouss each 30 s.

We tested three spatial grid resolutions for the tsunami propagation model GloBouss 4100 s after failure. The first wave amplitude of the second finest resolution deviated 2% from the finest resolution, and the coarsest resolution deviated by 8% from the finest resolution. Since the finest resolution was feasible, we applied that one. The CFL number in GloBouss was chosen as 0.8.

Kinematics

As we run our models in a depth-averaged regime, we divide the slump mass into vertical columns with length of one cell size. The height difference of the slump surface H in each column at two adjacent time steps serves as input for the vertical velocity calculation. With that velocity we calculate the vertical acceleration a_z in each column. It should be noted that resulting vertical velocities are half a time step behind the time of the surface heights of the next time loop, and the vertical accelerations are half a time step behind the velocities. The actual calculation of the vertical acceleration is a central derivative:

$$a_z^{(n)} = \frac{H^{(n+1)} - 2H^{(n)} + H^{(n-1)}}{\Delta t^2} \quad \text{for} \quad n \geq 2 \qquad \text{(A2)}$$

The first calculated acceleration refers to $t^{(n=\,1)} = 0.25\Delta t$, which means that the time interval for the calculations Δn is not constant for $t < \Delta t$.

Another kinematic quantity is the bed-parallel acceleration $a_{||}$, evaluated at the same time. In order to do so, we need to average the bed-parallel velocity $u_{||}$ between two time steps and then evaluate the time derivative:

$$a_{||}^{(n)} = \frac{\dfrac{v_{||}^{(n+1)} + v_{||}^{(n)}}{2} - \dfrac{v_{||}^{(n)} + v_{||}^{(n-1)}}{2}}{\Delta t} \quad \text{for} \quad n \geq 2 \qquad \text{(A3)}$$

The same exception for the first calculation step $n = 1$ applies here.

A third quantity is the angular momentum $\vec{L}$ of the entire slump mass, which is defined as $d\vec{L} = m(\vec{r} \times \vec{v})$ where m is the mass of a vertical column, $\vec{r}$ the position vector, and $\vec{v}$ the velocity vector. Each quantity is time dependent. The position vector ranges from the dynamic centre of mass to the average centre of a vertical column between two time steps. Position and velocity vectors are both split into horizontal and vertical components, r_x and r_z, v_x and v_z, respectively. The vertical velocity component corresponds to the one from the calculations above, but we approximate the horizontal velocity component v_x with the bed-parallel velocity $v_{||}$, as the bed is nearly horizontal. Maximum bed slope angle is 5.25°. Equation (A4) shows the calculation for the total angular momentum, which is a sum of all angular momenta for each vertical column.

$$\begin{aligned} \vec{L}^{(n-\frac{1}{2})} &= \sum_{n=1}^{n_{\text{end}}} m \frac{\vec{r}^{(n)} + \vec{r}^{(n-1)}}{2} \times \frac{\vec{u}^{(n)} + \vec{u}^{(n-1)}}{2} \\ &= \sum_{n=1}^{n_{\text{end}}} m\left(r_x^{(n-\frac{1}{2})} v_z^{(n-\frac{1}{2})} - r_z^{(n-\frac{1}{2})} v_{||}^{(n-\frac{1}{2})}\right) \end{aligned} \qquad \text{(A4)}$$

For the analysis in this study, we use maximum values of all times of each quantity described above.

References

Bardet, J.-P., Synolakis, C., Davis, H., Imamura, F. and Okal, E. 2003. Landslide tsunamis: recent findings and research directions. *Pure and Applied Geophysics*, **160**, 1793–1809, https://doi.org/10.1007/s00024-003-2406-0

Berger, M.J., George, D.L., LeVeque, R.J. and Mandli, K.T. 2011. The GeoClaw software for depth-averaged flows with adaptive refinement. *Advances in Water Resources*, **34**, 1195–1206, https://doi.org/10.1016/j.advwatres.2011.02.016

Courant, R., Friedrichs, K. and Lewy, H. 1967. On the partial difference equations of mathematical physics. *IBM Journal of Research and Development*, **11**, 215–234, https://doi.org/10.1147/rd.112.0215

Fine, I., Rabinovich, A., Bornhold, B., Thomson, R. and Kulikov, E. 2005. The Grand Banks landslide-generated tsunami of November 18, 1929: preliminary analysis and numerical modeling. *Marine Geology*, **215**, 45–57, https://doi.org/10.1016/j.margeo.2004.11.007

George, D., Iverson, R. and Cannon, C. 2017. New methodology for computing tsunami generation by subaerial landslides: application to the 2015 Tyndall Glacier landslide, Alaska. *Geophysical Research Letters*, **44**, 7276–7284, https://doi.org/10.1002/2017GL074341

Grilli, S.T., Tappin, D.R. *et al.* 2019. Modelling of the tsunami from the December 22, 2018 lateral collapse of Anak Krakatau volcano in the Sunda Straits, Indonesia. *Scientific Reports*, **9**, 11946–11946, https://doi.org/10.1038/s41598-019-48327-6

Gylfadóttir, S.S., Kim, J. *et al.* 2017. The 2014 Lake Askja rockslide-induced tsunami: optimization of numerical tsunami model using observed data. *Journal of Geophysical Research: Oceans*, **122**, 4110–4122, https://doi.org/10.1002/2016JC012496

Hammack, J.L. 1973. A note on tsunamis: their generation and propagation in an ocean of uniform depth. *Journal of Fluid Mechanics*, **60**, 769–799, https://doi.org/10.1017/S0022112073000479

Harbitz, C., Løvholt, F. and Bungum, H. 2014. Submarine landslide tsunamis: how extreme and how likely? *Natural Hazards*, **72**, 1341–1374, https://doi.org/10.1007/s11069-013-0681-3

Haugen, K.B., Løvholt, F. and Harbitz, C.B. 2005. Fundamental mechanisms for tsunami generation by submarine mass flows in idealised geometries. *Marine and Petroleum Geology*, **22**, 209–217, https://doi.org/10.1016/j.marpetgeo.2004.10.016

Heezen, B.C. and Ewing, W.M. 1952. Turbidity currents and submarine slumps, and the 1929 Grand Banks [Newfoundland] earthquake. *American Journal of Science*, **250**, 849–873, https://doi.org/10.2475/ajs.250.12.849

Heezen, B.C., Ericson, D. and Ewing, M. 1954. Further evidence for a turbidity current following the 1929 Grand Banks earthquake. *Deep Sea Research (1953)*, **1**, 193–202, https://doi.org/10.1016/0146-6313(54)90001-5

Huang, X. and Garcia, M.H. 1998. A Herschel–Bulkley model for mud flow down a slope. *Journal of Fluid Mechanics*, **374**, 305–333, https://doi.org/10.1017/S0022112098002845

Imran, J., Parker, G., Locat, J. and Lee, H. 2001. 1D numerical model of muddy subaqueous and subaerial debris flows. *Journal of Hydraulic Engineering*, **127**, 959–968, https://doi.org/10.1061/(ASCE)0733-9429(2001)127:11(959)

Kajiura, K. 1963. The leading wave of a tsunami. *Bulletin of the Earthquake Research Institute*, **41**, 535–571.

Kim, J., Løvholt, F., Issler, D. and Forsberg, C.F. 2019. Landslide material control on tsunami genesis: the Storegga Slide and tsunami (8,100 Years BP). *Journal of Geophysical Research: Oceans*, **124**, 3607–3627, https://doi.org/10.1029/2018JC014893

Løvholt, F., Harbitz, C.B. and Haugen, K.B. 2005. A parametric study of tsunamis generated by submarine slides in the Ormen Lange/Storegga area off western Norway. *Marine Petroleum and Geology*, **22**, 219–231, https://doi.org/10.1016/j.marpetgeo.2004.10.017

Løvholt, F., Pedersen, G. and Gisler, G. 2008. Oceanic propagation of a potential tsunami from the La Palma Island. *Journal of Geophysical Research: Oceans*, **113**, C09026, https://doi.org/10.1029/2007JC004603

Løvholt, F., Pedersen, G.K. and Glimsdal, S. 2010. Coupling of dispersive tsunami propagation and shallow water coastal response *Open Oceanography Journal, Caribbean Waves Special Issue*, **4**, 71–82, https://doi.org/10.2174/1874252101004010071

Løvholt, F., Pedersen, G., Harbitz, C.B., Glimsdal, S. and Kim, J. 2015. On the characteristics of landslide tsunamis. *Philosophical Transactions of the Royal Society A*, **373**, 20140376, https://doi.org/10.1098/rsta.2014.0376

Løvholt, F., Bondevik, S., Laberg, J.S., Kim, J. and Boylan, N. 2017. Some giant submarine landslides do not produce large tsunamis. *Geophysical Research Letters*, **44**, 8463–8472, https://doi.org/10.1002/2017GL074062

Løvholt, F., Schulten, I., Mosher, D., Harbitz, C. and Krastel, S. 2019. Modelling the 1929 Grand Banks slump and landslide tsunami. *Geological Society, London, Special Publications*, **477**, 315–331, https://doi.org/10.1144/SP477.28

Mandli, K.T., Ahmadia, A.J. *et al.* 2016. Clawpack: building an open source ecosystem for solving hyperbolic PDEs. *PeerJ Computer Science*, **2**, e68, https://doi.org/10.7717/peerj-cs.68

Mesinger, F. and Arakawa, A. 1976. Numerical methods used in atmospheric models. *GARP Publication Series*, **17**.

Mosher, D.C. and Piper, D.J. 2007. Analysis of multibeam seafloor imagery of the Laurentian Fan and the 1929 Grand Banks landslide area. *In*: Lykousis, V., Sakellariou, D. and Locat, J. (eds) *Submarine Mass Movements and their Consequences*. Advances in Natural and Technological Hazards Research, **27**. Springer, Dordrecht, 77–88, https://doi.org/10.1007/978-1-4020-6512-5_9

Paris, A., Okal, E.A., Guérin, C., Heinrich, P., Schindelé, F. and Hébert, H. 2019. Numerical modeling of the June 17, 2017 landslide and tsunami events in Karrat Fjord, West Greenland. *Pure and Applied Geophysics*, **176**, 3035–3057, https://doi.org/10.1007/s00024-019-02123-5

Pedersen, G. and Løvholt, F. 2008. Documentation of a global Boussinesq solver. *Preprint series. Mechanics and Applied Mathematics*, http://urn.nb.no/URN:NBN:no-23418

Piper, D.J., Cochonat, P. and Morrison, M.L. 1999. The sequence of events around the epicentre of the 1929 Grand Banks earthquake: initiation of debris flows and turbidity current inferred from sidescan sonar. *Sedimentology*, **46**, 79–97, https://doi.org/10.1046/j.1365-3091.1999.00204.x

Ren, Z., Zhao, X. and Liu, H. 2019. Numerical study of the landslide tsunami in the South China Sea using Herschel-Bulkley rheological theory. *Physics of Fluids*, **31**, 056601, https://doi.org/10.1063/1.5087245

Schambach, L., Grilli, S.T., Kirby, J.T. and Shi, F. 2018. Landslide tsunami hazard along the upper US East Coast: effects of slide deformation, bottom friction, and frequency dispersion. *Pure and Applied Geophysics*, **176**, 3059–3098, https://doi.org/10.1007/s00024-018-1978-7

Schulten, I., Mosher, D.C., Krastel, S., Piper, D.J. and Kienast, M. 2019*a*. Surficial sediment failures due to the 1929 Grand Banks Earthquake, St Pierre Slope. *Geological Society, London, Special Publications*, **477**, 583–596, https://doi.org/10.1144/SP477.25

Schulten, I., Mosher, D.C., Piper, D.J. and Krastel, S. 2019*b*. A massive slump on the St. Pierre Slope, a new perspective on the 1929 Grand Banks submarine landslide. *Journal of Geophysical Research: Solid Earth*, **124**, 7538–7561, https://doi.org/10.1029/2018JB017066

Sepúlveda, S.A. and Serey, A. 2009. Tsunamigenic, earthquake-triggered rock slope failures during the April 21 2007 Aisén earthquake, southern Chile (45.5° S). *Andean Geology*, **36**, 131–136.

Synolakis, C.E., Bardet, J.-P. *et al.* 2002. The slump origin of the 1998 Papua New Guinea tsunami *Proceedings of the Royal Society of London. Series A: Mathematical, Physical and Engineering Sciences*, **458**, 763–789, www.jstor.org/stable/3067390

Tappin, D. 2010. Mass transport events and their tsunami hazard. *In*: Mosher, D.C., Shipp, R.C., Moscardelli, L., Chaytor, J.D., Baxter, C.D.P., Lee, H.J. and Urgeles, R. (eds) *Submarine Mass Movements and Their Consequences*. Advances in Natural and Technological Hazards Research, **28**. Springer, 667–684.

Tappin, D.R., Watts, P. and Grilli, S.T. 2008. The Papua New Guinea tsunami of 17 July 1998: anatomy of a catastrophic event. *Natural Hazards and Earth System Science*, **8**, 243–266, https://doi.org/10.5194/nhess-8-243-2008

Tinti, S., Bortolucci, E. and Chiavettieri, C. 2001. Tsunami excitation by submarine slides in shallow-water approximation. *Pure and Applied Geophysics*, **158**, 759–797, https://doi.org/10.1007/PL00001203

Vanneste, M., Løvholt, F. *et al.* 2019. A novel quasi-3D landslide dynamics model: from theory to applications and risk assessment. Paper OTC-29363-MS, presented at the Offshore Technology Conference, 6–9 May, Houston, TX, https://doi.org/10.4043/29363-MS

Wang, J., Ward, S.N. and Xiao, L. 2015. Numerical simulation of the December 4, 2007 landslide-generated tsunami in Chehalis Lake, Canada. *Geophysical Journal International*, **201**, 372–376, https://doi.org/10.1093/gji/ggv026

Ward, S.N. 2001. Landslide tsunami. *Journal of Geophysical Research*, **6**, 11201–11215, https://doi.org/10.1029/2000JB900450

Watts, P. 2000. Tsunami features of solid block underwater landslides. *Journal of Waterway, Port, Coastal, and Ocean Engineering*, **126**, 144–152, https://doi.org/10.1061/(ASCE)0733-950X(2000)126:3(144)

Yavari-Ramshe, S. and Ataie-Ashtiani, B. 2016. Numerical modeling of subaerial and submarine landslide-generated tsunami waves-recent advances and future challenges. *Landslides*, **13**, 1325–1368, https://doi.org/10.1007/s10346-016-0734-2

Yavari-Ramshe, S. and Ataie-Ashtiani, B. 2019. On the effects of landslide deformability and initial submergence on landslide-generated waves. *Landslides*, **16**, 37–53, https://doi.org/10.1007/s10346-018-1061-6

Yeh, H., Imamura, F., Synolakis, C., Tsuji, Y., Liu, P. and Shi, S. 1993. The Flores Island tsunamis. *Eos, Transactions American Geophysical Union*, **74**, 369–373, https://doi.org/10.1029/93EO00381

Yudhicara, B.P., Bani, P. and Darmawan, A. 2015. Geothermal system as the cause of the 1979 landslide tsunami in Lembata Island, Indonesia. *Indonesian Journal on Geoscience*, **2**, 91–99, https://doi.org/10.17014/ijog.2.2.91-99

A scenario-based assessment of the tsunami hazard in Palermo, northern Sicily, and the southern Tyrrhenian Sea

Jack Dignan[1], Aaron Micallef[2], Christof Mueller[3], Attilio Sulli[4*], Elisabetta Zizzo[4] and Daniele Spatola[2]

[1]School of the Environment, Geography and Geosciences, University of Portsmouth, PO1 2UP, UK

[2]Marine Geology and Seafloor Surveying, University of Malta, MSD 2080, Malta

[3]GNS Science, Lower Hutt, 5010, New Zealand

[4]Department of Earth and Marine Sciences, University of Palermo, 90123, Italy

JD, 0000-0002-6765-0361; AS, 0000-0002-7705-3632; EZ, 0000-0002-2436-3469; DS, 0000-0001-6219-7062
*Correspondence: attilio.sulli@unipa.it

Abstract: Palermo is a populous city situated on the northern coast of Sicily, bordered by the Tyrrhenian Sea. This central part of the Mediterranean Sea features dramatic bathymetry, numerous subaqueous landslides and is also the epicentre to many subaqueous earthquakes. As such, the region is an area prone to tsunamis. This investigation uses the Cornell Multi-Grid Coupled Tsunami (COMCOT) tsunami modelling package to simulate five near-field landslides, and five near-field earthquakes regarded as worst-case credible scenarios for Palermo. The seismic simulations produced waves on a very small scale, the largest being *c.* 5 cm at its maximum height, and none of the earthquake-generated tsunami waves produced any measurable inundation. The landslide simulations produced larger waves ranging from 1.9 to 12 m in maximum height, two of which resulted in inundation in areas surrounding the Port of Palermo. Sensitivity analysis identified that fault width and dislocation as well as landslide-specific gravity did have significant influence over maximum wave height, inundation and maximum run-up wave height. There are methodological issues limiting the extent to which this study forms a comprehensive tsunami hazard assessment of Palermo, such as gaps in bathymetric data, computational restrictions and lack of a probabilistic element. These issues are counteracted by the fact that this study does serve as a robust first step in identifying that landslides in the region may produce larger tsunami waves than earthquakes, and that the directionality of mass movement is critical in landslide-driven tsunami propagation in the southern Tyrrhenian region.

Tsunamis are defined as a series of gravity-driven water waves created by an impulsive disturbance to the water body (Cox 1963; Parls *et al.* 2014; Sambah and Miura 2019). Whilst a relatively infrequent geohazard, they have potential for widespread and devastating impact; for example the infamous 1755 Lisbon tsunami, the 2004 Indonesian Boxing Day tsunami and the 2011 Tōhoku tsunami, leading to the catastrophic deaths of approximately 20 000, 200 000 and 20 000 people, respectively (Pelinovsky and Poplavsky 1996; Kathiresan and Rajendran 2005; Ruffman 2006; Docey *et al.* 2007; Nakahara and Ichikawa 2013). On a global scale, the majority of these tsunamis are generated by megathrust earthquakes at subducting plate margins, followed in frequency by subaqueous mass movements (Watts *et al.* 2005; Tappin 2010). However, tsunami waves produced by smaller causative forces still pose a danger to coastal vessels, poorly constructed coastal buildings, coastal infrastructure (transport networks, nuclear plants, etc.) and are able to inundate entire communities in lower-lying favourable topographic regions (Papadopoulos and Imamura 2001).

Tsunami generation can be explained with relative simplicity; in order for a tsunami wave to generate, water displacement is required. This displacement can take the form of seismically generated faults breaching the seabed, subaqueous mass movement (i.e. landsliding, canyon collapse, etc.), trench rupture from megathrust earthquakes along subduction zones, coastal volcano flank collapse or subaqueous eruptions (Grilli *et al.* 2019; IRIS 2019; Paris and Ulvrova 2019). The vast majority of tsunamis (*c.* 80%) are generated by seafloor seismic rupture, followed by landsliding and volcanism, in that order (International Tsunami Information Centre 2018).

Across countries bordering the Mediterranean Sea (especially in Italy, France and Spain), many populous hubs are located along coastal regions, thus presenting significant vulnerability to coastal hazards. Therefore, countries like these, having

From: Georgiopoulou, A., Amy, L. A., Benetti, S., Chaytor, J. D., Clare, M. A., Gamboa, D., Haughton, P. D. W., Moernaut, J. and Mountjoy, J. J. (eds) 2020. *Subaqueous Mass Movements and their Consequences: Advances in Process Understanding, Monitoring and Hazard Assessments*. Geological Society, London, Special Publications, **500**, 63–80.
First published online March 31, 2020, https://doi.org/10.1144/SP500-2019-181

vulnerabilities spatially concentrated in coastal regions, require extensive assessment and quantification of the hazard posed by events such as tsunamis. The city of Palermo, with a population of more than 660 000, is the most populated city in Sicily, and borders with the tectonically active Tyrrhenian Sea, making it predisposed to tsunami hazard.

The coastal regions of Sicily and southern Italy have historically fallen victim to a number of significant tsunami events, generated within the Tyrrhenian Basin and surrounding Mediterranean Sea. In 1908, a 7.1 MW earthquake in the Messina Straits generated one of the deadliest tsunamis in Sicilian history, destroying coastal towns and killing *c.* 60 000 people (Omori 1910). It is also reported that a tsunami of similar severity hit Messina coastlines around AD 1169. In more recent history, multiple tsunamis are reported to have been generated by landslides around Stromboli in the Aeolian Islands. Events in 1916, 1919, 1930, 1944 and 2002 produced waves affecting the north of Sicily and southern Italy, with assigned intensity values of 3 to 4 (Maramai *et al.* 2005; Tinti *et al.* 2005). Due to the low frequency of these more significant tsunami events, research has focused on understanding and recreating historical events, and less on attempting to quantify the more general tsunami hazard posed to infrastructural and population hubs in Sicily.

There is little time between a tsunami trigger and the resulting waves reaching coastlines in confined seas such as the Tyrrhenian; as such there is not enough time to run complex models to accurately forecast tsunami evolution, making probabilistic approaches a popular method for providing insights into likely scenarios. A variety of methodologies have been adopted to reduce this time lag; namely, using probabilistic-based scenario modelling (Tonini *et al.* 2011), correlational analysis of tsunami source and run-up properties (Sulli *et al.* 2018) or through mass-scenario modelling, all aiming to model tsunamis ahead of their generation.

An emerging technique for assessing tsunami hazard is the use of Probabilistic Tsunami Hazard Assessment (PTHA), which models and quantifies the probability that a tsunami of any given magnitude or source will occur. PTHA has also been implemented globally to assess the coastal tsunami hazard in a manner which also incorporates probability, thus making it an adaptable and relevant methodology (Gonzalez *et al.* 2014; Horspool *et al.* 2014; Davies and Griffin 2018). Often, PTHA analysis concentrates on quantifying the probability of an event, rather than the modelling of them, and so does not fully analyse local influences and potential exacerbating factors of tsunami events revealed through numerical modelling (Anita *et al.* 2012).

However, before using PTHA analysis, in order to understand the scale of the hazard, often the first stage is to undertake an analysis of 'worst-case scenarios'. This can be through the use of historical records to find some of the events in recorded history that have had the largest impact (e.g. Farreras and Sanchez 1991), by using some of the parameters of the worst-possible scenarios, or by running simulations for credible worst-case tsunami scenarios (known as Worst-Case Credible Tsunami Scenario Analysis – WCTSA). WCTSA generally uses instrumental or historical records to reconstruct worst-case scenarios of causative events within the realms of credible probability, in order to understand the scale of the impacts such events would have and is a popular first stage to tsunami assessment in literature (Tonini *et al.* 2011; Harbitz *et al.* 2012; Jelinek *et al.* 2012).

Anita *et al.* (2012) quantified the probability of tsunami hazard focused on Catania and the Messina Strait using a PTHA, providing a foundational understanding of the southern Italian tsunami hazard. Other research analyses far-field causative events from the Hellenic Arc and Messina Strait regions and their effect on western Sicily and Catania (Tonini *et al.* 2011), continuing to skirt the impact on the eastern Sicilian coast. Despite significant (albeit lower than eastern Sicily) probability of tsunami occurrence along the northern Sicilian coastline, few studies target the area for extensive tsunami hazard assessment.

Therefore, there is a gap in the understanding of the tsunami hazard to northern Sicily, particularly its capital city, Palermo. Palermo is the fifth most populated city in Italy (Istat 2011), and therefore holds significant potential for huge loss of life should an extreme disaster such as a tsunami occur. Furthermore, much focus of previous hazard assessment of these areas considers only far-field, catastrophic events which escape instrumental record and, as such, the modelling of these is somewhat speculative. However, research in this area conducted by Sulli *et al.* (2018) has assessed hazard posed by near-field tsunamigenic landslides within the Gulf of Palermo, by calculating run-up values of tsunamis generated by a range of historical mass movements on a theoretical basis.

As mentioned, the vast majority of tsunamis are caused by seismic rupture or mass movement and, as such, it stands to reason that the first stages of tsunami hazard assessment focus on these mechanisms. The first stage will be to understand the worst-case scenarios of both seismic and landslide hazard in the surrounding areas of Palermo, with the goal being to aid hazard zonation in the region and taking the first step in the process of forming a complete tsunami hazard assessment.

This research will use a WCTSA methodology to make use of local high-resolution and instrumental data to understand the potential impact that near-field tsunami generation would have on the city of Palermo. This research will form an important 'first step' in the process of fully understanding the tsunami hazard in Palermo and can be used to inform evacuation efforts when little time is available for real-time modelling, which is almost always the case for proximal generation.

Regional setting

Within the central sector of the Mediterranean basin lies the Tyrrhenian Sea, bounded by Corsica, Sardinia, the Italian Peninsula and Sicily, as well as containing several small islands, such as Ustica, Elba and the Aeolian Archipelago. The Tyrrhenian Sea is a back-arc basin, formed due to the Ionian slab rollback towards the SE (Sartori 2003). Subduction and significant faulting occur along the entire western seaboard of the Tyrrhenian Sea, the Italian Peninsula and Sicily (Dilek 2006). The plates converge producing a belt of volcanism within subaerial Sicily, Italy and the Aeolian Islands, which remain extremely active at present and pose significant hazard to surrounding areas. Much like global tsunamigenic event ratios, 89% of Mediterranean tsunami records have a seismic origin (either subaqueous or, in some cases, terrestrial), 9% were due to gravity-initiated subaqueous slope failure and 5% caused by volcanism (Maramai *et al.* 2014). The clustered distribution of these causative events has been spatially categorized, with the highest potential tsunamigenic zones situated in the central Mediterranean regions surrounding the Hellenic Arc, followed by the Sicilian coastlines (Papadopoulos *et al.* 2014).

Due to the both global and Mediterranean trends, as well as the complexity introduced by modelling volcanic tsunamis through subaqueous explosivity and flank collapse, this study will focus on assessing the more probable impact caused by subaqueous landslide- and seismically-triggered tsunamis.

Seismicity in the Tyrrhenian Basin

Understanding the distribution and size of seismic events within the study area is important before selecting the scenarios to be modelled. Within the Tyrrhenian Basin, the vast majority of seismic events have occurred along the curved coastline of northern Sicily and SW Italy, with the exception of a few events towards the centre of the basin (Fig. 1). In addition to understanding the spatial distribution of subaqueous seismicity, one must also acknowledge the magnitude distribution. Especially when simulating using a WCTSA methodology, the scenarios must be credible, and defining the bounds of credibility relies on having reliable historical records. Through the use of reliable instrumental records, it was established that the largest seismic events that have occurred since instrumental records began, in 1985,

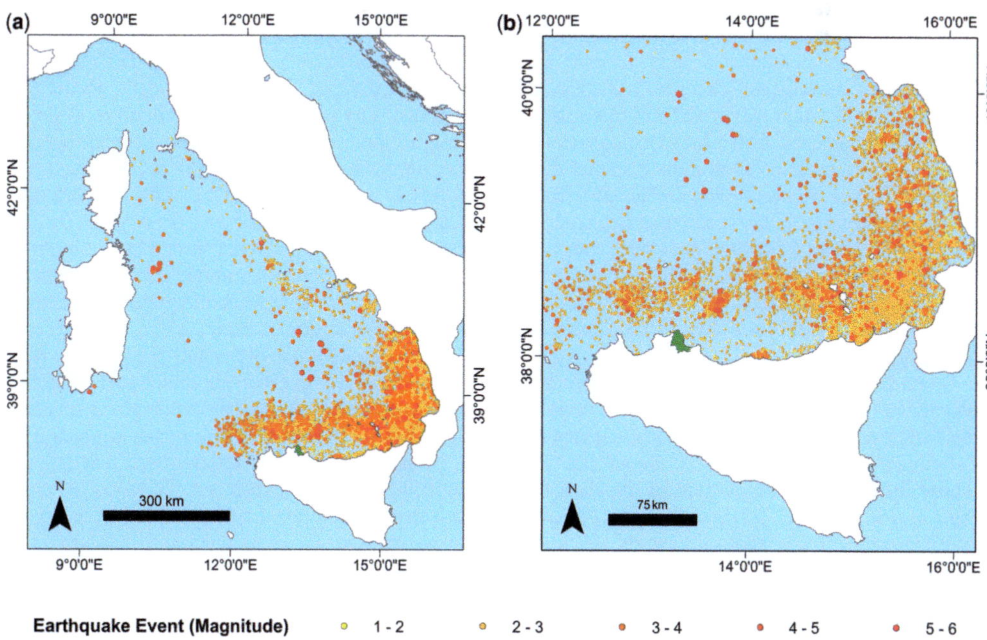

Fig. 1. Spatial distribution and magnitudes of all subaqueous seismic events within (**a**) the Tyrrhenian Basin and (**b**) northern Sicily between 1985 and 2018 as held on the INGV seismic database (2019), with Palermo highlighted in green (GADM 2018).

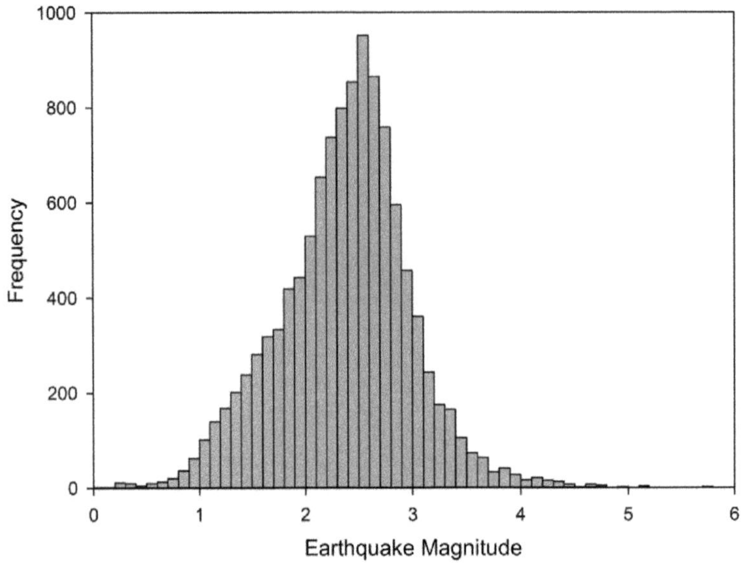

Fig. 2. Frequency–magnitude distribution of all subaqueous seismic events within the Tyrrhenian Basin between 1985 and 2018 as held on the INGV seismic database (2019).

measured between 4 and 6 in event magnitude (Fig. 2).

Subaqueous landsliding in the Tyrrhenian Basin

Although less comprehensive than seismic catalogues, mass movements cited in literature and those which fall within surveyed bathymetric zones have been collated by EMODnet Geology (2019), providing a resource for approximate spatial distribution and sizes of these events. The distribution of mass movements in the Tyrrhenian Sea is almost entirely in the coastal regions, with very few occurring in the central basin (Fig. 3). This would suggest that the majority of these events occur as a result of collapse of continental slope and the receding of the continental shelf.

When analysing the size of the mass movements recorded in the EMODnet Geology catalogue alongside their frequency, it is apparent that the majority of slides are very small in area and, as such, are likely to pose the most probable hazard in the short term (Fig. 4) (Sulli *et al.* 2018; EMODnet Geology 2019). However, larger slides, in excess of 200 km^2 in area, do occur. This is a point, however, where one must recognize the unreliability of these data due to their incompleteness; the delineation of landslides is a subjective process that primarily makes use of knowledge of subaqueous morphometric features. It is not unreasonable to suggest that the delineation of one slide is, in fact, a number of smaller ones, or that marine debris overlying a landslide scarp prevents identification in bathymetric surveys.

Data and methodology

A number of open source and proprietary tsunami modelling software packages exist, each aiming to either improve or develop the output accuracy or computational efficiency of simulations. Many of these computer programs modelling tsunami generation, propagation and inundation focus on an aspect of the tsunami lifespan, aiming to increase accuracy in a given stage (Hwang and Divoky 1970; Zhou *et al.* 2011). Other programs place a focus on providing hazard assessment through rapid computation and therefore are able to be integrated into tsunami early warning systems (Tkalich *et al.* 2007; Gusman and Tanioka 2014; Behrens and Dias 2015). However, with the advances in computational power and knowledge, the creation of programs able to combine these stages with reasonable accuracy is on the rise.

Some examples of widely used packages to model tsunami life cycles are TUNAMI (Tohoku University's Numerical Analysis Model for Investigation of Near-Field Tsunamis) (Imamura *et al.* 2006), TUNA (Tsunami Analysis) (Teh *et al.* 2006; Koh *et al.* 2007), MOST (Method of Splitting Tsunami) (Titov and Gonzalez 1997) and COMCOT (Cornell Multi-Grid Coupled Tsunami model) (Wang and Power 2011). Due to its ability to produce outputs relevant to this study and its ability to run simulations based upon multiple types of generation

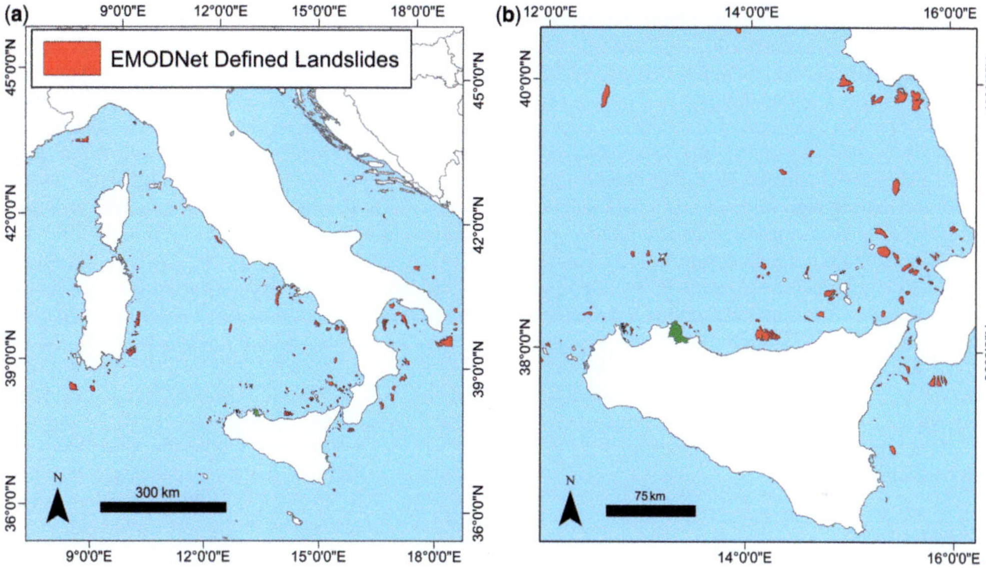

Fig. 3. Spatial distribution of subaqueous mass movements, delineated with red polygons, within (**a**) the Tyrrhenian Basin and (**b**) northern Sicily, as part of the EMODnet Geology (2019) catalogue, with Palermo highlighted in green (GADM 2018).

mechanisms, and validated use in studies around the world (Wang and Liu 2006; Liu *et al.* 2009; Wu and Huang 2009; Fraser *et al.* 2014), COMCOT was used in this scenario-based assessment.

COMCOT modelling package

COMCOT uses explicit staggered leap-frog finite difference schemes to solve shallow-water equations in both Cartesian and Spherical coordinates (Cho 1995; Wang and Power 2011). The model uses two-way dynamic links between multiple grids across multiple levels, varying in spatial resolution to allow for coarse modelling over large sections of open ocean, and higher resolutions closer to the area of interest, thus maintaining computational efficiency. COMCOT also simulates non-linear processes such as run-up and inundation. Examples of

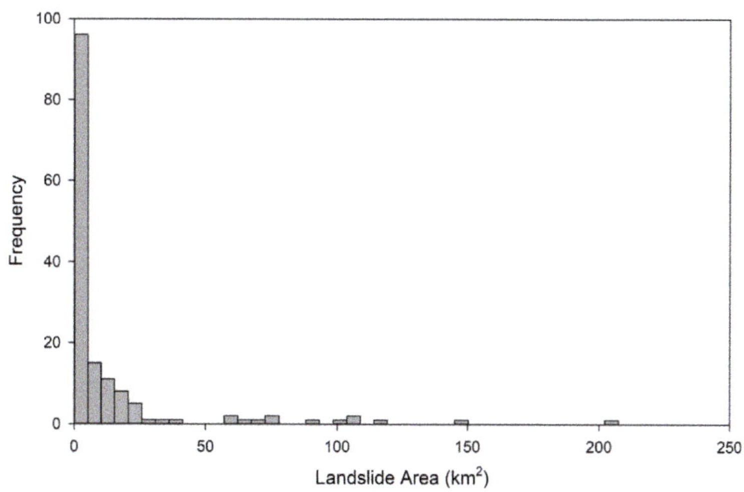

Fig. 4. Area–frequency distribution of subaqueous mass movements, featuring on the EMODnet Geology (2019) catalogue, within the Tyrrhenian Basin.

potential sources capable of being modelled in COMCOT are submarine fault plane rupture (single or multiple), submarine landslides and debris flows, custom transient seafloor motion, as well as custom water surface displacement or boundary waves. We invite the reader to explore the references given above if they are interested in further details regarding the modelling algorithms used by COMCOT.

COMCOT uses control files containing input parameters for both seismic rupture and landslide deformation calculation to generate an initial condition, as well as layer information to form the nested grid setup. COMCOT then propagates the water waves across the series of nested grids and, if selected in the control file, uses shallow-water equations to calculate run-up and inundations in any of the chosen grid layers. A flow chart for COMCOT can be seen in Figure 5.

Bathymetric grids

In the case of this study, the area of interest is the Gulf of Palermo; therefore, the smallest and highest resolution grid will be in the nearshore and coastal terrestrial areas of Palermo. Nesting the high-resolution Palermo grid will be a coarser resolution grid with an extent covering the majority of the Tyrrhenian Sea, focusing on the SE corner. This set-up allows for the wave propagation distance in deeper waters and distal coastlines to be appreciated.

The outermost grid layer covers the SE Tyrrhenian Sea specifically. These data were downloaded from EMODnet and resampled to have a spatial resolution of *c.* 70 m (EMODnet Bathymetry Consortium 2018). The inner Palermo grid was generated from bathymetric data acquired between 2001 and 2010 by the Marine GeoGroup at the University of Palermo as part of CARG (National Official Geological Cartography) and MAGIC (Marine Geohazards along the Italian Coasts) projects from various oceanographic cruises (DFP04, R/V *Universitatis*; Patti2003, R/V *Thetis*; PUMA2009, R/V *Universitatis*; MACS2010, R/V *Universitatis*) (Sulli *et al.* 2018). This grid includes both the aforementioned bathymetric data at a mean spatial resolution of *c.* 23 m and terrestrial digital elevation model (DEM) data resampled at the same resolution.

Seismic events

To effectively simulate worst-case scenarios of tsunami hazard to northern Sicily, the selection of seismic events to simulate was of critical importance. A selection of five significant seismic events (referred

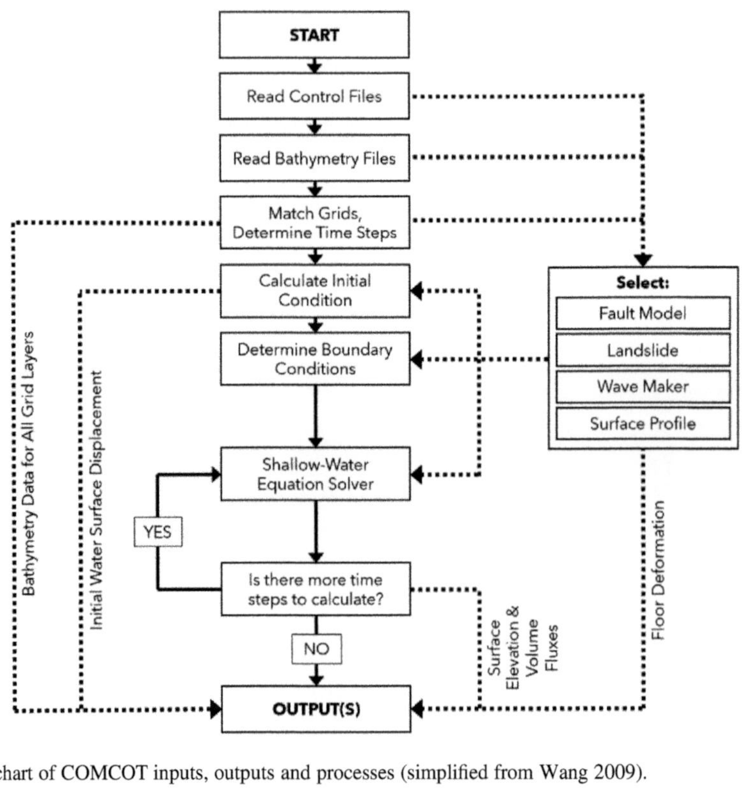

Fig. 5. Flow chart of COMCOT inputs, outputs and processes (simplified from Wang 2009).

Table 1. *Parameters for seismically generated tsunami simulations*

Parameter	Earthquake ID				
	EQ_01	EQ_02	EQ_03	EQ_04	EQ_05
Earthquake date*	20/06/1998	09/08/1985	13/04/2012	20/12/2015	06/09/2002
Magnitude*	M_D 4.3	M_D 4.0	M_W 4.3	M_W 4.2	M_L 5.6
Focal depth (km)*	10.0	9.5	9.2	22.7	5.0
Length of fault Plane (m)	2113	705	476.1	1129.2	2217.5
Width of fault Plane (m)†	2305	1109	854	1581	2381
Dislocation of fault plane (m)‡	0.662	1.212	1.504	0.935	0.645
Strike angle*§	69	44	124	93	37
Dip angle*§	22	90	55	84	42
Slip/rake angle*§	76	90	−81	−170	64

*Istituto Nazionale di Geofisica e Vulcanologia (2019).
†Scaled using Leonard (2010).
‡Scaled using Wells and Coppersmith (1994).
§Pondrelli (2002).

to as EQ_01 to EQ_05) were identified and their parameters derived from earthquake catalogues and literature where relevant (Table 1).

It is worth noting the spatial distribution of the selected worst-case scenarios, in that they are spread around the surrounding coastline in order to assess the extent of refraction around the Capo Gallo headland to the west and Capo Zafferano to the east (Fig. 6).

Key inputs for COMCOT simulations based on seismic tsunamigenic events include fault width (defined in this study as the fault damage zone – Fig. 7) and fault dislocation. For the selected events, this information was not available through either literature or earthquake catalogues, and so was calculated using scaling relationships of known variables. Fault width and average dislocation were scaled from the known fault length, using correlation-derived relationships (Wells and Coppersmith 1994; Leonard 2010) widely used in seismic modelling (Fig. 8). Also, EQ_02 does not have a published dip or slip value. As the methodology is focused

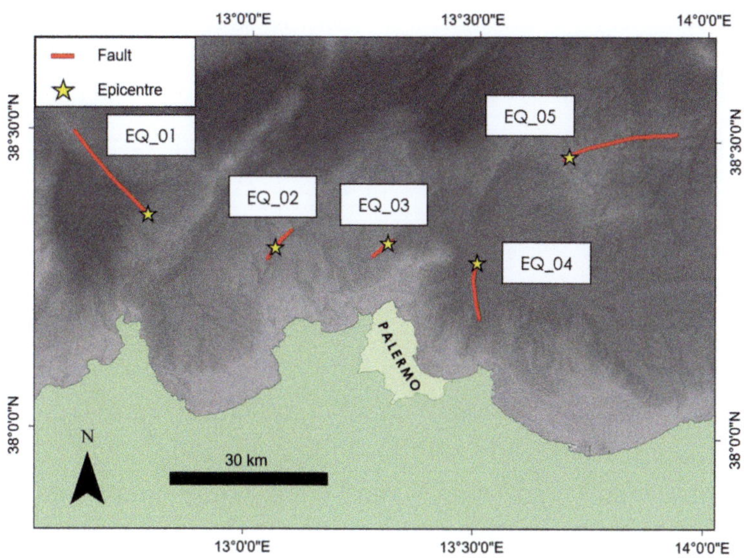

Fig. 6. Spatial distribution of the seismic events from the INGV seismic database (2019) (yellow stars) and their respective faults obtained by using the classical evolutionary models of faulting and the scale relationship between width of the damage zone and displacement, as well as deformational mechanism and involved lithologies (red lines) for the simulated scenarios, with their relative position to the northern Sicilian coastline and the city of Palermo (GADM 2018), underlain by EMODnet bathymetry.

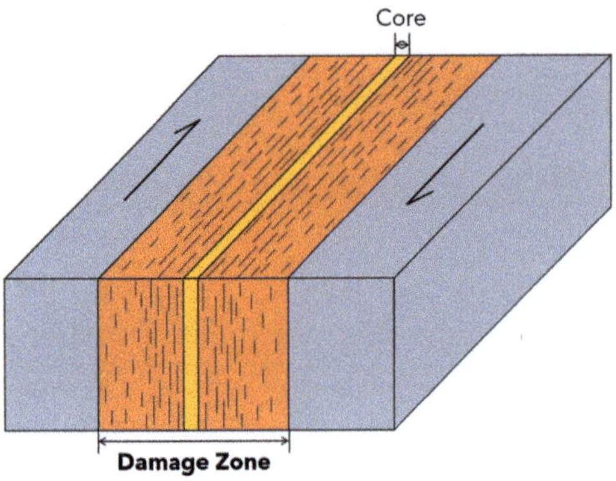

Fig. 7. Diagram of fault width as used for this study, with the damage zone identified being the measured fault width input for COMCOT.

on credible worst-case scenarios, the worst-case scenario – to induce maximum water displacement – would be to have a dip and slip value of 90° and, as such, these were the parameters used to fill in these gaps in published knowledge. A dip/slip value of 90° has been commonly used as a value in previous tsunami modelling works on the basis that it focuses all dislocation on the vertical axis, thus producing a worst-case wave (Tanioka and Satake 1996; Kowalik *et al.* 2005).

Mass movement events

Several canyon-head submarine landslides have already been identified and parameterized along the continental slope and within the Gulf of Palermo and have had their theoretical run-up and wave heights calculated (Sulli *et al.* 2018). Based on having considerable calculated run-up in previous works, a selection of five submarine landslides were identified for modelling (referred to as LS_01 through LS_05), with many of their morphometric characteristics and parameters derived from the aforementioned literature (Table 2). These characteristics are fed into COMCOT as an input to the initial condition calculation, which also calculates factors such as acceleration and velocity based upon relationships and equations developed by Enet and Grilli (2007).

The selected submarine landslides identified across the Palermo Gulf are mainly localized along the upper continental slope. The headwalls of the scars are generally linear to arcuate, and steep (12–18°) (Fig. 9). These mapped scars are connected to the headwall of the submarine canyons dominating the seafloor morphology. All the physical parameters of the selected submarine landslides are tabulated in Table 2.

These slides were selected to give a range in size, directionality and proximity to the Palermo coast, in order to identify key factors influencing tsunami risk by Palermo Gulf landslides. The chosen slides also provide the ability to identify the impact local submarine features are likely to have upon wave propagation, such as other submarine canyons and the La Barra High (Fig. 9).

The Sulli *et al.* (2018) paper is aimed to compare different algorithms used to calculate expected run-up heights, for only landslide-induced anomalous waves. Here we tested both landslide-generated

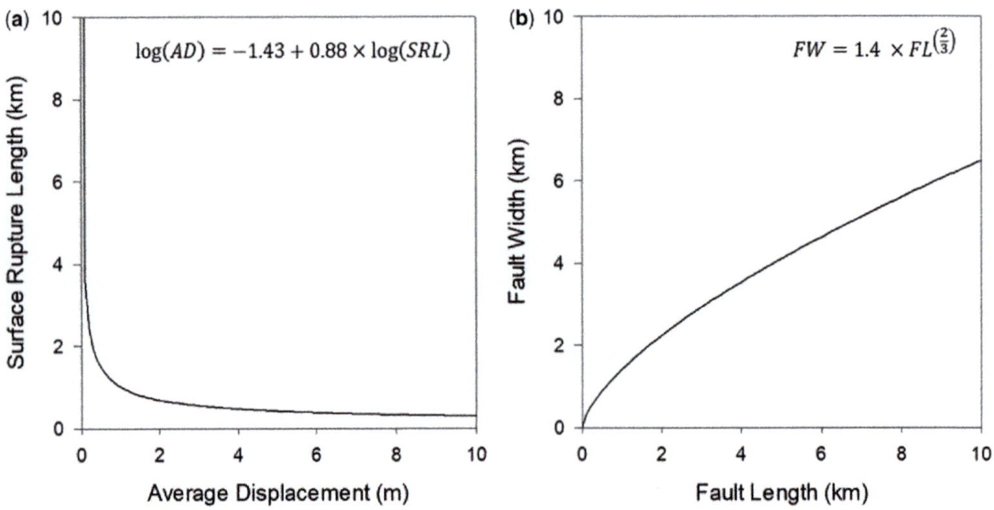

Fig. 8. Scaling relationships used to derive (**a**) average fault displacement from surface rupture length (Wells and Coppersmith 1994) and (**b**) fault width from fault length (Leonard 2010).

Table 2. *General parameters for mass movement simulations*

Parameter	Landslide ID				
	LS_01	LS_02	LS_03	LS_04	LS_05
ID in Sulli *et al.* (2018)	PMO-10	PMO-05	PMO-03	PMO-02	PMO-01
Slope angle (θ)	13°	13°	16°	18°	12°
Length (m)	900	1200	1600	800	1055
Width (m)	800	770	7000	750	1250
Thickness (m)	20	11	10	10	6
Detachment area (km^2)	1.22	2.22	2.36	0.96	3.04
Distance from coast (km)	3.50	6.50	7.30	5.97	2.50
Crown depth (m)	140	383	540	331	118

Based on Sulli *et al.* (2018)

and earthquake-generated anomalous waves. Sulli *et al.* (2018) tested the selected algorithms on real, historical tsunami events, and suggested which of them were most suitable, pointing out an empirical formula to obtain an expeditious method to evaluate run-up heights. The main advance is that in this study we propose a numerical method to obtain a continuous model of wave propagation and inundation, whilst in the previous study they provided only values at the source point and at the coast, giving also an assessment of the propagation time, which is essential for tsunami risk.

Similar to the seismic simulations mentioned above, there is some uncertainty surrounding modelling parameters for landsliding simulations also. The exact geology of the area is unknown and therefore the specific density (ratio between mass and volume) of the selected mass movement material is unknown. To overcome this, a value of 1.7 was used, as has been cited in literature as within the average range of marine sediment specific gravity (Tenzer and Gladkikh 2014), and very close to a moderately compacted sampled bulk sediment density of 1.8 g cm^{-3} as sampled in the southern Tyrrhenian Sea (Sammartini *et al.* 2019).

Sensitivity analysis

By using a 25% error range for density and dislocation, and 10% for fault width (the input parameters that introduced uncertainty), sensitivity analysis was used to re-run the simulations with the error extremes to identify any differences in run-up and inundation between the altered parameter simulation and the used parameter simulation. A smaller error range (10%) was used for fault width due to an increased level of certainty in this variable.

A recreation of simulations EQ_05 and LS_05, both producing a clear and widely propagating wave, took place with the altering of the aforementioned

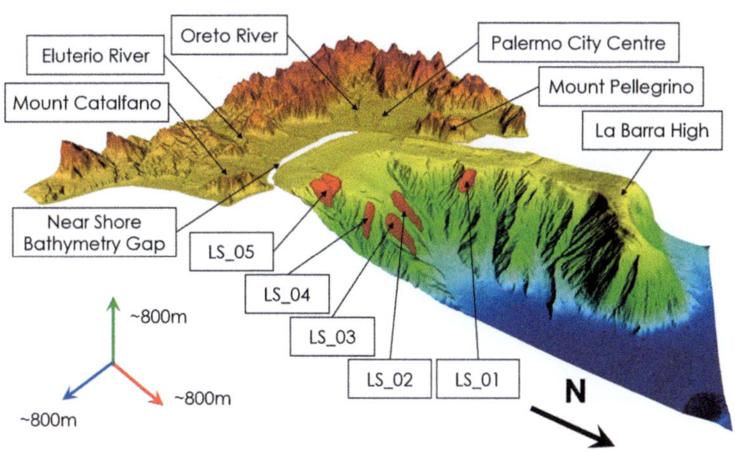

Fig. 9. 3D annotated diagram of high-resolution grid highlighting offshore and onshore features relative to Palermo, with mass movements modelled in this study delineated with red polygons. 3× vertical exaggeration.

Table 3. *Results generated by tsunamigenic seismic event simulations*

	EQ_01	EQ_02	EQ_03	EQ_04	EQ_05
Time to landfall (s)	0	0	10	0	20
Maximum inundation (m)	0	0	0	0	0
Maximum wave height (m)	0.0278	0.0140	0.0065	0.0021	0.0510
Maximum coastal wave height (m)	*c.* 0	*c.* 0	*c.* 0	*c.* 0	*c.* 0

key variables. Differences in modelled inundation, wave height and arrival time were analysed to quantify the uncertainty present within the simulations in this study. The percentage alterations given these two scenarios were: a fault width of 1785.75 and 2976.25 m, a fault dislocation of 0.484 and 0.8062 m and specific density of 1.275 and 2.125 g cm^{-3}.

Results

For each simulation, four key variables were extracted: the time between the tsunamigenic event and the first wave peak to reach the nearest coastline, maximum wave height within the computational area, maximum inundation distance and the maximum wave height along the coastline. These measures quantify the key information commonly desired by emergency planners to understand tsunami impact, and coastal wave height (also referred to as run-up) being a calculated parameter in Sulli *et al.* (2018), allowing for direct comparison.

Seismic events

The tallest wave height originating from a seismic source was 0.051 m, from a 5.6 M_L event, EQ_05 (Table 3), with other simulations producing even lower amplitude waves. Smaller waves range from 0.014 m (EQ_02) to 0.0021 m (EQ_04) in wave amplitude. These very low values are smaller than waves created by other marine disturbances such as the wake of a passing ship, or those created by the dominant northerly prevailing winds (Windfinder 2019). Due to the extremely small waves produced by these scenarios, the wave energy had dissipated before reaching the coastline and therefore no measurable wave height or inundation was possible. EQ_01, 02 and 04 all produced tsunamis making landfall within the first timestep; however, EQ_03 and 05 took 10 and 20 s, respectively to reach land.

Mass movement events

The same measures were used to quantify and compare the landslide scenarios (see Table 4). These results showed that the waves created by mass movement produce much larger waves than the seismically generated tsunamis above, which although moving slower through the water, produce more significant wave heights towards the coast, and therefore inundation. The largest tsunami wave produces a maximum height of 12.08 m (LS_02), with the second largest reaching 11.83 m (LS_01). Smaller waves produced by the sliding masses reached 5.13 m (LS_03), 2.37 m (LS_04) and 1.92 m (LS_05) in wave height, all of which were larger than those produced by the seismic events in this study. The fastest moving wave (LS_02) reached land within 10 s; however, the shorter tsunami waves had travel times of up to 200 s. Upon reaching the coastline, two of the tsunami waves produced measurable wave heights: LS_02 (3.22 m) and LS_05 (0.05 m). This coastal wave height allowed run-up and inundation to take place for these two events, with inundation distances reaching 528.12 m (LS_02) and 17.69 m (LS_05).

An important feature to note is the directionality of the landslide-generated tsunamis. The displacement forces the water to dissipate its energy in the direction of the mass movement and this shows in the simulation outputs. Each of the resultant tsunami waves propagated out in the direction in which the movement flowed. Using LS_04 as an example, the movement of mass is from the SW, moving to

Table 4. *Results generated by tsunamigenic mass movement simulations*

	LS_01	LS_02	LS_03	LS_04	LS_05
Time to landfall (s)	20	10	80	200	200
Maximum inundation (m)	0	528.12	0	0	17.69
Maximum wave height (m)	11.833	12.079	5.136	2.375	1.920
Maximum coastal wave height (m)	*c.* 0	3.221	*c.* 0	*c.* 0	0.055

Table 5. *Table of sensitivity simulation results demonstrating impact of change in uncertain parameters*

	Specific density		Fault width		Fault dislocation	
	25% increase	25% decrease	10% increase	10% decrease	25% increase	25% decrease
Time to landfall (s)	200	200	3200	3200	3200	3200
Maximum inundation (m)	0	0	0	0	0	0
Maximum wave height (m)	2.982	1.017	0.034	0.028	0.039	0.023
Maximum coastal wave height (m)	*c.* 0	*c.* 0	*c.* 0	*c.* 0	*c.* 0	*c.* 0

the NE and the direction of propagation follows the direction of mass movement.

Sensitivity analysis

For the sensitivity analysis, both the raw data outputs (Table 5) and the percentage change from the original simulation (Table 6) were calculated. Whilst there was no change in speed of wave progression from time step to time step, there was significant change in other categories caused by the changes in uncertain variables used across all sensitivity simulations.

For the changes in specific density, the increase and decrease caused inundation and coastal wave height to reduce to an infinitesimal number. However, fluctuations did occur in the maximum wave height after increase and decrease of specific density: increase in sliding mass density caused the wave height to increase from 1.92 to 2.98 m (an increase of 55%), and a decrease in density reduced the wave height from 1.92 to 1.01 m (a decrease of 47%).

Changes in landslide specific density influenced tsunami dynamics in somewhat unpredictable ways; an increase in density would likely cause more of a solid block slide, and it could be assumed that there would be a larger wave. These results support this hypothesis; however, a larger wave would normally cause larger inundation and larger coastal wave height, which is not the case in these results.

Looking at the sensitivity results derived from earthquake-based simulations (Tables 5 and 6), there are some equally unusual results. An increase in fault width and dislocation would likely increase the fault slip surface, thus water displacement, and therefore produce a larger wave, yet both an increase and decrease in these parameters result in smaller wave heights, based on the modelled scenario.

Discussion

Many assume that only large earthquakes create large tsunamis and this is an opinion that has also been published in peer-reviewed journals and publications (e.g. Scheidegger 1975; Dawson *et al.* 1988). However, recent studies have shown submarine mass movements are also capable of producing tsunamis of large amplitude and run-up (e.g. Watts *et al.* 2003). Using a numerical modelling approach, we show that through the simulations in this study and in the area of Palermo, landslides could be responsible for larger tsunamis than proximal seismic events. To this effect, the results of the smaller earthquake simulations in this study have shown the potential to create a tsunami wave, albeit small. Furthermore, all landslide-based simulations produced significantly higher waves than those generated by earthquakes (Tables 3 and 4). Therefore, this shows that smaller earthquakes do hold tsunamigenic potential under the proviso that tsunamis are simply gravity-driven displacement waves. However, this does raise questions as to what wave height (or other measure) constitutes a tsunami.

Table 6. *Table of sensitivity simulation results – percentage change from original simulation*

	Specific density		Fault width		Fault dislocation	
	25% increase	25% decrease	10% increase	10% decrease	25% increase	25% decrease
Time to landfall (s)						
Maximum inundation (m)	▼ 100	▼ 100				
Maximum wave height (m)	▲ 55	▼ 47	▼ 33	▼ 45	▼ 24	▼ 55
Maximum coastal wave height (m)	▼ 100	▼ 100				

When observing the landslide-generated tsunamis, the results follow the hypothesis that in the Palermo region, submarine landsliding may hold more of a tsunami hazard than seismically triggered tsunamis. Relative to other EMODnet-listed Tyrrhenian landslides (Fig. 3), the ones modelled were amongst but not the largest, nor were any of them angled as such to produce a strong wave in the direction of Palermo. Yet, LS_02 and LS_05 managed to produce a wave with large enough amplitude to inundate inland. The inundation focused around the port area of Palermo and the area surrounding it (Fig. 10). Given that the Port of Palermo is one of the major ports for passenger traffic in the whole of the Mediterranean and is quoted to be a driving force of Sicily's economy and history (Fabbri 2000), having this area susceptible to tsunami hazard is of concern.

Each of the landslide-generated tsunami waves reached maximum heights of at least *c.* 2 m. Although some of the smaller waves (LS_01, 03 and 04) did not inundate the coast, likely due to the directionality of the wave's energy, the smallest wave was created by LS_05, at maximum height of 1.92 m, and was one of the two which caused inundation. As mentioned above, inundation focused on the low-lying port area and urban areas directly to the north yet is significantly reduced as the terrain steepens inshore and further north, highlighting the influence topography has on inundation paths. This slide, 0.008 km^3 in cubic volume, was not the largest in terms of volume and therefore this demonstrates there is more to tsunami hazard from landslides than just the mass of the slide. Suggested hypotheses for this phenomenon are that the directionality of the sliding mass is one of the major influences for resultant run-up and propagation.

The recent study conducted by Sulli *et al.* (2018) was used to compare and validate the results of this study. This study included the five landslides used within this study and many of the parameters for the simulations were taken from this research. As such, it is appropriate to compare the theoretical calculations from Sulli *et al.* (2018) with those found within the more complex simulations run within this investigation. The run-up was calculated by Sulli *et al.* (2018) and, comparing this with the 'coastal wave height' in this investigation, we found that in all cases the simulations produced consistently lower values than the theoretical influences. The biggest notable difference would be that we found the coastal wave height of LS_01 to be infinitesimal; however, the theoretical approach calculated up to 21 m of run-up, using equations from Bryant (2014). This confirms the complexity of tsunami generation, propagation and inundation not being able to be accurately defined using theoretical and relatively simple equations.

The results demonstrate the influence landslide-driven tsunamis can have in tsunami hazard assessment. It can be said that landslides pose a more significant risk to Palermo than earthquakes from the 1985–2018 period. A working hypothesis is that the landslides posed a more significant hazard; however, the majority of their energy was focused away from the coastline, suggesting it would have had much more severe impact had they been facing the Gulf of Palermo. Therefore, it would perhaps be prudent to assess the hazard further by considering landslides further afield on the other side of the

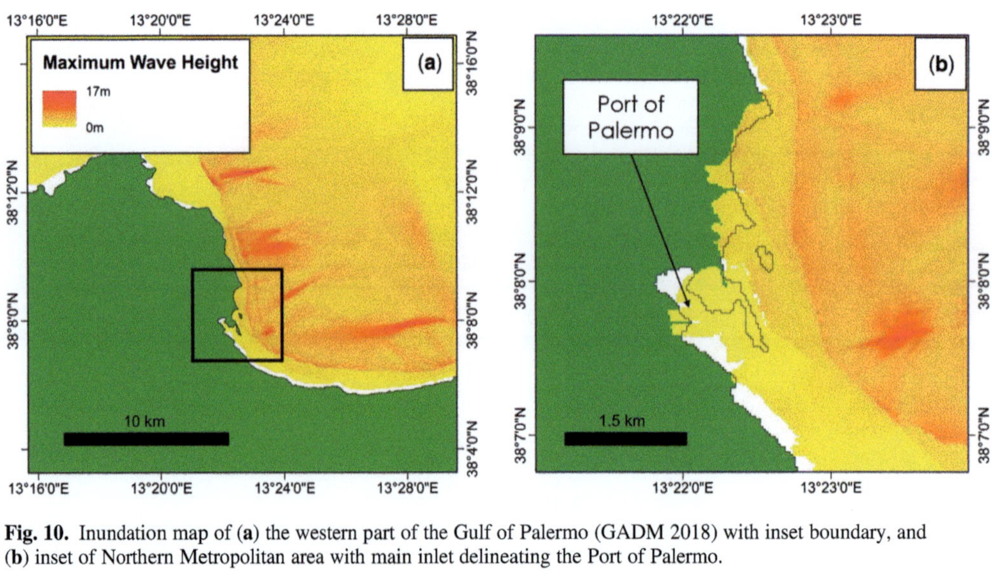

Fig. 10. Inundation map of (**a**) the western part of the Gulf of Palermo (GADM 2018) with inset boundary, and (**b**) inset of Northern Metropolitan area with main inlet delineating the Port of Palermo.

Tyrrhenian Basin (Fig. 3), facing Palermo, to assess what impact a stronger landslide-generated wave striking this coast would have. It is worth mentioning, however, that significant tsunamigenic landslides occur much less frequently than seismically generated tsunamis, rendering them less of a hazard for most regions (Harbitz *et al.* 2013). Furthermore, it would be wise to include the hazard posed by volcanic-driven tsunami, either through flank collapse or submarine volcanism, in this highly volcanic region. With the key active volcanoes of Stromboli (part of the nearby Aeolian Islands) and Vesuvius bordering the southern Tyrrhenian Basin having shown tsunamigenic potential in recorded history, they are often considered in tsunami hazard assessments of Italian seas as posing significant risk (Tinti 1991; Milia *et al.* 2003; Tinti *et al.* 2006).

Whilst these results are specific to the Palermo Gulf, this inlet is largely similar to others that board the Tyrrhenian Basin and, in fact, seas across the world: it has nearshore bathymetry of slowly sloping continental shelf, before a steep continental slope, usually with a number of submarine canyons carved into it. This provides some scope to apply the landslide results to other coastal bays such as, perhaps, offshore Naples, yet we are unable to say to what extent the Palermo Gulf is representative of all medium-sized bays with canyons in their continental shelves in a quantitative manner and therefore caution must be taken before generalizations such as these can be considered.

Uncertainty associated with unknown variables

The most disruptive variable change in our modelling was the specific density of the landsliding mass, which caused both inundation and coastal wave height to drop to zero when increased or decreased. A 25% increase in specific density figures increased the maximum wave height by 55%, while decreasing density by 25% caused a reduction of 47%. So, the 25% alteration in specific density caused an output change as significant as *c.* 50% from the original value of 12.079 m. The link between mass specific density and tsunami generation is not well researched and so it is difficult to comment on whether these findings can be replicated in other studies; however, based on the larger waves produced by denser masses, one can infer that a denser moving mass would likely hold its shape enough to cause more violent water displacement than a more viscous flow caused by a less dense mass. This relationship between block slide density, wave generation and generated wave height is something which has been documented in research, although generally within laboratory conditions, such as in the use of a wave tank (Watts 2000).

Wells and Coppersmith (1994) used a selection of 421 historical earthquakes to form their catalogue; 244 of these were deemed reliable enough to use to form empirical relationships. Looking at the individual scaling correlation used in this study, the relationship between rupture length and displacement, only 66 earthquakes were used to derive this correlation. The correlation coefficient of this relationship was 0.71 with a standard deviation of 0.36 (Wells and Coppersmith 1994). In addition, the rupture length range to derive these correlations was between 3.8 and 432 km, well beyond the scope of the smaller earthquakes simulated within this study, which had a rupture length range of just 854–2381 m. It is also worth drawing attention to the fact that a small 10% decrease in fault width caused the maximum wave height to drop by 45%, showing it to be an extremely sensitive variable; this link can be further validated by the knowledge that fault dislocation is a uniform measure, not changing as fault length does, therefore highlighting the sensitivity of fault length and width alone. This leads to the questioning of the reliability of the scaling used. It can therefore be said that the scaling used for these smaller earthquakes was not totally suitable and might be more accurate for larger magnitude events, upon which the relationship itself was founded. However, the relationships outlined by Wells and Coppersmith (1994) are broadly well established and still provide a reliable approximation of these relationships.

As well as the uncertainty addressed above through sensitivity analysis, further epistemic uncertainty is introduced through the incompleteness of input data available. This modelling does not consider smaller submarine landslide events which are not accounted for in databases. This error, however, can be counteracted by the fact that the landslides used were relatively small within the catalogue base and the smaller of them did not produce a wave large enough to inundate or produce measurable coastal waves, and so smaller landslides would not have contributed to an increase in the hazard assessment to Palermo. Perhaps more relevant epistemic uncertainty introduced into this study is the centre of landsliding mass input within the COMCOT set-up process. The selection of this point was calculated based on the centre of the remaining scarp, which could be improved by digitally recreating the pre-slide state to identify the exact centre of sliding mass. The scale of the slides in this study means, however, that the error introduced here is unlikely to have measurably impacted the generated outputs.

Further aleatory uncertainty is introduced when assuming that the scarps of the slides were generated by an *en-masse* event, and not a multi-stage retrogressive failure. Breaking the slides down into smaller multi-stage events would likely produce significantly smaller tsunamis and pose less of a tsunami hazard to Palermo, yet the nature of this study is a 'worst-case scenario' approach and we have modelled on the worst-case basis that these events occurred *en masse*.

Bathymetric data gaps

In offshore regions, the bathymetric data are interpolated in order to form a single image without gaps in data coverage, and COMCOT's nested grid set-up ensures that the grid boundaries do not also result in data gaps. However, the two sections of high-resolution data, the Gulf of Palermo bathymetry and the coastal high-resolution DEM, are missing a strip of data in some of the coastal zone. This 'white strip' is the result of shipborne multibeam sounder not reaching the shallower waters due to the draught of the vessel being restricted from being so close to shore, and the aerial light detection and radar (LiDAR) imagery not penetrating the coastal waters. These areas are also often cropped out and cleaned during image processing to remove turbulence and backscatter caused by coastal water debris (e.g. litter, seaweed, etc.), or wave action in the nearshore area. If one was to interpolate over this data gap, simply using a straight line, it would create ideal and favourable conditions for run-up, allowing the wave to build in size and approach the shore in a virtually idealistic manner, resulting in much larger wave height and inundation measurements to be produced by the simulations. Other techniques of interpolation do exist in order to create a more realistic interpolated 'filler' layer; however, coastal zones, especially in the intertidal region, are notoriously unique and vary significantly in a matter of 20 m. Using generalized computational methods in order to fill this data gap introduces significant error to an already sensitive and yet critical output variable. Even slight changes to submarine obstacles and points of possible friction, diffraction and energy dissipation can vary the resultant inundation to the point where incorrect maximum inundation contours are concluded upon.

Whilst not filling this gap does arguably also introduce error, it does not attempt to fill this gap with idealized interpolated slopes and thus modelling based on assumed data fraught with issues as outlined above. The ideal solution would be to improve the completeness of data either through the relatively expensive use of a smaller vessel fitted with side-scanning equipment to map this coastal area, or to carry out aerial surveys with higher-powered LiDAR scanners capable of penetrating the water to depths of *c.* 50 m in ideal conditions (Klemas 2011), a penetration depth well in exceedance of what is required to fill such data gaps. Being able to use this coastal bathymetry data to fill these gaps would undoubtedly improve the reliability of the results of coastal tsunami impact from the simulations.

Bathymetric influences on wave propagation

When observing both the wave propagation as the tsunami reaches coastal waters, and the maximum wave height outputs, there are clear disturbances to the wave front caused by bathymetric variation (Fig. 11). The major bathymetric shallow both close to the Gulf of Palermo and covered by the high-resolution bathymetry is the La Barra High. As the front of the tsunami wave approaches this region, the wave base is deeper than the water column itself, the wave slows and increases in height. This can be seen both in the progressing wave front slowing down and 'denting' the circular propagating wave in place of the La Barra High, as well as the offshore higher maximum wave heights in this zone and across other bathymetric features of the Tyrrhenian Basin.

Computational restrictions

In order to conduct the simulations within this study, significant computing power and time was required. The simulations took, on average, around 80 h to run from start to end – a simulated time period of approximately *c.* 45 mins – not including the time taken to convert file types and analyse the results. This demonstrates the true computational power required to run complex simulations such as these, and the impracticality of running them on a real-time basis.

COMCOT simulations are not practical to run in real time or implement into real-time early warning systems due to the fact that the wave itself will travel at speeds far in excess of the speed it can be modelled. Equally, these simulations cannot be easily run on a simple computer or mobile device without significant memory capacity and processing power.

However, for the purposes of scenario-based simulations not carried out in real time, it is well suited. For the purposes of producing sophisticated tsunami models, such as TsuCAT (Tsunami Coastal Assessment Tool), which produce extensive catalogues of simulations carried out in advance for almost every seismic eventuality, and when an event occurs, the results can simply be scaled to meet the inputs, producing a reasonable forecast almost instantly (Center for Tsunami Research 2019). The high levels of precision and accuracy COMCOT provides is extremely valuable; however, methodologies must be critically produced in order to ensure it is used within the correct and appropriate circumstances for the required use.

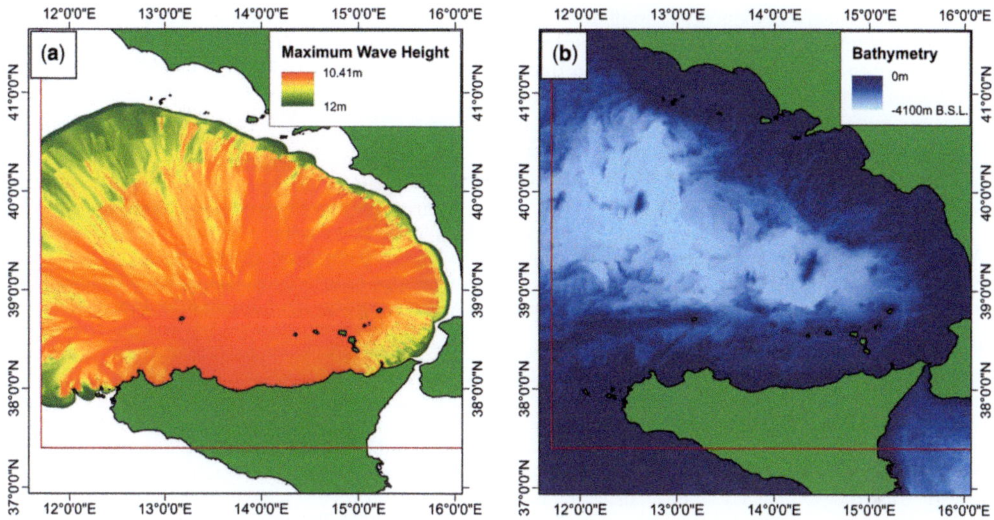

Fig. 11. (**a**) Maximum wave height from simulation LS_02 and (**b**) Tyrrhenian Basin bathymetry, with computational extent delineated with red border lines (GADM 2018).

Conclusions

From this limited study alone, we cannot derive a fully comprehensive and quantifiable tsunami hazard for Palermo and, as such, further research is required. However, the findings have enabled two main conclusions to be drawn on the basis of these results. Tsunami hazard, based on the simulations in this study, is much more significant when generated by landslides than seismic events. Two of the landslide-based scenarios simulated in this study resulted in inundation, and all produced maximum wave heights of more than 1 m. Compared with the seismic simulations, the maximum wave heights did not exceed 5.1 cm across all scenarios. This is likely attributable to the fact that the severity of seismic activity within the near-field area of Palermo is relatively low compared with more significant seismically triggered tsunamis of recent history. However, it can be concluded that submarine mass movements provide highly significant tsunamigenic potential and should be seriously and thoroughly considered as part of tsunami hazard analysis.

The point of largest inundation is shown to be around the Port of Palermo, caused by a landslide-driven tsunami in the Gulf of Palermo (LS_02). Based on these simulations, whilst inundation of the port is low in probability, the reliance on this port to support the local economy means inundation in this area is likely to cause significant and severe disruption. Therefore, it is suggested that tsunami impact mitigation strategies within this port area should be considered. Throughout the landslide model outputs, directionality of mass movement played a key role in focusing energy away from the Palermo coastline and is a key control in where and how tsunami waves propagate from source, thus influencing the hazard posed in this area.

Across both seismic and mass movement tsunami scenarios, the wave propagation and run-up is highly sensitive to bathymetric relief, especially in the shallower coastal waters and around the shallow La Barra High. These shallower waters are where there is limited (and in some cases no) available high-resolution bathymetric data. Future efforts to improve high-resolution bathymetry in the nearshore and intertidal zone would allow more robust tsunami modelling in the future.

The use of WCTSA as a methodology in this study provided a good introduction and first steps in assessing the tsunami hazard in Palermo, identifying areas at particular risk.

This research has started the process in assessing the tsunami hazard in Palermo, northern Sicily and more generally, the southern Tyrrhenian Sea, yet further investigation is required to progress knowledge and quality of assessment. Suggested research aims for the next steps in this process are as follows:

(1) integrate more distal landslide sources into the assessment of Palermo, assessing the impact directionality of mass movement has on tsunami hazard, and its relative influence over other factors such as distance from coastline, volume and specific density of sliding mass;
(2) integrate volcanic influences on tsunami hazard assessment of Palermo; in particular, the impact a large flank collapse of one of the

Aeolian Islands or explosive submarine volcanism would have on areas such as Palermo and Italian coasts bordering the Tyrrhenian Basin;
(3) carry out a comprehensive set of simulations on a larger sample size of both seismic activity and submarine landsliding, in order to perform correlational analysis of the influence particular parameters have on tsunami generation.

Acknowledgements We are grateful to the anonymous reviewers and to the editor for the helpful constructive suggestions that improved significantly the manuscript.

Funding This research received no specific grant from any funding agency in the public, commercial or not-for-profit sectors.

Author contributions **JD**: conceptualization (equal), data curation (equal), formal analysis (lead), investigation (lead), methodology (lead), writing – original draft (lead), writing – review and editing (lead); **AM**: conceptualization (equal), formal analysis (equal), methodology (equal), project administration (equal), supervision (equal), validation (equal), writing – review and editing (equal); **CM**: conceptualization (equal), formal analysis (equal), methodology (lead), software (lead), supervision (equal), validation (equal), writing – review and editing (equal); **AS**: conceptualization (equal), data curation (equal), resources (equal), validation (equal), writing – review and editing (equal); **EZ**: data curation (equal), investigation (equal), methodology (equal), validation (equal), writing – review and editing (equal); **DS**: data curation (equal), formal analysis (equal), investigation (equal), methodology (equal), writing – original draft (supporting), writing – review and editing (supporting).

References

Anita, G., Sandri, L. *et al.* 2012. Probabilistic tsunami hazard assessment for Messina Strait Area (Sicily, Italy). *Natural Hazards*, **64**, 329–358, https://doi.org/10.1007/s11069-012-0246-x

Behrens, J. and Dias, F. 2015. New computational methods in tsunami science. *Philosophical Transactions of the Royal Society A: Mathematical, Physical and Engineering Sciences*, **373**, https://doi.org/10.1098/rsta.2014.0382

Bryant, E.A. 2014. *Tsunami, the Underrated Hazard.* Springer, Switzerland.

Center for Tsunami Research 2019. *TsuCAT User's Manual.* NOAA, Seattle.

Cho, Y.-S. 1995. *Numerical Simulations of Tsunami and Runup.* PhD thesis, Cornell University.

Cox, D.C. 1963. *Status of Tsunami Knowledge.* International Union of Geodesy and Geophysics, Paris.

Davies, G. and Griffin, J. 2018. *The 2018 Australian Probabilistic Tsunami Hazard Assessment.* Geoscience Australia, Australian Government, Canberra.

Dawson, A.G., Long, D. and Smith, D.E. 1988. The Storegga Slides: evidence from Eastern Scotland for a possible tsunami. *Marine Geology*, **82**, 271–276, https://doi.org/10.1016/0025-3227(88)90146-6

Dilek, Y. 2006. Collision tectonics of the Mediterranean region: causes and consequences. *Geological Society of America, Special Paper*, **409**, 1–13.

Docey, S., Gorokhovich, Y. *et al.* 2007. Tsunami mortality estimates and vulnerability mapping in Aceh, Indonesia. *American Journal of Public Health*, **97**, S146, https://doi.org/10.2105/AJPH.2006.095240

EMODnet Bathymetry Consortium 2018. EMODnet Digital Bathymetry (DTM) [Online], http://www.emodnet-bathymetry.eu [last accessed 20 June 2019].

EMODnet Geology 2019. Events & Probabilities. [Online], https://www.emodnet-geology.eu/data-products/geological-events-and-probabilities/ [last accessed 6 September 2019].

Enet, F. and Grilli, S. 2007. Experimental study of tsunami generation by three-dimensional rigid underwater landslides. *Journal of Waterway, Port, Coastal, and Ocean Engineering*, **133**, 442–454, https://doi.org/10.1061/(ASCE)0733-950X(2007)133:6(442)

Fabbri, P. 2000. *Palermo and Monreale (Gold Guides to Popular European Cities).* Bonechi, Italy.

Farreras, S.F. and Sanchez, A.J. 1991. The tsunami threat on the Mexican west coast: a historical analysis and recommendations for hazard mitigation. *Natural Hazards*, **4**, 301–316, https://doi.org/10.1007/BF00162795

Fraser, S.A., Power, W. *et al.* 2014. Tsunami inundation in Napier, New Zealand, due to local earthquake sources. *Natural Hazards*, **70**, 415–445, https://doi.org/10.1007/s11069-013-0820-x

GADM 2018. GADM Maps – Version 3.6 [Online], https://gadm.org/download_country_v3.html [last accessed 6 April 2019].

Gonzalez, F.I., LaVeque, R.J. *et al.* 2014. *Probabilistic Tsunami Hazard Assessment (PTHA) for Crescent City, CA.* Department of Earth and Space Sciences Faculty Research, Washington.

Grilli, S.T., Tappin, D.R. *et al.* 2019. Modelling of the tsunami from the December 22, 2018 lateral collapse of Anak Krakatau volcano in the Sunda Straits, Indonesia. *Scientific Reports*, **9**, 11946, https://doi.org/10.1038/s41598-019-48327-6

Gusman, A.R. and Tanioka, Y. 2014. W phase inversion and tsunami inundation modelling for tsunami early warning: case study for the 2011 Tohoku event. *Pure and Applied Geophysics*, **171**, 1409–1422, https://doi.org/10.1007/s00024-013-0680-z

Harbitz, C.B., Glimsdal, S. *et al.* 2012. Tsunami hazard in the Caribbean: regional exposure derived from credible worst-case scenarios. *Continental Shelf Research*, **38**, 1–23, https://doi.org/10.1016/j.csr.2012.02.006

Harbitz, C.B., Lovholt, F. and Bungum, H. 2013. Submarine landslide tsunamis: how extreme and how likely? *Natural Hazards*, **72**, 1341–1374, https://doi.org/10.1007/s11069-013-0681-3

Horspool, N., Pranantyo, I.R. *et al.* 2014. A probabilistic tsunami hazard assessment for Indonesia. *Natural Hazards and Earth System Sciences*, **14**, 3105–3122, https://doi.org/10.5194/nhess-14-3105-2014

Hwang, L.S. and Divoky, D. 1970. Tsunami generation. *Journal of Geophysical Research*, **75**, 6802–6817, https://doi.org/10.1029/JC075i033p06802

Imamura, F., Yalciner, A.C. and Ozyurt, G. 2006. Tsunami Modelling Manual (TUNAMI Model), s.l.: UNESCO IOC International Training Course on Tsunami Numerical Modelling.

International Tsunami Information Centre 2018. *Tsunami Sources 1610 B.C. to A.D. 2017 from Earthquakes, Volcanic Eruptions, and Other Causes*. International Tsunami Information Centre, Honolulu.

Istituto Nazionale di Geofisica e Vulcanologia 2019. ISIDe: Italian Seismological Instrumental and Parametric Database, https://cnt.rm.ingv.it/iside

IRIS 2019. Subduction Zone: Tsunamis Generated by Megathrust Earthquakes [Online], https://www.iris.edu/hq/inclass/animation/subduction_zone_tsunamis_generated_by_megathrust_earthquakes [last accessed 30 August 2019].

Istat 2011. Geo Demo [Online], http://demo.istat.it/pop2011/index3_e.html [last accessed 6 September 2019].

Jelinek, R., Krausmann, E. *et al.* 2012. Approaches for tsunami risk assessment and application to the city of Cadiz, Spain. *Natural Hazards*, **60**, 273–293, https://doi.org/10.1007/s11069-011-0009-0

Kathiresan, K. and Rajendran, N. 2005. Coastal mangrove forests mitigated tsunami. *Estuarine, Coastal and Shelf Science*, **65**, 601–606, https://doi.org/10.1016/j.ecss.2005.06.022

Klemas, V. 2011. Beach profiling and LiDAR bathymetry: an overview with case studies. *Journal of Coastal Research*, **27**, 1019–1028, https://doi.org/10.2112/JCOASTRES-D-11-00017.1

Koh, H.L., Teh, S.Y. and Izani, A.M.I. 2007. *Tsunami Mitigation Management Technology. Special Feature: Natural Disaster Management Technologies*, The United Nations Asian and Pacific Centre for Transfer of Technology (UN-APCTT) Nov–Dec, 47–54.

Kowalik, Z., Knight, W., Logan, T. and Whitmore, P. 2005. Numerical modelling of the global tsunami: Indonesian tsunami of 26 December 2004. *Science of Tsunami Hazards*, **23**, 40–56.

Leonard, M. 2010. Earthquake fault scaling: self-consistent relating of rupture length, width, average displacement, and moment release. *Bulletin of the Seismological Society of America*, **100**, 1971–1988, https://doi.org/10.1785/0120090189

Liu, P.L.F., Wang, X. and Salisbury, A.J. 2009. Tsunami hazard and early warning system in South China Sea. *Journal of Asian Earth Sciences*, **36**, 2–12, https://doi.org/10.1016/j.jseaes.2008.12.010

Maramai, A., Graziani, L. and Tinti, S. 2005. Tsunamis in the Aeolian islands (Southern Italy): a review. *Marine Geology*, **215**, 11–21, https://doi.org/10.1016/j.margeo.2004.03.018

Maramai, A., Brizuela, B. and Graziani, L. 2014. The Euro-Mediterranean tsunami catalogue. *Annals of Geophysics*, **57**, S0435, https://doi.org/10.4401/ag-6437

Milia, A., Torrente, M. and Zuppetta, A. 2003. Offshore debris avalanches at Somma-Vesuvius volcano (Italy) implications for hazard evaluation. *Journal of the Geological Society, London*, **160**, 309–317, https://doi.org/10.1144/0016-764902-045

Nakahara, S. and Ichikawa, M. 2013. Mortality in the 2011 tsunami in Japan. *Journal of Epidemiology*, **23**, 70–73, https://doi.org/10.2188/jea.JE20120114

Omori, E. 1910. Preliminary report on the Messina-Reggio earthquake of Dec. 28 1908. *Bulletin of Imperial Earthquake Investigation Committee*, **3**, 37–45.

Papadopoulos, G.A. and Imamura, F. 2001. *A Proposal for a New Tsunami Intensity Scale*. The Abdus Salam International Centre for Theoretical Physics, Seattle.

Papadopoulos, G.A., Gràcia, E. *et al.* 2014. Historical and pre-historical tsunamis in the Mediterranean and its connected seas: geological signatures, generation mechanisms and coastal impacts. *Marine Geology*, **354**, 81–109, https://doi.org/10.1016/j.margeo.2014.04.014

Paris, R. and Ulvrova, M. 2019. Tsunamis generated by subaqueous volcanic explosions in Taal Caldera Lake, Philippines. *Bulletin of Volcanology*, **81**, 1–14.

Parls, R., Switzer, A.D. *et al.* 2014. Volcanic tsunami: a review of source mechanisms, past events and hazards in Southeast Asia (Indonesia, Philippines, Papua New Guinea). *Natural Hazards*, **70**, 447–470, https://doi.org/10.1007/s11069-013-0822-8

Pelinovsky, E. and Poplavsky, A. 1996. Simplified model of tsunami generation by submarine landslides. *Physics and Chemistry of the Earth*, **21**, 13–17, https://doi.org/10.1016/S0079-1946(97)00003-7

Pondrelli, S. 2002. European–Mediterranean Regional Centroid-Moment Tensors Catalog (RCMT) [Dataset].

Ruffman, A. 2006. Documentation of the farfield parameters of the November 1 1755 'Lisbon' tsunami along the shores of the western Atlantic Ocean. *International Tsunami Society Third Tsunami Symposium*, 23–25 May 2006, Honolulu, HI.

Sambah, A.B. and Miura, F. 2019. Geo spatial analysis for tsunami risk mapping, advanced remote sensing technology for synthetic aperture radar applications, tsunami disasters and infrastructure. *In*: Marghany, M. (ed.) *Advanced Remote Sensing Technology for Synthetic Aperture Radar Applications, Tsunami Disasters, and Infrastructure*. IntechOpen, 1–21, https://doi.org/10.5772/intechopen.82665

Sammartini, M., Camerlenghi, A. *et al.* 2019. Open-slope, translational submarine landslide in a tectonically active volcanic continental margin (Licosa Submarine Mass landslide, Southern Tyrrhenian Sea). *Geological Society, London, Special Publications*, **477**, 133–153, https://doi.org/10.1144/SP477.34

Sartori, R. 2003. The Tyrrhenian back-arc basin and subduction of the Ionian lithosphere. *Episodes*, **26**, 217–221, https://doi.org/10.18814/epiiugs/2003/v26i3/011

Scheidegger, A.E. 1975. *Physical Aspects of Natural Catastrophes*, 1st edn. Elsevier, Amsterdam.

Sulli, A., Zizzo, E. and Albano, L. 2018. Comparing methods for computation of run-up heights of landslide generated tsunami in the Northern Sicily continental margin. *Geo-Marine Letters*, **38**, 439–455, https://doi.org/10.1007/s00367-018-0544-8

Tanioka, Y. and Satake, K. 1996. Tsunami generation by horizontal displacement of ocean bottom. *Geophysical Research Letters*, **23**, 861–864, https://doi.org/10.1029/96GL00736

Tappin, D. 2010. Mass transport events and their tsunami hazard. *In*: Mosher, D.C. *et al.* (eds) *Submarine Mass*

Movements and Their Consequences. Springer, Dordrecht, 667–684.

Teh, S.Y., Koh, H.L. and Izani, A.M.I. 2006. A model investigation on tsunami propagation in Malaysian and Thailand coastal water. Association of Engineering Education in Southeast and East Asia and the Pacific (AEESEAP). *Journal of Engineering Education*, **31**, 7–14.

Tenzer, R. and Gladkikh, V. 2014. Assessment of density variations of marine sediments with ocean and sediment depths. *The Scientific World Journal*, **2014**, 1–9, https://doi.org/10.1155/2014/823296

Tinti, S. 1991. Assessment of tsunami hazard in the Italian seas. *Natural Hazards*, **4**, 267–283, https://doi.org/10.1007/BF00162792

Tinti, S., Armigliato, A. *et al.* 2005. Assessing the hazard related to tsunamis of tectonic origin: a hybrid statistical deterministic method applied to Southern Italy coasts. *Journal of Earthquake Technology*, **42**, 189–201.

Tinti, S., Pagnoni, G. and Zaniboni, F. 2006. The landslides and tsunamis of the 30th of December 2002 in Stromboli analysed through numerical simulations. *Bulletin of Volcanology*, **68**, 462–479, https://doi.org/10.1007/s00445-005-0022-9

Titov, V.V. and Gonzalez, F.I. 1997. *Implementation and Testing of the Method of Splitting Tsunami (MOST) Model*. NOAA/Pacific Marine Environmental Laboratory, Seattle.

Tkalich, P., Dao, M.H. and Chan, E.S. 2007. Tsunami propagation modelling and forecasting for Early Warning System. *Journal of Earthquake and Tsunami*, **1**, 87–98, https://doi.org/10.1142/S1793431107000067

Tonini, R., Armigliato, A. *et al.* 2011. Tsunami hazard for the city of Catania, eastern Sicily, Italy, assessed by means of Worst-case Credible Tsunami Scenario Analysis (WCTSA). *Natural Hazards and Earth System Sciences*, **11**, 1217–1323, https://doi.org/10.5194/nhess-11-1217-2011

Wang, X. 2009. User manual for COMCOT version 1.7 (First Draft).

Wang, X. and Liu, P.L.-F. 2006. An analysis of 2004 Sumatra earthquake fault plane mechanisms and Indian Ocean tsunami. *Journal of Hydraulic Research*, **44**, 147–154, https://doi.org/10.1080/00221686.2006.9521671

Wang, X. and Power, W. 2011. *COMCOT: A tsunami generation, propagation and run-up model*. GNS Science Report 2011/43, Lower Hutt.

Watts, P. 2000. Tsunami features of solid block underwater landslides. *Journal of Waterway, Port, Coastal and Ocean Engineering*, **126**, 144–152, https://doi.org/10.1061/(ASCE)0733-950X(2000)126:3(144)

Watts, P., Grilli, S.T. *et al.* 2003. Landslide tsunami case studies using a Boussinesq model and a fully nonlinear tsunami generation model. *Natural Hazards and Earth System Sciences*, **3**, 391–402, https://doi.org/10.5194/nhess-3-391-2003

Watts, P., Grilli, S.T., Tappin, D.R. and Fryer, G.J. 2005. Tsunami generation by submarine mass failure. II: predictive equations and case studies. *Journal of Waterway, Port, Coastal and Ocean Engineering*, **131**, 298–310, https://doi.org/10.1061/(ASCE)0733-950X(2005)131:6(298)

Wells, D.L. and Coppersmith, K.J. 1994. New empirical relationships among magnitude, rupture length, rupture width, rupture area, and surface development. *Bulletin of the Seismological Society of America*, **84**, 974–1002.

Windfinder 2019. Windfinder Palermo City Statistics [Online], https://www.windfinder.com/windstatistics/palermo_city [last accessed 9 September 2019].

Wu, T.-R. and Huang, H.-C. 2009. Modelling tsunami hazards from Manila trench to Taiwan. *Journal of Asian Earth Sciences*, **36**, 21–28, https://doi.org/10.1016/j.jseaes.2008.12.006

Zhou, H., Moore, C.W., Wei, Y. and Titov, V.V. 2011. A nested-grid Boussinesq-type approach to modelling dispersive propagation and runup of landslide-generated tsunamis. *Natural Hazards and Earth System Sciences*, **11**, 2677–2697, https://doi.org/10.5194/nhess-11-2677-2011

A workflow for the rapid assessment of the landslide-tsunami hazard in peri-alpine lakes

Michael Strupler[1]*, Flavio S. Anselmetti[2], Michael Hilbe[2], Katrina Kremer[1] and Stefan Wiemer[1]

[1]Swiss Seismological Service (SED) at ETH Zurich, Sonneggstrasse 5, 8092 Zurich, Switzerland

[2]Institute of Geological Sciences and Oeschger Centre for Climate Change Research, University of Bern, Baltzerstrasse 1+3, 3012 Bern, Switzerland

MS, 0000-0003-4237-3844

*Correspondence: michael.strupler@sed.ethz.ch

Abstract: Although lake tsunamis constitute a rare peril, they have repeatedly occurred in peri-alpine lakes in the past. There are several documented historical examples of landslide-triggered tsunamis in Swiss lakes. However, fundamental information and workflows to rapidly quantify the lacustrine tsunami hazard for multiple lakes are missing so far. The fact that the shorelines of major peri-alpine lakes are densely populated underlines the need for a hazard assessment. Detailed assessments require high-resolution geophysical, geotechnical and sedimentological data, and considerable computation time. Due to the involved data acquisition and calculation efforts, such assessments are mainly conducted as detailed case studies for single lakes. We present a workflow for a rapid first-order estimation of the landslide-triggered tsunami hazard along the shores of peri-alpine lakes. A crucial step is the identification of potential tsunami sources. Unstable slopes are mapped automatically, based on parameters that are derived from past studies. Such parameters include the bathymetry and derived parameters, and type, characteristics and thickness of the sediments. Wave amplitudes are estimated with existing predictive equations, based on the constructed maps of potentially unstable slopes. The results may be used for focusing more detailed, lake-specific tsunami-hazard assessments in respective areas.

Supplementary material: Forecasted peak ground accelerations, sedimentation model, further details for three lakes and a list of R libraries are available at https://doi.org/10.6084/m9.figshare.c.4782807

Tsunamis are waves that are generated by impulsive disturbances of huge water masses (e.g. De Blasio 2011). Although tsunamis are often generated directly by earthquakes, landslides that occur in the subaerial or subaqueous environment can also generate tsunamis. In this article, we focus on tsunamis that are caused by underwater landslides. Such landslides can be triggered by earthquakes, sediment overload, wave loading, fluid flow or human activity (Locat and Lee 2002). A favourable topography, the necessary amount of sediment and a trigger mechanism are needed for the occurrence of subaqueous landslides (Nadim *et al.* 1996). Among well-investigated subaqueous landslide tsunamis are the AD 1998 Papua New Guinea tsunami (Tappin *et al.* 2008), AD 1929 Grand banks tsunami (Fine *et al.* 2005; Ten Brink *et al.* 2009), and 8150 BP Storegga tsunami (Harbitz 1992; Bondevik *et al.* 2005; Løvholt *et al.* 2005); for more examples, see Okal (2019). Recently, the societal awareness and interest in landslide-tsunamis has arisen, due to the immediate sharing of photos and videos documenting an event via social networks (e.g. https://www.youtube.com/watch?v=61ItBglP-YM or https://twitter.com/gerrys/status/1045716841509703680).

Although many landslide-tsunamis are documented for marine slopes, they can also occur in lakes. On peri-alpine lakes, such tsunamis have been documented in early historical sources, such as for a large event on Lake Geneva in AD 563 (Tauredunum event, see chronicle of Marius of Avenches as summarized in Schoeneich *et al.* (2015) and Favrod (1991)) or for an event on Lake Lucerne in AD 1601 (Cysat 1969). Traces of mass movements that are dated to the described events (Schnellmann *et al.* 2002; Kremer *et al.* 2012) support these occurrences from a geoscientific point of view. Many of these gravitational underwater mass movements have been assigned to earthquakes as triggers, as multiple landslides occur coevally on a basin-wide scale (e.g. Siegenthaler *et al.* 1987; Schnellmann *et al.* 2002; Strasser *et al.* 2013; Kremer *et al.* 2015). Numerical simulations of past

From: Georgiopoulou, A., Amy, L. A., Benetti, S., Chaytor, J. D., Clare, M. A., Gamboa, D., Haughton, P. D. W., Moernaut, J. and Mountjoy, J. J. (eds) 2020. *Subaqueous Mass Movements and their Consequences: Advances in Process Understanding, Monitoring and Hazard Assessments*. Geological Society, London, Special Publications, **500**, 81–95.
First published online March 18, 2020, https://doi.org/10.1144/SP500-2019-166

and potential future tsunamis caused by underwater landslides have been conducted for Lakes Lucerne (Hilbe and Anselmetti 2015*b*), Geneva (Kremer *et al.* 2012, 2015) and Zurich (Strupler *et al.* 2018*c*), and suggest that tsunamis can occur on the investigated lakes. Documented evidence of sublacustrine landslides worldwide in more than 170 lakes (Sammartini *et al.* 2019) underlines the need for assessing the threat of such potential tsunamis. The assessment of the tsunami potential on peri-alpine lakes is crucial, especially as most of them have densely populated and often unprotected shorelines. Despite known examples of past tsunamis on peri-alpine lakes, to our knowledge no integrated approach to quickly assess the tsunami hazard caused by subaqueous landslides for multiple lakes exists to date.

The main goal of this study is to develop a workflow for the rapid first-order estimation of the subaqueous landslide-tsunami hazard, based on simple parameters that are typically known for all lakes or assumed from studies on other lakes. The results are expected to provide a rough estimate of the consequences of subaqueous landslides in Swiss peri-alpine lakes that are chosen as a case study for similar lacustrine settings worldwide. Also, the results should highlight hotspots that need more in-depth investigations through follow-up high-resolution geophysical, sedimentological, geotechnical and modelling data.

In particular, the following research questions are addressed:

(1) Which slopes of a peri-alpine lake are prone to subaqueous landslides?
(2) What are the expected consequences of potential subaqueous landslides?

Previous studies

Limnogeological research conducted on a plenitude of peri-alpine lakes during the past decades has shown that many of these lakes in Switzerland, France, Germany, Italy and Austria show similar characteristics, also due to the common deglaciation history. Typical characteristics of peri-alpine lakes include the fjord-type bathymetry with a glacially overdeepened elongated shape, a similar Late Glacial to Holocene sediment succession and the vicinity to the Alps. An overview on large and deep peri-alpine lakes can be found in Tolotti *et al.* (2018).

Large amounts of geomorphic, geophysical and sedimentological information already exist for key peri-alpine lakes. Examples of such high-resolution data include multibeam digital bathymetric models (DBMs) with grid resolutions of 1–2 m paired with high-resolution seismic reflection data for peri-alpine lakes, which are located either entirely in Switzerland (Lakes Aegeri, Biel, Murten, Neuchatel, Thun, Lucerne, Zug and Zurich; Hilbe *et al.* 2011; Strupler *et al.* 2015; Reusch 2016; Fabbri *et al.* 2017), partially in Switzerland (Lakes Constance and Geneva; Sastre *et al.* 2010; Corella *et al.* 2014; Wessels *et al.* 2015), or bordering Switzerland (Lakes Como and Le Bourget; Van Rensbergen *et al.* 1999; Fanetti *et al.* 2008; Ledoux *et al.* 2009). Moreover, the sediments of these lakes have been characterized through sediment cores – so that stratigraphic and sedimentological datasets are available. Despite the amounts of data available, at present, high-resolution data do not exist for all Swiss peri-alpine lakes (e.g. only low-resolution bathymetric data from single-beam surveys or even from historical soundings exist for smaller lakes; considerable amount of sediment-mechanical data are only available for Lakes Lucerne and Zurich; Strasser *et al.* (2007, 2011) and Strupler *et al.* (2017)). For this reason, assumptions of common characteristics of Swiss peri-alpine lakes have to be made. Findings from limnogeological research on geomorphology, lithostratigraphy, sediment-mechanical characteristics and slope stability on peri-alpine lakes constitute the basis of assumptions made in our workflow.

- Previous studies on submerged slope stability and subaqueous landslides in peri-alpine lakes (e.g. Schnellmann *et al.* 2006; Hilbe *et al.* 2011; Strasser *et al.* 2011; Strupler *et al.* 2018*a*, *b*) showed that most of the documented slides are translational slides on a distinct glide plane. Slope gradients of 10–25° favour landsliding, as they are gentle enough for considerable sediment accumulation yet steep enough to provide the necessary shear stress needed for slope failure. In contrast, slopes with gradients smaller than 5° largely remained stable.
- Ground shaking can trigger landslides both in the subaerial and subaqueous environment (e.g. Schnellmann *et al.* 2002, 2006; Strasser *et al.* 2011, 2013; Kremer *et al.* 2015; Reusch *et al.* 2016; Cauzzi *et al.* 2018). In Switzerland, earthquakes with M_W 6 or larger are expected to occur every 100–150 years (Wiemer *et al.* 2016). A probabilistic seismic hazard model for Switzerland (SuiHaz15) provides estimates of horizontal ground motions for different return periods (RP; see Supplementary material Fig. ESM.1) (Wiemer *et al.* 2016).
- Typically, in the sedimentological succession on the non-deltaic slopes of deglaciated peri-alpine lakes, glacial sediments (tills) that contain unsorted, poorly rounded clasts, are followed by glacio-lacustrine sediments (i.e. Late Glacial clays, which originate from a pro-glacial environments and are relatively dense (*c.* 1600 kg m^{-3}),

and of authigenic lacustrine sedimentation during the Holocene (e.g. Kelts 1978; Strasser *et al.* 2007, 2011; Strupler 2012; Strupler *et al.* 2017).

- Strasser *et al.* (2007, 2011) and Strupler *et al.* (2017) investigated the sediment-mechanical properties in peri-alpine lakes Zurich and Lucerne. A weak layer has been identified in a sedimentary unit consisting of Late Glacial clays. To date, in the case of yet unfailed slopes in the investigated peri-alpine lakes, this potential weak layer is covered by *c* .3–10 m postglacial and potentially mobile sediment. The bulk density (ρ_{bulk}) of this potentially mobile sediment package generally increases with depth from *c.* 1300 kg m^{-3} at the lake floor to values exceeding *c.* 1600 kg m^{-3} in Late Glacial sediments. The undrained peak shear strength (s_u) at the weak layer amounts to *c.* 8 kPa that remains relatively constant with depth. Therefore, slope sequences with a thicker Holocene drape put additional stress on the slopes, which makes them less stable.

A first step for assessing the landslide-tsunami potential consists of assessing the slope-failure susceptibility. Slope stabilities are often assessed with limit-equilibrium equations, which assume that a slope becomes unstable if the downward-driving forces (e.g. gravity forces and external loading) exceed the resisting forces (i.e. s_u) that act on a slope. Basin-wide slope-stability assessments were conducted for Lake Lucerne by Strasser *et al.* (2011) and Lake Zurich by Strupler *et al.* (2018*b*). Such detailed analyses require high-resolution stratigraphic profiles of sediment-mechanical characteristics including for instance ρ_{bulk} and s_u values.

Hilbe and Anselmetti (2015*b*) combined publicly available codes for modelling mass movements, wave generation, propagation and inundation to simulate historically reported tsunamis in Lake Lucerne. Landslides were modelled with MassMov2d (Beguería *et al.* 2009) as single-phase, fluid-like mass movements propagating on the lake floor. The output of the landslide propagation model was fed into GeoClaw (Berger *et al.* 2011) to simulate tsunami generation, propagation and run-up. GeoClaw uses a finite volume method to solve the nonlinear shallow-water equations (George and LeVeque 2006).

Predictive equations for a quick assessment of the tsunami amplitude and runup (e.g. Pelinovsky and Poplavsky 1996; Grilli and Watts 2005; Watts *et al.* 2005; Ward 2011) have been shown to provide similar results for case studies as those obtained by numerical models. Their aim is to estimate a tsunami amplitude based on only a few input parameters, which mainly includes the geometric and topographic characteristics of a landslide. Such predictive equations have been applied for submarine landslide-tsunami case studies in Papua New Guinea, Alaska, Turkey (Watts *et al.* 2005), the eastern Australian margin (Clarke *et al.* 2014; Talukder and Völker 2014) and for Lake Ohrid (Macedonia/Albania) (Lindhorst *et al.* 2014).

Research approach

Using the example of Lake Lucerne, we present a workflow for the rapid estimation of the landslide-triggered tsunami potential in peri-alpine lakes. It consists of (1) an automatic identification of the location and estimation of the characteristics of potential subaqueous landslides (based on bathymetric data, forecasted peak ground accelerations (PGA; Wiemer *et al.* 2016) and assumptions on sediment characteristics that rely on findings from previous studies; cf. section 'Previous studies' above); and (2) estimates of the characteristics of tsunamis caused by the potential landslides based on existing empirical equations. The workflow gives an overview of the tsunami hazard, while requiring little input data and computation time. Results are compared to amplitudes from scenario-based numerical simulations in Lake Lucerne (Hilbe and Anselmetti 2015*b*). Furthermore, distributions of characteristic tsunami amplitudes of three selected peri-alpine lakes are compared.

Methodology

All calculations and illustrations are carried out with the free software 'R' (R Core Team 2018). The library 'RPyGeo' (Brenning *et al.* 2018) was used to execute commands in the commercial software ArcGIS (ESRI, Inc.). A complete list of used R libraries can be found in the Supplementary material (Table ESM.3).

Mapping of potential subaqueous landslide evacuation zones and their characteristics

Input datasets. We use a generic approach that should be valid for various glacially excavated, peri-mountainous lakes. As input datasets for our workflow, a digital bathymetric model (DBM), stratigraphic and geotechnical information (ρ_{bulk} and s_u), and earthquake-shaking (PGA) maps for different return periods are required. Additionally, the basin-wide thickness estimations of the potentially mobile sediment need to be available.

In the most basic form, a seismic reflection dataset, sediment cores and information on the seismic hazard are sufficient to extract the datasets needed in our approach. The seismic reflection data can be used to extract a simple DBM and to get information on the seismic stratigraphy. Sediment cores give information on the lithostratigraphy that can be correlated to seismic reflection data. Also, sedimentation rates and sediment-mechanical characteristics can be extracted directly from sediment cores. High-resolution bathymetric data and *in-situ* geotechnical measurements improve the quality of the input data.

The following paragraphs describe the input datasets used to automatically identify the location and the characteristics of the potential subaqueous landslides on the slopes of the three example lakes, Lucerne, Zug and Walensee, investigated in this article.

- *Bathymetric data:* For a few lakes in Switzerland, high-resolution DBMs exist. For other peri-alpine lakes, bathymetries are interpolated from Swiss-MapVector25 (SMV25; Swisstopo) isobaths with a scale of 1:25 000. The workflow accepts any spatial resolution of the input DBM. In the following we choose a grid resolution of 10 m as it constitutes a reasonable compromise between the two resolutions of the high-resolution DBMs and the interpolated ones from SMV25. As near-shore zones are not mapped in most of the available high-resolution DBMs, these zones are added from interpolated SMV25 datasets. Figure 1a shows example bathymetric data from Lake Lucerne used as input in the workflow. Shoreline data were extracted from the SMV25 dataset. From the bathymetric dataset, slope and aspect maps are derived (Fig. 1b).
- *Thickness of the potentially mobile sediment (T_{PMS}):* Strupler *et al.* (2018*b*) constructed a sedimentation model for Holocene sediments in Lake Zurich that calculates the authigenic sedimentation rate and thus thickness of the Holocene drape at any non-deltaic location in the lake as a function of water depth and slope gradient. In general, the greater the water depth, the more sedimentation, and the steeper a slope, the less sedimentation occurs (Supplementary material; Table ESM.1 and equation ESM.1). Based on maps of water depth, slope gradients and the sedimentation model from Strupler *et al.* (2018*b*), a map representing potentially mobile sediment thicknesses on the slopes is constructed. In general, sediment thickness increases with increasing water depth and decreases with greater slope angles. The model was calibrated for Lake Zurich, but is assumed to be valid for all peri-alpine lakes here. Details on the sedimentation model can be found in the electronic supplementary material (equation ESM.1 and Table ESM.1).
- *Earthquake-shaking information:* PGA values for mean return periods of 475 and 2475 years (Fig. ESM.1) from Wiemer *et al.* (2016) are interpolated across Switzerland and resampled to the same grid size as the lake bathymetries.

Sensitivity test of potentially unstable zones based on slope gradient, thickness of the potentially mobile sediment and PGA values. Due to uncertainties involved in geotechnical parameters, we do not explicitly calculate the stabilities of the subaqueous slopes. Instead, we define potentially unstable zones based on the slope geometry, thickness of the potentially mobile sediment and expected peak ground accelerations. We assume that the observations from previous studies (type of sliding, drape thickness and sediment-mechanical characteristics) are also valid for other peri-alpine lakes, due to their similar deglaciation and sedimentation history.

To test which combinations of slope gradients (range: 5–25° in steps of 5°), drape thickness (range: 1–10 m in steps of 1 m) and peak ground accelerations (range: 0–2 m s^{-2} in steps of 0.1 m s^{-2}) favour slope instabilities, safety factors (FS) using a limit equilibrium equation on an infinite slope are calculated after equation (1) for all 1050 combinations of these three input vectors. Boundary conditions were set based on findings from previous studies. For slope gradients, a lower threshold was set to slope gradients of 5°. A thick sediment drape causes great gravitational forces, which may also lead to landslides on relatively gentle slopes. The upper slope threshold of 25° was selected, as on slopes steeper than 25° no considerable sediment accumulation occurred in peri-alpine lakes (e.g. Schnellmann *et al.* 2006; Strasser *et al.* 2011; Strupler *et al.* 2018*b*).

$$FS = \frac{s_u}{\left(\int_0^z \gamma' dz\right) \times \sin(\alpha) \times \cos(\alpha) + k \times \mathrm{PGA} \times \left(\int_0^z \gamma' + \gamma_w dz\right) \times \cos^2(\alpha)} \quad (1)$$

where z is slide thickness (measured vertical to lake surface, in m), s_u is undrained shear strength (Pa), γ' is submerged unit weight of the potentially mobile sediment (N m^{-3}), γ_w is unit weight of water (N m^{-3}), α is slope gradient, PGA is peak ground acceleration (m s^{-2}) and k is pseudostatic coefficient = 0.2.

The sediment-mechanical characteristics used in equation (1) are estimated in a simplified form from data of Lakes Zurich and Lucerne (Strasser *et al.* 2011; Strupler *et al.* 2017, 2018*b*): ρ_{bulk} is assumed to be 1300 kg m^{-3} at the top of the Holocene drape and 1600 kg m^{-3} at the transition between the Late Glacial clays and the base of the Holocene drape, which marks the potential glide plane. We assume that ρ_{bulk} constantly increases with depth, interpolating the values between the top of the potentially mobile sediment drape and the glide plane. ρ_{bulk} is thus not linearly dependent on the sub-bottom depth. Instead our approach uses

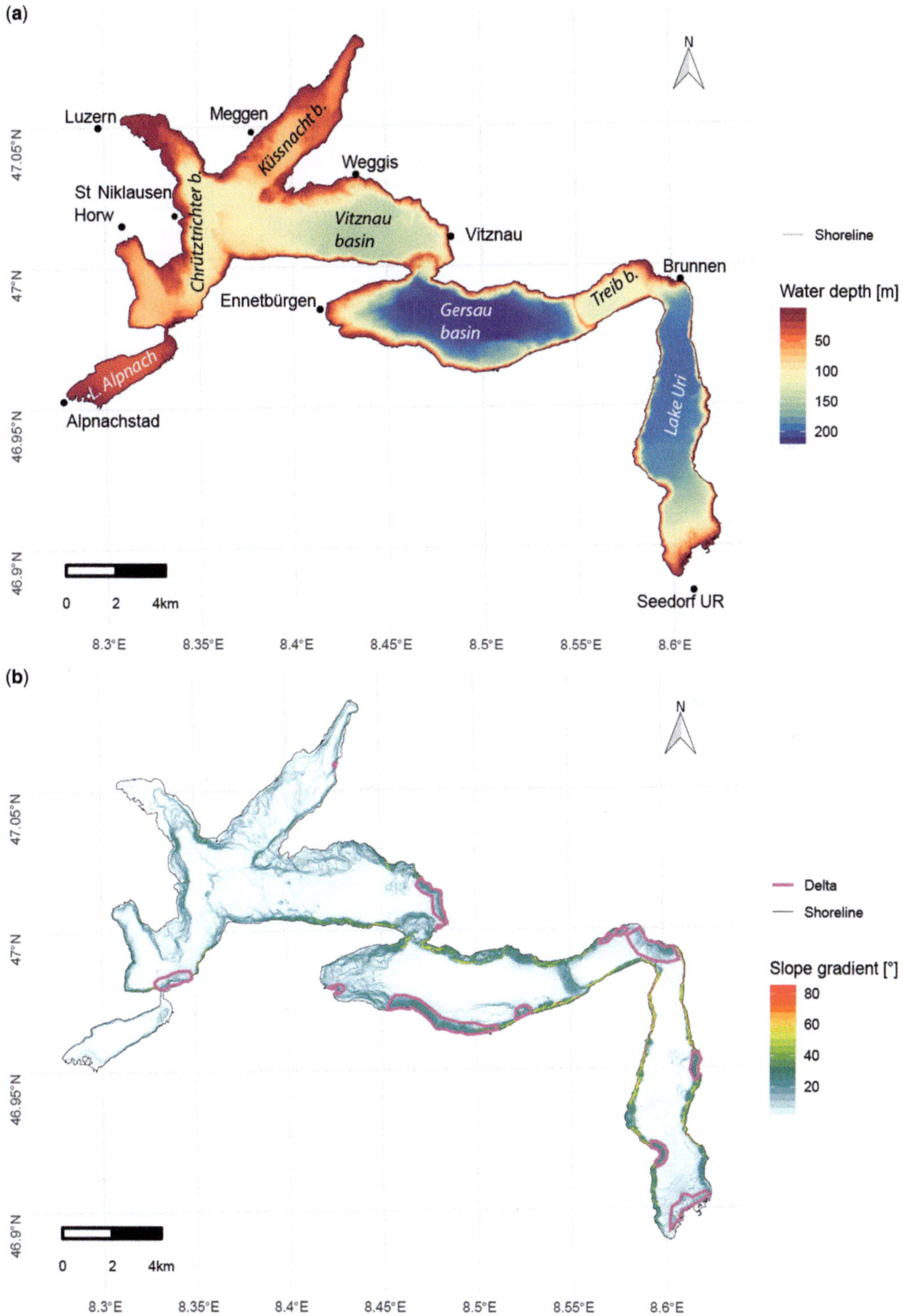

Fig. 1. (**a**) Bathymetry and (**b**) slope gradients of Lake Lucerne.

Table 1. *Combinations of slope gradient, thickness of potentially mobile sediment and PGAs leading to unstable slopes*

PGA ranges ($m\ s^{-2}$)	Slope ranges (°)	Drape thickness equal or greater than (m)
0.0–0.1	5–10	8
	10–15	6
	15–25	5
0.1–0.2	5–10	7
	10–20	5
	20–25	4
0.2–0.3	5–10	6
	10–15	5
	15–25	4
0.3–0.4	5–10	5
	10–25	4
0.4–0.5	5–15	4
	15–25	3
0.5–0.6	5–10	4
	10–25	3
0.6–0.9	10–25	3
0.9–1.0	5–20	3
	20–25	2
1.0–1.1	5–10	3
	10–25	2
1.1–2.0	5–25	1

'stretching' and 'shrinking' of the ρ_{bulk} profile to local sediment-thickness conditions (Strupler *et al.* 2017). s_u at the potential glide plane is assumed to be constant at *c.*8 kPa (Strasser *et al.* 2007, 2011; Strupler *et al.* 2017).

Based on the sensitivity test, we assume that zones of the slopes with the combinations of peak ground accelerations, slope gradients and potentially mobile sediment thickness shown in Table 1 are prone to failure.

Automatic identification of unstable zones. Potentially unstable zones are mapped automatically, based on input maps of slope gradients, potentially mobile sediment drape and peak ground accelerations for different return periods, and the identified combinations of these input data leading to instabilities (Table 1). The result of the mapping of potentially unstable zones is a binary grid indicating zones of potential landslides and stable zones.

For identifying the extent of individual potential landslide zones, it is assumed that single landslides have a similar aspect. Therefore, an aspect map is created (with 8 sub(directions); i.e. north, NE, east, …) and filtered (focal mean of 9 × 9 cells) to avoid that potential landslides located exactly on a class border between two different aspect classes are split into separate landslide patches. The binary landslide grid is then masked by the filtered aspect map and converted to polygons, representing individual potential landslide scars. Island polygons $<2500\ m^2$ (i.e. small, potentially stable zones) within the landslide scar polygons are removed. Slide polygons that represent multiple large areas that are combined by only a small band (i.e. <50 m) are automatically separated into individual patches.

For reasons of homogeneous data treatment (i.e. for many lakes, no information on subaqueous landslides exist), our approach assumes that all slopes are covered by a yet unfailed Holocene drape. Currently active deltas (Fig. 1b) are manually mapped (by visual interpretation based on bathymetric and isobaths data) and are excluded in our approach, as our understanding of the mechanical and dynamic delta characteristics is not sufficient to date.

At each node of the potential landslide polygons, elevation values from the bathymetric dataset, as well as slope gradients are queried. This allows the assessment of the characteristics of the potential landslides (Fig. 2). The following characteristics are calculated for each landslide polygon using the terminology of Clare *et al.* (2018):

- evacuation height (H_0): maximum – minimum elevation of each slide scar;
- mean gradient of slide scar (mean S_e): zonal mean of slope gradients for each polygon;
- evacuated length (length of the slide scar) (L_e): calculated from H_0 and sin(mean S_e) of each polygon;
- submergence depth at centre of mass motion (d): retrieved by querying the elevation values from the bathymetric dataset for the centre of the landslide scar polygons;
- slide thickness (T): zonal mean of T_{PMS} that was estimated with the sedimentation model of Strupler *et al.* (2018*b*).

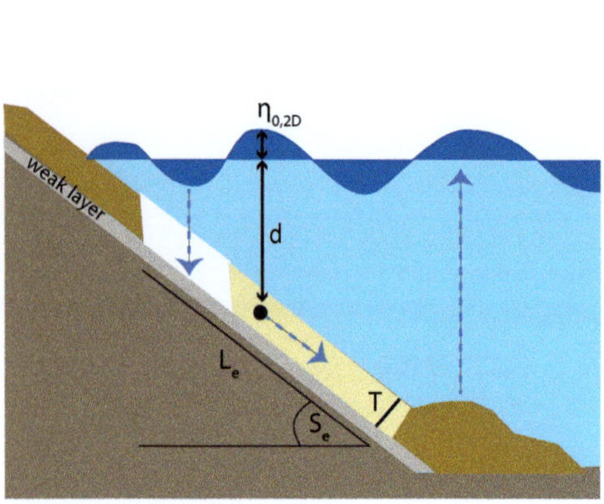

Fig. 2. Illustration of the parameters used for estimating $\eta_{0,2D}$.

Estimation of tsunami amplitudes

Due to its simplicity and due to the fact that it has already been used by various landslide-studies to get a first overview (cf. section 'Previous studies' above), we assess the size of potential tsunamis caused by each potential landslide with the characteristic tsunami amplitude after Watts *et al.* (2005), for a 2D case (equation 2). The equation is derived from numerical simulations and uses as input only slope gradient, slide thickness, slide length and depth of submergence. In our workflow, these input parameters are taken from the automatically mapped potential subaqueous landslide scars. We assume that the mass movements are translational slides occurring as one sliding block and that retrogressive behaviour does not occur. Equation (2) is valid only in the vicinity of tsunami sources, hence it does not consider amplitude changes during wave propagation.

The characteristic tsunami amplitude $\eta_{0,2D}$ is defined as the maximum tsunami amplitude above the initial mass-failure location (Watts *et al.* 2005; equation 2; Fig. 2):

$$\eta_{0,2D} \simeq 0.0286T\,(1 - 0.750 \times \sin S_e) \times \left(\frac{L_e \times \sin S_e}{d}\right)^{1.25} \quad (2)$$

Equation (2) is valid for a reference specific density = 1850 kg m^{-3}. The density of the potentially mobile sediment in peri-alpine lakes (*c.* 1300–1600 kg m^{-3}) is slightly lower, which may lead to small errors in absolute $\eta_{0,2D}$ values.

Results

Results of our workflow presented here are for the example of Lake Lucerne.

Modelled thickness of the potentially mobile sediment drape

The estimated thickness of the potentially mobile sediment drape shows values between *c.* 2 and *c.* 7 m (Fig. 3).

Location and characteristics of potential subaqueous landslide scars

Figure ESM.3 (in the Supplementary material) shows the locations of potential landslides and the expected volume (estimated by multiplication of slide thickness with slide area) for RP = 475 and

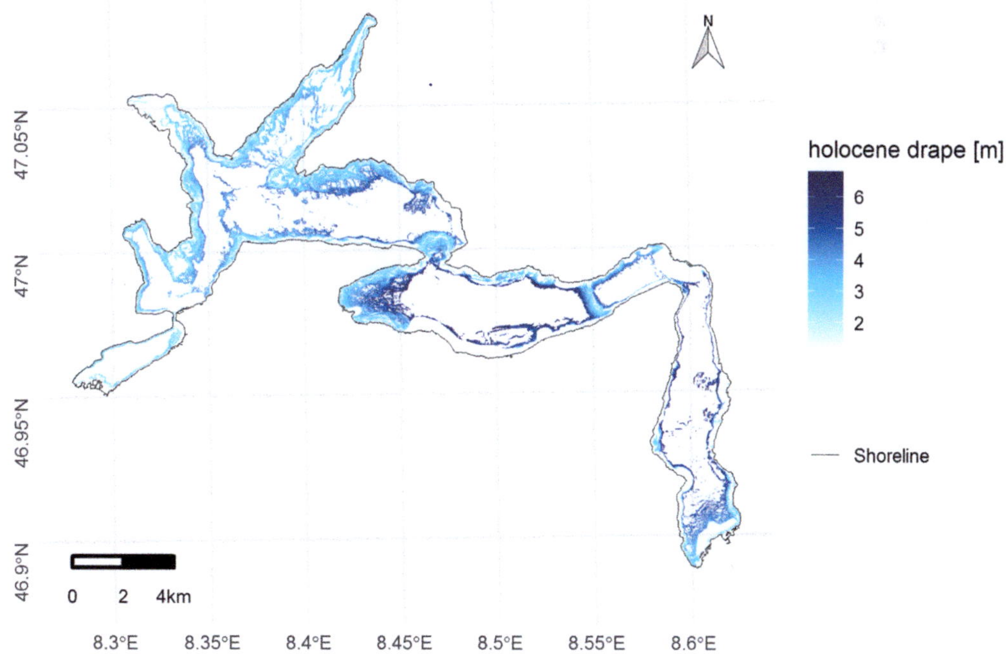

Fig. 3. Thickness of the Holocene sediments draping the slopes of Lake Lucerne estimated with the sedimentation model of Strupler *et al.* (2018*b*). Only zones within the defined boundary conditions (slope gradients between 5 and 25°, deltas excluded) are considered.

2475 years. Although, for the estimation of $\eta_{0,2D}$ only the slide thickness (and not the volume) is needed, we plot the volume to compare the data to documented landslide volumes for other peri-alpine lakes.

In general, for earthquake accelerations forecasted for larger RPs, larger slide areas and volumes are estimated. For an RP of 475 years, the 'Chrütztrichter', 'Vitznau' and the 'Gersau' basin (Fig. 1a) are mainly prone to landslides. The largest single landslide (volume *c.* 2×10^6 m^3) is expected to occur on the 'Chindli moraine'; separating the Gersau basin from the Treib basin. For an RP of 2475 years, all basins except for the Alpnach basin are prone to subaqueous landslides. The landslides with the greatest volumes (up to *c.* 5×10^6 m^3) are forecasted offshore Ennetbürgen, in the Gersau basin.

Histograms of T, L_e, d and S_e (RPs stacked) show distributions of the landslide characteristics used for the tsunami estimation (Fig. 4). For earthquakes with an RP = 475 years, slopes with a minimum thickness T of *c.* 4.5 m can be mobilized, while for an RP = 2475 years, slides can also occur on slopes with potentially mobile sediment thicknesses as low as 3.5 m (Fig. 4a). Most potential slides have an estimated L_e of *c.* 100–250 m. However, L_e can be as long as *c.* 600 m (Fig. 4b). For an RP = 475 years, the minimum submergence depth of the centre of the mass movement d is at 70 m below lake level, whereas for an RP = 2475 years, d can be located at a depth <10 m below lake level (Fig. 4c). Most of the potential slides are located on slopes between 10 and 15° (Fig. 4d).

Estimated characteristic tsunami amplitude for different return periods

Figure 5 shows the characteristic wave amplitude $\eta_{0,2D}$ caused by each single landslide for earthquakes with mean return periods of 475 (Fig. 5a) and 2475 years (Fig. 5b), according to equation (2). For an RP = 475 years, $\eta_{0,2D}$ expected to be caused by individual landslides are in the range of *c.* 0.002 to *c.* 0.11 m, and for an RP = 2475, in the range of *c.* 0.002 to *c.* 0.41 m. See also Figure 6.

$\eta_{0,2D}$*: distributions among selected peri-alpine lakes*

Distributions of forecasted $\eta_{0,2D}$ for three peri-alpine Swiss lakes (Lakes Lucerne, Walensee and Zug) are represented as violin plots (Fig. 6). For an RP = 475 years, maximum $\eta_{0,2D}$ for individual forecasted slides in all the lakes are in the range of *c.* 0.1 to *c.* 0.2 m. The largest number of potential slides (reflected by the width of the violin plots) can be found in Lake Lucerne ($n = 104$), followed by Lake Zug ($n = 82$) and Lake Walensee ($n = 30$).

For an RP = 2475 years, max $\eta_{0,2D}$ is largest for Lakes Lucerne and Zug (*c.* 0.4 m), followed by Walensee (*c.* 0.3 m). The largest number of potential tsunamigenic landslides can be found in Lake Lucerne ($n = 333$), followed by Lakes Zug ($n = 141$) and Walensee ($n = 52$). Distributions of $\eta_{0,2D}$ for Lakes Lucerne and Zug have maxima at *c.* 0.1 m. For Lake Walensee, the distribution of characteristic wave amplitudes is nearly uniform.

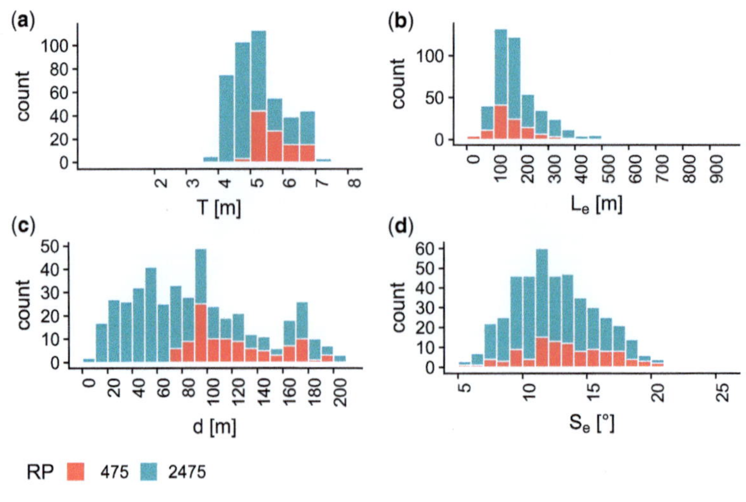

Fig. 4. Frequency distributions of LS characteristics in Lake Lucerne for various return periods. (**a**) Thickness of the LS (T); (**b**) slide length (L_e); (**c**) submergence depth of LS centre point; (**d**) mean slope gradient of LS (S_e).

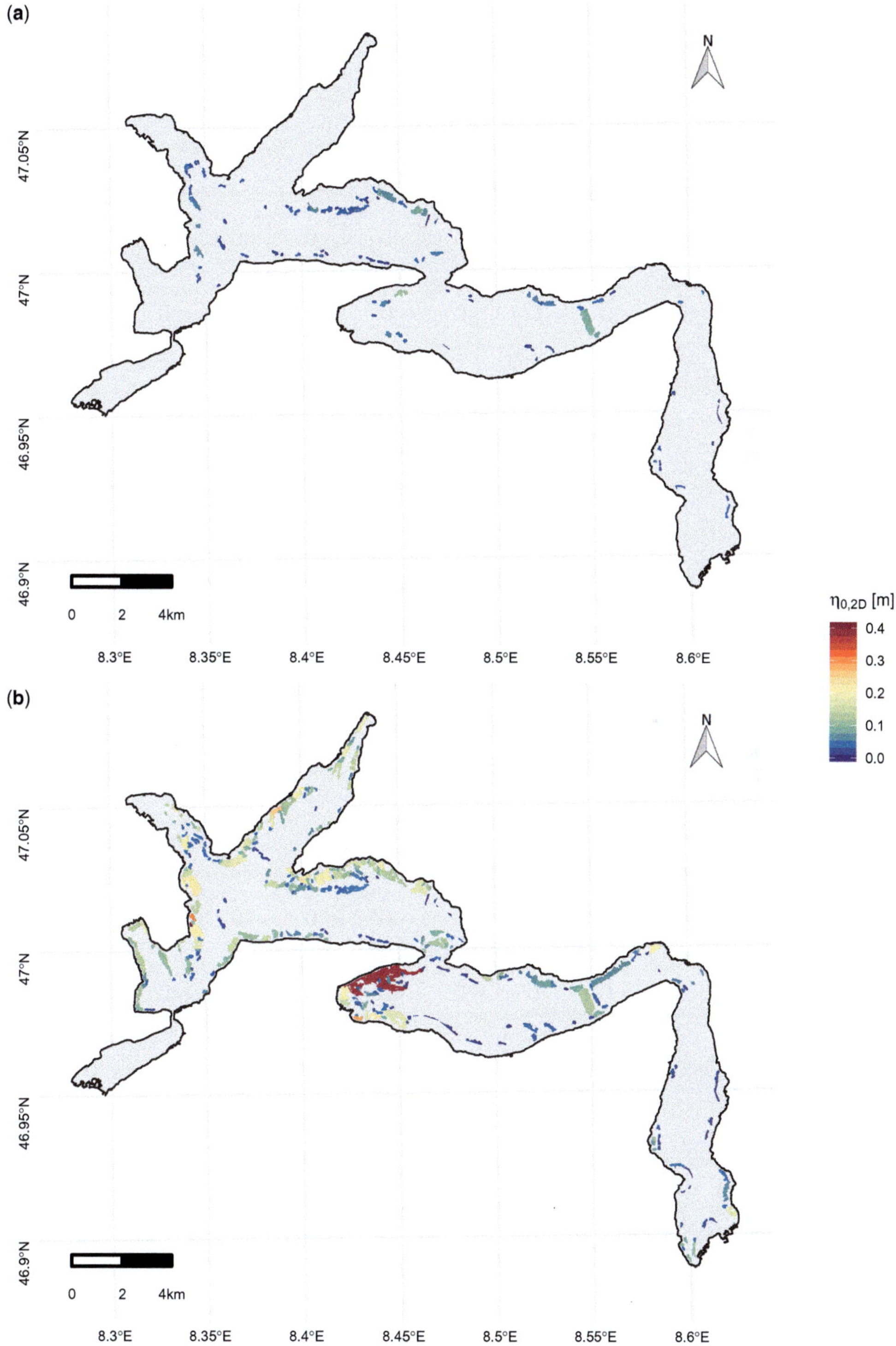

Fig. 5. Characteristic tsunami height for each landslide in Lake Lucerne for a mean RP of (**a**) 475 and (**b**) 2475 years.

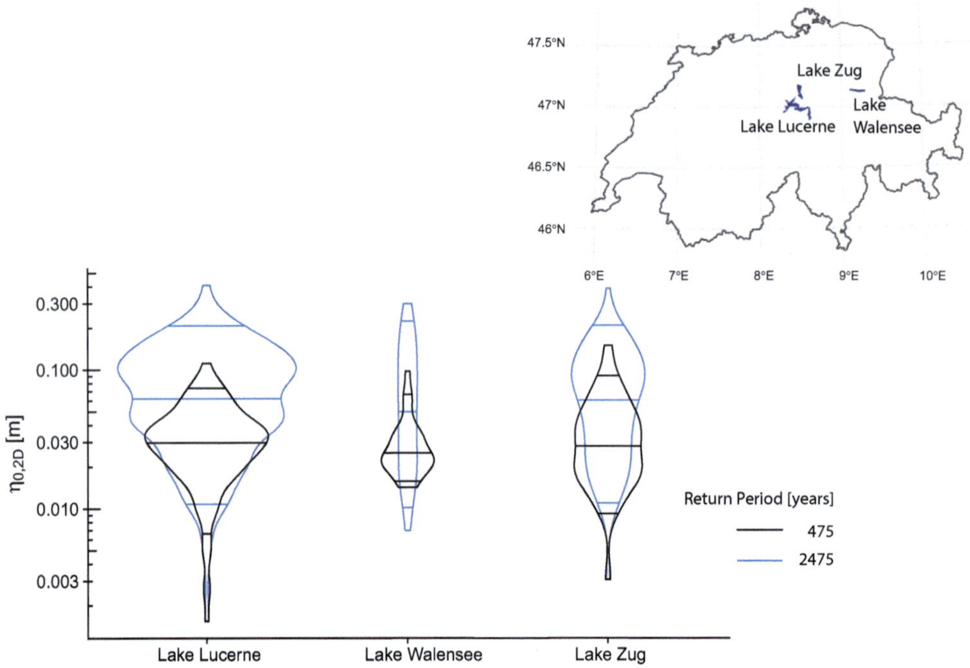

Fig. 6. Violin plots showing the distributions of the characteristic amplitude of tsunamis generated by earthquake shaking with an RP of 475 (black) and 2475 years (blue). The maximum width of the violin plots is proportional to the number of landslides. Horizontal lines show median, 5th and 95th quantiles.

Discussion

Quality and limitations of the proposed methodology

The workflow presented here allows for rapid first-order estimations of the location of potential landslides and their consequences on lakes with peri-alpine characteristics. Due to the generic approach used, the results are not to be interpreted as exact for every single landslide and the potential tsunami produced by it. Rather, the results are intended to give a rapid overview of locations within a lake with a greater potential for tsunamis.

As the assessment of the landslide-tsunami hazard is due to the large number of complex parameters involved, many uncertainties exist. Regarding the location of potential subaqueous landslides, the main uncertainties are related to the slope-stability assumptions, which are affected by uncertainties in the local drape thickness, geotechnical and geomorphic characteristics, PGA and the model used to assess slope stability.

- Without sampling geophysical and sedimentological data from each lake, we can only make assumptions of the potentially mobile sediment thickness. A simple comparison of the sedimentation model applied in this study to measured Holocene sediment thicknesses from sediment cores by Strasser *et al.* (2007) shows that the potentially mobile sediment modelled with the approach of Strupler *et al.* (2018*b*) does not show a systematic over- or underestimation (Table ESM.2 and Fig. ESM.2).
- The geotechnical assumptions taken here are strong simplifications. Also, the spatial variability of geotechnical properties (e.g. Strupler *et al.* 2017) is ignored.
- The used grid-resolution of the DBM influences the geomorphic characteristics (i.e. larger cell size results in a 'smoother' DBM).

Regarding the estimation of the tsunami amplitudes, the main uncertainties are related to the actual location of the estimated subaqueous landslides and their characteristics such as the slide thickness T, actual submergence depth of the centre of mass movement d, length L_e and mean gradient S_e of the slide scar. Also, uncertainties exist concerning the actual mechanism of the landslides (e.g. single- or multi-stage landslides, retrogressive behaviour, cohesion of slide material, etc.). Although retrogressive slides generally reduce the tsunamis' height due to the element-wise sliding procedure, interference of individual waves produced by each slide element may also increase the aggregated tsunami wave heights (e.g. Bondevik *et al.* 2005; Haugen *et al.* 2005; Løvholt *et al.* 2005; Harbitz *et al.* 2014).

Equation (2) shows reasonable results in the marine realm, where water depths are generally great (Watts *et al.* 2005). A disadvantage of applying the equation to relatively shallow lakes is that the equation uses only submergence depth at the centre of mass motion as an indicator, but not the maximum lake depth, which could (in case of very shallow lakes) limit the development of necessary slide accelerations.

Some uncertainties were mitigated by setting boundary conditions and stating the assumptions that need to be made for the results to be valid. For example, we assume landslides to occur as one landslide-block, resulting in rather conservative estimates. Also, delta failures are excluded as tsunami sources, as our understanding of delta-failures as tsunami source is not advanced enough to date.

The workflow presented here estimates $\eta_{0,2D}$ for each single potential landslide, but does not consider potential wave interference that may occur as a result of multiple landslides occurring coevally. This can be considered a major drawback of the rapid tsunami assessment, as earthquakes are known to often trigger multiple, coeval landslides on a basin-wide scale (e.g. Jibson 1996). Also, edge effects and potential effects of the local bathymetry on the tsunami wave, such as funnelling, are neglected. For the consideration of such effects, numerical modelling needs to be conducted, which is not the scope of our simple rapid assessment.

Location and characteristics of potential subaqueous landslides (Lake Lucerne)

Modelled slide thicknesses are in the range of the values estimated by Hilbe and Anselmetti (2015*b*). who estimated landslide volumes between *c.* 5.5 × 10^6 m^{-3} and 20.8 × 10^6 m^{-3}, whereas our estimated volumes of single potential landslides range between *c.* 4 × 10^4 m^{-3} and *c.* 5 × 10^6 m^{-3}. However, a technical report by Hilbe and Anselmetti (2015*a*) assumes landslide volumes in the range of 1.7 × 10^5 m^{-3} to 5.4 × 10^6 m^{-3}.

The generally larger slide areas and volumes estimated for earthquakes with greater accelerations assigned to greater RPs, are explained by the fact that according to the model, stronger PGAs mobilize many zones that show relatively gentle slope angles (5–10°). They would be stable for earthquakes with weaker PGAs, as forecasted for smaller RPs (Fig. ESM.3). This finding may be related to the fjord-like morphology of the lake basin: visual interpretation shows that many nearshore zones are relatively gently dipping down to a sudden transition of steeper gradients. Such relatively gentle zones close to the lake surface may be stable enough for weaker earthquakes but lose stability for stronger earthquakes.

Visual comparison shows that the zones identified as susceptible to landslides by Strasser *et al.* (2011) and Hilbe and Anselmetti (2015*a*) are contained in our automatically mapped potential landslide polygons. Due to reasons of comparability between single lakes with different information available, we do not exclude areas of already failed zones, so that our results show slightly more potential landslide zones. We are aware that during the earthquake events of AD 1601 and 2220 BP, large parts of the slopes lost most of the sediment cover and are thus relatively stable to date (Strasser *et al.* 2011). This is a compromise that has to be made in our generic approach. The prediction of individual landslide locations and their characteristics is still a major challenge in our field (Harbitz *et al.* 2014).

Tsunami amplitudes (Lake Lucerne)

The estimated tsunami amplitudes $\eta_{0,2D}$ caused by each individual potential landslide should not be considered as absolute values. Rather, these values give a first-order information of tsunami potential of each landslide.

Free surface waves modelled with GeoClaw by Hilbe and Anselmetti (2015*b*) range from *c.* 3 to 10 m, thus much larger than estimated here ($\eta_{0,2D}$ *c.* 0.4 m). However, our workflow estimates $\eta_{0,2D}$ above the centre of each individual potential landslide. This implies that wave propagation and thus the effect of wave interference is not considered. An earthquake triggering basin-wide subaqueous landslides may thus cause tsunami with larger wave amplitudes if these effects are considered. Another reason for differing wave amplitudes modelled by Hilbe and Anselmetti (2015*b*) and estimated here may be related to the different assumptions of slide mechanisms, that is fluid-like slides (Hilbe and Anselmetti 2015*b*) v. block slides (this paper). For Lake Lucerne, according to our applied method, these slope patches are mainly located offshore of the villages of Ennetbürgen ($\eta_{0,2D}$ *c.* 0.4 m), St Niklausen ($\eta_{0,2D}$ *c.* 0.4 m), Meggen ($\eta_{0,2D}$ *c.* 0.25 m) and Weggis ($\eta_{0,2D}$ *c.* 0.2 m). Our results show that for stronger ground shaking, more landslides are expected to occur, and therefore the aggregated tsunami amplitude may be greater in that case.

As a simple rule of thumb, the run-up close to the landslide source equals *c.* 1 to 2 times $\eta_{0,2D}$ (Abe 1995; Watts *et al.* 2005). However, in reality, tsunami amplitude gets smaller during propagation of the wave to the shoreline due to geometric spreading and dispersion effects, and, at the same time, shoaling effects cause an amplitude gain, both having an influence on the run-up. The historically reported run-up (*c.* 4 m) (Cysat 1969; Schnellmann *et al.* 2002) and the run-up estimated by Hilbe and Anselmetti (2015*b*) are higher than our results. However, it is documented that GeoClaw often overestimates amplitude and run-up in the near-source area of a tsunami (e.g. Hilbe and Anselmetti 2015*b*). This has also been found in a comparison between interpolated field measurements and calculation with GeoClaw in a study area in Japan, where run-up was documented as clearly overestimated (Montoya and Lynett 2017).

Comparison of subaqueous-landslide tsunami potential on three selected peri-alpine lakes

Figure 6 allows for a quick comparison of the tsunami potential among investigated lakes. Whereas considering only single values, such as the maximum $\eta_{0,2D}$ estimated for each lake, the consideration of the whole distribution allows a better understanding of this tsunami metric. The central basins of Lake Lucerne seem to favour tsunamigenic landsliding, mainly due to the many slopes with ideal preconditions such as the location below lake level of patches with an ideal slope gradient and sediment thickness. From the lakes investigated here, the ones that show many potential landslides causing the highest $\eta_{0,2D}$ values in each distribution are considered as having the relatively highest tsunami potential, because in an earthquake event, chances are higher for triggering such a potential landslide if more candidates are around.

Conclusions and outlook

A workflow for the rapid first-order estimation of tsunamis triggered by sublacustrine landslides that conducts a series of geospatial operations is presented. As a first step, potential subaqueous landslides expected for various earthquake-shaking scenarios (i.e. PGAs for return periods of 475 and 2475 years) are automatically identified and characterized. As a second step, the characteristic tsunami amplitudes are estimated with empirical equations. We present a cost- and time-efficient method that relies on geological common sense and results in a rapid assessment.

The approach proposed here allows for a rapid, first-order identification of potentially tsunamigenic landslides. Due to the generic approach assuming the same geotechnical properties and sedimentation model for and within all peri-alpine lakes, results from our workflow may not be correct for each single slope. Nevertheless, the results give a first overview of the tsunami potential on the investigated lakes. However, our workflow will never be able to reach the same quality as numerical modelling, as simple assumptions are taken and very important steps, such as wave propagation and inundation effects, are not considered. Therefore, we recommend the use of the workflow proposed here to identify parts of submerged slopes that need more in-depth attention. The frequency distributions of potential landslide characteristics may be compared to distributions of mapped, already failed landslides. To this end, a dataset with mapped landslides of various peri-alpine lakes would be ideal. Finally, the investigated lakes may be classified accordingly into different hazard levels, based on their hazard metrics: the results may be used for focusing more detailed, lake-specific tsunami-hazard assessments in the respective areas. Also, future work may include tsunamis generated by delta collapses and subaerial mass movements in the workflow.

The approach presented here may be transferred to other lakes worldwide, given that (1) bathymetric data are available, (2) assumptions on the stratigraphic location of the potential glide plane exist, (3) information on the characteristics of the potentially mobile sediment (i.e. thickness, ρ_{bulk}, and s_u) is available, and (4) estimations or assumptions of forecasted PGA values exist. We suggest that the workflow proposed here may be applied with relatively little effort on other peri-alpine, European lakes, because bathymetric and sedimentological datasets are available for selected Austrian, French and Italian peri-alpine lakes (e.g. Van Rensbergen *et al.* 1999, Fanetti *et al.* 2008; Ledoux *et al.* 2009; Daxer *et al.* 2018). Seismic hazard information for Europe is available via the 'SHARE' (Seismic Hazard Harmonization in Europe) project (Giardini *et al.* 2014). The approach may also be transferred to other de-glaciated lakes in mountainous areas, such as on Alaskan lakes on the Kenai Peninsula where reflection seismic data and sediment cores are available (e.g. Praet *et al.* 2017; Boes *et al.* 2018), as well as estimations on the seismic hazard (Petersen *et al.* 2010). Another setting to test our approach would be the deglaciated, peri-andine lakes in Patagonia with existing bathymetric, reflection seismic and sediment-core data (e.g. Moernaut *et al.* 2007, 2014; Heirman *et al.* 2011; Waldmann *et al.* 2011; Van Daele *et al.* 2014). Seismic hazard information for South America is available from Petersen *et al.* (2018).

Acknowledgements The handling editor, Jasper Moernaut, and the two reviewers David Tappin and Brian G. McAdoo are gratefully acknowledged for their constructive comments. SMV25 data are copyrighted by Swisstopo.

Funding The authors would like to thank the Federal Office for Environment (FOEN) for project funding (17.0092.PJ).

Author contributions **MS**: formal analysis (lead), methodology (lead), visualization (lead), writing – original draft (lead); **FSA**: funding acquisition (equal), writing – review & editing (equal); **MH**: writing – review & editing (equal); **KK**: writing – review & editing (equal); **SW**: funding acquisition (equal), writing – review & editing (equal).

References

Abe, K. 1995. Estimate of tsunami run-up heights from earthquake magnitudes. *In*: Tsuchiya, Y. and Shuto, N. (eds) *Tsunami: Progress in Prediction, Disaster Prevention and Warning*. Springer, 21–35, https://doi.org/10.1007/978-94-015-8565-1_2

Beguería, S., Van Asch, T.W.J., Malet, J.-P. and Gröndahl, S. 2009. A GIS-based numerical model for simulating the kinematics of mud and debris flows over complex terrain. *Natural Hazards and Earth System Science*, **9**, 1897–1909, https://doi.org/10.5194/nhess-9-1897-2009

Berger, M.J., George, D.L., LeVeque, R.J. and Mandli, K.T. 2011. The GeoClaw software for depth-averaged flows with adaptive refinement. *Advances in Water Resources*, **34**, 1195–1206, https://doi.org/10.1016/j.advwatres.2011.02.016

Boes, E., Van Daele, M. *et al.* 2018. Varve formation during the past three centuries in three large proglacial lakes in south-central Alaska. *Geological Society of America Bulletin*, **130**, 757–774, https://doi.org/10.1130/B31792.1

Bondevik, S., Løvholt, F. *et al.* 2005. The Storegga Slide tsunami: comparing field observations with numerical simulations. *Marine and Petroleum Geology*, **22**, 195–208, https://doi.org/10.1016/j.marpetgeo.2004.10.003

Brenning, A., Polakowski, F. and Becker, M. 2018. RPyGeo: ArcGIS geoprocessing via python, Version 1.0.0, https://github.com/fapola/RPyGeo

Cauzzi, C., Fäh, D. *et al.* 2018. ShakeMap-based prediction of earthquake-induced mass movements in Switzerland calibrated on historical observations. *Natural Hazards*, **92**, 1211–1235, https://doi.org/10.1007/s11069-018-3248-5

Clare, M., Chaytor, J. *et al.* 2018. A consistent global approach for the morphometric characterization of subaqueous landslides. *Geological Society, London, Special Publications*, **477**, 455–477, https://doi.org/10.1144/SP477.15

Clarke, S., Hubble, T. *et al.* 2014. Morphology of Australia's Eastern Continental Slope and related tsunami hazard. *In*: Krastel, S., Behrmann, J.-H. *et al.* (eds) *Submarine Mass Movements and their Consequences: 6th International Symposium*, Springer, 529–538.

Corella, J.P., Arantegui, A. *et al.* 2014. Sediment dynamics in the subaquatic channel of the Rhone delta (Lake Geneva, France/Switzerland). *Aquatic Sciences*, **76**, 73–87, https://doi.org/10.1007/s00027-013-0309-4

Cysat, R. 1969. Collectanea Chronica und denkwürdige Sachen pro Chronica Lucernensi et Helvetiae. *In*: Schmid, J. (ed.) *Quellen und Forschungen zur Kulturgeschichte von Luzern und der Innerschweiz*. Diebold Schilling Verlag, Luzern, 882–888.

Daxer, C., Moernaut, J. *et al.* 2018. Late Glacial and Holocene sedimentary infill of Lake Mondsee (Eastern Alps, Austria) and historical rockfall activity revealed by reflection seismics and sediment core analysis. *Austrian Journal of Earth Sciences*, **111**, https://doi.org/10.17738/ajes.2018.0008

De Blasio, F.V. 2011. Subaqueous Landslides. *In*: *Introduction to the Physics of Landslides*. Springer Netherlands, Dordrecht, 295–351.

Fabbri, S.C., Herwegh, M. *et al.* 2017. Combining amphibious geomorphology with subsurface geophysical and geological data: a neotectonic study at the front of the Alps (Bernese Alps, Switzerland). *Quaternary International*, **451**, 101–113, https://doi.org/10.1016/j.quaint.2017.01.033

Fanetti, D., Anselmetti, F.S. *et al.* 2008. Megaturbidite deposits in the Holocene basin fill of Lake Como (Southern Alps, Italy). *Palaeogeography, Palaeoclimatology, Palaeoecology*, **259**, 323–340, https://doi.org/10.1016/j.palaeo.2007.10.014

Favrod, J. 1991. *La Chronique de Marius d'Avenches (455-581): Texte, traduction et commentaire. Section d'histoire, Faculté des lettres*. Université de Lausanne, Lausanne.

Fine, I.V., Rabinovich, A.B. *et al.* 2005. The Grand Banks landslide-generated tsunami of November 18, 1929: preliminary analysis and numerical modeling. *Marine Geology*, **215**, 45–57, https://doi.org/10.1016/j.margeo.2004.11.007

George, D. and LeVeque, R.J. 2006. Finite volume methods and adaptive refinement for global tsunami propagation and local inundation. *Science of Tsunami Hazards*, **24**, 319–328.

Giardini, D., Wössner, J. and Danciu, L. 2014. Mapping Europe's seismic hazard. *EOS*, **95**, 261–262, https://doi.org/10.1002/2014EO290001

Grilli, S.T. and Watts, P. 2005. Tsunami generation by submarine mass failure. I: modeling, experimental validation, and sensitivity analyses. *Journal of Waterway, Port, Coastal, and Ocean Engineering*, **131**, 283–297, https://doi.org/10.1061/(ASCE)0733-950X(2005)131:6(283)

Harbitz, C.B. 1992. Model simulations of tsunamis generated by the Storegga Slides. *Marine Geology*, **105**, 1–21, https://doi.org/10.1016/0025-3227(92)90178-K

Harbitz, C.B., Løvholt, F. and Bungum, H. 2014. Submarine landslide tsunamis: how extreme and how likely? *Natural Hazards*, **72**, https://doi.org/10.1007/s11069-013-0681-3

Haugen, K.B., Løvholt, F. and Harbitz, C.B. 2005. Fundamental mechanisms for tsunami generation by submarine mass flows in idealised geometries. *Marine and Petroleum Geology*, **22**, 209–217, https://doi.org/10.1016/j.marpetgeo.2004.10.016

Heirman, K., De Batist, M. *et al.* 2011. Detailed seismic stratigraphy of Lago Puyehue: implications for the mode and timing of glacier retreat in the Chilean Lake District. *Journal of Quaternary Science*, **26**, 665–674, https://doi.org/10.1002/jqs.1491

Hilbe, M. and Anselmetti, F. 2015*a*. Bericht Gefahrenpotential erdbebeninduzierter Flutwellen im Vierwaldstättersee. Technical Report. Institute of Geological Sciences and Oeschger Centre for Climate Change Research, University of Bern, Bern, Switzerland.

Hilbe, M. and Anselmetti, F.S. 2015*b*. Mass movement-induced Tsunami Hazard on Perialpine Lake Lucerne (Switzerland): scenarios and numerical experiments. *Pure and Applied Geophysics*, **172**, 545–568, https://doi.org/10.1007/s00024-014-0907-7

Hilbe, M., Anselmetti, F.S. *et al.* 2011. Subaqueous morphology of Lake Lucerne (Central Switzerland): implications for mass movements and glacial history. *Swiss Journal of Geosciences*, **104**, 425–443, https://doi.org/10.1007/s00015-011-0083-z

Jibson, R.W. 1996. Use of landslides for paleoseismic analysis. *Engineering Geology*, **43**, 291–323, https://doi.org/10.1016/S0013-7952(96)00039-7

Kelts, K. 1978. *Geological and sedimentary evolution of Lakes Zurich and Zug, Switzerland*. PhD thesis, ETH Zurich

Kremer, K., Simpson, G. and Girardclos, S. 2012. Giant Lake Geneva tsunami in ad 563. *Nature Geoscience*, **5**, 756–757, https://doi.org/10.1038/ngeo1618

Kremer, K., Hilbe, M. *et al.* 2015. Reconstructing 4000 years of mass movement and tsunami history in a deep peri-Alpine lake (Lake Geneva, France–Switzerland). *Sedimentology*, **62**, 1305–1327, https://doi.org/10.1111/sed.12190

Ledoux, G., Lajeunesse, P., Chapron, E. and St-Onge, G. 2009. Multibeam bathymetry investigations of mass movements in Lake Le Bourget (NW Alps, France) using a portable platform. *In*: Mosher, D.C., Shipp, R.C., Moscardelli, L., Chaytor, J.D., Baxter, C.D.P., Lee, H.J. and Urgeles, R. (eds) *Submarine Mass Movements and Their Consequences,* Springer, 423–434, https://doi.org/10.1007/978-90-481-3071-9_35

Lindhorst, K., Krastel, S., Papenberg, C. and Heidarzadeh, M. 2014. Modeling submarine landslide-generated waves in Lake Ohrid, Macedonia/Albania. *In*: Krastel, S., Behrmann, J.-H., *et al.* (eds) *Submarine Mass Movements and Their Consequences: 6th International Symposium*, Springer, 497–506.

Locat, J. and Lee, H.J. 2002. Submarine landslides: advances and challenges. *Canadian Geotechnical Journal*, **39**, 193–212, https://doi.org/10.1139/t01-089

Løvholt, F., Harbitz, C.B. and Haugen, K.B. 2005. A parametric study of tsunamis generated by submarine slides in the Ormen Lange/Storegga area off western Norway. *Marine and Petroleum Geology*, **22**, 219–231, https://doi.org/10.1016/j.marpetgeo.2004.10.017

Moernaut, J., De Batist, M. *et al.* 2007. Giant earthquakes in South-Central Chile revealed by Holocene mass-wasting events in Lake Puyehue. *Sedimentary Geology*, **195**, 239–256, https://doi.org/10.1016/j.sedgeo.2006.08.005

Moernaut, J., Van Daele, M. *et al.* 2014. Lacustrine turbidites as a tool for quantitative earthquake reconstruction: new evidence for a variable rupture mode in south central Chile. *Journal of Geophysical Research: Solid Earth*, **119**, 1607–1633, https://doi.org/10.1002/2013JB010738

Montoya, L.H. and Lynett, P. 2017. Quantifying numerical model accuracy and variability. *Coastal Engineering Proceedings*, **1**, 12, https://doi.org/10.9753/icce.v35.currents.12

Nadim, F., Kalsnes, B. and Eide, A. 1996. Analysis of submarine slope stability under seismic action. *In*: Senneset, K. (ed.) *Landslides*. Balkema, Rotterdam, 561–565.

Okal, E.A. 2019. Twenty-five years of progress in the science of "geological" tsunamis following the 1992 Nicaragua and Flores events. *Pure and Applied Geophysics*, **176**, 2771–2793, https://doi.org/10.1007/s00024-019-02244-x

Pelinovsky, E. and Poplavsky, A. 1996. Simplified model of tsunami generation by submarine landslides. *Physics and Chemistry of the Earth*, **21**, 13–17, https://doi.org/10.1016/S0079-1946(97)00003-7

Petersen, M.D., Frankel, A.D. *et al.* 2010. *Documentation for the 2008 update of the United States national seismic hazard maps*. Earthquake Research: Background and Select Reports **107–234**.

Petersen, M.D., Harmsen, S.C. *et al.* 2018. Seismic hazard, risk, and design for South America. *Bulletin of the Seismological Society of America*, **108**, 781–800, https://doi.org/10.1785/0120170002

Praet, N., Moernaut, J. *et al.* 2017. Paleoseismic potential of sublacustrine landslide records in a high-seismicity setting (south-central Alaska). *Marine Geology*, **384**, 103–119, https://doi.org/10.1016/j.margeo.2016.05.004

R Core Team 2018. *R: A language and environment for statistical computing*. R Foundation for Statistical Computing, Vienna, Austria, https://www.R-project.org/

Reusch, A. 2016. *Sublacustrine paleoseismology and fluid flow in the Western Swiss Molasse Basin: New constraints from the sedimentary archive of Lake Neuchâtel – Mass-transport deposits, subsurface sediment mobilization and geomorphology*. PhD Thesis, ETH Zurich, https://doi.org/10.3929/ethz-a-010638214

Reusch, A., Moernaut, J., Anselmetti, F.S. and Strasser, M. 2016. Sediment mobilization deposits from episodic subsurface fluid flow-A new tool to reveal long-term earthquake records? *Geology*, **44**, 243–246, https://doi.org/10.1130/G37410.1

Sammartini, M., Moernaut, J. *et al.* (2019). An atlas of mass-transport deposits in lakes. *In*: Ogata, K., Festa, A. and Pini, G.A. (eds) *Submarine Landslides: Subaqueous Mass Transport Deposits from Outcrops to Seismic Profiles*. AGU, 201–226.

Sastre, V., Loizeau, J.L. *et al.* 2010. Morphology and recent history of the Rhone River Delta in Lake Geneva (Switzerland). *Swiss Journal of Geosciences*, **103**, 33–42, https://doi.org/10.1007/s00015-010-0006-4

Schnellmann, M., Anselmetti, F.S. *et al.* 2002. Prehistoric earthquake history revealed by lacustrine slump deposits. *Geology*, **30**, 1131–1134, https://doi.org/10.1130/0091-7613(2002)030<1131:PEHRBL>2.0.CO;2

Schnellmann, M., Anselmetti, F.S., Giardini, D. and McKenzie, J.A. 2006. 15,000 Years of mass-movement history in Lake Lucerne: implications for seismic and tsunami hazards. *Eclogae Geologicae Helvetiae*, **99**, 409–428, https://doi.org/10.1007/s00015-006-1196-7

Schoeneich, P., Weidmann, M. and Blomjous, C. 2015. L' énigme du Tauredunum enfin résolue? *In*: Reynard, E., Evéquoz-Dayen, M. and Borel, G. (eds) *Le rhône, entre nature et société*. Cahiers de vallesia, Sion, 153–174.

Siegenthaler, C., Finger, W., Kelts, K. and Wang, S. 1987. Earthquake and seiche deposits in Lake Lucerne, Switzerland. *Eclogae Geologicae Helvetiae*, **80**, 241–260

Strasser, M., Stegmann, S. *et al.* 2007. Quantifying subaqueous slope stability during seismic shaking: Lake Lucerne as model for ocean margins. *Marine Geology*, **240**, 77–97, https://doi.org/10.1016/j.margeo.2007.02.016

Strasser, M., Hilbe, M. and Anselmetti, F.S. 2011. Mapping basin-wide subaquatic slope failure susceptibility as a tool to assess regional seismic and tsunami hazards. *Marine Geophysical Research*, **32**, 331–347, https://doi.org/10.1007/s11001-010-9100-2

Strasser, M., Monecke, K., Schnellmann, M. and Anselmetti, F.S. 2013. Lake sediments as natural seismographs: a compiled record of Late Quaternary earthquakes in Central Switzerland and its implication for Alpine deformation. *Sedimentology*, **60**, 319–341, https://doi.org/10.1111/sed.12003

Strupler, M. 2012. *Mapping and interpreting subaquatic moraines in Lake Lucerne*. MSc thesis, University of Zurich.

Strupler, M., Hilbe, M., Anselmetti, F.S., Strasser, M. 2015. Das neue Tiefenmodell des Zürichsees: Hochauflösende Darstellung der geomorphodynamischen Ereignisse im tiefen Seebecken. *Swiss Bulletin für angewandte Geologie*, **20**, 71–83, https://doi.org/10.5169/seals-632526

Strupler, M., Hilbe, M. *et al.* 2017. Probabilistic stability evaluation and seismic triggering scenarios of submerged slopes in Lake Zurich (Switzerland). *Geo-Marine Letters*, **37**, 241–258, https://doi.org/10.1007/s00367-017-0492-8

Strupler, M., Anselmetti, F.S., Hilbe, M. and Strasser, M. 2018*a*. Quantitative characterization of subaqueous landslides in Lake Zurich (Switzerland) based on a high-resolution bathymetric dataset. *Geological Society, London, Special Publications*, **477**, 399–412, https://doi.org/10.1144/SP477.7

Strupler, M., Danciu, L. *et al.* 2018*b*. A subaqueous hazard map for earthquake-triggered landslides in Lake Zurich, Switzerland. *Natural Hazards*, **90**, 51–78, https://doi.org/10.1007/s11069-017-3032-y

Strupler, M., Hilbe, M. *et al.* 2018*c*. Subaqueous landslide-triggered tsunami hazard for Lake Zurich, Switzerland. *Swiss Journal of Geosciences*, **5**, 353–371, https://doi.org/10.1007/s00015-018-0308-5

Talukder, A.R. and Völker, D. 2014. The Tsunami Generation Potential of Shovel and Bulli Slides in the Continental Margin SE Australia. *In*: Krastel, S., Behrmann, J.-H. *et al.* (eds) *Submarine Mass Movements and their Consequences: 6th International Symposium*, Springer, 539–548.

Tappin, D.R., Watts, P. and Grilli, S.T. 2008. The Papua New Guinea tsunami of 17 July 1998: anatomy of a catastrophic event. *Natural Hazards and Earth System Sciences*, **8**, 243–266, https://doi.org/10.5194/nhess-8-243-2008

Ten Brink, U.S., Lee, H.J., Geist, E.L. and Twichell, D. 2009. Assessment of tsunami hazard to the U.S. East Coast using relationships between submarine landslides and earthquakes. *Marine Geology*, **264**, 65–73, https://doi.org/10.1016/j.margeo.2008.05.011

Tolotti, M., Dubois, N., Straile, D. and Lami, A. 2018. *Large and Deep Perialpine Lakes: A paleolimnological perspective for the advance of ecosystem science*. Springer.

Van Daele, M., Moernaut, J. *et al.* 2014. The 600 yr eruptive history of Villarrica Volcano (Chile) revealed by annually laminated lake sediments. *Geological Society of America Bulletin*, **126**, 481–498, https://doi.org/10.1130/B30798.1

Van Rensbergen, P., De Batist, M., Beck, C. and Chapron, E. 1999. High-resolution seismic stratigraphy of glacial to interglacial fill of a deep glacigenic lake: Lake Le Bourget, Northwestern Alps, France. *Sedimentary Geology*, **128**, 99–129, https://doi.org/10.1016/S0037-0738(99)00064-0

Waldmann, N., Anselmetti, F.S. *et al.* 2011. Holocene mass-wasting events in Lago Fagnano, Tierra del Fuego (54°S): implications for paleoseismicity of the Magallanes-Fagnano transform fault. *Basin Research*, **23**, 171–190, https://doi.org/10.1111/j.1365-2117.2010.00489.x

Ward, S.N. 2011. Tsunami. *In*: Gupta, H.K. (ed.) *Encyclopedia of Solid Earth Geophysics*. Springer Netherlands, Dordrecht, 1473–1493.

Watts, P., Grilli, S.T., Tappin, D.R. and Fryer, G.J. 2005. Tsunami generation by submarine mass failure. II: predictive equations and case studies. *Journal of Waterway, Port, Coastal, and Ocean Engineering*, **131**, 298–310, https://doi.org/10.1061/(asce)0733-950x(2005)131:6(298)

Wessels, M., Anselmetti, F. *et al.* 2015. Bathymetry of Lake Constance: A high-resolution survey in a large, deep lake *zfv - Zeitschrift fur Geodasie, Geoinformation und Landmanagement*, **140**, 203–210, https://doi.org/10.12902/zfv-0079-2015

Wiemer, S., Danciu, L. *et al.* 2016. *Seismic Hazard Model 2015 for Switzerland*. Swiss Seismological Service (SED) at ETH Zurich, 1–163, https://doi.org/10.12686/a2

Towards a national-scale assessment of the subaqueous mass movement hazard in Canada

D. Gwyn Lintern[1]*, Jessica Rutherford[1], Philip R. Hill[1], Calvin Campbell[2] and Alexandre Normandeau[2]

[1]Geological Survey of Canada, PO Box 6000, 9860 West Saanich Rd, Sidney, BC, V8L4B2 Canada

[2]Geological Survey of Canada, 1 Challenger Drive, PO Box 1006, Dartmouth, NS, B2Y 4A2 Canada

DGL, 0000-0003-1057-2670; PRH, 0000-0001-7444-4884
*Correspondence: gwyn.lintern@canada.ca

Abstract: Characterized by an active margin to the west, passive margins to the east and north, and numerous fjords and estuaries, the seafloor of Canada is prone to subaqueous landslides. The Geological Survey of Canada (GSC) facilitates government response in times of crisis by providing timely and concise information to Canadians, and informs the strategies to address natural hazards. Thus, the GSC is conducting a national assessment of the subaqueous landslide hazard. This paper reviews dozens of major subaqueous mass movement deposits with an emphasis on recent publications and summarizes the attempt to produce a national database. The types range from ephemeral turbidity current deposits to very large deposits (>100 km^3). To date, 1266 deposits are identified with many more expected as mapping progresses. This work is important as it will feed into the larger national tsunami strategy, and is a step forward for the national government to manage the risk. Canada is among the first countries to enter its entire database using the consistent morphometric characterization recommended by members of the UNESCO IGCP-640 (S4SLIDE) Community.

Supplementary material: The 1266 subaqueous slide locations documented in the geodatabase discussed in this paper are available at: https://doi.org/10.6084/m9.figshare.c.4847649. The catalogue is a work in progress and, to date, includes only previously published deposits/events.

Subaqueous mass movements are considered a danger to life and infrastructure throughout the world. UNESCO working groups IGCP-511, IGCP-585 and most recently IGCP-640 were established to assess the geohazard from subaqueous landslides. IGCP-640 (S4SLIDE) describes the issue facing worldwide scientists and managers:

> …probabilistic approaches exist to determine seismic hazard based on extensive earthquake catalogs from instrumental records (e.g. Frankel *et al.* 1996). Our knowledge of subaqueous landslide occurrence, however, is comparatively crude, as we lack historical record, age control and extensive geologic mapping control. The offshore catalogue of slope failures and hence recurrence rates is largely incomplete (Maslin *et al.* 2004; Camerlenghi *et al.* 2010) and therefore our ability to estimate the hazard from subaqueous slope failures, including potential tsunamigenesis, is rather low.

The Geological Survey of Canada (GSC), part of Natural Resources Canada, plays a prominent role in developing the geoscience knowledge base to inform Canadian government policies and actions related to geohazards and climate change. As a result, GSC researchers study the underlying causes and impacts of geohazards (including earthquakes, landslides, tsunamis and space weather), and the quantification of hazard and risk from these natural phenomena across Canada.

In Canada, the hazard that subaqueous mass movements represent to the public and infrastructure is largely unquantified. The goals of a National Assessment of Subaqueous Landslide Activity are to: (i) create the most up-to-date national scale inventory of subaqueous mass movement deposits; (ii) to evaluate the appropriate technologies and methods for improving subaqueous landslide hazard estimates; and (iii) to provide recommendations for future research and for integrating subaqueous landslide hazard into the national risk profile for natural hazards.

Three outcomes of this work are important. First, the study will provide a comprehensive basis for evaluating the risk to coastal communities and populations. Second, the results will be available for use by members of the public, stakeholders, planners and regulators to determine the level of planning, design and mitigation which might be needed in any given infrastructure project. Third, potential mitigation

From: Georgiopoulou, A., Amy, L. A., Benetti, S., Chaytor, J. D., Clare, M. A., Gamboa, D., Haughton, P. D. W., Moernaut, J. and Mountjoy, J. J. (eds) 2020. *Subaqueous Mass Movements and their Consequences: Advances in Process Understanding, Monitoring and Hazard Assessments.* Geological Society, London, Special Publications, **500**, 97–113.
First published online May 11, 2020, https://doi.org/10.1144/SP500-2019-206

measures can be developed to reduce the risks to people and infrastructure. This work represents the first stage in the development of a national scale probabilistic tool that will likely take years to build.

Research on Canadian subaqueous mass movements

Probably the most famous subaqueous landslide and turbidity current of all time occurred off eastern Canada following the 1929 Grand Banks earthquake (Heezen *et al.* 1954). Canadian research interest in subaqueous mass movements has been ongoing ever since. Before multibeam sonar technology was developed, subaqueous mass movements had been identified on the continental margin of the east coast (Heezen *et al.* 1954; Hill 1983; Piper *et al.* 1988), in fjords of the west coast (Prior *et al.* 1982) and in the Beaufort Sea (Hill *et al.* 1982) using single beam sonar, high-resolution reflection seismic profiles and sidescan sonar mosaics. Since the advent of multibeam sonar technology, the number of subaqueous slides identified in Canadian waters has proliferated and the geographical distribution extended. Several reviews of this research have been completed over the years. Locat and Lee (2002) used several Canadian examples in a paper that identified numerous challenges, including the need for hazard assessment, and led to the ongoing series of International Symposium of Subaqueous Mass Movements and their Consequences (IGCP-511, IGCP-585 and IGCP-640). Mosher (2008 and 2009) reviewed a number of well-documented examples of submarine mass movements in Canadian waters and highlighted the degree of the country's exposure to these events and associated tsunamis. This exposure was taken into account in the first national tsunami hazard assessment conducted by Leonard *et al.* (2012), which again underscores that a comprehensive database of submarine mass movements, on which could be built a more quantitative national hazard and risk assessment, would be in the nation's interest.

There have been many studies of subaqueous mass movements in Canadian waters since the Mosher (2008 and 2009) cross-country status reports, and they are distributed geographically across all the major marine areas of the country. The following sections summarize, by region, some of the main advances towards quantitative assessment of the subaqueous slide and mass movements hazard since that date.

Pacific Margin

Conway *et al.* (2012) produced the first comprehensive archive of submarine failures in coastal British Columbia (BC) in the form of a Geographical Information System (GIS), which allows each gravity-driven event deposit to be queried for information such as morphology, classification, area, possible age and other details. A subsample of these data is presented for Howe Sound in Figure 1. Comprehensive review of available multibeam sonar data led to the identification of dozens of previously unreported deposits. Deposits were identified in all fiords and inlets that were examined including Bute, Knight, Toba and Jervis Inlets and Howe Sound. These fjords were extensively studied in recent years, serving as natural laboratories for the study of turbidity currents originating from river discharge and submarine sliding of fjord-head delta deposits (Hughes Clarke 2016; Stacey and Hill 2016; Gales *et al.* 2018; Stacey *et al.* 2018; Fig. 2). Therefore, in these areas, the researchers have been able to follow the transformation from river discharge or mass transport deposits to turbidity current deposits found in the geological record (Hage *et al.* 2018; Vendettuoli *et al.* 2019).

Howe Sound

Howe Sound provides an example of the variety of mass movement styles in BC fjords. Submarine slides in Howe Sound are mostly located seaward of mountainous river valleys (Fig. 1), the fjord-head Squamish delta and on submerged Quaternary deposits. The fjord sidewalls are typically characterized by steep fan deltas and intermittent subaerial debris flows, which directly feed submarine flows. One of the deadliest terrestrial debris flows in Canadian history occurred at the mouth of one such creek at Britannia Beach and an extensive subaqueous landslide deposit can be observed in the adjacent fjord. Anthropogenic influence is also significant at Britannia Beach, where a mine site disposed of mine tailings for decades into Howe Sound (Jackson *et al.* 2008). The subaqueous delta of the Squamish River has grown into the fiord and has developed an extensive channel system. Several older slides are apparent along the eastern edge of the Porteau Cove terminal moraine and Keats Island. These likely occurred shortly after deglaciation when supporting ice receded, leaving over-steepened coarse sediment piles susceptible to sliding.

Mass movement and turbidity current events on the Squamish delta front have been examined in great detail (Hughes Clarke *et al.* 2012, 2013). A training dyke, installed in 1971 to create the Squamish Port facility, has concentrated the full Squamish River discharge into a 500 m section of the fjord-head shoreline on the NW side of the fjord. The training of the river and removal of the natural ability for the river to avulse (switch course) is similar to the situations at the Fraser Delta and at other important developed areas. Hughes Clarke *et al.* (2012)

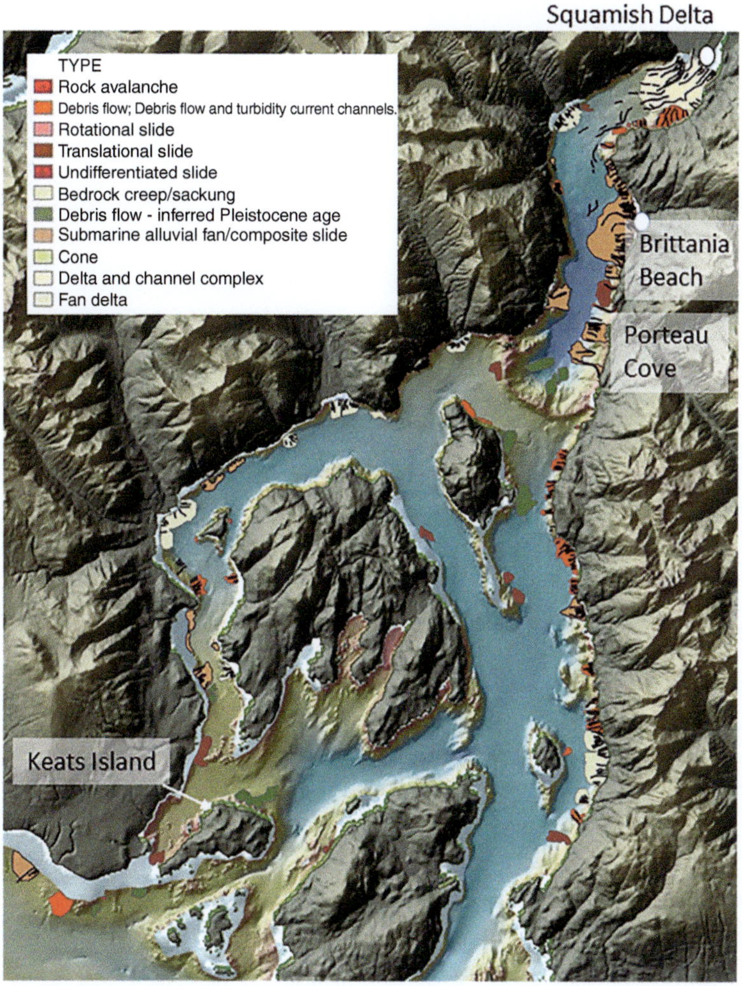

Fig. 1. Results of multibeam bathymetry analysis for Howe Sound, BC. The colours indicate different types of failures identified by Conway *et al.* (2012).

provide important insight into small scale but frequent failure modes of a delta, which initiate dozens of turbidity current events per year during the spring flood season. These failure modes include slope over-steepening of the delta front and settling river plumes (Hizzett *et al.* 2018; Hage *et al.* 2019).

Douglas Channel and Kitimat Arm

Significant mass transport bodies were also identified in focused studies of Douglas Channel (Shaw and Lintern 2016), Kitimat Arm (Shaw *et al.* 2017; Stacey *et al.* 2018). In Kitimat Arm, detailed analysis of the morphology, seismic stratigraphy and chronology showed that the classic Kitimat slide, mapped by Prior *et al.* (1982) and thought to have triggered the 1974 and 1975 tsunamis, was formed from several separate events, at least 1974 and 1975, but also a large slide, which occurred at *c.* 620 cal years BP (Stacey *et al.* 2019). An older, buried slide deposit with an age of *c.* 2.59 cal ka BP was identified, providing constraints on the natural return frequency of submarine slides and slide-generated tsunamis at this fjord-head.

An important outcome of Conway *et al.*'s (2012) inventory was the identification of two very large landslide features in Douglas Channel. These are the largest features seen in all of the BC inlets. Both occur on the east side of Douglas Channel, and each is in the order of 0.03 km^3. They are

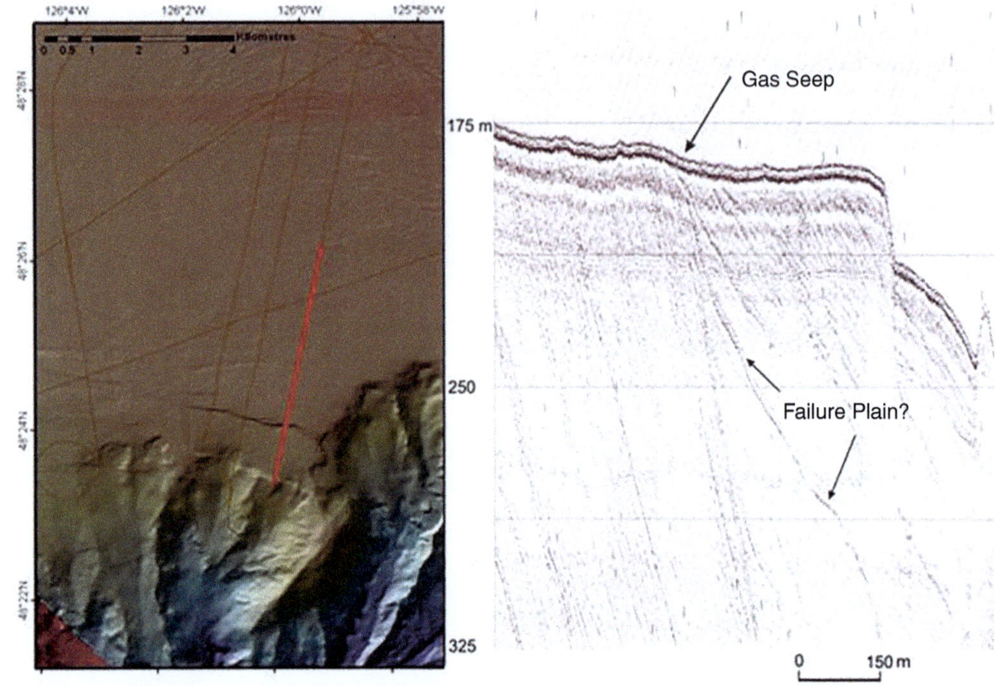

Fig. 2. Cross-section through a section of the shelf offshore of Vancouver Island. A gas seep reveals a possible failure plane in the bedrock.

interpreted to be rotational slides (sackungens) formed by the failure of scallop-shaped blocks that have fallen off the steep-sided fjord walls (Conway and Barrie 2018). The core of these blocks is probably Cretaceous diorite and, being a rigid block, has the potential to have caused tsunamis of up to 40 m in amplitude, or 80 m from trough to peak (Thomson *et al.* 2012). The run-up on the shoreline would likely have been greater, but was not modelled. The nature of rigid block tsunamis is that they generally have short wavelengths, which means that they lose (dissipate) energy rapidly as they travel. The model indicates that these huge tsunamis at the source are reduced to just a few tens of centimetres by the time they reach the head of the inlet at Kitimat. The ocean currents created by these events would be on the order of 15 m s^{-1} or 30 nautical miles per hour near the source in Douglas Channel, and similarly would diminish to tenths of m s^{-1} by at the entrance to Kitimat Arm, 10 to 15 minutes after the initial failure (Thomson *et al.* 2012).

Fraser River Delta

The Fraser River Delta is a strategically important location for Canada, being adjacent to the city of Vancouver and key infrastructure including an international airport and major coal and container port. The delta front is characterized by small to medium (0.01 km^3) retrogressive slides at the top of an active submarine channel and minor gullies on the adjacent slope in front of the most active distributary channel, Main Channel (McKenna *et al.* 1992; Hill 2012). Recent analysis of bed deposits on the levees of the active submarine channel has refined understanding of the nature of turbidity current activity in the channel (Stacey 2014). Slide-generated flows with a thickness greater than 10 m overspill the main channel on average every 10 to 15 years. Lower density flows occur more frequently and appear to be related to large flood years in the Fraser River.

Unconfined turbidity currents have also been observed on the open slope of the delta through the Ocean Networks Canada cabled observatory (Lintern and Hill 2010; Ayranci *et al.* 2012; Lintern *et al.* 2016). These flows occur almost every year during the river flood period; recent research has focused on the triggering mechanism, which likely relates to a combination of high sediment influx from the river, sediment resuspension by tidal currents and potentially groundwater flow and pore pressure effects.

Pacific continental slope

The continental slope off the west coast of Vancouver Island is relatively poorly covered by multibeam sonar. However, those areas surveyed have high potential for tsunamigenic slope failures, preconditioned by the presence of shallow gas or gas hydrates and triggered by megathrust earthquakes (Riedel *et al.* 2018). The work to map these areas is ongoing and it is anticipated that more submarine landslides will be discovered. A recent example is shown in Figure 2, where a gas seep and possible deep-seated failure plane are visible in seismic profiles at the shelf edge. The plane delimits a wedge of material 300 m wide at the surface and 150 m deep, in relatively well-consolidated sediment and rock, which has the potential to fail in this seismogenic zone.

Seafloor data from an area slightly further offshore, at the toe of the slope, reveal eight subaqueous mass movement deposits (Naegeli 2010). Scholz *et al.* (2016) determined that these features were likely early to mid-Holocene in age and that anticipated earthquake accelerations would not be sufficient to cause slope instability without significant preconditioning through excess pore pressure. Hamilton *et al.* (2015) have investigated turbidite beds associated with one of these slides, the Slip Stream Slide (formerly Humpback Slide), and determined that between 10.8 and 6.6 cal ka BP, the recurrence interval for the beds and inferred megathrust earthquakes was 460 ± 140 years. This rate is compatible with other estimates of palaeoseismic megathrust earthquake occurrence rates along the subduction zone.

Farther north, Greene *et al.* (2018) and Brothers *et al.* (2019) mapped the Queen Charlotte Fault, a seismically very active transform fault that shows abundant evidence for fluid migration as well as volcanic activity. The fault defines the trend of the continental slope on this part of the margin. Except for one large translational slide (greater than 13 km^2), mass movements on the upper slope, above the fault trace are relatively smaller (most less than 2 km^2) and the upper slope is characterized by canyons and gullies. Downslope of the fault trace on the margins of a terrace feature, there are several larger (up to 18 km^2) square-headed block slides.

Atlantic margin

Piper *et al.* (2003) reviewed the most important submarine landslides on the southeastern coast of Canada and their possible triggering mechanisms. Recent (post-2009) research on the Atlantic margin has mostly focused on large (>10 km^3) Quaternary and pre-Quaternary mass transport deposits revealed in petroleum exploration 2D and 3D seismic reflection data, as well as late Quaternary and Holocene deposits based on high-resolution seismic reflection data, multibeam bathymetry and sediment cores. Several studies have benefited from access to evenly spaced grids of industry 2D and 3D seismic reflection data, with a focus on deposits larger than 100 km^3. In addition, seabed surface renders from 3D seismic reflection data often provide superior resolution of seabed morphology compared to multibeam bathymetry in deep water (Mosher *et al.* 2006). The coverage by modern multibeam bathymetry surveys has increased substantially in the last 10 years, especially in deep water.

Offshore Nova Scotia

Recent research on subaqueous landslide deposits offshore Nova Scotia has mainly concentrated on the Scotian Slope. Studies that utilize industry 2D and 3D seismic reflection data provide new insights into failure processes such as their distribution along the margin (Campbell *et al.* 2015), erosional effects (Mosher and Campbell 2010; Deptuck and Campbell 2012; Campbell and Mosher 2016) and tsunami potential (Mosher 2009). Offshore exploration wells provide ground truth and age control for these studies, albeit the samples generally comprise drill cuttings and sidewall cores, and biostratigraphic age resolution is limited compared to radiometric methods.

Late Pleistocene glacial process have a significant influence on the modern seabed morphology of the NW Atlantic margin (Mosher *et al.* 2017; Normandeau *et al.* 2019*a*). On the Scotian Slope, a number of studies have looked at subaqueous landslide processes that were prevalent during deglaciation of the margin (e.g. Huppertz *et al.* 2010; Piper *et al.* 2012, 2016). Although these studies examine deglacial processes not currently active on the Atlantic margin, they are good analogies for what may happen in the future in Arctic settings.

Study of Holocene submarine landslides may hold the most relevance in terms of assessing modern slope stability and understanding future hazard. Recent discoveries of large (>14 000 km^2), previously unidentified Holocene-age submarine slides on the Scotian Slope (Normandeau *et al.* 2019*b*, *c*), have led to a re-evaluation of the submarine landslide hazard on the Atlantic margin. At least four landslides, including the most recent 1929 Grand Banks event, have occurred on this part of the margin within the last 4 ka, establishing a shorter return period than previously understood, i.e. approximately 1 ka. MacKillop *et al.* (2019) suggest that cyclical loading (e.g. earthquakes) or preconditioning factors (e.g. elevation of pore pressures) are critical for slope instability on the Scotian Slope. In the light of this new information, a re-evaluation of the rest of the margin is warranted.

Offshore Newfoundland and Labrador

Research has been conducted near the 1929 Grand Banks earthquake area. Giles *et al.* (2010) showed in 3D seismic reflection data that very large submarine landslides occurred in the region throughout the Cenozoic. Several publications were focused on the 1929 event specifically. Stevenson *et al.* (2018) reconstructed sediment concentration of the 1929 flow event using two distinct methods and showed that the concentration was orders of magnitude higher than observed elsewhere in smaller flows. Schulten *et al.* (2019*a*, *b*) put forward that the 1929 submarine landslide consisted of two failure mechanisms; massive slumping along deep faults and décollements followed by widespread surficial sediment failure. Løvholt *et al.* (2019) used the new failure mechanism to refine numerical models of the 1929 tsunami.

Along the continental slope off the SE Grand Banks, Rashid *et al.* (2017, 2019) examined the interplay of the Labrador Current and mass movement, in turn presenting a late Pleistocene to present chronology of subaqueous landslides and new ideas on how ice-rafting deposits and contourites impact the geotechnical parameters of the seabed.

Flemish Pass is a large saddle, separating the Grand banks from Flemish Cap and is an area of active hydrocarbon exploration. Using 3D seismic reflection data, Hawken (2017) examined the interaction between gas hydrate bottom-simulating reflectors (BSRs) and mass movement on Sackville Spur, a large contourite drift in northern Flemish Pass. She noted that the onset of mass movement deposition in the Pleistocene correlates to decreased sediment drift development and increased deposition in the base of Flemish Pass. There is no direct spatial association between the BSRs and mass-transport deposits (MTDs), although the onset of pockmarks and failure scarps are stratigraphically equivalent, suggesting hydrate dissociation and sediment failure may have a common triggering mechanism. In another study, Cameron *et al.* (2014) describe a large retrogressive slide complex with associated mass transport deposits on the northern flank of Flemish Cap. This multiple failure complex extends about 65 km along slope and individual arcuate failures are up to 10 km wide with scarps over 70 m high. Run-out distances extend up to 20 km downslope. Detailed sedimentological, geotechnical and geochronological studies of the slide complex and adjacent slope established that the failures occurred in the late Pleistocene and that the present slope appears to be stable. Recurrence intervals in this part of Flemish Pass are approximately 10 ka and most likely correspond to the recurrence of triggering earthquakes. Migration of deep fluids along the Cumberland Belt Fault Zone may have preconditioned the region and contributed to past failures.

In Orphan Basin, Li *et al.* (2012*a*) examined Pleistocene turbidites and canyon formation, showing that the origin of the canyon is synchronous with the onset of cross-shelf glaciation and that the occurrence of muddy turbidity flows is correlative with the movement of the ice stream. Using 3D seismic reflection data, Li *et al.* (2012*b*) described fundamental differences between glacigenic and blocky mass transport deposits; whereas blocky mass transport deposits have characteristic parallel to subparallel basal grooves and chaotic internal acoustic facies, glacigenic debris flows show a smooth basal surface and transparent internal acoustic character. Piper *et al.* (2019) reported that four large failures occurred since 85 ka and 18 failures were large enough to be recognized in seismic reflection profiles and/or cores. They postulated that the widespread mass movements were earthquake triggered and that on such a margin with modest sedimentation rates and low seismicity, mass movement recurrence provides a robust estimator of past seismicity. Further north on the NE Newfoundland shelf and slope, Shaw *et al.* (2012) and Roger *et al.* (2013) examined glacial-related mass movement processes and trough mouth fan formation.

St Lawrence Estuary

There has been considerable effort over the last 20 years to map and characterize submarine mass movements, canyons and gravity flow systems in the St Lawrence Estuary (e.g. St-Onge *et al.* 2004; Cauchon-Voyer *et al.* 2008, 2011; Gagné *et al.* 2009; Normandeau *et al.* 2015; Pinet *et al.* 2015; Poncet *et al.* 2010). Pinet *et al.* (2015) completed a comprehensive analysis of submarine slide morphology based on multibeam bathymetry. They documented 94 mass movements in this relatively small area, with a median area of 3.3 km^2 up to a maximum of 174 km^2. These deposits are preferentially located along the topographic break at the margin of the glacially incised Laurentian Channel. They consist of lateral spreading, slumps, debris flows and blocky submarine landslides (Campbell *et al.* 2008). One of the largest landslides in the St Lawrence Estuary was well studied by Cauchon-Voyer *et al.* (2008, 2011) in the Betsiamites area and had multiple phases of activity at 7.25 cal ka BP, AD 1663 and AD 1860 or AD 1870. These events are hypothesized to have been triggered by earthquakes originating from the Charlevoix-Kamouraska Seismic Zone. In addition, Turmel *et al.* (2019) showed that the 7.25 cal ka BP event might have produced a 1.5 m tsunami wave, which is considered low risk to coastal population since the tidal range is 5 m. St-Onge *et al.* (2004) suggested a reduced frequency of earthquakes in the Saguenay Fjord based on reduced evidence of landslide deposits after 4 ka BP. Although they are hypothesized to be earthquake-triggered, the ages of most of these slope failures are unknown and Pinet *et al.* (2015) state that the potential relationship with seismicity is difficult to establish for most of them without more detailed work. The preliminary hypothesis is that the recent ones may have been triggered by the M_w7.2 1663 earthquakes (Cauchon-Voyer *et al.* 2008) but further work is required to confirm their ages. There are few data to support suggestions of preconditioning of the sediments by high sedimentation rates and/or shallow gas (Duchesne *et al.* 2003; Cauchon-Voyer *et al.* 2008). Poncet *et al.* (2010) showed that one of the landslides in the upper St Lawrence Estuary had potentially triggered a tsunami although no physical evidence has been discovered so far.

The St Lawrence Estuary is also characterized by a significant number of small submarine canyon and channel systems, mostly on the north shore of the estuary (Normandeau *et al.* 2015). These channels and canyons were formed by turbidity currents and were mostly active during deglacial times (Normandeau *et al.* 2017). Most of them originate from glacier or river sediment supply in the past, are now detached from these sources and so are no longer active. Canyons supplied by longshore drift and triggered by intense storms are episodically active (Gagné *et al.* 2009, Normandeau *et al.* 2013, 2015), but, curiously, the most active canyons in the estuary are those that are sediment-starved but located on the steepest slopes (Normandeau *et al.* 2014). Recent direct monitoring of turbidity currents in this system revealed that storms trigger the turbidity currents by suspending small volumes of sediment in the canyon heads (Normandeau *et al.* 2020).

The results of this research effort have advanced our knowledge and understanding of the potential hazard posed by submarine slides and mass movements in the St Lawrence Estuary but further work on timing, preconditioning and trigger processes is required to quantify the hazard.

Baffin Bay and Baffin Island Fjords

The seismic hazard in Baffin Bay is high, especially in northern Baffin Bay where one of the largest passive margin earthquakes ever recorded occurred in in 1933 (M7.3; Bent 2002). Consequently, the potential submarine mass movement hazard on the Canadian continental margin of Baffin Bay can also be considered high until more understanding of the occurrence and timing of recent events is obtained. The difficulty of accessing this northern region has hindered systematic mapping of the margin and both multibeam and sub-bottom data coverage are sparse offshore. However, mapping of the fjords has revealed numerous submarine landslides and evidence of gravity flow processes (Hughes Clarke *et al.* 2015; Brouard and Lajeunesse 2019*a*, *b*; Normandeau *et al.* 2019*d*).

The Pleistocene history of the margin is characterized by episodes of shelf-crossing glacial ice advance, leaving complex ice proximal deposits including extensive glacigenic debris flow deposits (Campbell and Bennett 2013; Bennett *et al.* 2014; Jenner *et al.* 2018). Whereas the acoustic characteristics of these glacigenic debris flow deposits can be distinguished from non-glacigenic mass flow deposits, the complexity of the late Quaternary stratigraphy requires high-resolution surveys and detailed analysis to make possible a rigorous assessment of the submarine mass movement hazard. For example, the seafloor northern Baffin Bay in the vicinity of the 1933 earthquake has been examined and, whereas glacigenic debris flow deposits are widespread (Li *et al.* 2011), evidence of modern slope failure is absent (Bennett *et al.* 2014). Further south, off Clyde River and Qikiqtarjuaq, submarine mass movement deposits were reported by Aksu and Hiscott (1989) and Hiscott and Aksu (1994), which, upon further examination, are likely glacigenic in origin based on their acoustic character and location seaward of transverse troughs.

Some fresh slide scars and mass flow complexes have been identified on the continental slope and on the margins of some shelf-crossing glacial troughs (Campbell and Bennett 2013). In addition, mass movement and gravity flow complexes have been widely observed in multibeam surveys of fjords and fjord-head deltas of Baffin Island (Hughes Clarke *et al.* 2015; Broom 2019; Brouard and Lajeunesse 2019*a*, *b*; Normandeau *et al.* 2019*d*). Of particular note, at least 243 submarine slides were observed on the margins of Frobisher Bay, within 20 km of the town of Iqaluit, capital of Nunavut (Deering *et al.* 2015, 2018, 2019; Todd *et al.* 2016). Broom (2019) showed that Holocene submarine mass movement recurrence in a large basin north of the Hamlet of Pond Inlet is once every 800 years. Gravity flow processes present on delta fronts are associated with glacial retreat. A predictive framework of mass movement on fjord deltas for the entire eastern Baffin Island has been developed and shows that about 43% of the river mouths in Baffin are likely to have active gravity flow processes on their delta (Normandeau and Dietrich 2019). Therefore, the fjords and nearshore environments of Baffin Island appear to be particularly active areas for slope failures and work is underway to understand their distribution, triggers and recurrence.

Beaufort Sea

The presence of significant mass wasting on the continental slope of the Beaufort Sea has been recognized since the 1980s (Hill *et al.* 1982). The Beaufort Sea is dominated by sediment supply from the Mackenzie River and is therefore characterized by locally high postglacial sedimentation rates (Hill *et al.* 1992). The seismic hazard is considered moderate (Adams and Atkinson 2003) with the two largest recorded events a M6.5 earthquake in 1920 and a M5 event in 1975 (Hyndman *et al.* 2005).

Recent detailed studies of the shelf-edge failure identified by Hill *et al.* (1982) shows that it is a complex feature consisting of at least five phases (Cameron and King 2019), initiating along a single deep failure plane, followed by subsequent, smaller retrogressive failures. It is contiguous with, and is cut by, a second slide complex that originates in mid-slope. This failure complex has at least four phases and has an estimated volume of 40 km^3. A second, larger failure complex with a minimum volume of 100 km^3 has been identified along slope in a mid-slope position (Cameron and King 2019). This failure complex also has a minimum of four phases. It has proven difficult to establish the age and timing of the individual failure events, but based on ^{14}C and ^{210}Pb chronology, they can all be constrained to the last 1 ka.

The mid-slope location of two of these failure complexes is unusual, but is likely to be closely related to preconditioning by migrating fluids. Recent work has provided strong evidence for the presence of thick but degrading permafrost and gas hydrate beneath the continental shelf and for significant migration of gas towards the continental slope (Paull *et al.* 2007, 2015; Riedel *et al.* 2017). Both the Beaufort shelf and slope are characterized by pingo-like features and active mud volcanoes associated with significant venting of fluids (Paull *et al.* 2007, 2015). Whereas the composition of the fluids indicates multiple sources at depth, it is highly likely that the Beaufort shelf edge and slope sediments are preconditioned for failure by these migrating fluids. These factors suggest that the submarine mass movement hazard is considerable in this area and further work is necessary to better constrain recurrence intervals.

Development of a national-scale database for hazard assessment

A fundamental research question that needs to be answered when discussing a national-scale assessment of subaqueous mass movements is their geographical distribution. Therefore, it is necessary to develop a national database to provide a framework for more formal hazard and risk assessment. The following outlines the steps taken by the GSC to create the Canadian National Subaqueous Mass Movement database.

A literature review and inventory of GIS datasets on subaqueous mass movements were initially carried out to determine how to proceed in the creation of the national database for Canada. A Canadian national subaqueous landslide metrics spreadsheet based on Clare *et al.* (2018), outlining the main morphometric parameters, was completed and imported into an ArcGIS geodatabase file. All previous shapefiles and tables containing information on failures around Canada was then populated into the new database. At this time, the deposits for the rest of Canada were identified through already published databases or from published literature. Shapefiles (polygons and point layers) were merged from three main sources: (i) the BC Submarine Slope Failure Atlas in ArcGIS format (Conway *et al.* 2013); (ii) the dataset pertaining to the Douglas Channel and Kitimat Fiords of BC, published in Shaw and Lintern (2016); and (iii) the GSC (2015) unpublished dataset for the BC north coast. The resulting maps are shown in Figures 3, 4 and 5. Due to the previously published GIS databases being from the west coast of Canada, the data are skewed towards that coast; however, similar database efforts are underway on the east coast.

The ArcGIS map document is set up to easily navigate the geographical regions of Canada – Pacific Coast, Atlantic Coast, western Arctic, eastern Arctic and the Great Lakes and St Lawrence River and Seaway. In the figures, 'major' failure events were classified based on whether they had been thoroughly assessed in a publication and whether the deposit had been identified in the literature as being a potential hazard to infrastructure or populations. The preliminary database was then converted into a Google Earth-compatible format for easy sharing with the public and includes morphometric information. The database allows each defined mass movement to be queried for information such as the geographical region of Canada, slide classification, possible age, the area and volume of slide, hazard notes and various other slope stability descriptors. The database includes 60 different parameters documenting subaqueous slope stability, including all 54 parameters suggested by Clare *et al.* (2018), plus 6 additional parameters: geographical region, slide definition/category, scientist who defined the slide, area of the polygon, hazard information and comments.

A total of 1266 subaqueous slide locations have been documented in the geodatabase at this time (see Supplementary material), with 18 significant slides flagged in the map document. The significant slides were flagged because they have been cited

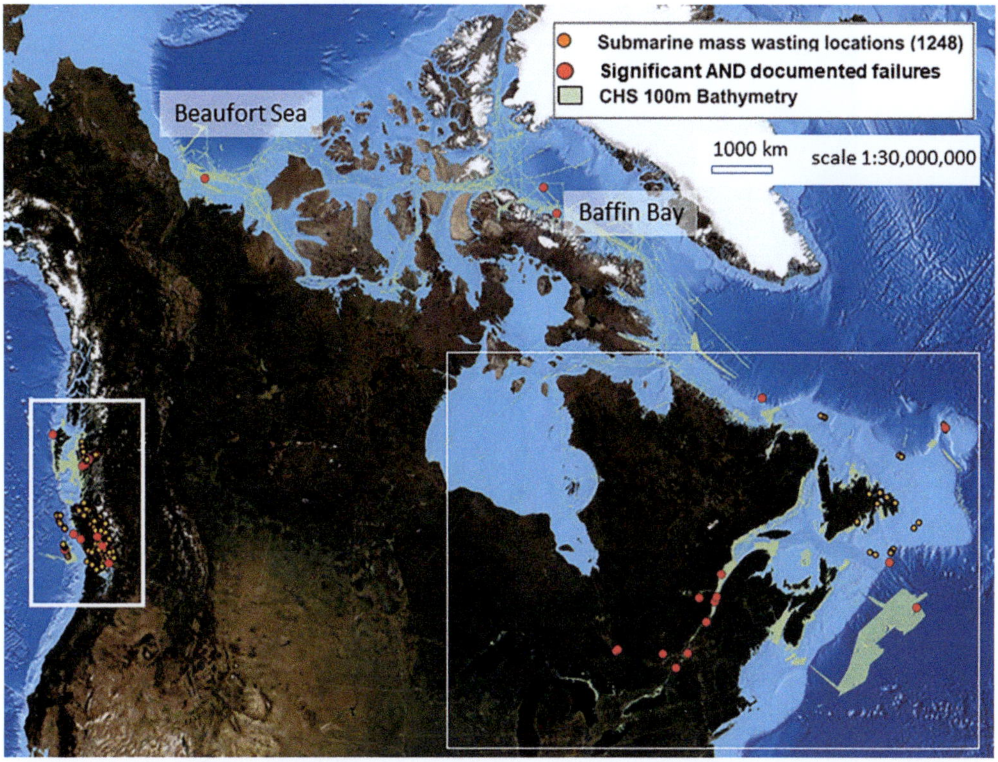

Fig. 3. Map of Canada showing current state of the database entries. There are 1266 mass transport deposits in the database. These are mainly clustered around areas where multibeam data exist, and often where targeted submarine landslide work was conducted. Significant failures, identified with the larger red dots, indicate those which are cited by researchers as posing a hazard.

by researchers as causing or posing significant hazards. Some issues of inclusion remain to be resolved. For example, Shaw and Lintern (2016) and Conway *et al.* (2013) classified slide location based on the geomorphology of the seascape, and have included cones, deltas, fan deltas and channel complexes, which are indirect indicators of slope failure. In addition, the issue of age of submarine landslides will need to be addressed. On the East Coast, 100 ka old landslides are observed on multibeam bathymetry but the same age landslides on the west coast would be buried under tens of metres of sediment. Therefore, defining a standard definition of landslides to be included in the database will need to be addressed. For example, would the giant 50 Ma-old Montagnais landslides be included in the database? Since the visible submarine landslides on the different coast vary considerably, including a homogenous definition of submarine landslides will be part of the challenge.

This exercise has revealed a great deal of useful information about levels of effort required to fully map and catalogue the submarine landslides of Canada's coast. The collection of high-resolution seismic reflection, sediment cores, and particularly multibeam bathymetry has revolutionized the field, and has led to many hundreds of new discoveries. Figures 3, 4 and 5 indicate where the Canadian Hydrographic Service has collected multibeam as of 2018, although we acknowledge that other partners from industry and academia have data in other locations. Nonetheless, qualitatively at least, the relationship between data density and number of deposits identified is very strong, as would be expected. Figure 4, on the west coast, illustrates this overlap well. The figure also clearly demonstrates how prone the steep-walled inlets (fjords) of BC are to this hazard. On the East Coast there are large areas of seafloor where multibeam is available and researchers have identified hundreds more deposits that will eventually feed into the database.

Case study

By its nature, the discovery of landslide deposits is a perpetual work in progress. To give an indication of

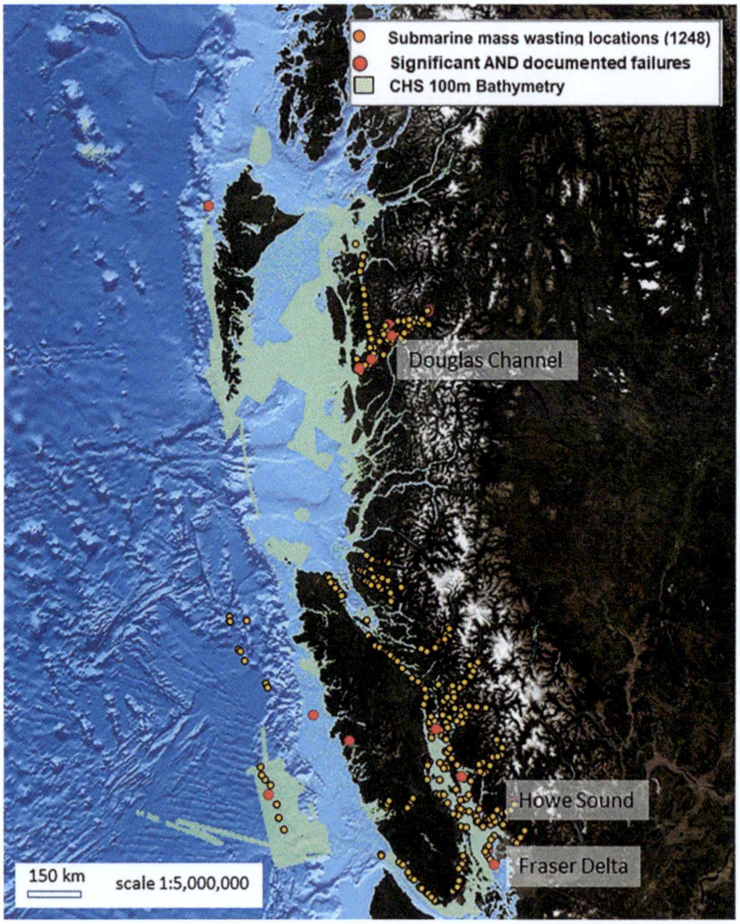

Fig. 4. Map of the west coast of Canada showing current state of the database entries.

effort required, and how technologies have evolved, we describe the case history of Kitimat Arm in Douglas Channel. This is a small and shallow area, relative to the coast of Canada, but with multiple attempts over decades to quantify the hazard. In 1975, a tsunami occurred in Kitimat Arm which triggered investigations. Fortunately, there was an eyewitness who, while placing a boulder on the seabed to build a crib wharf, saw the boulder and the whole seafloor disappear beneath his working barge. Had it not been for this, investigators would hardly have known where to start to look for the source of the wave. A series of investigations by consultants and scientists identified events in 1952, 1968, 1971, 1974 and 1975. So, a very specific lead allowed direct observation of several events in a confined area. This was adequate to determine the cause of the specific 1975 tsunami; however, it left questions about the hazard in this and other fjords.

The next attempt at the Douglas Channel was by Conway *et al.* (2013) during a coast-wide assessment of multibeam data which had been collected up to 2011. The data for Kitimat Arm consisted of perhaps two days of ship time, and of the wider Douglas Channel perhaps a week of ship time. These were not targeted specifically at this hazard. This surficial assessment revealed a further six recent mass transport deposits in Kitimat Arm, and over 80 outside the Arm in the larger Douglas Channel. So, the multibeam assessment certainly provided much information that submarine mass movements are common in the area during recent times. In terms of the hazard, it provided basic geometry or size of recent failures.

From 2013 to 2018, the area was targeted to obtain frequency and magnitude information about mass movements, to inform risk assessment with respect to public safety and a developing coast (Lintern *et al.* 2019). The area was resurveyed using a sub-bottom profiler and cored to determine the age

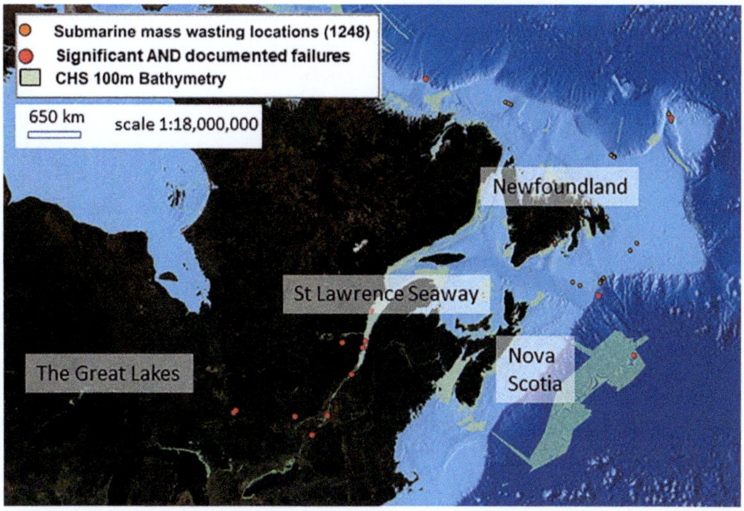

Fig. 5. Map of the east coast of Canada and the Great Lakes and St Lawrence Seaway showing current state of the database entries.

of failure. This led to the discovery of a further 25 buried deposits (Shaw and Lintern 2016; Shaw *et al.* 2017; Stacey *et al.* 2020). It provided researchers with information to calculate frequency and magnitude of mass transport events, and thus links could be made to natural triggers. Importantly though, the sub-bottom data indicated specific areas where landslides repeat throughout history, and Kitimat Delta (the source of the 1975 tsunami) has failed several times throughout the Holocene (Stacey *et al.* 2018).

As this effort is directed towards advising GSC managers on steps forward for the country's coastlines, it is worth summarizing the effort to examine this one stretch of coastline. It would take a modern ship with multibeam, piston coring and sub-bottom capabilities a maximum of two weeks to determine this high level of information about the hazard in waterways totalling 250 km long and an average of 3 km wide (750 km^2). This area indicated over 80 MTDs, 6 of which were investigated to be a present hazard to coastal infrastructure. In addition to the ship work, it is estimated that obtaining this information required, in total, three months of laboratory work (core lab and geotechnical lab), six months of spatial computing (GIS, seismic interpretation), one year of writing results, two months of public consultation and many months of planning in between. In the current configuration a capable scientific research vessel is used for coring, instrument deployment (cameras and other ground-truthing methods) during the day and multibeam and sub-bottom surveys run simultaneously at night. The surveys from the night prior are often processed by hydrographers and ready to plan the next day's science. Thus, it is a very efficient manner of conducting the work. It is tempting to think that some of this work could be transferred to autonomous underwater vehicle (AUV) surveys. In Canada, however, the large distances of uninhabited coastline would require a ship as a base for the AUV operations, and a hydrography team to be on hand regardless. Having an AUV team on many ships would come at the expense of berths for coring and other geoscience expertise. Nevertheless, the GSC's new marine geoscience programme is now commencing the use of AUVs for seabed mapping.

Lessons learnt and future endeavours

The database is, of course, far from complete because of the immense offshore land of Canada. More work is needed to ensure all mass movements have been documented with parameter fields filled and the validation of location of data and their references. This database is presently highly biased towards Pacific coastal data where high coastal populations, availability of several smaller multibeam vessels and infrastructure projects have prompted systematic seafloor interpretations. There are over 1200 slide areas documented in this dataset; only a few have descriptive features beyond two-dimensional geometry. Completion of the national database will require researchers to migrate their classifications into the fields provided by the specified international nomenclature. This has been done for some, but certainly not all the published deposits.

A national database such as this one relies heavily on available interpreted datasets, especially multibeam bathymetry and high-resolution seismic reflection profiles. The current status of this database clearly shows the critical importance of multibeam bathymetry for more mapping along the coasts of Canada, such as the more systematic interpretations done for coastal areas of BC. Many regions of Canada, especially on the Labrador and Arctic margins have poor data density, preventing a thorough assessment of submarine landslide hazard. Since Canada has one of the largest offshore areas in the world, mapping its complete extent will be a challenging endeavour, but one that needs to be addressed if a serious and nationwide database of submarine landslide hazards is to inform public policies. Until that time, data density will be considered an important qualifier in this national assessment.

In addition to the parameters outlined in Clare *et al.* (2018), there are some additional features required by the database worth noting. While a Google Earth version has been created for its easier viewing interface, and text files for sharing, it should be noted that any changes to the master Arc database would presently need to be changed into these other formats. As this is a lengthy process requiring supervision, it would likely be most efficiently done intermittently through a formal versioning process.

In making this start, a GIS methodology has been developed for measuring and calculating parameters and has generated a level of enthusiasm among Canadian contributors. Further discussion with the international community is underway for development of this database. Several discussion points have come up during population of the database. Specifically, clarification on a slide ID naming system is required to ensure consistency nationally and internationally. Without this clarification, populating the database will become confusing and has the potential for duplication. Furthermore, discussion with the international community is required on the naming system for the 'Parent_ID' field and what should be classified as a failure or potential failure and it should be included in the database. For example, should deltas, cones and channel complexes be included as secondary indicators of mass movements?

Consideration must eventually be given to what, how and when to make the database more publicly available for use in risk assessment and hazard mitigation. We are currently considering how the extensive results in Kitimat and other Canadian areas translate to similar areas.

Conclusions

This paper summarizes the current state of knowledge of subaqueous mass movement deposit distribution in Canada's offshore and coastal areas. Information has been gathered from all three coasts, although much more work is warranted to have a complete database. This review has shown that subaqueous landslides are present on all three coasts of Canada and have been observed everywhere multibeam bathymetry has been collected. New information published since Mosher (2008, 2009) shows that an increasing number of young landslides have been reported and new discoveries will likely be made as mapping efforts continue. For the database, we have chosen to follow the recommendations of the IGCP-585 – E-Marshal: Earth's Continental MARgins: aSsessing the geoHazard from submarine Landslides and described by Clare *et al.* (2018). Some of the main conclusions from this effort are as follows.

(1) Multibeam mapping of Canadian continental shelves, estuaries and fjords has multiplied by ten times or more the number of new subaqueous mass movements that have been identified, classified and precisely geo-positioned.
(2) Important progress has been made in the development of a National Subaqueous Mass Movements Geodatabase for Canada. The Canadian database is structured to include the 54 parameters developed by IGCP-585 with the addition of six new parameters (geographical region, slide definition/category, scientist who defined the slide, area of the polygon, hazard information and comments) that provide useful information from a hazard assessment point of view.
(3) The use of GIS technology to produce a geodatabase that is integrated into an existing digital elevation model (DEM) database allows for both search-ability and immediate visualization of the morphological expression of the mass movement by users. In Canada, this has been greatly facilitated by the fact that most data holdings lie in government institutions. However, university researchers and private stakeholders also hold data that should be integrated into the database.
(4) Much additional work needs to be done to populate the database. It is estimated that less than 50% of mass movements identified to date have been captured in the database. In addition, there are considerable amounts of multibeam data that have not yet been examined, but also vast areas of the Canadian submarine landmass that have not been surveyed with modern multibeam systems.
(5) Nevertheless, with continued infilling of data gaps, the National Subaqueous Mass Movements Geodatabase will provide a key tool for emergency managers to quantify the risk to communities and infrastructure posed by subaqueous mass movements and related tsunamis.

Acknowledgements The authors would like to thank the reviewers for their improvements to this paper. Early database preparation was completed by Tanner Dobson.

Funding This research received support from the Submarine Landslides and Tsunami activity of the Public Safety Geoscience Program, Geological Survey of Canada.

Author contributions **GL**: writing – original draft (lead); **JR**: data curation (lead), writing – original draft (equal); **PRH**: writing – original draft (equal); **CC**: writing – original draft (equal); **AN**: writing – original draft (equal).

References

Adams, J. and Atkinson, G. 2003. Development of seismic hazard maps for the proposed 2005 edition of the National Building Code of Canada. *Canadian Journal of Civil Engineering*, **30**, 255–271, https://doi.org/10.1139/102-070

Aksu, A.E., Hiscott, R.N. 1989. Slides and debris flows on the high-latitude continental slopes of Baffin Bay. *Geology*, **17**, 885–888, https://doi.org/10.1130/0091-7613(1989)017<0885:SADFOT>2.3.CO;2

Ayranci, K., Lintern, G., Hill, P.R. and Dashtgard, S.E. 2012. Tide-supported gravity flows on the upper delta front, Fraser River delta, Canada. *Marine Geology*, **326–328**, 166–170, https://doi.org/10.1016/j.margeo.2012.09.007

Bennett, R., Campbell, D.C., Furze, M.F. and Haggart, J.W. 2014. The shallow stratigraphy and geohazards of the Northeast Baffin Shelf and Lancaster Sound. *Bulletin of Canadian Petroleum Geology*, **62**, 217–231, https://doi.org/10.2113/gscpgbull.62.4.217

Bent, A.L. 2002. The 1933 $M_s = 7.3$ Baffin Bay earthquake: strike-slip faulting along the northeastern Canadian passive margin. *Geophysical Journal International*, **150**, 724–736, https://doi.org/10.1046/j.1365-246X.2002.01722.x

Broom, L.A. 2019. *Postglacial chronology and geohazards of Pond Inlet and Eclipse Sound, northeastern Baffin Island, Nunavut.* MSc thesis, Dalhousie University.

Brothers, D.S., Andrews, B.D. *et al.* 2019. Slope failure and mass transport processes along the Queen Charlotte Fault, Southeastern Alaska. *Geological Society, London, Special Publications*, **477**, 69–83, https://doi.org/10.1144/SP477.30

Brouard, E. and Lajeunesse, P. 2019*a*. Glacial to postglacial submarine landform assemblages in fiords of northeastern Baffin Island. *Geomorphology*, **330**, 40–56, https://doi.org/10.1016/j.geomorph.2019.01.007

Brouard, E. and Lajeunesse, P. 2019*b*. Submarine geomorphology of the northeastern Baffin Island fiords and cross-shelf troughs. *Journal of Maps*, **15**, 662–676, https://doi.org/10.1080/17445647.2019.1647302

Camerlenghi, A., Urgeles, R. and Fantoni, L. 2010. A database on submarine landslides of the Mediterranean Sea. *In*: Mosher, D.C., Shipp, R.C. *et al.* (eds) *Submarine Mass Movements and Their Consequences*. Advances in Natural and Technological Hazards Research, **28**. Springer, Dordrecht, https://doi.org/10.1007/978-90-481-3071-9_41

Cameron, G.D.M. and King, E.L. 2019. *Mass-Failure Complexes on the Central Beaufort Slope, Offshore Northwest Territories*. Geological Survey of Canada, Open File 8356, https://doi.org/10.4095/314644

Cameron, G.D.M., Piper, D.J.W. and Mackillop, K. 2014. *Sediment Failures in Northern Flemish Pass*. Geological Survey of Canada, Open File 7566, https://doi.org/10.4095/293680

Campbell, D.C. and Bennett, J.R. 2013. Preliminary results from recent investigations of marine geological hazards in Baffin Bay, Nunavut and Greenland. *In*: *Summary of Activities 2013*. Canada-Nunavut Geoscience Office, 121–128, (ESS Cont.# 20130281)

Campbell, D.C. and Mosher, D.C. 2016. Geophysical evidence for widespread Cenozoic bottom current activity from the continental margin of Nova Scotia, Canada. *Marine Geology*, **378**, 237–260. https://doi.org/10.1016/j.margeo.2015.10.005

Campbell, D.C., Duchesne, M. and Bolduc, A. 2008. Geomorphological and geophysical evidence of Holocene seafloor instability on the southern slope of the Lower St. Lawrence Estuary, Québec. In: *Proceedings of the 4th Canadian Conference on Geohazards: From Causes to Management*, Québec City, May, 367–374.

Campbell, D.C., Shimeld, J., Deptuck, M.E. and Mosher, D.C. 2015. Seismic stratigraphic framework and depositional history of a large Upper Cretaceous and Cenozoic depocenter off southwest Nova Scotia, Canada. *Marine and Petroleum Geology*, **65**, 22–42, https://doi.org/10.1016/j.marpetgeo.2015.03.016

Cauchon-Voyer, G., Locat, J. and St-Onge, G. 2008. Late-Quaternary morpho-sedimentology and submarine mass movements of the Betsiamites area, Lower St. Lawrence Estuary, Quebec, Canada. *Marine Geology*, **251**, 233–252, https://doi.org/10.1016/j.margeo.2008.03.003

Cauchon-Voyer, G., Locat, J., Leroueil, S., St-Onge, G. and Demers, D. 2011. Large-scale subaerial and submarine Holocene and recent mass movements in the Betsiamites area, Quebec, Canada. *Engineering Geology*, **121**, 28–45, https://doi.org/10.1016/j.enggeo.2011.04.011

Clare, M., Chaytor, J. *et al.* 2018. A consistent global approach for the morphometric characterization of subaqueous landslides. *Geological Society, London, Special Publications*, **477**, 455–477, https://doi.org/10.1144/SP477.15

Conway, K.W. and Barrie, V.J. 2018. Large bedrock slope failures in a British Columbia, Canada fjord: first documented submarine sackungen. *Geo-Marine Letters*, **38**, 195–209, https://doi.org/10.1007/s00367-018-0533-y

Conway, K.W., Barrie, J.V. and Thomson, R.E. 2012. Submarine slope failures and tsunami hazard in coastal British Columbia: Douglas Channel and Kitimat Arm. *Geological Survey of Canada, Current Research 2012*, **10**, 17, https://doi.org/10.4095/291732

Conway, K.W., Kung, R.B., Barrie, J.V., Hill, P.R. and Lintern, D.G. 2013. *A Preliminary Assessment of the Occurrence of Submarine Slope Failures in Coastal British Columbia by Analysis of Swath Multibeam*

Bathymetric Data Collected 2001–2011. Geological Survey of Canada, Open File 7348, 2013, https://doi.org/10.4095/292370

Deering, R., Bell, T., Forbes, D.L., Edinger, E. and Campbell, D.C. 2015. Morphology and setting of two submarine slope failures, inner Frobisher Bay, Baffin Island, Nunavut. *In*: *Poster Abstracts, ArticNet Annual Scientific Meeting*, 7–11 December, Vancouver, British Columbia.

Deering, R., Misiuk, B. *et al.* 2018. Characterization of the seabed and postglacial sediments of inner Frobisher Bay, Baffin Island, Nunavut. *In*: *Summary of Activities 2018*. Canada-Nunavut Geoscience Office, 139–152.

Deering, R., Bell, T., Forbes, D.L., Campbell, C. and Edinger, E. 2019. Morphological characterization of submarine slope failures in a semi-enclosed fjord, Frobisher Bay, eastern Canadian Arctic. *Geological Society, London, Special Publications*, **477**, 367–376, https://doi.org/10.1144/SP477.35

Deptuck, M.E. and Campbell, D.C. 2012. Widespread erosion and mass failure from the ~51 Ma Montagnais marine bolide impact off southwestern Nova Scotia, Canada. *Canadian Journal of Earth Sciences*, **49**, 1567–1594, https://doi.org/10.1139/e2012-075

Duchesne, M.J., Long, B.F., Urgeles, R. and Locat, J. 2003. New evidence of slope instability in the Outardes Bay delta area, Quebec, Canada. *Geo-Marine Letters*, **22**, 233–242, https://doi.org/10.1007/s00367-003-0121-6

Frankel, A., Mueller, C. *et al.* 1996. *National Seismic Hazard Maps: Documentation June 1996*. US Geological Survey, Open-File Rept, **96–532**.

Gagné, H, Lajeunesse, P, St-Onge, G and Bolduc, A. 2009. Recent transfer of coastal sediments to the Laurentian Channel, Lower St. Lawrence Estuary (Eastern Canada), through submarine canyon and fan systems. *Geo-Marine Letters*, **29**, 191–200, https://doi.org/10.1007/s00367-009-0138-6

Gales, J., Talling, P., Cartigny, M., Clarke, J., Lintern, G., Stacey, C. and Clare, M. 2018. What controls submarine channel development and the morphology of deltas entering deep-water fjords? *Earth Surface Processes and Landforms*, **44**, https://doi.org/10.1002/esp.4515

Giles, M.K., Mosher, D.C., Piper, D.J.W. and Wach, G.D. 2010. Mass transport deposits on the southwestern Newfoundland Slope. *In*: *Submarine Mass Movements and Their Consequences*. Springer, Dordrecht, 657–665.

Greene, G.H., Barrie, V.J. *et al.* 2018. Slope failure and mass transport processes along the Queen Charlotte Fault Zone, western British Columbia. *Geological Society, London, Special Publications*, **477**, 85–106, https://doi.org/10.1144/SP477.31

Hage, S., Cartigny, M.J. *et al.* 2018. How to recognize crescentic bedforms formed by supercritical turbidity currents in the geologic record: insights from active submarine channels. *Geology*, **46**, 563–566, https://doi.org/10.1130/G40095.1

Hage, S., Cartigny, M.J.B. *et al.* 2019. Direct monitoring reveals initiation of turbidity currents from extremely dilute river plumes. *Geophysical Research Letters*, **46**, 11310–11320, https://doi.org/10.1029/2019GL084526

Hamilton, T.S., Enkin, R.J., Riedel, M., Rogers, G.C., Pohlman, J.W. and Benway, H.M. 2015. Slipstream: an early Holocene slump and turbidite record from the frontal ridge of the Cascadia accretionary wedge off western Canada and paleoseismic implications. *Canadian Journal of Earth Sciences*, **52**, 405–430, https://doi.org/10.1139/cjes-2014-0131

Hawken, J.E. 2017. *Origin and Stratigraphic Setting of the Sackville Spur Bottom Simulating Reflector, Offshore Newfoundland*. MSc thesis, Dalhousie University.

Heezen, B.C., Ericson, D.B. and Ewing, M. 1954. Further evidence for a turbidity current following the 1929 Grand Banks earthquake. *Deep Sea Research (1953)*, **1**, 193–202, https://doi.org/10.1016/0146-6313(54)90001-5

Hill, P.R. 1983. Detailed morphology of a small area on the Nova Scotian continental slope. *Marine Geology*, **53**, 55–76, https://doi.org/10.1016/0025-3227(83)90034-8

Hill, P.R. 2012. Changes in submarine channel morphology and strata development from repeat multibeam surveys in the Fraser River delta, western Canada. *In*: Li, M.Z., Sherwood, C.R. and Hill, P.R. (eds) *Sediments, Morphology and Sedimentary Processes on Continental Shelves*. International Association of Sedimentologists Special Publication, **44**, 47–70.

Hill, P.R., Moran, K.M. and Blasco, S.M. 1982. Creep deformation of slope sediments in the Canadian Beaufort Sea. *Geo-Marine Letters*, **2**, 163–170, https://doi.org/10.1007/BF02462758

Hill, P.R., Blasco, S.M., Harper, J.R. and Fissel, D.B. 1992. Sedimentation on the Canadian Beaufort Shelf. *Continental Shelf Research*, **11**, 821–842, https://doi.org/10.1016/0278-4343(91)90081-G

Hiscott, R.N. and Aksu, A.E. 1994. Submarine debris flows and continental slope evolution in front of Quaternary ice sheets, Baffin Bay, Canadian Arctic. *AAPG Bulletin*, **78**, 445–460.

Hizzett, J.L., Hughes Clarke, J.E., Sumner, E.J., Cartigny, M.J.B., Talling, P.J. and Clare, M.A. 2018. Which triggers produce the most erosive, frequent, and longest runout turbidity currents on deltas? *Geophysical Research Letters*, **45**, 855–863, https://doi.org/10.1002/2017GL075751

Hughes Clarke, J.E. 2016. First wide-angle view of channelized turbidity currents links migrating cyclic steps to flow characteristics. *Nature Communication*, **7**, 11896, https://doi.org/10.1038/ncomms11896

Hughes Clarke, J.E., Brucker, S. *et al.* 2012. The Squamish ProDelta: monitoring Active Landslides and Turbidity Currents. *Proceedings of the Canadian Hydrographic Conference 2012*, Niagara Falls, Canada 15–17 May 2012.

Hughes Clarke, J.E., Marques, C.R.V. and Pratomo, D. 2013. Imaging active mass-wasting and sediment flows on a fjord delta, Squamish, British Columbia. *In*: Krastel, S., Behrmann, J.H. *et al.* (eds) *Submarine Mass Movements and Their Consequences 6th International Symposium*. Springer, 249–260.

Hughes Clarke, J.E., Muggah, J., Renoud, W., Bell, T., Forbes, D.L., Cowan, B. and Kennedy, J. 2015. Reconnaissance sea-bed mapping around Hall and Cumberland peninsulas, Nunavut: opening up southern Baffin Island to nearshore geological investigations. *In*: *Summary of Activities 2014*. Canada-Nunavut Geoscience Office, 133–144.

Huppertz, T.J., Piper, D.J.W., Mosher, D.C. and Jenner, K. 2010. The significance of mass-transport deposits for the evolution of a proglacial continental slope. *In*: Moscher, D.C., Shipp, R.C. *et al.* (eds) *Submarine Mass Movements and Their Consequences*. Advances in Natural and Technological Hazards Research, **28**. Springer, Dordrecht, 631–641, https://doi.org/10.1007/978-90-481-3071-9_51

Hyndman, R.D., Cassidy, J.F., Adams, J., Rogers, G.C. and Mazzotti, S. 2005. Earthquakes and seismic hazard in the Yukon-Beaufort-Mackenzie. *Canadian Society of Exploration Geophysicists Recorder*, **May 2005**, 32–66.

Jackson, L.E., Jr, Hermanns, R.L., Jermyn, C.E., Conway, K. and Kung, R. 2008. *Annotated Images of Submarine Landslides and Related Features Generated from Swath Multibeam Bathymetry, Howe Sound, British Columbia*. Geological Survey of Canada, **Open file 5662**, 1 CDROM.

Jenner, K.A., Campbell, D.C. and Piper, D.J.W. 2018. Along-slope variations in sediment lithofacies and depositional processes since the Last Glacial Maximum on the northeast Baffin margin, Canada. *Marine Geology*, **405**, 92–107, https://doi.org/10.1016/j.margeo.2018.07.012 (NRCan Cont.# 20170321)

Leonard, L.J., Rogers, G.C. and Mazzotti, S. 2012. *A Preliminary Tsunami Hazard Assessment of the Canadian Coastline*. Geological Survey of Canada, **Open File 7201, 126**, https://doi.org/10.4095/292067

Li, G., Piper, D.J.W. and Campbell, D.C. 2011. The Quaternary Lancaster Sound trough-mouth fan, NW Baffin Bay. *Journal of Quaternary Science*, **26**, 511–522, https://doi.org/10.1002/jqs.1479

Li, G., Piper, D.J., Campbell, D.C. and Mosher, D. 2012*a*. Turbidite deposition and the development of canyons through time on an intermittently glaciated continental margin: the Bonanza Canyon system, offshore eastern Canada. *Marine and Petroleum Geology*, **29**, 90–103, https://doi.org/10.1016/j.marpetgeo.2011.08.018

Li, G., Campbell, D.C., Mosher, D. and Piper, D.J. 2012*b*. Comparison of Quaternary glaciogenic debris flows with blocky mass-transport deposits in Orphan Basin, offshore eastern Canada. *In*: Yamada, Y., Kawamura, K. *et al.* (eds) *Submarine Mass Movements and Their Consequences*. Advances in Natural and Technological Hazards Research, **31**. Springer, Dordrecht, 723–733, https://doi.org/10.1007/978-94-007-2162-3_64

Lintern, D.G. and Hill, P.R. 2010. An underwater laboratory at the Fraser River Delta. *Eos Transactions of the AGU*, **91**, 333–334, https://doi.org/10.1029/2010EO380001

Lintern, D.G., Hill, P.R. and Stacey, C. 2016. Powerful unconfined turbidity current captured by cabled observatory on the Fraser River delta slope, British Columbia, Canada. *Sedimentology*, **63**, 1041–1064, https://doi.org/10.1111/sed.12262

Lintern, D.G., Blais-Stevens, A. *et al.* 2019. Providing multidisciplinary scientific advice for coastal planning in Kitimat Arm, British Columbia. *Geological Society, London, Special Publications*, **477**, 567–581, https://doi.org/10.1144/SP477.40

Locat, J. and Lee, H.J. 2002. Submarine landslides: advances and challenges. *Canadian Geotechnical Journal*, **39**, 193–212, https://doi.org/10.1139/t01-089

Løvholt, F., Schulten, I., Mosher, D., Harbitz, C. and Krastel, S. 2019. Modelling the 1929 Grand Banks slump and landslide tsunami. *Geological Society, London, Special Publications*, **477**, 315–331, https://doi.org/10.1144/SP477.28

Mackillop, K., Fenton, G., Mosher, D., Latour, V. and Mitchelmore, P. 2019. Assessing Submarine Slope Stability through Deterministic and Probabilistic Approaches: a Case Study on the West-Central Scotia Slope. *Geosciences*, **9**, 18, https://doi.org/10.3390/geosciences9010018

Maslin, M., Owen, M., Day, S. and Long, D. 2004. Linking continental-slope failures and climate change: testing the clathrate gun hypothesis. *Geology*, **32**, 53–56, https://doi.org/10.1130/G20114.1

Mckenna, G.T., Luternauer, J.L. and Kostaschuk, R.A. 1992. Large-scale mass-wasting events on the Fraser River delta front near Sand Heads, British Columbia. *Canadian Geotechnical Journal*, **29**, 151–156, https://doi.org/10.1139/t92-016

Mosher, D.C. 2008. Submarine mass movements in Canada: Geohazards with far-reaching implications. *In*: Locat, J., Perret, D., Turmel, D. and Demers, D.S.L. (eds) *Proceedings of the 4th Canadian Conference on Geohazards: From Causes to Management*. Laval University Press, Québec, 55–62.

Mosher, D.C. 2009. International Year of Planet Earth 7. Oceans: submarine Landslides and Consequent Tsunamis in Canada. *Geoscience Canada*, **36**.

Mosher, D.C. and Campbell, D.C. 2010. The Barrington submarine landslide, western Scotian slope. *In*: *The Importance of Mass-Transport Deposits in Deepwater Settings*. SEPM Special Publications, **91**.

Mosher, D.C., Bigg, S. and Lapierre, A. 2006. 3D seismic v. multibeam sonar seafloor surface renderings for geohazard assessment: case examples from the central Scotian Slope. *The Leading Edge*, **25**, 1484–1494, https://doi.org/10.1190/1.2405334

Mosher, D.C., Campbell, D.C., Gardner, J.V., Piper, D.J.W., Chaytor, J.D. and Rebesco, M. 2017. The role of deep-water sedimentary processes in shaping a continental margin: the Northwest Atlantic. *Marine Geology*, **393**, 245–259, https://doi.org/10.1016/j.margeo.2017.08.018

Naegeli, K. 2010. *Internship Report July-October 2010*. Natural Resources Canada, Pacific Geological Center, Sidney.

Normandeau, A. and Dietrich, P. 2019. *Distribution of Potentially Active Turbidity Currents in Eastern Baffin Island Fiords*. Geological Survey of Canada, Open File 8512, 2019, 1 sheet, https://doi.org/10.4095/314733

Normandeau, A., Lajeunesse, P. and St-Onge, G. 2013. Shallow-water longshore drift-fed submarine fan deposition (Moisie River Delta, Eastern Canada). *Geo-Marine Letters*, **33**, 391–403, https://doi.org/10.1007/s00367-013-0336-0

Normandeau, A., Lajeunesse, P., St-Onge, G., Bourgault, D., Drouin, S.S.O., Senneville, S. and Bélanger, S. 2014. Morphodynamics in sediment-starved inner-shelf submarine canyons (Lower St. Lawrence Estuary, Eastern Canada). *Marine Geology*, **357**, 243–255, https://doi.org/10.1016/j.margeo.2014.08.012

Normandeau, A., Lajeunesse, P. and St-Onge, G. 2015. Submarine canyons and channels in the Lower

St. Lawrence Estuary (Eastern Canada): morphology, classification and recent sediment dynamics. *Geomorphology*, **241**, 1–18, https://doi.org/10.1016/j.geomorph.2015.03.023

Normandeau, A., Dietrich, P., Lajeunesse, P., St-Onge, G., Ghienne, J.F., Duchesne, M.J. and Francus, P. 2017. Timing and controls on the delivery of coarse sediment to deltas and submarine fans on a formerly glaciated coast and shelf. *GSA Bulletin*, **129**, 1424–1441, https://doi.org/10.1130/B31678.1

Normandeau, A., Bourgault, D., Neumeier, U., Lajeunesse, P., St-Onge, G., Gostiaux, L. and Chavanne, C. 2020. Storm-induced turbidity currents on a sediment-starved shelf: insight from direct monitoring and repeat seabed mapping of upslope migrating bedforms. *Sedimentology*, **67**, 1045–1068, https://doi.org/10.1111/sed.12673

Normandeau, A., Campbell, D.C. and Cartigny, M.J.B. 2019*a*. The influence of turbidity currents and contour currents on the distribution of deep-water sediment waves offshore eastern Canada. *Sedimentology*, **66**, 1746–1767, https://doi.org/10.1111/sed.12557.

Normandeau, A., Campbell, C.D., Piper, D. and Jenner, K.A. 2019*b*. Are submarine landslides an underestimated hazard on the western North Atlantic passive margin? *Geology*, **47**, 848–852, https://doi.org/10.1130/G46201.1

Normandeau, A., Campbell, C.D., Piper, D. and Jenner, K.A. 2019*c*. New evidence for a major Quaternary submarine landslide on the external western levee of Laurentian Fan. *Geological Society, London, Special Publications*, **477**, 377–387, https://doi.org/10.1144/SP477.14

Normandeau, A., Dietrich, P., Hughes Clarke, J., Van Wychen, W., Lajeunesse, P., Burgess, D. and Ghienne, J.F. 2019*d*. Retreat pattern of glaciers controls the occurrence of turbidity currents on high-latitude fjord deltas (eastern Baffin Island). *Journal of Geophysical Research: Earth Surface*, **124**, 1559–1571, https://doi.org/10.1029/2018JF004970

Paull, C.K., Ussler, W. *et al.* 2007. Origin of pingo-like features on the Beaufort Sea Shelf and their possible relationship to decomposing methane gas hydrates. *Geophysical Research Letters*, **34**, https://doi.org/10.1029/2006GL027977

Paull, C., Dallimore, S.R. *et al.* 2015. Slope edge deformation and permafrost dynamics along the Arctic shelf edge, Beaufort Sea, Canada. Abstract OS22B-07 presented at 2015 Fall Meeting, AGU, 14–18 December, San Francisco, California.

Pinet, N., Brake, V., Campbell, C. and Duchesne, M.J. 2015. Geomorphological characteristics and variability of Holocene mass-transport complexes, St. Lawrence River Estuary, Canada. *Geomorphology*, **228**, 286–302, https://doi.org/10.1016/j.geomorph.2014.09.008

Piper, D.J.W., Shor, A.N. and Clarke, J.E.H. 1988. The 1929 'Grand Banks' earthquake, slump, and turbidity currents. *Geological Society of America, Special Papers*, **229**, 77–92.

Piper, D.J.W., Mosher, D.C., Gauley, B.-J., Jenner, K. and Campbell, D.C. 2003. The chronology and recurrence of submarine mass movements on the continental slope off southeastern Canada. *In*: Locat, J. and Mienert, J. (eds) *Submarine Mass Movements and their Consequences*. Kluwer Academic, The Netherlands, 299–306.

Piper, D.J., Deptuck, M.E., Mosher, D.C., Hughes Clarke, J.E. and Migeon, S. 2012. Erosional and depositional features of glacial meltwater discharges on the eastern Canadian continental margin. *SEPM Special Publications*, **99**, 61–80.

Piper, D.J.W., Campbell, D.C. and Mosher, D.C. 2016. Mid-latitude complex trough-mouth fans, Laurentian and Northeast fans, eastern Canada. *Geological Society, London, Memoirs*, **46**, 363–364, https://doi.org/10.1144/M46.93

Piper, D.J., Tripsanas, E., Mosher, D.C. and Mackillop, K. 2019. Paleoseismicity of the continental margin of eastern Canada: rare regional failures and associated turbidites in Orphan Basin. *Geosphere*, **15**, 85–107, https://doi.org/10.1130/GES02001.1

Poncet, R., Campbell, C., Dias, F., Locat, J. and Mosher, D. 2010. A study of the tsunami effects of two landslides in the St. Lawrence estuary. *In*: Moscher, D.C., Shipp, R.C. *et al.* (eds) *Submarine Mass Movements and Their Consequences*. Advances in Natural and Technological Hazards Research, **28**. Springer, Dordrecht, 755–764.

Prior, D.B., Bornhold, B.D., Coleman, J.M. and Bryant, W.R. 1982. Morphology of a submarine slide, Kitimat Arm, British Columbia. *Geology*, **10**, 588–592, https://doi.org/10.1130/0091-7613(1982)10<588:MOASSK>2.0.CO;2

Rashid, H., Mackillop, K., Sherwin, J., Piper, D.J.W., Marche, B. and Vermooten, M. 2017. Slope instability on a shallow contourite-dominated continental margin, southeastern Grand Banks, eastern Canada. *Marine Geology*, **393**, 203–215, https://doi.org/10.1016/j.margeo.2017.01.001

Rashid, H., Piper, D.J.W., Mackillop, K., Ouellette, D., Vermooten, M., Muñoz, A. and Jiménez, P. 2019. Dynamics of sediments on a glacially influenced, sediment starved, current-swept continental margin: the SE Grand Banks Slope off Newfoundland. *Marine Geology*, **408**, 67–86, https://doi.org/10.1016/j.margeo.2018.11.012

Riedel, M., Brent, T.A., Taylor, G., Taylor, A.E., Hong, J.-K., Jin, Y.-K. and Dallimore, S.R. 2017. Evidence for gas hydrate occurrences in the Canadian Arctic Beaufort Sea within permafrost-associated shelf and deep-water marine environments. *Marine and Petroleum Geology*, **81**, 66–78, https://doi.org/10.1016/j.marpetgeo.2016.12.027

Riedel, M., Côté, M.M., Urlaub, M., Geersen, J., Scholz, N.A., Naegeli, K. and Spence, G.D. 2018. Slope failures along the deformation front of the Cascadia margin: linking slide morphology to subduction zone parameters. *Geological Society, London, Special Publications*, **477**, 47–67, https://doi.org/10.1144/SP477.33

Roger, J., Saint-Ange, F., Lajeunesse, P., Duchesne, M.J. and St-Onge, G. 2013. Late Quaternary glacial history and meltwater discharges along the Northeastern Newfoundland Shelf. *Canadian Journal of Earth Sciences*, **50**, 1178–1194, https://doi.org/10.1139/cjes-2013-0096

Scholz, N.A., Riedel, M., Urlaub, M., Spence, G.D. and Hyndman, R. 2016. Submarine landslides offshore Vancouver Island along the northern Cascadia margin,

British Columbia: why preconditioning is likely required to trigger slope failure. *Geo-Marine Letters*, **36**, 323–337, https://doi.org/10.1007/s00367-016-0452-8

Schulten, I., Mosher, D.C., Krastel, S., Piper, D.J. and Kienast, M. 2019*a*. Surficial sediment failures due to the 1929 Grand Banks Earthquake, St Pierre Slope. *Geological Society, London, Special Publications*, **477**, 583–596, https://doi.org/10.1144/SP477.25

Schulten, I., Mosher, D.C., Piper, D.J. and Krastel, S. 2019*b*. A massive slump on the St. Pierre Slope, a new perspective on the 1929 Grand Banks submarine landslide. *Journal of Geophysical Research: Solid Earth*, **124**, 7538–7561 https://doi.org/10.1029/2018JB017066

Shaw, J. and Lintern, G. 2016. *Marine geology, geomorphology of the Kitimat Fiord System, British Columbia, parts of NTS 103-A, NTS 103-H and NTS 103-I*. Geological Survey of Canada, Canadian Geoscience Map 275, 2016, 1 sheet, https://doi.org/10.4095/297482

Shaw, J., Piper, D.J., Skulski, T., Lamplugh, M.J., Craft, A. and Roy, A. 2012. New evidence for widespread mass transport on the Northeast Newfoundland Shelf revealed by Olex single-beam echo sounding. *Geo-Marine Letters*, **32**, 5–15, https://doi.org/10.1007/s00367-011-0233-3

Shaw, J., Stacey, C., Wu, Y. and Lintern, G.D. 2017. Anatomy of the Kitimat fiord system, British Columbia. *Geomorphology*, **293**, 108–129, https://doi.org/10.1016/j.geomorph.2017.04.043

Stacey, C.D. 2014. *Frequency and initiation mechanisms of submarine slides on the Fraser Delta front*. MSc thesis, University of Victoria, Canada

Stacey, C.D. and Hill, P.R. 2016. Cyclic steps on a glacifluvial delta, Howe Sound, British Columbia. *Geological Society, London, Memoirs*, **46**, 93–94, https://doi.org/10.1144/M46.71

Stacey, C.D., Lintern, D.G. and Enkin, R. 2018. Multifaceted re-analysis of the enigmatic Kitimat slide complex, Canada. *Sedimentary Geology*, **369**, https://doi.org/10.1016/j.sedgeo.2018.01.006

Stacey, C.D., Hill, P., Talling, P., Enkin, R., Clarke, J. and Lintern, G. 2019. How turbidity current frequency and character varies down a fjord-delta system: combining direct monitoring, deposits and seismic data. *Sedimentology*, **66**, https://doi.org/10.1111/sed.12488

Stacey, C.D., Lintern, D.G., Shaw, J. and Conway, K.W. 2020. Slope stability hazard in a fjord environment: Douglas Channel, Canada. *Geological Society, London, Special Publications*, **500**, https://doi.org/10.1144/SP500-2019-191

Stevenson, C.J., Feldens, P. *et al.* 2018. Reconstructing the sediment concentration of a giant submarine gravity flow. *Nature Communications*, **9**, 2616, https://doi.org/10.1038/s41467-018-05042-6

St-Onge, G., Mulder, T., Piper, D.J., Hillaire-Marcel, C. and Stoner, J.S. 2004. Earthquake and flood-induced turbidites in the Saguenay Fjord (Québec): a Holocene paleoseismicity record. *Quaternary Science Reviews*, **23**, 283–294, https://doi.org/10.1016/j.quascirev.2003.03.001

Todd, B.J., Shaw, J., Campbell, D.C. and Mate, D.J. 2016. Preliminary interpretation of the marine geology of Frobisher Bay, Baffin Island, Nunavut. *In*: *Summary of Activities 2016*. Canada-Nunavut Geoscience Office, 61–66.

Thomson, R.E., Fine, I.V., Krassovski, M.V., Cherniawsky, J.Y., Conway, K. and Wills, P. 2012. *Numerical Simulation of Tsunamis Generated by Submarine Slope Failures in Douglas Channel, British Columbia*. Department of Fisheries and Oceans, Canadian science advisory research document 2012/115.

Turmel, D., Locat, J., Leblanc, J. and Cauchon-Voyer, G. 2019. Tsunami modelling of the 7250 cal years BP Betsiamites submarine landslide. *Geological Society, London, Special Publications*, **477**, 293–301, https://doi.org/10.1144/SP477.9

Vendettuoli, D., Clare, M.A. *et al.* 2019. Daily bathymetric surveys document how stratigraphy is built and its extreme incompleteness in submarine channels. *Earth and Planetary Science Letters*, **515**, 231–247, https://doi.org/10.1016/j.epsl.2019.03.033

Structural constraints on the subduction of mass-transport deposits in convergent margins

Jacob Geersen[1]*, Andrea Festa[2] and Francesca Remitti[3]

[1]GEOMAR Helmholtz Centre for Ocean Research Kiel, Wischhofstrasse 1–3, 24148 Kiel, Germany

[2]Dipartimento di Scienze della Terra, Università degli Studi di Torino, Via Valperga Caluso 35, 10125 Torino, Italy

[3]Dipartimento di Scienze Chimiche e Geologiche, Università di Modena e Reggio Emilia, via Campi 103, 41125 Modena, Italy

JG, 0000-0001-5625-493X; AF, 0000-0001-5325-0263; FR, 0000-0002-0720-1927
*Correspondence: jgeersen@geomar.de

Abstract: The subduction of large and heterogeneous mass-transport deposits (MTDs) is discussed to modify the structure and physical state of the plate boundary and therewith exert an influence on seismicity in convergent margins. Understanding which subduction-zone architectures and structural boundary conditions favour the subduction of MTDs, primarily deposited in oceanic trenches, is therefore highly significant. We use bathymetric and seismic reflection data from modern convergent margins to show that a large landslide volume and long runout, in concert with thin trench sediments, increase the chances for an MTD to become subducted. In regions where the plate boundary develops within the upper plate or at its base (non-accretionary margins), and in little-sedimented trenches (sediment thickness <2 km), an MTD has the highest potential to become subducted, particularly when characterized by a long runout. On the contrary, in the case of a heavily sedimented trench (sediment thickness >4 km) and short runout, an MTD will only be subducted if the thickness of subducting sediments is higher than the thickness of sediments under the MTD. The results allow identification of convergent margins where MTDs are preferentially subducted and thus potentially alter plate-boundary seismicity.

Submarine landslides occur on all types of continental margins and geological provinces in the oceans where inclined seafloors are present (Fig. 1). They include the largest sedimentary movements on Earth, forming different types of mass-transport deposits (MTDs). Such MTDs cover areas of hundreds to thousands of square kilometres in the oceans and range from tens to hundreds, and up to thousands of metres in thickness (Moscardelli and Wood 2016 and references therein). The Ruatoria Debris Avalanche off the coast of New Zealand (Fig. 2a), for example, moved around 2000 km^3 of sediment and basement rocks (thickness of the MTD up to *c.* 2000 m; Collot *et al.* 2001). The Storrega Slide off Norway affected around 3000 km^3 of slope material (Haflidason *et al.* 2004). MTDs related to large landslides (>1000 km^2) can have runout distances of some hundreds of kilometres (Urgeles and Camerlenghi 2013; Moscardelli and Wood 2016).

Along convergent margins, submarine landslides can mobilize both the sediments that drape the forearc slope and the underlying basement rock of the upper plate. As a consequence, their MTDs often consist of debris-blocky flows (Fig. 2a) with kilometre-sized megablocks embedded in a mud-rich to debrite matrix (Collot *et al.* 2001). Field studies on exhumed ancient analogues of convergent margins also documented mud-rich MTDs with block-in-matrix deposits, such as sedimentary mélanges or olistostromes (Pini 1999; Remitti *et al.* 2011; Pini *et al.* 2012; Festa *et al.* 2016; Ogata *et al.* 2019, 2020). They are commonly characterized by centimetre- to metre-sized lithic blocks that are randomly distributed in a clay- or shale-rich matrix, which may sustain bedding packages that locally reach tens of metres or up to kilometres in size (floaters or over-sized blocks; e.g. Festa *et al.* 2016, 2019).

Based on the observation that earthquake rupture during the moment magnitude (M_w) 9.5 Great Chile (1960) and M_w 8.8 Maule (2010) earthquakes decreased across the region where the MTDs of large submarine slope failures are currently subducted (Fig. 3b), Geersen *et al.* (2013) suggested that submarine landslides may affect seismicity at the plate boundary. In their model, the subduction of a section of chaotic MTDs, up to 700 m in thickness and including kilometre-scale blocks of upper-plate basement rocks, contributes to the development of a rough, structurally and lithologically complex plate boundary (Fig. 2c–e). Such a complex plate

From: Georgiopoulou, A., Amy, L. A., Benetti, S., Chaytor, J. D., Clare, M. A., Gamboa, D., Haughton, P. D. W., Moernaut, J. and Mountjoy, J. J. (eds) 2020. *Subaqueous Mass Movements and their Consequences: Advances in Process Understanding, Monitoring and Hazard Assessments.* Geological Society, London, Special Publications, **500**, 115–128.
First published online March 13, 2020, https://doi.org/10.1144/SP500-2019-174

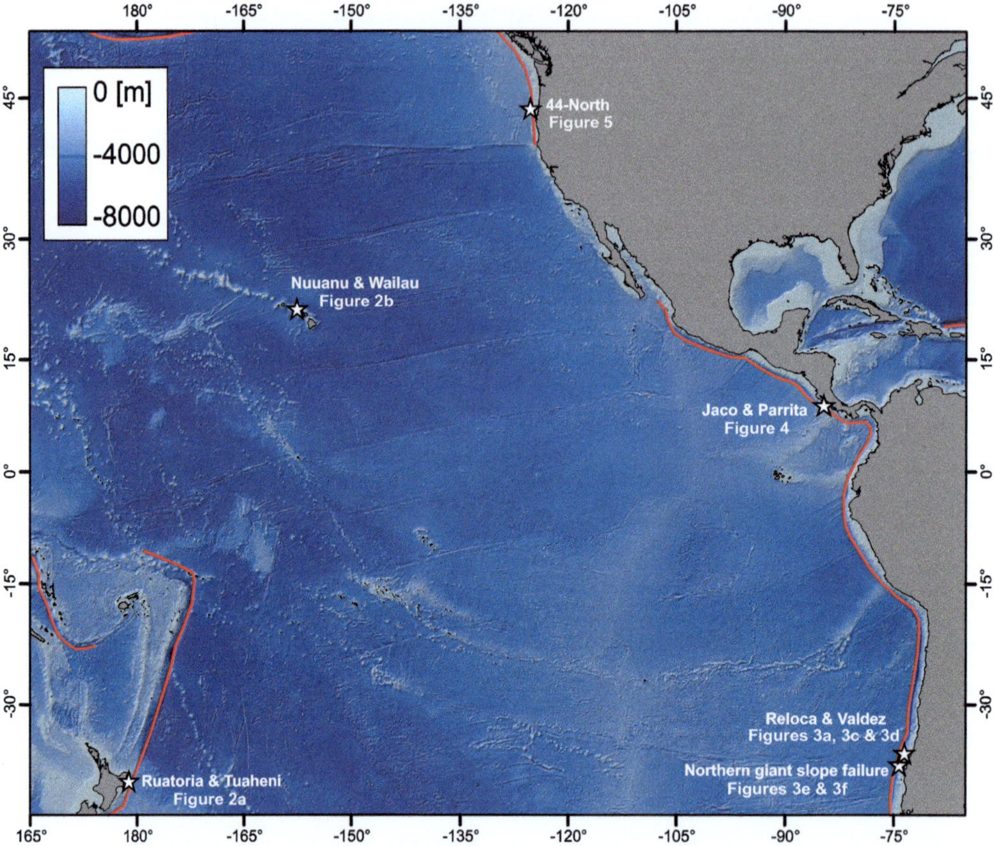

Fig. 1. Bathymetric map of the Pacific Ocean. White stars indicate submarine landslides shown in Figures 2–5. All slides occurred along convergent margins (red lines) except for Nuuanu and Wailau, which affected the Hawaiian Islands of Oahu and Molokai.

boundary may hamper the development of a spatially continuous and homogeneous shear zone as it might be required for large earthquake rupture propagation (Wang and Bilek 2011, 2014; Bilek and Lay 2018; Geersen 2019). This interpretation agrees with observations from exhumed subduction zones, where mud-rich MTDs with a chaotic internal fabric have been documented and thought to be responsible for modifying the structure and physical state of the ancient plate boundary (Festa *et al.* 2018). The scattered distribution of several MTDs may favour the development of an anastomosing fault network within the subduction plate boundary, and their heterogeneous (often less permeable) internal composition and block-in-matrix fabric with respect to the surrounding normal sediments is expected to influence the redistribution of overpressured fluids and shear strain (Fig. 2c–e). The mixture of clay-rich (incompetent) matrix and competent blocks exhibits both discrete and distributed deformation with minimum interaction between adjacent blocks (Festa *et al.* 2018). This scenario creates unstable sliding material enclosed by frictionally stable sliding or conditionally stable matrix, similar to that proposed for tectonic mélanges (Fagereng and Sibson 2010; Fagereng 2011; Beall *et al.* 2019). Therefore, large ruptures may not nucleate, but the development of several micro slip surfaces can trigger repetitive slip with micro-displacements, favouring the slow end-member of the spectrum of fault slip behaviour (e.g. tsunami earthquakes, slow slip events, very low frequency earthquakes, non-volcanic tremors, aseismic creep).

Considering the potential influence of subducting MTDs on the seismic behaviour of the plate boundary, it is important to better understand which subduction-zone architectures and structural boundary conditions favour the subduction of MTDs. In this work we use multibeam bathymetric and seismic reflection data to show field examples of MTDs in different convergent margins, representing a wide spectrum of landslide sizes and subduction-zone

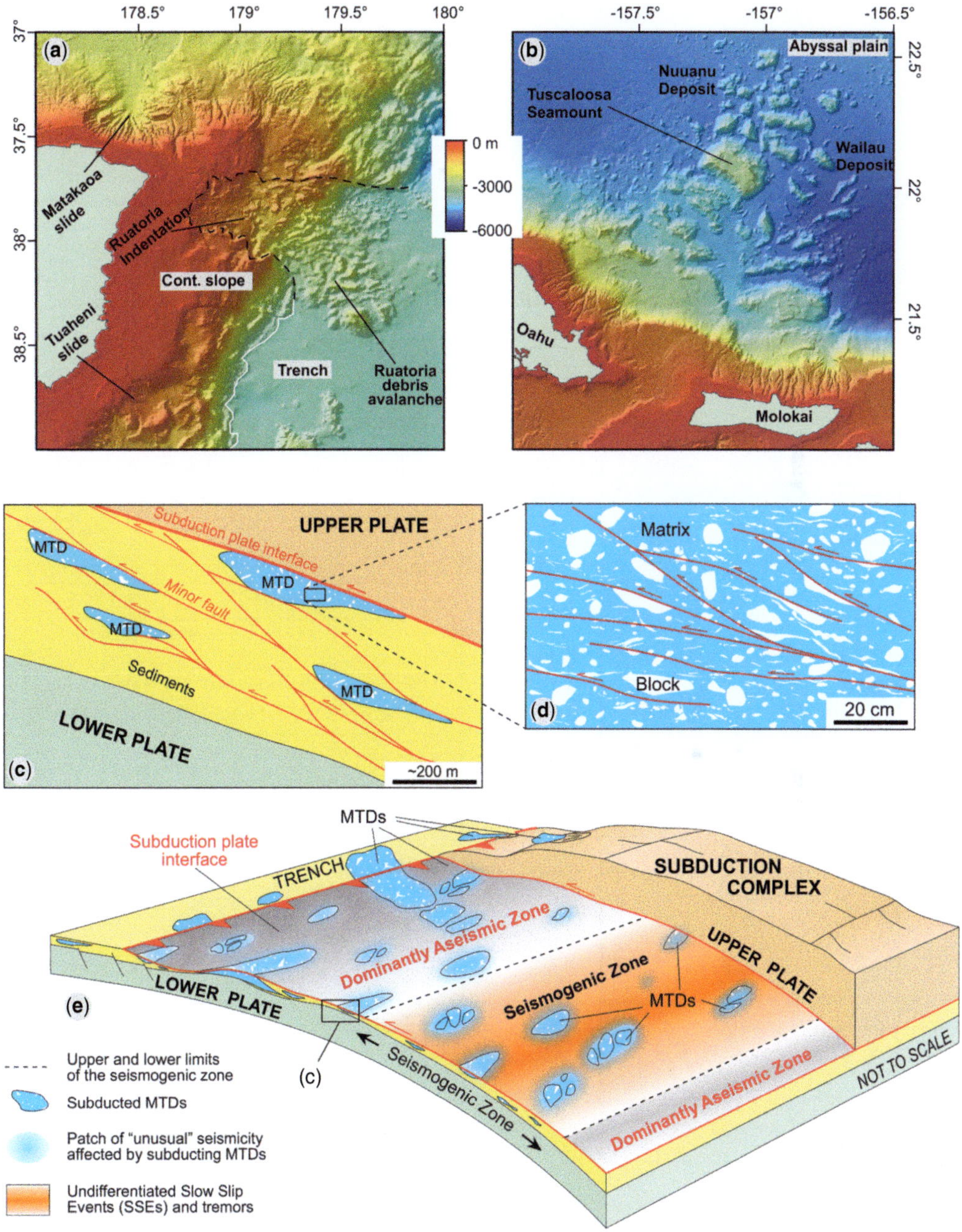

Fig. 2. (**a**) Bathymetric map of the Ruatoria Slide and debris avalanche off the North Island of New Zealand. (**b**) Bathymetric map of the Nuuanu Slide and Wailau Slide which affected the Hawaiian Islands of Oahu and Molokai. (**c**) Conceptual model on how subducting MTDs may affect the structural development of the plate boundary (modified from Festa *et al.* 2018). The subduction of several scattered distributed MTDs favours the formation of a heterogeneous fault network. This may support different seismic behaviours ranging from interseismic locking to aseismic creep. (**d**) Detail of the internal mesoscale block-in-matrix fabric of an MTD subducted to the up-dip limit of a shallow seismogenic zone. Note the inferred strain partitioning along some anastomosing shear surfaces around blocks and clasts, causing limited interaction between them. (**e**) Schematic view of a subduction plate boundary affected by the subduction of MTDs (modified from Festa *et al.* 2018).

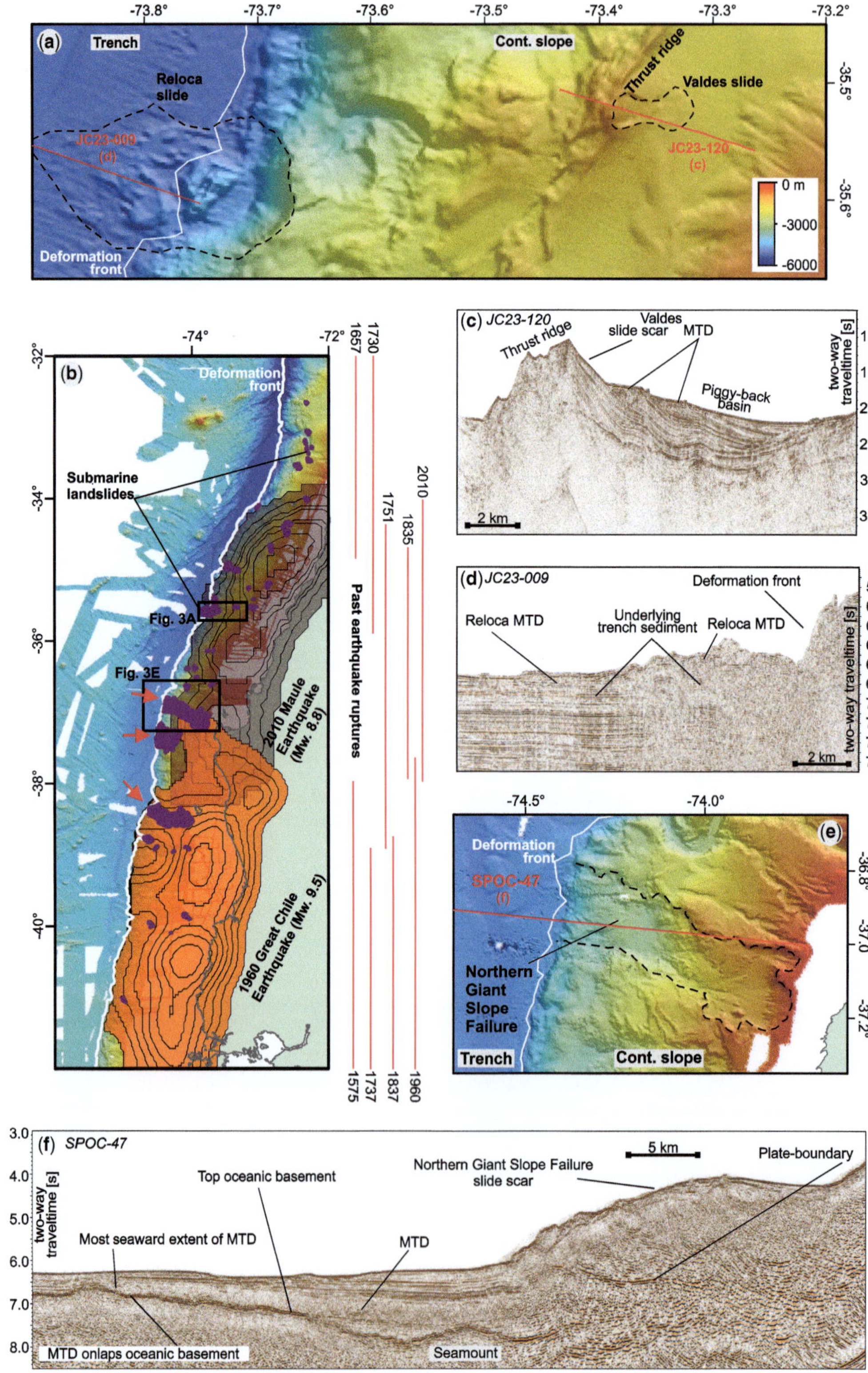
(a)
Trench
Cont. slope
Reloca slide
Thrust ridge
Valdes slide
JC23-009 (d)
JC23-120 (c)
Deformation front
0 m
-3000
-6000
(b)
Deformation front
Submarine landslides
Fig. 3A
Fig. 3E
2010 Maule Earthquake (Mw. 8.8)
1960 Great Chile Earthquake (Mw. 9.5)
Past earthquake ruptures
(c) JC23-120
Thrust ridge
Valdes slide scar
MTD
Piggy-back basin
2 km
two-way traveltime [s]
(d) JC23-009
Deformation front
Reloca MTD
Underlying trench sediment
Reloca MTD
2 km
two-way traveltime [s]
(e)
Deformation front
SPOC-47 (f)
Northern Giant Slope Failure
Trench
Cont. slope
(f) SPOC-47
5 km
Plate-boundary
Northern Giant Slope Failure slide scar
Top oceanic basement
Most seaward extent of MTD
MTD
MTD onlaps oceanic basement
Seamount
two-way traveltime [s]

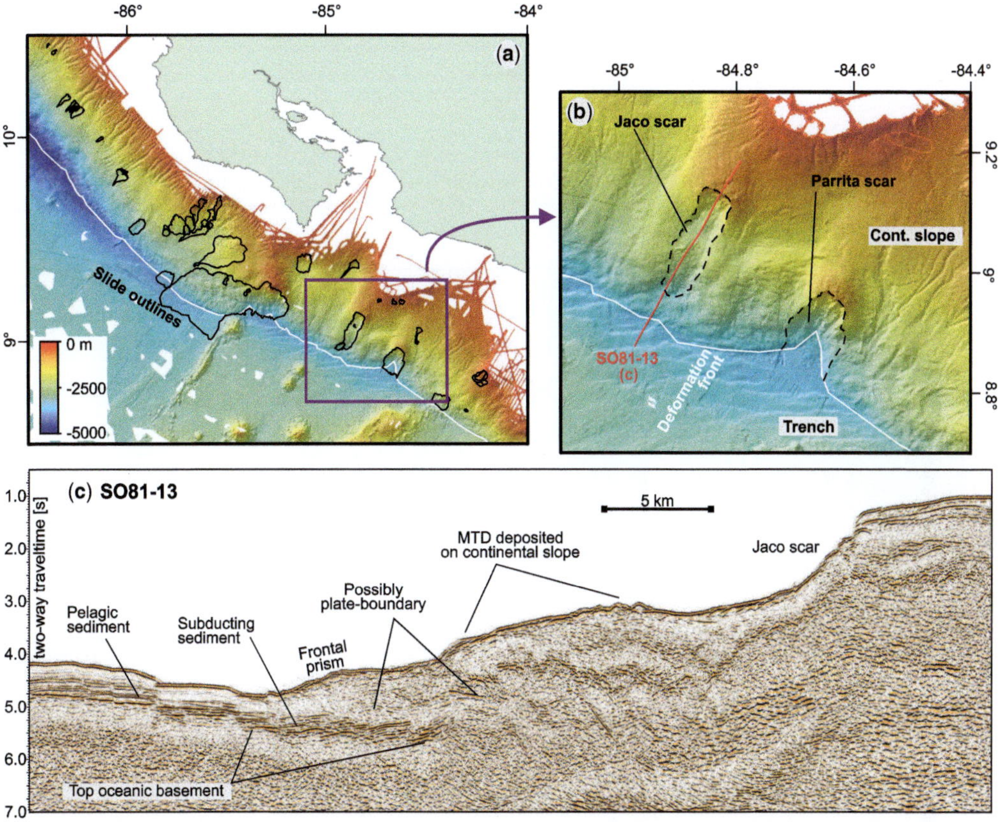

Fig. 4. (**a**) Overview bathymetric map of the convergent margin of Costa Rica with submarine landslides encircled by black lines (Harders *et al.* 2011). (**b**) Bathymetric map of Jaco Scar and Parrita Scar. Red line indicates the seismic profile shown in (c). (**c**) Seismic line SO81-13 across the trench and marine forearc of Central America at the location of Jaco Scar.

settings across both erosional and accretionary margins (Figs 2–5). We use the different cases to discuss the structural parameters that govern a possible subduction of MTDs. Finally we evaluate which of the global subduction zones are more likely to be affected by subducting MTDs. We consider only the chaotic part of an MTD that does not disintegrate into a turbidity current. Turbidites do not represent a chaotic deposit and often form the bulk of a sedimentary trench-fill.

MTDs in convergent margins

We first evaluate the structural boundary conditions, which control whether an MTD derived from the

Fig. 3. (**a**) Bathymetric map of the Reloca Slide and the Valdes Slide at the convergent margin of Central Chile. Red lines indicate seismic profiles shown in c and d. (**b**) Overview bathymetric map of the Central Chilean margin with the rupture areas of the 1960 Great Chile Earthquake (Mw. 9.5) and the 2010 Maule Earthquake (Mw. 8.8) indicated by orange (1960) and grey (2010) shaded areas (Moreno *et al.* 2009, 2012). Slip contour lines are given at intervals of 5 m (1960) and 2 m (2010). The purple polygons mark the source regions of submarine landslides (Völker *et al.* 2012) and the three red arrows point at the giant slope failures (Geersen *et al.* 2011*a*, 2013) which coincide with the region where earthquake rupture decreased. Red lines on the right side of the map mark the latitudinal extent of past megathrust earthquakes (Contreras-Reyes and Carrizo 2011). (**c**) Seismic line JC23-120 across the Valdes Slide which occurred on the landward inclined side of a thrust ridge. (**d**) Seismic line JC23-009 showing the MTD associated with the Reloca slide at the lowermost continental slope and trench of Central Chile. (**e**) Bathymetric map of the Northern Giant Slope Failure. Red line indicates the seismic profile shown in Figures 3F. (**f**) Seismic line SPOC-47 highlighting the up to 500 m thick MTD associated which the Northern Giant Slope Failure.

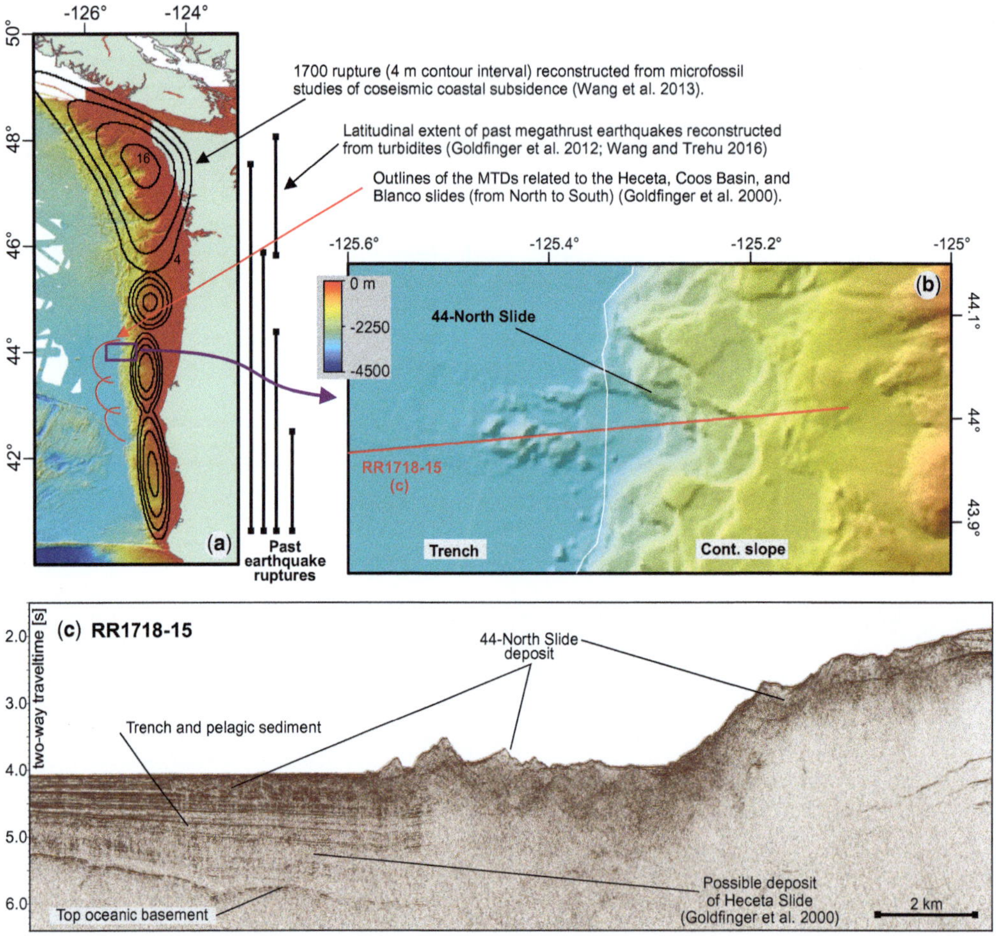

Fig. 5. (**a**) Overview bathymetric map of the Cascadia margin. Slip contour lines of the 1700 earthquake (M_w. *c*. 9) are given at 4 m intervals (Wang *et al.* 2013). Red semi-circles in the trench mark the latitudinal extend of the MTDs associated with the Blanco, Coos Basin and Heceta slides which are buried in the trench (Goldfinger *et al.* 2000). Black lines on the right side of the map mark the latitudinal extent of past megathrust earthquakes reconstructed from turbidites (Goldfinger *et al.* 2012; Wang and Trehu 2016). (**b**) Bathymetric map of the 44-North Slide. Red line indicates the seismic profile shown in c. (**c**) Seismic line RR1718-15 across the 44-North Slide (compare Tominaga *et al.* 2017; Lenz *et al.* 2019).

marine forearc is deposited on the continental slope or in the trench. For the latter we subsequently distinguish between little-sedimented trenches (sediment thickness <2 km), where the likelihood that MTDs can reach the plate boundary is high, and moderately (sediment thickness 2–4 km) to heavily (sediment thickness >4 km) sedimented trenches, where the probability is low. Subsequently we evaluate additional structural domains in the ocean basins (other than continental margins) where submarine landslides can cause the deposition of MTDs on oceanic plates before reaching a subduction zone.

Examples from MTDs on continental slopes

MTDs that become mobilized by a submarine landslide in a convergent margin do not automatically evacuate into the adjacent trench. Instead they can also be deposited on the forearc slope or within forearc basins. Figures 3a and c show the Valdes Slide off the coast of Central Chile as an example of a submarine landslide deposited in the marine forearc (Völker *et al.* 2012; Weiss *et al.* 2013). The Valdes Slide occurred at a water depth of around 1000 m on the landward inclined side of a thrust ridge and the mobilized sediments were deposited in the adjacent piggy-back basin (Geersen *et al.* 2011*b*; Völker *et al.* 2012). Another submarine landslide that occurred at the upper-continental slope (*c.* 500 m water depth) and did not evacuate into the trench is the Tuaheni Slide at the Hikurangi margin, New Zealand (Mountjoy *et al.* 2009; Fig. 2a). Factors that support the deposition of an MTD within a marine forearc (instead of being evacuated into the trench) include:

(1) a complex margin structure and surface morphology that blocks downslope-directed sediment transport (e.g. thrust ridges) and/or provides effective sediment traps (e.g. forearc basins);
(2) a low inclination angle of the continental slope that reduces the released energy and runout distance of the mobilized material;
(3) a shallow water depth of the submarine landslide, which increases the distance to the trench and the possibility that the mobilized material is blocked or trapped on the way to the trench;
(4) a small volume of mobilized material, which again increases the possibility that the mobilized material is blocked or trapped on the way to the trench.

The first two parameters are frequently met at accretionary margins which develop adjacent to heavily sedimented trenches. Here the transfer of sediments from the subducting to the overriding plate leads to compressional tectonics and the development of thrust ridges and adjacent piggy-back basins (Moore and Karig 1976; von Huene and Scholl 1991). Folding and thrusting associated with the compressional tectonic regime further reduce the inclination angle of the continental slope.

In contrast, non-accretionary margins, which develop adjacent to trenches with limited sediment accumulation, usually host steep continental slopes with upper-plate faulting dominated by extensional tectonics often expressed by margin-parallel normal fault scarps at the seafloor (von Huene and Scholl 1991; Clift and Vannucchi 2004). The large-scale structure and surface morphology of non-accretionary margins therefore support the unimpeded down-slope transport of slide material and possible deposition into the adjacent trench. Moreover, frontal tectonic erosion and subsequent underthrusting of the eroded material increase the possibility of MTD subduction.

Examples from MTDs in little-sedimented trenches

Figure 4b shows the bathymetric signature of the Jaco and Parrita scars, which both resulted from repeated submarine landsliding at an uplifted and oversteepened continental slope in the wake of subducting seamounts (von Huene *et al.* 2000; Harders *et al.* 2011). Seismic line SO81-13 (Fig. 4c), which runs perpendicular to the margin across Jaco scar and the adjacent trench, shows *c.* 300 m of layered and reflective pelagic sediment within the trench axis overlain by a *c.* 400 m thick unit of less reflective and partly chaotic sediment. The seismic image does not allow one to conclude whether the less reflective character of the shallow sediments results from small-scale faulting or if they correspond to the most recent and distal MTD related to submarine landsliding within Jaco scar. The pelagic sediment section seems to be completely subducted (also compare von Huene *et al.* 2000). Due to the chaotic nature of the shallower sediments it remains unclear how much of them are incorporated into a frontal prism and how much are subducted.

Ongoing and long-lasting subduction of Jaco slide deposits is, however, suggested by mass-balance calculations for the margin segment affected by the subducting seamount. Based on reconstruction of the undisturbed geometry of the continental slope (i.e. if not affected by a subducting seamount), Hühnerbach *et al.* (2005) suggested that around 20–23 km^3 of the slide MTD were deposited in the trench and subsequently subducted. In contrast, about 80 km^3 (75–80% of the total material involved in landsliding within the scar) were deposited on the continental slope and did not reach the trench (Hühnerbach *et al.* 2005). The latter is supported by chaotic seismic sequences in the shallow subsurface of the marine forearc which have been interpreted as MTDs (von Huene *et al.* 2000). Similar volumes of MTDs have likely been subducted at other locations along the margin (e.g. Parrita scar, Tarcoles scar, Nicoya slump) where seamount subduction resulted in large-scale instability of the continental slope (von Huene *et al.* 2000, 2004; Hühnerbach *et al.* 2005; Harders *et al.* 2011).

In summary, along little-sedimented trenches the chances for an MTD to become subducted are particularly high. They are usually located in front of non-accretionary margins where the plate boundary develops in the upper plate or at the top of the subducting plate. Due to the lack of significant amounts of trench sediment an MTD sourced from the marine forearc is deposited on top of (or close to) the pelagic sediments. In such a setting, the only possibility to not subduct the MTD (or at least a fraction of it) is when the plate boundary develops on top of (or close to) the oceanic basement. However, oceanic basement usually exhibits a rough morphology, including first-order structural features such as spreading-related abyssal hills, bending-related horst-and-graben, basement ridges, seamounts and oceanic fracture zones (e.g. Geersen 2019). A plate boundary that strictly follows a rough basement topography and that avoids any subducted sediment might be difficult to develop kinetically.

Examples from MTDs in moderately to heavily sedimented trenches

Figure 3f shows the seismic signature of the MTD related to the Northern Giant Slope Failure (for bathymetric expression see Fig. 3e), which occurred off the coast of Central Chile around 0.4 Ma (Geersen *et al.* 2011*a*; Völker *et al.* 2012). The MTD can be traced as a chaotic seismic unit up to 500 m thick about 30 km seaward of the deformation front. As the oceanic Nazca Plate has moved about 25 km since the slide occurred, the original extent of the MTD at the time of the slope failure was possibly an additional 55 km seaward of the deformation front and the thickness of the already subducted MTD likely largely exceeded 500 m. At *c.* 15 km from the deformation front, the MTD is located directly above the top of the oceanic basement while the maximum sediment thickness at the trench depression, including the MTD, ranges about 1.5–2.0 km. A high-amplitude seismic reflection in the marine forearc, which likely represents the shallow plate boundary, indicates that the MTD is currently subducted.

Another giant active margin landslide, which occurred in a comparable geological setting (accretionary margin, partly sediment-filled trench), is the Ruatoria Debris Avalanche at the Hikurangi margin (New Zealand) (Fig. 2a; Collot *et al.* 2001; Barnes *et al.* 2018). The maximum thickness of the MTD related to the slope failure ranges around 2000 m (Collot *et al.* 2001) on an average trench sediment thickness of *c.* 3 km (see Scholl *et al.* 2015). The Ruatoria Debris Avalanche moved *c.* 2000 km^3 of slope sediments and basement rocks (Collot *et al.* 2001), which have been recycled back into the subduction zone (Barnes *et al.* 2018).

The previously discussed giant active margin landslides deposited material over tens of kilometres seaward of the deformation front. However, the MTDs of many small- and medium-sized landslides that occur in comparable geological settings are often spatially confined to the region close to the deformation front. The Reloca Slide off Central Chile (Fig. 3a) affected around 24 km^3 of the lowermost continental slope (Völker *et al.* 2009) and the maximum runout distance of the MTD is 18 km seaward of the deformation front. As a consequence, the MTD of the Reloca Slide is confined to shallow (max. 200 m) depths within the trench sediment section (Fig. 3d; also compare Völker *et al.* 2009). Due to the presence of underlying trench and pelagic sediment *c.* 2300 m thick (Völker *et al.* 2013), the only structural framework that would lead to a subduction of the Reloca Slide MTD is a very thick (>2100 m) section of subducting sediment. Another example of a lower-slope failure with limited runout (*c.* 20–30 km) is the 44-North Slide, offshore Oregon (Tominaga *et al.* 2017; Fig. 5). The associated MTD covers an area of *c.* 100 km^2, reaching a thickness of up to 275 m (Lenz *et al.* 2019). More than 2000 m of underlying trench and pelagic sediments and a décollement position close to the oceanic basement (MacKay 1995) make it unlikely that the 44-North Slide alters the characteristics of the plate boundary. However, another chaotic seismic unit is visible in the trench closely above the oceanic basement. This unit, which may correspond to the MTD of the huge Heceta Slide (Goldfinger *et al.* 2000), is likely subducted together with the oceanic plate.

Generally, the chances for an MTD to become subducted in heavily sedimented trenches decrease with increasing thickness of the underlying trench sediments. Usually these are accretionary margins where the plate boundary develops within, or at the base of the trench sedimentary cover. For MTDs located at the seafloor or at shallow depth within the trench sediment section (e.g. Reloca Slide, Fig. 3d) there is a high chance that the plate boundary will form somewhere below and that the MTD will be incorporated into the accretionary prism.

However, large MTDs with a long runout (e.g. the Giant Slope Failure off Chile or the Ruataroria Debris Avalanche off New Zealand) can be deposited seaward of the point where the trench sediments onlap the pelagic sediments on the outer trench slope (e.g. Fig. 3f). In this case they are deposited on top of the pelagic sediment section. On their journey back towards the subduction zone their relative depth within the trench stratigraphy increases continuously. In general, the further the runout distance of the MTD, the deeper it will be located within the trench sediment section when arriving at the deformation front and the larger its chances to become subducted. Yet it has to be noted that in some heavily sedimented trenches, the thick sediments extend far on to the oceanic plate. In the case of Makran about 5 km of sediment rest on top of the oceanic basement at a distance of *c.* 100 km from the deformation front (Fruehn *et al.* 1997). Off northernmost Sumatra the thickness of the Bengal and Nicobar Fan sediments is *c.* 3 km at the same distance (Geersen *et al.* 2015). In these cases even a very long runout of an MTD would not significantly increase its chances to become subducted.

Other sources for MDTs in the ocean basins

In the previous sections we discussed the recycling of MTDs sourced from the marine forearc. MTDs that enter a subduction zone can, however, also be sourced from other geological domains within the oceanic basins. Figure 2b shows the deposits of the giant Nuuanu and Wailau Landslides off the Hawaiian Islands. The MTDs, which are up to 900 m thick, resulted from the collapse of Koolau

Volcano on Oahu (Nuuanu) and East Molokai Volcano (Wailau) (Moore *et al.* 1989, 1994; Garcia *et al.* 2006). The Nuuanu landslide occurred in multiple phases, with the initial slope failure between 1.7 and 2.6 Ma (Moore *et al.* 1994; Garcia *et al.* 2006). The runout of the slide debris is more than 200 km, and the slide mass includes huge individual blocks such as the Tuscaloosa Seamount, which is *c.* 30 km long, 17 km wide and more than 2 km tall. The sliding material likely eroded the pelagic sediment of the Pacific Plate and the MTD now sits above the Cretaceous oceanic basement. Although the MTDs are currently located about 6000 km (*c.* 67 Ma) away from the Izu-Bonin trench they will likely enter a subduction zone and become subducted at some point in their geological future. Other examples of MTDs within the ocean basins include submarine slope failures at the Owen Fracture Zone in the Indian Ocean (30–220 m thick MTDs; Rodriguez *et al.* 2012) or the Lomonosov Ridge in the Arctic Ocean (*c.* 200 m thick MTDs; Kristoffersen *et al.* 2007).

MTDs that are deposited on the oceanic plate before arriving at a convergent margin are at the same stratigraphic level as the pelagic sediments. Located deep within the trench stratigraphy, the chances are high that they will be part of the subducting sediment section. This even accounts for subduction zones adjacent to heavily sedimented trenches (case c in Fig. 6e). Although they may not be as frequent as submarine landslides at continental margins, the examples discussed above indicate the possibility that MTDs sourced from within the ocean basins are subducted.

Discussion

The combined analysis of bathymetric images and several seismic lines allows development of a conceptual model (Fig. 6) on how the (1) location of the plate boundary décollement, (2) runout distance of the MTD, (3) thickness of the sedimentary trench-fill and (4) subducting sediment thickness control whether an MTD is accreted to the upper plate or subducted with the down-going plate (see section 'Conceptual model on MTD subduction in convergent margins'). The knowledge gained can then be used to identify convergent margins at which MTDs are more or less likely to be subducted (see section 'Global evaluation of convergent margins affected by MTD subduction').

Conceptual model on MTD subduction in convergent margins

The cases discussed indicate that MTDs are likely subducted in several subduction zones. However, the subduction of an MTD does not automatically imply that it exerts an influence on plate boundary seismicity. This will happen only if the subducted MTD interacts with the plate boundary shear zone and represents a significant portion of it (Festa *et al.* 2018). Direct measurements of plate boundary thicknesses between 0 and 15 km depths in modern and ancient convergent margins show that the total thickness of the shear zone increases to 100–350 m at 1–2 km below seafloor, and this thickness is maintained to a depth of *c.* 15 km (Rowe *et al.* 2013). Therefore, MTD thicknesses of up to some hundreds to thousands of metres, as observed off Central Chile (Geersen *et al.* 2011*a*; Fig. 3f), Cascadia (Goldfinger *et al.* 2000) and New Zealand (Collot *et al.* 2001; Barnes *et al.* 2018), as well as in ancient convergent margins (Festa *et al.* 2016, 2018), are prone to potentially modify the structure and physical state of the shallow plate boundary.

The chances for an MTD to become subducted are always high in subduction zones where the plate boundary develops within the upper plate or at its base (non-accretionary margins) (Fig. 6c). In little-sedimented trenches an MTD is often located directly on top or closely above the oceanic basement. This implies that even for thin subducting sediment sections, the chances are high that the MTD is subducted and possibly interacts with the plate boundary shear zone (Fig. 6c).

A long runout distance of the MTD increases its chances to get subducted if the trench is filled with sediment (case c in Fig. 6d & e). The further away from the deformation front the mobilized material is deposited the thinner is usually the underlying sediment section. On its way back towards the deformation front (i.e. while riding passively on the oceanic plate), the relative depth of the MTD within the trench stratigraphy increases successively. If deposited beyond (seaward of) the onlap point of the trench sediments on the pelagic sediments (case c in Fig. 6d), the MTD will be located directly above the pelagic sediments.

In heavily sedimented trenches, or in cases where the MTD covers only a short runout distance, the section of sediment under the MTD is usually thick. In these cases, an MTD will only be subducted if the thickness of the subducting sediments exceeds the thickness of the sediments under the MTD. The likelihood of the interaction with the plate-boundary shear-zone depends on the ratio between MTD thickness and trench sediment thickness (Fig. 6d & e).

An MTD which is sourced from somewhere within an ocean basin (e.g. seamount, oceanic island, oceanic fracture-zone, basement ridge) is usually deposited on the oceanic plate before the trench sediments start to accumulate (Fig. 6a). This implies that its chances to become subducted are high due to its location directly on top of or closely above the

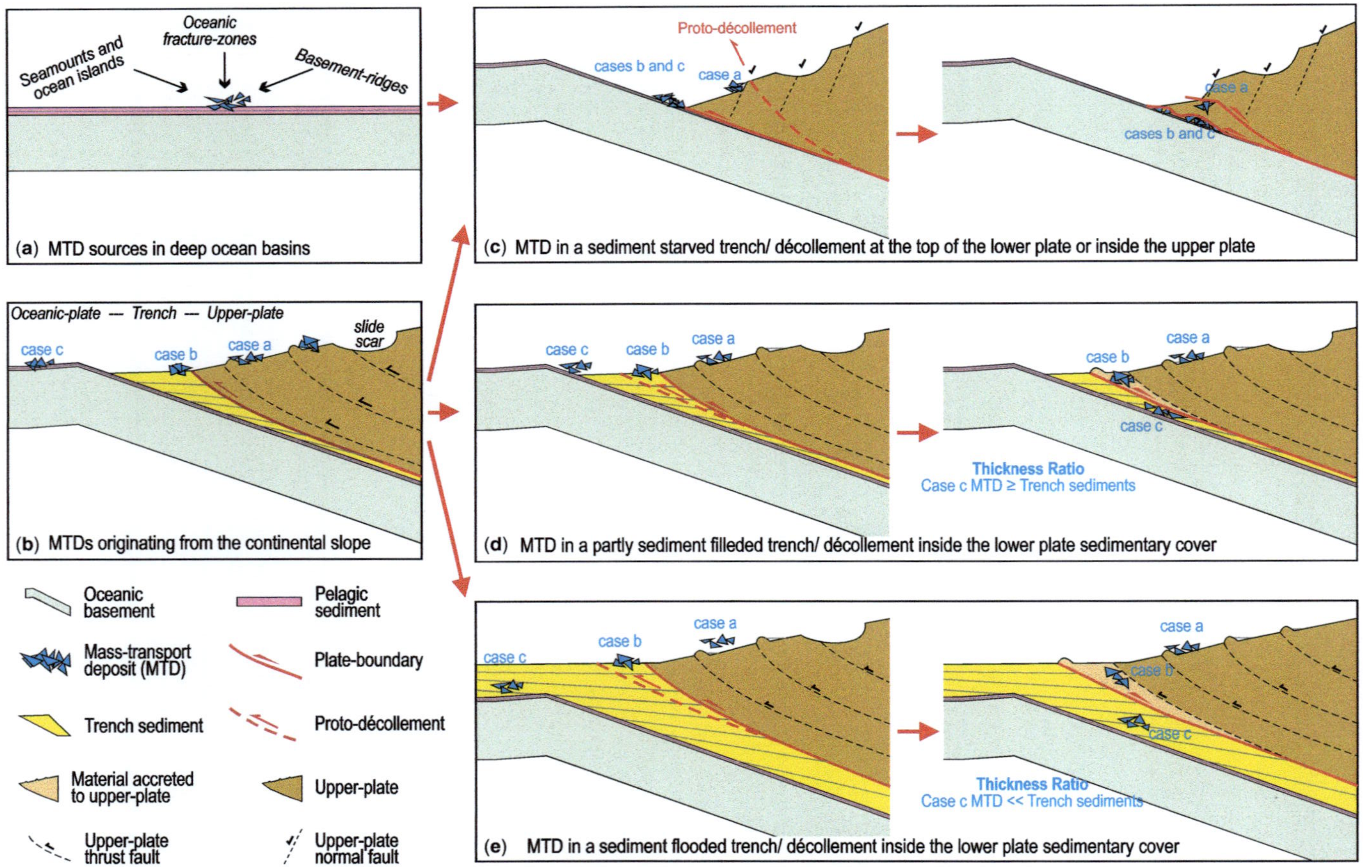

Fig. 6. Conceptual model of the structural controls on MTD subduction in convergent margins. (**a**) MTDs sourced from geological domains within the deep ocean basins are usually deposited on top of the pelagic sediments or the oceanic basement. When arriving at a subduction zone the MTDs are located deep within the trench stratigraphy and thus likely subducted. (**b**) Locations where MTDs sourced from the continental slope might be deposited. (**c**) Recycling and subduction of MTDs in an erosive margin adjacent to a sediment-starved trench. (**d**) Recycling and subduction of MTDs in an accretionary margin adjacent to a partly sediment-filled trench. (**e**) Recycling and subduction of MTDs in an accretionary margin adjacent to a sediment-flooded trench.

oceanic basement. However, increasing the thickness of the trench sediments decreases the likelihood that the MTD will interact with the plate boundary shear zone (case c in Fig. 6d & e).

Global evaluation of convergent margins affected by MTD subduction

Figure 7 summarizes trench sediment thickness adjacent to the global subduction zones (modified from Festa *et al.* 2018). The values need to be considered as an average over a large margin segment and do not capture short wavelength variations as, for example, induced by topographic features (seamounts, basement ridges, fracture zones) on the oceanic plates. Figure 3f shows how a seamount on the oceanic plate locally reduces sediment thickness in the Central Chilean trench.

The length of all subduction zone segments shown in Figure 7 is 49 270 km. From these about 8% are located adjacent to the heavily sedimented trenches (sediment thickness >4 km) in the Mediterranean, off northern Sumatra to Myanmar, and off Iran and Pakistan (Makran). Based on the conditions discussed above we conclude that subduction of MTDs originating from the continental slope is unlikely in these subduction zones.

Another 23% of global subduction zones are linked to oceanic trenches hosting 2–4 km thick sediments. The Northern Giant Slope Failure (Central Chile, Fig. 3e–f) generated a *c.* 500 m thick MTD within a 1.5–2 km thick trench sediment section. From seismic reflection data there is evidence that the décollement of Central Chile develops at shallow depth within the trench sediments and that *c.* 2 km of trench sediments are subducted with the downgoing plate (Olsen *et al.* 2017; also compare Fig. 3f). The deposits of the Northern Giant Slope Failure, as well as other large Central Chilean landslides (Fig. 3b, also compare Geersen *et al.* 2011*a*; Völker *et al.* 2012), are thus likely included in the subducting sediment section. Other examples for large MTDs (some hundreds of metres thick) within 2.0–4.0 km thick trench sediments include the deposits of the Blanco, Coos Basin and Heceta slides from the Cascadia subduction zone (Fig. 5a, also compare Goldfinger *et al.* 2000). As is the case in Central Chile, a subducting sediment section (up to 1.7 km thick) of Cascadia (Han *et al.* 2017) possibly allows the subduction of at least a significant fraction of these MTDs.

From the remaining subduction zones, 17% are located adjacent to oceanic trenches with 1–2 km of sediment and 52% develop in places where less than 1 km of sediment is found in the trench depression. For these subduction zones, the incorporation of MTDs in the subducting sediment section might be a common phenomenon, at least in places where MTDs are located in the trench. Frontal tectonic erosion of an upper plate made at least partly of MTDs (Fig. 6c) may add to the overall volume of subducting MTDs in these settings.

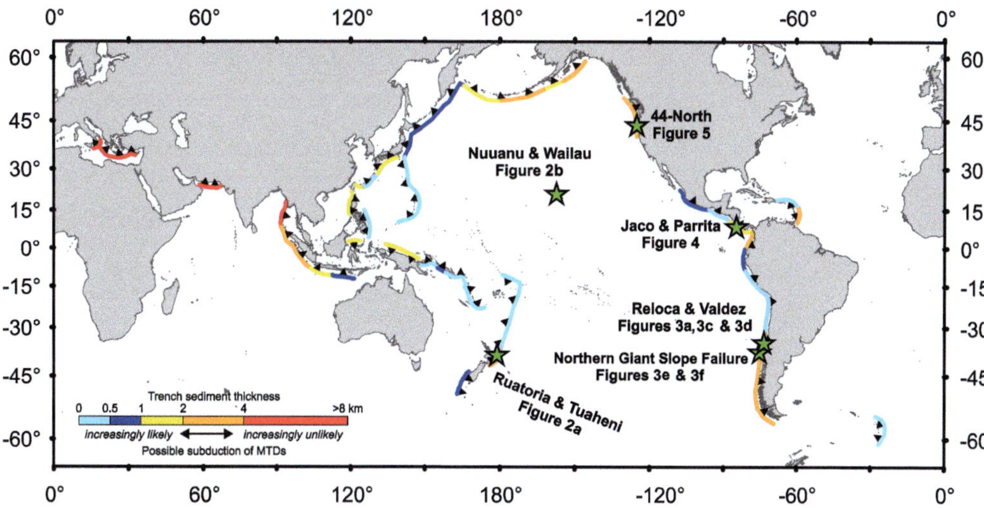

Fig. 7. Trench sediment thickness for global subduction-zones. The length of all subduction-zone segments shown in Figure 5 is 49 270 km. From these about 8% are located adjacent to trenches with >4 km of sediment fill, 23% are located adjacent to trenches with 2–4 km sediment fill, 17% are located adjacent to trenches with 1–2 km sediment fill, and 52% are located adjacent to trenches with <1 km sediment fill. Green stars indicate the locations of submarine landslides shown in Figures 2–5.

Conclusions

It has been suggested that subducting MTDs in convergent margins can alter plate boundary seismicity (Sage *et al.* 2006; Geersen *et al.* 2013; Festa *et al.* 2018). By reviewing marine geophysical images of MTDs of different sizes in modern subduction zones we evaluate the structural boundary conditions that control whether an MTD in a trench is subducted, and therefore may potentially exert an influence on the structure and physical state of the plate boundary. The likelihood that an MTD in a trench is subducted increases with (a) increasing size of the MTD, (b) decreasing thickness of the trench sediments, (c) increasing runout distance of the MTDs and (d) increasing thickness of the subducting sediment section. At the present state of knowledge, however, the formulation of a concrete cause-and-effect relationship between a subducting MTD and a distinct earthquake pattern proves difficult. A better understanding of the mechanical influence of a subducting MTD on the physical state of the plate boundary is usually hampered by the inaccessibility of subduction plate boundaries and challenges associated with detailed seismic imaging of plate boundary structure and composition under the marine forearc. Advances in marine geophysical imaging and detailed field studies on exhumed plate boundaries, possibly in concert with numerical modelling of plate boundary deformation, are required to overcome this shortcoming.

Acknowledgements We extend our sincere thanks to G.F. Moore and J. Chaytor for their constructive and thorough reviews, from which we have benefited greatly in revising our manuscript, and to the Editor D. Gamboa for his careful editorial handling.

Funding This work was supported by research grants from the Italian Ministry of Education, University and Research ('Finanziamento annuale individuale delle attività base di ricerca' 2017) and the Università di Torino (Ricerca Locale 'ex 60%' 2017–19) to AF, and Università di Modena e Reggio Emilia (FAR2017) to FR.

Author contributions **JG**: conceptualization (lead), formal analysis (lead), investigation (lead), methodology (lead), visualization (lead), writing – original draft (lead); **AF**: conceptualization (supporting), formal analysis (supporting), investigation (supporting), methodology (supporting), visualization (supporting), writing – original draft (supporting); **FR**: conceptualization (supporting), formal analysis (supporting), investigation (supporting), methodology (supporting), visualization (supporting), writing – original draft (supporting).

References

Barnes, P., Gamboa, D., Bell, R.E., Moore, G.F., Mountjoy, J.J., Paganoni, M. and Rabinowitz, H. 2018. Revisiting the giant Ruatoria Debris Flow on the Hikurangi Margin, New Zealand: results from IODP Expeditions 372 and 375, Site U1520. *In*: *AGU Fall Meeting Abstracts*, 10–14 December 2018, Washington, D.C., USA.

Beall, A., Fagereng, Å. and Ellis, S. 2019. Strength of strained two-phase mixtures: application to rapid creep and stress amplification in subduction zone mélange. *Geophysical Research Letters*, **46**, 169–178, https://doi.org/10.1029/2018GL081252

Bilek, S.L. and Lay, T. 2018. Subduction zone megathrust earthquakes. *Geosphere*, **14**, https://doi.org/10.1130/GES01608.1

Clift, P. and Vannucchi, P. 2004. Controls on tectonic accretion versus erosion in subduction zones: implications for the origin and recycling of the continental crust. *Reviews of Geophysics*, **42**, https://doi.org/10.1029/2003RG000127

Collot, J.Y., Lewis, K., Lamarche, G. and Lallemand, S. 2001. The giant Ruatoria debris avalanche on the northern Hikurangi margin, New Zealand: result of oblique seamount subduction. *Journal of Geophysical Research: Solid Earth*, **106**, 19271–19297, https://doi.org/10.1029/2001JB900004

Contreras-Reyes, E. and Carrizo, D. 2011. Control of high oceanic features and subduction channel on earthquake ruptures along the Chile–Peru subduction zone. *Physics of the Earth and Planetary Interiors*, **186**, 49–58, https://doi.org/10.1016/j.pepi.2011.03.002

Fagereng, Å. 2011. Geology of the seismogenic subduction thrust interface. *Geological Society, London, Special Publications*, **359**, 55–76, https://doi.org/10.1144/SP359.4

Fagereng, Å. and Sibson, R.H. 2010. Melange rheology and seismic style. *Geology*, **38**, 751–754, https://doi.org/10.1130/G30868.1

Festa, A., Ogata, K., Pini, G.A., Dilek, Y. and Alonso, J.L. 2016. Origin and significance of olistostromes in the evolution of orogenic belts: a global synthesis. *Gondwana Research*, **39**, 180–203, https://doi.org/10.1016/j.gr.2016.08.002

Festa, A., Dilek, Y., Mittempergher, S., Ogata, K., Pini, G.A. and Remitti, F. 2018. Does subduction of mass transport deposits (MTDs) control seismic behavior of shallow-level megathrusts at convergent margins? *Gondwana Research*, **60**, 186–193, https://doi.org/10.1016/j.gr.2018.05.002

Festa, A., Pini, G.A., Ogata, K. and Dilek, Y. 2019. Diagnostic features and field-criteria in recognition of tectonic, sedimentary and diapiric mélanges in orogenic belts and exhumed subduction–accretion complexes. *Gondwana Research*, **74**, 7–30, https://doi.org/10.1016/j.gr.2019.01.003

Fruehn, J., White, R.S. and Minshull, T.A. 1997. Internal deformation and compaction of the Makran accretionary wedge. *Terra Nova*, **9**, 101–104, https://doi.org/10.1046/j.1365-3121.1997.d01-13.x

Garcia, M.O., Sherman, S.B., Moore, G.F., Goll, R., Popova-Goll, I., Natland, J.H. and Acton, G. 2006.

Frequent landslides from Koolau Volcano: results from ODP Hole 1223A. *Journal of Volcanology and Geothermal Research*, **151**, 251–268, https://doi.org/10.1016/j.jvolgeores.2005.07.035

Geersen, J. 2019. Sediment-starved trenches and rough subducting-plates are conducive to tsunami earthquakes. *Tectonophysics*, **762**, 28–44, https://doi.org/10.1016/j.tecto.2019.04.024

Geersen, J., Völker, D., Behrmann, J.H., Reichert, C. and Krastel, S. 2011*a*. Pleistocene giant slope failures offshore Arauco peninsula, southern Chile. *Journal of the Geological Society, London*, **168**, 1237–1248, https://doi.org/10.1144/0016-76492011-027

Geersen, J., Behrmann, J.H., Völker, D., Krastel, S., Ranero, C.R., Diaz-Naveas, J. and Weinrebe, W. 2011*b*. Active tectonics of the South Chilean marine fore arc (35 S–40 S). *Tectonics*, **30**, https://doi.org/10.1029/2010TC002777

Geersen, J., Völker, D., Behrmann, J.H., Kläschen, D., Weinrebe, W., Krastel-Gudegast, S. and Reichert, C. 2013. Seismic rupture during the 1960 Great Chile and the 2010 Maule earthquakes limited by a giant Pleistocene submarine slope failure. *Terra Nova*, **25**, 472–477, https://doi.org/10.1111/ter.12060

Geersen, J., Bull, J.M., McNeill, L.C., Henstock, T.J., Gaedicke, C., Chamot-Rooke, N. and Delescluse, M. 2015. Pervasive deformation of an oceanic plate and relationship to large> Mw 8 intraplate earthquakes: the northern Wharton Basin, Indian Ocean. *Geology*, **43**, 359–362, https://doi.org/10.1130/G36446.1

Goldfinger, C., Kulm, L.D., McNeill, L.C. and Watts, P. 2000. Super-scale failure of the southern Oregon Cascadia margin. *Pure and Applied Geophysics*, **157**, 1189–1226, https://doi.org/10.1007/s0002400 50023

Goldfinger, C., Nelson, C.H. *et al.* 2012. *Turbidite event history: Methods and implications for Holocene paleoseismicity of the Cascadia subduction zone*. US Geological Survey Professional Papers, **1661-F**.

Haflidason, H., Sejrup, H.P. *et al.* 2004. The Storegga Slide: architecture, geometry and slide development. *Marine Geology*, **213**, 201–234, https://doi.org/10.1016/j.margeo.2004.10.007

Han, S., Bangs, N.L., Carbotte, S.M., Saffer, D.M. and Gibson, J.C. 2017. Links between sediment consolidation and Cascadia megathrust slip behaviour. *Nature Geoscience*, **10**, 954, https://doi.org/10.1038/s41561-017-0007-2

Harders, R., Ranero, C.R., Weinrebe, W. and Behrmann, J.H. 2011. Submarine slope failures along the convergent continental margin of the Middle America Trench. *Geochemistry, Geophysics, Geosystems*, **12**, https://doi.org/10.1029/2010GC003401

Hühnerbach, V., Masson, D.G., Bohrmann, G., Bull, J.M. and Weinrebe, W. 2005. Deformation and submarine landsliding caused by seamount subduction beneath the Costa Rica continental margin – new insights from high-resolution sidescan sonar data. *Geological Society, London, Special Publications*, **244**, 195–205, https://doi.org/10.1144/GSL.SP.2005.244.01.12

Kristoffersen, Y., Coakley, B.J., Hall, J.K. and Edwards, M. 2007. Mass wasting on the submarine Lomonosov Ridge, central Arctic Ocean. *Marine Geology*, **243**, 132–142, https://doi.org/10.1016/j.margeo.2007.04.012

Lenz, B., Sawyer, D., Phrampus, B., Davenport, K. and Long, A. 2019. Seismic imaging of seafloor deformation induced by impact from large submarine landslide blocks, offshore Oregon. *Geosciences*, **9**, 10, https://doi.org/10.3390/geosciences9010010

MacKay, M.E. 1995. Structural variation and landward vergence at the toe of the Oregon accretionary prism. *Tectonics*, **14**, 1309–1320, https://doi.org/10.1029/95TC02320

Moore, G.F. and Karig, D.E. 1976. Development of sedimentary basins on the lower trench slope. *Geology*, **4**, 693–697, https://doi.org/10.1130/0091-7613(1976)4<693:DOSBOT>2.0.CO;2

Moore, J.G., Clague, D.A., Holcomb, R.T., Lipman, P.W., Normark, W.R. and Torresan, M.E. 1989. Prodigious submarine landslides on the Hawaiian Ridge. *Journal of Geophysical Research: Solid Earth*, **94**, 17465–17484, https://doi.org/10.1029/JB094iB12p17465

Moore, J.G., Normark, W.R. and Holcomb, R.T. 1994. Giant Hawaiian landslides. *Annual Review of Earth and Planetary Sciences*, **22**, 119–144, https://doi.org/10.1146/annurev.ea.22.050194.001003

Moreno, M.S., Bolte, J., Klotz, J. and Melnick, D. 2009. Impact of megathrust geometry on inversion of coseismic slip from geodetic data: application to the 1960 Chile earthquake. *Geophysical Research Letters*, **36**, https://doi.org/10.1029/2009GL039276

Moreno, M., Melnick, D., Rosenau, M., Baez, J., Klotz, J., Oncken, O. and Socquet, A. 2012. Toward understanding tectonic control on the Mw 8.8 2010 Maule Chile earthquake. *Earth and Planetary Science Letters*, **321**, 152–165, https://doi.org/10.1016/j.epsl.2012.01.006

Moscardelli, L. and Wood, L. 2016. Morphometry of mass-transport deposits as a predictive tool. *GSA Bulletin*, **128**, 47–80, https://doi.org/10.1130/B31221.1

Mountjoy, J.J., McKean, J., Barnes, P.M. and Pettinga, J.R. 2009. Terrestrial-style slow-moving earthflow kinematics in a submarine landslide complex. *Marine Geology*, **267**, 114–127, https://doi.org/10.1016/j.margeo.2009.09.007

Ogata, K., Festa, A., Pini, G.A., Pogačnik, Ž. and Lucente, C.C. 2019. Substrate deformation and incorporation in sedimentary mélanges (olistostromes): examples from the northern Apennines (Italy) and northwestern Dinarides (Slovenia). *Gondwana Research*, **74**, 101–125, https://doi.org/10.1016/j.gr.2019.03.001

Ogata, K., Festa, A., Pini, G.A. and Alonso, J.L. 2020. Submarine landslide deposits in orogenic belts: olistostromes and sedimentary mélanges. *In*: Ogata, K., Festa, A. and Pini, G.A. (eds) *Submarine Landslides: Subaqueous Mass Transport Deposits from Outcrop to Seismic Profiles*. American Geophysical Union, Geophysical Monograph **247**, 3–26.

Olsen, K., Bangs, N.L., Arnulf, A.F., Trehu, A.M. and Contreras Reyes, E. 2017. Development of a shallow décollement along the south-central Chile Margin from 2D seismic reflection data. *In*: *AGU Fall Meeting Abstracts*, December.

Pini, G.A. 1999. *Tectonosomes and olistostromes in the Argille Scagliose of the Northern Apennines, Italy*. Geological Society of America Special Papers, **335**.

Pini, G.A., Ogata, K., Camerlenghi, A., Festa, A., Lucente, C.C. and Codegone, G. 2012. Sedimentary mélanges and fossil mass-transport complexes: a key for better understanding submarine mass movements? *In*: *Submarine Mass Movements and Their Consequences*. Springer, Dordrecht, 585–594.

Remitti, F., Vannucchi, P., Bettelli, G., Fantoni, L., Panini, F. and Vescovi, P. 2011. Tectonic and sedimentary evolution of the frontal part of an ancient subduction complex at the transition from accretion to erosion: the case of the Ligurian wedge of the northern Apennines, Italy. *GSA Bulletin*, **123**, 51–70, https://doi.org/10.1130/B30065.1

Rodriguez, M., Fournier, M., Chamot-Rooke, N., Huchon, P., Zaragosi, S. and Rabaute, A. 2012. Mass wasting processes along the Owen Ridge (northwest Indian Ocean). *Marine Geology*, **326**, 80–100, https://doi.org/10.1016/j.margeo.2012.08.008

Rowe, C.D., Moore, J.C., Remitti, F. and the IODP Expedition 343 Science Party 2013. The thickness of subduction plate boundary faults fromthe seafloor into the seismogenic zone. *Geology*, **41**, 991–994.

Sage, F., Collot, J.Y. and Ranero, C.R. 2006. Interplate patchiness and subduction-erosion mechanisms: evidence from depth-migrated seismic images at the central Ecuador convergent margin. *Geology*, **34**, 997–1000, https://doi.org/10.1130/G22790A.1

Scholl, D.W., Kirby, S.H., von Huene, R., Ryan, H., Wells, R.E. and Geist, E.L. 2015. Great (≥Mw8.0) megathrust earthquakes and the subduction of excess sediment and bathymetrically smooth seafloor. *Geosphere*, **11**, 236–265, https://doi.org/10.1130/GES01079.1

Tominaga, M., Trehu, A. and Lyle, M. 2017. Multichannel seismic shot data from the Seismic Early Career Chief Scientist Training Cruise 2017, Cascadia Margin, acquired during R/V Roger Revelle expedition RR1718 (2017). Interdisciplinary Earth Data Alliance (IEDA), https://doi.org/10.1594/IEDA/324504

Urgeles, R. and Camerlenghi, A. 2013. Submarine landslides of the Mediterranean Sea: trigger mechanisms, dynamics, and frequency-magnitude distribution. *Journal of Geophysical Research: Earth Surface*, **118**, 2600–2618, https://doi.org/10.1002/2013JF002720

Völker, D., Weinrebe, R.W., Behrmann, J.H., Bialas, J. and Klaeschen, D. 2009. Mass wasting at the base of the south central Chilean continental margin: the Reloca Slide. *Advances in Geosciences*, **22**, 155–167, https://doi.org/10.5194/adgeo-22-155-2009

Völker, D., Geersen, J., Behrmann, J.H. and Weinrebe, W.R. 2012. Submarine mass wasting off Southern Central Chile: distribution and possible mechanisms of slope failure at an active continental margin. *In*: *Submarine Mass Movements and Their Consequences*. Springer, Dordrecht, 379–389.

Völker, D., Geersen, J., Contreras-Reyes, E. and Reichert, C. 2013. Sedimentary fill of the Chile Trench (32–46° S): volumetric distribution and causal factors. *Journal of the Geological Society, London*, **170**, 723–736, https://doi.org/10.1144/jgs2012-119

von Huene, R. and Scholl, D.W. 1991. Observations at convergent margins concerning sediment subduction, subduction erosion, and the growth of continental crust. *Reviews of Geophysics*, **29**, 279–316, https://doi.org/10.1029/91RG00969

von Huene, R., Ranero, C.R., Weinrebe, W. and Hinz, K. 2000. Quaternary convergent margin tectonics of Costa Rica, segmentation of the Cocos Plate, and Central American volcanism. *Tectonics*, **19**, 314–334, https://doi.org/10.1029/1999TC001143

von Huene, R., Ranero, C.R. and Watts, P. 2004. Tsunamigenic slope failure along the Middle America Trench in two tectonic settings. *Marine Geology*, **203**, 303–317, https://doi.org/10.1016/S0025-3227(03)00312-8

Wang, K. and Bilek, S.L. 2011. Do subducting seamounts generate or stop large earthquakes? *Geology*, **39**, 819–822, https://doi.org/10.1130/G31856.1

Wang, K. and Bilek, S.L. 2014. Invited review paper: fault creep caused by subduction of rough seafloor relief. *Tectonophysics*, **610**, 1–24, https://doi.org/10.1016/j.tecto.2013.11.024

Wang, K. and Tréhu, A.M. 2016. Invited review paper: some outstanding issues in the study of great megathrust earthquakes – The Cascadia example. *Journal of Geodynamics*, **98**, 1–18, https://doi.org/10.1016/j.jog.2016.03.010

Wang, P.-L., Engelhart, S.E., Wang, K., Hawkes, A.D., Horton, B.P., Nelson, A.R. and Witter, R.C. 2013. Heterogeneous rupture in the great Cascadia earthquake of 1700 inferred from coastal subsidence estimates. *Journal of Geophysical Research*, **118**, 1–14, https://doi.org/10.1002/jgrb.50101

Weiss, R., Krastel, S., Anasetti, A. and Wünnemann, K. 2013. Constraining the characteristics of tsunami waves from deformable submarine slides. *Geophysical Journal International*, **194**, 316–321, https://doi.org/10.1093/gji/ggt094

Evaluating the sealing potential of young and thin mass-transport deposits: Lake Villarrica, Chile

Jasper Moernaut[1]*, Gauvain Wiemer[2], Achim Kopf[2] and Michael Strasser[1]

[1]Institute of Geology, University of Innsbruck, Innsbruck, Austria

[2]MARUM – Centre for Marine Environmental Sciences, University of Bremen, Bremen, Germany

JM, 0000-0003-3164-4274; MS, 0000-0003-3991-2405
*Correspondence: Jasper.Moernaut@uibk.ac.at

Abstract: Subaqueous mass-transport deposits (MTDs) can be important elements in hydrocarbon systems, forming potential reservoirs or seals. Most research has targeted outcrops or moderately to deeply buried MTDs and, therefore, the petrophysical properties of near-seafloor MTDs, and their influence in the trapping and release of shallow fluids, is poorly studied. Here, we investigate shallow MTDs in Lake Villarrica (Chile) by combining sub-bottom profiles, free-fall penetrometer data, pore pressure dissipation tests and geotechnical properties of sediment cores. Low undrained shear strength under a surficial MTD indicates underconsolidation caused by sudden loading and rapid sealing. Larger, buried MTDs show acoustic signatures of free gas at their base, indicating effective sealing. This is supported by degassing core gaps just below MTDs and by excess pore pressure ratios *c.* 30–70% within MTDs. Acoustic windows below rafted blocks suggest local fluid escape. MTDs exhibit elevated undrained shear strength and reduced porosity compared to surrounding sediments, but are comparable to upslope source sequences. This suggests that MTD sealing capacity in Villarrica relates to the apparently overconsolidated nature of the slope sequence, leaving a minor role for shear densification. This study shows that shallow MTDs can form a relatively rapid seal for fluid migration, locally degraded by rafted blocks.

Subaqueous mass-transport deposits (MTDs) can form economically relevant stratigraphic units with respect to hydrocarbon systems. In some cases MTDs act as potential oil and gas reservoirs because mass movements can transport sand-grained material into a mud-dominated basin sequence (Meckel 2011). In contrast, most muddy MTDs exhibit a high sealing competence because shear compaction and clay fabric alignment within the moving mass reduce porosity and permeability (Dugan 2012; Cardona *et al.* 2016; Sun *et al.* 2018). Consequently, MTDs in sediment cores often stand out in terms of measured physical properties (e.g. bulk density and shear strength) compared to surrounding sequences (Sawyer *et al.* 2009; Strasser *et al.* 2012). Due to their possible heterogeneous nature, MTDs can also locally compromise seal integrity, providing effective pathways for fluid escape (Gamboa and Alves 2015). Moreover, the morphological changes associated with slope failures and MTDs control the spatial distribution of overlying depositional units with reservoir capacity, such as those formed by sandy turbidite systems (Kneller *et al.* 2016).

Apart from their economic significance, gas trapping at the base of and within MTDs may also affect the stability of the MTD itself (Sun *et al.* 2017). For example, the Tuaheni landslide complex offshore New Zealand displays a variety of indicators for fluid migration and free gas accumulations, the presence of which may generate overpressure and facilitate a repeated slow creeping process of the landslide material (Gross *et al.* 2018). Other MTDs also act as barriers for gas migration and provide localized pathways for gas escape, creating fluid escape tunnels and pockmarks on the seabed (Bøe *et al.* 2012; Riboulot *et al.* 2013; Eng and Tsuji 2019).

In addition to trapping of migrating gas from below, pore fluid overpressure related to MTDs may also be generated by the mass movement process itself. The sudden loading of water-rich near-surface sediments immediately enhances pore pressure, which – due to the lower permeability of the MTD – may be preserved over long time periods and burial history (Diviacco *et al.* 2006; Dugan 2012). Another source of excess fluids can be the MTD itself as loose sediments tend to contract during shearing, inducing overpressures (Stoecklin *et al.* 2017). Products of MTD overpressure can be found on reflection seismic profiles in the form of acoustic wipe-outs in the sequences overlying the MTDs (Chapron *et al.* 2004; Diviacco *et al.* 2006; Praet *et al.* 2017), which – in some cases – can be attributed to past seismic shaking and may be used as a complementary palaeoseismic indicator (Moernaut *et al.* 2009, 2019).

Most research on MTD architecture has been conducted at outcrops or in moderately to deeply

From: Georgiopoulou, A., Amy, L. A., Benetti, S., Chaytor, J. D., Clare, M. A., Gamboa, D., Haughton, P. D. W., Moernaut, J. and Mountjoy, J. J. (eds) 2020. *Subaqueous Mass Movements and their Consequences: Advances in Process Understanding, Monitoring and Hazard Assessments*. Geological Society, London, Special Publications, **500**, 129–146.
First published online March 13, 2020, https://doi.org/10.1144/SP500-2019-155

buried MTDs (tens to hundreds of metres depth), some directly relevant to hydrocarbon systems, leaving many questions open about the role of MTDs on fluid migration during their early burial history (upper few metres). It is difficult to acquire such information from interpreting older, deeply buried MTDs, due to the tendency of losing diagnostic features with increasing consolidation (Alves *et al.* 2014). Here, we present a multi-method pilot study on three young earthquake-triggered MTDs in the glacigenic Lake Villarrica (Chile). We evaluate the potential of young MTDs to act as a seal for fluid migration, discuss which factors may compromise this capacity and outline the potential and pitfalls of investigating young MTDs in stratigraphic sequences.

Setting and previous research on landslides in Lake Villarrica

Lake Villarrica (39° 15′ S; 72° 02′ W in south-central Chile) is located in the South American subduction zone, which has hosted several great to giant megathrust earthquakes in recent times, such as in AD 1960 (M_w 9.5) along the Valdivia segment and in AD 2010 (M_w 8.8) along the Maule segment (Fig. 1a). Historical and palaeoseismic records suggest that events similar to the AD 1960 one at the Valdivia segment reoccur about every three centuries in a quasi-periodic manner (Cisternas *et al.* 2005; Moernaut *et al.* 2018). During such events, strong and prolonged ground shaking affects lacustrine slope sediments in the region, generating subaqueous landslides and surficial remobilization (i.e. removal of a few centimetres of thick veneer), leading to a repeated deposition of MTDs and turbidites in the lacustrine sequences (Van Daele *et al.* 2015; Moernaut *et al.* 2019).

Lake Villarrica is a large (21 × 9 km) and deep (167 m) glacigenic lake located at the western piedmont (at 214 m above sea-level (asl)) of the Andean Cordillera (Fig. 1b). The lake basin is comprised of a deep central basin and a shallower area with pronounced morphological variability in its southwestern part. This area yields several rock basement peaks, submerged moraine ridges and depositional

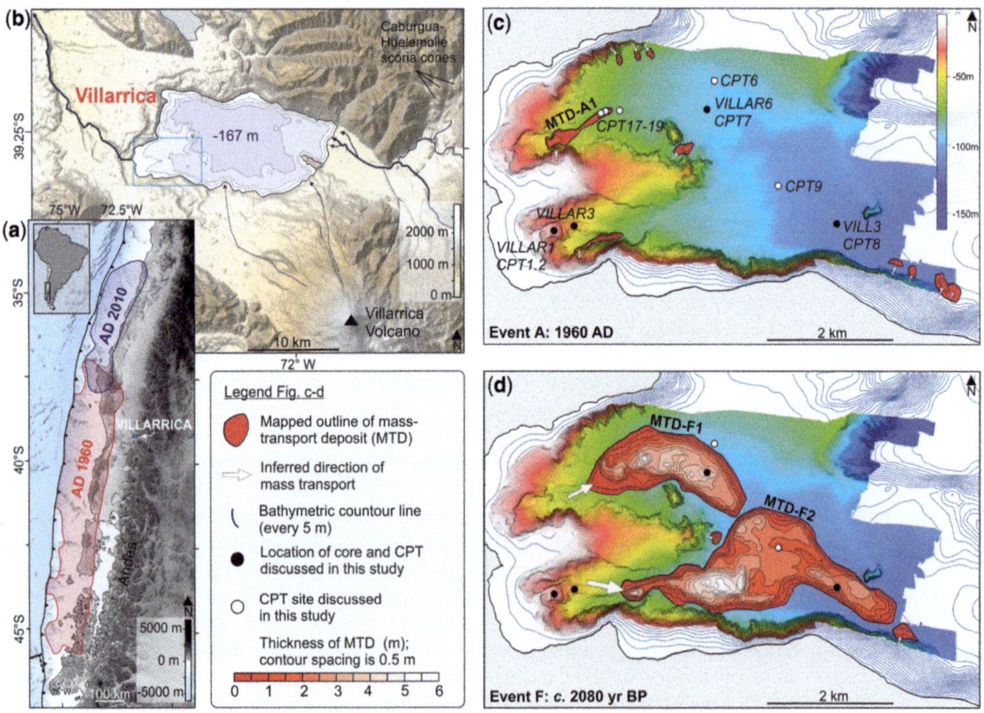

Fig. 1. Setting of Lake Villarrica and the studied mass-transport deposits MTD-A, -F1 and -F2. (**a**) South-central Chile showing the rupture area (>5 m slip) of the AD 1960 (M_w 9.5) and AD 2010 (M_w 8.8) megathrust earthquakes. (**b**) Local setting and bathymetry of Lake Villarrica, illustrating the study area in the southwestern part of the lake (blue rectangle), river network (arrows) and volcanoes. (**c–d**) Distribution and thickness of MTDs triggered by an earthquake in AD 1960 (event A) and in *c.* 2080 years BP (event F). Location of sediment cores and CPT sites discussed in this study are indicated as circles (see legend).

sub-basins (Fig. 1). The main fluvial input occurs at the lake's eastern shore. In the southwestern part, the acoustic basement (i.e. glacial till or bedrock) is covered by a thick and complex sedimentary sequence of up to about 120 m thick. Most of it was deposited in a glacier-proximal setting during the early stage of deglaciation, evidenced by the typical clastic varves encountered at the bottom of some sediment cores. Since *c.* 12.7 ka BP, a continuous drape of Holocene hemipelagic sediments has developed, reaching a thickness of 10–20 m (Moernaut *et al.* 2017). These background sediments consist mainly of diatomaceous mud with small amounts of organic matter, volcanic ash and terrigenous clays/silts. Tephra layers, lahar deposits and earthquake-triggered muddy turbidites frequently intercalate the basin sequence. Subaqueous slope stability has been extensively investigated by targeting an intact sedimentary platform on the upper slope, almost entirely bordered by slide scarps (Wiemer *et al.* 2015; Moernaut *et al.* 2017; location VILLAR1 on Fig. 1c). These studies focused on the *in situ* characterization of undrained shear strength and pore pressure, and an evaluation of the static and dynamic geotechnical properties of the main lithologies in the sequence. Considerable excess pore pressure ratios (up to 80–90% of the overburden stress) and weak, underconsolidated sediments were found, especially near areas of focused fluid escape visible on sub-bottom profiles. Such overpressure may have facilitated earthquake-triggered failure on the 2–4° dipping slopes. Moreover, the high permeability of a fine, sandy, tephra layer and the progressive failure behaviour of fine, silty, glacio-lacustrine sediments in the lower sequence may have strongly influenced the development of slope failures.

Seismic-stratigraphic analysis of the last 6 kyr of the sequence documented 11 stratigraphic levels (event A–M) on which multiple MTDs are located (Moernaut *et al.* 2019). Such coeval slope failure and MTD deposition was interpreted as a diagnostic signature of strong earthquake shaking. Voluminous MTDs ($1 - 6 \times 10^6$ m^3) were induced by three prehistoric earthquakes, whereas the giant AD 1960 earthquake induced only a few small MTDs ($<0.13 \times 10^6$ m^3). Most (small) landslides took place on slopes of 5–20°, whereas the few large (potentially tsunamigenic) landslides occurred exclusively on slopes <4°. In the present study, we focus on the characterization of one small MTD associated with event A (AD 1960) and two large MTDs associated with event F (*c.* 2080 years BP) (Fig. 1c, d).

Methods

The previous studies on subaqueous slope stability in Lake Villarrica (Wiemer *et al.* 2015; Moernaut *et al.* 2017) deployed the MARUM Free-Fall Piezocone Penetrometer (CPTU; Stegmann *et al.* 2006) and a percussion-driven piston corer from an anchored platform in December 2011. During the same fieldwork campaign, we also used these systems on selected buried mass-transport deposits, the results of which are presented in this study. The field data acquisition and processing were executed following the approaches outlined in the published slope stability studies and we present only the most relevant parameters here.

Bathymetric data were acquired using an R2Sonic 2024 multibeam system (200–400 kHz) in January 2013, using real-time kinematic (RTK) corrected GPS with horizontal accuracy of a decimetre (OmniSTAR G2). Pulse frequency and swath angle was adjusted on-the-fly depending on bottom morphology and water depth. Data processing – including corrections for vessel motion and orientation, sound velocity, sensor offsets and spikes – and visualization were performed using CARIS HIPS and SIPS 7.1 and standard GIS software. The presented bathymetric maps have a grid cell size of 3 m.

Dense networks of sub-bottom profiler data were acquired in 2001, 2007 and 2011 (Moernaut *et al.* 2019). A Kongsberg Geopulse sub-bottom profiler (3.5 kHz, 'pinger') was deployed from a towed cataraft system. Navigation and positioning were carried out using standard GPS, and data logging by an iXblue Delph Seismic system. Water and shallow subsurface depths were calculated using a constant acoustic velocity of 1500 m s^{-1}. Seismic-stratigraphic interpretation was performed using IHS Markit Kingdom Suite (v. 2018) after applying a frequency band-pass filter. Mass-transport deposits were identified based on their chaotic to transparent seismic facies and intercalated lens-shaped geometry, typical for frontally emergent landslides in lakes (Sammartini *et al.* 2019). Very strong reflections overlying zones of acoustic turbidity and blanking are interpreted as an indicator of free gas in the sediments, reflecting, scattering and attenuating the acoustic energy (e.g. Missiaen *et al.* 2002; Tóth *et al.* 2014).

Composite sediment cores were constructed by combining two overlapping drives of successive 3 m-long sections. Seven sites were cored, six of which have been presented in previous studies (Wiemer *et al.* 2015; Moernaut *et al.* 2017, 2019). Here, we introduce core site VILLAR6, located in the large, buried, MTD-F1 (Fig. 1c). The closed core was scanned with a Geotek multi-sensor core logger to obtain γ-density (i.e. bulk density) of the sediments. The split core was photographed and described sedimentologically. On a few selected sections, medical X-ray computed tomography (X-CT) scans were taken with a Siemens Somatom Definition Flash (Medical University Innsbruck), yielding a 0.20 × 0.20 × 0.30 mm voxel resolution. Natural water content was obtained by oven drying (>24 h at 60°) and solid volume by a helium pycnometer in order to derive grain density, bulk density and porosity. Undrained shear strength was measured using a motorized Wykeham-Farrance vane shear apparatus (rotation rate = 90° min^{-1}). Vane shear tests were also performed on core VILL3, located in the large buried MTD-F2 (Fig. 1c, d; Moernaut *et al.* 2019).

The Free-Fall Piezocone Penetrometer (CPTU) was used in a 5–8 m length configuration from an anchored platform (sites CPT1.2; CPT6–9) and in a 3 m configuration from a non-anchored small research vessel (sites CPT 17–19) (Fig. 1a). Only sites CPT1.2 and CPT6 have been studied previously (Moernaut *et al.* 2017). The CPTU measures cone resistance (q_c), sleeve friction (f_s) and pore pressure (u). The undrained shear strength (s_u) was derived as follows:

$$s_u = \frac{(q_{t,\text{quasi-stat}} - \sigma_{vo})}{N_{kt}} \tag{1}$$

where $q_{t,\text{quasi-stat}}$ is the pore pressure- and strain rate-corrected cone resistance, which is relative comparable to standard pushed CPT cone resistance; σ_{vo} is the total overburden stress, N_{kt} is an empirically determined cone factor (Lunne *et al.* 1997), here set to 14 (see Wiemer *et al.* 2015). The total overburden stress (σ_{vo}) was estimated as follows:

$$\sigma_{vo} = \gamma z \tag{2}$$

where γ is the unit weight of the sediments and z is the sediment depth. γ was taken as 13.9 kN m^{-3} based on the mean value from bulk density data. For illustrative clarity, we present smoothed s_u plots by applying a moving average of seven data points. More details on MARUM CPTU data processing, including the strain-rate correction for the dynamic (i.e. variable penetration rate) deployments, can be found in Steiner *et al.* (2014, 2015).

Ambient pore pressure at the deepest point of CPTU penetration at four sites was estimated using dissipation tests, monitoring the temporal evolution of the pore pressure signal after probe penetration. The test duration (t) was 40–45 min and the ambient pore pressure was estimated by extrapolating the incomplete dissipation curve towards 'infinity', using the 'inverse time' approach (Fig. 2). This

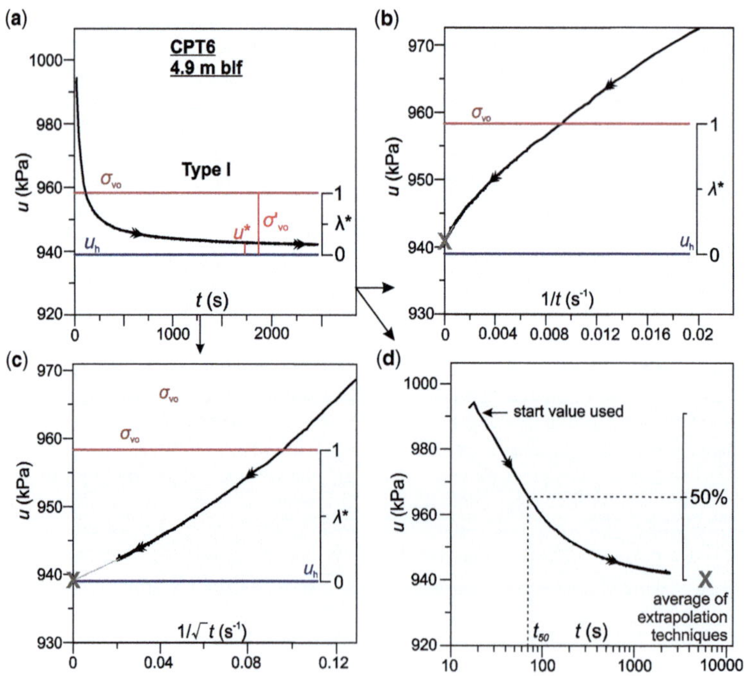

Fig. 2. (**a**) Example of a pore pressure dissipation test at site CPT6. After probe insertion, the pore pressure decreases towards a value between the hydrostatic pressure (u_h) and the overburden stress (σ_{vo}). (**b–c**) Extrapolation techniques $1/t$ and $1/\sqrt{t}$. The grey cross is the estimated ambient pore pressure. (**d**) Determination of t_{50} based on the pore pressure value at the start of the monotonic dissipation curve and the (extrapolated) ambient pore pressure. The grey cross is the average value obtained from both extrapolation techniques ($1/t$ and $1/\sqrt{t}$). m blf, metres below lake floor.

strategy includes the $1/t$ method and $1/\sqrt{t}$ method, which typically lead to a slight overestimation and underestimation of the ambient pore pressure, respectively (Moernaut *et al.* 2017). The dissipation plots of all four sites are classified as type 1, following the scheme of Sully *et al.* (1999), presenting a monotonous decrease of pore pressure immediately after the probe reached its final depth. Such a dissipation style is particularly suited to reliable extrapolation and estimation of ambient pore pressure (Moernaut *et al.* 2017). Overpressure (u^*) was calculated by subtracting the hydrostatic pressure (u_h) from the (extrapolated) ambient pore pressure. The overpressure ratio (λ^*) (Flemings *et al.* 2008) represents the amount of excess pore pressure relative to the hydrostatic vertical effective stress (σ'_{vo}), which is the difference between the overburden stress (σ_{vo}) and the hydrostatic pressure (u_h). We present both the maximum and minimum limits of the overpressure ratio based on the $1/t$ and $1/\sqrt{t}$ techniques, respectively. For a relative comparison of the hydraulic conductivity between the four test levels, we determined t_{50}, which is the time for 50% of pore pressure dissipation from the starting peak immediately after insertion until the (extrapolated) ambient value (Fig. 2d). Using t_{50}, we obtained a first-order estimation of the horizontal hydraulic conductivity (K_h) based on the simple empirical relationship presented in Parez and Fauriel (1988) given a filter element at the 'u2' (shoulder behind cone) position:

$$K_h \approx \left(\frac{1}{251\, t_{50}}\right)^{1.25} \quad (3)$$

with K_h in cm s^{-1} and t_{50} in s.

Small near-surface MTD-A1 (AD 1960) without basal gas horizon

Slope failure occurred during the AD 1960 earthquake on the northern slope of a buried submerged moraine (Moernaut *et al.* 2017). MTD-A1 is relatively small (0.127×10^6 m^3) compared to other MTDs in the lake. It has an elongated geometry because the mass movement propagated in a narrow morphological depression produced by the erosive character of a previous large-scale landslide (MTD-F1; Moernaut *et al.* 2019). Two CPTU drops in the distal part of MTD-A1 (CPT17–18) and one drop in front of it (reference site CPT19) allow one to compare the s_u evolution in this very young, near-lakefloor, MTD with an undisturbed hemipelagic sequence (Fig. 3). No sediment cores or dissipation tests are available on these sites. However, based on the sedimentation rate ($1-2$ mm a^{-1}) from short cores in this area of the lake (Moernaut *et al.* 2014), we estimate about $5-10$ cm of post-MTD drape.

The CPTU-derived undrained shear strength (s_u) of the reference site CPT19 shows a relatively steady, near-linear, increase with burial depth (Fig. 3c), as can be expected for an increasing consolidation (and density) with effective overburden stress. At a depth of 1.4 m, s_u reaches a value of 4.2 kPa. In contrast, the s_u of sites CPT17 and CPT18 shows highly elevated values within the MTD, with an average of 5.5 kPa (range 1.5–13.2 kPa) and 5.9 kPa (range 2.6–16.3 kPa), respectively. Below the MTD at a depth of 1.4 m, s_u reaches only 1.3 kPa, *c.* 30% of the value at the reference site (4.2 kPa) and similar to the s_u of the surface sediment overlying the MTD. We therefore consider the background sediments underlying the MTD-A1 to be much weaker and underconsolidated relative to what is expected for their burial depth.

Large MTD-F1 and MTD-F2 (*c.* 2.1 ka BP) with basal gas horizon

Rafted blocks and a thin basal shear surface

Two large slope failures occurred during an earthquake at *c.* 2.1 ka BP affecting slopes north and south of the submerged moraine in the southwestern end of Lake Villarrica (Fig. 1d; Moernaut *et al.* 2019). MTD-F1 and MTD-F2 yield a sediment volume of 3.62×10^6 m^3 and 6.29×10^6 m^3, respectively. The largest part of these slope failures occurred on rather gentle slopes of 2–4°. The source areas exhibit irregular headscarps (Fig. 4a, b) which are associated with large composite pockmarks, suggesting an important role of focused fluid escape on the development of these slope failures (Moernaut *et al.* 2017).

For both landslides, seismic-stratigraphic mapping shows that the basal shear surface (BSS) is located in relatively deep stratigraphic levels in the uppermost source area (Fig. 4a), i.e. within the glacio-lacustrine sediment units. Downslope, it ramps up to the R2 horizon, which forms the main BSS, and is associated with a sandy tephra layer in the early Holocene (Wiemer *et al.* 2015). In the basin plain, the BSS ramps up to several younger stratigraphic levels, forming a relatively narrow erosive slot. More distally, the BSS ramps up to the contemporaneous lake floor, allowing free unconfined propagation (Fig. 4a).

The large MTDs display numerous block-like features, producing topographic highs that are still visible on the present-day lake floor morphology (Fig. 4b), even though they are buried by several metres of hemipelagic sediments (Figs 4c, 5c–f). Typically, these blocks are slightly elongated and

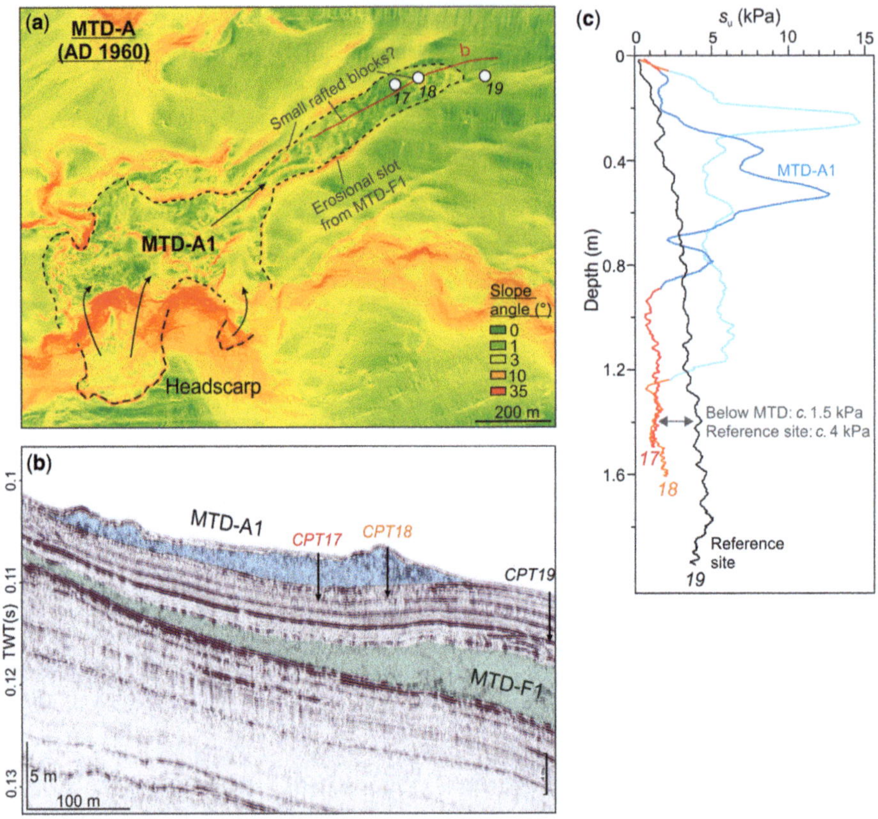

Fig. 3. (**a**) Slope gradient map indicating the headscarp and MTD studied from event A (AD 1960). MTD-A1 contains small rafted blocks and is emplaced within an erosional slot created by the previous larger landslide event F1. CPT sites are indicated by circles. The location of the sub-bottom profile shown in (b) is indicated by a red line. (**b**) Sub-bottom profile showing MTD-A1 and MTD-F1 and projected reach of the CPT drops; TWT, two-way travel time. (**c**) *In situ* s_u derived from the CPT tests. The reference site shows a gentle increase in s_u with depth, whereas data in MTD-A1 (in blue) show much higher values. Sediments below MTD-A1 have much lower s_u than comparable depths at the reference site.

have a diameter of about 40 m. Relative to the host MTD, they protrude about 1 m as measured on present-day bathymetry and about 1.5–2 m as measured on sub-bottom profiles. Several blocks contain a well-stratified seismic facies with continuous reflections (Fig. 4c). The similar seismic facies succession in stable upslope sequences (Fig. 4c) suggests that the blocks are, in fact, 'rafted blocks' which originated at the collapsing upper slopes and were transported for several kilometres without strongly disturbing their internal stratigraphy. For MTD-F1, the rafted blocks are situated 0.4–1 km downslope of the final frontal ramp, which results in an estimated travel distance of about 2.5 km (2.1–3.4 km). For MTD-F2, the rafted blocks are located about 0.2–0.8 km from the final ramp, resulting in an estimated travel distance of about 3 km (2.2–4.5 km). The mapped rafted blocks have a different morphological and reflection seismic expression than the buried sediment volcanoes located on top of the studied MTD (Fig. 5f; see Moernaut *et al.* 2009); the sediment volcanoes lack continuous internal reflections and generally have smaller dimensions and reduced morphological expression.

Sediment cores in the intact upper slope ('unfailed platform'), failed slope and base of the rafted block allow a precise identification of the stratigraphic level that acted as a BSS during event F2 (Fig. 4d). On the slope, the main BSS is located *c.* 1 cm above a sandy tephra layer ('R2' on Fig. 4d). Remarkably, the base of the rafted block in the basin – corresponding to the base of MTD-F1 – exhibits a comparable sediment stratigraphy to the one on the intact upper slope platform. A diagnostic unit of multiple thin tephra laminae in both cores acts as a key correlation marker, which

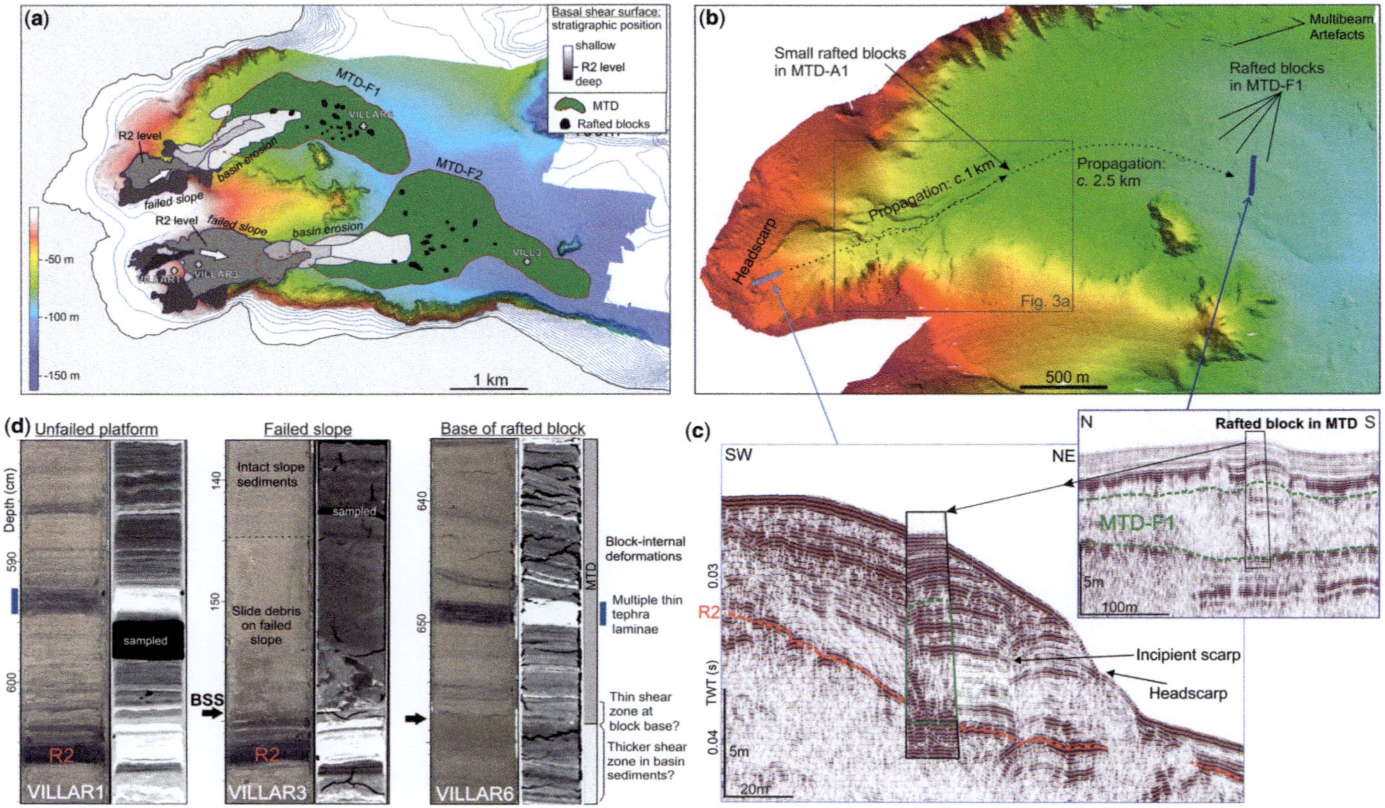

Fig. 4. (**a**) Distribution of MTD-F1 and -F2 (red outline), with rafted blocks (in black) and the relative stratigraphic position of the basal shear surface (BSS) (in grey-scale, based on stratigraphic mapping of Moernaut *et al.* 2019). The frontally emergent part of the MTD (in green) is associated with free landslide propagation over the contemporaneous lake floor. (**b**) Present-day lake floor morphology showing several buried rafted blocks. (**c**) The seismic facies sequence in the blocks can be traced back to the unfailed slope sequence upslope of the headscarp, indicating limited internal deformation of the rafted block. (**d**) Comparison of sediment cores in the intact unfailed platform, failed slopes and rafted block (location shown on (a)) based on macroscopic observations on the split core surface and on X-CT data. The BSS developed just above the R2 tephra layer (Moernaut *et al.* 2017). Remarkably, the base of the rafted block matches well to the unfailed sequence above the BSS, indicating a thin discrete BSS at the base of the rafted block. As for the R2 tephra, the 'multiple thin tephra laminae' correspond to eruptions of the Caburgua-Huelemolle scoria cones at *c.* 10.3–10.5 ka BP (Fontijn *et al.* 2016), located in Villarricás catchment.

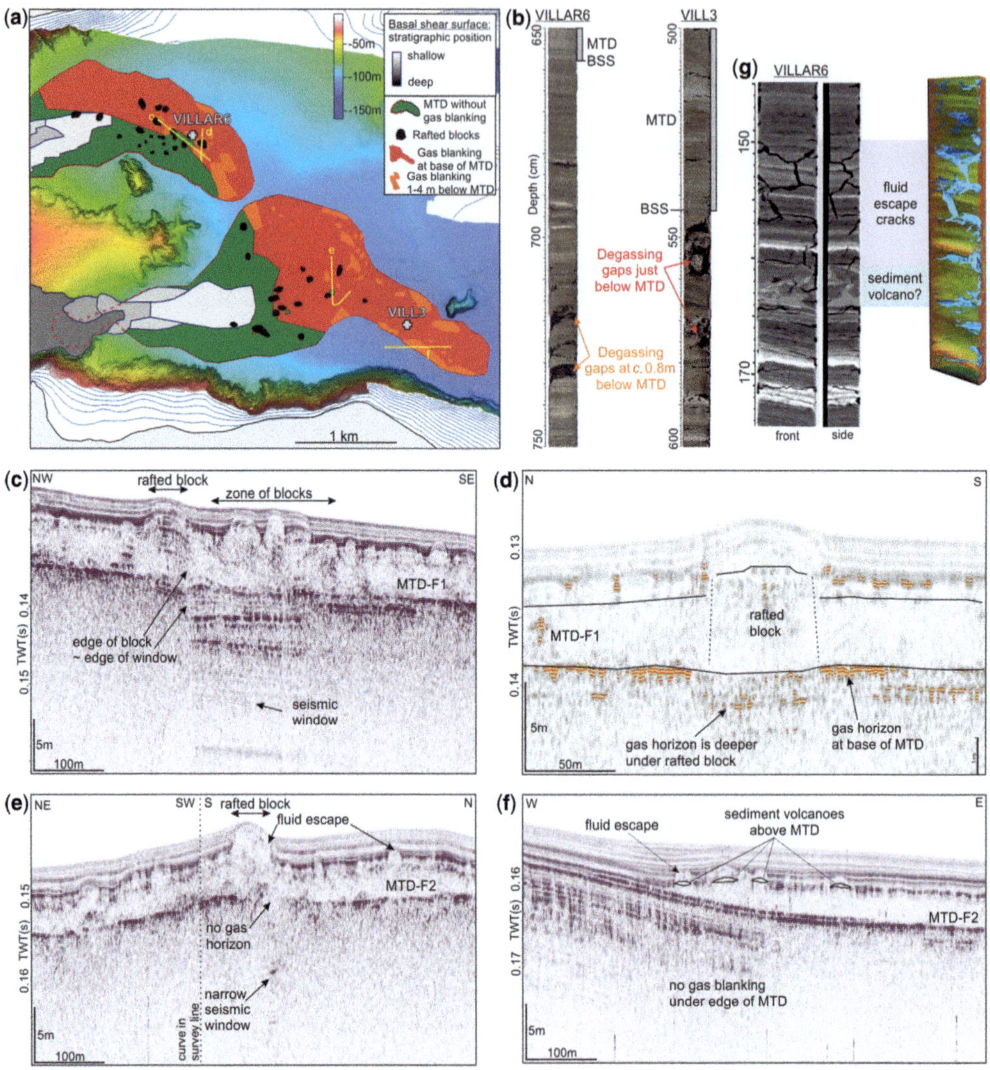

Fig. 5. (**a**) Distribution of MTD-F1 and -F2 (close-up of Fig. 4a), showing the gas blanking at the base of the MTDs (red), or at 1–4 m below them (orange). (**b**) Sediment cores showing degassing gaps immediately below the MTD (VILL3) or at *c.* 0.8 m below the MTD. (**c–f**) Sub-bottom profiles showing MTD-F1 and -F2, rafted blocks, gas horizons below the MTDs and seismic windows; (d) has a different colour scaling to highlight the strong amplitude (red/yellow) related to the gas horizon. Location of profiles shown on (a). (**g**) X-CT data of a selected portion of VILLAR6, located just above MTD-F1 (see also Fig. 6). In greyscale (white: highest CT density): front and side view showing horizontally laminated background sediments and volcanic layers, a chaotic deposit of a presumed sediment volcano (Moernaut *et al.* 2009) and a network of fluid-escape cracks above the latter. In colour: 3D view of the X-CT data (red: highest CT-density; dark green: low CT-density; blue: 3D structure of the air-filled cracks yielding the lowest CT-density).

shows that the shear zone at the base of the rafted block is very thin (maximum *c.* 2 cm), allowing the base of the failed sequence to be transported over several kilometres without significant deformation. X-CT data and visual observations show that the overridden basin sediments below the rafted block yield a somewhat thicker shear zone of about 10 cm. Above the unit of thin tephra laminae in the rafted block, the sequences of the intact platform and the MTD block do not correlate one to one, suggesting some block-internal deformation and displacement of relatively intact sequences.

Gas horizon below MTD: fluid-escape features above

Sub-bottom profiles show the presence of a gas horizon near the base of MTD-F1 and MTD-F2. This horizon covers about half of the MTD extent (Fig. 5a), has enhanced reflection amplitude and overlies areas of acoustic turbidity and blanking (Fig. 5c–f). The gas horizon mostly occurs directly at the base of the MTD, but in some areas it is located about 1–4 m below the MTD base (orange colour on Fig. 5a). A few narrow seismic windows exist, where seismic penetration reaches much deeper, i.e. to the same depths (about 30–40 m) as obtained in the Villarricás basin sediments not affected by MTDs. Several of such seismic windows and deeper gas horizons are located in zones of rafted blocks and near block edges (Fig. 5c–e), which indicate reduced gas accumulation in these areas. Moreover, the gas horizon is often deeper or absent near the edges of the MTD (Fig. 5f). Large gaps in the (otherwise continuous) sediment cores VILLAR6 and VILL3 can be regarded as an independent indicator for the presence of free gas below the MTD. The gaps in VILL3 occur immediately below the base of the MTD, whereas in VILLAR6 the gaps are *c.* 0.8 m lower. This relative difference in stratigraphic depth is in line with the seismic-stratigraphic mapping where the gas horizon is located directly below the MTD at VILL3 and somewhat deeper (*c.* 1–1.5 m) under the MTD at VILLAR6 (Fig. 5a).

Fluid-escape features are present above a large part of the MTDs (Fig. 5e, f). They consist of vertical acoustic wipe-out areas cross-cutting the seismic reflectors above the MTDs and are mainly restricted to the first *c.* 0.5–1 m overlying the MTDs. They occur above the edges of rafted blocks, above sediment volcanoes, but also above other parts of the MTD. In sediment core VILLAR6, a 10 cm thick network of sub-vertical and horizontal cracks overly the sediment volcano (Fig. 5g) without offsetting the sediment laminations. These sub-vertical cracks are interpreted as hydrofractures and have not been observed in other parts of VILLAR6. As is the case in the formation of sediment volcanoes, the hydrofractures are hypothesized to be caused by earthquake-triggered fluid escape from overpressured MTDs and/or the immediately underlying sequences (Moernaut *et al.* 2009).

Sediment cores: s_u *and porosity*

Core VILLAR6 shows the sedimentary facies of the rafted block in MTD-F1. It consists of near-horizontally laminated sequences with minor signs of disturbance, except for some gently dipping laminations (Fig. 4d) and a discrete steep-dipping fault offsetting strata for at least 0.5 m (Fig. 6b). A *c.* 5 cm thick turbidite overlies the MTD. The lower half of the MTD has slightly elevated γ-density values (*c.* 1.4–1.5 g cm^{-3}), whereas the sediments below the MTD show the lowest γ-density values ($<$ 1.3 g cm^{-3}), which – to some extent – relate to the presence of degassing cracks and gaps (Fig. 6a). Core-derived s_u in the MTD is slightly higher than in the surrounding sequences, albeit with considerable scatter. The porosity of the MTD is rather constant and clearly lower than in the surrounding units (about 77% v. 80–90%). However, the MTD porosity is rather similar to the porosity of the intact sequence of VILLAR1 (upslope unfailed platform; Wiemer *et al.* 2015), which is representative for the failed slope sequences on the upper slope (Fig. 6c).

Core VILL3 shows the sedimentary facies of MTD-F2 (Fig. 7), which is overlain by a small sediment volcano according to seismic-stratigraphic mapping (Moernaut *et al.* 2019). The MTD consists of faintly laminated sediments with slightly inclined sections, small-scale faulting and folding. Cracks and gaps related to gas expansion occur throughout the MTD and directly below it, producing very low γ-density values ($<$0.7 g cm^{-3}). The MTD exhibits slightly higher s_u values (5–7 kPa) than the surrounding sediments. Remarkably, there is a slightly decreasing trend downcore for s_u and γ-density, whereas the *in situ* s_u increases with depth, as can be expected for increasing consolidation with burial depth. Therefore, we think that the downward decreasing trends (in core s_u and γ-density) may be related to strong degassing of the sediment core during its acquisition, and thus do not represent the properties of the sequence in its natural conditions.

In situ undrained shear strength, excess pore pressure and hydraulic conductivity of MTDs

With the CPTU, we targeted a reference basin site (CPT6) next to MTD-F1 (Fig. 8a). As expected for an intact sequence of slow hemipelagic sedimentation, s_u steadily increases with depth and virtually no excess pore pressure was found (λ^*: 0–0.1 at 4.9 m depth) (Fig. 2; Table 1). Moreover, CPTU drops and dissipation tests were performed on three sites targeting different features of the MTDs: (1) rafted block with fluid-escape features (VILLAR6; CPT7); (2) MTD overlain by sediment volcano and fluid-escape features (VILL3; CPT8); and (3) MTD without fluid-escape features.

The average s_u in the MTD is highest in the rafted block and lowest under the sediment volcano (16.1 v. 7.3 kPa; Table 1; Fig. 8b). Compared to the immediately overlying sequences, s_u in the MTD is significantly higher (Fig. 8a, b), which is most pronounced for the block site (CPT7). Compared to the source sequence (CPT1.2; Wiemer *et al.* 2015), the rafted

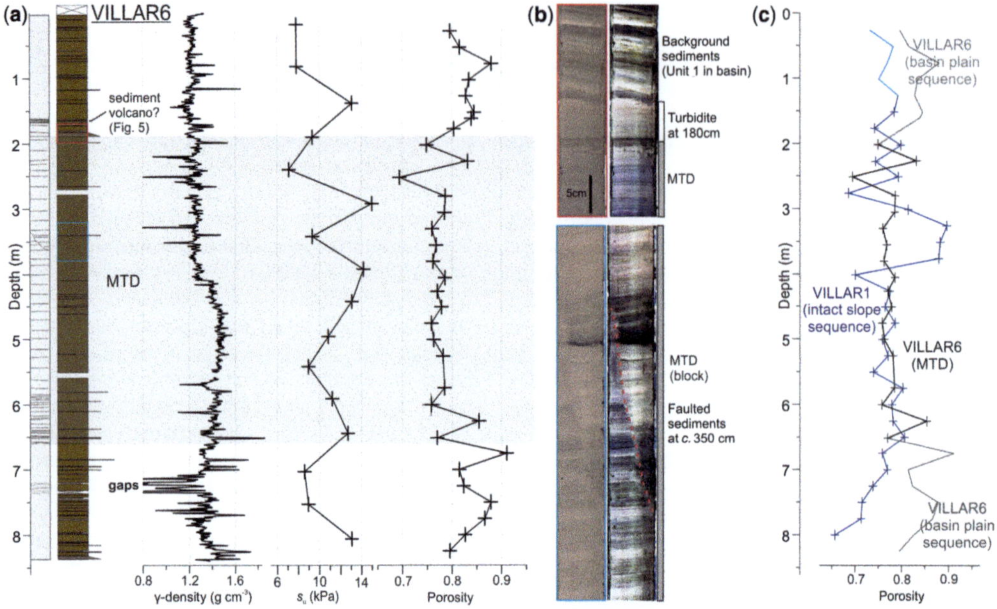

Fig. 6. (**a**) Core VILLAR6 taken within a rafted block, showing the structures, lithology and measured physical properties. Large parts of the MTD have higher γ-density and s_u and decreased porosity compared to the surrounding sediments. Density peaks relate to volcanic layers (tephra, lahar deposits). (**b**) Core pictures (original and colour-enhanced) of MTD material showing a turbidite overlying the MTD and a fault offsetting relatively undisturbed (horizontally laminated) sequences. (**c**) Comparison of porosity between cores VILLAR1 and VILLAR6, taken in an intact source slope sequence and rafted block, respectively. Parts of the curve are irrelevant for this comparison (no crosses) as these represent post- or pre-event sequences. VILLAR1 data after Wiemer *et al.* (2015).

block has a very similar average s_u (16.1 v. 16.2 kPa), whereas the other MTD sites exhibit significantly lower average s_u values (Fig. 8b, c; 13.5 and 7.3 kPa).

The excess pore pressure ratio (λ^*) is highest under the sediment volcano (CPT8; 0.62–0.66) and lowest where no fluid-escape features are present (CPT9; 0.2–0.44). The rafted block site presents intermediate overpressure values (0.32–0.54). Based on the dissipation t_{50} value, the reference site (CPT6) shows the highest horizontal hydraulic conductivity (K_h: 4.94×10^{-6} cm s^{-1}), whereas the MTD sites yield significantly lower values of 15–67% of the reference site (Table 1).

Discussion

Data quality

The multibeam bathymetry data and sub-bottom profiles are considered to be of good quality and suited for the mapping of landslide features such as rafted blocks. Imperfect corrections for the vessel motion generated some striped artefacts on the multibeam maps (Figs 1 & 3), but do not impact the operator interpretation and mapping. Resolution of the sub-bottom profiles is good enough for mapping of blocks larger than 10 m in diameter as the vertical resolution for a 3.5 kHz system is 10–20 cm and the horizontal resolution (Fresnel zone) at 100 m depth is *c.* 9.2 m. However, noise in the data due to the vessel's engine and waves complicates detailed mapping of fluid-escape features.

The calculation of *in situ* s_u based on the CPTU data relies on some empirically derived parameters, such as the cone factor, which are dependent on the sediment composition and behaviour of the sediments upon CPT impact (see discussion in Steiner *et al.* 2014, 2015). Therefore, we use the calculated s_u values for relative trend comparisons between sites and do not put much emphasis on the absolute values.

Estimation of the *in situ* excess pore pressure is challenging because most dissipation profiles are incomplete and extrapolation techniques are needed. To account for this, we adopt two approaches that are known to respectively under- and overestimate the ambient pore pressure and we consider both values as delimiting the real value. Moreover, due to the small subsurface depth of our tests, measurement uncertainties in the estimation of the hydrostatic pressure at the lake bottom and the overburden stress at the measurement depth become relevant. To

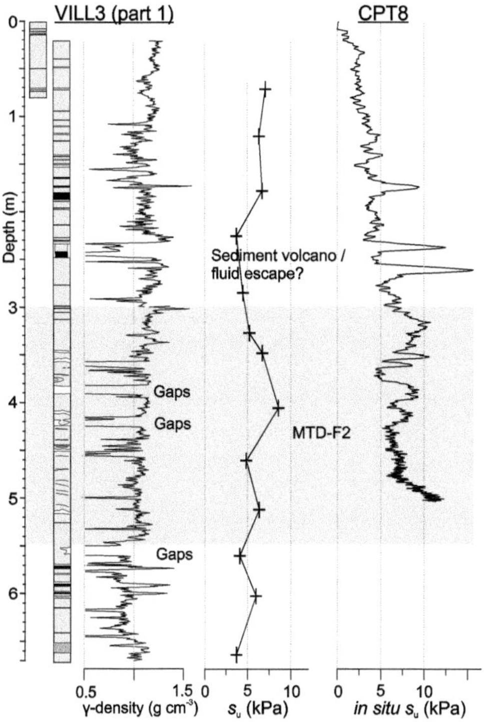

Fig. 7. Core VILL3 taken in the distal part of VILL3, showing the structures (Moernaut *et al.* 2019), γ-density, vane shear-derived s_u, and compared to the *in situ* s_u of CPT8 at the same site. In the sediment volcano, MTD and below the MTD, many degassing gaps and cracks occur, producing negative excursions in the γ-density data. Core-derived s_u shows a slight decrease with depth, whereas *in situ* s_u shows a clear increase with depth and rather constant values in the MTD. Peaks in *in situ* s_u relate to volcanic layers, such as tephra (black units on log) and lahar deposits (white units).

reliably estimate the hydrostatic pressure, we averaged the bottom hydrostatic pressure readings from several drops at the same site. For the overburden stress estimation, we do not expect that the use of a constant unit weight has a significant influence as the background sediments overlying the MTDs are very similar (Fig. 8b; Moernaut *et al.* 2019).

More important quality issues concern the lab measurements on sediment cores. In particular, the presence of degassing cracks has a profound influence on γ-density, vane shear-derived s_u and pycnometer-derived porosity. In VILLAR6, most of the core was intact and measurements are considered reliable, except for the sediments underlying the MTD. In VILL3, gas cracks are a major issue in several parts of the core, probably explaining the large discrepancy between the *in situ* s_u and the lab-derived s_u trends (Fig. 7). Other core quality issues may relate to the effect of the hammer-driven piston coring. This coring method may induce compaction of the uppermost (1 m?) part of the soft sediments, and lab data from this section should be treated with caution. Other coring deformation (i.e. u-shaped laminae) is present in VILL3 but not in VILLAR6, as the latter was taken with an improved piston system.

The Holocene sediments in Lake Villarrica consist of diatomaceous mud/ooze. The presence of diatoms has a profound effect on all standard geotechnical index properties that involve water content (e.g. bulk density, porosity, Atterberg limits etc.) because the hollow diatom frustules can hold large amounts of intra-frustule water (Wiemer *et al.* 2015, 2017). Consequently, bulk density is typically low; water content and derived porosity values are high. Changes in these index values downcore may relate to consolidation state, but can also represent a change in the abundance (and species) of diatoms. Therefore, bulk density and porosity in diatom-rich sediments are poor indicators for sediment consolidation or permeability, the proper assessment of which requires different techniques.

Mechanisms for MTD-related overpressure and free gas

There are no signs of free gas on sub-bottom profiles for the small AD 1960 MTD-A1 (Fig. 3), whereas the large MTD-F1 and -F2 show prominent basal gas horizons. We therefore discuss these as separate case studies.

For MTD-A1, we used s_u as a first-order indicator of sediment consolidation. The relatively low s_u values under the MTD are interpreted as follows. The sediments below the MTD were not capable of consolidating to a state that would be expected from their burial depth, but retained the low s_u (and probably density) of the near-surface sediments. The sudden loading by the MTD emplacement generated excess pore pressure in the underlying sediment. The elapsed 51 years since MTD emplacement were too short to allow a significant upward dissipation of these excess fluids through the MTD. In other words, the MTD can be regarded as a potentially rapidly forming seal for the excess fluids generated in the overridden sequences, regardless of the small thickness (*c.* 1 m) and dimensions of the MTD. The same process can be invoked for all muddy MTDs and predicts the presence of excess pore pressure and underconsolidated sediments below shallow MTDs. Extra complexity to this model is added by the influence of moving MTDs on the underlying sediments, potentially producing shear deformation and compaction. However, our observation of very low s_u below MTD-A1 indicates that this process was not significant there. Moreover, the MTD itself may also contribute to excess pore

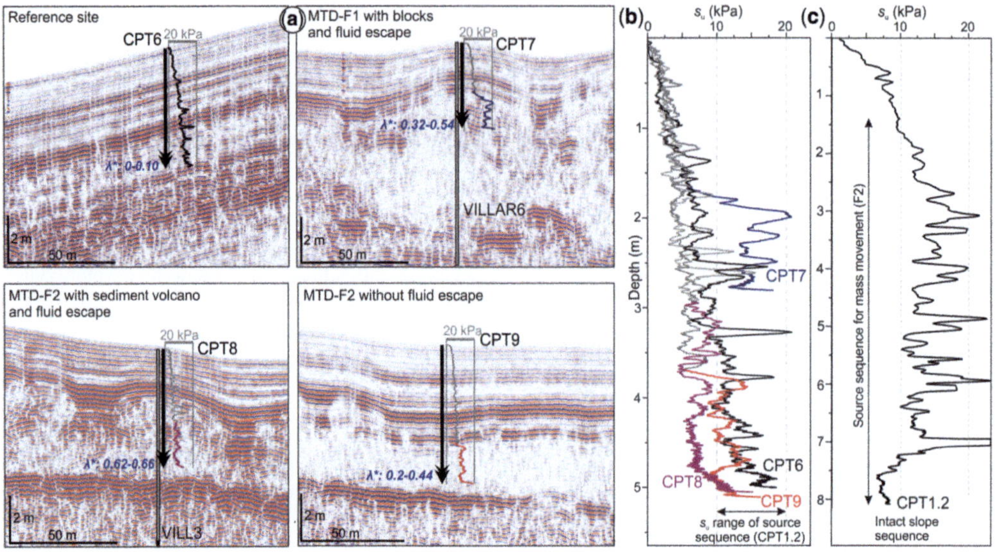

Fig. 8. (**a**) CPT sites showing s_u with depth and overpressure ratio (λ^*) at the deepest penetration point. Data within the MTD are colour-coded for each site. (**b**) s_u data plotted together showing that the MTD material can be stronger (CPT7), equally strong (CPT9) or weaker (CPT8) than background sediments (CPT6) at comparable depth. (**c**) s_u data from the intact upslope source sequence (CPT1.2; Wiemer *et al.* 2015) showing mostly higher values than the basin reference site CPT6 (shown in b).

pressure development as loose sediments tend to contract during shearing. If the muddy MTD has relatively low permeability and/or is capped by a finer-grained turbidite, it requires significant time or an external trigger mechanism for expelling most excess pore water. Water expulsion can form fluid-escape features – such as acoustic wipe-outs and sediment volcanoes – which are typically observed in the sequences overlying muddy MTDs (Diviacco *et al.* 2006; Moernaut *et al.* 2009; Praet *et al.* 2017).

For MTD-F1 and -F2, excess pore pressure due to MTD emplacement is expected, but the presence of free gas at the base of the MTDs may also play a crucial role in the development of the measured excess pore pressure values in these MTDs. Free gas in the sedimentary sequences of Lake Villarrica is observed in the deep basin, is of biogenic origin and relates to the relatively high organic matter content of the surface sediments originating by fluvial input, coastal erosion and biogenic productivity (Moernaut *et al.* 2017). Large amounts of freshly deposited organic matter can degrade by oxygenation at the lake bottom in this monomictic lake. However, the sudden emplacement of an MTD seals the organic matter from oxygen exposure and microbial activity in the anoxic sediment sequence leads to organic matter degradation and methanogenesis. This methane is trapped at the base of the MTD forming the typical gas horizon and acoustic turbidity observed on sub-bottom profiles. It is possible that increasing amounts of free gas take in more and more pore space, resulting in the development of excess pore pressure in the overlying MTD units, as long as these have a relatively low permeability and/or are capped by a low-permeability layer. Relatively lower hydraulic conductivity values derived from dissipation tests in the MTD compared to the background reference site support the hypothesis of a relatively low-permeability MTD. The presence of a gas horizon under MTD-F1 and -F2 depends (among others) on high organic matter accumulation on the lake bottom and on an effective sealing by the MTD. Due to the absence of inflowing rivers in the western part of the study area, we expect lower influx of organic matter there, which may explain the absence of free gas indicators in the western slopes of the lake, under MTD-A1 and in the western parts of MTD-F1 and F2(Figs 3b & 5a; Moernaut *et al.* 2017). An alternative hypothesis for the absence of free gas under MTD-A1 is that the deposition of MTD-A1 (in AD 1960) may have occurred too recently for sufficient degradation of organic matter and accumulation of free gas to have developed.

Sealing character of young MTDs

The studied MTDs are embedded in relatively weaker, water-rich basin sediments, as indicated by

Table 1. *CPTU pore pressure dissipation data at different sites*

Position	Depth (m)	Time (min)	Type	Measured u_h (kPa)	Measured σ_{vo} (kPa)	Measured u^* (kPa)	Measured λ^*	Extrapolated via 1/t u^* (kPa)	Extrapolated via 1/t λ^*	Extrapolated via $1/\sqrt{t}$ u^* (kPa)	Extrapolated via $1/\sqrt{t}$ λ^*	t_{50} (s)	K_h (cm s^{-1})	Lithology at dissipation	Average s_u of MTD (kPa)
CPT6	4.9	41	I	939.00	958.30	3.00	0.16	1.80	0.10	0.00	0.00	70	4.94×10^{-6}	U1	NA
CPT7	2.85	41	I	936.96	948.07	7.84	0.71	6.04	0.54	3.54	0.32	201	1.32×10^{-6}	MTD-F1 blocks	16.1
CPT8	5.05	45	I	1106.54	1126.24	14.46	0.73	13.06	0.66	12.16	0.62	96	3.33×10^{-6}	MTD-F2 + sediment volcano	7.3
CPT9	5.2	40	I	1016.01	1036.29	16.99	0.84	8.99	0.44	3.99	0.20	306	0.78×10^{-6}	MTD-F2 no fluid-escapes	13.5

u_h = hydrostatic pressure; σ_{vo} = overburden stress; u^* = overpressure; λ^* = overpressure ratio. t_{50} = time to reach 50% of dissipation. K_h = horizontal hydraulic conductivity derived from t_{50}. u_h is calculated for fresh water at the penetration depth reached by the probe. σ_{vo} is derived from the average bulk density on sediment cores.

the *in situ* undrained shear strength data (Fig. 8) and – possibly – the reduced MTD porosity in VILLAR6. Several indicators, such as gas horizons, underconsolidation of underlying sediments and excess pore pressure, suggest that the MTDs relatively rapidly form seals for fluid migration. However, fluid-escape features in the first 0.5–1 m overlying the MTDs indicate some fluid leakage shortly after MTD deposition (*c.* 400–800 years; Moernaut *et al.* 2019). In case these fluids are sourced from the overridden basin sediments, this would imply poor MTD sealing. In contrast, if the MTD itself is the source of excess fluids, this process would enhance the consolidation and seal competence of the MTD. Based on marine MTD investigations, it is often proposed that shear deformation and densification during mass movements play a key role in creating the sealing facies of muddy MTDs (e.g. Dugan 2012). Compared to its source sequence, a MTD should have reduced porosity and increased density and undrained shear strength. However, in our case study, the MTD porosity in VILLAR6 is very similar to the source sequence of VILLAR1 (Fig. 6c) and the *in situ* undrained shear strength is similar to or lower in the MTD (CPT7-9) than in the source sequence CPT1.2 (Fig. 8c). This suggests that shear densification is not the dominant process behind the apparent sealing character of the studied MTDs in Villarrica. Wiemer *et al.* (2015) found that upslope source sequences are strongly overconsolidated (i.e. overconsolidation ratio of 2–8) and, thus, we explain the outstanding MTD physical properties by the more consolidated nature of the source sequence, which became preserved partly during mass transport, and which contrasts to the weaker host sediments in the basin. This inference has important implications for marine studies where elevated density and s_u in MTDs is often attributed to shear densification without fully considering the physical properties of the source sequences upslope. However, with an increasing run-out distance, such a source signature may get lost and shear-induced porosity loss may become the dominant process (e.g. Dugan 2012).

Rafted blocks affecting seal integrity

Numerous rafted blocks are present within MTD-F1 and -F2. These blocks travelled about 2.1–4.5 km without significant remoulding of the original stratigraphic sequence. We propose that the enhanced resistance to deformation and disintegration of the blocks relates to the overconsolidated nature of the upper slope sequences. By comparing cores in the intact sequence, failed sequences and rafted block we constrained the BSS within the block base (at VILLAR6) to a thin interval of a maximum thickness of *c.* 2 cm. The feasibility for such a thin

BSS and intact block base may be related to the presence of the sandy tephra (R2) just below the BSS. Seismic shaking of such granular material may cause densification and excess pore pressure development that weakens the overlying hemipelagic sediments by effective stress reduction (e.g. Alves and Lourenço 2010; Wiemer *et al.* 2015). This allows a sharp detachment of the sliding block and, given the laterally continuous nature of the tephra layer, this process may have occurred over a large part of the failing slope. Liquefaction of the R2 tephra did not take place as this would have remobilized it together with the sliding mass, which is inconsistent with our observation of an intact R2 tephra on the failed slope (Fig. 4d) (see discussion in Moernaut *et al.* 2019).

The shear zone within the overridden distal basin sediments is also very thin (*c.* 10 cm at VILLAR6), leading to a ratio of *c.* 39 for the rafted block height (*c.* 4.7 m) over the total basal shear zone thickness (*c.* 12 cm). This value is much higher than the reported average ratio of 5–10 based on a compilation of marine and outcrop studies (Alves and Lourenço 2010). Variability in this ratio may relate to the nature of the overridden sequence (stratification, stiffness etc.) and the weight, flow regime and mode of transport of the mass movements (Alves and Lourenço 2010; Laberg *et al.* 2017; Sobiesiak *et al.* 2018). For the case of Lake Villarrica, we hypothesize that significant erosion, shearing and transport of basin sediments in the proximal part ('basin erosion' on Fig. 4a) may have broken up diatom frustules, releasing intra-frustule water into the base of the moving mass. This may facilitate 'shear wetting', allowing the mass movement to detach from the substrate and propagate as a 'free-slip' flow (see Sobiesiak *et al.* 2018).

The rafted blocks are exclusively emplaced in the frontally emergent parts of the MTDs, where the landslide masses propagated over the contemporaneous lake bottom. The erosional slots are devoid of blocks (Fig. 4a). We interpret this block distribution as a consequence of the slowing down of the propagating masses once they emerged over the frontal ramp and out of the narrow erosional slots. The sudden lack of lateral confinement leads to lateral spreading and slowing down of the moving mass. Consequently, most rafted blocks halted within a rather short distance from the frontal ramp and no blocks reached the more distal parts of the MTD. This block distribution may have important implications for the sealing integrity of the MTDs. Previous marine studies inferred that rafted blocks can form significant fluid migration pathways when they are embedded in rather low permeability sequences (Alves 2015; Gamboa and Alves 2015). Fluid flow can be facilitated due to the discontinuity formed by the block edges, the topographic expression of the blocks focusing fluids towards their top, or by faults, fractures and sheared beds within the blocks. In Lake Villarrica, we interpret a similar influence of rafted blocks on the MTD seal integrity, by the observations of (1) seismic windows or lowered gas horizons under areas of rafted blocks, especially near block edges (Fig. 5c–e), (2) the abundance of fluid-escape features in the sequences overlying rafted blocks, (3) distinct thrusts and sheared beds within the sediment core of the rafted block (VILLAR6, Fig. 6), all of which potentially facilitate fluid migration. The sediment volcanoes mapped on sub-bottom profiles seem to affect sealing integrity in a much reduced manner, given the smaller fluid-escape features on top of them and the presence of the largest overpressure ratio (λ^*: 0.62–0.66) in the underlying MTD (site CPT8; Fig. 8a). The reduced gas blanking at the very edges of the MTD (Fig. 5a, f) suggests that MTD edges have a poor sealing potential, possibly due to lateral leaching of free gas into undisturbed sequences where no sealing units are present. In summary, we propose that the better sealing facies can be found in the most distal part of the MTD, out of reach of rafted blocks and where shear compaction may have been more significant due to the long run-out and largest disintegration of the sliding mass. However, additional measurements are needed to constrain the significance of shear compaction for these MTDs.

Significance of investigating young MTDs in lakes

The typically heterogeneous character of MTDs stresses the need for high-resolution mapping of MTD geometry and its internal architecture. In lacustrine settings, this can be obtained in a relative fast and cost-effective manner. Aiming at young and thin MTDs, one can sample the entire MTD and some underlying sediments with conventional piston coring techniques and characterize the *in situ* undrained shear strength and pore pressure regime with free-fall CPTU measurements. Such CPTU tests (including dissipation tests) can be deployed from anchored platforms or from dynamically positioned vessels, involving tens of tests per day. Low-cost logistics allow for repeated surveys after, for example, earthquakes, opening perspectives to map migrating gas fronts, fluid-escape features, new landslides and so on. As the target features are often rather small, there can be great potential in the application of innovative 3D sub-bottom profiler techniques (e.g. Vardy *et al.* 2017).

A striking enigma in marine studies is the preservation of reduced porosity of many muddy MTDs compared to their surrounding sequences, regardless of their burial depth. This contradicts traditional soil mechanics models, which predict that deeply buried MTDs would not stand out in terms of consolidation state and physical properties, because the effective vertical stress imposed on the buried MTD is much higher than the stress these sediments experienced on the pre-failure slope and by shearing during the mass movement process (Dugan 2012). Possible explanations relate to a different micro-fabric of MTDs, where lateral shear may have induced high clay particle alignment and thus smaller pore-throat size and better sealing capacity (Cardona *et al.* 2016), a process that seems to be especially relevant in the lower part of MTDs (Kanamatsu *et al.* 2014). In contrast, for some MTDs, the effect of shear compaction on porosity seems to be overruled by overburden stress and the MTD sediments tend to a normal consolidation state (e.g. Sun *et al.* 2018). For deeply buried MTDs, it is difficult to assess the evolution of their sealing capacity, since early depositional features may be obliterated due to sediment compaction and cementation (Alves *et al.* 2014). Therefore, detailed micro-fabric studies on samples from near-surface MTDs may be crucial to better understand the outstanding physical properties of many muddy MTDs.

Conclusions

By combining results from seismic stratigraphic mapping, standard geotechnical measurements on sediment cores and *in situ* penetrometer (CPTU) data, we characterized the physical properties of three young MTDs in Lake Villarrica (Chile) and discussed their potential role as sealing facies for subsurface fluid migration. We present the following conclusions.

- Thin MTDs (1–5 m) can rapidly form seals for fluid migration. Sealing capacity does not require thick MTDs and is not necessarily a feature obtained during deep burial (hundreds to thousands of metres) and associated consolidation of the MTD.
- In Lake Villarrica, the sealing character of the studied MTDs seems to be related to the overconsolidated state of the source slope sediments that were emplaced within soft, water-rich, basin sediments. Shear densification, often invoked for marine MTDs, does not seem to be the dominant process.
- The rapid loading by the emplacement of MTDs can lead to excess pore pressure and/or gas accumulation in the directly underlying sequences, without the requirement of fluid transfer from deeper sedimentary units. Free (biogenic) gas production is caused by the trapping of bottom organic matter and its subsequent methanogenesis, leading to enhanced reflection amplitude and acoustic turbidity directly below MTDs.
- Repeated strong earthquake shaking and excess pore pressure within or below the Villarrica MTDs may have induced fluid-escape fractures and sediment volcanoes in the sequences directly overlying the MTDs. Thus, some of the excess fluids escaped shortly after deposition of the MTDs and – depending on the source unit of these excess fluids (MTD or underlying sequences) – this may imply a reduced MTD sealing potential in the very early burial stage (<1 m).
- Rafted blocks can form pathways of fluid leakage and compromise seal integrity, which in our study is indicated by subdued gas horizons under MTD blocks and significant fluid-escape features above them. Potential fluid pathways are the block edges and intra-block faults, fractures and sheared beds. The development of rafted blocks was facilitated by the overconsolidation of the source slope sequence and excess pore pressure development in the continuous, sandy tephra layer below the basal shear surface.
- Our first results using standard analyses highlight the need for more detailed, quantitative research approaches on the sediment cores, involving the characterization of the MTD micro-fabric, compressibility, permeability and inter-particle porosity, for example. Typical challenges in lacustrine settings may be the strong influence of the diatoms' intra-frustule water on standard geotechnical measurements and – due to the typical high organic matter content in lake sediment – the presence of degassing-related gaps and cracks, inhibiting reliable measurements of permeability and porosity in the lab. Therefore, a special effort should be made to obtain high-quality *in situ* data with dense spatial coverage.

This pilot study highlights the potential of investigating young lacustrine MTDs for a better understanding of the role of submarine MTDs on subsurface fluid migration, especially regarding their early-emplacement characteristics and subsequent evolution during burial. Even though young lacustrine MTDs are not economically relevant themselves, their study can help assess the economic value of large marine MTDs in terms of hydrocarbon systems and the drilling hazards related to potentially overpressured MTDs.

Acknowledgements IHS Markit Kingdom and OmniSTAR are acknowledged for their educational grant programmes providing seismic interpretation software and high-accuracy satellite positioning data, respectively. We thank the numerous researchers and technicians making the field surveys possible: Marc De Batist, François Charlet, Anna Reusch, Bruno Ladrón de Guevara, Philipp Kempf,

Nina Stark, Robert Brümmer, Roberto Urrutia, Mario Pino, Koen De Rycker, Alejandro Peña, Manuel Novoa and Francisco Martin. We thank survey company BENTOS (Santiago, Chile) for the multibeam data acquisition and the Renard Centre of Marine Geology (Ghent University) for making available the sub-bottom profiler and piston coring system. Ariana Molenaar is acknowledged for the 3D visualization of the X-CT data and Markus Erhardt for the X-CT scanning. We thank Tiago M. Alves and Greg Moore for constructive reviews of an earlier version of this manuscript.

Funding This research was funded by the Swiss National Science Foundation (grant 133481), a research grant of the Research Foundation Flanders (FWO-Vlaanderen) and the MARUM – Center for Marine Environmental Sciences via the Deutsche Forschungsgemeinschaft (Bonn, Germany).

Author contributions **JM**: conceptualization (lead), funding acquisition (supporting), investigation (lead), methodology (equal), project administration (supporting), visualization (lead), writing – original draft (lead), writing – review & editing (lead); **GW**: investigation (equal), methodology (equal); **AK**: funding acquisition (supporting), resources (lead), writing – review & editing (supporting); **MS**: funding acquisition (supporting), investigation (supporting), project administration (supporting), resources (supporting), writing – review & editing (supporting).

References

Alves, T.M. 2015. Submarine slide blocks and associated soft-sediment deformation in deep-water basins: a review. *Marine and Petroleum Geology*, **67**, 262–285, https://doi.org/10.1016/j.marpetgeo.2015.05.010

Alves, T.M. and Lourenço, S.D.N. 2010. Geomorphologic features related to gravitational collapse: submarine landsliding to lateral spreading on a Late Miocene–Quaternary slope (SE Crete, eastern Mediterranean). *Geomorphology*, **123**, 13–33, https://doi.org/10.1016/j.geomorph.2010.04.030

Alves, T.M., Kurtev, K., Moore, G.F. and Strasser, M. 2014. Assessing the internal character, reservoir potential, and seal competence of mass-transport deposits using seismic texture: a geophysical and petrophysical approach. *American Association of Petroleum Geologists Bulletin*, **98**, 793–824, https://doi.org/10.1306/09121313117

Bøe, R., Bellec, V.K., Rise, L., Buhl-Mortensen, L., Chand, S. and Thorsnes, T. 2012. Catastrophic fluid escape venting-tunnels and related features associated with large submarine slides on the continental rise off Vesterålene-Troms, North Norway. *Marine and Petroleum Geology*, **38**, 95–103, https://doi.org/10.1016/j.marpetgeo.2012.08.008

Cardona, S., Wood, L.J., Day-Stirrat, R.J. and Moscardelli, L. 2016. Fabric development and pore-throat reduction in a mass-transport deposit in the Jubilee Gas Field, eastern Gulf of Mexico: consequences for the sealing capacity of MTDs. *In*: Lamarche, G. *et al.* (eds) *Submarine Mass Movements and their Consequences*. Advances in Natural and Technological Hazards Research, **41**. Springer, 27–37, https://doi.org/10.1007/978-3-319-20979-1_3

Chapron, E., Van Rensbergen, P., De Batist, M., Beck, C. and Henriet, J.-P. 2004. Fluid-escape features as a precursor of a large sublacustrine sediment slide in Lake Le Bourget, NW Alps, France. *Terra Nova*, **16**, 305–311, https://doi.org/10.1111/j.1365-3121.2004.00566.x

Cisternas, M., Atwater, B.F. *et al.* 2005. Predecessors of the giant 1960 Chile earthquake. *Nature*, **437**, 404–407, https://doi.org/10.1038/nature03943

Diviacco, P., Rebesco, M. and Camerlenghi, A. 2006. Late Pliocene mega debris flow deposit and related fluid escapes identified on the Antarctic Peninsula continental margin by seismic reflection data analysis. *Marine Geophysical Researches*, **27**, 109–128, https://doi.org/10.1007/s11001-005-3136-8

Dugan, B. 2012. Petrophysical and consolidation behavior of mass transport deposits from the northern Gulf of Mexico, IODP Expedition 308. *Marine Geology*, **315–318**, 98–107, https://doi.org/10.1016/j.margeo.2012.05.001

Eng, C. and Tsuji, T. 2019. Influence of faults and slumping on hydrocarbon migration inferred from 3D seismic attributes: Sanriku-Oki forearc basin, northeast Japan. *Marine and Petroleum Geology*, **99**, 175–189, https://doi.org/10.1016/j.marpetgeo.2018.10.013

Flemings, P.B., Long, H. *et al.* 2008. Pore pressure penetrometers document high overpressure near the seafloor where multiple submarine landslides have occurred on the continental slope, offshore Louisiana, Gulf of Mexico. *Earth and Planetary Science Letters*, **269**, 309–325, https://doi.org/10.1016/j.epsl.2007.12.005

Fontijn, K., Rawson, H. *et al.* 2016. Synchronisation of sedimentary records using tephra: a postglacial tephrochronological model for the Chilean Lake District. *Quaternary Science Reviews*, **137**, 234–254, https://doi.org/10.1016/j.quascirev.2016.02.015

Gamboa, D. and Alves, T.M. 2015. Three-dimensional fault meshes and multi-layer shear in mass-transport blocks: implications for fluid flow on continental margins. *Tectonophysics*, **647–648**, 21–32, https://doi.org/10.1016/j.tecto.2015.02.007

Gross, F., Mountjoy, J.J. *et al.* 2018. Free gas distribution and basal shear zone development in a subaqueous landslide – Insight from 3D seismic imaging of the Tuaheni Landslide Complex, New Zealand. *Earth and Planetary Science Letters*, **502**, 231–243, https://doi.org/10.1016/j.epsl.2018.09.002

Kanamatsu, T., Kawamura, K., Strasser, M., Novak, B. and Kitamura, Y. 2014. Flow dynamics of Nankai Trough submarine landslide inferred from internal deformation using magnetic fabric. *Geochemistry, Geophysics, Geosystems*, **15**, 4079–4092, https://doi.org/10.1002/2014GC005409

Kneller, B., Dykstra, M., Fairweather, L. and Miliana, P. 2016. Mass-transport and slope accommodation: implications for turbidite sandstone reservoir. *American Association of Petroleum Geologists Bulletin*, **100**, 213–235, https://doi.org/10.1306/09011514210

Laberg, J.S., Strasser, M., Alves, T.M., Gao, S., Kawamura, K., Kopf, A. and Moore, G.F. 2017. Internal deformation of a muddy gravity flow and its interaction with the seafloor (site C0018 of IODP Expedition 333, Nankai Trough, SE Japan). *Landslides*, **14**, 849–860, https://doi.org/10.1007/s10346-016-0766-7

Lunne, T., Robertson, P.K. and Powell, J.J.M. 1997. Cone penetration testing in geotechnical practice. *Soil Mechanics and Foundation Engineering*, **46**, 237, https://doi.org/10.1007/s11204-010-9072-x

Meckel, L.D. 2011. Reservoir characteristics and classification of sand-prone submarine mass-transport deposits. *SEPM Special Publications*, **96**, 423–452, https://doi.org/10.2110/sepmsp.096.423

Missiaen, T., Murphy, S., Loncke, L. and Henriet, J.-P. 2002. Very high-resolution seismic mapping of shallow gas in the Belgian coastal zone. *Continental Shelf Research*, **22**, 2291–2301, https://doi.org/10.1016/S0278-4343(02)00056-0

Moernaut, J., De Batist, M., Heirman, K., Van Daele, M., Pino, M., Brummer, R. and Urrutia, R. 2009. Fluidization of buried mass-wasting deposits in lake sediments and its relevance for paleoseismology: results from a reflection seismic study of lakes Villarrica and Calafquén (South-Central Chile). *Sedimentary Geology*, **213**, 121–135, https://doi.org/10.1016/j.sedgeo.2008.12.002

Moernaut, J., Van Daele, M. *et al.* 2014. Lacustrine turbidites as a tool for quantitative earthquake reconstruction: new evidence for a variable rupture mode in south central Chile. *Journal of Geophysical Research – Solid Earth*, **119**, 1607–1633, https://doi.org/10.1002/2013JB010738

Moernaut, J., Wiemer, G. *et al.* 2017. The influence of overpressure and focused fluid flow on subaquatic slope stability in a formerly glaciated basin: Lake Villarrica (South-Central Chile). *Marine Geology*, **383**, 35–54, https://doi.org/10.1016/j.margeo.2016.11.012

Moernaut, J., Van Daele, M. *et al.* 2018. Larger earthquakes recur more periodically: new insights in the megathrust earthquake cycle from lacustrine turbidite records in south-central Chile. *Earth and Planetary Science Letters*, **481**, 9–19, https://doi.org/10.1016/j.epsl.2017.10.016

Moernaut, J., Van Daele, M. *et al.* 2019. The subaqueous landslide cycle in south-central Chilean lakes: the role of tephra, slope gradient and repeated seismic shaking. *Sedimentary Geology*, **381**, 84–105, https://doi.org/10.1016/j.sedgeo.2019.01.002

Parez, L. and Fauriel, R. 1988. Le piézocône. Améliorations apportées à la reconnaissance de sols. *Revue Française de Géotechnique*, **44**, 13–27, https://doi.org/10.1051/geotech/1988044013

Praet, N., Moernaut, J. *et al.* 2017. Paleoseismic potential of sublacustrine landslide records in a high-seismicity setting (south-central Alaska). *Marine Geology*, **384**, 103–119, https://doi.org/10.1016/j.margeo.2016.05.004

Riboulot, V., Cattaneo, A., Sultan, N., Carziglia, S., Ker, S., Imbert, P. and Voisset, M. 2013. Sea-level change and free gas occurrence influencing a submarine landslide and pockmark formation and distribution in Deepwater Nigeria. *Earth and Planetary Science Letters*, **375**, 78–91, https://doi.org/10.1016/j.epsl.2013.05.013

Sammartini, M., Moernaut, J., Anselmetti, F.S., Hilbe, M., Lindhorst, K., Praet, N. and Strasser, M. 2019. An atlas of mass-transport deposits in lakes. *In*: Ogata, K., Festa, A. and Pini, A. (eds) *Submarine Landslides: Subaqueous Mass Transport Deposits from Outcrops to Seismic Profiles*. AGU Geophysical Monograph Series. Wiley, 201–226, https://doi.org/10.1002/9781119500513.ch13

Sawyer, D.E., Flemings, P.B., Dugan, B. and Germaine, J.T. 2009. Retrogressive failures recorded in mass transport deposits in the Ursa Basin, Northern Gulf of Mexico. *Journal of Geophysical Research*, **114**, B10102, https://doi.org/10.1029/2008JB006159

Sobiesiak, M.S., Kneller, B., Alsop, G.I. and Milana, J.P. 2018. Styles of basal interaction beneath mass transport deposits. *Marine and Petroleum Geology*, **98**, 629–639, https://doi.org/10.1016/j.marpetgeo.2018.08.028

Stegmann, S., Moerz, T. and Kopf, A.J. 2006. Initial results of a new Free-Fall Cone Penetrometer (FF-CPT) for geotechnical in situ characterisation of soft marine sediments. *Norwegian Journal of Geology*, **86**, 199–208.

Steiner, A., Kopf, A.J., L'Heureux, J.-S., Kreiter, S., Stegmann, S., Haflidason, H. and Moerz, T. 2014. In-situ dynamic piezocone penetrometer tests in natural clayey sediments – a reappraisal of strain-rate corrections. *Canadian Geotechnical Journal*, **51**, 272–288, https://doi.org/10.1139/cgj-2013-0048

Steiner, A., Kopf, A.J., Henry, P., Stegmann, S., Apprioual, R. and Pelleau, P. 2015. Cone penetration testing to assess slope stability in the 1979 Nice landslide area (Ligurian Margin, SE France). *Marine Geology*, **369**, 162–181, https://doi.org/10.1016/j.margeo.2015.08.008

Stoecklin, A., Friedli, B. and Puzrin, A.M. 2017. Sedimentation as a control for large submarine landslides: mechanical modelling and analysis of the Santa Barbara basin. *Journal of Geophysical Research: Solid Earth*, **122**, 8645–8663, https://doi.org/10.1002/2017JB014752

Strasser, M., Henry, P., Kanamatsu, T., Thu, M.K. and Moore, G.F. and IODP Expedition 333 Scientists 2012. Scientific drilling of mass-transport deposits in the Nankai accretionary wedge: first results from IODP Expedition 333. *Advances in Natural and Technological Hazards Research*, **31**, 671–681, https://doi.org/10.1007/978-94-007-2162-3_60

Sully, J.P., Robertson, P.K., Campanella, R.G. and Woeller, D.J. 1999. An approach to evaluation of field CPTU dissipation data in overconsolidated fine-grained soils. *Canadian Geotechnical Journal*, **36**, 369–381, https://doi.org/10.1139/t98-105

Sun, Q., Alves, T.M., Xie, X., He, J., Li, W. and Ni, X. 2017. Free gas accumulations in basal shear zones of mass-transport deposits (Pearl River Mouth Basin, South China Sea): an important geohazard on continental slope basins. *Marine and Petroleum Geology*, **81**, 17–32, https://doi.org/10.1016/j.marpetgeo.2016.12.029

Sun, Q., Alves, T.M., Lu, X., Chen, C. and Xie, X. 2018. True volumes of slope failure estimated from a Quaternary mass-transport deposit in the northern South China Sea. *Geophysical Research Letters*, **45**, 2642–2651, https://doi.org/10.1002/2017GL076484

Tóth, Z., Spieß, V. and Jensen, J. 2014. Seismo-acoustic signatures of shallow free gas in the Bornholm Basin, Baltic Sea. *Continental Shelf Research*, **88**, 228–239, https://doi.org/10.1016/j.csr.2014.08.007

Van Daele, M., Moernaut, J. *et al.* 2015. A comparison of the sedimentary records of the 1960 and 2010 great Chilean earthquakes in 17 lakes: implications for quantitative lacustrine paleoseismology. *Sedimentology*, **62**, 1466–1496, https://doi.org/10.1111/sed.12193

Vardy, M.E., Vanneste, M., Henstock, T.J., Clare, M.A., Forsberg, C.F. and Provenzano, G. 2017. State-of-the-art remote characterization of shallow marine sediments: the road to a fully integrated solution. *Near Surface Geophysics*, **15**, 387–402, https://doi.org/10.3997/1873-0604.2017024

Wiemer, G., Moernaut, J. *et al.* 2015. The role of sediment composition and behavior under dynamic loading conditions on slope failure initiation: a study of a subaqueous landslide in earthquake-prone South-Central Chile. *International Journal of Earth Sciences*, **104**, 1439–1457, https://doi.org/10.1007/s00531-015-1144-8

Wiemer, G., Dziadek, R. and Kopf, A. 2017. The enigmatic consolidation of diatomaceous sediment. *Marine Geology*, **385**, 173–184, https://doi.org/10.1016/j.margeo.2017.01.006

Influence of mass transport deposit (MTD) surface topography on deep-water deposition: an example from a predominantly fine-grained continental margin, New Zealand

Suzanne Bull[1]*, Greg H. Browne[1], Malcolm J. Arnot[1] and Lorna J. Strachan[2]

[1]GNS Science, 1 Fairway Drive, Lower Hutt, New Zealand, 5010

[2]University of Auckland, 23 Symonds Street, Auckland, New Zealand, 1010

SB, 0000-0003-0909-1415; LJS, 0000-0001-9522-1583

*Correspondence: s.bull@gns.cri.nz

Abstract: Three-dimensional (3D) seismic data reveal the complex interplay between the surface topography of a *c.* 4405 km^3 mass transport deposit (MTD) and overlying sedimentary packages over approximately the last two million years. The data image part of the Pleistocene to recent shelf to slope to basin-floor Giant Foresets Formation in offshore western New Zealand. The MTD created substantive topographic relief and rugosity at the contemporaneous seabed, formed by the presence of a shallow basal detachment surface, and very large (up to 200 m high) intact slide blocks, respectively. Sediments were initially deflected away from high-relief MTD topography and confined in low areas. With time, the MTD was progressively healed by a series of broadly offset-stacked and increasingly unconfined packages comprised of many channel bodies and their distributary complexes. Positive topography formed by the channels and their distributary complexes further modified the seafloor and influenced the location of subsequent sediment deposition. Channel sinuosity increased over time, interpreted as the result of topographic healing and reduced seafloor gradients. The rate of sediment supply is likely to have been non-uniform, reflecting tectonic pulses across the region. Sediments were routed into deep water via slope-confined channels that originated shortly before emplacement of the MTD.

Episodic sedimentation processes occur as a continuum ranging from subaqueous mass-wasting events to dilute turbidity currents (Stow *et al.* 1996). All of these processes are dependent upon gravity acting upon material stored on, or moving down, a slope and are the primary means by which sediments are transferred into deep-water settings (Einsele *et al.* 1991; Stow *et al.* 1996). Long-term sedimentary archives from deep-water settings can yield valuable information relating to tectonic changes, climatic variability and geohazard occurrence over a range of timescales. Such information underpins robust future projections critical to hazard mitigation and planning for sustainable and resilient societies. However, unravelling the narrative contained within deep-water strata is challenging. Accessing long-term and complete deep-water sedimentary records is difficult and there is a general paucity of such data (Allin *et al.* 2018). In addition, the sensitivity of deep-water sedimentary flow-routing to evolving seafloor topography during deposition introduces complex feedback behaviour (Burgess *et al.* 2019). This has the effect of 'breaking up' external signals and making them difficult to distinguish from autogenic controls (Paola 2017).

Mass transport deposits (MTDs) are ubiquitous in modern and ancient subaqueous slopes and can involve vast volumes of sediment, such that they have been argued to be the principal sediment transport mechanism operating on continental margins (Nardin *et al.* 1979). MTDs can have a significant impact on seafloor topography and therefore the configuration of subsequent sedimentary deposits. Specifically, MTD occurrence has been shown to create significant bathymetric depressions by evacuation of slope sediments (Hackbarth and Shew 1994; Schultz *et al.* 2005; Qin *et al.* 2017). Conversely, substantial bathymetric highs created by the accumulation of remobilized sediments can divert subsequent sediment flows, leading to the generation of new sediment routes (Hansen *et al.* 2013; Corella *et al.* 2016; Masalimova *et al.* 2016). The typically irregular topography of MTDs can locally capture and focus sediment flows (Kastens and Shor 1985; Deptuck *et al.* 2003; Pickering and Corregidor 2005; Gee *et al.* 2007; Kneller *et al.* 2016; Ward *et al.* 2018; Zhao *et al.* 2019) or facilitate localized ponding (Pickering and Corregidor 2005; Schultz *et al.* 2005; Moscardelli *et al.* 2006; Moscardelli and Wood 2008; Armitage *et al.* 2009; Jackson

From: Georgiopoulou, A., Amy, L. A., Benetti, S., Chaytor, J. D., Clare, M. A., Gamboa, D., Haughton, P. D. W., Moernaut, J. and Mountjoy, J. J. (eds) 2020. *Subaqueous Mass Movements and their Consequences: Advances in Process Understanding, Monitoring and Hazard Assessments*. Geological Society, London, Special Publications, **500**, 147–171.
First published online April 28, 2020, https://doi.org/10.1144/SP500-2019-192

and Johnson 2009; Alves 2010; Olafiranye *et al.* 2013). The volume of localized accommodation space is typically controlled by the internal MTD structure and in particular by the presence and distribution of large slide blocks, which can be thicker than the enclosing MTD (McAdoo *et al.* 2000; Pickering and Corregidor 2005; Armitage *et al.* 2009; Alves 2010; Gamboa *et al.* 2010; Olafiranye *et al.* 2013; Ortiz-Karpf *et al.* 2015; Kneller *et al.* 2016; Ward *et al.* 2018). Furthermore, once-buried lithological contrasts between slide blocks, the surrounding MTD matrix and the overburden may create differential compaction, which can result in the persistence of seafloor rugosity long after emplacement of the MTD (Alves and Cartwright 2010; Ward *et al.* 2018).

Here we describe the interactions between a *c.* 4000 km^3 Pleistocene-age MTD (first described by Omeru 2014) from the Deepwater Taranaki Basin (Fig. 1), offshore western New Zealand, and subsequent sedimentary flow processes that have been deposited over the last 2 million years, up to and including the present day. The succession we focus on has accumulated within the rapidly evolving Australia–Pacific plate boundary zone and throughout a time period that has seen dramatic changes in both local and global climatic conditions. The excellent spatial resolution (7.5 m vertically) of the 3D seismic volumes (Fig. 1), upon which this analysis is primarily based, allows detailed examination of the MTD surface morphology and stratal geometries of the subsequently deposited units. We

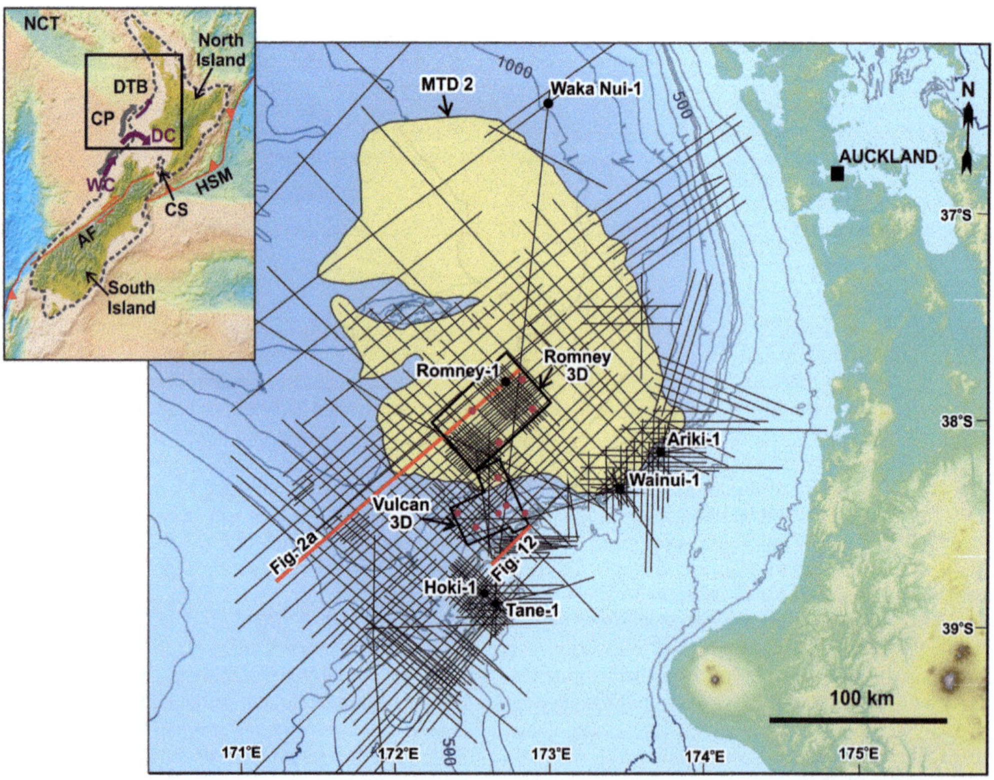

Fig. 1. Bathymetric map of the NW New Zealand region. Bathymetric contours are in metres below sea-level and are drawn every 100 m. 3D seismic data are indicated by black polygons and 2D seismic data by black lines. Petroleum wells used in this study are shown by black dots. Red dots indicate locations of surficial sediment sample stations from the nzSEABED database (Bostock *et al.* 2019*a*, *b*). The outline of MTD 2 (labelled) is based on mapping across the 2D seismic data. Inset: Regional setting of the study area, including the Deepwater Taranaki Basin (black rectangle). The outline of the New Zealand landmass at *c.* 20 ka is indicated by the grey dashed boundary (after Lorrey *et al.* 2012). Historical oceanic currents are shown by grey arrows (after Lorrey *et al.* 2012), and modern-day currents by purple arrows (after Carter *et al.* 1998). NCT, New Caledonia Trough; DTB, Deepwater Taranaki Basin; CP, Challenger Plateau; AF, Alpine Fault; CS, Cook Strait; HSM, Hikurangi Subduction margin; WC, West coast current; DC, D'Urville current.

describe basin-floor depositional units fed by a series of slope-confined channels and their downslope continuations as deep-water channels. We address the ongoing interplay between downslope processes that have shaped the recent and present-day seafloor. Specifically, we seek to answer the following key questions: (1) what impact does the emplacement of such volumetrically significant MTDs have on the ongoing deposition of relatively smaller volume, higher frequency sediment gravity flows? (2) Is it possible to identify external signals from the observed stratal patterns? (3) What can our results tell us about the conditions under which MTDs have typically occurred in the area and the potential for future reoccurrence?

Geological setting and development

The Deepwater Taranaki Basin is the NW extension of the petroleum-producing Taranaki Basin, and itself extends to the NW beneath the present-day New Caledonia Trough (Fig. 1). The Deepwater Taranaki Basin originated in the mid-Cretaceous as the first of two temporally distinct rifting phases that record the Gondwana break-up (Strogen *et al.* 2017). The 'early syn-rift' succession (*c.* 105–95 Ma) (Fig. 2a) is inferred to be mainly terrestrial, while the 'late syn-rift' (*c.* 95–83 Ma) (Fig. 2a) strata is dominantly marine transgressive (Strogen *et al.* 2017). Up to 10 km of sediment has accumulated above these rift sequences (for details see Uruski *et al.* 2002; Uruski 2008, 2010; Baur *et al.* 2014; Arnot *et al.* 2016) including a progradational Neogene package (Fig. 2a). The Neogene succession is characterized by high clastic input related to the onset of subduction along the Hikurangi margin further to the east (Fig. 1) that subsequently evolved into the modern-day plate boundary (Stern and Davey 1989; King and Thrasher 1992, 1996; Holt and Stern 1994; Kamp *et al.* 2004; Pulford and Stern 2004; Nicol *et al.* 2005, 2007; Stagpoole and Nicol 2008; Trewick and Bland 2012).

Throughout the Miocene, large mounded channel and turbidite systems formed a broad compensational stacking pattern (Uruski 2008; Baur *et al.* 2014) succeeded by the Giant Foresets Formation (GFF). The GFF is a late-early Pliocene to recent succession of shelf to slope to basin-floor fine-grained muds, silts and sands (Beggs 1990; Hansen and Kamp 2004*a*, *b*) (Fig. 2b). Seismic reflection profiles show that the GFF locally reaches thicknesses of over 2 km and is characterized by high-relief clinoforms (up to 1500 m thick; Salazar *et al.* 2015) that offlap in a basinward direction. Very rapid shelf-wedge deposition is attributed to increased sediment supply related to uplift of the ancestral Southern Alps and central North Island hinterland, both in response to increasing convergence along the plate boundary from the Pliocene to the present day (Kamp *et al.* 2004; Hansen and Kamp 2006). Derived sediments were transported north and westward by longshore drift and oceanic current action (Hansen and Kamp 2002; Anell and Mitkandel 2015; Salazar *et al.* 2015) at estimated sedimentation rates of up to 450 m Ma^{-1} (ARCO 1992).

As a result of high Neogene sediment supply and deposition, the modern-day continental shelf and slope of the Deepwater Taranaki Basin is a major sediment storage area, with a broad curvilinear shelf that is up to 112 km across in the study area (Fig. 1). The shelf edge occurs in *c.* 200 m water depth and, along with the slope, is heavily incised by canyons (Uruski 2008; Li *et al.* 2017). The canyons, which we hereafter refer to as 'slope channels', continue downslope into the deeper water of the basin floor (i.e. >1500 m) as submarine channels and fans (Uruski 2008). The basin floor slopes at <1.0° and is locally smooth to slightly rugose, forming a broad bathymetric depression which continues to the New Caledonia Trough to the NW (Uruski 2008; Rad 2015) (Fig. 1).

In this study we focus on Pleistocene–recent sediments and also consider contemporary sedimentation. MTDs locally comprise *c.* 50% of the Pleistocene–recent succession (Omeru 2014). At the Romney-1 well location (Fig. 1) sediments of this age are predicted to be clays and silts with sand interbeds, along with MTDs consisting of claystones and sandstones (no lithological samples younger than Miocene-age were recovered during drilling of Romney-1; Rad 2015). The most recent surficial sediments of the continental shelf are predominantly muddy in coastal regions (<60%) and sand-dominated in the mid-shelf area with some gravel (up to 80% sand and up to 40% gravel; Bostock *et al.* 2019*b*). Moving into deeper water, the continental slope and deep basin are blanketed by mud, most of which is terrigenous in origin, sourced from the rivers along the west coast of South Island and, to a lesser extent, from the North Island (Bostock *et al.* 2019*a*, *b* and references therein). On the slope, the muds are mixed with calcareous macro- and microfossils and carbonate-bearing sediments (Bostock *et al.* 2019*a*). The latter are considered to be relict from previous lower sea-levels (Carter and Heath 1975; Gillespie *et al.* 1998), while high mud contents generally reflect deep water and low current speeds (Bostock *et al.* 2019*a*).

Data sources and methods

The primary dataset for this study is the Romney 3D survey (Fig. 1), acquired in 2011. It images an 1826 km^2 area of the continental rise from water

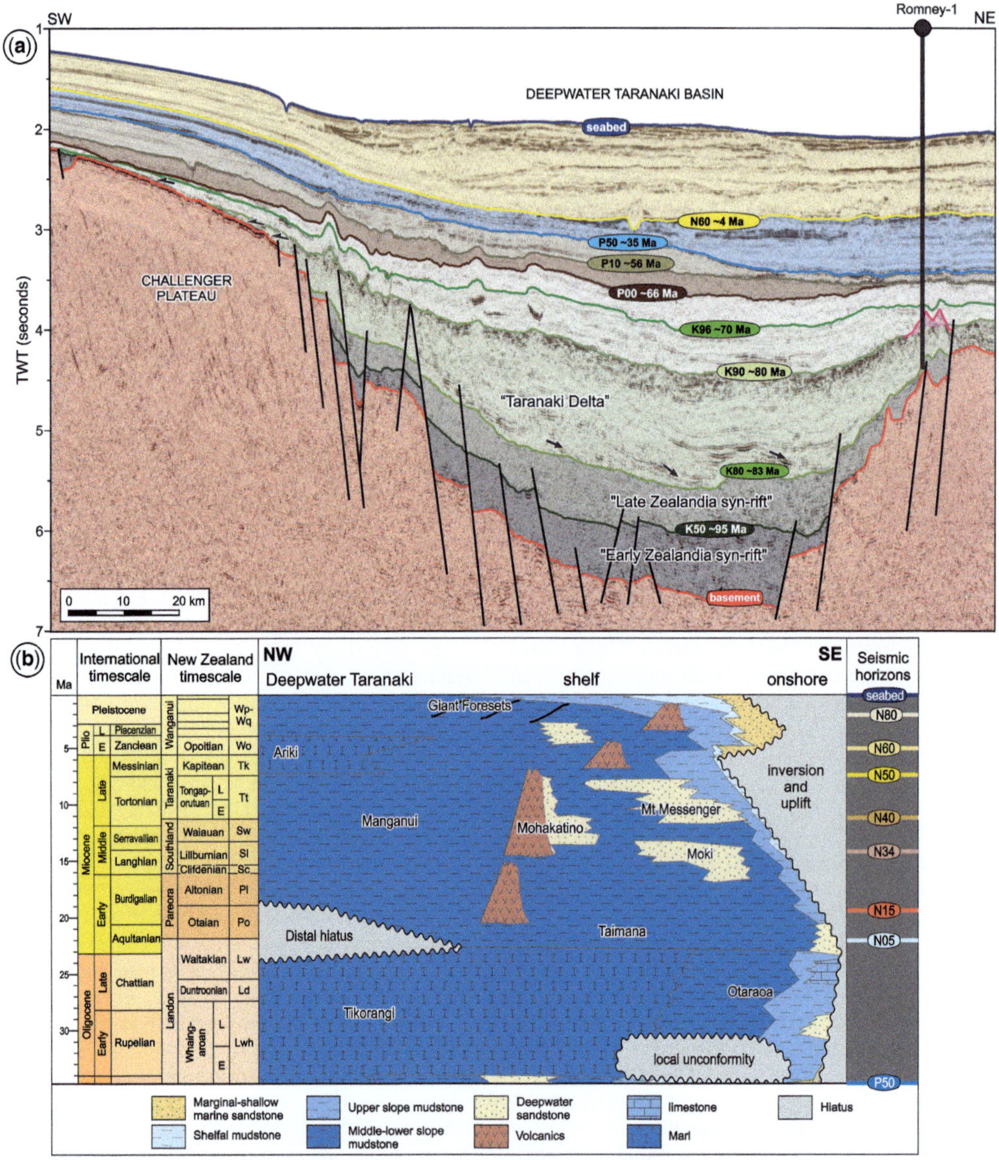

Fig. 2. (**a**) Interpreted regional cross-section through the Deepwater Taranaki Basin (2D seismic Line dtb01-18) (modified after Strogen *et al.* 2017). Regional seismic horizons shown are taken from Arnot *et al.* (2016). Please note orientation is normal to the trend of the Deepwater Taranaki Basin. Location shown in Figure 1. (**b**) Chronostratigraphic panel showing Neogene formations of the Taranaki Basin and key stratigraphic facies and trends during the Neogene (taken from Bull *et al.* 2016) including the Giant Foresets Formation upon which this study focuses. Seismic horizons shown are taken from Arnot *et al.* (2016), with the exception of N80 (*c.* 2 Ma) which is based on Bull *et al.* (2016). Horizon names are based on the convention set out in Strogen and King (2014).

depths of 1340–1550 m (Fig. 1). The data were post-stack time-migrated and zero-phased, meaning the geophysical interface that has produced the reflection is located at the positive peak or negative trough of the wavelet. Because the data are normal polarity according to the convention adopted by the Society of Exploration Geophysicists (SEG), an increase in acoustic impedance (i.e. at the seabed) is represented by a positive number and a peak in the seismic record. The main frequency is 50 Hz, yielding a vertical resolution of *c.* 7.5 m at the seabed using a seawater velocity of 1500 m s^{-1}. The sample rate is

2 ms and the final bin spacing of the seismic grid is 25 × 25 m. We also utilize the adjacent Vulcan 3D survey, which images 1086 km^2 of the continental slope and rise, from water depths of 175–1360 m (Fig. 1). The survey is pre-stack time-migrated and processed to near zero-phase characteristics and displayed as SEG normal polarity. The dominant frequency of the data is 50 Hz, the sample rate is 2 ms, and the final bin spacing of the seismic grid is 25 × 25 m.

Where possible, we also draw upon an extensive two-dimensional (2D) seismic grid derived from a variety of surveys of differing vintages and data characteristics (Fig. 1), and a 250 m gridded regional bathymetry dataset published by the National Institute of Water and Atmospheric Research Ltd (NIWA 2016). All seismic data used in this study are publicly available through the Ministry of Business, Innovation and Employment (MBIE) as the New Zealand Exploration Data Pack. The regional seismic stratigraphic framework for our detailed analysis is provided by recent regional mapping studies covering the offshore NW region of New Zealand (Arnot *et al.* 2016). Up to 20 regionally extensive Cretaceous–Cenozoic horizons have been mapped to define regionally extensive, time-equivalent depositional surfaces and are named using the Zealandia-wide scheme of Strogen and King (2014) (Fig. 2). The MTDs in the broader north Taranaki offshore area have been previously described by Uruski (2008) and Rad (2015) and studied by Omeru (2014), Rusconi (2017) and Panpichityota *et al.* (2018). Omeru *et al.* (2016) and Omeru and Cartwright (2019) focused on the two most recent MTDs, which they termed MTD 1 and MTD 2. Onshore outcropping equivalent MTDs have been studied from the late Miocene succession cropping out within the north Taranaki coastal section of North Island (e.g. King *et al.* 2011; Sharman *et al.* 2015; Masalimova *et al.* 2016; Bull *et al.* 2019). The present study focuses on MTD 2 of Omeru *et al.* (2016) and Omeru and Cartwright (2019), and we adopt this name for continuity. MTD 2 is the most recent of six large-scale MTDs in the Deepwater Taranaki Basin (Omeru 2014). It has an estimated age of *c.* 1.8 Ma based on the micropaleontological studies of Hayward (1984) and Hansen and Kamp (2006). It has an area of 26 470 km^2 and estimated volume of 4405 km^3 (based on analysis carried out as part of the present study), and was directed to the SW (Omeru 2014; Omeru and Cartwright 2019). The basal detachment surface varies from concordant to erosive, exhibiting a step-like series of down-cutting ramps (Omeru 2014) (Fig. 3). The upper surface is represented by a high-amplitude reflection and its topography varies significantly from smooth to highly rugose (Omeru 2014) (Fig. 3). High rugosity is associated with the presence of large intact slide blocks within the MTD (Omeru 2014; Omeru and Cartwright 2019).

Detailed analysis presented herein is based on integrated 3D seismic interpretation techniques including horizon interpretation, construction of time structure and isochron (time–thickness) maps, volume-based seismic attribute extraction, time-slice interpretation and 3D visualization using Paradigm™ interpretation software. Where necessary, horizons have been merged to form continuous surfaces for the sake of creating isochron maps, with the youngest stratigraphic surface taking priority over the older surface. No lithologic or well data are available to tie directly to the 3D seismic data. Therefore, estimated seismic velocities of 1500 m s^{-1} for seawater and 1900 m s^{-1} for the shallow subsurface sediments have been used when converting from time to the depth domain, based on velocity data taken from the key offset wells (Fig. 1).

Data analysis

To provide context, we have mapped MTD 2 to the fullest possible extent using the regional grid of 2D seismic data available for this study (Fig. 1). MTD 2 was identified by recognition of a disrupted and chaotic seismic facies unit representing the failed mass, defined at its base and top by continuous, undeformed surfaces (the basal surface and top surfaces, respectively) (Bull *et al.* 2009). We note that our MTD 2 outline extends beyond the original outline mapped by Omeru (2014). Following emplacement of MTD 2, the seafloor began to heal with the progressive deposition of background sedimentation and the influx of sediment gravity flows. Five seismic horizons have been interpreted to subdivide the overburden succession above the MTD, and are assumed to represent Pleistocene–recent deposition, up to and including the present-day seafloor (horizons H1–H4 and Seafloor; Fig. 3a). The horizons divide the overburden into five seismic facies packages based on several parameters, including amplitude, geometry and continuity of significant reflections, which together describe seismic character. Based on these parameters and their stratal relationships with other seismic packages, we infer stratigraphic associations and depositional environments based to some extent on previous seismic studies of deep-water systems made elsewhere (e.g. Posamentier and Kolla 2003; Catterall *et al.* 2010; Ortiz-Karpf *et al.* 2015) and the geology of the study area (Uruski 2008; Omeru 2014; Rad 2015; Arnot *et al.* 2016; Omeru and Cartwright 2019). In the following sections we first describe pertinent observations from MTD 2, followed by each subsequent overburden package in stratigraphic order from oldest to youngest.

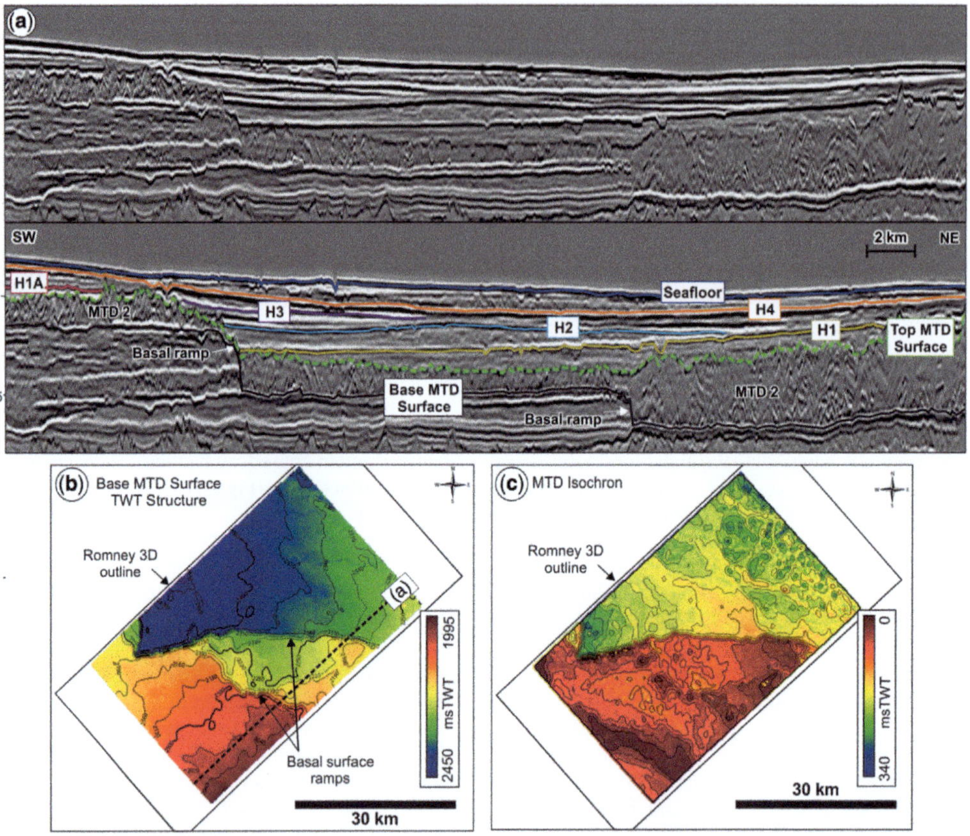

Fig. 3. (**a**) Uninterpreted (top) and interpreted (bottom) seismic traverse from the Romney 3D survey showing seismic character of MTD 2, overburden packages and mapped seismic horizons. Location shown in (b). (**b**) TWT structure map of the Base MTD Surface. Contours are every 20 ms. Location of Romney 3D survey is shown in Figure 1. (**c**) Isochron (TWT thickness) map of MTD 2. Contours are every 20 ms.

MTD 2

Using the Romney 3D survey, we have mapped part of the top surface of MTD 2 in detail (Top MTD Surface; Fig. 3a.). The basal surface of MTD 2 has also been mapped within the Romney 3D survey ('Base MTD Surface' on Fig. 3a, b) to allow consideration of major basal surface features and MTD thickness distribution (Fig. 3b, c). A comparison of the Base MTD Surface two-way-time (TWT) structure map and MTD thickness reveals that the presence of basal surface ramps corresponds to major changes in MTD thickness (Fig. 3).

The Top MTD Surface TWT structure map gives an approximation of the complex seafloor topography created following emplacement of MTD 2 (Fig. 4), which consists of variation in both the relief and texture of the surface. The top surface of MTD 2 exhibits relief, or the difference in height between the highest and lowest points, of up to 360 ms (300 m; Fig. 4). The surface texture of the MTD 2 varies between highly rugose and smooth, due to small-scale variations in surface height caused by the presence of large intact blocks deposited within MTD 2 (Fig. 3a). Many blocks exceed the general thickness of MTD 2 and create highly localized features of positive relief on a scale of tens of metres vertically and hundreds of metres horizontally.

The variations in each of these factors define three distinct zones within the Romney 3D area, which we term zones 1–3 (Fig. 4). Zone 1 has the highest relative elevation. Although this zone does not coincide with the thickest areas of MTD 2, the basal surface of MTD 2 is situated at a relatively shallow stratigraphic level (Fig. 3). Zone 1 also has a high level of rugosity, with numerous individual and highly localized features of high positive relief that correspond in seismic section to deformed but relatively intact blocks of up to 135 m in height and 560 m width (Fig. 3a). Individual blocks exceed

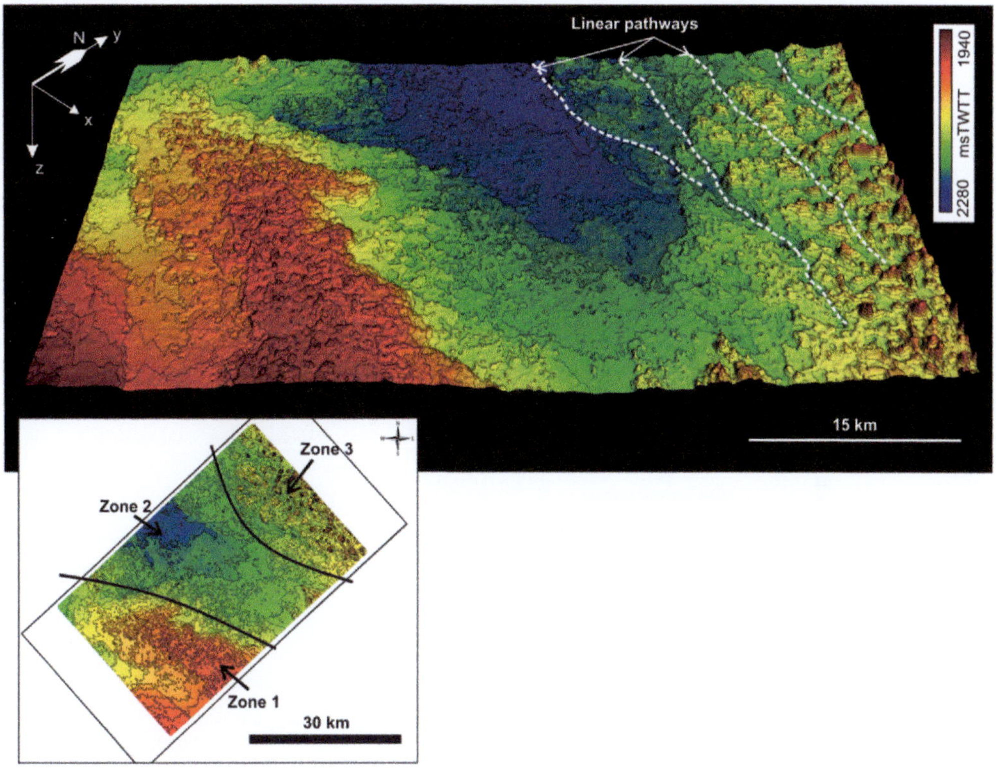

Fig. 4. TWT structure map of Top MTD Surface, shown in a 3D perspective view. Contours are every 20 ms. Inset: zones 1–3 of differing MTD Top Surface character.

the thickness of the surrounded MTD by <100 m (Fig. 3a). In the central and western zones of the Romney 3D area, zone 2 possesses pronounced lower topography (Fig. 4). In zone 2, MTD 2 is thick (145–180 m) with a smooth top surface and internally is seismically transparent, lacking continuous internal reflections (Omeru and Cartwright 2019) (Fig. 3). In the northern area of the Romney 3D survey, zone 3 possesses very high rugosity, created by very localized variations in relief linked to the presence of numerous very large intact blocks previously identified by Omeru (2014) and Omeru and Cartwright (2019) (Fig. 4). Blocks of up to 290 m high and 2800 m wide are noted (Figs 3 & 4), which exceed the general thickness of MTD 2 by up to 100 m and, in places, extrude above the subsequent sediment cover and are evident as morphological features on the present-day seabed (Omeru 2014). A small number of subtle, roughly linear pathways can be discerned threading through the area of large intact blocks in zone 3 (labelled on Fig. 4), suggesting that conduits for oceanic and/or turbidity currents formed and reworked the top of the MTD to a minor degree.

Overburden Package 1 (Top MTD Surface–H1)

This package lies directly on top of MTD 2 and is up to 120 ms (110 m) thick (Fig. 5a). Internally it comprises parallel to sub-parallel, continuous, low-amplitude, locally slightly deformed reflections which onlap and downlap onto the top of MTD 2 (Fig. 5b). Package 1 overlies zone 2, and also occupies more localized depocentres in zone 3, created by the presence of large slide blocks within the MTD (Figs 3, 4 & 5c). Channel bodies are interpreted from relatively high-amplitude reflector facies that can be flat, inclined or concave in cross-section, and in planform create a series of laterally continuous and straight-to-sinuous distributary channels (Mitchum *et al.* 1977; Posamentier and Kolla 2003) (Fig. 5b).

Channels present within Package 1 are identified from amplitude extractions and flattened time-slices (Fig. 5d). Multiple channels bodies *c.* 200–400 m wide travel to the west and NW, and exhibit sinuosity indexes (SI = channel thalweg length/straight-line distance) of up to 1.17. Conversely, at

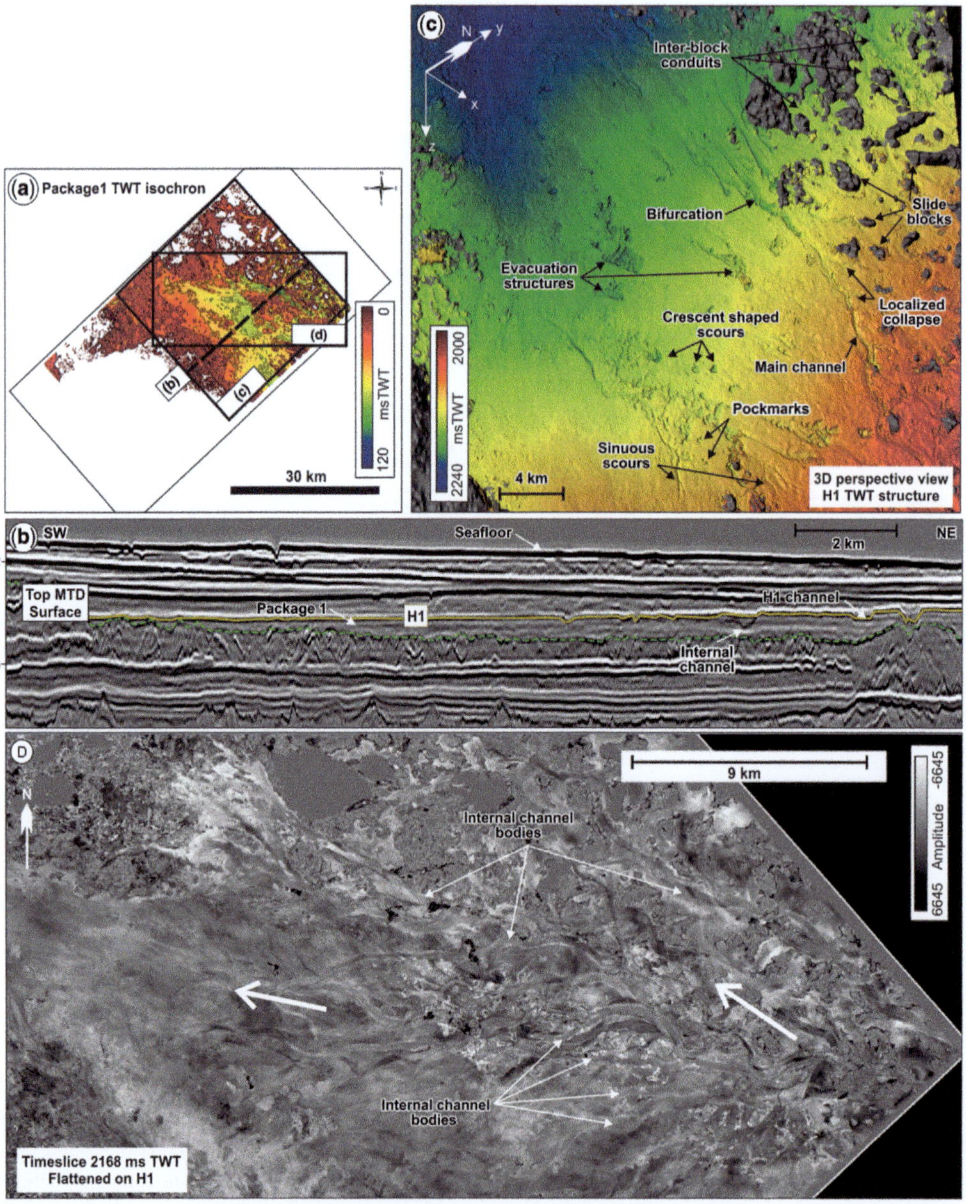

Fig. 5. (**a**) Isochron map showing TWT thickness and distribution of Package 1. (**b**) 3D seismic traverse from Romney 3D showing the character of Package 1. Location shown in (a). (**c**) TWT structure map of horizon H1, shown in a 3D perspective view. Location shown in (a). (**d**) Reflectivity time-slice 2168, flattened on horizon H1. Note the presence of numerous channels trending west and NW (main directions indicated by bold white arrows).

the top of Package 1, horizon H1 features a prominent single channel (labelled 'main channel' on Fig. 5c) that originates outside of the Romney 3D survey and is directed to the NW, varying between 280 and 330 m wide and 3 and 28 m deep. It has relatively low sinuosity (SI = 1.09) and varies in depth along its course where it interacts with buried slide blocks and localized collapse features, some of which appear to feed into the main channel body (Fig. 5c).

Along with the main channel, horizon H1 also features further channel-like features that occur in the areas between intact slide blocks present in MTD 2 (labelled 'inter-block conduits' on Fig. 5c).

These inter-block conduits are inferred to have been formed by localized focusing of sediment flows between intact blocks. Other features of H1 include localized and elongate-downslope depression features with rough interiors (in the order of 100–500 m wide and 1–2 km long; labelled 'evacuation structure' on Fig. 5c), and smoother, smaller crescent-shaped features (*c.* 500 m to 1 km across; labelled 'crescent-shaped scours' on Fig. 5c). The former are interpreted as localized surface failures or evacuation structures that could be related to fluid escape, and the latter as localized seafloor scouring by downslope-directed currents.

Several oval to circular pit-like depressions are noted from horizon H1 (labelled on Fig. 5c), and are up to 280 m across and 20 m deep. These depressions are interpreted as pockmarks, classically defined as shallow crater-like seabed depressions attributed to fluid expulsion from the underlying sedimentary column that dislodges the soft seafloor sediments into the water column (King and MacLean 1970). They can be linked to various controls including faults and fractures, lithology, sedimentation rate, seabed erosion, sea-level change, biotic activity and development of overpressure (Eichhubl *et al.* 2000; Dimitrov and Woodside 2003; Gay *et al.* 2007; Judd and Hovland 2007; Andresen and Huuse 2011; Chenrai and Huuse 2017). Comparable palaeo-fluid expulsion features have been described by Chenrai and Huuse (2017) from a similar base-of-slope setting in the nearby northern Taranaki Basin.

Overburden Package 1A (Top MTD Surface–H1A)

Package 1A also overlies the top of MTD 2 (Fig. 6a) but is only present in zone 1, where it reaches a maximum of 100 ms (90 m) thick (Fig. 6b). It does not overlie or overlap with the previous Package 1 (compare Figs 6b to 5a) and appears to be laterally confined by the high-relief topography of zone 1 (Fig. 4). Package 1A is defined at its base by the Top MTD Surface and its top by horizon H1A (Fig. 6a). Internally, Package 1A is characterized by low-amplitude, moderate to high continuity, locally disrupted, sub-parallel to gently inclined reflections (Fig. 6a). Seismic reflection profiles and attribute extractions reveal evidence of a single channel at the top of the package (H1A; Fig. 6c, d). The channel originates from a zone of sinuous scours (Fig. 6c) and continues for 20 km before moving outside of the area imaged by the Romney 3D survey. It is similar in geometry to the channel observed from the previous Package 1, having an SI of 1.06, and width and depth ranges from 145–185 to 5–26 m, respectively. This channel also interacts with intact slide blocks from the MTD below and, in several instances, bifurcates around the blocks before merging again to continue as a single channel (Fig. 6c, d). However, it is unclear whether both channel branches were active at the same time or if one was abandoned following establishment of the preferred route.

Overburden Package 2 (H1–H2)

Package 2 is defined at its base by horizon H1 and at its uppermost limit by horizon H2 (Fig. 7a). It is a maximum of 100 ms (90 m) thick (Fig. 7b) with parallel to sub-parallel, continuous and low to moderate amplitude reflectors (Fig. 7a). RMS amplitude extractions from within Package 2 reveal a single channel with flow directed to the NNE and a 'channelized domain' consisting of several locally clustered and overlapping individual channel bodies, with flow directed to the NW (Fig. 7c). Package 2 occupies zones 1 and 2 (Fig. 4), and both onlaps and downlaps the positive topography formed by MTD 2 (Fig. 7a, d). In zone 1, Package 2 is cut by a pair of younger sinuous channels that incise down from the present-day seafloor (labelled on Fig. 7d). The area of Package 2's greatest thickness is offset compared to the main thickness of Package 1 (compare to Fig. 5a), and this is interpreted to reflect broad compensational stacking above the older deposits (terminology after Slatt 2006).

The thickest area of Package 2 is composed in part by two lobe-like features, one of which has an elongate south–north trend and the other has a locus orientated to the NW (Fig. 7b). Seismic reflection profiles reveal bi-directional downlap in cross-sectional view (as defined by Mitchum 1985), which is diagnostic of the formation of a discrete submarine fan deposit, formed as a series of levee deposits adjacent to a centralized distributary channel (Fig. 7a). Levees typically form as the finer-grained portion of individual sediment gravity flows overtop the banks of distributary channels and spreads laterally beyond the channel margin (Weimer and Slatt 2004). A channel body is evident at the top of this package (horizon H2; Fig. 7a, d), first appearing close to the boundary between the high elevation topography of zone 1 and the lower topography of zone 2 (labelled 'channel emergence' on Fig. 7d). The channel continues to the NNE with a low sinuosity thalweg (SI = 1.07) and width and depth ranges of 80–285 and 3–18 m, respectively. It likely that this channel originates further updip from its emergence, either within or updip of zone 1, but is obscured by both the complex seismic character and relationship of the MTD with the overburden. We infer that this channel exploits a locally steep route linking the high elevation MTD topography in zone 1 to the lower topography of zone 2, where it debouches and forms the

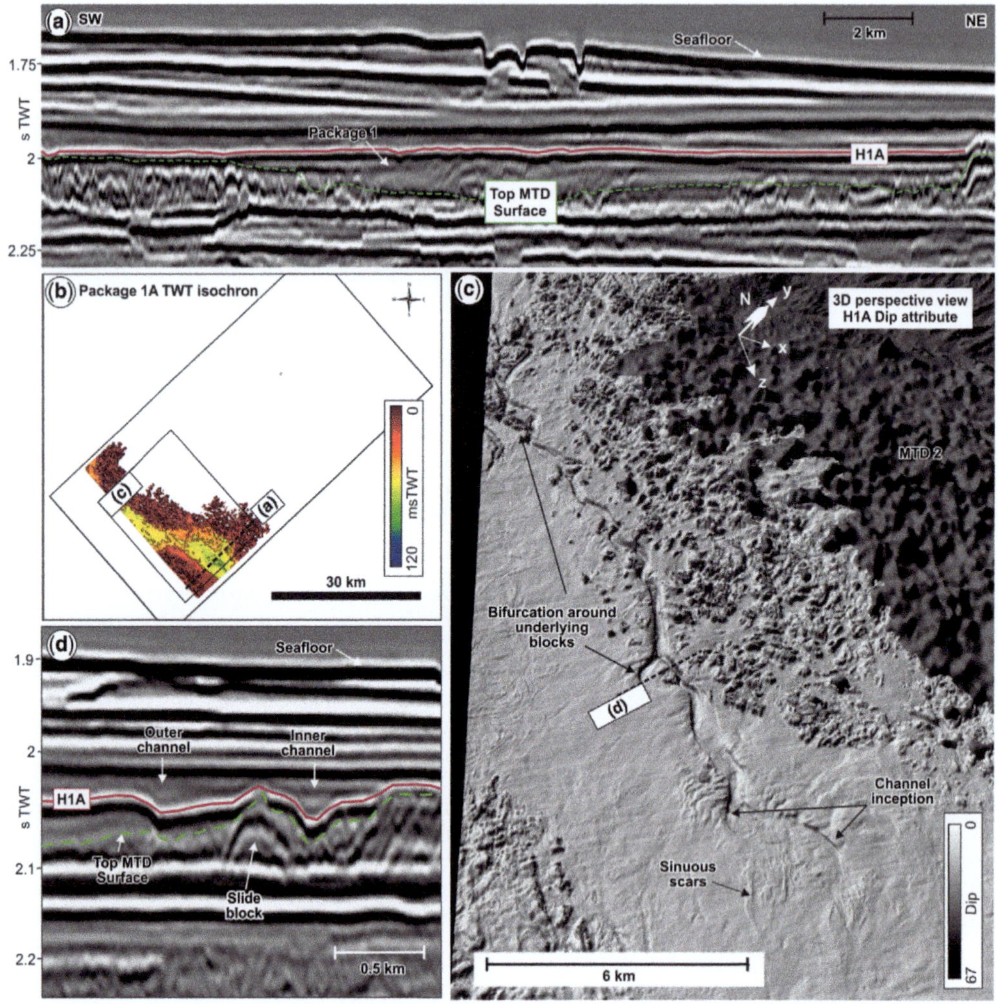

Fig. 6. (**a**) 3D seismic traverse from Romney 3D showing the character of Package 1A. Location shown in (b). (**b**) Isochron map of Package 1A. (**c**) Dip attribute from horizon H1A, shown as a 3D perspective view. Bold white arrows indicate main directional trends of the channels. Location shown in (b). (**d**) Romney 3D seismic traverse through main channel where it diverges around an intact slide block. Location shown in (c).

lobe-shaped deposit. We note that pockmarks, previously observed from horizon H1, are also present on horizon H2 (labelled on Fig. 7d).

Overburden Package 3 (H2–H3)

Package 3 is defined at its base by horizon H2 and at its top by horizon H3 (Fig. 8a). It is up to 60 ms (55 m) thick and is less extensive than the packages described previously (Fig. 8b). Package 3 is defined predominantly by one single high-amplitude laterally continuous reflection and, in part, infills topographic lows created by Package 2 (Fig. 8a). A dip attribute map of horizon H3 reveals further channel bodies including the main channel from the underlying Package 2. At horizon H3, the channel from Package 2 is partially infilled and creates negative topography (Fig. 8c). Two further channels pertaining to horizon H3 are identified (Fig. 8c) and indicate that the active thalwegs of the channel-levee complex, which was first identified in Package 2, have migrated to the north.

Overburden Package 4 (H3–H4)

Package 4 is defined by horizon H3 at its base and horizon H4 at its uppermost extent (Fig. 9a). Internally, Package 4 comprises generally

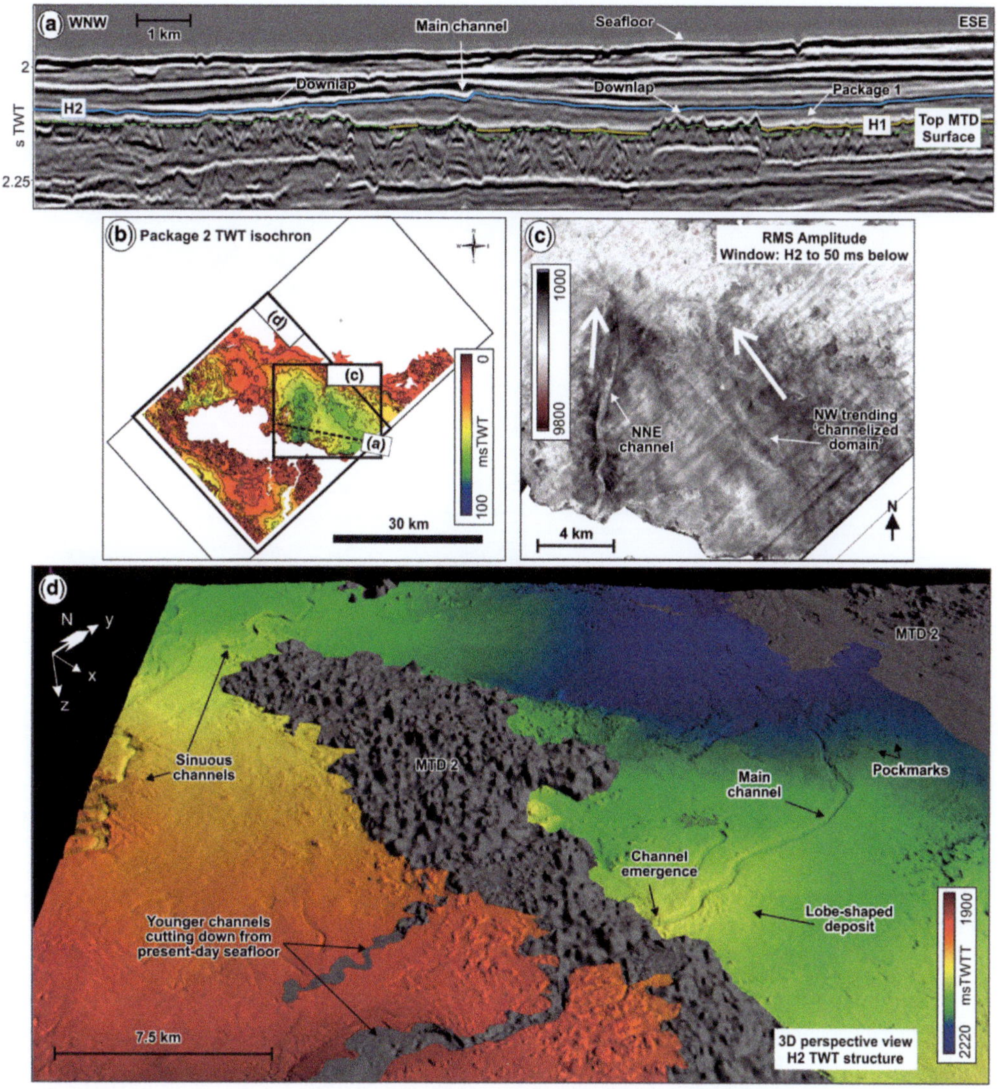

Fig. 7. (**a**) 3D seismic traverse from Romney 3D showing character and bi-directional downlap of Package 2. Location shown in (b). (**b**) Isochron map of Package 2. (**c**) RMS amplitude extraction over a window from horizon H2 to 50 ms below. Location shown in (b). (**d**) TWT structure map of horizon H2 shown as a 3D perspective view. Location shown in (b).

high-amplitude, continuous, parallel to sub-parallel reflections (Fig. 9a). Package 4 is laterally extensive, covering the entire Romney 3D area except for localized areas where intact blocks from the underlying MTD extrude (Fig. 9a). Package 4 is thin in most places (i.e. <30 ms/25 m), except for small areas within zones 1 and 2 where it is up to 120 ms (110 m) thick (Fig. 9b), and a larger area across zone 3, where it reaches up to 90 ms (80 m) thick. In zone 3, Package 4 is distributed around and between the large intact blocks protruding from the MTD (Fig. 9b). This distribution is interpreted as further compensational stacking and represents a significant period when sediment infilled a large area including between the largest of the intact blocks. Internally, RMS amplitude extractions reveal broad channelized domains with three different directional trends; NNE, NW and NE (indicated by the bold arrows on Fig. 9c). Although the channel bodies are subtle features observable only from the amplitude extractions, it is possible to measure the sinuosity index of several examples, which range from

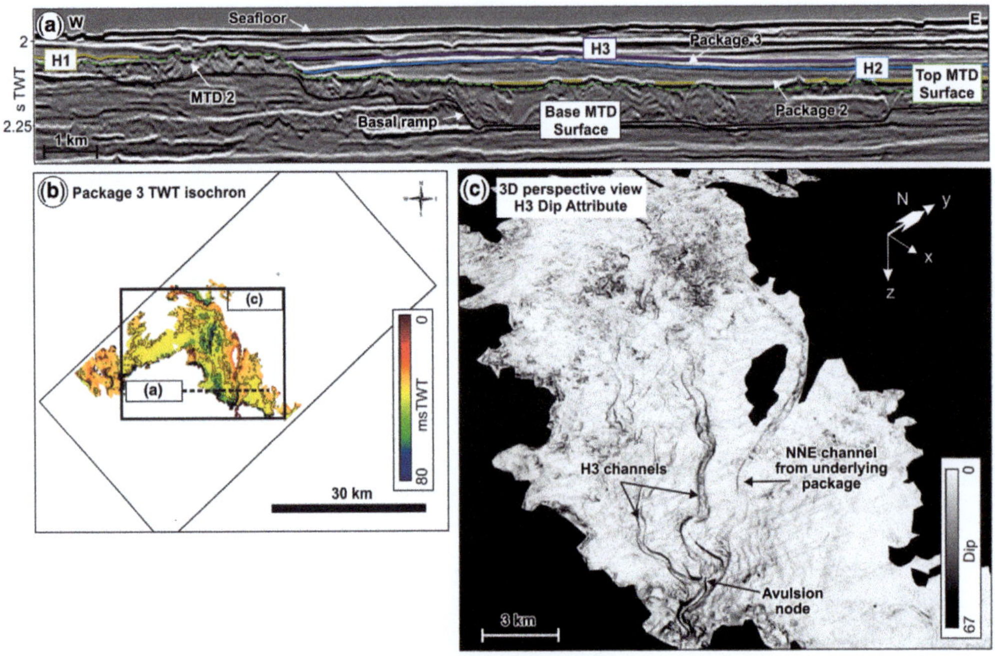

Fig. 8. (**a**) 3D seismic traverse from Romney 3D showing character and bi-directional downlap of Package 3. Location shown in (b). (**b**) Isochron map of Package 3. (**c**) Dip attribute map from horizon H3 shown as a 3D perspective view. Location shown in (b).

1.05–1.12 (Fig. 9c). The unconfined style of the internal channelized belts in Package 4 is reminiscent of those seen in Package 1 (Fig. 5d). Healing of the original MTD topography is now apparent (Fig. 5b, d). While the high topography of zone 1 is by this time overtopped by horizon H4, it retains a rugose character (Fig. 9d). H4 exhibits both linear and sinuous scouring (Fig. 9d) with linear scours mainly confined to inter-block areas in zone 3 (Fig. 9d). However, conversely to observations made from previous packages, it is notable that horizon H4 does not feature any channel bodies.

Overburden Package 5 (H4–present-day seafloor)

Package 5 is the most recently deposited, with its base defined by horizon H4 and its upper surface by the present-day seafloor (Fig. 10a). Internally, the unit is composed of moderate amplitude, continuous parallel to sub-parallel reflections (Fig. 10a). While Package 5 is continuous across the entirety of the Romney 3D area, its greatest thickness is deposited within zone 2 (up to 65 m) and it is very thin across zone 3 (<10 m) (Fig. 10b). Two lobe-shaped features are present in zone 2, forming distinct elements that are elongate from south to north (Fig. 10b). Bi-directional downlap is observed in seismic section (Fig. 10a), indicating the formation of distributary channel and levee deposits, which are offset from the channel-levee deposits identified in Packages 2 and 3 (see Figs 7b & 8b), thus continuing the pattern of offset stacking observed throughout this succession.

Numerous channel bodies are present internally within Package 5, evidenced by flattened time-slices from the reflectivity volume (Fig. 10c). The channels exhibit three main directional trends; to the NNW, north and NE (Fig. 10c, d). At the top of Package 5, singular channel bodies are observed at the present-day seafloor, labelled A to D on Figure 10d, and are traceable across the seafloor over substantial distances (tens of kilometres). Channels A and B flow around the high elevation MTD topography in zone 1 and have SIs of 1.17 and 1.10, respectively. They are 120–480 m wide and 13–26 m deep. Both channels diverge into two thalwegs (labelled on Fig. 10d) possibly due to localized MTD-related rugosity, and then converge further downslope. Channels A and B exhibit low-amplitude signatures (Fig. 10b), in contrast to the areas immediately surrounding the channel bodies, which exhibit high amplitudes distributed in a narrow zone along the channel lengths (Fig. 10b).

Channels C and D cut across zone 1 and exhibit high SIs of 1.8 and 1.22, respectively. Both channels

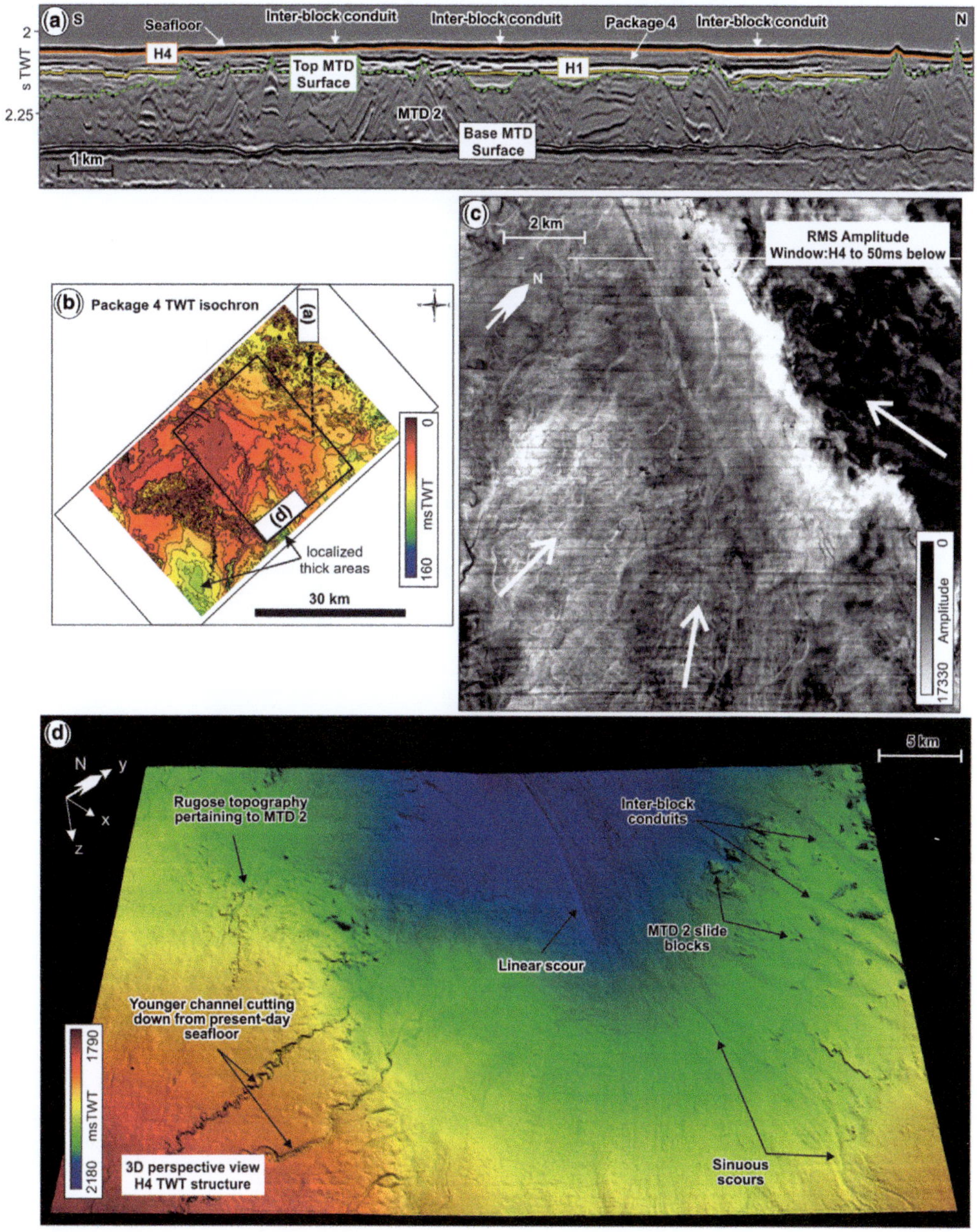

Fig. 9. (**a**) 3D seismic traverse from Romney 3D showing character of Package 4. Location shown in (b). (**b**) Isochron map of Package 4. (**c**) RMS amplitude extraction over a window from horizon H4 to 50 ms below. Directional trends of channels indicated by the bold white arrows. Location shown in (b). (**d**) TWT structure map of horizon H4 across entire Romney 3D area, shown as a 3D perspective view.

incise through Package 5 and the underlying packages (this is seen in Figs 7d & 9d). Channel C is 120–200 m wide and up to 25 m deep. Channel D is 140–315 m wide and 35 m deep. Both Channels C and D avulse upon reaching the relatively lower-gradient seafloor of zone 2 (Fig. 10d). In zone 2, they diverge into two separate channels accompanied by paddle-shaped, high-amplitude zones interpreted as overbank deposits and crevasse splays (cf. Posamentier and Kolla 2003) (Fig. 10d).

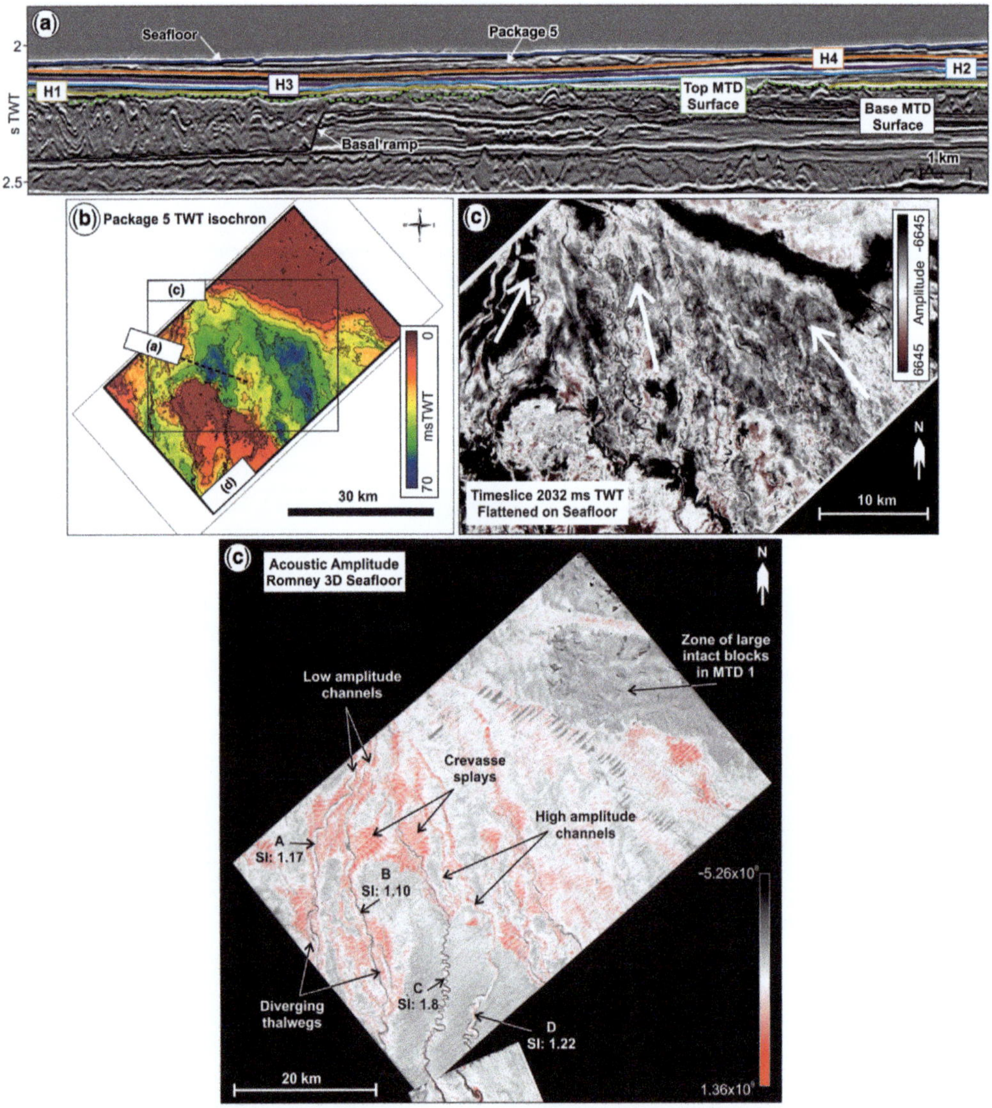

Fig. 10. (**a**) 3D seismic traverse from Romney 3D showing character of Package 5. Location shown in (b). (**b**) Isochron map of Package 5. (**c**) Reflectivity time-slice 2032, flattened on the seafloor. Note the presence of numerous channels trending to the NW and NNE (channel directions indicated by bold white arrows). Location shown in (b). (**d**) Seafloor amplitude map of Romney 3D area.

Crevasse splays typically occur at the outer bends of channels or where a channel begins to avulse, whereby the channel becomes infilled leading to the establishment of an alternative pathway (Weimer and Slatt 2004). Channels C and D both avulse into two separate paths; one exhibiting a low-amplitude signature and the other, a high-amplitude signature (Fig. 10d). This pattern may be indicative of one of the channel pathways becoming infilled with sediment and consequently being abandoned, while the alternative pathway represents the newly established and active route.

Channel C can be correlated directly into the adjacent Vulcan 3D survey (Fig. 1) and Channel B can also be correlated into the Vulcan survey by way of a 350 m jump across the data gap (Fig. 11a). A series of slope-confined channels incise the modern continental slope and some manifest in deeper water as the channels at the top of Package 5 (i.e. Channels B and C) (Fig. 11a). The channel heads originate

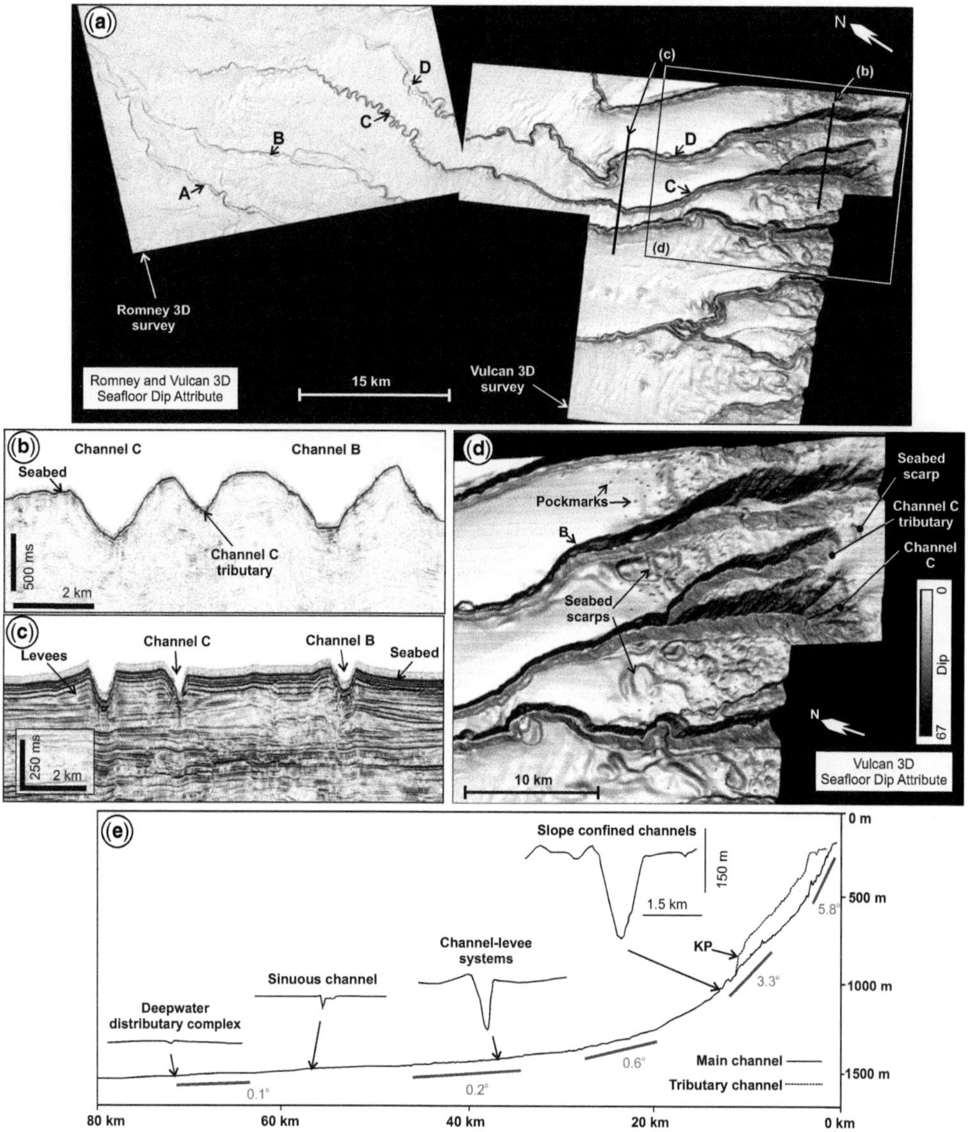

Fig. 11. (**a**) Seafloor-dip attribute map of Romney and Vulcan 3D surveys, showing linked slope-confined and deep-water submarine channels. (**b**) 3D seismic traverse from Vulcan 3D showing cross-sectional character of slope-channel heads. Location shown in (a). (**c**) 3D seismic traverse from Vulcan 3D showing cross-sectional character of mid-slope channel-levee systems. Location shown in (a). (**d**) Zoomed-in dip attribute map from inset area of Vulcan 3D survey. Note seabed scarps and pockmarks. Location shown in (a). (**e**) Channel profile along Channel C, showing cross-sectional profiles and slope gradients. KP, knick point.

at or below the shelf break in water depths between 175 and 210 m (Fig. 11a). In its head region Channel C exhibits a deeply incised V-shaped cross-section up to 385 m deep and 1.85 km wide (Fig. 11b). It is associated with a smaller tributary channel (labelled on Fig. 11d) and proceeds downslope with a relatively straight thalweg. On the lower slope, Channel C narrows and shallows gradually and develops levees (Fig. 11c). A marked increase in sinuosity is observed on the lower slope where Channel C is imaged by the Romney 3D survey (Fig. 11a).

Along-channel profiles for Channel C and its tributary are shown in Figure 11e, constructed by plotting distance v. water depth along the channel thalweg from its origin at the shelf edge to the point where it exits the NW boundary of the Romney 3D survey (Fig. 11a). We also show cross-sectional profiles at various points to illustrate its downslope morphological evolution from a 'slope-confined channel' through to channel-levee complexes, a highly sinuous channel and finally the deep-water distributary complex. It has been inferred that submarine channel systems tend towards a smooth axial profile, much like river systems, and do so by erosion and deposition that compensate for tectonic and slope failure-induced changes to the equilibrium profile or to base level (Lewis and Barnes 1999). The erosion rate is dependent on the resistance of seabed sediments to erosion, the nature of transported material and number or frequency of erosive downslope events (Pirmez *et al.* 2000; Piper and Normark 2009; Covault *et al.* 2011; Ratzov *et al.* 2012). Profiles therefore allow consideration of the balance of these competing factors and exhibit convex curvature when the rate of uplift is higher than the rate of erosion, and concave curvature when erosion exceeds uplift (Ratzov *et al.* 2012). Close to its origin in the upper slope region, there are sections of Channel C's profile which appear steep, convex and rugose (Fig. 11e). We also note a pronounced knickpoint (KP; Fig. 11e) where the tributary joins the main channel, which can be indicative of adjustment to a new base level (Tubau *et al.* 2015). Conversely, in the lower slope and basin-floor regions, Channel C exhibits a smooth, concave profile. Although more detailed analysis is required to fully assess relevant factors, this profile provides a snapshot of Channel C's present-day stratigraphic equilibrium and general maturity. The profile shown in Figure 11e suggests recent uplift is outpacing erosion on the upper continental slope, while the two factors are better countered on the lower slope and basin floor.

Discussion

In the following sections we summarize observations and discuss controls on the development of the overburden. We also consider the nature and evolution of the sediment feeder system, the present-day depositional setting of the study area and the implications of our findings for future occurrence of MTDs.

Pleistocene–recent stratigraphic evolution of the Deepwater Taranaki Basin

Emplacement of MTD 2 at *c.* 1.8 Ma created bathymetric relief of differing scales: (1) hundreds of metres vertically and tens of kilometres laterally, principally linked to the relatively shallow level of the basal surface in zone 1 (Figs 3 & 4); and (2) broad regions of pronounced rugosity of the order tens of metres vertically and hundreds of metres laterally, due to the presence of intact blocks within MTD 2 (zones 1 and 3) (Fig. 4). High relief and high rugosity may be superimposed, as seen in zone 1 (Fig. 4). Similarities are drawn with the study of Armitage *et al.* (2009), who identified three hierarchical tiers of MTD surface topography. We liken our observations to their 'Tier 3' MTD surface topography (hundreds of metres horizontally and thousands of metres laterally), and 'Tier 2' topography (tens of metres vertically and horizontally, formed by the presence of large intact blocks in the MTD) (Armitage *et al.* 2009). With reference to MTD surface topography, we identify three distinct stages of overburden deposition following emplacement of MTD 2:

1st stage: Early overburden packages are accommodated in, and laterally confined by, 'Tier 3' MTD topography. Overburden sediments are distributed across 'Tier 2' topography, but are locally confined to relatively small-scale depocentres. Internally within the overburden packages, 'channelized belts' develop, comprised of multiple discrete channel bodies exhibiting a range of flow directions. At the top of the packages, low sinuosity, singular channels develop which locally interact with 'Tier 2' topography created by intact slide blocks.

2nd stage: Overburden packages begin to form a broad compensational stacking pattern, with both tiers of MTD topography increasingly onlapped and downlapped. 'Tier 3' MTD topography is seen to create locally steeper routes for incoming sediment flows, which then form lobe-shaped deposits upon reaching lower and flatter areas (i.e. in zone 2). The resulting lobate features create their own positive topography leading to the next series of flow events being deposited in a different location. Channelized domains continue to occur within the overburden packages and singular channels occur at the top of the packages.

3rd stage: With the deposition of the most recent overburden packages, MTD topography is largely infilled and overtopped, with the exception of the largest intact blocks in zone 3. Depositional packages have evolved to become slightly thinner than earlier packages but are laterally unconfined and continuous across the study area. The pattern of broad offset stacking continues, along with numerous, increasingly sinuous channels within the packages. The 'Tier 3' topography of MTD 2 in zone 1 continues to exert an influence on present-day activity. Channels at the present-day seafloor cut across and incise zone 1, and are also diverted around it (Fig. 10d).

Controls on overburden development

It is clear that MTD 2 topography exerts a primary control on the evolution of the overburden, particularly in the earlier stages. Interestingly, the highest relief MTD topography results from the relatively shallow level of the detachment surface (Fig. 3b, c), and so in this case it can be argued that the geometry of the basal surface is also a key factor. The overall change from confined deposition of early overburden sediments towards more extensive, unconfined deposition over tens of kilometres laterally, following smoothing of MTD topography, has been observed from previous studies (Armitage *et al.* 2009). Related to this, we interpret the increased sinuosity of channels observed from Package 5 compared to earlier packages to be the result of generally reduced seafloor gradient (see Wynn *et al.* 2007) as MTD topography is progressively healed.

On a smaller scale, early overburden sediments pond between the intact blocks within zone 3, and are funnelled through conduits formed between them (Figs 3a, 5c & 9d). Blocks within the MTD determining the evolution and spatial deposition of depocentres have been previously noted by Schultz *et al.* (2005), Armitage *et al.* (2009) and Ward *et al.* (2018). We observe the confinement of sediment fairways to inter-block areas defined by roughly linear striations formed on the contemporaneous seafloor (Figs 4, 5c & 9d), which have previously been interpreted as evidence of sediment transport in a deep-water setting (Posamentier and Kolla 2003; Gee *et al.* 2005; Ward *et al.* 2018; Zhao *et al.* 2019).

Within the primary topographic constraints imposed by MTD 2, further complexity is introduced by lobe-switching processes typical of deep-water sediments in uncomplicated settings (Burgess *et al.* 2019). This process occurs repeatedly but irregularly over time as a consequence of feedback between flow-routing and the evolving depositional topography (Burgess *et al.* 2019). This process is observed in Package 2 where the topography of zone 1 creates localized seafloor gradients which divert sediment flows into the low areas of zone 2 (Fig. 7d) resulting in the formation of lobe-shaped deposits (Fig. 7b). These deposits possess positive topography that presently measures up to 90 m thick and kilometres across (Fig. 7b). Thus, these deposits further modify the seafloor and combine with the MTD topography to influence the depositional location of subsequent flows comprising Package 3 (compare Fig. 7b with Fig. 8b).

In the present study, an interesting observation is the identification of channelized belts within many of the overburden packages, comprising multiple channel bodies which exhibit differing directional trends (Figs 5d, 7c, 9c & 10c). This is in contrast to the bounding horizons at the top of the packages, which with the exception of Package 4, exhibit a small number of singular channels (Figs 5c, 6c, 7d, 8c & 10d). A possible explanation is non-uniform rates of sedimentary supply, with changing rates of sediment input and deposition over time due to extrinsic factors. Distinction of autogenic v. allogenic signals is important for understanding long-term frequencies and triggering of episodic sedimentation processes. This is often made difficult by the described autogenic behaviours, which have the effect of breaking down external signals (Burgess *et al.* 2019). We infer that many channel bodies develop to collectively form channelized domains in order to shed sediment load and possibly represent a waning phase of deposition. Conversely, the single channel marks renewed sediment input, and/or an increased rate of deposition. Although we are not able to provide vigorous age constraint or provenance data for our overburden packages, it is possible that the sediment transported within the channels described here were in part the eroded products of tectonic pulses in the continental hinterland of the North and South islands of New Zealand (e.g. Tippett and Kamp 1993; Browne 2004; Kamp *et al.* 2004; Pulford and Stern 2004; Nicol 2011; Trewick and Bland 2012; Bland *et al.* 2019).

Sediment feeder system

The presence of large singular channels at the top of overburden packages may be indicative of a sizeable point source located updip of the study area, or represent one channel element that forms part of a broader sediment input zone. In the present day, the slope channels imaged by the Vulcan 3D survey are confined to the continental slope, with heads that terminate at or below the shelf break (Fig. 11a). These so-called ‘slope-confined’ or ‘blind’ channels are typically isolated from shelf-originating downslope erosive flows or retrogressive slope failure mechanisms and lack a direct connection to modern terrestrial fluvial systems (Lastras *et al.* 2011; Brothers *et al.* 2013; Huang *et al.* 2014). Their evolution is likely influenced by a variety of external factors, including climate, sea-level, tectonics and proximity to terrestrial and shelfal sediment supply (Covault *et al.* 2007; Lamb and Mohrig 2009; Mountjoy *et al.* 2009).

2D seismic reflection profiles across the shelfal area, updip of the modern-day seafloor channels, reveal that the channels extend up the slope into a series of slope-confined channels feeding off the shelf break (Fig. 12). Horizon N80 (Strogen and King 2014; Bull *et al.* 2016) approximates to 2 Ma and is extended from the southern Taranaki area by correlation to biostratigraphic age markers in key offset wells (Fig. 1). The channels originate close

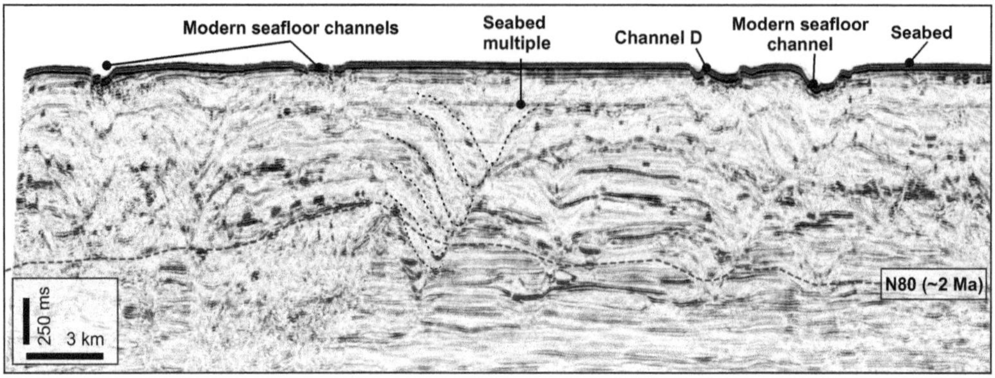

Fig. 12. 2D seismic profile slightly updip of present-day shelf break showing slope channels and *c.* 2 Ma horizon. Location shown in Figure 1.

to this marker, suggesting that they have been temporally persistent (i.e. originated at *c.* 2 Ma) and may have formed a more regionally extensive system of point sources as has been suggested for older late Miocene successions of north Taranaki (King *et al.* 2007). Given their origin at *c.* 2 Ma, the channels may have originated during relatively lower sea-levels and may previously have been connected to fluvial river systems that carried sediment across what is now the shelf (cf. Lewis and Barnes 1999). Studies in older late Miocene sedimentary systems from the broader Taranaki region suggest that submarine channels feeding basin-floor fan systems were more active during lowstands (King *et al.* 2007) and a similar situation may have existed for the younger strata of our study area. It is noted that the first appearance of the channels at *c.* 2 Ma occurs at a similar time to a further MTD in the study area (MTD 1 of Omeru *et al.* 2016; Omeru and Cartwright 2019; labelled on Fig. 1), which precedes the emplacement of MTD 2. This observation raises questions regarding the interplay between the slope-confined channels and the occurrence of MTDs, and further analysis of the extensive subsurface dataset available from the study area is recommended to address them.

There is evidence that slope failure on a smaller scale plays a role in slope-channel development in the study area. Enlargement by retrogressive headwall failure is indicated by the presence of an arcuate seafloor scarp just updip of the tributary of Channel C (labelled on Fig. 11d). In addition, localized step-like seafloor scarps (labelled on Fig. 11d) are interpreted as small rotational failures. Furthermore, both the slope-confined channels and localized step-like scarps occur in close association with seafloor pockmarks that appear in a zone across the mid-slope, down to water depths of 780 m (Fig. 11d). Focused fluid flow on active margins has been proposed as a mechanism for slope failure and gully system initiation by McAdoo *et al.* (1997), Orange *et al.* (1997), Lewis *et al.* (1998) and Elger *et al.* (2018). Pockmarks have also been noted to occur above abandoned and buried slope-channel axes from other locations (Jobe *et al.* 2011), but we do not make the same connection here as the pockmarks appear to be limited to a zone along, rather than down, the slope (Fig. 11d).

Contemporary oceanographic setting, present-day sedimentation and implications for future MTD occurrence

Discovery of a complex and dynamic succession of deep-water sedimentary packages linked to the shelf via numerous seafloor channels over the last *c.* 2 myr, and following the emplacement of MTD 2, invites further enquiry into the rate and mode of sediment delivery in the present day. It is well established that the majority of the late-early Pliocene to recent sediments underlying the present-day shelf and slope accumulated very rapidly (Beggs 1990; Hansen and Kamp 2004*a*, *b*; Salazar *et al.* 2015) through transport of sediments from the south and east by shore-parallel and other currents (Hansen and Kamp 2002; Anell and Mitkandel 2015; Salazar *et al.* 2015). MTD 2 emplacement is linked to periods of very high sediment supply (Omeru 2014) but there have been few studies to document the present-day sedimentation regime.

The nzSEABED surficial sediment database (Bostock *et al.* 2019*a*, *b*), a compilation of seafloor sediment samples from the oceans around New Zealand, provides some insight into modern grain-size distributions. Sample stations plotted directly onto seafloor amplitude maps from the Vulcan and Romney 3D surveys (Fig. 13) show that despite the

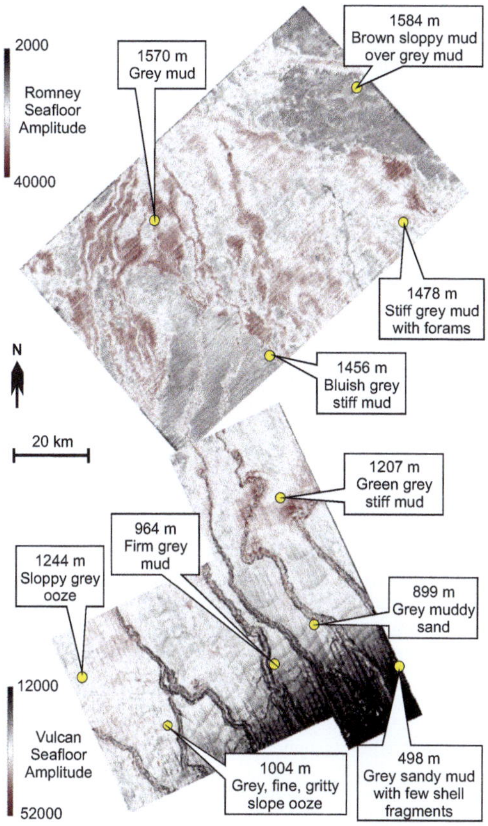

Fig. 13. Romney and Vulcan 3D survey seafloor amplitude maps shown with nzSEABED surficial sediment sample stations from nzSEABED database (Bostock *et al.* 2019*a*, *b*), annotated with sample station water depths and sediment descriptions.

complex seismic amplitude distribution of the seafloor reflection, surficial sediment samples indicate a generally muddy environment on the basin floor, with coarser material in shallower water on the continental slope and shelf. Thin Holocene pelagic/hemipelagic drapes have been described from large, modern, deep-water channel systems elsewhere, such as the Amazon and Indus fans (Wynn *et al.* 2007). We postulate that given the *c.* 7.5 m vertical resolution of the 3D seismic data at the seafloor, both a recent mud drape and lithological differences related to earlier active deep-water deposition are both likely to be captured within the first (water column–sediment interface) wavelet.

Recent and dramatic changes in sea-level and physical oceanography of the broader NW New Zealand region are linked to Pleistocene glaciations that affected New Zealand (Suggate 1990). Prior to the Last Glacial Maximum (LGM) at *c.* 30–20 ka (Clark *et al.* 2009; Golledge *et al.* 2012), for example, sea-level was markedly lower than at present and reconstructions of palaeo-circulations around the New Zealand landmass from *c.* 21 ka show substantial changes in oceanic currents in the western New Zealand region compared to the present day (Lorrey *et al.* 2012). Of significance to this study, a major oceanic current at LGM travelled NE along the west coast of South Island, similar to the modern-day west coast current (Carter *et al.* 1998) and continued north into the northern Taranaki Basin and Deepwater Taranaki Basin (Fig. 1; inset). This mirrors conditions invoked to explain rapid deposition of the GFF beginning in early-late Pliocene times (Hansen and Kamp 2002; Anell and Mitkandel 2015; Salazar *et al.* 2015) (Fig. 1, inset). In the present day, the west coast current feeds the D'Urville current (Carter *et al.* 1998; Chiswell *et al.* 2015), which swings from NE to SE around the northern South Island and is diverted through the Cook Strait (Carter *et al.* 1998; Chiswell *et al.* 2015) (Fig. 1; inset). It is proposed that the Cook Strait opened and closed over many eustatic sea-level cycles linked to glacial and interglacial periods (Lewis *et al.* 1994; Alloway *et al.* 2007). A relatively weaker remnant of the west coast current continues north to the northern Taranaki area, where present-day currents are reported as weak and variable (Chiswell *et al.* 2015; Rad 2015) (Fig. 1; inset). It is therefore inferred that sediment reaching the Deepwater Taranaki Basin in the present day is primarily derived from rivers of western North Island (Bostock *et al.* 2019*a*, *b*).

Classical models of deep-water submarine fan deposition predict that during relative sea-level highstands, such as in the present day, there will be a relative paucity of sediment delivery to the slope and basin floor (Posamentier and Walker 2006). Many examples of modern passive-margin fans are currently experiencing much reduced levels of activity as a result (Wynn *et al.* 2007). However, there is growing evidence to suggest that deep-water sediment delivery is poorly correlated to sea-level and rates of sea-level change, which cannot account for the multiple factors involved and their interactions (Covault and Graham 2010; Clare *et al.* 2014; Allin *et al.* 2018; Harris *et al.* 2018). While it seems unlikely that in the modern highstand sediment can be transported across the broad shelf to the heads of the slope-confined channels, it is possible that the channels may still be active but limited to relatively infrequent seismic-triggered events (i.e. Mountjoy *et al.* 2018) and significant storm events (Carter and Heath 1975).

Understanding the long-term controls on the frequency, magnitude and triggering of seafloor instability on the scale observed from the study area (Omeru 2014) is of vital importance for geohazard assessment. In our study area, MTD occurrence has been linked to periods of relatively high sediment supply (Omeru 2014). However, our understanding is limited by relatively coarse age control in the shallow succession, inability to identify specific triggering mechanisms and lack of knowledge regarding present-day slope conditions. Furthermore, the timing of MTD occurrence has been shown to be independent of temporally non-random processes, such as glacial cycles, and failure may occur thousands of years after peak sedimentation has occurred (Clare *et al.* 2014). Further investigation will be required to develop a robust age-frequency model for MTDs in the study area and understand the likelihood of future seafloor instability in the region.

Conclusions

Deep-water sedimentation patterns are highly sensitive to seabed topography and have a complex interplay with MTD 2 in the Deepwater Taranaki Basin, western New Zealand.

Emplacement of MTD 2 at *c.* 1.8 Ma created high-relief topography in the order of hundreds of metres vertically and tens of kilometres laterally, principally linked to the relatively shallow level of the basal surface. In addition, broad areas of high rugosity created by the presence of intact blocks preserved within MTD 2 introduced relief of tens of metres vertically and hundreds of metres laterally.

Subsequent sediment flows were initially deflected away from the high MTD topography and/or confined in low areas, while, more locally, sediment was both trapped within, and routed through, 'inter-block areas' between the upper protrusions of large intact slide blocks.

In time, sediment flows exploited new routes across high MTD topography and classical autogenic lobe-switching behaviour was re-established. A broad pattern of offset-stacked overburden packages emerged, together with an overall change to unconfined overburden distribution as MTD topography was healed.

Sediment input into the study area may have been non-uniform, reflecting tectonic pulses that shaped the continental hinterland of Zealandia. Eroded sediments were routed into the deep water by long-lived point sources, most likely slope channels that are presently confined to the continental slope.

It is likely that present-day sediment delivery into deep water in the study area has been strongly influenced by recent climatic variations (beginning in the Pleistocene), but it is as yet unclear what influence such changes may have on the future stability of the continental margin in this region.

Acknowledgements We thank New Zealand Petroleum and Minerals (Ministry of Business, Innovation and Employment, MBIE) for providing the seismic data used in this study and Paradigm for provision of SeisEarth interpretation software. NIWA provided bathymetric data and surficial sediment data. We thank several colleagues for valuable discussions of various aspects of this work, in particular Glenn Thrasher, Tusar Sahoo, Paul Viskovic and Nicholas Thompson. We thank Kyle Bland and Karsten Kroeger (GNS Science) for their comments which improved this manuscript. We gratefully acknowledge reviews by Glenn Sharman and Michael Clare, which improved the quality of the final manuscript.

Funding This research was supported by Strategic Science Investment Funding from MBIE to GNS Science under the Sedimentary Basin Research and Te Riu-a-Māui/Understanding Zealandia programmes (to SB, GB, MA).

Author contributions **SB**: conceptualization (lead), formal analysis (lead), investigation (lead), methodology (lead), validation (lead), visualization (lead), writing – original draft (lead), writing – review & editing (lead); **GHB**: conceptualization (supporting), formal analysis (supporting), investigation (supporting), methodology (supporting), writing – original draft (supporting), writing – review & editing (supporting); **MJA**: formal analysis (supporting), investigation (supporting); **LJS**: validation (supporting), writing – original draft (supporting), writing – review & editing (supporting).

References

Allin, J.R., Hunt, J.E., Clare, M.A. and Talling, P.J. 2018. Eustatic sea-level controls on the flushing of a shelf-incising submarine canyon. *GSA Bulletin*, **130**, 222–237, https://doi.org/10.1130/B31658.1

Alloway, B.V., Lowe, D.J. *et al.* NZ-INTIMATE members 2007. Towards a climate event stratigraphy for New Zealand over the past 30,000 years (NZ-INTIMATE project). *Journal of Quaternary Science*, **22**, 9–35, https://doi.org/10.1002/jqs.1079

Alves, T.M. 2010. 3D Seismic examples of differential compaction in mass-transport deposits and their effect on post-failure strata. *Marine Geology*, **271**, 212–224, https://doi.org/10.1016/j.margeo.2010.02.014

Alves, T.M. and Cartwright, J.A. 2010. The effect of mass-transport deposits on the younger slope morphology, offshore Brazil. *Marine and Petroleum Geology*, **27**, 2027–2036, https://doi.org/10.1016/j.marpetgeo.2010.05.006

Andresen, K.J. and Huuse, M. 2011. 'Bulls-eye' pockmarks and polygonal faulting in the Lower Congo Basin: relative timing and implications for fluid expulsion during shallow burial. *Marine Geology*, **279**, 111–127, https://doi.org/10.1016/j.margeo.2010.10.016

Anell, I. and Mitkandel, I. 2015. The quantifiable clinothem – types, shapes and geometric relationships in the Plio-Pleistocene Giant Foresets Formation, Taranaki Basin, New Zealand. *Basin Research*, **29**, 277–297, https://doi.org/10.1111/bre.12149

ARCO 1992. *Arawa-1 Final Well Report PPL38436*. Ministry of Commerce New Zealand, Unpublished Petroleum Report, **PR1824**.

Armitage, D.A., Romans, B.W., Covault, J.A. and Graham, S.A. 2009. The influence of mass-transport-deposit surface topography on the evolution of turbidite architecture: the Sierra Contreras, Tres Pasos Formation (Cretaceous), Southern Chile. *Journal of Sedimentary Research*, **79**, 287–301, https://doi.org/10.2110/jsr.2009.035

Arnot, M.J., Bland, K.J. *et al.* (compilers) 2016. *Atlas of Petroleum Prospectivity, Northwest Province: ArcGIS geodatabase and technical report*. GNS Science Data Series 23b. 5 ArcGIS projects, 1 ArcGIS geodatabase, and technical report.

Baur, J., Sutherland, R. and Stern, T. 2014. Anomalous passive subsidence of deep-water sedimentary basins: a prearc basin example, southern New Caledonia Trough and Taranaki Basin, New Zealand. *Basin Research*, **26**, 242–268, https://doi.org/10.1111/bre.12030

Beggs, J.M. 1990. Seismic Stratigraphy of the Plio-Pleistocene Giant Foresets Formation, Western Platform, Taranaki Basin. *1989 New Zealand Oil Exploration Conference Proceedings*, 201–207.

Bland, K.J., Nicol, A., Kamp, P.J.J. and Nelson, C.S. 2019. Stratigraphic constraints on the late Miocene–Pleistocene evolution of the North Island Fault System and axial ranges in the central Hikurangi subduction margin, New Zealand. *New Zealand Journal of Geology and Geophysics*, **62**, 248–272, https://doi.org/10.1080/00288306.2018.1545675

Bostock, H., Jenkins, C. *et al.* 2019*a*. Distribution of surficial sediments in the ocean around New Zealand/Aotearoa. Part A: continental slope and deep ocean. *New Zealand Journal of Geology and Geophysics*, **62**, 1–23, https://doi.org/10.1080/00288306.2018.1523198

Bostock, H., Jenkins, C. *et al.* 2019*b*. Distribution of surficial sediments in the ocean around New Zealand/Aotearoa. Part B: continental shelf. *New Zealand Journal of Geology and Geophysics*, **62**, 24–45, https://doi.org/10.1080/00288306.2018.1523199

Brothers, D.S., ten Brink, U., Andrews, B.D. and Chaytor, J.D. 2013. Geomorphic process fingerprints in submarine canyons. *Marine Geology*, **337**, 53–66, https://doi.org/10.1016/j.margeo.2013.01.005

Browne, G.H. 2004. Late Neogene sedimentation adjacent to the tectonically evolving North Island axial ranges: insights from Kuripapango, western Hawke's Bay. *New Zealand Journal of Geology and Geophysics*, **47**, 663–674, https://doi.org/10.1080/00288306.2004.9515082

Bull, S., Cartwright, J.A. and Huuse, M. 2009. A review of kinematic indicators from mass-transport complexes using 3D seismic data. *Marine and Petroleum Geology*, **26**, 1132–1151, https://doi.org/10.1016/j.marpetgeo.2008.09.011

Bull, S., Hill, M., Strogen, D.P., Arnot, M.J., Seebeck, H., Kroeger, K.F. and Zhu, H. 2016. *Seismic Reflection Interpretation, Static Modelling and Velocity Modelling of the Southern Taranaki Basin (4D Taranaki Project) (91 p)*. GNS Science Report, **2015/02**.

Bull, S., Arnot, M., Browne, G., Crundwell, M., Nicol, A. and Strachan, L. 2019. Neogene and Quaternary mass transport deposits from the Northern Taranaki Basin (North Island, New Zealand): morphologies, transportation processes, and depositional controls. *In*: Ogata, K., Festa, A. and Pini, G.A. (eds) *Submarine Landslides: Subaqueous Mass Transport Deposits from Outcrops to Seismic Profiles*. American Geophysical Union, Geophysical Monograph **247**, https://doi.org/10.1002/9781119500513.ch11

Burgess, P.M., Masiero, I., Tocy, S.C. and Duller, R.A. 2019. A big fan of signals? Exploring autogenic and allogenic processes and product in a numerical stratigraphic forward model of submarine-fan development. *Journal of Sedimentary Research*, **89**, 1–12, https://doi.org/10.2110/jsr.2019.3

Carter, L. and Heath, R.A. 1975. Role of circulation, tides and waves in the transport of bottom shelf sediment on the New Zealand continental shelf. *New Zealand Journal of Marine and Freshwater Research*, **9**, 423–448, https://doi.org/10.1080/00288330.1975.9515579

Carter, L., Garlic, R.D., Sutton, P., Chiswell, S., Oien, N.A. and Stanton, B.R. 1998. *Ocean Circulation New Zealand*. NIWA Chart Miscellaneous Series, **76**. NIWA, Wellington, New Zealand.

Catterall, V., Redfern, J., Gawthorpe, R., Hansen, D. and Thomas, M. 2010. Architectural style and quantification of a submarine channel-levee system located in a structurally complex area: offshore Nile delta. *Journal of Sedimentary Research*, **80**, 991–1017, https://doi.org/10.2110/jsr.2010.084

Chenrai, P. and Huuse, M. 2017. Pockmark formation by porewater expulsion during rapid progradation in the offshore Taranaki Basin, New Zealand. *Marine and Petroleum Geology*, **82**, 399–413, https://doi.org/10.1016/j.marpetgeo.2017.02.017

Chiswell, S.M., Bostock, H.C., Sutton, P.J.H. and Williams, M.J.M. 2015. Physical oceanography of the deep seas around New Zealand: a review. *New Zealand Journal of Marine and Freshwater Research*, **49**, 286–317, https://doi.org/10.1080/00288330.2014.992918

Clare, M.A., Talling, P.J., Challenor, P., Malgesini, G. and Hunt, J. 2014. Distal turbidites reveal a common distribution for large (>0.1 km^3) submarine landslide recurrence. *Geology*, **42**, 263–266, https://doi.org/10.1130/G35160.1

Clark, P.U., Dyke, A.S. *et al.* 2009. The last glacial maximum. *Science*, **325**, 710–714, https://doi.org/10.1126/science.1172873

Corella, J.P., Loizeau, J.L. *et al.* 2016. The role of mass-transport deposits and turbidites in shaping modern lacustrine deepwater channels. *Marine and Petroleum Geology*, **77**, 515–525, https://doi.org/10.1016/j.marpetgeo.2016.07.004

Covault, J.A. and Graham, S.A. 2010. Submarine fans at all sea-level stands: tectono-morphologic and climatic controls on terrigenous sediment delivery to the deep sea. *Geology*, **38**, 939–942, https://doi.org/10.1130/G31081.1

Covault, J.A., Normark, W.R., Romans, B.W. and Graham, S.A. 2007. Highstand fans in the California borderland: the overlooked deep-water depositional systems. *Geology*, **35**, 783–786, https://doi.org/10.1130/G23800A.1

Covault, J.A., Fildani, A., Romans, B.W. and McHargue, T. 2011. The natural range of submarine canyon-and-channel longitudinal profiles. *Geosphere*, **7**, 313–332, https://doi.org/10.1130/GES00610.1

Deptuck, M.E., Steffens, G.S., Barton, M. and Pirmez, C. 2003. Architecture and evolution of upper fan channel-belts on the Niger Delta slope and in the Arabian Sea. *Marine and Petroleum Geology*, **20**, 649–676, https://doi.org/10.1016/j.marpetgeo.2003.01.004

Dimitrov, L. and Woodside, J.M. 2003. Deep sea pockmark environments in the eastern Mediterranean. *Marine Geology*, **195**, 263–276, https://doi.org/10.1016/S0025-3227(02)00692-8

Eichhubl, P., Greene, H.G., Naehr, T. and Maher, N. 2000. Structural control of fluid flow: offshore fluid seepage in the Santa Barbara Basin, California. *Journal of Geochemical Exploration*, **69/70**, 545–549, https://doi.org/10.1016/S0375-6742(00)00107-2

Einsele, G., Ricken, W. and Seilacher, A. (eds) 1991. *Cycles and Events in Stratigraphy*. Springer-Verlag, Berlin.

Elger, J., Berndt, C., Rüpke, L., Krastel, S., Gross, F. and Geissler, W.H. 2018. Submarine slope failures due to pipe structure formation. *Nature Communications*, **9**, 715, https://doi.org/10.1038/s41467-018-03176-1

Gamboa, D., Alves, T., Cartwright, J. and Terrinha, P. 2010. MTD distribution on a 'passive' continental margin: the Espírito Santo Basin (SE Brazil) during the Palaeogene. *Marine and Petroleum Geology*, **27**, 1311–1324, https://doi.org/10.1016/j.marpetgeo.2010.05.008

Gay, A., Lopez, M., Berndt, C. and Séranne, M. 2007. Geological controls on focused fluid flow associated with seafloor seeps in the Lower Congo Basin. *Marine Geology*, **244**, 68–92, https://doi.org/10.1016/j.margeo.2007.06.003

Gee, M.J.R., Gawthorpe, R.L. and Friedmann, S.J. 2005. Giant striations at the base of a submarine landslide. *Marine Geology*, **214**, 287–294, https://doi.org/10.1016/j.margeo.2004.09.003

Gee, M.J.R., Gawthorpe, R.L., Bakke, K. and Friedmann, S.J. 2007. Seismic geomorphology and evolution of submarine channels from the Angolan continental margin. *Journal of Sedimentary Research*, **77**, 433–446, https://doi.org/10.2110/jsr.2007.042

Gillespie, J.L., Nelson, C.S. and Nodder, S.D. 1998. Post-glacial sea-level control and sequence stratigraphy of carbonate–terrigenous sediments, *Wanganui shelf, New Zealand. Sedimentary Geology*, **122**, 245–266, https://doi.org/10.1016/S0037-0738(98)00109-2

Golledge, N.R., Mackintosh, A.N. *et al.* 2012. Last Glacial Maximum climate in New Zealand inferred from a modelled Southern Alps icefield. *Quaternary Science Reviews*, **46**, 30–45, https://doi.org/10.1016/j.quascirev.2012.05.004

Hackbarth, C.J. and Shew, R.D. 1994. Morphology and stratigraphy of a mid-Pleistocene turbidite leveed channel from seismic, core, and log data. *In*: Bouma, A.H., Weimer, P. and Perkins, B. (eds) *Submarine Fans and Turbidite Systems: SEPM*, Gulf Coast Section, 15th Annual Research Conference, 127–133.

Hansen, R.J. and Kamp, P.J. 2002. Evolution of the Giant Forests Formation, Northern Taranaki Basin, New Zealand. *New Zealand Petroleum Conference*. Crown Minerals, Ministry of Economic Development, Wellington, 419–449.

Hansen, R.J. and Kamp, P.J.J. 2004*a*. Late Miocene to early Pliocene stratigraphic record in northern Taranaki Basin: condensed sedimentation ahead of Northern Graben extension and progradation of the modern continental margin. *New Zealand Journal of Geology and Geophysics*, **47**, 645–662, https://doi.org/10.1080/00288306.2004.9515081

Hansen, R.J. and Kamp, P.J.J. 2004*b*. Rapid progradation of the Pliocene–Pleistocene continental margin, northern Taranaki Basin, New Zealand, and implications. *Proceedings of the NZ Petroleum Conference*, Auckland, New Zealand, 2004.

Hansen, R.J. and Kamp, P.J. 2006. An integrated biostratigraphy and seismic stratigraphy for the late Neogene continental margin succession in northern Taranaki Basin, New Zealand. *New Zealand Journal of Geology and Geophysics*, **49**, 39–56, https://doi.org/10.1080/00288306.2006.9515146

Hansen, L., L'Heureux, J.S. *et al.* 2013. The effect of mass-wasting on the stratigraphic architecture of a fjord-valley fill: correlation of onshore, shear wave seismics and marine seismic data at Trondheim, Norway. *Sedimentary Geology*, **289**, 1–18, https://doi.org/10.1016/j.sedgeo.2013.02.002

Harris, A.D., Baumgardner, S.E., Sun, T. and Granjeon, D. 2018. A poor relationship between sea-level and deep-water sand delivery. *Sedimentary Geology*, **370**, 42–51, https://doi.org/10.1016/j.sedgeo.2018.04.002

Hayward, B.W. 1984. *Foraminiferal Biostratigraphy of Wainui-1 Offshore Well, West Taranaki, New Zealand Geological Survey*. Department of Scientific and Industrial Research.

Holt, W.E. and Stern, T.A. 1994. Subduction, platform subsidence, and foreland thrust loading: the late Tertiary development of Taranaki Basin, New Zealand. *Tectonics*, **13**, 1068–1092, https://doi.org/10.1029/94TC00454

Huang, Z., Nichol, S.L., Harris, P.T. and Caley, M.J. 2014. Classification of submarine canyons of the Australian continental margin. *Marine Geology*, **357**, 362–383, https://doi.org/10.1016/j.margeo.2014.07.007

Jackson, C.A.I. and Johnson, H.D. 2009. Sustained turbidity currents and their interaction with debrite-related topography: Labuan Island, offshore Borneo, Malaysia. *Sedimentary Geology*, **219**, 77–96, https://doi.org/10.1016/j.sedgeo.2009.04.008

Jobe, Z.R., Lowe, D.R. and Uchytil, S.J. 2011. Two fundamentally different types of submarine canyons along the continental margin of Equatorial Guinea. *Marine and Petroleum Geology*, **28**, 843–860, https://doi.org/10.1016/j.marpetgeo.2010.07.012

Judd, A.G. and Hovland, M. 2007. *Seabed Fluid Flow: The Impact of Geology, Biology, and the Marine Environment*. Cambridge University Press, Cambridge, UK.

Kamp, P.J., Vonk, A.J., Bland, K.J., Hansen, R.J., Hendy, A.J., McIntyre, A.P. and Nelson, C.S. 2004. Neogene stratigraphic architecture and tectonic evolution of Wanganui, King Country, and eastern Taranaki Basins, New Zealand. *New Zealand Journal of Geology and Geophysics*, **47**, 625–644, https://doi.org/10.1080/00288306.2004.9515080

Kastens, K.A. and Shor, A.N. 1985. Depositional processes of a meandering channel on Mississippi Fan. *American Association of Petroleum Geologists Bulletin*, **69**, 190–202.

King, L.H. and MacLean, B. 1970. Pockmarks on the Scotian Shelf. *Geological Society of America Bulletin*, **81**, 3141–3148, https://doi.org/10.1130/0016-7606(1970)81[3141:POTSS]2.0.CO;2

King, P.R. and Thrasher, G.P. 1992. Post-Eocene development of the Taranaki Basin, New Zealand: convergent overprint of a passive margin. *American Association of Petroleum Geologists, Memoirs*, **53**, 93–118.

King, P.R. and Thrasher, G.P. 1996. *Cretaceous-Cenozoic geology and petroleum systems of the Taranaki Basin, New Zealand*. Institute of Geological & Nuclear Sciences Monograph **13**. Institute of Geological & Nuclear Sciences Ltd, Lower Hutt.

King, P.R., Browne, G.H., Arnot, M.J. and Crundwell, M.P. 2007. A 2-D oblique-dip outcrop transect through an entire third-order, progradational, deep-water clastic succession (Late Miocene Mount Messenger-Urenui Formations), Taranaki Basin, New Zealand. *AAPG Studies in Geology*, **56**, CD Rom, https://doi.org/10.1306/1240948St563302

King, P.R., Ilg, B.R., Arnot, M., Browne, G.H., Strachan, L.J., Crundwell, M.P. and Helle, K. 2011. Outcrop and seismic examples of mass-transport deposits from a late Miocene deep-water succession, Taranaki Basin, New Zealand. *SEPM Special Publications*, **96**, 311–350.

Kneller, B., Dykstra, M., Fairweather, L. and Milana, J.P. 2016. Mass-transport and slope accommodation: implications for turbidite sandstone reservoirs. *American Association of Petroleum Geologists Bulletin*, **100**, 213–235, https://doi.org/10.1306/09011514210

Lamb, M.P. and Mohrig, D. 2009. Do hyperpycnal-flow deposits record river-flood dynamics? *Geology*, **37**, 1067–1070, https://doi.org/10.1130/G30286A.1

Lastras, G., Acosta, J., Muñoz, A. and Canals, M. 2011. Submarine canyon formation and evolution in the Argentine Continental Margin between 44°30'S and 48°S. *Geomorphology*, **128**, 116–136, https://doi.org/10.1016/j.geomorph.2010.12.027

Lewis, K.B. and Barnes, P.M. 1999. Kaikoura Canyon, New Zealand: active conduit from near-shore sediment zones to trench-axis channel. *Marine Geology*, **162**, 39–69, https://doi.org/10.1016/S0025-3227(99)00075-4

Lewis, K.B., Carter, L. and Davey, F.J. 1994. The opening of the Cook Strait: interglacial tidal scour and aligning basins at a subduction to transform plate edge. *Marine Geology*, **116**, 293–312, https://doi.org/10.1016/0025-3227(94)90047-7

Lewis, K.B., Collot, J.Y. and Lallem, S.E. 1998. The dammed Hikurangi Trough: a channel-fed trench blocked by subducting seamounts and their wake avalanches (New Zealand–France GeodyNZ Project). *Basin Research*, **10**, 441–468, https://doi.org/10.1046/j.1365-2117.1998.00080.x

Li, Q., Yu, S., Wu, W., Tong, L. and Kang, H. 2017. Detection of a deep-channel in 3D seismic data using the sweetness attribute and seismic geomorphology: a case study from the Taranaki Basin, New Zealand. *New Zealand Journal of Geology and Geophysics*, **60**, 199–208, https://doi.org/10.1080/00288306.2017.1307230

Lorrey, A., Vandergoes, M. *et al.* 2012. Palaeocirculation across New Zealand during the last glacial maximum at ~21 ka. *Quaternary Science Reviews*, **36**, 189–213, https://doi.org/10.1016/j.quascirev.2011.09.025

Masalimova, L.U., Lowe, D.R., Sharman, G.R., King, P.R. and Arnot, M.J. 2016. Outcrop characterization of a submarine channel-lobe complex: the lower Mount Messenger Formation, Taranaki Basin, New Zealand. *Marine and Petroleum Geology*, **71**, 360–390, https://doi.org/10.1016/j.marpetgeo.2016.01.004

McAdoo, B.G., Orange, D.L., Screaton, E., Lee, H. and Kayen, R. 1997. Slope basins, headless canyons, and submarine palaeoseismology of the Cascadia accretionary complex. *Basin Research*, **9**, 313–324, https://doi.org/10.1046/j.1365-2117.1997.00049.x

McAdoo, B.G., Pratson, L.F. and Orange, D.L. 2000. Submarine landslide geomorphology, U.S. continental slope. *Marine Geology*, **169**, 103–136, https://doi.org/10.1016/S0025-3227(00)00050-5

Mitchum, R.M., Jr 1985. Seismic stratigraphic expression of submarine fans. *American Association of Petroleum Geologists, Memoirs*, **39**, 117–136.

Mitchum, R.M., Jr, Vail, P.R. and Thompson, S., III 1977. Seismic stratigraphy and global changes of sea level; Part 2, The depositional sequence as a basic unit for stratigraphic analysis. *American Association of Petroleum Geologists, Memoirs*, **26**, 53–62.

Moscardelli, L. and Wood, L. 2008. New classification system for mass transport complexes in offshore Trinidad. *Basin Research*, **20**, 73–98, https://doi.org/10.1111/j.1365-2117.2007.00340.x

Moscardelli, L., Wood, L. and Mann, P. 2006. Mass-transport complexes and associated processes in the offshore area of Trinidad and Venezuela. *American Association of Petroleum Geologists Bulletin*, **90**, 1059–1088, https://doi.org/10.1306/02210605052

Mountjoy, J.J., Barnes, P.M. and Pettinga, J.R. 2009. Morphostructure and evolution of submarine canyons across an active margin: cook Strait sector of the Hikurangi Margin, New Zealand. *Marine Geology*, **260**, 45–68, https://doi.org/10.1016/j.margeo.2009.01.006

Mountjoy, J., Howarth, J. *et al.* 2018. Earthquakes drive large-scale submarine canyon development and sediment supply to deep-ocean basins. *Science Advances*, **4**, eaar3748, https://doi.org/10.1126/sciadv.aar3748

Nardin, T., Hein, F.J., Gorsline, M.S. and Edwards, B.D. 1979. A review of mass movement processes, sediment and acoustic characteristics, and contrasts in slope and base-of-slope systems v. canyon-fan-basin floor systems. *Society of Economic Paleontologists and Mineralogist Special Publications*, **27**, 61–73.

Nicol, A. 2011. Landscape history of the Marlborough Sounds, New Zealand. *New Zealand Journal of Geology and Geophysics*, **54**, 195–208, https://doi.org/10.1080/00288306.2010.523079

Nicol, A., Walsh, J., Berryman, K. and Nodder, S. 2005. Growth of a normal fault by the accumulation of slip over millions of years. *Journal of Structural Geology*, **27**, 327–342, https://doi.org/10.1016/j.jsg.2004.09.002

Nicol, A., Mazengarb, C., Chanier, F., Rait, G., Uruski, C. and Wallace, L. 2007. Tectonic evolution of the active Hikurangi subduction margin, New Zealand, since the Oligocene. *Tectonics*, **269**, TC4002, https://doi.org/10.1029/2006TC002090

National Institute of Water and Atmospheric Research Ltd (NIWA) 2016. New Zealand Regional Bathymetry, https://www.niwa.co.nz>our-science>oceans>download-the-data

Olafiranye, K., Jackson, C.A.-L. and Hodgson, D.M. 2013. The role of tectonics and mass-transport complex emplacement on upper slope stratigraphic evolution: a 3D seismic case study from offshore Angola. *Marine and Petroleum Geology*, **44**, 196–216, https://doi.org/10.1016/j.marpetgeo.2013.02.016

Omeru, T. 2014. *Mass Transport Deposits: Implications for Reservoir Seals*. PhD thesis, Cardiff University, UK.

Omeru, T. and Cartwright, J.A. 2019. The efficacy of kinematic indicators in a complexly deformed Mass Transport Deposit: insights from the deepwater Taranaki Basin, New Zealand. *Marine and Petroleum Geology*, **106**, 74–87, https://doi.org/10.1016/j.marpetgeo.2019.04.037

Omeru, T., Cartwright, J.A. and Bull, S. 2016. Kinematics of submarine slope failures in the deepwater Taranaki Basin, New Zealand. *In*: Lamarche, J., Mountjoy, J. *et al.* (eds) *Submarine Mass Movements and their Consequences. Advances in Natural and Technological Hazards Research*, **41**. Springer, 61–70.

Orange, D.L., McAdoo, B.G. *et al.* 1997. Headless submarine canyons and fluid flow on the toe of the Cascadia accretionary complex. *Basin Research*, **9**, 303–312, https://doi.org/10.1046/j.1365-2117.1997.00045.x

Ortiz-Karpf, A., Hodgson, D.M. and McCaffrey, W.D. 2015. The role of mass-transport complexes in controlling channel avulsion and the subsequent sediment dispersal patterns on an active margin: the Magdalena Fan, offshore Colombia. *Marine and Petroleum Geology*, **64**, 58–75, https://doi.org/10.1016/j.marpetgeo.2015.01.005

Panpichityota, N., Morley, C.K. and Ghosh, J. 2018. Link between growth faulting and initiation of a mass transport deposit in the northern Taranaki Basin. *Basin Research*, **30**, 237–248, https://doi.org/10.1111/bre.12251

Paola, C. 2017. A mind of their own: recent advances in autogenic dynamics in rivers and deltas. *SEPM Special Publications*, **106**, 5–17, https://doi.org/10.2110/sepmsp.106.04

Pickering, K.T. and Corregidor, J. 2005. Mass transport complexes and tectonic control on confined basin-floor submarine fans, Middle Eocene, south Spanish Pyrenees. *Geological Society, London, Special Publications*, **244**, 51–74, https://doi.org/10.1144/GSL.SP.2005.244.01.04

Piper, D.J.W. and Normark, W.R. 2009. Processes that initiate turbidity currents and their influence on turbidites: a marine geology perspective. *Journal of Sedimentary Research*, **79**, 347–362, https://doi.org/10.2110/jsr.2009.046

Pirmez, C., Beaubouef, R.T., Friedmann, S.J. and Mohrig, D.C. 2000. Equilibrium profile and baselevel in submarine channels: Examples from late Pleistocene systems and implications for the architecture of deepwater reservoirs. *Gulf Coast Section SEPM (Society of Sedimentary Geology) 20th Annual Research Conference Proceedings*, Houston, Texas, 782–805.

Posamentier, H.W. and Kolla, V. 2003. Seismic Geomorphology and Stratigraphy of Depositional Elements in Deep-Water Settings. *Journal of Sedimentary Research*, **73**, 367–388, https://doi.org/10.1306/111302730367

Posamentier, H.W. and Walker, R.G. 2006. Deep-water turbidites and submarine fans. *SEPM Special Publications*, **84**, 399–520, https://doi.org/10.2110/pec.06.84.0399

Pulford, A. and Stern, T. 2004. Pliocene exhumation and landscape evolution of central North Island, New Zealand: the role of the upper mantle. *Journal of Geophysical Research Earth Surface*, **109**, F01016, https://doi.org/10.1029/2003JF000046

Qin, Y., Alves, T.M., Constantine, J. and Gamboa, D. 2017. The role of mass wasting in the progressive development of submarine channels (Espirito Santo Basin, SE Brazil). *Journal of Sedimentary Research*, **87**, 500–516, https://doi.org/10.2110/jsr.2017.18

Rad, F. 2015. *PEP 38451 Romney-1 Well Completion Report*. New Zealand Unpublished Petroleum Report, **PR495**.

Ratzov, G., Sosson, M., Collot, J.-Y. and Migeon, S. 2012. Late Quaternary geomorphologic evolution of submarine canyons as a marker of active deformation on convergent margins: the example of the South Colombian margin. *Marine Geology*, **315**, 77–97, https://doi.org/10.1016 /j.margeo.2012.05.005

Rusconi, F.J. 2017. *3D Seismic Interpretation of a Plio-Pleistocene Mass Transport Deposit in the Deepwater Taranaki Basin of New Zealand*. Theses and Dissertations, 2583. University of Arkansas, http://scholarworks.uark.edu/etd/2583

Salazar, M., Moscardelli, L. and Wood, L. 2015. Utilising clinoform architecture to understand the drivers of basin margin evolution: a case study in the Taranaki Basin, New Zealand. *Basin Research*, **28**, https://doi.org/10.1111/bre.12138

Schultz, M.R., Fildani, A., Cope, T.D. and Graham, S.A. 2005. Deposition and stratigraphic architecture of an outcropping ancient slope system: Tres Pasos Formation, Magallanes Basin, southern Chile. *Geological Society, London, Special Publications*, **244**, 27–50, https://doi.org/10.1144/GSL.SP.2005.244.01.03

Sharman, G.R., Graham, S.A., Masalimova, L.U., Shumaker, L.E. and King, P.R. 2015. Spatial patterns of deformation and paleoslope estimation within the marginal and central portions of a basin-floor mass-transport deposit, Taranaki Basin, New Zealand. *Geosphere*, **11**, 266–306, https://doi.org/10.1130/geos01126.1

Slatt, R.M. 2006. *Stratigraphic Reservoir Characterization for Petroleum Geologists, Geophysicists, and Engineers*. Elsevier.

Stagpoole, V.M. and Nicol, A. 2008. Regional structure and kinematic history of a large subduction back thrust: Taranaki Fault, New Zealand. *Journal of Geophysical Research*, **113**, https://doi.org/10.1029/2007JB005170

Stern, T.A. and Davey, F.J. 1989. Crustal structure and origin of basins formed behind the Hikurangi subduction zone, New Zealand. *In*: Price, R.A. (ed.) *Origin and Evolution of Sedimentary Basins and Their Energy and Mineral Resources*. American Geophysical Union Geophysical Monograph Series, **48**, 73–86.

Stow, D.A.V., Reading, H.G. and Collinson, J. 1996. Deep Seas. *In*: Reading, H.G. (ed.) *Sedimentary Environments and Facies* 3rd edn. Blackwell, Oxford, 395–453.

Strogen, D.P., King, P.R. 2014. *A New Zealandia-Wide Seismic Horizon Naming Scheme*. GNS Science Report, **2014/34**.

Strogen, D.P., Seebeck, H., Nicol, A. and King, P.R. 2017. Two-phase Cretaceous–Paleocene rifting in the Taranaki Basin region, New Zealand; implications for Gondwana break-up. *Journal of the Geological Society, London*, **174**, 929–946, https://doi.org/10.1144/jgs2016-160

Suggate, R.P. 1990. Late Pliocene and Quaternary glaciations of New Zealand. *Quaternary Science Reviews*, **9**, 175–197, https://doi.org/10.1016/0277-3791(90)90017-5

Tubau, X., Paull, C.K.P. *et al.* 2015. Submarine canyons of Santa Monica Bay, Southern California: variability in morphology and sedimentary processes. *Marine Geology*, **365**, 61–79, https://doi.org/10.1016/j.margeo.2015.04.004

Tippett, J.M. and Kamp, P.J.J. 1993. Fission track analysis of the Late Cenozoic vertical kinematics of continental Pacific crust, South Island, New Zealand. *Journal of Geophysical Research*, **98**, 16119–16148, https://doi.org/10.1029/92JB02115

Trewick, S.A. and Bland, K.J. 2012. Fire and slice: palaeogeography for biogeography at New Zealand's North Island/South Island juncture. *Journal of the Royal Society of New Zealand.*, **42**, 153–183, https://doi.org/10.1080/03036758.2010.549493

Uruski, C. 2008. Deepwater Taranaki, New Zealand: structural development and petroleum potential. *Exploration Geophysics*, **39**, 94–107, https://doi.org/10.1071/EG08013

Uruski, C. 2010. New Zealand's deepwater frontier. *Marine and Petroleum Geology*, **27**, 2005–2026, https://doi.org/10.1016/j.marpetgeo.2010.05.010

Uruski, C.I., Stagpoole, V., Isaac, M.J., King, P.R. and Maslen, G. 2002. *Seismic interpretation Report – Astrolabe survey, Taranaki Basin, New Zealand Institute of Geological & Nuclear Sciences Confidential Client Report 2002/70*, New Zealand Open-file Petroleum Report 3072, Crown Minerals, Wellington.

Ward, N.I.P., Alves, T.M. and Blenkinsop, T.G. 2018. Submarine sediment routing over a blocky mass-transport deposit in the Espırito Santo Basin, SE Brazil. *Basin Research*, **30**, 816–834, https://doi.org/10.1111/bre.12282

Weimer, P. and Slatt, R.M. 2004. *Petroleum Systems of Deepwater Settings*. Distinguished Instructor Series, Issue 7. Society of Exploration Geophysicists.

Wynn, R.B., Cronin, B.T. and Peakall, J. 2007. Sinuous deep-water channels: genesis, geometry and architecture. *Marine and Petroleum Geology*, **24**, 341–387, https://doi.org/10.1016/j.marpetgeo.2007.06.001

Zhao, X., Qi, K., Patacci, M., Tan, C. and Xie, T. 2019. Submarine channel network evolution above an extensive mass-transport complex: a 3D seismic case study from the Niger delta continental slope. *Marine and Petroleum Geology*, **104**, 231–248, https://doi.org/10.1016/j.marpetgeo.2019.03.029

A multi-disciplinary investigation of the AFEN Slide: the relationship between contourites and submarine landslides

Ricarda Gatter[1]*, Michael A. Clare[2], James E. Hunt[2], Millie Watts[2], B. N. Madhusudhan[3], Peter J. Talling[4] and Katrin Huhn[1]

[1]MARUM – Center for Marine Environmental Sciences, University of Bremen, Leobener Str. 8, 28359, Germany

[2]University of Southampton, Engineering, Southampton Boldrewood Innovation Campus, Burgess Road, Southampton SO16 7QF, UK

[3]National Oceanography Centre, University of Southampton Waterfront Campus, Southampton European Way, Southampton, SO14 3ZH, UK

[4]University of Durham, Departments of Earth Sciences and Geography, Durham, DH1 3LE, UK

RG, 0000-0001-6548-4534; BNM, 0000-0002-2570-5934
*Correspondence: rgatter@marum.de

Abstract: Contourite drifts are sediment deposits formed by ocean bottom currents on continental slopes worldwide. Although it has become increasingly apparent that contourites are often prone to slope failure, the physical controls on slope instability remain unclear. This study presents high-resolution sedimentological, geochemical and geotechnical analyses of sediments to better understand the physical controls on slope failure that occurred within a sheeted contourite drift within the Faroe–Shetland Channel. We aim to identify and characterize the failure plane of the late Quaternary landslide (the AFEN Slide), and explain its location within the sheeted drift stratigraphy. The analyses reveal abrupt lithological contrasts characterized by distinct changes in physical, geochemical and geotechnical properties. Our findings indicate that the AFEN Slide likely initiated along a distinct lithological interface, between overlying sandy contouritic sediments and softer underlying mud-rich sediments. These lithological contrasts are interpreted to relate to climatically controlled variations in sediment input and bottom current intensity. Similar lithological contrasts are likely to be common within contourite drifts at many other oceanic gateways worldwide; hence our findings are likely to apply more widely. As we demonstrate here, recognition of such contrasts requires multi-disciplinary data over the depth range of stratigraphy that is potentially prone to slope failure.

Thermohaline-driven ocean bottom currents create sedimentary accumulations called contourites that are found along the world's continental margins (e.g. McCave and Tucholke 1986; Rebesco and Stow 2001; Stow *et al.* 2002). Contourites can cover extremely large areas (from <100 km^2 to >100 000 km^2), forming a variety of depositional geometries that include elongated, mounded, sheeted, channelized and mixed drift systems (Faugères *et al.* 1999; Rebesco and Stow 2001; Stow *et al.* 2002; Faugères and Stow 2008). It has become increasingly apparent that contourite drifts are prone to slope instability (Laberg and Camerlenghi 2008), with submarine landslides recognized in a wide range of locations affected by bottom currents (Table 1).

The affinity of contourite drifts for slope failure can be linked in part to deposit morphology (Fig. 1, Table 1). In some locations, contour-parallel currents modify the continental slope profile, creating mounded accumulations of sediment that are thicker and steeper than those on slopes unaffected by bottom currents (Laberg and Camerlenghi 2008; Rebesco *et al.* 2014). Factors such as sediment supply, intensity and location of currents, and sea-level and climatic changes control the presence or absence, location, growth and morphology of contourites (Faugères and Stow 2008; Rebesco *et al.* 2014). A number of compound morphological effects have been implicated as preconditioning and/or triggering mechanisms for slope instability. These include (1) slope over-steepening due to rapid sediment accumulation (A, Fig. 1) or due to erosion by vigorous along-slope currents (B, Fig. 1) and (2) loading resulting from differential sediment accumulation (C, Fig. 1). These effects occur particularly where contourites form as mounded accumulations (Laberg and Camerlenghi 2008; Prieto *et al.* 2016; Miramontes *et al.* 2018). However, submarine landslides, some of which include the largest on our planet (e.g. Storegga; Bryn *et al.* 2005*a*), often occur

From: Georgiopoulou, A., Amy, L. A., Benetti, S., Chaytor, J. D., Clare, M. A., Gamboa, D., Haughton, P. D. W., Moernaut, J. and Mountjoy, J. J. (eds) 2020. *Subaqueous Mass Movements and their Consequences: Advances in Process Understanding, Monitoring and Hazard Assessments.* Geological Society, London, Special Publications, **500**, 173–193.
First published online April 28, 2020, https://doi.org/10.1144/SP500-2019-184

Table 1. *Examples of submarine landslides in contourites*

Slide name	Location	Setting	Slide volume (km^3)	Seabed gradient (°)	Sediment accumulation rate	Drift type	Main control	References
Hinlopen-Yermak Slide	Northern Svalbard margin, Arctic Ocean	Northern high-latitudes	1200–1350	<0.5		?	Lithological and geotechncial contrasts	Vanneste *et al.* (2006), Winkelmann *et al.* (2008)
Fram Slide Complex	Offshore northwest Svalbard, Arctic Ocean	Northern high-latitudes	*c.* 1470 (17 failures)	*c.* 1.5–4.5	3–19 cm ka^{-1}	Plastered drift	Toe erosion, morphology	Mattingsdal *et al.* (2014), Elger *et al.* (2017)
–	Lofoten Islands, offshore Norway, Norwegian Sea	Northern high-latitudes	<1–8.7 (individual landslides)	4–1	Up to 4 m ka^{-1}	Mounded, elongated drift (Lofoten drift)	Under-cutting	Laberg *et al.* (2001), Baeten *et al.* (2013, 2014)
Trænadjupet Slide	Offshore Norway, Norwegian Sea	Northern high-latitudes	*c.* 900	2.3–0.6	Up to 65 m ka^{-1}	Mounded, elongated drift (Nyk drift)	Weak layer	Laberg and Vorren (2000), Laberg *et al.* (2001, 2002, 2003)
Nyk Slide	Offshore Norway, Norwegian Sea	Northern high-latitudes			Up to 1.2 m ka^{-1}	Mounded, elongated drift (Nyk drift)	Weak layer	Laberg *et al.* (2001, 2002), Lindberg *et al.* (2004)
Sklinnadjuped Slide	Offshore Norway, Norwegian Sea	Northern high-latitudes			Up to 0.5 m ka^{-1}	Infilling drift (Sklinnadjuped drift)	Weak layer (?)	Laberg *et al.* (2001), Dahlgren *et al.* (2002)
Storegga Slide	Offshore Norway, Norwegian Sea	Northern high-latitudes	2400–3200	0.5–1.0		Mounded, elongated drift	Sensitive clay layer	Bryn *et al.* (2005*a*, *b*), Haflidason *et al.* (2005), Kvalstad *et al.* (2005)
Tampen Slide	Offshore Norway, Norwegian Sea	Northern high-latitudes				Mounded elongated drift (?)		Evans *et al.* (2005), Solheim *et al.* (2005)
Northern Faroe Slide Complex	Faroe Islands, offshore UK, Norwegian Sea	Northern high-latitudes			14–30 cm ka^{-1}	Mounded, elongated drift (Faroe drift)		Rasmussen *et al.* (1996, 1998), Van Weering *et al.* (1998), Kuijpers *et al.* (2001), Long *et al.* (2004)
AFEN Slide	Offshore UK, Faroe–Shetland Channel	Northern high-latitudes	*c.* 0.153 (all phases)	1–3	Up to 10 cm ka^{-1}	Sheeted to mounded drift (West Shetland drift)	Sandy layer (?)	Knutz and Cartwright (2004), Wilson *et al.* (2004)
Rockall Bank Slide Complex	Offshore Ireland, Rockall Trough	Northern high-latitudes	265–765	5–10	5–17.1 cm ka^{-1}	Elongated, mounded drift (Feni drift)	Weak layers	Van Weering and de Rijk (1991), Faugères *et al.* (1999), Georgiopoulou *et al.* (2013, 2019)
–	Offshore eastern Canada, North Atlantic	Northern mid-latitudes				Plastered drift (?)		Piper (2005)
–	Grand Banks, offshore eastern Canada, North Atlantic	Northern mid-latitudes		2	Up to 50 cm ka^{-1}	Plastered drift	Lithological and geotechnical contrasts	Rashid *et al.* (2017)

–	Pianosa Ridge, Mediterranean Sea	Northern mid-latitudes		3–10 (locally 20)	13 cm ka^{-1}	Plastered drift	Over-steepening	Miramontes *et al.* (2016, 2018)
–	Gela and south Adriatic Basin, Mediterranean Sea	Northern mid-latitudes	0.1–0.2 (individual mass transport deposits)	*c.* 3	22.5 cm ka^{-1}	Elongated and separated drifts	Mechanical boundary, clay layer	Minisini *et al.* (2007), Verdicchio and Trincardi (2008)
–	SW Mallorca Island, Mediterranean Sea	Northern mid-latitudes		1.3–2.9	5.8 cm ka^{-1} (?)	Mounded, elongated drifts		Lüdmann *et al.* (2012)
–	Alboran Sea, Mediterranean Sea	Northern mid-latitudes				Contourite despositional system		Ercilla *et al.* (2016)
–	Levant Basin, Mediterranean Sea	Northern mid-latitudes	Generally, <1 (individual landslides)	>4	25–130 cm ka^{-1}	Plastered drift	Over-steepening	Katz *et al.* (2015), Hübscher *et al.* (2016)
–	Bahamas Bank	Norther low-latitudes	2–20 (individual landslides)	*c.* 3		Plastered drift	Stratigraphic control (?)	Mulder *et al.* (2012); Principaud *et al.* (2015), Tournadour *et al.* (2015)
–	Offshore Uruguay	Southern mid-latitudes	<2 (individual landslides)	1–3	8–18 cm ka^{-1}	Contourite depositional system	Lithological control	Henkel *et al.* (2011), Krastel *et al.* (2011), Ai *et al.* (2014), Hernández-Molina *et al.* (2016)
–	Offshore Argentina	Southern mid-latitudes		3–7	Up to 1.6 m ka^{-1}	Contourite depositional system	Lithological control; over-steepening	Hernández-Molina *et al.* (2009), Ai *et al.* (2014), Krastel *et al.* (2011), Preu *et al.* (2013)
–	Offshore Antarctic Peninsula, Pacific Ocean	Southern low-latitudes		2–3	Decrease from 18 to *c.* 8 cm ka^{-1}	Mounded drifts	Under-cutting; weak layer	Iwai *et al.* (2002), Volpi *et al.* (2003, 2011)

Slide volume, seabed gradient and sediment accumulation rate are given where available. Main controls of slope failure are listed where they are known or discussed in the literature.

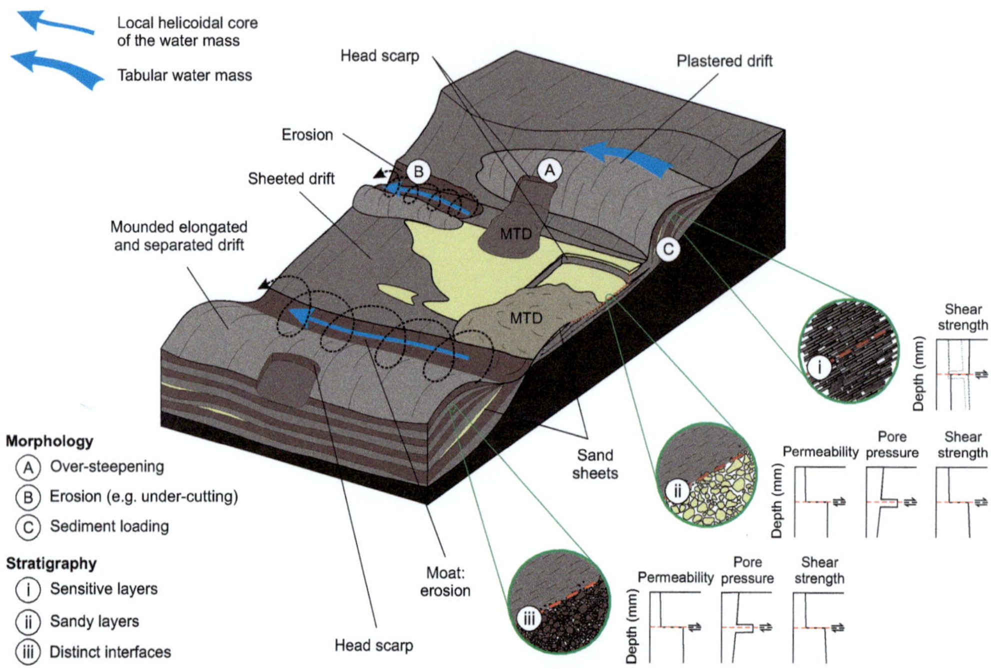

Fig. 1. Key characteristics of contourites that favour the formation of submarine landslides. Morphological controls: (A) over-steepening, (B) erosion, (C) sediment loading; stratigraphic controls: (i) laterally extensive sensitive clay layers that are prone to sudden strength loss, possible shear strength depth profiles are shown as black; dark grey, dashed and light grey, dotted lines; (ii) thick accumulation of sandy layers that can accommodate excess pore pressure due to high sedimentation rates; (iii) distinct lithological and/or geotechnical interfaces. Contourite depositional system adopted from Hernández-Molina *et al.* (2008).

within contourite drifts on very low angle (<2°) slopes (e.g. Hühnerbach *et al.* 2004). Another explanation for slope instability in contourite drifts, therefore, relates to specific compositional and geotechnical properties of contourites (Fig. 1, Table 1; Lindberg *et al.* 2004; Kvalstad *et al.* 2005). Plausible controls include prominent layers within the slope stratigraphy (Fig. 1) that may feature a lower peak or post-peak shear strength than over- and underlying strata, such as (i) laterally extensive (sometimes cm-thin) homogeneous layers of weaker, sensitive material that is prone to sudden strength loss (e.g. sensitive clay in the Storegga Slide, Norway – Kvalstad *et al.* 2005; sensitive zeolite layer in the N Tyrrhenian Sea – Miramontes *et al.* 2018), or (ii) thick accumulations of sandy material that is characterized by high sedimentation rates, promoting excess pore pressure (Laberg and Camerlenghi 2008; Ai *et al.* 2014). Another plausible control relates to lithological and/or geotechnical contrasts within a depositional sequence that may result from rapid changes in current regime, sediment input or type (e.g. Rashid *et al.* 2017; see (iii) on Fig. 1).

Detailed sedimentological and geotechnical studies of landslides within contourites are scarce (Baeten *et al.* 2013; Miramontes *et al.* 2018), and there is still much uncertainty as to which specific aspects act as the dominant control on slope instability. Many studies rely solely upon remote geophysical data for landslide characterization and, if sediment cores are acquired, they typically do not penetrate to the failure plane (which may be tens to hundreds of metres below the seafloor; Talling *et al.* 2014). Such cores also tend to focus on characterization of the failed landslide mass, rather than targeting sediments from adjacent undisturbed slopes. Targeting the undisturbed sediments of the adjacent slopes, including those stratigraphically equivalent to the failure plane of the landslide, however, is necessary in order to identify and characterize the material along which the landslide initiated, as these are usually removed or remoulded during failure. It is of critical importance to be able to identify sediments that are prone to failure in order to perform reliable slope stability assessments (L'Heureux *et al.* 2012; Vardy *et al.* 2012).

Aims

Here, we present a detailed characterization of a bedding-parallel, cohesive submarine landslide (called the AFEN Slide), which occurred within a low angle (<2.5°) laterally extensive sheeted contourite drift, based on physical, geochemical, sedimentological and geotechnical analyses. We focus on a sediment core targeted to sample the pre-landslide sedimentary sequence, including sediments that correlate stratigraphically with the failure plane located further upslope. Based on centimetre-resolution characterization of these deposits we address the following questions: first, what is the nature of the undisturbed sediment and do material heterogeneities explain the location of the failure plane? As many aspects of cohesive landslides appear to be scale invariant, this study of a relatively small landslide may provide key insights into our understanding of much larger ones (Micallef *et al.* 2008; Chaytor *et al.* 2009; Baeten *et al.* 2013; Casas *et al.* 2016; Kuhlmann *et al.* 2017; Clare *et al.* 2018). Second, what causes the observed heterogeneities within the stratigraphy? We explore how climatic changes and ocean circulation may play a key role in governing not just the failure plane depth, but also influence the timing of slope failure. Finally, we discuss the implications of climatically controlled sediment supply and deep ocean circulation for pre-conditioning slope instability in contourite depositional systems in oceanic gateways, which are narrow, deep passages connecting adjacent basins, elsewhere in the world.

Background

Regional setting

Geological and morphological setting. The study area lies on the eastern margin of the Faroe–Shetland Channel, which is located north of Scotland, extending over 400 km between the Wyville–Thomson Ridge and the Norwegian Basin (Fig. 2). The Faroe–Shetland Channel is a narrow basin, measuring 250 km at its widest in the NE and less than 130 km in the SW. The channel closely follows the trend of the regional NE–SW structural lineaments, and one of the NW–SE transfer zones (Victory Transfer Zone) passes close to the study area (Rumph *et al.* 1993; Wilson *et al.* 2004). The Faroe–Shetland Channel is the present-day expression of the Faroe–Shetland Basin that can be dated back to the Late Paleozoic (e.g. Rumph *et al.* 1993). Basin formation was probably initiated during the Devonian, while the main rift phase occurred during Cretaceous times (Dean *et al.* 1999; Roberts *et al.* 1999). Although localized extension continued until the early to mid-Paleocene (Smallwood and Gill 2002), more or less continuous post-rift subsidence predominated throughout the Cenozoic (Turner and Scrutton 1993). This subsidence was interrupted at various stages by contractional deformation (Ritchie *et al.* 2003, 2008; Johnson *et al.* 2005; Stoker *et al.* 2005) and regional uplift and tilting (Andersen *et al.* 2000; Smallwood and Gill 2002; Stoker *et al.* 2002, 2005). Following Late Paleocene uplift, the Faroe–Shetland Channel subsided about 2000 m, with present-day water depths of 1700 m in the NE and 1000 m in the SW, and slope angles between 1° and 3° flanking the eastern channel margin (Stoker *et al.* 1998; Andersen *et al.* 2000; Smallwood and Gill 2002). The channel forms an important oceanic gateway, exchanging water masses between the North Atlantic and the Norwegian Sea (Broecker and Denton 1990; Rahmstorf 2002) since at least the Early Oligocene (Davies *et al.* 2001).

Oceanography and palaeoceanography. In general, the present-day oceanography in the Faroe–Shetland Channel consists of warm surface water moving towards the NE, and cold bottom water, generating relatively strong, erosive bottom currents (with velocities in the range between <0.3 and >1.0 m s^{-1}; Masson *et al.* 2004), moving towards the SW (Fig. 2a; Saunders 1990; Turrell *et al.* 1999; Rasmussen *et al.* 2002). Five distinct water masses can be recognized based on their salinity and temperature characteristics (Turrell *et al.* 1999). Two distinct surface water masses transport warm water from the North Atlantic into the channel. North Atlantic Water (NAW) flows northward from the Rockall Trough (Turrell *et al.* 1999), while Modified North Atlantic Water (MNAW) flows clockwise around the Faroe Islands before turning northward in the Faroe–Shetland Channel (Saunders 1990). These surface waters typically occupy the upper 200–400 m of the water column (Turrell *et al.* 1999). Arctic Intermediate Water (AIM) flows anticlockwise along the southern edge of the Norwegian Basin and around the Faroe–Shetland Channel, typically between 400 and 600 m water depth (Blindheim 1990). At the base of the channel (usually below 600 m water depth), the Norwegian Sea Arctic Intermediate Water (NSAIW) and the Faroe–Shetland Channel Bottom Water (FSCBW) are funnelled along the Faroe–Shetland Channel towards the south (Turrell *et al.* 1999) and flow along the Faroe Bank Channel into the Atlantic (Saunders 1990). A small portion of the cold bottom water flows across the western end of the Wyville–Thomson Ridge south into the Rockall Trough (Stow and Holbrook 1984). The velocity of these water masses is variable, both across the channel and over time. The average along-slope velocities, mainly directed NE, of around 0.2–0.25 m s^{-1} were measured at around 500–700 m water depth (Van Raaphorst *et al.* 2001; Bonnin *et al.* 2002) and velocities over >1.0 m s^{-1} associated with SW-directed bottom currents were inferred from observed bedforms (Masson *et al.* 2004). Periodic

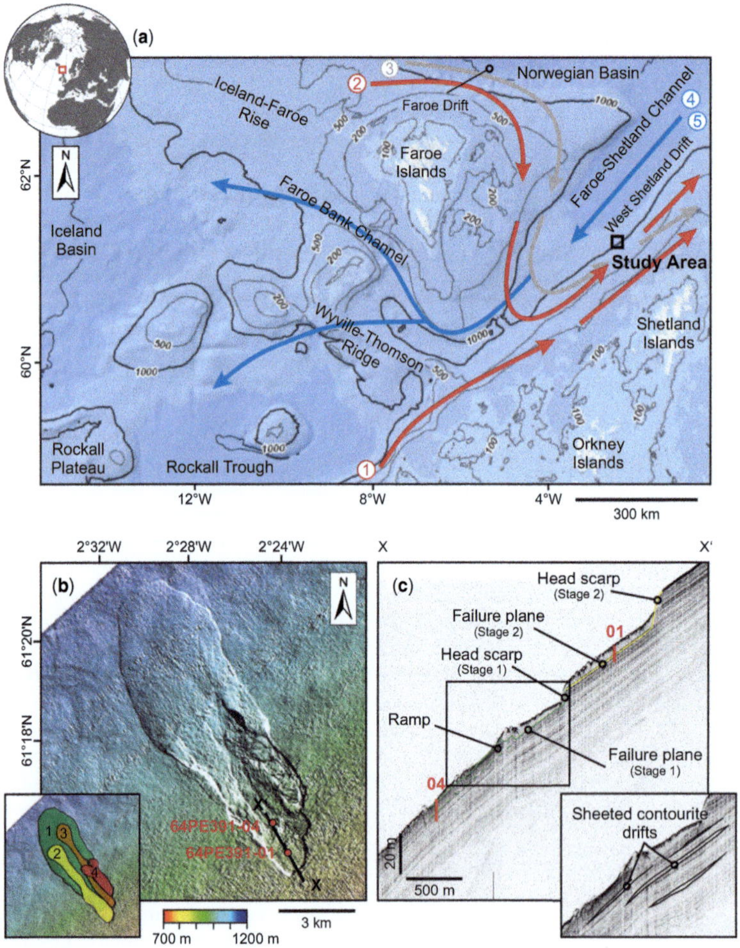

Fig. 2. (**a**) Schematic diagram of current regime in and around the Faroe–Shetland Channel. Arrows indicate the five main water masses: red 1, North Atlantic Water; red 2, Modified North Atlantic Water; grey 3, Arctic Intermediate Water; blue 4, Norwegian Sea Arctic Intermediate Water; blue 5, Faroe–Shetland Channel Bottom Water (after Turrell *et al.* 1999). The study area is outlined with a black square. (**b**) Outline of the AFEN Slide, showing piston core 64PE391-01 (61° 15′ 40.679″ N, 02° 23′ 42.899″ W; Madhusudhan *et al.* 2017) and Core 64PE391-04 (61° 16′ 17.651″ N, 02° 24′ 21.959″ W) as red circles. A black line illustrates the seismic line shown in (c). Inset image shows the four stages of the failure as interpreted by Wilson *et al.* (2004). Modified from Madhusudhan *et al.* (2017). (**c**) Seismic line across the AFEN Slide showing piston core 01 and 04. Inset image illustrates the distribution of sheeted contourite drifts in the area (after Wilson *et al.* 2004).

changes in salinity and temperature cause shifts of the boundaries between water masses on timescales from decades to hours (Turrell *et al.* 1999). Since the Last Glacial Maximum (LGM), when bottom and surface currents were weak, eight distinct changes in the surface and bottom current regime were identified, which are related to the changes in climatic conditions (Rasmussen *et al.* 2002). Climatic and palaeoceanographic changes also reportedly caused strong cyclical variation in sediment accumulation (with up to 30 cm ka^{-1} along the Faroe Drift and up to 10 cm ka^{-1} along the West Shetland Drift; Rasmussen *et al.* 1996, 1998; Knutz and Cartwright 2004; Nielsen *et al.* 2007).

Contourite deposits in the Faroe–Shetland Channel

The regional oceanography has controlled the depositional architecture of the slope sediments, creating elongated mounded contourite drifts at the base of the slope (to the NE of the AFEN Slide) and sheeted contourite drifts in the slide area (Long *et al.* 2004; Hohbein and Cartwright 2006). These sheeted drifts are characterized by parallel, laterally continuous reflectors on seismic profiles (Masson 2001). These reflectors can be traced over more than 50 km below the seafloor of the Faroe–Shetland Channel, which emphasizes the regional scale of bottom current activity and sheeted contourite drift accumulation (Stoker *et al.* 1998).

The AFEN Slide

The AFEN Slide was first identified in 1996, during an environmental survey for the Atlantic Frontiers Environmental Network in the region (Wilson *et al.* 2004). The slide is interpreted as a four-stage retrogressive landslide that occurred NW of the Shetland Islands (UK) at water depths of 830–1120 m on a slope varying from approximately 0.7° to about 2.5° (Wilson *et al.* 2003, 2004; Fig. 2b). The total length from the head scarp to the toe of the lobe is over 12 km, and the maximum width is around 4.5 km. The slide involved *c.* 200×10^6 m^3 of sediment and the slide debris has a maximum thickness of 20 m, averaging between 5 and 10 m (Wilson *et al.* 2004). Radiocarbon dating and biostratigraphy from the slide suggest that the first stage took place around 16–13 ka BP and the later retrogressive phases after 5.8 ka BP and prior to 2.8 ka BP (Wilson *et al.* 2004). Initial studies, based on high-resolution seismic data and sediment cores that did not penetrate the base of the slide, inferred that the failure plane comprised well-sorted contourite sands, which may liquefy during an earthquake (e.g. 10 000-year return period earthquake; Jackson *et al.* 2004). This hypothesis was supported by the presence of a buried slide, which appears to have occurred under similar physiographic conditions (Masson 2001; Wilson *et al.* 2003, 2004). Such well-sorted contourite sands were not found by Madhusudhan *et al.* (2017), who analysed a new sediment core (64PE391-01) that penetrated through the full extent of the deposits from the second stage of the landslide (Fig. 2c). Instead, they proposed progressive failure of geotechnically sensitive clays or liquefaction of silt layers. None of these previous cores sampled undisturbed material that corresponds stratigraphically with the failure plane.

Data and methods

Core 64PE391-04, which is the focus of this present study, was obtained during the RV *Pelagia* cruise 64PE391 in 2014 using a piston corer. The core was sampled within the AFEN Slide area, at a water depth of 945 m. It was targeted to sample undisturbed sediments, i.e. those characterized on seismic data by continuous reflectors and avoiding acoustically transparent, chaotic or disrupted seismic units and areas of hummocky seafloor texture likely indicative of slope failure (Shipp *et al.* 2011; Fig. 2). Figure 2 shows the location of core 64PE391-04 on the deep tow boomer seismic profile, which has a maximum theoretical vertical resolution of 0.5 m, with a penetration of 100 ms, and was obtained from the BGS 00/02 survey (Wilson *et al.* 2005). The core recovered 11.49 m of sediment in a 15 m core barrel and was kept in refrigerated storage at the British Ocean Sediment Core Facility (BOSCORF), UK, prior to study.

Physical properties analysis

A Geotek MSCL-S (Standard) multi-sensor core logger, based at BOSCORF, was used to measure P-wave velocity, gamma-ray bulk density, electrical resistivity, magnetic susceptibility and fractional porosity, which is derived from the measured sediment density at 1 cm intervals on split cores (Fig. 3). MSCL is a commonly used, non-destructive tool that allows the recognition of subtle changes in sediment physical properties. The data are commonly used for correlation between cores, and calibration of seismic data using P-wave velocity. Density serves as an effective proxy for changes in sediment lithology and is used for the calculation of fractional porosity (Gunn and Best 1998). Core images were obtained using the BOSCORF Geotek MSCL-CIS (Core Imaging System), which enables the acquisition of precise depth-registered images that can be correlated with the other datasets.

Geochemical analysis

Micro-XRF (X-ray fluorescence) core scanning was used to determine the geochemical composition of the sediment (ITRAX™ COX Ltd at BOSCORF; Croudace *et al.* 2006) at a spatial resolution of 1 cm. ITRAX scanning is a useful, rapid, non-destructive, high-resolution scanning technique that is widely used in earth and environmental sciences (Croudace and Rothwell 2015). This method enables the measurement of element intensities, such as Ca and Sr, which correlate well with the carbonate content, or Fe, Ti and K, which are related to the siliciclastic components, and vary directly with the terrigenous sediment input (e.g. Röhl and Abrams 2000; Hepp *et al.* 2006). ITRAX data represent a semi-quantitative analysis of the relative element abundances down-core. Data are expressed as counts per second (cps) and are presented as log ratios, which are accepted as a more accurate estimation of element concentrations. In addition, all XRF data are shown as log ratios of two elements, in order to show element concentrations more accurately

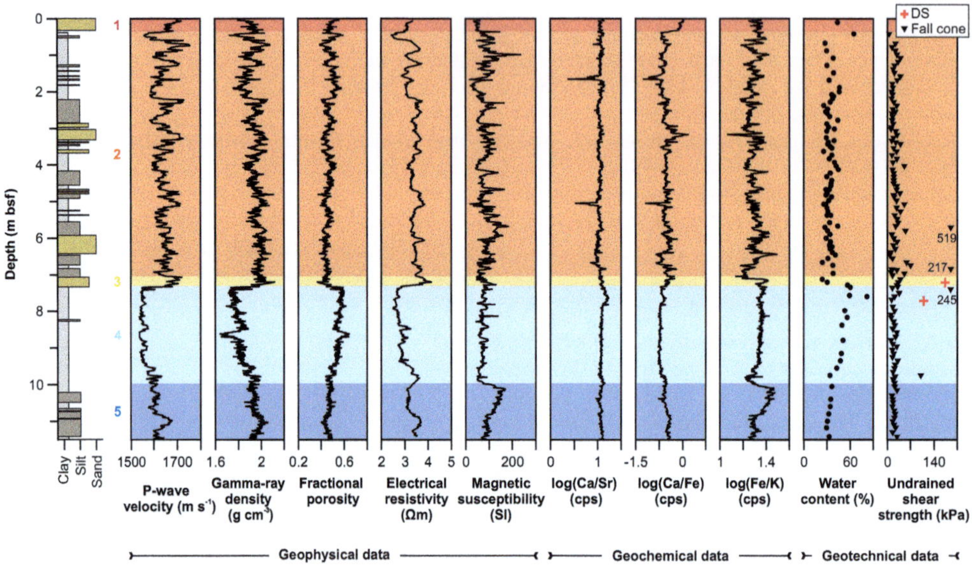

Fig. 3. Summary of sediment core analyses (64PE391-04), including visual sedimentary, physical properties (multi-sensor core logging) and geochemical (ITRAX X-ray fluorescence) core-log data, and geotechnical data (water content, drained and undrained shear strength). Units 1–5 are outlined.

and minimize matrix effects inherent to XRF (Weltje and Tjallingii 2008). Ca/Sr, Ca/Fe and Fe/K have been selected, as these element ratios have been shown to reflect changes in sea-level and temperature, sediment supply, and have been applied in climate studies (see Croudace and Rothwell 2015). In addition to geochemical composition, the ITRAX instrument provided X-radiographs. X-radiographs are digital images of the internal structure and physical property changes within a split core section that are obtained using optical and radiographic line cameras.

Grain-size distribution

Grain-size analysis was carried out at 10 cm depth intervals for sediments of Units 2, 3 and 4 (see results for definition), following the procedures in Rothwell and Rack (2006). The sediment was sieved to remove particles larger than 2 mm before the sample was dispersed in a 1 l mixing chamber by shaking it for 24 h. The dispersed sediment was circulated through a Malvern Mastersizer 3000 for 120 s over which time 12 measurements are taken and then averaged to obtain the grain-size distribution.

Geotechnical analyses

Water content and fall cone measurements were carried out at 10 cm intervals (BSI 1990; BSI 2004). Measurements of water content could be used as a first-order approximation of the sediment's shear strength and compressibility (i.e. higher water content is related to low shear strength and high compressibility). An 80 g 30° fall cone was used on the split cores, regardless of the grain size and whether the tested material was considered to be saturated or not. The undrained shear strength was calculated from the fall cone measurements assuming all tests were carried out on saturated clays. Subsamples were taken for subsequent direct shear and oedometric tests.

Static, drained shear test. Direct shear experiments were carried out to compare the drained shear strength of prominent layers, identified from downcore logging, grain-size distribution and standard geotechnical data. Cylindrical, undisturbed samples (*c.* 5 cm^2, 2.5 cm height) of intact samples were placed in the shear apparatus and consolidated via a vertical ram to *in situ* normal stress (σ_n). The sample was consolidated until the sample height was constant (or min. 24 h), so that the sample is assumed to be fully drained and the applied σ_n is approximately equal to the effective normal stress (σ'_n). The effective normal stress is the difference between the normal stress and the pore water pressure, u ($\sigma'_n = \sigma_n - u$; Terzaghi 1925). Shearing occurs on a predefined plane, perpendicular to the vertical ram that exerts the normal stress. The shear displacement for each experiment was 9.5 mm at a shear rate of 0.008 mm min^{-1}. This shear rate was slow enough to allow constant drainage during shearing (Deutsches

Institut für Normung 2002). Samples were taken from around 7 m core depth, which corresponds to around 18 m below seafloor (assuming around 10 m of sediment was removed during the failure). The samples were sheared at a normal stress of 170 kPa, simulating the effective hydrostatic vertical overburden stress (σ'_{v0}) acting at around 18 m below seafloor (m bsf) assuming an average sediment effective unit weight (γ') of 9.5 kN m^{-3}.

Oedometer test. One-dimension consolidation tests were performed on selected undisturbed core samples (*c.* 20 cm^2, 1.9 cm height) in order to measure and compare their permeability and consolidation parameters. The measured initial porosity (n), coefficient of compression (c_v) and permeability (k) can be used to make assumptions regarding the sediments' potential to build excess pore pressure. Incremental loading and unloading of 1 kPa to 7100 kPa stress were applied onto the sediment and the resulting displacement (change in volume) was measured. Each load was applied gradually and left until the displacement stabilized or primary consolidation was completed. Consolidation and permeability parameters were calculated from the settlement characteristics of the sediment using standard equations (Powrie 2013).

Data analysis

Physical and geochemical properties were compared using non-parametric tests that compare two unpaired groups of data and compute p-values testing the null hypothesis of two groups having the same distribution. The data were analysed for the discrepancy between the mean ranks of two groups (Mann–Whitney test) and for their varying cumulative distribution (Kolmogorov–Smirnov test) (Sheskin 2011). The significance level for both tests was set to 0.05 (Fisher 1926).

Results

Piston core 64PE391-04 was obtained about 750 m down-slope from where the sediment ramped up the failure plane on to the seabed (failure Stage 1, Wilson *et al.* 2004; Fig. 2c). The deep-tow boomer reflection seismic data indicate that the core penetrated the pre-landslide sediments, including those stratigraphically equivalent to the failure plane of the slide. Based on the newly obtained data, we identify five main lithological units within the sediment core, which we now characterize using results from visual sediment core logging, particle size distribution, X-ray scanning, and continuous physical properties (MSCL) and geochemical (micro-XRF) measurements (see summary in Figs 3 & 4). In addition, we present a geotechnical characterization of the recovered sediment based on water content and fall cone analyses, as well as direct shear (DS) and oedometer tests.

Visual sedimentary logging and grain-size analyses indicate that the general lithology is bioturbated silty clay to clayey silt with a number of sandy silt and silty sand layers, consistent with previous analyses of sediment cores from the area (Madhusudhan *et al.* 2017). Sandy layers are found only in the upper part of the core (above 7.3 m depth). The lithology in the lower part of the core is generally homogeneous with an absence of sand.

Multi-Sensor Core Logger (MSCL) data

Down-core logging data show an abrupt and distinct change in physical properties at around 7.3 m depth, as well as more subtle variations that enabled demarcation of the five sediment units (Fig. 3; Table 2). Unit 1 is largely indiscernible from Unit 2 based on physical properties, but does have much lower magnetic susceptibility. The sediments above the abrupt contact at 7.3 m (Units 2 and 3) are generally characterized by high relative P-wave velocities, gamma-ray densities, electrical resistivity, and low relative values of fractional porosity (on average under 0.5). Unit 3 shows the highest electrical resistivity and gamma-ray densities in the core; hence, it is demarcated as an individual unit, rather than being subsumed within Unit 2. In the sediments immediately below 7.3 m (Unit 4), the most marked step in physical properties is observed, including a reduction in gamma-ray density from 2.0 to 1.7 g cm^{-3}, and an increase in fractional porosity from approximately 0.45 to >0.55. Such a marked change was not observed in the magnetic susceptibility this side of the contact either; however, the signal is generally more erratic above and less variable below (Fig. 3). Below the contact at 7.3 m, P-wave velocity, gamma-ray density and electrical resistivity gradually increase down-core (inversely mirroring a steady decrease in fractional porosity) until the start of Unit 5, which is marked by a sharp increase in magnetic susceptibility (from <70 to >165 m^3 kg^{-1}), and a subtle increase in average P-wave velocity and gamma-ray density (Fig. 3).

Micro-X-ray fluorescence (XRF) data

Distinct changes in geochemistry are also observed from the micro-XRF analysis between the sediment units (Figs 3 & 5), which correspond to very similar depths ($\pm$0.3 m) where physical property changes are noted. The first-order observations are of: (i) a step in Fe/K, Ca/Fe and Ca/Sr elemental ratios between 7.1 and 7.3 m (i.e. straddling Unit 2/3/4 contacts); and (ii) a switch from more variable (noisy) elemental ratios above 7.1–7.3 m (Units 2

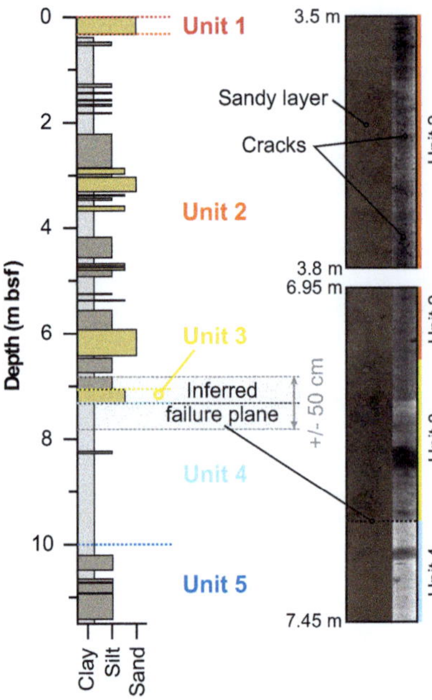

Fig. 4. Inferred location of the main failure plane based on down-core logging and deep-tow boomer reflection seismic data. Units 1–5 are outlined. Vertical error in the failure plane delineation, resulting from the vertical resolution of the seismic data is indicated by grey lines (±50 cm from the inferred failure plane). Core images and X-radiographs from the inferred failure plane and cracks in Unit 2 are also shown.

and 3), with centimetre-scale variations in geochemical composition, to less noisy ratios below (Unit 4). Below Unit 4, variations in elemental ratios are also observed, supporting the demarcation of Unit 5. Cross-plotting of the elemental ratios (Fig. 6) supports the demarcation of the five identified sediment units, as well as illustrating the range in variability between each unit (e.g. a large spread of values in Unit 2, compared to Unit 4).

Grain-size distribution

Figure 7 summarizes grain-size distribution data for core section 64PE391-04-D (6.5–7.7 m depth), which include sediments from Units 2, 3 and 4. The data illustrate the change in composition at around 7.3 m depth. Unit 4 (below 7.3 m depth) is characterized by a higher silt content in comparison to overlying sediments. Unit 3 is recognized as a sandy silt layer, and the sampled sediments of Unit 2 show a switch from sandy silt to clayey silt, which supports the distinct changes in lithology seen in the visual core log.

Geotechnical data

A distinct change in water content can be observed, which increases from around 30% to over 60% at 7.3 m depth (i.e. at the contact between Units 3 and 4; Fig. 3). Unit 1 has a slightly higher water content than Unit 2 (more or less constant 30%). Units 4 and 5 are characterized by decreasing water content. A distinct change in the undrained shear strength is not observed, although the scatter is greater in the upper part of the core (Units 2 and 3). Individual outliers (>100 kPa) are related to dropstones or mud clasts.

A summary of the key sample parameters and test results of the direct shear and oedometer tests are given in Table 3. The peak drained shear strength of Units 3 and 4 are shown in Figure 3 (indicated by red crosses). It can be seen that Unit 3 encompasses a higher peak shear strength (173 kPa) than Unit 4 (109 kPa). Typical porosity (n) v. applied normal stress (σ_n) is shown in Figure 8. It is apparent that porosity decreases with increasing normal stress and increases slightly during the rebound phase. Unit 3 has a lower initial porosity, and higher permeability (k) and compressibility (c_v) than Unit 4.

Discussion

The recovered slope sediment obtained from core 64PE391-04 is characterized by a distinct step change in both physical and geochemical properties between around 7.1 and 7.3 m depth, as well as a distinct high-density contrast at that depth that was recorded by X-ray imaging (Figs 3 & 4). These transitions are related to an abrupt change in lithology from a thick relatively homogeneous clayey silt, silty clay unit (Unit 4; Figs 3 & 5) to an overlying 25 cm-thick sandy silt layer (Unit 3; Figs 3 & 5). The depth of this distinct change matches well with the seismostratigraphic horizon that is equivalent to the main failure plane outlined in the deep-tow boomer reflection seismic data (assuming a seismic velocity of 1600 m s^{-1}; Wilson *et al.* 2004), which is supported by the available MSCL data.

The sediment above this distinct interface is characterized by slightly higher P-wave velocities and gamma-ray densities, as well as a lower fractional porosity than would be expected for continental slope sediments (Fig. 3; Hamilton 1970). Small cracks were recorded by X-ray imaging, but are limited to parts of Unit 2 (Fig. 4). These observations could be related to a slight compaction of the sediment, for example, due to compression by the partially confined landslide debris above the sediment ramp (Fig. 2c; e.g. Frey-Martínez *et al.* 2006; Principaud *et al.* 2015; Brooks *et al.* 2018), or to the around

Table 2. *Summary of sediment core's sedimentological, geophysical and geochemical characteristics*

Unit and depth range (m)	General sedimentological description	MSCL characterization	XRF characterization	Possible deposit interpretation
Unit 1 (0–0.33)	Muddy sand	Lower magnetic susceptibility, no distinct trends in other geophysical properties	>Ca/Fe; No distinct Ca/Sr or Fe/K trend	Recent current reworked deposits
Unit 2 (0.33–7.11)	Stratified unit, consisting of bioturbated clayey silt to silty clay and sandy silt to silty sand layers; drop stones in the upper part of the unit	Strong variations in P-wave velocity, gamma-ray density, fractional porosity and magnetic susceptibility; down-core increase in P-wave velocity and gamma-ray density, and decrease in fractional porosity	Strong variations especially in Ca/Fe	Post-glacial deposits, with variable pulses of sediment flux including melt-water plumes
Unit 3 (7.11– 7.32)	Sandy silt layer; mud clasts	High P-wave velocity and electrical resistivity	Increase in Ca/Sr; decrease in Ca/Fe; distinct increase in Fe/K	Sandy contourite, reworked from immediate post-glacial meltwater-derived sediments
Unit 4 (7.32–10.00)	Relatively homogeneous bioturbated silty clay to clayey silt; drop stones throughout the unit	Distinct and abrupt decrease in P-wave velocity, gamma-ray density and electrical resistivity, and increase in fractional porosity at contact with Unit 3; less variation in magnetic susceptibility	Relatively constant element ratios; higher average Ca/Sr (and peak); lower average Ca/Fe; higher average Fe/K	Steady glaciomarine deposition
Unit 5 (10.00–end)	Clayey silt to sandy silt	Distinct and abrupt increase in magnetic susceptibility at contact with Unit 4; slight increase in P-wave velocity and gamma-ray density	Slightly variations in Ca/Sr; increasing Ca/Fe; distinct increase in Fe/K;	Steady interstadial deposition

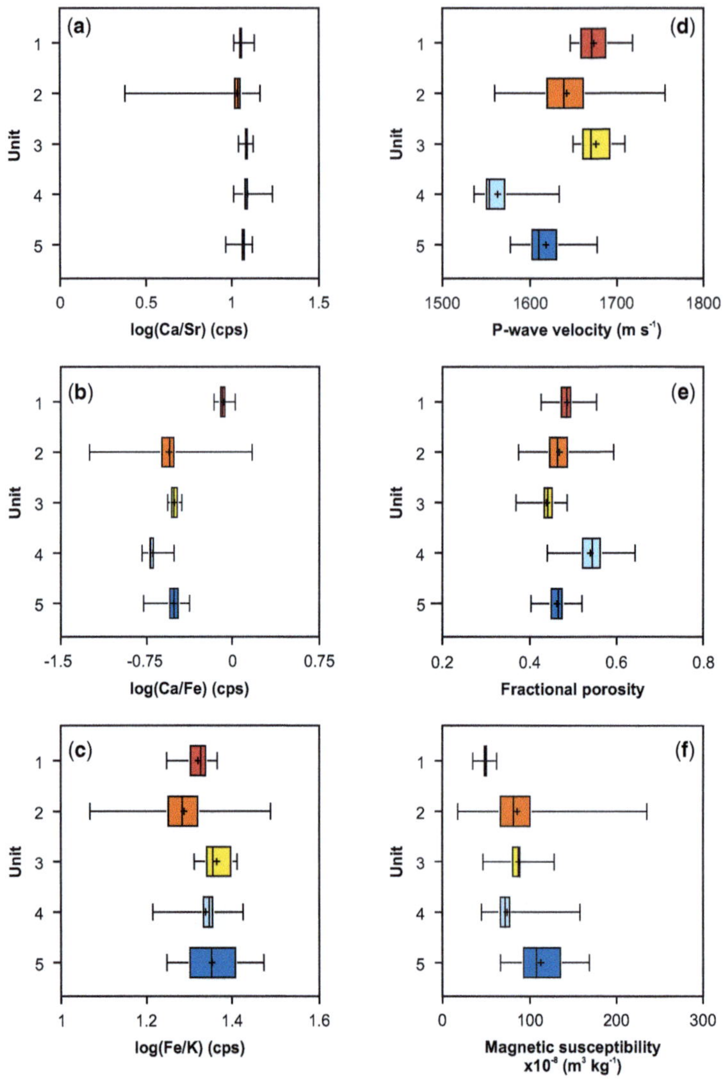

Fig. 5. Box-whisker plots showing the variation in element ratios (**a**) Ca/Sr, (**b**) Ca/Fe and (**c**) Fe/K, and (**d–f**) physical properties between Units 1–5. The lines of the box indicate the upper and lower quartiles and the median, lines extending parallel from the boxes indicate the maximum and minimum values and the cross illustrates the mean value.

10 m missing sediment sequence at the 64PE391-04 core location (Fig. 2), whose removal could have disturbed the slope sediments. The potential deformation, however, is not resolved in the seismic data, and the distinct change at around 7.1–7.3 m depth is not limited to the physical properties, but is also noted in the geochemical properties. We therefore infer that although the sediment might have been slightly deformed, it probably did not move (no sliding motion) and the stratigraphy was not altered.

Lithological contrasts appear to play a key role in dictating the location of the failure plane

Wilson *et al.* (2004) previously suggested that the AFEN Slide could have initiated along a sandy contouritic layer embedded within the slope stratigraphy, but were unable to sample deep enough to show its occurrence. Our deeper core now shows that this hypothesis may be plausible, given the presence of Unit 3. Although this unit was not identified

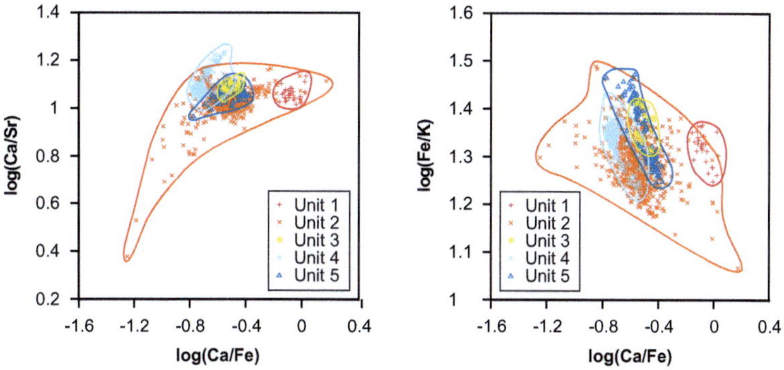

Fig. 6. ITRAX XRF composition of individual subunits compared as cross plots. Red crosses, Unit 1; orange crosses, Unit 2; yellow circles, Unit 3; light blue stars, Unit 4; dark blue triangles, Unit 5.

as a contourite in the seismic data (Fig. 2c, Wilson *et al.* 2004), we interpret it as a sheeted sandy contourite drift. This assumption is considered reasonable as the vertical resolution of the seismic data (0.5 m; Wilson *et al.* 2005) might be too low to register this 25 cm-thick layer. Furthermore, we also show that there is much greater lithological heterogeneity (based on physical properties and geochemistry) within these sheeted drifts than has been previously documented, aside from simply variations in grain size. Without detailed geochemical and physical properties data, this abrupt lithological change would not have been identified.

Abrupt lithological changes (such as between Units 3 and 4) may instead play a key role in defining the location of the failure plane. Unfortunately, the vertical resolution of the existing seismic data does not enable us to categorically determine whether the failure plane should correspond to the contact of Units 3/4 or 2/3. Although varying the assumed seismic velocity within reasonable ranges for sediments only results in a vertical offset of 0.5 m, the failure plane falls within the depth window that includes the interfaces between Units 2/3 and 3/4 (Fig. 4). Wilson *et al.* (2004) implicated sandy contouritic sediments as potential 'weak layers' (i.e. Units 2/3 scenario) because of their potential to generate excess pore pressures when bound by an overlying lower permeability unit. This is a reasonable suggestion; however, the fractional porosity data indicate that the sand-rich Unit 3 instead features slightly lower porosity than the overlying sediments, while the underlying mud-rich sediments (Unit 4) have an even higher porosity. This observation is supported by water content data, which show the highest values in the mud-rich Unit 4 and abruptly

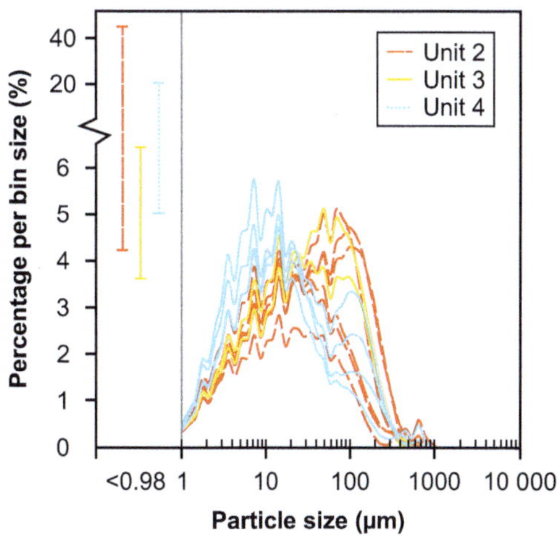

Fig. 7. Particle-size distribution data illustrated as percentage per bin. The data are binned as frequency per 1/4 phi unit. Particle size <0.98 is given as percentage range per Unit.

Table 3. *Key sample parameters and results from direct shear and oedometer tests*

Sample	Unit 3	Unit 4
LL (%)	26.5	56.1
PL (%)	–	25
γ' (kN m^{-3})	9.5	9.5
σ'_n (kPa)	170	170
τ_{peak} (kPa)	173	109
n	0.43	0.55
c_v (m^2 s^{-1})	5.2×10^{-4}	7.6×10^{-5}
k (m s^{-1})	4.3×10^{-7}	7.8×10^{-8}

LL, liquid limit; PL, plastic limit; γ', effective unit weight; σ'_n, effective normal stress; τ_{peak}, peak shear strength; n, porosity; c_v, compressibility; k, permeability.

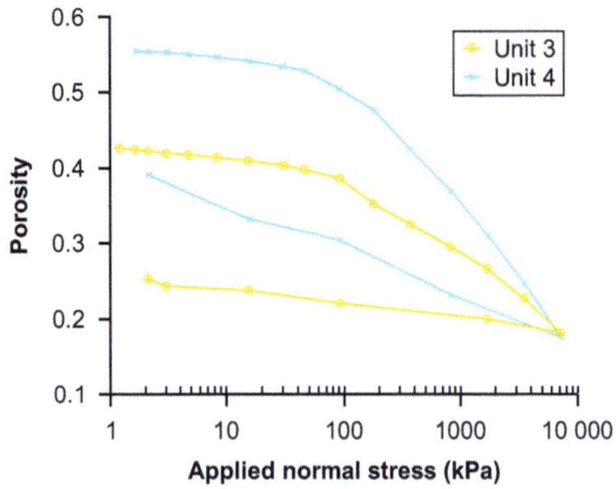

Fig. 8. Porosity (n) v. applied normal stress (σ_n) curves from one-dimensional consolidation tests.

decrease at the interface to Unit 3. Oedometer tests carried out on undisturbed samples from Units 3 and 4 reveal a higher initial porosity and lower compressibility of Unit 4. This relationship is in contrast to an established empirical relationship between coarser grain size and greater porosity (or larger pore size; Ren and Santamarina 2018). This apparent contradiction is explained by the presence of detrital clay that fills in pore spaces between sand grains (Unit 3), whereas the relatively open structure of the underlying muddier deposits (Unit 4) explains their higher relative porosity (Marion *et al.* 1992; Revil and Cathles III 1999). In contrast to porosity, however, permeability is found to be higher in the sand-rich sediments (Unit 3; Table 3). Considering the higher permeability and compressibility of Unit 3, it is possible for excess pore pressure to accumulate within the sandy contouritic sediments (e.g. during an earthquake). Although this observation would support the 'weak layer' hypothesis, it has to be noted that the water content is actually higher in Unit 4 and abruptly drops at the interface with Unit 3, instead of increasing within the layer.

Another noticeable observation is the difference in shear strength between Units 3 and 4. Both drained and undrained shear strength are lower in the mud-rich Unit 4, which can be related to the higher water content and the lack of sandy material within the unit. Taking all these observations into account, we suggest that it is possible that a failure plane could generate at an interface where sand overlies finer-grained cohesive sediments. The high water content and lower shear strength of the fine-grained material could allow the overlying sediment to slide on top of it. We are unable to be more absolute on the failure depth, but we have demonstrated that variability in sheeted drifts can also include abrupt whole-scale changes in sediment properties, as well as the presence of thin coarser units, which have traditionally been invoked to explain bedding-parallel failures in contourite sheeted drifts (Laberg and Camerlenghi 2008). Such variability may not necessarily be expected based on the available seismic data alone.

Climate change is a likely control on creating failure-prone lithological contrasts

Down-core changes in Ca/Sr ratios have been successfully related to variations in sea-level and water temperature (through integration with oxygen isotope curves and biostratigraphy), wherein high Ca/Sr ratios are indicative of ice-rafted debris and changes from colder to warmer conditions (e.g. Smith *et al.* 1979; Thomson *et al.* 2004; Hodell *et al.* 2008). High Fe/K ratios and low Ca/Fe, on the other hand, have been related to colder periods (Kuijpers *et al.* 2003; Perez *et al.* 2016). The increased Ca/Sr ratio above 7.6 m depth could therefore indicate a stronger meltwater flux, carrying ice-rafted debris into the channel, while the changes in Fe/K and Ca/Fe ratios at 7.1–7.3 m are also interpreted to indicate a switch from cold conditions (Unit 4) to warmer conditions (Units 2/3). This switch was coincident with a transition from finer-grained, stable sedimentation to a more variable regime with pulsed influxes of coarser material. Given the existing knowledge about the timing of the AFEN Slide (Unit 1 should post-date 2.8–5.8 ka BP, while the pre-failure sediments must be older than 16 ka BP; Wilson *et al.* 2004), this transition fits within a time window that includes the switch from the LGM (18 ka BP) to post-glacial conditions. Glacial conditions would have seen sediment largely locked up in ice sheets, while the melt-out during the immediate postglacial window involved pulses of fine- and coarser-grained sediment. The nearby Faroe–Shetland Channel is the main oceanic gateway between the North Atlantic and the Norwegian Sea (Broecker and Denton 1990; Rahmstorf 2002), where a direct relation exists between ocean circulation and climate. Rapid changes in the exchange of water masses between the NE Atlantic and the Norwegian Sea occurred following the LGM at 18 ka BP (Rasmussen *et al.* 2002), which would have compounded the abruptness of a switch in sediment transfer. We therefore suggest that the abrupt change in physical properties and geochemistry may relate to this climatic transition.

Previous studies have investigated the role of climate change on submarine landslides, primarily focusing on their timing. A number of early studies suggested that submarine landslides, particularly in higher latitudes, may be more likely during sea-level lowstands. Recent work, however, has suggested that there is no clear statistical relationship or at least that there are too few observations to be confident (e.g. Maslin *et al.* 2004; Brothers *et al.* 2013; Urlaub *et al.* 2013, 2014; Pope *et al.* 2015). Indeed, recent work has shown that such margins may feature many more late Holocene submarine landslides than previously thought (Normandeau *et al.* 2019). Proving a clear link between submarine landslides and sea-level or climate change is most likely complicated by a range of factors, including time lags in offshore sediment transport, residence times of excess pore pressures following periods of rapid sediment accumulation, local sea-level changes (e.g. isostatic rebound following glaciations) and other factors (Masson *et al.* 2006; Urgeles and Camerlenghi 2013; Talling *et al.* 2014). Whether climate change has played any role in the timing of the slope failures at AFEN remains unclear; however, it may have played a key role in one aspect: the location of the failure plane. Our data indicate that the slope failure most likely initiated along a distinct lithological interface that is interpreted to relate to a switch in depositional regime: from cold and uniform to warm and variable depositional conditions. The close connection between thermohaline circulation, sea-level and temperature, and sediment supply in this region may explain why the switch in deposition was so rapid.

Broader implications for slope instability in contourites at climatically influenced ocean gateways

The origin of distinct lithological interfaces may arise in a variety of ways, and may be very common in contouritic sediments near ocean gateways where climatic changes may affect bottom current intensity (and thus control the grain-size that is transported; Faugères and Mulder 2011), as well as the type of sediment that is distributed by bottom currents (e.g. terrestrial and biogenic fluxes may vary during different climatic windows; Faugères *et al.* 1993; Maldonado *et al.* 2005). Such effects can be felt at a variety of latitudes, ranging from tropical to polar settings (e.g. Kuijpers *et al.* 2001; Principaud *et al.* 2015; Elger *et al.* 2017). Climate may play a key role in dictating the location of potential failure planes. While many previous studies have invoked dominantly geometric controls on slope failure in contourite drifts, our study contributes to a growing literature base that indicates that lithological interfaces may explain the strong affinity of contourite deposits to slope instability. We posit that in low-angle, sheeted contourite drifts, such as AFEN, it is such material interfaces that are most important for preconditioning slopes to failure.

Conclusions

The integration of physical properties and geochemical core-log data, grain-size distribution, and geotechnical data indicate that the AFEN Slide initiated along a distinct lithological interface within the slope stratigraphy, which matches the depth of the failure plane obtained from seismic data. This lithological interface correlates with the base of a 25 cm sandy contourite layer, overlying a thick, relatively homogeneous silty clay unit. Based on this high-resolution multi-proxy analysis, it was possible to resolve small-scale material changes within the slope stratigraphy, which cannot be distinguished from seismic data alone (owing to the limited vertical resolution of 0.5 m). Integrating the core analyses with our knowledge about the current regime prevailing in the Faroe–Shetland Channel for the last 18 ka, it seems that climate change might precondition the location of failure initiation. This highlights the fact that in order to understand submarine landslide hazard, it is necessary to include information from all different scales, ranging from the small-scale high-resolution analysis of core material to the understanding of the regional oceanographic setting.

Acknowledgements The authors thank the British Geological Survey for the supply of the deep-tow boomer reflection seismic data and previous contributions (in particular David Tappin and David Long). Thanks are also given to the crew of the RV *Pelagia* for their efforts during data collection. We thank BOSCORF and its staff (S. MacLachan, M. Edwards and M. Charidemou) for their services in maintaining the cores and assisting with some of the analytical techniques, Achim Kopf for letting us use his geotechnical laboratory at MARUM to carry out the shear tests, and the National Infrastructure Laboratory, Southampton, for letting us conduct our oedometer tests. We acknowledge the constructive reviews by A. Cattaneo and U. Nicholson, and editor J. Mountjoy. Credit for the bathymetric metadata is given to ESRI, Garmin, GEBCO, NOAA NGDC, and other contributors.

Funding This project has received funding from the European Union's Horizon 2020 research and innovation programme under the Marie Skłodowska-Curie grant agreement No. 721403. We acknowledge research funding from the UK National Environmental Research Council (NE/K0008X/1) for sediment core collection. M. Clare was supported by the Climate Linked Atlantic Sector Science (CLASS) programme (Natural Environment Research Council Grant No. NE/R015953/1).

Author contributions **RG**: conceptualization (equal), data curation (supporting), investigation (equal), methodology (equal), validation (equal), visualization (lead), writing – original draft (lead), writing – review and editing (lead); **MAC**: conceptualization (equal), project administration (equal), supervision (lead), visualization (supporting), writing – original draft (equal), writing – review and editing (equal); **JEH**: data curation (equal), investigation (equal), methodology (equal), project administration (equal), supervision (supporting), validation (equal), writing – review and editing (supporting); **MW**: data curation (equal), investigation (equal), validation (equal), writing – review and editing (supporting); **BNM**: conceptualization (equal), investigation (equal), methodology (equal), supervision (equal), validation (equal), writing – review and editing (supporting); **PJT**: conceptualization (equal), funding acquisition (lead), project administration (equal), writing – review and editing (equal); **KH**: conceptualization (equal), funding acquisition (lead), supervision (equal), writing – review and editing (supporting).

References

Ai, F., Strasser, M., Preu, B., Hanebuth, T.J.J., Krastel, S. and Kopf, A. 2014. New constraints on oceanographic vs. seismic control on submarine landslides initiation: a geotechnical approach off Uruguay and northern Argentina. *Geo-Marine Letters*, **34**: 399–417, https://doi.org/10.1007/s00367-014-0373-3

Andersen, M.S., Nielsen, T., Sørensen, A.B., Boldreel, L.O. and Kuijpers, A. 2000. Cenozoic sediment distribution and tectonic movements in the Faroe region. *Global and Planetary Change*, **24**, 239–259, https://doi.org/10.1016/S0921-8181(00)00011-4

Baeten, N.J., Laberg, J.S. *et al.* 2013. Morphology and origin of smaller-scale mass movements on the continental slope off northern Norway. *Geomorphology*, **187**, 122–134, https://doi.org/10.1016/j.geomorph.2013.01.008

Baeten, N.J., Laberg, J.S. *et al.* 2014. Origin of shallow submarine mass movements and their glide planes – Sedimentological and geotechnical analyses from the continental slope of northern Norway. *Journal of Geophysical Research: Earth Surface*, **119**, 2335–2360, https://doi.org/10.1002/2013JF003068

Blindheim, J. 1990. Arctic intermediate water in the Norwegian Sea. *Deep-Sea Research*, **37**, 1475–1489, https://doi.org/10.1016/0198-0149(90)90138-L

Bonnin, J., van Raaphorst, W., Brummer, G.-J., van Haren, H. and Malschaert, H. 2002. Intense mid-slope resuspension of particulate matter in the Faeroe–Shetland Channel: short-term deployment of near-bottom sediment traps. *Deep-Sea Research I*, **49**, 1485–1505, https://doi.org/10.1016/S0967-0637(02)00030-4

Broecker, W.S. and Denton, G.H. 1990. The role of ocean–atmosphere reorganizations in glacial cycles. *Quaternary Science Reviews*, **9**, 305–341, https://doi.org/10.1016/0277-3791(90)90026-7

Brooks, H.L., Hodgson, D.M., Brunt, R.L., Peakall, J. and Flint, S.S. 2018. Exhumed lateral margins and increasing flow confinement of a submarine landslide complex. *Sedimentology*, **65**, 1067–1069, https://doi.org/10.1111/sed.12415

Brothers, D.S., Luttrell, K.M. and Chaytor, J.D., 2013. Sea-level–induced seismicity and submarine landslide occurrence. *Geology*, **41**, 979–982, https://doi.org/10.1130/G34410.1

Bryn, P., Berg, K., Forsberg, C.F., Solheim, A. and Kvalstad, T.J. 2005*a*. Explaining the Storegga Slide. *Marine and Petroleum Geology*, **22**, 11–19, https://doi.org/10.1016/j.marpetgeo.2004.12.003

Bryn, P., Berg, K., Stoker, M.S., Haflidason, H. and Solheim, A. 2005*b*. Contourites and their relevance for mass wasting along the Mid-Norwegian Margin. *Marine and Petroleum Geology*, **22**, 85–96, https://doi.org/10.1016/j.marpetgeo.2004.10.012

BSI 1990. Methods of test for soils for civil engineering purposes – Part 2: classification tests. *In*: *British Standard BS 1377–2*. British Standards Institution, London.

BSI 2004. Geotechnical investigation and testing – laboratory testing of soils. Part 6: fall cone tests. *In*: *DD CEN ISO/TS 17892-6*. British Standards Institution, London.

Casas, D., Chiocci, F., Casalbore, D., Ercilla, G. and De Urbina, J.O. 2016. Magnitude-frequency distribution of submarine landslides in the Gioia Basin (southern Tyrrhenian Sea). *Geo-Marine Letters*, **36**, 405–414, https://doi.org/10.1007/s00367-016-0458-2

Chaytor, J.D., Uri, S., Solow, A.R. and Andrews, B.D. 2009. Size distribution of submarine landslides along the US Atlantic margin. *Marine Geology*, **264**, 16–27, https://doi.org/10.1016/j.margeo.2008.08.007

Clare, M.A., Chaytor, J. *et al.* 2018. A consistent global approach for the morphometric characterization of subaqueous landslides. *Geological Society, London, Special Publications*, **477**, https://doi.org/10.1144/SP477.15

Croudace, I.W. and Rothwell, R.G. (eds) 2015. *Micro-XRF Studies of Sediment Cores. Applications of a Non-Destructive Tool for the Environmental Sciences*. Springer, Dordrecht.

Croudace, I.W., Rindby, A. and Rothwell, R.G. 2006. ITRAX: description and evaluation of a new multi-function X-ray core scanner. *Geological Society, London, Special Publications*, **267**, 51–63, https://doi.org/10.1144/GSL.SP.2006.267.01.04

Dahlgren, K.I.T., Vorren, T.O. and Laberg, J.S. 2002. Late Quaternary glacial development of the mid-Norwegian margin – 65 to 68°N. *Marine and Petroleum Geology*, **19**, 1089–1113, https://doi.org/10.1016/S0264-8172(03)00004-7

Davies, R.J., Cartwright, J.A., Pike, J. and Line, C. 2001. Early Oligocene initiation of North Atlantic deep water formation. *Nature*, **410**, 917–920, https://doi.org/10.1038/35073551

Dean, K., McLachlan, K. and Chambers, A. 1999. Rifting and the development of the Faeroe-Shetland Basin. *Geological Society, London, Petroleum Geology Conference Series*, **5**, 533–544, https://doi.org/10.1144/0050533

Deutsches Institut für Normung 2002. Baugrund, Untersuchung von Bodenproben – Bestimmung der Scherfestigkeit. *In*: *DIN-Norm 18137- 3*. Beuth, Berlin.

Elger, J., Berndt, C., Krastel, S., Piper, D.J.W., Gross, F. and Geissler, W.H. 2017. Chronology of the Fram Slide Complex offshore NW Svalbard and its implications for local and regional slope stability. *Marine Geology*, **393**, 141–155, https://doi.org/10.1016/j.margeo.2016.11.003

Ercilla, G., Juan, C. *et al.* 2016. Significance of bottom currents in deep-sea morphodynamics: an example from the Alboran Sea. *Marine Geology*, **378**, 157–170, https://doi.org/10.1016/j.margeo.2015.09.007

Evans, D., Harrison, Z. *et al.* 2005. Paleoslides and other mass failures of Pliocene to Pleistocene age along the Atlantic continental margin of NW Europe. *Marine and Petroleum Geology*, **22**, 1131–1148, https://doi.org/10.1016/j.marpetgeo.2005.01.010

Faugères, J.-C. and Mulder, T. 2011. Contour currents and contourite drifts. *Developments in Sedimentology*, **63**, 149–214, https://doi.org/10.1016/B978-0-444-53000-4.00003-2

Faugères, J.-C. and Stow, D.A.V. 2008. Contourite drifts: nature, evolution and controls. *Developments in Sedimentology*, **60**, 259–288, https://doi.org/10.1016/S0070-4571(08)10014-0

Faugères, J.-C., Mézerais, M.L. and Stow, D.A.V. 1993. Contourite drift types and their distribution in the North and South Atlantic Ocean basins. *Sedimentary Geology*, **82**, 189–203, https://doi.org/10.1016/0037-0738(93)90121-K

Faugères, J.-C., Stow, D.A.V., Imbert, P. and Viana, A. 1999. Seismic features diagnostic of contourite drifts. *Marine Geology*, **162**, 1–38, https://doi.org/10.1016/S0025-3227(99)00068-7

Fisher, R.A. 1926. The arrangement of field experiments. *Journal of the Ministry of Agriculture of Great Britain*, **33**, 503–513.

Frey-Martínez, J., Cartwright, J. and James, D. 2006. Frontally confined versus frontally emergent submarine landslides: a 3D seismic characterisation. *Marine and Petroleum Geology*, **23**, 585–604, https://doi.org/10.1016/j.marpetgeo.2006.04.002

Georgiopoulou, A., Shannon, P.M., Sacchetti, F., Haughton, P.D.W. and Benetti, S. 2013. Basement-controlled multiple slope collapses, Rockall Bank Slide Complex, NE Atlantic. *Marine Geology*, **336**, 198–214, https://doi.org/10.1016/j.margeo.2012.12.003

Georgiopoulou, A., Krastel, S. *et al.* 2019. On the timing and nature of the multiple phase of slope instability on Eastern Rockall Bank, Northeast Atlantic. *Geochemistry, Geophysics, Geosystems*, **20**, 594–613, https://doi.org/10.1029/2018GC007674

Gunn, D.E. and Best, A.I. 1998. A new automated nondestructive system for high resolution multi-sensor core logging of open sediment cores. *Geo-Marine Letters*, **18**, 70–77, https://doi.org/10.1007/s003670050054

Haflidason, H., Lien, R., Sejrup, H.P., Forsberg, C.F. and Bryn, P. 2005. The dating and morphometry of the Storegga Slide. *Marine and Petroleum Geology*, **22**, 123–136, https://doi.org/10.1016/j.marpetgeo.2004.10.008

Hamilton, E.L. 1970. Sound velocity and related properties of marine sediments, North Pacific. *Journal of Geophysical Research*, **75**, 4423–4446, https://doi.org/10.1029/JB075i023p04423

Henkel, S., Strasser, M. *et al.* 2011. An interdisciplinary investigation of a recent submarine mass transport deposit at the continental margin off Uruguay. *Geochemistry, Geophysics, Geosystems*, **12**, Q08009, https://doi.org/10.1029/2011GC003669

Hepp, D.A., Mörz, T. and Grützner, J. 2006. Pliocene glacial cyclicity in a deep-sea sediment drift (Antarctic Peninsula Pacific Margin). *Paleogeography, Paleoclimatology, Paleoecology*, **231**, 181–198, https://doi.org/10.1016/j.palaeo.2005.07.030

Hernández-Molina, F.J., Llave, E. and Stow, D.A.V. 2008. Continental slope contourites. *Developments in Sedimentology*, **60**, 379–408, https://doi.org/10.1016/S0070-4571(08)10019-X

Hernández-Molina, F.J., Paterlini, M., Violante, R., Marshall, P., de Isasi, M., Somoza, L. and Rebesco, M. 2009. Contourite depositional system on the Argentine Slope: an exceptional record of the influence of Antarctic water masses. *Geology*, **37**, 507–510, https://doi.org/10.1130/G25578A.1

Hernández-Molina, F.J., Soto, M. *et al.* 2016. A contourite depositional system along the Uruguayan continental margin: sedimentary, oceanographic and paleoceanographic implications. *Marine Geology*, **378**, 333–349, https://doi.org/10.1016/j.margeo.2015.10.008

Hodell, D.A., Channell, J.E.T., Curtis, J.H., Romero, O.E. and Röhl, U. 2008. Onset of "Hudson Strait" Heinrich events in the eastern North Atlantic at the end of the middle Pleistocene transition (~640 ka)?. *Paleoceanography*, **23**, PA4218, https://doi.org/10.1029/2008PA001591

Hohbein, M. and Cartwright, J. 2006. 3D seismic analysis of the West Shetland Drift system: implications for Late Neogene paleoceanography of the NE Atlantic. *Marine Geology*, **230**, 1–20, https://doi.org/10.1016/j.margeo.2006.03.009

Hübscher, C., Betzler, C. and Reiche, S. 2016. Seismo-stratigraphic evidences of deep base level control on middle to late Pleistocene drift evolution and mass wasting along southern Levant continental slope (Eastern Mediterranean). *Marine and Petroleum Geology*, **77**, 526–534, https://doi.org/10.1016/j.marpetgeo.2016.07.008

Hühnerbach, V., Masson, D.G. and partners of the COSTA-Project 2004. Landslides in the North Atlantic and its adjacent seas: an analysis of their morphology, setting and behaviour. *Marine Geology*, **213**, 343–362, https://doi.org/10.1016/j.margeo.2004.10.013

Iwai, M., Acton, G.D., Lazarus, D., Osterman, L.E. and Williams, T. 2002. Magnetobiochronologic synthesis of ODP Leg 178 rise sediments from the Pacific sector of the Southern Ocean: sites 1095, 1096, and 1101. *In*: Barker, P.F., Camerlenghi, A., Acton, G.D. and Ramsay, A.T.S. (eds) *Proceedings of ODP, Scientific Results*, **178**. Ocean Drilling Program, College Station, TX, 1–40.

Jackson, P.D., Gunn, D.A. and Long, D. 2004. Predicting variability in the stability of slope sediments due to earthquake ground motion in the AFEN area of the western UK continental shelf. *Marine Geology*, **213**, 363–378, https://doi.org/10.1016/j.margeo.2004.10.014

Johnson, H., Ritchie, J.D., Hitchen, K., Mcinroy, D.B. and Kimbell, G.S. 2005. Aspects of the Cenozoic deformational history of the Northeast Faroe–Shetland Basin, Wyville-Thomson Ridge and Hatton Bank areas. *Geological Society, London, Petroleum Geology Conference Series*, **6**, 993–1007, https://doi.org/10.1144/0060993

Katz, O., Reuven, E. and Aharonov, E. 2015. Submarine landslides and fault scarps along the eastern Mediterranean Israeli continental-slope. *Marine Geology*,

369, 100–115, https://doi.org/10.1016/j.margeo.2015.08.006

Knutz, P.C. and Cartwright,. 2004. 3D anatomy of late Neogene contourite drifts and associated mass flows in the Faroe–Shetland Basin. *Geological Society, London, Memoirs*, **29**, 63–71, https://doi.org/10.1144/GSL.MEM.2004.029.01.07

Krastel, S., Wefer, G. *et al.* 2011. Sediment dynamics and geohazards off Uruguay and the de la Plata River region (northern Argentina and Uruguay). *Geo-Marine Letters*, **31**, 271–283, https://doi.org/10.1007/s00367-011-0232-4

Kuhlmann, J., Asioli, A., Trincardi, F., Klügel, A. and Huhn, K. 2017. Landslide frequency and failure mechanisms at NE Gela Basin (Strait of Sicily). *Journal of Geophysical Research: Earth Surface*, **122**, 2223–2243, https://doi.org/10.1002/2017JF004251

Kuijpers, A., Nielsen, T., Akhmetzhanov, A., de Haas, H., Kenyon, N.H. and van Weering, T.C.E. 2001. Late Quaternary slope instability on the Faeroe margin: mass flow features and timing of events. *Geo-Marine Letters*, **20**, 149–159, https://doi.org/10.1007/s0036 70000053

Kuijpers, A., Troelstra, S.R. *et al.* 2003. Late Quaternary sedimentary processes and ocean circulation changes at the Southeast Greenland margin. *Marine Geology*, **195**, 109–129, https://doi.org/10.1016/S0025-3227(02)00684-9

Kvalstad, T.J., Andresen, L., Forsberg, C.F., Berg, K., Bryn, P. and Wangen, M. 2005. The Storegga slide: evolution of triggering sources and slide mechanics. *Marine and Petroleum Geology*, **22**, 245–256, https://doi.org/10.1016/j.marpetgeo.2004.10.019

Laberg, J.S. and Camerlenghi, A. 2008. The significance of contourites for submarine slope stability. *Developments in Sedimentology*, **60**, 537–556, https://doi.org/10.1016/S0070-4571(08)10025-5

Laberg, J.S. and Vorren, T.O. 2000. The Trænadjuped Slide, offshore Norway – morphology, evacuation and triggering mechanisms. *Marine Geology*, **171**, 95–114, https://doi.org/10.1016/S0025-3227(00)00112-2

Laberg, J.S., Dahlgren, T., Vorren, T.O., Haflidason, H. and Bryn, P. 2001. Seismic analyses of Cenozoic contourite drift development in the Northern Norwegian Sea. *Marine Geophysical Researches*, **22**, 401–416, https://doi.org/10.1023/A:1016347632294

Laberg, J.S., Vorren, T.O. *et al.* 2002. Late Quaternary paleoenvironment and chronology in the Trænadjuped Slide area offshore Norway. *Marine Geology*, **188**, 35–60, https://doi.org/10.1016/S0025-3227(02)00274-8

Laberg, J.S., Vorren, T.O. and Mienert, J. 2003. Preconditions leading to the Holocene Trænadjuped Slide offshore Norway. *In*: Locat, J. and Mienert, J. (eds) *Submarine Mass Movements and Their Consequences*. Advances in Natural and Technological Hazards Research, **19**, Springer, Dordrecht, 247–254, https://doi.org/10.1007/978-94-010-0093-2_28

L'Heureux, J.-S., Longva, O. *et al.* 2012. Identification of weak layers and their role for the stability at Finneidfjord, Northern Norway. *In*: Yamada, Y., Kawamura, K. *et al.* (eds) *Submarine Mass Movements and Their Consequences*. Advances in Natural and Technological Hazard Research, **31**, Springer, Dordrecht, 321–330, https://doi.org/10.1007/978-94-007-2162-3_29

Lindberg, B., Laberg, J.S. and Vorren, T.O. 2004. The Nyk Slide – morphology, progression, and age of a partly buried submarine slide offshore northern Norway. *Marine Geology*, **213**, 277–289, https://doi.org/10.1016/j.margeo.2004.10.010

Long, D., Bulat, J. and Stoker, M.S. 2004. Sea bed morphology of the Faroe-Shetland Channel derived from 3D seismic datasets. *Geological Society, London, Memoirs*, **29**, 53–61, https://doi.org/10.1144/GSL.MEM.2004.029.01.06

Lüdmann, T., Wiggershaus, S., Betzler, C. and Hübscher, C. 2012. Southwest Mallorca Island: a cool-water carbonate margin dominated by drift deposition associated with giant mass wasting. *Marine Geology*, **307-310**, 73–87, https://doi.org/10.1016/j.margeo.2011.09.008

Madhusudhan, B.N., Clare, M.A., Clayton, C.R.I. and Hunt, J.E. 2017. Geotechnical profiling of deep-ocean sediments at the AFEN submarine slide complex. *Quarterly Journal of Engineering Geology and Hydrology*, **50**, 148–157, https://doi.org/10.1144/qjegh2016-057

Maldonado, A., Barnolas, A. *et al.* 2005. Miocene to recent contourite drifts development in the northern Weddell Sea (Antarctica). *Global and Planetary Change*, **45**, 99–129, https://doi.org/10.1016/j.gloplacha.2004.09.013

Marion, D., Nur, A., Yin, H. and Han, D. 1992. Compressional velocity and porosity in sand-clay mixtures. *Geophysics*, **57**, 554–563, https://doi.org/10.1190/1.1443269

Maslin, M., Owen, M., Day, S. and Long, D. 2004. Linking continental-slope failures and climate change: testing the clathrate gun hypothesis. *Geology*, **32**, 53–56, https://doi.org/10.1130/G20114.1

Masson, D.G. 2001. Sedimentary processes shaping the eastern slope of the Faeroe-Shetland Channel. *Continental Shelf Research*, **21**, 825–857, https://doi.org/10.1016/S0278-4343(00)00115-1

Masson, D.G. Wynn, R.B. and Bett, B.J. 2004. Sedimentary environment of the Faroe-Shetland and Faroe Bank Channels, north-east Atlantic, and the use of bedforms as indicators of bottom current velocity in the deep ocean. *Sedimentology*, **51**, 1207–1241, https://doi.org/10.1111/j.1365-3091.2004.00668.x

Masson, D.G., Harbitz, C.B., Wynn, R.B., Pedersen, G. and Løvholt, F. 2006. Submarine landslides: processes, triggers and hazard prediction. *Philosophical Transactions: Mathematical, Physical and Engineering Sciences*, **354**, 2009–2039, https://doi.org/10.1098/rsta.2006.1810

Mattingsdal, R., Knies, J., Andreassen, K., Fabian, K., Husum, K., Grøsfjeld, K. and De Schepper, S. 2014. A new 6 Myr stratigraphic framework for the Atlantic-Arctic Gateway. *Quaternary Science Reviews*, **92**, 170–178, https://doi.org/10.1016/j.quascirev.2013.08.022

McCave, I.N. and Tucholke, B.E. 1986. Deep current controlled sedimentation in the western North Atlantic. *In*: Vogt, P.R. and Tucholke, B.E. (eds) *The Western North Atlantic Region*. Geology of North America, Volume **M**, 451–468.

Micallef, A., Berndt, C., Masson, D.G. and Stow, D.A. 2008. Scale invariant characteristics of the Storegga

Slide and implications for large-scale submarine mass movements. *Marine Geology*, **247**, 46–60, https://doi.org/10.1016/j.margeo.2007.08.003

Minisini, D., Trincardi, F., Asioli, A., Canu, M. and Foglini, F. 2007. Morphologic variability of exposed mass-transport deposits on the eastern slope of Gela Basin (Sicily channel). *Basin Research*, **19**, 217–240, https://doi.org/10.1111/j.1365-2117.2007.00324.x

Miramontes, E., Cattaneo, A. *et al.* 2016. The Pianosa Contourite Depositional System (Northern Tyrrhenian Sea): drift morphology and Plio-Quaternary stratigraphic evolution. *Marine Geology*, **378**, 20–42, https://doi.org/10.1016/j.margeo.2015.11.004

Miramontes, E., Garziglia, S., Sultan, N., Jouet, G. and Cattaneo, A. 2018. Morphological control of slope instability in contourites: a geotechnical approach. *Landslides*, **15**, 1085–1095, https://doi.org/10.1007/s10346-018-0956-6

Mulder, T., Ducassou, E. *et al.* 2012. New insights into the morphology and sedimentary processes along the western slope of Great Bahama Bank. *Geology*, **40**, 603–606, https://doi.org/10.1130/G32972.1

Nielsen, T., Rasmussen, T.L., Ceramicola, S. and Kuijpers, A. 2007. Quaternary sedimentation, margin architecture and ocean circulation variability around the Faroe Islands, North Atlantic. *Quaternary Science Reviews*, **26**, 1016–1036, https://doi.org/10.1016/j.quascirev.2006.12.005

Normandeau, A., Campbell, D.C., Piper, D.J. and Jenner, K.A. 2019. Are submarine landslides an underestimated hazard on the western North Atlantic passive margin? *Geology*, **47**, 848–852, https://doi.org/10.1130/G46201.1

Perez, L., Garcá-Rodríguez, F. and Hanebuth, T.J.J. 2016. Variability in terrigenous sediment supply offshore of the Río de la Plata (Uruguay) recording the continental climatic history over the past 1200 years. *Climate of the Past*, **12**, 623–634, https://doi.org/10.5194/cp-12-623-2016

Piper, D.J.W. 2005. Late Cenozoic evolution of the continental margin of eastern Canada. *Norwegian Journal of Geology*, **85**, 305–318.

Pope, E.L., Talling, P.J., Urlaub, M., Hunt, J.E., Clare, M.A. and Challenor, P. 2015. Are large submarine landslides temporally random or do uncertainties in available age constraints make it impossible to tell? *Marine Geology*, **369**, 19–33, https://doi.org/10.1016/j.margeo.2015.07.002

Powrie, W. 2013. *Soil Mechanics: Concepts and Applications*. 3rd edn. CRC Press, Taylor & Francis.

Preu, B., Hernández-Molina, F.J. *et al.* 2013. Morphosedimentary and hydrographic features of the northern Argentine margin: the interplay between erosive, depositional and gravitational processes and its conceptual implications. *Deep-Sea Research I*, **75**, 157–174, https://doi.org/10.1016/j.dsr.2012.12.013

Prieto, M.I., Moscardelli, L. and Wood, L.J. 2016. Exploring the influence of deepwater currents as potential triggers for slope instability. *In*: Lamarche, G., Mountjoy, J. *et al.* (eds) *Submarine Mass Movements and their Consequences*. Advances in Natural and Technological Hazards Research, **41**, Springer, Cham, 331–338, https://doi.org/10.1007/978-3-319-20979-1_33

Principaud, M., Mulder, T., Gillet, H. and Borgomano, J. 2015. Large-scale carbonate submarine mass-wasting along the northwestern slope of the Great Bank (Bahamas): morphology, architecture, and mechanisms. *Sedimentary Geology*, **317**, 27–42, https://doi.org/10.1016/j.sedgeo.2014.10.008

Rahmstorf, S. 2002. Ocean circulation and climate during the last 120,000 years. *Nature*, **419**, 207–214, https://doi.org/10.1038/nature01090

Rashid, H., MacKillop, K., Sherwin, J., Piper, D.J.W., Marche, B. and Vermooten, M. 2017. Slope instability on a shallow contourite-dominated continental margin, southeastern Grand Banks, eastern Canada. *Marine Geology*, **393**, 203–215, https://doi.org/10.1016/j.margeo.2017.01.001

Rasmussen, T.L., Thomsen, E., van Weering, T.C.E. and Labeyrie, L. 1996. Rapid changes in surface and deep water conditions at the Faeroe Margin during the last 58,000 years. *Paleoceanography*, **11**, 757–771, https://doi.org/10.1029/96PA02618

Rasmussen, T.L., Thomsen, E. and van Weering, T.C.E. 1998. Cyclic sedimentation of the Faeroe Drift 53-10 ka BP related to climatic variations. *Geological Society, London, Special Publications*, **129**, 255–267, https://doi.org/10.1144/GSL.SP.1998.129.01.16

Rasmussen, T.L., Bäckström, D. *et al.* 2002. The Faroe-Shetland Gateway: late Quaternary water mass exchange between the Nordic seas and the northeastern Atlantic. *Marine Geology*, **188**, 165–192, https://doi.org/10.1016/S0025-3227(02)00280-3

Rebesco, M. and Stow, D. 2001. Seismic expression of contourites and related deposits: a preface. *Marine Geophysical Researches*, **22**, 303–308, https://doi.org/10.1023/A:1016316913639

Rebesco, M., Hernández-Molina, F.J., Van Rooij, D. and Wåhlin, A. 2014. Contourites and associated sediments controlled by deep-water circulation processes: state-of-the-art and future considerations. *Marine Geology*, **352**, 111–154, https://doi.org/10.1016/j.margeo.2014.03.011

Ren, X.W. and Santamarina, J.C. 2018. The hydraulic conductivity of sediments: a pore size perspective. *Engineering Geology*, **233**, 48–54, https://doi.org/10.1016/j.enggeo.2017.11.022

Revil, A. and Cathles, III, L.M. 1999. Permeability of shaly sands. *Water Resource Research*, **35**, 651–662, https://doi.org/10.1029/98WR02700

Ritchie, J.D., Johnson, H. and Kimbell, G.S. 2003. The nature and age of Cenozoic contractional deformation within the NE Faroe-Shetland Basin. *Marine and Petroleum Geology*, **20**, 399–409, https://doi.org/10.1016/S0264-8172(03)00075-8

Ritchie, J.D., Johnson, H., Quinn, M.F. and Gatliff, R.W. 2008. The effects of Cenozoic compression within the Faroe-Shetland Basin and adjacent areas. *Geological Society, London, Special Publications*, **306**, 121–136, https://doi.org/10.1144/SP306.5

Roberts, D.G., Thompson, M., Mitchener, B., Hossack, J., Carmichael, S. and Bjørnseth, H.-M. 1999. Palaeozoic to Tertiary rift and basin dynamics: mid-Norway to the Bay of Biscay – a new context for hydrocarbon prospectivity in the deep water frontier. *Geological Society, London, Petroleum Geology Conference Series*, **5**, 7–40, https://doi.org/10.1144/0050007

Röhl, U. and Abrams, L.J. 2000. High-resolution, downhole and non-destructive core measurements from Sites 999 and 1001 in the Caribbean Sea: application to the Late Paleocene Thermal Maximum. *In*: Leckie, R.M., Sigurdsson, H., Acton, G.D. and Draper, G. (eds) *Proceedings of the ODP, Scientific Results*, **165**. Ocean Drilling Program, College Station, TX, 191–203.

Rothwell, R.G. and Rack, F.R. 2006. New techniques in sediment core analysis: an introduction. *Geological Society, London, Special Publications*, **267**, 1–29, https://doi.org/10.1144/GSL.SP.2006.267.01.01

Rumph, B., Reaves, C.M., Orange, V.G. and Robinson, D.L. 1993. Structuring and transfer zones in the Faeroe Basin in a regional tectonic context. *Geological Society, London, Petroleum Geology Conference Series*, **4**, 999–1009, https://doi.org/10.1144/0040999

Saunders, P.M. 1990. Cold outflow from the Faroe Bank Channel. *Journal of Physical Oceanography*, **20**, 29–43, https://doi.org/10.1175/1520-0485(1990)020<0029:COFTFB>2.0.CO;2

Sheskin, D.J. 2011. *Handbook of Parametric and Nonparametric Statistical Procedures*. 5th edn. CRC Press, Taylor & Francis.

Shipp, R.C., Weimer, P. and Posamentier, H.W. (eds) 2011. *Mass-Transport Deposits in Deepwater Settings*. SEPM Special Publications, **96**. SEPM Society for Sedimentary Geology, https://doi.org/10.2110/sepmsp.096

Smallwood, J.R. and Gill, C.E., 2002. The rise and fall of the Faroe-Shetland Basin: evidence from seismic mapping of the Balder Formation. *Journal of the Geological Society, London*, **159**, 627–630, https://doi.org/10.1144/0016-764902-064

Smith, S.V., Buddemeier, R.W., Redalje, R.C. and Houck, J.E., 1979. Strontium-calcium thermometry in coral skeletons. *Science*, **207**, 404–407, https://doi.org/10.1126/science.204.4391.404

Solheim, A., Berg, K., Forsberg, C.F. and Bryn, P. 2005. The Storegga Slide complex: repetitive large scale sliding with similar cause and development. *Marine and Petroleum Geology*, **22**, 97–107, https://doi.org/10.1016/j.marpetgeo.2004.10.013

Stoker, M.S., Akhurst, M.C., Howe, J.A. and Stow, D.A.V. 1998. Sediment drifts and contourites on the continental margin off northwest Britain. *Sedimentary Geology*, **15**, 33–51, https://doi.org/10.1016/S0037-0738(97)00086-9

Stoker, M.S., Nielsen, T., van Weering, T.C.E. and Kuijpers, A. 2002. Towards an understanding of the Neogene tectonostratigraphic framework of the NE Atlantic margin between Ireland and the Faroe Islands. *Marine Geology*, **188**, 233–248, https://doi.org/10.1016/S0025-3227(02)00282-7

Stoker, M.S., Hoult, R.J. *et al.* 2005. Sedimentary and oceanographic responses to early Neogene compression on the NW European margin. *Marine and Petroleum Geology*, **22**, 1031–1044, https://doi.org/10.1016/j.marpetgeo.2005.01.009

Stow, D.A.V. and Holbrook, J.A. 1984. Hatton Drift contourites, northeast Atlantic, Deep Sea Drilling Project Leg 81. *Initial Reports of the Deep Sea Drilling Project*, **81**. US Government Printing Office, Washington, DC, 695–699, https://doi.org/10.2973/dsdp.proc.81.125.1984

Stow, D.A.V., Faugères, J.C., Howe, J.A., Pudsey, C.J. and Viana, A.R. 2002. Bottom currents, contourites and deep-sea sediment drifts: current state-of-the-art. *Geological Society, London, Memoirs*, **22**, 7–20, https://doi.org/10.1144/GSL.MEM.2002.022.01.02

Talling, P.J., Clare, M.L., Urlaub, M., Pope, E., Hunt, J.E. and Watt, S.F. 2014. Large submarine landslides on continental slopes: geohazards, methane release, and climate change. *Oceanography*, **27**, 32–45, https://doi.org/10.5670/oceanog.2014.38

Terzaghi, K. 1925. *Erdbaumechanik auf Bodenphysikalischer Grundlage*. F. Deuticke, Leipzig.

Thomson, J., Crudeli, D., De Lange, G., Slomp, C.P., Erba, E., Corselli, C. and Calvert, S.E. 2004. *Florisphaera profunda* and the origin and diagenesis of carbonate phases in eastern Mediterranean sapropel units. *Paleoceanography*, **19**, PA3003, https://doi.org/10.1029/2003PA000976

Tournadour, E., Mulder, T., Borgomano, J., Hanquiez, V., Ducassou, E. and Gillet, H. 2015. Origin and architecture of a Mass Transport Complex on the northwest slope of Little Bahama Bank (Bahamas): relations between off-bank transport, bottom current sedimentation and submarine landslides. *Sedimentary Geology*, **317**, 9–26, https://doi.org/10.1016/j.sedgeo.2014.10.003

Turner, J.D. and Scrutton, R.A. 1993. Subsidence patterns in western margin basins: evidence from the Faeroe-Shetland Basin. *Geological Society, London, Petroleum Geology Conference Series*, **4**, 975–983, https://doi.org/10.1144/0040975

Turrell, W.R., Slesser, G., Adams, R.D., Payne, R. and Gillibrand, P.A. 1999. Decadal variability in the composition of Faroe Shetland Channel bottom water. *Deep-Sea Research I*, **46**, 1–25, https://doi.org/10.1016/S0967-0637(98)00067-3

Urgeles, R. and Camerlenghi, A. 2013. Submarine landslides of the Mediterranean Sea: trigger mechanisms, dynamics, and frequency-magnitude distribution. *Journal of Geophysical Research: Earth Surface*, **118**, 2600–2618, https://doi.org/10.1002/2013JF00 2720

Urlaub, M., Talling, P.J. and Masson, D.G. 2013. Timing and frequency of large submarine landslides: implications for understanding triggers and future geohazard. *Quaternary Science Reviews*, **72**, 63–82, https://doi.org/10.1016/j.quascirev.2013.04.020

Urlaub, M., Talling, P. and Clare, M. 2014. Sea-level-induced seismicity and submarine landslide occurrence: comment. *Geology*, **42**, 337, https://doi.org/10.1130/G35254C.1

Van Raaphorst, W., Malschaert, H., van Haren, H., Boer, W. and Brummer, G.-J. 2001. Cross-slope zonation of erosion and deposition in the Faeroe-Shetland Channel. *North Atlantic Ocean. Deep-Sea Research I*, **48**, 567–591, https://doi.org/10.1016/S0967-0637(00)00052-2

Van Weering, Tj.C.E. and de Rijk, S. 1991. Sedimentation and climate-induced sediments on Feni Ridge, Northeast Atlantic Ocean. *Marine Geology*, **101**, 49–69, https://doi.org/10.1016/0025-3227(91)90062-9

Van Weering, T.C.E., Nielsen, T., Kenyon, N.H., Akentieva, K. and Kuijpers, A.H. 1998. Sediments and sedimentation at the NE Faeroe continental margin; contourites and large-scale sliding. *Marine Geology*, **152**, 159–176, https://doi.org/10.1016/S0025-3227(98)00069-3

Vanneste, M., Mienert, J. and Bünz, S. 2006. The Hinlopen Slide: a giant, submarine slope failure on the northern Svalbard margin, Arctic Ocean. *Earth and Planetary Science Letters*, **245**, 373–388, https://doi.org/10.1016/j.epsl.2006.02.045

Vardy, M.E., L'Heureux, J.-S. *et al.* 2012. Multidisciplinary investigation of a shallow near-shore landslides, Finneidfjord, Norway. *Near Surface Geophysics*, **10**, 267–277, https://doi.org/10.3997/1873-2012022

Verdicchio, G. and Trincardi, F. 2008. Mediterranean shelf-edge muddy contourites: examples from the Gela and South Adriatic basins. *Geo-Marine Letters*, **28**, 137–151, https://doi.org/10.1007/s00367-007-0096-9

Volpi, V., Camerlenghi, A., Hillenbrand, C.D., Rebesco, M. and Ivaldi, R. 2003. Effects of biogenic silica on sediment compaction and slope stability on the Pacific margin of the Antarctic Peninsula. *Basin Research*, **15**, 339–363, https://doi.org/10.1046/j.1365-2117.2003.00210.x

Volpi, V., Amblas, D., Camerlenghi, A., Canals, M., Rebesco, M. and Urgeles, R. 2011. Late Neogene to recent seafloor instability on the deep pacific margin of the Antarctic Peninsula. *SEPM Special Publications*, **96**, 161–177, https://doi.org/10.2110/sepmsp.096.161

Weltje, G.J. and Tjallingii, R. 2008. Calibration of XRF core scanners for quantitative geochemical logging of sediment cores: theory and application. *Earth and Planetary Science Letters*, **274**, 423–438, https://doi.org/10.1016/j.epsl.2008.07.054

Wilson, C.K., Long, D. and Bulat, J. 2003. The Afen Slide – A multistage slope failure in the Faroe-Shetland Channel. *In*: Locat, J. and Mienert, J. (eds) *Submarine Mass Movements and Their Consequences*. Advances in Natural and Technological Hazards Research, **19**, Springer, Dordrecht, 317–324, https://doi.org/10.1007/978-94-010-0093-2_35

Wilson, C.K., Long, D. and Bulat, J. 2004. The morphology, setting and processes of the Afen Slide. *Marine Geology*, **213**, 149–167, https://doi.org/10.1016/j.margeo.2004.10.005

Wilson, C.K., Bulat, J. and Long, D. 2005. The Afen Slide. British Geological Survey Commissioned Report **CR/05/003**.

Winkelmann, D., Geissler, W., Schneider, J. and Stein, R. 2008. Dynamics and timing of the Hinlopen/Yermak Megaslide north of Spitsbergen, Arctic Ocean. *Marine Geology*, **250**, 34–50, https://doi.org/10.1016/j.margeo.2007.11.013

Indonesian Throughflow as a preconditioning mechanism for submarine landslides in the Makassar Strait

Rachel E. Brackenridge[1,2], Uisdean Nicholson[1]*, Benyamin Sapiie[3], Dorrik Stow[1] and Dave R. Tappin[4,5]

[1]School of Energy, Geoscience, Infrastructure and Society, Heriot-Watt University, Edinburgh EH14 4AS, UK

[2]School of Geosciences, University of Aberdeen, Aberdeen AB24 3FX, UK

[3]Faculty of Earth Sciences and Technology, Institut Teknologi Bandung, Indonesia

[4]British Geological Survey, Keyworth, Nottingham NG12 5GG, UK

[5]Department of Earth Sciences, University College London (UCL), London, UK

REB, 0000-0002-0572-314X; UN, 0000-0003-0746-8549;
DRT, 0000-0003-3186-8403
*Correspondence: U.Nicholson@hw.ac.uk

Abstract: The Makassar Strait is an important oceanic gateway, through which the main branch of the Indonesian Throughflow (ITF) transports water from the Pacific to the Indian Ocean. This study identifies a number of moderate (>10 km^3) to giant (up to 650 km^3) mass transport deposits within the Makassar North Basin Pleistocene–Recent section. The majority of submarine landslides that formed these deposits originated from the Mahakam pro-delta, with the largest skewed to the south. We see clear evidence for ocean-current erosion, lateral transport and contourite deposition across the upper slope. This suggests that the ITF is acting as an along-slope conveyor belt, transporting sediment to the south of the delta, where rapid sedimentation rates and slope oversteepening results in recurring submarine landslides. A frequency for the >100 km^3 failures is tentatively proposed at 0.5 Ma, with smaller events occurring at least every 160 ka. This area is therefore potentially prone to tsunamis generated from these submarine landslides. We identify a disparity between historical fault rupture-triggered tsunamis (located along the Palu-Koro fault zone) and the distribution of mass transport deposits in the subsurface. If these newly identified mass failures are tsunamigenic, they may represent a previously overlooked hazard in the region.

The Indonesian Archipelago is seismically active due to its location at the intersection of four major tectonic plates. Earthquakes, volcanic eruptions and tsunamis represent significant geological hazards across the island nation. Of these natural hazards, tsunamis pose a specific risk to the sustainability and resilience of coastal communities. There have been a number of devastating Indonesian tsunamis over the last 15 years, triggered by various mechanisms. The 2004 megathrust earthquake and tsunami offshore Sumatra resulted in over 220 000 fatalities across the Indian Ocean region, 165 000 of these on Sumatra, making it one of the worst natural disasters of the last 100 years (National Geophysical Data Center/World Data Service, NGDC/WDS). The mechanism of the Palu tsunami (September 2018) is still uncertain but is proposed as a result of a combination of earthquake-related seafloor rupture and subaerial/submarine landslides (Carvajal *et al.* 2019; Takagi *et al.* 2019). Two surges, with maximum wave heights of over 10 m (NGDC/WDS), were recorded. Over 4000 fatalities occurred as a result of tsunami surges and widespread onshore liquefaction due to seismic shaking (Sangadji 2019). The Palu event was closely followed in December 2018, by the Anak Krakatau tsunami, where flank collapse, probably triggered by the erupting volcano, resulted in tsunami run-up of 30 m and over 400 fatalities on the local coasts of Java and Sumatra (NGDC/WDS; Grilli *et al.* 2019). The high number of fatalities of these two events has been attributed to the limited understanding of tsunami generation from mechanisms other than fault rupture and the lack of an effective tsunami warning system for non-seismic events (Yalciner *et al.* 2018; Williams *et al.* 2019).

The Makassar Strait has the highest frequency of tsunamis in Indonesia (Prasetya *et al.* 2001). Historical records show that most are caused by earthquake-generated fault rupture of the seafloor, with the exception of the September 2018 Palu event, which probably had a landslide component (Jamelot *et al.* 2019). However, there are numerous other factors in the Strait that could make it susceptible to submarine landslide-triggered tsunamis,

From: Georgiopoulou, A., Amy, L. A., Benetti, S., Chaytor, J. D., Clare, M. A., Gamboa, D., Haughton, P. D. W., Moernaut, J. and Mountjoy, J. J. (eds) 2020. *Subaqueous Mass Movements and their Consequences: Advances in Process Understanding, Monitoring and Hazard Assessments.* Geological Society, London, Special Publications, **500**, 195–217.
First published online April 1, 2020, https://doi.org/10.1144/SP500-2019-171

including oversteepening of the continental slope due to carbonate growth and faulting, or sediment influx from the Mahakam Delta. The Strait is also the main channel for the Indonesian Throughflow (ITF), a strong current that transports 10–15 million Sverdrups (where 1 Sverdrup (Sv) = 1×10^6 m^3 s^{-1}) of water from the Pacific into the Indian Ocean (Kuhnt *et al.* 2004). Elsewhere, it has been shown that ocean currents and their associated deposits can precondition the slope for failure through rapid sediment deposition and erosion (Laberg and Camerlenghi 2008; Nicholson *et al.* 2020).

This study evaluates slope stability in the Makassar Strait, specifically the role of the ITF in preconditioning the slope for failure. Understanding the distribution of mass transport deposits (MTDs) that result from submarine landslides will allow us to identify specific regions of hazard and risk should these landslides be tsunamigenic. This has important implications for hazard mitigation and early warning systems across coastal regions of the Makassar Strait.

Geographical and oceanographic setting

Regional tectonic setting

The Makassar Strait separates the Indonesian islands of Kalimantan (Borneo) and Sulawesi in SE Asia (Fig. 1). It sits within a highly complex and dynamic plate tectonic setting (Daly *et al.* 1991), located at

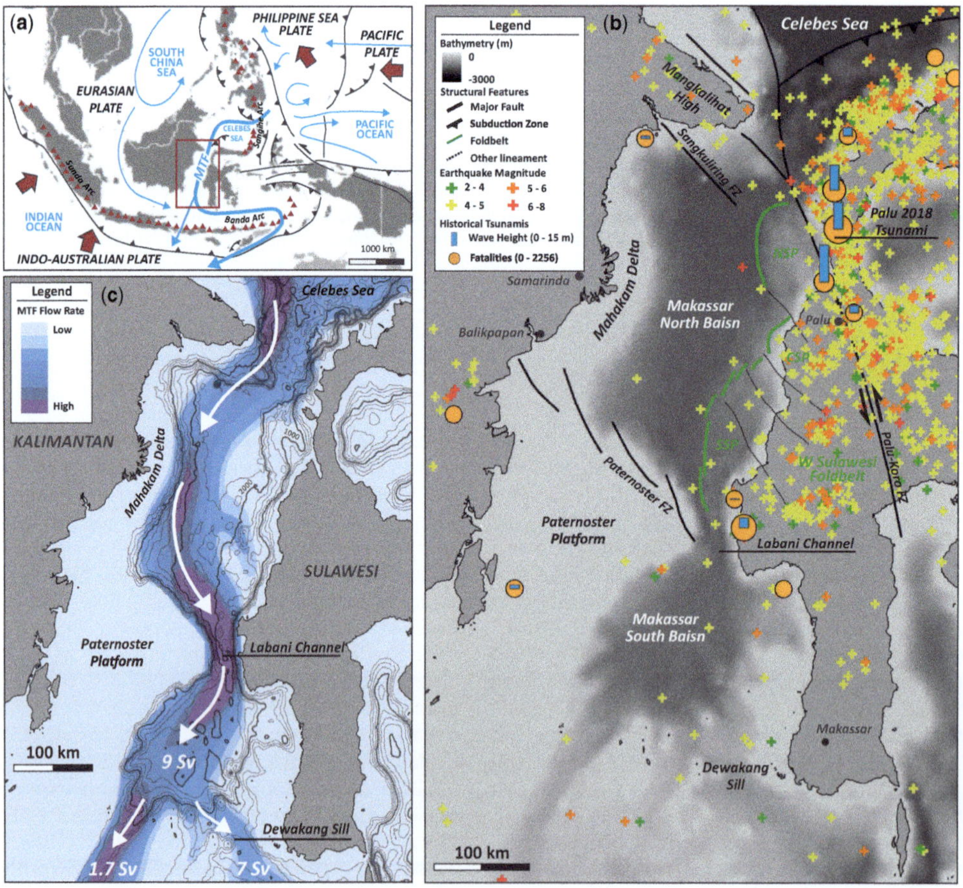

Fig. 1. (**a**) Regional tectonic and oceanographic setting. Study area indicated in red. MTF, Makassar Throughflow. (**b**) Structural features of the Makassar Strait from Puspita *et al.* (2005) (NSP, CSP, SSP refer to the Northern, Central and Southern structural provenances) and Cloke *et al.* (1999). SRTM30_PLUS Global Bathymetry Data from Becker *et al.* (2009). Historical earthquakes from the United States Geological Survey Earthquake Catalog (https://earthquake.usgs.gov/earthquakes). Historical tsunami data from the National Geophysical Data Center/World Data Service (NGDC/WDS) Global Historical Tsunami Database (NGDC/WDS). (**c**) Simulated daily average of Sv (Sv = 1×10^6 m^3 s^{-1}) for the main Makassar Throughflow jet between 52 and 420 m (Mayer and Damm 2012). Sv values from Kuhnt *et al.* (2004).

the intersection between four major lithospheric plates (Fig. 1a). The Indo-Australian Plate, located to the south of the Makassar Strait, is colliding with the Eurasian Plate in a northwards direction and is subducting below Sumatra and Java at a rate of *c.* 7 cm a^{-1} (Bergman *et al.* 1996; Koop *et al.* 2006). To the east, the large Pacific Plate interacts with the NW-moving (and clockwise-rotating) Philippine Sea Plate (Hall *et al.* 1995). As a result of plate interactions, a complex system of subduction, back-arc-thrusting, extension and major transform zones has developed (Hall 1997; Prasetya *et al.* 2001) (Fig. 1a). The region is extremely susceptible to seismic activity and volcanic arcs line the compressional plate boundaries, including the Sunda Arc and Banda Arc to the south and the Sangihe Arc to the NE (Katili 1975).

Due to its active plate tectonic setting, earthquakes are common across the Indonesian archipelago, with some associated with tsunamis. Along the Sunda Arc, historical tsunamis are generally generated by megathrust earthquakes. The mechanisms for these events are well understood and have been studied extensively following the devastating 2004 Indian Ocean tsunami (e.g. McCloskey *et al.* 2008; Okal and Synolakis 2008). Prasetya *et al.* (2001) show that the high frequency of historical tsunamis within the Makassar Strait, which is located away from any major subduction zone, is the result of shallow-depth earthquakes along the Palu-Koro transform fault zone (Fig. 1b).

Geological framework

The Makassar Strait forms a deep seaway that separates Kalimantan from Sulawesi. It varies in width from 100 to 200 km and is *c.* 600 km in length. It formed in the middle Eocene, when rifting and seafloor spreading was initiated (Hall *et al.* 2009). The onset of compression in the Late Miocene to Pliocene resulted in fold-and-thrust belt development along the margins of the Makassar Strait basins (Bergman *et al.* 1996). At present, far-field stresses and resulting faulting generate topographic features on the seabed, particularly in the east where Miocene–Pliocene compression has generated the West Sulawesi Fold Belt (Puspita *et al.* 2005). The Palu-Koro fault zone is a NNW–SSE-trending strike-slip fault and connects with the North Sulawesi Trench in the Celebes Sea. It crosses the Makassar Strait at its narrow northernmost opening, separating it from the Celebes Sea (Prasetya *et al.* 2001). The NW–SE-trending Eocene Paternoster fault zone acts as a topographic barrier separating the North and South Makassar basins (Fig. 1b).

The North Makassar Basin is north–south trending, 340 km long and 100 km wide, with water depths of 200–2000 m (Fig. 1b). The South Makassar Basin shows a similar range of water depths, is 300 km long, 100 km wide and NE–SW trending. The margins of the Makassar Strait show contrasting characteristics. To the west, the Paternoster Platform forms a wide shelf less than 200 m deep. This shelf has been periodically exposed during glacial conditions to form part of the Sundaland landmass, connecting Kalimantan, Java and Sumatra with the Asian continent (Bird *et al.* 2005; Hall 2009).

In the east of the Makassar Strait two fold belts are expressed as topographic features on the seabed (Fig. 1b). Puspita *et al.* (2005) define a Northern (NSP) and Southern (SSP) fold belt separated by a Central Structural Province (CSP). They note the different characteristics of the two fold belts, with the northern showing a steep western deformation front and a chaotic internal seismic character, whereas the southern shows well-developed thin-skinned fold-and-thrust deformation. These differences are attributed to interaction with basement structures and sediment supply, with the NSP likely consisting of muddier sediment (Puspita *et al.* 2005).

Studies of the active sedimentary systems within the Makassar Strait are limited but reveal a complex distribution of different depositional systems. Shallow-water sedimentation along the margins of the Makassar South Basin is largely carbonate dominated. The Paternoster Platform in the SW of the Strait forms a *c.* 40 000 km^2 medium- to coarse-grained carbonate sand sheet made up of bioclasts close to reefal build-ups, and benthic foraminifera in open marine regions (Burollet *et al.* 1986). The actively prograding Mahakam Delta is a major source of clastic sediment to the study area, with an estimated annual sediment discharge rate of 8×10^6 m^3 a^{-1} of clay, silt and sand-rich sediments (Roberts and Sydow 2003). Outboard and north of the delta front, carbonate deposition is found on the shelf edge, where a number of bioherm and shelf-edge build-ups are identified on bathymetry and seismic data (Roberts and Sydow 2003).

The continental slopes surrounding the Makassar Basins are locally fault-controlled (Guntoro 1999; Prasetya *et al.* 2001). Canyons are present on the steepest slopes, particularly in areas where sediment input into the basin is restricted. MTDs are identified at the foot of the slope and are generated by sediment failure (Saller and Dharmasamadhi 2012). Outboard of the Mahakam Delta, high sediment influx onto the slope has formed sinuous channel-levee complexes (Saller and Dharmasamadhi 2012).

The Makassar Basin floor is smooth, with no evidence of tectonic disturbance (Puspita *et al.* 2005). Seismic mapping at the base of the slope in front of the Mahakam Delta shows a large number of deep-water depositional features, including turbidite channels, levees and splays, as well as significant

MTDs (Posamentier and Meizarwin 2000). Sediment waves are a common feature, with Decker *et al.* (2004) attributing those along the Sulawesi Slope Apron to non-channelized hyperpycnal or turbidite flows triggered by storm events.

Oceanographic setting

The ITF is made up of numerous branches sourced from the Pacific Ocean. The main branch, the Makassar Throughflow (MTF), constitutes around 80% of water transfer (Fig. 1a) (Gordon *et al.* 2008) making the Makassar Strait the main location for water and heat exchange between the two oceans. The MTF exits the North Pacific between the Philippines and New Guinea and enters the Makassar Strait via the Celebes Sea (Tillinger 2011), where an average of 9.3 $\pm$ 2.5 Sv of water are transferred (Kuhnt *et al.* 2004) (Fig. 1c). It then splits into two branches, one of which flows through the Lombok Strait, between the islands of Bali and Lombok, and the other through the Timor Sea, between Timor and Australia (Fig. 1a, c) (Kuhnt *et al.* 2004). The Makassar Strait is the only low-latitude ocean gateway in the global ocean 'conveyor belt' of thermohaline circulation and is, therefore, a critical regulator of global oceanic energy exchange (Gordon and Fine 1996; Kuhnt *et al.* 2004; Rahmstorf 2007).

Within the Makassar Strait, moorings (Susanto and Gordon 2005) and modelling (Mayer and Damm 2012) show the MTF to act as a western boundary current along the Kalimantan continental slope. The flow of the surface 50 m is primarily wind-driven, with average velocities of 0.5 m s^{-1} (Gordon *et al.* 2008). Velocities increase with depth, reaching a maximum of around 1 m s^{-1} within a subsurface 'jet' extending down to 300 m, with a core corresponding to the thermocline, at about 100–150 m (Mayer and Damm 2012). Below 300 m, velocities reduce to under 30 cm s^{-1} (Tillinger 2011). The depth and speed of this high-velocity water mass varies seasonally in response to monsoonal forcing. It shallows and intensifies during the SE and NW monsoon seasons (March–April and August–September) (Gordon *et al.* 2008). Locally in the Makassar Strait, the MTF exerts a control on sedimentation. Seismic mapping of the Mahakam Delta by Roberts and Sydow (2003) shows the fine-grained prodelta (at *c.* 40 m water depth) is deflected to the south due to interaction with the MTF, which can reach velocities of 80 cm s^{-1} at the delta front.

Data and methodology

Bathymetric data

This study is based on three multibeam echosounder bathymetry surveys, acquired by Gardline Geosurvey on behalf of TGS, for their Indonesian Frontier Basins programme. The TGS_MakN, TGS_MakS and TGS_Pat surveys (Fig. 2) were acquired between December 2006 and February 2007 and cover around 60 000 km^2 in total. The multibeam echosounder data have been gridded at 25 $\times$ 25 m resolution, with features enhanced by a shaded relief map (0° azimuth, 45° angle). Backscatter data, at 5 m resolution, were also used to inform interpretation. These data cover the shelf edge, at around 200 m depth, to the basin floor, at over 3000 m water depth.

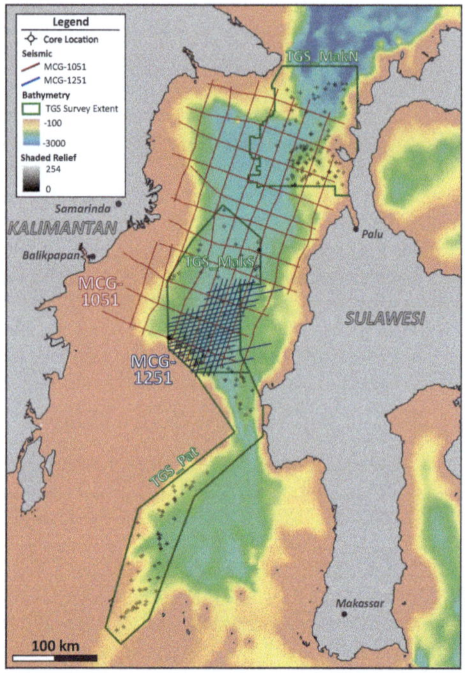

Fig. 2. Data used in the study. SRTM30_PLUS Global Bathymetry Data from Becker *et al.* (2009).

Additional bathymetric data, used for regional mapping (Fig. 2), include the Global Multi-Resolution Topography synthesis (Ryan *et al.* 2009) and the SRTM30_PLUS Global Bathymetry Data at 30 arc seconds resolution (Becker *et al.* 2009).

Slope gradient maps were extracted from bathymetric data and depth profiles generated to aid interpretation of physiographical domains. Structural lineations and depositional bodies were interpreted using bathymetric, slope and backscatter data to generate environment of deposition maps.

Seismic data

Two 2D seismic surveys from Multiclient Geophysical were interpreted in this study. The Makassar

Regional Super-Tie MCG2D Survey (MCG-1051) provides regional coverage of the Makassar North Basin (Fig. 2) with 21 2D pre-stack time migration (PSTM) seismic lines totalling *c.* 5000 km line length. The data were acquired in 2010 and processed by CGG Veritas in early 2011. The MCG-1051 survey was followed in 2012 by an infill survey (MCG-1251). The Makassar Infill 2D PSTM (Fig. 2) consists of 27 lines with a total of *c.* 2400 km line length. Both seismic surveys are high resolution in the upper few seconds below the seabed, which is the main section of interest in this study.

From the seismic data, we map Quaternary sediments deposited in the Makassar North Basin. Precise age constraints were not possible because of the lack of well control; however, a Top Pliocene horizon was interpreted based on del Negro *et al.* (2013). This allows us to date the age of the youngest sedimentary sequence at <2.6 Ma. Seismic data were interpreted using standard seismostratigraphic methods, and seismic facies (SF) were used to distinguish the different environments of deposition. Reflection terminations were used to identify key seismic stratigraphic surfaces, including tops and bases of individual MTDs. Due to the spacing of the 2D seismic data (17–40 km) only MTDs larger than *c.* 80 km^2 and 60 ms (*c.* 110–130 m thick) were mapped. Events smaller than 100 km^3 are under-represented due to limitations in spatial coverage of the 2D seismic survey, and the resolution for imaging thinner events. Seismic resolution deteriorates with depth; therefore, accurate mapping of MTDs was limited to the Pleistocene–Recent section. The volume of each MTD was calculated and converted to depth using a velocity of 2000 m s^{-1}, based on well velocity data gathered from MTDs of similar depositional character and burial depths in the Falkland Basin (Nicholson *et al.* 2020). An uncertainty range of 20% has been added to capture possible MTD velocities between 1800 and 2200 m s^{-1}. The total number of mapped landslides over the last *c.* 2.6 Ma was used to derive an average frequency for events of different volumes. As a result of the low seismic imaging resolution and line spacing, estimates of the frequency of smaller events (<100 km^3) are considered to be conservative.

Additional data

As part of the TGS-NOPEC Indonesian Frontier Basins programme, a number of additional datasets were acquired to complement the multibeam echosounder data used in the study. These data include gravity and magnetics acquired during the bathymetry survey, as well as regional gravity data compiled by Sandwell *et al.* (2014). Also, 234 piston cores were acquired within the region of the bathymetry survey (Fig. 2) in 2017 by TDI-Brooks International for geochemical analysis. From the cores over 600 samples were analysed for hydrocarbon indications. Shear strength measurements and basic sedimentary information were also available for each sample.

Historical earthquake data were compiled from the United States Geological Survey Earthquake Catalog (USGS; https://earthquake.usgs.gov/earthquakes, accessed May 2019). These records extend back to 1900 and list earthquake date, time, magnitude and epicentre depth. Historical tsunami data were downloaded from the National Geophysical Data Center/World Data Service Global Historical Tsunami Database (NGDC/WDS), which documents source location, date, time, event magnitude, maximum water height, total number of deaths, injuries and damage for each event.

Additional information was compiled from the literature where data coverage was unavailable for this study. Specifically, modelled routing of the MTF was georeferenced and digitized from Mayer and Damm (2012). Additional information about the features on the seafloor were gathered from studies by Fowler *et al.* (2004), Saller and Dharmasamadhi (2012) and Frederik *et al.* (2019).

Results and interpretation

Surface bathymetry interpretation

Physiographic domains. The main physiographic domains of the Northern Makassar Basin were mapped regionally on the SRTM30_PLUS Global Bathymetry Data (Becker *et al.* 2009). The continental shelf-slope break is defined by a change in slope gradient from less than 0.2° across the shelf, to slope gradients of 3–23° (Fig. 3). The shelf break occurs at *c.* 200 m water depth and varies significantly with distance from the coastline. To the west of the basin, a broad shelf of around 30–50 km extends out in front of the Mahakam Delta. This is significantly wider than the eastern shelf, which reaches a maximum width of 10–15 km and where a narrow, and locally very steep (up to 24°) continental slope grades to the basin floor (Fig. 3). The northern and southern limits of the basin are defined by structural highs (the Mangkalihat High and the Paternoster Platform respectively) which are bisected by deep, narrow channels. The deep Southern Makassar Basin is significantly smaller than the Northern Basin at approximately 180 km wide and 200 km in length (Fig. 3). This is a result of the extensive Paternoster Platform extending out from the Sulawesi Margin. Two channels connect the Southern Makassar Basin over the Dewakang Sill, to the Flores Sea in the South.

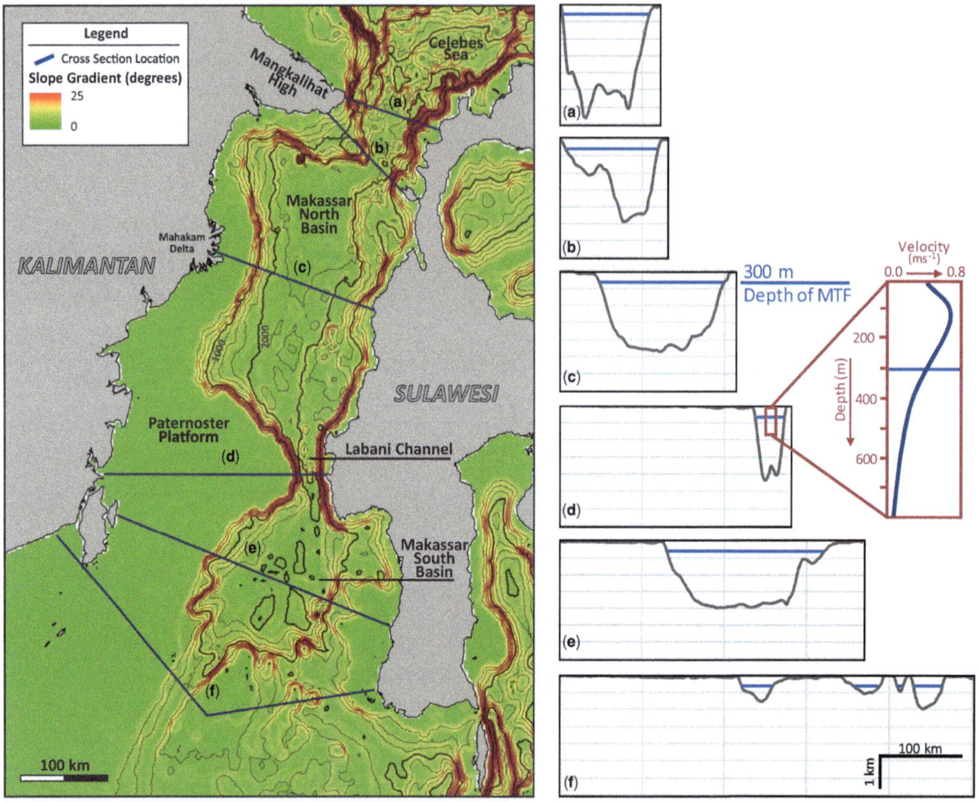

Fig. 3. Slope map extracted from the regional bathymetry RTM30_PLUS Global Bathymetry Data (Becker *et al.* 2009). Bathymetry contours at 250 m spacing). (**a–f**) Depth profiles across the Strait showing the main basin and channel features. Depth of Makassar Throughflow (MTF) indicated in blue. Velocity–depth profile shows average velocity profile for years 2004–09 (Susanto *et al.* 2012).

Structural features. The resolution of the SRTM30_PLUS Global Bathymetry Data (Becker *et al.* 2009) is insufficient to identify structural and sedimentary features in detail. However, the active tectonic nature of this region is evidenced by a number of features mapped on the high resolution TGS bathymetric surveys (Figs 4–6). Most notably, structural features include: (1) major strike-slip fault zones; and (2) fold-and-thrust complexes. The Palu-Koro fault zone is clearly imaged in the bathymetry data (Fig. 4). It extends from Palu Bay, Sulawesi, northwestwards to the northern edge of the Mangkalihat High. On the high-resolution TGS_MakN survey, the details of this fault zone can be mapped (Fig. 4). It is made up of multiple individual left-lateral strike-slip faults with both transtensional pull-apart and transpressional pop-up structures evident along the fault zone. The faults generate topography on the seabed of 250–300 m. The basin floor is modified in the east of the Makassar Strait where two fold-and-thrust belts are evident on the bathymetry, and a series of thrust-cored anticlines are clearly imaged on the seafloor (Fig. 4). Puspita *et al.* (2005) define two distinct fold-and-thrust belts named the Northern and Southern Structural Provinces (Fig. 1).

Sedimentary features. There is evidence for multiple downslope systems transporting sediment into the Makassar North Basin. From the backscatter data we interpret subtle sedimentary features, distinguishing bedrock or carbonates of high acoustic reflectivity (strongly negative values) from medium reflectivity sands and low reflectivity mud-dominated deposition (values close to zero) (Fig. 5). Dominant sedimentary features include: (1) slope canyons; (2) basin floor fans; (3) slope failure scarps; and (4) and deep-water turbidite channels.

The TGS_MakN bathymetric survey, on the eastern margin of the Makassar North Basin, images a deep-water channel in the northern fold belt (Fig. 4). The channel is sourced from the mouth of Palu Bay, which incises to a depth of 300 m in the upper slope, down to 600 m deep in its distal portion. The channel profile shows steep, erosive flanks,

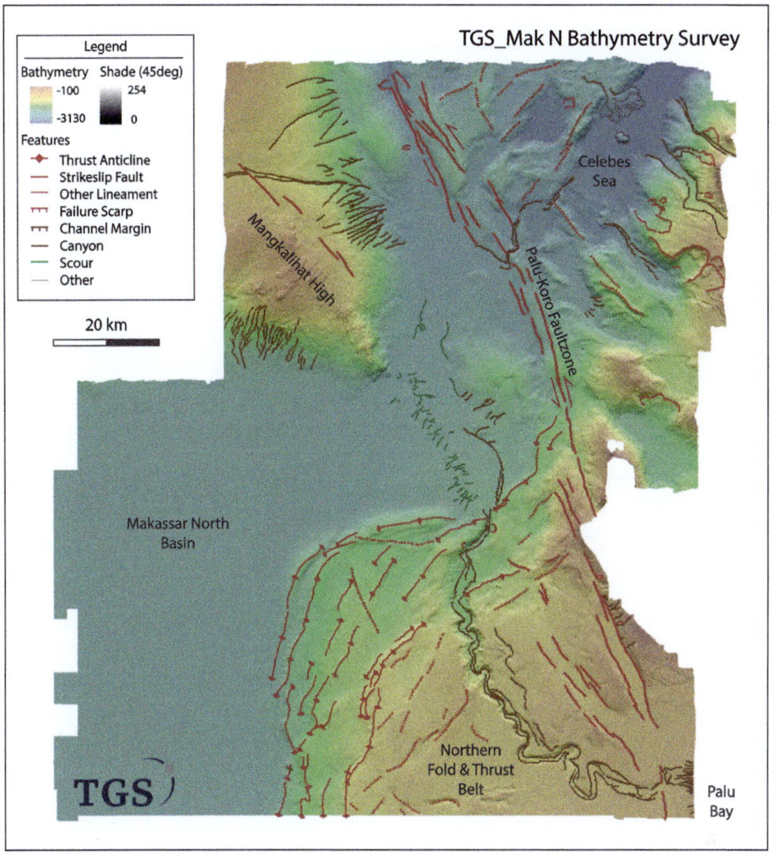

Fig. 4. Bathymetry data interpretation of the Makassar North TGS survey. Bathymetry data overlaying a shade map with interpretation of structural and sedimentary features. Survey location indicated on Figure 2.

bordering a flat channel axis. The channel shows a moderate sinuosity of 1.4, which is modified by thrust-cored anticlines in the outer part of the fold belt. There is no evidence for an associated depositional fan on the basin floor.

A downslope (turbidite) depositional fan is seen in the TGS_MakS bathymetric survey, where an extensive sediment wave field has developed along the eastern side of the basin. The Makassar South backscatter data image a small downslope channel *c.* 2.5 km wide, 200 m deep on the eastern continental slope, which feeds a sediment fan with low backscatter values compared to the surrounding basinal sediment (Fig. 5). The sediment wave field suggests periodic unconfined flow across the continental margin between the two fold-and-thrust belts. Wave crests are orientated parallel or slightly oblique to the continental slope. The sediment waves are sinuous with wave lengths of 0.5 to 1 km, and heights of 5–30 m. Backscatter data show higher values on the lee slide of waves.

Where sinuous channels and sediment waves dominate the eastern margin of the basin, v-shaped and straight canyons dominate on the northern, western and southern margins. High-resolution bathymetric data only cover a small portion of the continental slope of the Mangkalihat High and the Paternoster Platform. Both these continental margins show canyon systems of very different characteristics. The Mangkalihat High canyons are tightly spaced, straight, v-shaped downslope canyons that range between 40 and 80 m deep (Fig. 4). The TGS_MakS bathymetric data cover a number of canyons fed from the Paternoster Carbonate Platform. Here, four large v-shaped canyons (over 5 km wide, 400–700 m deep) are separated by slope-perpendicular 'spurs' (Fig. 5). This change in character may be due to clastic v. carbonate-dominated sedimentation and erosive processes in the north and south, respectively. We do not have access to high-resolution bathymetric data over the western continental slope of the Makassar North Basin;

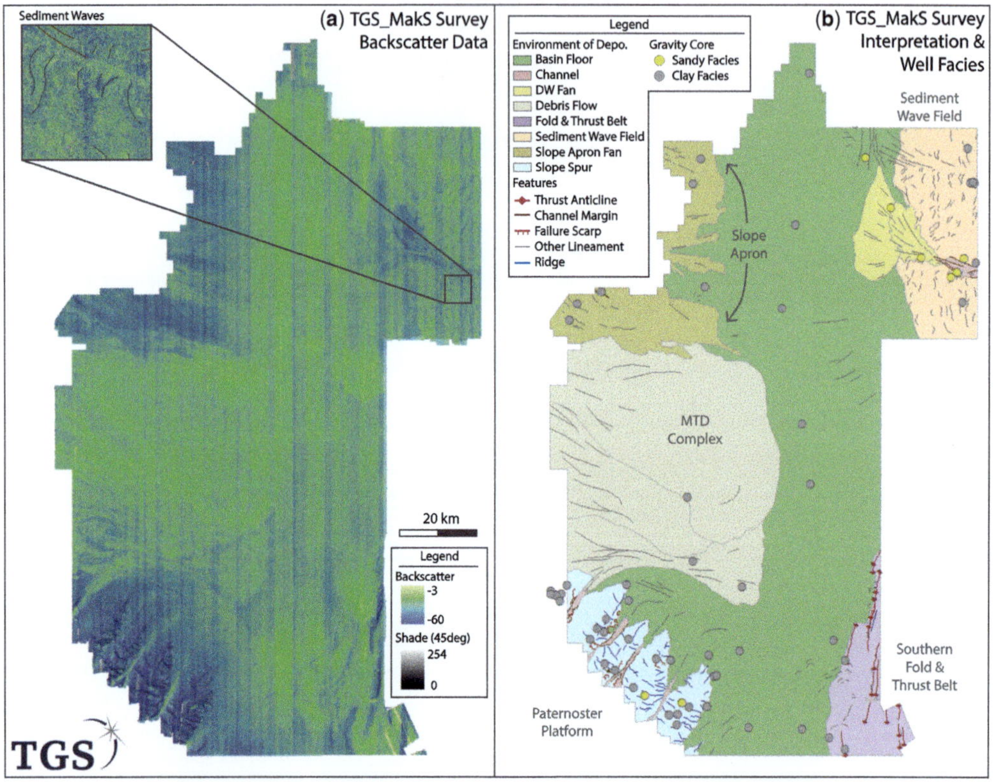

Fig. 5. Bathymetry data interpretation from the Makassar South TGS survey. (**a**) Backscatter overlaying a shade map. Highly negative values (indicated in blues) represent bed rock, carbonates or sands. Greens indicate mud-dominated deposition. Insert shows zoom-in of sediment wave crests. (**b**) Interpretation of sedimentary and structural features with seabed facies from piston cores. Survey location indicated on Figure 2.

however, published studies suggest that downslope canyons and associated slope apron fans are widespread (e.g. Fowler *et al.* 2004; Saller *et al.* 2008).

Small failure scarps (<5 km^2) are a common feature of the forelimbs of thrust belt anticlines (Fig. 6) and in the TGS_MakN bathymetric data to the north of the Palu-Koru Fault zone (Fig. 4). The backscatter imaging over the TGS_MakS bathymetric survey highlights a region of mass-transport deposition (Fig. 5). These deposits are characterized by low backscatter values, suggesting low acoustic reflectivity and therefore high mud content. Kinematic indicators, expressed as lineaments on the seafloor (Fig. 5b), suggest a flow direction from the west. Evidence for slope failure is also seen in the TGS_Pat bathymetric survey (Fig. 6). Large blocks (reaching 20–35 m in diameter) are distributed at the base of slope along the western margin of the Paternoster Platform in the Makassar South Basin (Fig. 6b).

The TGS_Pat bathymetric survey data image a number of features that we interpret as contourites. Immediately south of the Labani Channel, a scalloped, scour-like feature is imaged at 1300 m water depth (Fig. 6a). The feature is orientated perpendicular to the slope, with a steep lee and shallow stoss side; all characteristic of formation by alongslope erosion. To the south, there is an erosional terrace across the upper slope at 200 m water depth (Fig. 6b). This terrace is 700–800 m wide and is elongated alongslope for *c.* 10 km. It is formed along the north side of a large embayment. On the south side of this embayment, a mounded topographic feature around 500 m high is seen at the base of the slope. It is elongated alongslope, (Fig. 6a) and tentatively interpreted as a contourite mounded drift; however, further geophysical data are required to confirm this interpretation.

Subsurface seismic interpretation

Seismic tectonostratigraphy. The Top Pliocene was interpreted across the entire seismic survey and an isochron thickness map of the seabed to Pliocene

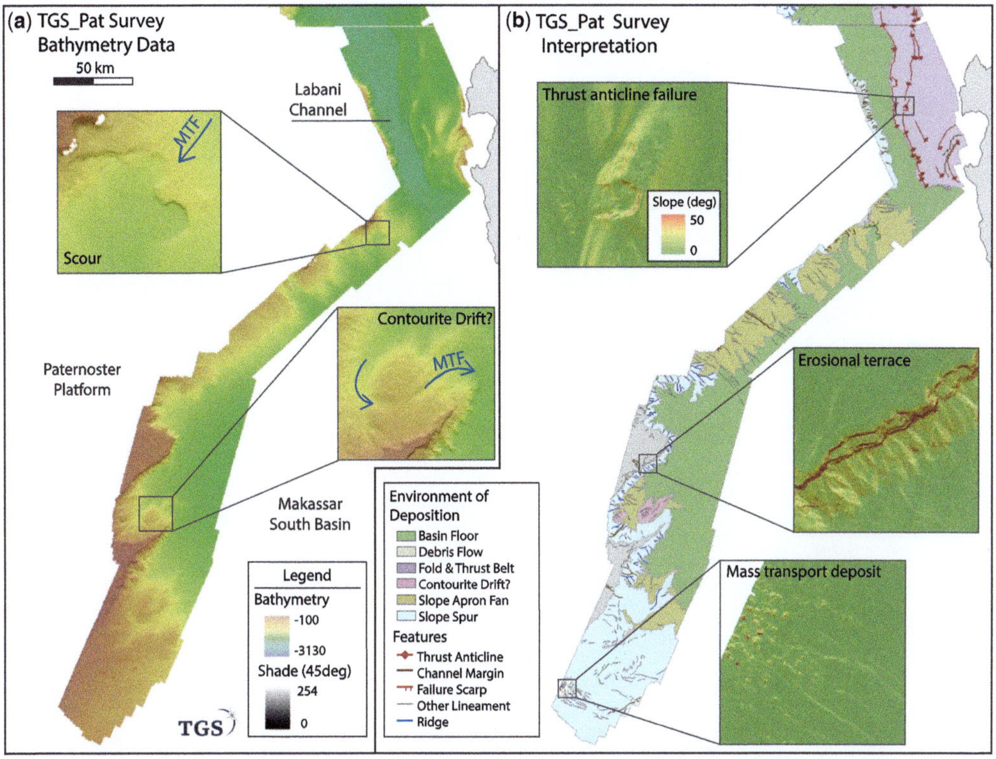

Fig. 6. Bathymetry data interpretation from the Paternoster TGS survey. (**a**) Bathymetry overlaying a shade map. MTF (inserts) indicates the direction of the Makassar Throughflow. (**b**) Interpretation of sedimentary and structural features. Inserts show slope gradient. Survey location indicated on Figure 2.

sequence generated to show depositional trends in Pleistocene–Recent sediments of the Makassar North Basin (Fig. 7). The isochron thickness map reveals a basin depocentre in the north, with a general thinning to the south. An additional depocentre is identified in the west on the edge of the data in front of the Mahakam Delta (Fig. 7).

Seismic facies and environment of deposition. SF interpretation identifies eight dominant facies (I–VIII) (Figs 8 & 9; Table 1). Each facies is characterized from its internal reflection character, continuity, amplitude and frequency, in addition to external morphology and/or bounding relationships (Table 1). The regional extent of each of the facies was mapped (Fig. 8). These observations were combined with those from the bathymetric mapping to interpret an environment of deposition for each facies (Table 1).

SF I. Two regions of polygonally faulted pelagic sediments (SF I) are located in the deepest part of the basin, directly in front of the two fold-and-thrust belts (Fig. 8). Here, high-amplitude, high-frequency reflections are offset due to densely spaced high-angle extensional faulting (Fig. 9 I). These faults are characteristic of sediment compaction and fluid expulsion from fine-grained biosiliceous sediments (e.g. Davies and Ireland 2011). These facies are associated with a prominent bottom-simulating reflector (BSR). Based on the polarity reversal of this BSR, and the polygonally faulted nature of the associated facies, we interpret this as a diagenetic Opal A/CT boundary zone within highly siliceous pelagic sediments (Lee *et al.* 2003). Polygonal faulting and a strong Opal A/CT transition suggest a high silica content, implying a lack of terrigenous sediments (turbidites, hemipelagic mud), allowing us to distinguish this facies from hemipelagites (see SF II).

SF II. High-amplitude, high-frequency, laterally continuous reflections with draping morphology are interpreted as hemipelagic facies (Fig. 9 II). SF II transitions towards the basin centre into SF I. Based on the lack of polygonal faulting and BSR, we interpret this facies as less biosiliceous, but still very fine-grained, hemipelagites.

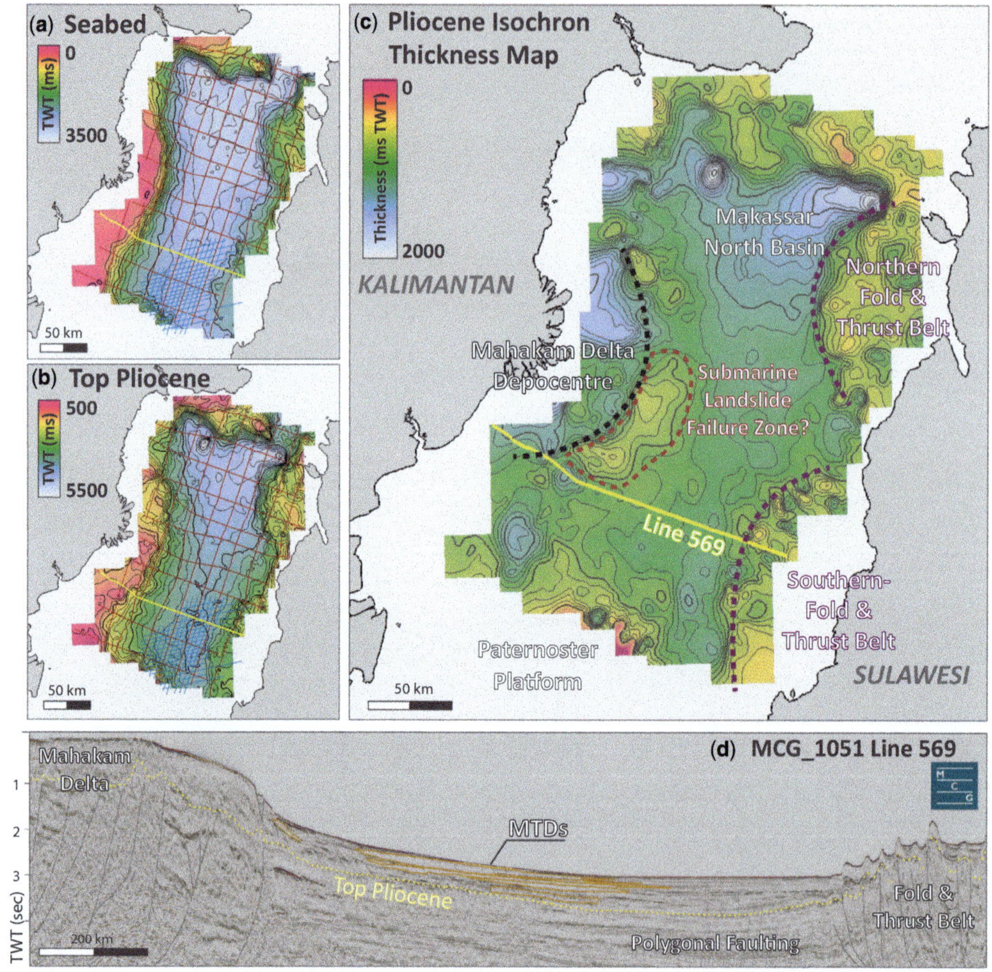

Fig. 7. (**a**)–(**c**) Seabed to Pliocene mapping and (**d**) example regional seismic line from the MCG-1051 survey.

SF III. Sediment waves are found locally on the western basin margin and an extensive sediment wave field dominates the eastern margin, between the two fold-and-thrust belts (Fig. 8). The waves are characterized by hummocky, subparallel, semi-continuous reflections (SF III) (Fig. 9 III). The package is 200–400 ms two-way travel time (TWT) in thickness and affects the seafloor bathymetry as seen in the Makassar North Basin (Fig. 5). The base of this seismic package is a high amplitude reflection showing opposite polarity to the seafloor. This is likely a BSR associated with gas hydrates that could have promoted gravitational movement (creep) downslope. The seabed topography generated by such a process appears to have promoted the growth of upslope-migrating sediment waves that have likely been deposited by unconfined downslope (turbiditic) processes.

SF IV. Dominating the Pleistocene basin fill are numerous MTDs, clearly identifiable from their semi-transparent and chaotic seismic signature (SF IV). The MTDs are characteristically deformed, chaotic, discontinuous and low amplitude (Fig. 9 IV). Rare internal reflections show extension in their proximal portions, and imbrication and compression in their distal parts (Fig. 10). Individual MTDs are mappable because they form lenticular bodies with erosional bases of negative acoustic impedance (AI), and a strong positive AI upper surface onto which younger stratigraphy onlaps (Fig. 10). MTDs thin and pinch out, upslope and laterally. Their downslope extent is either thinning (e.g. Fig. 10), or shows an abrupt, stepped change in facies. SF IV is predominantly distributed along the western side of the basin, with the largest events clustered in the SW (Fig. 8).

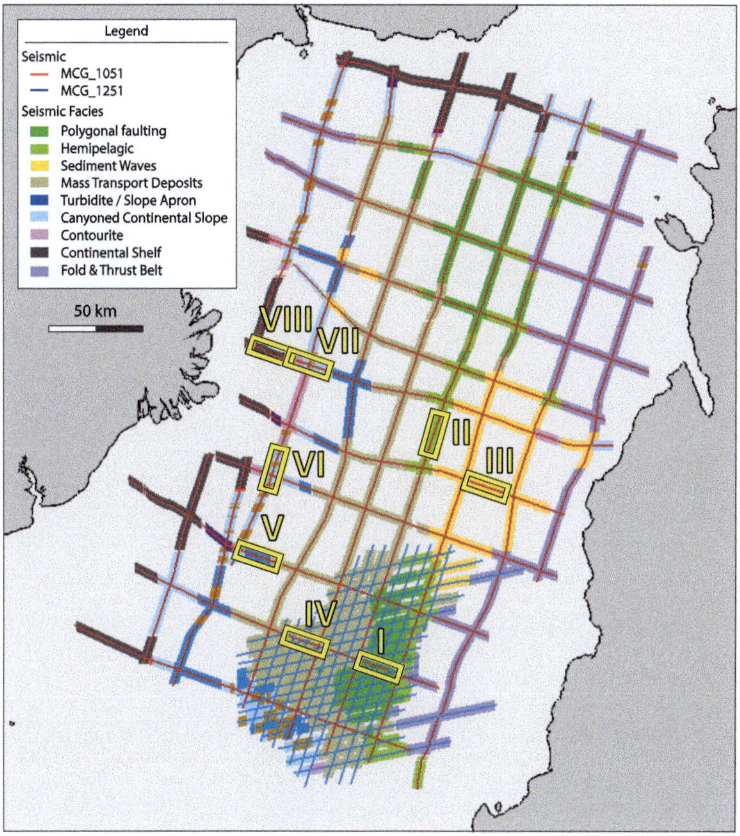

Fig. 8. Mapped seismic facies. Yellow boxes (I–VIII) indicate location of seismic sections shown in Figure 9.

SF V. Upslope of the MTDs and basin floor, SF V is characterized by subparallel and semi-continuous reflections (Fig. 9 V). Locally, this facies shows a mounded external morphology and thinning upslope. Reflections dip basinward but are at a lower angle than the seabed forming the upper continental slope. This facies is interpreted as a turbidite-dominated slope apron consisting of sediment fans and channel-levee systems.

SF VI. The upper slope shows varying seismic reflection amplitudes. Internally reflections are laterally semi-continuous, but commonly are truncated by downcutting, erosive canyons (Fig. 9 VI). Basinward-dipping reflections are seen in addition to mounded morphologies. Rare BSR reflections are also identified within SF VI.

SF VII. On the upper slope in the SW of the Makassar North Basin (Fig. 8), the SF is characterized by basinward-dipping, parallel, semi-continuous internal reflections (Fig. 9 VII), interpreted as contourites. The external morphology is distinctly convex up, with evidence of alongslope elongation (despite limited 2D data coverage). The upslope progradation of internal seismic reflections, combined with their external morphology, is typical of contourite-plastered drifts (Rebesco *et al.* 2014).

SF VIII. Characterized by horizontal to basinward-dipping, semi-parallel, discontinuous reflections, SF VIII is confined to the continental shelf (Fig. 9 VIII). Internal reflections are low amplitude, compared to the seabed reflection horizon (which locally is high amplitude), particularly in the north of the study area. Locally on the shelf or at the shelf break, isolated build-ups (Fig. 11a, b) are interpreted as biohermal carbonates (Roberts and Sydow 2003). These suggest that the locally high amplitude seabed seismic reflections are due to carbonate deposition.

Combining the observations and interpretations from the bathymetric and seismic mapping, an environment of deposition map for the Quaternary has been compiled (Fig. 12).

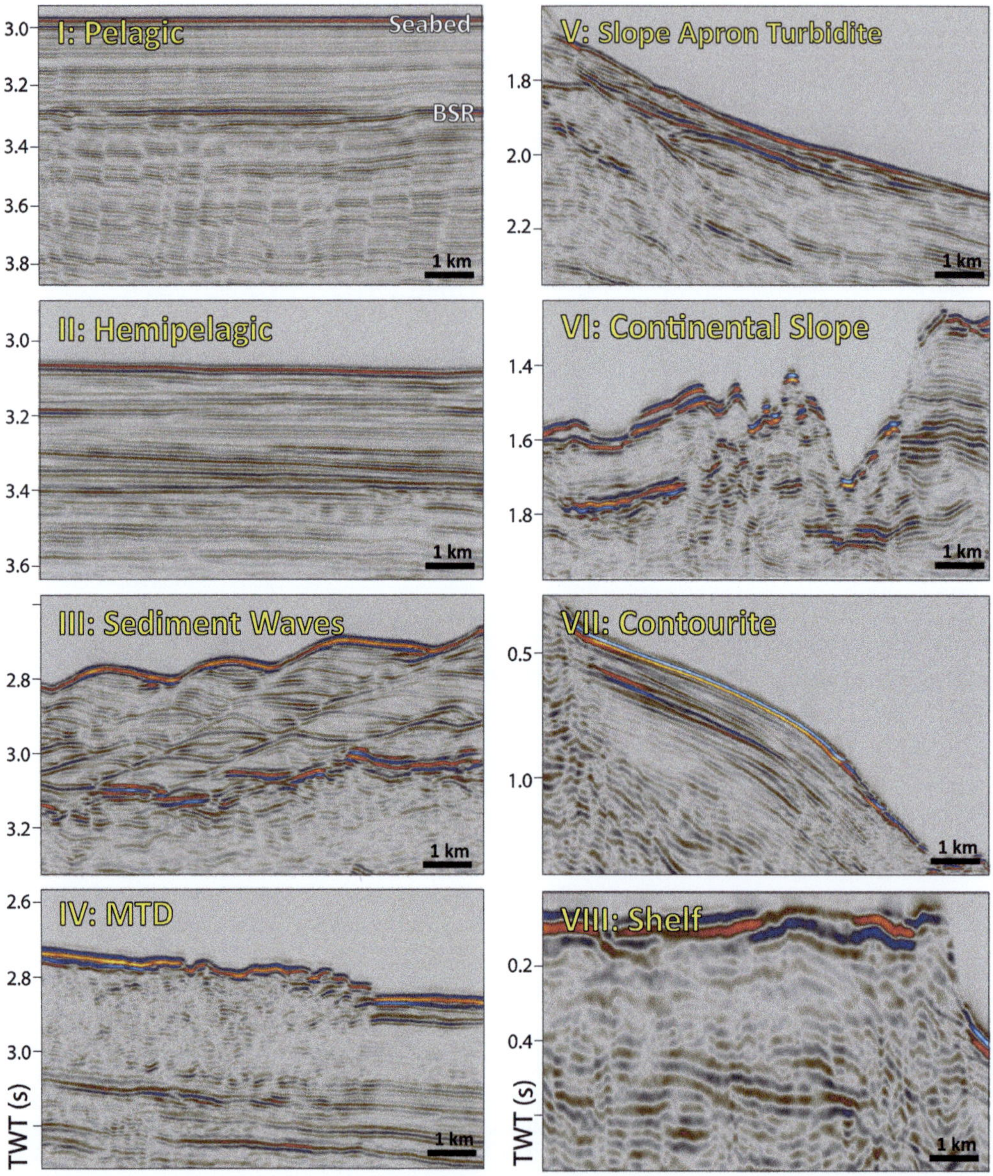

Fig. 9. Seismic facies of the Makassar Strait. Locations indicated on Figure 8.

MTDs. In total, 19 individual MTD deposits were mapped in detail (Table 2; Fig. 12), with volumes ranging from 5 km^3 to over 600 km^3 (Table 2). Five MTDs are over 100 km^3 in volume, with the three largest located at the base of the continental slope in the SW of the Makassar North Basin, south of the Mahakam Delta (Fig. 13). Mapping of internal and top-surface directional indicators (imbrications and lineations) reveals that the largest MTDs are sourced from the southwestern margin of the basin. Smaller (<50 km^3) events are sourced from the Mangkalihat High in the north and the Paternoster Platform in the south. No significant volume MTD events are identified in the fold-and-thrust belt – although numerous smaller events (<5 km^3) associated with thrust-cored anticlines are evident in the bathymetric data (e.g. Fig. 6b).

Based on their internal and external characteristics, the MTDs mapped in this study, are interpreted as deposited from translational submarine

Table 1. *Pliocene–Recent seismic facies (SF) across the Makassar Strait MCG surveys*

SF	Configuration and continuity	Amplitude and frequency	Bounding relationship	Depositional environment
I	Horizontal, parallel, discontinuous	High amplitude, high frequency	Fault-plane truncations. Laterally, merging with SF II	Pelagic with polygonal faulting
II	Horizontal, parallel, laterally continuous	High amplitude, high frequency	Laterally continuous, draping morphology	Hemipelagic
III	Hummocky, subparallel, semi-continuous	High amplitude, low frequency	Laterally continuous, onlapping upslope	Sediment waves
IV	Deformed – chaotic, discontinuous	Low amplitude, low frequency	Lenticular, basal erosional truncations and steps	MTDs
V	Dipping, subparallel, semi-continuous	Varied amplitude, mod. frequency	Localized mounded morphology. Thinning towards slope	Slope apron/ turbidite
VI	Dipping, subparallel, semi-continuous	Varied amplitude, low frequency	Canyon downcutting results in erosional truncation	Continental slope with canyons
VII	Dipping, parallel, semi-continuous	High amplitude, high frequency	Convex-upward lenticular. Elongate alongslope	Contourite
VIII	Horizontal, semi-parallel, discontinuous	Low amplitude, low frequency	High-amplitude top reflection with localized build-ups	Continental shelf

landslides. The isochron thickness map identifies a thin Pleistocene–Recent sequence in the upper slope which could be a possible failure zone (Fig. 7c). This suggests that the submarine landslides were initiated in the upper- to middle-continental slope between water depths of 500 and 1250 m.

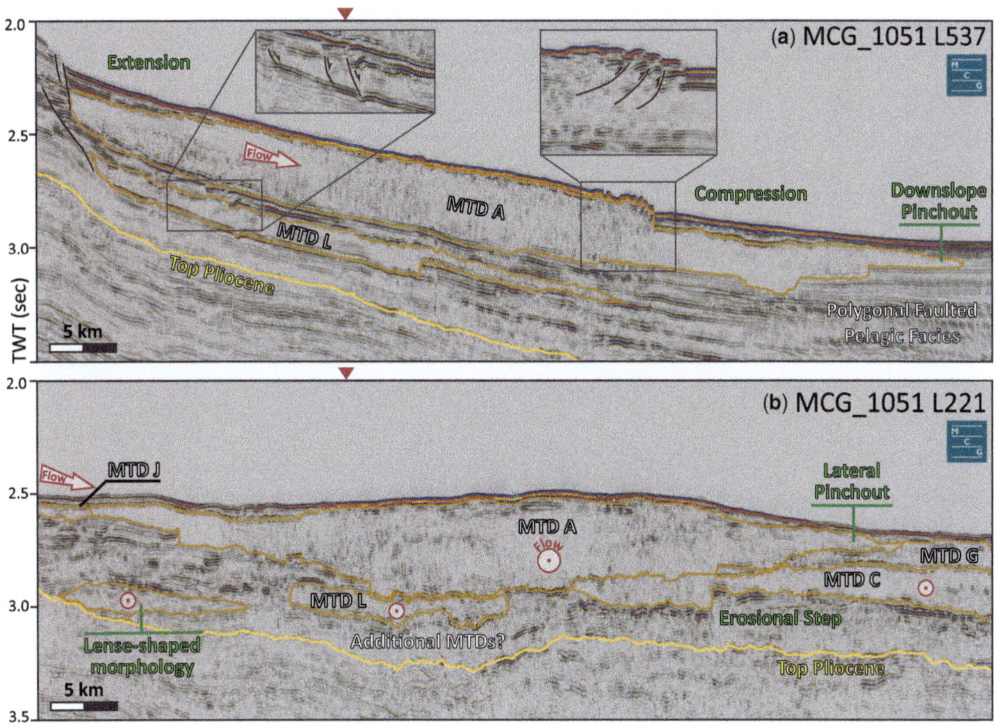

Fig. 10. Seismic character of MTDs in (**a**) dip and (**b**) strike. Location of section indicated in Figure 12. Red triangles indicate intersection. Flow direction in (b) towards the reader.

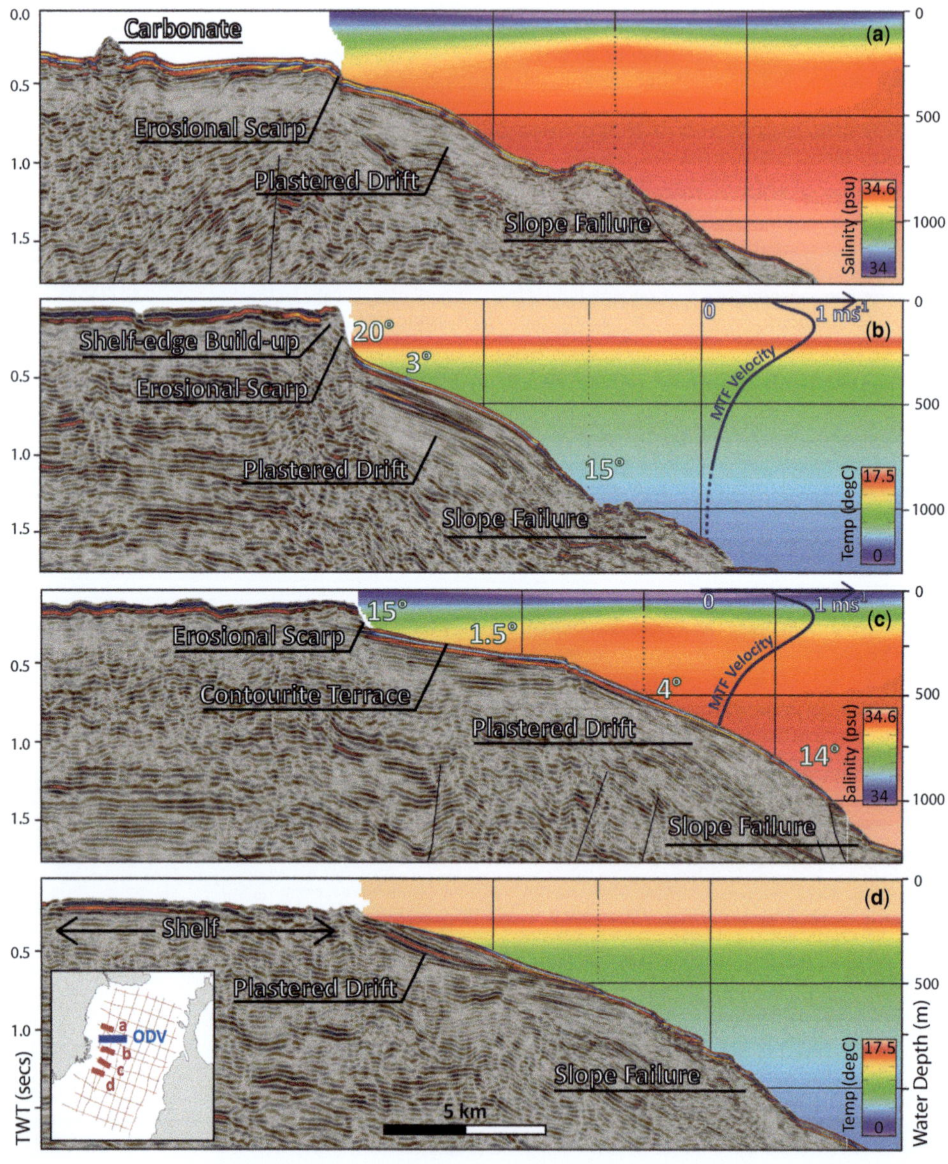

Fig. 11. Seismic character of the contourite depositional system across seismic line (**a**)–(**d**) indicated in inset map. Temperature and salinity data from Ocean Data View (ODV) section indicated (Schlitzer 2020). Slope gradients annotated in sections (b) and (c). Line locations indicated on Figure 12. Velocity–depth profile shows average velocity profile 2004–09 (Susanto *et al.* 2012).

Contourite features. On the upper slope, erosional and depositional contourite features are interpreted. Along the western edge of the Northern Basin, outboard of the shelf break there is a steep (15–20°) erosional escarpment that correlates well with the maximum velocity core of the MTF at 100–150 m water depth (Gordon *et al.* 2008). Locally, the base of this steep scarp is associated with a low gradient upper slope (1.5°) before a second shelf-slope break at 450 ms TWT (*c.* 300 m water depth). This is seen clearly on seismic line MCG_1051 L639 (Fig. 11b) where the 'step' in the upper slope corresponds to the average water depth of the base MTF at 300 m (Mayer and Damm 2012). It is therefore interpreted as a contourite terrace, formed by MTF erosion. The mechanisms for contourite terrace

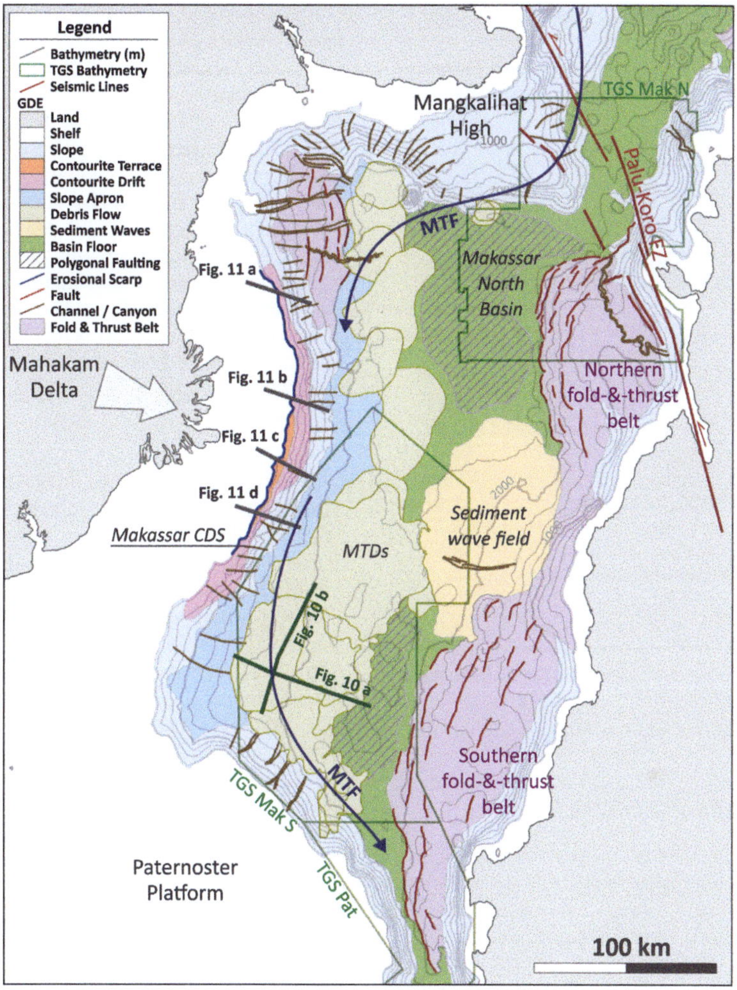

Fig. 12. Map of the environment of deposition of the Makassar North Basin. TGS bathymetry data extent indicates areas of high-resolution data. Additional information gathered from Fowler *et al.* (2004); Saller and Dharmasamadhi (2012); and Frederik *et al.* (2019). CDS, contourite depositional system.

formation are poorly understood (Hernández-Molina *et al.* 2008); however, the association between terrace depth and the base of the MTF suggests the possibility of enhanced current energy or internal wave-related erosion at the thermocline or pycnocline. The presence of an erosional scarp upslope of the terrace shows that this western boundary current is highly erosive in nature along much of the western margin of the Makassar North Basin. Downslope of the erosional features, SF VII shows a distinctly convex-upwards slope morphology, alongslope elongation and upslope progradation of internal reflections (Fig. 9) – all features diagnostic of contourites (Nielsen *et al.* 2008). We interpret these depositional features as plastered contourite drifts. In the Makassar South Basin, the Paternoster bathymetric data (Fig. 6) show a number of additional features that could represent contourites including: (1) an erosional scour orientated perpendicular to the Paternoster margin (Fig. 6a insert) – these scours are characterized by steep lee and shallow stoss sides, and are interpreted as formed by bottom current turbulence within the MTF on exiting the Labani Channel, similar to features described by Nicholson and Stow (2019); (2) a contourite terrace similar to that seen in the Makassar North Basin (Fig. 6b); and (3) a potential mounded drift within an embayment (Fig. 6a). However, additional seismic data are required to corroborate this interpretation.

Table 2. *Measurements from mapped MTDs and their calculated volumes*

MTD	Area (km^2)	Volume (TWT)	km^3 at 2000 m s^{-1}	Uncertainty (km^3) range ±
A	3410	616.47	616.47	123.29
B	938	53.88	53.88	10.78
C	3891	422.99	422.99	84.60
D	1482	110.60	110.60	22.12
E	1313	137.06	137.06	27.41
F	444	26.32	26.32	5.26
G	3959	212.19	212.19	42.44
H	230	41.87	41.87	8.37
I	193	5.85	5.85	1.17
J	882	50.57	50.57	10.11
K	249	12.71	12.71	2.54
L	660	58.64	58.64	11.73
M	523	36.81	36.81	7.36
N	865	57.56	57.56	11.51
O	963	75.64	75.64	15.13
P	659	42.48	42.48	8.50
Q	680	44.48	44.48	8.90
R	86	4.91	4.91	0.98
S	130	8.14	8.14	1.63

20% uncertainty range applied to allow for acoustic velocities between 1800 and 2200 m s^{-1}.

Discussion

Drivers for slope instability

There are a number of factors that contribute to instability of continental slopes: (1) seismic activity, including seismic loading and earthquake-related shaking; (2) slope oversteepening, due to erosion, carbonate build-up, faulting or diapirism; (3) loading, from water, sediments or ice – this can occur over various timescales, for example short-lived cyclic loading during storm events, or longer-term sea-level changes; (4) rapid accumulation and underconsolidation that can result in elevated pore pressures; and (5) gas hydrate disassociation (Locat and Lee 2002). All these processes act to either reduce the shear strength of continental slope sediments, or increase the stress acting upon them (Hampton *et al.* 1996). In addition, settings that result in steep seabed morphology such as fjords, volcanoes and carbonate build-ups are particularly susceptible to failure (Hampton *et al.* 1996). Here we discuss the possible drivers for slope failure in the Makassar Strait.

Carbonate build-up. Carbonate slopes are capable of reaching gradients that far exceed that of their siliciclastic counterparts (Schlager and Camber 1986). This ability to reach gradients of up to 29°, combined with their capacity to form a concave-upward slope morphology, makes them susceptible to mass failure (Hampton *et al.* 1996). Carbonates are present across the western margin of the Makassar Strait, both as isolated biohermal build-ups along the NW continental shelf, and more extensively across the Paternoster Platform in the SW. Based on kinematic indicators and gross morphology, a number of moderate size (<50 km^3) MTDs are sourced from the Paternoster Platform (Fig. 13). However, the largest MTDs are sourced from the western continental margin where clastic sedimentation from the Mahakam Delta dominates (Fig. 13), rather than the carbonate platform margin. It is therefore unlikely that oversteepening of carbonate slopes is a driver for large-scale submarine landslides in the Makassar North Basin.

Seismic activity. There is historical evidence for earthquake rupture of the seabed causing tsunamis (NGDC/WDS) and seismic activity is frequent across the Makassar Strait region (Fig. 1) (https://earthquake.usgs.gov/earthquakes), particularly in Sulawesi where earthquakes of up to M_w 8 are associated with the left lateral Palu-Koro fault zone (Fig. 1b). There are historical earthquakes along the eastern margin of the Makassar Strait; however, our mapping shows that the MTDs are sourced from the western margin. It is therefore unlikely that seismic activity and related seafloor fault rupture is the primary driver for these. Nevertheless, with the narrow Makassar Strait *c.* 300 km in width, it is possible that far-field seismic shaking is the *triggering* mechanism for submarine landslides of the Kalimantan continental margin. The distribution of the largest MTDs in the SW of the Makassar North Basin, away from historical earthquake epicentres, suggests that there is an additional *preconditioning* mechanism in this region for the generation of large-volume MTDs.

Deltaic deposition and sediment loading. A prominent sedimentary feature of the Kalimantan margin is the Mahakam Delta. With an estimated annual sediment discharge of 8×10^6 m^3 a^{-1} (Roberts and Sydow 2003), the Mahakam River provides significant volumes of sediment to this actively prograding delta. Delta fronts studied elsewhere (Hampton *et al.* 1996) are known to be vulnerable to collapse due to the interplay of various factors. For example, in the case of the Mississippi Delta, high sedimentation rates result in under-consolidated sediment accumulations and frequent failure (Coleman *et al.* 1993). Furthermore, these deltaic sediments are generally fine grained with high organic content and gas charge, which acts to reduce sediment shear strength. As a result, such sediments are preconditioned for failure.

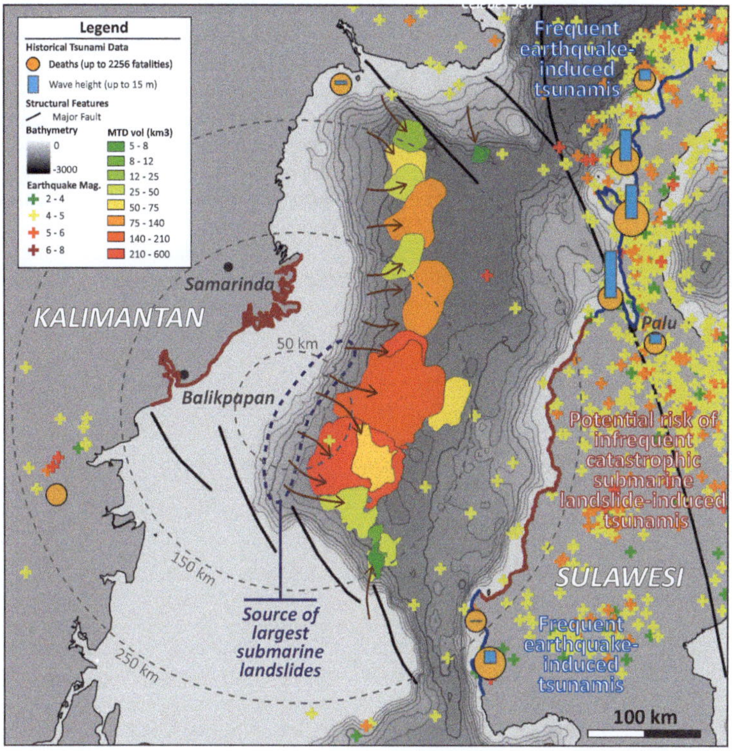

Fig. 13. Risk map showing the Makassar Throughflow. MTDs colour-coded with respect to their volume. Coastlines at risk of infrequent and catastrophic submarine landslide-induced tsunamis are indicated in red.

In the case of the Mahakam Delta, similar processes to those on the Mississippi Delta may promote mass failure of the continental slope. The Makassar Strait is a major oil and gas province, and sediments supplied to the delta are high in organic content (Saller *et al.* 2006) and therefore are likely to be low in shear strength. In addition, microbial consumption of organic matter and thermogenic maturation of kerogens is likely to result in methane production, which could reduce stability. Detailed interpretation of the seismic, bathymetry and gravity cores offshore the Mahakam Delta (Roberts and Sydow 2003) reveals that during Pleistocene lowstand conditions, delta growth occurred along the shelf-slope transition. This would have provided sediment directly to the upper slope, causing oversteepening and sediment loading, resulting in the slope being highly susceptible to failure. This situation explains why MTDs are distributed only along the western slope of the basin, unlike the eastern margin, where there is no comparable sediment influx. However, this observation does not explain the distribution of the largest mapped MTDs clustered to the SW of the Mahakam Delta (Fig. 13).

Slope failure preconditioning by the ITF. Mapping the present-day route of the MTF (Fig. 1c) shows that this western boundary current is interacting with the Kalimantan margin and the Mahakam Delta. Detailed mapping of the delta front shows the fine-grained prodelta (at *c.* 40 m water depth) is deflected to the south due to interaction with the MTF (Roberts and Sydow 2003). Regional gravity and depth to basement mapping (Cloke *et al.* 1999; Moss and Chambers 1999; Sandwell *et al.* 2014) demonstrate that the Mahakam depocentre has a marked asymmetry from north to south. Oceanographic measurements show the MTF current reaches 1 m s^{-1} within a subsurface 'jet' extending down to 300 m below the sea surface with a high velocity core at 100–150 m (Mayer and Damm 2012). Currents of this strength are sufficient to erode and transport sediments (Stow *et al.* 2008), as evidenced by the distribution of contourite erosional and depositional features across the upper slope of the Kalimantan margin, herein named the Makassar Contourite Depositional System (CDS) (Fig. 11).

Previous authors have highlighted that contourite deposition is often associated with slope instability along continental margins (Laberg and Camerlenghi 2008; Stow *et al.* 2012; Miramontes *et al.* 2018; Nicholson *et al.* 2020). Along the NW European margin, the giant (*c.* 3500 km^3) Storegga Slide is, in part, a result of the reduced shear strength within the basal failure plane comprising clay-rich contourites (Bryn *et al.* 2005). In the South Falkland Basin, an over-steepened and basinward-dipping contourite drift on the Burdwood Bank has regularly failed due to upper-slope deposition and lower-slope erosion within a contourite moat (Nicholson *et al.* 2020). Along the East Canadian margin, MTDs are associated with shallow-water contourites (Rashid *et al.* 2017). A geotechnical review of these sediments concludes that their high fluid content, together with the highly cyclical nature of the sediments resulting in failure surfaces throughout the succession, may promote liquefaction from seismic shaking (Rashid *et al.* 2017). In these three regions, the cause of sediment failure is attributed to the nature and morphology of contourite sediments.

Thus, three key factors make contourites susceptible to mass transport failure. Firstly, the anomalously high sedimentation rates of contourite drifts reaching 60 cm ka^{-1} (Howe *et al.* 1994) can lead to high fluid content and elevated pore pressures upon burial. Secondly, the fine muddy nature of contourite sediments makes then likely to have low internal shear strength (Laberg and Camerlenghi 2008). Finally, their convex-upwards external morphology can locally result in high gradients on the continental slope making failure more likely (Stow *et al.* 2012; Miramontes *et al.* 2018). Laberg and Camerlenghi (2008) note that there is often an association between gas hydrate accumulation and contourite deposition due to the elevated pore space in these rapidly deposited sediments and may be an additional factor making contourites susceptible to failure.

In the case of the Makassar CDS, the high velocity MTF jet correlates well with the formation of a steep (20°) erosional scarp and relatively flat (1.5°) contourite terrace on the upper slope directly to the north and east of the Mahakam Delta (Fig. 11). This is evidence for the erosion and removal of sediments from this area, which are then transported and deposited downslope and downcurrent. This winnowing of sediments from the north results in enhanced sedimentation on the slope to the SE of the delta, directly above the region where the largest Quaternary MTDs have been mapped (Figs 12 & 13). Contourite-plastered drifts are identified along the margin, generally directly below the base of the MTF at 300 m water depth (Fig. 11). These regions of enhanced sedimentation are vulnerable to failure due to high fluid content, the high organic content of the Mahakam Delta sediments, and their convex-upwards morphology which results in an oversteepened, basinward-dipping slope, of around 15°. We propose these regions as the main source of mass transport sediments to the deep basin. It is noted that the plastered contourite drifts identified in this study are significantly smaller than examples seen along other passive continental margins (Hernández-Molina *et al.* 2008). We propose this to be a result of regular removal of material from the drifts during submarine landslide events.

The MTF is likely to have varied significantly in response to Quaternary climatic fluctuations, which may further play a role in controlling slope instability. The resolution of the data in our study, and the lack of well control, do not allow detailed evaluation of the timescales and the control of sea-level or climate on triggering these events, but it is expected that climate exerts a strong control on slope stability by controlling the depth, velocity and nature of the MTF. The evidence from previous studies, however, is conflicting. During glacial lowstands, much of the Paternoster Platform and the continental shelf was exposed (Hall 2009; Roberts and Sydow 2003), and the width and cross-sectional area of the Makassar Strait was considerably reduced. As a result, the MTF may have initially been accelerated due to flow constriction (Hall 2009). However, ocean-current velocity proxies suggest that the MTF was significantly reduced, or even switched off during glacial conditions (Gingele *et al.* 2002; Holbourn *et al.* 2011), due to the reduced exchange between the Pacific and South China seas (Godfrey 1996) and reduced Atlantic meridional overturning (Hendrizan *et al.* 2017). Modelling of the response of ocean currents to fluctuating sea-level in SE Asia is further complicated by the effect of monsoonal conditions on the region. Although the interplay between sea-level-induced restriction of the MTF and changes in velocity profile are not well understood, it is probable that sea-level fluctuations resulting from Pleistocene glacial–interglacial cyclicity affected slope instability. Further work is required to constrain variations in the timing and frequency of slope failure events to reveal the drivers.

Tsunamigenic risk

MTDs of the scale and volume mapped in this study in the Quaternary section of the Makassar North Basin, by analogy with other locations, would be expected to generate hazardous tsunamis (e.g. Harbitz *et al.* 2006; Tappin *et al.* 2008, 2014; Parsons *et al.* 2014) (Table 3). Many past large volume landslide tsunamis, such as Storegga, are prehistoric, and are recognized and mapped from their palaeo-record preserved in sediments (e.g. Dawson *et al.* 1988; Bondevik *et al.* 1997). There is no historical evidence for tsunamis originating in the

Table 3. *Key characteristics of previous tsunamigenic submarine landslides*

Tsunami	MTD volume (km^3)	Water depth (m)	Initial acceleration ($m\ s^{-1}$)	Distance to shore (km)	Run up (m)	Reference
PNG	4	1200	0.47	25	10–15	Tappin *et al.* (2008)
Japan	500	4000–5500	0.37	150	40	Tappin *et al.* (2014)
Falklands	100	860	0.37	180	40	Nicholson *et al.* (2020)
Grand Banks*	200	500	unknown	340	13	Fine *et al.* (2005)
Makassar Strait	100–600 km^3	500–1000	unknown	E = 180; W = 100	unknown	

*Grand Banks reverse wave documented.

Makassar Strait, and no one has yet looked for evidence of prehistoric (Pleistocene–Recent) events. This leads us to conclude that either: (1) the submarine landslides that generated these MTDs were not tsunamigenic; or (2) tsunamis generated by submarine landslides are of much lower frequency than those generated by fault movements and have not occurred since historical records began. This needs to be tested by investigating evidence for palaeo-tsunamis in the surrounding coastal regions, although evidence is only likely to be preserved for Holocene–Recent events.

From our tentative chronology of the MTDs sourced from the Mahakam Delta, they have occurred frequently throughout the Quaternary. A conservative estimate of the frequency of submarine landslides >100 km^3 is approximately every 0.5 Ma. Events larger than 10 km^3 are conservatively estimated to have occurred every 160 ka; detailed mapping of high-resolution 3D seismic surveys would likely reveal multiple smaller events and a significantly higher frequency. It is, however, likely that the last large-scale failure event occurred prior to historical records. If the landslides were tsunamigenic, they would impact the local Sulawesi and Kalimantan coastlines which previously were not considered at risk of tsunamis. There are a number of towns along the Sulawesi coast, *c.* 180 km east and directly in front of the largest MTD deposits that could be affected by any forward-propagating wave. The distance is comparable to that travelled by previous tsunamis in Papua New Guinea (Tappin *et al.* 2008) and Japan (Tappin *et al.* 2014) (Table 3). The high-energy point source release of tsunamis from submarine landslides also generates a significant reverse wave, travelling in the opposite direction to slope failure. This reverse wave would travel only 100 km before reaching the low-lying cities of Balikpapan (population >850 000) and Samarinda (population >840 000). The destructive nature of such a reverse wave was demonstrated in the 1929 Grand Banks event, when a 13 m wave struck the Newfoundland Coast following a slope failure some 340 km offshore (Fine *et al.* 2005). With the exception of distance to shoreline, the characteristics of the Grand Banks slope failure (MTD volume and water depth) are similar to the MTDs mapped in the Makassar Strait. Therefore, if the Makassar Strait submarine landslides were capable of generating tsunami waves, it is expected that the back-wave could also form a tsunami risk, with local morphological features such as channels and estuaries, including Balikpapan Bay further amplifying the wave height.

Tsunamis generated by submarine landslides differ in character from those sourced from fault rupture of the seabed (Okal and Synolakis 2003). Unlike the linear wave formed by fault rupture (Aida 1978; Tanioka and Satake 1996), landslides generate a tsunami wave from a point source that radially spreads out in the direction of the submarine failure, with a negative reverse wave spreading in the opposite direction (Harbitz *et al.* 2006). The magnitude of the wave is much larger than those formed by fault rupture; however, rapid radial dampening limits the destructive distance of the tsunami (Okal and Synolakis 2003). The magnitude of a tsunami generated by a submarine landslide is strongly controlled by landslide volume, initial acceleration and the water depth of initial failure (Harbitz *et al.* 2006). Tsunami wave modelling is urgently required to test slope failure scenarios and identify any coastal regions at risk. This will allow for the potential risk to be quantified, and inform recommendations for mitigation measures to be put into effect and evacuation plans be implemented in high-risk regions.

Conclusions

This study identifies, for the first time, multiple large-volume MTDs generated by submarine landslides within the Quaternary section of the Makassar North Basin. The largest MTDs are clustered in the SW of the basin and are over 600 km^3 in volume.

Kinematic indicators demonstrate that the MTDs are sourced from the western margin of the Makassar Strait, south of the Mahakam Delta. Mapping along this western margin also reveals a contourite depositional system along the upper slope: the Makassar CDS.

We propose that the largest and most frequent submarine landslides in the Makassar North Basin are genetically related to these contourites. Sediments transported into the basin from the Mahakam Delta are redistributed by the MTF and deposited on the upper slope to the south of the delta, resulting in elevated sedimentation rates in this region. In combination with fine grain size, rapid burial promotes fluid retention and overpressure generation, increasing the probability of slope failure. The fine, muddy nature of the drifts makes them likely to have low shear strength, and their convex-upwards morphology results in locally high gradients on the upper continental slope. All of these factors make these deposits inherently unstable and prone to failure.

Based on analogue studies of MTDs elsewhere, it is probable that the submarine landslides that generated the mapped deposits were tsunamigenic. The wave generated from the submarine landslides mapped in this study could impact the Sulawesi and Kalimantan coastlines in regions not previously affected by historical events. It is therefore critical to understand the triggering mechanisms of these events and their hazard. Future work will aim to constrain the frequency, and the role of climate and sea-level on submarine landslide triggering. Tsunami wave modelling is required to test slope failure scenarios and identify coastal regions at highest risk.

Acknowledgements We would like to thank our partners at Bandung Institute of Technology, and the members of the Indonesian Marine Geological Institute and Geological Survey of Indonesia for their discussion and contribution to this research. We thank TGS and Multiclient Geophysical for permission to publish seismic and multibeam data respectively. D.R. Tappin publishes with the permission of the Executive Director of the BGS (United Kingdom Research and Innovation).

Funding This work was completed as part of a Scottish Funding Council (SFC) Global Challenges Research Fund (GCRF) pump-priming project at Heriot Watt University (PI Nicholson).

Author contributions **REB**: formal analysis (lead), investigation (lead), writing – original draft (lead); **UN**: conceptualization (lead), funding acquisition (lead), supervision (lead), writing – review & editing (lead); **BS**: funding acquisition (equal), validation (supporting); **DS**: conceptualization (supporting), funding acquisition (supporting), supervision (supporting), writing – review & editing (supporting); **DRT**: funding acquisition (equal), validation (supporting), writing – review & editing (supporting).

References

Aida, I. 1978. Reliability of a tsunami source model derived from fault parameters. *Journal of Physics of the Earth*, **26**, 57–73, https://doi.org/10.4294/jpe1952.26.57

Becker, J.J.D.T., Sandwell, W.H.F. *et al.* 2009. Global Bathymetry and Elevation Data at 30 Arc Seconds Resolution: SRTM30_PLUS. *Marine Geodesy*, **32**, 355–371, https://doi.org/10.1080/01490410903297766

Bergman, S.C., Coffield, D.Q., Talbot, J.P. and Garrard, R.A. 1996. Tertiary tectonic and magmatic evolution of western Sulawesi and the Makassar Strait, Indonesia: evidence for a Miocene continent-continent collision. *Geological Society, London, Special Publications*, **106**, 391–429, https://doi.org/10.1144/GSL.SP.1996.106.01.25

Bird, M.I., Taylor, D. and Hunt, C. 2005. Palaeoenvironments of insular Southeast Asia during the Last Glacial Period: a savanna corridor in Sundaland? *Quaternary Science Reviews*, **24**, 2228–2242, https://doi.org/10.1016/j.quascirev.2005.04.004

Bondevik, S., Svendsen, J.I., Johnsoen, G., Mangerud, J.A.N. and Kaland, P.E. 1997. The Storegga tsunami along the Norwegian coast, its age and run up. *Boreas*, **26**, 29–53, https://doi.org/10.1111/j.1502-3885.1997.tb00649.x

Bryn, P., Berg, K., Forsberg, C.F., Solheim, A. and Kvalstad, T.J. 2005. Explaining the Storegga slide. *Marine and Petroleum Geology*, **22**, 11–19, https://doi.org/10.1016/j.marpetgeo.2004.12.003

Burollet, P.F., Boichard, R., Lambert, B. and Villain, J.M. 1986. The Pater Noster Carbonate Platform. *Indonesian Petroleum Association 15th Annual Convention Proceedings*, **1**, 155–169.

Carvajal, M., Araya-Cornejo, C., Sepúlveda, I., Melnick, D. and Haase, J.S. 2019. Nearly instantaneous tsunamis following the Mw 7.5 2018 Palu earthquake. *Geophysical Research Letters*, **46**, 5117–5126, https://doi.org/10.1029/2019GL082578

Cloke, I.R., Milsom, J. and Blundell, D.J.B. 1999. Implications of gravity data from East Kalimantan and the Makassar Straits: a solution to the origin of the Makassar Straits? *Journal of Asian Earth Sciences*, **17**, 61–78, https://doi.org/10.1016/S0743-9547(98)00056-7

Coleman, J.M., Prior, D.B., Garrison, L.E. and Lee, H. 1993. Slope failures in an area of high sedimentation rate: Offshore Mississippi River delta. *In*: Schwab, W.C., Lee, H.J. and Twichell, D.C. (eds) *Submarine Landslides: Selected Studies in the US Exclusive Economic Zone*. US Geological Survey Bulletin, **2002**, 79–91.

Daly, M.C., Cooper, M.A., Wilson, I., Smith, D.G. and Hooper, B.G.D. 1991. Cenozoic plate tectonics and basin evolution in Indonesia. *Marine and Petroleum Geology*, **8**, 2–21, https://doi.org/10.1016/0264-8172(91)90041-X

Davies, R.J. and Ireland, M.T. 2011. Initiation and propagation of polygonal fault arrays by thermally triggered volume reduction reactions in siliceous sediment.

Marine Geology, **289**, 150–158, https://doi.org/10.1016/j.margeo.2011.05.005
Dawson, A.G., Long, D. and Smith, D.E. 1988. The Storegga slides: evidence from eastern Scotland for a possible tsunami. *Marine Geology*, **82**, 271–276, https://doi.org/10.1016/0025-3227(88)90146-6
Decker, J., Teas, P.A., Schneider, R.D., Saller, A.H. and Orange, D.L. 2004. Modern deep-sea sedimentation in the Makassar Strait: insights from high-resolution multibeam bathymetry and backscatter, sub-bottom profiles, and USBL-navigated cores. *In*: Noble, R.A., Argenton, A. and Caughey, C.A. (eds) *Deepwater and Frontier Exploration in Asia & Australasia*. Proceedings of the International Geoscience Conference, Indonesian Petroleum Association, Jakarta, DFE-04-PO-042, 377–387.
Del Negro, R., Castellano, P., Kuhfuss, A., Mattioni, L., Moss, J. and Vialla, M. 2013. Structural Styles and Petroleum System Modelling of the North Makassar Straits, Indonesia. *Proceedings of the Indonesian Petroleum Association Thirty-Seventh Annual Convention & Exhibition*, May 2013, Jakarta, IPA13-G-077.
Fine, I.V., Rabinovich, A.B., Bornhold, B.D., Thomson, R.E. and Kulikov, E.A. 2005. The Grand Banks landslide-generated tsunami of November 18, 1929: preliminary analysis and numerical modeling. *Marine Geology*, **215**, 45–57, https://doi.org/10.1016/j.margeo.2004.11.007
Fowler, J.N., Guritno, E. *et al.* 2004. Depositional architectures of recent deepwater deposits in the Kutei Basin, East Kalimantan. *Geological Society, London, Memoirs*, **29**, 25–34, https://doi.org/10.1144/GSL.MEM.2004.029.01.03
Frederik, M.C., Adhitama, R. *et al.* 2019. First results of a bathymetric survey of Palu Bay, Central Sulawesi, Indonesia following the Tsunamigenic Earthquake of 28 September 2018. *Pure and Applied Geophysics*, **176**, 3277–3290, https://doi.org/10.1007/s00024-019-02280-7
Gingele, F.X., De Deckker, P., Girault, A. and Guichard, F. 2002. History of the South Java Current over the past 80 ka. *Palaeogeography, Palaeoclimatology, Palaeoecology*, **183**, 247–260, https://doi.org/10.1016/S0031-0182(01)00489-8
Godfrey, J.S. 1996. The effect of the Indonesian throughflow on ocean circulation and heat exchange with the atmosphere: a review. *Journal of Geophysical Research: Oceans*, **101**, 12217–12237, https://doi.org/10.1029/95JC03860
Gordon, A.L. and Fine, R.A. 1996. Pathways of water between the Pacific and Indian oceans in the Indonesian seas. *Nature*, **379**, 146, https://doi.org/10.1038/379146a0
Gordon, A.L., Susanto, R.D., Field, A., Huber, B.A., Pranowo, W. and Wirasantosa, S. 2008. Makassar Strait throughflow, 2004 to 2006. *Geophysical Research Letters*, **35**, https://doi.org/10.1029/2008GL036372
Grilli, S.T., Tappin, D.R. *et al.* 2019. Modelling of the tsunami from the December 22, 2018 lateral collapse of Anak Krakatau volcano in the Sunda Straits, Indonesia. *Scientific Reports*, **9**, 1–13, https://doi.org/10.1038/s41598-019-48327-6
Guntoro, A. 1999. The formation of the Makassar Strait and the separation between SE Kalimantan and SW Sulawesi. *Journal of Asian Earth Sciences*, **17**, 79–98, https://doi.org/10.1016/S0743-9547(98)00037-3
Hall, R. 1997. Cenozoic plate tectonic reconstructions of SE Asia. *Geological Society, London, Special Publications*, **126**, 11–23, https://doi.org/10.1144/GSL.SP.1997.126.01.03
Hall, R. 2009. Southeast Asia's changing palaeogeography. *Blumea-Biodiversity, Evolution and Biogeography of Plants*, **54**, 148–161, https://doi.org/10.3767/000651909X475941
Hall, R., Ali, J.R., Anderson, C.D. and Baker, S.J. 1995. Origin and motion history of the Philippine Sea Plate. *Tectonophysics*, **251**, 229–250, https://doi.org/10.1016/0040-1951(95)00038-0
Hall, R., Cloke, I.R., Nur'aini, S., Puspita, S.D., Calvert, S.J. and Elders, C.F. 2009. The North Makassar Straits: what lies beneath? *Petroleum Geoscience*, **15**, 147–158, https://doi.org/10.1144/1354-079309-829
Hampton, M.A., Lee, H.J. and Locat, J. 1996. Submarine landslides. *Reviews of Geophysics*, **34**, 33–59, https://doi.org/10.1029/95RG03287
Harbitz, C.B., Løvholt, F., Pedersen, G. and Masson, D.G. 2006. Mechanisms of tsunami generation by submarine landslides: a short review. *Norwegian Journal of Geology/Norsk Geologisk Forening*, **86**, 255–265.
Hendrizan, M., Kuhnt, W. and Holbourn, A. 2017. Variability of Indonesian Throughflow and Borneo runoff during the last 14 kyr. *Paleoceanography*, **32**, 1054–1069, https://doi.org/10.1002/2016PA003030
Hernández-Molina, F.J., Llave, E. and Stow, D.A.V. 2008. Continental slope contourites. *Developments in Sedimentology*, **60**, 379–408, https://doi.org/10.1016/S0070-4571(08)10019-X
Holbourn, A., Kuhnt, W. and Xu, J. 2011. Indonesian Throughflow variability during the last 140 ka: the Timor outflow. *Geological Society, London, Special Publications*, **355**, 283–303, https://doi.org/10.1144/SP355.14
Howe, J.A., Stoker, M.S. and Stow, D.A.V. 1994. Late Cenozoic sediment drift complex, northeast Rockall Trough, North Atlantic. *Paleoceanography*, **9**, 989–999, https://doi.org/10.1029/94PA01440
Jamelot, A., Gailler, A., Heinrich, P., Vallage, A. and Champenois, J. 2019. Tsunami simulations of the Sulawesi M w 7.5 event: Comparison of seismic sources issued from a tsunami warning context v. post-event finite source. *Pure and Applied Geophysics*, **176**, 3351–3376, https://doi.org/10.1007/s00024-019-02274-5
Katili, J.A. 1975. Volcanism and plate tectonics in the Indonesian island arcs. *Tectonophysics*, **26**, 165–188, https://doi.org/10.1016/0040-1951(75)90088-8
Koop, H., Flueh, E.R., Petersen, C.J., Weinrebe, W. and Wittwer, A., Meramex Scientists 2006. The Java margin revisited: Evidence for the subduction erosion off Java. *Earth and Planetary Science Letters*, **242**, 130–142, https://doi.org/10.1016/j.epsl.2005.11.036
Kuhnt, W., Holbourn, A., Hall, R., Zuvela, M. and Käse, R. 2004. Neogene history of the Indonesian throughflow. Continent-Ocean Interactions within East Asian Marginal Seas. *Geophysical Monograph*, **149**, 299–320, https://doi.org/10.1029/149GM16
Laberg, J.S. and Camerlenghi, A. 2008. The significance of contourites for submarine slope stability. *Developments*

in Sedimentology, **60**, 537–556, https://doi.org/10.1016/S0070-4571(08)10025-5

Lee, G.H., Kim, H.J., Jou, H.T. and Cho, H.M. 2003. Opal-A/opal-CT phase boundary inferred from bottom-simulating reflectors in the southern South Korea Plateau, East Sea (Sea of Japan). *Geophysical Research Letters*, **30**, https://doi.org/10.1029/2003GL018670

Locat, J. and Lee, H.J. 2002. Submarine landslides: advances and challenges. *Canadian Geotechnical Journal*, **39**, 193–212, https://doi.org/10.1139/t01-089

Mayer, B. and Damm, P.E. 2012. The Makassar Strait throughflow and its jet. *Journal of Geophysical Research: Oceans*, **117**, https://doi.org/10.1029/2011JC007809

McCloskey, J., Antonioli, A. *et al.* 2008. Tsunami threat in the Indian Ocean from a future megathrust earthquake west of Sumatra. *Earth and Planetary Science Letters*, **265**, 61–81, https://doi.org/10.1016/j.epsl.2007.09.034

Miramontes, E., Garziglia, S., Sultan, N., Jouet, G. and Cattaneo, A. 2018. Morphological control of slope instability in contourites: a geotechnical approach. *Landslides*, **15**, 1085–1095, https://doi.org/10.1007/s10346-018-0956-6

Moss, S.J. and Chambers, J.L. 1999. Tertiary facies architecture in the Kutai basin, Kalimantan, Indonesia. *Journal of Asian Earth Sciences*, **17**, 157–181, https://doi.org/10.1016/S0743-9547(98)00035-X

NGDC/WDS National Geophysical Data Center/World Data Service: NCEI/WDS Global Historical Tsunami Database. NOAA National Centers for Environmental Information, https://doi.org/10.7289/V5PN93H7 [last accessed May 2019].

Nicholson, U. and Stow, D. 2019. Erosion and deposition beneath the Subantarctic Front since the Early Oligocene. *Scientific Reports*, **9**, https://doi.org/10.1038/s41598-019-45815-7

Nicholson, U., Libby, S., Tappin, D.R. and McCarthy, D. 2020. The Subantarctic Front as a sedimentary conveyor belt for tsunamigenic submarine landslides. *Marine Geology*, **424**, 106161, https://doi.org/10.1016/j.margeo.2020.106161

Nielsen, T., Knutz, P.C. and Kuijpers, A. 2008. Seismic expression of contourites depositional systems. *Developments in Sedimentology*, **60**, 379–408, https://doi.org/10.1016/S0070-4571(08)10019-X

Okal, E.A. and Synolakis, C.E. 2003. A theoretical comparison of tsunamis from dislocations and landslides. *Pure and Applied Geophysics*, **160**, 2177–2188, https://doi.org/10.1007/s00024-003-2425-x

Okal, E.A. and Synolakis, C.E. 2008. Far-field tsunami hazard from mega-thrust earthquakes in the Indian Ocean. *Geophysical Journal International*, **172**, 995–1015, https://doi.org/10.1111/j.1365-246X.2007.03674.x

Parsons, T., Geist, E.L. *et al.* 2014. Source and progression of a submarine landslide and tsunami: The 1964 Great Alaska earthquake at Valdez. *Journal of Geophysical Research: Solid Earth*, **119**, 8502–8516, https://doi.org/10.1002/2014JB011514

Posamentier, H.W. and Meizarwin, W.T. 2000. Deep water depositional systems–ultra-deep Makassar Strait, Indonesia. *GCSSEPM Foundation 20th Annual Research Conference Deep-Water Reservoirs of the World*, **3** (6), 806–816.

Prasetya, G.S., De Lange, W.P. and Healy, T.R. 2001. The Makassar Strait tsunamigenic region, Indonesia. *Natural Hazards*, **24**, 295–307, https://doi.org/10.1023/A:1012297413280

Puspita, S.D., Hall, R. and Elders, C.F. 2005. Structural styles of the offshore West Sulawesi fold belt, North Makassar Straits, Indonesia. *Proceedings of the Indonesian Petroleum Association Thirtieth Annual Convention & Exhibition*, August 2005, Jakarta, 519–542, IPA05-G-110.

Rahmstorf, S. 2007. Thermohaline circulation. *In*: Elias, S.A. (ed.) *Encyclopedia of Quaternary Sciences*, 739–750, Elsevier, https://doi.org/10.1016/B0-44-452747-8/00014-4

Rashid, H., MacKillop, K., Sherwin, J., Piper, D.J.W., Marche, B. and Vermooten, M. 2017. Slope instability on a shallow contourite-dominated continental margin, southeastern Grand Banks, eastern Canada. *Marine Geology*, **393**, 203–215, https://doi.org/10.1016/j.margeo.2017.01.001

Rebesco, M., Hernández-Molina, F.J., Van Rooij, D. and Wåhlin, A. 2014. Contourites and associated sediments controlled by deep-water circulation processes: state-of-the-art and future considerations. *Marine Geology*, **352**, 111–154, https://doi.org/10.1016/j.margeo.2014.03.011

Roberts, H.H. and Sydow, J. 2003. Late Quaternary stratigraphy and sedimentology of the offshore Mahakam delta, east Kalimantan (Indonesia). *SEPM Special Publication*, **76**, 125–145, https://doi.org/10.2110/pec.03.76.0125

Ryan, W.B.F., Scarbotte, S.M. *et al.* 2009. Global Multi-Resolution Topography synthesis. *Geochemistry Geophysics, Geosystems*, **10**, Q03014, https://doi.org/10.1029/2008GC002332

Saller, A. and Dharmasamadhi, I.N.W. 2012. Controls on the development of valleys, canyons, and unconfined channel–levee complexes on the Pleistocene Slope of East Kalimantan, Indonesia. *Marine and Petroleum Geology*, **29**, 15–34, https://doi.org/10.1016/j.marpetgeo.2011.09.002

Saller, A., Lin, R. and Dunham, J. 2006. Leaves in turbidite sands: The main source of oil and gas in the deep-water Kutei Basin, Indonesia. *AAPG Bulletin*, **90**, 1585–1608, https://doi.org/10.1306/04110605127

Saller, A., Werner, K., Sugiaman, F., Cebastiant, A., May, R., Glenn, D. and Barker, C. 2008. Characteristics of Pleistocene deep-water fan lobes and their application to an upper Miocene reservoir model, offshore East Kalimantan, Indonesia. *AAPG Bulletin*, **92**, 919–949, https://doi.org/10.1306/03310807110

Schlager, W. and Camber, O. 1986. Submarine slope angles, drowning unconformities, and self-erosion of limestone escarpments. *Geology*, **14**, 762–765, https://doi.org/10.1130/0091-7613(1986)14<762:SSADUA>2.0.CO;2

Sandwell, D.T., Müller, R.D., Smith, W.H.F., Garcia, E. and Francis, R. 2014. New global marine gravity model from CryoSat-2 and Jason-1 reveals buried tectonic structure. *Science*, **346**, 65–67, https://doi.org/10.1126/science.1258213

Sangadji, R. 2019. Central Sulawesi disasters killed 4340 people, final count reveals. *Jakarta Post*, retrieved 16 September 2019.

Schlitzer, R. 2020. Ocean Data View, ODV 5.2.1, http://odv.awi.de

Stow, D.A.V., Hunter, S., Wilkinson, D. and Hernández-Molina, F.J. 2008. The nature of contourite deposition. *Developments in Sedimentology*, **60**, 143–156, https://doi.org/10.1016/S0070-4571(08)10009-7

Stow, D.A.V., Brackenridge, R., Patel, U. and Toulmin, S. 2012. Geohazards and ocean hazards in deepwater: overview and methods of assessment. Paper OTC 23697, presented at the Offshore Technology Conference, Houston.

Susanto, R.D. and Gordon, A.L. 2005. Velocity and transport of the Makassar Strait throughflow. *Journal of Geophysical Research: Oceans*, **110**, https://doi.org/10.1029/2004JC002425

Susanto, R.D., Field, A., Gordon, A.L. and Adi, T.R. 2012. Variability of Indonesian throughflow within Makassar Strait, 2004–2009. *Journal of Geophysical Research: Oceans*, **117**, https://doi.org/10.1029/2012JC008096

Takagi, H., Pratama, M.B., Kurobe, S., Esteban, M., Aránguiz, R. and Ke, B. 2019. Analysis of generation and arrival time of landslide tsunami to Palu City due to the 2018 Sulawesi Earthquake. *Landslides*, **16**, 983–991, https://doi.org/10.1007/s10346-019-01166-y

Tanioka, Y. and Satake, K. 1996. Tsunami generation by horizontal displacement of ocean bottom. *Geophysical Research Letters*, **23**, 861–864, https://doi.org/10.1029/96GL00736

Tappin, D.R., Watts, P. and Grilli, S.T. 2008. The Papua New Guinea tsunami of 17 July 1998: anatomy of a catastrophic event. *Natural Hazards and Earth System Science*, **8**, 243–266, https://doi.org/10.5194/nhess-8-243-2008

Tappin, D.R., Grilli, S.T. *et al.* 2014. Did a submarine landslide contribute to the 2011 Tohoku tsunami? *Marine Geology*, **357**, 344–361, https://doi.org/10.1016/j.margeo.2014.09.043

Tillinger, D. 2011. Physical oceanography of the present day Indonesian Throughflow. *Geological Society, London, Special Publications*, **355**, 267–281, https://doi.org/10.1144/SP355.13

Williams, R., Rowley, P. and Garthwaite, M.C. 2019. Reconstructing the Anak Krakatau flank collapse that caused the December 2018 Indonesian tsunami. *Geology*. **47**, 973–976, https://doi.org/10.1130/G46517.1

Yalciner, A.C., Hidayat, R. *et al.* 2018. Field Survey on the Coastal Impacts of the September 28, 2018 Palu, Indonesia Tsunami. *AGU Fall Meeting Abstracts*, 10–14 December 2018, Washington D.C. Walter E Washington Convention Center.

Morphological signature of gully development by rapid slide retrogression in a layered coarse-grained delta foreslope

Jacques Locat[1]*, Ali Azizian[2], Jim Stronach[2], Aurélien Hospital[2], Chris Young[2], Dominique Turmel[1] and Andrew Bevan[3]

[1]Laboratoire d'études sur les risques naturels (LERN), Department of Geology and Geological Engineering, Université Laval, Pavillon Adrien-Pouliot 1065, Avenue de la Médecine, Québec, QC G1V 0A6, Canada

[2]Air, Coastal and Lake Engineering Department, Tetra Tech Canada, Vancouver, BC V6C 1N5, Canada

[3]500 Queen Street, Brisbane, Queensland, 4001, Australia

AH, 0000-0001-5367-5405
*Correspondence: jacques.locat@ggl.ulaval.ca

Abstract: Coarse-grained deltas are often characterized by steep foreslopes (often more than 10°) that are traversed by delta-front channels. The channels thus erode into relatively steeply inclined bedding. In this context, the slopes flanking the channels can be steeper than the friction angle since they include a component of dip related to the delta-front slope as well as the channel-related erosion slope. In this study, part of the Busu River delta (Papua New Guinea) was imaged using a high-resolution multibeam bathymetry survey over an area where the angle of the slopes flanking the channels locally reaches 50°. A detailed analysis of the delta slope morphology has revealed an additional source of instability due to erosion within the main channels. In some places, erosion cuts into the channel flank forming a local knickpoint inclined in a direction approaching that of the bed dip. The cut can then initiate breaching or static liquefaction failure from that point up to the crest of the interfluve resulting in a V-shaped gully.

Knowledge concerning channel and gully system development on submarine slopes has been developed over many years (e.g. Rebesco *et al.* 2014; Amblas *et al.* 2018). Submarine erosion networks can have a range of morphologies and complexities. Often they include canyons fed by channels that connect upslope with several orders of gully. Shumaker *et al.* (2017) considered gullies as relatively straight, regularly spaced erosional features that are distinguished from submarine channels by their width/depth ratio: ranging from 3.5 to 5 for gullies, and from 17 to 29.5 for channels. In this study, submarine gullies are distinguished following the definition of Amblas *et al.* (2018) and Clare *et al.* (2018) as small-scale linear erosional features with less than 10 m of erosional relief and less than 200 m long. As will be seen below, the gullies studied here fall within the range of aspect ratios reported by Shumaker *et al.* (2017).

It is recognized that channel and canyon development may involve significant retrogressive erosion (Baztan *et al.* 2005) by landsliding, and that some channels can be formed by upslope migration of knickpoints governed by both hydromechanical and geotechnical aspects (Turmel *et al.* 2011, 2015). Channels can also form by top-down processes, as discussed by Mauffrey *et al.* (2017).

Gully formation has been linked to various mechanisms and their origin remains poorly understood. Micallef and Mountjoy (2011) studied examples from the Cook Strait, New Zealand, and concluded that, given the location of the gullies near the crest of the slope, these were likely to have been formed by unconfined fluid or gravity flows accelerating downslope. Vachtman *et al.* (2013) studied upper slope gullies along the central USA Atlantic continental margin and also considered that they may have been formed by sedimentary flows spilling over the shelf edge. They also considered that gullies lower on the slope were mostly formed by retrograde erosion. Retrograde erosion has also been invoked by Nemec (1990) and Klaucke and Cochonat (1999). Sliding mechanisms (excavation by slumps and slides) have also been considered for gully formation by Pratson *et al.* (2007) and Gales *et al.* (2013). According to Amblas *et al.* (2018), these retrogressive slides could have been spontaneous.

The study reported here focuses on mass-wasting processes taking place in gullies on the coarse-grained Busu River delta front (Fig. 1), Papua New

From: Georgiopoulou, A., Amy, L. A., Benetti, S., Chaytor, J. D., Clare, M. A., Gamboa, D., Haughton, P. D. W., Moernaut, J. and Mountjoy, J. J. (eds) 2020. *Subaqueous Mass Movements and their Consequences: Advances in Process Understanding, Monitoring and Hazard Assessments*. Geological Society, London, Special Publications, **500**, 219–234.
First published online March 13, 2020, https://doi.org/10.1144/SP500-2019-159

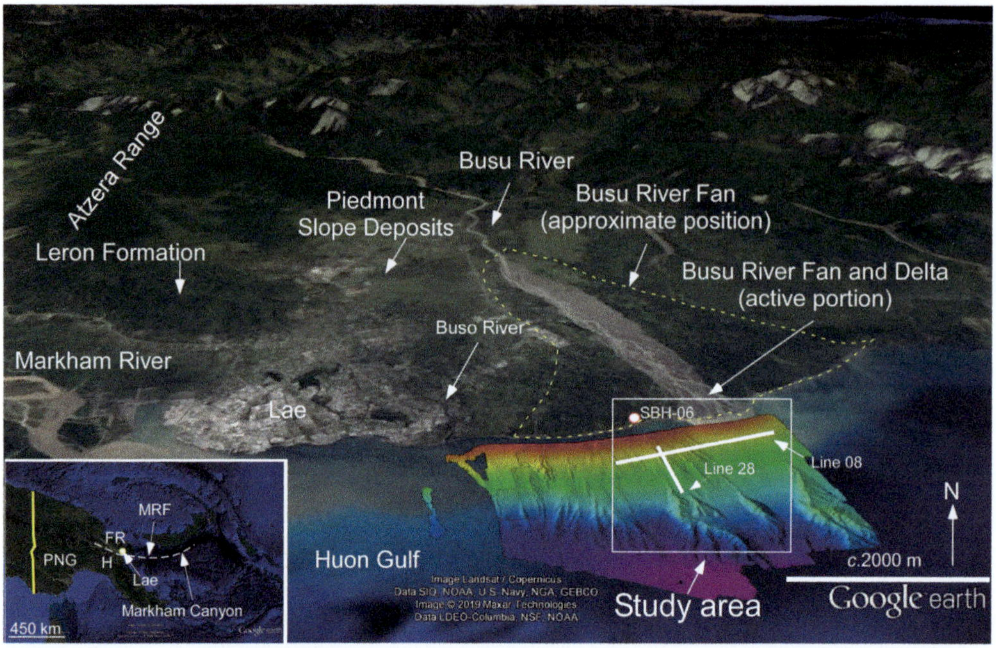

Fig. 1. The location of the study area on the Busu River delta front in Papua New Guinea (PNG) several kilometres to the east of the town of Lae. The location of borehole SBH-06, and seismic lines 08 and 28 referred to in the text are shown. The dotted line delineates the approximate location of the Busu River fan and delta. (images are from Google Earth with data from Maxar Technologies and Terrametrics). The Markham–Ramu Fault (MRF), the Finistere Range (FR), the New Guinea Highlands (H) and Markham Canyon are labelled on the inset image (bottom left).

Guinea. It was made possible by the availability of a very-high-resolution (0.5 m) multibeam survey of part of the Busu River delta using an autonomous underwater vehicle (AUV: Fig. 2b). This survey was carried out as part of a larger study related to sediment dynamics in the Huon Gulf. A sector close to the active delta was recognized as having fresh morphological evidence of channel erosion (channels 1–3 in Fig. 2b). The channels are flanked by slopes on which gullies are well expressed. One of these channels (channel 2) is active or recently active and may provide insight into the mechanism of gully development. The hypothesis is that local erosion at the margin of the channels may trigger breaching or static liquefaction of the slope, resulting in the rapid formation of the gully up to the crest of the interfluve (Berg *et al.* 2002). To that effect, this paper will examine the potential role of the inclined delta-front bedding on gully initiation along the margins of submarine channels.

Methodology

Fieldwork off the Busu River delta, on which this study builds, involved coring and cone penetration tests with pore pressure measurements (CPTu), together with seismic and multibeam surveys.

CPTu and PC coring

Samples were obtained using a 3 m Kullenberg piston corer (PC) with either a 3 m or 6 m barrel length. Cone penetration tests with pore pressure measurements (CPTu) were carried out using a dual-drive 50 kN unit with a 6 m mast. Piston cores were acquired within 5 m of the CPTu site. Data from a sonic borehole drilled near the beach were also used in this study (SBH-06: see Fig. 1).

Seismic survey

The seismic survey, carried out in 2017, consisted of a 2D ultrahigh-resolution (2DUHR) seismic profiler using a 10 cubic inch mini-sleeve gun as the energy source and a Teledyne 48 channel hydrophone streamer, a survey line spacing of about 200 m with a survey speed of 4 knots. Survey lines were run along and across the depth contour lines. Data were recorded in two-way travel time v. shot point number. The distance between shot

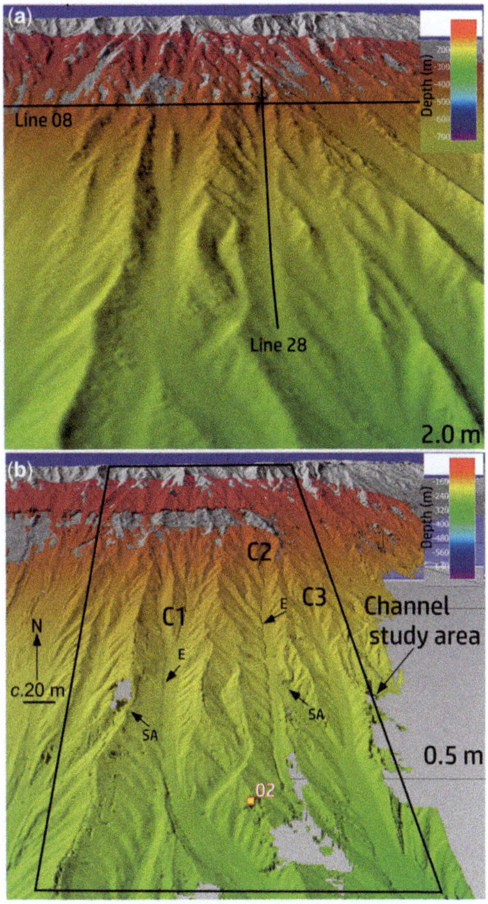

Fig. 2. Effect of the survey method on image resolution: (**a**) hull-mounted multibeam (2.0 m) v. (**b**) AUV-mounted multibeam system (0.5 m) running about 20–40 m above the seafloor. Water depth in the image varies from 100 to 250 m. In both cases, illumination is from the left (azimuth at 269° and sun angle at 68°). In (b) the channel study area is shown along with the location of channels 1, 2 and 3. The grey bathymetry at the top of the figure is not considered in the comparison of the resolution. E, an indication of local erosion; SA, survey artefact; 02, the location of sample 02.

points was about 6.5 m, and the speed of sound was taken as 1540 m s^{-1} in water and 1700 m s^{-1} in sediments.

Quality checks included the inspection of shot, near trace and auxiliary gathers to confirm that the guns and streamer were functioning correctly. Noise levels were monitored to determine the effect of sea state on data quality. The data were stacked to verify that the equipment configuration was suitable for the local geological conditions and to facilitate an initial geological interpretation.

Bathymetry surveys

The bathymetry of the study area was obtained using various multibeam echosounder (MBES) surveys that were undertaken independently. Three different techniques were employed: a nearshore hull-mounted survey (2012), an AUV survey (2016) and a hull-mounted survey simultaneously with the 2DUHR seismic survey (2017).

The 2012 shoreline survey from depths of 10–40 m was acquired using a hull-mounted SVS3 240 kHz multibeam system at a speed of 3.5–4 knots. The results were used to generate a 0.4 m gridded map with a 3.4 bin hits/m^2. A Kongsberg Munin® AUV was used in 2016 and equipped with a Kongsberg EM 2040 200–400 kHz MBES with a 1° × 1° resolution and a 120°–140° swath depth depending on frequency. The AUV was flown to about 20–40 m above the seafloor at a speed of 2–4.5 knots. Most of the AUV lines were collected in a direction perpendicular to the contour lines and at a frequency of 400 kHz to produce a map with a more or less constant 0.5 m resolution. Unless specified otherwise, all bathymetry profiles presented here were done using the 2016 AUV bathymetry survey.

The 2017 swath bathymetry was acquired using a hull-mounted Kongsberg EM 2040 shallow-water MBES to *c.* 150 m water depth and a hull-mounted Kongsberg EM 302 deep-water MBES where the water was >150 m deep. The EM 2040 was operated at a frequency of between 200 and 300 kHz. The EM 302 used at a frequency of 30 kHz. This set of bathymetry data were acquired at the same time as the 2DUHR seismic survey.

Figure 2 presents a comparison between the 2017 survey gridded at 2.0 m and the 2016 survey gridded at 0.5 m to illustrate the much higher resolution of the AUV survey, although there may be more artefacts due to the survey technique. All the multibeam images were analysed using Fledermaus software.

Regional settings

The Basu delta lies within the Huon Gulf area of Papua New Guinea (PNG: see Fig. 1), an area of complex geology and a tectonically active region with at least 25 M6 earthquakes occurring between 1900 and 2016 within 100 km of the study site (source: USGS). A review of various aspects of the Huon Gulf geology and geohazards can be found in Dow (1977: geology), Chappell and Shackleton (1986: sea-level changes), Liu *et al.* (1995: deltas in Huon Gulf), Buleka *et al.* (1999: geohazards) and Baldwin *et al.* (2012: tectonics).

The Busu River is one of the many rivers discharging sediment to the ocean on the north side of the Huon Gulf, with much of the sediment ultimately passing downslope into the Markham Canyon (Liu et al. 1995). The Busu River delta is located just east of the town of Lae. The study site is located north of the active Markham–Ramu Fault (MRF), which is driving uplift of the region north of the fault at rates of 2–7 mm a^{-1}. These high rates are likely to be responsible for the active mass-wasting processes in the region (Buleka *et al.* 1999).

Onshore, the morphology of the area consists of a low-relief hill flanked by two major rivers: the Markham River to the west and the Busu River to the east (Fig. 1). The main topography consists of the Atzera Range, a low-relief feature (<400 m) mostly made up of the Leron Formation. According to Dow (1977), the Leron Formation was deposited in a narrow arm of the sea between the rising New Guinea Highlands to the south and the Finisterre Range to the north, both separated by the Markham–Ramu Fault Zone. It consists of poorly indurated, mostly well-bedded sandstone, pebbly sandstone and conglomerate, which are about 1.8 myr old (Crook 1989).

The relatively flat alluvial plains and deltas were formed by the emplacement of the Piedmont slope deposits that overlie the older Leron Formation. Both formations consist of similar sediments but the Leron Formation is weakly cemented (Dow 1977). The Busu River erodes into the Leron Formation and also partly into the Piedmont slope deposits to form the modern Busu River fan delta (Fig. 1). Further offshore, large ridges, lower on the slope seaward of the Busu River delta (Fig. 1), also consist of the Leron Formation. These ridges are also intensively eroded by a complex system of gullies (but these are not considered in this study). From seismic lines obtained over these submarine ridges, the internal bedding appears more deformed closer to the Markham canyon.

The Busu River is a fast flowing river with a peak discharge of up to 25 000 m^3 s^{-1} which can carry large amounts of sediments so that its course can evolve rapidly with time (Buleka *et al.* 1999; Sekac and Jana 2013). The Busu River has a

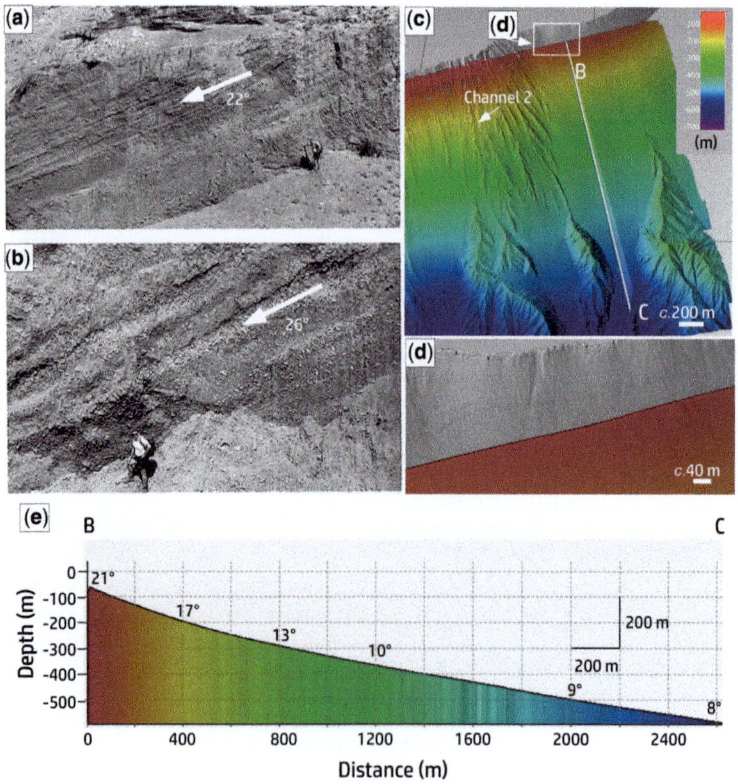

Fig. 3. The Busu River is a coarse-grained delta: (**a**, **b**) are adapted from Nemec (1990: licence No. 4691980646407) and illustrate steeply-dipping layers in coarse-grained delta-front deposits. (**c**) Morphology of the active portion of the Busu River delta using the 2017 survey with an almost featureless surface shown in (**d**) with a blow up of the 2012 shoreline survey in grey, and in (**e**) a regularly decreasing slope gradient from 21° at a depth of 60 m to 8° at a depth of about 600 m. The high-relief ridges on the lower part of the slope in (c) are the outcrop on the seafloor of the Leron Formation.

catchment area of 1311 km^2 (Samanta 2015) and the subaerial fan has an area of about 60 km^2. The alluvial plain transporting the sediments to the delta front extends over a distance of about 12 km with an average slope of 0.46° (Liu *et al.* 1995). The tidal range in this area varies from 0.4 to 1.0 m (Renagi *et al.* 2013). The actual active part of the fan delta is located almost in the middle of the fan and is about 3 km wide (Fig. 1). Sediments reaching the coastline accumulate on the delta front or are transported down into the Markham Canyon at a depth of about 700 m, or even deeper into the Salomon Trench at a depth of 5000 m.

The Busu River delta

Upon reaching the shoreline, the sediments enter the Huon Gulf to form a steep and featureless (Fig. 3c, d, e) delta front sloping from 25° near the shoreline to less than 8° at a water depth of 550 m. Such a foreslope angle is likely to reflect the angle of deposition of the coarse-grained sediments, as shown by Nemec (1990) (Fig. 3a, b), and is similar to the Type 1 delta proposed by Gales *et al.* (2013). A typical sample (Fig. 4) obtained at a depth of 138 m in borehole SBH-06 located just east of the active delta (Fig. 1) illustrates the sandy and gravelly nature of the sediments (also shown in Fig. 3a, b). It is believed that this is typical of what has been deposited over time by the Busu River. Occasionally, significant pulses of sediment input come from large landslides within the watershed generated by heavy rain or earthquakes (Buleka *et al.* 1999).

Seismic line 28 (Fig. 5a) shows the overall attitude of the beds, which are dipping at a similar angle to that of the delta front (Fig. 5b) while appearing subhorizontal in a strike section along the 240 m contour line, as shown on seismic line 08 in Figure 6a. The attitude of the bedding is similar to what Patruno and Helland-Hansen (2018) call a delta-scale clinoform. There is no evidence of major landslides in this part of the delta front. The seismic reflector highlighted in pink in Figures 5a and 6a is interpreted to be the contact between the underlying Leron Formation and the overlying more recent deltaic sediments. The green line in Figure 5a represents the bathymetry from the synchronous multibeam survey. The bathymetry across seismic line 08 is shown in Figure 6b to illustrate the morphology of the active part of the delta compared to that of the abandoned portion where channels have been eroded into recently deposited deltaic sediments. This is also evidenced by the morphology shown in seismic line 08 with channels cut into the deltaic sediments (Fig. 6a).

Channels

Except for the active portion of the Busu River delta, all along the coastline a series of sub-parallel channels (active and inactive) were carved into the deltaic sediments in a direction likely to be similar to that of the dip of the delta front and its clinoforms (Figs 1–3). The orientation of the long axis of the channels down to a depth of 500 m is more or less perpendicular to the coastline, and the relative height of the ridges between the channels within the study area, with respect to the floor of the channel, varies between 5 and 15 m (Figs 6 & 7d). Three channels at a water depth of less than 250 m were selected for the present study, with most of the analysis done around channel 2 (Figs 2 & 7–11). This channel appears to have been more recently active than the other two, in that there is little evidence of sediment draping the channel-floor morphology. There are no sand waves visible in any of the channels.

The slope of the channels, as seen on the slope map (Fig. 7b), are quite regular and vary between 15° and 16° in the portion investigated (see the slope of channel 2 in Fig. 7c). The upper parts of the channels are shown in Figure 8 where the coastal multibeam survey has been added to the hig- resolution multibeam survey (HRMS) to provide a detailed view of the morphology between water depths of 10 and 50 m (Fig. 8). Figure 8b presents the backscatter image of the HRMS for the same area as in Figure 8a.

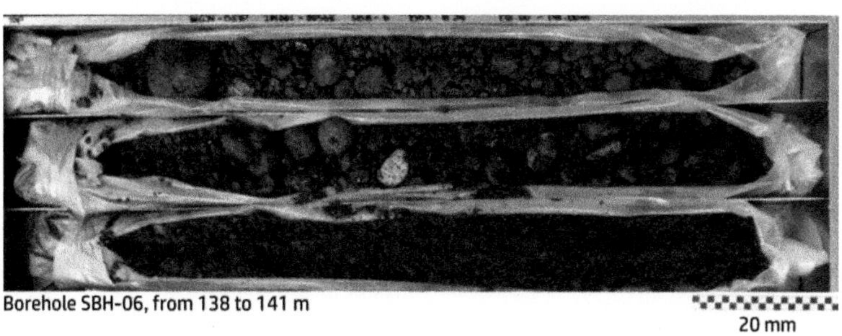

Fig. 4. Example of sandy gravel with well-rounded clasts typical of the of the Piedmont slope deposits forming the Busu River delta, and taken from a shoreline borehole (SBH-06: see Fig. 1) at depths of between 138 and 141 m.

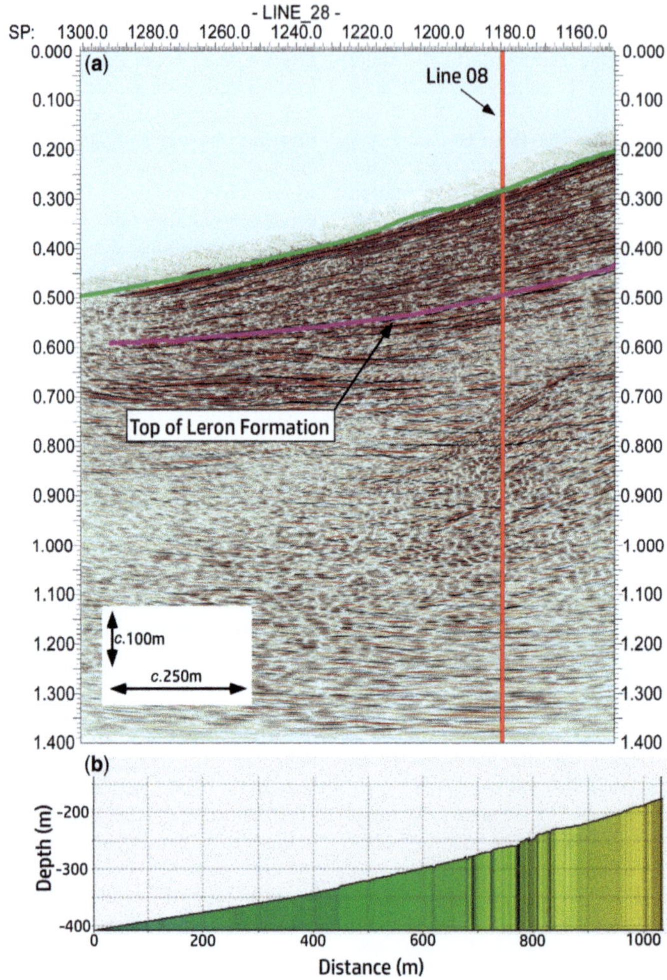

Fig. 5. (**a**) Seismic line 28 illustrating the dipping (12°–15°) delta-front clinoforms in recent sediments overlying the deformed Leron Formation. For the location, see Figures 1 and 2. Line 28 has been acquired approximately along the axis of channel 2 (Fig. 2a). The green line at the surface represents the bathymetry from the 2017 multibeam survey. The seismic profile and multibeam bathymetry show a fairly good overlap (airgun seismic line provided and interpreted by Neptune Marine Services; one shot point (sp) is equal to a horizontal distance of about 6.5 m). (**b**) The slope profile corresponding to the same profile as shown in (a).

The location of the head of channels 1 and 2 is indicated on this figure, while that of channel 3 lies at a greater depth (Fig. 7). For the backscatter image, the white tone represents coarse material and the darker tone represents finer material. Channel 1 may have a weak connection to the nearshore area, while channel 2 appears to have been cut off by the development of another channel (C in Fig. 8). The timing of the formation of channel C is not known but considering the highly dynamic environment, major sediment inputs and frequent earthquakes, changes could take place rapidly. If one compares the freshness of the gully system flanking channel 2 relative to channels 1 and 3, it appears that channel 2 has accumulated little fine-grained sediments since its closure by channel C.

An attempt was made to make a detailed comparison between the 2012 and 2016 bathymetry surveys but there were large uncertainties resulting from the differences in resolution of each survey and the survey methods. This was particularly an issue in high-relief areas, such as the study area. Thus, a detailed comparison could be done once similar survey methods were used. Nonetheless, a comparison of the two surveys (Fig. 8) shows that over the 4 years, the position of the channels have remained stable.

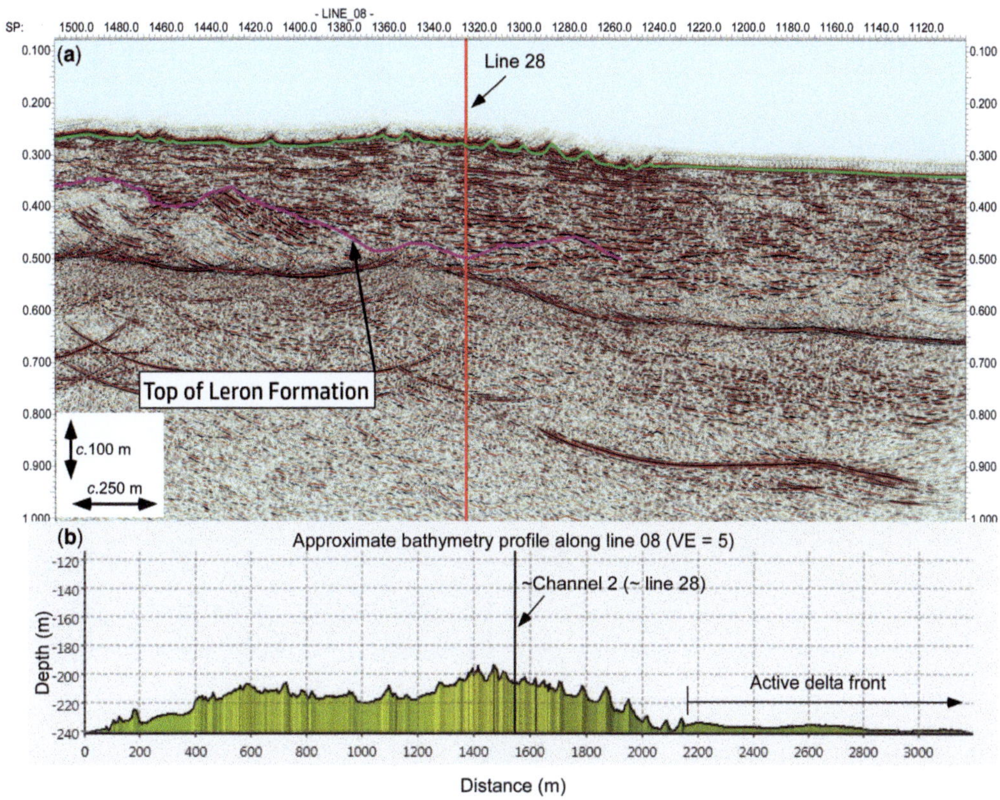

Fig. 6. (**a**) Seismic line 08 illustrating the subhorizontal layering of the sediments parallel to the slope in about 200 m water depth showing the Busu River delta sediments overlying the Leron Formation. See Figures 1 and 2a for the location. Note that the seismic line extends from the eroded (left) to the active (right) sectors of the Basu delta front. The green line at the surface represents the bathymetry from the multibeam survey. The seismic profile and multibeam bathymetry show a fairly good overlap (airgun seismic line provided and interpreted by Neptune Marine Services; one shot point (sp) is equal to a horizontal distance of about 6.5 m). (**b**) The across-slope profile along approximately the same profile as shown in (a).

Channel 1 has three major branches in its upper reaches whereas channels 2 and 3 each have two major branches (Fig. 7a). Channels in this part of the delta have widths of up to 10 m and erosional margins dipping between 25° and 30°. There are indications of recent erosion in channels 1 and 2, mostly along the west flanks of both channels (e.g. Figs 9 & 10). In some places, mostly along the western side of channel 2, the base of the slope descending onto the channel floor is characterized by several small erosional scarps, about 2 m high, with local slope angles reaching up to 54° (Figs 9 & 10).

Selected sites with CPTu and piston coring (PC) data are used to describe the nature of the seafloor in parts of channels 1 and 2. For these sites, CPTu and PC results are summarized in Table 1. Test hole locations are shown in Figures 2, 7 and 8. All samples collected on the floor of channel 2 consist of sandy gravel (samples 03, 05 and 09: Fig. 7b). Samples obtained on the lateral slopes above the channels typically contain fined-grained sediments up to 0.7 m in thickness resting on sand and gravel (sample 02 in Fig. 2b, and sample 06 in Fig. 8). The fine-grained sediment is attributed to hemipelagic sedimentation that is likely to have formed from hypopycnal plume sedimentation visible at the outlet of the Markham and Busu rivers (Fig. 1).

Gullies

Gullies related to channels 1–3 are identified in Figure 7a and, except for one gully in channel 2, there is only one order of branching here, indicating that the V-shaped gully system would correspond to the Type I system proposed by Vachtman *et al.* (2013). The gully thalwegs are mostly linear and their orientation is oblique to the main channel to

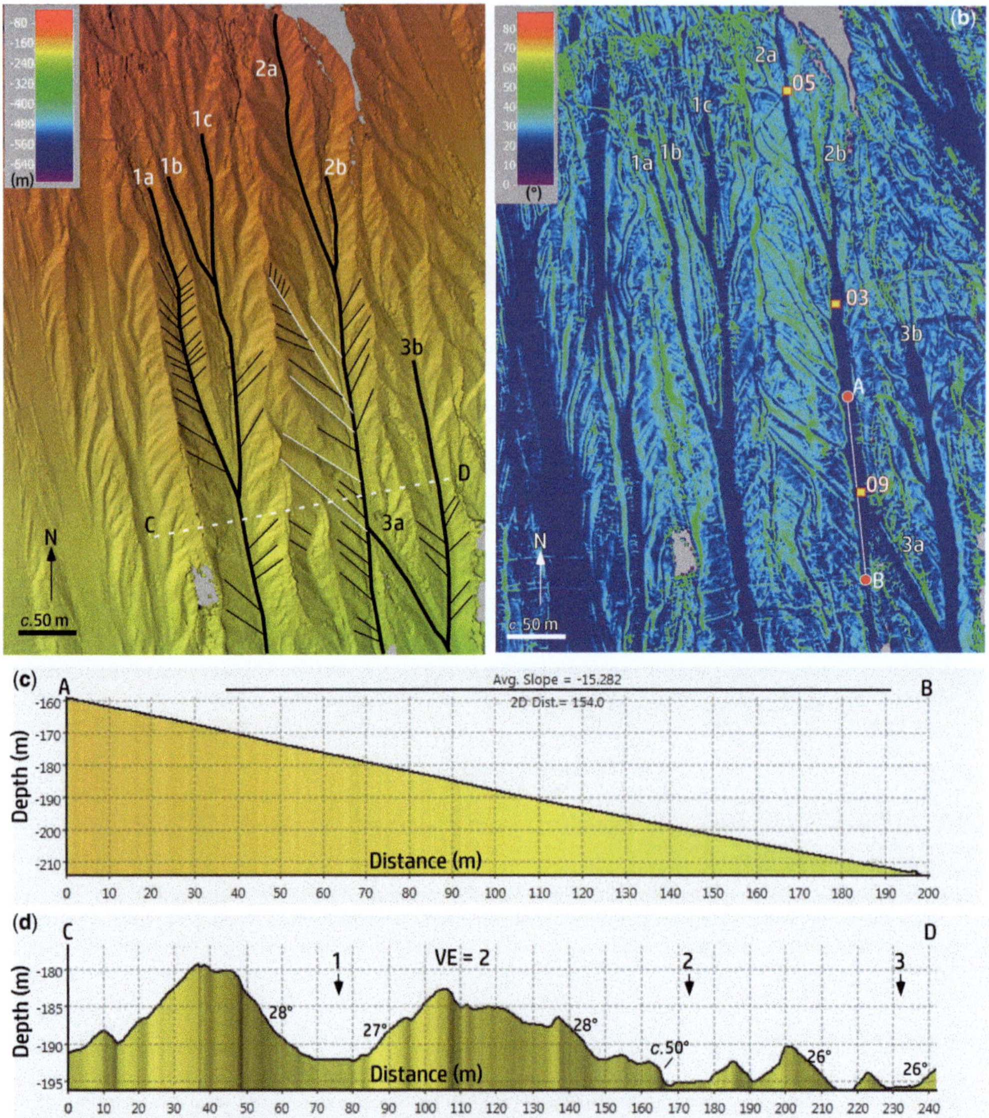

Fig. 7. Morphology of the inactive Basu delta front showing the relationship between channels and gullies. (**a**) Gully network system where the gullies (highlighted by fine black lines) have been extended to the thalweg of the channels (bold black lines). The white lines are the five gullies described in more detail in the text. (**b**) Slope map of the same area showing the very regular gradient of the channels, with angles varying between 15° and 16° as shown in (**c**) for the lower part of channel 2. (**d**) Cross-section along line C–D located in (a) showing the various channels in cross-section and the angles of the gullied slopes lateral to the channels.

which they connect, with intersection angles varying between 20° and 55°. Linear gullies can be expected because of the steepness of the slope (Clare *et al.* 2018). For channels 1, 2 and 3, the interfluves of the gullies are rounded, except for those along the west side of channel 2 which appear sharp (Figs 2b & 11).

Gullies along channel 2

For the purpose of this study, gullies associated with channel 2 are analysed in more detail in order to assess how the gullies may have formed (Figs 9–11). Erosion features, seen mostly along the western flank of channel 2, locally indent the channel margin,

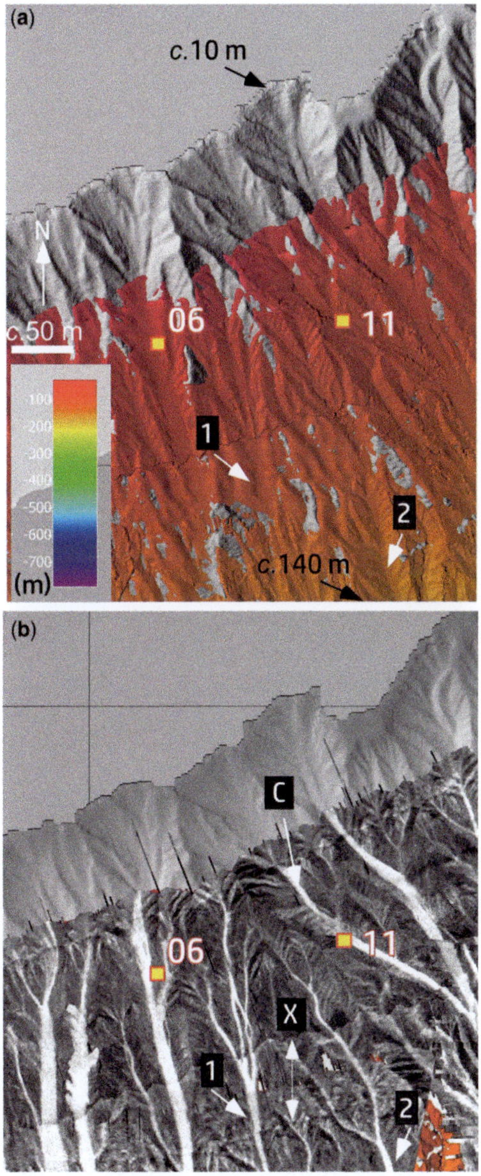

Fig. 8. (**a**, **b**) Path of channels 1 and 2 on the upper part of the delta foreslope. In (b), the backscatter data collected at the time of the AUV shows that channel 1 is still connected to the nearshore zone, while channel 2 has been cut off by the incision of channel C. The area in X is noisy. The bathymetry data in grey are from the 2012 nearshore bathymetry survey (10–40 m).

as shown in Figures 9 and 11. These notches are seen at the entrance of each gully and extend more or less from the flank of the channel to the thalweg of the gully, with lengths varying between 3 and 6 m. These erosion features are considered to be knickpoints.

Just upstream of the gully mouth, the channel flank can been seen to display small scarps about 2 m high with a local steep slope angle of up to *c.* 50° (see Fig. 10). These scarps are small and close to the limit of the resolution of the dataset but they are commonly seen along channel 2. However, caution is warranted concerning the slope angles in Figure 10 since they have been computed using only a few points. Each of the profiles shows that there may be shallow scouring at the toe of the scarps, as seen by a small depression relative to the floor of the channel.

To extend the analysis, five gullies were selected along the western flank of channel 2 (Figs 9–11). The associated interfluves are mostly sharp with an incision depth ranging from 3 to 7 m and small toe scarps (knickpoints) at the gully mouths with heights of less than 1 m. The gulley surfaces are smooth, suggesting that they are fully developed. Their thalwegs make an angle varying from 17° to 23° with the channel they connect to, except for gully 1 which presents an S-shape planform that is seen occasionally in this area (Fig. 11a). For gullies 2–5, the thalweg length varies from 58 to 115 m, and the slope angle perpendicular to the thalweg direction is between 30° and 40°; the slope facing towards the NE is always steeper by a few degrees, reaching nearly 40° in places (Fig. 7b). The width/depth ratio of gullies 2–5 varies from 3.2 to 3.8, which is within the range of aspect ratios reported by Shumaker *et al.* (2017) for gullies elsewhere.

The excavated volumes of channels 2–5 were estimated assuming that the dip of the slope in a direction perpendicular to the thalweg is the same on both sides, at 30°, and using a single cross-section as shown by the black dashed lines in Figure 9. Using this simplified wedge shape and the length of the thalweg for these gullies, the excavated volumes were estimated to be 4788, 1322, 7066 and 1949 m^3 for gullies 2, 3, 4 and 5, respectively. In addition, there is no evidence of sediment accumulation in the channels that could come from the gullies: that is, the sediments removed from the gullies had enough mobility to be transported to greater depths. Channel 4 has a much greater volume due to its wider south-facing slope (Fig. 9).

The knickpoints that may have triggered the development of the five gullies are detailed in Figure 11. This figure shows the orientation of the remnants of the five knickpoints for the selected gullies with respect to the axis of channel 2 (Fig. 11b–f). The angle between the knickpoint scarp and the axis of the adjacent channel varies from 28° to 63°. The line following the knickpoint scarp has also been extended to the gully thalweg (dashed white line in Fig. 11) where they meet close to 90° to each other.

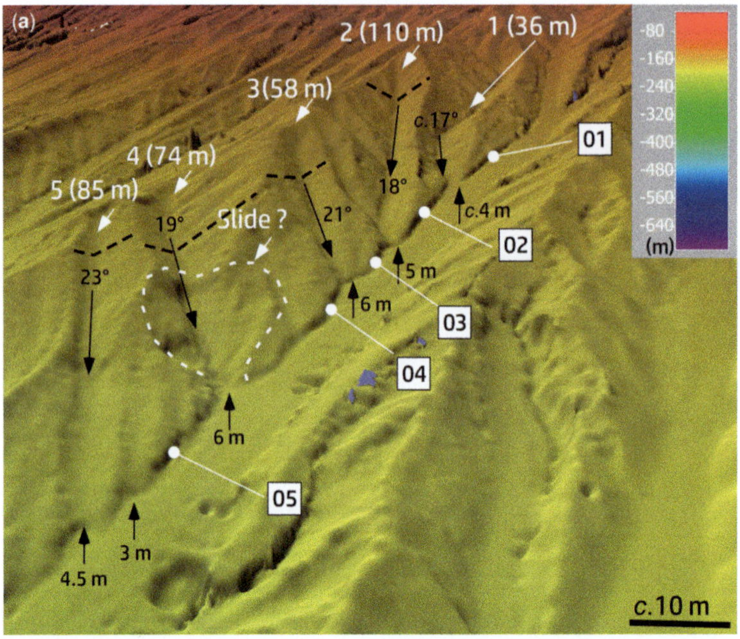

Fig. 9. Morphometric analysis of five gullies along channel 2 showing their inclinations and lengths, and also highlighting the position and height of the knickpoints located at the toe of the gullies (see the text for more detail). Dashed black lines are approximate positions of cross-sections used for gully volume computation. The numbers in boxes refer to profiles in Figure 10.

From the geometry of the various lines shown in Figure 11, an estimation of the relative orientation of the various scarp slopes relative to the delta-front bedding is possible and is presented in Figure 12. It is considered here that the bedding plane dip direction can be estimated from the depth contour lines within the study area, with dip directions ranging from 150° to 170° (Fig. 12). The channel 2 slope direction, taken from the contour line orientation, is towards 168°. The dip direction of the western slope flanking channel 2 is towards 76°, which is nearly at 90° to the axis of the channel and the bedding dip direction. The range of dip directions for the knickpoint remnants at the toe of the gullies (shown in Fig. 11) is also provided so they can be compared to the delta-front bedding dip directions.

Interpretation

Before proposing an interpretation of a potential process for gully formation on a coarse-grained delta slope, the above morphological analysis enables the formulation of some observations or considerations.

- Deltaic sediments were deposited in such a way to form layers of sediment that are dipping at more or less the same angle as the modern delta slope (i.e. at angles of between 15° and 22°).
- Channels are orientated more or less in the same orientation as the dip of the delta-front clinoforms, so that the slopes flanking the channels generally dip in a direction nearly perpendicular to that of the bedding. For channel 2, the dip direction of the bedding is towards 168° (Fig. 11a).
- The notches seen at the foot of the gullied slope connected to the western margin of channel 2 are the result of channel lateral erosion forming a local knickpoint.
- Knickpoints scarps for gullies 1–5 (Fig. 11) are nearly perpendicular to the gully thalweg.
- Steep small scarp slopes (< 2 m) are mostly seen at the toe of the western flank of channel 2 (Fig. 10), and these slopes are dipping at about 50°.
- The erosion process generating the gullies reached the crest of the interchannel slope leaving a smooth surface, suggesting that they are fully developed.
- There are no traces of sediments originating from the erosion of the gullies on the floor of channel 2, and sand waves are absent in the channels.

Busu River sediments mostly consist of rounded sandy gravel (Fig. 4), and in these conditions the slope angle of the stratification could reach values of 35° (Nemec 1990). However, on the Busu River

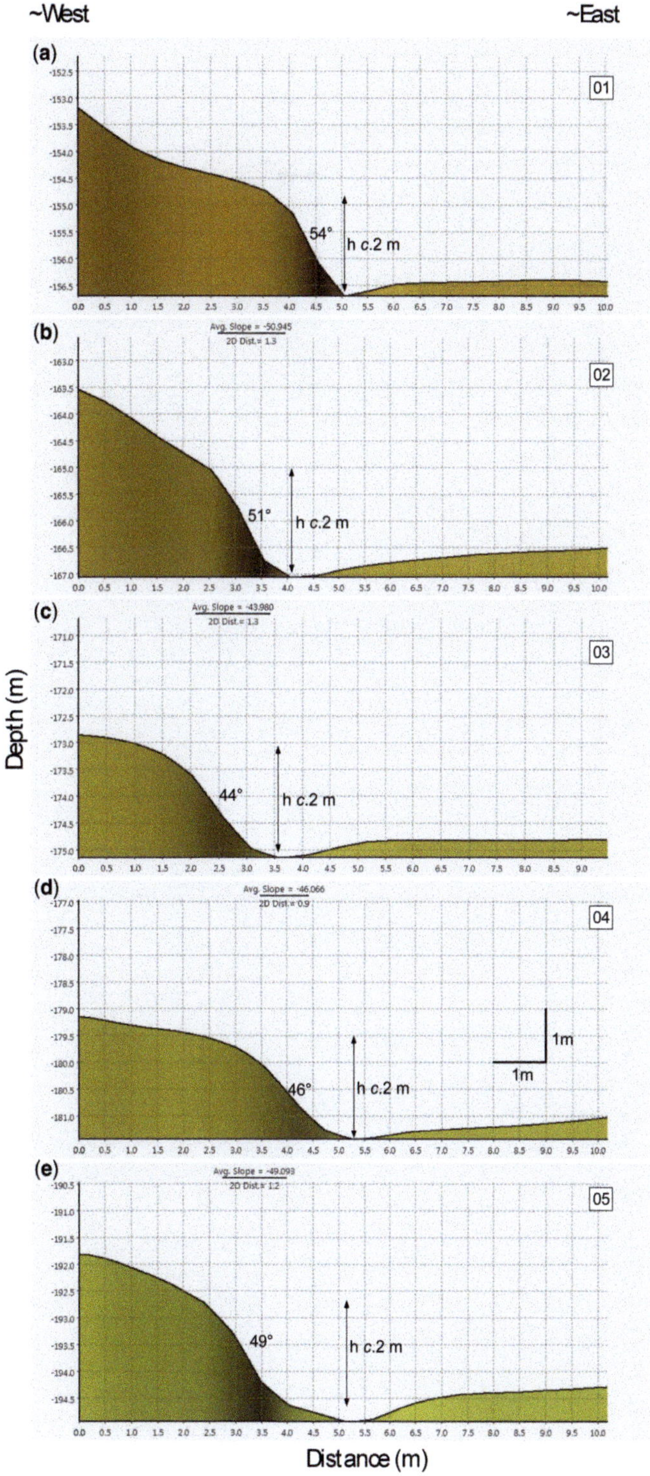

Fig. 10. Selected profiles, from west to east, along the western margin of channel 2 annotated with maximum slope angles (see Fig. 9 for the locations).

Table 1. *CPT and PC sample descriptions at selected sites*

Site no.	CPT		PC	
	Latitude, longitude *z* (m)	Depth (m): description	Latitude, longitude *z* (m)	Depth (m): description
02	9254849.0° N, 504962.0° E 228.4	0.0–1.24: gravel with traces of clay and sand	9254847.4° N, 504958.6° E 245.5	0.0–0.20: clay, very soft, likely to be stopped by an underlying gravel
03	9255175.4° N, 504921.7° E 143.0	0.0–0.74: clay, with soft sand 0.74–1.64: gravel	9255172.8° N, 504930.3° E 143.0	0.0–0.55: clay, very soft 0.55–1.68: gravel
05	9255304.4° N, 504869.5° E 106.0	0.0–0.44: clay, very soft 0.44–1.04: gravel	9255299.5° N, 504871.0° E 103.9	0.0–0.10: gravel and traces of sand and clay, the gravel is angular to subrounded
06	9255456.5° N, 504701.3° E 45.9	0.0–0.67: clay, very soft with gravel 0.67–1.30: gravel	9255457.6° N, 504705.3° E 40.6	0.0–0.50: clay, very soft with traces of gravel (the core head was damaged)
09	9255016.9° N, 504949.5° E 192.1	0.0–0.47: clay, with sand lenses 0.47–1.46: gravel	9255019.6° N, 504952.7° E 195.6	0.0–0.20: sand fine or coarse with clay 0.20–0.90: gravel
11	9255479.0° N, 504840.3° E 53.2	0.0–0.23: gravel 0.25–0.80: clay 0.80–1.76: gravel with a 4 cm clay layer	9255478.7° N, 504839.4° E 53.5	0.0–0.6: gravel

Depth z is in metres below seafloor.

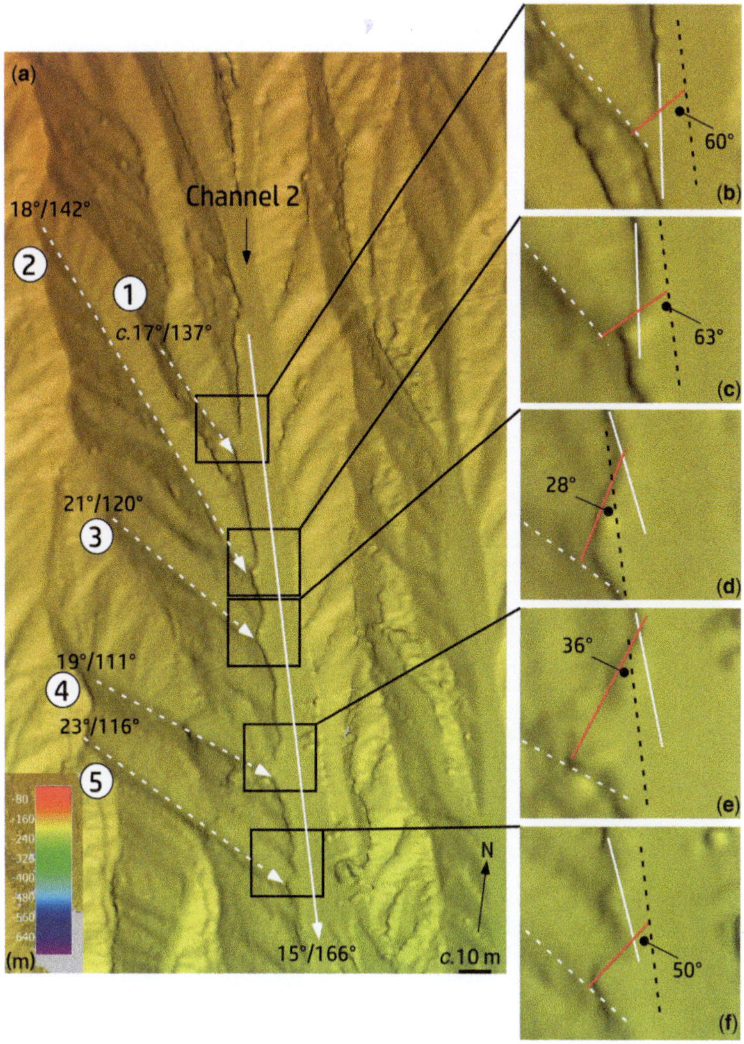

Fig. 11. Analysis of knickpoints along part of the western margin of channel 2. In (**a**) the dashed white lines represent the positions of the thalweg of the gullies, and the values in degrees are the dip and dip direction of the thalweg. The white solid line is the axis of channel 2 and represents the estimated dip direction of the bedding planes. (**b**)–(**f**) present an enlargement of the knickpoints (red lines show their strike) to illustrate the approximate difference in strike orientation (in degrees) relative to that of the general channel margin orientation (dashed black line). The white lines in (**b**)–(**f**) represent the extension of the local direction of the margin of channel 2. The dashed black lines represent the orientation of the channel axis shown in (a). Note the orientation of the north (N) in (a).

delta, the maximum values seen on the active delta front are about 25°. This is below the lower limit of the friction angle of a sandy gravel mixture, which can vary from 32° to 55° depending on the amount of gravel (Simoni and Houlsby 2006) and relative density (compaction). Even though the dip of the stratification is quite high, it was formed by deposition, not by erosion (Locat 2018). The sedimentation process results in the accumulation of layers with different grain sizes, which, by their variability, can act as a preconditioning factor that could generate a weak layer, particularly under seismic loading (Locat *et al.* 2014).

Slopes flanking the channels and in gullies are considered to be eroded slopes with slope angles reaching values of up to 38°, which could reflect the average friction angle of the Busu River delta sediments (Locat 2018).

From the above, sediments forming the delta can be considered to dip in a direction similar to that of

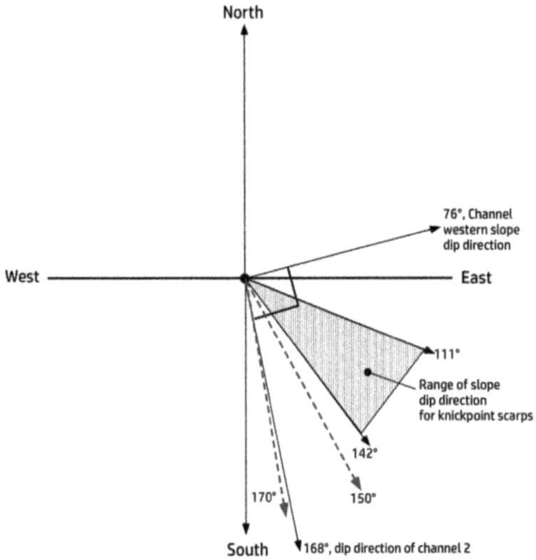

Fig. 12. Summary of dip orientations for channel 2 and its flanking gullied slope. The western slope dip direction is taken at 90° to the direction of the western margin of channel 2, although this may include a component of the delta-front slope gradient. Dashed red arrows represent the range of bedding dip directions estimated from the modern slope orientation within the study area, and vary between 150° and 170°. The dip direction of the floor of channel 2 is towards 168°. Also shown here is the range of knickpoint scarp dip directions (ranging from 111° to 142°) taken from Figure 11. See the text for discussion.

channel 2 (i.e. towards 168°). This is a simplification made for comparison since, in the study area, channel azimuths vary from 138° to 170° (Fig. 12). Any shift in the position of the active delta may also result in a different bedding orientation. The dip of the bedding planes is also considered to be close to the slope of the channel in this part of the foreslope (i.e. *c.* 15°). The sediment mixture is considered to have an average friction angle of 38°.

An important assumption introduced here is that the layering developed by the sedimentation process on the delta foreslope creates bedding planes with dip directions similar to the dip direction of the delta foreslope. The presence of such bedding planes creates a strength anisotropy (Fu and Dafalias 2011), so that eroded scarps with a slope in the same direction as the bedding would be less stable than those at 90° to it (e.g. along the channel flank). These conditions would apply on either side of the channel.

Local deviation by erosion into the western flank of channel 2 initiates the formation of a knickpoint. As the knickpoint expands into the western flank, the dip direction of the knickpoint wall gets closer to that of the dip direction of the bedding, which is assumed to be similar to that of the main channel axis (Fig. 11). It is not clear what could initiate the notch but one could consider that as the dip direction of the bedding plane may vary along the flank of the channel, there could be some parts of the flank that are more prone to erosion than others, thus favouring the initiation of a notch to form a local knickpoint.

Such a scenario is illustrated in Figure 12 and can be explained as follows. The erosion of the channel occurs in the same direction as the bedding, so the slopes formed by the channel erosion are in a direction perpendicular to that of the bedding inclination, which enables a steeper local slope angle to be sustained due to the sediment fabric. These local slope angles can be as high as 50° but only for a small slope height (not exceeding 2 m here).

The notch resulting in the formation of a knickpoint will have a slope direction closer to the bedding dip direction. At some critical slope height (not known here), erosion into the slope will favour a failure along the bedding plane that will initiate breaching or static liquefaction (or dynamic liquefaction if an earthquake triggers the slide) of the sediments, enabling a rapid upward migration of a wedge failure up to the crest of the interchannel slope in a manner similar to that which has been proposed by Berg *et al.* (2002). The overall failure is rapid enough to ensure that all the excavated material flows out of the gully into the channel, where it is then transported to greater depths.

Results from the above analysis suggest another mechanism for gully formation in granular subaqueous sediments. Linear V-shape gullies in coarse-grained sediments may form as a result of rapid upslope excavation (involving breaching or static liquefaction) from the site of a knickpoint or scour along a channel margin, aided by the favourable orientation of the steep delta-front bedding and associated weak layers left by seaward delta progadation.

Concluding remarks

The availability of a very-high-resolution multibeam survey has provided an opportunity to analyse the detailed morphology of a small channel network on the foreslope of the coarse-grained delta of the Busu River. The analysis indicates that the formation of a knickpoint along the flank of the channel can lead to conditions with a lower stability because the dip direction of the slope of the knickpoint approaches that of the bedding. The less favourable orientation of the bedding and associated heterogeneity along the flank of the channel could play a role in localizing scouring.

There are still many questions which cannot be answered at this point. For example, what caused the initiation of the notch and the development of a knickpoint has not been investigated but is likely to result from hydrodynamic processes which can be quite complex (Rebesco *et al.* 2014).

The actual sediment is also made of layers of sand and gravel, and one could ask whether such layering does indeed produce a series of weak layers (Locat *et al.* 2014), particularly for the case of earthquake loading where sand and gravel could liquefy (Bezerra *et al.* 2005). The morphology of the gullies presents a clear V-shape and the absence of remaining sediments within the V-shape area is seen as an indication that the failure occurred rapidly and progressed upslope, reaching the crest of the interfluve between adjacent channels. On the other hand, the orientation of the gullies does not appear to follow the apparent dip of the bedding: that is, the slopes within the gullies are much steeper than the bedding plane dip (i.e. at more than 30°). Still, slopes on either side of the gullies are not at the same angle and appear significantly steeper on the north-facing slopes: that is, on a slope with a dip opposite to the bedding plane dip direction. This may indicate some influence of the bedding dip direction on the final shape of the gully. With the slope of the gullies being between 30° and 40°, the friction angle is likely to still play a role on the shape of the gully. This may suggest that the bedding may only play a role in the initiation of the failure, while the friction angle will be the main limiting condition for breaching or static liquefaction. This is still an interesting element for future consideration.

Regarding the bedding dip direction, it is close to that of the channel in the study area. However, when looking at the arcuate shape of the active delta, it is clear that there is a spread in bedding orientations along the shoreline and also in channel orientations (Fig. 12). In rock mechanics, bedding is often seen as a primary discontinuity along which planar failure takes place (Hoek and Bray 1981) and where the shear strength may be lowest (Fu and Dafalias 2011). Here, we propose that a similar role can be played by the bedding orientation in sediments with respect to erosion and initiation of breaching or static liquefaction.

Acknowledgements The results presented here were made possible by the efforts of many in order to carry out the fieldwork which involved scientists and personnel from Kongsberg (Craig Wallace), Neptune Marine Services, IHA, Advisian and Tetra Tech. The authors would also like to thanks P. Haughton, H. Lee, and R. Urgeles for providing a stimulating review of the manuscirpt.

Funding This research received no specific grant from any funding agency in the public, commercial, or not-for-profit sectors.

Author contributions **JL**: formal analysis (lead), validation (lead), visualization (lead), writing – original draft (lead), writing – review & editing (lead); **AA**: formal analysis (supporting), validation (supporting), writing – review & editing (supporting); **JS**: project administration (lead), validation (supporting); **AH**: investigation (supporting), writing – review & editing (supporting); **CY**: resources (supporting); **DT**: formal analysis (supporting), validation (supporting), writing – review & editing (equal); **AB**: funding acquisition (lead), project administration (lead), writing – review & editing (supporting), writing – review & editing (supporting).

References

Amblas, D., Ceramicola, S. *et al.* 2018. Submarine canyons and gullies. *In*: Micallef, A., Krastel, S. and Savini, A. (eds) *Submarine Geomorphology*. Springer Geology. Springer, Cham, Switzerland, 251–272, https://doi.org/10.1007/978-3-319-57852-1_14

Baldwin, S., Fitzgerald, P.G. and Webb, L.E. 2012. Tectonics of the New Guinea region. *Annual Review of Planetary Sciences*, **40**, 495–520, https://doi.org/10.1146/annurev-earth-040809-152540

Baztan, J., Borne, S. *et al.* 2005. Axial incision: The key to understand submarine canyon evolution (in the western Gulf of Lion). *Marine and Petroleum Geology*, **22**, 805–826, https://doi.org/10.1016/j.marpetgeo.2005.03.011

Berg, J.V.D., Van Gelder, A. and Mastbergen, D.R. 2002. The importance of breaching as a mechanism of subaqueous slope failure in fine sand. *Sedimentology*, **49**, 81–95, https://doi.org/10.1111/j.1525-139X.2006.00168.x-i1

Bezerra, F.H.R., Da Fonseca, V.P., Vita-Finzi, C., Lima-Filho, F.P. and Saadi, A. 2005. Liquefaction-induced structures in Quaternary alluvial gravels and gravelly sediments, NE Brazil. *Engineering Geology*, **76**, 191–208, https://doi.org/10.1016/j.enggeo.2004.07.007

Buleka, J., Prior, D. and Van Der Spek, A. 1999. *Coastplan: Geology and Natural Hazards of Lae City and Surroundings, Papua New Guinea*. Report for the Port Moresby/Bangkok.

Chappell, J. and Shackleton, N.J. 1986. Oxygen isotopes and sea level. *Nature*, **324**, 137–140, https://doi.org/10.1038/324137a0

Clare, M.A., Le Bas, T. *et al.* 2018. Complex and cascading triggering of submarine landslides and turbidity currents at volcanic islands revealed from integration of high-resolution onshore and offshore surveys. *Frontiers in Earth Science*, **6**, 223, https://doi.org/10.3389/feart.2018.00223

Crook, K.A.W. 1989. Quaternary uplift rates at a plate boundary, Lae urban area, Papua New Guinea. *Tectonophysics*, **163**, 105–118, https://doi.org/10.1016/0040-1951(89)90121-2

Dow, D.B. 1977. *A Geological Synthesis of Papua New Guinea*. Australian Department of Natural Resources, Bureau of Mineral Resources, Geology and Geophysics Bulletin, **201**.

Fu, P. and Dafalias, Y.F. 2011. Study of anisotropic shear strength of granular materials using DEM simulation. *International Journal for Numerical and Analytical*

Methods, **35**, 1098–1126, https://doi.org/10.1002/nag.945

Gales, J.A., Larter, R.D., Mitchell, N.C. and Dowdeswell, J.A. 2013. Geomorphic signature of Antarctic submarine gullies: Implications for continental slope processes. *Marine Geology*, **337**, 112–124, https://doi.org/10.1016/j.margeo.2013.02.003

Hoek, E. and Bray, J.W. 1981. *Rock Slope Engineering*. 3rd edn. Institute of Mining and Metallurgy, London.

Klaucke, I. and Cochonat, P. 1999. Analysis of past seafloor failures on the continental slope off Nice (SE France). *Geo-Marine Letters*, **19**, 245–253, https://doi.org/10.1007/s003670050115

Liu, K., Crook, K.A.W., Clarke, J.H. and Whitmore, G.P. 1995. Submarine features of modern open-sea fan deltas, Huon Peninsula, Papua New Guinea. *Sedimentary Geology*, **98**, 63–77, https://doi.org/10.1016/0037-0738(95)00027-6

Locat, J. 2018. Failure and post-failure analysis of submarine mass movements using geomorphology and geomechanical concepts. *Geological Society, London, Special Publications*, **477**, 333–351, https://doi.org/10.1144/SP477.27

Locat, J., Leroueil, S., Locat, A. and Lee, H. 2014. Weak layers: their definition and classification from a geotechnical perspective. *Natural and Technological Hazards Research*, **37**, 3–13, https://doi.org/10.1007/978-3-319-00972-8_1

Mauffrey, M.-A., Urgeles, R., Berné, S. and Canning, J. 2017. Development of submarine canyons after the Mid-Pleistocene Transition on the Ebro margin, NW Mediterranean: The role of fluvial connections. *Quaternary Science Reviews*, **158**, 77–93, https://doi.org/10.1016/j.quascirev.2017.01.006

Micallef, A. and Mountjoy, J.J. 2011. A topographic signature of a hydrodynamic origin for submarine gullies. *Geology*, **39**, 115–118, https://doi.org/10.1130/G31475.1

Nemec, W. 1990. Aspects of sediment movement on steep delta slopes. *International Association of Sedimentologists Special Publications*, **10**, 29–73.

Patruno, S. and Helland-Hansen, W. 2018. Clinoforms and clinoform systems: Review and dynamic classification scheme for shorelines, subaqueous deltas, shelf edges and continental margins. *Earth-Science Reviews*, **185**, 202–233, https://doi.org/10.1016/j.earscirev.2018.05.016

Pratson, L.F., Nittrouer, C.A. *et al.* 2007. Seascape evolution on clastic continental shelves and slopes. *International Association of Sedimentologists Special Publications*, **10**, 339–380, https://doi.org/10.1002/9781444304398.ch7

Rebesco, M., Hernandez-Molina, F.J., Van Rooij, D. and Wahlin, A. 2014. Contourites and associated sediments controlled deep-water circulation processes: State-of-the-art and future considerations. *Marine Geology*, **352**, 111–154, https://doi.org/10.1016/j.margeo.2014.03.011

Renagi, O., Heron, S.F. and Ridd, P.V. 2013. The river plume of the Southern Huon Peninsula, PNG. *In*: *Proceedings of the 7th Huon Seminar: Achieving Vision 2050 Through Higher Education, Research, Science & Technology, November 13–14, 2013, Papua New Guinea University of Technology, Lae, Papua New Guinea*. The Papua New Guinea University of Technology, Lae, PNG, HS72013-HS-03, 20–34.

Samanta, S. 2015. Geospatial data for surface runoff and transport capacity modeling. *International Journal of Remote Sensing*, **4**, 91–100.

Sekac, T. and Jana, S.K. 2013. Change detection of Busu River course in Papua New Guinea – Impact on local settlements using remote sensing and GIS technology. *International Journal of Scientific and Engineering Research*, **5**, 891–899.

Shumaker, L.E., Jobe, Z.R. and Graham, S.A. 2017. Evolution of submarine gullies on a prograding slope: Insights from 3D seismic reflection data. *Marine Geology*, **393**, 35–46, https://doi.org/10.1016/j.margeo.2016.06.006

Simoni, A. and Houlsby, G.T. 2006. The direct shear strength and dilatancy of sand–gravel mixtures. *Geotechnical and Geological Engineering*, **24**, 523–549, https://doi.org/10.1007/s10706-004-5832-6

Turmel, D., Locat, J. and Parker, G. 2011. Upstream migration of knickpoints: geotechnical considerations. *Advances in Natural and Technological Hazards Research*, **31**, 123–132, https://doi.org/10.1007/978-94-007-2162-3_11

Turmel, D., Locat, J. and Parker, G. 2015. Morphological evolution of a well-constrained, subaerial–subaqueous source to sink system: Wabush Lake. *Sedimentology*, **62**, 1636–1664, https://doi.org/10.1111/sed.12197

Vachtman, D., Mitchell, N.C. and Gawthorpe, R. 2013. Morphologic signatures in submarine canyons and gullies, central USA Atlantic continental margins. *Marine and Petroleum Geology*, **41**, 250–263, https://doi.org/10.1016/j.marpetgeo.2012.02.005

Morphology and spatio-temporal distribution of lacustrine mass-transport deposits in Wörthersee, Eastern Alps, Austria

Christoph Daxer[1]*, Maddalena Sammartini[1], Ariana Molenaar[1], Thomas Piechl[2], Michael Strasser[1] and Jasper Moernaut[1]

[1]Institute of Geology, University of Innsbruck, Innsbruck, Austria

[2]Amt der Kärntner Landesregierung, Klagenfurt, Austria

CD, 0000-0003-1276-4596; MSa, 0000-0003-1338-5475; MSt, 0000-0003-3991-2405; JM, 0000-0003-3164-4274

*Correspondence: Christoph.Daxer@uibk.ac.at

Abstract: In lakes, landslides can be studied in high resolution due to their accessibility and limited size. Here, we investigate mass-transport deposits in glacigenic Wörthersee (Eastern European Alps) by integration of seismic, sediment core and multibeam bathymetric data. Two outstanding landslide events were revealed: the first occurred in the Late Glacial, leading to multiple deposits of up to 15 m thickness; they consist of sandy turbidites and mudclast conglomerates, which are overlain by a 2.5 m thick megaturbidite. The extensive, likely earthquake-triggered failure linked to this event was preconditioned by rapid sedimentation of fine-grained glaciolacustrine sediments and associated build-up of excess pore pressure. The second event was presumably triggered by a major earthquake ($M_w \approx 7$) in AD 1348 and comprises a mass-transport complex and several landslides, which led to a *c.* 30 cm thick turbidite. In total, 62 landslides are imaged in the multibeam map, 6 of which are most likely human-induced. Some of these show horseshoe-type compressional ridges and frontal breaching, whereas others exhibit an extensive zone of rafted blocks. We attribute these morphological differences to four main factors: (1) slope gradient and changes therein; (2) preconditioning of the impacted zone; (3) volume of remobilized sediment; and (4) type of impactor.

Other than their often devastating consequences, including tsunamis (Bondevik *et al.* 2005; Tappin *et al.* 2008; Urgeles *et al.* 2019) or loss of infrastructure (Assier-Rzadkiewicz *et al.* 2000; Pope *et al.* 2017), the processes involved in initiation, propagation and emplacement of submarine landslides cannot be readily monitored (Masson *et al.* 2006). Therefore, we either rely on physical or numerical models (Rzadkiewicz *et al.* 1996; Enet *et al.* 2003; Bondevik *et al.* 2005; Sue *et al.* 2011) or detailed characterization of the landslide morphology in source area, translation area and deposition area (Hühnerbach and Masson 2004; Twichell *et al.* 2009; Urgeles and Camerlenghi 2013) in order to understand the fundamental mechanisms underlying submarine mass movements. Comparison of these processes in the oceanic (Owen *et al.* 2007; Krastel *et al.* 2014 and references therein) and lacustrine realm (Girardclos *et al.* 2007; Moernaut and De Batist 2011; Wiemer *et al.* 2015) indicates broad similarities, allowing us to use the smaller and more easily accessible lacustrine landslides as a natural laboratory for the ocean (Strasser *et al.* 2007; Sammartini *et al.* 2020).

Slope failures are often ultimately triggered by a single event, for example earthquakes (Piper *et al.* 1999), cyclones (Pope *et al.* 2017), or sudden gravitational loading (Assier-Rzadkiewicz *et al.* 2000). Nevertheless, sediment accumulation rates, deposition of lithologically varying sediments (L'Heureux *et al.* 2012) as well as the development of excess pore pressure (Talling *et al.* 2014) are important preconditioning factors, acting on longer timescales (Urgeles and Camerlenghi 2013). In order to statistically assess the influence of such time-dependent factors on slope stability and understand the hazard potential of slope failures, knowledge about the spatio-temporal variability of landslides in both marine and lacustrine environments is of paramount importance (Masson *et al.* 2006).

Glacigenic lakes in temperate climates offer the possibility of investigating the influence of changing sedimentary properties (from glacio-lacustrine to hemipelagic sediments) and – due to the small dimensions of most lakes – the critical stratigraphic levels can often be sampled via standard piston coring techniques. Moreover, accurate dating of landslide events reaching back into Late Glacial time is

From: Georgiopoulou, A., Amy, L. A., Benetti, S., Chaytor, J. D., Clare, M. A., Gamboa, D., Haughton, P. D. W., Moernaut, J. and Mountjoy, J. J. (eds) 2020. *Subaqueous Mass Movements and their Consequences: Advances in Process Understanding, Monitoring and Hazard Assessments*. Geological Society, London, Special Publications, **500**, 235–254.
First published online March 19, 2020, https://doi.org/10.1144/SP500-2019-179

possible due to the general abundance of organic macro-remains for ^{14}C dating and the possible presence of annual laminations (varves). Limited water depths allow for very high-resolution morphologic and seismic-stratigraphic mapping. In the present study, we describe the morphology and seismic-stratigraphic character of landslide deposits in an Austrian glacigenic lake, from very recent, human-induced events, to extensive basin collapses in the Late Glacial. We focus on addressing a number of research questions. (1) Why do some mass-transport deposits (MTDs) have numerous rafted blocks whereas others seem to form more coherent masses? (2) Which factors control the potential of landslides to deform and erode basin-plain sediments? (3) To what extent are the dimensions and morphology of landslide deposits influenced by the geotechnical properties of failed slope sequences?

Geological and geomorphological setting

Wörthersee (*See* = German for lake) is located in the Austrian state of Carinthia (Eastern European Alps), near the border of Italy and Slovenia (Fig. 1). The lake covers 19.4 km^2 and is embedded in a glacially overprinted landscape consisting of metamorphic crystalline units (phyllites and mica-schists; Homann 1962). The three main lake basins as well as their sub-basins (Kapuziner, Reifnitz and Klagenfurt basins) are separated by bedrock and/or moraine ridges and follow the trend imposed by a series of WSW–ENE-striking (transform) faults and their associated auxiliary faults (Kahler 1962; Anderle 1977; Pistotnik *et al.* 1980). Wörthersee reaches its maximum depth (84 m) in the western basin, which mainly exhibits steep (>15°) slopes. In the middle basin (max. depth 40 m) and the eastern basin (70 m), shallow (20 m) bays and nearshore terraces are common features. Most of the inflows to the lake are small rivulets entering the western lake basin, together draining a catchment area of 162 km^2. The main inflow enters the lake near Reifnitz at an average discharge of 0.63 m^3 s^{-1}. The only outlet, the artificially regulated Glanfurt, leaves the lake in the very east near Klagenfurt. According to the Carinthian Institute for Lake Research (KIS), the lake level dropped severely around AD 1770, when construction activity at the Glanfurt was completed (KIS 2019). Human influence is also clearly visible along the shoreline, of which only 23% remains in a natural to near-natural state. The remaining 77% have either been reinforced (47%) or altered by human activity (30%; Schulz *et al.* 2008). On 11 December 2013, a landslide happened at a construction site located near Bad Saag at the northern shore of the western lake basin (Fig. 1). Dug-out material, as well as three containers, was placed close to the shoreline, which then collapsed. The resulting landslide transported the containers and 'several thousands of m^3' (APA/derStandard 2013) of onshore material down into the lake basin, fortunately causing no fatalities.

The study area is one of Austria's seismically most active regions and has been affected by several damaging earthquakes (M_w>6) in AD 1348, AD 1511, AD 1690 and AD 1976, with the AD 1348 event being the strongest historical earthquake (M_w≈7) documented in the entire Alps (Stucchi *et al.* 2013). Due to its location at the southeastern border of the Alps, deglaciation in the area surrounding Wörthersee occurred earlier than in other peri-alpine lakes (Seguinot *et al.* 2018). This is supported by radiocarbon ages from sediments in the nearby Jeserzer See (*c.* 2 km north of Velden; see Fig. 1), where two (bulk-) samples taken at the base of a 7.26 m long sediment core yielded ages of 17 140–17 825 cal years BP and 17 581–19 366 cal years BP, respectively (Schultze 1984; Schmidt *et al.* 2012).

Material and methods

Multibeam bathymetry

Bathymetric data were acquired in 2017 using a high-resolution SeaBat T50-P multibeam echo-sounder combined with an AsteRx-U MARINE GNSS Heading System. Data processing was conducted with Qinsy software. The resulting bathymetric maps used in this study have a cell size of 2 m. To enhance the topography, the maps showing the detailed landslide morphologies are presented as a combination of topographic openness (Yokoyama *et al.* 2002), analytical hill-shade and colour slices from the digital elevation model generated in SAGA GIS (Conrad *et al.* 2015) and combined in Adobe Illustrator software. Description and interpretation of landslide morphometric parameters were carried out in a standard GIS framework (ArcGIS 10.5.1) following Clare *et al.* (2018).

Seismic reflection data

High-resolution seismic data were acquired during three separate surveys in 2017 and 2019 using two different systems: a 3.5 kHz Geopulse pinger source (theoretical vertical resolution: *c.* 10 cm; 203 km of survey lines) and an Innomar SES-2000 light sub-bottom profiler (100 kHz primary frequency, 8 kHz secondary low frequency; theoretical vertical resolution: *c.* 5 cm; 121 km of survey lines; Fig. 1b). Positioning was carried out with stand-alone GPS. The data were processed in the IHS Markit Kingdom (v. 2018) Seismic and Geological Interpretation software applying a bandpass filter for the 3.5 kHz data

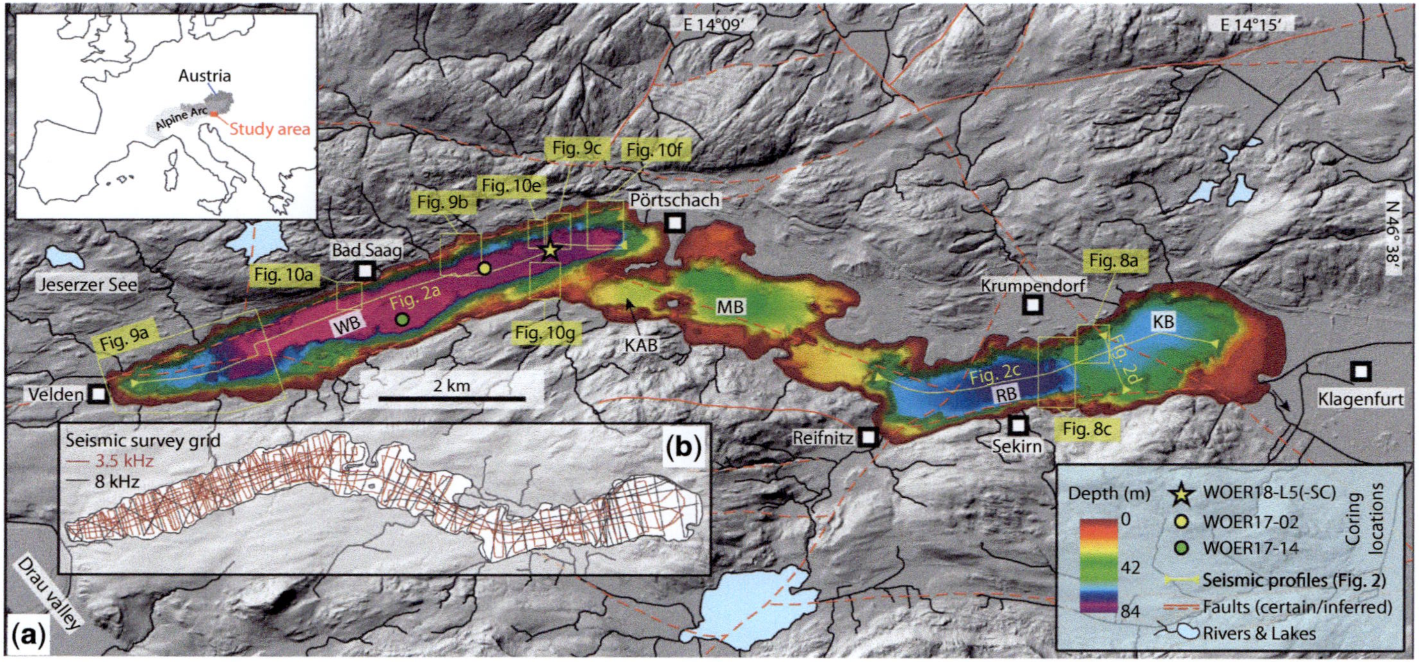

Fig. 1. (**a**) Bathymetric map of Wörthersee, located at the southeastern rim of the European Alps, Austria (see inset). The lake is divided into three main basins: the western basin (WB), the middle basin (MB; including the Kapuziner basin, KAB) and the eastern basin, consisting of the Reifnitz sub-basin (RB) and the Klagenfurt sub-basin (KB). Red lines indicate the main faults (dashed where inferred; according to Austrian Geological maps: Kahler 1962; Anderle 1977). (**b**) Seismic survey grid showing profiles acquired using a 3.5 kHz source (red lines) and a *c.* 8 kHz source (black lines).

(lower cut 2 kHz, upper cut 6 kHz). Sedimentary deposits related to landsliding were distinguished from regular background sedimentation – exhibiting continuous reflections – by their irregular to chaotic seismic facies and intercalated geometry, which in turn differs from the mappable transparent seismic intervals and ponding geometry attributed to megaturbidites (*sensu* Schnellmann *et al.* 2002). Such deposits are herein interchangeably also referred to as MTDs if they constitute singular, unidirectional events. For synchronous, overlapping and multidirectional event deposits, where individual landslides cannot be distinguished in the seismic data (but might be discernible in the multibeam data), the term mass-transport complex (MTC) is used. Event horizons are defined as the seismic-stratigraphic horizons related to the pinch-out points of MTDs and are potentially overlain by the associated megaturbidites. These horizons were mapped over the whole lake to determine synchronicity of individual landslides (Strasser *et al.* 2013). If steep slope angles and non-sedimentation prohibited mapping of seismic event horizons over lake basin boundaries, correlation polygons were used to determine correlative seismic-stratigraphic horizons. For the time to depth conversions, a constant velocity of 1500 m s^{-1} for both water and sediment was assumed. Errors arising from changes in velocity are negligible (*c.* 5 cm ms^{-1}) and therefore irrelevant for the purpose of this study. Volumetric analyses were carried out in Surfer (Golden Software).

Coring campaigns and sediment core analyses

From 2017 to 2019, short (*c.* 1.5 m) sediment cores were retrieved using a percussion gravity corer. In September 2018, a *c.* 11 m long gravity piston core (Kullenberg type; Kelts *et al.* 1986) was retrieved in the deepest part of the western basin. X-ray computed tomography (CT) scans were carried out on whole round sediment cores at the Medical University of Innsbruck, using a Siemens SOMATOM Definition AS. For data visualization, the software VGSTUDIO (v. 3.3) as well as FIJI (Schindelin *et al.* 2012) were used. γ-density and magnetic susceptibility (MS) were measured at 0.5 cm and 0.2 cm vertical resolution, respectively, using a GEOTEK Multi-Sensor Core Logger (MSCL) and a Bartington MS2E point sensor at the Austrian Core Facility for scientific core logging and scanning (Institute of Geology, Innsbruck). The grain size of untreated samples was determined with a Malvern Mastersizer 3000 laser diffractometer, applying 1 min ultrasonication at 40% strength before measurement. Smear slide analyses were carried out following the protocol defined by Schnurrenberger *et al.* (2003).

Chronology

AMS radiocarbon dating was conducted on terrestrial plant macro-remains at the Ion beam Physics Laboratory of ETH Zürich (Table 1). Radionuclide activity measurements (^{210}Pb and ^{137}Cs) were carried out at EAWAG (Dübendorf, Switzerland) using *CANBERRA* and *Princeton Gamma-Tech* germanium well detectors.

Results

Seismic landslide stratigraphy

The sedimentary infill of Wörthersee can be generally divided into two main seismostratigraphic units separated by a high amplitude reflection package, which is present in all lake basins at about 5.5 ms (in the middle basin) to 8.5 ms (in the western basin) sediment depth (Figs 2 & 3). The upper seismic unit 1 (SU-1) consists of a draping sedimentary succession, showing parallel, low- to intermediate-amplitude reflections. The reflections are generally continuous but partly interrupted by higher amplitudes where MTDs are present. The distinctive reflection patterns in SU-1 are similar in all lake basins and allow for a detailed seismic stratigraphic correlation (Fig. 3b). The underlying seismic unit 2 (SU-2) constitutes a sequence of up to 43 ms (*c.* 32 m) thick. The reflection pattern, which is frequently disturbed by vertical acoustic wipe-outs, commonly onlaps on to the acoustic basement, therefore exhibiting a ponding geometry (Figs 2 & 3a). Due to varying seismic facies, SU-2 is further distinguished into two time-equivalent seismic subunits: seismic unit 2a (SU-2a), present only in the western basin, shows low- to intermediate-amplitude reflections and intercalated, laterally fluctuating and partly truncated high-amplitude packages; seismic unit 2b (SU-2b), constituting the majority of sedimentary infill in the middle and eastern basins, is characterized by closely spaced intermediate- to high-amplitude reflections and little lateral variation.

Of several seismic-stratigraphic MTD event horizons, which can be identified especially in the western basin, three horizons are particularly prominent and are the subject of the present study: event horizon A (EH-A) constitutes the stratigraphically highest event horizon, event horizons B (EH-B) and C (EH-C) stand out due to their many time-correlative large-volume MTDs.

In terms of MTD volume, EH-A constitutes a minor stratigraphic event horizon (Fig. 2a; volume = *c.* 0.08×10^6 m^3) consisting of six individual landslides, of which four are in the western basin (see Fig. 4 for locations). These MTDs are overlain by a thin (<1 ms) sedimentary drape, thus representing rather recent MTDs.

Table 1. *Radiocarbon ages from organic macro-remains*

Sediment core	Sediment core depth (cm)	Section	Sediment core section depth (cm)	Lab no.	Conventional radiocarbon age (^{14}C years BP)	2σ calibration (cal years BP)	2σ calibration (BC/AD)	Dated material
WOER17-02	32.5–35.5	–	32.5–35.5	ETH-88627	388 ± 30		1442–1632	*Betula* fruits, plant macro remains
WOER17-02	46–51	–	46–51	ETH-85092	519 ± 23		1331–1441	*Betula* fruits, plant macro remains
WOER17-02	129–131	–	129–131	ETH-85093	2562 ± 24		−804–588	leaf
WOER18-L5	222	F	8.5–9.5	ETH-96546	6938 ± 25	7832–7693		plant macro remains
WOER18-L5	344	F	127–128	ETH-96545	5273 ± 24	6179–5944		plant macro remains
WOER18-L5	452	E	96–97	ETH-96547	9952 ± 28	11600–11 251		plant macro remains
WOER18-L5	524	D	10–11	ETH-96548	11 785 ± 27	13 733–13 480		plant macro remains

EH-B is overlain by *c.* 2.6 ms of sediment in the basins. It comprises several landslides, including a MTC in the very western area of the lake near the village of Velden ('Velden MTC'; Fig. 2a). Compressional ridges are present in the seismic data as sub-vertical high-amplitude reflections, indicating amalgamation of individual landslides. These ridges, as well as the generally irregular surface of the MTC, account for a hummocky lake floor. Due to limited acoustic penetration in the central part of the Velden MTC, the absolute maximum thickness remains uncertain. In areas that have been sufficiently mapped, however, thickness reaches up to 15 ms (Fig. 4). Apart from the Velden MTC, seven smaller MTDs are also attributed to EH-B in the western basin, five in the middle and eight in the eastern basin, respectively. These MTDs show well-developed lateral or frontal pinch-out geometries and a related, distally fading high-amplitude reflection, that extends over the basin plain (Fig. 2c). Overall, the MTDs amount to a volume of 5.6×10^6 m^3, covering an area of 1.7×10^6 m^2. In the western basin and in small areas between compressional ridges of the Velden MTC, a megaturbidite with a maximum thickness of 1.5 ms is present. It covers an area of *c.* 0.6×10^6 m^2 and amounts to a volume of *c.* 0.44×10^6 m^3 (Fig. 4).

EH-C occurs within the uppermost part of seismic unit 2 and is overlain by about 10 ms (7.5 m) of sediment in the basins. The respective MTDs cover almost entirely all lake basins, as well as the nearshore terraces in both the middle and eastern basin. In the western basin, the deposits reach a maximum thickness of 20 ms. Near the village of Velden, the overlying Velden MTC of EH-B, as well as the thick deposits of EH-C itself, strongly attenuate the seismic signal (Fig. 2a), making mapping of the base of EH-C difficult to impossible. In the western basin, the deposits thin out towards the present-day depocentre in the east (Fig. 2a). East of a delta area influenced by gas blanking, where older MTDs underlie EH-C (Fig. 2b), the landslide deposit fills in an undulating palaeobathymetry, levelling the lake floor towards its present-day, nearly flat morphology. Near the coring site WOER18-L5, individual MTDs are imaged by transparent to chaotic seismic facies (Fig. 2a, b). These bodies pinch out into a high-amplitude reflection package. No clear landslide scar can be identified from the seismic data in the western basin. In contrast, slide scars of well-defined individual MTDs exist in the middle and eastern basin. There, individual landslides occur both on sedimentary sequences covering the lateral slopes of the basin (Fig. 2d), as well as along an intra-basinal ridge separating the Reifnitz basin from the Klagenfurt basin (Fig. 2c). Other than the deposits of EH-B, which originate exclusively on steep (>15°) slopes, event C also exhibits

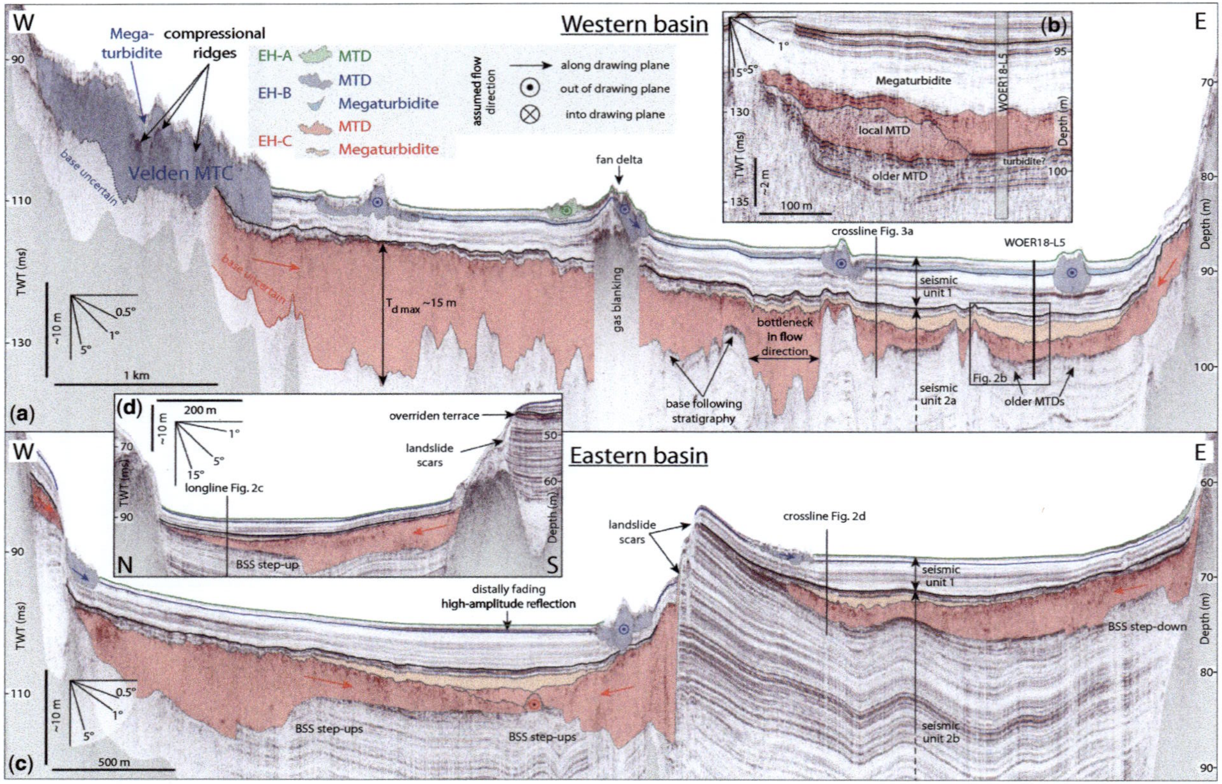

Fig. 2. 3.5 kHz reflection seismic profiles along the main lake axis (a, b, c) and perpendicular to it (d). See Figure 1a for location of seismic lines. (**a**) Seismic profile along the western basin. Three main event horizons are distinguished (EH-A, EH-B and EH-C). EH-A is in the stratigraphically highest position, only overlain by a thin (max. 0.5 ms) reflection package. EH-B and EH-C show extensive landsliding, especially in the west, with the MTDs of EH-C covering almost the whole lake basin and being overlain by a *c.* 3 m thick megaturbidite (light orange). (**b**) Close-up of (a) at the long coring site WOER18-L5. The MTD associated with EH-C is underlain by older MTDs and exhibits lateral variation in seismic facies, indicating smaller, local MTDs. (**c**) Seismic profile along the eastern basin. As in the western basin, EH-C covers most of the lake basin, but individual landslides can be more easily distinguished due to step-ups of the basal shear surface (BSS) and distinct changes in seismic facies where landslides intersect. At the ridge separating the Reifnitz basin from the Klagenfurt basin, a draped failure scar of EH-C is clearly visible. (**d**) Profile perpendicular to the eastern basin's main axis. A prominent failure scar can be traced along a terrace overridden by a landslide of EH-C.

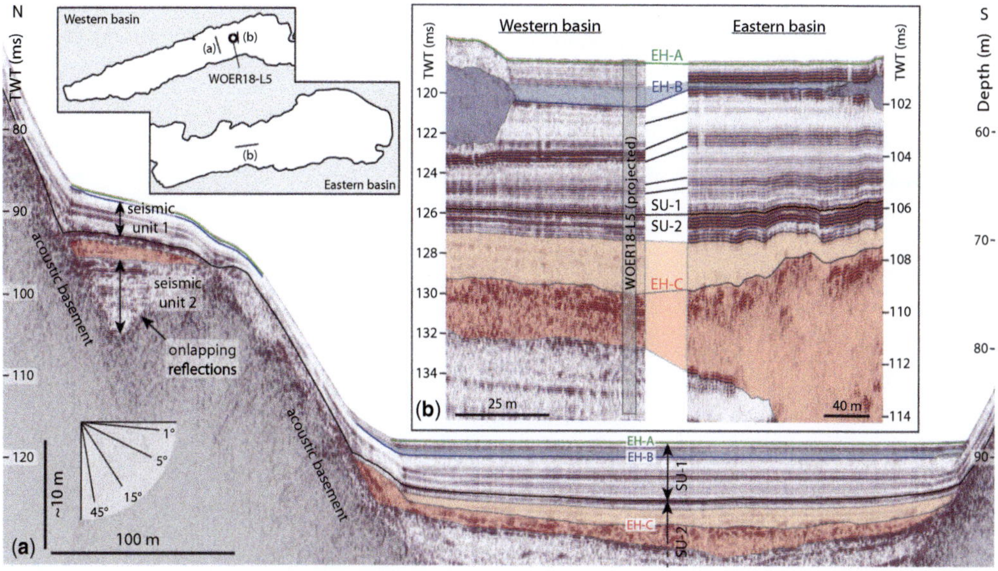

Fig. 3. (**a**) 3.5 kHz reflection seismic profile perpendicular to the lake axis in the western basin. The reflections of seismic unit 1 form a drape which also covers slopes as steep as 20°. Seismic unit 2 shows reflections onlapping on to the acoustic basement, indicating a ponding sedimentation. Colour coding as in Figure 2. (**b**) Correlation of 8 kHz reflection seismic profiles in the deepest part of the western and eastern basins, respectively. Note the different vertical axes, indicating higher sedimentation rates in the western basin.

failure on slopes as gentle as 3°, for instance at the eastern end of the western basin near Pörtschach. Overall, the EH-C MTDs amount to a volume of at least *c.* 28.7 × 10^6 m^3, covering an area of *c.* 6.6 × 10^6 m^2 and exhibiting an average thickness of *c.* 4.3 m. In all basins, the MTDs are overlain by a megaturbidite, with low-amplitude internal reflectivity and ponding geometries. This megaturbidite is up to 2.6 m thick (Fig. 4; overall volume *c.* 1.9 × 10^6 m^3) and flattens the rough MTD surface to some extent. The total deposits attributed to EH-C (MTDs and megaturbidites) amount to *c.* 30.6 × 10^6 m^3.

Sediment core data

Lithostratigraphy. The general lithostratigraphic succession of Wörthersee recovered at coring location WOER18-L5 (see Fig. 1 for coring location) can be subdivided into three main lithostratigraphic units from top to bottom (Fig. 5). Lithostratigraphic unit I (LU-I; 0–0.5 m core depth) consists of annually-laminated (varved) sediments. Based on varve thickness, presence of authigenic calcite, organic content and colour, this unit can be further subdivided into two subunits: unit 1a (LU-Ia; varve thickness >1 mm; abundant authigenic calcite, little organic content) and unit Ib (LU-Ib; varve thickness <1 mm; hardly any authigenic calcite, increased organic content).

Lithostratigraphic unit IIa (LU-IIa) is composed of dark, non-laminated, organic-rich (gyttja-like) sediments (0.5–1.75 m). Unit IIb (LU-IIb) consists of well-laminated, calcite-rich deposits (like LU-Ia), and interbedded parts of non-laminated dark organic mud (1.75–5.1 m) The transition between the two subunits is gradual. Several degassing cracks and low-density values (*c.* 1–1.2 g cm^{-3}) further reflect the organic content of these units. Magnetic susceptibility is generally low (5–10 SI × 10^{-5}) with a few spikes up to 35 SI × 10^{-5}.

Lithological unit III (LU-III) consists of entirely different, light grey sediments laminated at a millimetre-scale. At the boundary of LU-IIb to LU-III (5.1 m core depth), both density and magnetic susceptibility increase rapidly. Density rises from *c.* 1.25 to 1.5 g cm^{-3} and reaches values as high as 2 g cm^{-3} further down-core. Magnetic susceptibility increases to values of about 25 SI × 10^{-5}.

The sedimentary changes indicated by optical and physical properties are further corroborated by grain-size measurements and smear slide analysis (Fig. 6): LU-IIa and LU-IIb range from coarse to very coarse silt and have a clay content below 1%, whereas LU-III consists of fine to medium silt and has a clay content of 5–10%. Under the microscope, LU-II has a rather heterogenic appearance, mainly consisting of diatoms, amorphous organic matter and calcite crystals. LU-III, on the other hand, consists almost exclusively of

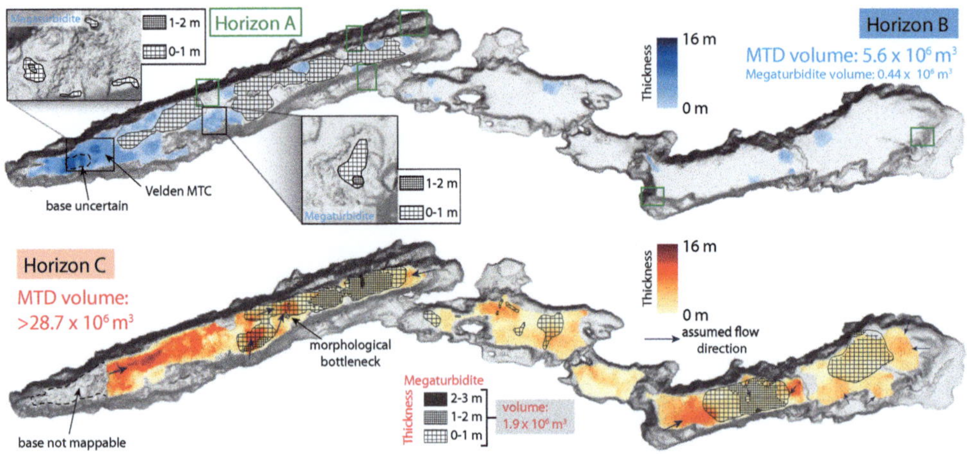

Fig. 4. Thickness maps of landslide events related to EH-B and EH-C. For both event horizons, mapping of the landslide base was restricted in the very west due to indistinct changes in seismic facies and limited penetration of the seismic signal. Blue arrows indicate inferred flow direction of individual landslides attributed to event horizon EH-C. To ensure comparability, the same scale (0–16 m) was chosen, although the maximum thickness of the MTDs associated with EH-B is only *c.* 12 m.

silt- to clay-sized particles and has a low amount of organic matter.

Identifying and dating of landslide events in cores. Within the uppermost lithological unit, LU-Ia, two graded beds of about 3 cm thickness, separated by a bright lamina, are recovered (Fig. 7). They contain large organic macro remains and are interpreted as turbidites resulting from near-shore landslides. The short-lived radionuclide data from core WOER18-L5-SC proves the annual character of the calcite-rich laminae: *c.* 22 $\pm$ 1 laminae can be counted by the unaided eye between the ^{137}Cs-peaks of AD 1986 and AD 1963. The two turbidites of this core thus are dated to about AD 1987 but occurred a year apart.

A more prominent turbidite of *c.* 30 cm thickness occurs in cores WOER17-14, WOER17-02 and WOER18-L5-SC, at *c.* 50 cm core depth, and therefore can be traced all over the western lake basin. Its lithofacies varies from a coarse, sandy base (*c.* 5 cm in core WOER17-02), showing multiple pulses, to a fine-grained, homogeneous grey sediment with abundant organic matter at the top. Radiocarbon dating of terrestrial plant material in laminated background sediment of LU-Ib revealed an age range of AD 1331–1441 (2σ range) just 3 cm above this turbidite.

The base of the most prominent, 3.7 m thick MTD, recovered from the long core WOER18-L5 between 5.3 and 9 m core depth and within the uppermost part of LU-III, consists of rounded to subangular gravels ranging from 0.5 cm to about 3 cm in size, which are interspersed in a silty matrix (Fig. 5). Physical properties exhibit a sharp density-contrast to the underlying silts. The biggest gravel clasts are present at the top of the *c.* 16 cm thick basal interval of the MTD, which is overlain by a *c.* 23 cm thick, sandy turbidite showing three fining-upward cycles. The central part of the MTD is composed of a 1 m thick mudclast conglomerate: soft clasts of 4–5 cm in diameter, composed of LU-III lithology are disseminated in a fine, dense matrix, and are clearly visible in X-CT data and false-colour optical images (Fig. 5b). The top of the deposit is represented by a 2.3 m thick megaturbidite deposit, which is not entirely homogeneous but shows fluctuations in both density and magnetic susceptibility.

A radiocarbon date about 18 cm above the megaturbidite indicates an age of 13 480–13 733 years BP, revealing a Late Glacial age for LU-III.

Core-to-core and core-to-seismic correlation. Lithostratigraphical correlation of the prominent marker layers and the different lithological units enables robust core-to-core correlation across the basin (Fig. 5). Comparison of the seismic reflection pattern to the γ-density and CT scans (radiodensity) allows for the precise correlation of long core WOER18-L5 (and short core WOER18-L5-SC) to the seismic stratigraphy (Fig. 5). The MTD associated with EH-C clearly correlates to the 3.7 m thick deposit ranging from 5.3 to 9 m core depth, with the basal gravel layer exhibiting a strong seismic reflector.

The 30 cm thick turbidite in the core correlates to the seismically resolved megaturbidite associated with EH-B, whereas the core-to-seismic and

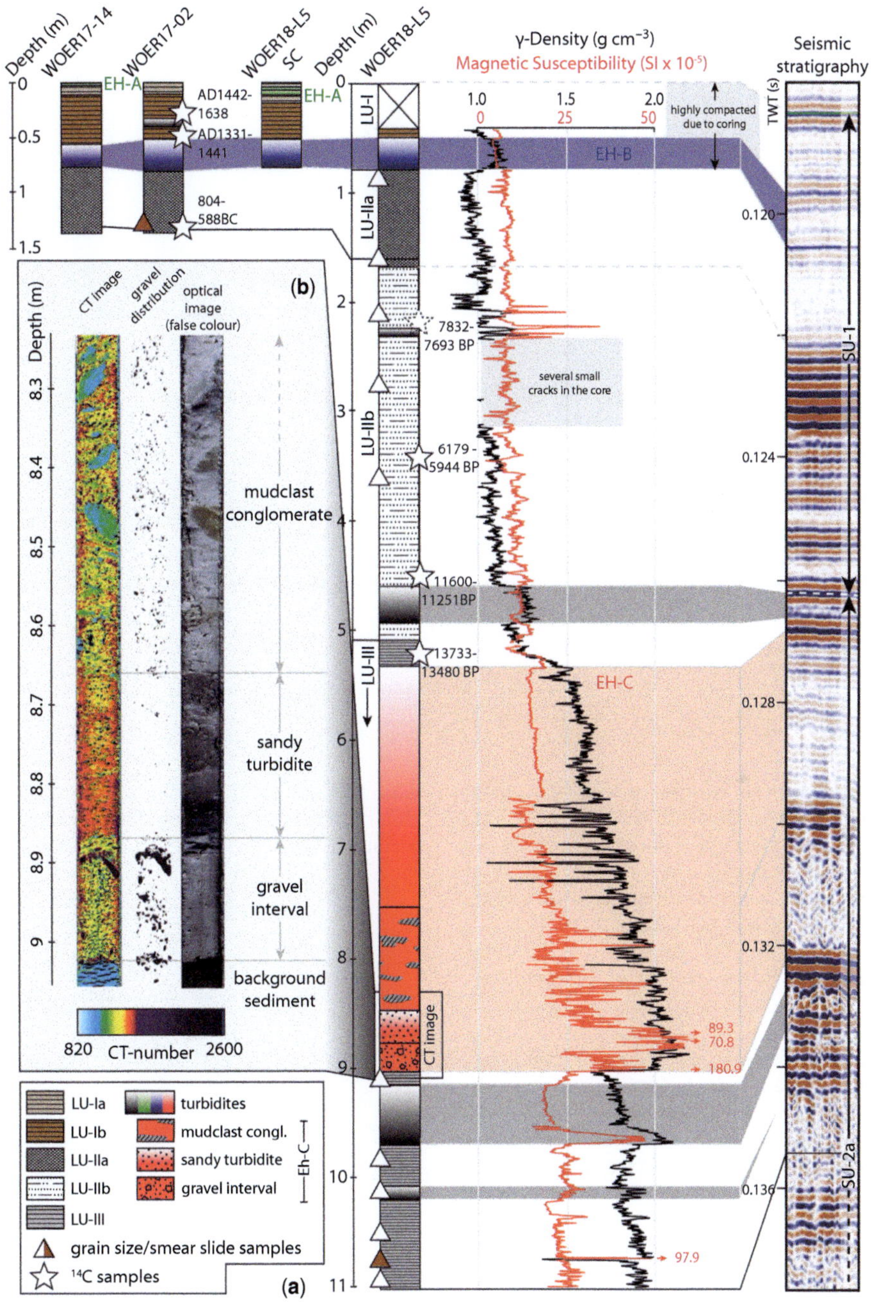

Fig. 5. (**a**) Seismic to core correlation, geophysical data and lithology of long core WOER18-L5 and core to core correlation of short cores WOER17-14, WOER17-02 and WOER18-L5-SC. Small turbidites (<10 cm) correlating to EH-A are present in cores WOER17-14 and WOER18-L5-SC, but too small to be visible in the core log. The blue interval corresponds to MTDs or megaturbidites of EH-B, the red interval to those of EH-C. White stars denote radiocarbon samples, white triangles grain-size samples and brown triangles indicate smear slide samples. (**b**) CT data showing the base of the MTD related to EH-C and the distribution of gravel in the sediment (extracted from the CT data). Note the similar radiodensity of mudclasts to background sediment below the base. The optical image has been modified (histogram equalization) to enhance contrast.

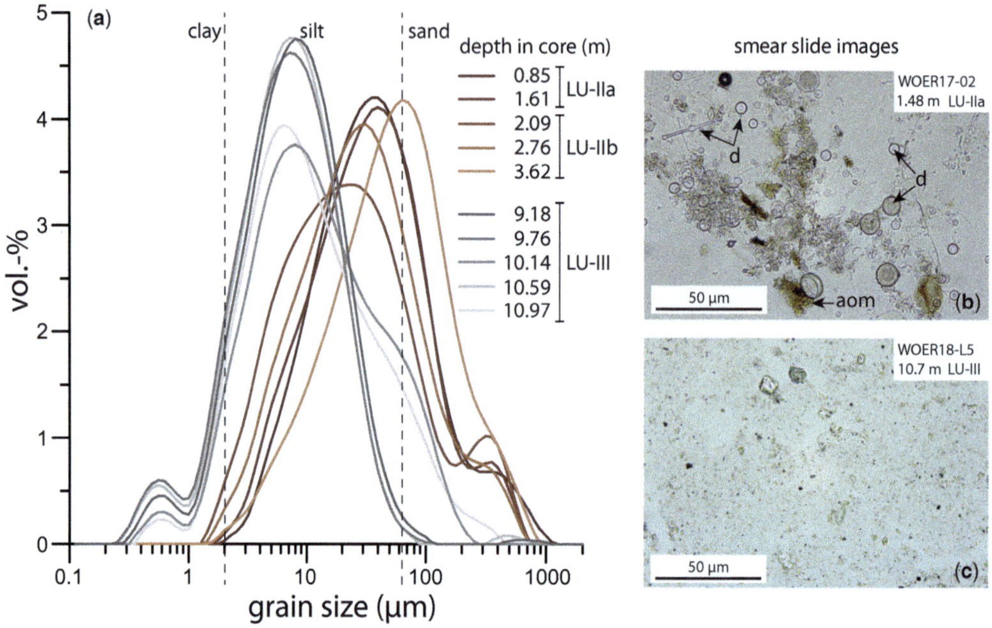

Fig. 6. (**a**) Grain-size measurements on core WOER18-L5 (see Fig. 5a for sampling locations). (**b**) Representative smear slide image of LU-IIa, consisting of amorphous organic matter (aom) and diatoms (d). (**c**) Representative smear slide image of silt- to clay-rich sediments (LU-III).

core-to-core correlation for the uppermost part of the sequence (which is subject to strong compaction in the cores) reveals that EH-A does not link to one single simultaneous landslide event, but is actually the result of several, non-concurrent landslides that all took place after about AD 1950.

Detailed landslide morphology: integrating multibeam bathymetry and seismic data

In total, 62 individual landslides (including landslides that are part of the Velden MTC) are observable in the multibeam bathymetric map. Most of these can be attributed to either EH-A or EH-B, as inferred from core-to-seismic correlation and basin-wide event horizon mapping (Figs 8–10). The deposits of EH-C are covered by about 5 m or more of sediment that smoothened the hummocky MTD surface morphology. Therefore, there is hardly any morphological expression left of the MTDs related to EH-C in today's bathymetry; only their slide scars are still partly visible in the multibeam bathymetric map (Fig. 8). Moreover, as the MTDs of EH-C extend over the entire basin plains, it is often unclear how many individual slope segments failed. In the following subsections, some landslides exhibiting peculiar morphological features will be discussed and characterized in detail. The most important morphometrical parameters are listed in Table 2.

Eastern basin. In the eastern basin, three landslides differing strongly in their morphology occur in close vicinity to each other (Fig. 8): the Krumpendorf East landslide (KEL), exhibiting strong disintegration and

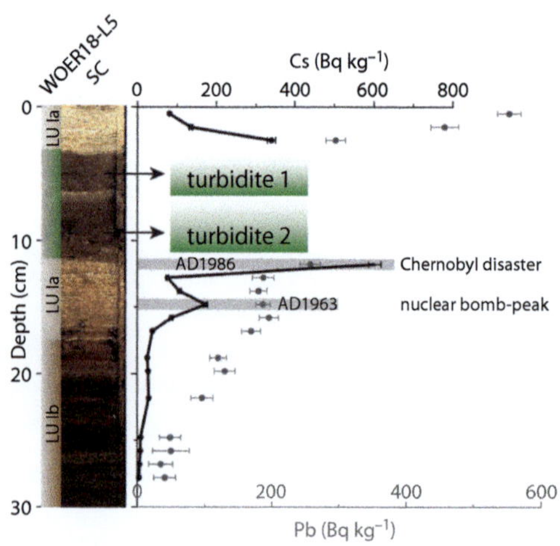

Fig. 7. Optical image and radionuclide data of short core WOER18-L5-SC. No radionuclide samples were collected in the turbidites (3–11.5 cm core depth). Note the bright layer separating the two turbidites at about 7 cm core depth, indicating that the two related landslides occurred a year apart.

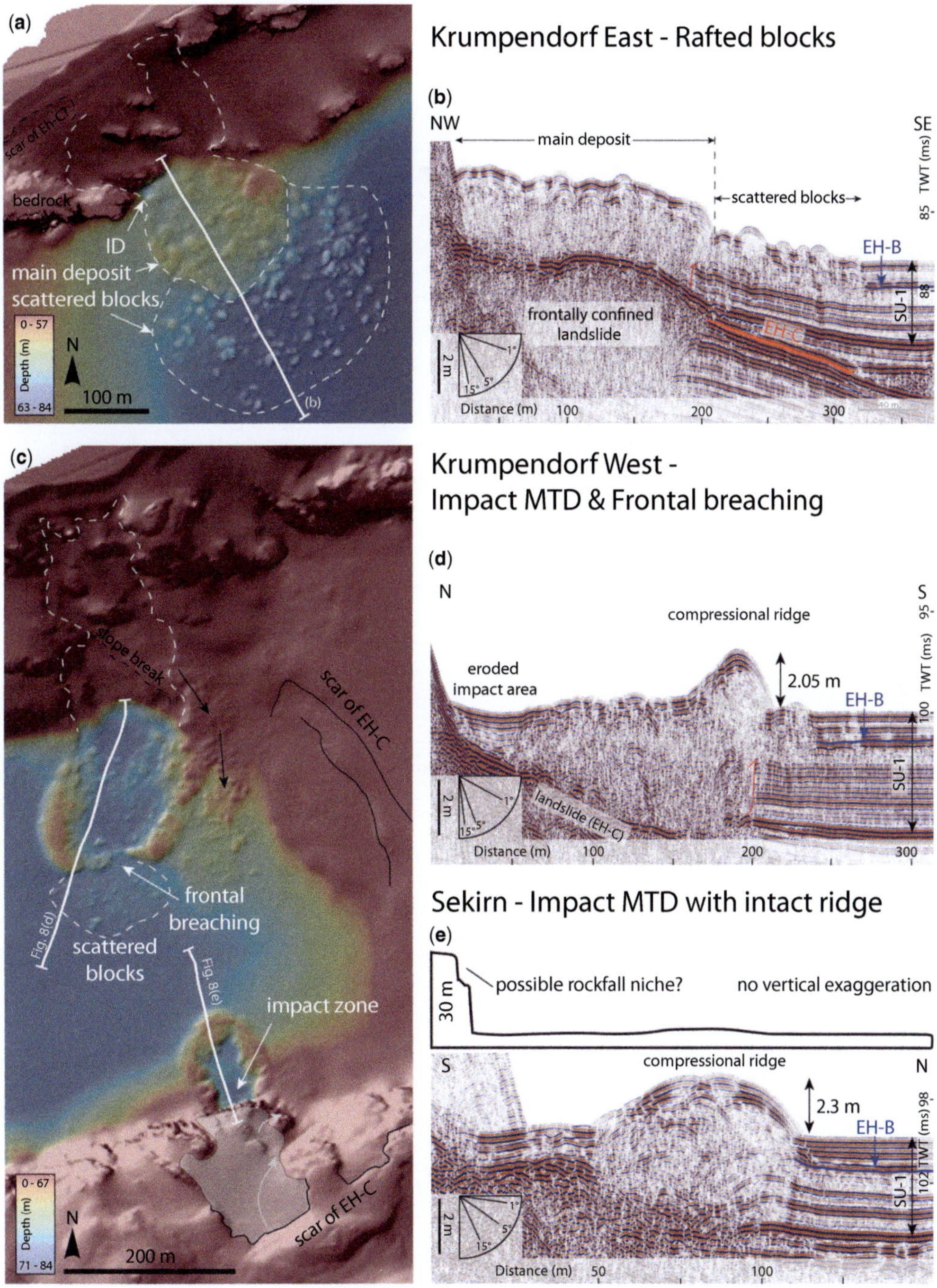

Fig. 8. Landslide morphologies in the eastern basin (see Fig. 1 for locations). (**a**) Morphology of Krumpendorf East landslide (KEL). The bright linear features on the slopes indicate bedrock ridges crossing the evacuation zone. ID, impact depression. (**b**) Longitudinal seismic profile through KEL, showing a frontally confined landslide (part of EH-C) below. (**c**) Morphology of landslides Krumpendorf West (KWL) in the north and Sekirn (SL) in the south. Black arrows indicate the flow path of landslide mass that travelled along the gentle slope and did not impact on the basin plain. (**d**) Longitudinal seismic profile through landslide Krumpendorf West. (**e**) Longitudinal seismic profile through landslide Sekirn.

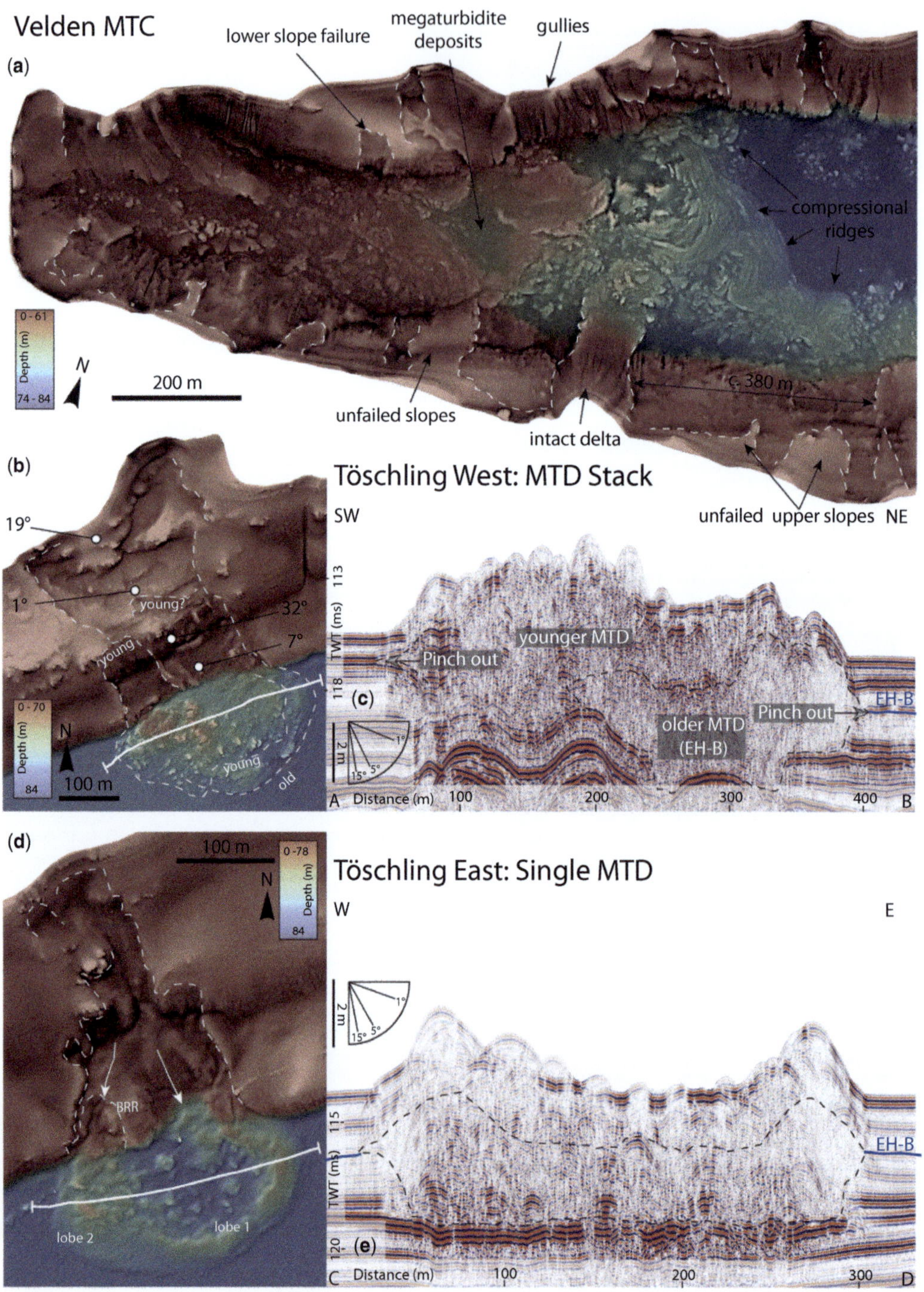

Fig. 9. Landslide morphologies in the western basin (see Fig. 1 for location). (**a**) Morphology of the Velden MTC. (**b**) Morphology of landslide Töschling West. Note the staircase-type slide scars. (**c**) Seismic profile perpendicular to landslide Töschling West. An older MTD (related to EH-B) is overlain by a younger MTD, pinching out higher in the seismic stratigraphy. (**d**) Morphology of landslide Töschling East. BRR, bedrock ridge. (**e**) Seismic profile perpendicular to landslide Töschling East.

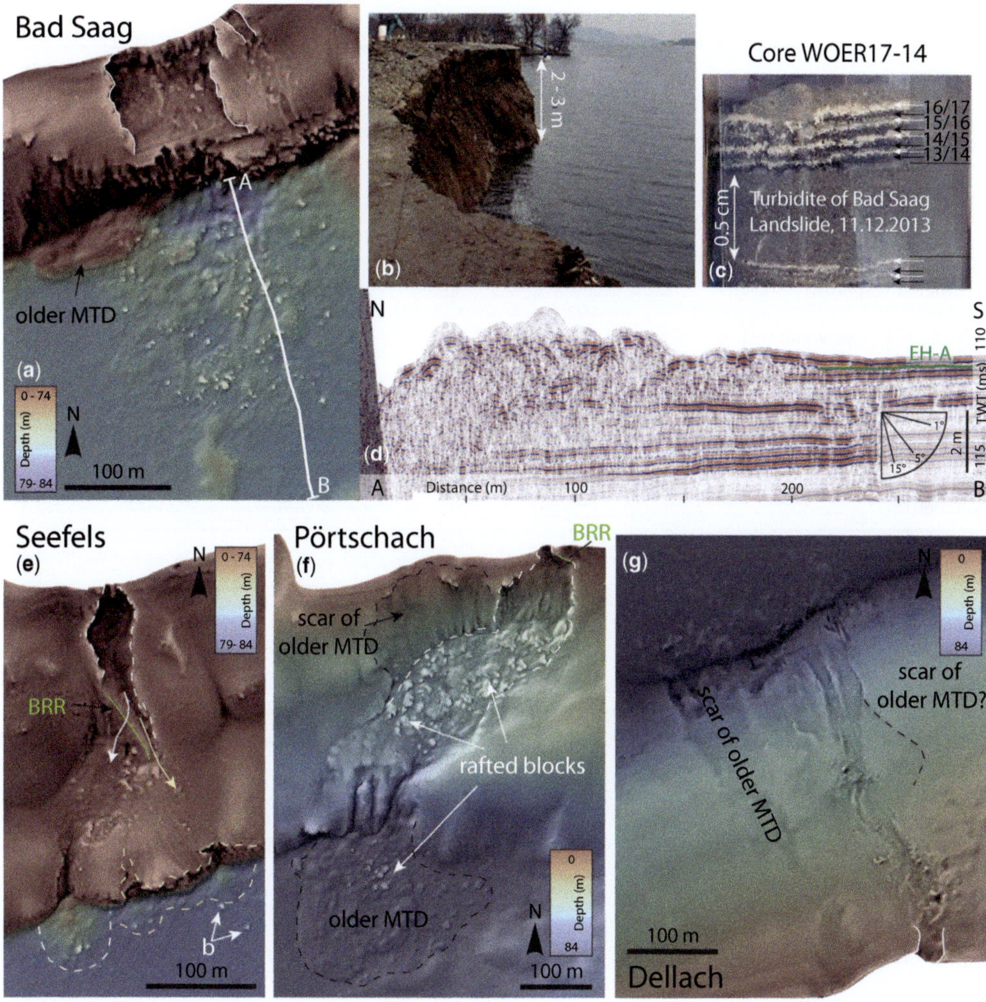

Fig. 10. Landslide morphologies of human-induced landslides (see Fig. 1 for location). (**a**) Morphology of Bad Saag landslide (BSL). (**b**) Photograph of the onshore headscarp of BSL (courtesy of Christoph Posch, ÖBB). (**c**) Photograph of core WOER17-14 directly after acquisition (see Fig. 1 for location). The annual layers on top of the turbidite related to the Bad Saag landslide are clearly visible. (**d**) Longitudinal seismic profile through the Bad Saag landslide. (**e**) Morphology of Seefels landslide (SEL). White arrow indicates main flow path, yellow arrow the assumed flow path of the outlier blocks (b). BRR, bedrock ridge. (**f**) Morphology of Pörtschach landslide (PL). Note the older slide scars and deposits related to EH-B. (**g**) Morphology of landslide Dellach. Next to the recent landslide, the slide scar of an older event can easily be recognized.

a multitude of blocks; the Krumpendorf West landslide (KWL), with a horseshoe morphology showing frontal breaching; and the Sekirn landslide (SL), showing a horseshoe morphology with intact compressional ridges.

The main deposit of the KEL (Fig. 8a) is characterized by a concentric morphology, outlined by compressional ridges, and an extensive zone of rafted blocks surrounded by landslide matrix (Fig. 8a, b). These blocks spread over an area of 49 200 m^2, extending about 140 m out of the main depositional zone. Only a small impact depression (0.5 m max. depth) is present adjacent to an upslope bedrock ridge, which protrudes about 10 m from the basin plain. Several of these bedrock ridges cross the evacuation zone, leading to steps (max. 45°) in the slide scar, which otherwise shows a slope gradient of 10–13° and gently levels out into the basin plain. The main MTD is underlain by an older (EH-C) MTD that shows frontal thrusting, but does not ramp up from its initial BSS (frontally confined landslide; Moernaut and De Batist 2011),

Table 2. *Morphometric data of selected landslides*

	KEL	KWL	SL	TEL	BSL
Depth below lake floor (m)	0.675	0.9	0.9	1.35	0.3
Total length L_t (m)	266	425	121	403	385
Deposit length L_d (m)	158	189	112	139.6	227
Evacuated length (m)	108	236	9	263.4	158
Maximum length rafted blocks (m)	143	67	0	0	0
Scar perimeter length (m)	475	721	–	673	348
Sidewall height* (m)	*c.* 1.6	1.5–2	–	3.1	3.5
Evacuation height (m)	38	50.5	max. 30	59.5	81
Scar width (m)	194	137	–	167	106
Maximum deposit width (m)	178	245	104	291	200.5
Maximum deposit thickness (m)	2.7	4.9	4.7	4.4	4.5
Total height drop (m)	43.1	54.2	31	60	82.6
Slope gradient (unfailed)	10°	16°	73°	19.2°	18.1°
Slope gradient at the toe	0.2°	0.1°	0.1°	0.1°	0°
Slope gradient change at toe of slope	10°	10–17°	73°	11°	36°
Area (m^2)	71 205	56 300	11 030	32 810	43 730
Volume (m^3)	79 034	96 309	26 896	92 600	68 885

Morphometric data were measured according to Clare *et al.* (2018). KEL, Krumpendorf East landslide; KWL, Krumpendorf West landslide; SL, Sekirn landslide; TEL, Töschling East landslide; BSL, Bad Saag landslide.
*Sidewall height was measured instead of the headscarp height, because no well-developed headscarps are visible in the bathymetric data.

leading to a topographic bulge in the subsurface. The BSS of the KEL follows the stratigraphic level defined by the top of this frontally confined landslide in the subsurface of the main deposit and ramps up in the stratigraphy by about 1 m where scattered blocks are present.

The BSS of the KWL (Fig. 8c, d) follows the top of an MTD (EH-C) as well, before ramping up about 2 m in the stratigraphy, but in contrast to the KEL, this BSS gently dips basinward. The KWL shows a large impact depression (max. depth 0.6 m) and high (up to 2.5 m) lateral compressional ridges (Fig. 8c, d). Some scattered blocks are present in front of a 56 m wide frontal breaching zone, spreading over an area of about 9500 m^2. The evacuated length is considerably longer than that of the KEL (236 m v. 108 m) and the maximum deposit thickness is almost double that of the KEL (4.9 m v. 2.7 m). Some of the landslide mass spilled laterally to the east (marked with black arrows in Fig. 8c). There, the sediments moved along the gentle slope, progressively disintegrating to form a frontally emergent MTD.

The SL (Fig. 8c, e) differs considerably from the KEL and KWL, because it exhibits an intact, 2.3 m high frontal compressional ridge and a small impact depression (0.5 m compared to the present-day lake floor), giving it a perfect horseshoe morphology. Flow-path analysis, as well as considerations of slide scar roughness and limited lateral extent of the SL, indicate that the upslope area (light grey in Fig. 8c) cannot constitute the evacuated zone of the SL, because (1) there is a significant misfit between the large excavation volume and the small impact zone, (2) bedrock ridges cause a positive morphology in a north–south direction, diverting a potential sediment flow to the NE (grey arrow in Fig. 8c) and (3) the area has a smooth surface, indicating a significant sediment drape, which is not visible in the scars of the coeval KEL and KWL. Instead, in the 30 m high cliff bordering the impact zone, a small niche is present. This niche might either have been created during a rockfall from the cliff or host enough space for sediment to accommodate and fail. However, considering the narrow bedrock ridges in the eastern basin, which do not show a substantial sedimentary drape, the latter explanation seems unlikely.

Western basin. The Velden MTC (Fig. 9a), which has already been introduced in the section on seismic landslide stratigraphy, defines the structure of the lake bottom in the very west, spreading over an area of *c.* 0.7×10^6 m^2. Several amalgamated MTDs contribute to the complex morphology consisting of multiple frontal and lateral compressional ridges, rafted blocks, as well as smooth surfaces in local depocentres, where megaturbidites pond depressions in underlying rough MTC surfaces. Some slope segments of up to 380 m in length failed almost entirely, leaving only small areas of unfailed upper slope sediments behind, whereas other slopes, including deltaic bodies, remained intact. The sidewalls and headscarps of most landslides are about 4 m high.

The Töschling West landslide (TWL) shows a highly complex evacuation area, with multiple downstepping, staircase-type scars along a terraced slope (Fig. 9b). The slope gradient changes drastically from the nearshore area (19°), to the terrace (1°) and along the lower slope (32–7°). The depositional area is characterized by a large, semi-circular compression ridge covering an area of 46×10^3 m^2 without an apparent frontal breaching zone. Large (up to 80 m wide) blocks, probably originating from lower slope failures, are dispersed on the landslide lobe. The deposits pinch out at two different stratigraphical intervals (Fig. 9c), indicating two superimposed MTDs ('MTD-stack'), the older (TWL-I) of which correlates stratigraphically to EH-B. Part of the younger MTD (TWL-II) levels out the impact depression of the older landslide.

The big lobe (lobe 1 in Fig. 9d) of the Töschling East landslide (TEL) on the other hand, has a well-developed, 1.5 m high compressional ridge and an internal depression zone. Together with the smaller lobe, which has a blocky appearance, the deposit constitutes a single MTD, covering an area of 30×10^3 m^2. The presence of two depositional lobes is due to a bedrock ridge present in the lower slope, splitting the flow path. The scar width of the TEL increases from *c.* 70 m at the upper slope to about 160 m at the lower slope (Fig. 9d).

Anthropogenic landslides. The Bad Saag landslide (BSL; Fig. 10a–d) of 2013, already introduced above, led to the failure of several thousand cubic metres of onshore sediment along a 106 m broad scar with 4.5 m high sidewalls. Impact on the basin plain led to a 2 m deep depressional zone. The sediments spread over an area of 44×10^3 m^2 and led to a turbiditic current, which can be recovered as a 1 cm thick turbidite about 650 m away from the main depositional area (Fig. 10c; see Fig. 1 for location of core WOER17-14).

The Seefels landslide (SEL) originates in a very narrow (10 m) onshore area and shows a downslope-diverging slide scar, reaching a maximum width of 45 m (Fig. 10e). Most of the dispersed landslide mass was either deposited on the lower slope or at the toe of a small gully, covering an area of *c.* 3.5×10^3 m^2. Some of the remobilized sediment might have triggered minor failures at the very base of the slope (zone indicated by yellow dashed lines in Fig. 10e).

The deposits of the Pörtschach landslide (PL; Fig. 10f) are dispersed over an area of 28×10^3 m^2 along a terraced slope. They show heavy disintegration and follow a slide scar of an older MTD, related to EH-B. Only a small number of blocks made it into the deep basin. The slide scar can be traced onshore, originating in a narrow area (*c.* 15 m wide) and spreading downslope along a morphology defined by bedrock ridges.

The Dellach landslide (DL; Fig. 10g) only eroded the uppermost metre of sediment, contrasting with the deep (4–5 m) landslide scars of most landslides in Wörthersee, including the BSL, SEL and PL. In the nearshore area, the narrow scar (20 m) is blurred by a rough lake-floor morphology showing several small-scale irregularities. Downslope, the slide scar branches out into individual sub-scars.

Discussion

Preconditioning and trigger of landslides in Wörthersee

Several preconditioning factors can increase the landslide-susceptibility of slopes, for instance (1) excess pore pressure and underconsolidation due to rapid deposition of sediments (Prior *et al.* 1982), (2) weak layers (Locat *et al.* 2014 and references therein) or (3) oversteepened slopes (Piper and McCall 2003). The ultimate slope failure is caused by the combination of these prerequisites with short-term triggering mechanisms, such as: (a) loading of slopes, either associated with human activity (Assier-Rzadkiewicz *et al.* 2000), subaerial rockfalls (Kremer *et al.* 2015; Daxer *et al.* 2018) or heavy rainfall (Longva *et al.* 2003); (b) lake level fluctuations (Anselmetti *et al.* 2009); and, most importantly, (c) earthquakes (Schnellmann *et al.* 2002; Simonneau *et al.* 2013; Van Daele *et al.* 2015; Normandeau *et al.* 2016).

Based on our estimates and the drastic lithological change from LU-III to LU-II (i.e. from light-grey, well-laminated, fine-grained sediments to coarser deposits with high amounts of authigenic calcite and organic matter), as well as the characteristic change from ponding to draping geometries across the boundary of SU-2 to SU-1 (described in many other formerly glaciated peri-alpine lakes, e.g. Daxer *et al.* 2018 and references therein), we interpret the transition from LU-III to LU-II (and the correlating changeover from SU-2 to SU-1; Fig. 5) as the Late Pleistocene to Holocene transition, when the lake switched from a glacial meltwater-influenced lake to a peri-alpine lake dominated by hemipelagic sedimentation. The sedimentation rates during the deposition of SU-2 (thickness >43 ms) exceeded 4–5 mm a^{-1}, assuming glacier-free lake basins at 20 kaBP and a seismic velocity of 1500 m s^{-1}. In the Holocene, the sedimentation rates decrease considerably to about 0.5 mm a^{-1}. This lithological succession closely resembles that of many other previously glaciated depositional basins exhibiting large-scale landsliding, be it fjords (Aarseth *et al.* 1989; Hjelstuen *et al.* 2009; Hjelstuen and Brendryen 2014) or lakes (Chapron *et al.* 1996; Reusch *et al.* 2016; Strupler *et al.* 2017). It has been shown that high sedimentation rates (>1 mm a^{-1}) and low permeability can inhibit a normal consolidation process during sedimentation, leading to the generation of shallow pore fluid overpressure (Dugan and Germaine 2008; Dugan and Sheahan 2012). Underconsolidation and a reduced shear strength due to overpressure (e.g. Stigall and Dugan 2010; Wiemer *et al.* 2015) in the Late Glacial LU-III may be one of the reasons why EH-B and EH-C differ so strongly in remobilized volume (1:5.5), source area of failure (steep slopes v. slope & intra-basin) and consequently flow evolution (megaturbidite v. mass flow/megaturbidite). Because the sidewall height of most landslides visible in the multibeam bathymetry is in the range of 1.5–4.5 m and sedimentation rates are generally lower on slopes (Blais and Kalff 1995), failure of these MTDs most likely occurred in Late Glacial clays or at the transition to LU-II. This is similar to findings in Lake Lucerne (Strasser *et al.* 2007) and Lake Zurich (Strupler *et al.* 2017), indicating further dependence of slope initiation on lithological variability and the possible presence of predefined weak layers in the LU-III stratigraphy.

Coevality (within the limits of seismic resolution) of landslides with different source areas, as seen in EH-B and EH-C, has proven to be a good indicator of a seismic trigger mechanism (Schnellmann *et al.* 2002; Praet *et al.* 2017; Moernaut *et al.* 2019). As shown by a radiocarbon age (AD 1331–1441), the AD 1348 earthquake constitutes a likely trigger of the multiple landslides related to EH-B. This earthquake is the largest seismic event historically recorded in the entire Alps (Stucchi *et al.* 2013) and led to massive rockslides at the nearby Dobratsch mountain (Lenhardt 2007). The reported intensities (EMS-98) of this earthquake were around VIII–IX at the lake site (Hammerl 1994; Stucchi *et al.* 2013). Compared with studies on lacustrine palaeoseismology, this is well above the empirical intensity threshold (seismic intensities $\geq$ VI) for seismically triggered landslides (Monecke *et al.* 2004; Van Daele *et al.* 2015). Therefore, a causal relationship between the AD 1348 earthquake and EH-B is very likely. Similar assumptions can be made for EH-C dated to *c.* 13.7 cal ka BP, which is represented by synchronous landslides on an even larger scale. The fact that all lake basins (and sub-basins) failed at about the same time can only be explained by either a very local major seismic event or a sudden drop in lake level. The lithological change from fine-grained, bright grey, millimetre-scale laminae (LU-III) to darker coloured, more organically rich sediments (LU-II) a few centimetres above the turbidite deposit related to EH-C indicates a major change in lake system dynamics and sedimentation after the landslide event. A concomitant drop in lake level, however, is unlikely due to the lack of visible erosional surfaces in the seismic data at shallow water depths, thus favouring an earthquake trigger.

Human-induced landslides

Since the youngest mass movements cannot be attributed to the AD 1976 Friuli earthquake (M_w *c.* 6.5), which was the most recent major ($M_w > 6$) earthquake in the region, an anthropogenic trigger is a likely explanation for the Bad Saag, Seefels, Pörtschach and Dellach landslides. Indeed, for the BSL, loading of slopes during building activity is documented as the cause of failure initiation. In the 1980s, piers were built in the headscarp area of the SFL and the DL, probably leading to the small-scale slope failures visible in the bathymetry and accounting for the rough lake floor at the DL. These landslides are also a very likely source for the turbidites visible in core WOER18-L5-SC (Fig. 7). For the bigger Pörtschach landslide, the construction of a boathouse seems to be the triggering event. The slide scar developed from a small, point-source failure and propagated downslope along a bedrock ridge, leading to a larger-scale subaqueous failure. Similar observations have been made for the Oberrieden slide in Lake Zurich (Strupler *et al.* 2018). This MTD, as well as the well-studied anthropogenic Horgen slide, however, are much bigger in volume (Oberrieden: 1.49×10^6 m^3; Horgen: 1.29×10^6 m^3; Sammartini *et al.* 2020) than the human-induced landslides in Wörthersee. The development of large translational landslides from point sources in Wörthersee is largely hindered by bedrock ridges, which act as lateral boundaries to failure propagation. Just as for the BSL, the evacuation zones of the SEL, PL and DL extend onshore, making this the distinctive feature compared to the older non-anthropogenic (EH-B) events, where the headscarp is well below the water level (>15 m water depth).

Factors controlling landslide morphology

The morphology of landslides in Wörthersee seems to be mainly controlled by three driving forces: (1) slope gradient of the evacuation zone and changes therein; (2) preconditioning of the impacted area by previous landslides; and (3) the remobilized sediment volume.

The most significant difference between the two 'end-members' of landslide morphologies observed in the eastern basin, the heavily disintegrated KEL and the horseshoe-shaped SL, is the slope gradient (14° v. 73°). Steeper slope gradients generally cause higher impact velocities (depending on height drop) and, therefore, lead to stronger deformation of basin-plain sediments and frontal thrusts. In contrast to the deposits originating from very steep (>45°)

slopes in Aysén fjord (Lastras *et al.* 2013; Van Daele *et al.* 2013), however, the SL does not show a distal unconfined part. This might be explained by (1) the drastic change in slope gradient at the toe of slope, which does not allowing the kinetic energy of the (possibly very small) impactor to be transferred into lateral movement of basin-plain sediment (Mazzanti and De Blasio 2010), (2) different types of impactors, i.e. onshore debris entrained in a mass-flow in Aysén v. bedrock material from an underwater cliff in Wörthersee, and (3) the much larger height drop at the Aysén fjord (>195 m, excluding the onshore part; Van Daele *et al.* 2013). The landslide morphology of the BSL supports the second explanation: similar to the Aysén mass-flow, coarse, already disintegrated material mixed with hemipelagic sediment impacted on the basin plain during the BSL, resulting in a deep depressional zone, but still managed to emerge distally for a considerable distance. The KWL, showing high compressional ridges as well as a few rafted blocks, represents an intermediate member of landslide morphology. Here, the change of slope gradient is not as extreme as for the SL, but greater than for the KEL (17° v. 10°).

Aside from the slope gradient, the variations in flow development and disintegration could also be linked to the morphology of the impacted zone: upon impact, both the KEL as well as the KWL remobilized basin-plain sediment along a shear surface developed at the lithological boundary between older MTDs time-correlative to EH-C and the overlying sedimentary sequence. Whereas at the location of KWL this shear surface is gently inclined along the path of travel and thus facilitated the remobilized mass to plough into the basin-plain sediment, the same surface on top of a frontally confined Late Glacial MTD exhibits a slightly positive upward bend at KEL, likely facilitating ramping up and disintegration.

Another type of preconditioning of the depositional area is seen at the Töschling West slides, where the first event, which probably resembled the single MTD of Töschling East, became overprinted by subsequent events. The impact depression of TWL-I was filled by TWL-II and the compressional ridge was partly buried by the lower slope failure.

Run-out length has generally been found to be proportional to failure volume (e.g. Gee *et al.* 1999; Moernaut and De Batist 2011). This explains the long run-out of the Velden MTC and of the landslides related to EH-C. For the latter, the high content of clay-sized minerals probably contributed to the high mobility (Longva *et al.* 2003), facilitating propagation even on an almost flat basin plain. For such massive landslides to happen along slopes or onshore, enough sediment potentially ready to fail needs to be present. In the Velden area, this is guaranteed by several inflows, leading to active recharge of slopes.

Conclusions

In the present study, we depicted the dimensions, morphology and unravelled the date of several landslides in glacigenic Wörthersee by means of reflection seismic mapping of MTDs, coring of MTDs and their respective turbidites, and analysis of multibeam bathymetric data. Based on these studies, we come to a number of general (1, 2) and site-specific (3–5) conclusions.

(1) Preconditioning of the landslide impact area by previous mass movements can have a strong influence on landslide morphology: the top of an older MTD can act as the basal shear surface, altering the propagation path of a landslide.
(2) Deformation and erosion of basin-plain sediments is dependent on (a) slope gradient of the slide scar and changes therein, (b) excavation height and (c) the type of impactor (hemipelagic sediment, (onshore) mass flows, hard rock).
(3) Human-induced landslides in Wörthersee were all initiated onshore, but downslope failure propagation partly resulted in a larger subaqueous landslide.
(4) Rapid sedimentation of fine-grained glaciolacustrine sediments in the Late Glacial allowed for extensive landsliding (*c.* 28.7×10^6 m^3) in Wörthersee. This is relevant for all types of glacigenic basins that show a similar lithological evolution.
(5) The AD 1348 earthquake probably triggered numerous landslides, including a large (*c.* 4.5×10^6 m^3) mass-transport complex in the western basin.

Acknowledgements We acknowledge ETH Zurich for access to the Kullenberg coring system and platform and we are very grateful to all the people who helped carrying out fieldwork and retrieving Kullenberg-cores on Wörthersee: Michael Hilbe, Reto Seifert and Patrick Oswald. The Research Department for Limnology (ILIM, Mondsee) and Hannes Höllerer are thanked for providing the research vessel *Luna* and logistical help, as is the Carinthian Institute of Lake Research (KIS). The multibeam data were kindly provided by the state of Carinthia (Land Kärnten, Abteilung Umwelt, Energie und Naturschutz). Further acknowledged are Irka Hajdas for radiocarbon analyses and Pascal Rünzi for both radionuclide dating and help during field work. We are very grateful to Markus Erhardt, Gerald Degenhart and Wolfgang Recheis for the use of the medical CT scanner at the Medical University of Innsbruck. We thank Marc De Batist, an anonymous reviewer and Jason Chaytor for their constructive comments on an earlier version of this manuscript.

Funding This project was funded by the Austrian Science Fund (FWF), project number P30285-N34, and the Austrian Academy of Sciences project 'S[4]SLIDE-Austria'. Maddalena Sammartini acknowledges funding from the European Union's Horizon 2020 research and innovation programme under the Marie-Skłodowska-Curie grant agreement No 721403. IHS Markit is thanked for its educational grant programme providing the KINGDOM seismic interpretation software.

Author contributions **CD**: conceptualization (lead), investigation (lead), writing – original draft (lead), writing – review & editing (lead); **MSa**: investigation (supporting); **AM**: investigation (supporting); **TP**: funding acquisition (supporting), resources (supporting); **MSt**: conceptualization (supporting), funding acquisition (equal), project administration (supporting), resources (equal), supervision (equal); **JM**: conceptualization (equal), funding acquisition (lead), investigation (supporting), methodology (supporting), project administration (lead), resources (lead), supervision (lead).

References

Aarseth, I., Lønne, Ø. and Giskeødegaard, O. 1989. Submarine slides in glaciomarine sediments in some western Norwegian fjords. *Marine Geology*, **88**, 1–21, https://doi.org/10.1016/0025-3227(89)90002-9

Anderle, N. 1977. *Geologische Karte der Republik Österreich 1:50000*, 201–210 Villach-Assling. 1.

Anselmetti, F.S., Ariztegui, D. *et al.* 2009. Environmental history of southern Patagonia unravelled by the seismic stratigraphy of Laguna Potrok Aike. *Sedimentology*, **56**, 873–892, https://doi.org/10.1111/j.1365-3091.2008.01002.x

APA/derStandard 2013. Kärnten: Erdrutsch bei ÖBB-Baustelle riss Straße weg, https://www.derstandard.at/story/1385170631020/erdrutsch-bei-oebb-baustelle-strasse-am-woerthersee-nordufer-weggerissen

Assier-Rzadkiewicz, S., Heinrich, P., Sabatier, P.C., Savoye, B. and Bourillet, J.F. 2000. Numerical modelling of a landslide-generated Tsunami: the 1979 nice event. *Pure and Applied Geophysics*, **157**, 1707–1727, https://doi.org/10.1007/PL00001057

Blais, J.M. and Kalff, J. 1995. The influence of lake morphometry on sediment focusing. *Limnology and Oceanography*, **40**, 582–588, https://doi.org/10.4319/lo.1995.40.3.0582

Bondevik, S., Løvholt, F., Harbitz, C., Mangerud, J., Dawson, A. and Svendsen, J.I. 2005. The Storegga Slide tsunami – comparing field observations with numerical simulations. *Marine and Petroleum Geology*, **22**, 195–208, https://doi.org/10.1016/j.marpetgeo.2004.10.003

Chapron, E., Van Rensbergen, P., Beck, C., De Batist, M. and Paillet, A. 1996. Lacustrine sedimentary records of brutal events in Lake Le Bourget (Northwestern Alps-Southern Jura). *Quaternaire*, **7**, 155–168, https://doi.org/10.3406/quate.1996.2068

Clare, M., Chaytor, J. *et al.* 2018. A consistent global approach for the morphometric characterization of subaqueous landslides. *Geological Society, London, Special Publications*, **477**, 455–477, https://doi.org/10.1144/SP477.15

Conrad, O., Bechtel, B. *et al.* 2015. System for automated geoscientific analyses (SAGA) v. 2.1.4. *Geoscientific Model Development*, **8**, 1991–2007, https://doi.org/10.5194/gmd-8-1991-2015

Daxer, C., Moernaut, J., Taylor, T., Haas, J.N. and Strasser, M. 2018. Late Glacial and Holocene sedimentary infill of Lake Mondsee (Eastern Alps, Austria) and historical rockfall activity revealed by reflection seismics and sediment core analysis. *Austrian Journal of Earth Sciences*, **111**, 111–134, https://doi.org/10.17738/ajes.2018.0008

Dugan, B. and Germaine, J.T. 2008. Near-seafloor overpressure in the deepwater Mississippi Canyon, northern Gulf of Mexico. *Geophysical Research Letters*, **35**, https://doi.org/10.1029/2007GL032275

Dugan, B. and Sheahan, T.C. 2012. Offshore sediment overpressures of passive margins: mechanisms, measurement, and models. *Reviews of Geophysics*, **50**, https://doi.org/10.1029/2011rg000379

Enet, F., Grilli, S.T. and Watts, P. 2003. Laboratory experiments for Tsunamis generated by underwater landslides: comparison with numerical modeling. *Proceedings of the Thirteenth (2003) International Offshore and Polar Engineering Conference*, ISOPE03, Honolulu, USA, May 2003, 372–379.

Gee, M.J.R., Masson, D.G., Watts, A.B. and Allen, P.A. 1999. The Saharan debris flow: an insight into the mechanics of long runout submarine debris flows. *Sedimentology*, **46**, 317–335, https://doi.org/10.1046/j.1365-3091.1999.00215.x

Girardclos, S., Schmidt, O.T., Sturm, M., Ariztegui, D., Pugin, A. and Anselmetti, F.S. 2007. The 1996 AD delta collapse and large turbidite in Lake Brienz. *Marine Geology*, **241**, 137–154, https://doi.org/10.1016/j.margeo.2007.03.011

Hammerl, C. 1994. The earthquake of January 25th, 1348, discussion of sources. *Historical Investigation of European Earthquakes. Materials of the CEC Project "Review of Historical Seismicity in Europe"*, **2**, 225–240.

Hjelstuen, B.O. and Brendryen, J. 2014. Submarine mass movements and trigger mechanisms in Byfjorden, western Norway. *In*: *Advances in Natural and Technological Hazards Research.* Springer, Netherlands, 351–359, https://doi.org/10.1007/978-3-319-00972-8_31

Hjelstuen, B.O., Haflidason, H., Sejrup, H.P. and Lyså, A. 2009. Sedimentary processes and depositional environments in glaciated fjord systems – evidence from Nordfjord, Norway. *Marine Geology*, **258**, 88–99, https://doi.org/10.1016/j.margeo.2008.11.010

Homann, O. 1962. Die geologisch-petrographischen Verhältnisse im Raume Ossiachersee - Wörthersee (südlich Feldkirchen zwischen Klagenfurt und Villach). *Jahrbuch der Geologischen Bundesanstalt*, **105**, 243–272.

Hühnerbach, V. and Masson, D.G. 2004. Landslides in the North Atlantic and its adjacent seas: an analysis of their morphology, setting and behaviour. *Marine Geology*, **213**, 343–362, https://doi.org/10.1016/j.margeo.2004.10.013

Kahler, F. 1962. *Geologische Karte der Umgebung von Klagenfurt 1:50000*. 1.

Kelts, K., Briegel, U., Ghilardi, K. and Hsu, K. 1986. The limnogeology-ETH coring system. *Swiss Journal of Hydrology*, **48**, 104–115, https://doi.org/10.1007/BF02544119

KIS 2019. Wörthersee. Seenseite, http://www.kis.ktn.gv.at/188551_DE-Kaerntner_Seen-Seenseite.?seeid=46

Krastel, S., Behrmann, J.-H. *et al.* (eds) 2014. *Submarine Mass Movements and Their Consequences*. Advances in Natural and Technological Hazards Research, **37**. Springer, Cham, https://doi.org/10.1007/978-3-319-00972-8

Kremer, K., Corella, J.P., Adatte, T., Garnier, E., Zenhäusern, G. and Girardclos, S. 2015. Origin of turbidites in deep Lake Geneva (France-Switzerland) in the last 1500 years. *Journal of Sedimentary Research*, **85**, 1455–1465, https://doi.org/10.2110/jsr.2015.92

Lastras, G., Amblas, D. *et al.* 2013. Landslides cause tsunami waves: insights from aysén fjord, Chile. *Eos*, **94**, 297–298, https://doi.org/10.1002/2013EO340002

Lenhardt, W.A. 2007. Earthquake-triggered landslides in Austria–Dobratsch revisited. *Jahrbuch der Geologischen Bundesanstalt*, **147**, 193–199.

L'Heureux, J.-S., Longva, O. *et al.* 2012. Identification of weak layers and their role for the stability of slopes at Finneidfjord, Northern Norway. *Submarine Mass Movements and Their Consequences*. Dordrecht, Springer Netherlands, 321–330, https://doi.org/10.1007/978-94-007-2162-3_29

Locat, J., Leroueil, S., Locat, A. and Lee, H. 2014. Weak layers: their definition and classification from a geotechnical perspective. *Submarine Mass Movements and Their Consequences, 6th International Symposium*. Kluwer Academic, 3–12, https://doi.org/10.1007/978-3-319-00972-8_1

Longva, O., Janbu, N., Blikra, L.H. and Bøe, R. 2003. The 1996 Finneidfjord Slide; seafloor failure and slide dynamics. *In*: *Advances in Natural and Technological Hazards Research*, **19**, 531–538, https://doi.org/10.1007/978-94-010-0093-2_58

Masson, D.G., Harbitz, C.B., Wynn, R.B., Pedersen, G. and Løvholt, F. 2006. Submarine landslides: processes, triggers and hazard prediction. *Philosophical Transactions of the Royal Society A: Mathematical, Physical and Engineering Sciences*, **364**, 2009–2039, https://doi.org/10.1098/rsta.2006.1810

Mazzanti, P. and De Blasio, F.V. 2010. Peculiar morphologies of subaqueous landslide deposits and their relationship to flow dynamics. *Submarine Mass Movements and Their Consequences – 4th International Symposium*, 141–151, https://doi.org/10.1007/978-90-481-3071-9

Moernaut, J. and De Batist, M. 2011. Frontal emplacement and mobility of sublacustrine landslides: results from morphometric and seismostratigraphic analysis. *Marine Geology*, **285**, 29–45, https://doi.org/10.1016/j.margeo.2011.05.001

Moernaut, J., Van Daele, M. *et al.* 2019. The subaqueous landslide cycle in south-central Chilean lakes: the role of tephra, slope gradient and repeated seismic shaking. *Sedimentary Geology*, **381**, 84–105, https://doi.org/10.1016/j.sedgeo.2019.01.002

Monecke, K., Anselmetti, F.S., Becker, A., Sturm, M. and Giardini, D. 2004. The record of historic earthquakes in lake sediments of Central Switzerland. *Tectonophysics*, **394**, 21–40, https://doi.org/10.1016/j.tecto.2004.07.053

Normandeau, A., Lamoureux, S.F., Lajeunesse, P. and Francus, P. 2016. Sediment dynamics in paired High Arctic lakes revealed from high-resolution swath bathymetry and acoustic stratigraphy surveys. *Journal of Geophysical Research: Earth Surface*, **121**, 1676–1696, https://doi.org/10.1002/2016JF003873

Owen, M., Day, S. and Maslin, M. 2007. Late Pleistocene submarine mass movements: occurrence and causes. *Quaternary Science Reviews*, **26**, 958–978, https://doi.org/10.1016/j.quascirev.2006.12.011

Piper, D.J.W. and McCall, C. 2003. A synthesis of the distribution of submarine mass movements on the eastern Canadian margin. *In*: Locat, J., Mienert, J. and Boisvert, L. (eds) *Submarine Mass Movements and Their Consequences: 1st International Symposium*. Dordrecht, Netherlands, 291–298, https://doi.org/10.1007/978-94-010-0093-2_32

Piper, D.J.W., Cochonat, P. and Morrison, M.L. 1999. The sequence of events around the epicentre of the 1929 Grand Banks earthquake: initiation of debris flows and turbidity current inferred from sidescan sonar. *Sedimentology*, **46**, 79–97, https://doi.org/10.1046/j.1365-3091.1999.00204.x

Pistotnik, J., Daurer, A. *et al.* 1980. Die östlichen Zentralalpen (südlich der Hohen Tauern und östlich der Katschberg-Linie). *In*: *Der Geologische Aufbau Österreichs*. Springer, Vienna, 348–404, https://doi.org/10.1007/978-3-7091-3744-4_19

Pope, E.L., Talling, P.J., Carter, L., Clare, M.A. and Hunt, J.E. 2017. Damaging sediment density flows triggered by tropical cyclones. *Earth and Planetary Science Letters*, **458**, 161–169, https://doi.org/10.1016/j.epsl.2016.10.046

Praet, N., Moernaut, J. *et al.* 2017. Paleoseismic potential of sublacustrine landslide records in a high-seismicity setting (south-central Alaska). *Marine Geology*, **384**, 103–119, https://doi.org/10.1016/j.margeo.2016.05.004

Prior, D.B., Coleman, J.M. and Bornhold, B.D. 1982. Results of a known seafloor instability event. *Geo-Marine Letters*, **2**, 117–122, https://doi.org/10.1007/BF02462751

Reusch, A., Moernaut, J., Anselmetti, F.S. and Strasser, M. 2016. Sediment mobilization deposits from episodic subsurface fluid flow – A new tool to reveal long-term earthquake records? *Geology*, **44**, 243–246, https://doi.org/10.1130/G37410.1

Rzadkiewicz, S.A., Mariotti, C. and Heinrich, P. 1996. Modelling of submarine landslides and generated water waves. *Physics and Chemistry of the Earth*, **21**, 7–12, https://doi.org/10.1016/S0079-1946(97)00002-5

Sammartini, M., Moernaut, J., Anselmetti, F.S., Hilbe, M., Lindhorst, K., Praet, N. and Strasser, M. 2020. An atlas of mass-transport deposits in lakes. *In*: Ogata, K., Festa, A. and Pini, G.A. (eds) *Submarine Landslides: Subaqueous Mass Transport Deposits from Outcrops to Seismic Profiles*. American Geophysical Union.

Schindelin, J., Arganda-Carreras, I. *et al.* 2012. Fiji: an open-source platform for biological-image analysis. *Nature Methods*, **9**, 676–682, https://doi.org/10.1038/nmeth.2019

Schmidt, R., Weckström, K., Lauterbach, S., Tessadri, R. and Huber, K. 2012. North Atlantic climate impact on early late-glacial climate oscillations in the southeastern Alps inferred from a multi-proxy lake sediment record. *Journal of Quaternary Science*, **27**, 40–50, https://doi.org/10.1002/jqs.1505

Schnellmann, M., Anselmetti, F.S., Giardini, D., McKenzie, J.A. and Ward, S.N. 2002. Prehistoric earthquake history revealed by lacustrine slump deposits. *Geology*, **30**, 1131, https://doi.org/10.1130/0091-7613(2002)030<1131:PEHRBL>2.0.CO;2

Schnurrenberger, D., Russell, J. and Kelts, K. 2003. Classification of lacustrine sediments based on sedimentary components. *Journal of Paleolimnology*, **29**, 141–154, https://doi.org/10.1023/A:1023270324800

Schultze, E. 1984. Neue Erkenntnisse zur spät-und früh-postglazialen Vegetations-und Klimaentwicklung im Klagenfurter Becken. *Carinthia II*, **174**, 261–266.

Schulz, L., Schönhuber, M., Santner, M. and Swaton, T. 2008. Erhebung des Ist-Zustandes der Uferverbauung und Seeeinbauten des Wörthersees und des Millstätter Sees. *Publikationen des Kärntner Instituts für Seenforschung*, **4**, 1–13.

Seguinot, J., Ivy-Ochs, S., Jouvet, G., Huss, M., Funk, M. and Preusser, F. 2018. Modelling last glacial cycle ice dynamics in the Alps. *The Cryosphere*, **12**, 3265–3285, https://doi.org/10.5194/tc-12-3265-2018

Simonneau, A., Chapron, E. *et al.* 2013. Mass-movement and flood-induced deposits in Lake Ledro, southern Alps, Italy: implications for Holocene palaeohydrology and natural hazards. *Climate of the Past*, **9**, 825–840, https://doi.org/10.5194/cp-9-825-2013

Stigall, J. and Dugan, B. 2010. Overpressure and earthquake initiated slope failure in the Ursa region, northern Gulf of Mexico. *Journal of Geophysical Research*, **115**, 1–11, https://doi.org/10.1029/2009jb006848

Strasser, M., Stegmann, S., Bussmann, F., Anselmetti, F.S., Rick, B. and Kopf, A. 2007. Quantifying subaqueous slope stability during seismic shaking: Lake Lucerne as model for ocean margins. *Marine Geology*, **240**, 77–97, https://doi.org/10.1016/j.margeo.2007.02.016

Strasser, M., Monecke, K., Schnellmann, M. and Anselmetti, F.S. 2013. Lake sediments as natural seismographs: a compiled record of Late Quaternary earthquakes in Central Switzerland and its implication for Alpine deformation. *Sedimentology*, **60**, 319–341, https://doi.org/10.1111/sed.12003

Strupler, M., Hilbe, M., Anselmetti, F.S., Kopf, A.J., Fleischmann, T. and Strasser, M. 2017. Probabilistic stability evaluation and seismic triggering scenarios of submerged slopes in Lake Zurich (Switzerland). *Geo-Marine Letters*, **37**, 241–258, https://doi.org/10.1007/s00367-017-0492-8

Strupler, M., Anselmetti, F.S., Hilbe, M. and Strasser, M. 2018. Quantitative characterization of subaqueous landslides in Lake Zurich (Switzerland) based on a high-resolution bathymetric dataset. *Geological Society, London, Special Publications*, **477**, 399–412, https://doi.org/10.1144/SP477.7

Stucchi, M., Rovida, A. *et al.* 2013. The SHARE European Earthquake Catalogue (SHEEC) 1000–1899. *Journal of Seismology*, **17**, 523–544, https://doi.org/10.1007/s10950-012-9335-2

Sue, L.P., Nokes, R.I., Davidson, M.J., Nokes, R.I. and Davidson, M.J. 2011. Tsunami generation by submarine landslides: comparison of physical and numerical models. *Environmental Fluid Mechanics*, **11**, 133–165, https://doi.org/10.1007/s10652-010-9205-9

Talling, P.J., Clare, M., Urlaub, M., Pope, E., Hunt, J.E. and Watt, S.F.L. 2014. Large submarine landslides on continental slopes: geohazards, methane release, and climate change. *Oceanography*, **27**, 32–45, https://doi.org/10.5670/oceanog.2014.38

Tappin, D.R., Watts, P. and Grilli, S.T. 2008. The Papua New Guinea tsunami of 17 July 1998: anatomy of a catastrophic event. *Hazards Earth Systems Science*, **8**, 243–266, https://doi.org/10.5194/nhess-8-243-2008

Twichell, D.C., Chaytor, J.D., ten Brink, U.S. and Buczkowski, B. 2009. Morphology of late Quaternary submarine landslides along the U.S. Atlantic continental margin. *Marine Geology*, **264**, 4–15, https://doi.org/10.1016/j.margeo.2009.01.009

Urgeles, R. and Camerlenghi, A. 2013. Submarine landslides of the Mediterranean Sea: trigger mechanisms, dynamics, and frequency-magnitude distribution. *Journal of Geophysical Research: Earth Surface*, **118**, 2600–2618, https://doi.org/10.1002/2013JF002720

Urgeles, R., Bahk, J.-J. *et al.* 2019. Tsunami hazard from submarine landslides: scenario-based assessment in the Ulleung Basin, East Sea (Japan Sea). *Geosciences Journal*, **23**, 439–460, https://doi.org/10.1007/s12303-018-0044-x

Van Daele, M., Versteeg, W., Pino, M., Urrutia, R. and De Batist, M. 2013. Widespread deformation of basin-plain sediments in aysén fjord (chile) due to impact by earthquake-triggered, onshore-generated mass movements. *Marine Geology*, **337**, 67–79, https://doi.org/10.1016/j.margeo.2013.01.006

Van Daele, M., and Moernaut, J. *et al.* 2015. A comparison of the sedimentary records of the 1960 and 2010 great Chilean earthquakes in 17 lakes: implications for quantitative lacustrine palaeoseismology. *Sedimentology*, **62**, https://doi.org/10.1111/sed.12193

Wiemer, G., Moernaut, J. *et al.* 2015. The role of sediment composition and behavior under dynamic loading conditions on slope failure initiation: a study of a subaqueous landslide in earthquake-prone South-Central Chile. *International Journal of Earth Sciences*, **104**, 1439–1457, https://doi.org/10.1007/s00531-015-1144-8

Yokoyama, R., Shirasawa, M. and Pike, R.J. 2002. Visualizing topography by openness A new application of image processing to DEMs. *Photogrammetric Engineering and Remote Sensing*, **68**, 257–265.

A numerical investigation of excess pore pressures and continental slope stability in response to ice-sheet dynamics

Morelia Urlaub[1]*, Isabel Kratzke[1] and Berit Oline Hjelstuen[2]
[1]GEOMAR Helmholtz Centre for Ocean Research Kiel, Kiel, Germany
[2]Department of Earth Science, University of Bergen, Allegt 41, 5007 Bergen, Norway
MU, 0000-0002-1116-636X
*Correspondence: murlaub@geomar.de

Abstract: Submarine landslides are common at glaciated continental margins. The onset of large-scale landslides coincides with the initiation of Northern Hemisphere glaciations in the Quaternary. This implies that processes related to glacial cycling provide favourable conditions for submarine landslides at high-latitude margins. Potential processes include glacial deposition patterns and enhanced seismicity. It is also possible that advances and retreats of ice sheets, a highly dynamic process in geological terms, makes slopes discernible to failure by modifying the stress regime. Here, we quantify this effect using 2D finite element modelling of a glaciated continental margin. Different model runs investigate the pore-pressure development in homogeneous, as well as layered, slopes during glaciation when loaded by an ice stream with one or more ice advances. Ice streams cause significant variations in excess pore pressure in the very shallow sediment sequences at the continental shelf. However, lateral fluid flow is not efficient enough to increase pore pressures significantly at the slope, where large-scale submarine slides are observed. Hence, while ice-sheet dynamics appear to favour the occurrence of shallow slides close to the shelf edge, ice sheets seem to be irrelevant for the generation of large-scale submarine landslides at the continental slope.

The NE Atlantic margin, from Ireland to Svalbard (Fig. 1), has experienced about 30 large submarine landslides throughout the last 2.7 myr (e.g. Vorren *et al.* 1998; Evans *et al.* 2005; Solheim *et al.* 2005; Hjelstuen *et al.* 2007). These submarine-slope failures represent a geohazard as they can cause damage to seabed installations, infrastructure and people. Mass failures may also generate devastating tsunamis, as was the case *c.* 8200 years ago when the Storegga Slide, offshore NW Norway (Fig. 1), generated a tsunami that impacted large parts of the North Atlantic region (Bondevik *et al.* 2005; Haflidason *et al.* 2005; Wagner *et al.* 2007; Fruergaard *et al.* 2015). For geohazard assessments, it is essential to understand what causes slope failure. Numerous hypotheses have been put forward as to what mechanisms could cause submarine landslides (e.g. Leynaud *et al.* 2009), which, however, are difficult to test in the real world mainly due to the inaccessibility of landslide-prone areas in the deep sea.

Remarkably, the largest slides occur on continental slopes with gradients of 2° or less (Hühnerbach *et al.* 2004). The failure of slopes with such low inclinations is difficult to explain, as this is well below typical friction angles for any type of sediment. To cause failure at such low slope angles, pore pressure must greatly exceed the hydrostatic pressure. A simple infinite slope calculation shows that for a $<2°$ slope, pore pressure must take up more than 94% of the lithostatic stress to destabilize sediment with an angle of repose of 30° (Urlaub *et al.* 2015).

Interestingly, the geological record of large submarine landslides along the NE Atlantic margin seems to be initiated in the Plio-Pleistocene (e.g. Solheim *et al.* 2005), and is thus coincident with the initiation of the Northern Hemisphere glaciations (Sejrup *et al.* 2005; Knies *et al.* 2009). Hence, a connection between submarine slides and glacial processes is very likely, meaning that the factors that cause large-scale instability are probably linked to glacial dynamics. Moving into a glacial–interglacial world invokes fundamental changes in the nature of deposited sediments, as well as the style and rate of deposition. During glaciations, thick packages of glacimarine and hemipelagic sediments are deposited. In addition, ice sheets transport glacial till to the shelf edge, which is then redistributed down slope towards the deep-sea basin as glacigenic debris flows (GDF) (Laberg and Vorren 1995; King *et al.* 1998). In contrast, slow deposition of purely marine hemipelagic sediments occurs during interglacials. In recent studies, Bellwald *et al.* (2019) and Llopart *et al.* (2019) numerically analysed how these sedimentary processes affect the pore pressure regime in the North Sea and Storfjorden Trough Mouth Fans (TMFs) (Fig. 1), respectively. They conclude

From: Georgiopoulou, A., Amy, L. A., Benetti, S., Chaytor, J. D., Clare, M. A., Gamboa, D., Haughton, P. D. W., Moernaut, J. and Mountjoy, J. J. (eds) 2020. *Subaqueous Mass Movements and their Consequences: Advances in Process Understanding, Monitoring and Hazard Assessments*. Geological Society, London, Special Publications, **500**, 255–266.
First published online March 13, 2020, https://doi.org/10.1144/SP500-2019-185

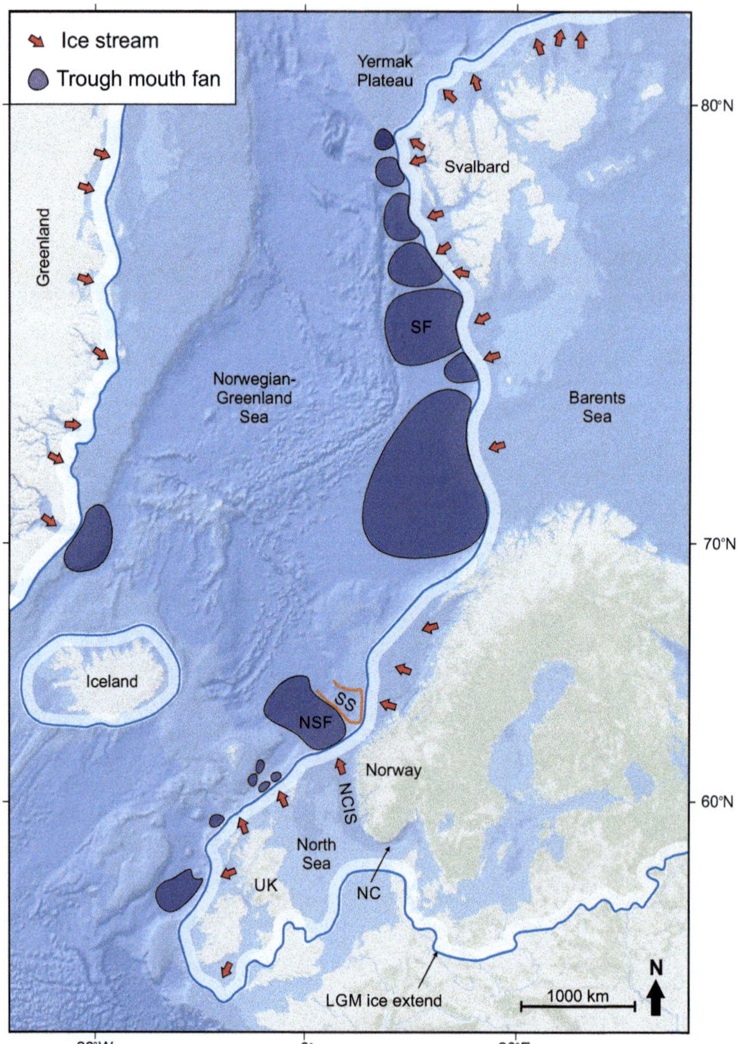

Fig. 1. Maximum ice extent (blue line) during the Last Glacial Maximum (LGM) including locations of trough mouth fans, commonly representing regions with recurrent submarine slides and ice streams following Sejrup *et al.* (2018). NCIS, Norwegian Channel Ice Stream; NC, Norwegian Channel; SF, Storfjorden Trough Mouth Fan; NSF, North Sea Trough Mouth Fan; SS, Storegga Slide.

that rapid deposition during glacials results in high excess pore pressures, but that an additional trigger is required to generate deep-seated slope failures.

In addition to changes in sedimentary processes, loading and unloading by thick ice sheets leads to elastic responses of the lithosphere (isostatic rebound) accompanied by increased levels of seismicity. The latter could account for an additional trigger (Bellwald *et al.* 2019) but is difficult to test conclusively. However, loading of the shelf by an ice sheet can also cause excess pore pressures in the sediments and thus decrease the effective stress, if dissipation of excess pore fluid is prevented. It is also possible that excess pore pressure underneath the ice induces fluid flow into ice-free areas (i.e. the continental slope), where effective stresses are reduced.

Therefore, the aim of this contribution is to test the hypothesis that ice loading causes pore pressure build-up in shelf sediments, which leads to lateral fluid flow towards the continental slope, where it might reduce slope stability. We use numerical models to quantify the amount of excess pore pressure that is generated by ice loading and lateral fluid flow and assess the effect on slope stability. Our simulations are set up to represent ice conditions of the Norwegian Channel Ice Stream (NCIS) in the North Sea, and the slope geometry and prevailing sediment types at the northern North Sea margin (Fig. 1). The northern North Sea margin is a well-studied region, where previous studies have resolved the spatial and temporal regional ice-sheet development. Moreover, several shallow boreholes and sediment cores in the Storegga Slide and Storfjorden TMF region (Fig. 1) (Kvalstad *et al.* 2005; Llopart *et al.* 2019) provide *in situ* information on sediment properties.

Ice-sheet dynamics

The Northern Hemisphere glaciations initiated at *c.* 2.74 Ma (Kleiven *et al.* 2002; Knies *et al.* 2009, 2014) and represent a time period characterized by repeated ice-sheet advances to the continental shelf edges of the northern North Atlantic. During maximum ice stages the Eurasian Ice Sheets were dissected by numerous ice streams (Fig. 1) (Ottesen *et al.* 2005), representing arteries of higher ice fluctuations (e.g. Stokes and Clark 2001). Ice streams are typically warm based (Anandakrishnan *et al.* 1998; De Angelis and Kleman 2008), which essentially means that their bases are impervious. The ice streams are normally restricted to cross-shelf troughs and are capable of transporting huge quantities of sediments (e.g. Nygård *et al.* 2007) to the continental slope, where TMF systems build up (Vorren *et al.* 1998). Commonly, these TMF systems (Fig. 1) are focus regions for recurrent submarine slide events (Nygård *et al.* 2005; Solheim *et al.* 2005; Hjelstuen *et al.* 2007).

It is suggested that ice sheets have reached the shelf edge offshore Norway in five marine isotope stages (MIS) during the last 0.5 myr of the Pleistocene time period (Sejrup *et al.* 2005). During the Last Glacial Maximum (LGM; MIS 2) the entire Barents Sea, Scandinavia and the North Sea were covered by an extensive ice sheet for the last time (Fig. 1). Nygård *et al.* (2005) have inferred that smaller fluctuations of stadial–interstadial phases occurred within MIS 2, 6, 8, 10 and 12 (Fig. 2a). The ice sheet extended to the shelf edge at least four times during MIS 2, three or more times during MIS 6, probably once during MIS 8 and at least once during MIS 10 and 12. Becker *et al.* (2018) have also reported a high variability of the NCIS during the LGM and confirm at least four distinct advances of the NCIS to the shelf edge.

Modelling approach

The model represents a generic glaciated continental-slope setting, based on the NCIS area. A load, variable in time and space, simulates ice-stream dynamics. The models do not consider sedimentation in order to isolate the effect of ice dynamics. Fully coupled 2D pore fluid diffusion and stress is solved to simulate flow of a single fluid phase through porous media. The porous medium is modelled by attaching the finite element mesh to the solid phase. The fluid can flow through the mesh and is governed by Darcy's law. The mechanical part of the model is based on the effective stress principle, where the sediments are fully saturated.

Model set-up

The geometry of the finite element model displays a simplified, but typical glaciated continental margin (Fig. 3). The model is 120 km long and includes a continental shelf (20 km, no inclination), a continental slope (80 km, 2° inclination), and a deep-sea basin setting (20 km, no inclination). Although the focus is on the upper 500 m below the sea floor, the model area is larger, with a depth of 4.8 km at the landward boundary and 2 km at the seaward boundary, to avoid boundary effects. The modelled slope is subdivided into several layers with different material properties. At the landward boundary the upper layer has a thickness of 100 m, the second layer 20 m, the third layer 100 m, the fourth layer 300 m, the fifth layer 500 m and the deepest layer 3780 m. The first five layers thin out to the half thickness in the direction of the seaward boundary to simulate deposition of sediments originating from the continent.

Boundary conditions

The base is fixed against movements in both horizontal and vertical directions. The lateral boundaries are free to move vertically but are fixed in the horizontal direction. The seafloor is free to move in both horizontal and vertical directions. The base of the model is impermeable due to the highly lithified sediments at this depth. The seaward lateral boundary is permeable, and a hydrostatic pore-pressure boundary condition is implied. The landward lateral boundary is impermeable. The seafloor of the slope and the deep-sea basin are permeable. The shelf is permeable in ice-free periods but is impermeable when an ice stream covers the respective area, because ice streams are warm based (Anandakrishnan *et al.* 1998; De Angelis and Kleman 2008).

Mesh

The finite element mesh is finest within the upper 200 m, with 20 m long and wide elements. Between depths of 200 and 1020 m, the elements increase in size up to 80 m in length and width. At the deeper

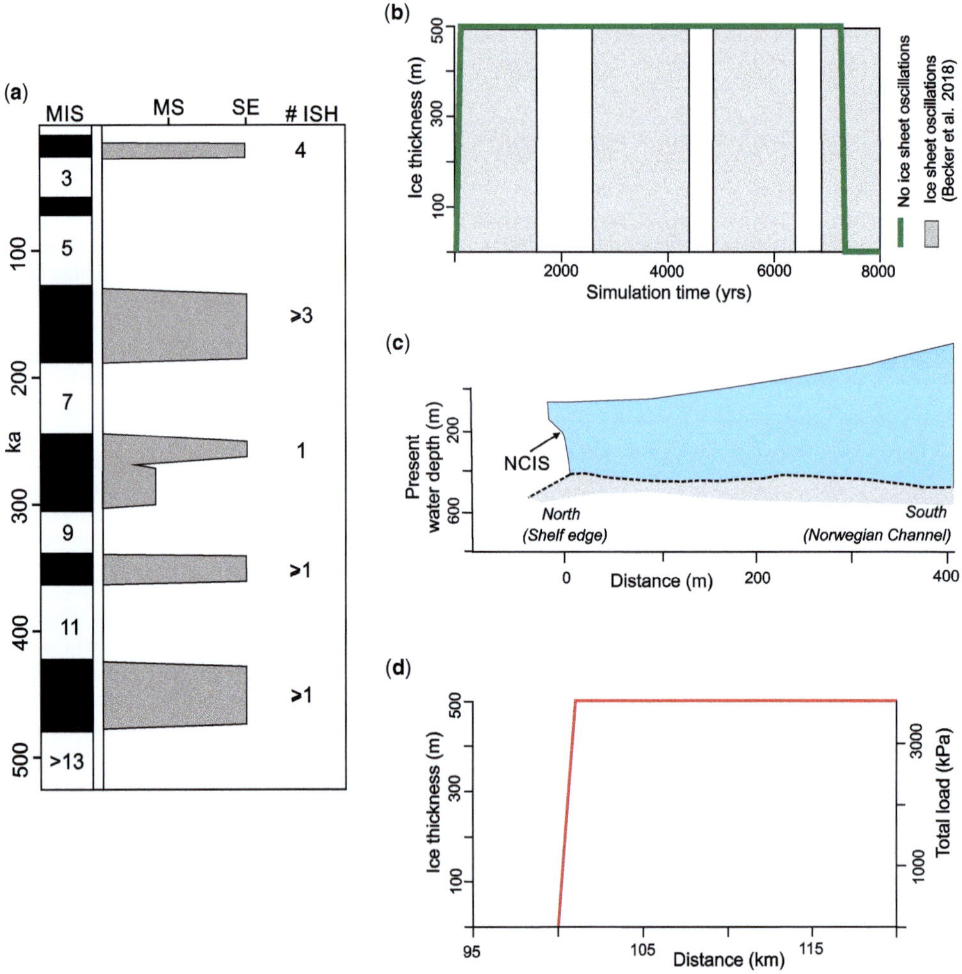

Fig. 2. Temporal and spatial ice-sheet characteristics from field observations and its simulation. (**a**) Assumed shelf edge glaciations of the Norwegian Channel Ice Stream (NCIS) during the last 0.5 myr (Nygård *et al.* 2005). MIS, marine isotope stage; MS, mid shelf; SE, shelf edge; ISH, ice-sheet oscillations. (**b**) Variation of ice-stream thickness over time as used in the models. (**c**) Proposed longitudinal NCIS profile (Sejrup *et al.* 2003). (**d**) Corresponding spatial load distribution in the models.

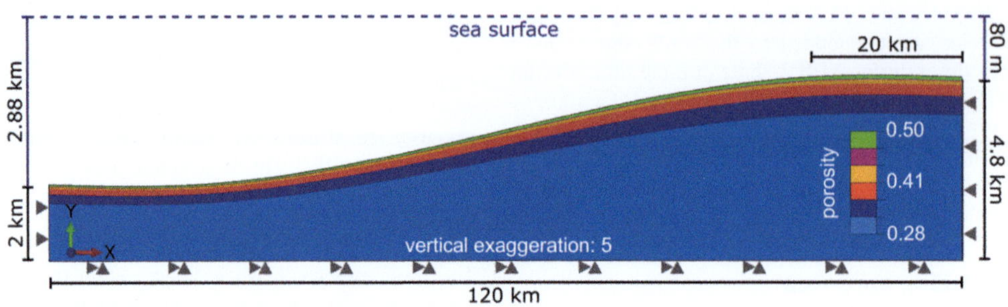

Fig. 3. 2D model geometry of all models. Colours show initial porosity of the reference model according to consolidation stress under gravity loading for normally consolidated materials.

part, which is outside of the interest area, the element size increases to 1000 m at the basal boundary of the model. The total number of elements is 135 625.

Loading

Gravity ($g = 9.81$ m s^{-2}) is taken into account by a body force load based on the sediment's density. We simulate ice as a load with varying spatial and temporal distribution (Fig. 2). The maximum ice-stream thickness is 500 m. The resulting load is calculated from the ice-stream thickness (500 m), the ice density (917 kg m^{-3}) and the gravity (9.81 m s^{-2}) and yields about 4500 kPa. As an ice stream displaces seawater, the maximum load on the continental shelf is the difference between the load of the ice stream and the load of the seawater. The present-day shelf edge in the study area has a water depth of about 200 m (Sejrup *et al.* 2003). During the LGM, the sea-level was *c.* 120 m lower than at present (Lambeck *et al.* 2014). Thus, the previous shelf edge in the study area was located in a water depth of about 80 m. The seawater load is calculated from the water depth (80 m), the salt-water density (1025 kg m^{-3}) and the gravity (9.81 m s^{-2}), which yields about 800 kPa. Therefore, a maximum load of 3700 kPa on the seabed is applied.

The spatial evolution of the load is distributed along the continental shelf from the landward boundary ($x = 120$ km) to the edge of the shelf ($x = 100$ km) and is defined as $f(x) = \tanh(0.01 \times (x - 100\,000))$ (Fig. 2d). The ice-sheet build-up rate is assumed in the range of <200 m a^{-1} (Patton *et al.* 2016) and the retreat rate is 450 m a^{-1} (Morén *et al.* 2018). The lateral advance of the ice sheet across the 20 km wide shelf thus takes 100 years for the ice-stream build-up and 45 years for the ice-stream retreat. The total duration of the NCIS during the last glacial period (MIS 2) is about 7400 years long (Becker *et al.* 2018). In our reference model, we simulate that the ice stream remains at the edge of the shelf at its maximum position for a period of 7250 years (Fig. 2b, no ice-sheet oscillations). In another model ('dynamic ice stream') we simulate the ice stream moving back and forth, as suggested by Becker *et al.* (2018) (Fig. 2b, four ice-sheet oscillations). After the ice-stream retreat another time step with ice-free conditions of 2000 years is added. The total simulation time is thus 9400 years.

Material

The models include up to three different materials representing typical sediment types at glaciated margins (Hjelstuen *et al.* 2004; Sejrup *et al.* 2004), glacimarine sediments, GDFs and meltwater plumites. The corresponding hydromechanical properties (Table 1) are adopted from measurements on samples from the Storegga Slide region (Kvalstad *et al.* 2005; Bellwald *et al.* 2019) and the Storfjorden TMF (Fig. 1) (Llopart *et al.* 2019). All sediments show a non-linear porous elastoplastic behaviour, which is described by the Modified Cam Clay constitutive model. Permeability for glacimarine sediments and GDFs is a function of void ratio and is updated at each numerical increment. All materials have a Poisson's ratio of 0.3 and a critical effective friction angle of 30°. The specific weight of seawater is 10.25 kN m^{-3}.

Initial conditions

At the start of simulations, pore pressures are hydrostatic. Void ratios are assigned according to consolidation stress under gravity loading for normally consolidated materials.

Results

We present three simulations and evaluate their results in terms of excess pore pressure, fluid flow and overpressure ratio. The latter is defined as the ratio of excess pore pressure (p_e) to vertical effective stress (σ_v) under hydrostatic conditions (with p_h as hydrostatic pore pressure): $p_e/(\sigma_v - p_h)$. The overpressure ratio is a useful estimate of the impact of excess pore pressure on the stability because it is independent of the prevailing lithostatic stress. An overpressure ratio of zero represents hydrostatic conditions, while unity indicates that the pore fluid carries the entire overburden and no stabilizing effective stresses exist.

Table 1. *Mechanical properties required for the Modified Cam Clay constitutive model for the three sediment types used in this study: void ratio at 1 kPa (e_0), logarithmic hardening modulus (or compressibility) (λ) and the isotropic swelling index (κ). Density (ρ) and permeability (k) are also given*

Type	e_0	λ	κ	ρ (kg m^{-3})	k (m s^{-1})	Reference
Glacimarine	3.5	0.25	0.025	1800	$\log(k) = 7.6 \times e/(1+e) - 13.8$	Kvalstad *et al.* (2005)
Glacial debris	1.2	0.09	0.009	2000	$\log(k) = 11.5 \times e/(1+e) - 14.6$	Kvalstad *et al.* (2005)
Plumite	1.67	0.15	0.022	2000	$\log(k) = (e - 4.2)/0.35$	Llopart *et al.* (2019)

Reference model

The first simulation is a reference model made up of a homogeneous body of glacimarine sediment loaded by an ice stream with a long residence time at the shelf edge (no oscillations, Fig. 2b). After the onset of ice loading, excess pore pressure increases at the shelf below the seafloor–ice interface (Fig. 4a). Due to the load on the shelf and the impermeable seabed, the pore-fluid flow indicated by black arrows is directed downwards. At the shelf edge the direction of fluid flow changes seawards. The slope is ice-free and thus has a permeable seafloor allowing the pore fluid to escape upwards. When the ice thickness is at a maximum (Fig. 4b), p_e reaches a peak of 3700 kPa on the shelf, and maintains similar, high levels for the entire time of ice loading at depth >500 m. Once the ice stream retreats after 7370 years (Fig. 4c), the direction of pore-fluid flow changes to upwards on the shelf and dominantly downwards along the slope. However, it cannot flow back as fast as the load disappears so that negative excess pore pressures occur near the seafloor along the slope. Just after all the ice has disappeared (Fig. 4d), the pore fluid still escapes upwards along the shelf and downwards along the slope in order to balance pore pressures. Over time and as ice-free conditions persist, excess pore pressures slowly approach zero (not shown here).

The ice load directly affects stress conditions in the shelf sediments. In the initial ice advance phase excess pore pressure rises rapidly in alignment with total pressure to almost 3700 kPa, while the effective stress rises by less than 400 kPa (Fig. 4e). Hence, the ice load is carried almost entirely by the pore fluid. Rapid negative vertical displacement documents sediment compression by the load. As the ice persists (100–7350 years), drainage sets on and leads to decreasing excess pore pressure and rising effective stresses. The sediment consolidates under the load, also reflected by a further downwards displacement. The following ice retreat marks significant drops in total pressure, effective stress and excess pore pressure. The latter turns negative because of the sediment's rapid increase in volume (also expressed in the rapid upwards displacement) in response to unloading. As pore fluid flows back, the stress conditions approach those at the start of the analysis. This is not the case for vertical displacement. The previous upward displacement attributed to load compression is recovered. However, the volume loss due to consolidation is irreversible and the sediment is now overconsolidated. The ice cover does not extend to the slope and therefore the transient stress changes are different (Fig. 4f). However, due to proximity of the large ice load (<5 km), load spread induces a change in total pressure of 23 kPa when the ice appears. In the same period, excess pore pressure rises to 21 kPa, along with a relative decrease in effective stress by the same amount and an uplift of almost 1 m. The uplift is a partial consequence of volume increase as well as load spread, which induces a heave in adjacent sediments. During the duration of the ice cover, pore-fluid discharges, which causes a slow decrease in excess pore pressure as well as an increase in effective stress, and the sediment consolidates by reducing its void ratio and, hence, volume. This effect is also notable in a subtle downwards displacement. Load reduction due to ice retreat ($t = 7350$ years) leads to an instantaneous increase in effective stress, decrease in excess pore pressure and rapid downward displacement. The previous upward displacement attributed to heave is recovered. However, the volume loss due to consolidation processes is irreversible and the sediment is overconsolidated.

The ice load is capable of generating excess pore-pressure ratios of almost 1 in very shallow sediments at the shelf during the period of maximum extent and thickness (100–7350 years, Fig. 4g). During the ice retreat (7370 years) these values decrease again and turn to zero or even negative overpressure ratios when all ice has disappeared (7400 years). During the final ice-free state, the shelf slowly approaches hydrostatic conditions again (9400 years). The development of the overpressure ratio at the slope (Fig. 4h) is different to the shelf. First, overpressure ratios are significantly lower. Second, the overpressure ratio is almost constant with depth. This is in contrast to the exponential decrease from high surface near to low-deep values at the shelf. During the ice-stream build-up (10–100 years) the overpressure ratio increases linearly over the entire depth up to 0.017. While the ice remains at the shelf (100–7350 years), the overpressure ratio slightly decreases and continues to decrease during ice retreat (7370–7400 years) to negative values of −0.016 near the seafloor. During times of no ice cover (7400–9400), values rise again and approach zero over the entire depth.

Material contrasts

The second simulation includes three layers of different sediment types loaded by an ice stream without oscillations. The uppermost 100 m thick layer is made of GDFs, followed by melt-water plumites at 100–120 m below seafloor and glacimarine sediment at depth >120 m. At the time of ice increase the total pressure and vertical effective stress at the shelf inside the uppermost glacial debris layer rise equally to 3700 kPa, suggesting that the entire load is taken up by the pore fluid. The overpressure ratio reaches 0.7. During constant ice cover, a subtle decrease of excess pore pressure and effective stress of about 100 kPa are observed. This is different to

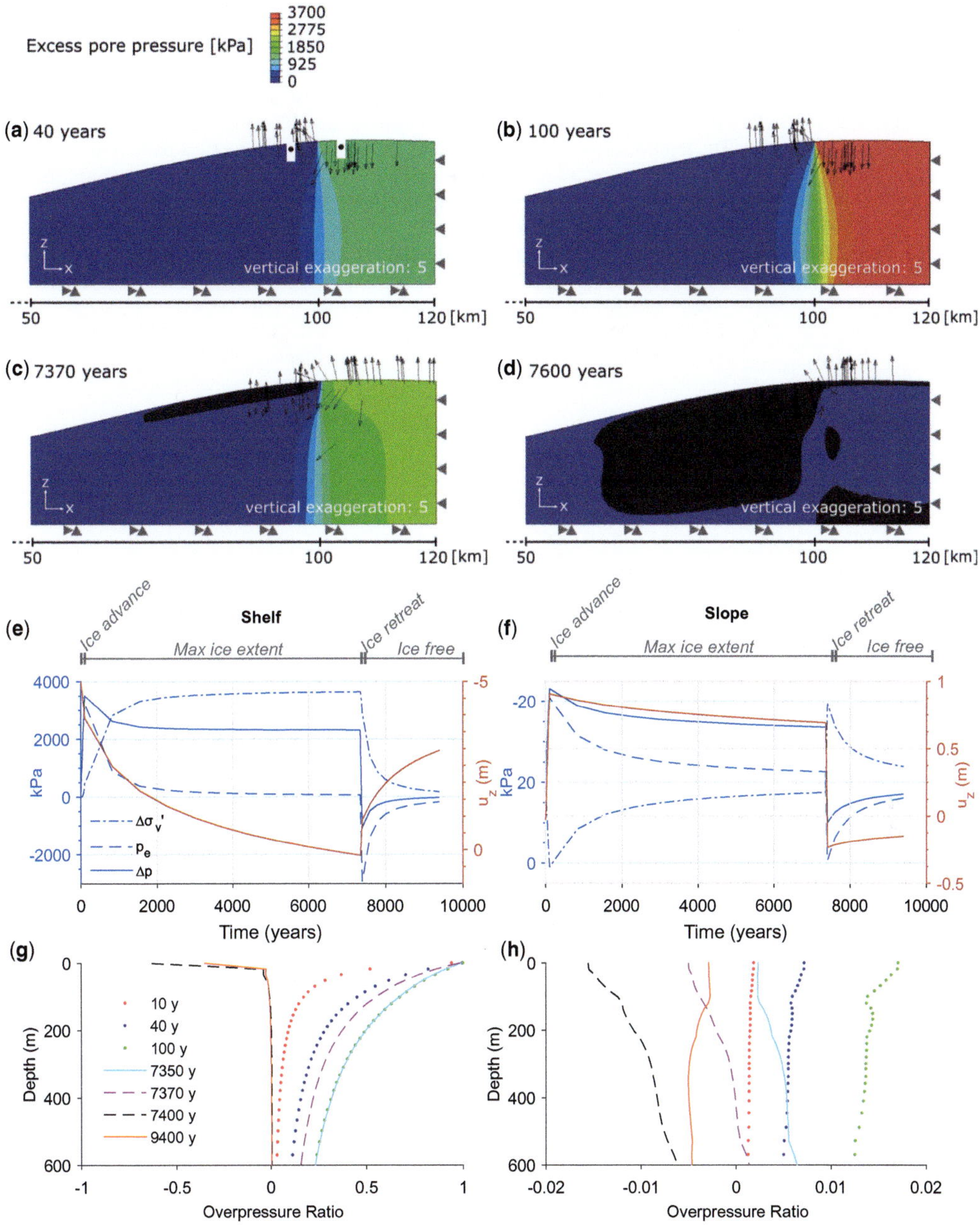

Fig. 4. Reference model results. Filled contour plots show excess pore pressure with vectors indicating the direction of fluid flow during (**a**) the ice build-up phase, (**b**) the maximum ice extent, (**c**) ice retreat and (**d**) ice-free conditions. Vectors represent the direction of fluid flow. The middle panels show the temporal evolution of relative changes in total pressure (Δp) and vertical effective stress ($\Delta\sigma_v'$), excess pore pressure (p_e) and vertical displacement (u_z) for (**e**) one element at the shelf ($x = 105$ km) and (**f**) one element at the upper slope ($x = 95$ km), both at 90 m below seafloor. Black dots in (a) mark the element locations. The lower panels are vertical profiles of overpressure ratio at (**g**) the shelf and (**h**) the continental slope. White bars in (a) mark their specific locations. Colours indicate different model times with dotted lines as ice build-up phase, solid line as the end of maximum ice extent, and dashed lines as retreat and ice-free phases.

the reference model, where excess pore fluid can drain and pore pressure drops to almost hydrostatic (Fig. 4e). When the ice sheet disappears, effective stress and excess pore pressure decrease by the load of the ice (3700 kPa), resulting in negative values in ice-free conditions. The temporal evolution of the stress conditions at the upper slope of a layered model show a similar pattern as in a homogeneous model (compare Figs 5b & 4f, respectively). Nevertheless, the values of excess pore pressure and overpressure ratio in the layered model are almost twice that of a homogeneous one.

On the shelf (Fig. 5c) the overpressure ratio increases to almost one as the ice sheet reaches its maximum extent. The values then decrease during ice retreat in an almost identical pattern as in the reference model (Fig. 4g). At the slope (Fig. 5d), the overpressure ratio increases especially in the first layer of glacial debris flow and reaches almost 0.035 after 100 years. This is twice as high as in

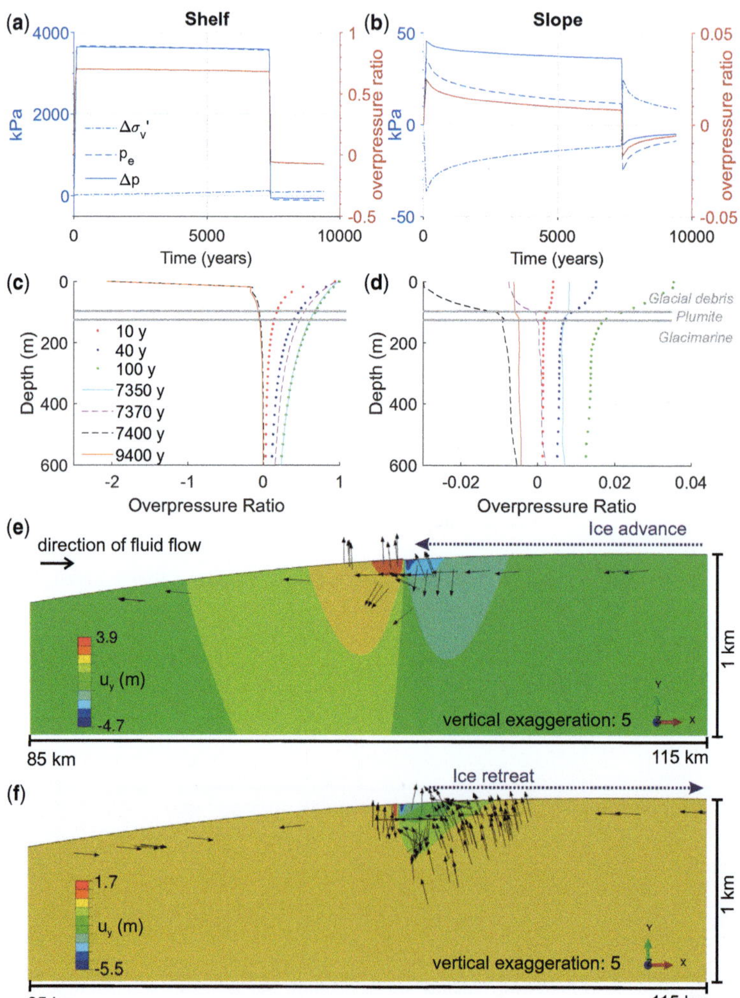

Fig. 5. Results of model run with different materials (layering). Temporal evolution of relative changes in total pressure (Δp) and vertical effective stress ($\Delta\sigma_v'$), excess pore pressure (p_e) and overpressure ratio for (**a**) one element at the shelf ($x = 105$ km) and (**b**) one element at the upper slope ($x = 95$ km), both in the glacial debris layer at 90 m below seafloor (see Fig. 4a for locations). Overpressure ratio at several time steps for vertical profiles at (**c**) the shelf and (**d**) the continental slope (locations in Fig. 4a). Dotted lines represent ice build-up phase, solid line the end of maximum ice extent, and dashed lines the retreat and ice-free phases. Lower panels show the direction of fluid flow ($>1\ e^{-12}$ m s^{-1}) at model times of (**e**) 100 years and (**f**) 7400 years. Focused flow takes place inside the plumite layer.

the reference model (Fig. 4h). The overpressure ratio shows a steep decrease in the plumite layer down to magnitudes in the glacimarine layer that are comparable to the reference model. The fluid flow also displays a particular pattern (Fig. 5e & f), likely governed by the permeability contrasts of the different materials. Pore-fluid flow is particularly high inside the meltwater plumite layer owing to its comparatively high permeability. During the ice stream build-up (Fig. 5e), fluid escapes away from the shelf towards the slope. The direction does not change until the ice stream retreats (Fig. 5f), then the flow paths along the slope are directed towards the shelf. Vertical displacement is negative (downwards) in areas covered by ice, reaching a maximum subsidence of −4.7 m (Fig. 5e). There is an abrupt change at the ice edge ($x = 100$ km) to localized upward displacement, forming a bulge with maximum vertical displacement up to 3.9 m.

Dynamic ice stream

The third model scenario is a body of homogeneous glacimarine sediment that is loaded by a highly dynamic ice stream. The ice stream reaches the shelf edge four times within one glacial period (Fig. 2b). The effect of repeated loading and unloading by a highly dynamic ice stream on slope stability can best be visualized by showing the temporal evolution of the overpressure ratio at one point at the upper slope ($x = 95$ km) at a depth of 100 m (Fig. 6). The results clearly reflect the loading and unloading cycles. The overpressure ratio increases rapidly as the ice stream builds up and reaches its maximum value of 0.014 after 100 years at the end of the first ice-stream build-up. The next three maximum peaks, each at the end of an ice-stream build-up, are lower than the first peak with the values of 0.012, 0.010 and 0.011. We explain this with the material being overconsolidated after the first ice advance, and with the different durations of ice-covered and ice-free periods. While the ice stream remains at the shelf edge at its maximum position, the overpressure ratios decrease slightly over time. A rapid decline follows as the ice retreats. The four minimum peaks at the end of each retreat have values of −0.008, −0.010, −0.0114 and −0.0112. During ice-free conditions, the overpressure ratio increases slightly, and the material can recover until the next advance. Due to a longer interstadial phase between 5200 and 6000 years, the material has more time to recover. Therefore, the last ice-stream build-up has a slightly higher impact on the element and indicates a higher peak than the third stadial phase.

Discussion

Ice loading is capable of generating very high overpressure ratios. However, these are limited to regions directly covered by the ice (the continental shelf), where the impermeable ice hampers fluid discharge at the seafloor. The continental slope, where large landslides have been mapped, is not affected by ice coverage and the perturbations in stress conditions are small. The maximum overpressure ratio obtained in our models is 0.035. It takes an overpressure ratio larger than 0.9 to cause direct failure of a typical continental slope setting with a 2° slope and a sediment with an effective friction angle of 30° (Urlaub *et al.* 2015). Hence, when considering the magnitude of overpressure ratio that would be required for failure, the contribution of ice loading will be almost negligible. Moreover, a negative impact on stability occurs only during a short time interval when the ice sheet/ice stream has just reached its maximum extent. After one or several cycles of ice loading the slope is in an

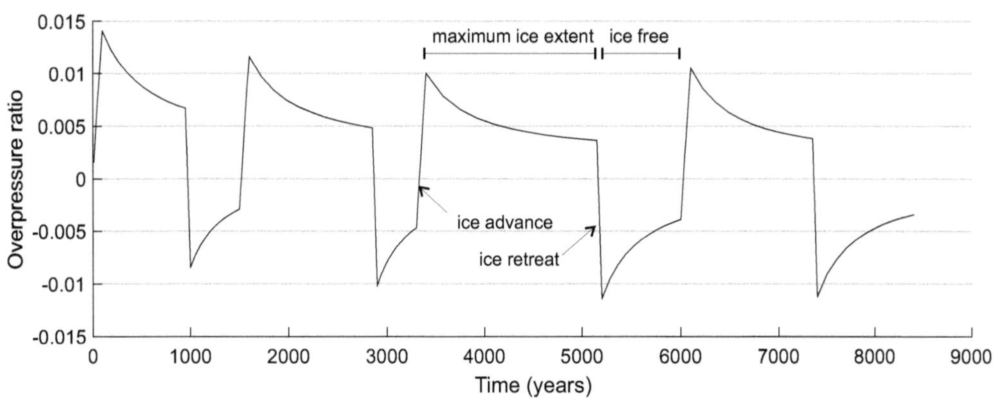

Fig. 6. Results of model run with four ice advances during one glaciation. Temporal evolution of overpressure ratio at one element at the slope ($x = 95$ km) in about 100 m depth (see black dot in Fig. 4a for location).

overconsolidated state and consequently has gained additional strength.

The reasons for changes in stress conditions at the slope in response to ice loading are twofold. First, spreading of the large and rapid ice loading invokes an instantaneous increase in excess pore pressure in the absence of a total stress increase. Second, high excess pore pressures prevail directly underneath the ice and excess pore fluid flows away and towards the slope, where it causes an increase in the overpressure ratio. This process is particularly effective if a high permeable layer allows comparably rapid discharge, as in our layered model. Here, the maximum overpressure ratio is twice that of the homogeneous model (Fig. 5b, d). Excess pore fluid resulting from the focused fluid flow inside the higher permeable plumite layer (Fig. 5e) might explain the higher overpressure ratio.

The weight of the ice causes the shelf to subside and the slope sediments to form a bulge (comparable to crustal glacial isostatic adjustment, where the peripheral area is uplifted during glaciations). This leads to local increases in slope angle. In our reference model the slope angle increase is almost threefold at the seaward-facing side of the uplifting bulge (Fig. 7). This steepening is locally confined to the shelf edge and is therefore unlikely to be the source of large deep-seated submarine landslides. However, it provides a viable mechanism for smaller failures directly at the ice edge during periods of maximum ice extent.

The model geometry represents a simplified synthetic and geological setting. The model neglects sediment deposition, which is an important process at glaciated margins and acts as an additional source of excess pore pressure at depth (Bellwald *et al.* 2019; Llopart *et al.* 2019). The way in which the spatial advance and retreat of the ice stream is simulated adds another simplification that may result in an overestimation of near-surface pore pressures during the short periods of advance and retreat (100 and 45 years, respectively). The results presented here should therefore be understood in a qualitative way and the provided estimates of pore pressure levels considered as approximations.

Total stress decrease when ice sheets retreat could lead to excess pore pressure when a compressible pore fluid, such as free gas, is present. However, the models do not consider the presence of gas because of a lack of constraints on quantities, locations and types of gas. Moreover, while implementing a compressible pore fluid is straightforward, a realistic description of the sensitivity of the gas to changes in boundary conditions and related feedback processes is not easily possible.

Conclusions

The onset of large-scale submarine landslides at glaciated margins coincides with the initiation of Northern Hemisphere glaciations in the Quaternary. Consequently, direct or indirect effects of glaciations must have caused these landslides. In this study we used numerical models to test the hypothesis that advancing and retreating ice sheets can reduce the stability of continental slopes by causing excess pore pressures, thereby promoting the occurrence of large submarine landslides widely observed at continental slopes. The models show a significant loss in stability due to elevated pore pressures only in shallow sediments on the shelf (the area directly covered by ice), which is essentially flat. Lateral fluid flow from the shelf into the slope occurs within a layer of high permeability but cannot account for sufficient high pore pressures to cause a significant reduction in slope stability. In contrast, repeated ice loading and unloading make the shelf sediments and those affected by load spread stronger due to

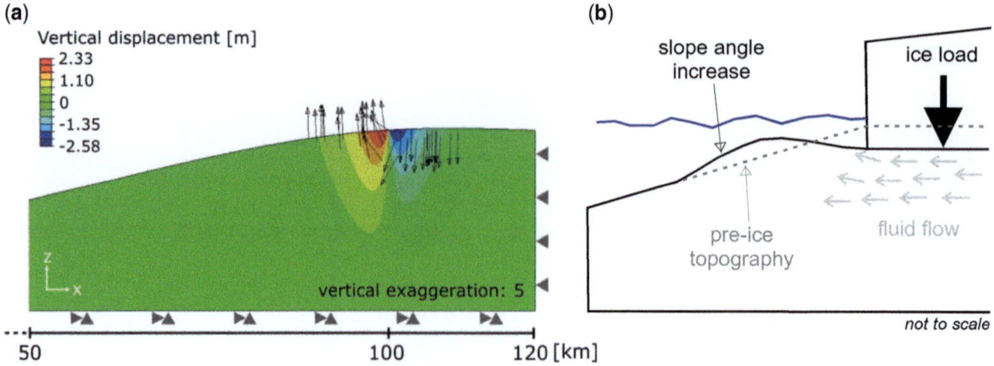

Fig. 7. (**a**) Vertical displacement at the time of maximum ice extent (reference model) showing significant subsidence below the ice load at the shelf and uplift seawards of the load. (**b**) Sketch of the impact of ice on slope topography and fluid flow.

overconsolidation. A further effect of ice loading is a local increase in slope angle near the shelf edge during periods of maximum ice extent. Localized oversteepening could be the reason for local and small-scale slope failures at the shelf edge.

Acknowledgements We appreciate the careful review of this manuscript by the Editor, Tom Arne Rydningen and Jaume Llopart. We thank Hans Petter Sejrup and Atle Nesje for helpful discussions.

Funding This research was completed as part of the German-Norwegian collaborative research support scheme (DAAD project 57245348 / Norwegian Research Council project 254789).

Author contributions **MU**: conceptualization (lead), funding acquisition (lead), investigation (equal), methodology (equal), visualization (equal), writing – original draft (lead), writing – review & editing (lead); **IK**: conceptualization (supporting), formal analysis (lead), investigation (equal), methodology (equal), writing – original draft (supporting), writing – review & editing (supporting); **BOH**: conceptualization (equal), funding acquisition (equal), investigation (supporting), supervision (supporting), writing – original draft (supporting), writing – review & editing (supporting).

References

Anandakrishnan, S., Blankenship, D.D., Alley, R.B. and Stoffa, P.L. 1998. Influence of subglacial geology on the position of a West Antarctic ice stream from seismic observations. *Nature*, **394**, 62–66, https://doi.org/10.1038/27889

Becker, L.W., Sejrup, H.P., Hjelstuen, B.O., Haflidason, H. and Dokken, T.M. 2018. Ocean-ice sheet interaction along the SE Nordic Seas margin from 35 to 15 ka BP. *Marine Geology*, **402**, 99–117, https://doi.org/10.1016/j.margeo.2017.09.003

Bellwald, B., Urlaub, M., Hjelstuen, B.O., Sejrup, H.P., Sørensen, M.B., Forsberg, C.F. and Vanneste, M. 2019. NE Atlantic continental slope stability from a numerical modeling perspective. *Quaternary Science Reviews*, **203**, 248–265, https://doi.org/10.1016/j.quascirev.2018.11.019

Bondevik, S., Mangerud, J., Dawson, S., Dawson, A. and Lohne, O. 2005. Evidence for three North Sea tsunamis at the Shetland Islands between 8000 and 1500 years ago. *Quaternary Science Reviews*, **24**, 1757–1775, https://doi.org/10.1016/j.quascirev.2004.10.018

De Angelis, H.D. and Kleman, J. 2008. Palaeo-ice-stream onsets: examples from the north-eastern Laurentide Ice Sheet. *Earth Surface Processes and Landforms*, **33**, 560–572, https://doi.org/10.1002/esp.1663

Evans, D., Harrison, Z. *et al.* 2005. Paleoslides and other mass failures of Pliocene to Pleistocene age along the Atlantic continental margin of NW Europe. *Marine and Petroleum Geology*, **22**, 1131–1148, https://doi.org/10.1016/j.marpetgeo.2005.01.010

Fruergaard, M., Piasecki, S., Johannessen, P.N., Noe-Nygaard, N., Andersen, T.J., Pejrup, M. and Nielsen, L.H. 2015. Tsunami propagation over a wide, shallow continental shelf caused by the Storegga slide, southeastern North Sea, Denmark. *Geology*, **43**, 1047–1050, https://doi.org/10.1130/G37151.1

Haflidason, H., Lien, R., Sejrup, H.P., Forsberg, C.F. and Bryn, P. 2005. The dating and morphometry of the Storegga Slide. *Marine and Petroleum Geology*, **22**, 123–136, https://doi.org/10.1016/j.marpetgeo.2004.10.008

Hjelstuen, B.O., Sejrup, H.P., Haflidason, H., Nygård, A., Berstad, I.M. and Knorr, G. 2004. Late Quaternary seismic stratigraphy and geological development of the south Vøring margin, Norwegian Sea. *Quaternary Science Reviews*, **23**, 1847–1865, https://doi.org/10.1016/j.quascirev.2004.03.005

Hjelstuen, B.O., Eldholm, O. and Faleide, J.I. 2007. Recurrent Pleistocene mega-failures on the SW Barents Sea margin. *Earth and Planetary Science Letters*, **258**, 605–618, https://doi.org/10.1016/j.epsl.2007.04.025

Hühnerbach, V., Masson, D.G. and partners of the COSTA-Project 2004. Landslides in the North Atlantic and its adjacent seas: an analysis of their morphology, setting and behaviour. *Marine Geology*, **213**, 343–362, https://doi.org/10.1016/j.margeo.2004.10.013

King, E.L., Haflidason, H., Sejrup, H.P. and Løvelie, R. 1998. Glacigenic debris flow on the North Sea Trough Mouth Fan during ice stream maxima. *Marine Geology*, **152**, 217–246, https://doi.org/10.1016/S0025-3227(98)00072-3

Kleiven, H.F., Jansen, E., Fronval, T. and Smith, T.M. 2002. Intensification of Northern Hemisphere glaciations in the circum Atlantic region (3.5–2.4 Ma)–ice-rafted detritus evidence *Palaeogeography, Palaeoclimatology, Palaeoecology*, **184**, 213–223, https://doi.org/10.1016/S0031-0182(01)00407-2

Knies, J., Matthiessen, J. *et al.* 2009. The Plio-Pleistocene glaciations of the Barents Sea-Svalbard region: a new model based on revised chronostratigraphy. *Quaternary Science Reviews*, **28**, 812–829, https://doi.org/10.1016/j.quascirev.2008.12.002

Knies, J., Mattingsdal, R. *et al.* 2014. Effect of early Pliocene uplift on late Pliocene cooling in the Arctic–Atlantic gateway. *Earth and Planetary Science Letters*, **387**, 132–144, https://doi.org/10.1016/j.epsl.2013.11.007

Kvalstad, T.J., Nadim, F., Kaynia, A.M., Mokkelbost, K.H. and Bryn, P. 2005. Soil conditions and slope stability in the Ormen Lange area. *Marine and Petroleum Geology*, **22**, 299–310, https://doi.org/10.1016/j.marpetgeo.2004.10.021

Laberg, J.S. and Vorren, T.O. 1995. Late Weichselian submarine debris flow deposits on the Bear Island Trough mouth fan. *Marine Geology*, **127**, 45–72, https://doi.org/10.1016/0025-3227(95)00055-4

Lambeck, K., Rouby, H., Purcell, A., Sun, Y. and Sambridge, M. 2014. Sea level and global ice volumes from the Last Glacial Maximum to the Holocene. *Proceedings of the National Academy of Sciences*, **111**, 15296–15303, https://doi.org/10.1073/pnas.1411762111

Leynaud, D., Mienert, J. and Vanneste, M. 2009. Submarine mass movements on glaciated and non-glaciated European continental margins: a review of triggering mechanisms and preconditions to failure. *Marine and Petroleum Geology*, **26**, 618–632, https://doi.org/10.1016/j.marpetgeo.2008.02.008

Llopart, J., Urgeles, R., Forsberg, C.F., Camerlenghi, A., Vanneste, M., Rebesco, M. and Lantzsch, H. 2019. Fluid flow and pore pressure development throughout the evolution of a trough mouth fan, western Barents Sea. *Basin Research*, **31**, 487–513, https://doi.org/10.1111/bre.12331

Morén, B.M., Sejrup, H.P., Hjelstuen, B.O., Borge, M.V. and Schäuble, C. 2018. The last deglaciation of the Norwegian Channel–geomorphology, stratigraphy and radiocarbon dating. *Boreas*, **47**, 347–366, https://doi.org/10.1111/bor.12272

Nygård, A., Sejrup, H.P., Haflidason, H. and Bryn, P. 2005. The glacial North Sea Fan, southern Norwegian Margin: architecture and evolution from the upper continental slope to the deep-sea basin. *Marine and Petroleum Geology*, **22**, 71–84, https://doi.org/10.1016/j.marpetgeo.2004.12.001

Nygård, A., Sejrup, H.P., Haflidason, H., Lekens, W.A.H., Clark, C.D. and Bigg, G.R. 2007. Extreme sediment and ice discharge from marine-based ice streams: new evidence from the North Sea. *Geology*, **35**, 395–398, https://doi.org/10.1130/G23364A.1

Ottesen, D., Dowdeswell, J.A. and Rise, L. 2005. Submarine landforms and the reconstruction of fast-flowing ice streams within a large Quaternary ice sheet: the 2500-km-long Norwegian–Svalbard margin (57°–80° N). *Geological Society of America Bulletin*, **117**, 1033–1050, https://doi.org/10.1130/B25577.1

Patton, H., Hubbard, A., Andreassen, K., Winsborrow, M. and Stroeven, A.P. 2016. The build-up, configuration, and dynamical sensitivity of the Eurasian ice-sheet complex to Late Weichselian climatic and oceanic forcing. *Quaternary Science Reviews*, **153**, 97–121, https://doi.org/10.1016/j.quascirev.2016.10.009

Sejrup, H.P., Larsen, E. *et al.* 2003. Configuration, history and impact of the Norwegian Channel Ice Stream. *Boreas*, **32**, 18–36, https://doi.org/10.1080/0300948031 0001029

Sejrup, H.P., Haflidason, H., Hjelstuen, B.O., Nygård, A., Bryn, P. and Lien, R. 2004. Pleistocene development of the SE Nordic Seas Margin. *Marine Geology*, **213**, 169–200, https://doi.org/10.1016/j.margeo.2004.10.006

Sejrup, H.P., Hjelstuen, B.O. *et al.* 2005. Pleistocene glacial history of the NW European continental margin. *Marine and Petroleum Geology*, **22**, 1111–1129, https://doi.org/10.1016/j.marpetgeo.2004.09.007

Sejrup, H.P., Bentley, M., Hjelstuen, B.O. and O Cofaigh, C. 2018. Geological evolution and processes of the glaciated North Atlantic margins. *Marine Geology*, **402**, 1–4, https://doi.org/10.1016/j.margeo.2018.06.009

Solheim, A., Berg, K., Forsberg, C.F. and Bryn, P. 2005. The Storegga Slide complex: repetitive large scale sliding with similar cause and development. *Marine and Petroleum Geology*, **22**, 97–107, https://doi.org/10.1016/j.marpetgeo.2004.10.013

Stokes, C.R. and Clark, C.D. 2001. Palaeo-ice streams. *Quaternary Science Reviews*, **20**, 1437–1457, https://doi.org/10.1016/S0277-3791(01)00003-8

Urlaub, M., Talling, P.J., Zervos, A. and Masson, D. 2015. What causes large submarine landslides on low gradient (<2°) continental slopes with slow (~0.15 m/kyr) sediment accumulation? *Journal of Geophysical Research: Solid Earth*, **120**, 6722–6739, https://doi.org/10.1002/2015JB012347

Vorren, T.O., Laberg, J.S. *et al.* 1998. The Norwegian-Greenland Sea continental margins: morphology and late Quaternary sedimentary processes and environment. *Quaternary Science Reviews*, **17**, 273–302, https://doi.org/10.1016/S0277-3791(97)00072-3

Wagner, B., Bennike, O., Klug, M. and Cremer, H. 2007. First indication of Storegga tsunami deposits from East Greenland. *Journal of Quaternary Science*, **22**, 321–325, https://doi.org/10.1002/jqs.1064

Impact of sea-level fluctuations on the sedimentation patterns of the SE African margin: implications for slope instability

Aaron Micallef[1,2]*, Aggeliki Georgiopoulou[3], Andrew Green[4] and Vittorio Maselli[5]

[1]Marine Geology & Seafloor Surveying, University of Malta, MSD 1805, Malta

[2]Helmholtz Centre for Ocean Research, GEOMAR, Kiel, 24148, Germany

[3]School of Environment and Technology, University of Brighton, BN2 4GJ, UK

[4]Geological Sciences, University of KwaZulu-Natal, X54001, South Africa

[5]Department of Earth and Environmental Sciences, Dalhousie University, B3H 4R2, Canada

AM, 0000-0002-9330-0648; AGr, 0000-0001-9438-1315
*Correspondence: aaron.micallef@um.edu.mt

Abstract: The sheared-passive margin offshore Durban (South Africa) is characterized by a narrow continental shelf and steep slope hosting numerous submarine canyons. Supply of sediment to the margin is predominantly terrigenous, dominated by discharge from several short but fast-flowing rivers. International Ocean Discovery Program Expedition 361 provides a unique opportunity to investigate the role of sea-level fluctuations on the sedimentation patterns and slope instability along the South African margin. We analysed >300 sediment samples and downcore variations in *P*-wave, magnetic susceptibility, bioturbation intensity and bulk density from site U1474, as well as regional seismic reflection profiles to: (1) document an increase in sand input since the Mid-Pliocene; (2) associate this change to a drop in sea-level and extension of subaerial drainage systems towards the shelf-edge; (3) demonstrate that slope instability has played a key role in the evolution of the South Africa margin facing the Natal Valley. Furthermore, we highlight how the widespread occurrence of failure events reflects the tectonic control on the morphology of the shelf and slope, as well as bottom-current scour and instability of fan complexes. This information is important to improve hazard assessment in a populated coastal region with growing offshore hydrocarbon activities.

Borehole U1474, drilled in the Natal Valley during the International Ocean Discovery Program (IODP) Expedition 361 in 2016, provides a unique opportunity to investigate sedimentation patterns and slope instability in the SE African passive continental margin offshore Durban, KwaZulu-Natal, South Africa. The margin is relatively underexplored, yet it remains of interest in view of its unusual geological setting. It consists of one of the narrowest continental shelves in the world (Shepard 1963), which passes basinward into a steep slope that hosts large-scale mass-wasting deposits, including submarine canyons (Wiles *et al.* 2013), slumps (Dingle and Robson 1985) and big (up to 34 000 km^2) landslides (Martin 1984; Goodlad 1986). The area is considered susceptible to slope instability due to the preconditioning of the slope by weakly consolidated siltstone layers (Green *et al.* 2008), coupled to the large quantities of sediment delivered by numerous fast-flowing rivers from the hinterland (Martin 1987) and undercutting by strong bottom-current activity (Ben-Avraham *et al.* 1994; Niemi *et al.* 2000; Wiles *et al.* 2013). The hazard associated with such processes needs to be quantified in consideration of the proximity to coastal cities and undersea cable infrastructure (Chattopadhyaya 2018), as well as the recently approved oil and gas exploration activities in the area (Myeni 2019). Here we combine borehole data from IODP Site U1474 with seismic reflection profiles to understand which processes control sediment supply in the southern Natal Valley, to determine the causes of slope instability and to evaluate how mass-wasting processes affect the evolution of the SE African margin facing the Indian Ocean.

Regional setting

The Natal Valley developed during the Early Cretaceous (Goodlad *et al.* 1982; Watkeys and Sokoutis 1998) following the break-up of Gondwana. The Natal Valley is a sediment-filled basin where open marine sedimentation has been recorded since the Late Cretaceous (Davison and Steel 2017). Uplift

From: Georgiopoulou, A., Amy, L. A., Benetti, S., Chaytor, J. D., Clare, M. A., Gamboa, D., Haughton, P. D. W., Moernaut, J. and Mountjoy, J. J. (eds) 2020. *Subaqueous Mass Movements and their Consequences: Advances in Process Understanding, Monitoring and Hazard Assessments*. Geological Society, London, Special Publications, **500**, 267–276.
First published online March 13, 2020, https://doi.org/10.1144/SP500-2019-172

of the hinterland started during the Oligocene and triggered an increase in the rate of sediment supply (Said *et al.* 2015), favouring the formation of thick sequences of deep-water deposits, at places eroded and re-worked by ocean bottom circulation (Niemi *et al.* 2000; Wiles *et al.* 2013). The Natal Valley is bordered to the west by the Agulhas–Falkland Fracture Zone (Fig. 1), which represents a 1200 km long right-lateral strike-slip fault zone that evolved between the South American and African plates during the Early Cretaceous break-up. To the north and east, the Natal Valley is confined by the Mozambique Ridge (Fig. 1), a large igneous province of oceanic origin (Sinha *et al.* 1981; Fischer *et al.* 2017).

The South African margin facing the Natal Valley (Fig. 1) is characterized by a narrow continental shelf, up to 10 km wide, which abruptly ends on a steep escarpment associated with the Agulhas–Falkland Fracture Zone (Ben-Avraham *et al.* 1993). The Natal Valley is a NE–SW-elongated bathymetric low that reaches a modern water depth of 4000 m. Southwestward, the Natal Valley opens and deepens into the Transkei Basin (Fig. 1), while the Agulhas Plateau rises to 2000 m from the abyssal plain further to the SW (Fig. 1). The western slope of the Natal Valley hosts a number of submarine canyons that have variable depth and width, and that often indent the shelf. Several submarine landslides have also been recognized along the South African margin, including the Agulhas Slump, presumably the world's largest submarine landslide (Dingle 1977).

Sediments are delivered to the Natal Valley by numerous steep and fast-flowing rivers, such as the Tugela River, which are sourced from an area of high topographic relief (>3000 m) that is currently characterized by a sub-tropical high-rainfall climate. The main focus of sedimentation occurs within the Tugela Cone, an asymmetrical delta-fan complex seaward of the Tugela River (Goodlad *et al.* 1982). Additional clastic input is derived from the Limpopo and Zambezi rivers located further to the north, as well as by deep-water ocean circulation, which creates a series of elongated sediment drifts along the slopes bordering the Natal Valley, the Agulhas Plateau and the Mozambique Ridge (Niemi *et al.* 2000).

The modern circulation of the south Indian Ocean is characterized by two main currents (Schott *et al.* 2009). The Agulhas Current is a warm and saline surface current that represents the western boundary current of the Indian Ocean and affects the entire water column (Lutjeharms 2006). The current flows along the continental shelf of Mozambique and South Africa, reaching velocities of up to 1.5 m s^{-1} (Bryden *et al.* 2005), and contributing to the dispersal of surface sediments. When reaching the Atlantic Ocean, the Agulhas Current retroflects, flowing back to the south Indian Ocean (Toole and Warren 1993; Lutjeharms 2006). The Agulhas Undercurrent carries the North Atlantic Deep Water and flows northeastwards along the slope at water depths of 1000–2900 m and 11–60 km offshore Durban (Lutjeharms and Ansorge 2001; Beal 2009). IODP Site U1474 is located in an area where the Agulhas Current is particularly stable due to the confinement generated by the steep slope (Simon *et al.* 2015). The IODP site is not affected by the Agulhas Undercurrent.

Data and methods

Seismic reflection data

The seismic reflection data presented here comprise legacy, reprocessed, migrated-stack 2D seismic reflection profiles obtained for petroleum exploration during the mid-1970s (Hicks and Green 2016). The single-channel 2D seismic profiles were obtained from the Petroleum Agency of South Africa (PASA) in SEG-Y format and imported into IHS Global Inc. Kingdom Advanced V2015.0. Digital well shoot/velocity data from the Jc-D1 well were utilized by PASA to define two-way time v. measured depth (m) in order to tie well data to the seismic data (Hicks and Green 2016).

Borehole data

IODP Site U1474 is located in the northern portion of the Natal Valley (latitude: 31° 13.00′ S, longitude: 31° 32.71′ E, Fig. 1 (Hall *et al.* 2017)) at a water depth of 3045 m below sea-level; 317 sediment samples were obtained from holes A, B and F, which were processed with a Malvern Mastersizer 3000™ to determine grain-size distribution. Downcore variations in bioturbation intensity were measured during the expedition, whereas changes in *P*-wave velocity, magnetic susceptibility and bulk density were determined using a multi-sensor core logger with a Bartington Instruments MS2C system and a GRA bulk densitometer (Hall *et al.* 2017).

Results

Seismic reflection data

Two principal seismic facies can be interpreted from the seismic reflection profiles (Figs 2 & 3).

- Facies A (light grey in Figs 2c & 3c): high amplitude, sub-parallel, continuous reflectors. In view of their correlation with drift deposits, formed by the northward-flowing Agulhas Undercurrent at IODP Site U1474 (Niemi *et al.* 2000; Hall *et al.* 2017) and their seismic geometry (most notably the continuous and flat-lying nature of the reflectors, which bear strong similarity

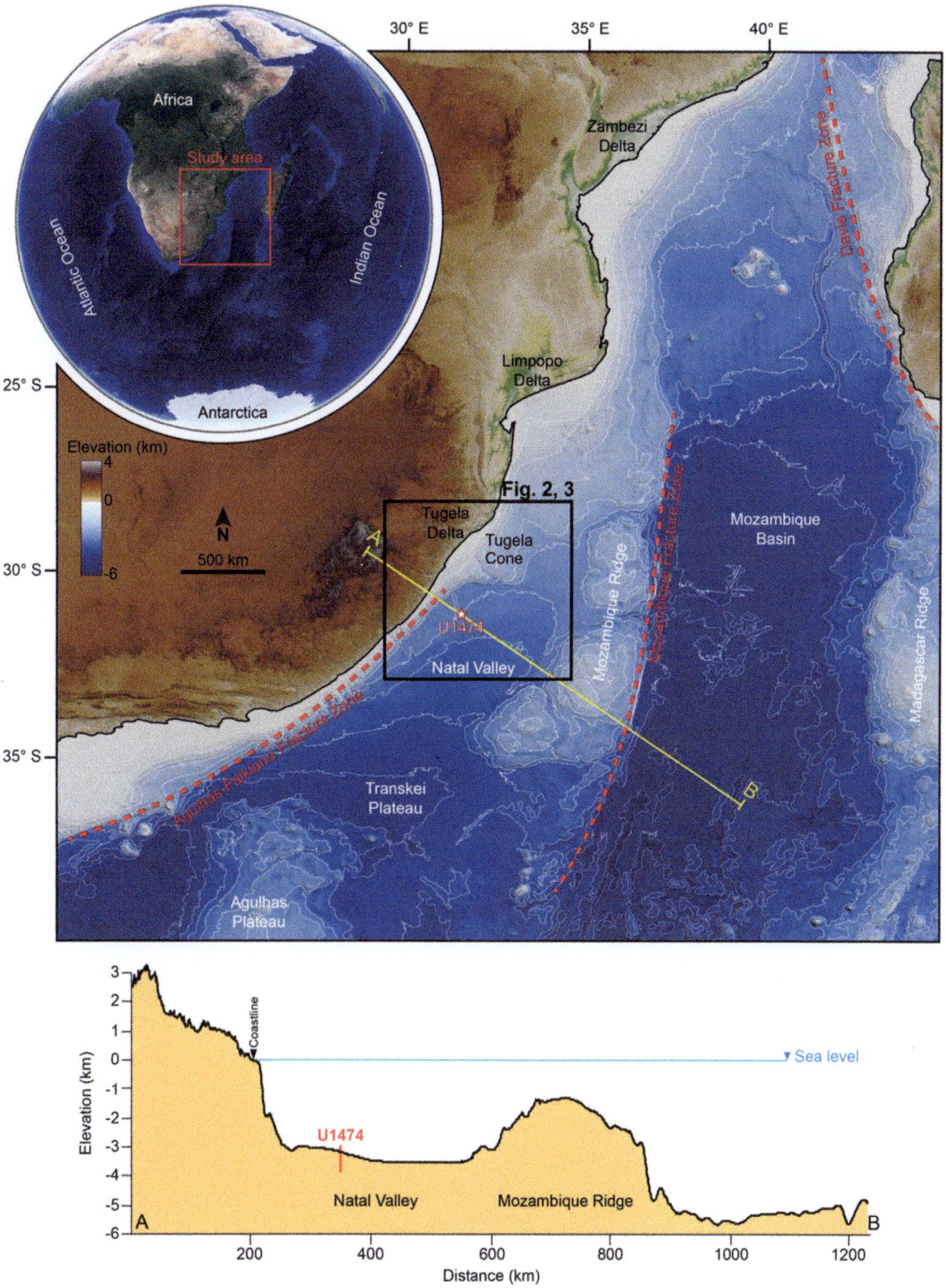

Fig. 1. Location map and regional physiography. Bathymetric data from Dorschel *et al.* (2018). Red star denotes location of IODP Site U1474.

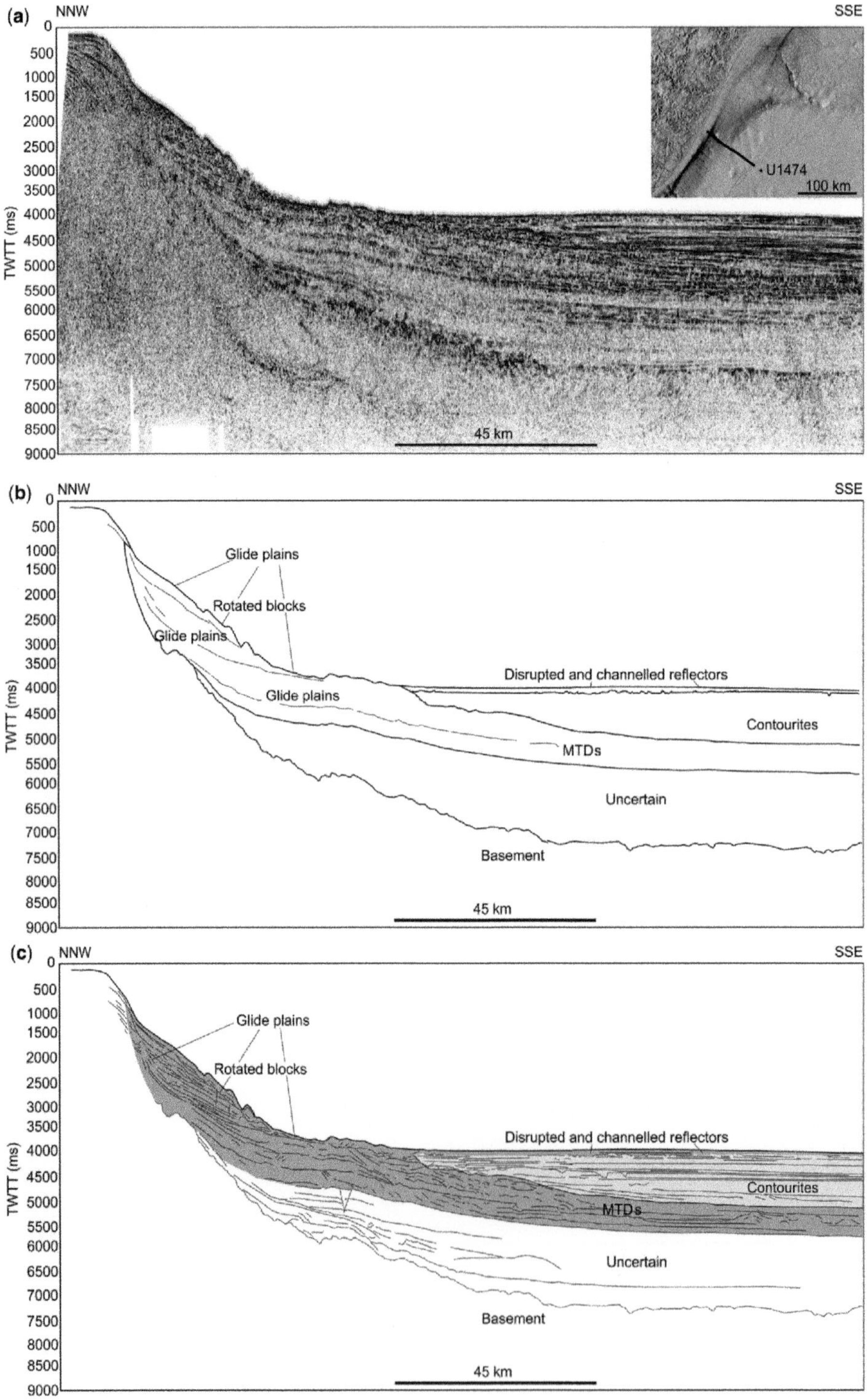

Fig. 2. (**a**) Seismic reflection profile P1 and (**b, c**) its interpretation. See Figure 1 for the location of the inset. Bathymetric data used in the inset sourced from Dorschel *et al.* (2018). TWTT, two-way travel time; MTD, mass transport deposit.

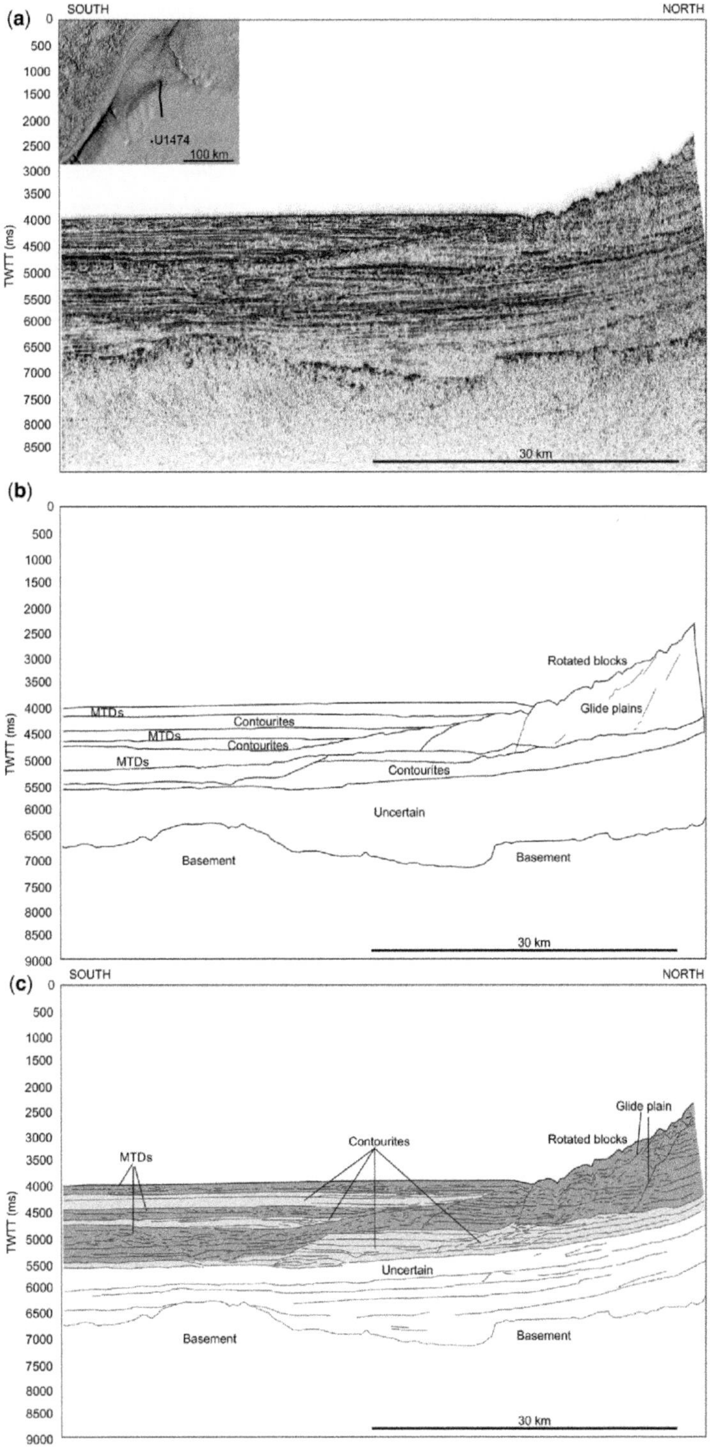

Fig. 3. (**a**) Seismic reflection profile P2 and (**b**, **c**) its interpretation. See Figure 1 for the location of the inset. Bathymetric data used in the inset sourced from Dorschel *et al.* (2018). TWTT, two-way travel time; MTD, mass transport deposit.

to those recognized as contouritic sheeted drift deposits (Nielsen *et al.* 2008; Rebesco *et al.* 2014)), we interpret facies A as contouritic drift deposits.

- Facies B (dark grey in Figs 2c & 3c): acoustically chaotic facies with prominent high amplitude reflectors occurring in association with upslope rotated blocks, which we interpret as mass transport deposits (MTD) (Dingle 1977).

Profile P1 is perpendicular to the shoreline and located to the SW of the Tugela Cone (Figs 1 & 2). The continental slope and rise consist of facies B, which is up to 1 s two-way travel time (TWTT) thick. The high amplitude reflectors correspond to glide plains and allow identification of a number of rotated blocks (Fig. 2). The abyssal plain consists of facies A, up to 1 s TWTT in thickness, which intersects IODP Site U1474 and which is underlain by facies B (Fig. 2). The base of facies A corresponds to the O reflector of Niemi *et al.* (2000).

Profile P2 is located south of, and partly intersects the southern edge of, the Tugela Cone (Figs 1 & 3). The continental slope comprises up to 1.5 s TWTT of facies B, which extend into the abyssal plain. Here, facies B is overlain by an alternating sequence of facies A with facies B (Fig. 3). The top of the sequences consists of facies B.

Borehole data

Sediments at IODP Site U1474 are 60% terrigenous (predominantly clay) and 40% biogenic (predominantly nannofossils) (Hall *et al.* 2017). We identify four units based on the physical logs, grain-size distribution and bioturbation intensity (Fig. 4).

(1) Unit 1 comprises the top 3 m. It is marked by a sharp drop and then an increase in *P*-wave velocity, similar to the magnetic susceptibility, and a peak in the D10 grain-size fraction (Fig. 4). Bioturbation intensity is moderate. Unit 1 is likely the Holocene cover and is the equivalent of unit 1 of the shipboard description (Hall *et al.* 2017).

(2) Unit 2 is observable between 3 and 33 m below seafloor (bsf). Using the age model, we

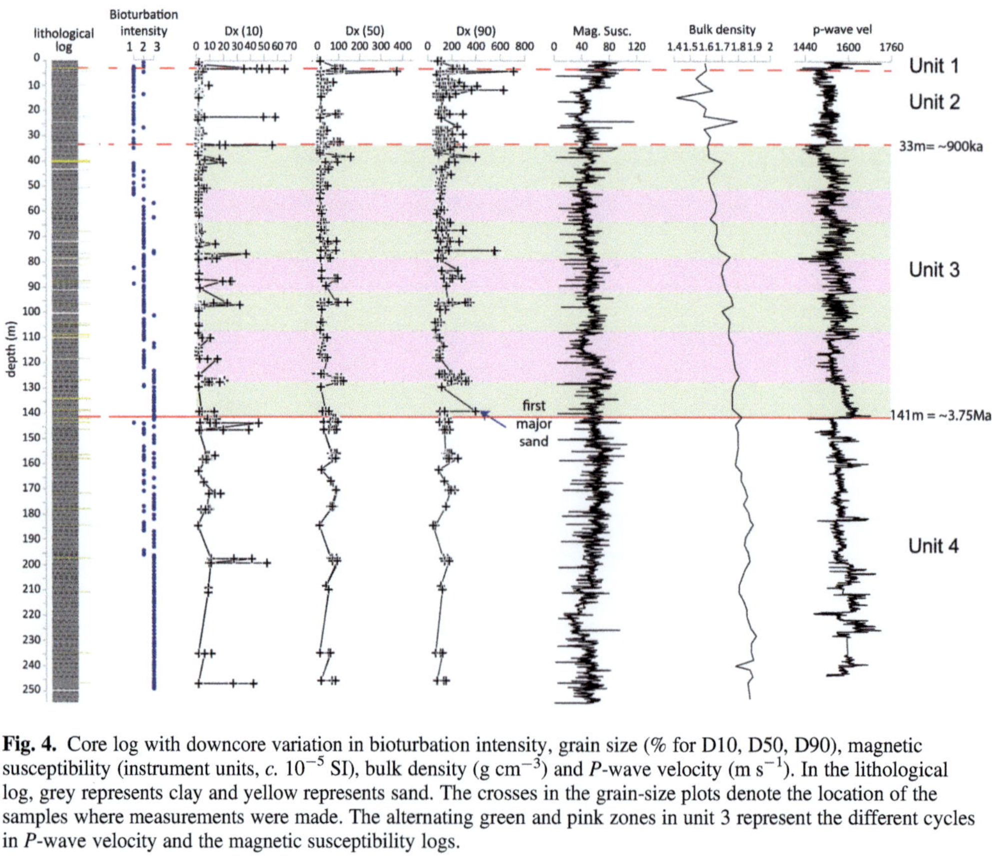

Fig. 4. Core log with downcore variation in bioturbation intensity, grain size (% for D10, D50, D90), magnetic susceptibility (instrument units, *c.* 10^{-5} SI), bulk density (g cm^{-3}) and *P*-wave velocity (m s^{-1}). In the lithological log, grey represents clay and yellow represents sand. The crosses in the grain-size plots denote the location of the samples where measurements were made. The alternating green and pink zones in unit 3 represent the different cycles in *P*-wave velocity and the magnetic susceptibility logs.

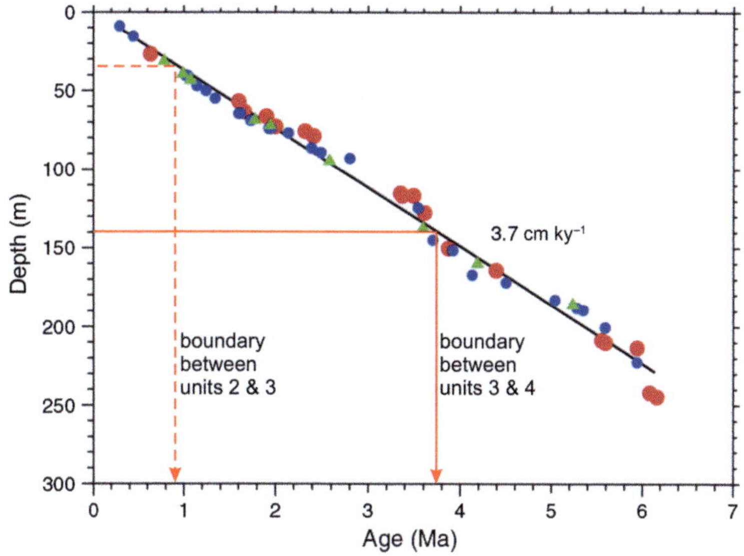

Fig. 5. Age–depth relationships for IODP Site U1474 based on planktonic foraminifera (red circles), calcareous nannofossils (blue circles) and magnetostratigraphy (green triangles) (adapted from Hall *et al.* 2017; credit, IODP). Orange lines correspond to the age–depth line for the boundary between units 2 and 3 (dashed), and between units 3 and 4 (solid).

determine the base of the unit to have an age of 900 ka (Fig. 5), which corresponds to the Late Pleistocene. In unit 2 there is an overall grain-size increase (seen in the high D90 grain sizes), erratic bulk density, and consistent P-wave velocity (Fig. 4). This interval is weakly bioturbated.

(3) Unit 3 covers the depths 33–141 m bsf and its base has an age of 3.75 Ma (Fig. 5), which corresponds to the Mid-Pliocene (P reflector of Niemi *et al.* (2000)). In this unit we observe the most frequent sand input. Bioturbation is moderate to high, and there is an apparent cyclicity, particularly in the *P*-wave velocity and the magnetic susceptibility logs (Fig. 4). It is likely that the cyclicity is not restricted to unit 3, as similar patterns in these two logs can be seen deeper in the stratigraphy. However, the cyclicity in unit 3 is more pronounced, possibly due to the presence of more sand units that amplify the signal in *P*-wave velocity and the magnetic susceptibility as they respond to the presence of (terrigenous) ferro-manganese minerals. Based on the age model, the cycles appear to have a duration of *c.* 400 ka (Fig. 4).

(4) Unit 4 extends from 141 m bsf to the base of the core at 254 m bsf. It is characterized by thinner and less frequent sand beds, and increased bioturbation that becomes more intense with depth. Unit 4 is a finer-grained sequence compared to unit 3 (Fig. 4).

Discussion and conclusions

Source of sediment in IODP Site U1474

Sediment recovered from IODP Site U1474 comprises a combination of terrigenous and hemipelagic sediments. The terrigenous clays are likely sourced by the main drainage systems of the KwaZulu-Natal coast (Martin 1987; Ziegler *et al.* 2013; Simon *et al.* 2015). Low rates of sedimentation characterize much of the sequence (3.7 cm ka^{-1} (Hall *et al.* 2017)) and the occurrence of thin sand beds points to a sediment supply from the distal part of the slope turbidite system of the northern Natal Valley. However, since IODP Site U1474 is located on a drift deposit (Figs 2 & 3), we cannot exclude that coarse-grained layers may be related to bottom-current activity.

Control of sand input to drift sequence in IODP Site U1474

The major drift bodies in the vicinity of the Natal Valley started to develop at the Eocene–Oligocene boundary, in correspondence with the onset of cold, abyssal current circulation (Hall *et al.* 2017). In IODP Site U1474, a higher input of sand is observed from the Mid-Pliocene. In southern Africa, this period corresponds to a significant drop in relative sea-level from >300 m down to −200 m compared to present sea-level (Siesser and Dingle 1981). From the Mid-Pliocene onwards, sea-level fluctuated between −200 and 0 m (Siesser and Dingle 1981). We therefore suggest that the higher sand input from the Mid-Pliocene onwards is a result of the extension of subaerial drainage systems across the continental shelf and upper slope, and the discharge of terrigenous sediments directly on to the slope, particularly via submarine canyons. In view of this, we infer that the sand intervals are more likely to be turbidites than contourite structures. We interpret the *c.* 400 ka cyclicity in the *P*-wave velocity and the magnetic susceptibility logs as a sedimentary signature of Milankovitch-type eccentricity of the Earth's orbit (Hays *et al.* 1976).

Causes of margin-scale slope instability

The seismic reflection profiles show that slope instability has played a key role in the evolution of the South African margin facing the Natal Valley. We attribute the widespread occurrence of slope failure signatures in the continental slope, rise and abyssal plain to the following factors.

(1) Shelf width: the shelf is relatively narrow and shallow. It is considered a morphologically inherited feature of the initial shearing phase during margin development (Martin 1984). Fluvial sediments delivered here were not stored on the shelf, but readily moved offshore, especially during sea-level lowstands. The 100 m deep shelf break is always shallower than the Pleistocene lowstands (Compton 2011), and hence no accommodation space was available for lowstand deltas, resulting in strong loading of the upper slope during glaciations, and their subsequent failure. In addition, in view of the narrow shelf, it is likely that rivers could have delivered sediments directly to the continental slope in places, supplying turbidity currents.

(2) Steep continental slope and bottom-current activity: the continental slope, due to preservation of a Miocene/Pliocene shelf-edge wedge, is over-steepened in the upper portions (Green *et al.* 2007). Any undercutting of the lower continental slope would quickly evolve into retrogressive failures, which we observe in profiles P1 and P2 (Figs 2 & 3). We suggest that undercutting of the lower continental slope by vigorous bottom-current activity, particularly the Agulhas Undercurrent, is an important cause of slope instability. Wiles *et al.* (2013) have reported similar scouring by North Atlantic Deep Water in the lower Tugela Cone.

(3) Uplift: starting from the Oligocene, hinterland uplift enhanced slope instability and shedding of large quantities of sediment to the slope (Said *et al.* 2015; Hicks and Green 2017). Seismic triggers might also have been important during this time considering that the study site has been recently affected by earthquakes up to 5 M_L in magnitude (Wiles *et al.* 2014).

(4) Fan complex instability: profile P2 shows that, in comparison to profile P1, MTDs are more common and extensive in the vicinity of the Tugela Cone, which indicates that rapid accumulation of sediments in this fan complex led to widespread mass wasting until recently. Hicks and Green (2017) document continued mass wasting and canyon development just north of the study site, focused along the Tugela Cone through Oligocene to Pleistocene times.

Future work

We have demonstrated that sea-level fluctuations and slope instability play an important role in the evolution of the sheared-passive margin offshore Durban. Investigation of the margin with higher-resolution geophysical techniques will allow us to better understand the style, spatial distribution, preconditioning and triggering factors of slope failures, and to quantify the hazard that they pose.

Acknowledgements We thank Jason Chaytor, Samantha Clarke and an anonymous reviewer for their insightful comments.

Funding This research received no specific grant from any funding agency in the public, commercial, or not-for-profit sectors.

Author contributions **AM**: investigation (lead), writing – original draft (lead); **AGe**: formal analysis (lead), writing – review & editing (lead); **AGr**: formal analysis (lead), writing – review & editing (lead); **VM**: formal analysis (supporting), writing – review & editing (lead).

References

Beal, L.M. 2009. A time series of Agulhas Undercurrent transport. *Journal of Physical Oceanography*, **39**, 2436–2450, https://doi.org/10.1175/2009JPO4195.1

Ben-Avraham, Z., Hartnady, C.J.H. and Malan, J.A. 1993. Early tectonic extension between the Agulhas bank and the Falkland Plateau due to rotation of the Lafonia microplate. *Earth and Planetary Science Letters*, **117**, 43–58, https://doi.org/10.1016/0012-821X(93)90116-Q

Ben-Avraham, Z., Niemi, T.M. and Hartnady, C.J. 1994. Mid-Tertiary changes in deep ocean circulation patterns in the Natal Valley and Transkei Basin, Southwest Indian Ocean. *Earth and Planetary Science Letters*, **121**, 639–646, https://doi.org/10.1016/0012-821X(94)90097-3

Bryden, H.L., Beal, L.M. and Duncan, L.M. 2005. Structure and transport of the Agulhas Current and its temporal variability. *Journal of Oceanography*, **61**, 479–492, https://doi.org/10.1007/s10872-005-0057-8

Chattopadhyaya, S. 2018. How is digital economy of Africa enabled by subsea cable connectivity, https://www.submarinenetworks.com/en/insights/how-is-digital-economy-of-africa-enabled-by-subsea-cable-connectivity, 2019.

Compton, J.S. 2011. Pleistocene sea-level fluctuations and human evolution on the southern coastal plain of South Africa. *Quaternary Science Reviews*, **30**, 506–527, https://doi.org/10.1016/j.quascirev.2010.12.012

Davison, I. and Steel, I. 2017. Geology and hydrocarbon potential of the East African continental margin: a review. *Petroleum Geoscience*, **24**, 57–91, https://doi.org/10.1144/petgeo2017-028

Dingle, R.V. 1977. The anatomy of a large submarine slump on a sheared continental margin (SE Africa). *Journal of the Geological Society, London*, **134**, 293–310, https://doi.org/10.1144/gsjgs.134.3.0293

Dingle, R.V. and Robson, S. 1985. Slumps, canyons and related features on the continental margin off East London, SE Africa (SW Indian Ocean). *Marine Geology*, **67**, 37–54, https://doi.org/10.1016/0025-3227(85)90147-1

Dorschel, B., Jensen, L. *et al.* 2018. The Southwest Indian Ocean Bathymetric Compilation (swIOBC). *Geochemistry, Geophysics, Geosystems*, **19**, 968–976, https://doi.org/10.1002/2017GC007274

Fischer, M.D., Uenzelmann-Neben, G., Jacques, G. and Werner, R. 2017. The Mozambique Ridge: a document of massive multistage magmatism. *Geophysical Journal International*, **208**, 449–467, https://doi.org/10.1093/gji/ggw403

Goodlad, S.W. 1986. *Tectonic and sedimentary history of the mid-Natal Valley (SW Indian Ocean)*. PhD thesis, University of Cape Town.

Goodlad, S.W., Martin, A.K. and Hartnady, C.J.H. 1982. Mesozoic magnetic anomalies in the southern Natal Valley. *Nature*, **295**, 686–688, https://doi.org/10.1038/295686a0

Green, A.N., Goff, J.A. and Uken, R. 2007. Geomorphological evidence for upslope canyon-forming processes on the northern KwaZulu-Natal shelf, SW Indian Ocean, South Africa. *Geo-Marine Letters*, **27**, 399– 409, https://doi.org/10.1007/s00367-007-0082-2

Green, A.N., Ovechkina, M. and Uken, R. 2008. Nannofossil age constraints on shelf-edge wedge development: implications for continental margin dynamics, northern KwaZulu-Natal, South Africa. *Continental Shelf Research*, **28**, 2442–2449, https://doi.org/10.1016/j.csr.2008.06.007

Hall, I.R., Hemming, S.R. *et al.* 2017. Site U1474. *In*: Hall, I.R., Hemming, S.R., LeVay, L.J. and the Expedition 361 Scientists (eds) *Proceedings of the International Ocean Discovery Program 361*. International Ocean Discovery Program, College Station, https://doi.org/10.14379/iodp.proc.361.103.2017

Hays, J.D., Imbrie, J. and Shackleton, N.J. 1976. Variations in the Earth's orbit: pacemaker of the ice ages. *Science*, **194**, 1121–1132, https://doi.org/10.1126/science.194.4270.1121

Hicks, N. and Green, A.N. 2016. Sedimentology and depositional architecture of a submarine delta-fan complex in the Durban Basin, South Africa. *Marine and Petroleum Geology*, **78**, 390–404, https://doi.org/10.1016/j.marpetgeo.2016.09.032

Hicks, N. and Green, A.N. 2017. A Mid-Miocene erosional unconformity from the Durban Basin, SE African margin: a combination of global eustatic sea level change, epeirogenic uplift, and ocean current initiation. *Marine and Petroleum Geology*, **86**, 798–811, https://doi.org/10.1016/j.marpetgeo.2017.06.037

Lutjeharms, J.R.E. 2006. *The Agulhas Current*. Springer-Verlag, Berlin.

Lutjeharms, J.R.E. and Ansorge, I.J. 2001. The Agulhas return current. *Journal of Marine Systems*, **30**, 115–138, https://doi.org/10.1016/S0924-7963(01)00041-0

Martin, A.K. 1984. *Plate tectonic status and sedimentary basin in-fill of the Natal Valley (SW Indian Ocean)*, PhD thesis, University of Cape Town.

Martin, A.K. 1987. Comparison of sedimentation rates in the Natal Valley, south-west Indian Ocean, with modern sediment yields in east coast rivers of Southern Africa. *South African Journal of Science*, **83**, 716–724.

Myeni, G. 2019. Offshore oil exploration gets green light, https://zululandobserver.co.za/199022/offshore-oil-exploration-gets-green-light/, 2019.

Nielsen, T., Knutz, P.C. and Kuijpers, A. 2008. Seismic expression of contourite depositional systems. *Developments in Sedimentology*, **60**, 301–321, https://doi.org/10.1016/S0070-4571(08)10016-4

Niemi, T.M., Ben-Avraham, Z., Hartnady, C.J.H. and Reznikov, M. 2000. Post-Eocene seismic stratigraphy of the deep ocean basin adjacent to the southeast African continental margin: a record of geostrophic bottom current systems. *Marine Geology*, **162**, 237–258, https://doi.org/10.1016/S0025-3227(99)00062-6

Rebesco, M., Hernández-Molina, F.J., Van Rooij, D. and Wåhlin, A. 2014. Contourites and associated sediments controlled by deep-water circulation processes: state-of-the-art and future considerations. *Marine Geology*, **352**, 111–154, https://doi.org/10.1016/j.margeo.2014.03.011

Said, A., Moder, C., Clark, S. and Ghorbal, B. 2015. Cretaceous–Cenozoic sedimentary budgets of the Southern Mozambique Basin: implications for uplift history of the South African Plateau. *Journal of African Earth Science*, **109**, 1–10, https://doi.org/10.1016/j.jafrearsci.2015.05.007

Schott, F.A., Xie, S.P. and McCreary, J.P. 2009. Indian Ocean circulation and climate variability. *Reviews of Geophysics*, **47**, RG1002, https://doi.org/10.1029/2007RG000245

Shepard, F.P. 1963. *Submarine Geology*. 2nd edn. Harper and Row, London.

Siesser, W.G. and Dingle, R.V. 1981. Tertiary sea-level movements around southern Africa. *Journal of Geology*, **89**, 523–536, https://doi.org/10.1086/628618

Simon, M.H., Ziegler, M., Bosmans, J., Barker, S., Reason, C.J.C. and Hall, I.R. 2015. Eastern South African hydroclimate over the past 270,000 years. *Scientific Reports*, **5**, 18153, https://doi.org/10.1038/srep18153

Sinha, M.C., Louden, K.E. and Parsons, B. 1981. The crustal structure of the Madagascar Ridge. *Geophysical Journal International*, **66**, 351–377, https://doi.org/10.1111/j.1365-246X.1981.tb05960.x

Toole, J.M. and Warren, B.A. 1993. A hydrographic section across the subtropical South Indian Ocean. *Deep-Sea Research Part I*, **40**, 1973–2019, https://doi.org/10.1016/0967-0637(93)90042-2

Watkeys, M.K. and Sokoutis, D. 1998. Transtension in southeastern Africa associated with Gondwana break-up. *Geological Society, London, Special Publications*, **135**, 203–214, https://doi.org/10.1144/GSL.SP.1998.135.01.13

Wiles, E., Green, A., Watkeys, M., Jokat, W. and Krocker, R. 2013. The evolution of the Tugela canyon and submarine fan: a complex interaction between margin erosion and bottom current sweeping, southwest Indian Ocean, South Africa. *Marine and Petroleum Geology*, **44**, 60–70.

Wiles, E., Green, A., Watkeys, M., Jokat, W. and Krocker, R. 2014. Anomalous seafloor mounds in the northern natal valley, southwest Indian Ocean: implications for the east African rift system. *Tectonophysics*, **630**, 300–312, https://doi.org/10.1016/j.tecto.2014.05.030

Ziegler, M., Simon, M.H., Hall, I.R., Barker, S., Stringer, C. and Zahn, R. 2013. Development of Middle Stone Age innovation linked to rapid climate change. *Nature Communications*, **4**, 1905, https://doi.org/10.1038/ncomms2897

Geomechanical behaviour of gassy soils and implications for submarine slope stability: a literature analysis

P. Kaminski[1]*, M. Urlaub[2], J. Grabe[1] and C. Berndt[2]

[1]Hamburg University of Technology (TUHH), Institute for Geotechnical Engineering and Construction Management, Harburger Schlossstrasse 20, 21079 Hamburg, Germany

[2]GEOMAR Helmholtz Centre for Ocean Research, Wischhofstrasse 1-3, 24148 Kiel, Germany

PK, 0000-0002-2106-2660; MU, 0000-0002-1116-636X; JG, 0000-0001-8112-7131; CB, 0000-0001-5055-0180

*Correspondence: pauline.kaminski@tuhh.de

Abstract: Submarine slope failures pose a direct threat to seafloor installations and coastal communities. Here, we evaluate the influence of free gas on the soil's shear strength and submarine slope failures in areas with gassy soils based on an extensive literature review. We identify two potential destabilization mechanisms: gas bubbles in the pore space lead to a reduced shear strength of the soil and/or gas induces excess pore pressures that consequently reduce the effective stress in the soil. Our evaluation of the reported mechanical and hydraulic behaviour of gassy sediments indicates that the unfavourable impact of entrapped gas on a soil's shearing resistance is not sufficient to trigger large-scale slope failures. Liquefaction failure due to high gas pressures is, however, a viable scenario in coarse-grained soils. Transferring the gas influence on the soil mechanical behaviour to constitutive models is identified as the most important prerequisite for a successful future analysis of slope stability.

Submarine slope failures are a common feature encountered at continental margins worldwide. The consequences of the slide events are potentially detrimental for coastal communities and offshore infrastructure (Locat and Lee 2002). It is hence of great interest to understand the processes that lead to and accompany submarine landsliding. The triggering mechanisms for large slide events on very low gradient slopes (1–2°) have been especially inscrutable, partly due to the lack of on-site studies (Urlaub *et al.* 2012). Based on the observation that several submarine landslides coincide with the occurrence and migration of free marine gas in the soil (Bünz *et al.* 2005; Lafuerza *et al.* 2012), gas is considered a conceivable trigger mechanism (Elger *et al.* 2018). A predominantly negative impact of free gas on the soil's mechanical behaviour is suspected. A precise assessment of the bearing capacity of gas-containing soils is, however, difficult from a geotechnical point of view, as the investigation of gassy soils is not yet very advanced due to the low relevance for engineering applications and the complexity in the required experimental testing procedures.

Thus, the influence of free gas on marine soils requires further investigation in order to assess the feasibility of gas acting as a trigger mechanism. We developed two working hypotheses for possible gas-induced failure mechanisms: either a reduction of the soil's shearing resistance due to enclosed gas bubbles in the soil, or the build-up of weakening excess pore pressures due to the acting gas pressure. Previous findings on the influence of gas have to a certain extent been contradictory (Wheeler 1988*b*; Hong *et al.* 2018), making a more detailed analysis necessary in order to evaluate the introduced hypotheses. Here, we gather the existing knowledge in the field of gassy soils from an extensive literature review. Particularly, we peruse the main characteristics of a gassy soil and the associated alteration of its geomechanical properties. The implications of entrapped gas bubbles for the overall soil structure can be severe. Hence, we review the processes of gas bubble growth, gas migration and gas-induced soil fracturing. Based on the current state of knowledge, we provide a further development of the existing approaches for gas-induced slope failure mechanisms. Consequently, on the basis of the scientific evidence previously published, we assess the feasibility of enclosed free gas in the soil as a potential trigger mechanism for submarine slope failures.

Continuously, the term 'soil' is used as a collective term for all loose rocks above the bedrock and thereby comprises all marine sediments (Press and Siever 1998).

From: Georgiopoulou, A., Amy, L. A., Benetti, S., Chaytor, J. D., Clare, M. A., Gamboa, D., Haughton, P. D. W., Moernaut, J. and Mountjoy, J. J. (eds) 2020. *Subaqueous Mass Movements and their Consequences: Advances in Process Understanding, Monitoring and Hazard Assessments*. Geological Society, London, Special Publications, **500**, 277–288.
First published online March 13, 2020, https://doi.org/10.1144/SP500-2019-149

General characteristics of gassy soils

In the marine environment gases mainly exist in three different states: dissolved in the water phase, as gas hydrate, or as free gas in the pore space. While a gas in its dissolved form does not influence the mechanical behaviour of a soil, the formation or dissociation of hydrates and entrapped gas bubbles have a major impact on soil characteristics (Sills and Wheeler 1992; Jayasinghe 2013). Hydrates are only present within the gas hydrate stability zone, but marine gases from thermogenic or biogenic origin can occur in a wide variety of settings (Judd and Hovland 2009).

Conceptual delimitation of gassy soils

Gassy soil is a partially saturated soil whose void space is occupied by a continuous water phase and a discontinuous gas phase in the form of gas bubbles. The minimum degree of saturation required for the occurrence of a discontinuous gas phase is given with $S_r \geq 85$–90%; that is the gas content of the pore fluid is ≤ 10–15% (Nageswaran 1983; Fredlund and Rahardjo 1993; Grozic *et al.* 1999; Wang *et al.* 2018). Hence, capillary cohesion, which is often associated with unsaturated soils with low degrees of saturation, cannot develop (Fredlund and Rahardjo 1993). The characteristics of gassy soils, and thus also the mechanical behaviour, is influenced by several factors. For example, grain size, soil structure and load history, as well as the amount of gas, and the specific pressure regime play a role. In order to evaluate the role of soil texture, not enough research results have been published in this area so far. Above all, the grain size of a soil determines the structure of a gassy soil (Sills and Wheeler 1992). Consequently, this results in a fundamentally different soil behaviour depending on the grain size. A differentiated consideration of different soil types is therefore indispensable and hence well-sorted soils are presupposed in the following.

Fine-grained soils

By geotechnical definition fine-grained, cohesive soils show grain sizes below a diameter of 0.063 mm and, accordingly, small natural pore spaces through which the gas bubbles cannot move. It cannot be ruled out that similar characteristics may occur in soils with slightly larger grain diameters and that a transition zone may occur. In order for the gas to invade the void space, the water phase has to be displaced by the gas phase. This process requires the gas pressure in the gas phase to exceed the capillary pressure acting in the capillary tubes between the grains. The necessary gas pressures usually exceed the soil's strength. Consequently, the gas causes local damage in the soil structure and forms large voids within the fully saturated soil matrix. These large gas voids are isolated, fixed with respect to the grain structure and randomly distributed within the saturated soil matrix (Wheeler 1986; Mabrouk and Rowe 2011). Hence, the two pore fluids do not mix but are spatially separated and thus the water phase can still be considered incompressible.

The size of the formed bubbles depends on a variety of boundary conditions, such as the material properties of soil, pore water and gas, the thermodynamic conditions, and the stress state. Scans from experiments conducted with fine-grained gassy soil show average bubble diameters of 0.1–1 mm (Wheeler 1988*a*). A schematic sketch of a fine-grained, gassy soil is shown in Figure 1a. In order to fully understand the mechanical impact of the gas bubbles on the overall soil behaviour it is essential to assess the interdependencies of the different influencing factors. A dominating parameter is the gas solubility, which is directly dependent on the acting gas pressure, the global pressure regime, temperature and salinity – which in turn will also adapt to a change in any other boundary condition.

Additionally, the gas pressure in neighbouring bubbles is not necessarily identical (Wheeler 1988*a*; Sills *et al.* 1991). In the ambition to create equilibrium, the gas pressures in adjoining bubbles adapt. Because the gas voids are fixed in the soil matrix, a gas exchange between two formed bubbles can only take place in the dissolved form by diffusive or advective transport in the pore water.

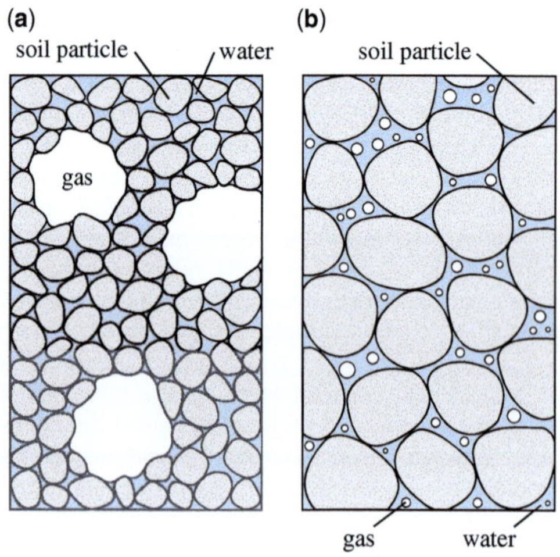

Fig. 1. Appearance of gassy soils according to Mabrouk and Rowe (2011): (**a**) fine-grained, gassy soil; (**b**) coarse-grained, gassy soil. Modified from Mabrouk and Rowe (2011, fig. 2), © 2011, with permission from Elsevier.

The existence of gas voids within the saturated soil matrix affects the stress state in the soil considerably. Knowledge about visual appearance of the gas voids is a requirement in order to evaluate the impact on the stress distribution. However, the description of the void's size and shape is not trivial, as it does not correlate directly with the internal gas pressure in the voids, the radii of the menisci, or the acting surface tension between the fluid phases (Wheeler 1988*a*; Sills *et al.* 1991). At the interface between the gas and the saturated soil matrix the pore water forms menisci bridging the spaces between the soil particles. The curvature radii of the menisci depend on the unique position of the solids, the surface tension, and the acting gas and water pressures (u_g and u_w; see Fig. 2). In the case of a high gas pressure compared to the pore water pressure, the menisci will have a concave shape to the gas phase and small radii. When the pressure difference $u_g - u_w$ reaches a critical upper boundary, gas encroachment to the saturated soil matrix occurs, that is the gas invades the pore space until reaching the limiting capillary pressure. In the opposite case of low gas pressures and comparably high pore water pressures in the surrounding saturated soil matrix, the radius of curvature will equal the radius of the gas void. A further decrease in the pressure difference $u_g - u_w$ leads to a flooding of the void with pore water.

The stress distribution in a gassy, fine-grained soil is comparable to a metal sheet containing a hole. The local stress state in the vicinity of the bubble is increased compared to the global stress state. There are localized effects of pore pressure build up and consolidation, as well as deformations and corresponding stress changes around the gas voids. Consequently, the pore pressure and stress distribution in the soil are potentially highly non-uniform on a small scale.

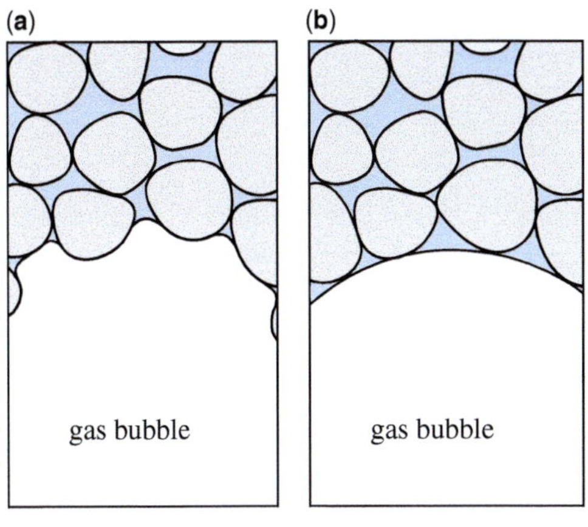

Fig. 2. Maximum and minimum expansion of a large gas bubble in a fine-grained saturated soil matrix (Wheeler 1988*a*): (**a**) maximum expansion ($u_g > u_w$); (**b**) minimum expansion ($u_g < u_w$).

As a consequence, Terzaghi (1943)'s approach of effective stresses ($\sigma' = \sigma - u$) is not applicable to cohesive, gassy soil in its entirety, but only for the water-saturated soil matrix excluding the gas voids. Providing a mathematical description for the stress state of the solid and the liquid phase, Wheeler (1986) introduced the parameter of operative stress σ'' ($\sigma'' = \sigma - u_w$). Local non-uniformities in the stress distribution around the gas voids are neglected in this approach as the current state of research does not allow a more sophisticated description. The quantification of the gas-induced stress changes is one of the major research objectives in the field of gassy soils.

To account for the significant changes in the soil's characteristics more accurately, the parameters of the void ratio of the saturated soil matrix, e_m, and the volume fraction of bubbles, *f*, are also introduced (see equations 1 and 2, respectively) as the conventional parameters of void space, *e*, and degree of saturation, S_r, do not provide an unobjectionable description for gassy soil anymore either (Wheeler 1986):

$$e_m = S_r \times e \quad (1)$$

$$f = \frac{(1 - S_r) \times e}{1 + e} \quad (2)$$

The void ratio of the saturated soil matrix describes solely the water-filled pore space, while the volume fraction of bubbles defines the gas-filled voids.

Coarse-grained soils

Sandy soils with grain fractions of diameters larger than 0.063 mm are considered coarse. In consequence of the comparably large grain size, the magnitude of the void space is correspondingly high. Therefore, in coarse-grained soils gas bubbles can be accommodated in the void space without structural changes in the grain matrix. Consequently, the existence of a gas phase in sands mostly changes the compressibility of the pore fluid (Sills and Wheeler 1992; Grozic *et al.* 1999). A characteristic coarse-grained, gassy soil is shown in Figure 1b. Because no structural changes in the grain matrix are required to host a gas phase in a sandy soil, the formation of gas bubbles requires less energy in coarse soil (Mabrouk and Rowe 2011).

When the pore fluids reach equilibrium, the pressure inside a gas bubble can be assumed equal to the pressure in the surrounding pore water ($u_g = u_w$). Hence, the effective stress approach ($\sigma' = \sigma - u$) is valid for gassy sands (Fredlund and Rahardjo 1993). This has also been experimentally verified by Finno *et al.* (2017).

Direct effect of gas on the geomechanical properties

The gas entrapped in a soil is a compressible medium. Consequently, considerable changes in the engineering properties of the gas–soil continuum can be expected. In fine-grained soils the gas alters the soil structure. Therefore, the soil supposedly does not behave like a saturated soil with a supplementary compressible pore fluid, but rather shows distinct properties. The precise expression of the alterations depends strongly on the type of gas, and on the influencing parameters solubility, temperature and stress state. Additionally, the drainage conditions are a relevant factor: drained conditions are characterized by the dissipation of excess pore pressures due to the drainage of the pore fluid from the pore space while under undrained conditions excess pore pressures build up as the pore fluid cannot escape the pore space. Whether drained or undrained conditions prevail depends on, amongst others, the load velocity and the soil permeability.

Compressibility and expansivity

The compressibility of fine-grained soils, irrespective of their degree of saturation, is characterized by their time-dependent consolidation behaviour. Under gravitational loading a low gas content prevents complete consolidation. When reaching critical gas pressures, higher gas contents can lead to fracturing of the saturated soil matrix. The fractures form a path for easy gas and pore water escape and consequently enhance the consolidation process. The impact of gas on self-weight consolidation can hence be estimated through the critical pressure for sediment fracturing. Before the onset of fracturing, a gassy soil under self-weight is prone to underconsolidation, while the formation of fractures decreases the time span until normal consolidation is reached (Sills and Gonzalez 2001). The consolidation behaviour under compressive loading is described by the double compressibility model, which was developed by Thomas (1987) based on a series of experiments. Under loading, gassy soil undergoes deformation due to the applied total stress and due to the consolidation stress. Both induced deformations happen independently.

Under undrained conditions the soil matrix will not change its volume, in contrast to the gas voids. In order to take on the volume change of the gas voids the soil matrix deforms at constant volume. The amount of stress, which is transferred to the gas void during the deformation process, is influenced by the shearing resistance of the saturated soil matrix (Thomas 1987).

Under drained conditions a saturated soil changes its volume. This change of void ratio depends on the effective stress. In a gassy soil, the void ratio of the saturated soil matrix depends on the operative stress. For isotropic loading (a soil cube under uniform loading from all coordinate directions) or earth pressure at rest (under the natural earth pressure from the surrounding soil) it can be assumed that the relationships for saturated soils and the saturated soil matrix of gassy soils are equal, that is the saturated soil matrix behaves identical to a saturated soil. The volume change behaviour of the gas voids, however, relates to the total stress state (Thomas 1987).

Consequently, the double compressibility of a gassy soil can be summarized by introducing two modes of deformation.

(1) Changes in total stress induces compression and dissolution of gas, and a deformation of the soil matrix around the voids due to local shearing. This is generally encountered during the initial undrained state after load application. Under unloading conditions, that is a decrease in total stress, the gas volume expands and a corresponding change in the stress state around the gas cavities takes place. Between the removal of load and changes in the gas regime a time difference in the range of minutes to hours occurs in fine-grained soils. The reasons for the temporal offset are not understood, yet (Thomas 1987).

(2) Changes in the operative stress cause an overall drainage and a subsequent compression of the saturated soil matrix (Thomas 1987). The independent development of the volume fraction of gas bubbles f (deformation mode 1) and of the matrix void ratio e_m (deformation mode 2) under compressive loading over time is shown in Figure 3. Under fully saturated

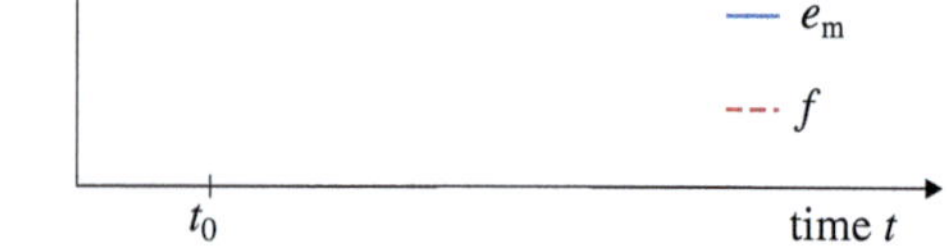

Fig. 3. The development of the volume fraction of bubbles f and the void ratio of the saturated soil matrix e_m over time following load application (Wheeler 1986; Thomas 1987).

conditions the development of the stress state during the consolidation process can be calculated based on the approach by Terzaghi (1943). The quantification of the gas impact is, however, difficult based on existing knowledge. Concerning the compressive behaviour of coarse-grained soils, no definite proposition on the geomechanical properties can be made due to a lack of research efforts on the topic.

Liquefaction potential

The unloading of coarse-grained soils with a gas-saturated pore fluid leads to gas exsolution which starts as soon as the total stress reaches the solubility limit. The exsolution process keeps the pore pressures of both fluids on a constant level while the total stress is further reduced. This leads to a significant reduction in effective stress. The pore pressure only starts to decrease when the effective stress becomes very small, that is when the compressibility of the grain skeleton equals that of the pore fluid mixture (Sobkowicz and Morgenstern 1984; Amaratunga and Grozic 2009). This process is depicted in Figure 4.

Once the effective stress, and thus the grain-to-grain contact forces approach zero the soil matrix liquefies. Because of the decreasing permeability and the accompanying drainage restriction, the pore pressure build-up and liquefaction potential rise in loose fine sands (Amaratunga and Grozic 2009).

Shear strength

For fine-grained, gassy soils a variety of triaxial test series with different boundary conditions have been conducted and published in the past. The most detailed works on the undrained shear strength of gassy, fine-grained soils were published by Nageswaran (1983), Wheeler (1986, 1988*a*, *b*) and Sultan *et al.* (2012), and can be summarized in a detailed conceptual model.

Shear loading induces a deformation of the soil. According to Wheeler (1988*a*) the deformation of the saturated soil matrix is directly coupled to changes in the gas void pressure. This has a major impact on the general shearing behaviour of a soil, as the soil experiences shrinkage or growth of the cavities, bubble flooding and plastic yielding of the surrounding saturated soil matrix caused by stress concentrations around the voids. These processes have a major impact on the shearing behaviour of the soil (Wheeler 1988*a*, *b*). It is important, therefore, to consider these processes when assessing the detrimental or beneficial effect of gas on the shear strength.

Shrinkage of the gas cavities leads to an overall decrease of the volume fraction of bubbles, *f*. Consequently, the required stress to cause further deformation increases, resulting in a strain-hardening behaviour of the soil. According to Boyle's gas law the gas pressure rises proportionally to a decrease in volume. Hence, a potential movement of gas into solution due to an increase in gas pressure by a decrease in gas volume during the process of cavity shrinkage reduces the bubble stiffness and thereby leads to a reduction in the strain-hardening effect. Vice versa, cavity growth increases *f*, decreases the required stress for further deformation, and results in a strain-softening response, which could be reduced by an increase in bubble stiffness by gas exsolution processes (Wheeler 1988*a*, *b*).

The pore water pressure in the saturated soil matrix increases significantly during undrained shearing, while the gas pressure within the cavities is expected to stay constant, being solely dependent on the total stress. The pressure difference ($u_g - u_w$) is hence likely to decrease to negative values and might thereby reach a range of values critical for bubble flooding. The widespread flooding of gas

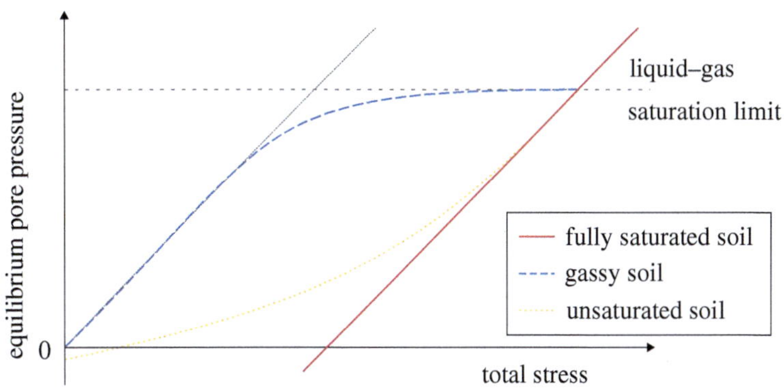

Fig. 4. Equilibrium pore pressure of the pore fluids during undrained isotropic unloading in dependence of the total stress state based on Sobkowicz and Morgenstern (1984).

bubbles strengthens the surrounding soil matrix by decreasing e_m, thereby, altering the shear strength beneficially. When the stage of complete bubble flooding is reached, the examined soil is fully saturated but contains water-filled cavities. As long as gas fills the voids, the cavity roofs are supported by the tension forces of the menisci. In a fully saturated state, the support is lost. Hence, cavity collapse and a resulting detrimental impact on the shear strength is likely (Wheeler 1988*a*, *b*). This process can, for example, be observed in the test results of Sultan *et al.* (2012) where the gas-caused damage in the soil structure results in a detrimental development of the sample's shearing resistance. Mostly, an increase in the amount of gas within the sample intensifies the detrimental impact of the gas (Grozic *et al.* 2005; Sultan *et al.* 2012). The influence of the bubble size on the shear strength has not been investigated, yet.

During triaxial testing of a coarse-grained gassy soil the gas leads to a soil response between the drained and the undrained response of the same soil in a fully saturated state. With regard to the shear strength this is either beneficial or detrimental, depending on the mechanical behaviour of the sand in question (e.g. dilative or contractive response, respectively). When all the gas is resoluted during the shearing process, a fully saturated soil behaviour can be expected (Rad *et al.* 1994; Grozic *et al.* 1999; Fourie *et al.* 2001; Vega-Posada *et al.* 2014).

Concluding, it can be said that from the current state of knowledge an unambiguous statement regarding the influence on the shear strength cannot be made – especially for fine-grained soils. Several processes with a significant influence on the shearing behaviour could be identified. However, the relevant small-scale processes are difficult to observe and hence challenging to describe mathematically. Also, the actual soil response depends strongly on the prevailing boundary conditions of the test series and the amount of studies is not sufficient in order to establish reliable empirical correlations. It has thus not been possible to quantify the impact of gas on a soil's shearing behaviour, yet. Hence, a conclusive and reliable relation between gas content and bearing capacity cannot be formulated based on the current state of knowledge. An overview on the qualitative shearing response of gassy soil is given in Figure 5.

Indirect effect of gas on the soil structure

Especially in fine-grained soil the expansion and potential migration of gas bubbles leads to damage in the soil structure, which changes the mechanical behaviour of the soil. Both mechanisms change the stress state in the immediate environment of the gas bubble (Wheeler 1986; Sultan *et al.* 2012). Because of the size of the pore space, coarse-grained soils are not affected by these processes (Grozic *et al.* 1999).

Bubble growth

The magnitude of bubble growth is strongly dependent on the availability and the production rate of the gas, and therefore relates to the individual boundary conditions. The growth of a gas bubble will affect the surrounding soil, because the increasing bubble displaces pore water and/or applies forces to the grain matrix. As explained above, the characteristics of the formed gas cavities in a fine-grained soil strongly depend on the applied total stress. The approach developed by Wheeler (1990) assumes a homogeneous soil matrix and spherical gas bubbles. Thereby, a symmetrical stress distribution around the gas cavity may be anticipated. When the gas pressure increases compared to the total stress – for example due to ongoing bacterial degradation of organic material – the cavity expands. Until the hypothetical yield criterion of twice the undrained shear strength ($2c_u$) is reached, linear elastic soil behaviour is assumed. The resultant bubble growth is expected to induce plastic yielding in the saturated soil matrix adjoining the gas void. When yielded, Wheeler (1990) suggests a perfectly plastic soil matrix behaviour, that is all deformations are irreversible and an increase or decrease in soil stiffness during yielding is neglected. The yielded zones spread out from the bubble in a spherical manner. A potential plastic collapse mechanism is formed once the yielded zones around several cavities connect.

The onset of yielding is defined by the constitutive description of the saturated soil and the applied yield criterion. Moreover, a certain difference in gas pressure and mean total stress in the soil matrix limits the process of cavity expansion and contraction, because the processes of bubble flooding and gas encroachment to the soil matrix have to be considered as well (see Fig. 2). While higher shear strengths and lower volume fractions of bubbles favour gas encroachment and bubble flooding, in soils with lower shear strength the gas cavities are more prone to expansion and contraction.

Bubble migration

The movement of gas in a soil matrix can occur as discrete bubbles through the soil matrix or along, for example, gas-induced fractures. Apart from these gas-induced fracture structures, existing geological features, such as faults or erosional surfaces, can represent a possible low-resistance migration path for free gas (Judd and Hovland 2009; Lafuerza *et al.* 2012).

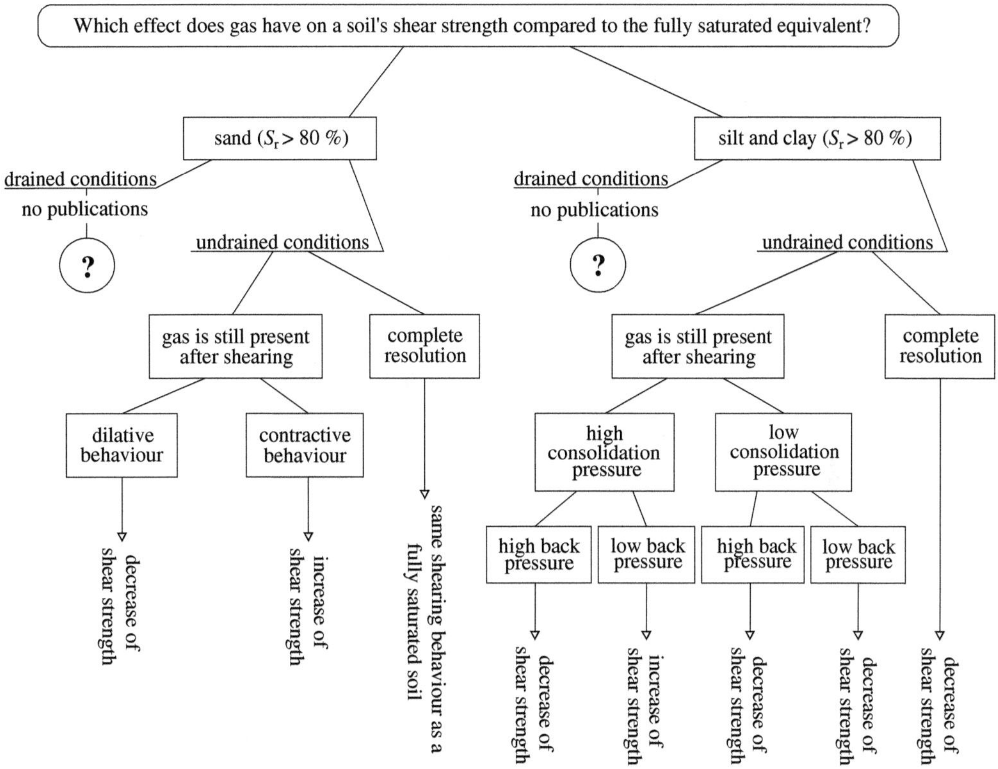

Fig. 5. The impact of gas on a soil's shear strength based on investigations concerning coarse-grained soils (Sobkowicz and Morgenstern 1984; Sills and Wheeler 1992; Rad *et al.* 1994; Grozic *et al.* 1999, 2000; Fourie *et al.* 2001; Grozic 2013; Amaratunga and Grozic 2009; Rebata-Landa and Santamarina 2012; Vega-Posada *et al.* 2014; Wang *et al.* 2018) and fine-grained soils (Nageswaran 1983; Wheeler 1986, 1988*a*, *b*; Sills *et al.* 1991; Sills and Wheeler 1992; Sills and Gonzalez 2001; Sultan *et al.* 2012; Hong *et al.* 2018).

In coarse-grained soil layers the migration of discrete bubbles through the comparably large pore space happens either by advective transport with the groundwater flow or is buoyancy driven (Judd and Hovland 2009). The description of the ascent of a discrete bubble in a fine-grained soil matrix is not so trivial due to several processes which influence the controlling boundary conditions. These include the change of the thermodynamic conditions along the path of ascent, the change of the bubble velocity in a vertical direction according to the change in bubble radius, processes of gas expansion and gas dissolution, and possibly the formation of hydrate when the conditions are suitable (Li and Huang 2016). However, these processes become irrelevant when the nature of the soil matrix inhibits the bubble migration. Movement of bubbles that are larger than the particles making up the soil matrix is only possible if the driving buoyancy force is higher than the soil's shearing resistance. For the calculation of the counteracting resisting force no analytical solution exists. Wheeler (1990) suggests two estimating approaches to analyse this scenario. Firstly, an equilibrium of forces consisting of the buoyancy of the bubble and the resistance of the soil is proposed. Secondly, based on the assumption of continuous cavity expansion in the upper part of the bubble and a corresponding continuous cavity contraction in the lower part of the bubble, another approach to define the resistance of the soil to gas migration was developed (Wheeler *et al.* 1990). A similar approach was published by van Kesteren and van Kessel (2002).

From the three introduced approaches the critical bubble radius for bubble ascent $r_{b,crit}$ through a saturated, fine-grained soil matrix can be derived. For a generic soil with a unit weight of saturated soil matrix of $\gamma_m = 20$ kN m^{-3}, the critical bubble radius depends on the undrained shear strength of the saturated soil matrix $c_{u,m}$ (see Fig. 6). Hereby, the unit weight of the gas is assumed $\gamma_g = 0$ kN m^{-3}. The depth scale is based on a typical development of $c_{u,m}$ with depth for a normally or slightly overconsolidated soil of 1 kN m^{-1} below the seafloor

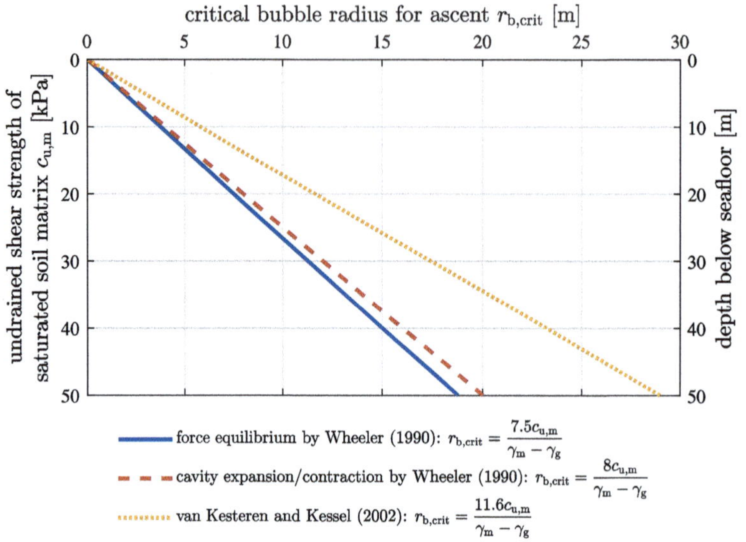

Fig. 6. Critical bubble radius for ascension $r_{b,crit}$ in a saturated soil depending on the undrained shear strength $c_{u,m}$ of the saturated soil matrix.

(Randolph and Gourvenec 2011), and merely serves as a reference.

It is obvious that bubble migration through the saturated soil matrix is only a realistic scenario for very low shear strengths. In some marine sediments these might occur in very shallow sediment layers close to the seafloor due to high water content (Baudet and Ho 2004; Madhusudhan *et al.* 2017). However, especially in sediments with high organic carbon content the formation of crusts with increased shear strengths due to interparticle bonding and cementing effects in the upper layers was observed (Busch and Keller 1982; Dean 2010). Shallow gas bubble migration in corresponding regions hence becomes less likely. Furthermore, due to the depth dependency of $c_{u,m}$ the resistance to bubble ascent increases with increasing depth below the seafloor. In the depth ranges where slide planes of submarine landslides are located (50–250 m below seafloor) bubble migration can therefore be ruled out. As buoyancy – the main driving force for bubble migration – is directed upwards, horizontal gas migration is even less likely. Because fine-grained soils are poor hosts for gas migration processes they often act as a natural barrier. Movement of the gas phase is thus restricted to layers with coarser grain sizes or to faults and fractures.

As the migration of discrete bubbles through the saturated sediment matrix was ruled out as a potential transport mechanism for free gas in marine soils, sediment fracturing induced by high gas pressure and a subsequent movement of gas through the fracture become more likely. As fractures provide low-resistance drainage paths for both pore fluids, gas and water, fracturing potentially describes a mechanism which strengthens rather than weakens the soil, as a dissipation of excess pore pressure becomes possible (Sills and Gonzalez 2001). Under these circumstances fracturing is a spatially restricted, beneficial feature in a gassy soil (see Fig. 7). Unfortunately, preferential conditions for the onset of fracturing could not be quantified so far. Many authors have hypothesized about the possibility of gas-induced fracturing of the soil matrix: the approaches include the application of a yield criterion and soil failure in the vicinity of the gas cavities, linear elastic fracture mechanics, or local liquefaction due to excess pore pressure build up. Additionally, techniques such as computed tomography scanning, X-ray density measurements and mercury intrusion porosimetry were applied in several test series on different soils and conditions (Wheeler 1990; Sills and Gonzalez 2001; van Kesteren and van Kessel 2002; Elger *et al.* 2018; Johnson *et al.* 2018). However, the development of a satisfying mathematical approach, which can be verified by experimental testing, is still pending. In the offshore oil industry approaches for hydraulic fracturing have been proven satisfying for engineering applications, which involve the fracturing of rock formations under high fluid pressures. Here, the mathematical approach for the description of the fracture behaviour presupposes a linear elastic material behaviour. The onset of fracturing is defined by the Mohr–Coulomb failure criterion (Valkó and Economides 1995). Whether this approach is a reliable

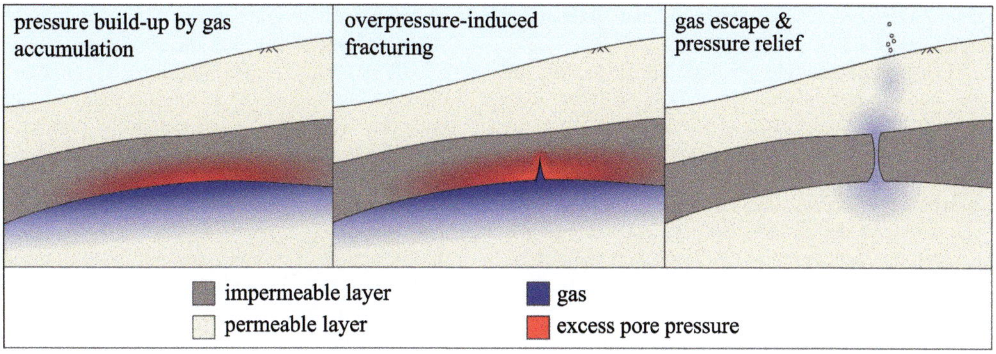

Fig. 7. Schematic process of a large-scale gas-induced fracture in a fine-grained soil layer initiating the dissipation of excess pore pressures by creating a low-resistance drainage path.

description of fracturing induced by a gaseous instead of a liquid fluid has not yet been proven. The development of a reliable mathematical description of gas-induced fracture behaviour of granular materials would contribute significantly to a better understanding of the mechanical behaviour of gassy soils and geological processes such as piping, which describes a vertically directed erosion process. It is therefore highly relevant for further research efforts in this field.

Discussion

Through an extensive literature study we analysed the mechanical behaviour of gas-bearing soil with regard to potential causes of slope instabilities at continental margins. Even though the characteristics of the *in-situ* shear stress during a slope failure event are most similar to shear in a direct simple shear test, the effect of gas on the shear strength has so far mostly been investigated in triaxial tests. Hence, the conclusions drawn from the published triaxial tests are applied in the evaluation of slope stability. A gassy soil's mechanical behaviour is influenced by many factors, and small changes in the boundary conditions may possibly lead to a fundamentally different soil response: detrimental to the shearing resistance, no change compared to the fully saturated equivalent, or beneficial to the shearing resistance. The establishment of a coherent relationship represents the current objective of research in the field of gassy soils. Based on the current state of knowledge, Figure 5 summarizes the effects of different influencing factors on the shear strength of a soil. The impact of gas on the shear strength is unfavourable for most boundary conditions. Because the favourable boundary conditions are not identical to those that are expected at the seafloor, we conclude that occluded gas in the soil matrix has a predominantly negative impact on the shearing resistance of marine soils. Because of the small slope angles, at which most failures occur, the driving forces of the slope are far lower than the frictional forces in the potential glide plane. Consequently, to overcome the resisting forces a vast shear strength decrease is required to induce slope failure. As a shear strength decrease in a realistic range cannot overcome the resisting forces a much more far-reaching change in the soil structure must take place to produce slope failure. Hence, a reduction of the shear strength is not a realistic trigger mechanism. A shear strength reduction due to occluded gas bubbles can therefore rather be classified as a preconditioning factor than as a trigger mechanism.

A significant decrease in effective stresses can lead to liquefaction failure of the soil. Coarse-grained soils show an unambiguous reduction of effective stresses when gas exsolutes and expands in the pore fluid, as shown in Figure 4. The exsolution process is controlled by the total stress. Potential geological processes that lead to a reduction in total stress are, for instance, changes in sea-level, slope failures, or other erosive processes. Gas-induced liquefaction failure of a slope in an area of shallow water depths under tidal influence, and hence under total stress variation, has been identified as a viable trigger mechanism by Clare *et al.* (2016). The destructive potential of liquefaction failure is therefore not to be overlooked. Under naturally occurring conditions, it is also likely that an ongoing gas production, due to biological activity, upward migration of thermogenically formed gases, or hydrate dissociation, triggers gas exsolution and expansion. This scenario, however, has not been investigated yet. Nevertheless, increasing pore pressures induced by gas under constant total stress conditions resulting in an effective stress decrease represents a logical assumption. Restricting the validity to coarse-grained soil, we can thus confirm the hypothesis of gas-induced excess pore pressures and subsequent liquefaction.

The processes in fine-grained soils are more complex and are not only influenced by the stress state but also by other factors, such as shear strength of the saturated soil and the degree of saturation. It seems that increasing gas pressures do not lead to decreasing effective stresses in fine-grained soil. The extent to which increasing gas pressures have a negative impact on the soil structure cannot be conclusively clarified based on the current state of research.

It is difficult to determine if an increase in gas pressures and its consequences can trigger submarine slope failures. Extensive liquefaction of soil surely provides a viable trigger mechanism. It is, however, questionable whether the occurring gas pressures reach sufficiently high magnitudes on a larger scale in order to trigger a liquefaction failure event. Rapid unloading due to slides triggered in a different way and a subsequent liquefaction proposes a realistic scenario for the retrogressive advance of a slide complex, but not for the trigger itself.

Based on the equivocal results of the evaluation, it must be questioned whether processes on a small-scale soil mechanical level are generally capable of triggering submarine slope failures. Therefore, we propose that large-scale processes, for example uplift of large sediment packages or piping mechanisms, are more likely trigger mechanisms. Nevertheless, a further investigation of the problem would be useful in order to reliably exclude gassy soil as a triggering mechanism. To conclusively clarify the question at hand for complex stratigraphic sequences and the involved feedback processes, an evaluation with the help of numerical models is indispensable. In this regard, the development of appropriate constitutive models is a prerequisite.

Conclusion

Based on the literature research carried out, the following findings concerning the general mechanical behaviour of gassy soils can be derived:

- Terzaghi (1943)'s principle of effective stresses is applicable for coarse-grained gassy soils but not for fine-grained gassy soils in their entirety;
- increasing gas pressures lead to a decrease in effective stresses and thus potentially to liquefaction failure in coarse-grained soils. In fine-grained soils increasing gas pressures do not result in an effective stress reduction.

With respect to the generation of slope failures the following findings are of importance:

- small changes in the boundary conditions potentially lead to a fundamentally different response of gassy soil ranging from detrimental, to no effect, to beneficial to the shearing resistance;
- for the common stress conditions at continental margins the impact of gas on the shear strength is mostly unfavourable but not strong enough to present a viable trigger mechanism for slope failures. Processes on a small-scale soil mechanical level do not show a lot of potential to work as a large-scale trigger mechanism;
- the migration of discrete gas bubbles through an intact fine-grained saturated soil matrix is not possible. Hence, fines act as gas barriers and migration is spatially restricted to sandy soil layers, faults and fractures.

A more detailed understanding of the gas-influenced soil mechanical behaviour is indispensable to develop constitutive models for gassy soils on the basis of which numerical stability analyses can be performed.

Acknowledgements This work is part of a co-operative project at Hamburg University of Technology and GEOMAR Helmholtz Centre for Ocean Research.

Funding The work is funded by grant GR 1024/35-1 from the Deutsche Forschungsgemeinschaft to Jürgen Grabe.

Author contributions **PK**: conceptualization (lead), methodology (lead), writing – original draft (lead); **MU**: methodology (equal), writing – review & editing (lead); **JG**: methodology (supporting), writing – original draft (supporting), writing – review & editing (supporting); **CB**: writing – original draft (supporting), writing – review & editing (supporting).

References

Amaratunga, A. and Grozic, J. 2009. On the undrained unloading behaviour of gassy sands. *Canadian Geotechnical Journal*, **46**, 1267–1276, https://doi.org/10.1139/T09-056

Baudet, B.A. and Ho, E.W. 2004. On the behaviour of deep-ocean sediments. *Géotechnique*, **54**, 571–580, https://doi.org/10.1680/geot.2004.54.9.571

Bünz, S., Mienert, J., Vanneste, M. and Andreassen, K. 2005. Gas hydrates at the Storegga Slide: constraints from an analysis of multicomponent, wide-angle seismic data. *Geophysics*, **70**, B19–B34, https://doi.org/10.1190/1.2073887

Busch, W.H. and Keller, G.H. 1982. Consolidation characteristics of sediments from the Peru-Chile continental margin and implications for past sediment instability. *Marine Geology*, **45**, 17–39, https://doi.org/10.1016/0025-3227(82)90178-5

Clare, M.A., Hughes Clarke, J.E., Talling, P.J., Cartigny, M.J.B. and Pratomo, D.G. 2016. Preconditioning and triggering of offshore slope failures and turbidity currents revealed by most detailed monitoring yet at a

fjord-head delta. *Earth and Planetary Science Letters*, **450**, 208–220, https://doi.org/10.1016/j.epsl.2016.06.021
Dean, E.T.R. (ed.) 2010. *Offshore Geotechnical Engineering: Principles and Practice*. Telford, London.
Elger, J., Berndt, C., Rüpke, L., Krastel, S., Gross, F. and Geissler, W. 2018. Submarine slope failures due to pipe structure formation. *Nature Communications*, **9**, 1–6, https://doi.org/10.1038/s41467-018-03176-1
Finno, R.J., Zhang, Y. and Buscarnera, G. 2017. Experimental validation of Terzaghi's effective stress principle for gassy sand. *Journal of Geotechnical and Geoenvironmental Engineering*, **143**, 1–10, https://doi.org/10.1061/(ASCE)GT.1943-5606.0001797
Fourie, A.B., Hofmann, B.A., Mikula, R.J., Lord, E.R. and Robertson, P.K. 2001. Partially saturated tailings sand below the phreatic surface. *Géotechnique*, **51**, 577–585, https://doi.org/10.1680/geot.2001.51.7.577
Fredlund, D.G. and Rahardjo, H. 1993. *Soil Mechanics for Unsaturated Soils*. John Wiley & Sons, Inc., New York.
Grozic, J.L. 2013. Liquefaction potential of gassy marine sands. *In*: Locat, J., Mienert, J. and Boisvert, L. (eds) *Submarine Mass Movements and Their Consequences*. Springer Netherlands, Dordrecht, 37–45, https://doi.org/10.1007/978-94-010-0093-2_5
Grozic, J., Robertson, P. and Morgenstern, N. 1999. The behavior of loose gassy sand. *Canadian Geotechnical Journal*, **36**, 482–492, https://doi.org/10.1139/t99-007
Grozic, J., Robertson, P. and Morgenstern, N. 2000. Cyclic liquefaction of loose gassy sand. *Canadian Geotechnical Journal*, **37**, 843–856, https://doi.org/10.1139/t00-008
Grozic, J.L., Nadim, F. and Kvalstad, T.J. 2005. On the undrained shear strength of gassy clays. *Computers and Geotechnics*, **32**, 483–490, https://doi.org/10.1016/j.compgeo.2005.10.002
Hong, Y., Wang, L.Z. and Yang, B. 2018. Undrained shear behaviour of gassy clay with varying initial pore water pressures. *Proceedings of China-Europe Conference on Geotechnical Engineering*, **1**, 524–528, https://doi.org/10.1007/978-3-319-97112-4
Jayasinghe, A.G. 2013. *Triaxial Compression Strength of Methane Hydrate-Bearing Sediments*. PhD thesis, University of Calgary, https://doi.org/10.11575/PRISM/28526
Johnson, M., Peakall, J., Jia, X., Fairweather, M., Harbottle, D., Biggs, S. and Hunter, T.N. 2018. Enhanced gas migration through permeable bubble networks within consolidated soft sediments. *American Institute of Chemical Engineering Journal*, **64**, 4131–4147, https://doi.org/10.1002/aic.16223
Judd, A. and Hovland, M. 2009. *Seabed Fluid Flow: The impact on geology, biology and the marine environment*. Cambridge University Press, New York.
Lafuerza, S., Sultan, N. *et al.* 2012. Failure mechanisms of Ana Slide from geotechnical evidence, Eivissa Channel, Western Mediterranean Sea. *Marine Geology*, **307–310**, 1–21, https://doi.org/10.1016/j.margeo.2012.02.010
Li, C. and Huang, T. 2016. Simulation of gas bubbles with gas hydrates rising in deep water. *Ocean Engineering*, **112**, 16–24, https://doi.org/10.1016/j.oceaneng.2015.12.002
Locat, J. and Lee, H. 2002. Submarine landslides: advances and challenges. *Canadian Geotechnical Journal*, **39**, 193–212, https://doi.org/10.1139/t01-089
Mabrouk, A. and Rowe, R.K. 2011. Effect of gassy sand lenses on a deep excavation in a clayey soil. *Engineering Geology*, **122**, 292–302, https://doi.org/10.1016/j.enggeo.2011.06.009
Madhusudhan, B.N., Clare, M.A., Clayton, C.R. and Hunt, J.E. 2017. Geotechnical profiling of deep-ocean sediments at the AFEN submarine slide complex. *Quarterly Journal of Engineering Geology and Hydrogeology*, **50**, 148–157, https://doi.org/10.1144/qjegh2016-057
Nageswaran, S. 1983. *Effect of gas bubbles on the seabed behaviour*. PhD thesis, University of Oxford.
Press, F. and Siever, R. 1998. *Earth*. 4th edn. Freeman, New York.
Rad, N.S., Vianna, A.J.D. and Berre, T. 1994. Gas in soils II: effect of gas on undrained static and cyclic strength of sand. *Journal of Geotechnical Engineering*, **120**, 716–736, https://doi.org/10.1061/(ASCE)0733-9410(1994)120:4(716)
Randolph, M. and Gourvenec, S. 2011. *Offshore Geotechnical Engineering*. 1st edn. CRC Press, London, https://doi.org/10.1201/9781315272474
Rebata-Landa, V. and Santamarina, J.C. 2012. Mechanical effects of biogenic nitrogen gas bubbles in soils. *Journal of Geotechnical and Geoenvironmental Engineering*, **138**, 128–137, https://doi.org/10.1061/(ASCE)GT.1943-5606.0000571
Sills, G.C. and Gonzalez, R. 2001. Consolidation of naturally gassy soft soil. *Géotechnique*, **51**, 629–639, https://doi.org/10.1680/geot.2001.51.7.629
Sills, G.C. and Wheeler, S.J. 1992. The significance of gas for offshore operations. *Continental Shelf Research*, **12**, 1239–1250, https://doi.org/10.1016/0278-4343(92)90083-V
Sills, G.C., Wheeler, S.J., Thomas, S.D. and Gardner, J.N. 1991. Behaviour of offshore soils containing gas bubbles. *Géotechnique*, **41**, 227–241, https://doi.org/10.1680/geot.1991.41.2.227
Sobkowicz, J.C. and Morgenstern, N.R. 1984. The undrained equilibrium behaviour of gassy sediments. *Canadian Geotechnical Journal*, **21**, 439–448, https://doi.org/10.1139/t84-048
Sultan, N., de Gennaro, V. and Puech, A. 2012. Mechanical behaviour of gas-charged marine plastic sediments. *Géotechnique*, **62**, 751–766, https://doi.org/10.1680/geot.12.OG.002
Terzaghi, K. 1943. *Theoretical Soil Mechanics*. 1st edn. John Wiley & Sons, Inc., New York.
Thomas, S.D. 1987. *The consolidation behaviour of gassy soils*. PhD thesis, University of Oxford.
Urlaub, M., Zervos, A., Talling, P.J., Masson, D.G. and Clayton, C.I. 2012. How do 2° slopes fail in areas of slow sedimentation? A sensitivity study on the influence of accumulation rate and permeability on submarine slope stability. *In*: Yamada, Y., Kawamura, K. *et al.* (eds) *Submarine Mass Movements and Their Consequences*. Springer, Dordrecht, 277–287, https://doi.org/10.1007/978-94-007-2162-3
Valkó, P. and Economides, M. 1995. *Hydraulic Fracture Mechanics*. John Wiley & Sons, Chichester.
van Kesteren, W. and van Kessel, T. 2002. Gas bubble nucleation and growth in cohesive sediments.

Proceedings in Marine Science, **5**, 329–341, https://doi.org/10.1016/S1568-2692(02)80025-0

Vega-Posada, C.A., Finno, R.J. and Zapata-Medina, D.G. 2014. Effect of gas on the mechanical behavior of medium-dense sands. *Journal of Geotechnical and Geoenvironmental Engineering*, **140**, 1–10, https://doi.org/10.1061/(ASCE)GT.1943-5606.0001163

Wang, Y., Kong, L., Wang, Y., Wang, M. and Wang, M. 2018. Liquefaction response of loose gassy marine sand sediments under cyclic loading. *Bulletin of Engineering Geology and the Environment*, **77**, 963–976, https://doi.org/10.1007/s10064-017-1164-7

Wheeler, S.J. 1986. *The stress-strain behaviour of soils containing gas bubbles*. PhD thesis, University of Oxford.

Wheeler, S.J. 1988*a*. A conceptual model for soils containing gas bubbles. *Géotechnique*, **38**, 389–397, https://doi.org/10.1680/geot.1988.38.3.389

Wheeler, S.J. 1988*b*. The undrained shear strength of soils containing large gas bubbles. *Géotechnique*, **38**, 399–413, https://doi.org/10.1680/geot.1988.38.3.399

Wheeler, S.J. 1990. Movement of large gas bubbles in unsaturated fine-grained sediments. *Marine Georesources & Geotechnology*, **9**, 113–129, https://doi.org/10.1080/10641199009388234

Wheeler, S.J., Sham, W.K. and Thomas, S.D. 1990. Gas pressure in unsaturated offshore soils. *Canadian Geotechnical Journal*, **27**, 79–89, https://doi.org/10.1139/t90-008

From gravity cores to overpressure history: the importance of measured sediment physical properties in hydrogeological models

Davide Mencaroni[1]*, Jaume Llopart[1], Roger Urgeles[1], Sara Lafuerza[2], Eulàlia Gràcia[1], Anne Le Friant[3] and Morelia Urlaub[4]

[1]Institut de Ciències del Mar (CSIC), 08003 Barcelona, Spain

[2]Sorbonne Université, Paris, 75006 France

[3]Institut de Physique du Globe de Paris, 75005 Paris, France

[4]GEOMAR Helmholtz Centre for Ocean Research Kiel, 24148 Kiel, Germany

DM, 0000-0002-4736-7526; JL, 0000-0002-9495-1194; RU, 0000-0001-9438-1753; SL, 0000-0003-2126-6505; EG, 0000-0001-9311-3108; ALF, 0000-0001-8901-2515; MU, 0000-0002-1116-636X
*Correspondence: mencaroni@icm.csic.es

Abstract: The development of overpressure in continental margins is typically evaluated with hydrogeological models. Such approaches are used to both identify fluid flow patterns and to evaluate the development of high pore pressures within layers with particular physical properties that may promote slope instability. In some instances, these models are defined with sediment properties based on facies characterization and proxy values of porosity; permeability or compressibility are derived from the existing literature as direct measurements are rarely available. This study uses finite-element models to quantify the differences in computed overpressure generated by fine-grained hemipelagic sediments from the Gulf of Cadiz, offshore Martinique and the Gulf of Mexico, and their consequences in terms of submarine slope stability. By comparing our simulation results with *in situ* pore pressure data measured in the Gulf of Mexico, we demonstrate that physical properties measured on volcanic-influenced hemipelagic sediments underestimate the computed stability of a submarine slope. Physical properties measured on sediments from the study area are key to improving the reliability and accuracy of overpressure models, and when that information is unavailable, literature data from samples with similar lithologies, composition and depositional settings enable better assessment of the overpressure role as a pre-conditioning factor in submarine landslide initiation.

Overpressure development in marine sediments is a major pre-conditioning factor that favours slope instability by decreasing the amount of gravitational driving stresses required to mobilize sediments (Dugan and Sheahan 2012). If the pressure of the water contained in the pore space exceeds the hydrostatic condition, failures can be initiated in low-angle submarine slopes by a relatively minor trigger such as a small magnitude earthquake (Urlaub *et al.* 2015). Generation of overpressure is commonly associated with rapid sedimentation (Dickinson 1953; Gibson 1958) and with the physical properties of the sediments, such as porosity, hydraulic conductivity and compressibility. The occurrence of overpressure within sediments can be studied directly through *in situ* measurements with piezocone tests (Lunne *et al.* 1997) and pore pressure penetrometers (Flemings *et al.* 2008), or indirectly from borehole logging measurements, such as sonic, resistivity and density logs (Mouchet and Mitchelle 1989), and from velocity analysis derived from seismic reflection data. Finally, overpressure development and evolution can be evaluated using hydrogeological models. The application at the basin scale of those models, in some cases, aims to provide a broad understanding of the conditions that may have generated previous slope instability (Gutierrez and Wangen 2005; Bellwald *et al.* 2019; Llopart *et al.* 2019). In many cases, however, sediment samples from the study area are not available for geotechnical testing. In such instances, porosity, permeability and compressibility values are typically derived from the literature based on their expected lithologies (Urlaub *et al.* 2015), depositional environment (Bellwald *et al.* 2019) or geographical area. On the other hand, measuring sediments' physical properties implies a series of inevitable errors and uncertainties, which affect the reliability of models, such as the presence of test errors, an insufficient number of tests and the inherent variability of

From: Georgiopoulou, A., Amy, L. A., Benetti, S., Chaytor, J. D., Clare, M. A., Gamboa, D., Haughton, P. D. W., Moernaut, J. and Mountjoy, J. J. (eds) 2020. *Subaqueous Mass Movements and their Consequences: Advances in Process Understanding, Monitoring and Hazard Assessments.* Geological Society, London, Special Publications, **500**, 289–300.
First published online May 22, 2020, https://doi.org/10.1144/SP500-2019-176

soil properties (Phoon and Kulhawy 1999; Zhang et al. 2012).

This study aims to evaluate the suitability of hydrogeological models based on literature-derived physical properties of sediments collected from similar depositional environments and with comparable grain-size distributions. To do this, we built a series of finite-element models that simulate the development of overpressure on a hemipelagic unit rapidly buried by a series of mass wasting events. Using the same depositional scenario, we defined the hemipelagic layer with initial porosity, hydraulic conductivity and specific storage measured on comparable sediments with similar grain-size distribution, but collected in three different depositional settings: Alentejo Basin (Gulf of Cadiz, SW Iberian margin), Grenada Basin (offshore Martinique, Lesser Antilles) and Ursa Basin (Gulf of Mexico). We further used the actual sedimentary history at Ursa Basin, where overpressure is groundtruthed with piezometer measurements, and tested the influence of using actual physical properties from the area v. physical properties taken from other depositional environments of similar grain size.

Depositional settings

We based this study on mechanical properties from 18 hemipelagic sediment samples collected from three different oceanic regions, characterized by different depositional settings, and all affected by well-documented subaqueous slope instabilities (Fig. 1a). The Alentejo Basin is located in the northern sector of the Gulf of Cadiz, SW offshore Portugal (Fig. 1b). The area is characterized by an extensive contourite depositional system (Hernández-Molina et al. 2015), generated by the Mediterranean Outflow Water, which distributes sediments entering the gulf mainly through the Guadalquivir river (Mulder et al. 2003). The Grenada Basin (Fig. 1c), offshore Martinique, is mainly composed of volcanic ashes, hemipelagic sediments and volcanoclastic turbidite deposits (Lafuerza et al. 2014) due to the activity of the surrounding volcanoes and absence of important sediment inputs from rivers. The sediments in the deep part of the basin accumulate at sedimentation rates up to 20 cm a^{-1} (Le Friant et al. 2015). The third area is the Ursa Basin (Fig. 1d), in the Gulf of Mexico. The region is known for the very high terrigenous sedimentation, entering the basin

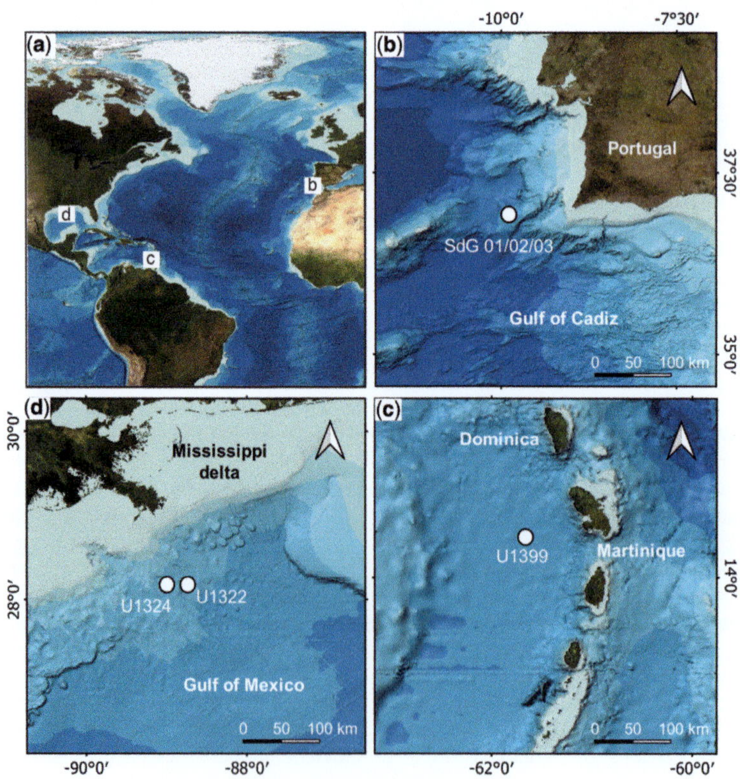

Fig. 1. (**a**) Location of the areas where the sediments considered in this study were collected. Detailed location of the cores in the (**b**) Alentejo Basin, (**c**) Grenada Basin and (**d**) Ursa Basin.

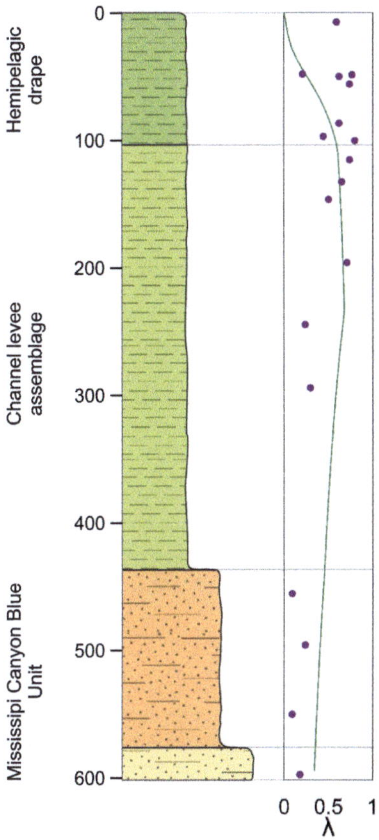

Fig. 2. Site U1324 stratigraphic model adopted for the 1D simulation with corresponding overpressure: *in situ* piezometer overpressure measurements (purple dots) (Expedition 308 Scientists 2005); reference 1D modelling result (green line) (Urgeles *et al.* 2010). Properties for overpressure models in Figure 7 are only changed for the upper hemipelagic drape.

from the Mississippi River, taking place upon a salt substrate (Worrall and Snelson 1989). The slope is characterized by well-documented overpressure due to sedimentation rates exceeding 25 m ka^{-1} during some intervals in the Pleistocene (Flemings *et al.* 2006; Sawyer *et al.* 2007). IODP Expedition 308 performed *in situ* pore pressure measurements using the DVTPP and T2P piezoprobes (see Flemings *et al.* 2008; Long *et al.* 2008 for further details on the measurements), and documented high overpressure associated mainly with the high sedimentation rates in the area (Expedition 308 Scientists 2005). Previous studies (Urgeles *et al.* 2007, 2010) modelled the stratigraphic evolution of the margin considering the physical properties of the different units in the Ursa Basin, with the aim of understanding the present-day pore pressure condition revealed by the *in situ* measurements. Despite the significant scatter observed in the direct overpressure measurement, especially within the uppermost hemipelagic drape, 1D simulations reproduced the general vertical overpressure distribution at site U1324 (Fig. 2) (for more information about previous overpressure modelling, see Urgeles *et al.* 2010).

Methods

Sediments from the Alentejo Basin (six samples) have been collected using a gravity corer during the INSIGHT_Leg1 (2018) cruise in the Gulf of Cadiz. Initial void ratio (e_0), compression index (C_c) and initial hydraulic conductivity (k_0) in these samples have been derived with stepped loading (Rowe and Barden type of cell) or Constant Rate of Strain (CRS) oedometer tests (Fig. 3a and Table 1). Samples from the Grenada Basin (three samples) and from the Ursa Basin (nine samples) were recovered from wells drilled during two separate IODP expeditions. The samples from the Grenada Basin used in this study were drilled at well U1399 during the IODP expedition 340 (Le Friant *et al.* 2015) and their compressibility/permeability properties measured with 1D consolidation tests (Fig. 3b) (Lafuerza *et al.* 2014). Sediments from the Ursa Basin were taken from wells U1322 and U1324, drilled during the IODP Expedition 308 (Expedition 308 Scientists 2005; Flemings *et al.* 2006). Their physical properties were also analysed using a combination of incremental loading and CRS oedometer tests (Fig. 3c) (Urgeles *et al.* 2007, 2010; Long *et al.* 2008; Stigall and Dugan 2010).

In our basin-modelling approach, we input the physical properties of the sediments at their initial depositional conditions, to simulate the pre-failure overpressure behaviour of the sediment composing the stratigraphic column. Therefore, input data such as porosity, hydraulic conductivity and specific storage are extrapolated to 1 kPa along the virgin consolidation line (e.g. Llopart *et al.* 2019). All of the samples were composed of hemipelagic sediments, collected outside mass transport deposit areas: for this reason, we assumed they did not undergo heavy deformation or mixing processes and the extrapolation of their physical properties to 1 kPa can be considered as a fair reproduction of their behaviour at depositional conditions.

Grain-size distribution for all samples was measured using laser diffractometers. We adopted the grain-size classification proposed by Wentworth (1922) to derive the percentages of sand (2 mm–62.5 μm), silt (62.5–4 μm) and clay (<4 μm) measured in the sediments. The three-component textural classification for muddy sediments proposed by Flemming (2000) was used to subdivide the analysed sediments into four classes based on their sand, silt

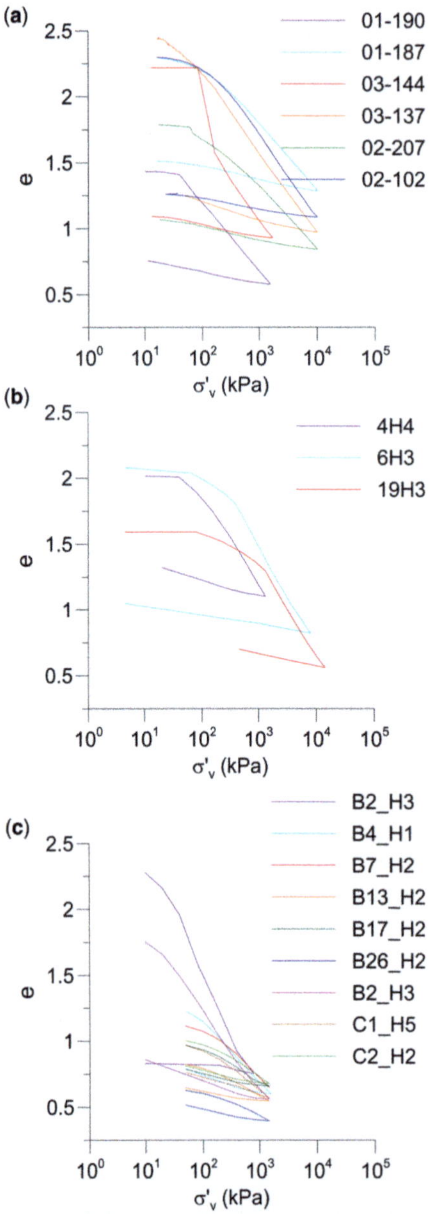

Fig. 3. Consolidation curves showing void ratio e reduction while increasing uniaxial loading σ'_v. Each plot refers to one of the analysed areas: (**a**) Alentejo Basin; (**b**) Grenada Basin (Lafuerza *et al.* 2014); (**c**) Ursa Basin (Urgeles *et al.* 2010).

and clay content: D-II (very silty, slightly sandy mud), D-III (silty, slightly sandy mud), E-II (slightly clayey silt) and E-III (clayey silt) (Fig. 4).

Overpressure development simulations were performed using the finite element software BASIN (Bitzer 1996, 1999). Modelling the stratigraphic and hydrodynamic evolution of a 2D section, BASIN allows the calculation of non-equilibrium compaction and overpressure generation on sediments by coupling compaction and 2D fluid flow. The consolidation model incorporates porosity-dependent sediment compressibility through equation (1) (Bitzer 1996):

$$\left(\frac{\partial}{\partial x}\frac{k_{x(\phi)}\partial p}{\partial x}\right) + \left(\frac{\partial}{\partial x}\frac{k_{z(\phi)}\partial p}{\partial z}\right) = \frac{(1-\phi)\rho g \alpha_{(\phi)}\partial p}{\partial t} \quad (1)$$

where $k_{x(\phi)}$ is the porosity-dependent hydraulic conductivity in the x-direction, $\alpha_{(\phi)}$ is the porosity-dependent sediment compressibility, p the fluid pressure, ϕ the porosity, ρ is the sediment bulk density, g is the gravity constant, t is the time (s) and z is depth (m). Sediment compressibility in BASIN is calculated from the specific storage (S_s), which is defined as the volume of water removed from a unit volume of a confined aquifer with an increase in the vertical stress. It relates to the more commonly used compression index by the relation (Jorgensen 1980):

$$S_s = \frac{0.434 C_c \gamma_w}{\sigma'_v(1+e_0)} \quad (2)$$

where C_c is the compression index, γ_w is the specific weight of water, σ'_v is the effective vertical stress and e_0 is the void ratio at depositional conditions (1 kPa).

In the first instance, we built a series of 1D models to simulate the deposition of a 500 m thick hemipelagic unit with a sedimentation rate of 30 cm a^{-1}, followed by a series of turbidite events that deposit 250 m of sand-rich material during 1000 years. Initial thickness of the units, sedimentation rates and boundary conditions, as well as physical properties of the turbidite unit, are kept constant through the different scenarios. Since no samples associated with turbidites were available in our database, we defined its physical properties with values taken from the literature (Reed *et al.* 2002). The only variables that change between the different models are initial porosity, initial hydraulic conductivity and initial specific storage of the hemipelagic layer. In the first set of analyses (Scenario A), we defined the properties of the hemipelagic unit by averaging arithmetically the values of the physical properties for each study area. In the second series of tests (Scenario B), we assigned physical properties to each hemipelagic layer based on the average specific storage, porosity and hydraulic conductivity of the samples from each grain-size class. The BASIN input parameters used for the simulation are provided in Table 2.

Secondly, to evaluate the reliability of our results and their significance in terms of submarine slope stability, we applied our approach on a case study where *in situ* overpressure measurements have been collected. To do so, we considered the results

Table 1. *Compiled physical and grain-size data from the Gulf of Cadiz, Martinique and the Gulf of Mexico*

Area	Sample	Original depth (mbsf)	Sand (%)	Silt (%)	Clay (%)	Texture class*	ϕ_0	C_c	k_0
Alentejo Basin	SdG-01_190	1.9	4.93	57.36	37.71	E-III	0.69	0.52	1.57×10^{-8}
	SdG-01_187	1.8	3.57	62.58	33.85	E-III	0.76	0.50	2.45×10^{-8}
	SdG-03_144	1.4	0.7	69.89	29.41	E-III	0.74	0.48	5.25×10^{-8}
	SdG-03_137	1.3	1.66	70.32	28.28	E-III	0.77	0.62	6.22×10^{-8}
	SdG-02_207	2.0	8.49	66.23	25.28	D-III	0.72	0.45	2.31×10^{-8}
	SdG-02_102	1.0	0.24	61.25	38.51	E-III	0.78	0.65	7.87×10^{-8}
Ursa Basin	308_1322_B2_H3	8.0	0	57.8	42.2	E-III	0.77	0.92	5.70×10^{-10}
	308_1322_B4_H1	24.0	0	68	32	E-III	0.71	0.57	1.12×10^{-8}
	308_1322_B7_H2	54.0	8.2	62.2	29.6	D-III	0.66	0.41	3.48×10^{-8}
	308_1322_B13_H2	111.0	0	68.1	31.9	E-III	0.56	0.23	1.06×10^{-8}
	308_1322_B17_H2	144.0	0	61.69	38.4	E-III	0.59	0.25	3.08×10^{-8}
	308_1322_B26_H2	212.0	0.63	73.37	26	E-III	0.52	0.22	4.41×10^{-8}
	308_1324_B2_H3	8.0	2.65	72.4	24.4	E-II	0.71	0.58	2.32×10^{-8}
	308_1324_C1_H5	57.0	0	74.3	25.7	E-III	0.63	0.35	8.87×10^{-9}
	308_1324_C2_H2	103.0	0	79.8	20.2	E-II	0.61	0.28	5.70×10^{-10}
Grenada Basin	340-1399B_4H4	30.5	10.5	75.8	13.7	D-II	0.78	0.69	1.55×10^{-9}
	340-1399B_6H3	46.5	3.9	76.1	20.1	E-II	0.79	0.75	1.32×10^{-9}
	340-1399B_19H3	138.9	5.8	83.1	11.1	D-II	0.77	0.67	1.41×10^{-9}

ϕ_0, initial porosity; C_c, compression index; k_0, initial hydraulic conductivity (m s^{-1}).
*Flemming (2000).

of site U1324 drilled in the Ursa Basin during IODP Expedition 308.

Using BASIN, we reproduced the stratigraphic architecture, sedimentation rate, boundary conditions and lithologies of a 1D overpressure model at site U1324 from Urgeles *et al.* (2010) (Fig. 2). Then, we substituted the original physical properties associated with the top hemipelagic drape with those measured from the sediments collected in the Alentejo and Grenada basins (Table 2). By doing this,

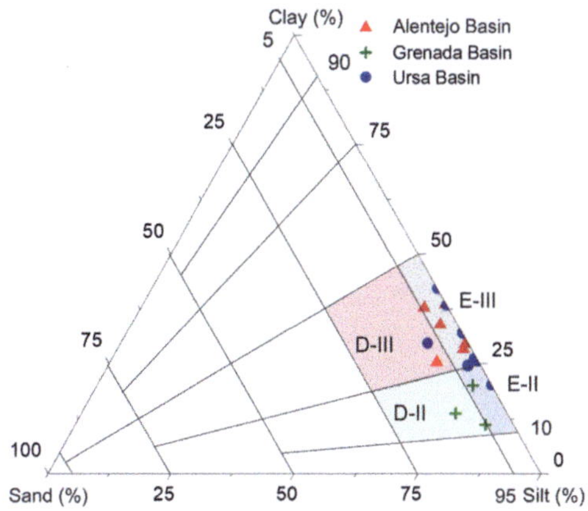

Fig. 4. Grain-size distribution of the sediments analysed for this study. The internal subdivision refers to the classification for muddy marine sediments proposed by (Flemming 2000) and adopted in this work.

we aim to evaluate the differences in overpressure generated by defining hydrogeological models with physical properties measured from hemipelagic sediments collected in different areas, and compare them with the reference calibrated results from Urgeles *et al.* (2010). At site U1324, 612 m of sedimentary sequence deposited between the Pleistocene and the Holocene (Expedition 308 Scientists 2005; Flemings *et al.* 2008) were drilled. From bottom to top, the main units composing the Ursa Basin consist of: (1) the sand-dominated lower Mississippi Canyon Blue Unit (Late Pleistocene) (Sawyer *et al.* 2007), (2) a mud-dominated channel levee assemblage and (3) a mud drape deposited during the last *c.* 20 ka (Behrmann *et al.* 2006; Flemings *et al.* 2008). The overpressure evolution of the basin has been modelled and is modelled in this study since deposition of the sand-rich Blue unit (*c.* 100 ka ago), followed by the silt lithologies associated with the filling of the Ursa Canyon, and by the mud-dominated sediments from the channel levee assemblage and hemipelagic sediments (Fig. 2) (Urgeles *et al.* 2007, 2010).

This study also refers to pore pressure in terms of the overpressure ratio (λ) (Flemings *et al.* 2008), defined as:

$$\lambda = \frac{(p - p_h)}{(\sigma_v - p_h)} \tag{3}$$

where p is pore pressure, p_h is the hydrostatic pressure and σ_v is the lithostatic or total vertical stress.

Table 2. *Parameters used for the simulations in BASIN*

	Samples	Initial porosity	Initial specific storage (m^{-1})	Initial hydraulic conductivity ($m\ s^{-1}$)
Turbidite layer	(Urgeles *et al.* 2010)	0.5	0.001	1×10^{-6}
Hemipelagic	Alentejo Basin	0.74	0.025	4.28×10^{-8}
Scenario A – areas	Grenada Basin	0.78	0.005	1.43×10^{-9}
	Ursa Basin	0.64	0.023	2.71×10^{-8}
Hemipelagic	D-II	0.77	0.003	1.48×10^{-9}
Scenario B – classes	D-III	0.69	0.014	2.90×10^{-8}
	E-II	0.70	0.017	1.04×10^{-7}
	E-III	0.69	0.027	3.09×10^{-8}

The physical properties associated with the turbidity layer were used in all the simulations, while the hemipelagic changed.

A λ value of 0 means that pore pressure is hydrostatic, while a value of 1 means that the overburden is fully sustained by the pore water.

Results

Physical properties

The grain-size distribution of sediments collected in each study area is relatively homogeneous although each location has a distinctly different composition. Whilst all samples analysed are silt dominated, the percentages vary significantly between 57.36 and 83.1%. Clay contents also ranged between 11.1 and 42.2%, and sand-sized grains comprised the lowest fraction at each site (0 to 10.5%) (Fig. 4). The Ursa Basin has the lowest sand content, with 8 out of 9 samples corresponding to the 'E' class of Flemming's classification (sand <5%). Within the Grenada Basin, 2 out of 3 samples tested show a sand content greater than 5% and a clay content less than 25%. Sediments from the Alentejo Basin are similar to the Ursa Basin: sand content is less than 5% for 5 out of 6 samples, with clay values between 25 and 35%. Therefore, group D-II in Flemming's classification is represented by only two samples from the Ursa Basin.

We used ternary contour diagrams to visualize the different textural composition of sediments in relation to their compression index (Fig. 5a), initial porosity (Fig. 5b) and initial hydraulic conductivity (Fig. 5c). Results show a broadly heterogeneous relationship between grain size and physical properties. The compression index (Fig. 5a) ranges between 0.22 and 0.92. The highest C_c values are related to an increase in silt percentage and tend to decrease in sediments with some sand content. Samples from the Grenada Basin, with higher silt content, show a compression index ranging between 0.67 and 0.75, while the lowest values are found in sediments from the Ursa Basin, where most of the samples show C_c values below 0.65. An exception is represented

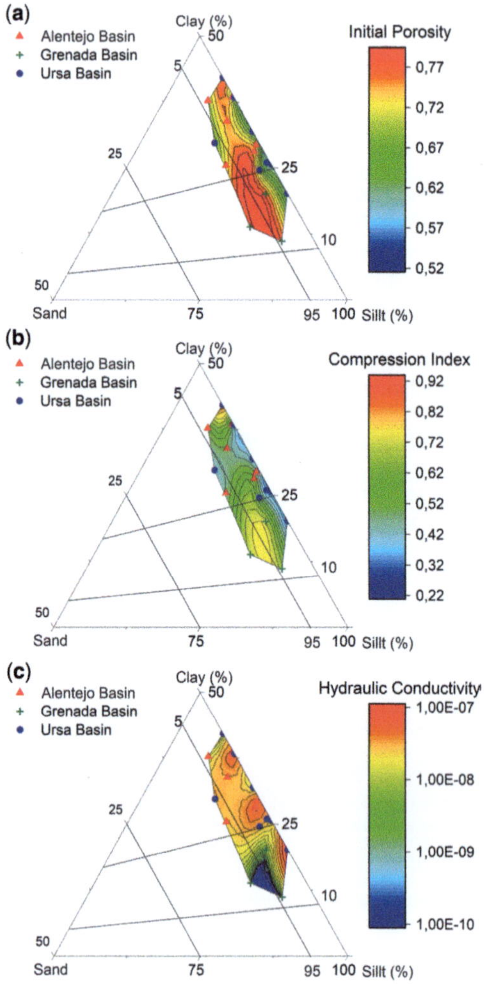

Fig. 5. Ternary contour diagrams comparing sediment composition (% sand, silt and clay) with (**a**) initial porosity, (**b**) compression index and (**c**) hydraulic conductivity. Sand axes have been reduced to the range 50–100% to facilitate visualization.

by sample 308-1322B_H08, which has higher clay content (42.2%) and compression index (0.92).

A similar trend is illustrated in the initial porosity diagram (Fig. 5b), with porosity values ranging between 0.52 and 0.79. The higher values are reached in sediments with higher silt content while the lower values are linked to higher sand content. However, sample SdG_02_102 differs from this trend, showing a high initial porosity (0.78) despite a relatively low silt content (61%).

Values of hydraulic conductivity at deposition (1 kPa) vary within two orders of magnitude (10^{-8} to 10^{-10} m s^{-1}). The Alentejo Basin samples produce the highest values while the lowest values correspond to samples from the Grenada Basin (Fig. 5c).

Our first modelling stage (Scenario A) (Fig. 6a) consisted of three separate models, where the physical properties of the hemipelagic layer described above were averaged for each study area (Table 2). Fine-grained sediments from the Alentejo Basin and from the Ursa Basin show a comparable range of overpressure development, reaching a maximum overpressure ratio (λ) of 0.18 at 343 mbsf for the Alentejo Basin model and 0.25 at 341 mbsf for the Ursa Basin (i.e. immediately below the turbidite unit). The computed overpressure ratio from the Grenada Basin model shows higher values, with λ up to 0.56 at 329 mbsf.

The second modelling scenario (Scenario B) consisted of defining the hemipelagic units with physical properties averaged from their grain-size distribution classes (Fig. 6b). The minimum peak value ($\lambda \sim 0.06$) is observed in the model defined with properties from group E-II (sand <5%, clay <25%). Models from groups E-III and D-III show comparable maximum overpressure ratios (0.24 and 0.19), while the maximum values ($\lambda \sim 0.53$) are reached by the model defined with the properties averaged from the samples in group D-II (sand >5%, clay <25%).

Figure 7 shows the present-day overpressure resulting at site U1324 from our 1D simulations.

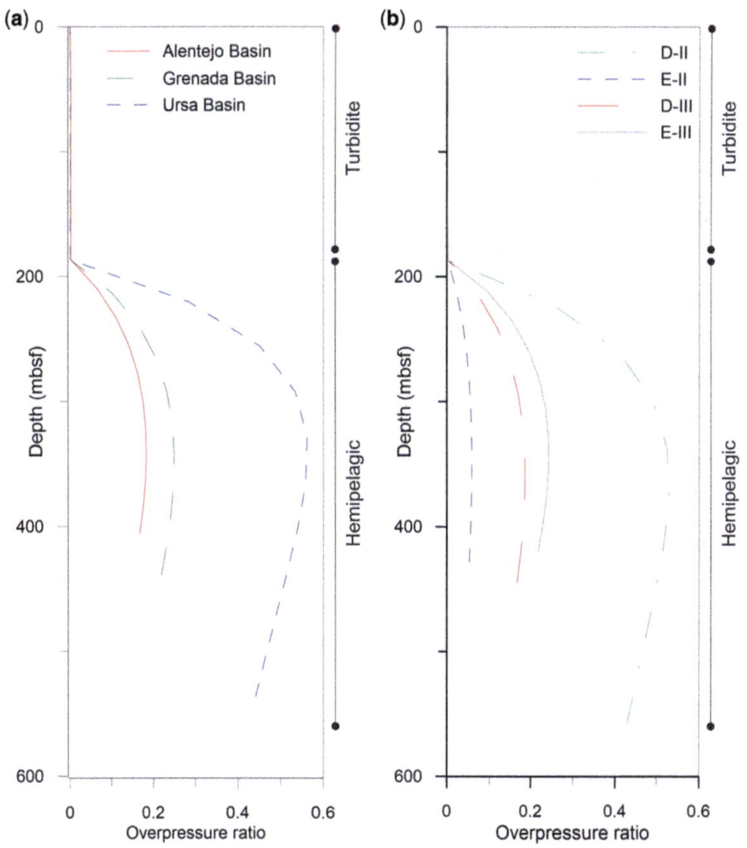

Fig. 6. Overpressure generated at the end of the simulations v. computed final depths. The plots refer to (**a**) Scenario A, where physical properties of the hemipelagic units are averaged from each study area and (**b**) Scenario B, where the properties of the hemipelagic layer are averaged from each of the four represented textural classes.

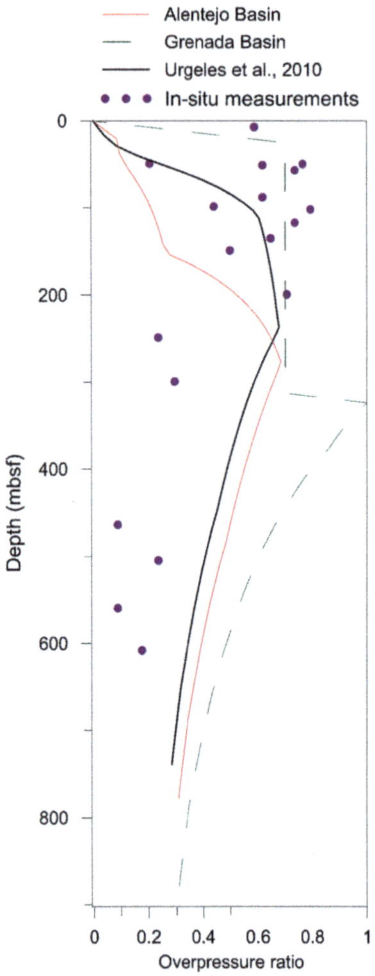

Fig. 7. Overpressure generated at site U1324 v. the computed final depth for the (1) *in situ* piezometer measurements (purple dots); reference overpressure modelled at the same site by Urgeles *et al.* (2010) (black line); overpressure resulting by defining the hemipelagic units with properties collected from the other study areas (red and green lines).

Changes in the physical properties of the hemipelagic unit, at the top of the stratigraphy, did not allow any of our scenarios to fit perfectly either with the reference 1D model (Urgeles *et al.* 2010) or the *in situ* overpressure measurements (Flemings *et al.* 2008). The model defined with hydraulic conductivity, specific storage and initial porosity from the Alentejo Basin underestimates the overpressure within the hemipelagic layer, mainly because of its higher permeability compared with the hemipelagic sediment from the Ursa Basin (Table 2). Finally, the model defined with physical properties from the Grenada Basin provides a very different output compared with the other models (Fig. 7). Even if it reproduces the measured *in situ* overpressures from the hemipelagic layer more effectively, the hydrodynamics of the model are split into two separate systems: in the lower part of the stratigraphy the overpressure increases constantly from the bottom to the top, reaching a maximum overpressure value of 1 at the base of the muddy sediments. Above this point, the overpressure drops, keeping a constant value of 0.7 up to the seafloor.

Discussion

In the first part of our study, two different scenarios were used to analyse the influence of variable grain-size composition of hemipelagic sediments in overpressure development. In Scenario (A), the hemipelagic layer was defined using physical properties collected from specific study areas and resulted in much higher overpressure generation for sediments of the Grenada Basin compared to the other two areas considered. In Scenario (B), the samples with relatively high silt and low clay content and sand content higher than 5% (sediment class D-II) generated the highest overpressure. This sediment class includes only two samples, both collected in the Grenada Basin. In this case, the results obviously coincide with those of samples for the Grenada Basin in Scenario A. Our results therefore seem to show that overpressure generation is rather related to the sediment composition and depositional setting instead of grain-size distribution. This is in agreement with previous studies showing that there is no direct correlation between grain-size distribution and sediment physical properties (Lafuerza *et al.* 2014) in marine muddy sediments.

However, when considering each depositional environment (e.g. river influenced), the grain-size distribution has a major influence on overpressure development. Thus, sediments with low clay content (E-II) and samples with similar clay content but higher sand content (D-III) generated lower overpressure (Figs 4 and 6a). Nonetheless, factors not considered in this study, such as sorting and particle rounding, might have played a significant role that should be investigated in future work.

Comparing the overpressure results with the physical properties used for the simulations indicates that the development of overpressure is associated with low initial hydraulic conductivities and specific storage (Table 2). Such differences in the physical properties between the Grenada Basin and the other two areas could reflect substantial diversity in sediment composition and mineralogy. This may, in turn, be directly linked to the depositional environment: (a) the hemipelagic sediments in the Alentejo and Ursa basins are largely fluvially derived from the Guadalquivir and Mississippi rivers, respectively; (b) volcanically derived sediments from the Grenada Basin are influenced by volcanic ash released by the extensive series of volcanos in the Lesser Antilles volcanic arc (Wiemer and Kopf 2015).

Comparing the 1D consolidation behaviour for volcanically derived hemipelagic sediments to the river-fed samples, it is possible to make assumptions about the reason for such notable differences in physical properties (Fig. 3). The Grenada Basin sediments show lower compressibility at low consolidation stress and a transition to high compressibility at higher vertical stress (Fig. 3b). Similar behaviour known as structuration has been documented especially for fossiliferous, cemented or highly bioturbated sediments (Locat *et al.* 2003; Spinelli *et al.* 2007): in these instances, sediments are capable of holding water in rigid structures, preventing compaction during the early stages of consolidation or, in the case of high degrees of cementation, also at high burial depths. Physical properties of the sediments from the Grenada Basin, with higher compression index and lower specific storage and hydraulic conductivity, suggest that a similar mechanism could occur in the volcanic ash-rich hemipelagic sediments from the Lesser Antilles volcanic arc. Volcanic ashes, in fact, may generate the same type of structures that prevent porewater being released into the system at low consolidation stresses.

Structuration in the Grenada Basin sediments retains the water in the pore space, reducing the specific storage (which is, by definition, the volume of water removed from a unit volume of a confined aquifer) and decreasing the permeability, since the porewater gets trapped in the rigid structures. Those physical properties represent the ideal conditions for the development of overpressure in marine sediments, and this could explain why the Grenada Basin sediments (Scenario A) develop higher overpressure (near lithostatic), which leads to a lower vertical displacement in the model.

It must be also mentioned that defining hydrogeological models with marine sediments that show structuration might question the hydrogeological model results. 'Delayed consolidation' (Locat *et al.* 2003), as seen in the Grenada Basin hemipelagic sediments (Fig. 3b), implies two different consolidation responses within the same sediment sample. Considering that the compression index is usually input as one representative value for each lithology, modelling such variation can represent a challenge. This is the reason why it is important to recognize structuration phenomena in the analysed samples and to consider their consequences while evaluating simulation results.

This behaviour is well exemplified when modelling the Ursa Basin sedimentation history with physical properties from the Grenada Basin (Fig. 7). The lower hydraulic conductivity and specific storage of the sediments from the Grenada Basin, mainly due to the structuring phenomena previously discussed, make the hemipelagic unit act as a perfectly impermeable seal in the case of the very high sedimentation rates of the Ursa Basin. The porewater from the lower formations, characterized by coarser grain size and higher permeability, flows upwards and accumulates at the base of the seal, reaching extreme overpressure values up to 1 (Fig. 7). Nonetheless, this trend does not find correspondence in the *in situ* measurement, where overpressure reaches its maximum within the hemipelagic layer and decreases constantly with depth, indicating a continuity between the porewater flow in the lower part of the stratigraphy and in the uppermost hemipelagic layer.

Significance for submarine slope stability

The development of overpressure in a submarine slope affects the magnitude of the events that can trigger a failure (e.g. earthquakes), or, in case of very high values and/or steep slope, can lead to slope instability with no external trigger (Stigall and Dugan 2010; Urlaub *et al.* 2015). A preliminary quantification of the consequences from the different overpressure resulting from our 1D models at site U1324 can be made by calculating the factor of safety (FoS) on a hypothetical 2° submarine slope using equation (4) (Flemings *et al.* 2008):

$$\mathrm{FoS} = \frac{\tan \phi_\mathrm{f}}{\sin \theta \cos \theta}(\cos^2 \theta - \lambda) \qquad (4)$$

where θ is the surface slope, λ the overpressure ratio and θ_f the friction angle of 28° as it was measured on the hemipelagic sediments in the Ursa Basin (Urgeles *et al.* 2007). FoS greater than 1 indicates a stable slope, while values lower than 1 indicate slope failure.

The FoS was calculated on the sediment deposited starting from 40 ka ago, corresponding to the beginning of the mud-dominated channel levee assemblage, until the present day (Fig. 8). As the FoS decreases with increasing overpressure, the model generated defining the hemipelagic units with the physical properties from the Alentejo and the original Ursa Basin reference model shows a decrease in the factor of safety until 20 ka. At this point, the hemipelagic sediment starts depositing and porewater migrates upwards with overpressure dissipating towards the seafloor (Fig. 8). The FoS for these two models reaches a minimum of around 3, increasing up to 5 at the present day. In contrast,

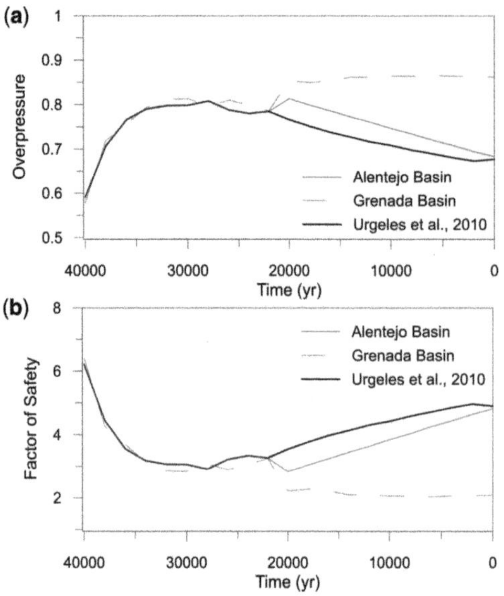

Fig. 8. (**a**) Overpressure evolution during the last 40 ka and (**b**) computed factor of safety for all models at the location of IODP Site U1324. The graph refers to the base of the fine-grained channel levee assemblage unit, which started depositing around 40 ka ago.

the model where the hemipelagic layer is defined with the physical properties from the Grenada Basin shows an increase in overpressure values also after 20 ka. This is because the deposition and compaction of the hemipelagic sediments creates a seal that prevents the water from the lower part of the stratigraphy flowing upwards. In this case, the FoS reaches values of around 2, which propagate almost constantly from 20 ka until the present day (Fig. 8).

At the end of the simulations, none of the models results in a FoS low enough to induce the slope to fail (<1) in the considered stratigraphic level. Nevertheless, 1D sedimentation models at site U1324 show a completely different overpressure development history when the model is defined with the hemipelagic sediments from the Grenada Basin, resulting in a considerable underestimation of the FoS at the base of the first deposited fine sediment unit.

Our modelling results show that physical properties derived from only the lithological characterization or the grain-size distribution available in the literature may produce inaccurate overpressure models, with consequent misleading slope stability estimations. On the other hand, models based on physical properties derived from similar grain size, depositional environment and expected compositional nature of sediments, such as the ones from the Alentejo Basin and the Ursa Basin, provided similar overpressure development results. This indicates that, in the absence of direct measurements, literature-derived physical properties based on grain-size characteristics and depositional setting are expected to produce more accurate overpressure development for stability analysis of submarine slopes. However, care should be taken in settings where the presence of microfossils such as diatoms, initial cementation or volcanic glass particles could provide sediment structuration.

Conclusions

Using hydrogeological modelling we show how the sediment characteristics in some depositional environments have a significant influence on the amount of overpressure generated in marine hemipelagic sediments. We demonstrate how fine-grained marine sediments collected from different sedimentary environments (volcanic-influenced v. river-dominated) can result in differences in overpressure in excess of one order of magnitude and discuss the consequences of those discrepancies for submarine slope stability assessment. Phenomena, such as structuration of marine sediments, need to be carefully considered when carrying out basin hydrogeological models. Our results support the idea that, to achieve accurate results, besides accurate stratigraphic architecture and age of units, hydrogeological models should be defined with physical properties measured from samples collected in the study area. In instances where such an approach is not feasible, assigning an initial void ratio, hydraulic conductivity and specific storage extracted from the literature for samples with similar expected lithologies, composition and depositional settings are likely to provide the most accurate estimates. Given that the data are generally difficult to find within the scientific literature, a global open database for published physical properties is paramount to increase the accuracy of basin hydrogeological models in the future.

Acknowledgements This research used samples and/or data provided by the International Ocean Discovery Program (IODP). We also thank Brandon Dugan and another reviewer for their comments that significantly improved our manuscript.

Funding Davide Mencaroni was supported by a Marie Skłodowska-Curie Doctoral Fellowship through the SLATE Innovative Training Network within the European Union Framework Programme for Research and Innovation Horizon 2020 under Grant Agreement No. 721403. The Spanish 'Ministerio de Economia y Competitividad' and the European Regional Development Fund through grant CTM2015-70155-R (project INSIGHT) are also acknowledged.

Author contributions **DM**: conceptualization (lead), data curation (lead), formal analysis (lead), writing – original draft (lead), writing – review & editing (lead); **JL**: data curation (supporting), formal analysis (equal), methodology (lead), supervision (lead), validation (equal), visualization (equal), writing – original draft (supporting), writing – review & editing (supporting); **RU**: conceptualization (supporting), data curation (supporting), formal analysis (supporting), methodology (equal), supervision (equal), visualization (supporting), writing – original draft (supporting), writing – review & editing (equal); **SL**: data curation (supporting), resources (supporting); **EG**: resources (supporting); **ALF**: resources (supporting); **MU**: conceptualization (supporting), methodology (supporting).

References

Behrmann, J.H., Flemings, P.B. *et al.* 2006. Rapid sedimentation, overpressure, and focused fluid flow, Gulf of Mexico continental margin. *Scientific Drilling*, **1**, 12–17, https://doi.org/10.2204/iodp.sd.3.03.2006

Bellwald, B., Urlaub, M., Hjelstuen, B.O., Sejrup, H.P., Sørensen, M.B., Forsberg, C.F. and Vanneste, M. 2019. NE Atlantic continental slope stability from a numerical modeling perspective. *Quaternary Science Reviews*, **203**, 248–265, https://doi.org/10.1016/j.quascirev.2018.11.019

Bitzer, K. 1996. Modeling consolidation sedimentary and fluid basins flow. *Computers and Geosciences*, **22**, 467–478, https://doi.org/10.1016/0098-3004(95)00113-1

Bitzer, K. 1999. Two-dimensional simulation of clastic and carbonate sedimentation, consolidation, subsidence, fluid flow, heat flow and solute transport during the formation of sedimentary basins. *Computers and Geosciences*, **25**, 431–447, https://doi.org/10.1016/S0098-3004(98)00147-2

Dickinson, G. 1953. Geological aspects of abnormal reservoir pressures in Gulf Coast Louisiana. *AAPG Bulletin*, **37**, 410–432.

Dugan, B. and Sheahan, T.C. 2012. Offshore sediment overpressures of passive margins: mechanisms, measurement, and models. *Reviews of Geophysics*, **50**, 271–276, https://doi.org/10.1029/2011RG000379

Expedition 308 Scientists 2005. *Overpressure and fluid flow processes in the deepwater Gulf of Mexico: slope stability, seeps, and shallow-water flow*. Integrated Ocean Drilling Program Expedition 308 Scientific Prospectus. IODP Publications.

Flemings, P., Behrmann, J. and John, C. and Expedition 308 Scientists 2006. *Proceedings of the Ocean Drilling Program*. IODP Management International, Inc., College Station TX.

Flemings, P.B., Long, H., Dugan, B., Germaine, J., John, C.M., Behrmann, J.H. and Sawyer, D. 2008. Pore pressure penetrometers document high overpressure near the seafloor where multiple submarine landslides have occurred on the continental slope, offshore Louisiana, Gulf of Mexico. *Earth and Planetary Science Letters*, **274**, 269–283, https://doi.org/10.1016/j.epsl.2008.06.027

Flemming, B.W. 2000. A revised textural classification of gravel-free muddy sediments on the basis of ternary diagrams. *Continental Shelf Research*, **20**, 1125–1137, https://doi.org/10.1016/S0278-4343(00)00015-7

Gibson, R.E. 1958. The progress of consolidation in a clay layer increasing in thickness with time. *Géotechnique*, **8**, 171–182, https://doi.org/10.1680/geot.1958.8.4.171

Gutierrez, M. and Wangen, M 2005. Modeling of compaction and overpressuring in sedimentary basins. *Marine and Petroleum Geology*, **22**, 351–363, https://doi.org/10.1016/j.marpetgeo.2005.01.003

Hernández-Molina, F.J., Sierro, F.J. *et al.* 2015. Evolution of the gulf of Cadiz margin and southwest Portugal contourite depositional system: Tectonic, sedimentary and paleoceanographic implications from IODP expedition 339. *Marine Geology*, **377**, 7–39, https://doi.org/10.1016/j.margeo.2015.09.013

Jorgensen, D.G. 1980. *Relationships between Basic Soils-Engineering Equations and Basic Ground-Water Flow Equations*. Geological Survey Water-Supply Paper 2064. US Department of the Interior.

Lafuerza, S., Le Friant, A. *et al.* 2014. Geomechanical characterization of submarine volcano-flank sediments, Martinique, Lesser Antilles Arc. *In*: Krastel, S., Behrmann, J.-H. *et al.* (eds) *Submarine Mass Movements and their Consequences: 6th International Symposium.* Springer, Cham, 73–81, https://doi.org/10.1007/978-3-319-00972-8_7

Le Friant, A., Ishizuka, O. *et al.* 2015. Submarine record of volcanic island construction and collapse in the Lesser Antilles arc: first scientific drilling of submarine volcanic island landslides by IODP Expedition 340. *Geochemistry Geophysics Geosystems*, **16**, 420–442, https://doi.org/10.1002/2014GC005684.Key

Llopart, J., Urgeles, R. *et al.* 2019. Fluid flow and pore pressure development throughout the evolution of a trough mouth fan, western Barents Sea. *Basin Research*, **31**, 487–513, https://doi.org/10.1111/bre.12331

Locat, J., Tanaka, H., Tan, T., Dasari, G.R. and Lee, H.J. 2003. Natural soils: geotechnical behavior and geological knowledge. *In*: Tan, T., Phoon, K., Hight, D. and Leroueil, S. (eds) *Characterisation and Engineering Properties of Natural Soils.* Swets & Zeitlinger, The Netherlands, 3–28.

Long, H., Flemings, P.B., Germaine, J.T., Saffer, D.M. and Dugan, B. 2008. Data report: consolidation characteristics of sediments from IODP Expedition 308, Ursa Basin, Gulf of Mexico. *Proceedings of the Integrated Ocean Drilling Program*, **308**, https://doi.org/10.2204/iodp.proc.308.204.2008

Lunne, T., Robertson, P.K. and Powell, J.J.M. 1997. *Cone Penetration Testing in Geotechnical Practice*. Spon-Press, London.

Mouchet, J.P. and Mitchelle, A. 1989. *Abnormal Pressures While Drilling – Origins, Prediction, Detection, Evaluation. Manuals Techniques 2*. Elf Aquita, Boussens.

Mulder, T., Voisset, M. *et al.* 2003. The Gulf of Cadiz: an unstable giant contouritic levee. *Geo-Marine Letters*, **23**, 7–18, https://doi.org/10.1007/s00367-003-0119-0

Phoon, K.-K. and Kulhawy, F.H. 1999. Characterization of geotechnical variability. *Canadian Geotechnical Journal*, **36**, 612–624, https://doi.org/10.1139/t99-038

Reed, A.H., Briggs, K.B. and Lavoie, D.L. 2002. Porometric properties of siliciclastic marine sand: a comparison of traditional laboratory measurements with image analysis and effective medium modeling. *IEEE Journal of Oceanic Engineering*, **27**, 581–592, https://doi.org/10.1109/JOE.2002.1040940

Sawyer, D.E., Flemings, P.B., Shipp, R.C. and Winker, C.D. 2007. Seismic geomorphology, lithology, and evolution of the late Pleistocene Mars–Ursa Turbidite region, Mississippi Canyon Area, Northern Gulf of Mexico. *AAPG Bulletin*, **91**, 215–234, https://doi.org/10.1306/08290605190

Spinelli, G.A., Mozley, P.S., Tobin, H.J., Underwood, M.B., Hoffman, N.W. and Bellew, G.M. 2007. Diagenesis, sediment strength, and pore collapse in sediment approaching the Nankai Trough subduction zone. *Geological Society of America Bulletin*, **119**, 377–390, https://doi.org/10.1130/B25920.1

Stigall, J. and Dugan, B. 2010. Overpressure and earthquake initiated slope failure in the Ursa region, northern Gulf of Mexico. *Journal of Geophysical Research*, **115**, B04101, https://doi.org/10.1029/2009JB006848

Urgeles, R., Locat, J. and Dugan, B. 2007. Recursive failure of the Gulf of Mexico continental slope: timing and causes. *In*: Lykousis, V., Sakellariou, D. and Locat, J. (eds) *Submarine Mass Movements and Their Consequences*. Springer, Dordrecht, 209–219, https://doi.org/10.1007/978-1-4020-6512-5_22

Urgeles, R., Locat, J., Sawyer, D.E., Flemings, P.B., Dugan, B. and Binh, N.T.T. 2010. History of pore pressure build up and slope instability in mud-dominated sediments of Ursa Basin, Gulf of Mexico Continental Slope. *In*: Mosher, D.C., Shipp, R.C., Moscardelli, L., Chaytor, J.D., Baxter, C.D.P., Lee, H.J. and Urgeles, R. (eds) *Submarine Mass Movements and Their Consequences*. Springer, Dordrecht, 179–190, https://doi.org/10.1007/978-90-481-3071-9_15

Urlaub, M., Talling, P.J., Zervos, A. and Masson, D.G. 2015. What causes large submarine landslides on low gradient sediment accumulation? *Journal of Geophysical Research: Solid Earth*, **120**, 1–18, https://doi.org/10.1002/2015JB012347.

Wentworth, C.K. 1922. A scale of grade and class terms for clastic sediments. *The Journal of Geology*, **30**, 377–392, https://doi.org/10.1086/622910

Wiemer, G. and Kopf, A. 2015. Altered marine tephra deposits as potential slope failure planes? *Geo-Marine Letters*, **35**, 305–314, https://doi.org/10.1007/s00367-015-0408-4

Worrall, D.M. and Snelson, S. 1989. Evolution of the northern Gulf of Mexico, with emphasis on Cenozoic growth faulting and the role of salt. *In*: Bally, A.W. and Paler, A.R. (eds) *The Geology of North America – An Overview*, Vol. **A**. Geological Society of America, Boulder, CO.

Zhang, J., Tang, W.H., Zhang, L.M. and Huang, H.W. 2012. Characterising geotechnical model uncertainty by hybrid Markov Chain Monte Carlo simulation. *Computers and Geotechnics*, **43**, 26–36, https://doi.org/10.1016/j.compgeo.2012.02.002

The influence of clay content on submarine slope failure: insights from laboratory experiments and numerical models

M. M. W. Silver[1]* and B. Dugan[1,2]

[1]Hydrologic Science and Engineering Program, Colorado School of Mines, Golden, CO 80401, USA

[2]Department of Geophysics, Colorado School of Mines, Golden, CO 80401, USA

MMWS, 0000-0001-5093-6876; BD, 0000-0002-2555-6430
*Correspondence: mmsilver@mymail.mines.edu

Abstract: Submarine slope failures pose risks to coastlines because they can damage infrastructure and generate tsunamis. Passive margin slope failures represent the largest mass failures on Earth, yet we know little about their dynamics. While numerous studies characterize the lithology, structure, seismic attributes and geometry of failure deposits, we lack direct observations of failure evolution. Thus, we lack insight into the relationships between initial conditions, slope failure initiation and evolution, and final deposits. To investigate submarine slope failure dynamics in relation to initial conditions and to observe failure processes we performed physical experiments in a benchtop flume and produced numerical models. Submarine slope failures were induced under controlled pore pressure within sand–clay mixtures (0–5 wt% clay). Increased clay content corresponded to increased cohesion and pore pressure required for failure. Subsurface fractures and tensile cracks were only generated in experiments containing clay. Falling head tests showed a log-linear relation between hydraulic conductivity and clay content, which we used in our numerical models. Models of our experiments effectively simulate overpressure (pressure in excess of hydrostatic) and failure potential for (non)cohesive sediment mixtures. Overall our work shows the importance of clay in reducing permeability and increasing cohesion to create different failure modes due to overpressure.

Submarine slope failures along continental margins pose risks to coastal communities because they can damage infrastructure and generate tsunamis. Mass transport deposits, which are archives of slope failures, have been widely studied because they record slope failure events on passive, active and transform margins. Submarine slope failures can travel large distances at high speeds, even on low-angle (<2°) slopes (Hühnerbach and Masson 2004). A submarine slope failure initiated by the 2018 Palu, Indonesia earthquake (magnitude (M_L) 7.5) produced a tsunami that killed 1944 people (Muhari *et al.* 2018). The 2006 Java tsunami that claimed 600 lives is also thought to have been the product of an earthquake-triggered submarine slope failure (Fritz *et al.* 2007). The 1929 Grand Banks slope failure was triggered by a M_L 7.2 earthquake and transported 200 km^3 of mud and sand a distance of up to 1000 km (Fine *et al.* 2005). The mass failure severed 12 telegraph cables, and 28 people were killed from the produced tsunami (Fine *et al.* 2005). Although earthquake activity is low along the US Atlantic margin, the 1929 Grand Banks failure indicates that submarine slope failures in the region continue to occur, and thus are a potential hazard (ten Brink *et al.* 2014). Infrastructure along the US Atlantic margin presents a potential tsunami flooding/damage vulnerability in the form of densely populated areas, port facilities, industrial sites and ten coastal nuclear power plants (ten Brink *et al.* 2014).

Despite the risks associated with submarine slope failures on passive margins, little is known about their initiation and dynamics. Along active margins, frequent earthquakes prevent thick sediment accumulation and seismic shaking may strengthen sediment, decreasing failure potential and size (Sawyer and Devore 2015). Submarine slope failures on passive margins have longer recurrence intervals than on active margins, which facilitates thick sediment accumulation before failure and yields much larger failures (ten Brink *et al.* 2016). While numerous studies characterize slope failure deposits in terms of lithology, structure, seismic attributes, geometry and rheology (McAdoo *et al.* 2000; Gazioğlu *et al.* 2005; Locat *et al.* 2010; Dan *et al.* 2010; Lucchi *et al.* 2012; Idarraga-García and Vargas 2013), we lack direct observations of slope failure initiation

From: Georgiopoulou, A., Amy, L. A., Benetti, S., Chaytor, J. D., Clare, M. A., Gamboa, D., Haughton, P. D. W., Moernaut, J. and Mountjoy, J. J. (eds) 2020. *Subaqueous Mass Movements and their Consequences: Advances in Process Understanding, Monitoring and Hazard Assessments*. Geological Society, London, Special Publications, **500**, 301–309.
First published online March 26, 2020, https://doi.org/10.1144/SP500-2019-186

and evolution. As a result, there is limited insight into the relationship between initial conditions, preconditioning, slope failure initiation and failure behaviour.

In this paper, we expand the knowledge of submarine slope failure processes through a suite of physical laboratory experiments. We investigated failure mechanics and dynamics as influenced by pore pressure, cohesion and clay content in a benchtop flume in the Geomechanics Laboratory at the Colorado School of Mines and developed numerical models that quantify the observed processes. In order to establish a reproducible foundation for understanding pore pressure, cohesion and clay content effects on slope stability, high sand to clay ratios (0–5 wt% clay) were selected for our initial experiments. We focused on a mechanistic understanding of dynamic processes rather than a specific geological study area. This experimental and modelling approach builds upon previous experiments that looked at the role of yield strength to assess failure morphology (Sawyer *et al.* 2012) by investigating how pre-failure geometry, clay content and pore pressure affect slope failure dynamics, including sediment remobilization.

Methods

Flume experiments

Subaqueous slope failures were induced within the flume under controlled pore pressure boundary conditions with varied mixtures of commercially available fine-grained sand (sub-angular quartz; 87% SiO_2 $D_{50} = 195$ μm) and clay (dioctahedral smectite; 63% SiO_2 and 21% Al_2O_3; $D_{90} = 44$ μm) constrained to a constant pre-failure slope geometry (Fig. 1). A wedge-shaped geometry was selected which measured 30.3 cm long and 6.8 cm tall, with a top horizontal surface of 10.7 cm that ultimately tapered down to the base of the flume at an angle of 19.4° (Fig. 1). Experiments utilizing 2% clay used a slight variation of this geometry (wedge extended to 34.8 cm length with a top horizontal plane 15.2 cm long with unchanged height and slope).

Sand–clay mixtures (wt%) investigated were: (1) 100% fine-grained sand (FG sand); (2) 98% FG sand + 2% clay; (3) 96% FG sand + 4% clay; and (4) 95% FG sand + 5% clay. A minimum of three experiment runs were performed for each mixture. For sand–clay mixtures, sand and clay were gently mixed with a counter-top mechanical mixer for 5 min. Sediment (sand or sand–clay mixtures) was then poured into the flume and shaped with a template to provide a consistent geometry between experiments. The flume was filled with tap water through the effluent fitting until the flume water level reached a height *c.* 0.5 cm above the downstream weir (Fig. 1). The system was then left undisturbed until the water level over the sediment equilibrated with the water level in Reservoir 2 (R2; Fig. 1). The height of Reservoir 1 (R1) was adjusted to raise the water level of R2 without inducing slope failure. The flow control inlet valve was then opened by a quarter turn (*c.* 90°) and water was permitted to flow from R1 into R2, through the pebble layer and overlying wire mesh, and into the sand or sand–clay mixture.

Next R1 was raised to increase pore pressure at the pebble–sediment interface. The pressure was increased by raising R1 in 1 cm increments, with pauses of 3–5 min after each increase, until failure was observed. When slope failure occurred, the time of the event, the water levels in R2 before and after the event, and the geomorphological characteristics of the failure event were recorded.

Photogrammetry was utilized to assist experiment analyses. High-resolution photographs and videos were taken using two GoPro HERO7 Black cameras, capturing cross-section and overhead views of each experiment. Photogrammetry allowed for analysis of failure and deposit geomorphology.

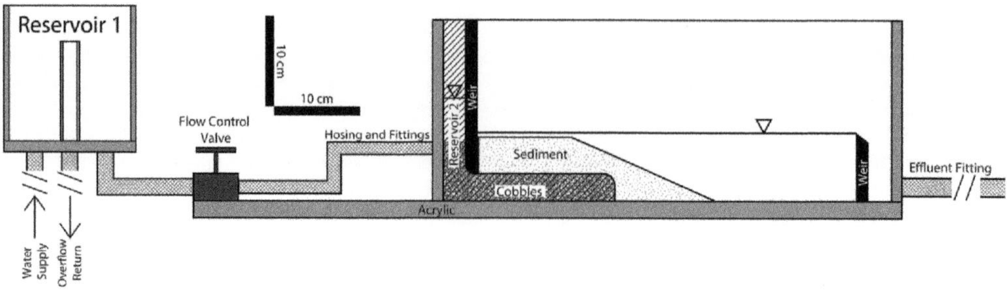

Fig. 1. Schematic diagram of the benchtop flume in the Geomechanics Laboratory at the Colorado School of Mines. The area labelled 'sediment' represents the typical experiment slope geometry. R2 is denoted with diagonal (45°) hatching.

Permeability/hydraulic conductivity and porosity tests

Falling head tests were performed for each sediment mixture to constrain hydraulic conductivity. Sand–clay mixtures were prepared as previously described and arranged in triplicate as 12–13 cm long specimens (4 cm in diameter) in 25 cm long apparatuses. Tap water was chosen as the permeant to match our flume experiment conditions. Specimens were capped with a porous stone (*c.* 0.5 cm thick; top cap) and permeable membrane (base). Assemblies were submerged in water to depths sufficient to produce flow across the specimens and facilitate air removal and saturation (Berthila *et al.* 2010). Permeant liquid flowed from the base of the specimen upwards until the specimen was completely saturated. The specimen was then left in this state for a minimum of 24 h. The sample assemblies were then removed from the water and suspended, and water was added to the top of the specimens. The change in water level (head) over time was recorded to determine hydraulic conductivity (equation 1; Berthila *et al.* 2010). Falling-head experiments using prepared assemblies were run in triplicate (i.e. nine runs per sediment mixture).

$$K = \frac{\ln(h_1/h_2) \times L}{\Delta t} \tag{1}$$

where K is hydraulic conductivity (m s^{-1}), L is the length of the specimen (m), Δt is the interval of time (s) over which flow occurred, h_1 is head at t_1 (m) and h_2 is head at t_2 (m).

Sediment mixture bulk density was measured in triplicate and used to determine porosity to inform numerical modelling (equation 2; Fetter 2001),

$$n = 100\left(1 - \frac{\rho_b}{\rho_s}\right) \tag{2}$$

where n is porosity (%), ρ_b is bulk density (g cm^{-3}) and ρ_s is the density of dry solids (g cm^{-3}).

Numerical modelling

To quantify the experimentally observed failure processes, we developed groundwater flow models that simulated transient fluid flow through the experimental system and performed simple stability calculations based on the infinite slope approximation as a first-order characterization for how changes in overpressure, or pressure in excess of hydrostatic, correlate with locations of failure in our experimental set-up. Groundwater flow modelling used a simplified version of the groundwater flow equation,

$$\frac{S_s}{K}\frac{\partial h}{\partial t} = \frac{\partial^2 h}{\partial x^2} + \frac{\partial^2 h}{\partial y^2} \tag{3}$$

where S_s is the specific storage and h is the head. We assumed that all sediments are isotropic and that S_s was 0.981 m^{-1}. For the pebbles, we assumed a hydraulic conductivity of 9.81×10^{-2} m s^{-1}. For the sand and sand–clay mixtures, we used the average hydraulic conductivity values from the falling head experiments.

We used the spatially and temporally varying head data from the fluid flow models to estimate local stability based on an infinite slope factor of safety (FS) analysis,

$$FS = \frac{c + (\sigma'_{vh}\cos^2\theta - \rho g h^*)\tan\phi_f}{\sigma'_{vh}\cos\theta\sin\theta} \tag{4}$$

where c is cohesion, σ'_{vh} is the vertical effective stress assuming hydrostatic conditions, g, is acceleration due to gravity, h^*, is the head above hydrostatic conditions, θ is the seafloor slope (assumed constant and equal to 19°, similar to that of the flume experiments) and ϕ_f is the internal angle of friction. For cohesion, we assumed 10 Pa for every wt% clay (e.g. 4% clay equates to 40 Pa cohesion). This assumption allowed us to investigate how failure initiates and evolves with low values of cohesion that increase slightly with clay content. These values are within the range of cohesion for sand–clay mixtures based on the Unified Soil Classification System. For friction angle we assume 35° for 100% FG sand and decreased friction angle by 0.1° for each wt% clay added (e.g. 4% clay equates to ϕ_f of 34.6°). This approach provides a first-order estimate of where failure might occur and allows us to relate observed failures to fluid flow processes.

Results

Flume experiments

Slope failures initiated after sufficient water pressure was supplied through R2 and the pressure front migrated into the sediment above the pebbles. The water level within R2 required to initiate slope failure was 9.2 ± 0.2 cm (±1 SD) in sediment consisting of 100% FG sand (0% clay), 11.0 ± 0.4 cm in 98 wt% FG sand + 2 wt% clay, 11.5 ± 0.1 cm in 96% FG sand + 4% clay, and 12.2 ± 0.4 cm in 95% FG sand + 5% clay (Table 1). In seven experiments, we documented a decrease in the water level in R2 after a slope failure initiation; these systems then equilibrated at steady state. These steady-state R2 water levels increased with clay content from

Table 1. *Water levels in R2 during experiments (in cm ± 1 SD) as a function of clay and FG sand content*

Sediment composition (wt%)	R2 water level prior to slope failure (cm)	Stable R2 water level following slope failure (cm)	Number of data points
100% FG sand	9.2 ± 0.2	9.2 ± 0.2	3
98% FG sand + 2% smectite (clay)	11.0 ± 0.4	10.6 ± 0.7	3
96% FG sand + 4% smectite (clay)	11.5 ± 0.1	10.9 ± 0	2
95% FG sand + 5% smectite (clay)	12.2 ± 0.4	8.9 ± 0.01	2

Column 3 presents steady-state R2 water level following slope failure initiation. The water level in R2 reached equilibrium with the water level within the experiment section once the influent valve was closed.

9.2 ± 0.2 cm (0% clay) to 10.9 ± 0 cm (4% clay) but decreased to their lowest point in 5% clay experiments (8.9 ± 0.01 cm; Table 1).

Geomorphic characteristics of slope failures varied with clay content. All experiments displayed some form of surficial eruption involving localized, vertical displacement of sediments ('sand volcanoes' Table 2; Fig. 2). Extensional surface cracking and suspension of clay particles was observed in all experiments involving clay. In experiments that had extensional surface cracking and surficial eruptions, extensional cracks were observed before surficial eruptions. Formation of elongated subsurface fissures along the flume–wall interface ('sidewall subsurface piping/fissures') was observed in one out of three of the 2% clay experiments, two out of three of the 4% clay experiments and all of the 5% clay experiments. Formation of subsurface fissures was observed after the formation of extensional cracking, but prior to surficial eruption. Each experiment resulted in downslope sediment transport, including but not limited to suspension of clay particles. No massive, blocky failures were observed.

Hydraulic conductivity and porosity

The hydraulic conductivity of sediment consisting of 100 wt% FG sand was $1.7 \times 10^{-4} \pm 1.6 \times 10^{-5}$ m s^{-1} (±1 SD); 98% FG sand + 2% clay was $5.4 \times 10^{-7} \pm 2.3 \times 10^{-7}$ m s^{-1}; 96% FG sand + 4% clay was 9.9×10^{-9} m s^{-1} $\pm 2.3 \times 10^{-9}$ m s^{-1}; and 95% FG sand + 5% clay was $7.8 \times 10^{-9} \pm 3.9 \times 10^{-9}$ m s^{-1}.

A best-fit linear regression was performed on $\log_{10}$ (hydraulic conductivity) as a function of clay content data for 40 experimental data points (Fig. 3). The regression from 0 to 5% clay yielded ($R^2 = 0.94$)

$$\log_{10}(K) = -0.8784cc - 4.1366 \quad (5)$$

where cc is clay content (fraction; i.e. 5 wt% = 0.05). Extrapolation of this regression to clay concentrations exceeding 5 wt%, or mixtures not comparable to those described here, should be verified with additional experiments.

The porosity of 100 wt% FG sand was 47.2 ± 0.7% (±1 SD); 98% FG sand + 2% clay was 46.7 ± 1.9%; 96% FG sand + 4% clay was 49.0 ± 1.6%; and 95% FG sand + 5% clay was 48.7 ± 1.9%.

Numerical modelling

Numerical simulations quantified the fluid flow, overpressure and failure potential for our flume

Table 2. *Geomorphology of slope failures as a function of clay and FG sand content*

Sediment composition (wt%)	Surface eruption	Sidewall subsurface piping/fissures	Tension cracks	Suspension of clay particles	Downslope sediment transport
100% FG sand	3/3	0/3	0/3	N/A	3/3
98% FG sand + 2% smectite (clay)	3/3	1/3	3/3	3/3	3/3
96% FG sand + 4% smectite (clay)	3/3	2/3	3/3	3/3	3/3
95% FG sand + 5% smectite (clay)	2/2	2/2	2/2	2/2	2/2
Example photo figures	2a & 2b	2b	2a	2b	2b

Experiments that had slope failures unexpectedly during experiment set-up were excluded.

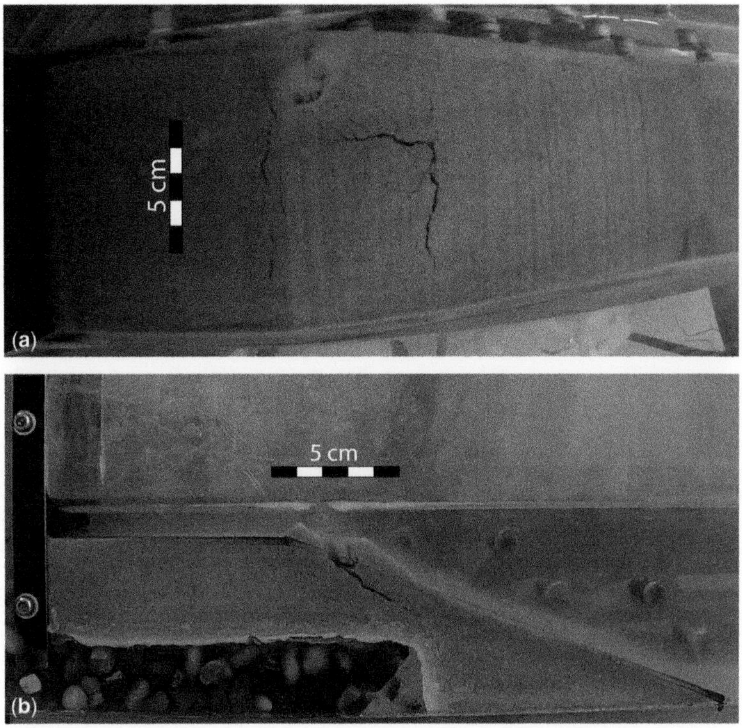

Fig. 2. (**a**) Overhead view of 4% clay experiment 2 showing four clustered sand volcanoes and extensional cracking. Surficial extensional cracking permitted focused water discharge. (**b**) Cross-section view of 4% clay experiment 3 displaying a sand volcano, subsurface piping/fissure, suspension of sediment particles and downslope sediment transport. Note: (a) and (b) are not to the same scale.

experiments. While our models are transient, we present results once the overpressure has equilibrated in the pebble layer, which allows us to determine failure potential when the entire high permeability zone is at its maximum overpressure. The numerical model of the 100% FG sand system documents a rapid (order of seconds) propagation of the overpressure from R2 through the sand. The maximum overpressure (pressure in excess of hydrostatic) at this point is 128 Pa in the pebble layer (Fig. 4a). The overlying sediment has low overpressure. The

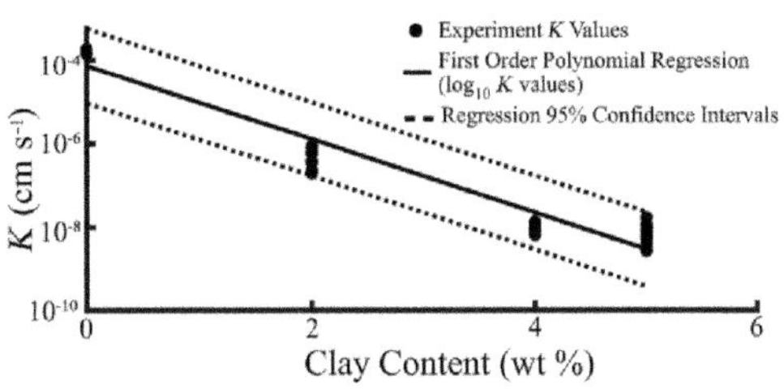

Fig. 3. Hydraulic conductivity as a function of clay determined by falling head method.

stability analysis predicts FS = 1 near the top of the sand (Fig. 4a). This is consistent with experimentally observed sand volcanoes at the surface of the sand (Table 2).

The numerical model of the 98% FG sand system documents a slower propagation (order of hundreds of seconds) of the overpressure from R2 into the sand–clay mixture. The maximum overpressure is 304 Pa in the pebble layer (Fig. 4b). The overlying sediment has moderate-to-low overpressure. The stability analysis predicts FS = 1 near the top of the sand–clay mixture and near the upper boundary of the pebble layer (Fig. 4b). This is consistent with experimentally observed sand volcanoes at the surface and fissures and cracks within the sand layer (Table 2).

The numerical models of the 96% FG sand and 95% FG sand are similar and document an even slower propagation (order of thousands of seconds) of the overpressure from R2 into the sand–clay mixture. The maximum overpressure is 372 Pa in the pebble layer of the 96% FG sand system (Fig. 4c) and is 442 Pa in the pebble layer of the 95% FG sand system (Fig. 4d). The stability analysis predicts

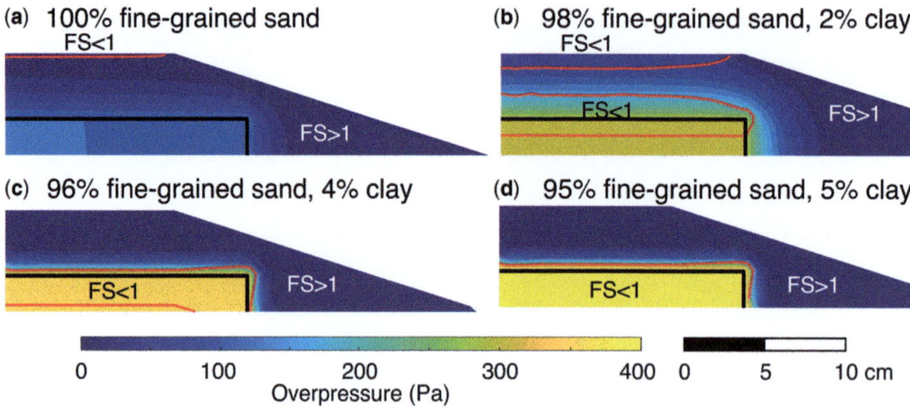

Fig. 4. Example results from numerical model of fluid flow and slope stability analysis for our flume experiments. For all plots, overpressure (or pressure in excess of hydrostatic) data are colour contoured at the same level (0–400 Pa), the thick red line is the factor of safety (FS) value of 1, and the thick black line is the boundary between the pebbles and the sediment (sand or sand–clay mixture). Zones where FS<1 and where FS>1 are labelled to show zones of instability (FS<1) and stability (FS>1). (**a**) Overpressure in the model of the 100% fine-grained (FG) sand experiment 1 s after raising the head in R2 to 9.2 cm. Maximum overpressure is 128 Pa. FS = 1 contour is localized near the top of the sediment between horizontal distance 0 and 10.5 cm, indicating that failure could initiate anywhere near the surface above that FS contour. (**b**) Overpressure in the model of the 2% clay experiment 500 s after raising the head in R2 to 11.0 cm. Maximum overpressure is 304 Pa. Two FS = 1 contours exist; one encompasses the boundary between the pebble layer and the sand–clay layer and extends to a horizontal distance of 16 cm and one is near the top of the sand–clay layer. This indicates that failure could initiate anywhere near the boundary of the pebble layer and the sand–clay layer or near the top of the sand–clay layer. Note: this model used a geometry based on the slightly different sediment geometry used in the 2% clay flume experiment. (**c**) Overpressure in the model of the 4% clay experiment 2500 s after raising the head in R2 to 11.5 cm. Maximum overpressure is 374 Pa. FS = 1 contour is near the base of the pebble layer and just above the pebble layer in the sand–clay layer, indicating that failure could initiate anywhere near the boundary of the pebble layer and the sand–clay layer. (**d**) Overpressure in the model of the 5% clay experiment 2500 s after raising the head in R2 to 11.5 cm. Maximum overpressure is 442 Pa. FS = 1 contour is in the sand–clay mixture parallel to the pebble layer, indicating that failure could initiate anywhere near the boundary of the pebble layer and the sand–clay layer. There is no vertical exaggeration.

FS = 1 near the boundaries of the pebble layer (Fig. 4c, d). This is consistent with experimentally observed sand volcanoes at the surface and fissures and cracks within the sand layer (Table 2).

Discussion

The water level of R2 controlled the slope failure, with increased R2 water levels prior to slope failure initiation directly correlating with increasing clay content (Table 1). As sand is non-cohesive and clay is cohesive, we interpret the increased cohesion with increasing clay content as increased resistance to failure. Decreases in R2 water levels immediately after slope failure initiation were likely a result of focused fluid flow through high permeability conduits produced by the failure. This was directly observed in cases where surface eruptions or subsurface fissures were produced in contact with the flume sidewall (e.g. Fig. 2b). The R2 water levels post-failure in relation to the water level within the flume experiment section may reflect differences in overall system permeability facilitated by differing scales of channellization following slope failure.

The lack of observed subsurface sidewall fissure formation and tensile cracking in sediment slopes of 100% FG sand indicated the importance of clay in their formation. We believe that clay particle size decreased permeability (Fig. 3), allowing the build-up of the higher pore pressures required to overcome the clay particle cohesive forces allowing failure. Once pore pressures exceeded the value to create failure, a subsurface fissure formed rapidly. Surficial extensional cracking was interpreted to be the result of sediment expansion produced by increasing pore pressure (Terzaghi *et al.* 1996). We note that subsurface sidewall fissures were present in experiments that contained clay (Table 2; Fig. 2b), but were not the origin of the failures. In many cases, these failures initiated within the sedimentary wedge and migrated to the sidewall. In addition, early deformation in the experiments included tension cracks on the surface that were not associated with sidewalls (Fig. 2a). Based on these observations, we surmise that these types of failures were controlled by water–sediment interactions and were not influenced by boundary effects (e.g. fluid migration along the sidewall).

Suspension of clay particles was anticipated for experiments involving sand–clay mixtures and is indicative of an active flow system where fluid expulsion resulted in the winnowing of clay particles. In most cases, the violent nature of slope failure initiation (surface eruption or sidewall piping) was enough to suspend clay particles within the top *c.* 1.5 cm of the flume experiment water column. Once suspended, clay particles travelled downslope due to gravity (towards the downstream weir) and remained in suspension for 12–24 h.

Slump failures, here defined as any cohesive, blocky failure or sedimentary folding, were not observed in any successful experiments. In some experiments, slumping was witnessed during experiment set-up, but this was prior to full saturation or total slope submergence. These experiments were ended and discredited as failed preparations, and they did not reflect the failure of a subaqueous slope under controlled pore pressure conditions. The lack of slumps in experiments may be the result of insufficient clay content (e.g. Sawyer *et al.* 2012) within the sediment and/or pore pressure being the only failure-inducing mechanism. Increased clay content is expected to increase cohesion within the slope and may allow for cohesive blocks of material to fail, rather than the observed channellization. The angle of the experiment slope was also considered as a potential cause of the lack of slumping. However, in natural marine environments, slump-like failures have been documented on slopes as low as 2° (Hühnerbach and Masson 2004). We therefore do not think the angle of our experiment slope (*c.* 19°) is the cause of this trend. A third effect that may increase slump-like behaviour is the rate of applied perturbations (e.g. pore pressure), which we did not explore in these experiments.

The sand volcano and tension crack geomorphological features observed in experiments resemble features observed in environments with greater clay content. For example, mud volcanoes have been observed in clay- and silt-dominated environments, such as the accretionary wedge offshore Costa Rica consisting of a carbonate-rich section overlain by hemipelagic sediment (clay- and silt-dominated; Shipley *et al.* 1990) or seaward of the Barbados accretionary complex (Westbrook and Smith 1983). Dambreak experiments performed by Sawyer *et al.* (2012), using mixtures of kaolin clay and silica silt, show fracture features in their source area resembling our observed experimental extensional crack features.

Surface eruption was the final stage of slope failure observed in all experiments. Formation of surficial extension cracks and subsurface cavities prior to surficial eruption was observed in some experiments. The observation of surface eruption as the most severe form of slope failure, combined with a lack of failure features post-eruption, indicates that eruption decreased excess pore pressure and increased local permeability, establishing a new stable system. Our documentation of progressive failure processes provides new insights into the temporal evolution of the final slope failure deposits.

Our numerical models correlate well with the general observations from our flume experiments. The models show that the decrease in permeability from the addition of clay impacts on the magnitude and distribution of the overpressure field (Fig. 4). This change in the overpressure field produces a change in the locus of failure (Fig. 4), with higher permeability, non-cohesive systems favouring shallow failure through surface eruption and lower permeability, cohesive systems evolving to deeper and more brittle failure modes (Table 2). In pure FG sand experiments, there were no failure processes or geomorphological features observed, suggesting deep sourcing for surface eruption. Clay experiments, however, displayed clear evidence for deeper sourcing for surface eruptions; in some instances, the source could be directly observed to extend to the pebble boundary. We interpret this to mean that small amounts of clay provide a decrease in permeability and cohesion that creates a change in the failure mode and the depth at which failure initiates.

While the qualitative link between our numerical models and physical experiments gives us confidence in understanding the basic processes and dynamics, our models do not provide exact replicates of our experiments. Specifically, our models do not predict the exact time or exact location of failure. In experiments with pure FG sand, surface eruptions nucleated at the location of greatest overpressure gradient, downslope of the model FS$<$1 boundary. We anticipate that this is a result of our models being homogeneous and isotropic, whereas the flume experiments have heterogeneity based on set-up and small variations in grain size and sorting. These heterogeneities will affect the permeability distribution and thus the pressure and instability distributions. More advanced models with stochastic heterogeneities could be developed to explore the importance of heterogeneity in localization of failure. We did not, however, address that issue as this research was focused on characterizing physical experiments and developing models that link the physical processes with the observed behaviour.

Our laboratory and experimental results show that for a fixed stratigraphic geometry composed of high-permeability pebbles overlain by sand–clay mixtures (up to 5% clay) the permeability, cohesion and failure angle influence the onset of failure and the type of failure. As clay content and cohesion increase and permeability decreases, we observe

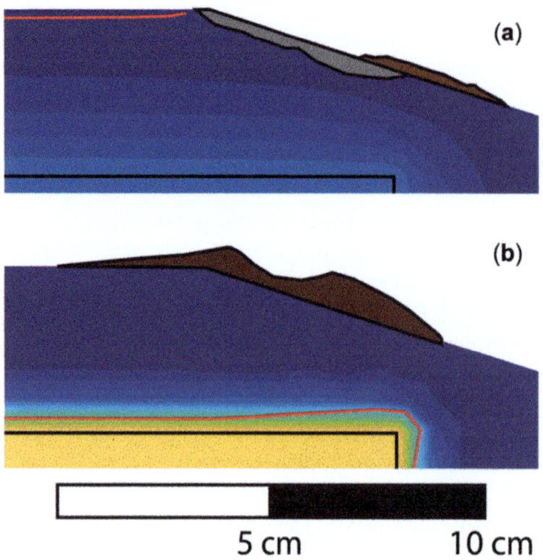

Fig. 5. Example results from numerical model of fluid flow and slope stability analysis from Figure 4 expanded and overlain with observed slope mass loss (in grey) or deposition (in brown) for (**a**) 100% FG sand experiment 4 and (**b**) 4% clay experiment 3. There is no vertical exaggeration.

that higher overpressure is required to induce failure and that failure transitions from surface eruption (100% FG sand) to surface eruption plus fissures, tension cracks and suspension of clay particles (2–5% clay). Our models of the physical experiments based on measured permeability, geometry and known boundary heads successfully predict the location of failures and the timing of failures to the zeroth order (e.g. Fig. 5). Combined, this documents a necessity to understand *in situ* properties to better understand the potential for failure location and type. Our modelling results, however, show that the geometry of the pebble layer and the overlying sand–clay mixtures influence the distribution of overpressure, which in turn affects the localization of failure. This result will motivate future studies on the influence stratigraphic architecture has on overpressure distribution and failure localization.

Conclusions

Submarine slope failures pose risks to coastal communities because of their ability to damage infrastructure and generate tsunamis. Slope failures on passive margins represent some of the largest mass failures on Earth, yet little is known about their dynamics. While numerous studies characterize slope failure deposits in terms of lithology, structure, seismic attributes and geometry, we lack direct observations of slope failure initiation and evolution. As a result, there is limited insight into the relationships between initial conditions, preconditioning, slope failure initiation and slope failure evolution.

We performed a suite of physical experiments in a benchtop flume to investigate submarine slope failure dynamics in relation to initial conditions, and numerical models were developed to quantify the observed processes. Submarine slope failures were induced under controlled pore pressure with varied mixtures of sand and clay (0–5 wt% clay) constrained to a constant pre-failure slope geometry.

Increased clay concentrations were observed to directly relate to an increase in pore pressure required to induce slope failure, attributed to clay particle cohesion and decreased hydraulic conductivity. Formation of elongated subsurface cavities along the flume–wall interface and surficial tensile cracking was only observed in experiments containing clays, indicating the importance of the presence of clay in their formation. No slump/block failure was observed in experiments. Insufficient clay content and a slow rate of pore pressure increase are considered to be the causes. Numerical models of our physical experiments using measured hydraulic properties are effective at simulating the distribution of overpressure and locations of potential failure for both cohesive and non-cohesive sediments and modes of failure. The model results demonstrate the importance clay has for permeability and cohesion, which helps define where failure conditions exist and when they occur. The correlation between the experiments and models gives strength to the interrogation of processes that models provide, and suggests that the upscaling of these models to natural systems can be effective at evaluating the potential for failure location and mode for known geometries and pressure sources.

Acknowledgements This research benefited from discussions with C. Paola and A. Abeyta. Comments by D. Sawyer and one anonymous reviewer helped improve the structure and content of the manuscript. Please contact the authors for original data and/or digital model results.

Funding This research was supported by the National Science Foundation (grant NSF-OCE 1753680 to BD).

Author contributions **MMWS**: formal analysis (equal), investigation (equal), methodology (supporting), software (equal), visualization (equal), writing – original draft (lead), writing – review & editing (equal); **BD**: conceptualization (lead), formal analysis (equal), funding acquisition (lead), investigation (equal), methodology (lead), project administration (lead), resources (lead), software (equal), supervision (lead), validation (lead), visualization (equal), writing – original draft (supporting), writing – review & editing (equal).

References

Berthila, E.V.P., Badillo, E.J., Rodríguez, A.R. and ASTM D5084-10. 2010. *Standard Test Methods for Measurement of Hydraulic Conductivity of Saturated Porous Materials Using a Flexible Wall Permeameter*. ASTM International, West Conshohocken, PA, https://doi.org/10.1520/D5084-10

ten Brink, U.S., Chaytor, J.D., Geist, E.L., Brothers, D.S. and Andrews, B.D. 2014. Assessment of tsunami hazard to the U.S. Atlantic margin. *Marine Geology*, **353**, 31–54, https://doi.org/10.1016/j.margeo.2014.02.011

ten Brink, U.S., Andrews, B.D. and Miller, N.C. 2016. Seismicity and sedimentation rate effects on submarine slope stability. *Geology*, **44**, 563–566, https://doi.org/10.1130/G37866.1

Dan, G., Sultan, N., Cattaneo, A., Everchère, J.D. and Yelles, K. 2010. Mass-transport deposits on the Algerian margin (Algiers area): morphology, lithology and sedimentary processes. *In*: Mosher, D.C. *et al.* (eds) *Submarine Mass Movements and Their Consequences*. Advances in Natural and Technological Hazards Research, **28**. Springer, 527–539, https://doi.org/10.1007/978-90-481-3071-9_43

Fetter, C.W. 2001. Chapter 3: Properties of Aquifers. *Applied Hydrogeology*, 4th edn. Prentice Hall, Upper Saddle River, NJ.

Fine, I.V., Rabinovich, A.B., Bornhold, B.D., Thomson, R.E. and Kulikov, E.A. 2005. The Grand Banks landslide-generated tsunami of November 18, 1929: preliminary analysis and numerical modeling. *Marine Geology*, **215**, 45–57, https://doi.org/10.1016/j.margeo.2004.11.007

Fritz, H.M., Kongko, W. *et al.* 2007. Extreme runup from the 17 July 2006 Java tsunami. *Geophysical Research Letters*, **34**, L12602, https://doi.org/10.1029/2007GL029404

Gazioğlu, C., Yücel, Z.Y. and Doğan, E. 2005. Morphological features of major submarine landslides of Marmara Sea using multibeam data. *Journal of Coastal Research*, **214**, 664–673, https://doi.org/10.2112/03-0060.1

Hühnerbach, V. and Masson, D.G. 2004. Landslides in the North Atlantic and its adjacent seas: an analysis of their morphology, setting and behaviour. *Marine Geology*, **213**, 343–362, https://doi.org/10.1016/j.margeo.2004.10.013

Idarraga-García, J. and Vargas, C.A. 2013. Morphological expression of submarine landslides in the accretionary prism of the Caribbean continental margin of Colombia. *In*: Krastel, S. *et al.* (eds) *Submarine Mass Movements and Their Consequences*. Advances in Natural and Technological Hazards Research, **37**. Springer, 391–401, https://doi.org/10.1007/978-3-319-00972-8_35

Locat, J., ten Brink, U.S. and Chaytor, J.D. 2010. The Block composite submarine landslide, Southern New England slope, U.S.A.: a morphological analysis. *In*: Mosher, D.C. *et al.* (eds) *Submarine Mass Movements and Their Consequences*. Advances in Natural and Technological Hazards Research, **28**. Springer, 267–277, https://doi.org/10.1007/978-90-481-3071-9_22

Lucchi, R.G., Pedrosa, M.T., Camerlenghi, A., Urgeles, R., De Mol, B. and Rebesco, M. 2012. Recent submarine landslides on the continental slope of Storfjorden and Kveithola Trough-Mouth Fans (North West Barents Sea). *In*: Yamada, Y. *et al.* (eds) *Submarine Mass Movements and Their Consequences*. Advances in Natural and Technological Hazards Research, **31**. Springer, 735–745, https://doi.org/10.1007/978-94-007-2162-3_65

McAdoo, B.G., Pratson, L.F. and Orange, D.L. 2000. Submarine landslide geomorphology, US continental slope. *Marine Geology*, **169**, 103–136, https://doi.org/10.1016/S0025-3227(00)00050-5

Muhari, A., Imamura, F., Arikawa, T., Hakim, A.R. and Afriyanto, B. 2018. Solving the puzzle of the September 2018 Palu, Indonesia, tsunami mystery: clues from the tsunami waveform and the initial field survey data. *Journal of Disaster Research*, **13**, sc20181108, https://doi.org/10.20965/jdr.2018.sc20181108

Sawyer, D.E. and Devore, J.R. 2015. Elevated shear strength of sediments on active margins: evidence for seismic strengthening. *Geophysical Research Letters*, **42**, 10216–10221, https://doi.org/10.1002/2015GL066603

Sawyer, D.E., Flemings, P.B., Buttles, J. and Mohrig, D. 2012. Mudflow transport behavior and deposit morphology: role of shear stress to yield strength ratio in subaqueous experiments. *Marine Geology*, **307–310**, 28–39, https://doi.org/10.1016/j.margeo.2012.01.009

Shipley, T.H., Stoffa, P.L. and Dean, D.F. 1990. Underthrust sediments, fluid migration paths, and mud volcanoes associated with the accretionary wedge off Costa Rica: Middle America Trench. *Journal of Geophysical Research*, **95**, 8743–8752, https://doi.org/10.1029/JB095iB06p08743

Terzaghi, K., Peck, R.B. and Mesri, G. 1996. *Soil Mechanics in Engineering Practice*. Wiley, New York

Westbrook, G.K. and Smith, M.J. 1983. Long decollements and mud volcanoes: evidence from the Barbados Ridge Complex for the role of high pore-fluid pressure in the development of an accretionary complex. *Geology*, **11**, 279–283, https://doi.org/10.1130/0091-7613(1983)11<279:LDAMVE>2.0.CO;2

Subduction of an extinct rift and its role in the formation of submarine landslides in NW South America

Carlos A. Vargas[1]*, Gustavo A. Gutiérrez[2] and Gustavo A. Sarmiento[1]

[1]Department of Geosciences, Universidad Nacional de Colombia at Bogotá, Colombia

[2]CIOH, National Hydrographic Service, Cartagena, Colombia

CAV, 0000-0002-5027-9519

*Correspondence: cavargasj@unal.edu.co

Abstract: On the eastern margin of the Panama Basin, the Nazca oceanic plate converges towards the continental plate of South America at approximately 53 mm a^{-1}. Subduction processes are accompanied by the presence of anomalous bathymetric elements including the Sandra Ridge. This east–west-orientated ridge is catalogued as an aborted rift derived from a magmatic spreading axis that was active between 12 and 9 Ma. Seismic activity within this structure is considered evidence of fault reactivation and tectonism. Once the structure reached the subduction trench several submarine landslides were triggered. Run-out lengths of these submarine landslides are perpendicular to the convergence of the structure with some units spreading and forming a wide fan that reaches tens of kilometres to the north and south of the trench. The area affected by the three main landslides varies between 130 and 300 km^2 approximately, with relatively superficial earthquakes (<33 km) and with magnitudes that reach up to M_w 7.2. The morphology of the landslides suggests a retrogressive nature with younger events proximal to shore. This paper presents estimates of the age of these landslides and discusses sources of uncertainty regarding these times of occurrence.

Supplementary material: A graphical compilation of the 12 profiles analysed is available at https://doi.org/10.6084/m9.figshare.c.4960862

Submarine canyons are generated by erosion of the continental slope and by the imprint of repeated gravity-driven processes. The recurrence of events and level of activity within submarine canyons can be linked to tectonic setting. For instance, Harris and Whiteway (2011) reported that active continental margins contain approximately 15% more canyons than their passive margin counterparts. According to these authors, canyons are steeper, shorter, more dendritic and more closely spaced on active tectonic margins. It has also been reported that the volume of sediment remobilization in canyons within passive margins is larger than in active settings but the frequency of events is higher in the latter (see Masson *et al.* 2010). Submarine landslides that develop within canyons in active tectonic margins tend to present a wider morphological character (see, for example, Idárraga-García *et al.* 2019). This variability is due to changes linked to rates of erosion/accretion bearing to subduction and differences in relief of the underthrusting plate (see, e.g. Harders *et al.* 2011). Hence, the tsunamigenic potential of submarine landslides within active tectonic margins is more susceptible to the dynamic processes associated with the converging plate setting. A review of worldwide scenarios associated with submarine landslides in passive and active margins can be found in Harris and Whiteway (2011), and Kawamura *et al.* (2014).

Geological setting

The northwestern region of South America is a natural tectonic laboratory that covers >3000 km in length where two lithospheric plates of oceanic origin, the Caribbean and Nazca plates, collide against the South American continental plate (Salazar and Vargas 2015). In this region (Fig. 1), the low-angle subduction of the Caribbean Plate under South America (Vargas 2019) is associated with the riding of the Panama Island arc (Salazar *et al.* 2017). Here, the magmatic arc is extinct, the seismic activity is low and smaller submarine landslides are the product of gravitational instabilities on the continental slope (Idarraga-Garcia and Vargas 2014; Vargas and Idárraga-Garcia 2014). In contrast, the subduction of the Nazca Plate beneath South America offers a panorama of greater seismic activity (Taboada *et al.* 2000; Pedraza *et al.* 2007; Vargas *et al.* 2007) where larger submarine landslides are developed within the forearc basin. Consistent with this greater tectonic activity, the tsunamigenic potential within the Nazca/South America subduction zone is more significant (Vargas *et al.* 2012).

Estimates of seismotectonic deformation rates and the presence of strike-slip seismic events have suggested that the area of the extinct oceanic rift known as Sandra Ridge (Fig. 2a) and its projection

From: Georgiopoulou, A., Amy, L. A., Benetti, S., Chaytor, J. D., Clare, M. A., Gamboa, D., Haughton, P. D. W., Moernaut, J. and Mountjoy, J. J. (eds) 2020. *Subaqueous Mass Movements and their Consequences: Advances in Process Understanding, Monitoring and Hazard Assessments*. Geological Society, London, Special Publications, **500**, 311–322.
First published online May 12, 2020, https://doi.org/10.1144/SP500-2019-189

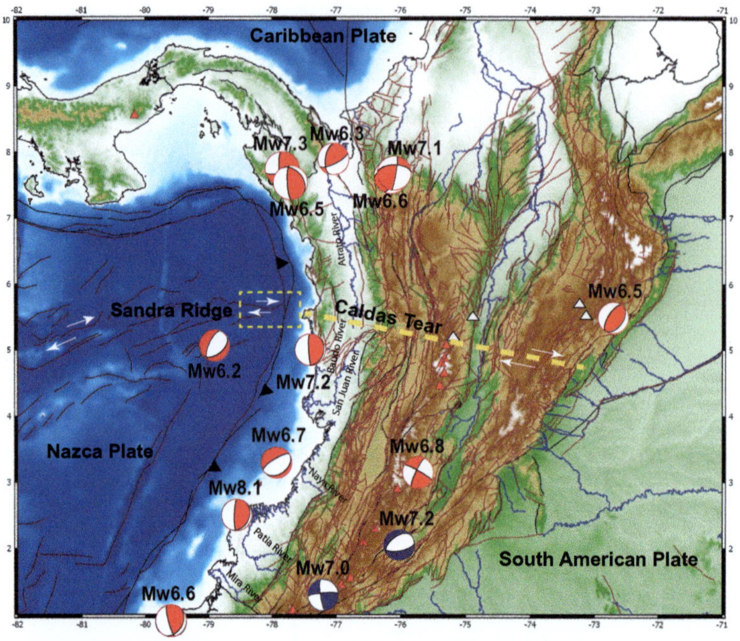

Fig. 1. Location of the main topographic and bathymetric characteristics in the tectonic convergence zone of the NW of South America. A bathymetric feature associated with an old oceanic spread centre, known as Sandra Ridge, can be seen north of the Nazca Plate. This structure is collinear with the lithospheric tear known as the Caldas Tear (Vargas and Mann 2013). The main rivers of the region have been highlighted with blue lines. Brown lines correspond to faults on the onshore reported from official geological cartography (Gómez *et al.* 2015). Offshore faults are presented in black lines. Beach balls correspond to the focal mechanisms of the events that have occurred in the region during recent decades and reported by the Global CMT (Ekström *et al.* 2012). The thin dashed yellow square corresponds to the study zone where the scarp fault dating estimations presented in this work are located. White arrows suggest right-lateral kinematics that connect the weakness zone associated with the Sandra Ridge and the Caldas Tear.

with the continental Caldas Tear may correspond to the boundary between the Caribbean and Nazca plates (Salazar and Vargas 2015). The presence of adakitic rocks (de Boer *et al.* 1988; Aguillón-Robles *et al.* 2001; Thorkelson and Breitsprecher 2005; Castillo 2006; Chadwick *et al.* 2009; Eyuboglu *et al.* 2011) in continental areas on the trend Sandra Ridge–Caldas Tear is evidence of the subduction of the extinct rift that nowadays constitutes a slab window and that has been documented in northwestern South America (see, e.g. Abratis and Worner 2001; Vargas and Mann 2013; Vargas *et al.* 2018). Beyond this hypothesis, the seismic activity in the Colombia–Ecuador trench suggests that the Sandra Ridge structure subducts under the South American continental plate, accompanied by energetic earthquakes (Fig. 1). Lateral differences in the thermal regime, age and thickness of the subducted plate, generate variations in the seismotectonic behaviour, as well as elastic deformation, which promote control mechanisms on the distribution of modern sediments along the trench (Gutsher *et al.* 1999; Collot *et al.* 2002). This process facilitates the development of typical forearc basins, e.g. Tumaco and Choco basins (Cediel *et al.* 2003; Witt *et al.* 2006; Barrero *et al.* 2007), as well as emplacement of anomalous depocentres along the trench.

The area proximal to the trench, the Sandra Ridge, is related to a series of submarine landslides that overlap each other and form an active depocentre (Fig. 2b). We have detected at least three large recent landslides (SR-North, SR-Central and SR-South). Boundaries between each one of them are difficult to establish, but reflect a high dynamism in the slope, near to the subduction trench. We have followed the proposal of Clare *et al.* (2018) to systematize some of the most relevant morphological characteristics of these identified landslides (Table 1). In general terms, we account that the high slopes of the margin, the occurrence of gas hydrates with potential of dissociation (Vargas 2012), the seismic activity and the processes of continental weathering, erosion and sedimentary loads produced by a rainfall between 2000 and 12 700 mm a^{-1} under tropical conditions (Restrepo and Lopez 2008) generate favourable scenarios for the development of these underwater landslides.

On the base of the previous observations, it seems reasonable that in slab window environments, and in the vicinity of the subduction trench, relatively young sedimentary basins may appear, which are filled with sediments from continental areas and where the material is partially incorporated into the subduction process. In this paper we present an interesting example of submarine landslides that feed a sedimentary basin associated with a slab window process. We will try to describe the current processes that trigger these landslides, as well as possible ages that allow us to project recurrences and the convergence process of the rift/ridge involved.

Data and methods

In this work, we used a bathymetric dataset acquired in several campaigns by the R/V *ARC Providencia*, a research ship of the Maritime Authority of Colombia (DIMAR). A swath-bathymetry-multibeam-echosounder of low-frequency system 'ATLAS HIDROSWEEP MD2' was used during data acquisition (Fig. 2a). This system uses an array of 395 beams to achieve an angular cover of 152°. Each beam has a transmission aperture of 1.3°. The system has a dome formed by two transmitters and two receivers, which operate at a frequency of 30 kHz at a depth range between 10 and 5000 m. The vertical accuracy of measurements at 30 m ranges between 0.36 to 0.24 m, and at a depth of 3000 m the precision is between 6.3 and 5.70 m.

In order to get a general view of the area, our dataset was combined with bathymetric records of the Expedition DANA02RR (Lonsdale 2005), which are stored in the Global Multi-Resolution Topography Data, immersed in the Marine Geoscience Data System – Geomapapp (Ryan *et al.* 2009). A digital terrain model was interpolated with a resolution of *c.* 60 × 60 m, which allowed estimating bathymetric profiles on pronounced slope breaks possibly associated with fault scarps related to the Sandra Ridge.

The Sandra Ridge is a structure that has been associated with the occurrence of recent submarine landslides, due to this association a quantitative geomorphological scarp analysis was undertaken to constrain age ranges for these events. The objective was to understand the evolution of this active margin through time, as well as the rate of growth of the associated sedimentary basin within the trench and the tsunamigenic potential associated with the recurrence of submarine landslides. Dating may inform whether recent earthquakes in the study zone were related to landslides detected on the sea bottom (Fig. 2b). This work followed Hanks (2000) by applying the diffusion equation under transport-limited conditions. Equation (1) is a second-order partial differential equation that relates the change in elevation H to time t using the proportionality constant κ and the distance x along a profile:

$$\frac{\partial H}{\partial t} = \frac{\partial^2 H}{\partial x^2} \qquad (1)$$

In this formula the diffusivity κ [L^2T^{-1}] varies with climate, scarp material and scarp height. The slope aspect is considered constant over the timescales investigated. Scarp height reduction is expressed by Hanks (2000) as:

$$\left.\frac{\partial u}{\partial x}\right|_{x=0} - b = \frac{a}{\sqrt{\pi t}} \equiv \tan\theta_s - \tan\theta_f \qquad (2)$$

In this reduction, $\theta_s = \tan^{-1}\left(\left.\frac{\partial u}{\partial x}\right|_{x=0}\right)$ represents the maximum slope-scarp angle, $\theta_f = \tan^{-1} b$ is the fan slope angle, and a is the effective half-vertical displacement.

We analysed 12 bathymetric profiles perpendicular to slope breaks of the Sandra Ridge and submarine landslides (see location on Fig. 2c), which were compared to theoretical shapes calculated by numerical solutions of equation (1). A least-squares fit was reached by iterative tuning of measurements of the effective displacement ($2a$) and the angles θ_s and θ_f. The constant κ and variable t were merged for calculating values that best fitted the data. One problem with this dating approach is that the profiles are not entirely dependent on κ or t, but rather their product. It means that the same profile would result for $\kappa = 1\ m^2\ a^{-1}$ and $t = 10$ years or $\kappa = 10\ m^2\ a^{-1}$ and $t = 1$ year. This methodology has been used in submarine scenarios (see e.g. Mitchell 1996; McAdoo and Simpson 2005; Vargas *et al.* 2012; Idarraga-Garcia and Vargas 2014). However, the use of this method in submarine environments has not usually been incorporated due to the limited knowledge of climate, erosion and oceanographic conditions on the evolution of underwater scarps. Incorporation of well or outcrop data within the continental shelf and onshore areas, respectively, and that could potentially contain tsunamigenic deposits, might help constrain age estimates; however, such information is not available as of today. Estimates presented in this work are the first approach to the dating of submarine landslides in this region of the world.

Finally, we used two seismic lines migrated in time and acquired by the National Hydrocarbon Agency of Colombia in 2005 to support our interpretations (ISPBR-2005-102053 and ISPBR-2005-202020).

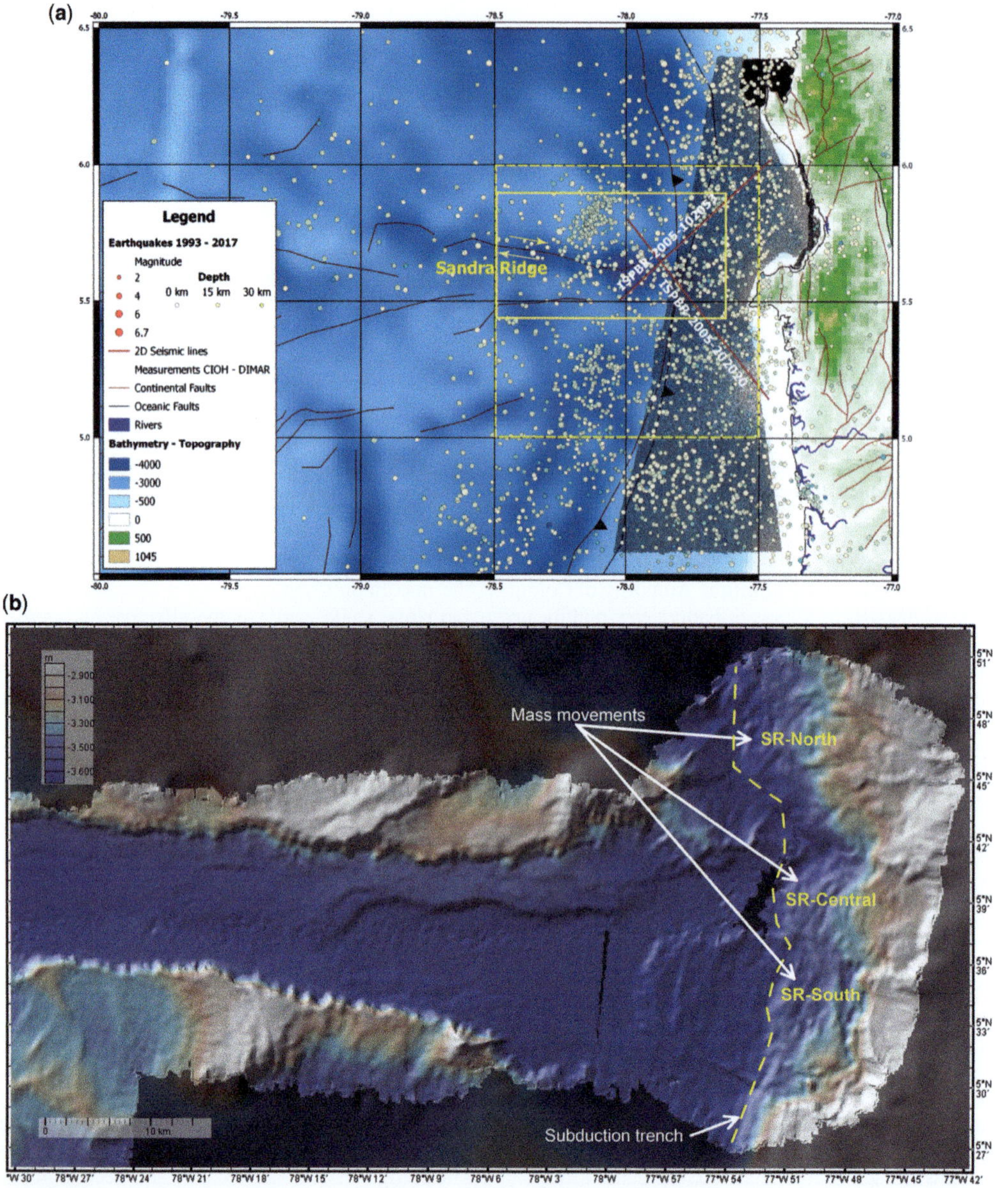

Fig. 2. Bathymetric expression of the Sandra Ridge. (**a**) The solid yellow line corresponds to (b); the dashed yellow line corresponds to (c) and the digital terrain model presented in Figure 4b. The dense network of black dots located east of Sandra Ridge corresponds to bathymetric measurements made by DIMAR through its Research Center of Oceanography and Hydrography. The red lines correspond to the seismic lines migrated in time and acquired by the National Hydrocarbons Agency of Colombia. Seismic events in the area at various depths have been represented with white dots (0–15 km), yellow dots (15–30 km) and green dots (>30 km). These seismological data were acquired by the Geological Survey of Colombia and can be accessed via the website https://www2.sgc.gov.co/sgc/sismos/Paginas/consultaexperta-rsnc.aspx. Black lines with indentation symbols represent the trench. Yellow arrows suggest the kinematics of the main faults that cross the Sandra Ridge. (**b**) Bathymetric detail of the Sandra Ridge in its proximity with the subduction trench. Image shows several landslides near to the subduction trench. Image provided by the Global Multi-Resolution Topography Data, immersed in the Marine Geoscience Data System – Geomapapp (Ryan *et al.* 2009).

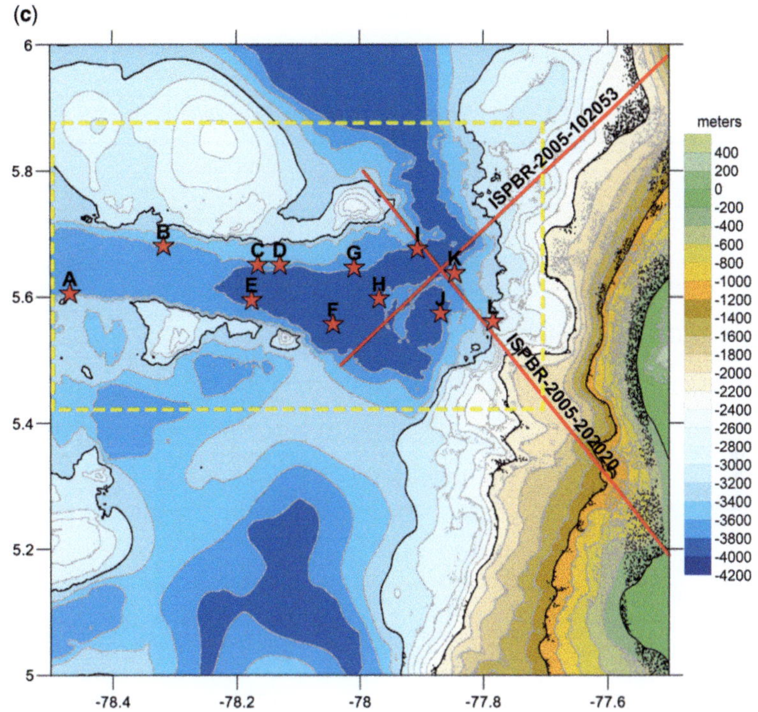

Fig. 2. *Continued.* (**c**) Location of places with estimated ages of fault scarps (red stars). Red lines correspond to position of the seismic lines of Figure 3. Dashed yellow rectangle corresponds to the area presented in Figure 4b.

This information was provided by the Geological Survey of Colombia in graphical format.

Results

Seismic lines showcase the variability in slope angles within the margin, as well as the presence of ridges that are the expression of tectonic convergence within the margin (Fig. 3). The presence of chaotic reflectors and rupture surfaces in the intra-ridge basins are indicative of the occurrence of submarine landslides sourced from continental areas that have reached the slope region. Triggering mechanisms are presumably linked to oceanic currents and seismotectonic activity (see Figs 2c & 3a). These sediments infill the trench generated by the crustal bulge as it approaches the subduction channel. It looks like the trench-related basin is later consumed during the subduction process (Fig. 3). It is also remarkable that this small basin presents a bulge in its central area with small scarp faces, probably related to the flexion of the crust just before subduction initiation. The general development of submarine landslides in this region appears to be a complicated process that includes tectonic and oceanographic triggered sedimentary process that are responsible for transporting sediments downslope, as well as tectonic processes that condition the tortuosity of the seafloor and the subsequent subduction of sedimentary material that reaches the trench.

Figure 4a illustrates the bathymetric profile A, located *c.* 70 km away from the trench, and perpendicular to the trend of the extinct spread centre (Fig. 4b). Almost all red points are inside theoretical profiles built with κt values between 30 000 and 300 000 m^2 (Fig. 4a). On the head and toe of the fault scarp, there are observations that fall outside the theoretical profiles, suggesting that the real profile may represent a complex process with successive scarps constructed downslope. Near the head scarp there is another scarp-slope change with respect to the surrounding fan (Fig. 4b). We also note that on the slump toe there is a protuberance that resembles a displacement of material from upper ridges. These slope breaks suggest other episodes of landslides and highlight a permanent process of geomorphological evolution in the study area. We tested extreme values of κt and found a geomorphological age of *c.* 45 000 m^2 for this profile. The Supplementary material presents a graphical compilation of the 12 profiles analysed and Table 2 resumes the

Table 1. *Compilation of some features of the submarine landslides identified in this work*

Feature	Submarine landslide		
	1	2	3
Parent ID	Sandra Ridge	Sandra Ridge	Sandra Ridge
Name	SR-SL-1	SR-SL-2	SR-SL-3
Aliases	SR-North	SR-Central	SR-South
Frontal confinement	Frontally confined	Frontally confined?	Frontally unconfined
Attachment	Attached	Attached	Attached
Object type	Mass-transport complex	Mass-transport complex	Mass-transport complex
Depth below seafloor (m)	3100	3000	2800
Latitude and longitude (WGS)	5.7743 N/77.8195 W	5.6471 N/77.8019 W	5.5602 N/77.8834 W
Total length (m)	18 470	16 000	21 840
Deposit length (m)	14 000	10 950	20 410
Evacuated length (m)	4000	5000	1500
Length metadata	Apparent measurement	Apparent measurement	Apparent measurement
Scar perimeter length (m)	71 760	51 560	68 280
Evacuation height (m)	150	520	895
Scar width (m)	15.9	8.8	14.4
Scar surface nature	Concave	Concave	Concave
Maximum deposit width (m)	14	9.2	14.6
Maximum deposit thickness (ms)	1450	900	990
Maximum unconfined deposit (ms)	450	550	550
Thickness metadata			
Total height drop (m)	1200	1400	1500
Slope gradient (°)	11	11	11
Slope gradient metadata (km at the south)	35	22	9
Slope gradient of the head scarp (°)	9?	14	22
Slope gradient of the head scarp metadata	5.7 N/77.7 W	5.6 N/77.7 W	5.5 N/77.8 W
Slope gradient at the toe (°)	2.3	2.3	0.3
Slope gradient at the toe metadata	5.75 N/77.85 W	5.65 N/77.85 W	5.6 N/78.05 W
Basal surface type	Rugose	Rugose	Rugose
Upper surface type	Rugose	Rugose	Rugose

Metrics and metadata suggested by Clare *et al.* (2018). This metric approach should be revisited in the future with a more trustworthy dataset orientated to these purposes.

parameters measured and estimations of geomorphological ages. In general terms, we found κt values ranging between *c.* 27 000 and *c.* 170 000 m^2. Our measurements show slopes with low gradients ($\theta_f = 0.71 \pm 0.63°$; $\theta_s = 14.3 \pm 9.9°$), which are consistent with other measurements in the region (e.g. Vargas *et al.* 2012). Values of the vertical effective displacement ($a = 108 \pm 59.2$ m) promote high values of κt.

We found the highest values of κt associated with profiles D, E and I and the lowest around the profile K. At least for this last place, we infer that no consolidated material that falls in the basin near to the trench is responsible for this behaviour. In contrast, we do not discard that the highest value around profile I corresponds to a scarp on crustal material, which is often highly resistant to erosive processes.

Discussion and conclusions

According to Restrepo and Kjerfve (2003), few rivers flow into the Pacific margin of Colombia, which despite their short length, have the highest rates of discharge per kilometre in South America due to a relatively constant temperature, high rainfall rates and high humidity. The rivers closest to the study area are the Baudó, Mira, Patia and San Juan rivers, which receive on average *c.* 5600 mm of rainfall annually (see Table 3). Despite the small basin areas involved (*c.* 13×10^3 km^2), these rivers discharge a relatively high volume of freshwater and sediments (Restrepo and Lopez 2008). These rivers also exhibit significant discharge variability as a result of ENSO, tectonic activity, wave energy, tides and sedimentary progradation on a narrow

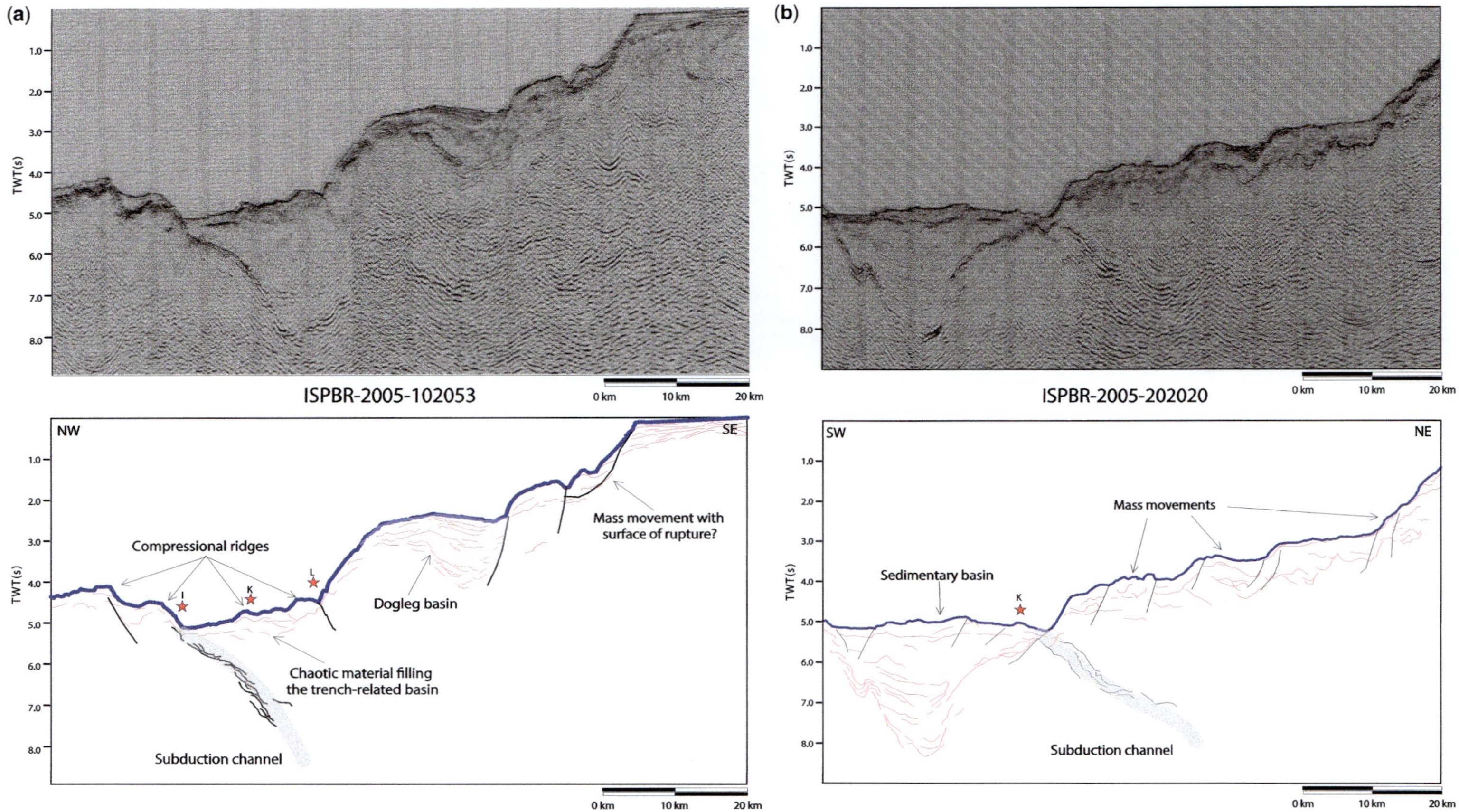

Fig. 3. Seismic lines (**a**) ISPBR-2005-102053 and (**b**) ISPBR-2005-202020. Upper panels present the information without interpretation; lower panels are interpreted. Red stars represent the approximate location of profiles used for dating scarps. Red lines represent reflectors associated with sedimentary structures; black lines represent faults; blue line corresponds to the ocean floor.

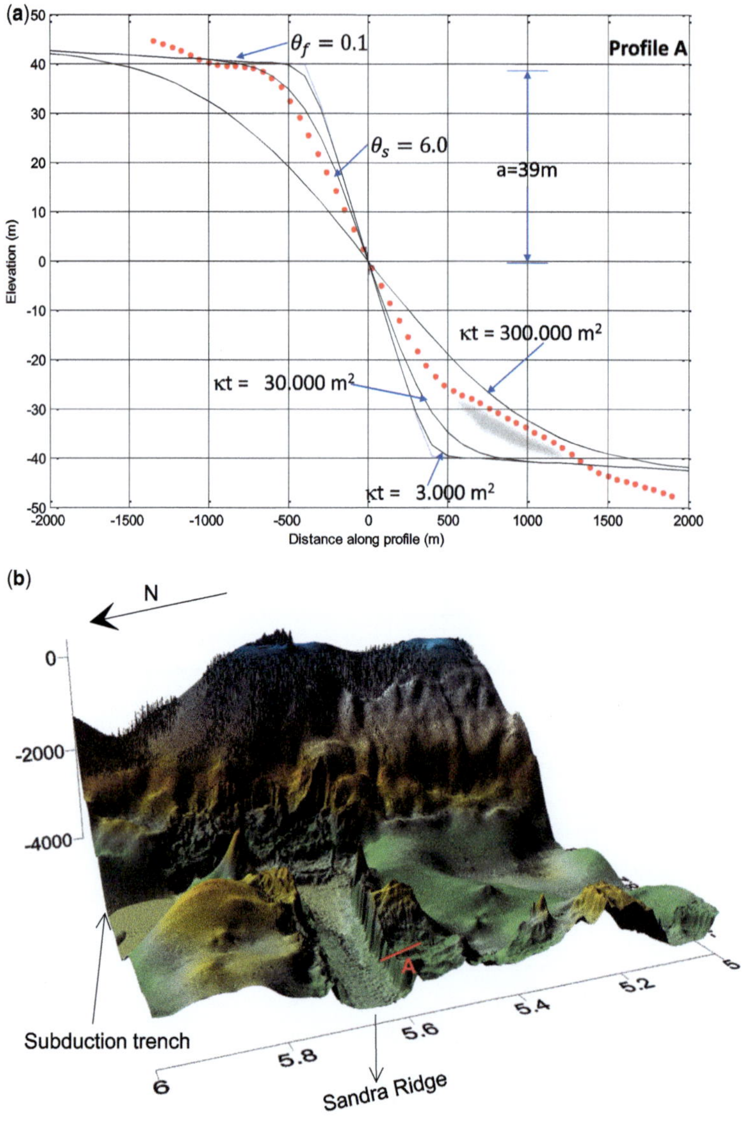

Fig. 4. (**a**) Profile located in position A (Fig. 2c). Red points correspond to real bathymetric measurements. Black lines represent different theoretical profiles built using the parameters presented in Table 2 and different values of κt. Near to the toe is a grey polygon that represents a debris zone, which may contribute to another landslide later. All profiles analysed can be observed in the figure in the Supplementary material. (**b**) Digital terrain model where it is possible to observe the location of Profile A and bathymetric textures derived from different sources of data. We observe details of the Sandra Ridge and the chaotic texture related to different landslides near to the subduction trench.

shelf bordering a deep trench (Restrepo *et al.* 2002). This scenario allows us to infer that towards the study area there is a significant amount of sedimentary load that spreads out downslope and reaches a broad area beyond the subduction trench.

We tested different κ values reported by Mitchell (1996), McAdoo and Simpson (2005), Vargas *et al.* (2012) and Vargas and Idárraga-Garcia (2014) to calculate relative ages associated with the formation of the fault scarps reported in Table 2. Estimations of the age of the landslides analysed suggest high contrasting values of age. Nevertheless, due to the high seismotectonic activity, the recurrence period, distance to seismic sources and history of recent energetic events involved in the region, we discarded

Table 2. *Parameters and results of the geomorphological scarp dating analyses in 12 bathymetric profiles located along the Sandra Ridge (Fig. 2c)*

	a (m)	Θ_s (°)	Θ_f (°)	κt (m^2)	κ (m^2y)	t (years)	Longitude (°)	Latitude (°)	Depth (m)
Profile A	39	6	0.1	45 319	0.007[1]	6 474 191	−78.17 559 209	5.59 460 615	3835
	39	6	0.1	45 319	0.015[2]	3 021 289			
	39	6	0.1	45 319	1.33[3-4]	34 075			
	39	6	0.1	45 319	8.9[3-4]	5092			
	39	6	0.1	45 319	35.7[3-4]	1295			
	39	6	0.1	45 319	133[3-4]	341			
Profile B	20	3	0.26	55 563	0.007[1]	7 937 563	−78.0 434 982	5.55 620 777	3891
	20	3	0.26	55 563	0.015[2]	3 704 196			
	20	3	0.26	55 563	1.33[3-4]	41 777			
	20	3	0.26	55 563	8.9[3-4]	6243			
	20	3	0.26	55 563	35.7[3-4]	1588			
	20	3	0.26	55 563	133[3-4]	418			
Profile C	80	10	0.22	68 472	0.007[1]	9 781 785	−77.96 838 044	5.597 149 238	3864
	80	10	0.22	68 472	0.015[2]	4 564 833			
	80	10	0.22	68 472	1.33[3-4]	51 483			
	80	10	0.22	68 472	8.9[3-4]	7694			
	80	10	0.22	68 472	35.7[3-4]	1956			
	80	10	0.22	68 472	133[3-4]	515			
Profile D	100	11	1.5	112 519	0.007[1]	16 074 190	−77.84 522 877	5.636 816 154	3663
	100	11	1.5	112 519	0.015[2]	7 501 288			
	100	11	1.5	112 519	1.33[3-4]	84 601			
	100	11	1.5	112 519	8.9[3-4]	12 643			
	100	11	1.5	112 519	35.7[3-4]	3215			
	100	11	1.5	112 519	133[3-4]	846			
Profile E	150	12	0.4	169 469	0.007[1]	24 209 926	−77.86 822 338	5.57 324 629	3764
	150	12	0.4	169 469	0.015[2]	11 297 965			
	150	12	0.4	169 469	1.33[3-4]	127 421			
	150	12	0.4	169 469	8.9[3-4]	19 042			
	150	12	0.4	169 469	35.7[3-4]	4842			
	150	12	0.4	169 469	133[3-4]	1274			
Profile F	190	30	2.2	39 563	0.007[1]	5 651 814	−77.78 416 348	5.560 277 042	3218
	190	30	2.2	39 563	0.015[2]	2 637 513			
	190	30	2.2	39 563	1.33[3-4]	29 746			
	190	30	2.2	39 563	8.9[3-4]	4445			
	190	30	2.2	39 563	35.7[3-4]	1130			
	190	30	2.2	39 563	133[3-4]	297			
Profile G	80	14	0.6	35 707	0.007[1]	5 101 067	−77.90 527 151	5.675 717 349	3821
	80	14	0.6	35 707	0.015[2]	2 380 498			
	80	14	0.6	35 707	1.33[3-4]	26 848			
	80	14	0.6	35 707	8.9[3-4]	4012			
	80	14	0.6	35 707	35.7[3-4]	1020			
	80	14	0.6	35 707	133[3-4]	268			
Profile H	110	17	0.8	45 244	0.007[1]	6 463 455	−78.0 090 051	5.645 969 429	3782
	110	17	0.8	45 244	0.015[2]	3 016 279			
	110	17	0.8	45 244	1.33[3-4]	34 018			
	110	17	0.8	45 244	8.9[3-4]	5084			
	110	17	0.8	45 244	35.7[3-4]	1293			
	110	17	0.8	45 244	133[3-4]	340			
Profile I	160	16	0.4	104 113	0.007[1]	14 873 341	−78.12 960 177	5.65 054 603	3668
	160	16	0.4	104 113	0.015[2]	6 940 892			
	160	16	0.4	104 113	1.33[3-4]	78 281			
	160	16	0.4	104 113	8.9[3-4]	11 698			
	160	16	0.4	104 113	35.7[3-4]	2975			
	160	16	0.4	104 113	133[3-4]	783			
Profile J	115	16	1.42	61 346	0.007[1]	8 763 729	−78.1 653 727	5.650 546 045	3696
	115	16	1.42	61 346	0.015[2]	4 089 740			

(*Continued*)

Table 2. *Continued.*

	a (m)	Θ_s (°)	Θ_f (°)	κt (m^2)	κ (m^{2}y)	t (years)	Longitude (°)	Latitude (°)	Depth (m)
	115	16	1.42	61 346	1.33[3-4]	46 125			
	115	16	1.42	61 346	8.9[3-4]	6893			
	115	16	1.42	61 346	35.7[3-4]	1753			
	115	16	1.42	61 346	133[3-4]	461			
Profile K	40	8	0.2	27 115	0.007[1]	3 873 584	−78.31 739 544	5.680 293 764	3615
	40	8	0.2	27 115	0.015[2]	1 807 672			
	40	8	0.2	27 115	1.33[3-4]	20 387			
	40	8	0.2	27 115	8.9[3-4]	3047			
	40	8	0.2	27 115	35.7[3-4]	775			
	40	8	0.2	27 115	133[3-4]	204			
Profile L	215	29	0.4	49 117	0.007[1]	7 016 722	−78.46 865 236	5.605 286	3498
	215	29	0.4	49 117	0.015[2]	3 274 470			
	215	29	0.4	49 117	1.33[3-4]	36 930			
	215	29	0.4	49 117	8.9[3-4]	5519			
	215	29	0.4	49 117	35.7[3-4]	1403			
	215	29	0.4	49 117	133[3-4]	369			

κ values taken from Mitchell (1996)[1], McAdoo and Simpson (2005)[2], Vargas *et al.* (2012)[3-4], and Vargas and Idárraga-Garcia (2014)[3-4].

Table 3. *Some features of the hydrographic basins related to the main rivers that flow out on the Pacific Ocean coast of Colombia*

River	Annual rainfall (mm)	Basin area ($\times 10^3$ km^2)	Water discharge (km^3 a^{-1})	Sediment flux (t km^{-2} a^{-1})
Baudó	6373	5.4	782	–
Mira	5546	4.8	839	1025
Patia	2821	23.7	1291	980
San Juan	7277	16.4	2550	1150

From Restrepo and Lopez (2008).

κ values less than 1.33 m^{2}y. Kogan and Bendick (2011), based on observations in the Hebgen Lake (Montana) and other worldwide data, have suggested that scarps require a finite period of mass failure to reach the initial conditions for hillslope diffusion, so the age of features less than 1 ka old cannot properly be estimated with methods based only on the linear diffusion equation. According to Vargas *et al.* (2012) and Dumont *et al.* (2006), a large event with regional impact in the south of the study area happened around 3 ka BP. If we assume this event as responsible for the generation of some fault scarps, we may identify the best candidate of κ ranging between 1.33 m^{2}y and 35.7 m^{2}y. Unfortunately, there is no information available in the study zone associated with other dating approaches to confirm these assumptions.

Due to the high complexity of the problem in submarine conditions, as well as the high rates of sediment flux, high seismotectonic activity and the lack of information regarding deep oceanographic conditions, the dating approach used is difficult to constrain. In addition, some of the profiles analysed correspond to a segment of contiguous and complicated landscape curvatures that suggest a complex history of events (Fig. 4a & b). We also remark that using the same range of κ (1.33–35.7 m^{2}y), there are contrasting ages regarding sites D, E and I. We hypothesize that although there is a large blanket of sediments covering the study area, there are outcrops of volcanic rocks (Fig. 4b) related to the old spread centre that may distort scarp dating, whose method was originally proposed on sedimentary environments. Regardless, the estimations of this work constitute one of the first attempts to constrain the time of development of submarine landslides associated with the collision and subduction processes of a submarine ridge in the frame of a slab window.

Acknowledgements The authors would like to express their appreciation to the Maritime Authority of Colombia (DIMAR), the Research Center of Oceanography and Hydrography (CIOH), and the Geological Survey of Colombia (SGC) for permitting the use of the data. This paper used data from the Global Multi-Resolution Topography (GMRT) by means of the application Geomapapp. We thank Dr Lorena Moscardelli and an anonymous reviewer for their constructive comments which improved the manuscript. Our gratitude also goes to Dr Joshu Mountjoy, one of the volume editors of *Subaqueous Mass Movements and their Consequences*.

Funding The authors thank the Universidad Nacional de Colombia for the partial funding of this research.

Author contributions **CAV**: conceptualization (lead), formal analysis (lead), investigation (lead), methodology (lead), project administration (lead), software (lead), validation (lead), visualization (lead), writing – original draft (lead), writing – review & editing (lead); **GAG**: data curation (lead), funding acquisition (lead), writing – review & editing (equal); **GAS**: formal analysis (equal), writing – original draft (supporting), writing – review & editing (equal).

References

Abratis, M. and Worner, G. 2001. Ridge collision, slab-window formation, and the flux of Pacific asthenosphere into the Caribbean realm. *Geology*, **29**, 127–130, https://doi.org/10.1130/0091-7613(2001)029<0127:RCSWFA>2.0.CO;2

Aguillón-Robles, A., Calmus, T. *et al.* 2001. Late Miocene adakites and Nb–enriched basalts from Vizcaino Peninsula, Mexico: indicators of East Pacific Rise subduction below southern Baja California? *Geology*, **29**, 531–534, https://doi.org/10.1130/0091-7613(2001)029<0531:LMAANE>2.0.CO;2

Barrero, D., Pardo, A., Vargas, C.A. and Martinez, J.F. 2007. *Colombian Sedimentary Basins*, 5th edn. Agencia Nacional de Hidrocarburos.

Castillo, P.R. 2006. An overview of adakite petrogenesis. *Chinese Science Bulletin*, **51**, 257–268, https://doi.org/10.1007/s11434-006-0257-7

Cediel, F., Shaw, R.P. and Caceres, C. 2003. Tectonic assembly of the northern Andean block, in the circum-Gulf of Mexico and the Caribbean: hydrocarbon habitats, basin formation, and plate tectonics *AAPG Memoir*, **79**, 1–34.

Chadwick, J., Perfit, M., Mcinnes, B., Kamenov, G., Plank, T., Jonasson, I. and Chadwick, C. 2009. Arc lavas on both sides of a trench: slab window effects at the Solomon Islands triple junction, SW Pacific. *Earth and Planetary Science Letters*, **279**, 293–302, https://doi.org/10.1016/j.epsl.2009.01.001

Clare, M. *et al.* 2018. A consistent global approach for the morphometric characterization of subaqueous landslides. *Geological Society, London, Special Publications*, **477**, 455–477, https://doi.org/10.1144/SP477.15

Collot, J.Y., Charvis, P., Gutscher, M.A. and Operto, S. 2002. The Sisteur scientific party, Exploring the Ecuador–Colombia active margin and inter-plate seismogenic zone. *Eos*, **83**, 189–190.

de Boer, J.Z., Defant, M.J., Stewart, R.H., Restrepo, J.F., Clark, L.F. and Ramirez, A.H. 1988. Quaternary calc-alkaline volcanism in western Panama: Regional variation and implication for the plate tectonic framework. *Journal of South American Earth Sciences*, **1**, 275–293, https://doi.org/10.1016/0895-9811(88)90006-5

Dumont, J.F., Santana, E., Valdez, F., Tihay, J.P., Usselmann, P., Iturralde, D. and Navarette, E. 2006. Fan beheading and drainage diversion as evidence of a 3200–2800 BP earthquake event in the Esmeraldas–Tumaco seismic zone: a case study for the effects of great subduction earthquakes. *Geomorphology*, **74**, 100–123, https://doi.org/10.1016/j.geomorph.2005.07.011

Ekström, G., Nettles, M. and Dziewonski, A.M. 2012. The global CMT project 2004–2010: centroid-moment tensors for 13 017 earthquakes. *Physics of the Earth and Planetary Interiors*, **200–201**, 1–9, https://doi.org/10.1016/j.pepi.2012.04.002

Eyuboglu, Y., Chung, S.L., Santosh, M., Dudas, F.O. and Akaryah, E. 2011. Transition from shoshonitic to adakitic magmatism in the eastern Pontides, NE Turkey: implications for slab window melting. *Gondwana Research*, **19**, 413–429, https://doi.org/10.1016/j.gr.2010.07.006

Gómez, J., Montes, N.E., Nivia, Á. and Diederix, H. 2015. *Geologic Map of Colombia – 2015*. Scale 1:1 000 000. Servicio Geológico Colombiano, 2 sheets. Bogotá.

Gutsher, M.A., Malavieille, J., Lallemand, S. and Collot, J.Y. 1999. Tectonic segmentation of the North Andean margin: impact of the Carnegie Ridge collision. *Earth and Planetary Science Letters*, **168**, 255–270, https://doi.org/10.1016/S0012-821X(99)00060-6

Hanks, T.C. 2000. The age of scarplike landforms from diffusion-equation analysis. *In*: Nollet, J.S., Sowers, J.M. and Lettis, W.R. (eds) *Quaternary Geochronology: Methods and Applications*. American Geophysical Union.

Harders, R., Ranero, C.R., Weinrebe, W. and Behrmann, J.H. 2011. Submarine slope failures along the convergent continental margin of the Middle America Trench. *Geochemistry, Geophysics, Geosystems*, **12**, Q05S32, https://doi.org/10.1029/2010GC003401

Harris, P.T. and Whiteway, T. 2011. Global distribution of large submarine canyons: geomorphic differences between active and passive continental margins. *Marine Geology*, **285**, 69–86, https://doi.org/10.1016/j.margeo.2011.05.008

Idarraga-Garcia, J. and Vargas, C.A. 2014. Morphological expression of submarine landslides in the accretionary prism of the Caribbean continental margin of Colombia. *In*: Krastel, S. *et al.* (eds) *Submarine Mass Movements and their Consequences*. Advances in Natural and Technological Hazards Research, **37**. Springer, Cham, https://doi.org/10.1007/978-3-319-00972-8_35

Idárraga-García, J., Masson, D.G., García, J., Leon, H. and Vargas, C.A. 2019. Architecture and development of the Magdalena Submarine Fan (southwestern Caribbean). *Marine Geology*, **414**, 18–33, https://doi.org/10.1016/j.margeo.2019.05.005

Kawamura, K., Laberg, J.S. and Kanamatsu, T. 2014. Potential tsunamigenic submarine landslides in active margins. *Marine Geology*, **356**, 44–49, https://doi.org/10.1016/j.margeo.2014.03.007

Kogan, L. and Bendick, R. 2011. A Mass Failure Model for the Initial Degradation of Fault Scarps, with Application to the 1959 Scarps at Hebgen Lake, Montana. *Bulletin of the Seismological Society of America*, **101**, 68–78, https://doi.org/10.1785/0120100107

Lonsdale, P. 2005. Creation of the Cocos and Nazca plates by fission of the Farallon plate. *Tectonophysics*, **404**, 237–264, https://doi.org/10.1016/j.tecto.2005.05.011

Masson, D.G., Wynn, R.B. *et al.* (eds) *Submarine Mass Movements and Their Consequences*. Advances in Natural and Technological Hazards Research, **28**. Springer Science + Business Media.

McAdoo, B and Simpson, G. 2005. Morphometric dating of submarine landslide scarps. *Geophysical Research Abstracts*, **7**, EGU05-A-00629.

Mitchell, N.C. 1996. Creep in pelagic sediments and potential for morphologic dating of marine fault scarps. *Geophysical Research Letters*, **23**, 483–486, https://doi.org/10.1029/96GL00421

Pedraza, P., Vargas, C.A. and Monsalve, H. 2007. Geometric model of the Nazca Plate subduction in Southwest Colombia. *Earth Sciences Research Journal*, **11**, 117–130.

Restrepo, J.D. and Kjerfve, B. 2003. The Pacific and Caribbean rivers of Colombia: water discharge, sediment transport and dissolved loads *In*: Drude de Lacerda, L., Santelli, R.E., Duursma, E.K. and Abrão, J.J. (eds) *Environmental Geochemistry in Tropical and Subtropical Environments*. Springer, https://doi.org/10.1007/978-3-662-07060-4_14

Restrepo, J.D. and Lopez, S. 2008. Morphodynamics of the Pacific and Caribbean deltas of Colombia, South America. *Journal of South American Earth Sciences*, **25**, 1–21, https://doi.org/10.1016/j.jsames.2007.09.002

Restrepo, J.D., Kjerfve, B., Correa, I.D. and Gonzalez, J.L. 2002. Morphodynamics of a high discharge tropical delta, San Juan River, Pacific coast of Colombia. *Marine Geology*, **192**, 355–381, https://doi.org/10.1016/S0025-3227(02)00579-0

Ryan, W.B.F., Carbotte, S.M. *et al.* 2009. Global Multi-Resolution Topography (GMRT) synthesis data set. *Geochemistry, Geophysics, Geosystems*, **10**, Q03014, https://doi.org/10.1029/2008GC002332

Salazar, J.M. and Vargas, C.A. 2015. Fractal dimension and seismotectonic deformation rates along an inter-plate setting: seismic regime along the Caribbean plate boundary zone. *AAPG Memoir*, **108**, https://doi.org/10.1306/13531940M1083644

Salazar, J.M., Vargas, C.A. and Leon, H. 2017. Curie point depth in the SW Caribbean using the radially averaged spectra of magnetic anomalies. *Tectonophysics*, **694**, 400–413, https://doi.org/10.1016/j.tecto.2016.11.023

Taboada, A., Rivera, L.A., Fuenzalida, A., Cisternas, A., Philip, H., Bijwaard, H. and Olaya, J. 2000. Geodynamics of northern Andes: subduction and intra–continental deformation (Colombia). *Tectonics*, **19**, 787–813, https://doi.org/10.1029/2000TC900004

Thorkelson, D.J. and Breitsprecher, K. 2005. Partial melting of slab window margins: genesis of adakitic and non-adakitic magmas. *Lithos*, **79**, 25–41, https://doi.org/10.1016/j.lithos.2004.04.049

Vargas, C.A. 2012. Evaluating total yet-to-find hydrocarbon volume in Colombia. *Earth Science Research Journal*, **16**, 1–246.

Vargas, C.A. 2019. Subduction geometries in northwestern South America. *In*: Gómez, J. and Pinilla–Pachon, A.O. (eds) *The Geology of Colombia, Volume 4 Quaternary*. Servicio Geológico Colombiano, Publicaciones Geológicas Especiales, **38**, 445–495. Bogotá, https://doi.org/10.32685/pub.esp.38.2019.11

Vargas, C.A. and Idárraga-Garcia, J. 2014. Age estimation of submarine landslides in the La Aguja Submarine Canyon, Northwestern Colombia. *In*: Krastel, S. *et al.* (eds) *Submarine Mass Movements and their Consequences*. Advances in Natural and Technological Hazards Research, **37**, 629–638. Springer, Cham, https://doi.org/10.1007/978-3-319-00972-8_56

Vargas, C.A. and Mann, P. 2013. Tearing and breaking off of subducted slabs as the result of collision of the Panama Arc-Indenter with northwestern South America. *Bulletin of the Seismological Society of America*, **103**, 2025–2046, https://doi.org/10.1785/0120120328

Vargas, C.A., Pujades, L.G. and Montes, L. 2007. Seismic structure of South-Central Andes of Colombia by tomographic inversion. *Geofísica Internacional*, **46**, 117–127.

Vargas, C.A., Mann, P. and Gomez, C. 2012. Morphologic expression of accretionary processes and recent submarine landslides along the southwestern Pacific margin of Colombia. *In*: Yamada, Y. *et al.* (eds) *Submarine Mass Movements and their Consequences*. Advances in Natural and Technological Hazards Research, **31**, 365–377. Springer, Dordrecht, https://doi.org/10.1007/978-94-007-2162-3_33

Vargas, C.A., Pulido, J.E. and Hobbs, R.W. 2018. Thermal structure of the Panama Basin by analysis of seismic attenuation. *Tectonophysics*, **730**, 81–99, https://doi.org/10.1016/j.tecto.2018.02.017

Witt, C., Bourgois, J., Michaud, F., Ordonez, M., Jimenez, N. and Sosson, M. 2006. Development of the Gulf of Guayaquil (Ecuador) during the Quaternary as an effect of the North Andean block tectonic escape. *Tectonics*, **25**, TC3017, https://doi.org/10.1029/2004TC001723

Mass wasting on Alpha Ridge in the Arctic Ocean: new insights from multibeam bathymetry and sub-bottom profiler data

Kai Boggild[1]*, David C. Mosher[1], Paola Travaglini[2], Catalina Gebhardt[3] and Larry Mayer[4]

[1]Geological Survey of Canada, Dartmouth, NS, Canada

[2]Canadian Hydrographic Service, Dartmouth, NS, Canada

[3]Alfred Wegener Institute Helmholtz-Center for Polar and Marine Research, Bremerhaven, Germany

[4]Center for Coastal and Ocean Mapping/Joint Hydrographic Center, University of New Hampshire, Durham, NH, USA

KB, 0000-0003-0090-5400; CG, 0000-0002-3227-0676
*Correspondence: kai.boggild@canada.ca

Abstract: Marine geological and geophysical data from Alpha Ridge in the Arctic Ocean are sparse because of thick perennial sea-ice cover, which prevents access by most surface vessels. Rare seismic data in this area, acquired largely from drifting ice-camps, had shown the hemipelagic drape that covers most of the ridge is highly disrupted within a large (>90 000 km^2) south central region. Here, evidence of pronounced seafloor erosion and debris flows infilling seafloor lows was previously interpreted to be the result of a possible bolide impact. In recent years, several icebreaker expeditions have successfully acquired multibeam bathymetry and sub-bottom profiler data in the western segment of this region. Analysis of these data reveals a complex seafloor morphology characterized by ridges and troughs, angular blocks and escarpments as well as seismic facies characterized by hyperbolic seafloor reflections, and convoluted to incoherent and transparent sub-bottom reflectivity. These features are interpreted as evidence of sediment mass movement with varying degrees of lateral transport deformation. At least two episodes of failure are interpreted based on the presence of both buried and surficial mass-transport features. As multiple events are interpreted, seismicity is the most plausible trigger mechanism rather than bolide impact.

Alpha Ridge extends from the Canadian Arctic Archipelago landmass into the Amerasia Basin in the central Arctic Ocean between *c.* 90° and 150° W (Fig. 1). In this part of the Arctic, the wind-driven surface currents of the Beaufort Gyre and the Transpolar Drift converge to drive the Arctic icepack against the Canadian margin and northern Greenland, forming accumulations of the thickest sea-ice in the central Arctic Ocean (Rigor *et al.* 2002; Kwok 2018). This thick perennial sea-ice cover limits access by even the most capable ice-breakers and presents significant challenges to acquisition of geological and geophysical data (see Hutchinson *et al.* 2009; Armstrong *et al.* 2012; Mosher *et al.* 2013). As a result, much of Alpha Ridge remains relatively uncharted (see Fig. 1). In these regions, regional bathymetry is mostly constrained by single-beam measurements made from drifting ice-camps, submarines and widely spaced spot-soundings made from the sea-ice surface during aircraft-supported hydrographic surveys (Jakobsson *et al.* 2012). Rare seismic reflection data exist in these areas, most of which were acquired from ice-camps.

Previous studies using seismic reflection data from the crest of Alpha Ridge have shown a large region (>90 000 km^2) characterized by extensive erosion and mass wasting of the uppermost sedimentary cover (see Fig. 1; Kristoffersen *et al.* 2009). In this region, seismic data show truncated and disrupted reflections along seafloor highs and deposition of seismically chaotic sediment in intervening lows. The style of deformation and lack of modern seismicity were interpreted to suggest that this vast disturbed region was the result of a bolide impact (Kristoffersen *et al.* 2009). Beyond observations from 2D seismic and single-beam echosounder profiles, however, little is known about the high-resolution seabed morphology or near-surface stratigraphy in this region.

From: Georgiopoulou, A., Amy, L. A., Benetti, S., Chaytor, J. D., Clare, M. A., Gamboa, D., Haughton, P. D. W., Moernaut, J. and Mountjoy, J. J. (eds) 2020. *Subaqueous Mass Movements and their Consequences: Advances in Process Understanding, Monitoring and Hazard Assessments*. Geological Society, London, Special Publications, **500**, 323–340.
First published online May 13, 2020, https://doi.org/10.1144/SP500-2019-196

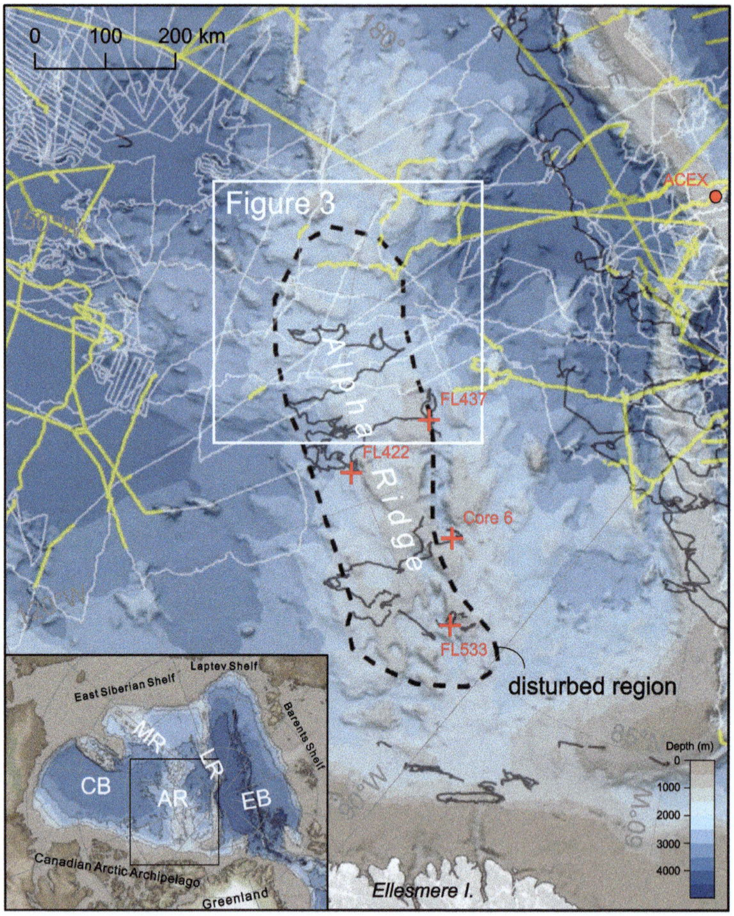

Fig. 1. Bathymetry of central Arctic Ocean (Jakobsson *et al.* 2012), showing the location of the Alpha Ridge (AR) and surrounding areas (inset: LR, Lomonosov Ridge; MR, Mendeleev Rise; EB, Eurasian Basin; CB, Canada Basin). The thick dashed line indicates the extent of disturbed seabed identified in Kristoffersen *et al.* (2009). Seismic reflection data from drifting ice-stations are shown as thin black lines. Thin white lines show the tracklines of icebreaker cruises with coincident sub-bottom profiler and multibeam bathymetry, thick yellow lines represent multichannel seismic reflection data. Red markers indicate core samples that recovered pre-Neogene-aged sediments.

In the past decade, several joint icebreaker expeditions to the high Arctic have successfully acquired multibeam bathymetry and sub-bottom profiler data over the western Alpha Ridge, including newly acquired data in the western segment of the area described by Kristoffersen *et al.* (2009). These high-resolution data demonstrate several complex sedimentary features at the seafloor that are hypothesized to be a result of conventional sediment mass failure. In this study we assess these features to (1) determine if these features are a result of mass movement; (2) characterize the styles of deformation that have occurred in this region; and (3) evaluate if they could be a result of a bolide impact.

Background and setting

Alpha Ridge is an expansive submarine high measuring more than 450 000 km^2 in area and reaching from abyssal depths in excess of 4000 m to less than 1400 m at its shallowest portions. The ridge is thought to have formed as a large igneous province between about 130 and 83 Ma (Døssing *et al.* 2013; Oakey and Saltus 2016; Petrov *et al.* 2016). Its surface morphology is rugose at regional (e.g. 100 km) scales, thought to be largely related to volcanism that formed its shallowest basement rocks (Jackson *et al.* 1985; Jokat *et al.* 2013). Sediment accumulation over Alpha Ridge is largely a consequence of hemipelagic and pelagic processes, beginning in the early Cenozoic, that were extremely slow (Mudie *et al.* 1986). These processes have blanketed the rugged basement of the ridge with *c.* 500 to more than 1200 m of sedimentary cover (Bruvoll *et al.* 2010; Evangelatos and Mosher 2016; Evangelatos *et al.* 2017). Throughout the Quaternary, perennial ice cover conditions over the ridge have resulted in sedimentation rates that are the slowest in the central Arctic Ocean (Polyak *et al.* 2009; Levitan 2015).

The earliest scientific expeditions to the Alpha Ridge were accomplished from ice-camps established on drifting ice-floes in the Arctic pack ice (see Weber and Roots 1990). Among them was the T3 Ice Island, from which science operations were conducted in the Arctic Ocean between 1962 and 1974. Nearly 1700 km of single-channel seismic reflection data and hundreds of short sediment cores were acquired along a clockwise drift-path that circled the Amerasian Basin and included Alpha Ridge (Hunkins and Tiemann 1977; Lachenbruch *et al.* 2019).

Hall (1970, 1979) first reported on seismic data from T3 and demonstrated that a vast area of seafloor along the crest of central Alpha Ridge was characterized by overlapping hyperbolic seafloor reflections. These hyperbolae measured *c.* 1 km in wavelength and up to 55 m in amplitude and were initially interpreted as sediment waves on the seafloor relict from a period of enhanced bottom circulation in the basin. In this region, hyperbolic seafloor reflections dominate the upper 100 m of sediment, beneath which up to 800 m of stratified sediment bury the underlying basement. Despite the thick sediment cover, three short (<5 m) sediment cores recovered from graben slopes in the area contained material dating to the Upper Cretaceous and early Cenozoic (Clark 1974; Clark *et al.* 1986). Additional Upper Cretaceous sediments were again recovered from the area during the Canadian Expedition to Study Alpha Ridge (CESAR) in 1983, which sampled nearly 2 m of biosiliceous ooze containing Maastrichtian–Danian microfossil assemblages at the base of ‘Core 6’ (Mudie and Blasco 1985). The recovery of these ages at such shallow depths suggests that younger sediment had been removed from these sample locations, possibly as a result of mass wasting (Clark 1974; Mudie and Blasco 1985). Except for drill cores recovered from neighbouring Lomonosov Ridge during the Arctic Coring Expedition (ACEX; Moran *et al.* 2006; Backman *et al.* 2008), these four samples on Alpha Ridge contain the oldest sediments recovered from the central Arctic Ocean and provide rare constraints concerning the pre-Neogene palaeoenvironment of the central Arctic Ocean (e.g. Mudie *et al.* 1986; Clark 1988; Davies *et al.* 2009).

Research icebreakers were unable to access Alpha Ridge until relatively recently. The 1998 PS51 and 2005 HOTRAX expeditions were the first to collect multibeam bathymetry and multichannel seismic reflection data over the ridge (Jokat 1999; Darby *et al.* 2005). Seismic data from these expeditions confirmed that much of the rugged basement morphology of Alpha Ridge and adjacent Mendeleev Rise is buried by a thick (*c.* 500 to >1200 m) drape of hemipelagic sediment (Jokat 2003; Bruvoll *et al.* 2010, 2012; Dove *et al.* 2010).

Over an area of the central Alpha Ridge, the coherent stratigraphy of the surrounding sedimentary cover is disrupted (see Fig. 1 for location). Seismic data in this region showed truncated sub-bottom reflections and acoustically incoherent deposits overlying stratified sediments (Kristoffersen *et al.* 2009; Bruvoll *et al.* 2010). Coincident multibeam bathymetry data in these areas also revealed several large sub-parallel ridges at the seafloor, which were interpreted as further evidence of giant sediment waves (Darby *et al.* 2005; Polyak and Jakobsson 2011). Using these results along with existing T3 seismic data, Kristoffersen *et al.* (2009) showed that the vast area of rough seabed reported by Hall (1970, 1979) was characterized by evidence of extensive seabed erosion and sediment mass transport. Owing to the depth-limited style of deformation of these features and apparent lack of seismicity in the modern observational record, Kristoffersen *et al.* (2009) proposed that this region may have been the site of a recent bolide impact. Given that no sediment cores recovered from the disturbed region contained evidence of failed sediments (Hall 1970; Kristoffersen *et al.* 2009), a tentative Plio-Pleistocene age was suggested for this impact event. Subsequent erosion from such an event may have exhumed older sediment in locations where pre-Neogene sediments were recovered by short cores at the seabed (Kristoffersen *et al.* 2009).

Methods and data

Data presented in this study are compiled from multidisciplinary icebreaker expeditions undertaken in the central Arctic Ocean between 1998 and 2016 involving Canada, the USA, Germany and Sweden (Table 1). These expeditions were completed aboard the research icebreakers CCGS *Louis S. St-Laurent* (*LSSL*), USCGC *Healy*, FS *Polarstern* and IB *Oden*. In this study, multibeam bathymetry and sub-bottom profiler data are used in combination to interpret the high-resolution geomorphology of the seabed and shallow subsurface. Sub-bottom profiles also provide first-order observations of seafloor morphology and roughness where multibeam bathymetry data cannot be used due to noise or gaps in coverage.

Table 1. *Data used in this study*

Platform	Sub-bottom profiler (SBP)	SBP frequency (kHz)	SBP beam-width (°)	Multibeam echosounder (MBES)	MBES frequency (kHz)	Expedition
IB *Oden*	Kongsberg SBP120	2.5–7 (chirp)	3	Kongsberg EM122	12	ODEN2016[1]
CCGS *Louis S. St-Laurent*	Knudsen 3260	2.3–5.3 (chirp)	*c.* 26–30	Kongsberg EM122	12	LSSL2015 LSSL2016[2]
USCGC *Healy*	Knudsen 320 B/R	3–6 (chirp)	26	Kongsberg EM122 (2011, 2015) Seabeam 2112 (2005)	12	HLY0503[3] HLY1102[4] HLY1502[5]
FS *Polarstern*	Atlas Parasound (DS-2, DS-3-P70)	4 (parametric)	4	Atlas HydroSweep (DS-2, DS-3)	15.5	PS51 (1998)[6] PS70 (2007)[7] PS78 (2011)[8]

[1]Gårdfeldt and Lindgren (2017); [2]Shimeld and Boggild (2017); [3]Coakley *et al.* (2005); [4]Mayer and Armstrong (2011); [5]Kadko and Landing (2015); [6]Jokat (1999); [7]Schauer (2008); [8]Schauer (2012).

Multibeam bathymetry

Multibeam bathymetry data from *LSSL*, *Oden* and *Healy* (since 2011) were collected using Kongsberg EM122 12 kHz echosounders. *Healy* multibeam bathymetry data from the 2005 HOTRAX survey were collected with a Seabeam EM2112 12 kHz echosounder. Bathymetry from *Polarstern* surveys were collected with an Atlas Hydrographic Hydro-Sweep 15.5 kHz echosounder (DS-2, DS-3). Data were both automatically and manually edited to remove potential errant soundings. Data from *LSSL*, *Oden* and *Healy* were gridded at a 25 m spacing, *Polarstern* data were analysed as a 100 m grid. ArcMap 10.2.2 was used to calculate seafloor gradients and vector ruggedness measure (VRM) from gridded multibeam data. VRM is a method of calculating surface ruggedness within a specified neighbourhood that captures variability of slope and aspect into a single measure (see Sappington *et al.* 2007).

Sub-bottom profiler data

Sub-bottom profiler data were acquired using a variety of hull-mounted echosounders. *LSSL* data were collected using a Knudsen 3260 chirp echosounder (2.3–5.3 kHz), *Healy* data were collected using a Knudsen 320B/R chirp echosounder (3–6 kHz), and *Oden* data were collected using a Kongsberg SBP120 chirp echosounder (2.5–7 kHz). *Polarstern* used an Atlas Hydrographic Deep Sea Parasound parametric echosounder to collect sub-bottom data (DS-2 during 1998; DS-3-P70 in 2007, 2011). Parasound sub-bottom profiles were collected using a frequency of 4 kHz formed through non-linear interference of 18 and 22 kHz primary frequencies. This parametric interference technique produces a narrow beam composite source wavelet. In general, penetration of sub-bottom profiler data included in this study does not exceed 100 m below the seafloor. Vertical resolution of these data is determined by the bandwidth of chirp systems (e.g. *LSSL*, *Healy*, *Oden*) and the secondary frequency of parametric systems (e.g. *Polarstern*). Vertical resolution of data compiled in this study is therefore between *c.* 30 and 150 cm.

At the water depths along central Alpha Ridge (i.e. >1800 m), differences in beam angle between sub-bottom profilers influence their ability to resolve the stratigraphy of near-seafloor sediments since much of the seafloor roughness in this area is within the beam pattern of the sub-bottom profiler. The minimum beam-width of the 16-element TR-109 transducer array installed on the *Healy* for example, is approximately 26°, translating to a beam pattern footprint of *c.* 0.72 km^2 at a water depth of 2000 m. The comparatively narrow 3° beam-width of the SBP120 system installed on *Oden* translates to a footprint of only *c.* 0.035 km^2 under the same conditions (a reduction of 95%), and thus this system is better able to resolve the horizontal detail where the seabed is roughest. The narrow 4° beam-width on the Parasound system aboard *Polarstern* also results in better resolution of the near-seabed stratigraphy.

Influence of sea-ice on acoustic data

In the heavy icepack of the central Arctic, icebreaking results in erratic ship motion and introduces substantial noise to hull-mounted echosounders. This noise often overpowers the reflected signal from the seabed. As a result, swath bathymetry and sub-bottom profiler data collected in these conditions

	Facies	Description
smooth seafloor	1	Smooth seafloor echo with coherent parallel subbottom reflections that drape the seafloor morphology
smooth seafloor	1/t	Smooth seafloor echo with coherent parallel subbottom reflections overlying (or interbedded with) acoustically transparent lens-shaped sediment
smooth seafloor	1/4	Smooth seafloor echo with coherent parallel subbottom reflections comformably overlying high-amplitude basal reflection
rugose seafloor	2	Rough seafloor echo (hyperbolae to irregular) with semi-coherent subbottom reflections in the uppermost 20 - 50 ms that maintain a consistent amplitude profile laterally
rugose seafloor	2/t	Rough seafloor echo (hyperbolae to irregular) with semi-coherent subbottom reflections that maintain a consistent amplitude profile, overlying acoustically transparent, lens-shaped deposits
rugose seafloor	3	Narrow overlapping hyperbolae (< ~ 1 km wide) on flat or gently-sloping seafloor. No apparent subbottom reflections.
rugose seafloor	3/4	Narrow overlapping hyperbolae (< ~ 1 km wide) or irregular seafloor echo, with acoustically transparent subbottom overlying smooth or irregular basal reflection
rugose seafloor	3/1	Narrow overlapping hyperbolae or irregular seafloor echo, with acoustically transparent subbottom overlying basal reflections with some stratification
steep / irregular	4	Irregular seafloor echo with absent or few incoherent subbottom reflections. May have isolated out of plane hyperbolae.
steep / irregular	5	Varying amplitude, broad (1-2 km width) or one-sided hyperbolic bottom echo with no subbottom reflections, or loss of signal occuring on steep bathymetric gradients.

Examples*

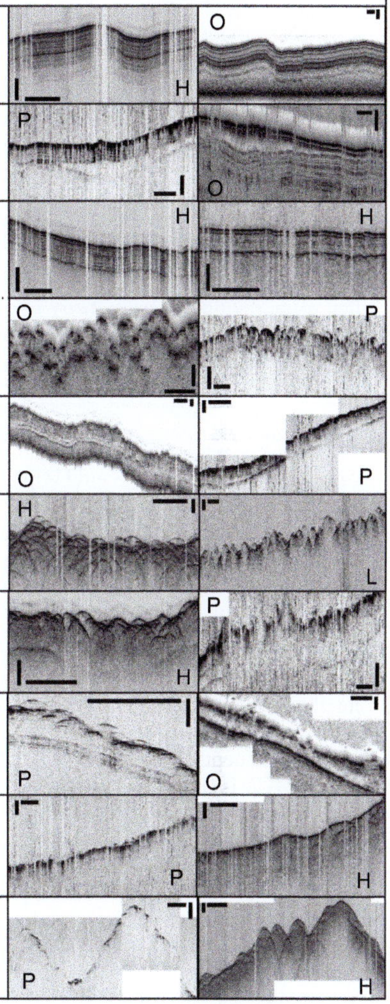

horizontal bar = 1 km, vertical bar = 40 ms, P: Polarstern, H: Healy, O: Oden, L: LSSL

Fig. 2. Sub-bottom profiler acoustic facies for the central Alpha Ridge.

suffer in quality. In most cases, surveys included in this study involved two icebreakers that operated simultaneously; one would act as a lead icebreaker while the second vessel would follow and acquire data in the track (see Armstrong *et al.* 2012). This configuration mitigates the effects of icebreaking on the trailing vessel; however, data quality and sounding density still vary significantly in these conditions. In general, swath widths of multibeam bathymetry data were limited to *c.* 3× water depth.

Several processing steps were taken to improve the quality of sub-bottom profiler data recorded in heavy ice conditions. These include: (1) correction for erratic ship motion by trimming excess data and geographical projection of data to a smoothed track; (2) speed correction to create a uniform aspect ratio in all records; and (3) improvement in the signal-to-noise ratio by amplitude clipping, gain normalization and band-pass filtering. Sub-bottom data from *Oden* have a 40 ms automatic gain control (AGC) applied. Where amplitude spikes were obvious, a horizontal despiking filter was also applied. Data were converted from SEG-Y to JPEG-2000 format (Courtney 2007) and displayed as trace envelope for interpretation. This format conversion uses wavelet compression which further removes white noise to improve the signal-to-noise ratio.

Results

The seabed morphology of Alpha Ridge is rugose at regional length scales (e.g. >100 km). At smaller length scales (e.g. between 1 and 5 km), compiled multibeam bathymetry and sub-bottom profiler data demonstrate features ranging from low rugosity seabed with stratified sub-bottom reflections, to high-rugosity seabed with acoustically transparent sub-bottom. To determine the extent of disturbed seabed, acoustic facies derived from sub-bottom profiler data are mapped. Following this, morphological analysis of features from the disturbed region is presented using multibeam bathymetry to describe the seafloor morphology and sub-bottom profiles to constrain the near-surface stratigraphy.

Acoustic facies

Acoustic facies relevant to central Alpha Ridge are shown in Figure 2. Facies categories are based on the character of the seafloor echo and sub-bottom reflections. A description of these facies is as follows:

- Facies 1 (F1) is characterized by a smooth seafloor with coherent parallel sub-bottom reflections. Variations of this facies are identified where it is interbedded with acoustically transparent units (F1/T) or conformably overlies a high-amplitude reflection at depth (F1/4).
- Facies 2 (F2) is characterized by a rough or hyperbolic seafloor echo with semi-coherent sub-bottom reflections in the near-surface (uppermost 20–50 ms). Facies 2/t (F2/T) has the same facies as F2, overlying acoustically transparent lens-shaped deposits.
- Facies 3 (F3) is characterized by a rough seafloor reflection consisting of narrow, overlapping hyperbolae and no sub-bottom reflections (or acoustically transparent). Variations of this facies are identified where it overlies a coherent basal reflection (F3/4) or stratified sub-bottom reflections (F3/1).
- Facies 4 (F4) is characterized by an irregular seafloor echo and few incoherent sub-bottom reflections.
- Facies 5 (F5) consists of broad hyperbolic seafloor echoes resulting from rapid changes in seafloor elevation along steep slopes.

The distribution of acoustic facies mapped on Alpha Ridge (Fig. 3) is grouped into two distinct regions: (1) an 'undisturbed region', predominantly characterized by facies with smooth seafloor echoes and stratified sub-bottom reflections (F1, F1/T); and (2) a 'disturbed region', consisting of mixed facies assemblages of varying seafloor rugosity and internal stratification (F2, F3, F1/4).

The undisturbed region comprises the periphery of central Alpha Ridge. Localized acoustically transparent lens-shaped deposits (F1/T) are observed interbedded with stratified sediments throughout this region; however, they are thin (<20 m thick) and have minimal seafloor expression. The disturbed region comprises the central Alpha Ridge region between water depths of 3000 and 1200 m. Local seafloor highs in the area are generally characterized by narrow hyperbolae (F3) or acoustically transparent sub-bottom reflection characteristics (F3/1 and F3/4). Valleys in this region are typically characterized by narrow hyperbolic seafloor echoes, with varying degrees of internal stratification (F2 and F3). Where seabed roughness is high, data acquired by broad-footprint sub-bottom profilers (e.g. *LSSL*, *Healy*) are unable to resolve the near-seabed stratigraphy (e.g. F2) so the ability to recognize the facies transition between F2 and F3 within the disturbed region is partially limited by horizontal resolution of the sub-bottom profiler system. F2 facies polygons were only mapped where there is interpreted to be evidence of stratification.

Morphological domains

While the coverage of acoustic data on central Alpha Ridge is far from complete, features identified from individual multibeam swaths and coincident sub-bottom profiles in this region are grouped into at least three broad morphological domains on length scales between 1 and 5 km: (1) **low rugosity** domains characterized by smooth seafloor and stratified sub-bottom; (2) **intermediate rugosity** domains characterized by hummocky and chaotic seafloor; and (1) **high rugosity** domains consisting of angular escarpments, ridges and blocks.

Low rugosity. Low rugosity domains are characterized by smooth seafloor and are typically underpinned by facies F1 and F1/T. This morphological domain is dominant for gentle slopes in the periphery of the study area (e.g. 'undisturbed region').

High rugosity. High rugosity domains are typically observed along several seafloor highs within the disturbed region. Characteristic features of this domain include angular escarpments, ridges and troughs with high-gradient flanks, and isolated blocks. Escarpments are up to 100 m in height and in some regions truncate smooth seafloor sediments. Figure 4 shows examples of two seafloor highs with angular escarpments at their crest. Downslope of these features several isolated blocks are visible in multibeam bathymetry. Sub-bottom profiles over these features reveal acoustically transparent facies at their crests overlying stratified sub-bottom reflections (F3/1).

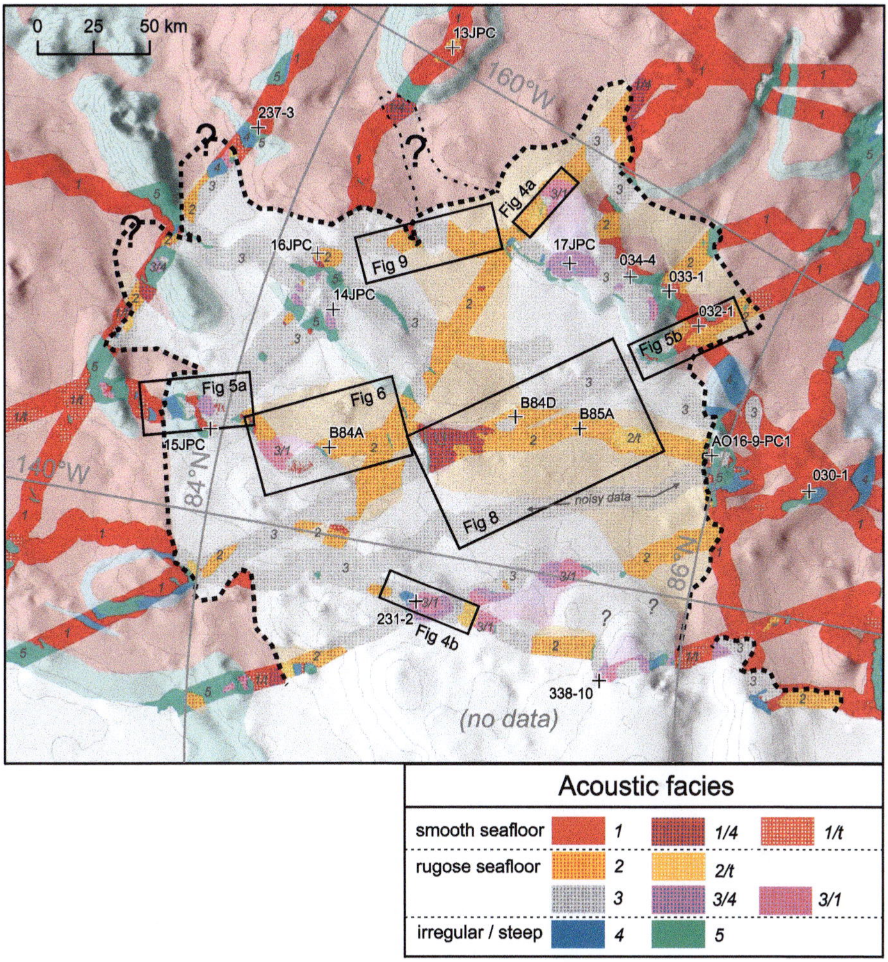

Fig. 3. Map of acoustic facies interpreted from sub-bottom profiler data from central Alpha Ridge (see Fig. 1 for location and Fig. 2 for descriptions). The hill-shaded basemap and contours are from Jakobsson *et al.* (2012). Areas within 5 km of sub-bottom profiler data are shown in bright colours to indicate positions of data control. Map polygons interpolated between data control are shown in muted colours to give regional context. The envelope of rugose seafloor features (i.e. disturbed region) is indicated by a dashed black line. Positions of selected short sediment cores are indicated by white crosses.

In some areas, ridges and troughs occur immediately downslope of amphitheatre-shaped escarpments (Fig. 5). These ridges have heights up to 50 m and widths between 200 and 500 m. Angular ridges and troughs are also observed along a near-level crest of a bathymetric high from the centre of the disturbed region (Fig. 6). These features are not linked to any visible escarpment and have blocky cross-sectional morphology with apparent flank gradients in excess of 30°. Ridges in this instance are 250 to 1000 m wide and are separated by NE-trending troughs that are 25 to 50 m deep. Sub-bottom profiler data demonstrate the ridges consist of coherently stratified sediment (F1) (Fig. 7). Immediately south of this region, the sloped seafloor (average gradient of *c.* 0.8°) is characterized by linearly orientated arrays of narrow blocks (<300 m wide). Sub-bottom profiler data over this terrain show acoustically transparent sub-bottom echo returns overlying a stratified basal reflection (F3/1) that is locally grooved.

Intermediate rugosity. Regions of intermediate rugosity are typically observed in valleys and low areas within the disturbed region. Seafloor morphology in these areas ranges from nearly smooth seafloor to hummocky terrain that includes subdued blocks and ridges (e.g. Figs 6, 8, 9). Sub-bottom profiler

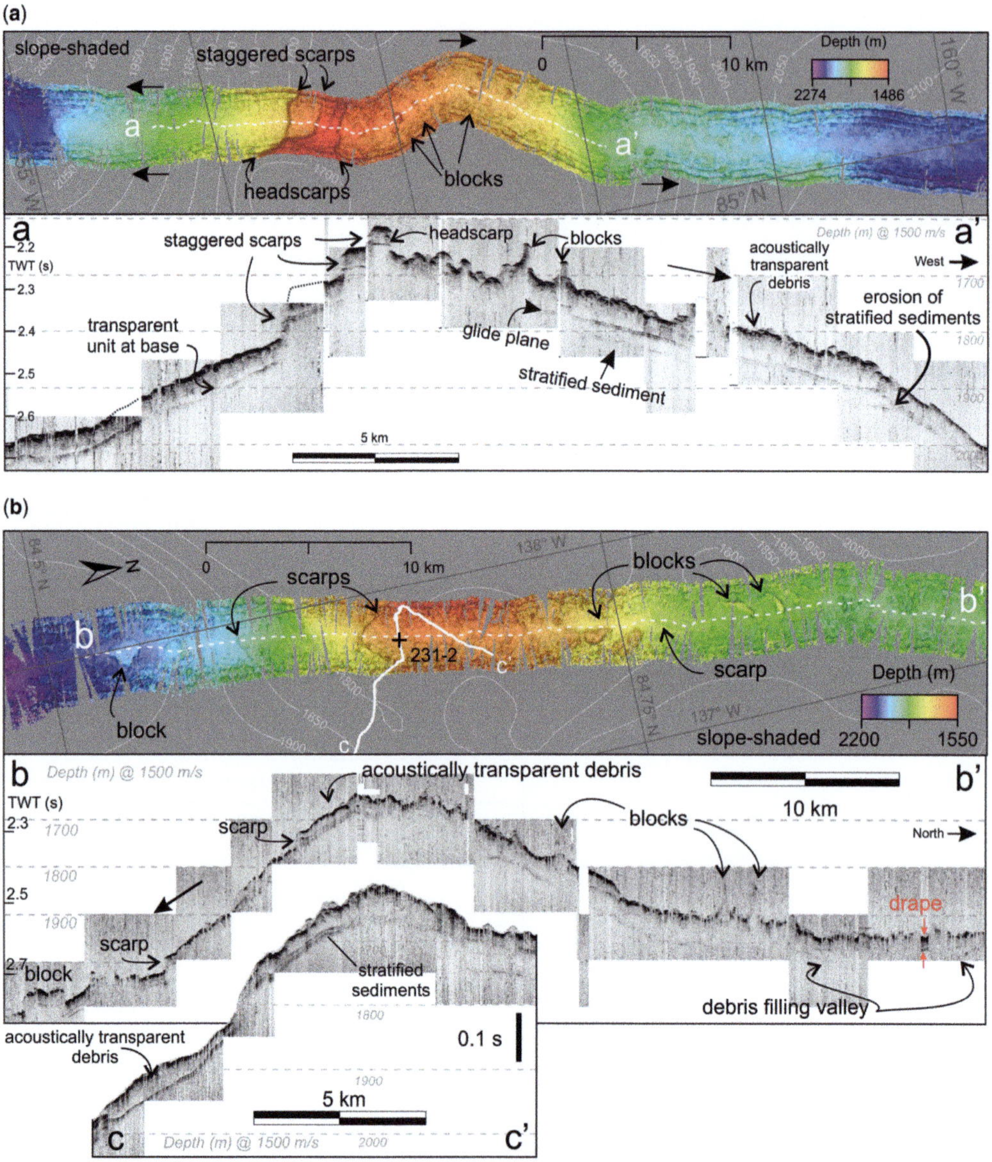

Fig. 4. High-rugosity features from local seafloor highs in the disturbed region (see Fig. 3 for locations). (**a**) Multibeam bathymetry (100 m grid) showing opposite-facing scarps along the crest of a local seafloor high. Sub-bottom profiler data (a–a′) from the crest demonstrate acoustically transparent surficial units overlying stratified sediment. Truncation (erosion) of underlying stratified sediments is clear on the western edge of this feature. (**b**) Slope-shaded bathymetry and sub-bottom profiles from southeastern Alpha Ridge showing similar seafloor features.

data in these areas typically show evidence of a rugose seafloor echo with semi-coherent stratification (F2) indicative of a surficial sedimentary drape unit.

Figure 8 shows an example of intermediate rugosity features from the centre of the disturbed region, where rugosity ranges from smooth seabed in the south to hummocky seafloor in the north. Hummocky morphology of the seabed in this region consists of convoluted ridges and narrow blocks (<300 m wide) with subdued rugosity. Seafloor ridges observed in this region are 300–1000 m wide and 25–50 m high, with gradients of *c.* 20°

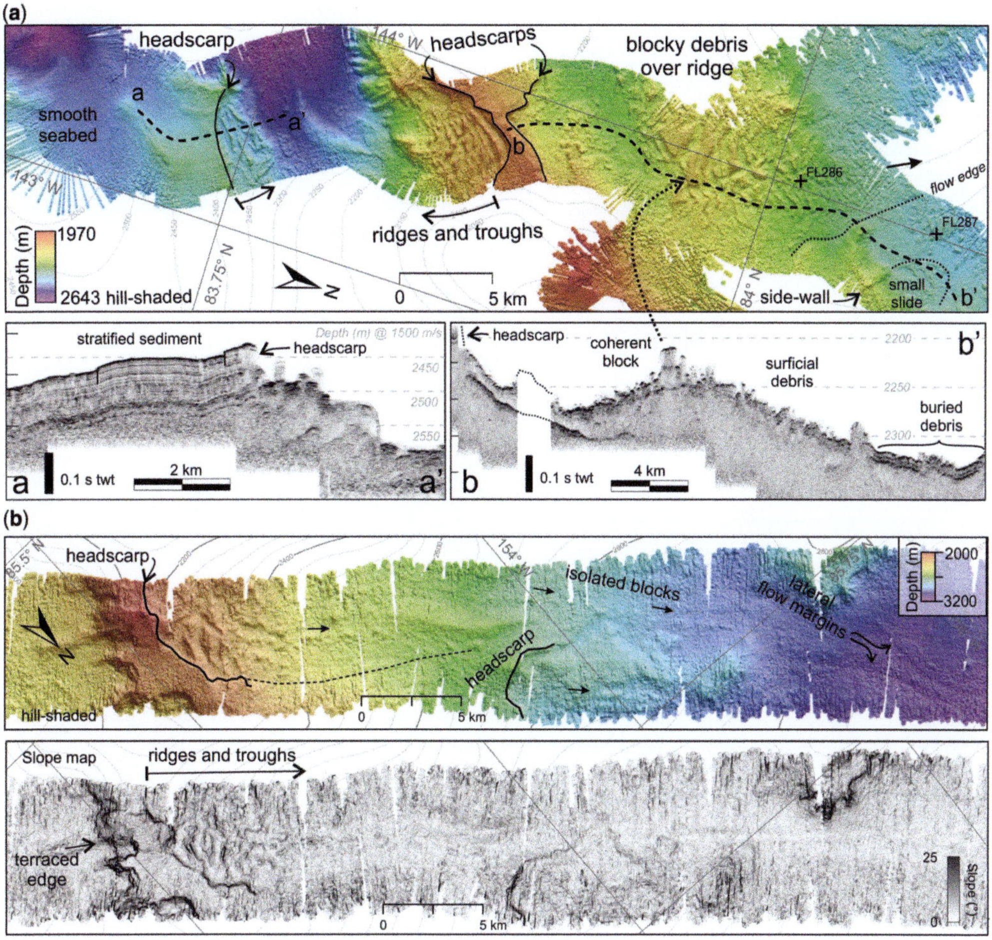

Fig. 5. Multibeam bathymetry (gridded at 25 m) and sub-bottom profiler data from Alpha Ridge showing angular escarpments and associated ridges and troughs from (**a**) the southern edge and (**b**) the northern edge of the disturbed region (see Fig. 3 for locations). The positions of short sediment cores are indicated by crosses.

along their flanks. Rare isolated large angular blocks (up to 2 km wide) are also apparent. Sub-bottom profiles show the seafloor in this region is blanketed by 30–40 ms of coherent sub-bottom reflections (F2, F1/4). This facies results in a muted seafloor rugosity. Several stacked acoustically transparent deposits are also imaged beneath this surficial unit (F2/T).

Intermediate rugosity features are observed in valleys where seismic data from the HOTRAX expedition show evidence of near-surface mass-transport deposits (Fig. 9). Chaotically orientated ridges in the north end of this valley range up to 50 m in height. In the southern end of the valley, undulating ridges orientated towards the NE ridges are 500–750 m wide and 10–25 m in height, with flanks that slope up to *c.* 12°. Longitudinal troughs in bathymetry trend orthogonally to the ridges. A sub-bottom profile over these features shows an intermittently visible coherent stratified unit at the seabed (F2); however, the rough seabed in this area mostly obscures these reflections (F3). A local seafloor high adjacent to these features is characterized by F1 facies that is truncated on either side by parallel escarpments, with gradients of *c.* 20°.

Discussion

Evidence of mass failure

It is interpreted that morphological domains along central Alpha Ridge are related to differing styles and ages of sediment mass transport.

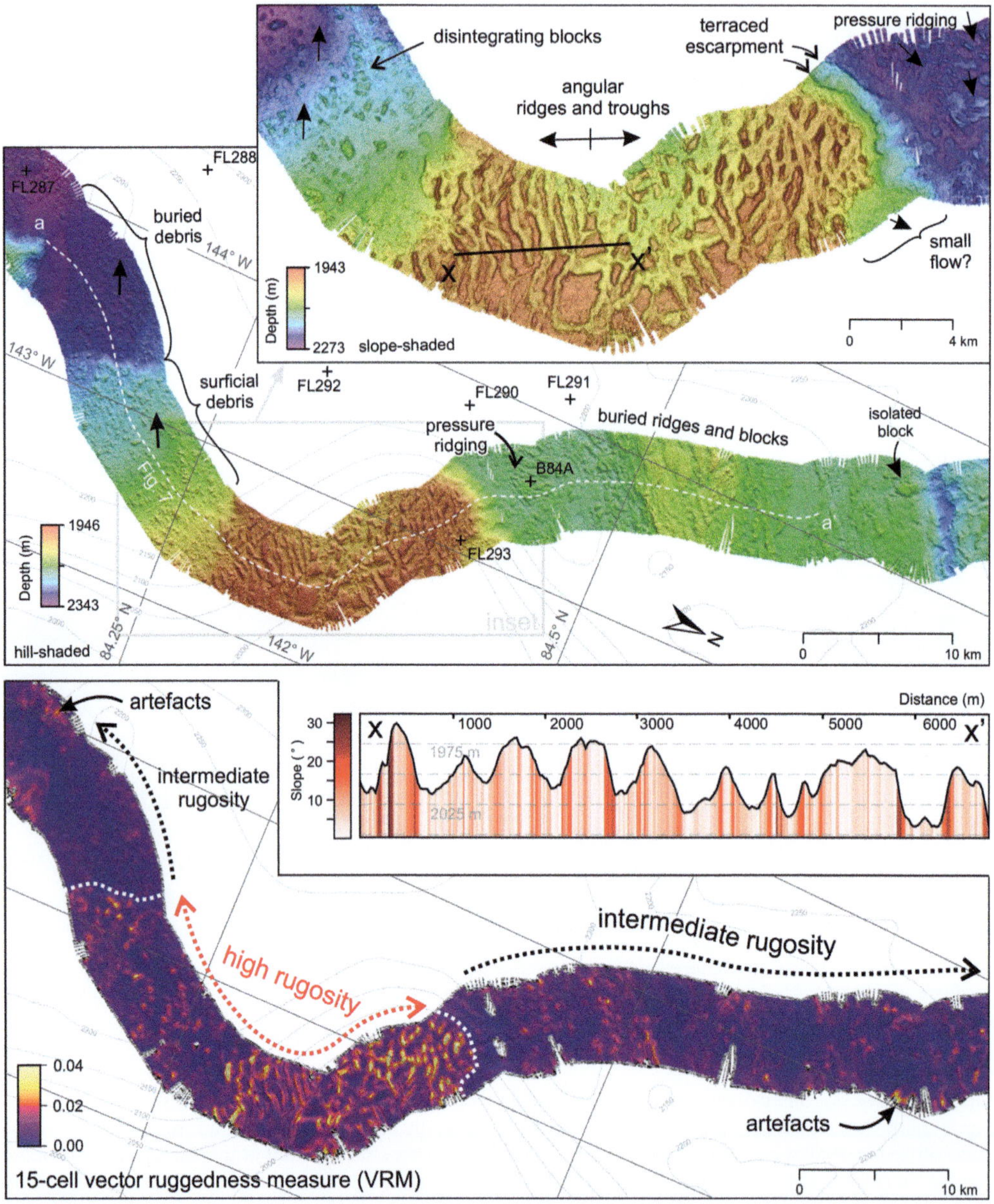

Fig. 6. Hill-shaded multibeam bathymetry (gridded at 25 m) showing angular ridges and troughs along a local bathymetric high from central Alpha Ridge (location in Fig. 3). The white dashed line (a–a′) shows location of sub-bottom profile shown in Figure 7. Upper right inset shows slope-shaded render, and location of bathymetric profile x–x′. Bottom panel shows a map of VRM calculated from bathymetry using a 15-cell wide (375 m) square neighbourhood. High rugosity and intermediate rugosity regions are indicated. Outer-beam artefacts result in anomalously high VRM values where indicated.

Low rugosity (smooth seafloor) regions dominate much of the western Alpha Ridge and Mendeleev Rise. These regions are underlain by stratified seismic reflections in sub-bottom records (F1, F1/T) and are interpreted as areas of undisturbed seafloor (e.g. Fig. 5a, profile a–a′). Sub-bottom facies and seafloor morphology in these regions are a result of pelagic and hemipelagic deposition that dominated

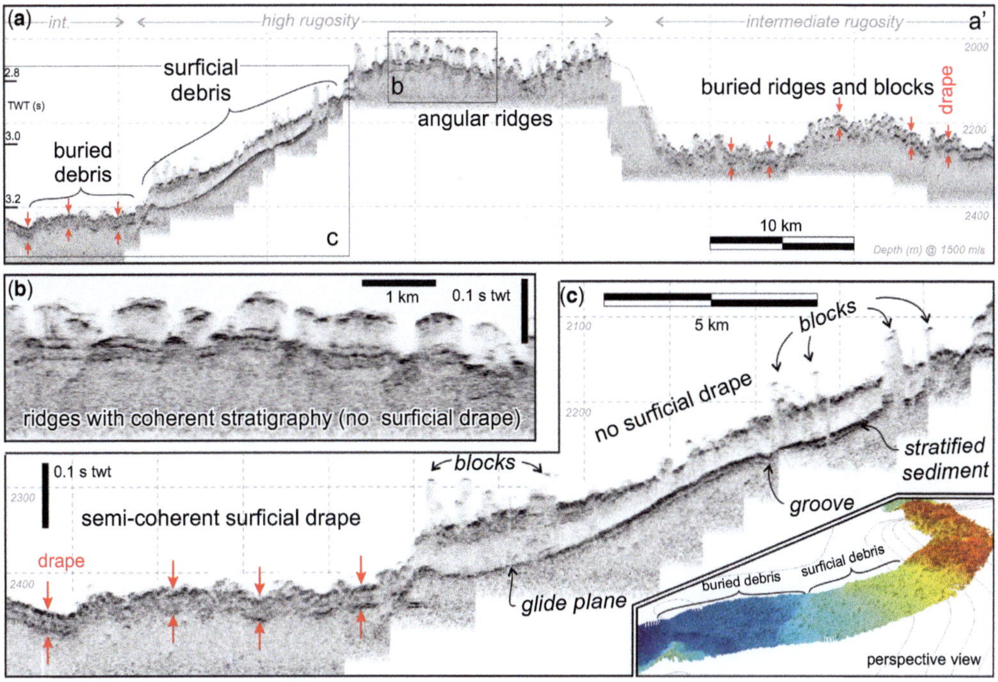

Fig. 7. (**a**) Sub-bottom profiler data over a local bathymetric high on central Alpha Ridge (location in Fig. 6) where high-rugosity features are observed. Enlarged insets show: (**b**) angular ridges at the crest underpinned by coherent stratified sub-bottom reflections; and (**c**) juxtaposition of buried debris with surficial debris. Perspective view of multibeam bathymetry over these features is shown in lower right.

the ridge's sedimentation history. These sediments are as much as 600 m thick over basement highs and yet reflect much of the underlying basement topography due to their draped nature (e.g. Jokat 2003; Bruvoll *et al.* 2010; Evangelatos and Mosher 2016; Evangelatos *et al.* 2017).

High rugosity regions are the result of a highly complex seafloor of escarpments, blocks and angular ridges. Where angular escarpments truncate or displace sub-bottom stratigraphy (Figs 4, 5 & 9), then these escarpments are interpreted as head scarps or side-wall scarps as a result of evacuation of sediment in a mass failure event (cf. McAdoo *et al.* 2000; Li *et al.* 2017). Ridges and troughs (e.g. Figs 5 & 6) are interpreted to result from an extensional domain of mass failure, where sediments are laterally displaced during initial downslope movement (cf. Micallef *et al.* 2007, 2016). The angular cross-sectional morphology and puzzle-fit geometry suggest they were formed by lateral spreading proximal to the site of failure initiation. Parallel stratified sub-bottom reflections that underpin some of these ridges suggest there was little to no deformation during spreading. Progressive transport leads to a domain dominated by disintegrating blocks, or isolated slump/raft blocks (cf. Nissen *et al.* 1999; Moscardelli *et al.* 2006; Hogan *et al.* 2013). Sediment with transparent or acoustically amorphous sub-bottom reflections (e.g. F3/1, F3/4) may represent either debris flow deposits (cf. Prior *et al.* 1984), or high deformation zones due to sliding of material, i.e. a décollement (cf. Posamentier and Martinsen 2011). Where acoustically transparent material overlies coherent stratified sediments (F3/1), failure is likely to have occurred along bedding-parallel glide planes (cf. Li *et al.* 2017).

Intermediate rugosity regions consist of morphological features that may be explained in at least two ways:

(1) *Mass failed sediment deposited within topographic valleys and low areas.* These accumulations may have incoherent ridging and complex terrain that appear as hyperbolic reflections as the seafloor return on sub-bottom profiler records (e.g. Fig. 9). The sub-bottom is not resolved or consists of incoherent or amorphous reflections (e.g. F3). These terrains are interpreted as debris flow deposits; the ridges represent the compressional domain of a debris flow (cf. Prior *et al.* 1984; Coleman and Prior 1988; Laberg and Vorren 2000; Bull

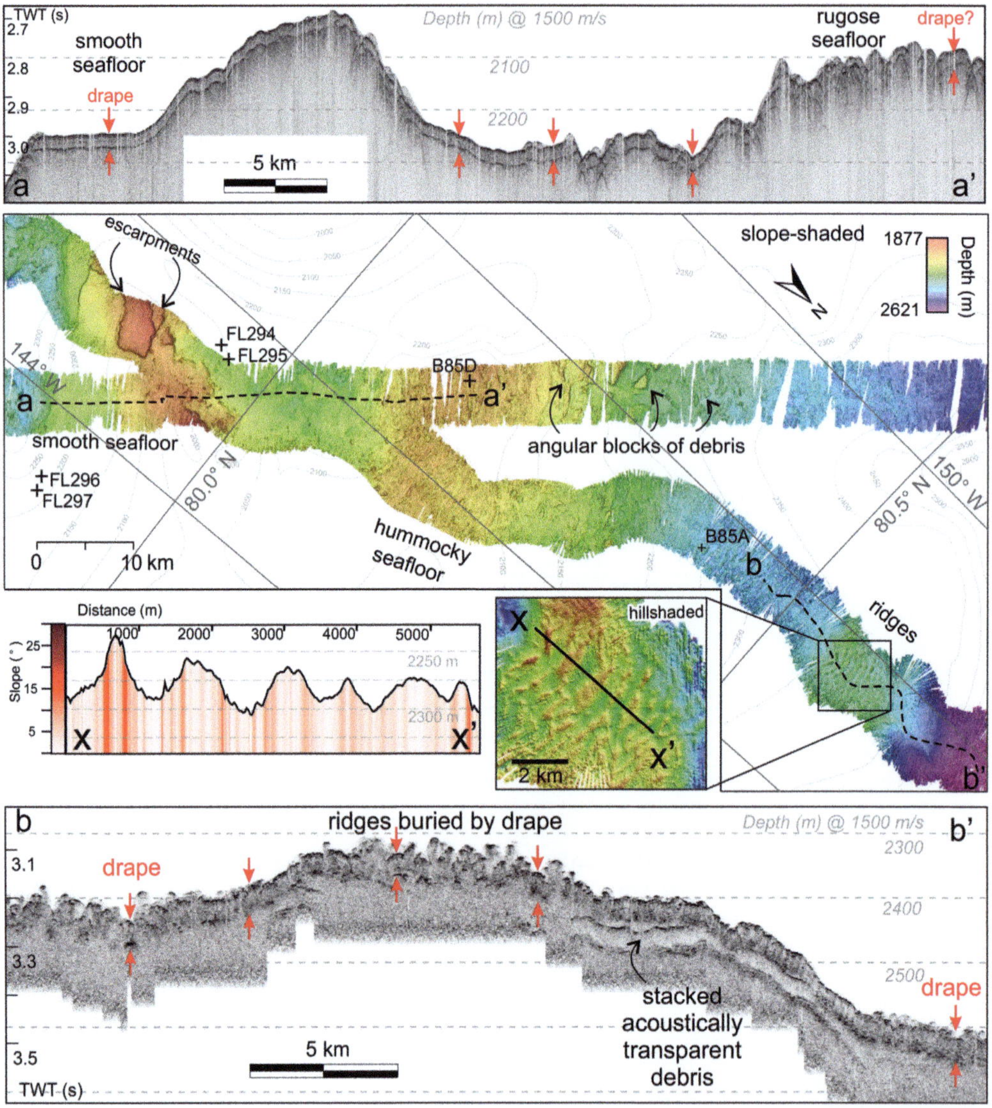

Fig. 8. Slope-shaded multibeam bathymetry (25 m grid) and sub-bottom profiler data from central Alpha Ridge where hummocky seafloor features are blanketed by *c.* 30–40 ms of coherent reflections (see location in Fig. 3). The locations of sediment cores are shown by crosses. Profile a–a′ demonstrates the consistent thickness of this coherent unit over a local high in bathymetry. This unit becomes more poorly defined towards the NW where the reflected signal is obscured by side-echoes of the rough seabed topography. Several angular blocks are indicated in multibeam bathymetry. Ridges in bathymetry are buried by *c.* 40 ms of semi-coherent reflections in sub-bottom profile (b–b′).

et al. 2009). The surface morphology of some of these deposits shows lineaments that indicate the direction of debris flow movement and shows boundaries between flow segments interpreted as longitudinal shear within different domains of the flow (cf. Masson *et al.* 2002; Bull *et al.* 2009).

(2) Complex topography generated by mass failure processes as described above, but a 30–40 ms thick (*c.* 22.5 to 30 m) sedimentary drape covers the features (F2). The sediment cover subdues the roughness of the original topography of the mass failure, leading to intermediate rugosity of the seabed.

The fact that some mass-transport deposits are deeply buried (>30 ms) and some appear at the seafloor indicates that at least two, and perhaps

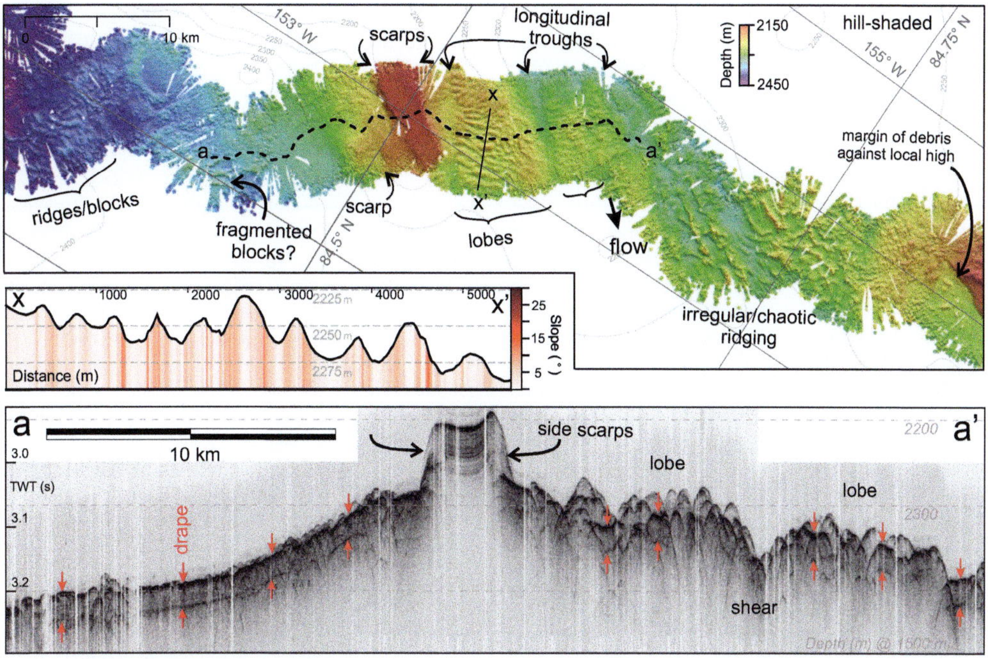

Fig. 9. Hill-shaded multibeam bathymetry (25 m grid) and sub-bottom profiler data from the HOTRAX expedition (see Fig. 3 for location) showing parallel escarpments truncating stratified sediments along a local bathymetric high and sub-parallel ridges in adjacent valleys. The valley to the north of the multibeam image corresponds to the location of the seismic profile 0531 (Coakley *et al.* 2005).

numerous, episodes of mass failure occurred within the disturbed region of Alpha Ridge.

Extent of mass failure

This study was limited to regions of multibeam bathymetric and sub-bottom profiler data on Alpha Ridge, which themselves provide only sparse coverage of the study area. Earlier studies, however, have shown similar sedimentary features that extend well beyond current data coverage (e.g. Kristoffersen *et al.* 2009; Fig. 1). Maps of hyperbolic echoes from precision depth recorder (PDR) data acquired from T3 demonstrate rough seabed features with varying rugosity that extend towards the Canadian margin (Hall 1979). Linear sedimentary features in this region with amplitudes up to 55 m and widths of up to 1 km were originally interpreted as sediment waves; however, new multibeam data led to their interpretation as ridges formed by sediment mass movement (e.g. Fig. 6). Similar features observed in single-beam echosounder data collected by nuclear submarines over eastern Alpha Ridge (Beal 1969) have comparable cross-sectional morphology, and are thus also likely related to mass failure.

Age of mass failures

The observation of (1) surficial mass-transport features with high rugosity and (2) buried mass-transport features with intermediate rugosity suggests at least two phases of failure have occurred in the disturbed region. A clear example of this is shown in Figure 6, where high rugosity features are along the crest and southern flank of a bathymetric high are juxtaposed to intermediate rugosity features in the nearby valleys. A sub-bottom profile (Fig. 7) reveals that intermediate rugosity features (interpreted as debris flows, pressure ridges and blocks) are buried by *c.* 25 m of conformal drape sediment (assuming 1500 m s^{-1} sedimentary velocity) whereas high rugosity features (interpreted as extensional ridges and blocky debris) are not buried by this unit.

The lack of visible drape covering the most recent failures in sub-bottom profile (e.g. Figs 5, 7 & 8) suggests they are possibly recent (i.e. Quaternary). Short (2–12 m) sediment cores recovered from these areas, however, contain no evidence of failed sediment (Hall 1979; Clark *et al.* 1980; Darby *et al.* 2005; Spielhagen and Gebhardt 2008; Kristoffersen *et al.* 2009; Matthiessen 2011) and many correlate over large distances (Clark *et al.* 1980; Sellén *et al.*

2010; Stein *et al.* 2010*a*). These facts suggest the shallowest coherent sediments in these regions are not well imaged. Based on core recoveries from these regions, the thickness of such a unit is estimated to be approximately 6–12 m and is likely to represent a minimum of 500–600 ka.

The oldest failed material in the disturbed region is buried by a laterally extensive conformal drape unit (e.g. F2). This unit is 22.5–30 m thick (assuming 1500 m s^{-1} sedimentary velocity) in most areas (e.g. Figs 5, 7–9). Currently, no existing sediment cores penetrate this unit to constrain the age. Three short (<3 m) sediment cores (B84A, B85A, B85D) were recovered from the top of this surficial unit where sub-bottom profiles show it is *c.* 25 m thick (locations in Figs 6, 8). Age models for these cores show uninterrupted sedimentation for the past *c.* 350–600 ka (Wang *et al.* 2018; Ye *et al.* 2019), suggesting the full thickness of this unit represents a substantially longer period of deposition (e.g. Pliocene in age). In absence of a comprehensive age model for this unit, a tentative age estimate is made by linear extrapolation of these and other Quaternary age models reported from Alpha Ridge. Most age models in this region result in average late-Quaternary sedimentation rates between 0.2 and 0.6 cm ka^{-1} (Polyak *et al.* 2009; Stein *et al.* 2010*b*; Wang *et al.* 2018, Ye *et al.* 2019). Therefore, if a representative sedimentation rate of 0.5 cm ka^{-1} is assumed for the 25 m of conformal drape (e.g. Fig. 8), then a minimum age of earliest Pliocene is estimated for the buried failed material.

These estimates are made with caution, however, since: (1) age controls have a high degree of uncertainty in this region of the Arctic Ocean (e.g. Sellén *et al.* 2009; Stein *et al.* 2010*b*; Hillaire-Marcel *et al.* 2017; Matthiessen *et al.* 2018); and (2) no Neogene sediments have been recovered from Alpha Ridge to confirm pre-Quaternary sedimentation rates. Pronounced palaeoenvironmental changes experienced by the central Arctic Ocean during this time (e.g. Stein *et al.* 2016; Dipre *et al.* 2018) likely influenced sedimentation rates over Alpha Ridge to some degree.

Cause of mass failure

Evaluating the cause of slope failure requires consideration of (1) preconditioning factors conducive to failure, and (2) triggering mechanisms. The paucity of core and seismic data preclude detailed assessment of slope stability in the region; however, several observations from acoustic data in this study are used to address these questions.

The observation of bed-parallel glide planes along seafloor highs (e.g. Fig. 4) suggests the possibility of failure along weak layers in the sedimentary succession. Weak layers consist of horizons with low-shear strength and may develop as a result of inherited or induced factors (Locat *et al.* 2014). Failure along weak layers typically results in lateral mass movement whereby translation of shallowest sediment is decoupled from the underlying sediment along a bed-parallel décollement (i.e. a shallow glide plane). This style of failure is often linked to extensional ridges and troughs formed by lateral spreading (e.g. Baeten *et al.* 2013), as identified on Alpha Ridge (e.g. Fig. 5). Failure of this sort may explain features previously identified in T3 seismic data where narrow seafloor hyperbolae overlie undisrupted stratified sediment (e.g. Hall 1970, 1979; Hunkins and Tiemann 1977). While the geotechnical basis of possible weak layers on Alpha Ridge is unknown without direct sample measurement, the variable depositional history of the ridge throughout the Cenozoic could play a role. Intervals of biosiliceous ooze and volcaniclastics (Clark 1974; Mudie *et al.* 1986) and grain-size variations related to glacial cycles (Polyak *et al.* 2009) suggest sufficient potential for the deposition of interbedded porous and impermeable layers. This stratigraphy establishes a framework of layers susceptible to reduced shear strength due to increased pore pressure during loading events, such as earthquakes. Weak layers are also interpreted to explain several failures near the ACEX site on neighbouring Lomonosov Ridge, where a biosiliceous-rich interval is likely to have acted as a failure plane (Kristoffersen *et al.* 2007).

The large areal extent of mass failure features and their occurrence over low-gradient and level seafloor suggests sediment mass failure on Alpha Ridge was likely triggered as the result of ground accelerations (seismicity or bolide impact), and not the result of localized slope undercutting or oversteepening. The closest known bolide impact crater to the disturbed features mapped by Kristoffersen *et al.* (2009) is located on NW Greenland, *c.* 800 km to the south, and is possibly Pleistocene in age (Kjær *et al.* 2018). Uncertain ages of both the bolide impact and the failed sediments on Alpha Ridge, however, make a direct link impossible. Unfailed sediment between the two areas suggests they are likely unrelated. A bolide impact trigger, while possible as a single event, seems unlikely given the observation of multiple failures in the same region over disparate periods of time. Multiple episodes of failure, as shown in this study, are most consistent with a recurring trigger mechanism such as earthquake loading during intermittent seismicity or volcanism.

Alpha Ridge is effectively aseismic during the modern observational period (Kristoffersen *et al.* 2009); however, intermittent seismicity during timescales suggested for the sediment failures shown here (i.e. Pliocene–Quaternary) is possible.

Seismicity due to glacial rebound can initiate submarine mass movements (e.g. Mosher *et al.* 2004; Piper *et al.* 2012). Direct glacial loading of Alpha Ridge, however, is not expected due to its great water depth and earthquakes generated along the glaciated Arctic Archipelago margin are likely too distant to be a plausible trigger. On longer timescales (i.e. Cenozoic), the potential for seismicity along Alpha Ridge is considerable. Prior to formation of the Eurasia Basin (i.e. >54 Ma) Alpha Ridge was believed to be emergent or within shallow water (Clark *et al.* 1986; Mudie *et al.* 1986). Late-stage seismicity related to subsidence of the ridge (e.g. Verzhbitskii *et al.* 2013) is thus a possibility. Basement faults are locally observed on Alpha Ridge (Hall 1979; Jokat 2003; Kristoffersen *et al.* 2009) and large faulted graben exist where basement has been sampled at the seafloor (e.g. Jackson *et al.* 1985). Numerous basement faults are also documented for Mendeleev Rise that may have resulted from late-stage tectonic activity between 22 and 14 Ma (Bruvoll *et al.* 2010). Episodic reactivation of existing basement structures during late-stage seismicity on Alpha Ridge (i.e. Pliocene–Quaternary) thus could provide sufficient ground acceleration to trigger multiple failures at the seafloor.

Another potential source of recent seismicity is late-Cenozoic volcanism along the Alpha Ridge. Radiometric ages of basement rocks sampled from the ridge confirm extrusive volcanism in the Upper Cretaceous (Jokat *et al.* 2013; Williamson *et al.* 2019); however, continued volcanism well into the Cenozoic is possible and may explain observations of several previous studies in the region. Pyroclastic bedrock dredged during CESAR, for example, contained Eocene pollen embedded in its matrix (Mudie *et al.* 1986), suggesting extrusive volcanism in shallow water along the ridge may have occurred as recently as *c.* 40 Ma (Van Wagoner *et al.* 1986). Eocene sediments recovered in cores FL-422 and Core 6 contain volcaniclastic sediments (Clark 1974; Bukry 1984; Mudie *et al.* 1986). Seismic reflection data from adjoining Mendeleev Rise demonstrate possible evidence of Miocene volcanic activity (Bruvoll *et al.* 2012). Interpreted mid-Cenozoic intrusion of volcanic sills in neighbouring Canada Basin (Shimeld *et al.* 2016) may also reflect late-stage magmatism in the area.

Conclusions

Recently acquired multibeam bathymetry and sub-bottom profiler data from Alpha Ridge show abundant evidence of sedimentary features formed by mass failure with varying degrees of lateral displacement. These features include, from minor displacement to maximum: (1) high-angle escarpments with truncated internal reflections interpreted as failure head scarps or side-wall scarps; (2) blocky ridges and troughs formed by extensional lateral spreading; (3) isolated blocks with semi-coherent internal stratigraphy interpreted as slump and slide blocks; (4) beds of acoustically transparent or incoherent reflections interpreted as debris flows; (5) undulating and contorted ridges in bathymetric lows, interpreted as compressional ridges within a debris flow; and (6) boundaries within debris flows interpreted as longitudinal shear zones. The extent of these features remains uncertain due to lack of data coverage; however, previous studies show it clearly extends further south towards the Canadian margin than currently mapped.

While some of these mass failure features appear near or on the surface, the few cores in the region suggest all areas are buried by at least a thin cover (metres thick) of pelagic/hemipelagic sediment. Sub-bottom profiler data clearly show seismic facies interpreted as mass failure deposits that are more deeply buried, suggestive of multiple episodes of mass failure. In this case, it seems improbable that mass failure is related to a bolide impact. Instead, it is proposed that ground accelerations related to seismicity are the likeliest trigger. This seismicity may be related to episodic basin subsidence or late-stage volcanism along Alpha Ridge.

Acknowledgements The authors would like to thank the commanding officers, crew and scientific personnel involved in the expeditions that acquired data used in this study. Comments and suggestions of two anonymous reviewers greatly improved this manuscript. We thank John Hall for useful discussions about the *Healy* and T3 datasets.

Funding This work was supported by the Geological Survey of Canada (Natural Resources Canada contribution number 20190297).

Author contributions **KB**: writing – original draft (lead); **DM**: writing – original draft (supporting); **PT**: data curation (equal); **CG**: data curation (equal); **LM**: data curation (equal).

References

Armstrong, A.A., Mayer, L.A. and Mosher, D.C. 2012. Gathering Multibeam Bathymetry Aboard Icebreakers. *Sea Technology*, **53**, 10–13.

Backman, J., Jakobsson, M. *et al.* 2008. Age model and core-seismic integration for the Cenozoic Arctic Coring Expedition sediments from the Lomonosov Ridge: ACEX age model. *Paleoceanography*, **23**, https://doi.org/10.1029/2007PA001476

Baeten, N.J., Laberg, J.S., Forwick, M., Vorren, T.O., Vanneste, M., Forsberg, C.F. and Ivanov, M. 2013.

Morphology and origin of smaller-scale mass movements on the continental slope off northern Norway. *Geomorphology*, **187**, 122–134. https://doi.org/10.1016/j.geomorph.2013.01.008

Beal, M.A. 1969. *Bathymetry and Structure of the Arctic Ocean*. PhD thesis, Oregon State University.

Bruvoll, V., Kristoffersen, Y., Coakley, B.J. and Hopper, J.R. 2010. Hemipelagic deposits on the Mendeleev and northwestern Alpha submarine Ridges in the Arctic Ocean: acoustic stratigraphy, depositional environment and an inter-ridge correlation calibrated by the ACEX results. *Marine Geophysical Researches*, **31**, 149–171, https://doi.org/10.1007/s11001-010-9094-9

Bruvoll, V., Kristoffersen, Y., Coakley, B.J., Hopper, J.R., Planke, S. and Kandilarov, A. 2012. The nature of the acoustic basement on Mendeleev and northwestern Alpha ridges, Arctic Ocean. *Tectonophysics*, **514–517**, 123–145, https://doi.org/10.1016/j.tecto.2011.10.015

Bukry, D. 1984. Paleogene paleoceanography of the Arctic Ocean is constrained by the middle or late Eocene age of USGS Core FL-422: evidence from silicoflagellates. *Geology*, **12**, 199–201, https://doi.org/10.1130/0091-7613(1984)12<199:PPOTAO>2.0.CO;2

Bull, S., Cartwright, J. and Huuse, M. 2009. A review of kinematic indicators from mass-transport complexes using 3D seismic data. *Marine and Petroleum Geology*, **26**, 1132–1151, https://doi.org/10.1016/j.marpetgeo.2008.09.011

Clark, D.L. 1974. Late Mesozoic and Early Cenozoic sediment cores from the Arctic Ocean. *Geology*, **2**, 41–44, https://doi.org/10.1130/0091-7613(1974)2<41:LMAECS>2.0.CO;2

Clark, D.L. 1988. Early history of the Arctic Ocean. *Paleoceanography*, **3**, 539–550 https://doi.org/10.1029/PA003i005p00539

Clark, D.L., Whitman, R.R., Morgan, K.A. and Mackey, S.D. (eds) 1980. *Stratigraphy and Glacial-Marine Sediments of the Amerasian Basin, Central Arctic Ocean*. Geological Society of America, Special Papers, **181**, https://doi.org/10.1130/SPE181

Clark, D.L., Byers, C.W. and Pratt, L.M. 1986. Cretaceous black mud from the central Arctic Ocean. *Paleoceanography*, **1**, 265–271, https://doi.org/10.1029/PA001i003p00265

Coakley, B.J., Kristoffersen, Y. and Hopper, J.R. 2005. *Cruise Report for Underway Geophysics Program HLY 05-03, 5 August 2005*; *Dutch Harbor, Alaska to 30 September, 2005;* Tromso, Norway.

Coleman, J.M. and Prior, D.B. 1988. Mass Wasting on Continental Margins. *Annual Review of Earth and Planetary Sciences*, **16**, 101–119, https://doi.org/10.1146/annurev.ea.16.050188.000533

Courtney, R. 2007. Storage and Dissemination of SEGY Data in JPEG2000 Format. *In*: *American Geophysical Union, Fall Meeting 2007*, 10–14 December, San Francisco, CA, Abstract Id. IN51B-0402.

Darby, D.A., Jakobsson, M. and Polyak, L. 2005. Icebreaker expedition collects key Arctic seafloor and ice data. *Eos, Transactions American Geophysical Union*, **86**, 549, https://doi.org/10.1029/2005EO520001

Davies, A., Kemp, A.E.S. and Pike, J. 2009. Late Cretaceous seasonal ocean variability from the Arctic. *Nature*, **460**, 254–258, https://doi.org/10.1038/nature08141

Dipre, G.R., Polyak, L. *et al.* 2018. Plio-Pleistocene sedimentary record from the Northwind Ridge: new insights into paleoclimatic evolution of the western Arctic Ocean for the last 5 Ma. *Arktos*, **4**, 24, https://doi.org/10.1007/s41063-018-0054-y

Døssing, A., Jackson, H.R., Matzka, J., Einarsson, I., Rasmussen, T.M., Olesen, A.V. and Brozena, J.M. 2013. On the origin of the Amerasia Basin and the High Arctic Large Igneous Province – Results of new aeromagnetic data. *Earth and Planetary Science Letters*, **363**, 219–230, https://doi.org/10.1016/j.epsl.2012.12.013

Dove, D., Coakley, B., Hopper, J. and Kristoffersen, Y. and HLY0503 Geophysics Team 2010. Bathymetry, controlled source seismic and gravity observations of the Mendeleev ridge; implications for ridge structure, origin, and regional tectonics: Mendeleev Ridge and regional tectonics. *Geophysical Journal International*, **183**, 481–502, https://doi.org/10.1111/j.1365-246X.2010.04746.x

Evangelatos, J. and Mosher, D.C. 2016. Seismic stratigraphy, structure and morphology of Makarov Basin and surrounding regions: tectonic implications. *Marine Geology*, **374**, 1–13, https://doi.org/10.1016/j.margeo.2016.01.013

Evangelatos, J., Funck, T. and Mosher, D.C. 2017. The sedimentary and crustal velocity structure of Makarov Basin and adjacent Alpha Ridge. *Tectonophysics*, **696**, 99–114, https://doi.org/10.1016/j.tecto.2016.12.026

Gårdfeldt, K. and Lindgren, Å. (eds) 2017. *SWEDARCTIC Arctic Ocean 2016 Cruise Report*. Swedish Polar Research Secretariat, **117**.

Hall, J.K. 1970. *Arctic Ocean Geophysical Studies: The Alpha Cordillera and Mendeleyev Ridge*. Lamont–Doherty Geological Observatory, Columbia University.

Hall, J.K. 1979. Sediment waves and other evidence of paleo-bottom currents at two locations in the deep Arctic Ocean. *Sedimentary Geology*, **23**, 269–299, https://doi.org/10.1016/0037-0738(79)90017-4

Hillaire-Marcel, C., Ghaleb, B. *et al.* 2017. A new chronology of late quaternary sequences from the Central Arctic Ocean Based on 'Extinction Ages' of their excesses in ^{231} Pa and ^{230} The Late Quaternary Arctic Ocean chronology. *Geochemistry, Geophysics, Geosystems*, **18**, 4573–4585, https://doi.org/10.1002/2017GC007050

Hogan, K.A., Dowdeswell, J.A. and Mienert, J. 2013. New insights into slide processes and seafloor geology revealed by side-scan imagery of the massive Hinlopen Slide, Arctic Ocean margin. *Geo-Marine Letters*, **33**, 325–343, https://doi.org/10.1007/s00367-013-0330-6

Hunkins, K. and Tiemann, W. 1977. *Geophysical Data Summary for Fletcher's Ice Island (T-3), 1962–1974*. Technical Report, Lamont-Doherty Survey of the World Ocean, **CU-177**.

Hutchinson, D.R., Jackson, H.R., Shimeld, J.W., Chapman, C.B., Childs, J.R., Funck, T. and Rowland, R.W. 2009. Acquiring Marine Data in the Canada Basin, Arctic Ocean. *Eos, Transactions American Geophysical Union*, **90**, 197, https://doi.org/10.1029/2009EO230001

Jackson, H.R., Mudie, P.J. and Blasco, S.M. (eds) 1985. *Initial Geological Report on CESAR – the Canadian Expedition to Study the Alpha Ridge*. Geological Survey of Canada, **Paper 84-22**, 3–10, https://doi.org/10.4095/120319

Jakobsson, M., Mayer, L. *et al.* 2012. The International Bathymetric Chart of the Arctic Ocean (IBCAO) Version 3.0: IBCAO VERSION 3.0. *Geophysical Research Letters*, **39**, https://doi.org/10.1029/2012GL052219

Jokat, W. 1999. *ARCTIC '98: The Expedition ARK-XIVII/1a of RV 'Polarstern' in 1998*. Alfred Wegener Institute, Reports on Polar and Marine Research.

Jokat, W. 2003. Seismic investigations along the western sector of Alpha Ridge, Central Arctic Ocean. *Geophysical Journal International*, **152**, 185–201, https://doi.org/10.1046/j.1365-246X.2003.01839.x

Jokat, W., Ickrath, M. and O'Connor, J. 2013. Seismic transect across the Lomonosov and Mendeleev Ridges: constraints on the geological evolution of the Amerasia Basin, Arctic Ocean. *Geophysical Research Letters*, **40**, 5047–5051, https://doi.org/10.1002/grl.50975

Kadko, D. and Landing, W. 2015. GEOTRACES Cruise Report: 09 August, 2015–12 October, 2015, https://doi.org/10.7284/906271

Kjær, K., Larsen, N.K. *et al.* 2018. A large impact crater beneath Hiawatha Glacier in Northwest Greenland. *Science Advances*, **4**, eaar8173, https://doi.org/10.1126/sciadv.aar8173

Kristoffersen, Y., Coakley, B.J., Hall, J.K. and Edwards, M. 2007. Mass wasting on the submarine Lomonosov Ridge, central Arctic Ocean. *Marine Geology*, **243**, 132–142, https://doi.org/10.1016/j.margeo.2007.04.012

Kristoffersen, Y., Hall, J.K., Hunkins, K., Ardai, J. and Hopper, J.R. 2009. Extensive local seabed disturbance, erosion and mass wasting on Alpha Ridge, Central Arctic Ocean: possible evidence for an extra-terrestrial impact? *Norwegian Journal of Geology*, **8**.

Kwok, R. 2018. Arctic sea ice thickness, volume, and multiyear ice coverage: losses and coupled variability (1958–2018). *Environmental Research Letters*, **13**, 105005, https://doi.org/10.1088/1748-9326/aae3ec

Laberg, J.S. and Vorren, T.O. 2000. The Trñnadjupet Slide, offshore Norway: morphology, evacuation and triggering mechanisms. *Marine Geology*, **171**, 95–114, https://doi.org/10.1016/S0025-3227(00)00112-2

Lachenbruch, A.H., Marshall, B.V. and Ruppel, C.D. 2019. *Post-Expedition Report for USGS T-3 Heat Flow Measurements in the High Arctic Ocean*. GUS eological Survey data release, https://doi.org/10.5066/P91XQ3IS

Levitan, M.A. 2015. Sedimentation rates in the Arctic Ocean during the last five marine isotope stages. *Oceanology*, **55**, 425–433, https://doi.org/10.1134/S000143701503011X

Li, W., Alves, T.M. *et al.* 2017. Morphology, age and sediment dynamics of the upper headwall of the Sahara Slide Complex, Northwest Africa: evidence for a large Late Holocene failure. *Marine Geology*, **393**, 109–123, https://doi.org/10.1016/j.margeo.2016.11.013

Locat, J., Leroueil, S., Locat, A. and Lee, H. 2014. Weak layers: their definition and classification from a geotechnical perspective. *In*: Krastel, S. *et al.* (eds) *Submarine Mass Movements and their Consequences*. Advances in Natural and Technological Hazards Research, **37**. Springer, Cham, 3–12, https://doi.org/10.1007/978-3-319-00972-8_1

Masson, D.G., Watts, A.B., Gee, M.J.R., Urgeles, R., Mitchell, N.C., Le Bas, T.P. and Canals, M. 2002. Slope failures on the flanks of the western Canary Islands. *Earth-Science Reviews*, **57**, 1–35, https://doi.org/10.1016/S0012-8252(01)00069-1

Matthiessen, J. 2011. Documentation of sediment core PS78/231-2. Alfred Wegener Institute – Polarstern core repository, PANGAEA, https://doi.org/10.1594/PANGAEA.771313

Matthiessen, J., Schreck, M., De Schepper, S., Zorzi, C. and de Vernal, A. 2018. Quaternary dinoflagellate cysts in the Arctic Ocean: potential and limitations for stratigraphy and paleoenvironmental reconstructions. *Quaternary Science Reviews*, **192**, 1–26, https://doi.org/10.1016/j.quascirev.2017.12.020

Mayer, L.A. and Armstrong, A.A. 2011. HLY1102 Cruise Report: US Law of the Sea cruise to map the foot of the slope and 2500-m isobaths of the US Arctic Ocean margin, https://ccom.unh.edu/sites/default/files/publications/Mayer_2011_cruise_report_HEALY1102.pdf

McAdoo, B.G., Pratson, L.F. and Orange, D.L. 2000. Submarine landslide geomorphology, US continental slope. *Marine Geology*, **169**, 103–136, https://doi.org/10.1016/S0025-3227(00)00050-5

Micallef, A., Masson, D.G., Berndt, C. and Stow, D.A.V. 2007. Morphology and mechanics of submarine spreading: a case study from the Storegga Slide. *Journal of Geophysical Research*, **112**, F03023, https://doi.org/10.1029/2006JF000739

Micallef, A., Masson, D.G., Berndt, C. and Stow, D.A.V. 2016. Submarine spreading in the Storegga Slide, Norwegian Sea. *Geological Society, London, Memoirs*, **46**, 411–412, https://doi.org/10.1144/M46.88

Moran, K., Backman, J. *et al.* 2006. The Cenozoic palaeoenvironment of the Arctic Ocean. *Nature*, **441**, 601–605, https://doi.org/10.1038/nature04800

Moscardelli, L., Wood, L. and Mann, P. 2006. Mass-transport complexes and associated processes in the offshore area of Trinidad and Venezuela. *American Association of Petroleum Geologists Bulletin*, **90**, 1059–1088, https://doi.org/10.1306/02210605052

Mosher, D.C., Piper, D.J.W., Campbell, D.C. and Jenner, K.A. 2004. Near-surface geology and sediment-failure geohazards of the central Scotian Slope. *American Association of Petroleum Geologists Bulletin*, **88**, 703–723, https://doi.org/10.1306/01260403084

Mosher, D.C., Chapman, C.B. *et al.* 2013. High Arctic marine geophysical data acquisition. *The Leading Edge*, **32**, 524–536, https://doi.org/10.1190/tle32050524.1

Mudie, P.J. and Blasco, S.M. 1985. Lithostratigraphy of the CESAR cores. *In*: *Initial Report on CESAR*. Geological Survey of Canada, **Paper 84-22**, 59–99.

Mudie, P.J., Stoffyn-Egli, P. and Van Wagoner, N.A. 1986. Geological constraints for tectonic models of the Alpha Ridge. *Journal of Geodynamics*, **6**, 215–236, https://doi.org/10.1016/0264-3707(86)90040-2

Nissen, S.E., Haskell, N.L., Steiner, C.T. and Coterill, K.L. 1999. Debris flow outrunner blocks, glide tracks, and pressure ridges identified on the Nigerian continental slope using 3-D seismic coherency. *The Leading Edge*, **18**, 595–599, https://doi.org/10.1190/1.1438343

Oakey, G.N. and Saltus, R.W. 2016. Geophysical analysis of the Alpha–Mendeleev ridge complex: characterization of the High Arctic Large Igneous Province. *Tectonøphysics*, **691**, 65–84, https://doi.org/10.1016/j.tecto.2016.08.005

Petrov, O., Morozov, A., Shokalsky, S., Kashubin, S., Artemieva, I.M., Sobolev, N. and Smelror, M. 2016. Crustal structure and tectonic model of the Arctic region. *Earth-Science Reviews*, **154**, 29–71, https://doi.org/10.1016/j.earscirev.2015.11.013

Piper, D.J.W., Mosher, D.C. and Campbell, D.C. 2012. Controls on the distribution of major types of submarine landslides. *In*: Stead, D. and Clague, J.J. (eds) *Landslides: Types, Mechanisms and Modeling*. Cambridge University Press, Cambridge, 95–107, https://doi.org/10.1017/CBO9780511740367.010

Polyak, L. and Jakobsson, M. 2011. Quaternary Sedimentation in the Arctic Ocean: recent advances and further challenges. *Oceanography*, **24**, 52–64, https://doi.org/10.5670/oceanog.2011.55

Polyak, L., Bischof, J. *et al.* 2009. Late Quaternary stratigraphy and sedimentation patterns in the western Arctic Ocean. *Global and Planetary Change*, **68**, 5–17, https://doi.org/10.1016/j.gloplacha.2009.03.014

Posamentier, H.W. and Martinsen, O.J. 2011. The character and genesis of submarine mass-transport deposits: insights from outcrop and 3d seismic data. *In*: Shipp, R.C., Weimer, P. and Posamentier, H.W. (eds) *Mass-Transport Deposits in Deepwater Settings*. SEPM Society for Sedimentary Geology, https://doi.org/10.2110/sepmsp.096.007

Prior, D.B., Bornhold, B.D. and Johns, M.W. 1984. Depositional characteristics of a submarine debris flow. *The Journal of Geology*, **92**, 707–727, https://doi.org/10.1086/628907

Rigor, I.G., Wallace, J. and Colony, R.L. 2002. Response of Sea Ice to the Arctic Oscillation. *Journal of Climate*, **15**, 16, https://doi.org/10.1175/1520-0442(2002)015<2648:ROSITT>2.0.CO;2

Sappington, J.M., Longshore, K.M. and Thompson, D.B. 2007. Quantifying landscape ruggedness for animal habitat analysis: a case study using bighorn sheep in the Mojave Desert. *Journal of Wildlife Management*, **71**, 1419–1426, https://doi.org/10.2193/2005-723

Schauer, U. 2008. *The Expedition of the Research Vessel 'Polarstern' to the Arctic in 2007 (ARK-XXII/2)*. Alfred Wegener Institute, Reports on Polar and Marine Research.

Schauer, U. 2012. *The Expedition of the Research Vessel 'Polarstern' to the Arctic in 2011 (ARK-XXVI/3–TransArc)*. Alfred Wegener Institute, Reports on Polar and Marine Research.

Sellén, E., Jakobsson, M., Frank, M. and Kubik, P.W. 2009. Pleistocene variations of beryllium isotopes in central Arctic Ocean sediment cores. *Global and Planetary Change*, **68**, 38–47, https://doi.org/10.1016/j.gloplacha.2009.03.024

Sellén, E., O'Regan, M. and Jakobsson, M. 2010. Spatial and temporal Arctic Ocean depositional regimes: a key to the evolution of ice drift and current patterns. *Quaternary Science Reviews*, **29**, 3644–3664, https://doi.org/10.1016/j.quascirev.2010.06.005

Shimeld, J. and Boggild, K. 2017. *Seismic Operations Report for the 2016 Canada-Sweden Polar Expedition in the Arctic Ocean*. Geological Survey of Canada, Open File 8239, https://doi.org/10.4095/300670

Shimeld, J., Li, Q., Chian, D., Lebedeva-Ivanova, N., Jackson, R., Mosher, D. and Hutchinson, D. 2016. Seismic velocities within the sedimentary succession of the Canada Basin and southern Alpha–Mendeleev Ridge, Arctic Ocean: evidence for accelerated porosity reduction? *Geophysical Journal International*, **204**, 1–20, https://doi.org/10.1093/gji/ggv416

Spielhagen, R.F. and Gebhardt, C. 2008. Documentation of sediment core PS70/338-10. Alfred Wegener Institute – Polarstern core repository, PANGAEA, https:// doi.org/10.1594/PANGAEA.680738

Stein, R., Matthiessen, J. and Niessen, F. 2010*a*. Re-coring at ice island t3 site of key core FL-224 (Nautilus Basin, Amerasian Arctic): sediment characteristics and stratigraphic framework. *Polarforschung*, **79**, 81–96.

Stein, R., Matthiessen, J., Niessen, F., Krylov, A., Nam, S. and Bazhenova, E. 2010*b*. Towards a better (litho-) stratigraphy and reconstruction of Quaternary paleoenvironment in the Amerasian Basin (Arctic Ocean). *Polarforschung*, **79**, 97–121.

Stein, R., Fahl, K. *et al.* 2016. Evidence for ice-free summers in the late Miocene central Arctic Ocean. *Nature Communications*, **7**, 11148, https://doi.org/10.1038/ncomms11148

Van Wagoner, N.A., Williamson, M.-C., Robinson, P.T. and Gibson, I.L. 1986. First samples of acoustic basement recovered from the Alpha Ridge, Arctic Ocean: new constraints for the origin of the ridge. *Journal of Geodynamics*, **6**, 177–196, https://doi.org/10.1016/0264-3707(86)90038-4

Verzhbitskii, E.V., Lobkovskii, L.I., Byakov, A.F. and Kononov, M.V. 2013. The origin and age of the Alpha–Mendeleev and Lomonosov ridges in the Amerasia Basin. *Oceanology*, **53**, 89–98, https://doi.org/10.1134/S0001437013010141

Wang, R., Polyak, L., Xiao, W., Wu, L., Zhang, T., Sun, Y. and Xu, X. 2018. Late-Middle Quaternary lithostratigraphy and sedimentation patterns on the Alpha Ridge, central Arctic Ocean: implications for Arctic climate variability on orbital time scales. *Quaternary Science Reviews*, **181**, 93–108, https://doi.org/10.1016/j.quascirev.2017.12.006

Weber, J.R. and Roots, E.F. 1990. Geophysical and geological exploration in the Arctic Ocean region: summary of the most important expeditions and surveys that have contributed to geophysical and geological knowledge of the Arctic Ocean. *Marine Geology*, **93**, 1–13, https://doi.org/10.1016/0025-3227(90)90075-U

Williamson, M.-C., Kellett, D. *et al.* 2019. Age and eruptive style of volcanic rocks dredged from the Alpha Ridge, Arctic Ocean. *In*: *Geophysical Research Abstracts – EGU General Assembly 2019*. Vienna, Austria.

Ye, L., März, C., Polyak, L., Yu, X. and Zhang, W. 2019. Dynamics of manganese and cerium enrichments in Arctic Ocean sediments: a case study from the Alpha Ridge. *Frontiers in Earth Science*, **6**, 236, https://doi.org/10.3389/feart.2018.00236

The Last Glacial Maximum Balearic Abyssal Plain megabed revisited

Antonio Cattaneo[1]*, Shray Badhani[1,2], Cristina Caradonna[3], Massimo Bellucci[2,3], Estelle Leroux[1], Nathalie Babonneau[2], Sébastien Garziglia[1], Jeffrey Poort[4], Grigorii G. Akhmanov[5], Germain Bayon[1], Bernard Dennielou[1], Gwenäel Jouet[1], Sébastien Migeon[6,7], Marina Rabineau[2], Laurence Droz[2] and Michael Clare[8]

[1]Unité géosciences marines, IFREMER, GM, F-29280 Plouzané, France

[2]UMR 6538, CNRS, Université de Brest, Université Bretagne-Sud, Géosciences Océan (LGO), F-29280 Plouzané, France

[3]Dipartimento di Matematica e Geoscienze, Università di Trieste, 34128 Trieste, Italy

[4]Sorbonne Université, CNRS-INSU, Institut des Sciences de la Terre Paris, ISTeP UMR 7193, F-75005 Paris, France

[5]Faculty of Geology, Lomonosov Moscow State University, 119992 Moscow, Russia

[6]Université Côte d'Azur, CNRS, OCA, IRD, Géoazur (UMR7329), 06560 Valbonne, France

[7]Sorbonne Université, UFR939, 06230 Villefranche-sur-Mer, France

[8]National Oceanography Centre, University of Southampton Waterfront Campus, Southampton SO14 3ZH, UK

AC, 0000-0001-5247-9138; SB, 0000-0001-8432-4042; MB, 0000-0001-6753-5649; EL, 0000-0003-3584-1179; NB, 0000-0003-1213-0059; SG, 0000-0003-1136-243X; JP, 0000-0001-5964-9697; GGA, 0000-0002-8884-2629; BD, 0000-0002-9528-2746; MR, 0000-0001-8420-1587; LD, 0000-0002-6694-1247

*Correspondence: antonio.cattaneo@ifremer.fr

Abstract: Megabeds are thick sedimentary layers extending over thousands of square kilometres in deep-sea basins and are thought to result from large slope failures triggered by major external events. Such deposits have been found in at least three areas of the Mediterranean Sea. Although their discovery dates back to the early 1980s, many questions remain concerning their initiation, source area, extent and the nature of their emplacement. One of the largest previously documented megabeds was emplaced during the Last Glacial Maximum across the Balearic Abyssal Plain, with a thickness of 8–10 m in water depths of up to 2800 m.

New 3.5 kHz sub-bottom profiles and sediment cores provide greater constraints on the lateral variability of the megabed and allow it to be mapped beyond previous estimates, with a revised areal extent of 90 000–100 000 km^2. The megabed terminations show a gradual pinchout to the west and an abrupt eastward termination against the steep Sardinia margin. The megabed presents, in seismic profiles and sediment cores, a tripartite subdivision, which most likely corresponds to the changes in flow regimes across the basin, with a central area of sandy facies and an erosional base oriented NNE–SSW; this allows renewed discussions about the sources and triggers of the megabed.

Anomalously thick (up to tens of metres), acoustically transparent sedimentary layers known as megabeds or megaturbidites are found in at least three deep basins of the Mediterranean Sea: the Balearic Abyssal Plain (Rothwell *et al.* 1998), the Ionian Sea (Cita and Aloisi 2000; Hieke *et al.* 2003) and the Herodotus Basin in the Eastern Mediterranean (Reeder *et al.* 2000). These deposits have volumes equivalent to some of the largest mass movements documented worldwide, often in excess of 100 km^3 (Fildes 2013 and references therein). The term megaturbidite indicates a turbidite (i.e. a fining-upward, sand/silt/clay bed accumulated from a waning, semi-laden turbulent flow; Mutti 1992) that has a thickness larger than the average surrounding turbidite beds above and below, a peculiar composition and a basinwide extent (Bouma 1987). Typically, megabeds are recognized in abyssal plain settings. In some cases, megabeds are capped by a thick (pluri-metric) accumulation of dominantly structureless, homogeneous

From: Georgiopoulou, A., Amy, L. A., Benetti, S., Chaytor, J. D., Clare, M. A., Gamboa, D., Haughton, P. D. W., Moernaut, J. and Mountjoy, J. J. (eds) 2020. *Subaqueous Mass Movements and their Consequences: Advances in Process Understanding, Monitoring and Hazard Assessments*. Geological Society, London, Special Publications, **500**, 341–357.
First published online May 14, 2020, https://doi.org/10.1144/SP500-2019-188

silty clay without microfossil content (also referred to as homogenite) (Kastens and Cita 1981; Cita *et al.* 1984). In this study, we use the term 'megabed', because the terms 'megaturbidite' and 'homogenite', though describing sedimentary facies, are also genetically connoted and correspond to the HmTu (homogenite–turbidite) couplet of genetically linked deposits (e.g. Campos *et al.* 2013). The reason to study megabeds is two-fold. Firstly, in spite of their massive appearance, detailed information on the sedimentary structures is scarce in the literature, inhibiting a clear interpretation of the associated sedimentary processes at the origin of these anomalously large event deposits. Secondly, they may be diagnostic of major catastrophic events, as their triggering is usually attributed to major external events, including large earthquakes and tsunamis, or they may themselves have triggered tsunamis (e.g. Talling *et al.* 2014). The emplacement of such a megabed could impact important deep-sea benthic communities as a result of a widespread and rapid burial of the seafloor over hundreds of square kilometres as well as the prolonged (several months' long) effects of turbid waters during the settling of suspended fine sediments. Following the 12 January 2010 Haiti earthquake (M_w 7.0), a 600 m-thick nepheloid layer took two months to settle after the initial event (McHugh *et al.* 2011).

As megabeds typically form widespread, low-relief deposits that blanket broad basin topography, their identification from seafloor data is challenging. As a result, their identification is usually based on the recognition of extensive acoustically transparent layers in sub-bottom profiles – a diagnostic echo-facies signature that results from the generally homogeneous nature of the deposits (Damuth 1980; Cita *et al.* 1984). Several hypotheses have been proposed that associate their mechanisms of origin with catastrophic high magnitude events (e.g. volcanic activity, earthquakes and tsunamis). Recent studies have proposed scenarios where multiple sedimentary processes contribute to the emplacement of megabeds (e.g. Mulder *et al.* 2009; San Pedro *et al.* 2017) or have resulted in the revision of previous age estimates for event timing. For instance, more accurate age dating linked the Augias megabed in the Ionian Sea to the large historical earthquake of AD 365 on the island of Crete (Vigliotti 2008; Polonia *et al.* 2013; San Pedro *et al.* 2017), rather than to the previously assumed trigger – the Santorini volcanic eruption (3.5 ka BP; Cita *et al.* 1984). Given their prodigious volumes, it is remarkable that there are still so many unanswered questions about the nature and processes related to these megabeds.

Although the discovery of megabeds dates back to the early 1980s (Kastens and Cita 1981), there are still questions concerning megabeds that relate to (1) determining the precise recurrence times of these deposits when present in superimposed layers, with important implications for the reconstruction of past catastrophic events; (2) the physical and mechanical characteristics of these deposits to more strictly constrain the sedimentary processes that led to these exceptional sediment accumulation events and their impact on benthic communities; and (3) the sedimentary and geochemical sources of the sediment that formed the megabeds. The Balearic Abyssal Plain (BAP) megabed, the largest megabed in the Mediterranean, is a perfect example of a puzzling deposit. While its age is well constrained by multiple radiocarbon dates and biostratigraphy (Rothwell *et al.* 1998; Hoogakker *et al.* 2004), the initiation processes and source areas remain far from clear, even on a regional basis (Rothwell *et al.* 1998, 2000; Droz *et al.* 2006). It is remarkable that a deposit with an estimated volume of 500–600 km^3 has no obvious source, especially given the extent of marine geological studies around the Western Mediterranean in recent years (e.g. Badhani *et al.* 2019; Dennielou *et al.* 2019). In order to understand the hazard posed by these large sediment transport events, it is important to address these uncertainties. Where did they come from? How did they initiate? What is their full spatial extent? What process(es) explain their emplacement and are multiple phases of activity involved? These questions are particularly important given the dense coastal populations that border confined Mediterranean basins, such as the Balearic Abyssal Plain, and because megabed emplacement is not a one-off event in these settings. Stacked megabeds have been observed in seismic data, indicating that these are recurrent events both in Quaternary sediment (e.g. in the Ionian Abyssal Plain, Hieke *et al.* 2003; in the Marmara Sea, Beck *et al.* 2007) and in ancient deposits (e.g. in the Northern Apennines, Amy and Talling 2006; in Argentina, Fallgatter *et al.* 2016).

Sub-bottom seismic profile datasets are important for near-surface structural and sedimentological analysis. This is why we present a basinwide analysis of the newly available sub-bottom profile data and a comparison with, in part, already published analysis of sediment cores that have sampled the BAP megabed. We present an extensive (though not comprehensive, because the vast area of the BAP is scarcely covered by geophysical surveys, see Fig. 1) view of the geometry, terminations, bounding surfaces and internal geometries of the megabed. We also set up the first steps to match seismic data to core data in order to reconstruct the sedimentary processes at the origin of the megabed, based on integrated and complementary new evidence. The aim of this study is to highlight details in the seismic images of the BAP megabed that unveil aspects of the sedimentary processes at its origin and source; in particular, we present detailed diagnostic features

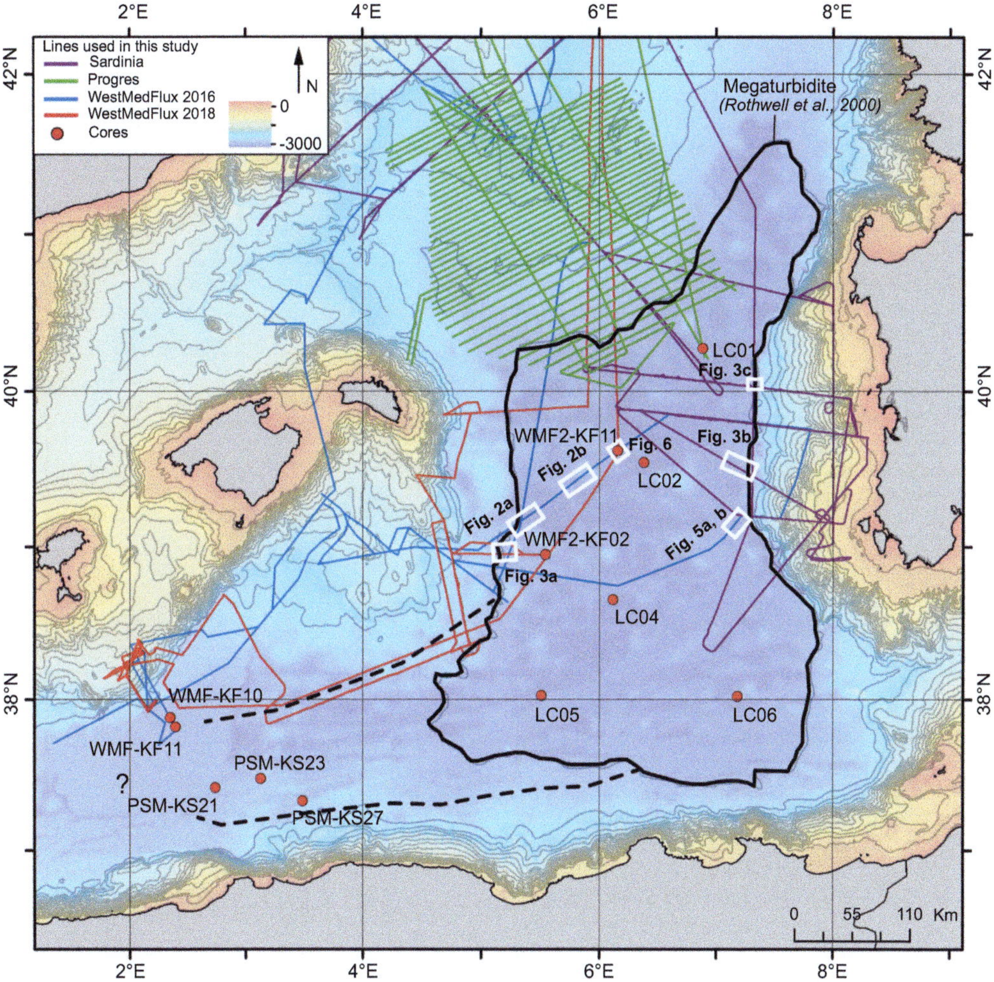

Fig. 1. General bathymetric map of the Balearic Abyssal Plain (BAP) with the location of seismic surveys (coloured lines; see the graphic legend to identify the name of the cruise) and sediment cores (red dots) mentioned or presented in this study. The continuous black line represents the indicative areal coverage of the BAP megabed as proposed by Rothwell *et al.* (2000). The white boxes show the location of the seismic profiles presented in Figures 2–6. The dashed black line shows the proposed extent of the megabed based on sediment core evidence from Babonneau *et al.* (2017) and this study.

of the sedimentary processes, including evidence of localized basal erosion, considerations on the nature and internal variability of the megabed indicative of possible flow transformations and some results on the mechanical properties of the homogenite part.

Background

The Western Mediterranean Basin formed behind a Miocene volcanic arc with the clockwise rotation of the Corsica–Sardinia block and the opening of the Provençal Basin (Réhault *et al.* 1984). It is the youngest part of the Mediterranean formed from the late Oligocene to the present (Mauffret *et al.* 2004). The BAP is the largest abyssal plain of the Mediterranean with an area of about 77 000 km^2 and may be identified by the 2800 m isobath (Rothwell *et al.* 1998, 2000). It is surrounded by the passive margins of the Gulf of Lions and the Ligurian Sea with the Rhone and Var fans to the north, the Algerian margin to the south, the Balearic margin and the Valencia fan to the west and the Sardinian–Corsican margin to the east; the continental shelves vary from the 200 km-wide Gulf of Lions to less than 50 km along Corsica, Sardinia and North Africa (Rothwell *et al.* 1998; Hoogakker *et al.* 2004; Zúñiga *et al.* 2007). Sources for turbidity currents have been documented from several of these continental margins and include earthquake-triggered events associated with tsunami waves, such as the 2003 Boumerdès event (Cattaneo *et al.* 2012).

The most striking morphological feature of the BAP is the presence of a thick Messinian evaporite layer, generating salt walls and diapirs at the seafloor (Stanley *et al.* 1974). Another key feature of the BAP is a thick layer identified in the abyssal plain by its acoustically transparent signature in sub-bottom profiles, known as a 'megaturbidite' (Rothwell *et al.* 1998), which extends over the entire abyssal plain and reaches a thickness of 8–10 m in the centre of the basin (Rothwell *et al.* 1998, 2000). The longest core records of the BAP megabed are the giant Calypso piston cores recovered during the PALAEOFLUX campaign in 1995 (Rothwell and Balut 1995; Table 1). These piston cores of up to 32 m in length were recovered along a NNE–SSW transect and were located about 120 km apart from each other (Fig. 1). They were dated to reconstruct the stratigraphic history of the accumulation, which accounts for up to 150 kyr and is constituted by about 90% of turbiditic successions and only 10% of hemipelagites (Rothwell *et al.* 2000, 2006; Hoogakker *et al.* 2004; Clare *et al.* 2014). Turbidite bases towards the northern part of the BAP are thicker and coarser in grain size, indicating a more likely provenance from the north, according to Hoogakker *et al.* (2004). The attribution of the age of the megabed is based on ten radiocarbon ages above and below the megabed in cores MD95-LC01, -LCO2, -LC04, -LC05 and -LC06 (Rothwell *et al.* 1998). Dennielou *et al.* (2019) performed a recalibration of the five dates on top of the megabed following Reimer *et al.* (2013) and obtained a 2-sigma age of between 20.3 and 20.9 cal ka BP with a median probability of 20.6 cal ka BP, placing it within the Last Glacial Maximum.

Data and methods

Sub-bottom profiles

All available sub-bottom data within the interest area were integrated in a single Kingdom database for interpretation, including, in particular, the newly acquired data during the WestMedFlux surveys (Poort and Lucazeau 2016; Poort and Gorini 2018; Figs 2–5) but also the datasets from the PROGRES (Droz 2003) and SARDINIA (Aslanian and Olivet 2006; Fig. 3) campaigns. Some profiles from the AM-MED survey (Rabineau *et al.* 2013) were also

Table 1. *Sediment cores in the Balearic Abyssal Plain with entire or partial recovery of the megabed, including the coordinates, water depth, basin sector, distribution of sediment facies within the megabed and megabed thickness*

Core name	Lat. N	Lon. E	Water depth (m)	Location within the BAP	Megabed sampled (Y/N)	Top megabed (clay) (mbsf)	Top silt of megabed (mbsf)	Top sand of megabed (mbsf)	Base megabed (mbsf)	Total megabed thickness (m)
MD81-LC01	40° 15.82′	6° 53.23′	2845	Abyssal Plain	Y	14.5	19.7	21.8	22.8	8.3
MD81-LC02	39° 31.88′	6° 22.58′	2860	Abyssal Plain	Y	8.7	15.0	/	16.4	7.7
MD81-LC04	38° 39.01′	6° 06.64′	2855	Abyssal Plain	Y	6.2	14.4	/	16.8	10.6
MD81-LC05	38° 00.92′	5° 29.95′	2845	Abyssal Plain	Y	6.0	10.2	10.9	11.4	5.4
MD81-LC06	38° 00.66′	7° 11.09′	2845	Abyssal Plain	Y	14.5	19.0	19.2	20.2	5.7
PSM-KS21	37° 09.953′	2° 42.127′	2779	Algerian Margin	Y	8.8	/	/	8.9	0.1
PSM-KS23	37° 27.093′	2° 59.526′	2775	Algerian Margin	Y	4.9	/	/	5.2	0.3
PSM-KS27	37° 18.102′	3° 24.317′	2791	Algerian Margin	Y	10.5	/	/	11.5	1.0
WMF2-KF02	38° 56.766′	5° 32.448′	2848	East of Minorca	Y	4.6	?	?	>6.99	>2.39
WMF2-KF11	39° 37.168′	6° 09.609′	2848	East of Minorca	Y	2.28	?	?	>5.92	>3.64
WMF-KF10	37° 52.797′	2° 20.883′	2795	South of Majorca	?	3.9?	?	?	4.3	maybe 0.4
WMF-KF11	37° 49.109′	2° 23.578′	2796	South of Majorca	?	3.0?	?	?	3.8	maybe 0.8

BAP, Balearic Abyssal Plain. Cores MD81- come from campaign PALEOFLUX (Rothwell and Balut 1995), cores PSM- come from campaign PRISME (Sultan 2007) and cores WMF- and WMF2- come from campaigns WestMedFlux (Poort and Lucazeau 2016) and WestMedFlux2 (Poort and Gorini 2018), respectively.

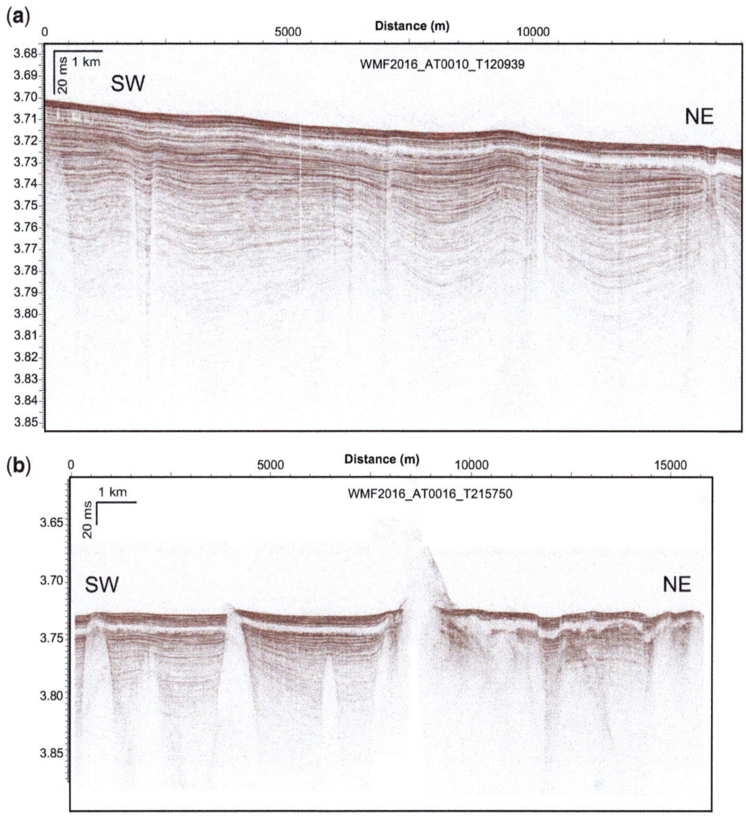

Fig. 2. Sub-bottom profiles from campaign WestMedFlux (Poort and Lucazeau 2016) with images of the BAP megabed identifiable as an acoustically transparent layer: (**a**) profile WMF2016-AT010_T120939 shows the acoustically transparent unit of the BAP megabed less than 10 m below the seafloor, with gentle folds following the geometry of underlying strata and a gradual pinchout of the unit towards the SW; (**b**) profile WMF2016-AT0016_T215750 shows the presence of faults displacing the megabed in the central part of the basin.

integrated. Successive processing steps were applied onboard: correlation (signal compression), correction of the spherical divergence, compensation of the movements of the ship (pitch, roll and heave), calculation of the location of the source and receptors and correction for the delay. The import of sub-bottom data in the Kingdom database was performed in amplitude, and the envelope was computed for each SEG-Y file. The locations of the cores were also loaded into the database so that they could be correlated with the seismic data (Table 1).

Sediment cores

Sediment cores reaching the top or the entire BAP megabed are summarized in Table 1, including cores presented in the literature and in this study (Figs 6 and 7). In most cases, the identification of the BAP megabed is based on two criteria for cores in the central part of the abyssal plain: (1) the identification of a turbidite layer outpacing in thickness the encasing successions of alternating hemipelagites and thin-bedded turbidites; and (2) the attribution to a comparable age range as that reported by Rothwell *et al.* (1998) and recalibrated by Dennielou *et al.* (2019). Grey, homogeneous mud beds identified in cores PSM-KS21 (*c.* 8.8 mbsf (metres below seafloor)), PSM-KS23 (*c.* 5 mbsf) and PSM-KS27 (*c.* 11 mbsf) are correlated to the BAP megabed mainly based on age attribution (Table 2) but also on sedimentary facies characteristics, mainly the sediment colour (different from the adjacent beds) and the homogeneous structureless texture (Fig. 8).

Sediment core WMF2-KF11 was positioned on seismic profile WMF2016-AT006 by adopting a velocity of 1529 ms^{-1} based on an averaged value of the Multi Sensor Core Logger data (P-wave velocity) measured on the whole core (Fig. 6).

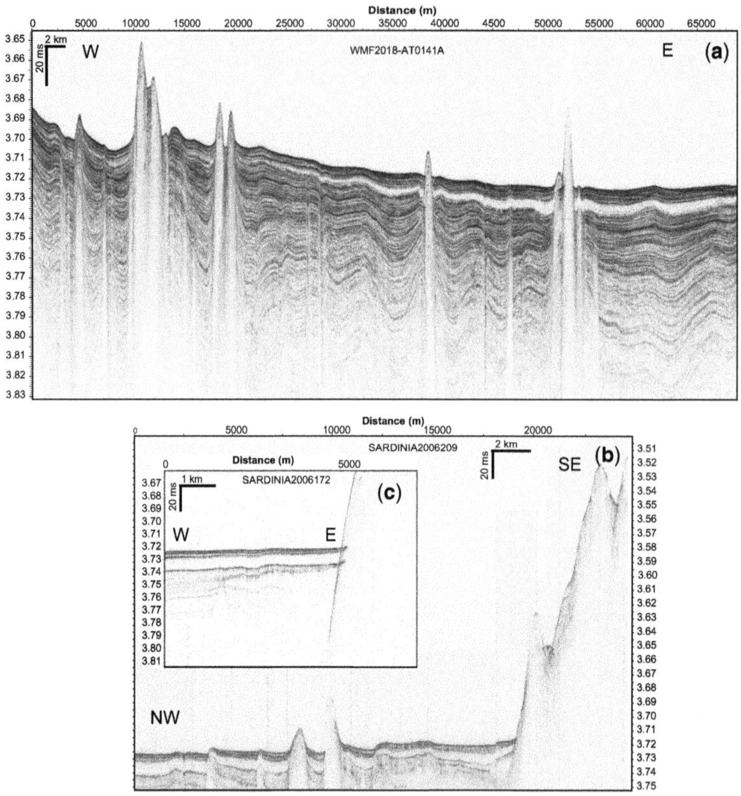

Fig. 3. (**a**) Profile WMF2018-AT141A with evidence of the pinchout of the BAP megabed towards the west beyond the salt domes. The lateral termination of the megabed seems to reach the offset of 15 000 m, but it actually goes further west (offset 1000 m). (**b**) Profile SARDINIA2006209 with abrupt eastward termination of the BAP megabed on the steep Sardinia margin testified by diffraction hyperbolae and thickness variability due to the presence of salt domes towards the NW. (**c**) Profile SARDINIA2006172 with a detail of the abrupt eastward termination against the Sardinia margin.

Radiocarbon dates on core WMF2-KF11 were obtained by accelerating mass spectrometry (AMS) at Beta Analytic laboratory and calibrated with the MARINE 13 database (Reimer *et al.* 2013) (Table 2).

Geotechnical analyses were performed on core WMF2-KF11 (Fig. 9). They included measurements on the freshly opened core (water content, vane shear, P-wave velocity with a celerimeter). Values of compressional wave velocities, undrained shear strength (peak and residual) and water content were measured on split sections of core WMF2-KF11 using devices and approaches as reported in Miramontes *et al.* (2018). Four oedometer tests with incremental loading were also carried out on preserved samples from the same core, two within the megabed and two in overlying sediments, to determine the recompression (Cr), compression (Cc) and swelling (Cs) indices as well as the yield stress ratio (YSR) of the sediments.

Results

Expression and variability of the BAP megabed in sub-bottom profiles

With the available bathymetric and seismic reflection datasets, it is difficult to have an exhaustive view of the lateral variability of the BAP megabed due to the lack of complete coverage of the abyssal plain. The BAP is characterized by variable morphologies and structures linked to the halokinesis of Messinian evaporites, creating a highly variable seafloor landscape. To account for such variability, we present here some evidence of the expression of the megabed in distinct sectors of the abyssal plain, including megabed terminations (Fig. 3), erosion at the base of the megabed (Fig. 4) and internal subdivisions (Figs 5–7). The BAP megabed is identifiable as an acoustically transparent layer all over the area outlined by a black line in Figure 1 and corresponds to

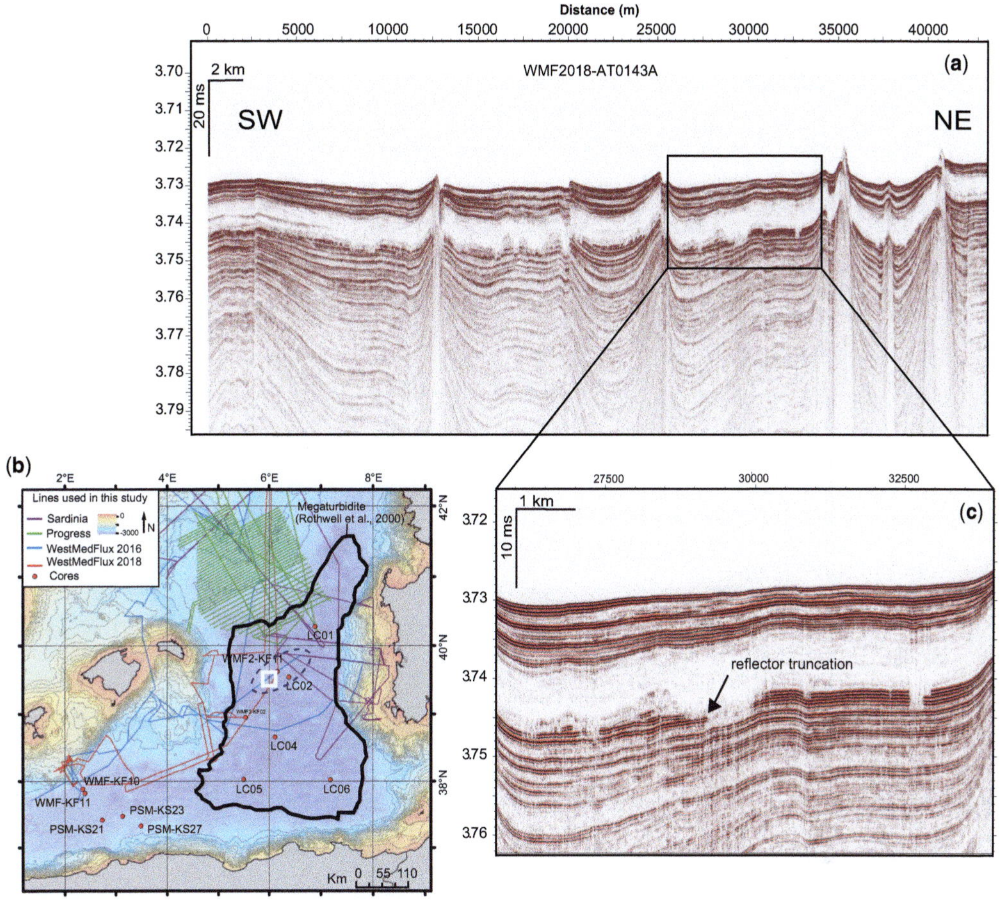

Fig. 4. Sub-bottom profile (SBP) WMF2018-AT143A in a sector where the BAP megabed is relatively isopach, folded by salt diapirs and cut by vertical faults showing metric displacement (**a**). (**b**) Map with location of the seismic profile (white square). The dashed blue line shows the area with the occurrence of seismic reflector truncations at the base of the megabed. The pop-up image (**c**) shows evidence of erosion at the base of the BAP megabed with apparent truncation of seismic reflectors.

the limits of this layer according to Rothwell *et al.* (1998, 2000), i.e. to the 2800 m isobath. Sub-bottom profile WMF2016-AT010 shows the laterally continuous, acoustically transparent unit of the BAP megabed less than 10 m below the seafloor, with gentle folds following the geometry of underlying strata and a gradual pinchout of the unit towards the west (Fig. 2a). In sub-bottom profile WMF2018-AT141A, however, the same acoustically transparent layer still presents a good degree of lateral continuity but with an accentuated effect of deformation due to the presence of salt domes; it is also possible to lose the lateral continuity of the layer and find it beyond the highest salt domes (see Fig. 3a towards the west). In Figure 2b, sub-bottom profile WMF2016-AT0016_T215750 shows the presence of faults in the central part of the basin, which are most likely linked to salt dome movements, displacing the megabed by several metres either in distinct locations (Fig. 2b, towards the west) or in sets of narrowly spaced faults (Fig. 2b, towards the east). By appreciating the geometry of the BAP megabed and that of the underlying strata, it is possible to deduce that the halokinetic deformation postdates the emplacement of the megabed.

Another characteristic feature of sedimentary bodies, and of their expression as seismic units, is the geometry of pinchout terminations. Two contrasting styles of megabed termination exist on two opposite sides of the abyssal plain (Fig. 3). Towards the west, the pinchout of the BAP megabed seems to reach the offset of 15 000 m, but it actually goes gradually further west beyond an area with salt domes (offset 1000 m in Fig. 3a). Towards the

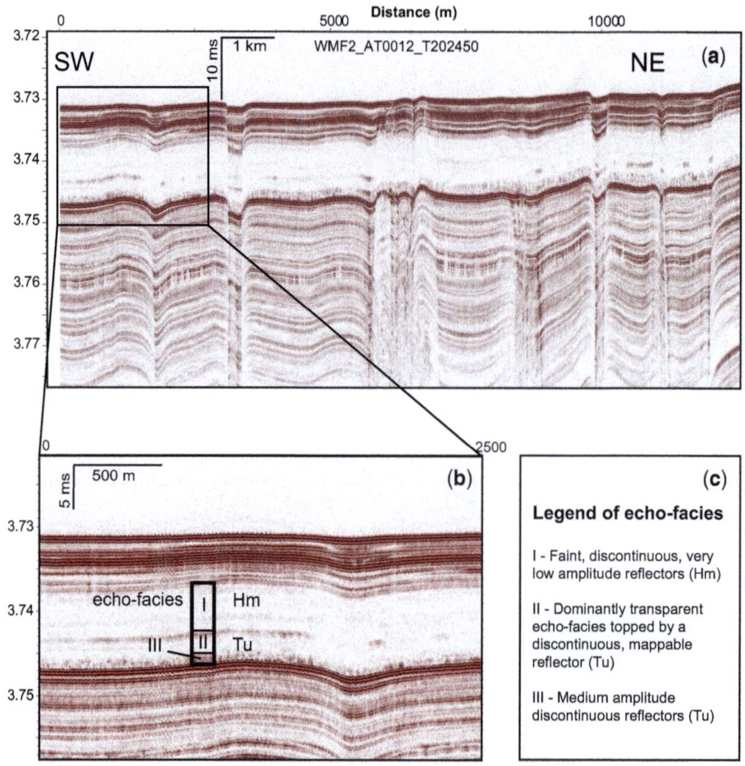

Fig. 5. (**a**) Profile WMF2_AT0012_T202450 where the BAP megabed is dissected by several vertical faults (location in Fig. 1). (**b**) The pop-up image shows three distinct echo-facies within the BAP megabed: I, faint discontinuous and very low amplitude reflectors echo-facies; II, dominantly transparent echo-facies topped by a discontinuous but mappable reflector; and III, basal unit with a discontinuous base and top showing medium-amplitude discontinuous reflectors. (**c**) Legend of the echo-facies.

east, the termination of the acoustically transparent unit is against the Sardinia margin, which is relatively steep with a slope angle exceeding 11° (Deiana *et al.* 2016). The termination is abrupt, passing from about 10 m in thickness to zero, and masked by the hyperbolae of the Sardinia margin (Fig. 3b, c). In this latter case, the termination is reminiscent of the first identifications of 'homogenites' in the Eastern Mediterranean by Kastens and Cita (1981), who described these deposits as lying flat within each basin with abrupt terminations against the steep basin walls. The megabed thickness does not decrease gradually towards the west; instead, values irregularly increase towards the centre of the basin due to deformation by salt domes (Fig. 3b).

The contact of the megabed with the underlying strata is conformable in the majority of the analysed seismic lines (Figs 2 and 6), but it is irregular in a part of the central area of the BAP (Fig. 4). A diagnostic indicator of the processes generating turbidite beds is the erosional nature of the basal surfaces of the beds. Figure 4a, b presents evidence of seismic reflectors truncated below the base of the megabed; we interpret these features, present mostly in the area within the dashed blue line in Figure 4c, as localized erosional features. An alternative interpretation might suggest that these features have resulted from fluid escape, but this is unlikely because of the lack of evidence of vertical anomalies in the seismic profiles that could be identified as fluid escape chimneys. Other sub-bottom profiles in this central-western area of the abyssal plain show similar features.

Far from being fully homogeneous, the dominantly acoustically transparent megabed also presents a certain variability in its acoustic response. We summarize this variability in three slightly differentiated seismic echo-facies (Damuth 1980; Fig. 5). Echo-facies I, below the top of the megabed, presents faint discontinuous and very low amplitude reflectors. Echo-facies II is dominantly transparent, but it is topped by a discontinuous seismic reflector mappable in several areas where the megabed is relatively thick. Finally, echo-facies III is characterized by a discontinuous base and top showing

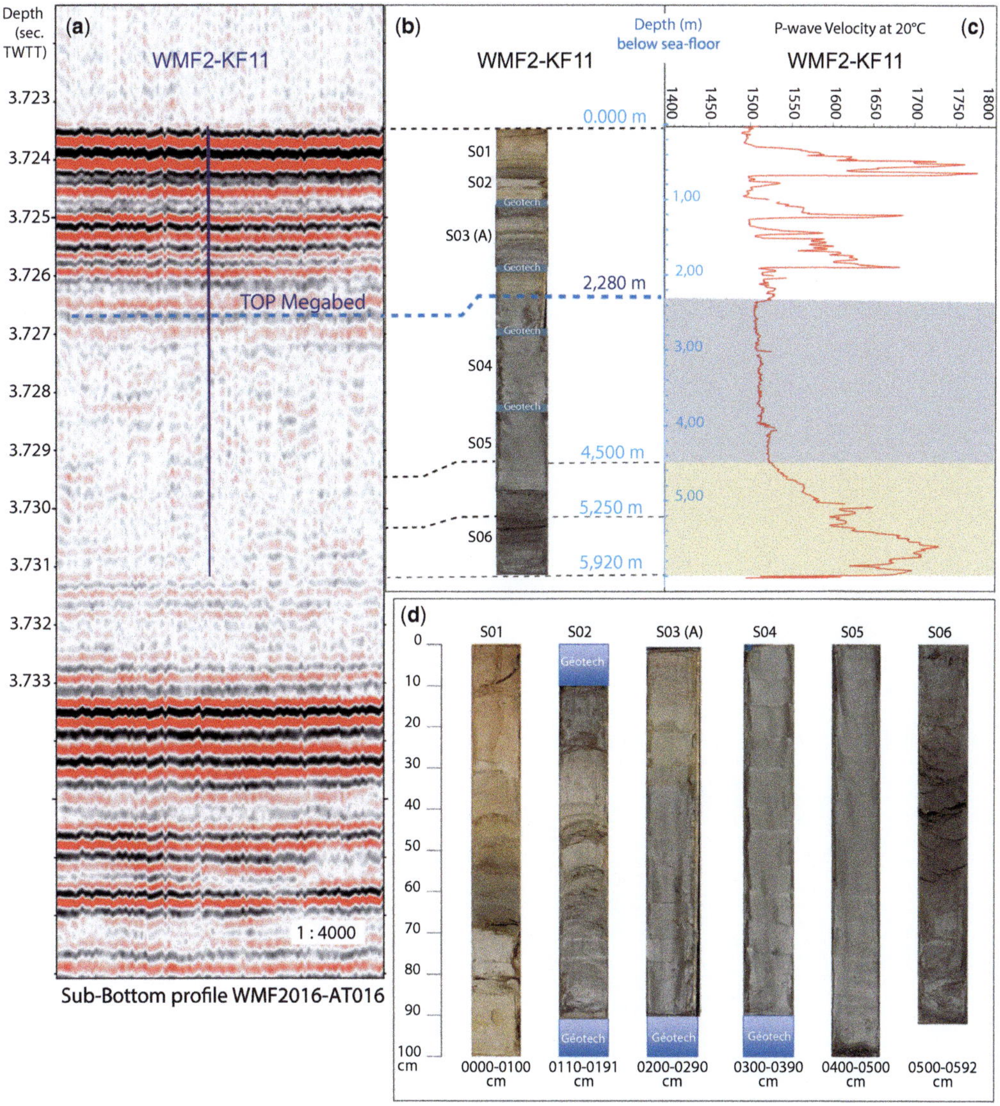

Fig. 6. (**a**) Portion of SBP WMF2016-AT016 at the site of sediment core WMF2-KF11 (location in Fig. 1). The core identifier for WMF2-KF11 is http://igsn.org/BFBGX-134138. (**b**, **d**) Photographs of core WMF2-KF11. Note the four intervals in blue where whole core samples for oedometer tests were reserved: two within the homogeneous clay of the megabed and two in overlying layers characterized by alternating thin beds of sand and silt with dominant hemipelagic silty clay (the latter two samples were from fine-grained sediment). (**c**) P-wave velocity log of the core showing clear identification of three lithological units.

medium-amplitude discontinuous reflectors; it also occurs in association with local erosional features (Fig. 4b). Echo-facies III is present at the base of the megabed, where it reaches a thickness exceeding *c.* 10 m in the central area of the abyssal plain (Fig. 4c). The three echo-facies cannot be identified thoroughly over the entire extent of the megabed.

Sediment character of the BAP megabed

The lithology and sedimentary structures of the BAP megabed from the literature were compared with the new observations on core WMF2-KF11. Table 1 shows the identification of the BAP megabed, its grain-size divisions and thickness in

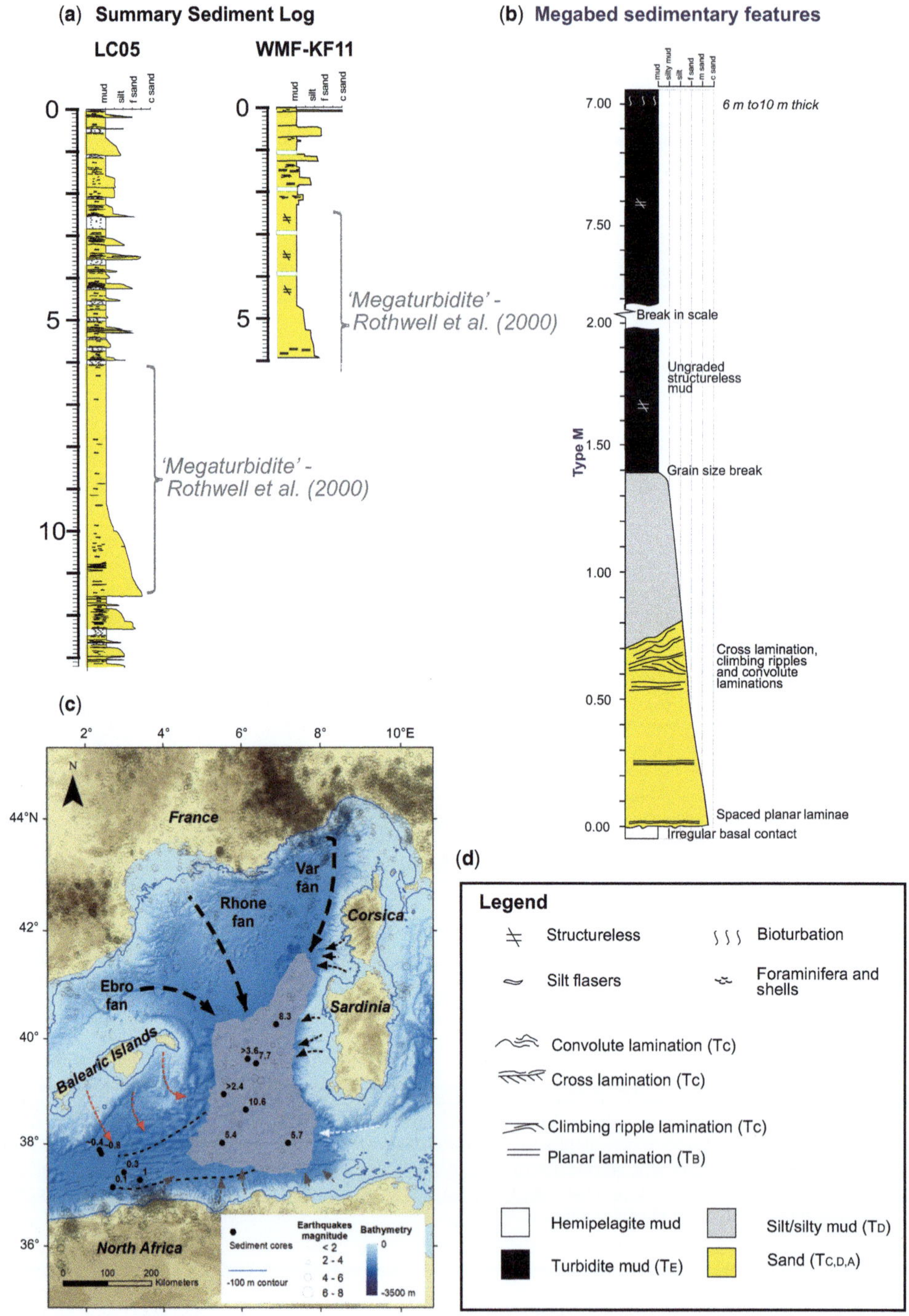

Fig. 7. (**a**) Lithological log of sediment cores LC05 (only the upper *c.* 13 m, modified from Clare *et al.* 2014) and WMF2-KF11. Note that the majority of the sedimentary succession is represented by turbidites. (**b**) Synthetic sedimentary structures of the megabed. (**c**) Map showing the possible sources of the megabed (arrows, modified from Clare *et al.* 2014); thick black arrows from the major Western Mediterraneans fans, black arrows from Corsica and Sardinia margins, grey arrows from the Algerian margin, red arrows from the Balearic Islands and white arrow from the Tyrrhenian Sea. Map also shows the thickness value of the megabed (from Table 1) and the earthquake epicentres of $M_w > 2$ from the EMSC Catalogue (http://www.emsc-csem.org).

Table 2. *List of radiocarbon ages*

Core	Sample interval (cm)	Foram. species	Lab. code	Age (^{14}C years)	Error ±	Age min. (cal years BP)	Age max. (cal years BP)	Source
PSM-KS21	837–838	*Globigerinoides ruber*	Poz-68081	16 790	90	19 383	19 881	Babonneau *et al.* (2017)
PSM-KS23	423–424.5	*Globigerinoides ruber*	Poz-68119	15 270	80	17 806	18 521	Babonneau *et al.* (2017)
WMF2-KF11	1.51–1.52	*Globigerina bulloides*	Beta-508507	10 530	30	11 250	12 031	This study
WMF2-KF11	2.02–2.03	*Globigerina bulloides*	Beta-508508	14 450	50	16 595	17 385	This study
WMF2-KF11	2.10–2.11	*Globigerina bulloides*	Beta-508509	17 030	60	19 652	20 325	This study

Lab. code 'Poz' indicates measurements by Poznan Radiocarbon Laboratory (Poland), while 'Beta' is for Beta Analytic. Age min. and age max. correspond to the calibrated ages (2σ; cal years BP) calculated by the CALIB program with the calibration curve Marine 13 (Reimer *et al.* 2013), with a reservoir age of 400 years.

ten cores: five MD81 cores published in Rothwell *et al.* (1998), three PSM cores presented in Babonneau *et al.* (2017) and two WMF2 cores presented in this study (Poort and Gorini 2018). The presence of the BAP megabed is also suggested in two additional WMF cores (Poort and Lucazeau 2016). In general, cores located within the limits identified by Rothwell *et al.* (1998) and reaching the megabed present an overall bipartite lithology with a normally graded, sandy to silty base capped by a homogeneous clay unit that can be several metres thick. The megabed can thus be described as a HmTu couplet, since it is composed of two distinct components (Campos *et al.* 2013; see also McHugh *et al.* 2006).

Sediment core WMF2-KF11 was collected at 2848 m water depth in the central part of the Balearic Abyssal Plain. It is 5.92 m long and presents brownish-yellow silty clay with foraminifera and shell fragments in the upper *c.* 30 cm, gradually passing to alternating very thin to medium thick, normally graded sandstones and olive-grey silty clays down to 2.28 mbsf (Figs 6b, d and 7). This interval is interpreted as alternating turbidites and hemipelagites. The rest of the core down to 5.92 mbsf is composed of homogeneous dark grey silty clay (2.28–4.50 mbsf), dark grey clayey silt with sub-millimetric laminae (4.50–5.25 mbsf) and dark grey structureless, possibly normally graded, silt. The entire interval from 2.28 to 5.92 mbsf is interpreted as part of the BAP megabed and includes the Hm division (2.28–4.50 mbsf) and part of the Tu division (*sensu* Campos *et al.* 2013). By using a conversion velocity of 1529 m s^{-1} based on Multi Sensor Core Logger (MSCL) data (Fig. 6c), the core was projected on the sub-bottom profile WMF2016-AT016 (Fig. 6a). It is possible to tentatively correlate the three echo-facies to sedimentary facies: echo-facies III would correspond to sandy turbidite facies, echo-facies II to silty turbidite facies, and echo-facies I to homogenite facies.

The same tripartition of sedimentary facies within the megabed in core WMF2-KF11 is also visible in core LC05, which is from the southwestern sector of the BAP (Fig. 7). The megabed presents a sharp and irregular (eroded) base and an upper gradational and bioturbated contact. Its grain size is coarse sand to clay with positive grading, although negative grading may exist in the coarse fraction at the base. There is an overall upward-fining trend to the homogeneous mud unit at the top, with a slight grain-size break at the transition from silt to mud (Fig. 7b). The sedimentary structures include a mostly massive basal division with faint planar lamination, which may include planar intrabed contacts at the base and a transition into cross-bedded, climbing ripples and convolute laminae (Fig. 7b).

It is likely that the megabed represents a ponded megaturbidite, which resulted from a steady, long-lived turbidity current as suggested in the existing literature (e.g. Rothwell *et al.* 1998, 2000; Clare *et al.* 2014). Planar laminae formed as a result of grain flow at the head of the turbidity current. The rest of the flow settled out due to low density turbulent suspension, and/or gel strength and/or excess pore pressure.

A transect of sediment cores offshore the Algerian margin (PSM3-KS21, PSM3-KS23 and PSM3-KS27; Fig. 8), fully described in Babonneau *et al.* (2017, fig. 7), shows a sedimentary succession of fining-upward, sandy/silty layers (turbidites) alternating with thin mud layers (hemipelagites). A mud layer is distinctive in photographs due to its grey colour being different from that of other encasing layers. The age constraints above and below the grey bed (Babonneau *et al.* 2017; Table 2) suggest that it is possible to identify it as the distal expression of the BAP megabed. This identification of the BAP megabed in distal cores goes up to 200 km beyond the pinchout of the megabed, as is visible in seismic lines, and suggests that other cores, such as cores WMF-KF10 and -KF11 south of the Balearic Islands, could also contain the same layer (Fig. 1; Table 1).

Mechanical properties of the BAP megabed in core WMF2-KF11

A marked change in both density and V_p profiles occurs at 2.28 m in core WMF2-KF11 (Fig. 9). This change is reflected by YSRs higher than 1 above 2.28 m and lower than 1 below 2.28 m. However, apart from the recompression indices, the other indices determined from oedometer tests do not show a clear change; this contrasts with the values of undrained shear strength (S_u), which decrease abruptly below 2.28 m (Fig. 9). Comparisons with the trends generally exhibited by normally consolidated sediments ($S_u = 0.2\,\sigma'v$ or $0.3\,\sigma'v$; e.g. Garziglia *et al.* 2010) can be reconciled with YSR values to indicate that below 2.28 m the sediment is underconsolidated.

Discussion

Overall distribution of the BAP megabed and inference on sources

The BAP megabed is reported in the literature as extending in the BAP identified by the 2800 m isobath over a surface of 60 000 km^2 for a volume

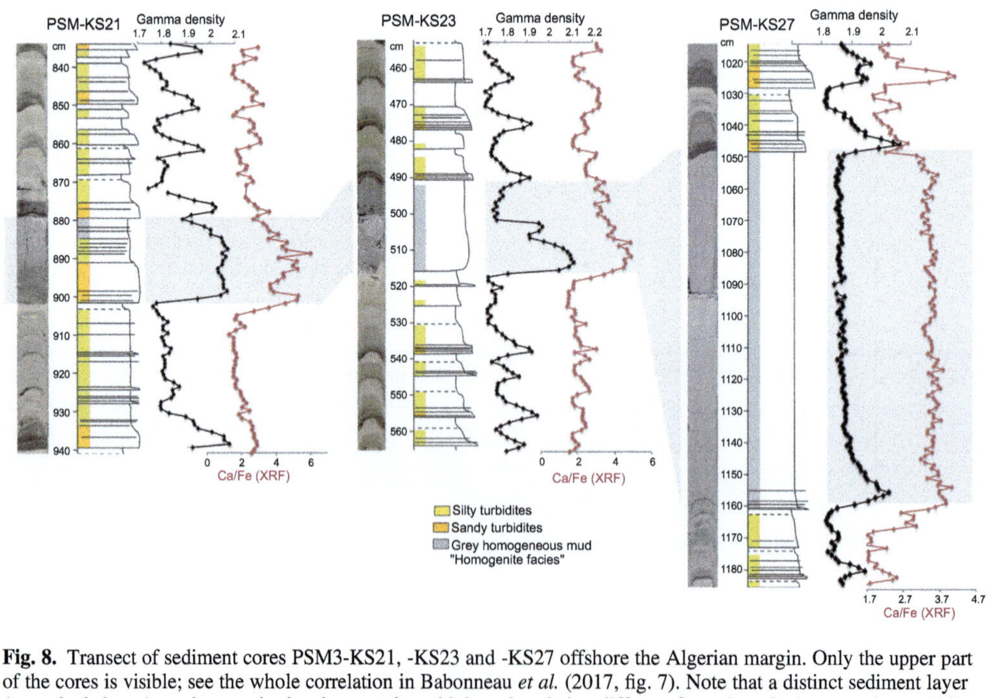

Fig. 8. Transect of sediment cores PSM3-KS21, -KS23 and -KS27 offshore the Algerian margin. Only the upper part of the cores is visible; see the whole correlation in Babonneau *et al.* (2017, fig. 7). Note that a distinct sediment layer (grey shaded area) can be seen in the photographs, with its colour being different from that of other encasing turbidity layers; its age is compatible with that of the BAP megabed (Table 2). The core identifiers are PSM-KS21 (http://igsn.org/BFBGX-86346), PSM-KS23 (http://igsn.org/BFBGX-86348 and PSM-KS27 (http://igsn.org/BFBGX-86352).

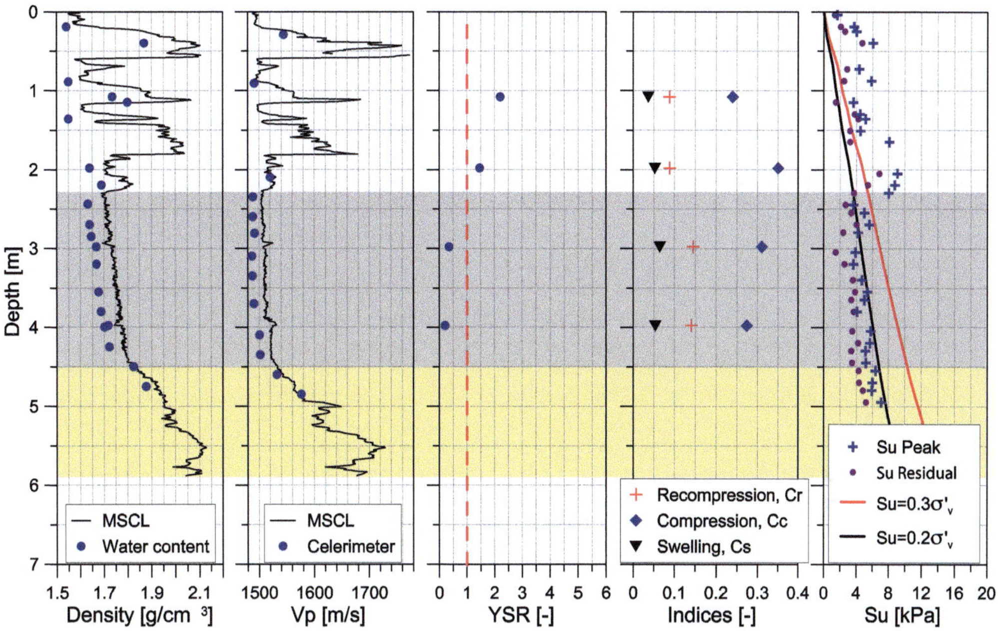

Fig. 9. Physical and geotechnical parameters measured in core WMF2-KF11. From left to right: density from Multi Sensor Core Logger (MSCL) data (black line) and from water content on discrete samples (blue dots); P-wave velocity from MSCL data (black line) and discrete measurements on half core via celerimeter (blue dots); derived parameters from oedometer tests – yield stress ratios (YSRs) and recompression, compression and swelling indices; and undrained shear strength (Su) measured by the Vane test in the half core – normally consolidated sediment would fall within the black and red lines. Note the abrupt change of several parameters at 2.28 mbsf. The grey shading corresponds to the depth of the homogenite part of the BAP megabed (2.28–4.50 mbsf), whereas the yellow shading is the silty/sandy portion of the megabed sampled by the core.

of 500 km^3 (Rothwell *et al.* 1998) and, successively, over a surface of 77 000 km^2 for a volume of 600 km^3 (Rothwell *et al.* 2000; Hoogakker *et al.* 2004). We find, thanks to the identification of a distal correlation of the megabed to the west (Fig. 8), that the megabed extends 200 km further along the central Algerian basin (Fig. 1, dashed lines). The resulting extent would thus be up to 90 000–100 000 km^2, while the total volume would not be affected much by this extension and would be up to 640 km^3 because of the limited thickness of this part of the megabed (<1 m; Fig. 8).

Beyond its considerable size, the importance of the information from the analysis of the BAP megabed consists in its exceptional state of preservation and basinwide extent. The BAP megabed has preserved much detail that is missing in many other more ancient settings, but unfortunately the extent of the dataset is limited and uneven. Core data (Table 1) show that the megabed thickness is at its maximum in the deepest central part of the BAP (core LC04). The contrasting styles of the megabed terminations, with a gradual pinchout towards the west and an abrupt termination with significant thickness to the east, could be related to the distance from the source or to an enhanced effect of ponding with possibly oblique reflection against the steep Sardinia margin (*sensu* Kneller *et al.* 1991).

As in the case of several ancient megabeds in outcrops, the provenance problem cannot be solved simply by a single argument. The sediment provenance of the Oligocene Numidian megabed has been debated and attributed to European sources based on zircon analysis (Fildes *et al.* 2010), to African provenance (Thomas *et al.* 2010) or, again, African based on a series of argumentations, including palaeogeographical reconstructions (Stow *et al.* 2010). The case of the BAP megabed could be theoretically simpler, because palaeogeography during the Last Glacial Maximum can be inferred with a higher degree of certainty and corresponds to the present setting of the megabed with a sea-level lowered by about 120 m during the Last Glacial Maximum (Fig. 7c, modified from Clare *et al.* 2014). Nevertheless, the solution is not easy and several independent bodies of evidence have to be explored, possibly with the support of future geochemical investigations. The location of historical seismicity, concentrated in three sectors (Ligurian, Pyrenean and Algerian margins), could suggest possible source areas of mobilized sediment following major earthquakes (Fig. 7c).

Internal structure of BAP megabed and inferred sedimentary processes

Hoogakker *et al.* (2004) and Rothwell *et al.* (2006) found that over the last 130 kyr over 90% of the sediment accumulation in the BAP was composed of turbidites and that the overall amount of erosion was negligible, allowing a rather continuous record of hemipelagites. By analysing the sub-bottom profiles in detail, we found that the base of the BAP megabed could be, in part, an exception to this general rule, and we think that this fact could be interesting for reconstructions of the sedimentary processes. Even though the interpretation of the erosional base of turbidite beds is far from straightforward (Eggenhuisen *et al.* 2011), the evidence of seismic reflector truncations and, in particular, the location of zones of enhanced basal erosion could suggest areas of the BAP impacted by stronger turbidity flows in the centre of the BAP (Fig. 4). The distribution of the erosional areas in available profiles suggests that the area of erosion is oriented roughly north–south or NE–SW; it is possible that this orientation represents a direction of axial transport of the main initial flow in the deepest part of the basin.

The notion of turbidity currents running very long distances down to the abyssal plains has been demonstrated in several cases, both in recent and older case studies (e.g. Zuffa *et al.* 2000; Clare *et al.* 2014). Zúñiga *et al.* (2007) demonstrated that turbidity currents originating from the continental margins surrounding the BAP can reach the centre of the basin also during sea-level highstands as has happened in the present one. An implication in the long distance of a turbiditic flow is the series of flow transformations that could occur along the turbidity flow path (Haughton *et al.* 2009; Talling 2013; Fallgatter *et al.* 2016). The tripartite structure of the megabed found in seismic profiles and in sediment cores might reflect flow transformations, even if, given the scarce spatial coverage of the data, it is difficult to reconstruct a continuous scenario of flow evolution. We agree with former interpretations of sediment cores (e.g. Rothwell *et al.* 2000; Clare *et al.* 2014) in considering the megabed as a single event, and we think that the presence/absence of tripartition in echo-facies could be due to changes in flow velocity possibly linked to seafloor irregularities, such as the presence of salt domes.

Finally, the homogenite shows peculiar physical and geotechnical properties, in particular the consolidation state (underconsolidated). The megabed appears to be composed of sediment that has remained underconsolidated since its deposition. Water incorporation during the flow may explain the underconsolidated state of the sediments during tens of years after rapid deposition. However, since burial has ensued over the last *c.* 20 kyr, one might expect this characteristic signature to have been erased upon consolidation, unless the transport and depositional processes gave the deposit an atypical structure (a combination of particle arrangement and bounding). This underconsolidation could be suggestive of an extremely rapid and water-saturated deposition of the megabed, with the clay on top acting as a seal, and could further support the hypothesis of a relatively rapid accumulation from a single event. *In situ* tests, such as the cone penetration test (undrained) (CPTU) with measurement of excess pore pressure, could thus help confirm this observation.

Conclusions

A large deposit sitting on the Balearic Abyssal Plain – originally mapped using the extent of the acoustically transparent reflection – has been redefined and remapped to discover that it is larger than previously thought. The Balearic Plain megabed is one of the most known examples of megabeds in abyssal plains and an analogue to megabeds studied in the outcrop of ancient rocks. It occupies the whole abyssal plain over an extent of at least 77 000 km^2, but likely up to 90 000–100 000 km^2, with a thickness of 8–10 m in the centre of the basin and a volume of up to 640 km^3 that is comparable to giant landslide deposits. The megabed is affected by faults and halokinetic deformation (salt domes) suggesting recent tectonic activity (post-Last Glacial Maximum) based on the age of the deformed megabed.

The internal geometry of the BAP megabed presents three faint but distinctive echo-facies from bottom to top, which correspond tentatively to the three sedimentary facies identified in cores possibly linked to flow transformations within a single event bed. The analysis of new data in the area evidenced some potentially diagnostic details of its geophysical appearance – in particular, the presence of an abrupt pinchout on the eastern Sardinian margin contrasting with the gradual pinchout towards the west. This suggests that there could have been some reflected turbidite pattern against the Sardinia margin.

In the central area of the abyssal plain, the basal layer of the BAP megabed shows evidence of the erosional truncation of underlying deposits, which might indicate enhanced erosion of turbiditic flows and/or proximity of sedimentary sources; however, a composite source from several margins cannot be discarded, due to the present state of knowledge.

Further work on the geochemistry of sediment cores could help to identify the source areas and, thus, the geographical location of the predisposing mechanisms that influenced the deposition of the Western Mediterranean megabed and will further enhance the understanding of geohazards in the Mediterranean region.

Acknowledgements We thank the captains, crew and science parties of campaigns WESTMEDFLUX (R/V *L'Atalante* 2016, https://doi.org/10.17600/16000600), WESTMEDFLUX2 (R/V *L'Atalante* 2018, https://doi.org/10.17600/18000402), AM-MED (R/V *Le Suroit*, 2013, https://doi.org/10.17600/13020050), PRISME (R/V *L'Atalante* 2007, https://doi.org/10.17600/7010090), SARDINIA (R/V *L'Atalante*, 2006, https://doi.org/10.17600/6010150 and PROGRES (R/V *Le Suroit* 2003, https://doi.org/10.17600/3020080). We thank core repositories BOSCORF (NOC, Southampton, UK) and CREAM (IFREMER, Géosciences Marines, Brest, France) and their staff for support in core maintenance/handling and assistance with analytical techniques. This contribution would not have been possible without the valuable suggestions and encouragement of reviewers C. McHugh and A. Fildani and the scientific guidance and perseverance of Guest Editor A. Georgiopoulou.

Funding This project has received funding from the European Union's Horizon 2020 research and innovation programme under the Marie Skłodowska-Curie grant agreement no. 721403 (ITN SLATE project, PhD of SB).

Author contributions **AC**: conceptualization (lead), methodology (equal), resources (lead), supervision (lead), writing – original draft (lead), writing – review & editing (lead); **SB**: conceptualization (equal), formal analysis (equal), methodology (equal), software (lead), writing – review & editing (equal); **CC**: data curation (equal), formal analysis (equal), methodology (supporting), software (equal), writing – review & editing (supporting); **MB**: formal analysis (equal), supervision (equal), writing – review & editing (supporting); **EL**: conceptualization (equal), formal analysis (equal), funding acquisition (equal), software (equal), supervision (lead), writing – original draft (equal), writing – review & editing (equal); **NB**: conceptualization (equal), formal analysis (equal), investigation (equal), writing – original draft (supporting), writing – review & editing (equal); **SG**: data curation (equal), methodology (lead), software (equal), writing – original draft (equal), writing – review & editing (equal); **JP**: conceptualization (equal), funding acquisition (lead), project administration (lead), supervision (equal), writing – original draft (supporting), writing – review & editing (supporting); **GGA**: conceptualization (supporting), data curation (supporting), investigation (supporting), methodology (equal), software (equal), visualization (lead); **GB**: conceptualization (equal), formal analysis (supporting), methodology (equal), supervision (equal), writing – review & editing (supporting); **BD**: conceptualization (equal), methodology (equal), supervision (equal), writing – review & editing (equal); **GJ**: conceptualization (supporting), writing – review & editing (supporting); **SM**: conceptualization (equal), data curation (equal), funding acquisition (equal), investigation (lead), project administration (equal), resources (equal), supervision (equal), writing – original draft (supporting), writing – review & editing (supporting); **MR**: conceptualization (supporting), formal analysis (supporting), funding acquisition (equal), resources (supporting), supervision (supporting), writing – review & editing (supporting); **LD**: conceptualization (equal), formal analysis (equal), funding acquisition (equal), investigation (equal), supervision (lead), writing – review & editing (supporting); **MC**: conceptualization (equal), data curation (lead), investigation (equal), methodology (equal), writing – original draft (supporting), writing – review & editing (equal).

References

Amy, L.A. and Talling, P. 2006. Anatomy of turbidites and linked debrites based on long distance (120–30 km) bed correlation, Marnoso Arenacea Formation, Northern Apennines, Italy. *Sedimentology*, **53**, 161–212, https://doi.org/10.1111/j.1365-3091.2005.00756.x

Aslanian, D. and Olivet, J.L. 2006. SARDINIA cruise, RV L'Atalante, https://doi.org/10.17600/6010150

Babonneau, N., Cattaneo, A., Ratzov, G., Déverchère, J., Yelles-Chaouche, A., Lateb, T. and Si Bachir, R. 2017. Turbidite chronostratigraphy off Algiers, central Algerian margin: a key for reconstructing Holocene paleoearthquake cycles. *Marine Geology*, **384**, 63–80, https://doi.org/10.1016/j.margeo.2016.10.017

Badhani, S., Cattaneo, A. *et al.* 2019. Morphology of retrogressive failures in the Eastern Rhône interfluve during the last glacial maximum (Gulf of Lions, Western Mediterranean). *Geomorphology*, **351**, https://doi.org/10.1016/j.geomorph.2019.106894

Beck, C., Mercier de Lépinay, B. and the MARMACORE Cruise Party 2007. Late Quaternary co-seismic sedimentation in the Sea of Marmara's deep basins. *Sedimentary Geology*, **199**, 69–85, https://doi.org/10.1016/j.sedgeo.2005.12.031

Bouma, A.H. 1987. Megaturbidite: an acceptable term? *Geo-Marine Letters*, **7**, 63–67, https://doi.org/10.1007/BF02237985

Campos, C., Beck, C., Crouzet, C., Carrillo, E., Van Welden, A. and Tripsanas, E. 2013. Late Quaternary paleoseismic sedimentary archive from deep central Gulf of Corinth: time distribution of inferred earthquake-induced layers. *Annals of Geophysics*, **56**, S0670, https://doi.org/10.4401/ag-6226

Cattaneo, A., Babonneau, N. *et al.* 2012. Searching for the seafloor signature of the 21 May 2003 Boumerdès earthquake offshore central Algeria. *Natural Hazards and Earth System Sciences*, **12**, 2159–2172, https://doi.org/10.5194/nhess-12-2159-2012

Cita, M.B. and Aloisi, G. 2000. Deep-sea tsunami deposits triggered by the explosion of Santorini (3500 y BP), Eastern Mediterranean. *Sedimentary Geology*, **135**, 181–203, https://doi.org/10.1016/S0037-0738(00)00071-3

Cita, M.B., Camerlenghi, A., Kastens, K.A. and McCoy, F.W. 1984. New findings of Bronze Age homogenites in the Ionian Sea: geodynamic implications for the Mediterranean. *Marine Geology*, **55**, 47–62, https://doi.org/10.1016/0025-3227(84)90132-4

Clare, M.A., Talling, P.J., Challenor, P., Malgesini, G. and Hunt, J. 2014. Distal turbidites reveal a common distribution for large (>0.1 km^3) submarine landslide recurrence. *Geology*, **42**, 263–266, https://doi.org/10.1130/G35160.1

Damuth, J.E.D. 1980. Use of high-frequency (3.5–12 kHz) echograms in the study of near-bottom sedimentation processes in the deep-sea: a review. *Marine Geology*, **38**, 51–75, https://doi.org/10.1016/0025-3227(80)90051-1

Deiana, G., Meleddu, A., Paliaga, E.M., Todde, S. and Orrù, P.E. 2016. Continental slope geomorphology: landslides and pockforms of Southern Sardinian margin (Italy). *Geografia Fisica e Dinamica Quaternaria*, **39**, 129–136, https://doi.org/10.4461/GFDQ.2016.39.12

Dennielou, B., Jégou, I. *et al.* 2019. Major modification of sediment routing by a large Mass Transport Deposit in the Gulf of Lions (Western Mediterranean). *Marine Geology*, **411**, 1–20, https://doi.org/10.1016/j.margeo.2019.01.011

Droz, L. 2003. PROGRES cruise, RV Le Suroît, https://doi.org/10.17600/3020080

Droz, L., dos Reis, A.T., Rabineau, M., Berné, S. and Bellaiche, G. 2006. Quaternary turbidite systems on the northern margins of the Balearic Basin (Western Mediterranean): a synthesis. *Geo-Marine Letters*, **26**, 347–359, https://doi.org/10.1007/s00367-006-0044-0

Eggenhuisen, J.T., McCaffrey, W.D., Haughton, P.D.W. and Butler, R.W.H. 2011. Shallow erosion beneath turbidity currents and its impact on the architectural development of turbidite sheet systems. *Sedimentology*, **58**, 936–959, https://doi.org/10.1111/j.1365-3091.2010.01190.x

Fallgatter, C., Kneller, B., Paim, P.S.G. and Milana, J.P. 2016. Transformation, partitioning and flow-deposit interactions during the run-out of megaflows. *Sedimentology*, **64**, 359–387, https://doi.org/10.1111/sed.12307

Fildes, C.T. 2013. *Megabeds: Emplacement Mechanics of Large-Volume Event Beds*. PhD thesis, University of Auckland, https://researchspace.auckland.ac.nz/handle/2292/22136

Fildes, C., Stow, D., Riahi, S., Soussi, M., Patel, U., Milton, A.J. and Marsh, S. 2010. European provenance of the Numidian Flysch in northern Tunisia. *Terra Nova*, **22**, 94–102, https://doi.org/10.1111/j.1365-3121.2009.00921.x

Garziglia, S., Sultan, N. *et al.* 2010. Identification of shear zones and their causal mechanisms using a combination of cone penetration tests and seismic data in the Eastern Niger delta. *In*: Mosher, D. *et al.* (eds) *Submarine Mass Movements and their Consequences*. Advances in Natural and Technological Hazards Research, **28**. Springer, 55–65, https://doi.org/10.1007/978-90-481-3071-9_5

Haughton, P., Davis, C., McCaffrey, W. and Barker, S. 2009. Hybrid sediment gravity flow deposits – classification, origin and significance. *Marine and Petroleum Geology*, **26**, 1900–1918, https://doi.org/10.1016/j.marpetgeo.2009.02.012

Hieke, W., Hirschleber, H.B. and Dehghani, G.A. 2003. The Ionian Abyssal Plain (central Mediterranean Sea): morphology, subbottom structures and geodynamic history – an inventory. *Marine Geophysical Researches*, **24**, 279–310, https://doi.org/10.1007/s11001-004-2173-z

Hoogakker, B.A.A., Rothwell, R.G., Rohling, E.J., Paterne, M., Stow, D.A.V., Herrle, J.O. and Clayton, T. 2004. Variations in terrigenous dilution in western Mediterranean Sea pelagic sediments in response to climate change during the last glacial cycle. *Marine Geology*, **211**, 21–43, https://doi.org/10.1016/j.margeo.2004.07.005

Kastens, K. and Cita, M.B. 1981. Tsunami-induced sediment transport in the abyssal Mediterranean Sea. *Geological Society of America Bulletin*, **92**, 845–857, https://doi.org/10.1130/0016-7606(1981)92<845:TSTITA>2.0.CO;2

Kneller, B., Edwards, D., McCaffrey, W. and Moore, R. 1991. Oblique reflection of turbidity currents. *Geology*, **19**, 250–252, https://doi.org/10.1130/0091-7613(1991)019<0250:OROTC>2.3.CO;2

Mauffret, A., Frizon de Lamotte, D., Lallemant, S., Gorini, C. and Maillard, A. 2004. E–W opening of the Algerian Basin (Western Mediterranean). *Terra Nova*, **16**, 257–264, https://doi.org/10.1111/j.1365-3121.2004.00559.x

McHugh, C.M.G., Seeber, L. *et al.* 2006. Submarine earthquake geology along the North Anatolia fault in the Marmara Sea, Turkey: a model for transform basin sedimentation. *Earth and Planetary Sciences*, **248**, 661–684, https://doi.org/10.1016/j.epsl.2006.05.038

McHugh, C.M., Seeber, L. *et al.* 2011. Offshore sedimentary effects of the 12 January 2010 Haiti earthquake. *Geology*, **39**, 723–726, https://doi.org/10.1130/G31815.1

Miramontes, E., Garziglia, S., Sultan, N., Jouet, G. and Cattaneo, A. 2018. Morphological control of slope instability in contourites: a geotechnical approach. *Landslides*, **15**, 1085–1095, https://doi.org/10.1007/s10346-018-0956-6

Mulder, T., Zaragosi, S., Razin, P., Grelaud, C., Lanfumey, V. and Bavoil, F. 2009. A new conceptual model for the deposition process of homogenite: application to a cretaceous megaturbidite of the western Pyrenees (Basque region, SW France). *Sedimentary Geology*, **222**, 263–273, https://doi.org/10.1016/j.sedgeo.2009.09.013

Mutti, E. 1992. *Turbidite Sandstones*. AGIP – Istituto di Geologia, Università di Parma.

Polonia, A., Bonatti, E., Camerlenghi, A., Lucchi, R.G., Panieri, G. and Gasperini, L. 2013. Mediterranean megaturbidite triggered by the AD 365 Crete earthquake and tsunami. *Scientific Reports*, **3**, 1285, https://doi.org/10.1038/srep01285

Poort, J. and Lucazeau, F. 2016. WESTMEDFLUX cruise, RV L'Atalante, https://doi.org/10.17600/16000600

Poort, J. and Gorini, C. 2018. WESTMEDFLUX-2 cruise, RV L'Atalante, https://doi.org/10.17600/18000402

Rabineau, M., Droz, L. and Aslanian, D. 2013. AM-MED-1 Cruise, RV Le Suroît, https://doi.org/10.17600/18000821

Reeder, M.S., Rothwell, R.G. and Stow, D.A.V. 2000. Influence of sea level and basin physiography on emplacement of the late Pleistocene Herodotus Basin Megaturbidite, SE Mediterranean Sea. *Marine and Petroleum Geology*, **17**, 199–218, https://doi.org/10.1016/S0264-8172(99)00048-3

Réhault, J.P., Boillot, G. and Mauffret, A. 1984. The Western Mediterranean Basin geological evolution. *Marine Geology*, **55**, 447–477, https://doi.org/10.1016/0025-3227(84)90081-1

Reimer, P.J., Bard, E. *et al.* 2013. IntCal13 and Marine13 radiocarbon age calibration curves 0–50 000 years cal BP. *Radiocarbon*, **55**, 1869–1887, https://doi.org/10.2458/azu_js_rc.55.16947

Rothwell, R.G. and Balut, Y. 1995. PALAEOFLUX cruise, RV *Marion Dufresne*, https://doi.org/10.17600/95200090

Rothwell, R.G., Thomson, J. and Kähler, G. 1998. Low-sea-level emplacement of a very large Late Pleistocene 'megaturbidite' in the western Mediterranean Sea. *Nature*, **392**, 377–380, https://doi.org/10.1038/32871

Rothwell, R.G., Reeder, M.S., Anastasakis, G., Stow, D.A.W., Thomson, J. and Kähler, G. 2000. Low sea-level stand emplacement of megaturbidites in the western and eastern Mediterranean Sea. *Sedimentary Geology*, **135**, 75–88, https://doi.org/10.1016/S0037-0738(00)00064-6

Rothwell, R.G., Hoogakker, B., Thomson, J., Croudace, I.W. and Frenz, M. 2006. Turbidite emplacement on the southern Balearic Abyssal Plain (western Mediterranean Sea) during Marine Isotope Stages 1–3: an application of ITRAX XRF scanning of sediment cores to lithostratigraphic analysis. *Geological Society, London, Special Publications*, **267**, 79–98, https://doi.org/10.1144/GSL.SP.2006.267.01.06

San Pedro, L., Babonneau, N., Gutscher, M.A. and Cattaneo, A. 2017. Origin and chronology of the Augias deposit in the Ionian Sea (Central Mediterranean Sea), based on new regional sedimentological data. *Marine Geology*, **384**, 199–213, https://doi.org/10.1016/j.margeo.2016.05.005

Stanley, D.J., McCoy, F.W. and Diester-Haass, L. 1974. Balearic Abyssal Plain: an example of modern basin plain deformation by salt tectonism. *Marine Geology*, **17**, 183–200, https://doi.org/10.1016/0025-3227(74)90080-2

Stow, D., Fildes, C., Riahi, S., Soussi, M., Patel, U., Milton, J.A. and Marsh, S. 2010. Reply to comment on 'European provenance of the Numidian Flysch in northern Tunisia'. *Terra Nova*, **22**, 504–505, https://doi.org/10.1111/j.1365-3121.2010.00966.x

Sultan, N. 2007. PRISME cruise, RV *L'Atalante*, https://doi.org/10.17600/7010090

Talling, P.J. 2013. Hybrid submarine flows comprising turbidity current and cohesive debris flow: deposits, theoretical and experimental analyses, and generalized models. *Geosphere*, **9**, 460–488, https://doi.org/10.1130/GES00793.1

Talling, P.J., Clare, M., Urlaub, M., Pope, E., Hunt, J.E. and Watt, S.F.L. 2014. Large submarine landslides on continental slopes: geohazards, methane release, and climate change. *Oceanography*, **27**, 32–45, https://doi.org/10.5670/oceanog.2014.38

Thomas, M.F.H., Bodin, S., Redfern, J. and Irving, D.H.B. 2010. A constrained African craton source for the Cenozoic Numidian Flysch: implications for the palaeogeography of the western Mediterranean basin. *Earth-Science Reviews*, **101**, 1–23, https://doi.org/10.1016/j.earscirev.2010.03.003

Vigliotti, L. 2008. Comment on 'Lost tsunami' by Maria Teresa Pareschi *et al. Geophysical Research Letters*, **35**, L02608, https://doi.org/10.1029/2007GL031155

Zuffa, G.G., Normark, W.R., Serra, F. and Brunner, C.A. 2000. Turbidite megabeds in an oceanic rift valley recording jökulhlaups of late Pleistocene glacial lakes of the Western United States. *Journal of Geology*, **108**, 253–274, https://doi.org/10.1086/314404

Zúñiga, D., García-Orellana, J. *et al.* 2007. Late Holocene fine-grained sediments of the Balearic Abyssal Plain, Western Mediterranean Sea. *Marine Geology*, **237**, 25–36, https://doi.org/10.1016/j.margeo.2006.10.034

Integrated geophysical, sedimentological and geotechnical investigation of submarine landslides in the Gulf of Lions (Western Mediterranean)

Shray Badhani[1,2]*, Antonio Cattaneo[1], Stefano Collico[3], Roger Urgeles[4], Bernard Dennielou[1], Estelle Leroux[1], Florent Colin[1,2], Sebastien Garziglia[1], Marina Rabineau[2] and Laurence Droz[2]

[1]IFREMER, Unité de Recherche Géosciences Marines, Centre de Bretagne, 1625 Route de Sainte-Anne, 29280 Plouzané, France

[2]Université de Brest, CNRS, Université Bretagne-Sud, Laboratoire Géosciences Océan, UMR6538 – IUEM, rue Dumont d'Urville, F-29280 Plouzané, France

[3]Igeotest, Carrer Holanda, Num 14 – Poligon Recinte Firal, 17600 Figueres, Girona, Spain

[4]Institut de Ciències del Mar, Pg. Marítim de la Barceloneta, 37-49, 08003 Barcelona, Spain

SB, 0000-0001-8432-4042; AC, 0000-0001-5247-9138; SC, 0000-0003-2834-9813; RU, 0000-0001-9438-1753; BD, 0000-0002-9528-2746; EL, 0000-0003-3584-1179; FC, 0000-0002-7594-8491; SG, 0000-0003-1136-243X; MR, 0000-0001-8420-1587; LD, 0000-0002-6694-1247

*Correspondence: shray.badhani@ifremer.fr

Abstract: The Gulf of Lions presents recurring mass-transport deposits (MTDs) within the Plio-Quaternary sediments, suggesting a long history of mass movements. The two large, surficial MTDs are located on the eastern and western levee of the Rhone canyon over an area exceeding 6000 km^2 and volumes exceeding 100 km^3. Both MTDs were emplaced 21 ka ago (peak of the Last Glacial Maximum), suggesting a common trigger. Here, we present a multidisciplinary high-resolution geophysical, sedimentological and *in-situ* geotechnical study of the source and deposit areas of both MTDs to characterize distinct expressions of sediment deformation as well as their spatial and chronological distributions. We show the internal structure of mass movements and resulting MTDs with unprecedented details that were previously represented in the conventional seismic data as transparent and chaotic facies. The combination of multidisciplinary approaches shows new insights into the nature of basal surfaces of the slope failures. In particular, we show that the basal surfaces of the failures consist of clay-rich material contrasting with the overlying turbiditic deposits, suggesting that a strong lithological heterogeneity exists within the strata. We suggest that this change in lithology between clay-rich sediments and turbiditic sequences most likely controls the localization of weak layers and landslide basal surfaces.

Supplementary material: Sediment cores are curated at IFREMER core repository in Plouzané (France); core data related to this article are available at: http://igsn.org/BFBGX-87813 (core PSM3-CS049: IGSN BFBGX-87813); http://igsn.org/BFBGX-87814 (core PSM3-CS050: IGSN BFBGX-87814); http://igsn.org/BFBGX-87815 (core PSM3-CS051: IGSN BFBGX-87815)

Submarine landslides have been widely recognized on passive and active continental margins, in a wide range of environments, including submarine canyons (Sultan *et al.* 2007*a*; Micallef *et al.* 2012), carbonate platforms (Principaud *et al.* 2014), siliciclastic platforms (Silva *et al.* 2010) and volcanic islands (Moore *et al.* 1989; Urgeles *et al.* 1997; Le Friant *et al.* 2015). These landforms have a profound effect on the shaping of margin architecture (Posamentier and Kolla 2003), providing useful insights into the evolution of continental margins, along with providing constraints on palaeoseismicity (St-Onge *et al.* 2004; Brooks 2016; Moernaut *et al.* 2017; Praet *et al.* 2017). Submarine landslides are regarded as important marine geological processes given their widespread occurrences and potentially large volumes (>10^3 km^3; Hampton *et al.* 1996; Masson *et al.* 2006; Talling *et al.* 2014), and are able to redistribute significant amounts of sediments from the shelf or the upper slopes of the margins to the lower slopes or the surrounding deep basin. Moreover, submarine landslides can be two to

From: Georgiopoulou, A., Amy, L. A., Benetti, S., Chaytor, J. D., Clare, M. A., Gamboa, D., Haughton, P. D. W., Moernaut, J. and Mountjoy, J. J. (eds) 2020. *Subaqueous Mass Movements and their Consequences: Advances in Process Understanding, Monitoring and Hazard Assessments.* Geological Society, London, Special Publications, **500**, 359–376.
First published online May 14, 2020, https://doi.org/10.1144/SP500-2019-175

three orders of magnitude larger, in terms of volume, than their terrestrial counterparts (Hampton *et al.* 1996; Masson *et al.* 2006), and therefore threaten offshore infrastructure. Submarine landslides also hold a high hazard potential for tsunamigenic risks, which can be devastating to increasingly populated coastlines (Tappin *et al.* 1999; Bryn *et al.* 2003, 2005; Kvalstad *et al.* 2005; Heidarzadeh *et al.* 2018; Carvajal *et al.* 2019; Sassa and Takagawa 2019).

Owing to their locations in deep-water environments, direct observation of submarine landslides and their deposits is difficult. Therefore, geophysical techniques are used to observe submarine landslides, such as multibeam bathymetry and reflection seismic, and often sediment cores are collected to ground truth geophysical data. In reflection seismic data, submarine landslide deposits are generally referred to as mass-transport deposits (MTDs) and mass transport complexes (Alves 2015). MTDs in the seismic data are often defined by acoustically transparent or chaotic facies (Damuth 1980). However, outcrop examples often show a complex internal geometry including seismic-scale folding and faulting, contorted lithologies and slide blocks within the MTDs that can be used as kinematic indicators (Ogata *et al.* 2014, 2016).

Often the failure surface (i.e. the basal surface or glide plane) over which the mass movement of sediments occurs develops at the base of a weak layer and follows stratigraphic horizons (Baeten *et al.* 2014). Landslide basal surfaces are identified on seismic profiles by high-amplitude reflections (Baeten *et al.* 2014). In order to understand mass wasting processes and to correctly perform geohazard assessments, identification and analysis of weak layers and basal surfaces is imperative (Lastras *et al.* 2004; Kvalstad *et al.* 2005; L'Heureux *et al.* 2012; Locat *et al.* 2014). Owing to alteration in physical properties during failure or the occurrence of thick post-failure sedimentary sequences, weak layers and basal surfaces are rarely penetrated using current coring techniques. Therefore, only a few studies have reported information about failure-related weak layers and basal surfaces of submarine landslides (e.g. Kvalstad *et al.* 2005; Dan *et al.* 2007; L'Heureux *et al.* 2012; Baeten *et al.* 2014; Locat *et al.* 2014; Miramontes *et al.* 2018; Urlaub *et al.* 2018).

Submarine landslides in the Gulf of Lions (GoL) are widespread from the upper slope to the deep basin, within the canyon flanks and in the interfluves of major canyons (Fig. 1). The two main submarine landslides present in the GoL are the Eastern Rhône Interfluve Slide (ERIS, Badhani *et al.* 2019) and an unnamed slide complex on the western side of the Petit Rhone Canyon (PRC, Dennielou *et al.* 2019) that we refer to as the Western Slide in this paper. Their resulting MTDs, the Rhone Eastern MTD (REMTD) and the Rhone Western MTD (RWMTD) have previously been described in detail in many studies (Droz and Bellaiche 1985; Torres *et al.* 1997; Droz *et al.* 2001, 2006; Bonnel *et al.* 2005; Dennielou *et al.* 2019). However, owing to the lack of a high-resolution multidisciplinary dataset, such as high-resolution seismic, sediment cores and *in-situ* geotechnical measurements, a detailed analysis of weak layers and preconditioning factors was never performed. The overarching aim of this study was to analyse for the first time a suite of multidisciplinary datasets to better understand the distribution and characteristics of seafloor instabilities in the GoL. Specifically, the objectives of this study were: (1) to use deep-towed multichannel seismic data to image the internal structure of the mass-wasting products, previously imaged as seismically transparent or chaotic intervals in conventional seismic data; and (2) to use very high-resolution seismic, sedimentological data and *in-situ* geotechnical measurements to analyse the basal surfaces of the slope failures in the GoL.

Geological setting

The GoL (Fig. 1) is a SW–NE-orientated passive margin that is described by a wide continental shelf of *c.* 70 km, which was subaerially exposed during the Last Glacial Maximum (LGM) about 21 ka ago (Rabineau *et al.* 2005; Berné *et al.* 2007; Jouet 2007; Bassetti *et al.* 2008). Sedimentary succession composing the distal Rhone Sedimentary System originates from 15 distinct submarine canyons. The major canyon feeding the Rhone Sedimentary System is the PRC, located in the central part of the study area (Fig. 1), which is fed by Alpine sediments through the Rhone River and its tributaries.

During the Messinian Salinity Crisis (*c.* 5.33 Ma ago; Lofi *et al.* 2003; Roveri *et al.* 2008), a major drop in the Mediterranean sea-level led to a basin-wide accumulation of a kilometre-scale-thick evaporite succession (Hsu 1973). This changed the sedimentary architecture of the GoL drastically. Since then, syn-sedimentary salt tectonics and gravity-driven deformation of Pliocene and Quaternary strata mainly control the evolution of the deeper part of the GoL (dos Reis *et al.* 2005). As a result of the gravity-driven deformation above the salt layer, which acts as a 'décollement' surface, several extensional areas developed, where listric faults and grabens predominate in the upper and lower slopes (Fig. 1), and formed compressional areas in the deep basin driven by salt diapirs and domes (dos Reis *et al.* 2005). Subsequent glacio-eustatic sea-level fluctuations related to the Milankovitch cycles mostly control the Pleistocene sedimentation in the GoL (Rabineau *et al.* 1998, 2005; Tesson and Gensous 1998; Bassetti *et al.* 2008). After the

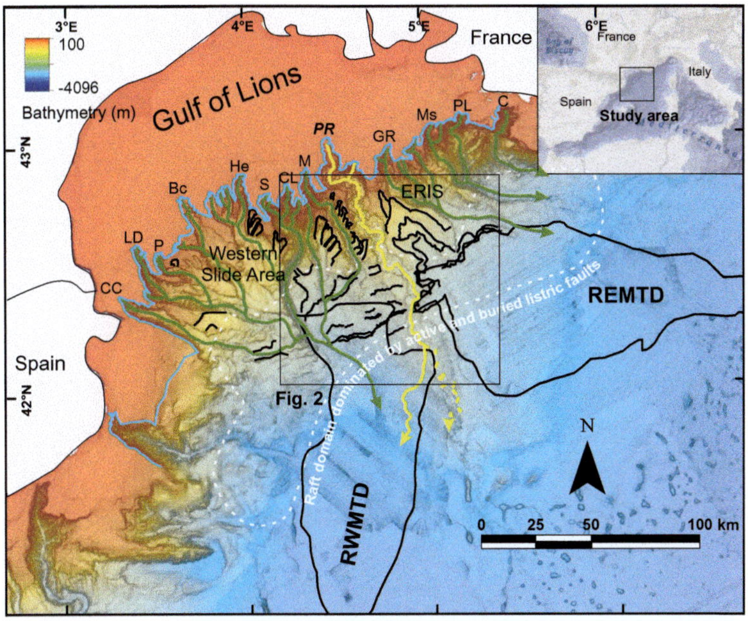

Fig. 1. EMODnet bathymetric map of the GoL margin highlighting the distribution of main sedimentary and mass movement features (black lines, headwall of submarine landslides and mass transport deposits; yellow line, present-day thalweg of the Petit Rhone Canyon; green lines, thalwegs of the main submarine canyons; blue line, shelf break; white dashed polygon, raft (salt) domain dominated by listric faults). Abbreviations for the GoL canyons (Berné *et al.* 2002), from west to east: CC, Cap de Creus; LD, Lacaze Duthier; P, Pruvost; Bc, Bourcart; He, Herault; S, Sète; CL, Catherine Laurence; M, Marti; PR, Petit-Rhone; GR, Grand-Rhone; Ms, Marseille; PL, Planier; C, Cassidaigne. REMTD, Rhone Eastern Mass Transport Deposit; RWMTD, Rhone Western Mass Transport Deposit. Limits of sedimentary bodies are from Droz *et al.* (2006). Raft domain dominated by active and buried listric faults is from dos Reis *et al.* (2005).

Mid-Pleistocene, sedimentation within the GoL mainly consisted of terrigenous sediments from the Alpine rivers (mainly Rhone River) delivered to the basin through turbiditic systems. The lower slope of the GoL (1500–2000 m water depth) is mainly fed by the PRC (Lombo Tombo *et al.* 2015). Sedimentation rates in the lower slopes during the glacial and deglacial periods were reported to be of the order of 1–3 m ka^{-1} (Sierro *et al.* 2009; Dennielou *et al.* 2019). During the interglacial periods, the sedimentation rates were much lower, falling in the range of 0.05–0.10 m ka^{-1} (Sierro *et al.* 2009; Lombo Tombo *et al.* 2015; Dennielou *et al.* 2019).

Within the GoL, immediately at the foot of the slope, two large submarine landslides, the ERIS and the Western Slide, have been recognized and investigated by Bellaiche *et al.* (1986), Droz *et al.* (2006), Dennielou *et al.* (2019) and Badhani *et al.* (2019). The ERIS extends from the shelf break (*c.* 120 m) to the water depth of 1500 m on the eastern side of the PRC (Badhani *et al.* 2019), and the Western Slide complex extends from 1800 to 2300 m on the western side of the PRC (Dennielou *et al.* 2019). The ERIS consists of successive headwall scarps of 30–120 m height aligned in a staircase style and the Western landslide consists of several scarps parallel to the slope whose heights reach 50–100 m (Fig. 2). The resulting deposits, the REMTD and RWMTD, have surface areas exceeding 6500 km^2, volumes greater than 100 km^3 and an extremely long run out of more than 200 km (Droz *et al.* 2006). Both landslides occurred during the peak of the LGM (21 ka cal BP), suggesting a common trigger behind the failure (Badhani *et al.* 2019; Dennielou *et al.* 2019). Badhani *et al.* (2019) showed repeated slope failures above high-amplitude horizons at different stratigraphic levels, indicating control of specific sediment type on the failure.

Material and methods

Geophysical data

To characterize the morphology of the study area, we use an openly available European Marine Observation and Data Network (EMODnet) 115 m resolution multibeam bathymetry (Fig. 1; EMODnet Bathymetry Consortium 2018; http://www.emodnet-bathy

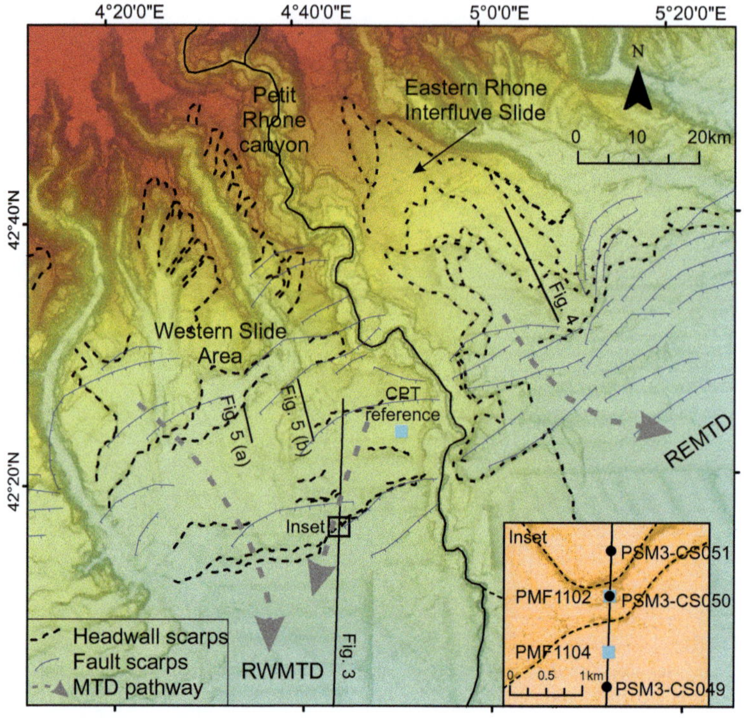

Fig. 2. High-resolution multibeam bathymetric map of the source area of the landslides (see Fig. 1 for location). Landslide headwalls are highlighted with dashed black lines and active faults related to Messinian salt are highlighted with blue lines. Solid black lines represent locations of the seismic profiles shown in Figures 3, 4 and 5a, b. In the inset, black circles and light blue rectangles respectively represent the locations of the sediment cores and the *in-situ* geotechnical measurements acquired during PRISME3 cruise (see Figs 7 & 9). REMTD, Rhone Eastern Mass Transport Deposit; RWMTD, Rhone Western Mass Transport Deposit.

metry.eu) and a high-resolution (HR) bathymetry with 25 m resolution acquired during several French cruises (for details see Dennielou *et al.* 2019). The 2D multichannel seismic (MCS) reflection data (Figs 3–5) used in this study were acquired during two surveys, PRISME2 onboard R/V *L'Atalante* (Cattaneo 2013*a*) and AM-MED onboard R/V *Le Suroit* (Rabineau *et al.* 2013). The processing sequence for the 2D HR MCS data includes quality control, filtering, binning (6.25 m), band-pass filtering (15–30–375–400 Hz), seawater velocity (1500 m s^{-1}) normal move-out, stacking and seawater velocity post-stack Kirchoff time migration.

Two-dimensional deep-towed multichannel very high-resolution seismic data (Fig. 6) were acquired for the first time in the GoL during the PRISME2 campaign (Cattaneo 2013*a*) using Ifremer's in-house acquisition system SYSIF (Système Sismique de Fond; Ker *et al.* 2010; Marsset *et al.* 2014). It consists of an open-cavity acoustic source and a specially developed streamer that is towed at a constant altitude above the seafloor (50–100 m). The source consists of a Janus–Helmholtz transducer that emits a 100 ms sweep signal in the 220–1050 Hz frequency bandwidth. The multichannel streamer is composed of 52 digital hydrophones spaced by 2 m (offsets 10–112 m). The processing of the data includes a reconstruction of the variable streamer geometry, a deconvolution of the seismic signal by the source signature and a Kirchhoff pre-stack depth migration. SYSIF can obtain a maximum vertical resolution of 0.6 m and up to 1.5 m horizontally, which allows us to image mass-movement features in unprecedented detail.

Sediment cores

Lithological characterization and age estimation of sedimentary facies were based on three sediment cores – PSM3-CS049, 50 and 51 – collected during cruise PRISME 3 (Cattaneo 2013*b*) on the GoL slope using a Calypso Piston corer. These cores respectively penetrated 9.20, 11.26 and 22 65 m deep (Fig. 7), and were corrected for true depth using Ifremer software CINEMA[2] based on accelerometer measurements performed during coring. Identification of lithofacies is based on the visual description, which was performed partially onboard and partially

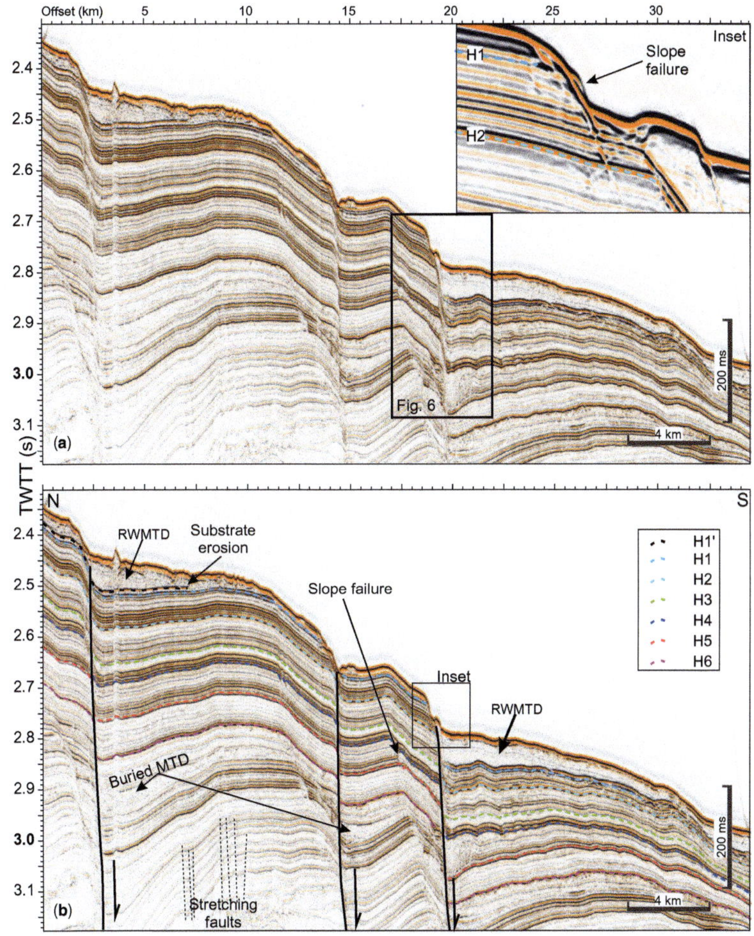

Fig. 3. North–south-oriented seismic profile crossing the western slide area showing the internal architecture of the study area. Solid black lines represent prominent faults imaged in the seismic profile. The seismic stratigraphy of the study area is highlighted by seven key high-amplitude reflectors (H1–H6). See Figure 6 for a focus on the central part of the profile (offset 17–21 km). RWMTD, Rhone Western Mass Transport Deposit.

at Ifremer–Brest. Geotek® Multi-Sensor Core Logger was used for gamma-density and p-wave velocity (V_p) measurements at 1 cm intervals (Fig. 7). Bulk X-ray fluorescence was measured with the Avaatech X-ray fluorescence core scanner on split cores at 1 cm intervals (Fig. 7). Grain-size analysis was performed on core PSM3-CS050 using a laser diffractometer device at CSIC-ICM, Barcelona in 2019 (Fig. 8a). Vane shear tests on sediment core PSM3-CS050 at 10 cm interval were performed at Ifremer–Brest (Fig. 8b).

Three radiocarbon dates (Fig. 7) were determined by accelerator mass spectrometry on core PSM3-CS049. Dates were obtained from handpicked monospecific planktonic foraminifera (*Globigerina bulloides*) extracted from the hemipelagic sediments and were analysed at Beta Analytics, London, UK. Conventional ^{14}C ages were converted into calibrated calendar ages with CALIB V7.1 (Stuiver *et al.* 2017) using the Marine13 curve for calibration (Reimer *et al.* 2013) with a reservoir correction (ΔR) value of 48 ± 101 years and reported in this study with an error of two standard deviations (Table 1).

In-situ geotechnical measurements

Five *in-situ* geotechnical measurements were carried out using the Penfeld seabed rig with a standard CPTu cone developed at Ifremer (Sultan *et al.* 2007*b*; Garziglia *et al.* 2010), of which two were used in this study (Fig. 9). The Penfeld penetrometer provides a continuous measurement of the tip resistance (q_c), sleeve friction (f_s) and induced excess pore pressure (Δ_{u2}) at a standard rate of 2 cm s^{-1} down to 30 m below the seafloor (mbsf). The

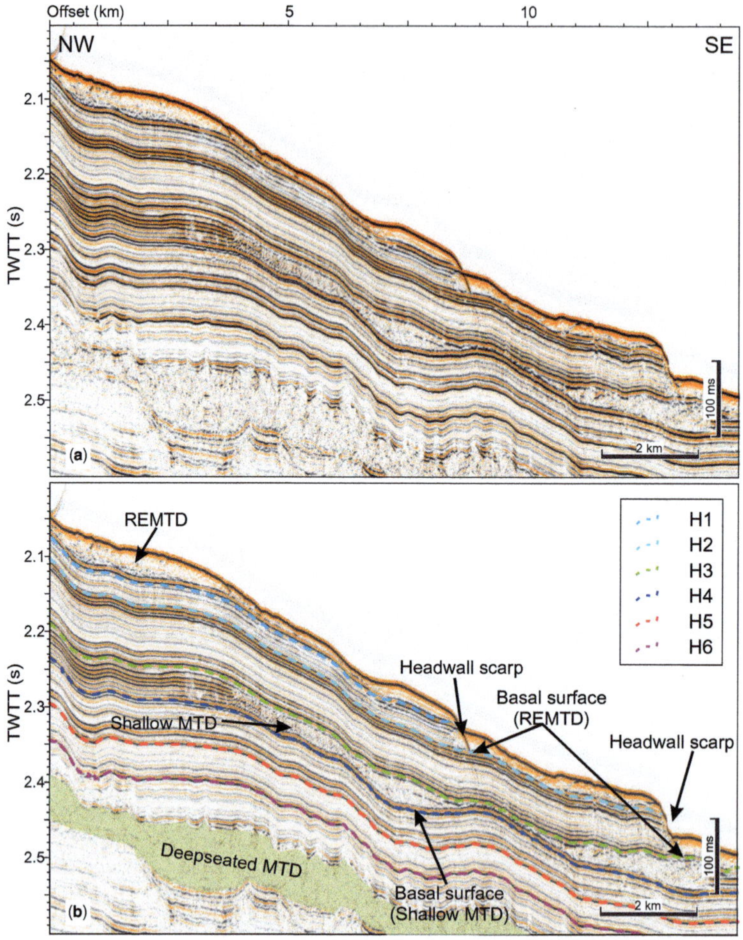

Fig. 4. NNW–SSE-oriented seismic profile AM-MED018 crossing the eastern part of the ERIS. It illustrates two headwall scarps (HS2 and HS3) showing particularly sharp truncation and a transparent unit interpreted as a buried MTD. Headwall scarps of the landslides are represented by sharp truncations of well-stratified sediments disappearing downslope. The basal surface of the MTD corresponds to the prolongation of a sharp contact between underlying low amplitude reflectors and overlying high amplitude reflectors. REMTD, Rhone Eastern Mass Transport Deposit.

corrected tip resistance (q_t) is calculated using the measured excess pore pressure (Δ_{u2}) values.

Values of peak undrained shear strength (S_u) were calculated from:

$$S_u = \frac{(q_t - \sigma_{v0})}{N_{kt}} \quad (1)$$

where σ_{v0} is the vertical effective stress and N_{kt} is a cone factor typically varying from 10 to 20. Following Low *et al.* (2010), a minimum and a maximum value of S_u were calculated using a N_{kt} value of 10 and 20.

Sensitivity values were derived using:

$$St = \frac{S_u}{f_s} \quad (2)$$

where f_s is sleeve friction. The overconsolidation ratio (OCR) can be calculated as the ratio between the preconsolidation stress (σ'_p) and the vertical effective stress (σ'_v).

$$\text{OCR} = \frac{\sigma'_p}{\sigma'_v} \quad (3)$$

Results

Morphology and internal structure of seafloor instabilities in the GoL

The GoL slope extends from a water depth of 500 to 1200 m. The slope is mainly characterized by three types of geomorphological features: (1) sinuous elongated valleys that incise the seafloor

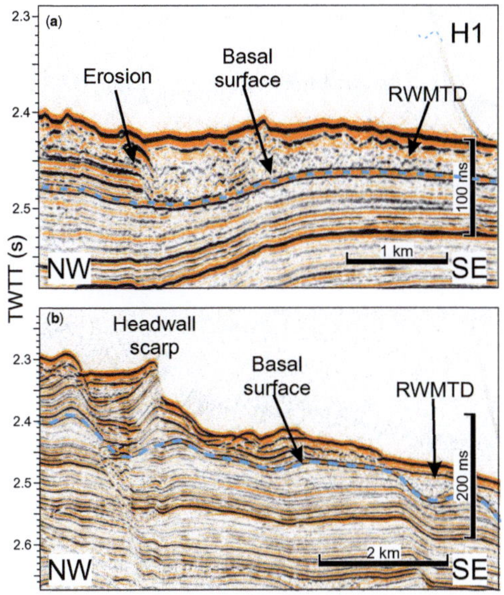

Fig. 5. (**a**, **b**) Seismic profiles showing examples of headwall scarps and other erosional features within the western slide area, similar to the ERIS; the basal surface of the landslides coincides with the transition between underlying low-amplitude and overlying high-amplitude reflectors. RWMTD, Rhone Western Mass Transport Deposit.

down to several hundred metres, interpreted as submarine canyons; (2) irregular, arcuate, curved breaks in the bathymetry, interpreted as submarine landslide headwall scarps; and (3) strike-parallel linear breaks in the bathymetry, interpreted as faults (Fig. 2).

The subsurface architecture of the study area is mainly characterized by parallel, well-stratified reflections related to the Petit Rhone channel levee complex (Fig. 3). Several steep-dipping, low-amplitude anomalies interrupt the seismic reflections of the well-stratified sediments and are interpreted as normal listric faults (Figs 3, 6). The lack of penetration of our seismic data until the base of faults prevents us from clearly imaging the origin of the faults. Nevertheless, dos Reis *et al.* (2005) demonstrated that a majority of listric faults in the GoL are rooted in the Messinian strata and are attributed to the gliding of post-Messinian sediments over the evaporite décollement surface. The architecture of the entire GoL slope is characterized by a morphology that is controlled by structural movements owing to these listric faults. At times these faults reach the seafloor, suggesting an ongoing fault activity that contributes to continuous steepening of the slope (e.g. Figs 2, 3 & 6). In particular, the southernmost fault in the seismic profile PSM2-117A shows a displacement of more than 150 ms, creating a small-scale failure on the seafloor, exposing the strata until Horizon H2′ (Figs 3 & 6).

Seismic profile AM-MED18 partially shows the internal structure of the ERIS (Fig. 4). Part of the source area shows two morphological steps of 30 and 70 ms height, which correspond to headwall scarps of the slide complex (Fig. 4). The basal surfaces of two failure steps of the ERIS imaged in the seismic data lie above the high-amplitude reflectors H2 and H3 (Fig. 4). Additionally, we observe a small MTD (shallow MTD, Fig. 4) between horizons H3 and H4, similar to the ERIS; the basal surface of the shallow MTD lies on top of the high-amplitude reflector H4 (Fig. 4).

The source area of the RWMTD is unfortunately not fully covered by our seismic data. Nevertheless, we observe a similar trend of failures in the area covered within the Western Slide area. Seismic data show that, similar to the ERIS, the basal shear surface of the RWMTD also lies above the high-amplitude reflector H1 (Fig. 5a, b). The RWMTD propagated translationally above the high-amplitude reflector H1′ (marked with a black dashed line in Fig. 3) in the proximal part. After the erosion of the basal surface H1′, further downslope the MTD propagated above reflector H1 (Fig. 3). The seismic profile PSM2-117A is located on the main pathway of the MTD (Figs 2, 3). However, between offset 10 and 20 km, where the slope angles become steeper, the RWMTD gradually thins to less than 5 ms in thickness. In places, especially between 15 and 20 km offset (Fig. 3), the RWMTD completely disappears, most likely owing to steeper slope angles preventing deposition of failed sediments. Nevertheless, since the seismic profile is located at the main pathway of the MTD, it is likely that a significant amount of MTD propagated over horizon H1.

We observe a small-scale failure at 150 msbsf (ms below seafloor) between horizons H4 and H5 (Fig. 3). In the conventional HR seismic data, the source area of this failure is defined by chaotic reflections typical of slump facies (Fig. 3). However, in the deep-towed seismic profile, the internal structure of the failure is imaged with unprecedented detail (Fig. 6b, e). The source area of the failure is characterized by a sharp truncation of well-stratified sediments of around 30 ms thickness and translation of slide blocks with well-preserved internal reflections. It is interesting to note that the basal surface of this failure also occurs above the high-amplitude reflector H5 (Fig. 6b). The deposit area of the failure seems to be buttressed against the pre-existing strata, partially faulting and folding the strata. The resulting MTD from this failure seems to be of frontally emergent type that overflowed the pre-existing strata after generating a compressional zone by folding and faulting the strata (Fig. 6e). The source and the

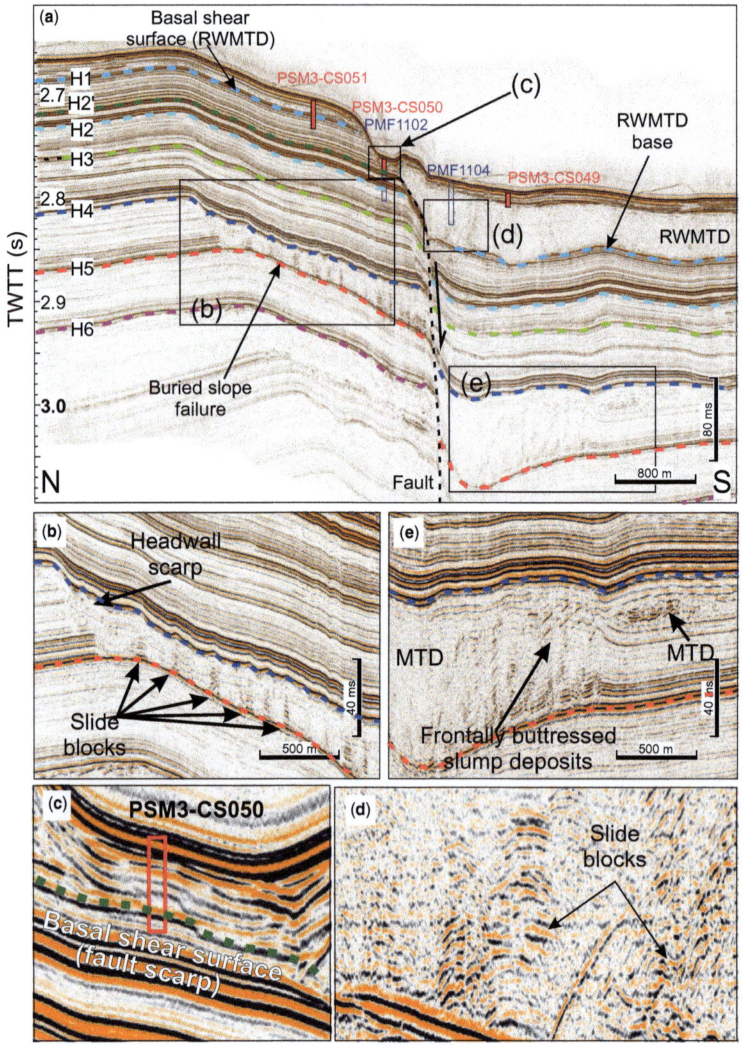

Fig. 6. (**a**) North–south-oriented very high-resolution seismic profile corresponding to a focus of Figure 3 (see location on Fig. 1). The internal architecture of the study area is dominated by well-stratified seismic reflections. Note the presence of internal reflections within the transparent Rhone Western Mass Transport Deposit (RWMTD) facies. Red and blue rectangles show the location of sediment cores and *in-situ* measurements, respectively. (**b**) Close-up of the small slope failure shows *c.* 30 ms-high slide blocks with intact internal reflections. Note that the seismic facies in this area was previously interpreted as chaotic on the conventional high-resolution seismic data. (**c**) Zoom at the fault scarp showing the location of core PSM3-CS050. (**d**) Zoom of the RWMTD at the hanging wall showing slide blocks within the MTD. (**e**) Close-up of the deposit area of the slope failure, which is identified by frontal buttressing of the slide deposit against the unfailed material, as shown by folded strata in the frontal part.

deposit area of the buried slope failure are vertically separated by 200 ms, most likely owing to fault displacement (Figs 3 & 6a).

The internal structure of the REMTD and RWMTD is defined by transparent acoustic facies in conventional HR seismic profiles (Figs 3 & 4). Also in the deep-towed seismic profile, the REMTD shows transparent facies (Fig. 6a). However, about 1 km south of the fault scarp, within the hanging wall, we observe some kilometre-scale irregular blocks of high-amplitude internal reflections within the MTD (Fig. 6d). These reflections do not seem to be associated with the processing artefacts, as they tend to mimic the seafloor. Therefore, we attribute these internal reflections to the slide blocks within the REMTD (Fig. 6d).

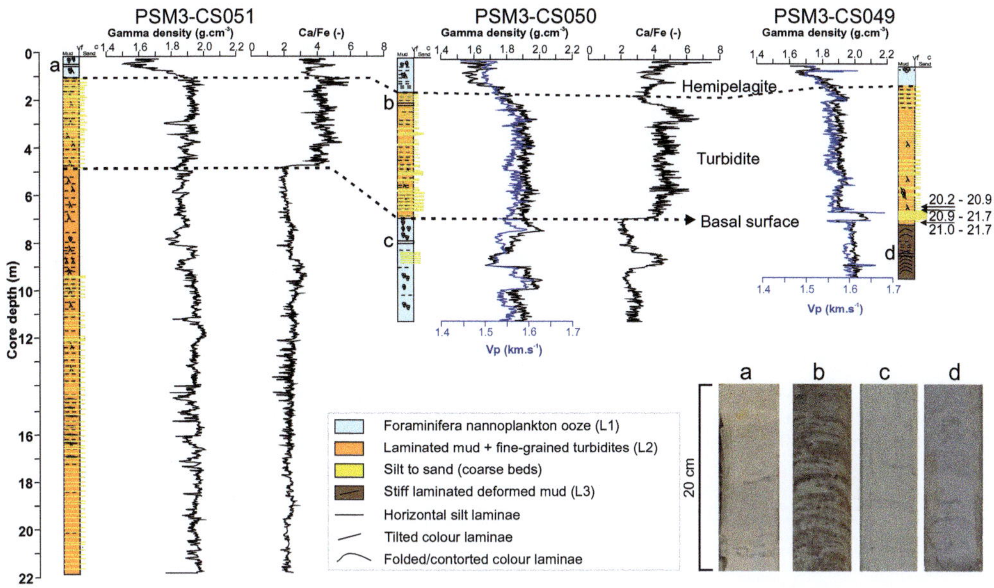

Fig. 7. Gamma density, V_p, Ca/Fe ratio, sediment photos and lithofacies of sediment cores (from left to right, PSM3-CS051, -50 and -49) collected in the western slide area and the RWMTD. See Figure 6 for core locations.

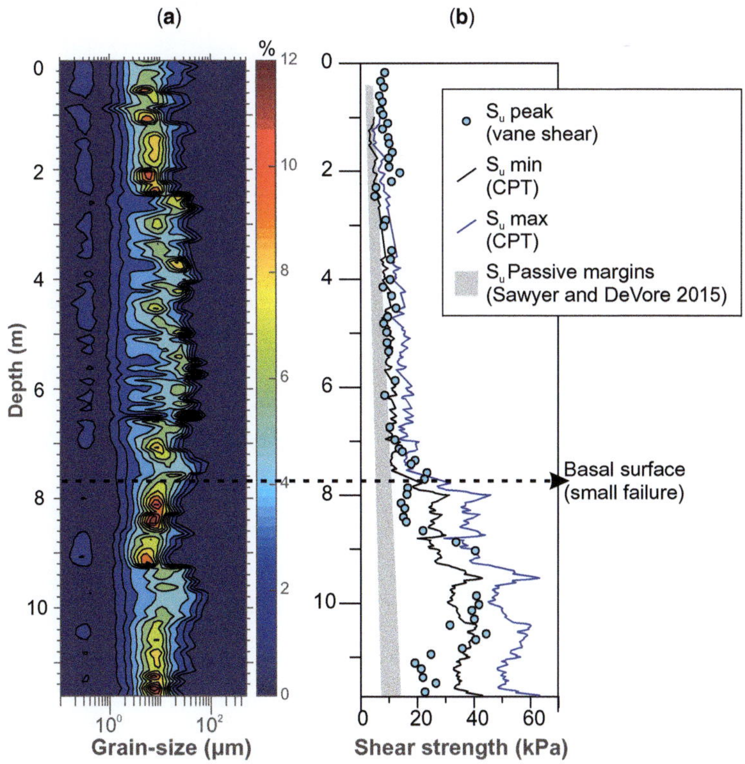

Fig. 8. (**a**) Grain-size distribution and (**b**) vane shear strength measurements of core PSM3-CS050.

Table 1. *Radiocarbon dating carried out on sediment core PSM3-CS049*

Core-section-depth (m)	Laboratory reference no.	^{14}C age (years BP)	Calibrated age	Sampled material
PSM3-CS049-S07-6.12	510580	17 540 ± 50	20 935–20 295	*Globigerina bulloides*
PSM3-CS049-S07-6.25	508505	18 100 ± 60	21 723–20 974	*Globigerina bulloides*
PSM3-CS049-S07-6.64	508506	18 150 ± 60	21 785–21 038	*Globigerina bulloides*

Sediment cores

Three sediment cores were recovered on the western side of the Rhone submarine valley in the turbiditic levees (Figs 6 & 7). Cores PSM3-CS049 and PSM3-CS051 were recovered downslope and upslope the fault scarp, respectively. Core PSM3-CS050 was strategically retrieved at the fault scarp in order to sample the basal surface in the area of sediment removal owing to fault reactivation (Fig. 6c). Lithological observations on the sediment cores may be classified into three lithofacies (L1–L3; Fig. 7):

- lithofacies L1 is composed of light grey foraminifera and calcareous nanoplankton oozes that correspond to hemipelagic–pelagic sedimentation

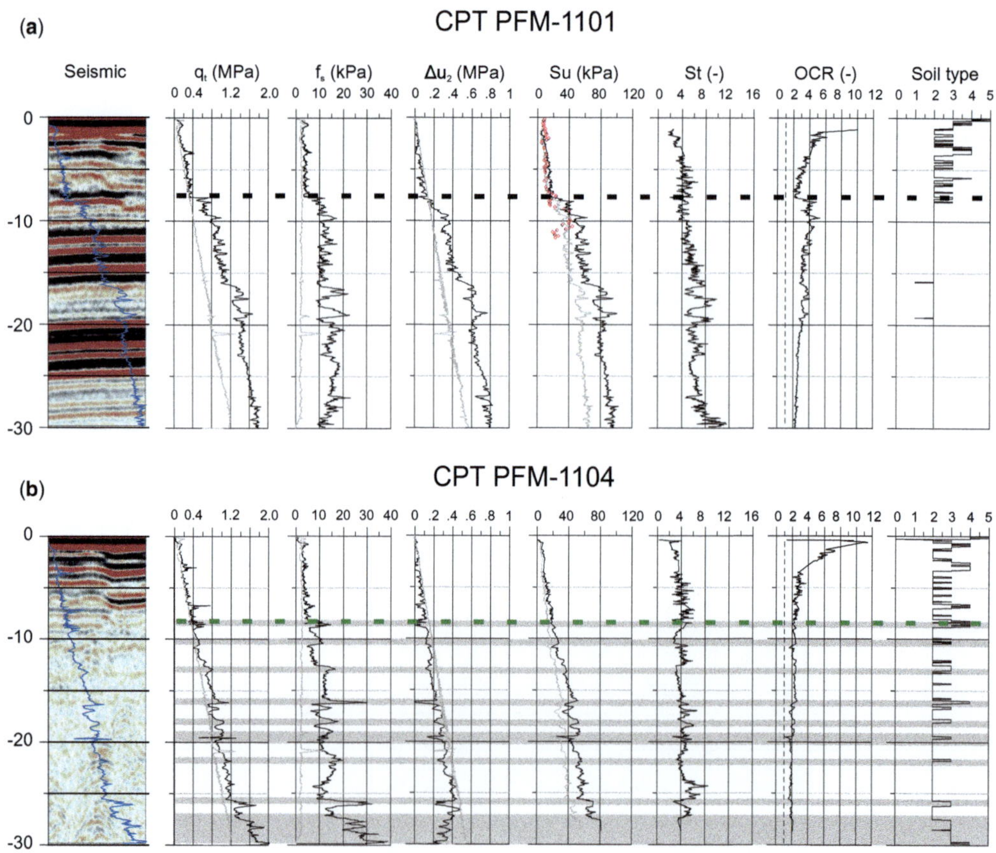

Fig. 9. (**a, b**) From left to right: corrected cone resistance (q_t) and reference q_t (grey line), sleeve friction (f_s), excess pore pressure (Δ_{u2}), peak undrained shear strength (S_u) from CPTu (minimum, grey lines and maximum, black lines) and from laboratory vane shear test on core PSM3-CS050 (red diamonds only for (a), undrained shear strength (S_u), sensitivity (*St*) and overconsolidation ratio (OCR) and soil type from CPTu. Soil type from CPTu data is derived from Q_t/F_r, where soil type 1 = sensitive, fine-grained, 2 = organic soils-peats, 3 = clay to silty clay, 4 = clayey silt to silty clay and 5 = silty sand to sandy silt (after Robertson 1990). The dashed black line represents the basal surface of the small failure and the green dashed line represents the RWMTD top. See Figure 6 for CPTu locations.

during the sea-level highstands (Badhani *et al.* 2019; Dennielou *et al.* 2019).
- lithofacies L2 is composed of laminated clay with silt to very fine sand laminae interpreted as turbidites deposited by spillover from the Rhone valley (Dennielou *et al.* 2019).
- lithofacies L3 is composed of stiff mud of generally higher bulk density than the background values (1.85–1.90 g cm^{-3}). Based on the similarities observed in other studies in the same area (e.g. Bonnel *et al.* 2005; Lombo Tombo *et al.* 2015; Dennielou *et al.* 2019, 2006), we associate the lithofacies L3 with mass transport deposits. These lithofacies at times show colour banding corresponding to laminae and a few silt layers characteristic of the Rhodanian fan lithofacies (Lombo Tombo *et al.* 2015; Dennielou *et al.* 2019).

Core PSM3-CS051

Core PSM3-CS051 is 22.65 m long and was collected at the footwall of the fault at a water depth of 2033 m (Fig. 7). The core is located in the area of a steep slope where we observe an absence of MTD in the seismic profile (Fig. 3). The substrate erosion (horizon H1′) upslope allowed penetration of high-amplitude reflector H1 (Figs 3 & 6). The uppermost 0.85 m of the core corresponds to lithofacies L1, showing lower gamma density and Ca/Fe ratio than average values. Below 0.85 m until the bottom of the core, the core displays lithofacies L2, typically consisting of frequent alternating silty turbidites. The gamma density and Ca/Fe ratio values are above average until 4.8 m (Fig. 7). Below 4.8 m, we observe a sharp decrease in Ca/Fe ratio, with values remaining constant until the bottom of the core, whereas the gamma density values are close to average (1.8 g cm^{-3}). We observe an interval of fine-grained material of dark grey colour with less frequent turbidites between 4.8 and 9.5 m. Upon correlation with seismic data (Fig. 6), we observe that this interval lies at the basal surface H1.

Core PSM3-CS050

Core PSM3-CS050 (Fig. 7) is 11.26 m long and was collected at 2073 m water depth, at the fault scarp where the high-amplitude reflector H2 has been exposed owing to removal of material related to faulting activity (Figs 3 & 6). The uppermost 1.50 m of the core is composed of lithofacies L1 and the gamma density and V_p values within this interval are lower than the background values (1.8 g cm^{-3} and 1550 m s^{-1}, respectively). This unit shows Ca/Fe ratios in the range of 3–5 (Fig. 7). From 1.5 to 7 m, the core is composed of lithofacies L2, where higher gamma density, V_p and Ca/Fe ratios values than the background values are observed (Fig. 7). From 7 m to the bottom of the core, it presents light greyish clayey material typical of lithofacies L1. Between 9.5 and 10.25 m, we observe few thin-bedded turbidites (Fig. 7), which are consistent with the higher gamma density and V_p values. Otherwise, within this interval gamma density, V_p and Ca/Fe ratio values are below average.

Grain-size distribution (Fig. 8a) shows the presence of higher fine-grained material within the upper 1.5 m. Between 1.5 and 7 m, the grain-size distribution curve underlines a distribution of clayey to fine sand material in mostly equal percentages. Below 7 m, the grain-size distribution plot shows a higher percentage of clay-sized particles, except between 9.5 and 10.25 m, where we observe a broad distribution of particle size from clay to silt, most likely owing to the presence of thin-bedded turbidites (Fig. 7).

The peak shear strength (Fig. 8b) obtained from vane shear tests tends to increase linearly with depth, from 8 kPa at the seafloor to 20 kPa down to the base of the core. Peaks in shear strength at 7.5 and 8.6 m interrupt this trend, as the shear strength increases to 23 and 44 kPa, respectively. When compared with the global shear strength trends for passive margins (Sawyer and DeVore 2015), shear strengths measured on core PSM3-CS050 generally fall at the higher end of the plot, especially within the peaks in the shear strength at 7.5 m and between 8.6 and 10.5 m (Fig. 8b).

Core PSM3-CS049

Core PSM3-CS049 is 9.2 m long and was collected at 2096 m water depth at the footwall of the fault within the RWMTD (Fig. 2). The uppermost 0.80 m of sediments shows clay-rich material of low gamma density and V_p (Fig. 7), which relate to lithofacies L1. Between 0.80 and 6.75 m, we observe frequent silty turbidites embedded within hemipelagic sediments, which we consider as corresponding to lithofacies L2. The gamma density and V_p values in this interval are higher compared with lithofacies L1. Between 6.35 and 6.65 m, we observe a fine to medium sand layer most likely related to an event bed such as a coarse-grained turbidite after the main failure event. This sand layer coincides with an interval of high values of gamma density (2.05 g cm^{-3}) and V_p (>1.7 m s^{-1}, Fig. 7). Below 6.75 m to the bottom of the core, we observe folded coloured bands most likely associated with the RWMTD. We thus attribute this sediment interval to lithofacies L3. The gamma density and V_p values within this interval are higher than in lithofacies L2. Radiocarbon dating at the base of lithofacies L2 in core PSM3-CS049 yields a 2σ age between 21 and 21.7 cal ka BP, with an average median age of 21.4 cal ka BP (at the peak of the LGM, Fig. 7; Table 1).

In-situ *geotechnical measurements*

CPTu PMF-1102. CPTu PMF-1102 (Fig. 9a) was performed in the area of sediment removal owing to fault activity and coincides with core PSM3-CS050 (Figs 6 & 7). The uppermost 7.5 m of sediments shows a quasi-linear increase in the corrected tip resistance (q_t), sleeve friction (f_s) and induced excess pore pressure (Δ_{u2}), which correlates with values of reference CPTu measurements performed away from the failure area (grey curve in Fig. 9). This sediment interval is described by a predominance of turbidites intercalated with hemipelagic sediments (Fig. 7). Below 7.5 m, we observe a sharp increase in q_t, f_s and Δ_{u2} values, as well as an increase in the shear strength profile (Fig. 9a). Interestingly, below 7.5 m, until the bottom of the measurements, sediments are classified as fine-grained sediments, which is in agreement with the grain-size distribution curve of coinciding core PSM3-CS050 (Fig. 8a). Also, a peak in the OCR profile is observed at the same depth, indicating that sediments are overconsolidated. Below 7.5 m, OCR ratios decrease gradually, but remain above the value of 1, indicating that sediments are slightly overconsolidated. Shear strength values also show increased values at the basal surface and there seems to be a good fit between the values obtained from CPT and laboratory vane shear test. The entire sensitivity profile is dominated by low to moderately sensitive sediments. In particular, at the basal surface sensitivity values decrease to a value of 2.

CPTu PMF-1104. CPTu PMF-1104 (Fig. 9b) was performed within the RWMTD. Similar to CPTu PMF-1102, the uppermost 6.5 m of the profile is characterized by undeformed sediments, which show a quasi-linear increase in q_t, f_s and Δ_{u2} values, typical for undisturbed seafloor sediments. Below 6.5 m, and extending to the bottom of the profile, no clear trend can be seen. This could be attributed to the presence of silty layers within the MTD, which is evident with the drop of pore pressure where we observe peaks in the cone tip resistance (represented by grey intervals in Fig. 9b). We observe a quasi-linear increment in shear strength and sensitivity, while OCR values remain mostly constant until the bottom of the profile. In general, the RWMTD is slightly overconsolidated and shows a normal sensitivity and shear strength as compared with the post-slide sediments.

Discussion

Internal structure of mass movement products in the GoL

Our data highlight recurring slope failures at different stratigraphic levels within the Plio-Quaternary sequence (Figs 3–6), suggesting a long history of mass movements in the GoL (Badhani *et al.* 2019; Dennielou *et al.* 2019). Previous studies claimed a seismically homogeneous character of the surficial MTDs (Badhani *et al.* 2019; Dennielou *et al.* 2019). Newly collected deep-towed seismic data show few internal structures in the proximal part of the MTD at the hanging wall of the fault (Fig. 6). Coinciding *in-situ* measurements within the RWMTD confirm several closely spaced sediment layers with high tip resistance and coarse-grained beds suggested by negative peaks in pore pressure between 20 and 30 m (Fig. 9b), which contrasts with homogenized material such as debris flow deposits that show an overall constant CPT profile with small-scale irregularities (Garziglia *et al.* 2010). In addition, few contorted stratified layers have been observed in core PSM3-CS049 (Fig. 7). This suggests that RWMTD is not very homogeneous as previously thought based on transparent seismic facies and that the degree of sediment disintegration in the proximal area is most likely lower than previously considered (Badhani *et al.* 2019; Dennielou *et al.* 2019). One can argue that the degree of sediment disintegration increases as the MTD propagates further downslope, such as in the case of the Big'95 debris flow (Lastras *et al.* 2005). However, for the RWMTD, this could not be demonstrated owing to a lack of ground truth information in the distal part of the MTD.

In addition, in the subsurface we were able to image a small-scale landslide with intact slide blocks of 30 ms height, which kept their stratigraphy after the sliding (Fig. 6b) and were represented as chaotic reflections in the conventional HR seismic data (Fig. 3). The intact slide blocks within the failure area suggest less deformation of strata during a local failure related to slope steepening owing to halokinesis (dos Reis *et al.* 2005) that did not progress into a full dynamic slide (Alves 2015). Therefore, we suggest that the chaotic reflections in the conventional seismic data can be imaged if the source and streamer are towed close to the seafloor in order to avoid diffractions from scattered blocks within the MTD.

Based on our observation, we suggest that the chaotic reflections in the MTDs (e.g. Figs 3 & 6b, c) and transparent reflections in the debris flow deposits (e.g. RWMTD; Figs 3 & 6a) are generated owing to a lack of resolution of the conventional seismic data, which is not able to resolve scattered blocks embedded within the transported mass. This is important because with the previously available conventional seismic data most of the kinematic indicators (Frey-Martínez *et al.* 2006; Bull *et al.* 2009; Moscardelli and Wood 2016) of the sliding are ignored or poorly understood. Therefore, a better resolution of seismic data, such as deep-towed seismic data, especially when coupled with *in-situ* measurements as presented in this study, will allow us to better understand large-scale, complex events in order to constrain seafloor instabilities in deep-water environments such as the GoL.

Sedimentological, geotechnical and geophysical properties of basal surfaces

With the help of combined geophysical, sedimentological and *in-situ* geotechnical properties of the GoL slope, we characterized the lithofacies within the study area. Mainly two types of genetically different sediments exist in the study area: (1) hemipelagic to pelagic fine-grained clay-rich sediments likely representing a sea-level highstand or interglacial period; and (2) silty turbiditic sediments interbedded with hemipelagic intervals probably representing sea-level lowstands or glacial and deglacial periods. The lithofacies L1 is characterized by fine-grained material of low Ca/Fe ratios (Fig. 7). This lithofacies was most likely deposited after the Holocene and represents present-day conditions (Dennielou *et al.* 2019). The glacial and deglacial conditions in the study area could be represented by the frequent occurrence of silty turbidites within lithofacies L2, showing that the river sources rapidly deposited these sediments close to the present-day shelf break. This lithofacies is represented by high Ca/Fe ratios in cores PSM3-CS050 and 51 (Fig. 7). Radiocarbon dating on top of the MTD within the turbiditic sediments shows that these lithofacies L1 and L2 (Fig. 7) were deposited after the failure event during the LGM (Fig. 7). Most slope failures in the GoL occur bedding parallel over high-amplitude reflectors, which are sometimes interpreted as impressions of condensed sections formed during the sea-level highstands in the upper slope setting (Sierro *et al.* 2009). High-resolution seismic data clearly show that on the western side of the PRC the basal surface of the failure at the fault scarp lies below these units at a depth of *c.* 7.5 m (Fig. 6). At the basal surface in core PSM3-CS050, we observe sediment type similar to that within lithofacies L1 (clay-rich hemipelagic sediments). This lithofacies also shows more clayey sediments than the overlying turbiditic sequence (Fig. 8a), suggesting a less energetic environment or a time interval characterized by small-volume turbidity currents that could be ascribed to highstand conditions. Also, the basal shear surface of the RWMTD in core PSM3-CS051 lies below lithofacies L1 and L2, although on a younger stratigraphic level (Fig. 7). Visual description of core PSM3-CS051 shows that the basal surface of the RWMTD is in a sediment interval with fewer turbidites if compared with overlying and underlying intervals (below 4.8 m, Fig. 7).

Core PSM3-CS051 and -50 are located in distinct stratigraphic units. Core 51 penetrated the basal surface of the RWMTD and core 50 penetrated the basal surface of the small failure on the fault scarp (Figs 6 & 7). With the help of gamma density, V_p and Ca/Fe signatures of cores PSM3-CS050 and -51, we were able to correlate the basal surface of the RWMTD and the failure at the fault scarp (Fig. 7). This allowed us to extrapolate the detailed analysis on sediment core PSM3-CS050 to the Western Slide area. It is interesting to note that both basal surfaces consist of higher fractions of clay-rich sediments. Similarly, on the eastern side of the PRC, successive headwall scarps of the ERIS and the shallow MTD took place over high-amplitude reflectors (Fig. 4) that are interpreted to have deposited during highstands (Badhani *et al.* 2019). Sediment core PSM3-CS050 shows a good example where failure mostly affects turbiditic successions, even if the core is located in an area of limited sediment remobilization (Fig. 6) and is not representative of large-scale failures (e.g. REMTD and RWMTD). We observe a similar process of failure over the high-amplitude reflectors throughout the Plio-Quaternary strata in the GoL. Therefore, it is likely that the failure surfaces in the GoL are influenced by climatic forcing related to Milankovitch cyclicity tuning the number and volume of turbiditic flows in the Rhone system and forming surfaces of significantly different mechanical properties in the sedimentary column. The inference of increased turbiditic deposits during glacial and deglacial times is based on evidence from the nearby Var turbiditic system (Jorry *et al.* 2011). We suggest that this could also be extrapolated to the entire GoL where a majority of failures took place, bedding parallel over clay-rich sediments deposited during the highstand periods (Figs 3–6).

Many studies have pointed out the role of lithological heterogeneities in enhancing seafloor instabilities (e.g. Laberg and Camerlenghi 2008; L'Heureux *et al.* 2012). We point out that the interface between clay-rich hemipelagic successions and overlying turbiditic sequences is critical. *In-situ* geotechnical measurements show that the sediments below the basal surface of the failure on the fault scarp show higher shear strength and OCR, and lower sensitivity values than the background values (Fig. 9a, b). This is consistent with laboratory vane shear measurements for the basal surface of the failure at the fault scarp (Fig. 8). However, it should be noted that these elevated values of shear strength measurement could be due to the removal of sediments above the basal surface during the failure at the fault scarp. Nevertheless, several processes can induce slope instabilities where lithological heterogeneity within the strata exists such as the liquefaction potential of coarse-grained material and excess pore pressure generation in the presence of very high sedimentation rates such as in the GoL during several lowstands (Kettner and Syvitski 2009; Sierro *et al.* 2009; Lombo Tombo *et al.* 2015; Dennielou *et al.* 2019). Dan *et al.* (2009) showed that coarser-grained materials (such as turbidites) are prone to liquefaction even in cases of mild seismic loading. Lafuerza *et al.* (2014) and Hornbach *et al.* (2015) demonstrated that, in the sediment column, excess pore pressure resides in coarse-grained intervals while fine-grained intervals show very low permeability and act as seals generating weak layers at the boundary between coarse-grained and fine-grained sediments. This may potentially imply that also in deep-water environments such as the GoL, even if not significant, lithological heterogeneity exists and most likely controls the localization of potential weak layers. This may explain the recurrence of failures over clay-rich hemipelagic sediments identified at several stratigraphic levels in the GoL (Figs 3–6).

Variations in sediment grain size also occur in multiple settings around the globe, e.g. volcanic flanks (Lafuerza *et al.* 2014; Hornbach *et al.* 2015; Brunet *et al.* 2016), passive margins (Sawyer *et al.* 2009) and glacial margins (Laberg and Camerlenghi 2008). With the example provided by the GoL, we further support the evidence that lithological heterogeneity within the sediment successions is a key to localization of seafloor instability. The identification of this stratigraphic control as a major predisposing factor for seafloor instability allows better assessment by understanding the causes of recurrent failures.

Conclusions

The Gulf of Lions is a passive margin of about 200 km length and 70 km width with main sediment supply from the Rhone River supplying Alpine sediments to the Rhone delta. Previous studies highlighted two large surficial submarine landslides with resulting volumes exceeding 100 km^3. We presented here the internal structure of mass transport deposit and small-scale slope failures with unprecedented detail using newly available very high-resolution deep-towed seismic data for the first time in the GoL. We showed an example of failure that consisted of slid blocks and folded and faulted strata with preserved stratigraphy that were previously imaged as ‘transparent’ or ‘chaotic’ reflections in the conventional reflection seismic data. The combination of deep-towed seismic, sedimentological data and *in-situ* measurements allowed us to analyse the basal surface of the slope failures in greater detail. We showed that the basal surfaces of the recurring slope failures consist of clay-rich sediments as compared with turbiditic deposits that dominate the GoL margin. It is evident that the greater degree of lithological heterogeneity promotes slope failure in the GoL most likely related to higher liquefaction potential of coarser-grained material, excess pore pressure and maybe resulting variation in sediment strength. However, further analysis is necessary to confirm if these factors influence slope stability in the GoL.

Acknowledgements We thank the captains, crew and science parties of the PRISME-2 and PRISME-3 campaigns onboard R/V *L'Atalante* and R/V *Pourquoi pas?*, respectively for their support during the campaigns. Elena Martinez Planchart (ICM, Barcelona) is thanked for carrying out grain-size analysis on sediment core PSM3-CS050. We thank Nabil Sultan for discussions on CPTu data interpretation. We are also grateful to reviewers Samantha Clarke and Sara Lafuerza and editor Lawrence Amy for providing insightful reviews that significantly improved this manuscript.

Funding This work and the PhD of Shray Badhani are funded by the European Union's Horizon 2020 research and innovation programme under the Marie-Skłodowska-Curie grant via project ITN SLATE (grant agreement no. 721403).

Author contributions **SB:** conceptualization (lead), data curation (lead), formal analysis (lead), investigation (lead), visualization (lead), writing – original draft (lead), writing – review & editing (lead); **AC:** conceptualization (lead), methodology (equal), supervision (lead), writing – original draft (equal); **SC:** formal analysis (equal), methodology (equal), writing – original draft (equal), writing – review & editing (equal); **RU:** conceptualization (equal), methodology (equal), writing – original draft (equal), writing – review & editing (equal); **BD:** conceptualization (equal), methodology (equal), writing – original draft (equal), writing – review & editing (equal); **EL:** conceptualization (equal), writing – original draft (equal), writing – review & editing (equal); **FC:** conceptualization (equal), data curation (lead), writing – original draft (equal), writing – review & editing (equal); **SG:** data curation (equal), writing – original draft (equal), writing – review & editing (equal); **MR:** conceptualization (equal), writing – original draft (equal), writing – review & editing (equal); **LD:** conceptualization (equal), writing – original draft (equal), writing – review & editing (equal).

References

Alves, T.M. 2015. Submarine slide blocks and associated soft-sediment deformation in deep-water basins: a review. *Marine and Petroleum Geology*, **67**, 262–285, https://doi.org/10.1016/j.marpetgeo.2015.05.010

Badhani, S., Cattaneo, A. *et al.* 2019. Morphology of retrogressive failures in the Eastern Rhone Interfluve during the Last Glacial Maximum (Gulf of Lions, Western Mediterranean). *Geomorphology*, https://doi.org/10.1016/j.geomorph.2019.106894.

Baeten, N.J., Laberg, J.S. *et al.* 2014. Origin of shallow submarine mass movements and their glide planes – sedimentological and geotechnical analyses from the continental slope off northern Norway. *Journal of Geophysical Research: Earth Surface*, **119**, 2335–2360, https://doi.org/10.1002/2013JF003068

Bassetti, M.A., Berné, S. *et al.* 2008. The 100-ka and rapid sea level changes recorded by prograding shelf sand bodies in the Gulf of Lions (western Mediterranean Sea). *Geochemistry, Geophysics, Geosystems*, **9**, 11, https://doi.org/10.1029/2007GC001854

Bellaiche, G., Coutellier, V. and Droz, L. 1986. Seismic evidence of widespread mass transport deposits in the Rhône deep-sea fan: their role in the fan construction. *Marine Geology*, **71**, 327–340, https://doi.org/10.1016/0025-3227(86)90076-9

Berné, S., Satra, C. *et al.* 2002. *Carte morpho-bathymétrique du Golfe du Lion, notice explicative*. Ifremer, Brest, **48**.

Berné, S., Jouet, G., Bassetti, M.A., Dennielou, B. and Taviani, M. 2007. Late Glacial to Preboreal sea-level rise recorded by the Rhône deltaic system (NW Mediterranean). *Marine Geology*, **245**, 65–88, https://doi.org/10.1016/j.margeo.2007.07.006

Bonnel, C., Dennielou, B., Droz, L. and Mulder, T. 2005. Architecture and depositional pattern of the Rhône Neofan and recent gravity activity in the Gulf of Lions (western Mediterranean). *Marine and Petroleum Geology*, **22**, 827–843, https://doi.org/10.1016/j.marpetgeo.2005.03.003

Brooks, G.R. 2016. Evidence of late glacial paleoseismicity from submarine landslide deposits within Lac Dasserat, northwestern Quebec, Canada. *Quaternary Research*, **86**, 184–199, https://doi.org/10.1016/j.yqres.2016.06.005

Brunet, M., Le Friant, A. *et al.* 2016. Composition, geometry, and emplacement dynamics of a large volcanic island landslide offshore Martinique: from volcano flank-collapse to seafloor sediment failure? *Geochemistry, Geophysics, Geosystems*, **17**, 699–724, https://doi.org/10.1002/2015GC006034

Bryn, P., Solheim, A. *et al.* 2003. *The Storegga Slide Complex; Repeated Large Scale Sliding in Response to Climatic Cyclicity*. Springer, Dordrecht, 215–222, https://doi.org/10.1007/978-94-010-0093-2_24.

Bryn, P., Berg, K., Forsberg, C.F., Solheim, A. and Kvalstad, T.J. 2005. Explaining the Storegga Slide. *Marine and Petroleum Geology*, **22**, 11–19, https://doi.org/10.1016/j.marpetgeo.2004.12.003

Bull, S., Cartwright, J. and Huuse, M. 2009. A review of kinematic indicators from mass-transport complexes using 3D seismic data. *Marine and Petroleum Geology*, **26**, 1132–1151, https://doi.org/10.1016/j.marpetgeo.2008.09.011

Carvajal, M., Araya-Cornejo, C., Sepúlveda, I., Melnick, D. and Haase, J.S. 2019. Nearly instantaneous tsunamis following the *Mw* 7.5 2018 Palu Earthquake. *Geophysical Research Letters*, **46**, 2019GL082578, https://doi.org/10.1029/2019GL082578.

Cattaneo, A. 2013*a*. *PRISME 2 cruise, RV L'Atalante*, https://doi.org/10.17600/13010050

Cattaneo, A. 2013*b*. *PRISME 3 Cruise, RV Pourquoi Pas?*, https://doi.org/10.17600/13030060

Damuth, J.E. 1980. Use of high-frequency (3.5–12 kHz) echograms in the study of near-bottom sedimentation processes in the deep-sea: a review. *Marine Geology*, **38**, 51–75, https://doi.org/10.1016/0025-3227(80)90051-1

Dan, G., Sultan, N. and Savoye, B. 2007. The 1979 Nice harbour catastrophe revisited: trigger mechanism inferred from geotechnical measurements and numerical modelling. *Marine Geology*, **245**, 40–64, https://doi.org/10.1016/j.margeo.2007.06.011

Dan, G., Sultan, N., Savoye, B., Deverchere, J. and Yelles, K. 2009. Quantifying the role of sandy–silty sediments in generating slope failures during earthquakes: example from the Algerian margin. *International Journal of Earth Sciences*, **98**, 769–789, https://doi.org/10.1007/s00531-008-0373-5

Dennielou, B., Huchon, A., Beaudouin, C. and Berné, S. 2006. Vertical grain-size variability within a turbidite levee: autocyclicity or allocyclicity? A case study from the Rhône neofan, Gulf of Lions, Western Mediterranean. *Marine Geology*, **234**, 191–213, https://doi.org/10.1016/j.margeo.2006.09.019

Dennielou, B., Jégou, I. *et al.* 2019. Major modification of sediment routing by a large mass transport deposit in the Gulf of Lions (Western Mediterranean). *Marine Geology*, **411**, 1–20, https://doi.org/10.1016/j.margeo.2019.01.011

dos Reis, A.T., Gorini, C. and Mauffret, A. 2005. Implications of salt–sediment interactions on the architecture of the Gulf of Lions deep-water sedimentary systems – western Mediterranean Sea. *Marine and Petroleum Geology*, **22**, 713–746, https://doi.org/10.1016/j.marpetgeo.2005.03.006

Droz, L. and Bellaiche, G. 1985. Rhone deep-sea fan: morphostructure and growth pattern. *AAPG Bulletin*, **69**, 460–479.

Droz, L., Kergoat, R., Cochonat, P. and Berné, S. 2001. Recent sedimentary events in the western Gulf of Lions (Western Mediterranean). *Marine Geology*, **176**, 23–37, https://doi.org/10.1016/S0025-3227(01)00147-5

Droz, L., dos Reis, A.T., Rabineau, M., Berné, S. and Bellaiche, G. 2006. Quaternary turbidite systems on the northern margins of the Balearic Basin (Western Mediterranean): a synthesis. *Geo-Marine Letters*, **26**, 347–359, https://doi.org/10.1007/s00367-006-0044-0

EMODnet Bathymetry Consortium. 2018. EMODnet Digital Bathymetry (DTM 2018), EMODnet Bathymetry Consortium, http://doi.org/10.12770/18ff0d48-b203-4a65-94a9-5fd8b0ec35f6

Frey-Martínez, J., Cartwright, J. and James, D. 2006. Frontally confined versus frontally emergent submarine landslides: a 3D seismic characterisation. *Marine and Petroleum Geology*, **23**, 585–604, https://doi.org/10.1016/j.marpetgeo.2006.04.002

Garziglia, S., Sultan, N. *et al.* 2010. Identification of shear zones and their causal mechanisms using a combination of cone penetration tests and seismic data in the Eastern Niger Delta. *In*: Mosher, D.C. *et al.* (eds) *Submarine Mass Movements and Their Consequences.* Advances in Natural and Technological Hazards Research, **28**. Springer, Dordrecht, 55–65, https://doi.org/10.1007/978-90-481-3071-9_5

Hampton, M.A., Lee, H.J. and Locat, J. 1996. Submarine landslides. *Reviews of Geophysics*, **34**, 33–59, https://doi.org/10.1029/95RG03287

Heidarzadeh, M., Muhari, A. and Wijanarto, A.B. 2018. Insights on the source of the 28 September 2018 Sulawesi Tsunami, Indonesia based on spectral analyses and numerical simulations. *Pure and Applied Geophysics*, 1–19, https://doi.org/10.1007/s00024-018-2065-9.

Hornbach, M.J., Manga, M. *et al.* 2015. Permeability and pressure measurements in Lesser Antilles submarine slides: evidence for pressure-driven slow-slip failure. *Journal of Geophysical Research: Solid Earth*, **120**, 7986–8011, https://doi.org/10.1002/2015JB012061

Hsu, K.J. 1973. The origin of the Mediterranean evaporites. *Initial Reports of the Deep Sea Drilling Project*, **13**, 1203–1231.

Jorry, S.J., Jégou, I., Emmanuel, L., Silva Jacinto, R. and Savoye, B. 2011. Turbiditic levee deposition in response to climate changes: the Var Sedimentary Ridge (Ligurian Sea). *Marine Geology*, **279**, 148–161, https://doi.org/10.1016/j.margeo.2010.10.021

Jouet, G. 2007. *Enregistrements stratigraphiques des cycles climatiques et glacio-eustatiques du Quaternaire terminal – Modélisations de la marge continentale du Golfe du Lion.* Phd thesis, Université de Bretagne Occidentale, Brest, https://archimer.ifremer.fr/doc/00000/3692/

Ker, S., Marsset, B., Garziglia, S., Le Gonidec, Y., Gibert, D., Voisset, M. and Adamy, J. 2010. High-resolution seismic imaging in deep sea from a joint deep-towed/OBH reflection experiment: application to a Mass Transport Complex offshore Nigeria. *Geophysical Journal International*, **182**, 1524–1542, https://doi.org/10.1111/j.1365-246X.2010.04700.x

Kettner, A.J. and Syvitski, J.P.M. 2009. Fluvial responses to environmental perturbations in the Northern Mediterranean since the Last Glacial Maximum. *Quaternary Science Reviews*, **28**, 2386–2397, https://doi.org/10.1016/j.quascirev.2009.05.003

Kvalstad, T.J., Nadim, F., Kaynia, A.M., Mokkelbost, K.H. and Bryn, P. 2005. Soil conditions and slope stability in the Ormen Lange area. *Marine and Petroleum Geology*, **22**, 299–310, https://doi.org/10.1016/j.marpetgeo.2004.10.021

Laberg, J.S. and Camerlenghi, A. 2008. The significance of contourites for submarine slope stability. *Developments in Sedimentology*, **60**, 537–556, https://doi.org/10.1016/S0070-4571(08)10025-5

Lafuerza, S., Le Friant, A. *et al.* 2014. Geomechanical characterization of submarine volcano-flank sediments, Martinique, Lesser Antilles Arc. *In*: Krastel, S. *et al.* (eds) *Submarine Mass Movements and Their Consequences.* Advances in Natural and Technological Hazards Research, **37**. Springer, Cham, 73–81, https://doi.org/10.1007/978-3-319-00972-8_7

Lastras, G., Canals, M., Urgeles, R., Hughes-Clarke, J.E. and Acosta, J. 2004. Shallow slides and pockmark swarms in the Eivissa Channel, western Mediterranean Sea. *Sedimentology*, **51**, 837–850, https://doi.org/10.1111/j.1365-3091.2004.00654.x

Lastras, G., De Blasio, F.V., Canals, M. and Elverhoi, A. 2005. Conceptual and numerical modeling of the BIG'95 Debris Flow, Western Mediterranean Sea. *Journal of Sedimentary Research*, **75**, 784–797, https://doi.org/10.2110/jsr.2005.063

Le Friant, A., Ishizuka, O. *et al.* 2015. Submarine record of volcanic island construction and collapse in the Lesser Antilles arc: first scientific drilling of submarine volcanic island landslides by IODP Expedition 340. *Geochemistry, Geophysics, Geosystems*, **16**, 420–442, https://doi.org/10.1002/2014GC005652

L'Heureux, J.-S., Longva, O. *et al.* 2012. Identification of weak layers and their role for the stability of slopes at Finneidfjord, Northern Norway. *In*: Yamada, Y. *et al.* (eds) *Submarine Mass Movements and Their Consequences.* Advances in Natural and Technological Hazards Research, **31**. Springer, Dordrecht, 321–330, https://doi.org/10.1007/978-94-007-2162-3_29

Locat, J., Leroueil, S., Locat, A. and Lee, H. 2014. Weak layers: their definition and classification from a geotechnical perspective. *In*: Krastel, S. *et al.* (eds) *Submarine Mass Movements and Their Consequences.* Advances in Natural and Technological Hazards Research, **37**. Springer, Cham, 3–12, https://doi.org/10.1007/978-3-319-00972-8_1

Lofi, J., Rabineau, M. *et al.* 2003. Plio-Quaternary prograding clinoform wedges of the western Gulf of Lion continental margin (NW Mediterranean) after the Messinian Salinity Crisis. *Marine Geology*, **198**, 289–317, https://doi.org/10.1016/S0025-3227(03)00120-8

Lombo Tombo, S., Dennielou, B. *et al.* 2015. Sea-level control on turbidite activity in the Rhone canyon and the upper fan during the Last Glacial Maximum and Early deglacial. *Sedimentary Geology*, https://doi.org/10.1016/j.sedgeo.2015.04.009.

Low, H.E., Lunne, T., Andersen, K.H., Sjursen, M.A., Li, X. and Randolph, M.F. 2010. Estimation of intact and remoulded undrained shear strengths from penetration tests in soft clays. *Géotechnique*, **60**, 843, https://doi.org/10.1680/geot.9.P.017

Marsset, B., Menut, E. *et al.* 2014. Deep-towed high resolution multichannel seismic imaging. *Deep-Sea Research Part I: Oceanographic Research Papers*, **93**, 83–90, https://doi.org/10.1016/j.dsr.2014.07.013

Masson, D., Harbitz, C., Wynn, R., Pedersen, G. and Løvholt, F. 2006. Submarine landslides: processes, triggers and hazard prediction. *Philosophical Transactions of the Royal Society A: Mathematical, Physical and Engineering Sciences*, **364**, 2009–2039, https://doi.org/10.1098/rsta.2006.1810

Micallef, A., Mountjoy, J.J., Canals, M. and Lastras, G. 2012. Deep-seated bedrock landslides and submarine Canyon evolution in an active tectonic margin: Cook Strait, New Zealand. *In*: Yamada, Y. *et al.* (eds) *Submarine Mass Movements and Their Consequences.* Advances in Natural and Technological Hazards Research, **31**. Springer, Dordrecht, 201–212, https://doi.org/10.1007/978-94-007-2162-3_18

Miramontes, E., Sultan, N., Garziglia, S., Jouet, G., Pelleter, E. and Cattaneo, A. 2018. Altered volcanic deposits as basal failure surfaces of submarine landslides. *Geology*, **46**, 663–666, https://doi.org/10.1130/G40268.1

Moernaut, J., Van Daele, M. *et al.* 2017. Lacustrine turbidites produced by surficial slope sediment remobilization: a mechanism for continuous and sensitive turbidite paleoseismic records. *Marine Geology*, **384**, 159–176, https://doi.org/10.1016/j.margeo.2015.10.009

Moore, J.G., Clague, D.A., Holcomb, R.T., Lipman, P.W., Normark, W.R. and Torresan, M.E. 1989. Prodigious submarine landslides on the Hawaiian Ridge. *Journal*

of Geophysical Research, **94**, 17465, https://doi.org/10.1029/JB094iB12p17465
Moscardelli, L. and Wood, L. 2016. Morphometry of mass-transport deposits as a predictive tool. *Bulletin of the Geological Society of America*, **128**, 47–80, https://doi.org/10.1130/B31221.1.
Ogata, K., Pini, G.A. *et al.* 2014. High-resolution studies of mass transport deposits: outcrop perspective for understanding modern submarine slope failure and associated natural hazards. *In*: Lollino, G., Manconi, A., Locat, J., Huang, Y. and Canals Artigas, M. (eds) *Engineering Geology for Society and Territory*, **4**. Springer, Cham, 209–213, https://doi.org/10.1007/978-3-319-08660-6_40
Ogata, K., Pini, G.A., Festa, A., Pogačnik, Ž. and Lucente, C.C. 2016. Meso-scale kinematic indicators in exhumed mass transport deposits: definitions and implications. *In*: Lamarche, G. *et al.* (eds) *Submarine Mass Movements and their Consequences*. Advances in Natural and Technological Hazards Research, **41**. Springer, Cham, 461–468, https://doi.org/10.1007/978-3-319-20979-1_46
Posamentier, H.W. and Kolla, V. 2003. Seismic geomorphology and stratigraphy of depositional elements in deep-water settings. *Journal of Sedimentary Research*, **73**, 367–388, https://doi.org/10.1306/111302730367
Praet, N., Moernaut, J. *et al.* 2017. Paleoseismic potential of sublacustrine landslide records in a high-seismicity setting (south-central Alaska). *Marine Geology*, **384**, 103–119, https://doi.org/10.1016/j.margeo.2016.05.004
Principaud, M., Mulder, T., Gillet, H. and Borgomano, J. 2014. Large-scale carbonate submarine mass-wasting along the northwestern slope of the Great Bahama Bank (Bahamas): morphology, architecture, and mechanisms. *Sedimentary Geology*, **317**, 27–42, https://doi.org/10.1016/j.sedgeo.2014.10.008
Rabineau, M., Berné, S., Ledrezen, É., Lericolais, G., Marsset, T. and Rotunno, M. 1998. 3D architecture of lowstand and transgressive Quaternary sand bodies on the outer shelf of the Gulf of Lion, France. *Marine and Petroleum Geology*, **15**, 439–452, https://doi.org/10.1016/S0264-8172(98)00015-4
Rabineau, M., Berné, S. *et al.* 2005. Sedimentary sequences in the Gulf of Lion: a record of 100,000 years climatic cycles. *Marine and Petroleum Geology*, **22**, 775–804, https://doi.org/10.1016/j.marpetgeo.2005.03.010
Rabineau, M., Droz, L. and Aslanian, D. 2013. *AM-MED-1 Cruise, RV Le Suroît*, https://doi.org/10.17600/18000821
Reimer, P.J., Edouard Bard, B., Alex Bayliss, B., Warren Beck, B.J., Paul Blackwell, B.G. and Christopher Bronk Ramsey, B. 2013. IntCal13 and Marine13 radiocarbon age calibration curves 0–50,000 years cal bp. *Radiocarbon*, **55**, 1869–1887, https://doi.org/10.2458/azu_js_rc.55.16947
Robertson, P.K. 1990. Soil classification using the cone penetration test. *Canadian Geotechnical Journal*, **27**, 151–158, https://doi.org/10.1139/t90-014
Roveri, M., Lugli, S., Manzi, V. and Schreiber, B.C. 2008. The Messinian Sicilian stratigraphy revisited: new insights for the Messinian salinity crisis. *Terra Nova*, **20**, 483–488, https://doi.org/10.1111/j.1365-3121.2008.00842.x
Sassa, S. and Takagawa, T. 2019. Liquefied gravity flow-induced tsunami: first evidence and comparison from the 2018 Indonesia Sulawesi earthquake and tsunami disasters. *Landslides*, **16**, 195–200, https://doi.org/10.1007/s10346-018-1114-x
Sawyer, D.E. and DeVore, J.R. 2015. Elevated shear strength of sediments on active margins: evidence for seismic strengthening. *Geophysical Research Letters*, **42**, 10216–10221, https://doi.org/10.1002/2015GL066603
Sawyer, D.E., Flemings, P.B., Dugan, B. and Germaine, J.T. 2009. Retrogressive failures recorded in mass transport deposits in the Ursa Basin, Northern Gulf of Mexico, **114**, https://doi.org/10.1029/2008JB00 6159.
Sierro, F.J., Andersen, N. *et al.* 2009. Phase relationship between sea level and abrupt climate change. *Quaternary Science Reviews*, **28**, 2867–2881, https://doi.org/10.1016/j.quascirev.2009.07.019
Silva, C.G., Araújo, E., Reis, A.T., Perovano, R., Gorini, C., Vendeville, B.C. and Albuquerque, N. 2010. Megaslides in the Foz do Amazonas Basin, Brazilian Equatorial Margin. *In*: Mosher, D.C., Shipp, R.C. *et al.* (eds) *Submarine Mass Movements and their Consequences*. Advances in Natural and Technological Hazards Research, **28**. Springer, Dordrecht, 581–591, https://doi.org/10.1007/978-90-481-3071-9_47
St-Onge, G., Mulder, T., Piper, D.J.W., Hillaire-Marcel, C. and Stoner, J.S. 2004. Earthquake and flood-induced turbidites in the Saguenay Fjord (Québec): a Holocene paleoseismicity record. *Quaternary Science Reviews*, **23**, 283–294, https://doi.org/10.1016/j.quascirev.2003.03.001
Stuiver, M., Reimer, P.J. and Reimer, R.W. 2017. Calib 7.1 [WWW program] at http://calib.org, last accessed September 2019.
Sultan, N., Gaudin, M., Berne, S., Canals, M.M., Urgeles, R. and Lafuerza, S. 2007*a*. Analysis of slope failures in submarine canyon heads: an example from the Gulf of Lions. *Journal of Geophysical Research: Earth Surface*, **112**, F01009, https://doi.org/10.1029/2005JF000408
Sultan, N., Voisset, M., Marsset, B., Marsset, T., Cauquil, E. and Colliat, J.-L. 2007*b*. Potential role of compressional structures in generating submarine slope failures in the Niger Delta. *Marine Geology*, **237**, 169–190, https://doi.org/10.1016/j.margeo.2006.11.002
Talling, P., Clare, M., Urlaub, M., Pope, E., Hunt, J. and Watt, S. 2014. Large submarine landslides on continental slopes: geohazards, methane release, and climate change. *Oceanography*, **27**, 32–45, https://doi.org/10.5670/oceanog.2014.38
Tappin, D.R., Matsumoto, T. *et al.* 1999. Sediment slump likely caused 1998 Papua New Guinea tsunami. *Eos. Transactions of the American Geophysical Union*, **80**, 329, https://doi.org/10.1029/99EO00241
Tesson, M. and Gensous, B. 1998. L'enregistrement des cycles climatiques et eustatiques quaternaires de marges récentes du bassin Nord-Méditerranéen. *Comptes Rendus de l'Académie des Sciences – Series IIA – Earth and Planetary Science*, **326**, 121–127, https://doi.org/10.1016/S1251-8050(97)87456-3
Torres, J., Droz, L., Savoye, B., Terentieva, E., Cochonat, P., Kenyon, N.H. and Canals, M. 1997. Deep-sea avulsion and morphosedimentary evolution of the Rhône Fan Valley and Neofan during the Late Quaternary (north-western

Mediterranean Sea). *Sedimentology*, **44**, 457–477, https://doi.org/10.1046/j.1365-3091.1997.d01-36.x
Urgeles, R., Canals, M., Baraza, J., Alonso, B. and Masson, D. 1997. The most recent megalandslides of the Canary Islands: El Golfo debris avalanche and Canary debris flow, west El Hierro Island. *Journal of Geophysical Research: Solid Earth*, **102**, 20305–20323, https://doi.org/10.1029/97JB00649
Urlaub, M., Geersen, J., Krastel, S. and Schwenk, T. 2018. Diatom ooze: crucial for the generation of submarine mega-slides? *Geology*, **46**, 2–5, https://doi.org/10.1130/G39892.1

Characterization of recent deep-sea debrites in the eastern Mediterranean based on foraminiferal taphonomy

Oded Katz[1]*, Leeron Ashkenazi[1,2], Shani Sultan-Levi[1], Sigal Abramovich[2], Ahuva Almogi-Labin[1] and Orit Hyams-Kaphzan[1]

[1]Geological Survey of Israel, 32 Yesha'ayahu Leibowitz, Jerusalem 9692100, Israel

[2]Department of Geological and Environmental Science, Ben Gurion University of the Negev, Beer-Sheva 8410501, Israel

AA-L, 0000-0002-4082-7120

*Correspondence: odedk@gsi.gov.il

Abstract: It is hypothesized that submarine transport of sediments down a continental slope induces physical disintegration of pristine (non-broken) foraminiferal shells, and thus mass transport deposits should include a significant percentage of fragmented shells. To validate this hypothesis, we studied two gravity-cores from the eastern Mediterranean continental slope, offshore Israel: AM113 sampled within a landslide lobe at 848 m water depth, and AM015 located away from a landslide at 1080 m. At least one interval, *c.* 0.5 m thick, of heterogeneous sediments (i.e. debrite) was identified within each core. The timing of these debrites, based on biostratigraphy, oxygen isotopes and total organic carbon data, predates sapropel S1 in both cores and is contemporaneous (AM113) or slightly predates (AM015) the most recent deglaciation.

We found a noticeable increase in the fragmentation of benthic and planktic foraminiferal shells through the last deglaciation and up to the base of S1. This strongly fragmented sequence is located in the debrite of AM113 but overlays the debrite of AM015. Accordingly, we suggest two possible mechanisms for the increased fragmentation of foraminiferal shells in both cores: sediment transport and turbulence related to submarine mass-transport events, or geochemical changes in the lower water column properties at the transition from MIS-2 to the Holocene.

Supplementary material: Photo-mosaic of cores AM015 and AM113 are available at: https://doi.org/10.6084/m9.figshare.c.4862370

Sediment mass-transport events down a continental slope (e.g. turbidity currents, mud flows, landslides) are important mechanisms moving sediments from the shelf-break and upper continental slope towards the bathyal environments (Talling *et al.* 2012). These events also have the potential to damage submarine infrastructures, such as energy-related ones or communication lines (Parker *et al.* 2009; Pope *et al.* 2017). For these two important aspects, Late Quaternary submarine mass-transport events are extensively studied often utilizing high-resolution bathymetric data (e.g. ten Brink *et al.* 2006; López-Venegas *et al.* 2008) and shallow subsurface analysis using high-resolution seismics (e.g. Moore and Sawyer 2016; Georgiopoulou *et al.* 2019). A common approach to date and establish the recurrence interval of mass-transport events is by performing detailed sediment core analysis (Ducassou *et al.* 2007, 2013; Polonia *et al.* 2016; Georgiopoulou *et al.* 2019). Cores document the subsurface sediment, where mass-transport deposits (MTDs) are visually identified based on their deformed stratigraphy. Such stratigraphy is different from the homogeneous strata that are a result of the continuous deposition of hemipelagic and pelagic fine-grained material from the water column (Georgiopoulou *et al.* 2009). Other approaches to identify mass-transport deposits within cores are based on grain-size analysis and textural characteristics (e.g. Almagor and Schilman 1995; Ducassou *et al.* 2013) and on analyses of stable isotopes, radiometric ages and foraminiferal biostratigraphy (Ducassou *et al.* 2007).

The taphonomy of foraminiferal shells, especially their mode of preservation, can also be used as an independent tool to identify mass-transport deposits as well as characterize their related transport distance and transport-related processes (Ducassou *et al.* 2013). In different laboratory experiments it was found that long transport distances and high current velocities enhance shell fragmentation of larger benthic foraminifera (Beavington-Penney 2004; Beavington-Penney and Racey 2004; Ash-Mor *et al.* 2018). This process was also suggested based on marine sediment cores taken from the Red Sea (Ash-Mor *et al.* 2017). Nevertheless, although this method looks promising due to the fact that it

From: Georgiopoulou, A., Amy, L. A., Benetti, S., Chaytor, J. D., Clare, M. A., Gamboa, D., Haughton, P. D. W., Moernaut, J. and Mountjoy, J. J. (eds) 2020. *Subaqueous Mass Movements and their Consequences: Advances in Process Understanding, Monitoring and Hazard Assessments*. Geological Society, London, Special Publications, **500**, 377–391.
First published online May 11, 2020, https://doi.org/10.1144/SP500-2019-170

analyses the direct effect of sediment deformation, it has not been often used globally and never been used to study MTDs in the continental slope of the eastern Mediterranean.

Coping with the dearth of studies, in this research we focused on the type and timing of recent debrites along the eastern Mediterranean continental slope, offshore southern Israel. Our main goal is to develop a new tool to identify and analyse MTDs utilizing the taphonomic features of benthic and planktic foraminifera. We analysed two marine sediment cores and integrated the stratigraphic-sedimentological and the micropaleontological-taphonomic approaches to evaluate the nature of submarine mass-transport events.

The post-mortem change in the state of preservation of foraminiferal calcareous shells, referred to as taphonomy (Beavington-Penney 2004; Dessandier *et al.* 2018), is a result of various physical (e.g. fragmentation), chemical (e.g. dissolution) and/or biological (e.g. predation) processes that deteriorate whole foraminiferal shells into a highly fragmented matrix. In the current work we address the physical aspect of taphonomy, especially shell breakage. We quantify the amount of non-pristine (i.e. broken) benthic and planktic foraminiferal shells within the studied continental slope sediments. Transport of sediment downslope coupled with high turbidity can cause the fragmentation of shells (Duros *et al.* 2012; Ducassou *et al.* 2013). Hence, it is expected to find a larger percentage of fragmented shells in samples from disrupted and deformed sequences of the studied cores.

The study area

Along the eastern Mediterranean continental slope, Katz *et al.* (2015) mapped 447 small- to medium-scale landslides in water depths between 170 and 1000 m (Fig. 1). Their surface area ranges from 0.0024 to 91 km^2. Using the relief of the landslides' evacuation areas and the known sedimentation rates on the slopes, the authors suggested that these landslides are younger than 40 ka. This age estimation was later generally confirmed by Elad *et al.* (2018) and Katz *et al.* (2019) using sediment cores acquired within the landslide scars. The majority of the mapped landslides behaved retrogressively as they resulted from sequential slope failure events. In addition to the landslides, north–south-elongated step structures, recognized as surficial scarps of growth faults rooted in the underlying Messinian evaporites, are also observed along the studied continental slope (Garfunkel 1984; Gvirtzman *et al.* 2015; Katz *et al.* 2015; Elfassi *et al.* 2019). Some of them bisect the landslides and others are covered by the landslide deposits. For this paper, the studied marine sediment cores were sampled from the lower part of the continental slope, offshore southern Israel (Fig. 1). Additional details about the cores are given in Table 1.

Methods

Two sediment cores were studied: AM113 and AM015 (Figs 1, 2 & S1; Table 1). Both cores were sampled by 'Samedan Mediterranean Sea' in 2002 and were later archived in the Geological Survey of Israel repository. Core AM015 was sampled on the slope at 1080 m water depth, more than 2 km south of the closest mapped landslide (Fig. 1). Core AM113 was sampled within an elongated distal lobe of a mapped landslide at 848 m water depth, about 10 km downslope from its headscarp (Fig. 1). The lobe forms a depression a few metres deep in the seafloor, which apparently serves as a sediment conduit from the headscarp downwards.

In order to study the stratigraphy of the cores and to identify heterogeneous deformed sequences as opposed to more homogeneous hemipelagic/pelagic ones, we first scanned the cores using a medical X-ray computed tomography (CT) to create a high-resolution 3D tomography. CT scanning was carried out at the Carmel Medical Center, Haifa, using a Philips Brilliance ICT 128 scanner. The scanning parameters used were: an 80 kV electric potential and 800 mA electric charge, which provided 817 DLP (dose length product). Scanning resolution is 1 mm along the core and *c.* 0.2 mm in both lateral directions.

Cores were then sliced lengthwise, visually described, documented and sub-sampled throughout in 1 cm thick intervals. Samples were then freeze-dried for further analysis at the Geological Survey of Israel laboratories as described below.

The total organic carbon dry weight percentage (TOC wt%) of the bulk sediment was measured by the potassium dichromate method (Gaudette and Flight 1974) with a potentiometric detection threshold of 0.02%.

Micropaleontology

To study the foraminiferal assemblages at specific depth intervals along the cores, 2–4 g from each sediment sample were oven dried at 50°C, weighed and then washed through a 63 µm sieve. Later, specimens of benthic and planktic foraminifera were hand-picked from the >150 µm size-fraction using a stereo-microscope (Leica M205 C) and documented by a Leica DFC295 digital camera attached to it. A total of *c.* 300 individuals (combined benthic and planktic foraminifera) with complete shells (pristine) or partly fragmented (>50% of the shell is still preserved) were picked on micro-slides from

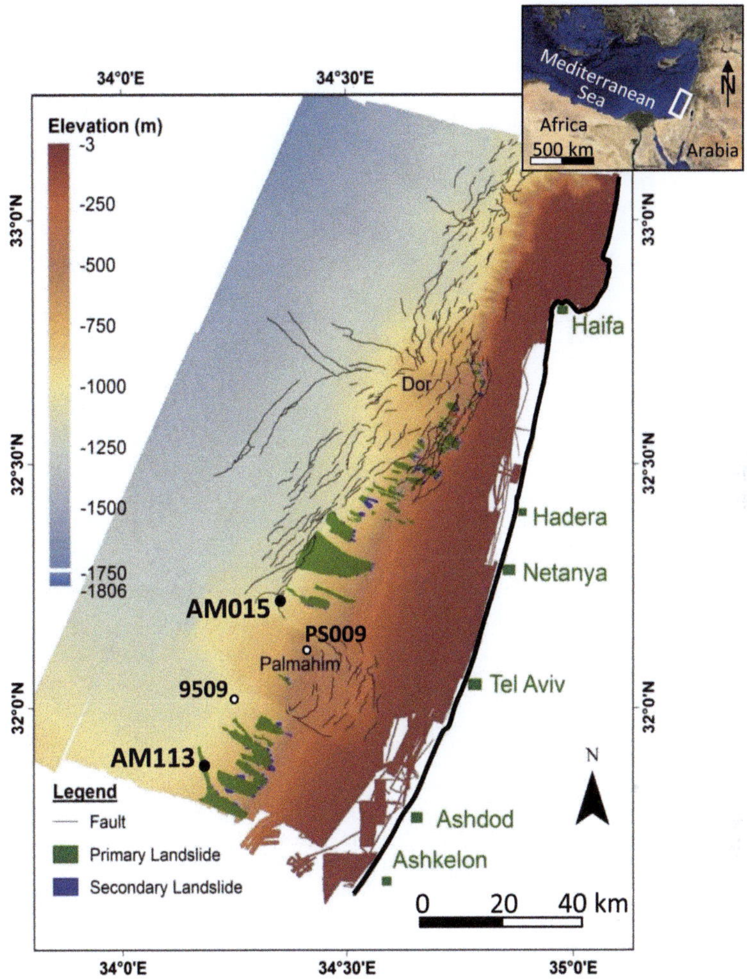

Fig. 1. Bathymetry of the eastern Mediterranean margin (after Sade *et al.* 2007; Tibor *et al.* 2013) showing the location of the study cores (AM015 at 1080 m water depth and AM113 at 848 m water depth). Submarine landslides and faults mapped along the Israeli offshore are also shown on top of the bathymetric grid (following Katz *et al.* 2015). The different generations of landslides are colour-coded. Global location of the Israeli offshore is marked in the inset. Dor and Palmahim are bathymetrical disturbances related to the Messinian evaporates (Garfunkel 1984). Other cores, 9509 (Almogi-Labin *et al.* 2009) and PS009 (Hennekam *et al.* 2014), discussed in the text, are marked. Major cities along the coastline are shown.

an entire aliquot. Foraminifera were identified to the lowest taxonomic level possible based on Parker (1962), Loeblich and Tappan (1987, 1994), Hemleben *et al.* (1989), Cimerman and Langer (1991), Hottinger *et al.* (1993), Reiss *et al.* (1999), Holbourn *et al.* (2013), and Schiebel and Hemleben (2017).

The distribution of both planktic foraminifera (PF) and benthic foraminifera (BF) was analysed

Table 1. *Details of the studied cores*

Core	Location (Fig. 1)	Water depth (m)	Core-length (m)	Remarks
AM015	32° 13′ 33.79″ N 34° 21′ 19.04″ E	1080	3.09	On an open slope
AM113	31° 51′ 05.73″ N 34° 11′ 33.56″ E	848	3.06	Within a landslide lobe

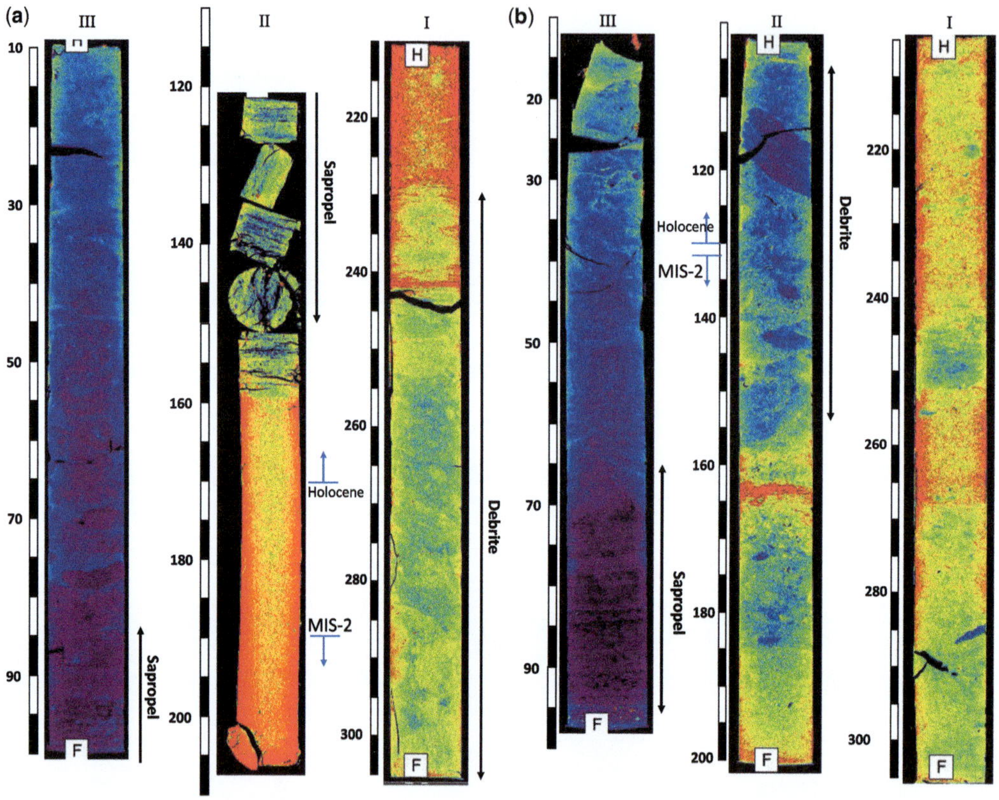

Fig. 2. CT scans of the studied cores: (**a**) AM015 and (**b**) AM113 presented in colour-coded CT-density. Dark colours (up to black) and bright colours (up to red) are relatively low and high density, respectively. Each core is cut to three *c.* 1 m long segments marked by I (the deepest segment), II and III (the shallow segment that includes the seafloor level at its top). Scale is in centimetres; for location of cores, see Figure 1.

quantitatively (i.e. individually counted) along the studied cores. Their abundances (pristine and partly fragmented) were normalized to 1 g of dry bulk sediment. Disappearance and re-appearance of key species since the Last Glacial Maximum (LGM) were used as an independent tool for dating the cores. In core AM113, *Globigerinoides ruber* pink was counted semi-quantitatively, in a small, reserved aliquot, because most of the sample had already been used for radiocarbon dating.

Stable isotopes

For measuring oxygen isotopic composition ($\delta^{18}O$) of the shells of the planktonic foraminifera *G. ruber* (*sensu stricto* form), 20–30 shells of *G. ruber* (250–350 µm size fraction) were handpicked from the same aliqot of the total picked foraminiferal assemblage. The picked specimens were rinsed three times with ultrapure water and ethanol (super special grade) and ultrasonicated to remove adherent clay particles. Oxygen isotopes were measured using a Finnigan Gas Bench II extraction system attached to a ThermoFinnigan Delta PLUS XP continuous flow mass spectrometer. A Carrara Marble standard ($\delta^{18}O = -3.85‰$) was measured every 8 samples. All $\delta^{18}O$ values obtained were calibrated against the internal lab standard, which was calibrated against the international standards NBS-19 and NBS-18. Analytical reproducibility is better than 0.1‰ for $\delta^{18}O$.

Taphonomy

We hypothesize that the transport of sediments results in progressive physical disintegration of pristine (non-broken) foraminiferal shells. For the purpose of the current analysis we define two consecutive disintegration states regarding the foraminiferal shell: (1) a fragmented shell where >50% of the shell is preserved (Fig. 3); and (2) a crushed shell where less than 50% of the pristine form is preserved (Fig. 4). We did not find evidence of other taphonomic features suggesting extensive

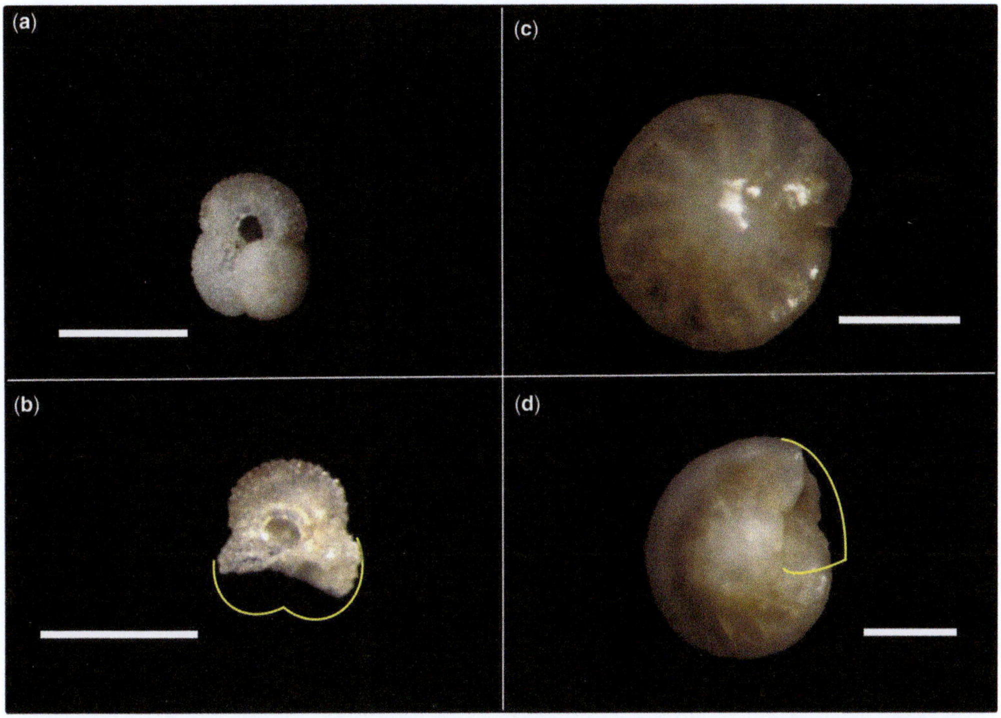

Fig. 3. Example of benthic (*Cibicidoides pachyderma*) and planktic (*Globigerinoides ruber*) foraminifera in a pristine state (**a**, **c**) and 'fragmented' (**b**, **d**) states. Specimens are from core AM015 at 256 cm core depth. Scale bar is 0.2 mm.

dissolution, reworking, shell-colouring or bioerosion, thus these processes are not included in the following analysis.

The percentage of fragmented shells out of the total number of picked shells found in the original sample (*c.* 300 pristine and fragmented specimens) is estimated for *G. ruber* and total BF separately. The reason for choosing *G. ruber* as a representative of the PF is due to its dominance throughout the entire studied cores and due to its high tendency to fragment (Kucera 2007). In the BF no dominant species is present, therefore we use the whole assemblage. The definition of a fragmented shell, preserved above 50%, prevents overcounting individuals. The fragmentation percentage of *G. ruber* in core AM113 could not be determined because the size of the aliquot left for this purpose was too small.

We then quantified three relative crushing grades of the foraminiferal shells, using a semi-qualitative procedure based on visual inspection of the residual sample (after picking of *c.* 300 pristine and fragmented shells from an aliqot used for the procedure described above). Grade 1: the amount of crushed shells in the residual material is negligible (less than *c.* 10%) relative to the non-foraminiferal material (e.g. pteropods, ostracods, echinoderm fragments, terrigenous grains) (Fig. 4a, b). This means that almost all foraminiferal shells in the original sample were pristine or fragmented and they were previously picked from the sample to estimate the fragmentation percentage. Grade 3: the amount of crushed foraminiferal shells in the residual sample relative to the non-foraminiferal material is significant (more than *c.* 50%; Fig. 4e, f). This means that a significant number of specimens in the original sample were crushed. 'Grade 2' is defined by the presence of a noticeable (*c.* 10–50%) amount of crushed foraminiferal shells in the residual sample relative to the non-foraminiferal material (Fig. 4c, d).

Results

Temporal setting of the cores

Core AM015. In order to have a general age constraint of the slope-instability events recorded in core AM015, we first identified the planktic foraminifera species. We focused on the species indicative of the Last Glacial period (Marine Isotopic Stage 2; MIS-2) that disappeared in the Mediterranean *c.* 14 ka BP (*Globorotalia scitula*; Jorissen *et al.*

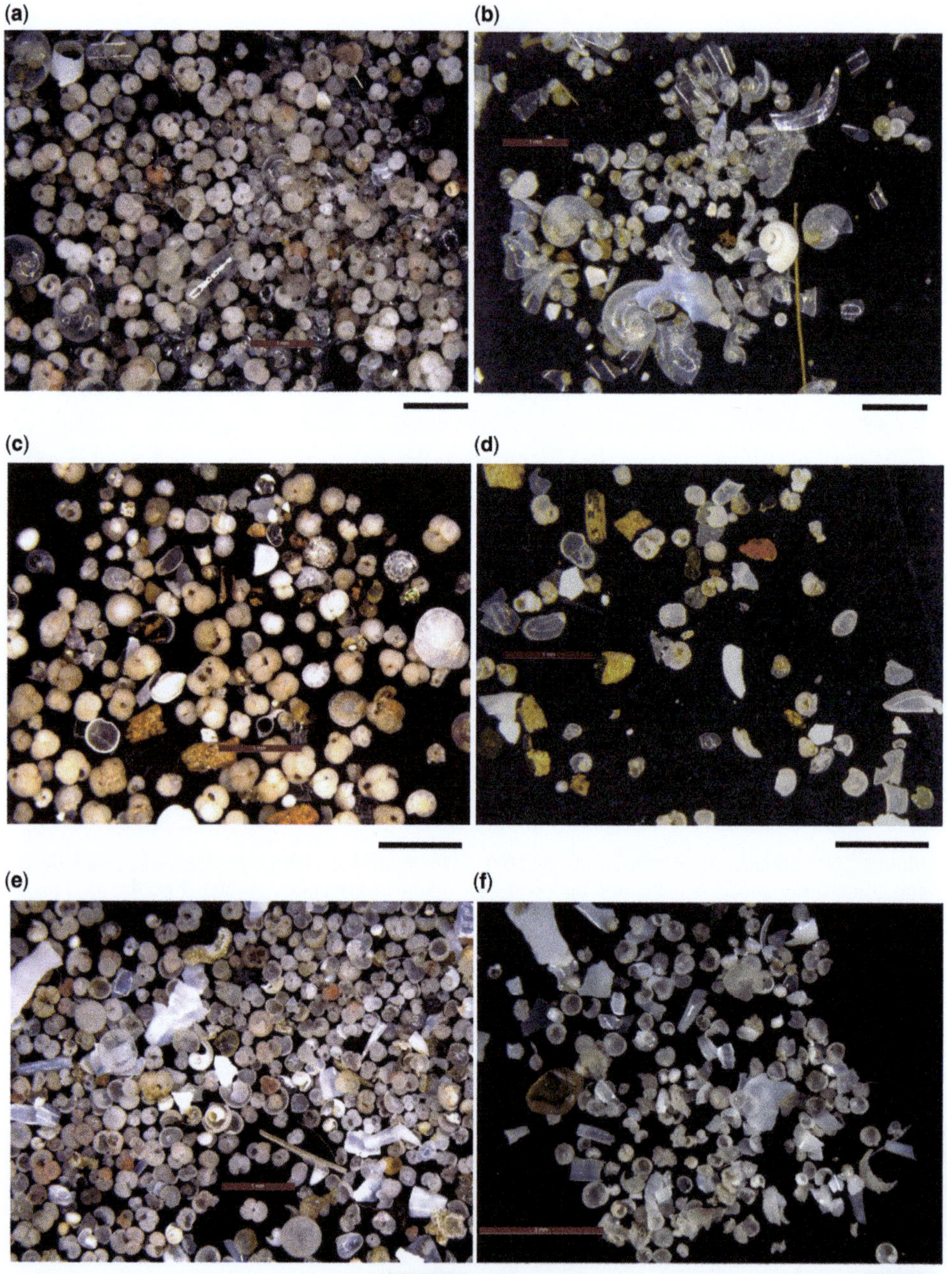

Fig. 4. Pictures depicting the definition of crushing grades of foraminiferal shells. Grade 1: (**a**) original material (before picking of *c.* 300 pristine and fragmented specimens out of an aliqot) composed of many pristine and well-preserved BF and PF shells; (**b**) residual material (after picking of *c.* 300 pristine and fragmented specimens) contains negligible amounts of crushed shells with well-preserved (transparent) and broken pteropod (aragonitic) shells. Grade 2: (**c**) original material composed of many pristine or fragmented BF and PF shells badly preserved; (**d**) residual material contains some crushed foraminiferal shells alongside badly preserved pteropods, ostracods, sea urchin spines and pyrite. Grade 3: (**e**) original material composed mainly of non-pristine PF, although well preserved (e.g. *Orbulina universa*); (**f**) residual material contains mostly crushed foraminiferal shells (well preserved) with some pteropod shell fragments and quartz. Pictures were taken from core AM015 at core depths of 94 cm (a–b), 221 cm (c–d) and 144 cm (e–f). Scale bar is 1 mm.

1993), and the one that thrives in warmer waters during the Holocene (*Globigerinoides ruber* pink, Pérez-Folgado *et al.* 2003; Mojtahid *et al.* 2015). We found that *G. ruber* pink is absent in the lower part of the core up to 185 cm below core top. *Globorotalia scitula* is observed in the lower part of the core up to 175 cm below core top (Fig. 5). Thus, a transition from MIS-2 to the Holocene is suggested by these observations to occur at a core depth around 180 cm below core top.

The $\delta^{18}O$ values of the PF *G. ruber* (white) were measured on six selected samples between 292 and 90 cm (Fig. 5). This interval contains the last appearance (LA) of *G. scitula* and the first appearance (FA) of *G. ruber* pink (Fig. 5). Between 292 and 233 cm, $\delta^{18}O$ values of *G. ruber* (white) are relatively high, varying between 1.66 and 2.41‰ and can be interpreted as representing the cooler glacial stage (Almogi-Labin *et al.* 2009). Between 155 and 90 cm, $\delta^{18}O$ values range between −0.1 and −1.07‰, which is indicative of a warmer period (Almogi-Labin *et al.* 2009). The shift to low interglacial values of the upper part of the core is in agreement with previous studies on other cores from the eastern Mediterranean. Hennekam *et al.* (2014) studied a core from a shallower water depth (532 m), located about 12 km SE of core AM015 (Fig. 1, core PS009) and found a similar shift from enriched to depleted $\delta^{18}O$ *G. ruber* values from 12.5 to 9.5 ka. Almogi-Labin *et al.* (2009) studied a core from 884 m water depth, midway between AM113 and AM015 (Fig. 1, core 9509) and reported a shift from high to low $\delta^{18}O$ *G. ruber* values occurring between 20 and 10 ka. Thus, the interval between 233 and 155 cm, hosting the change from high to low $\delta^{18}O$ *G. ruber* values, is assigned to the time interval between 12.5 and 9.5 ka BP, following Almogi-Labin *et al.* (2009) and Hennekam *et al.* (2014). It was not possible to measure $\delta^{18}O$ values within this *c.* 80 cm interval due to lack of *G. ruber* (white). Nevertheless, this interval hosts the transition from cold to warm-water foraminiferal species with LA of *G. scitula* at 175 cm (Fig. 5). Based on these observations, it is interpreted that

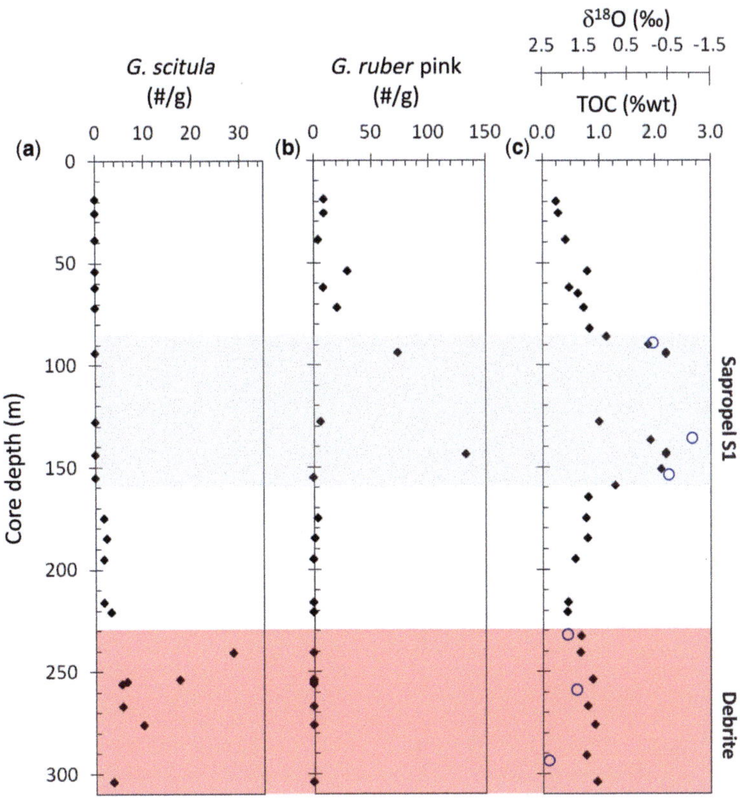

Fig. 5. Core AM015. (**a**, **b**) Abundance of PF: *G. scitula* (#/g dry sediment) and *G. ruber* pink (#/g dry sediment). (**c**) Percent of total organic carbon (TOC; black diamonds) and values of $\delta^{18}O$ (open blue circles). The S1 sequence along the core, based on the TOC values, and the debrite section along the core (Fig. 2a), are presented.

the upper part of core AM015 (the top 180 cm) corresponds to the Holocene and its lower part to the late MIS-2 (Last Glacial period).

The TOC in 31 selected samples along the core also provides a general indication of the strata age (Fig. 5). High TOC values (>2%) are characteristic of a sapropel period (Rohling and Hilgen 1991; Grant *et al.* 2016). In the eastern Mediterranean, the last sapropel event (S1) occurred between 10.5 and 6.1 ka BP (de Lange *et al.* 2008; Almogi-Labin *et al.* 2009; Mojtahid *et al.* 2019). In core AM015, TOC values are mostly below 0.7%, except in the interval between 165 and 82 cm, where they rise up to 2.2% (double peak). Hence, we interpret this section as representing sapropel S1, which is consistent with its stratigraphic position above the transition to the Holocene.

Core AM113. Globigerinoides ruber pink is absent at the base of the core and its FA is at 160 cm below core top (Fig. 6b). *Globorotalia scitula* was observed from the core bottom and its abundance declined from 170 cm up to 130 cm, indicating a transition from cooler to warmer sea surface temperatures (SST). In addition, the FA of *G. ruber* pink at 160 cm and the LA of *G. scitula* at 110 cm (Fig. 6) indicate a disturbed area with a mixture of these two species.

The $\delta^{18}O$ values of *G. ruber* (white) were measured in 12 selected samples along the interval where *G. scitula* declines and *G. ruber* pink appears (166–86 cm below core top; Fig. 6). The $\delta^{18}O$ values between 166 and 132 cm vary from 2.11 to 2.45‰, suggesting cooler SST and higher global ice volume. In contrast, the $\delta^{18}O$ values of the interval between 117 and 86 cm vary between 0.06 and −0.56‰, which are indicative of the early Holocene when global ice volume was lower and SST higher. Following the same reasoning described for core AM015, the core section at a depth of 233 to 117 cm, which hosts the transition of $\delta^{18}O$ values from enriched to depleted ones and a change in the distribution of some PF species (Fig. 6), is assigned to the deglaciation period.

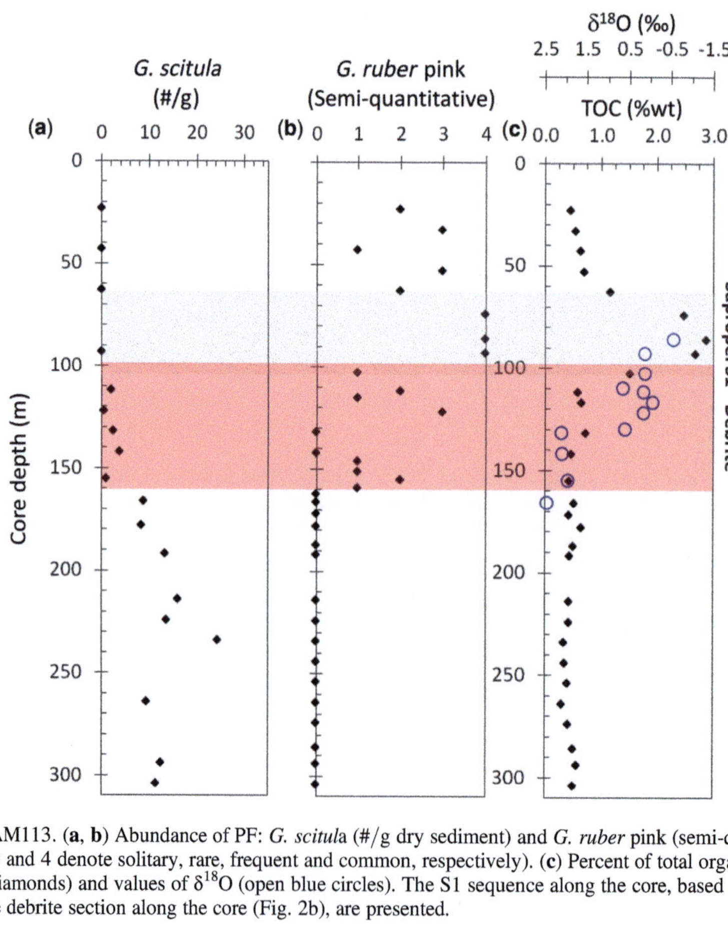

Fig. 6. Core AM113. (**a, b**) Abundance of PF: *G. scitula* (#/g dry sediment) and *G. ruber* pink (semi-quantitative: 0 is none, 1, 2, 3 and 4 denote solitary, rare, frequent and common, respectively). (**c**) Percent of total organic carbon (TOC; black diamonds) and values of $\delta^{18}O$ (open blue circles). The S1 sequence along the core, based on the TOC values, and the debrite section along the core (Fig. 2b), are presented.

TOC values of core AM013, measured on 29 samples, are typically lower than 0.7 wt%, except for the interval between 103 and 63 cm where they rise up to 2.9 wt% (one peak; Fig. 6), marking the position of sapropel S1. As for core AM015, its stratigraphic position above the transition to the Holocene agrees with its suggested younger age (10.5–6.1 ka BP).

Stratigraphy related to mass wasting

Core AM015. Based on the CT scanned core and the visual inspection, the sediments are homogeneous down to a depth of *c.* 50 cm (Figs 2a, S1). Further down the core, between 50 and 98 cm, the sediments include burrows and consist of abundant and well-preserved pteropods shells such as *Clio pyramidata* and *Styliola subula* and occasionally *Creseis acicula, Cavolinia gibbosa* and the heteropod *Atlanta* sp. This interval has the highest TOC values, typical of a sapropel event (Fig. 5). Between 98 and 150 cm the core is dry and cracked as a consequence of bad archiving. This section also shows high TOC values. From 160 to 240 cm sediments are homogeneous, with an S-shaped shear structure in the lower 10 cm. The lower part of the core (240–309 cm) consists of heterogeneous sediments with clastic appearance in the CT scan (Fig. 2a). This heterogeneous clastic appearance of the sediments is characteristic of debrites, i.e. a sediment gravity flow in which larger grains and/or clasts float within a matrix of cohesive fine-grained sediments (Moscardelli and Wood 2008; Talling *et al.* 2012). According to our age constraints, this event predates S1 and the transition to the Holocene (Figs 2 & 5).

Core AM113. Based on the CT scan and the visual inspection, the sediments are homogeneous from top down to a depth of 70 cm (Figs 2b, S1). Further down the core, between 70 and 97 cm the sediments are laminated and consist of abundant pteropods shells. In this interval TOC values are high, typical of a sapropel event (Fig. 6). The stratigraphy along the lower 2 m of the core alternates between sequences of heterogeneous v. homogeneous sediments. Between 97 and 185 cm the sediments are heterogeneous with dark clasts embedded in a lighter matrix. The clasts are less dense in the CT image (Fig. 2b). A coarse (and dense) 1 cm-thick layer appears at a depth of 163 cm (Fig. 2b) and consists of abundant foraminiferal shells and quartz grains. Between 185 and 268 cm the sediments are mostly homogeneous, with a clast-hosting interval between 245 and 253 cm. The lowermost part of the core, between 286 and 304 cm, consists of sediments with clastic appearance in the CT image. Based on these observations, a *c.* 0.5 m thick debrite is interpreted at the central part of core AM113. This event predates S1 and is contemporaneous to or slightly predates the transition to the Holocene (Figs 2b & 6).

Micropaleontology and taphonomic analysis

The next step of this study was to analyse the taphonomy of the PF and BF in order to detect a possible increase in breakage of the foraminiferal shells in association with the deformed sediments of the debrites. In core AM015 we found a relatively high abundance of BF (*c.* 100 specimens/g) in sediments of MIS-2, none in S1 sediments and low abundances in the Holocene sediments postdating S1 (Fig. 7). In contrast, PF show a relatively high abundance (>200 specimens/g) both in MIS-2 and Holocene sediments, including a peak of abundance in sediments of S1 (Fig. 7). The abundance pattern of BF along core AM113 is similar to that in AM015 (Fig. 8), with relatively high and low abundance in sediments of MIS-2 and the Holocene respectively, and none in S1. The abundance pattern of the PF along AM113 (Fig. 8) is similar to that of the BF in the lower part of the core, up to S1 (*c.* 100 cm). During and above S1 the maximal abundance of PF is up to 600 specimens/g.

The fragmentation percentage of BF in core AM015 shows relative low values (less than 20%) in MIS-2 sediments (Fig. 7). Then, it gradually increases to 40% at 155 cm below core top (base of S1). During the Holocene, BF show a relatively high fragmentation percentage, but their abundance is low (less than 10 specimens). Thus, the high fragmentation percentage in this interval might be misleading. The fragmentation percentage of the planktic *G. ruber* follows the BF trend, although with consistently lower values (Fig. 7). The similarity in fragmentation percentage trends of both BF and PF (up to the S1 level) suggests a mechanical process that affected the sediments after the deposition of the PF. The highest fragmentation percentage in this core is observed at the section overlying the debrite (195–155 cm below core top), where sediments are homogeneous (Fig. 2a).

In core AM113 the fragmentation percentage of the BF is relatively low (<35%) in the lower part of the core, up to 192 cm below core top. Values increase to 47% at 132 cm, at the transition from MIS-2 to the Holocene and base of S1 (Fig. 8) and decrease again towards the core top. The core section identified as a debrite (*c.* 160–97 cm; Fig. 2b) hosts the highest fragmentation percentage observed within this core. Overall, the BF shells in AM113 are more broken than the ones in AM015.

To further address the extent of the mechanical deterioration of the foraminiferal shells, 22 samples were examined along core AM015 to assess their crushing grade (Fig. 7). MIS-2 sediments are

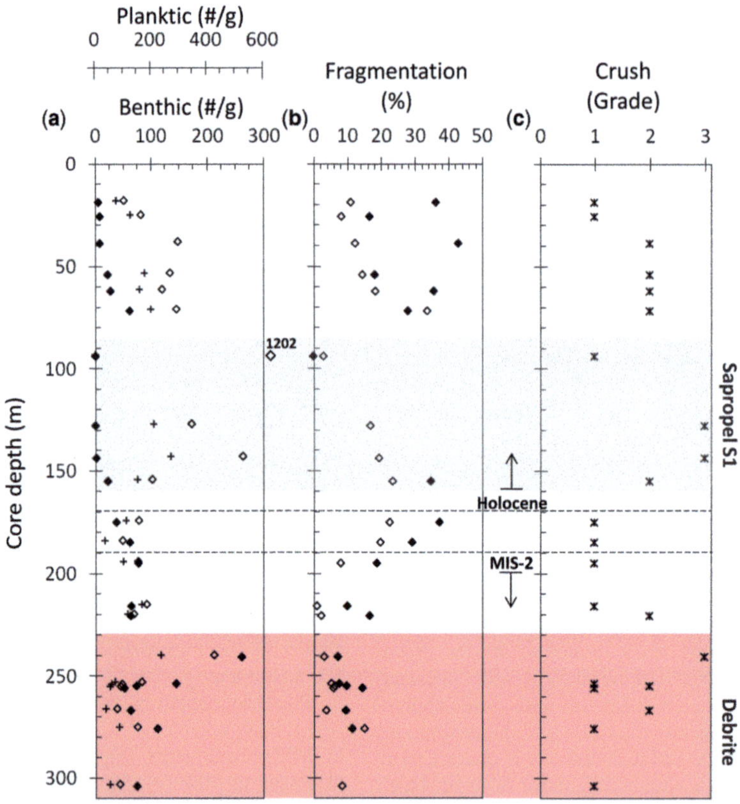

Fig. 7. Core AM015. (**a**) Abundance (#/g dry sediment) of total benthic foraminifera (closed diamonds), total planktic foraminifera (open diamonds) and G. *ruber* (plus signs). (**b**) Fragmentation percentage of benthic (closed diamond) and planktic (open diamond) foraminifera represented by G. *ruber*. (**c**) Crush grade. For definition of fragmentation percentage and crush grade, see 'Methods' and Figures 3 and 4, respectively. The MIS-2 to Holocene transition, based on $\delta^{18}O$ values and *G. scitula* v. *G. ruber* pink abundance (Fig. 5), the S1 sequence along the core, based on TOC values, and the debrite section along the core (Fig. 2), are presented.

generally only mildly crushed (grade 1), except for the interval between 266 and 216 cm below the core top, which shows the higher grades of 2 and 3 (at 241 cm). The lower part of this interval (up to 230 cm) is interpreted as a debrite based on its heterogeneous appearance on the CT data (Fig. 2). The Holocene sediments are more crushed, with grade 2 as a background and a grade 3 peak at S1 sediments, 144–128 cm below the core top. Along core AM113, 17 sediment samples were examined to assess their crushing grade. The sediment of AM113 is generally more crushed than AM015 and shows a general background of grade 2 (Fig. 8). Higher values of grade 3 are observed in MIS-2 sediments at 264 cm below the core top, and in the upper part of the core, 166–63 cm below the core top (debrite overlain by Holocene sediment).

In the debrite sequence of both cores the size distribution of foraminiferal shells was heterogeneous, implying negligible mechanical sorting.

Discussion

Timing of mass-transport processes offshore southern Israel

A mass-transport deposit, best described as a debrite, of at least 0.5 m thick was identified in both studied cores (Fig. 2). These deposits are much smaller than the larger-scale landslides mapped by Katz *et al.* (2015) and involve significantly lower sediment volume than the landslides that are typically tens of metres thick. The debrites apparently represent instability of the uppermost part of the slope in which unconsolidated sediment with probably high water content locally fluidized and moved downslope (Hyams-Kaphzan *et al.* 2018). The fact that debrites are found in a lobe of a larger-scale landslide (as in core AM113), as well as on the open slope where no landslide is mapped (as in core AM015), suggests that surficial debrites are a common process relevant

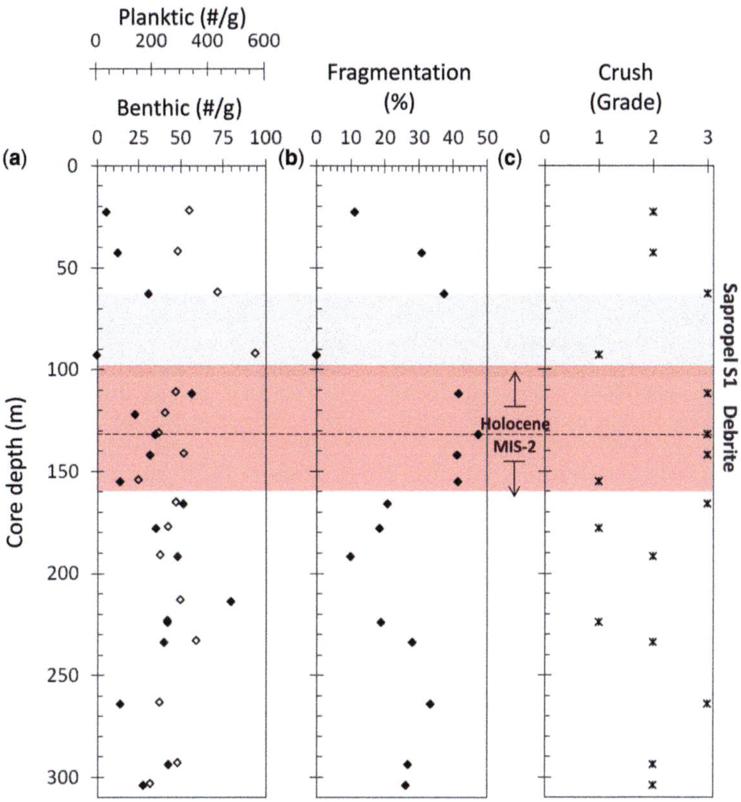

Fig. 8. Core AM113. (**a**) Abundance (#/g dry sediment) of total benthic and planktic foraminifera. (**b**) Fragmentation percentage of BF. (**c**) Crush grade. For definition of fragmentation percentage and crush grade, see 'Methods' and Figures 3 and 4, respectively. The MIS-2 to Holocene transition, based on $\delta^{18}O$ values and *G. scitula* v. *G. ruber* abundance (Fig. 5), the S1 sequence along the core, based on TOC values, and the debrite section along the core (Fig. 2), are presented.

to the entire studied slope in the Southeastern Levantine Basin.

The timing of the debrites, based on the LA and FA of selected planktic foraminifera, oxygen isotopes and TOC values (Figs 5–8), predates S1 in both cores and is contemporaneous (AM113) or slightly predates the transition to the Holocene (AM015). The timing of the debrite in AM113 also gives a minimal age of *c.* 8 ka BP for the hosting larger-scale landslide (Fig. 1), in agreement with radiocarbon ages for that landslide given by Elad *et al.* (2018). No instability is observed to postdate S1, i.e. younger than 8 ka BP. These findings are in general agreement with a suggested post-LGM increase in Mediterranean submarine landslide activity (e.g. Urgeles and Camerlenghi 2013) and an observed decrease in the global occurrence of submarine landslides since the mid-Holocene (Lee 2009; Korup 2012). The reason for this decrease might be the post-LGM widening of the continental shelves, increasing the distance between the sediment sources (e.g. river deltas, long-shore currents) and the continental slope off the shelf break as well a prevailing quasi-stable sea-level.

Estimated average sedimentation rates at the base of the continental slope for the mid-upper Holocene, postdating S1 (using 8.3 ka BP as the age of S1), are 10 cm ka^{-1} for AM113 and 15 cm ka^{-1} for AM015, which is in general agreement with other calculated rates from the eastern Mediterranean Sea at 800–1000 m water depth (Hamann *et al.* 2008; Langgut 2018).

Foraminifera taphonomy as a tool to evaluate mass-transport dynamics

We identified debrites within the two studied cores using a visual criterion of deformed and heterogeneous stratigraphy (Fig. 2). Our working hypothesis is that if foraminiferal shell breakage is the result of sediment deformation induced by transport and turbulence (following Beavington-Penney 2004; Ducassou *et al.* 2013; Ash-Mor *et al.* 2018), we should find a relatively high percentage of shell fragmentation followed by a high crushing grade in the debrites. The findings in AM113 do support our hypothesis. There, the highest fragmentation values (*c.* 40%) are located between 150 and 110 cm below the core top, coinciding with the deformed stratigraphy, i.e. debrite (Figs 2a & 8). Our observations suggest that the debrite sediments in AM113 experienced a possible long transport distance and significant turbulence, both promoting shell breakage. The core is located in a landslide lobe related to a headscarp on the slope below the shelf break, at *c.* 350 m water depth (Fig. 1), suggesting a possible transport distance of more than 10 km, extending from below the shelf break to the foot of the continental slope. A sediment source on the shelf is ruled out by the presence of only continental slope BF species found within the debrite: e.g. *Cibicidoides pacyderma, Hoeglundina elegans, Planulina ariminensis, Uvigerina* sp. 1, *Bolivina spathulata, Gyroidina orbicularis, Melonis affinis* (Hyams-Kaphzan *et al.* 2018). Occurrence of turbulence within the debrite of AM113 is supported by the mixing of both *G. ruber* pink and *G. scitula* (Fig. 6a, b; 160–110 cm below the core top), suggesting a mixture of late MIS-2 and early Holocene sediments.

In contrast to AM113, the fragmentation percentage in the debrite of AM015 (310 to 230 cm below the core top, Fig. 2a) is low, *c.* 10% (Fig. 7) and the BF and PF shells are abundant and well preserved. Thus, we suspect that the transport distance of the debrite in AM015 is shorter than that of AM113. This is also supported by the absence of any shallow-water shelf BF species in the debrite and lack of spatial relation to a scar of a landslide (Fig. 1). In addition, no mixed sediments containing both PF species *G. ruber* pink and *G. scitula* are observed (Fig. 5), ruling out high sediment turbulences within the AM015 debrite.

The fragmentation percentage in core AM015 increases remarkably (up to *c.* 40%) above the debrite interval (from 230 cm upwards), coinciding with the identified climatic transition from MIS-2 and the early Holocene sapropel formation in the eastern Mediterranean Sea (Figs 5 & 7). Former studies that dealt with BF successions across the late Quaternary Mediterranean sapropel (e.g. Jorissen 1999; Kuhnt *et al.* 2007; Schmiedl *et al.* 2010) showed a high spatio-temporal variability in deep-water oxygenation and biogeochemical processes at the seafloor during that time. Here we have focused on physical shell deterioration as a result of submarine landslides. Yet, bottom-water conditions at that time, progressing from oxic to dysoxic and anoxic conditions, culminating in the formation of sapropel S1, may also lead to poor preservation of calcareous foraminiferal shells. This trend of a sharp increase in fragmented shells during the late deglaciation period is observed in both our studied cores, regardless of the location of the debrite interval along the core (Figs 7 & 8). Further studies focusing on this phenomenon within this time interval are required.

The crushing grade of the foraminiferal shells in sediments from core AM113 is commonly higher than that of core AM015 (Figs 7 & 8), in accordance with the location of this core within a landslide lobe. The interval with the highest crushing grade in core AM113 (*c.* 160–110 cm below the core top) overlaps with the highest fragmentation percentage, both situated in the debrite. However, in core AM015 there is no clear correlation between the highest crushing grade and the highest fragmentation percentage. The highest crushing grade is found in the debrite interval (within MIS-2) and within S1 sediments and the highest fragmentation percentage is situated at the transition from MIS-2 to Holocene sediments. Further studies are required to improve understanding of the crushing mechanism and its application to MTD analysis.

To summarize, we think that our observational approach, which quantifies breakage of foraminiferal shells as a proxy for sediment deformation, provides solid results, thus showing trends that reflect natural processes affecting the sediments. At this preliminary stage of the study we are still assessing which process is the most important. Is it the extent of mechanical sediment deformation as shown in core AM113, which is located in a landslide lobe? Or is it the oceanographic and geochemical changes from the last deglaciation towards the early Holocene, potentially affecting the preservation of calcareous shells, as shown in core AM015, located outside a landslide.

Conclusions

In this work we used the taphonomic properties of benthic and planktonic foraminiferal shells in order to analyse two debrites identified in marine cores of the lower continental slope, offshore Israel. We quantified what we think are reliable physical breakage parameters (fragmentation and crushing) showing coherent trends along the cores. We link the decrease in the shells' state of preservation to two possible scenarios: (1) downslope transport and turbulence of the hosting sediments in the frame of submarine slope failure; or (2) geochemical changes in the lower water column at the transition from MIS-2 to Holocene. Both scenarios apparently can lead to a decreasing preservation state of the foraminiferal shells. However, at the current stage we cannot fully evaluate which process is more important.

The studied debrites were contemporaneous or slightly predate the transition to the Holocene, with no observed instability postdating sapropel S1. These findings are in general agreement with the observed global trends of a decrease in submarine mass wasting activity since the mid-Holocene.

Acknowledgements The authors thank the reviewers Ángel Puga Bernabéu, and an anonymous reviewer and the editor for their time, suggestions and comments that significantly focused and improved the quality of the manuscript. Support was provided to Leeron Ashkenazi by the Mediterranean Sea Research Center of Israel. The authors thank M. Kitin, T. Zilberman and R. Binstock (GSI) for overseeing lab work. A. Ayalon (GSI) is thanked for fruitful discussions. B. Cohen (GSI-publishing unit) is thanked for help in constructing the figures.

Funding This study was funded by the Israel Science Foundation (grant No. 954/15) to Oded Katz and by the Chief Scientist office of the Ministry of Energy, Israel (grant No. 76/17) to Oded Katz and Orit Hyams-Kaphzan.

Author contributions **OK:** formal analysis (equal), funding acquisition (lead), methodology (equal), writing – original draft (lead); **LA:** data curation (lead), formal analysis (equal), methodology (lead), writing – original draft (supporting); **SSL:** data curation (equal); **SA:** funding acquisition (equal), supervision (equal), writing – review & editing (supporting); **AAL:** methodology (supporting), supervision (equal), writing – review & editing (equal); **OHK:** conceptualization (equal), formal analysis (equal), funding acquisition (equal), supervision (equal), writing – original draft (equal).

References

Almagor, E. and Schilman, B. 1995. Sedimentary structures and sediment transport across the continental slope of Israel from piston core studies. *Sedimentology*, **42**, 575–592, https://doi.org/10.1111/j.1365-3091.1995.tb00393.x

Almogi-Labin, A., Bar-Matthews, M. *et al.* 2009. Climatic variability during the last ~90 ka of the southern and northern Levantine Basin as evident from marine records and speleothems. *Quaternary Science Reviews*, **28**, 2882–2896, https://doi.org/10.1016/j.quascirev.2009.07.017

Ash-Mor, A., Bookman, R., Kanari, M., Ben-Avraham, Z. and Almogi-Labin, A. 2017. Micropaleontological and taphonomic characteristics of mass transport deposits in the northern Gulf of Eilat/Aqaba, Red Sea. *Marine Geology*, **391**, 36–47, https://doi.org/10.1016/j.margeo.2017.07.009

Ash-Mor, A., Bookman, R., Bouchet, V., Seuront, L., Ben-Avraham, Z. and Almogi-Labin, A. 2018. Going with the flow: re-suspension and fragmentation of foraminifera shells caused by high velocity currents. *FORAMS2018* Abstract, Edinburgh, Scotland.

Beavington-Penney, S.J. 2004. Analysis of the effects of abrasion on the test of *Palaeonummulites venosus*: implications for the origin of nummulithoclastic sediments. *PALAIOS*, **19**, 143–155, https://doi.org/10.1669/0883-1351(2004)019<0143:AOTEOA>2.0.CO;2

Beavington-Penney, S.J. and Racey, A. 2004. Ecology of extant nummulitids and other larger benthic foraminifera: applications in palaeoenvironmental analysis. *Earth-Science Reviews*, **67**, 219–265, https://doi.org/10.1016/j.earscirev.2004.02.005

Cimerman, F. and Langer, M.R. 1991. *Mediterranean Foraminifera. Academia Scientarium et Aritium Slovenica, Dela, Opera*. **30**, Classis IV: Historia Naturalis.

de Lange, G.J., Thomson, J., Reitz, A., Slomp, C.P., Principato, M.S., Erba, E. and Corselli, C. 2008. Synchronous basin-wide formation and redox-controlled preservation of a Mediterranean sapropel. *Nature Geoscience*, **1**, 606–610, https://doi.org/10.1038/ngeo283

Dessandier, P.-A., Bonnin, J., Kim, J.-H. and Racine, C. 2018. Comparison of living and dead benthic foraminifera on the Portuguese margin: understanding the taphonomical processes. *Marine Micropaleontology*, **140**, 1–16, https://doi.org/10.1016/j.marmicro.2018.01.001

Ducassou, E., Capotondi, L. *et al.* 2007. Multiproxy Late Quaternary stratigraphy of the Nile deep-sea turbidite system – Towards a chronology of deep-sea terrigeneous systems. *Sedimentary Geology*, **200**, 1–13, https://doi.org/10.1016/j.sedgeo.2007.01.023

Ducassou, E., Migeon, S., Capotondi, L. and Mascle, J. 2013. Run-out distance and erosion of debris-flows in the Nile deep-sea fan system: evidence from lithofacies and micropalaeontological analyses. *Marine and Petroleum Geology*, **39**, 102–123, https://doi.org/10.1016/j.marpetgeo.2012.09.006

Duros, P., Fontanier, C., de Stigter, H.C., Cesbron, F., Metzger, E. and Jorissen, F.J. 2012. Live and dead benthic foraminiferal faunas from Whittard Canyon (NE Atlantic): focus on taphonomic processes and paleo-environmental applications. *Marine Micropaleontology*, **94–95**, 25–44, https://doi.org/10.1016/j.marmicro.2012.05.004

Elad, M., Bookman, R., Almogi-Labin, A. and Makovsky, Y. 2018. Insights on the evolution of a recent slide complex: A case study of the Goliath Slide, offshore southern Israel. *20th EGU General Assembly, EGU2018 Proceedings*, Vienna, Austria, p. 5331.

Elfassi, Y., Gvirtzman, Z., Katz, O. and Aharonov, E. 2019. Chronology of post-Messinian faulting along the Levant continental margin and its implications for salt tectonics. *Marine and Petroleum Geology*, **109**, 574–588, https://doi.org/10.1016/j.marpetgeo.2019.05.032

Garfunkel, Z. 1984. Large-scale submarine rotational slumps and growth faults in the eastern Mediterranean. *Marine Geology*, **55**, 305–324, https://doi.org/10.1016/0025-3227(84)90074-4

Gaudette, H.E. and Flight, W.R. 1974. An inexpensive titration method for the determination of organic carbon in Recent sediments. *Journal of Sedimentary Petrology*, **44**, 249–253.

Georgiopoulou, A., Wynn, R.B., Masson, D.G. and Frenz, M. 2009. Linked turbidite-debrite resulting from recent

Sahara Slide headwall reactivation. *Marine and Petroleum Geology*, **26**, 2021–2031, https://doi.org/10.1016/j.marpetgeo.2009.02.013

Georgiopoulou, A., Krastel, S. *et al.* 2019. On the timing and nature of the multiple phases of slope instability on eastern Rockall Bank, Northeast Atlantic. *Geochemistry, Geophysics, Geosystems*, **20**, 594–613, https://doi.org/10.1029/2018GC007674

Grant, K.M., Grimm, R., Mikolajewicz, U., Marino, G., Ziegler, M. and Rohling, E.J. 2016. The timing of Mediterranean sapropel deposition relative to insolation, sea-level and African monsoon changes. *Quaternary Science Reviews*, **140**, 125–141, https://doi.org/10.1016/j.quascirev.2016.03.026

Gvirtzman, Z., Reshef, M., Buch-Leviatan, O., Groves-Gidney, G., Karcz, Z., Makovsky, Y. and Ben-Avraham, Z. 2015. Bathymetry of the Levant basin: interaction of salt-tectonics and surficial mass movements. *Marine Geology*, **360**, 25–39, https://doi.org/10.1016/j.margeo.2014.12.001

Hamann, Y., Ehrmann, W., Schmiedl, G., Krüger, S., Stuut, J.B. and Kuhnt, T. 2008. Sedimentation processes in the eastern Mediterranean Sea during the Late Glacial and Holocene revealed by end-member modelling. *Marine Geology*, **248**, 97–114, https://doi.org/10.1016/j.margeo.2007.10.009

Hemleben, C., Spindler, M. and Anderson, O.R. 1989. Modern planktonic foraminifera. *Springer Science & Business Media*, **20**, 90–91, https://doi.org/10.2113/gsjfr.20.1.90

Hennekam, R., Jilbert, T., Schnetger, B. and de Lange, G.J. 2014. Solar forcing of Nile discharge and sapropel S1 formation in the early to middle Holocene eastern Mediterranean. *Paleoceanography*, **29**, 343–356, https://doi.org/10.1002/2013PA002553

Holbourn, A., Henderson, A.S. and MacLeod, N. 2013. *Atlas of Benthic Foraminifera*. Wiley-Blackwell, https://doi.org/10.1002/9781118452493

Hottinger, L., Halicz, E. and Reiss, Z. 1993. *Recent Foraminiferida from the Gulf of Aqaba, Red Sea*. Opera Sazu, Ljubljana, **33**.

Hyams-Kaphzan, O., Lubinevsky, H. *et al.* 2018. Live and dead deep-sea benthic foraminiferal macrofauna of the Levantine basin (SE Mediterranean) and their ecological characteristics. *Deep Sea Research Part I*, **136**, 72–83, https://doi.org/10.1016/j.dsr.2018.04.004

Jorissen, F.J. 1999. Benthic foraminiferal successions across Late Quaternary Mediterranean sapropels. *Marine Geology*, **153**, 91–101, https://doi.org/10.1016/S0025-3227(98)00088-7

Jorissen, F.J., Asioli, A. *et al.* 1993. Late Quaternary central Mediterranean biochronology. *Marine Micropaleontology*, **21**, 169–189, https://doi.org/10.1016/0377-8398(93)90014-O

Katz, O., Reuven, E. and Aharonov, E. 2015. Submarine landslides and fault scarps along the eastern Mediterranean Israeli continental-slope. *Marine Geology*, **369**, 100–115, https://doi.org/10.1016/j.margeo.2015.08.006

Katz, O., Hyams-kaphzan, O., Sultan-levi, S. and Almogi-labin, A. 2019. The dynamic and timing of mass wasting processes along the eastern Mediterranean continental slope, off the Israeli shores. *Geophysical Research Abstracts*, **21**, EGU2019-2539.

Korup, O. 2012. Earth's portfolio of extreme sediment transport events. *Earth-Science Reviews*, **112**, 115–125, https://doi.org/10.1016/j.earscirev.2012.02.006

Kucera, M. 2007. Chapter six planktonic foraminifera as tracers of past oceanic environments. *In*: Hillaire–Marcel, C. and De Vernal, A. (eds) *Developments in Marine Geology*. Elsevier, 213–262.

Kuhnt, T., Schmiedl, G., Ehrmann, W., Hamann, Y. and Hemleben, C. 2007. Deep-sea ecosystem variability of the Aegean Sea during the past 22 kyr as revealed by Benthic Foraminifera. *Marine Micropaleontology*, **64**, 141–162, https://doi.org/10.1016/j.marmicro.2007.04.003

Langgut, D. 2018. Late Quaternary Nile flows as recorded in the Levantine Basin: the palynological evidence. *Quaternary International*, **464**, 273–284, https://doi.org/10.1016/j.quaint.2017.07.006

Lee, H.J. 2009. Timing of occurrence of large submarine landslides on the Atlantic Ocean margin. *Marine Geology*, **264**, 53–64, https://doi.org/10.1016/j.margeo.2008.09.009

Loeblich, A.R. and Tappan, H. 1987. *Foraminiferal Genera and Their Classification*. Van Nostrand Reinhold, New York, **2v**.

Loeblich, A.R. and Tappan, H. 1994. Foraminifera of the Sahul shelf and Timor Sea. *Cushman Foundation for Foraminiferal Research, Special Publication*, **31**, 661

López-Venegas, A.M., ten Brink, U.S. and Geist, E.L. 2008. Submarine landslide as the source for the October 11, 1918 Mona Passage tsunami: observations and modeling. *Marine Geology*, **254**, 35–46, https://doi.org/10.1016/j.margeo.2008.05.001

Mojtahid, M., Manceau, R., Schiebel, R., Hennekam, R. and de Lange, G.J. 2015. Thirteen thousand years of southeastern Mediterranean climate variability inferred from an integrative planktic foraminiferal-based approach. *Paleoceanography*, **30**, 402–422, https://doi.org/10.1002/2014PA002705

Mojtahid, M., Hennekam, R. *et al.* 2019. Evaluation and application of foraminiferal element/calcium ratios: assessing riverine fluxes and environmental conditions during sapropel S1 in the Southeastern Mediterranean. *Marine Micropaleontology*, **153**, 101783, https://doi.org/10.1016/j.marmicro.2019.101783

Moore, Z.T. and Sawyer, D.E. 2016. Assessing post-failure mobility of submarine landslides from seismic geomorphology and physical properties of mass transport deposits: an example from seaward of the Kumano Basin, Nankai Trough, offshore Japan. *Marine Geology*, **374**, 73–84, https://doi.org/10.1016/j.margeo.2016.02.003

Moscardelli, L. and Wood, L. 2008. New classification system for mass transport complexes in offshore Trinidad. *Basin Research*, **20**, 73–98, https://doi.org/10.1111/j.1365-2117.2007.00340.x

Parker, E.J., Traverso, C., Del Giudice, T., Evans, T. and Moore, R. 2009. Geohazard risk assessment – vulnerability of subsea structures to geohazards – risk implications. *Offshore Technology Conference*, Houston, Texas, 10.

Parker, F.L. 1962. Planktonic foraminiferal species in Pacific sediments. *Micropaleontology*, **8**, 219–254, https://doi.org/10.2307/1484745

Pérez-Folgado, M., Sierro, F.J., Flores, J.A., Cacho, I., Grimalt, J.O., Zahn, R. and Shackleton, N., 2003. Western Mediterranean planktonic foraminifera events and millennial climatic variability during the last 70 kyr. *Marine Micropaleontology*, **48**, 49–70, https://doi.org/10.1016/S0377-8398(02)00160-3

Polonia, A., Vaiani, S.C. and de Lange, G.J. 2016. Did the AD 365 Crete earthquake/tsunami trigger synchronous giant turbidity currents in the Mediterranean Sea? *Geology*, **44**, 191–194, https://doi.org/10.1130/G37486.1

Pope, E.L., Talling, P.J. and Carter, L. 2017. Which earthquakes trigger damaging submarine mass movements: insights from a global record of submarine cable breaks? *Marine Geology*, **384**, 131–146, https://doi.org/10.1016/j.margeo.2016.01.009

Reiss, Z., Halicz, E. and Luz, B. 1999. Late-Holocene foraminifera from the SE Levantine basin. *Israel Journal of Earth Science*, **48**, 1–27.

Rohling, E.J. and Hilgen, F.J. 1991. The eastern Mediterranean climate at times of sapropel formation; a review. *Geologie en Mijnbouw*, **70**, 253–264.

Sade, R., Hall, J.K., Amit, G., Golan, A., Gur-Arieh, L. and Tibor, G. 2007. The Israel national bathymetric survey – a new look at the seafloor of Israel. *Israel Journal of Earth Sciences*, **55**, 185–187, https://doi.org/10.1560/IJES_55_3_185

Schiebel, R. and Hemleben, C. 2017. *Planktic Foraminifers in the Modern Ocean*. Springer.

Schmiedl, G., Kuhnt, T. *et al.* 2010. Climatic forcing of eastern Mediterranean deep-water formation and benthic ecosystems during the past 22 000 years. *Quaternary Science Reviews*, **29**, 3006–3020, https://doi.org/10.1016/j.quascirev.2010.07.002

ten Brink, U.S., Geist, E.L. and Andrews, B.D. 2006. Size distribution of submarine landslides and its implication to tsunami hazard in Puerto Rico. *Geophysical Research Letters*, **33**, https://doi.org/10.1029/2006GL026125

Talling, P.J., Mason, D.G., Sumner, E.J. and Malgesini, G. 2012. Subaqueous sediment density flows: depositional processes and deposit types. *Sedimentology*, **59**, 1937–2003, https://doi.org/10.1111/j.1365-3091.2012.01353.x

Tibor, G., Sade, R., Sade, H. and Hall, J.K. 2013. *Data collection and processing of multibeam data from the deep water offshore Israel*. IOLR report, **H-31/2013**.

Urgeles, R. and Camerlenghi, A. 2013. Submarine landslides of the Mediterranean Sea: trigger mechanisms, dynamics, and frequency-magnitude distribution. *Journal of Geophysical Research: Earth Surface*, **118**, 2600–2618, https://doi.org/10.1002/2013JF002720

Widespread mass-wasting processes off NE Sicily (Italy): insights from morpho-bathymetric analysis

Daniele Casalbore[1]*, Romano Clementucci[2], Alessandro Bosman[1], Francesco Latino Chiocci[3], Eleonora Martorelli[1] and Domenico Ridente[1]

[1]Institute of Environmental Geology and Geo-Engineering, Italian National Research Council (IGAG-CNR), Rome, Italy

[2]University Roma Tre, Rome, Italy

[3]Department of Earth Sciences, University Sapienza of Rome, Italy

DC, 0000-0002-0151-6901; RC, 0000-0003-2332-5812; AB, 0000-0002-1441-8016
*Correspondence: daniele.casalbore@uniroma1.it

Abstract: The NE Sicilian continental margin is largely affected by canyons and related landslide scars. Two main types of submarine canyons are recognizable: the first type carves the shelf up to depths <20 m, a few hundred metres from the coast, acting as a main collector for sediments transported by hyperpycnal flows and/or littoral drift. These canyons mostly have a V-shaped cross-section and are characterized by a strong axial incision, where a network of dendritic gullies carving the canyon flanks converges. The second type of canyon occurs where the shelf is wider, hindering the direct connection between the subaerial and submarine drainage system. This setting exhibits canyon heads mostly confined to the shelf break, characterized by a weaker axial incision of the canyon and U-shaped cross-section. A total of 280 landslide scars are recognized in the study area and these are divided into three groups according to their morphology and location. A morphometric analysis of these scars is performed to investigate which parameters might be key factors in controlling instability processes and how they correlate with each other. We also try to assess the possible tsunamigenic potential associated with these landslide events by coupling the morphometric analysis with semi-empirical relationships available in the literature.

In recent decades, interest in submarine landslides and their associated hazards has exponentially grown. The advances in seafloor mapping systems have allowed the mapping of several geomorphic features associated with such processes (Mitchell 2012). Submarine landslides are in fact ubiquitous features identified on the seafloor, with different spatial and temporal scales (Masson *et al.* 2006; Mountjoy and Micallef 2018). Until now, most of the attention has been drawn to large-scale landsliding processes that can mobilize hundreds or thousands of cubic kilometres of material and generate large tsunami waves (Harbitz *et al.* 2014 and reference therein). In contrast, the characterization of smaller landslides has received less attention, despite their being more widespread and frequent than the large-scale ones, especially in geologically active continental margins (Casalbore *et al.* 2012; Urgeles and Camerlenghi 2013). Geologically active continental margins are often characterized by a narrow or absent continental shelf and carved by shelf-indenting canyons, where retrogressive landslides can approach the coastline, representing a major hazard for coastal communities and offshore infrastructures (Chiocci and Casalbore 2017; Amblas *et al.* 2018 and reference therein). Examples of geohazards associated with slope failures at the heads of canyons and channels have been witnessed offshore Gioia Tauro in 1977 (Colantoni *et al.* 1992), Nice in 1979 (Assier-Rzadkieaicz *et al.* 2000) and Finneifjord in 1996 (Longva *et al.* 2003).

In this paper, we show a remarkable example of a continental margin largely carved by shelf-indenting canyons between Cape D'Orlando and the Gulf of Patti in NE Sicily (Fig. 1). Our attention is mainly focused on the morphometric characterization of landslide scars detected both at and within the canyon heads. This analysis allows us to better constrain the main morphometric features and spatial distribution of landsliding processes that contribute to the recent morphogenesis of the canyon as well as to roughly estimate the tsunamigenic potential above the barycentre of the landslide scar using predictive equations available in the literature (McAdoo and Watts 2004).

Geological setting

The study area is located in NE Sicily between the Gulf of Patti to the east and Cape D'Orlando to the west (Fig. 1). The onshore area is part of the Calabro-Peloritani arc, formed by a stack of crystalline rocks, overlain discontinuously by Mesozoic limestones

From: Georgiopoulou, A., Amy, L. A., Benetti, S., Chaytor, J. D., Clare, M. A., Gamboa, D., Haughton, P. D. W., Moernaut, J. and Mountjoy, J. J. (eds) 2020. *Subaqueous Mass Movements and their Consequences: Advances in Process Understanding, Monitoring and Hazard Assessments*. Geological Society, London, Special Publications, **500**, 393–403.
First published online May 15, 2020, https://doi.org/10.1144/SP500-2019-195

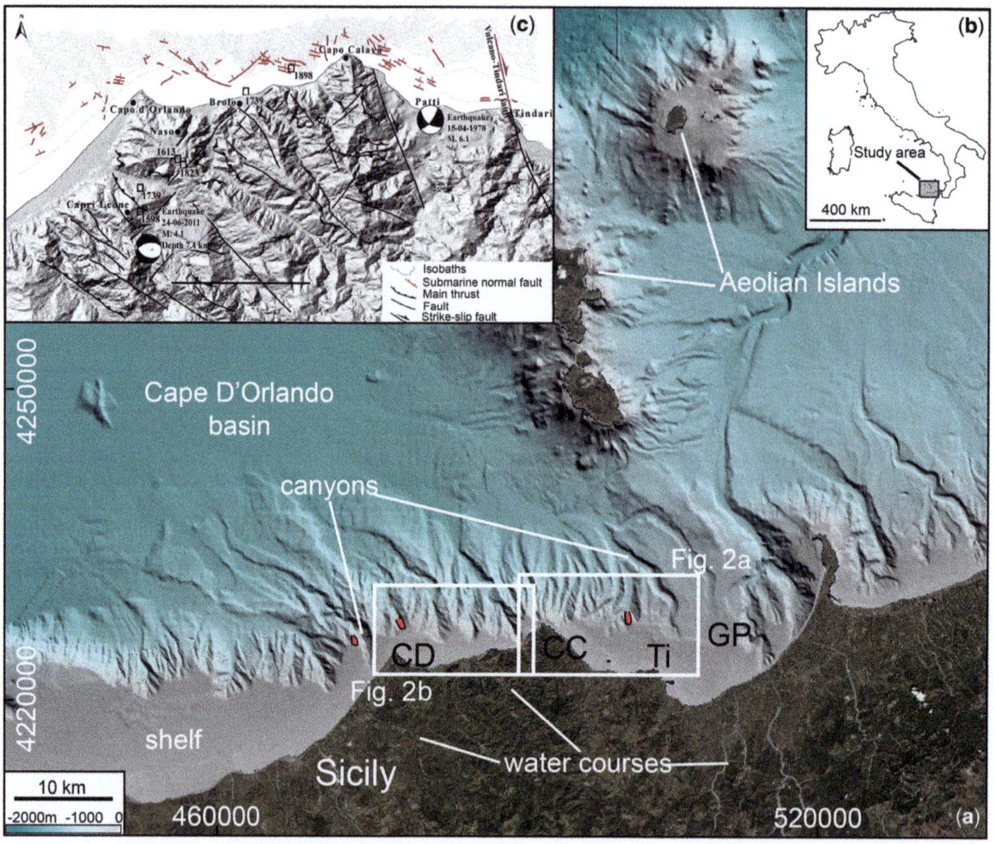

Fig. 1. (**a**) Shaded relief image (artificial sunlight from NE) of the NE Sicilian Margin (bathymetric data downloaded from https://www.emodnet-bathymetry.eu), with the indication of the three landslides scars (red polygons) previously identified by Sulli *et al.* (2018). CD, Cape D'Orlando; CC, Cape Calava; Ti, Tindari; GP, Gulf of Patti. (**b**) Location of the study area; (**c**) morpho-structural map showing the main structural features recognized both on land (black) and submarine areas (red), the location of the main historical earthquakes (black squares) and the focal mechanisms of the largest earthquakes (modified from Sulli *et al.* 2013).

and Cenozoic flysch deposits (Lentini *et al.* 2006). The Calabro-Peloritani arc has been affected by regional uplift at rates ranging from 0.8 to 1.6 mm a^{-1} since the Pleistocene (e.g. Ferranti *et al.* 2010), and amplified by local vertical movements along NE- and NW-trending faults (Catalano *et al.* 2003; Scicchitano *et al.* 2011). Comparison between onshore and offshore sectors (Sulli *et al.* 2013) also suggests that the activity of fault systems parallel to the coastline caused differential vertical movements between the uplifting coast (as witnessed by raised marine terraces) and the facing subsiding margin (as witnessed by the depth of a lowstand submarine depositional terrace). Seismicity is frequent in the study area, with over 1800 earthquakes of small to moderate magnitude ($1.0 \leq M_L \leq 4.8$) recorded between 1984 and 2014 and two main (M_w 6.1) events occurring in 1786 and 1978 in the Gulf of Patti (CPTI Working Group 2004; Chiarabba *et al.* 2005). Most of the seismicity mainly occurs along the NNW–SSE-trending seismic belt associated with the activity of a major fault system running across NE Sicily, i.e. the Vulcano–Tindari Fault System (Fig. 1c; Cultrera *et al.* 2017 and references therein). The main subaerial and submarine faults recognized by Sulli *et al.* (2013) between Capo D'Orlando and Tindari are shown in Figure 1c, where the epicentres of the main historical earthquakes (black squares) and the focal mechanism of the largest earthquakes are also indicated.

In the study area, the continental shelf is narrow (6 km at maximum) or almost absent, and the continental slope is steep (10°–20°) and carved by a dense network of submarine canyons, affecting *c.* 64% of the surface (Figs 1 & 2). Most of these canyons are spatially linked to a network of steep and short

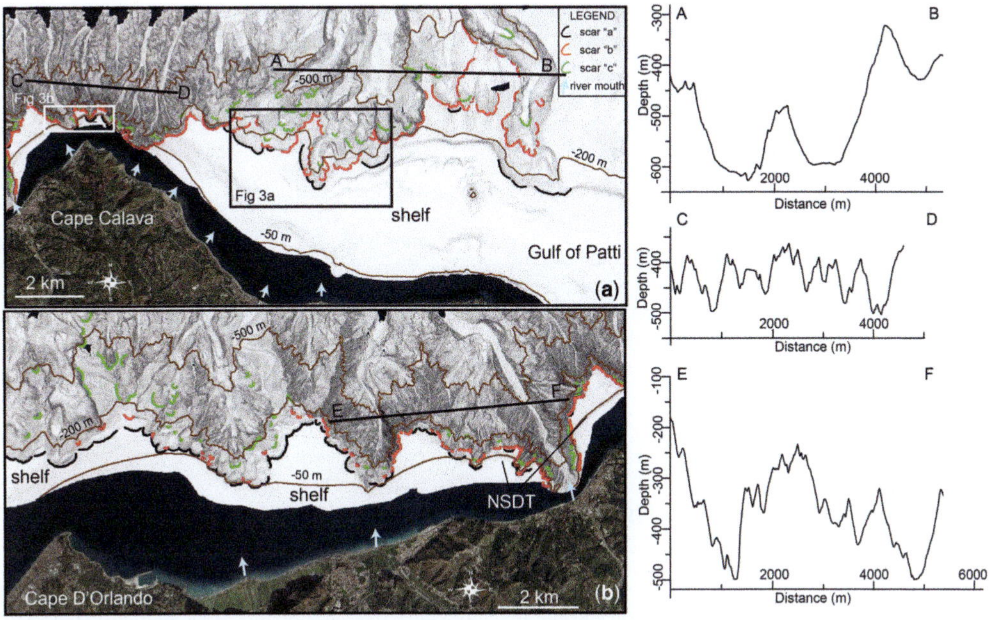

Fig. 2. Shaded relief maps (azimuthal artificial sunlight) of the study area divided into eastern (**a**) and western (**b**) sectors (location in Fig. 1). The three different groups of landslide scars (black, red and green lines) and the main river mouths (light blue arrows) recognized in the study area are also indicated. NSDT, near-shore submarine depositional terrace. On the right, three bathymetric profiles show the cross-section of some canyons; note that in the eastern sector the canyons are generally U-shaped (profile A–B), whereas in the central and western sectors the canyons are generally V-shaped (profiles C–D and E–F).

water courses (Fig. 1 and light-blue arrows in Fig. 2) draining the Peloritani Mountains and characterized by a torrential regime (locally named Fiumara; Sabato and Tropeano 2004). These water courses are able to transport a large amount of debris into the sea during flash-flood events, leading to the development of submarine deltaic systems where the continental shelf is present (Gamberi *et al.* 2014; Casalbore *et al.* 2017*a*) or directly funnelling material within canyon heads, where it can also trigger small, retrogressive slope failures (Casalbore *et al.* 2011*a*). Three larger submarine landslides have been also recognized on the continental slope, with a detachment surface of *c.* 0.5 km^2 (red polygons in Fig. 1a; Sulli *et al.* 2018). At the confluence of canyons/channel systems with the Cape D'Orlando Basin (Fig. 1), sedimentary gravity flows associated with flash-flood events and/or submarine slope failures feed turbidite fans (Gamberi *et al.* 2015).

Data and methods

In this paper, we mainly rely on morpho-bathymetric data acquired during two oceanographic cruises in August 2011 and January 2012 on board RV *Urania* belonging to the National Research Council (Italy). Data were collected at depths ranging from −20 to −1000 m using a Kongsberg EM710 multibeam system, operating at a frequency of 70/100 kHz. A differential GPS was used to position the hull-mounted multibeam system; the multibeam survey was carried out by performing the sensor patch test and collecting sound velocity casts of the water column with a CTD 911 probe daily. A sound speed sensor was also used to update in real time the sound velocity values close to the flat face of the multibeam transducer. Data processing was performed using Caris Hips & Sips. The processing workflow included: (1) navigation and tidal corrections; (2) sound velocity ray-tracing; (3) application of statistical and geometrical filters to remove coherent/incoherent noise; and (4) manual editing of outlier soundings (for details see Bosman *et al.* 2015). Data were gridded at different cell sizes to maximize resolution at different water depths, ranging from 1 m in shallow water (first −100 m) to 20 m in deeper water (down to −1000 m).

The obtained digital elevation models (DEM) were used to identify landslide scars having a minimum area of 50 m^2 and headscarp height of 3 m.

The main morphometric parameters of the landslide scar were measured by applying (where possible) the criteria proposed by Clare *et al.* (2018). In particular, we measured: (1) the minimum water depth of the landslide scar and the water depth at the landslide barycentre; (2) the area, width and length of the landslide scar; (3) the height and slope gradients of the main headscarp; and (4) the slope gradients of the landslide scar and adjacent slopes. Some of these parameters were also used to roughly compute the potential initial maximum tsunami amplitude above these failure sources using predictive equations available in the literature (Rahiman and Pettinga 2006):

$$\lambda = 3.87(bd/\sin\sigma)^{0.5} \quad (1)$$

$$A = 0.224\,T[w/(w+\lambda)] \times [(\sin\sigma)^{1.29} - 0.746(\sin\sigma)^{2.29} + 0.170(\sin\sigma)^{3.29}](b/d)^{1.25} \quad (2)$$

where λ and A are the tsunami wavelength and amplitude, b and w are the length and width of the scar, d is the water depth at the landslide barycentre, T is the headscarp height and σ is the mean slope of the failure scar. Finally, we used high-resolution seismic profiles acquired with the Geo Marine Survey Systems GeoSpark 1500 during a 2015 cruise to observe the relationship between landslides and local stratigraphy. Constant sound velocities in water and sediment (1500 m s^{-1}) were used for time–depth conversions.

Results

The study area can be divided into two sectors (the western and the eastern sectors in Figs 1 & 2) according to their morphological characteristics. The western sector between Cape D'Orlando and Cape Calavà is almost completely carved by a network of shelf-indenting canyons. These canyons show an overall V-shaped cross-section and are characterized by narrow (a few hundred metres wide at maximum) thalweg (profiles C–D and E–F in Fig. 2), where several gullies from the canyon flanks and head converge downslope, giving rise to an overall dendritic morphology. The canyon heads are commonly located very close to the coast (a few hundred metres at maximum), especially around Cape Calavà, where they lie at water depths of less than 20 m. This morphological setting leads to a very narrow shelf, where a terraced feature, with the outer edge located at depth of 25–35 m, is recognizable (NSDT in Fig. 3b). Towards Cape D'Orlando, the shelf becomes progressively wider and the canyon heads

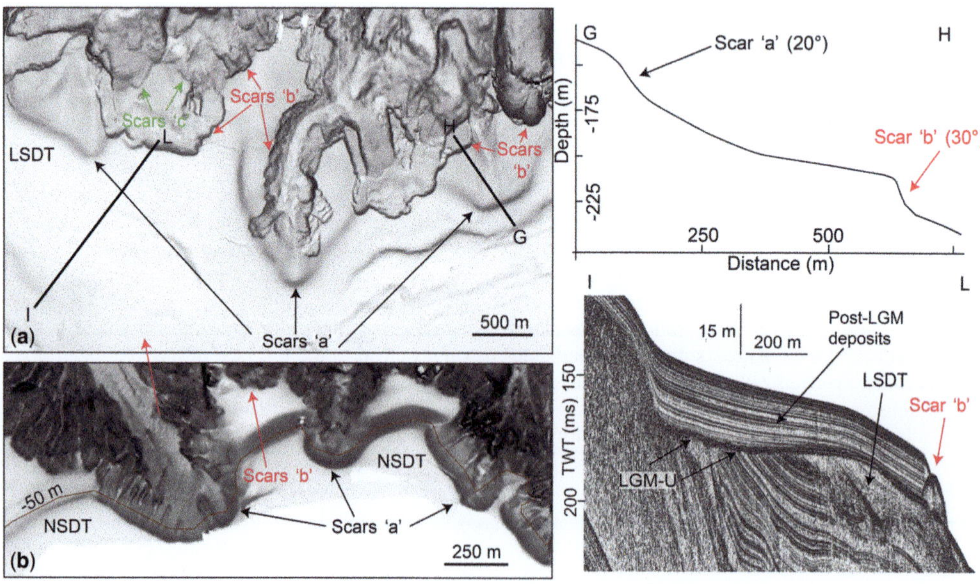

Fig. 3. Shaded relief map of high-resolution DEMs shows canyon heads located at the shelf edge (**a**) and proximal to the coastline (**b**), with the indication of the main groups of scars (see text for details and Fig. 2 for location). On the right, the bathymetric profile G–H shows the different morphology in terms of headscarp height and slope gradient between scars 'a' and 'b'. The Sparker profile I–L shows the erosive truncation of seismic reflectors associated with post-glacial deposits at the scar 'b'. LGM-U, Last Glacial Maximum unconformity; LSDT, lowstand submarine depositional terrace; NSDT, near-shore submarine depositional terrace.

are located at depths greater than 40 m, 1–2 km away from the coast.

The eastern sector, between Cape Calavà and the Gulf of Patti, is characterized by a progressive enlargement of the shelf eastwards, with the canyons mostly confined to the shelf edge at depths greater than 120 m, at a distance of >3 km from the coastline (Fig. 2a). These canyons are commonly wider (1–2 km) and show U-shaped cross-sections (profile A–B in Fig. 2) with respect to those observed in the previous sector. Gullies are almost absent from the canyon flanks, where landslide scars are instead predominant.

Distribution of landslide scars in the study area

A total of 280 landslide scars are recognized in the study area between −15 and −550 m, having lengths of *c.* 5–300 m, widths of *c.* 20–700 m and areas of *c.* 70–140.000 m^2 (Fig. 2; Table 1). The mapped landslides exhibit headscarps with heights of 3–54 m and slope gradients of 12–63°.

Landslide scars can be clustered into three main groups based on their morphology (the most important parameter) and location: the first group (scars 'a', black lines in Fig. 2) occurs between −15 and −250 m in the shallowest part of the canyon heads or just upslope of their main head. These landslide scars are characterized by an overall smooth and subdued morphology (Fig. 3), with headscarps having median values of 14 m and 21° for their height and slope gradients, respectively (Table 1). The landslide scars have a width/length ratio of 4×, affecting areas with median value of 9450 m^2; slope gradients of the landslide scars have a median value of 11° (Table 1).

The second group (scars 'b', red lines in Fig. 2) occurs between −17 and −334 m, just downslope of the scars 'a'. Compared with the latter, landslide scars 'b' have a higher morphological relief (Fig. 3), with headscarps having median values of 10 m and 34° for their height and slope gradients, respectively (Table 1). The landslide scars have a width/length ratio of *c.* 3, affecting smaller areas than type 'a', with a median value of 1500 m^2; slope gradients of the landslide scars show median values of *c.* 18° (Table 1). The seismic profile I–L (Fig. 3) clearly shows the erosive truncation of the post-glacial deposits owing to the scars 'b'.

The third group (scars 'c', green lines in Fig. 2) occurs between −76 and −527 m within the canyon

Table 1. *Descriptive statistics of the morphometric parameters measured both for the entire database and single groups of landslide scars*

	WD1 (m)	WD2 (m)	L (m)	W (m)	HH (m)	HS (°)	SS (°)	AS (°)	Area (m^2)
Total scars (280)									
Minimum	15	30	5	19	3	12	0	1	70
Maximum	527	555	305	692	54	63	39	27	138 000
Mean	146	169	45	142	13	32	17	7	7818
Median	106	129	33	105	11	32	16	6	2300
Standard deviation	104	107	38	125	7	10	7	5	15 977
Scars 'a' (48)									
Minimum	15	30	10	39	3	13	3	1	300
Maximum	253	273	243	692	45	36	23	11	138 000
Mean	77	100	66	259	15	22	11	4	17 749
Median	76	103	54	229	14	21	11	4	9450
Standard deviation	47	51	45	173	8	5	5	2	24 825
Scars 'b' (140)									
Minimum	17	32	5	22	4	14	4	1	70
Maximum	334	359	305	666	54	63	33	20	109 000
Mean	123	146	41	121	12	33	18	7	5914
Median	123	114	27	85	10	34	18	7	1500
Standard deviation	78	81	40	103	8	7	7	4	13 754
Scars 'c' (92)									
Minimum	76	92	9	19	5	12	0	2	100
Maximum	527	555	174	463	34	61	39	27	70 000
Mean	218	239	41	113	13	36	18	12	5534
Median	201	221	33	97	11	35	18	10	2250
Standard deviation	119	124	31	87	6	11	8	8	10 606

See text for details of scars 'a', 'b' and 'c'.
WD1, minimum water depth; WD2, water depth at the barycentre of the landslide scar; L, scar length; W, scar width; HH, headscarp height; HS, headscarp slope gradients; SS, scar slope gradients; AS, adjacent slope gradients.

Table 2. *Rank correlation coefficients (Spearman's rho) computed for the entire distribution of landslides (lower part, i.e. below the black box) and group of scars 'a' (upper part, i.e. above the black box)*

Variables	WD1 (m)	WD2 (m)	L (m)	W (m)	HH (m)	HS (°)	SS (°)	AS (°)	Area (m^2)
WD1 (m)		**0.94**	**0.45**	**0.52**	**0.30**	**−0.54**	**−0.58**	−0.11	**0.51**
WD2 (m)	**0.98**		**0.53**	**0.62**	**0.49**	**−0.40**	**−0.47**	−0.09	**0.60**
L (m)	**0.45**	**0.50**		**0.86**	**0.41**	−0.23	**−0.38**	−0.14	**0.95**
W (m)	**0.43**	**0.48**	**0.85**		**0.46**	−0.28	**−0.36**	−0.19	**0.96**
HH (m)	**0.23**	**0.30**	**0.44**	**0.53**		0.03	−0.13	−0.12	**0.45**
HS (°)	−0.07	−0.04	**−0.46**	**−0.45**	0.00		**0.64**	0.09	−0.28
SS (°)	**−0.30**	**−0.28**	**−0.58**	**−0.58**	**−0.19**	**0.69**		**0.30**	**−0.37**
AS (°)	0.05	0.03	**−0.28**	**−0.31**	**−0.19**	**0.34**	**0.32**		−0.16
Area (m^2)	**0.46**	**0.52**	**0.92**	**0.95**	**0.52**	**−0.45**	**−0.60**	**−0.29**	

Values in bold are different from 0 with a significance level $\alpha = 0.05$. Grey cells show the highest (>0.8) correlation between measured parameters.

head, showing a variable morphological relief, with headscarps having median values of 11 m and 35° for their height and slope gradients, respectively (Table 1). Landslide scars 'c' have a similar width/length ratio (3×) compared with landslide scars 'b', whereas the median value of the scar area (2250 m^2) is slightly higher than the previous case, but markedly lower than for landslides scars 'a'.

In order to evidence correlations between the measured morphometric parameters, rank correlation coefficients (Spearman's rho) were computed (Tables 2 & 3) and selected scatter plots are shown in Figure 4.

This statistical analysis shows that the width, length and area of landslide scars are positively correlated, with width and length having a correlation coefficient of 0.79 for scars 'b' and approximately 0.85 for the total distribution of scars, for scars 'a' and 'c' (Tables 2 & 3). The scatter plots of the width/length and area/length (Fig. 4a, b) indicate that a relatively tight correlation between these parameters is recognizable only for landslide scars affecting areas smaller than 20.000 m^2, whereas their distribution becomes more scattered for larger scar areas. A weaker positive correlation (0.4–0.5) is also recognized between scar area and headscarp height (Tables 2 & 3). The other measured parameters do not show a significant correlation (Tables 2 & 3), as also evidenced by scatter plots of the scar area against water depth or scar area against slope gradients of the landslide scars (Fig. 4). However, an overall decrease in the scar area with increasing slope gradients is observed for the different groups of landslides, with correlation coefficients ranging from −0.4 to −0.6 (Tables 2 & 3). In more detail, a slightly negative correlation (coefficient between −0.4 and −0.6) is observed between the scar slope against the length or width of the scars, whereas there is no correlation between the scar slope and the headscarp height (maximum coefficient of 0.2).

Table 3. *Rank correlation coefficients (Spearman's rho) computed for scars 'b' (lower part, i.e. below the black box) and scars 'c' (upper part, i.e. above the black box)*

Variables	WD1 (m)	WD2 (m)	L (m)	W (m)	HH (m)	HS (°)	SS (°)	AS (°)	Area (m^2)
WD1 (m)		**0.99**	**0.72**	**0.73**	**0.34**	**−0.55**	**−0.47**	**−0.42**	**0.76**
WD2 (m)	**0.97**		**0.75**	**0.76**	**0.41**	**−0.53**	**−0.46**	**−0.45**	**0.79**
L (m)	**0.60**	**0.64**		**0.87**	**0.48**	**−0.58**	**−0.57**	**−0.59**	**0.95**
W (m)	**0.69**	**0.76**	**0.79**		**0.56**	**−0.49**	**−0.49**	**−0.67**	**0.94**
HH (m)	**0.22**	**0.31**	**0.31**	**0.45**		0.02	−0.07	−0.29	**0.56**
HS (°)	**−0.20**	−0.14	**−0.32**	**−0.25**	0.10		**0.69**	**0.56**	**−0.54**
SS (°)	**−0.48**	**−0.44**	**−0.57**	**−0.58**	**−0.19**	**0.61**		**0.77**	**−0.56**
AS (°)	−0.08	−0.11	**−0.26**	**−0.37**	−0.13	0.13	**0.23**		**−0.69**
Area (m^2)	**0.64**	**0.71**	**0.87**	**0.94**	**0.44**	**−0.26**	**−0.59**	**−0.30**	

Values in bold are different from 0 with a significance level $\alpha = 0.05$. Grey cells show the highest (>0.8) correlation between measured parameters.

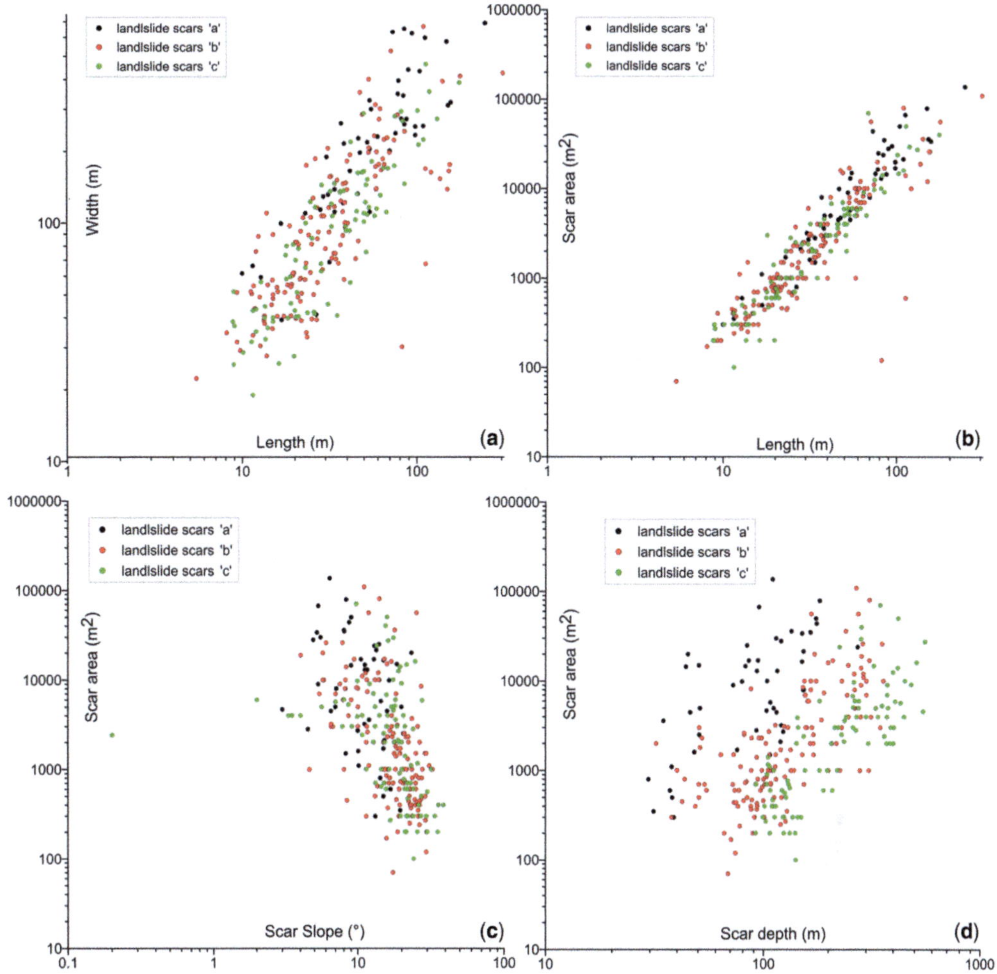

Fig. 4. Scatter plots of selected morphometric parameters measured for the three groups of landslide scars identified in the study area.

Discussions and conclusions

Recent morphological evolution of submarine canyons

The high-resolution morpho-bathymetric data allowed us to make a detailed characterization of the Sicilian continental margin between Cape D'Orlando and the Gulf of Patti, highlighting a key role of mass-wasting processes in the most recent evolution of this margin. The development of widespread instability processes can be related to the concomitant occurrence of several predisposing and triggering factors in the area, including the steepness (>10°) of the continental slope, recent tectonics (Fig. 1c) and associated vertical movements, high sedimentation rates and frequent seismicity (i.e. Sulli *et al.* 2013; Gamberi *et al.* 2015; Cultrera *et al.* 2017; Casalbore *et al.* 2017*a*). In addition, the width of the shelf seems to play a major role in controlling the most recent morphogenetic evolution of the canyons, as shelf-indenting canyons located on a narrow shelf (around Cape Calava) are commonly V-shaped and show a dendritic morphology with respect to the canyons confined to the shelf edge (eastward of Cape D'Orlando and in the Gulf of Patti). The lack of a well-developed shelf enables the connection between submarine and subaerial drainage systems. In such cases, canyon heads can act as the main collectors for sediments transported by flash flood-generated hyperpycnal flows or littoral drift (Chiocci and Casalbore 2011; Casalbore *et al.*

2011*a*; Gamberi *et al.* 2014; Pierdomenico *et al.* 2019). Sediments can be directly funnelled to higher depths during the main floods or remobilized through small failures owing to cyclical load by earthquakes or storm-waves (Locat and Lee 2002). In both cases, axial erosion associated with sedimentary gravity flows on the canyon floor may favour the development of retrogressive failures at the head and flanks of the canyons, thus producing a network of small gullies converging on the main canyon axis (Amblas *et al.* 2018 and references therein). The main role of axial erosion in driving the recent most morphological evolution of these canyons is testified by their narrow thalwegs and V-shaped cross-sections (Baztan *et al.* 2005). In contrast, the occurrence of a wider shelf hinders the direct connection between the subaerial and submarine system drainages (at least after the Last Glacial Maximum, LGM), thus representing a storage area for riverine and coastal sediments (Casalbore *et al.* 2017*a*). This setting leads to canyon heads mostly confined to the shelf break, characterized by lower sedimentary dynamics and axial incision of the canyon, in agreement with their more subdued morphology, lack of a well-developed network of gullies and the occurrence of wider and flat-bottomed thalwegs.

Landsliding processes at the canyon head and flanks

We have detected three main groups of landslide scars in both types of canyons, whose characterization can provide useful information for their stage of activity, failure and post-failure behaviour. In particular, the shallower landslide scars ‘a’ have an overall subdued morphology that can be related to their partial infilling. These landslide scars are often found at or close to the edge of terraced bodies on the shelf (Fig. 3) and interpreted as submarine depositional terraces (SDT), i.e. a sedimentary prograding wedge whose rollover depth is related to storm-wave base level, estimated around 25–30 m in the Southern Tyrrhenian Sea (Casalbore *et al.* 2017*b*). This means that the near-shore terraces found at 25–35 m (NSDT in Fig. 3b) are probably formed during the present-day highstand condition, whereas deeper terraces at the shelf edge (LSDT in Fig. 3a) were probably formed during the last lowstand (see also Sulli *et al.* 2013). The progradation of these terraced deposits within the landslide scars can be responsible for the scar’s subdued morphology (profile G–H in Fig. 3), thus the age of the landslide scars is older than that of the associated SDTs (i.e. pre-LGM in the case of lowstand terraces or more recently for the near-shore terraces). The landslide scars ‘b’ show instead a fresher-looking morphology, indicative of a younger erosive stage that is affecting the canyons. This is clearly recognizable in the seismic profile I–L (Fig. 3), where the slide scars ‘b’ cut both part of the lowstand terrace (LSDT) and the overlying (post-LGM) deposits. Landslide scars ‘c’ show a more variable morphological relief and they are often coaxial (occurring downslope, Fig. 3a) to the landslide scars ‘b’, suggesting a retrogressive evolution of slope failures from ‘c’ to b’. The variable morphological relief of these scars can be related to the fact that their headscarps can be partially reworked by material mobilized from successive and upslope slope failures.

The morphometric analysis of landslide scars shows that they are much wider (up to 695 m) than longer (up to 305 m), like semicircular or elliptical landslides scars recognized in other canyon heads elsewhere. However, fresh-looking landslide scars ‘b’ and ‘c’ are generally characterized by a lower width/length ratio with respect to the older and smoother landslide scars ‘a’ (Table 1; Fig. 4a). This could be interpreted as the result of two main processes: (1) landslide scars ‘b’ and ‘c’ tend to be more elongated because they are the result of an upslope retrogressive evolution, as also testified by their coaxial position and steeper gradients (Table 1); (2) the partial infilling of the landslide scars ‘a’ can partially modify their original shape and/or does not allow the discrimination of single slope failure events, so that their final shape is actually the coalescence of more landslide scars. This latter explanation is also supported by the overall larger areas characterizing landslide scars ‘a’ with respect to the types ‘b’ and ‘c’ (see median values in Table 1). Moreover, scatter plots show that the length, width and area of the different landslide scars have a relatively tight positive correlation up to certain threshold values (*c.* 200 m, 100 m and 20.000 m^2, respectively), whereas they become more scattered for higher values (Fig. 3a, b). This observation supports the fact that larger scars are actually the result of the coalescence of multiple slope failures that are no longer detectable because of successive erosive-depositional processes in shallow water or owing to the lack of multibeam data resolution in deeper sectors.

Morphometric analysis also shows the main role played by slope gradients (median values of 16° for the total distribution of landslide scars in Table 1) in controlling the development of slope failures in canyons, similarly to what was observed elsewhere (e.g. Green and Uken 2008; Goswami *et al.* 2014; Alberico *et al.* 2018; Casalbore *et al.* 2019). The high slope gradients coupled with the confined setting of canyons can possibly explain the smaller size of recognized landslide scars in comparison with the larger slope failures commonly recognized on the continental slopes (e.g. McAdoo *et al.* 2000; Huhnerbach and Masson 2004). A moderate negative correlation is in fact observed between landslide scar area against slope angle for the total distribution of landslides recognized in the study area (Table 2). In particular, slope gradients seem to influence the length and width of the scars (mainly for scars 'b' and 'c'), whereas poor correlation is present between scar slope and headscarp height (Tables 2 & 3). These results suggest that other factors, such as sedimentation rates, erosion and local geology, may have an important role in controlling the final shape and size of slope failures. Finally, steep gradients are also probably responsible for the disintegrative-like behaviour characterizing most of these landslides (*sensu* McAdoo *et al.* 2000), with scars having a smooth morphology and related landslide deposits not recognizable on the bathymetry. This could be related to the fact that landslide mass can either lose cohesion during failure (static liquefaction) or disintegrate downslope owing to the high slope gradients and confined setting, evolving into debris flows and/or turbidity currents (Mulder and Cochonat 1996; Locat and Lee 2002).

Preliminary hazard assessment associated with landsliding processes in the canyons

Because of the retrogressive erosion (i.e. the progressive landward migration of the mass-wasting processes) that commonly characterizes the evolution of canyon heads and the proximity of the canyon heads to the coast (especially around Cape Calava), the assessment of tsunamigenic potential associated with landsliding processes is an important issue to be addressed in our study area. Considering the high number of recognized landslide scars and the lack of some morphometric and geotechnical parameters useful for a full characterization of the landslide events, a proper modelling of landslide-generated tsunami propagation is beyond the aim of this paper. However, a first estimation of maximum tsunami amplitude above the barycentre of the landslide can be roughly made by applying the empirical relation available in the literature (McAdoo and Watts 2004). Despite the potential inaccuracy of using equations with parametric coefficients derived for a different part of the world and with different conditions, the same equations have already been used for computing the tsunamigenic potential of three larger submarine landslide scars in the study area (Sulli *et al.* 2018) as well as landslide scars in neighbouring areas (Aeolian Islands and Campanian Margin; Casalbore *et al.* 2011*b*, 2018; Alberico *et al.* 2018), thus making the obtained results comparable. This analysis shows that the recognized landslide scars were not able to generate significant tsunami waves, because estimated values were always lower than 1 m. However, considering the ongoing erosive activity of the studied canyons and the common retrogressive evolution at their heads, we computed the tsunamigenic potential associated with future slope failures at shallower depths. This computation was performed using the minimum depth (WD1 in Table 1) of the recognized landslide scars as a proxy of the landslide barycentre in the case of future retrogressive slope failures by assuming that the geometry and size of these slope failure are similar to those presently recognizable in the bathymetry. This analysis shows that some scars can potentially generate a maximum tsunami amplitude of 2 m, thus representing a local coastal hazard.

More generally, the geohazard associated with the shelf-indenting canyons developed in this area is very high, considering their proximity to the coast and the retrogressive evolution of slope failures along with the frequent occurrence of flash flood-generated hyperpycnal flows funnelled within the canyons that can directly impact coastal settlements and offshore infrastructures (i.e. subsea telecommunication cables; Carter *et al.* 2014; Pope *et al.* 2017).

Acknowledgements Special thanks are due to the crews of RV *Urania* and RV *Minerva Uno* (CNR) along with the participants of the marine surveys. We also gratefully acknowledge Rui Quartau, Jasper Moernaut and an anonymous reviewer for their comments and suggestions that improved the manuscript.

Funding This research was funded by the Italian Project MaGIC (Marine Geohazards along the Italian Coasts) and carried out in the framework of the flagship project RITMARE (Ricerca ITaliana per il MARE) Project.

Author contributions **DC**: conceptualization (lead), formal analysis (equal), methodology (lead), visualization (lead), writing – original draft (lead); **RC**: formal analysis (equal); **AB**: writing – review & editing (equal); **FLC**: funding acquisition (lead); **EM**: writing – review & editing (equal); **DR**: writing – review & editing (equal).

References

Alberico, I., Budillon, F., Casalbore, D., Di Fiore, V. and Iavarone, R. 2018. A critical review of potential tsunamigenic sources as first step towards the tsunami hazard assessment for the Napoli Gulf (Southern Italy) highly populated area. *Natural Hazards*, **92**, 43–76, https://doi.org/10.1007/s11069-018-3191-5

Amblas, D., Ceramicola, S. *et al.* 2018. Submarine canyons and gullies. *In*: Micallef, A., Krastel, S. and Savini, A. (eds) *Submarine Geomorphology*. Springer, Cham, 251–272.

Assier-Rzadkieaicz, S., Heinrich, P., Sabatier, P.C., Savoye, B. and Bourillet, J.F. 2000. Numerical modelling of a landslide-generated tsunami: the 1979 Nice event. *Pure and Applied Geophysics*, **157**, 1707–1727, https://doi.org/10.1007/PL00001057

Baztan, J., Berné, S. *et al.* 2005. Axial incision: the key to understand submarine canyon evolution (in the western Gulf of Lion). *Marine Petroleum Geology*, **22**, 805–826, https://doi.org/10.1016/j.marpetgeo.2005.03.011

Bosman, A., Casalbore, D., Anzidei, M., Muccini, F., Carmisciano, C. and Francesco Latino, C. 2015. The first ultra-high resolution Digital Terrain Model of the shallow-water sector around Lipari Island (Aeolian Islands, Italy). *Annals of Geophysics*, **58**, 1–11, https://doi.org/10.4401/ag-6746

Carter, L., Gavey, R., Talling, P.J. and Liu, J.T. 2014. Insights into submarine geohazards frombreaks in subsea telecommunication cables. *Oceanography*, **27**, 58–67, https://doi.org/10.5670/oceanog.2014.40

Casalbore, D., Chiocci, F.L., Scarascia Mugnozza, G., Tommasi, P. and Sposato, A. 2011*a*. Flash-flood hyperpycnal flows generating shallow-water landslides at Fiumara mouths in Western Messina Straits (Italy). *Marine Geophysical Research*, **32**, 257–271, https://doi.org/10.1007/s11001-011-9128-y

Casalbore, D., Romagnoli, C., Bosman, A. and Chiocci, F.L. 2011*b*. Potential tsunamigenic landslides at Stromboli Volcano (Italy): insights from marine DEM analysis. *Geomorphology*, **126**, 42–50, https://doi.org/10.1016/j.geomorph.2010.10.026

Casalbore, D., Bosman, A. and Chiocci, F.L. 2012. Study of recent small-scale landslides in geologically active marine areas through repeated multibeam surveys: examples from the southern Italy. *In*: Yamada, Y. *et al.* (eds) *Submarine Mass Movements and their Consequences*. Advances in Natural and Technological Hazards Research, **31**, Springer, Dordrecht, 573–582, https://doi.org/10.1007/978-94-007-2162-3_51

Casalbore, D., Ridente, D., Bosman, A. and Chiocci, F.L. 2017*a*. Depositional and erosional bedforms in Late Pleistocene–Holocene pro-delta deposits of the Gulf of Patti (southern Tyrrhenian margin, Italy). *Marine Geology*, **385**, 216–227, https://doi.org/10.1016/j.margeo.2017.01.007

Casalbore, D., Falese, F., Martorelli, E., Romagnoli, C. and Chiocci, F.L. 2017*b*. Submarine depositional terraces in the Tyrrhenian Sea as a proxy for Paleo-Sea level reconstruction: problems and perspective. *Quaternary International*, **439**, 169–180, https://doi.org/10.1016/j.quaint.2016.02.027

Casalbore, D., Romagnoli, C., Bosman, A., Anzidei, M. and Chiocci, F.L. 2018. Coastal hazard due to submarine canyons in active insular volcanoes: examples from Lipari Island (southern Tyrrhenian Sea). *Journal of Coastal Conservation*, **22**, 989–999, https://doi.org/10.1007/s11852-017-0549-x

Casalbore, D., Bosman, A., Casas, D., Chiocci, F., Martorelli, E. and Ridente, D. 2019. Morphological variability of submarine mass movements in the tectonically controlled Calabro–Tyrrhenian Continental Margin (southern Italy). *Geosciences*, **9**, 43, https://doi.org/10.3390/geosciences9010043

Catalano, S., Monaco, C., Tortorici, L. and De Guidi, G. 2003. Morphological evidence of Holocene coseismic deformation in the Taormina region (NE Sicily). *Journal of Geodynamics*, **36**, 193–211, https://doi.org/10.1016/S0264-3707(02)00035-2

Chiarabba, C., Jovane, L. and DiStefano, R. 2005. A new view of Italian seismicity using 20 years of instrumental recordings. *Tectonophysics*, **395**, 251–268, https://doi.org/10.1016/j.tecto.2004.09.013

Chiocci, F.L. and Casalbore, D. 2011. Submarine gullies on Italian upper slopes and their relationship with volcanic activity revisited 20 years after Bill Normark's pioneering work. *Geosphere*, **7**, 1284–1293, https://doi.org/10.1130/GES00633.1

Chiocci, F.L. and Casalbore, D. 2017. Unexpected fast rate of morphological evolution of geologically-active continental margins during Quaternary: examples from selected areas in the Italian seas. *Marine and Petroleum Geology*, **82**, 154–162, https://doi.org/10.1016/j.marpetgeo.2017.01.025

Clare, M., Chaytor, J. *et al.* 2018. A consistent global approach for the morphometric characterization of subaqueous landslides. *Geological Society, London, Special Publications*, **477**, 455–477, https://doi.org/10.1144/SP477.15

Colantoni, P., Gennesseaux, M., Vanney, J.R., Ulzega, A., Melegari, G. and Trombetta, A. 1992. Processi dinamici del canyon sottomarino di gioia tauro (Mare Tirreno). *Giornale di Geologia*, **3**, 199–213.

CPTI Working Group 2004. *Catalogo Parametrico dei Terremoti Italiani, version 2004 (CPTI04)*. INGV, Bologna, http://emidius.mi.ingv.it/CPTI

Cultrera, F., Barreca, G. *et al.* 2017. Active faulting and continental slope instability in the Gulf of Patti (Tyrrhenian side of NE Sicily, Italy): a field, marine and seismological joint analysis. *Natural Hazards*, **86**, 253–272, https://doi.org/10.1007/s11069-016-2547-y

Ferranti, L., Antonioli, F., Anzidei, M., Monaco, C. and Stocchi, P. 2010. The timescale and spatial extent of vertical tectonic motions in Italy: insights from relative sealevel changes studies. *Journal of the Virtual Explorer*, paper 30, https://doi.org/10.3809/jvirtex.2010.00255

Gamberi, F., Rovere, M., Mercorella, A., Leidi, E. and Dalla Valle, G. 2014. Geomorphology of the NE Sicily continental shelf controlled by tidal currents, canyon head incision and river-derived sediments. *Geomorphology*, **217**, 106–121, https://doi.org/10.1016/j.geomorph.2014.03.038

Gamberi, F., Rovere, M., Marani, M.P. and Dykstra, M. 2015. Modern submarine canyon feeder- system and deep-sea fan growth in a tectonically active margin (northern Sicily). *Geosphere*, **11**, 307–319, https://doi.org/10.1130/GES01030.1

Goswami, R., Mitchell, N.C., Argnani, A. and Brocklehurst, S.H. 2014. Geomorphology of the western Ionian Seabetween Sicily and Calabria, Italy. *Geomarine Letters*, **34**, 419–433, https://doi.org/10.1007/s00367-014-0374-2

Green, A. and Uken, R. 2008. Submarine landsliding and canyon evolution on the northern KwaZulu–Natal continental shelf, South Africa, SW Indian Ocean. *Marine Geology*, **254**, 152–170, https://doi.org/10.1016/j.margeo.2008.06.001

Harbitz, C.B., Løvholt, F. and Bungum, H. 2014. Submarine landslide tsunamis: how extreme and how likely?

Natural Hazards, **72**, 1341–1374, https://doi.org/10.1007/s11069-013-0681-3

Huhnerbach, V., Masson, D.G. and COSTA Project Partners 2004. Landslides in the North Atlantic and its adjacent seas: an analysis of their morphology, setting and behaviour. *Marine Geology*, **213**, 343–362, https://doi.org/10.1016/j.margeo.2004.10.013

Lentini, F., Carbone, S. and Guarnieri, P. 2006. Collisional and post-collisional tectonics of the Apenninic–Maghrebian Orogen (Southern Italy). *Geological Society of America, Special Papers*, **409**, 57–81.

Locat, J. and Lee, H.J. 2002. Submarine landslides: advances and challenges. *Canadian Geotechnical Journal*, **39**, 193–212, https://doi.org/10.1139/t01-089

Longva, O., Janbu, N., Blikra, L.H. and Bøe, R. 2003. The 1996 Finneidfjord slide: seafloor failure and slide dynamics. *In*: Locat, J., Mienert, J. and Boisvert, L. (eds) *Submarine Mass Movements and their Consequences*. Advances in Natural and Technological Hazards Research, **19**, Springer, Dordrecht, 531–538, https://doi.org/10.1007/978-94-010-0093-2_58

Masson, D.G., Harbitz, C.B., Wynn, R.B., Pedersen, G. and Løvholt, F. 2006. Submarine landslides: processes, triggers and hazard prediction. *Philosophical Transactions of the Royal Society A*, **364**, 2009–2039, https://doi.org/10.1098/rsta.2006.1810

McAdoo, B.G. and Watts, P. 2004. Tsunami hazard from submarine landslides on the Oregon continental slope. *Marine Geology*, **203**, 235–245, https://doi.org/10.1016/S0025-3227(03)00307-4

McAdoo, B.G., Pratson, L.F. and Orange, D.L. 2000. Submarine landslide geomorphology, US continental slope. *Marine Geology*, **169**, 103–136, https://doi.org/10.1016/S0025-3227(00)00050-5

Mitchell, N.C. 2012. Aspects of marine geoscience: a review and thoughts on potential for observing active processes and progress through collaboration between the ocean sciences. *Philosophical Transactions of the Royal Society, London. A: Mathematical, Physical and Engineering Sciences*, **370**, 5567–5612, https://doi.org/10.1098/rsta.2012.0395

Mountjoy, J. and Micallef, A. 2018. Submarine landslides. *In*: Micallef, A., Krastel, S. and Savini, A. (eds) *Submarine Geomorphology*. Springer, Cham, 235–250.

Mulder, T. and Cochonat, P. 1996. Classification of offshore mass movements. *Journal of Sedimentary Research*, **66**, 46–57, https://doi.org/10.1306/D42682AC-2B26-11D7-8648000102C1865D

Pierdomenico, M., Casalbore, D. and Chiocci, F.L. 2019. Massive benthic litter funnelled to deep sea by flash-flood generated hyperpycnal flows. *Scientific Reports*, **9**, 5330, https://doi.org/10.1038/s41598-019-41816-8

Pope, E.L., Talling, P.J., Carter, L., Clare, M.A. and Hunt, J.E. 2017. Damaging sediment density flows triggered by tropical cyclones. *Earth and Planetary Science Letters*, **458**, 161–169, https://doi.org/10.1016/j.epsl.2016.10.046

Rahiman, T.I and Pettinga, J.R. 2006. The offshore morpho-structure and tsunami sources of the Viti Levu Seismic Zone, southeast Viti Levu, Fiji. *Marine Geology*, **232**, 203–225, https://doi.org/10.1016/j.margeo.2006.07.007

Sabato, L. and Tropeano, M. 2004. Fiumara: a kind of high hazard river. *Physics and Chemistry of the Earth*, **29**, 707–715, https://doi.org/10.1016/j.pce.2004.03.008

Scicchitano, G., Spampinato, C.R., Ferranti, L., Antonioli, F., Monaco, C., Capano, M. and Lubritto, C. 2011. Uplifted Holocene shorelines at Capo Milazzo (NE Sicily, Italy): evidence of co-seismic and steady-state deformation. *Quaternary International*, **232**, 201–213, https://doi.org/10.1016/j.quaint.2010.06.028

Sulli, A., Lo Presti, V., Gasparo Morticelli, M. and Antonioli, F. 2013. Vertical movements in NE Sicily and its offshore: outcome of tectonic uplift during the last 125 ky. *Quaternary International*, **288**, 168–182, https://doi.org/10.1016/j.quaint.2012.01.021

Sulli, A., Zizzo, E. and Albano, L. 2018. Comparing methods for computation of run-up heights of landslide-generated tsunami in the Northern Sicily continental margin. *Geo-Marine Letters*, **38**, 439–455, https://doi.org/10.1007/s00367-018-0544-8

Urgeles, R. and Camerlenghi, A. 2013. Submarine landslides of the Mediterranean Sea: trigger mechanisms, dynamics, and frequency-magnitude distribution. *Journal of Geophysical Research – Earth Surface*, **118**, 2600–2618, https://doi.org/10.1002/2013JF002720

Geomorphology and event-stratigraphy of recent mass-movement processes in Lake Hallstatt (UNESCO World Heritage Cultural Landscape, Austria)

Michael Strasser[1]*, T. Berberich[1], S. Fabbri[2], M. Hilbe[2], J-J. S. Huang[1], S. Lauterbach[3], M. Ortler[1], H. Rechschreiter[4], A. Brauer[5], F. Anselmetti[2] and K. Kowarik[4]

[1]Department of Geology, University of Innsbruck, Innrain 52, 6020 Innsbruck, Austria

[2]Institute of Geological Sciences and Oeschger Centre for Climate Change Research, University of Bern, Baltzerstrasse 1+3, 3012 Bern, Switzerland

[3]Leibniz Laboratory for Radiometric Dating and Stable Isotope Research, Kiel University, Max-Eyth-Str. 11-13, 24118 Kiel, Germany

[4]Naturhistorisches Museum Wien, Burgring 7, 1010 Wien, Austria

[5]Section Climate Dynamics and Landscape Evolution, GFZ-German Research Centre for Geosciences, Telegrafenberg, 14473 Potsdam, Germany

MS, 0000-0003-3991-2405; SF, 0000-0002-0015-0442; J-JSH, 0000-0001-8815-3957; SL, 0000-0003-4131-0366; MO, 0000-0002-2732-9514

*Correspondence: michael.strasser@uibk.ac.at

Abstract: We report the first-ever basin-wide geomorphological characterization of a high-resolution bathymetry map in a steeply incised valley-occupying intra-mountainous lake in the Eastern Alps. The resulting new geomorphological map of Lake Hallstatt is then combined with high-resolution reflection seismic and sedimentary core analyses to document, characterize and date recent (<200 years) subaquatic landslides. The area is located in the UNESCO World Heritage Cultural Landscape of the Hallstatt–Dachstein area, famous for its well-documented medieval and prehistoric human settlement history. This allows us to calibrate observed mass-transport deposits (MTDs) in the high-resolution sedimentary archive (sedimentation rates *c.* 0.5 cm a^{-1}) of this deep lake dominated by clastic sedimentation.

The hydro-acoustic data document a multitude of different MTDs linked to rock falls, subaqueous slope failures and shore collapses. Sediment cores document laminated background sediments intercalated with distinct event deposits that can be linked to historically documented major flood events and moderately strong earthquakes. Our study suggests that the larger MTDs result from earthquake-triggered subaquatic delta slope instabilities and that the deeper subsurface provides evidence of even larger mass-movement processes, yet to be validated by longer cores. Thus, Lake Hallstatt is a potential natural laboratory for studying causes and consequences of subaquatic landslides in steeply incised intra-mountainous lakes and comparable fjord settings.

Studying subaquatic mass-movement processes in lakes as model for oceans has contributed to the advancement of our understanding of causes and consequences of submarine landslides over the past two decades (e.g. Girardclos *et al.* 2007; Strasser *et al.* 2007; Moernaut and De Batist 2011; Hilbe and Anselmetti 2014). Using common state-of-the-art research methodologies as applied in the oceanic realm to geophysically image and geologically sample mass-transport deposits (MTDs) and related turbidites, lacustrine subaquatic geomorphology combined with seismic and sedimentary event-stratigraphy can reveal detailed information about past mass-movement activity and related sedimentary processes. Key advantages of lakes are the comparably high sedimentation rates (and thus high resolution of the archive) and that events can often be calibrated with historical records to investigate the underlying causal mechanism (e.g. written records for floods, earthquakes or subaerial mass movements into the lake (Schnellmann *et al.* 2002; Gilli *et al.* 2013; Van Daele *et al.* 2013)). While there have been several studies documenting MTDs and inferred processes of subaquatic slope failures on sediment-covered lateral slopes in peri-alpine piedmont lakes worldwide (see summary in Sammartini *et al.* (2019) and references therein; often as models for translation submarine slides, e.g. Strasser *et al.* (2007)), detailed studies from steeply incised valley-occupying intra-mountainous lakes

From: Georgiopoulou, A., Amy, L. A., Benetti, S., Chaytor, J. D., Clare, M. A., Gamboa, D., Haughton, P. D. W., Moernaut, J. and Mountjoy, J. J. (eds) 2020. *Subaqueous Mass Movements and their Consequences: Advances in Process Understanding, Monitoring and Hazard Assessments*. Geological Society, London, Special Publications, **500**, 405–426.
First published online March 31, 2020, https://doi.org/10.1144/SP500-2019-178

are relatively few. Similar to fjords in mountainous coastline regions, sediment supply in such lakes is typically dominated by clastic sediment transport, while autochthonous biogenic production is comparably low (e.g. Girardclos *et al.* 2007; Bellwald *et al.* 2016). Furthermore, lateral slopes are steep and may not be covered by thick sedimentary sequences. Predominating mass-movement processes are thus expected to be linked to rockfalls from steep subaerial cliffs (e.g. Van Daele *et al.* 2013) or instabilities developing along subaqueous river and/or alluvial fan delta slopes (e.g. Hilbe and Anselmetti 2014). However, inferring delta slope stability and evaluating preconditioning v. a trigger mechanism for past mass movements from sedimentary event deposits is very challenging, mainly because sediment gravity flows may evolve from both high-sediment supply during extreme precipitation/run-off/sediment-discharge events, and/or from delta slope failures of various magnitudes and frequencies, while respective causal mechanisms are not mutually exclusive (e.g. Clare *et al.* 2016). Furthermore, intra-mountainous and high-relief fjord regions often lack long historical records that allow for calibration of observed MTDs and turbidites in offshore geophysical and geological data. Yet, recent studies that addressed subaqueous mass-movement in such settings (e.g. Hansen *et al.* 2016; Rapuc *et al.* 2018) document the potential of elaborating magnitude–frequency relationships over time periods long enough to cover the whole width of possible mass-movement processes for a given lake or fjord system.

Here we study Lake Hallstatt, located in the Hallstatt–Dachstein area in upper Austria, which is well known for its alpine landscape and well-documented medieval and prehistoric human settlement history (Kern *et al.* 2009). The cultural importance of underground salt mining in Hallstatt, which has been documented since the Middle Bronze Age (*c.* 3.5 ka BP; Kern *et al.* 2009), was recognized long ago by assigning the status of a UNESCO World Cultural Heritage Site to the Hallstatt area, particularly because of the wealth of archaeological artefacts from the Early Iron Age. Local mining activity is well documented for prehistoric times and known to have been repeatedly affected by large-scale catastrophic landslides (Rohn *et al.* 2005). The Hallstatt chronicle, spanning more than 500 years of written history, documents several catastrophic events, such as mass movements, heavy precipitation events with debris flows, shore collapses and earthquakes (Urstöger 2000). Therefore, Lake Hallstatt potentially provides an excellent natural laboratory for studying mass-movement processes and their manifestation in the geological record of the lacustrine sedimentary infill of this intra-mountainous lake, which until now has not been investigated by state-of-the-art offshore geophysics and sedimentary core analyses. The objectives of this study are (i) to characterize, map and date subaqueous landforms (by means of high-resolution multibeam bathymetry), MTDs and turbidites (by means of reflection seismics and sedimentary core analyses), (ii) to investigate and compare the resulting event-stratigraphy with historical records to calibrate observable parameters in offshore datasets for process-based understanding of mass-movement processes (e.g. distinguishing flood- from earthquake-generated turbidites) and (iii) to evaluate the potential of Lake Hallstatt as a natural laboratory for studying causes and consequences of subaquatic landslides in steeply incised valley-occupying intra-mountainous lakes and fjords.

Geological setting and previous research in Lake Hallstatt

Lake Hallstatt (47° 33′ N; 13° 39′ E; 5.9 km long, 8.55 km^2 area, maximum water depth 124 m, Fig. 1) is a holomictic alpine lake, with occasional meromixis beyond 100 m water depth (Ficker *et al.* 2011). It is located within the Northern Calcareous Alps in the Salzkammergut of upper Austria. The fjord-type lake is surrounded by steep limestone mountains near the glaciated Dachstein Massif in the south. The lake today consists of two sub-basins with a deeper southern basin (Obersee, 124 m deep) and shallower northern basin (Untersee, 45 m deep). Its overdeepened bedrock basin was formed by repeated glacial erosion, with the last cycle ending with the retreat of the Traun Glacier from the Hallstatt Basin at *c.* 16 cal ka BP (Van Husen 2004). The basin was subsequently infilled by Late Pleistocene and Holocene lacustrine sediments consisting predominantly of clastic carbonates that were eroded from the surrounding Triassic lagoonal limestones of the Dachstein Formation (Mandl *et al.* 2012). The Traun River is the main inflow in the SE (53% of total inflow into Lake Hallstatt, mean annual discharge 1970–2006 = 20.7 m^3 s^{-1}; Ficker *et al.* 2011). Other tributaries are the Waldbach in the SW, the Gosaubach in the NW and the Zlambach in the NE (Fig. 1). The Mühlbach, which forms the cone-shaped alluvial fan delta on which the historical village Hallstatt is located (Fig. 1), drains the high valley of the Hallstatt salt-mine area. This mine area, geologically located in the Upper Permian to Lower Triassic Haselgebirge Formation (e.g. Schorn and Neubauer 2014), has been exploited for rock salt since the Bronze Age (at least since 1500 BC; Kern *et al.* 2009).

The long history of the Hallstatt area documents several catastrophic events, such as mass movements, heavy precipitation events with debris flows, shore

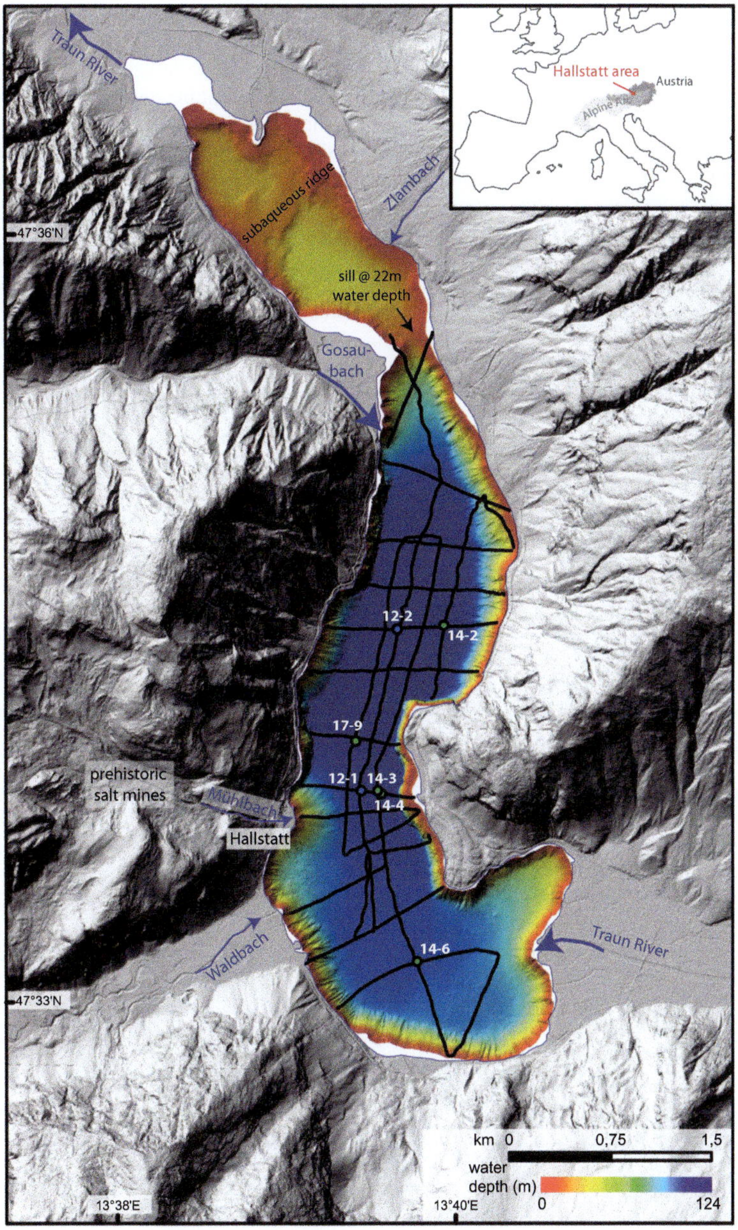

Fig. 1. Bathymetric map of Lake Hallstatt. Location of Lake Hallstatt in the Salzkammergut of upper Austria (upper right inset shows the Hallstatt area with respect to Europe, the Alpine Arc and Austria) and the new bathymetric map acquired with a Kongsberg EM2040 multibeam echo sounder system, embedded in a shaded relief map of the Hallstatt area, calculated from the 1 m digital terrain model of Airborne Laser Scanning (Light Detection And Ranging; LiDAR) data (Land-Oberösterreich 2019). The annotations locate tributary rivers and creeks, as well as local villages and valleys mentioned in the text. Short cores from the 2014 and 2017 coring campaigns and long cores from 2012 are depicted in green and blue, respectively. Black lines show the survey grid along which 3.5 kHz single channel reflection seismic profiles were acquired in 2014.

collapse and earthquakes (Urstöger 2000; Melzner 2017). Various historical information documents how society reacted to these geohazards with geoengineering in the catchments, for instance by routing the tributary rivers through lakeside villages (e.g. Pürstinger 2017). However, as of to date, there has been no research study that systematically uses state-of-the-art geophysical surveying and geological sampling techniques to investigate the sediment infill of Lake Hallstatt for event deposits related to these historically documented events around the lake. Only a pilot study by Heine (2017) has tested multibeam bathymetry in a small perimeter offshore the historical village of Hallstatt and imaged bathymetric features typical for subaquatic mass-movement processes. However, incomplete coverage of the lake and missing validation of interpreted features by coring has hampered process-based understanding of causes and consequences of mass-movement processes. The aim of this study is to overcome this limitation by applying the following state-of-the-art methodologies and research approach.

Material and methods

Geophysical data acquisition, processing and interpretation

High-resolution bathymetry of Lake Hallstatt was acquired in October 2017 using the Kongsberg EM2040 multibeam echo sounder system of the University of Bern (Hilbe and Anselmetti 2014) operated at 300 kHz in 1° by 1° beam-width configuration. Positioning was realized with a Leica GX1230 + GNSS receiver in combination with real-time kinematic positioning service provided by EPOSA.at. Sound correction was based on continuously monitored surface sound-velocity and vertical velocity profiles recorded at least daily using a Valeport MiniSVP probe. Speed of sound in the water column ranged from *c.* 1450 m s^{-1} at the surface to 1425 m s^{-1} in the deepest basin. The processed point cloud was rasterized with a cell size of 1 m with vertical accuracy on the order of a few decimetres (Hilbe and Anselmetti 2014). Geomorphological analyses were performed using ArcGIS. Furthermore, raster maps showing the backscatter intensity were produced to support of geological and sediment-dynamic interpretation. Imperfections in the image were corrected by the creation of a beam pattern file, which removes acoustic artefacts from the sonar data.

High-resolution single-channel seismic profiles were acquired in 2014 using a single channel GeoAcoustics 3.5 kHz 4-element pinger, providing a theoretical vertical subsurface resolution of *c.* 10 cm. Data were processed using a bandpass filter (lower cut 2 kHz, upper cut 6 kHz) in the IHS Kingdom® software (v. 2018). Seismic data interpretation focuses on the southern deep lake basin because of low penetration depths in the northern shallow basin. For time-to-depth conversions, a constant velocity of 1450 m s^{-1} for both water and sediment was assumed.

Coring campaigns

Sediment and lithostratigraphic analyses were performed on a multitude of sediment cores that were retrieved across the lake basin during three coring campaigns in 2012, 2014 and 2017 (Fig. 1). In 2012, three parallel piston-core sequences and one gravity core were recovered from a site located proximal to the subaquatic fan offshore the village of Hallstatt (site HAS_2012-1; *c.* 47° 33′ 50″ N, 13° 39′ 23″ E, *c.* 116 m water depth) by using a 90 mm diameter UWITEC piston corer system and a 90 mm diameter UWITEC gravity corer with additional hammer weight. In addition, two further piston-core sequences and one gravity core were obtained during the same coring campaign from a site located in the deepest part of the southern basin (site HAS_2012-2; *c.* 47° 34′ 28″ N, 13° 39′ 40″ E, *c.* 124 m water depth). All recovered core segments were split into halves, photographed, described lithologically and sub-sampled for further research at the GFZ Potsdam before storage at 4°C. Visual correlation of the overlapping, 2 m-long segments of the individual piston-core sequences and the gravity cores from each of the two coring sites via distinct macroscopic lithological marker layers resulted in a discontinuous composite sediment core (two core recovery gaps) of 816.0 cm length for site HAS_2012-1 and a continuous composite sediment core of 367.5 cm length for site HAS_2012-2. Here we report only results from the uppermost part of the composite sediment cores HAS_2012-1 (segments B1o, B1u, C2o and C2u) and HAS_2012-2 (segment D1o), which allow for the characterization and dating of recent subaquatic MTDs (see below). In the following text and in the figures the cores are labelled as 12-1 and 12-2.

In order to investigate the lithological characteristics and spatial distribution of the uppermost sedimentary sequences and individual event deposits across the entire deep basin of Lake Hallstatt, a total of fifteen 63 mm-diameter gravity cores of up to 1 m length were retrieved with a gravity short corer in 2014 and with a gravity short corer with hammer technique in 2017. Here we report results of the five most representative cores from these two campaigns (Fig. 1). The cores from the 2014 coring campaign were split into two halves and stored at 4°C at the University of Bern, before being transported to the Austrian Core Facility at the University of

Innsbruck where they were analysed together with the cores retrieved in 2017.

Sediment-core analysis

The split gravity cores from the 2014 and 2017 coring campaigns were photographed with the RGB-camera system attached to the COX Analytical Systems Itrax core scanner at the Austrian Core Facility, University of Innsbruck. Macroscopic visual description and microscopic smear-slide analyses following protocols defined by Schnurrenberger *et al.* (2003) were used to characterize general lithotypes and event layers in the sediment. A laser-diffraction particle analyser (Malvern Mastersizer 3000) was used to determine grain size of the major event deposits and adjacent 'background sediment'. Due to the predominantly clastic composition of the sediments, measurements were carried out on a bulk sediment mix averaged over a sampling interval of 5 mm. Average results of three successive measurements per sample (analytical error of <5%) are reported as relative fractions of clay, silt and sand, as well as D90 values derived from statistical analyses after Blott and Pye (2001).

In order to further characterize the sedimentary sequence and event deposits geochemically, and to guide advanced core-to-core correlation across the entire lake basin, X-ray fluorescence (XRF) core scanning was performed on the 2014 and 2017 gravity cores at 200 µm downcore resolution using a COX Analytical Systems Itrax XRF core scanner (Croudace *et al.* 2006) at the Austrian Core Facility, operated with a Molybdenum X-ray tube (voltage 30 kV; current 50 mA; exposure time 5 s). XRF results are given as counts and are considered semi-quantitative (Huang *et al.* 2016). In order to obtain a reliable, discernible dataset, a reduction of original data was performed by sorting out elements that showed low R^2 (<0.65) between two duplicate runs (Löwemark *et al.* 2019). Elements considered most reliable for scientific interpretation are K, Ca, Ti, Mn, Fe, Zn, Rb and Sr. In order to account for downcore variation in physical properties and matrix effects that introduce non-linear, close-sum and asymmetrical problems that deteriorate the integrity of the XRF core scanning data, centred log-ratio transformation was applied to the data (Weltje and Tjallingii 2008; Lee *et al.* 2019). With the main objective of using XRF data as a core correlation index, the centred log-ratio transformed data were analysed using principal component analyses (PCA) to identify the element ratio showing the maximum data variation across the analysed cores. PCA analysis reveal that principal component 1 (PC 1) can explain *c.* 41–48% data variation with Ca and Fe as the main end-member variables. Thus, we select ln (Ca/Fe) as the most powerful core correlation index for the Lake Hallstatt sediments and use this value hereafter to report the geochemical fingerprint of event deposits v. background sediment.

Chronology

To provide an absolute chronological control for the recent sediments of Lake Hallstatt, gamma spectrometric measurements were conducted on core segment D1o of composite sediment core HAS_2012-2. For this purpose, sediment samples (*c.* 5–10 g) were taken continuously at 1 cm resolution from the uppermost 38 cm of the core segment, freeze-dried, homogenized and subsequently analysed using a vertical-shaped Canberra Industries BE 3830 broad energy germanium (BEGe) detector (38 cm^2 active area, 70 mm diameter, ultra-low-background cryostat) that was mounted in a 6″ low-background Pb chamber and placed in an actively ventilated lab in the cellar of a concrete building at the GFZ Potsdam. The measurement time per sample was 1–5 days and the software Genie 2000 (Canberra Industries) was used for hardware control, data storage and spectrum analysis. Efficiency calibration is based on the measurement of milk powder (E2282, National Physical Laboratory, UK) with the same analytical setting. Here we report only results for the short-lived radionuclide ^{137}Cs (photo peak at 661.7 keV); the average counting uncertainty was <2.4% for samples exceeding 500 counts. Results are reported as mBq g^{-1} and are linked to known activity peaks related to maxima of radionuclide fallout reached after the Limited Test Ban Treaty for nuclear bomb tests in AD 1963 and the Chernobyl reactor accident in AD 1986 (e.g. Appleby 2000).

For estimating the chronology of the lower part of the sediment succession recovered in the short gravity cores, we also consider the results of accelerator mass spectrometry (AMS) ^{14}C dating of three samples of organic macro remains (Poz-50393, Poz-50390, Poz-50392) obtained at 191.0, 260.0, and 278.5 cm depth in composite sediment core HAS_2012-1. Samples were dated at the Poznań Radiocarbon Laboratory and conventional radiocarbon ages were calibrated using OxCal 4.2 (Bronk Ramsey 2009) with the IntCal13 calibration dataset (Reimer *et al.* 2013). AMS ^{14}C dating yielded undistinguishable 'modern' age ranges between 0 and 300 cal years BP (2σ) for all these samples, documenting very high sedimentation rates (Table 1; see more detail below).

Geomorphological description of the lake floor

The new bathymetric map of Lake Hallstatt (Fig. 1) reveals unprecedented details about the subaqueous geomorphology of this intra-mountainous lake.

Table 1. *Radiocarbon ages*

Sample code	Core section	Section depth (cm)	Composite depth (cm)	AMS ^{14}C age	Calibrated age (2σ cal years BP)	Sampled material
Poz-50393	HAS_2012-1 B1u	90.5	190.0	175 ± 25	0–289	organic macro remains
Poz-50390	HAS_2012-1 C2o	80.5	260.0	135 ± 30	8–278	organic macro remains
Poz-50392	HAS_2012-1 C2u	99	278.5	180 ± 30	0–297	organic macro remains

The southern deep basin has a north–south-trending elongated shape with >100 m high, steep lateral slopes (>20° up to nearly vertical cliffs) that are separated from the nearly flat basin floor (<1°) by a 100–200 m wide zone comprising the toe of slopes-to-basin transition. Several steeply fronted (20–30°) river deltas and alluvial fan deltas with convex-basinward cone shapes and rather constant slope profiles are imaged offshore the tributary rivers and creeks, respectively. The delta of the Gosaubach, entering the lake from the west, separates the deeper southern basin (max. water depth 124 m) from the shallower northern basin (max. water depth 45 m), forming a less than 100 m wide subaqueous sill at 22 m water depth (Fig. 1). The northern shallow basin is not entirely flat in its central part, but is separated by a SW–NE-trending ridge *c.* 15 m high above the flat basin floor (Fig. 1).

These bathymetric features define the four first-order elements of the new subaquatic landform map of Lake Hallstatt (Fig. 2). For classification we follow the concept of Harris *et al.* (2014), who introduced the seafloor geomorphic feature map for the geomorphology of the oceans. Yet, we use the term ‘subaquatic landforms’ instead of ‘submarine geomorphic features’, following terminology used by Micallef *et al.* (2018). To date there is no unifying classification scheme for geomorphic features or subaquatic landforms in lakes, so we freely introduce subclasses to further characterize differences within the first-order subaquatic landforms based on slope angle, lake-floor roughness, backscatter characteristics and occurrence of distinct elementary geomorphic features. In the following, these subaquatic landforms and their subclasses are described in more detail, with a special focus on morphological features indicative of subaquatic mass-movement processes in the southern Lake Hallstatt basin.

Subaquatic Landform 1 (SL-1 – basin floor). This category comprises the sediment-covered flat (<1°) lake floor located in the deepest parts of both the southern and the northern lake basins. In the southern deep basin, the flat lake floor exhibits two separate areas with different roughness and backscatter characteristics (geomorphic subunits SL-1a and SL-1b). While the lake floor of the deepest central basin is very smooth and shows low backscatter (white area in Fig. 3a; SL-1a), the slightly shallower basin floor towards the north and south (i.e. more proximal to the deltas of the Traun River and the Gosaubach) shows intermediate backscatter (Fig. 3a) and punctuated occurrence of several up to 1–2 m high and 10–20 m wide hummocks (Figs 3b & 4e; SL-1b). The hummocks have a rather smooth surface expression and exhibit slightly lower backscatter intensities than the surrounding lake floor. Their occurrence and shapes appear random, but there is a trend indicating that the hummocks become smaller and apparently smoother towards the deepest central basin. Thus, the boundary between SL-1a and SL-1b is transitional (dashed line in Fig. 2).

Subaquatic Landform 2 (SL-2 – lateral slopes). This category comprises the steep lateral slopes (>20°) and almost vertical subaquatic cliffs. Where slope angles are steeper than about 30°, the bathymetric data reveal high backscatter and rough surface textures that show comparable characteristics as the subaerial steep rock walls of outcropping well-bedded limestones of the Dachstein Formation (cf. onshore LiDAR map in Fig. 1 and subaquatic rock cliffs in Fig. 3b). In analogy, this characteristic geomorphic elementary feature (SL-2a) thus likely represents outcropping bedrock without lacustrine sediment cover. In contrast, many lateral slopes with typical slope angles of 20–26° exhibit less rough surface textures, and comparably lower backscatter intensities, and comprise geomorphic elements typical of erosion-like gullies and crests (e.g. Strupler *et al.* 2018; SL-2b). Backscatter is very high inside gullies located in subaerial steep graben and onshore talus slopes (Fig. 3a), likely indicating active sediment routing of detrital sand and gravel. Crests and surrounding slope areas show intermediate backscatter intensity, suggesting that the slope is covered by fine-grained sediments. Along the eastern and southern slopes, the crests and gullies do not reach up to the shore line, but

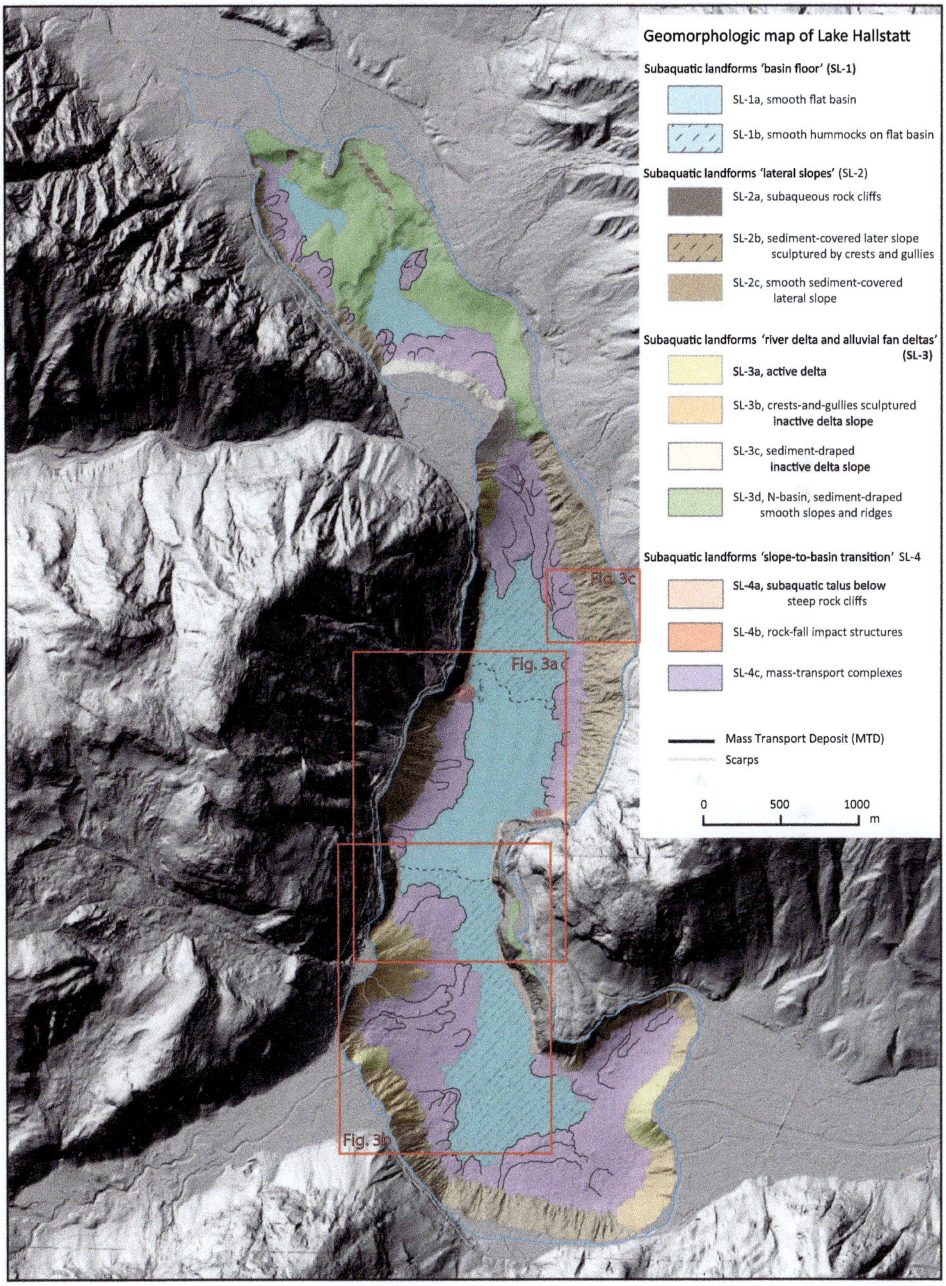

Fig. 2. New subaquatic landform map of Lake Hallstatt. The map shows the four main subaquatic landforms (SL, and their further subdivided subaquatic landform classes) resulting from our geomorphological analyses and interpretation of the new bathymetry map, based on slope angle, lake-floor roughness, backscatter characteristics and occurrence of distinct geomorphological features. Rectangles in red show locations of maps shown in Figure 3a–c.

SL-2b is separated from the uppermost slope area showing smoother surface textures (SL-2c) by few-metre high, coalescing concave-downward escarpments (Fig. 3c).

Several scarps on the slopes can be linked to lobe-shaped hummocky areas on the toe of the slope (Fig. 3c; see description of such 'mass-movement deposits' in SL-4c, below). The head scarps on the

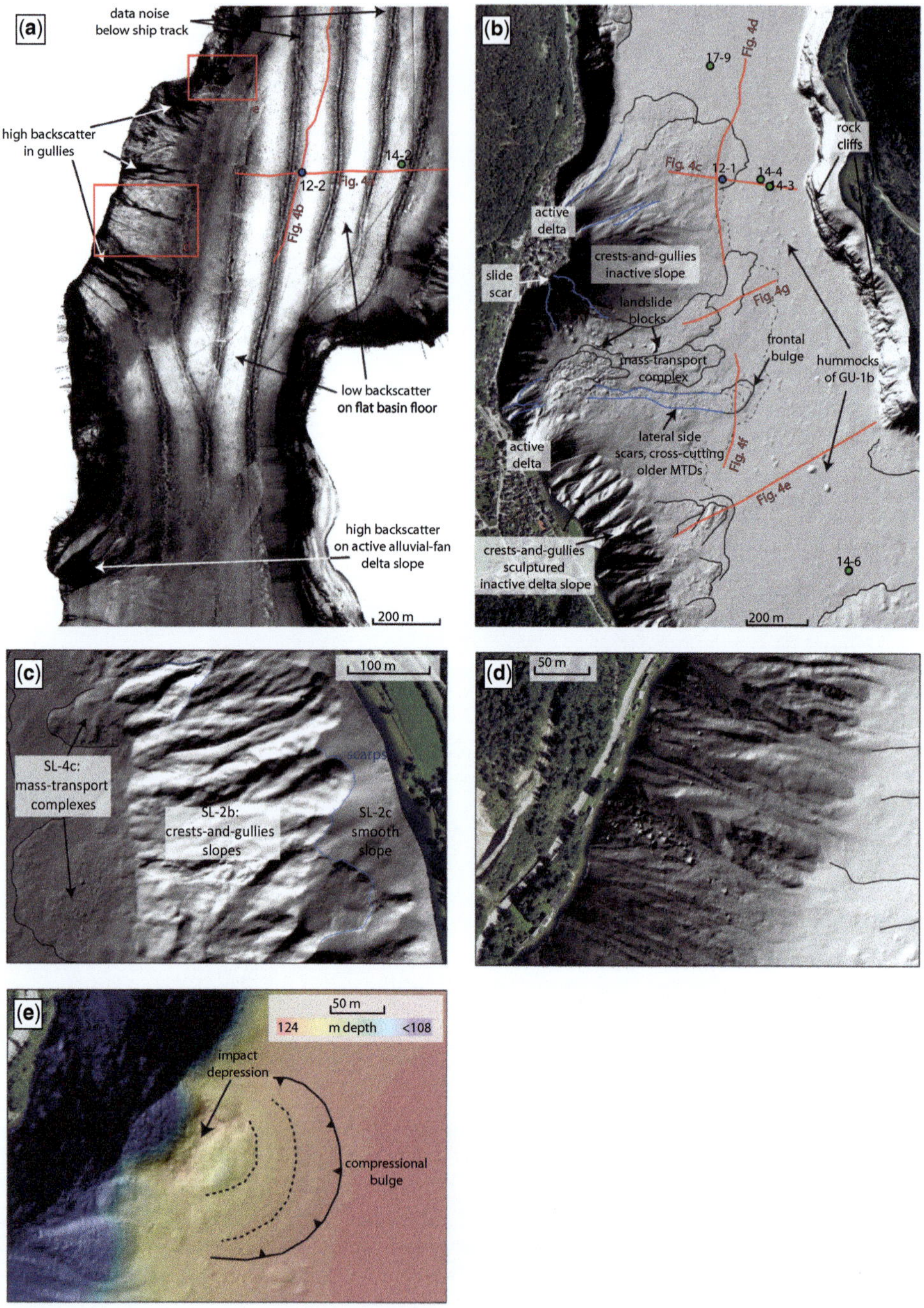

Fig. 3. Examples of geomorphic features of the lake floor of Lake Hallstatt. (**a**) Backscatter map showing low backscatter (white area) in the deepest central basin, while active deltas and gullies exhibit high backscatter (black). (**b–d**) Shaded relief from bathymetry: (**b**) offshore the active deltas of the Mühlbach and Waldbach (see Figs 1 & 2 for labelling of local tributaries and location of map, respectively); (**c**) along the eastern steep slope near the deepest part of Lake Hallstatt (see Fig. 2 for location), showing the differences between the gullies-and-crest sculptured (lower) slope (SL-2b) and the smooth upper slope (SL-2c); (**d**) along the western steep slope that contains blocks of a few metres in diameter. (**e**) Coloured bathymetric map overlain on shaded relief from bathymetry showing a prominent example of subaquatic landform SL-4b (rock-fall impact structure). Red lines in (a) and (b) show location of seismic profiles shown in Figure 4.

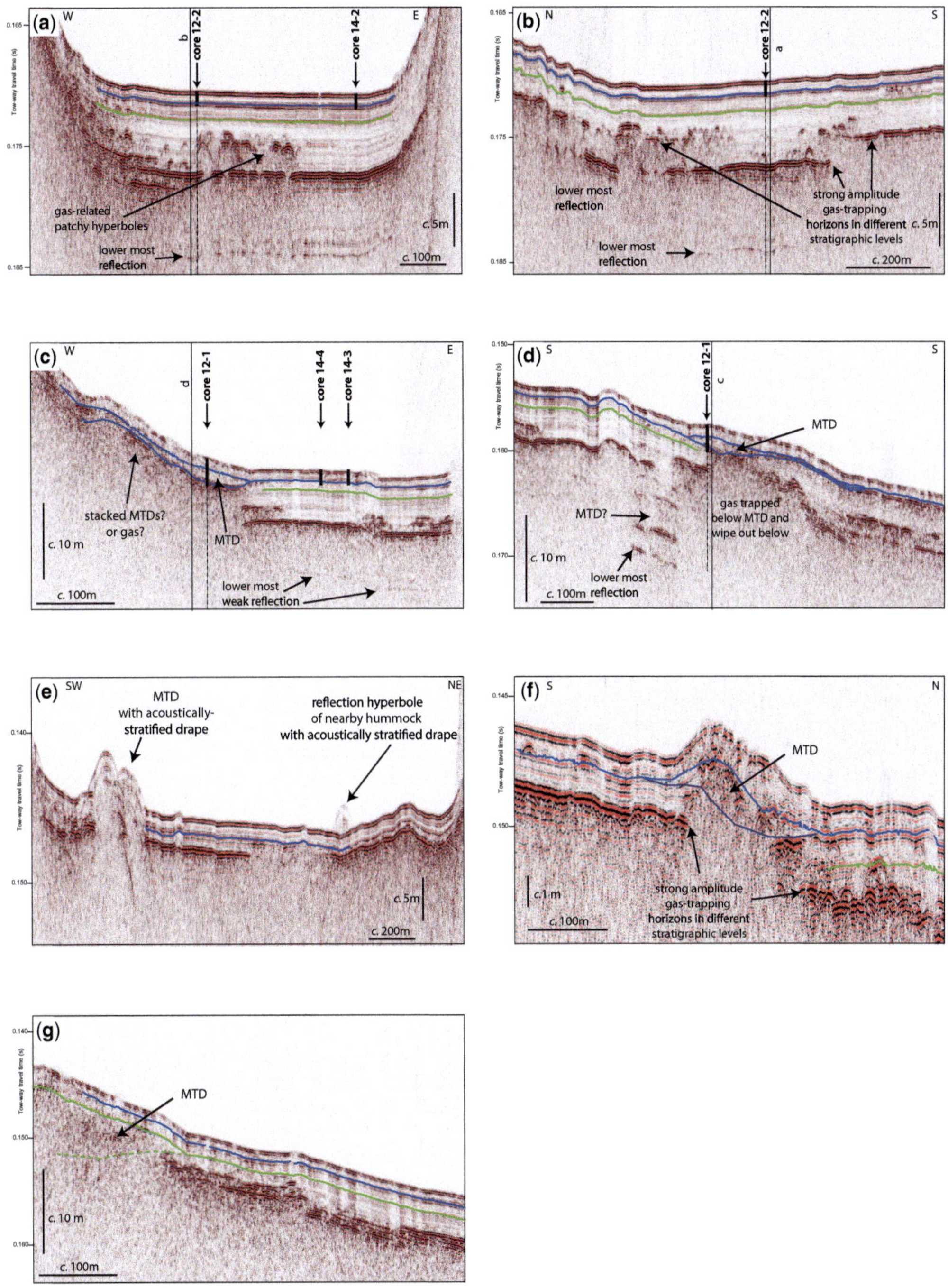

Fig. 4. Reflection seismic profiles showing typical subsurface characteristics of the uppermost sedimentary sequence of the southern deep Lake Hallstatt basin, as imaged by 3.5 kHz single-channel pinger data acquired in 2014. See Figure 3 for locations of seismic profiles. The blue and green lines show distinct seismic-stratigraphic horizons that can be mapped across the dataset and that correlate to mass-transport deposits (MTDs) linked to the AD 1895 earthquake (blue horizon) and – possibly – a historically reported shore collapse in AD 1808 (Urstöger 2000).

upper slope, however, are not sharp erosional features that would indicate very recent erosion; instead the lake floor across the scarps is rather smooth, suggesting lacustrine sedimentation draping these older geomorphic elements of subaquatic mass-movement processes (e.g. Strupler *et al.* 2018). Along the western slope, the slope of SL-2b, sculptured by crests and gullies, also contains blocks of a few metres in diameter (Fig. 3d), indicating that subaerial rockfall debris or waste from road construction are stable on this *c.* 30°-inclined sediment-covered lateral slope.

Subaquatic Landform 3 (SL-3 – deltas and subaquatic fans). Four different classes can be distinguished in the bathymetric data to describe the different morphometric and backscatter characteristics of Lake Hallstatt's deltas and subaquatic fans. Subunit SL-3a represents the active parts of the river and alluvial fan deltas, immediately offshore the river and creek mouths that today are mostly engineered. These depositional areas build cone-shaped positive reliefs and show downslope-decreasing slope angles with *c.* 30° in the proximal areas and flattening to 10° and less towards the basin. They exhibit the highest backscatter intensities where active sedimentation of coarser-grained sediment transported by the tributary rivers and creeks occurs (Fig. 3a). These active areas cover only a small part of the entire basinward-convex prograding delta of subaquatic fan systems, which, outside the currently active depositional areas, comprise comparable morphological characteristics to the slope of SL-2b, sculptured by crests and gullies (Fig. 3b). However, given their clear affinity to delta/fan sedimentation processes, they are assigned to SL-3b, and represent the currently inactive former delta lobes, which today experience hemipelagic-type sedimentation and are sculptured by gully erosion. Several scars exhibiting different seafloor roughness can be mapped, indicating delta-front slope instabilities and erosion by subaquatic mass-movement processes during different times in the past (Fig. 3b). Such processes are represented by scars with relatively smooth lake-floor roughness that are interpreted to be relatively older and covered by more recent sedimentation (cf. Strupler *et al.* 2018 and Fig. 4 and further description and interpretation below).

The larger delta of the Gosaubach shows some notable additional morphometric and geomorphologic features. The overall shape of the southeastern part of this delta does not exhibit the typical lakeward-convex curvature. Instead, it is characterized by a concave shape and rough lake-floor texture covered only by relatively small lakeward-convex cones of SL-3a. They occur at the location where the Gosaubach, which has been engineered since the late nineteenth century, enters the lake (Figs 1, 2). Towards the northern basin and the subaquatic sill, which separates the two main Lake Hallstatt basins, the typical appearance of the inactive delta slope, sculptured by crests and gullies, is significantly smoother compared to the inactive delta slope of the Traun River delta and lateral alluvial-fan deltas. This suggests that these slopes have not experienced erosion over longer time, but have been draped by settling of fine-grained lacustrine sediment particles (SL-3c).

Similarly, most of the gentle slopes of the offshore part of the Zlambach alluvial fan (Fig. 1), as well as the SW–NE-trending ridge of the northern shallow basin, exhibit very smooth surface textures, both categorized as SL-3d. Subtle diffuse morphological steps, with small downward-concave shapes, on slopes in the northern basin likely represent old lake-floor structures now mostly draped by Holocene lacustrine sedimentation.

Subaquatic Landform (SL-4 – slope-to-basin transition). This category comprises the gentle slopes and slope-break areas (10–1° inclination) defining the transition between the flat basin floor (SL-1) and the steep lateral slope and deltas (SL-2 and SL-3, respectively). This slope-to-basin transition unit exhibits the most diverse characteristics across the entire lake.

Located immediately below the subaquatic rock faces of SL-2a, the slope-to-basin transition is characterized by a sharp slope break only 20–30 m wide, with slope angles rapidly decreasing from 25 to 2°. The geomorphic elementary feature SL-4a is interpreted as subaquatic talus below the steep rock cliffs. In a few locations, the talus sediment morphology is interrupted by a depression, around which concentric bulges formed towards the basin (SL-4b). The most prominent example is located below the western slope (Fig. 3e). The depression is about 2 m deep, 40 m long and 15 m wide and is interpreted as an impact structure of a subaerial mass movement that resulted in deformation of basin-plain sediments to form the concentric bulges of the cone that extends about 75 m into the basin (cf. Van Daele *et al.* 2013; Daxer *et al.* 2020).

Apart from the rather few talus slope-break areas, most parts of the slope-to-basin transition zone are characterized by an uneven lake floor composed of several superimposed depositional lobes with hummocky surface texture (SL-4c), typical for subaquatic mass-transport complexes (MTC, e.g. Sammartini *et al.* 2019). The lobes are located downslope of scars (see SL-2b and SL-3b above), have positive relief and elongated shapes in the down-dip direction towards the deepest part of the lake (Fig. 3b) and/or may form frontal bulges against the basin sediments, typical for emergent v.

frontally confined lacustrine landslides, respectively (cf. Moernaut and De Batist 2011; Sammartini *et al.* 2019). The most prominent and distinct lobes also contain large blocks up to several tens of metres in width and up to 5 m in height (Fig. 3b). For many superimposed lobes, lateral boundaries and hummocky lake-floor areas, distinguishing individual deposits or uniquely identifying landslide blocks is not possible, mainly because the geomorphic structures are mostly smoothed and indistinct. This is likely due to the fact that younger sedimentation has draped the relatively older MTDs (Fig. 4e–g).

The sedimentary infill of Lake Hallstatt deep southern basin

Reflection-seismic data description and interpretation

Penetration of the seismic signal into the subsurface of Lake Hallstatt's sedimentary infill is limited by acoustic blanking, typical for sediments containing free gas (e.g. Tóth *et al.* 2014; Moernaut *et al.* 2020). In the northern basin, total wipe-out of seismic data immediately below the lake-bottom reflections precludes subsurface interpretation completely. Seismic data in the central part of the deep southern basin occasionally image up to 10 m subsurface depth (Fig. 4a, b), but mostly data are only coherent for the uppermost 4–5 m of the sedimentary infill, characterized by parallel and laterally continuous, low- to medium-amplitude reflections (Fig. 4). A strong reversed-polarity reflection at *c.* 5 m depth mostly limits further seismic penetration and suggests the occurrence of gas below this horizon (e.g. Tóth *et al.* 2014; Moernaut *et al.* 2020). Several patchy areas exist, where high-amplitude reflections with reversed polarity occur in shallower stratigraphic levels that are partly laterally continuous or hyperbolic and that show acoustic wipe-out below (Fig. 4). More proximal to the Gosaubach delta in the north and the Hallstatt alluvial fan and Traun River delta in the SW and SE, respectively, the high-amplitude reflection occurs successively in higher stratigraphic levels (near the SL-1 to SL-4c transition), until it coincides with the lake-floor reflection in proximal parts of the deltas (SL-4c to SL-3 transition). This reflection-seismic pattern is typical for distal-to-proximal transects of deltas that deposit higher amounts of organic material in proximal areas, which is decomposed to gas after deposition (e.g. Girardclos *et al.* 2007; Hilbe and Anselmetti 2014, Moernaut *et al.* 2020).

Within the acoustically parallel, well-stratified sequence of the deep basin, two horizons with slightly higher amplitudes at *c.* 1 and 2 m subsurface depth can be traced across the entire central part of the deep basin and also towards the south, where the lake-floor geomorphology is assigned to SL-1b and partly SL-4c (blue and green horizons in Fig. 4). The reflection pattern containing these two horizons can even be observed above interpreted landslide blocks (SL-4c) and smoothened hummocks of SL-1b (Fig. 4e), even though reflection profiles do not perfectly cross these features in a central position (cf. diffraction hyperboles in Fig. 4e). Therefore, seismic data suggest that these features are covered by at least 2 m of acoustically laminated lacustrine sediments.

The younger blue horizon stratigraphically correlates to the lateral pinchout points of an MTD imaged at the NE toe of the Hallstatt fan (Fig. 4c–d). In its most frontal part, the MTD shows a high-amplitude reflection at the base, overlain by a *c.* 1 m thick wedge-shaped body with internal chaotic acoustic facies. The lateral and frontal boundaries of this MTD imaged by seismic subsurface data nicely correlate with a prominent lobe with hummocky surface texture mapped in multibeam data (SL-4c, Fig. 3b). The MTD-overlying sequence shows more diffuse but still apparently parallel laterally continuous reflection patterns indicative of a rather undisturbed *c.* 1 m thick sequence deposited after the emplacement of the MTD. This overlying sequence can be traced further upslope towards more proximal areas of the SL-4 to SL-3 transition, but interpretation and separation of a single MTD body in the deeper subsurface becomes non-unique. Instead, the chaotic reflection pattern there might be indicative of several older amalgamated and superimposed sheet-like MTDs or, alternatively, of various degrees of gas content. A slope-parallel seismic profile crossing the distal part of this youngest MTD reveals clear evidence of accumulation of gas at the base of the MTD, while progressively deeper stratigraphic sequences are imaged below the areas not covered by the MTD (Fig. 4d). Moernaut *et al.* (2020) report similar occurrences of gas trapped below MTDs and interpret gas migration also from areas immediately adjacent to MTDs. Probably due to similar processes, seismic data reveal a small *c.* 100 m wide window that images the frontal part of the distal Hallstatt fan down to *c.* 10 m subsurface depth (Fig. 4d). In this seismic window, the sequence comprises a *c.* 4 m thick upper interval of acoustically parallel and laterally continuous reflections, overlying stacked bodies separated by high-amplitude reflections and internal laterally less continuous to chaotic seismic facies. This suggests the occurrence of older and thicker MTDs at the toe of the central part of Hallstatt fan that have been draped by normal lacustrine sedimentation. This is also suggested by the comparable smoother surface texture of this central part of the Hallstatt fan compared to the NE part affected by the youngest MTD (blue event) described above (Fig. 3b).

The stratigraphic position of MTDs within prominent MTCs in the southern part of the Hallstatt fan, characterized by large blocks and high lake-floor roughness (see above description of SL-4c and Fig. 3b), cannot be conclusively interpreted from reflection seismic profiles. This is likely due to the acoustic scattering of the seismic signal from this hummocky texture and limited seismic coverage in this area. A small wedge with a chaotic internal reflection pattern spatially linked – based on bathymetric interpretation – to the apparently youngest mass-flow with lateral slide scars and frontal compression bulges (Fig. 3b) might also correlate with the blue seismic-stratigraphic horizon (Fig. 4f). However, seismic interpretation remains non-unique because high-amplitude gas horizons in different stratigraphic levels do not allow one to track the seismic-stratigraphic horizons across the entire basin. Nevertheless, the blue horizon can be clearly tracked above the older surrounding MTDs, suggesting these landslides are older than the blue MTDs. The green horizon potentially could be the youngest horizon, immediately overlying the chaotic reflection pattern below the area mapped as an MTD in Figure 4g. However, seismic interpretation at the inferred pinchout point remains non-unique because of gas accumulation below a deeper high-amplitude reflection.

Sediment sequence and MTDs observed in cores

The uppermost metre of the sedimentary sequence in the central part of the deep basin, as recovered by cores 12-2 and 14-2, consists of millimetre- to sub-millimetre-scale laminated fine-grained carbonate mud with frequent intercalated centimetre-scale graded turbidites (Figs 5, 6). The sediments are predominantly composed of detrital silty carbonates, clays and a few diatom frustules with varying amounts of organic material. The fine-grained laminae show various types of colours and there are no systematically recurring couples or patterns that may suggest seasonality, even though the sediments reflect variable detrital input from the tributaries.

Composite sediment core 12-1, retrieved from the northeastern toe-of-slope of the Hallstatt fan within SL-4c, in an area where reflection-seismic data have imaged the MTDs correlated with the blue horizon (Fig. 4 c, d), shows a similar sequence of laminated carbonate mud from the core top down to 67 cm sediment depth (Fig. 5). The lamination is more diffuse and, similar to the other cores, the sequence also contains three prominent yellowish–light grey, 1–2 cm thick layers with clear grading-upward trends (Figs 5, 6). Below 67 cm, a 1.13 m thick MTD has been recovered (Fig. 5), composed from bottom to top of (i) a 41 cm thick interval of dark grey mixed mud and poorly sorted fine to coarse sand with matrix-supported texture, including common centimetre-scale soft mud clasts and gravels, grading upwards into (ii) a 19 cm thick interval of very dark grey mud with contorted and disrupted diffuse light grey laminae. The overlying sequence consists of (iii) a rather sharp MTD-internal boundary characterized by a 0.5 cm thick disrupted dark sand layer and (iv) an overlying 1–2 cm thick, light grey carbonate mud, which is partly intruded by the underlying sand (micro-flame structures; Fig. 5). The uppermost part (v) is a 45 cm thick mixed fine-grained carbonate mud overlain by (vi) a 2 cm thick dark grey layer with patchy occurrence of terrestrial organic material and sand at the top. These textures are typical for distal parts of subaquatic delta-slope failures, mixing granular (proximal) to cohesive (distal) sediments by rather cohesive mass-movement transport and deposition (Hilbe and Anselmetti 2014; Sammartini *et al.* 2019). These processes corroborate interpretation of multibeam bathymetry and reflection-seismic data interpretation described above. The rather sharp MTD internal boundary separating interval (ii) from (iii) might either indicate multiphase mass-flow pulses or MTD internal deformation during emplacement (e.g. Laberg *et al.* 2016).

Stratigraphic correlation and age control

In order to test for stratigraphic correlation between the core recovered from the MTD (12-1) and the basin core with age control (12-2), we further describe in more detail granulometry and geochemistry (i.e. XRF-based correlation proxy ln(Ca/Fe)) of potentially event-stratigraphic marker layers from short cores collected along a proximal-to-distal coring transect. Visual correlation between core 12-1 and the short core 14-3, about 200 m distant (Fig. 4c), is robust for the two prominent yellowish–light grey layers that occur 5 cm above the top of the MTD, and most reasonable for the *c.* 1 cm thick grey sand layer on top of the MTD (Fig. 6). The correlative turbidite layer in core 14-3 also exhibits a dark grey colour, shows pronounced normal grading from sand to silt and the ln(Ca/Fe) values from XRF core scanning are slightly lower than the background sediments between the event layers. Two other slightly coarser-grained dark layers below this turbidite in core 14-3 might potentially correlate with two sand layers below the base of the MTD in core 2012-2 (see base of section B1u in Fig. 5), but interpretation of the correlation is non-unique.

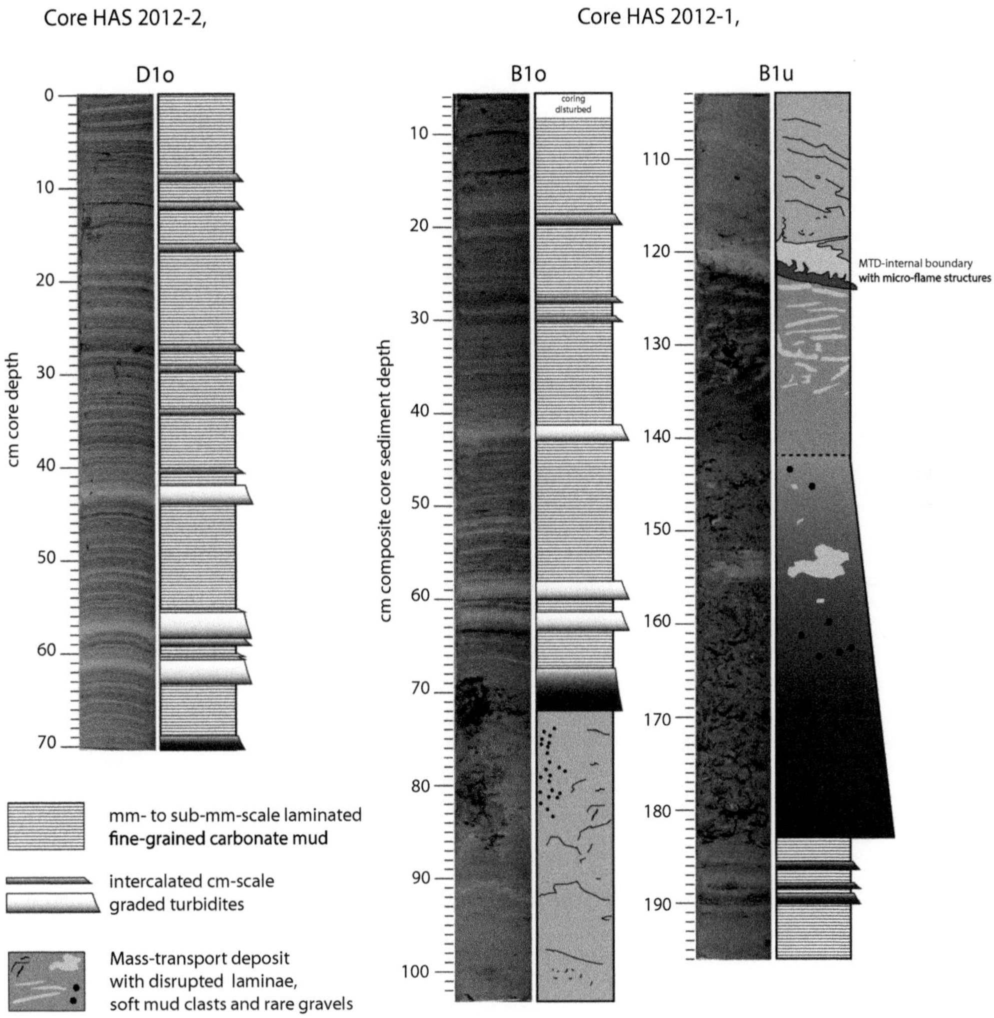

Fig. 5. Core photographs and lithologs of piston core segments HAS_2012-1 (B1o, B1u) and HAS_2012-2 (D1o). See Figures 1 and 4 for core locations (labelled as 12-1 and 12-2).

In contrast to the MTD-correlative turbidite in core 14-3, the two yellowish–light grey layers show a distinct ln(Ca/Fe) signal with very high values at the sharply defined base and decreasing values towards the top, mimicking both the clear fining-upward trend, as confirmed by granulometry, as well as likely decreasing calcium and inferred detrital carbonate contents. The two investigated yellowish–light grey event layers in core 14-2 from the depocentre of the deep basin, show overall very similar grain-size characteristics and ln(Ca/Fe) values, although the event deposits in this more distal position are internally more complex, showing internal lamination. The overall fining-upward texture and high ln(Ca/Fe) values compared to the background sediment corroborate stratigraphic correlation (Figs 6, 7). The character of the MTD-correlative turbidite, however, is distinctly different in this more distal position. The thickness of the layer has remained constant (*c.* 1 cm), but its internal character both in terms of the ln(Ca/Fe) proxy and grain size is best described as homogeneous. Apart from slight variations in the lowermost millimetres at the base of the event layer, ln (Ca/Fe) is entirely constant, with absolute values close to the average of the over- and underlying laminated background sediments that, however, show higher ln(Ca/Fe) variability. Grain size is constant in the lower part of the deposits, with absolute values slightly above the average grain size of the background sediment

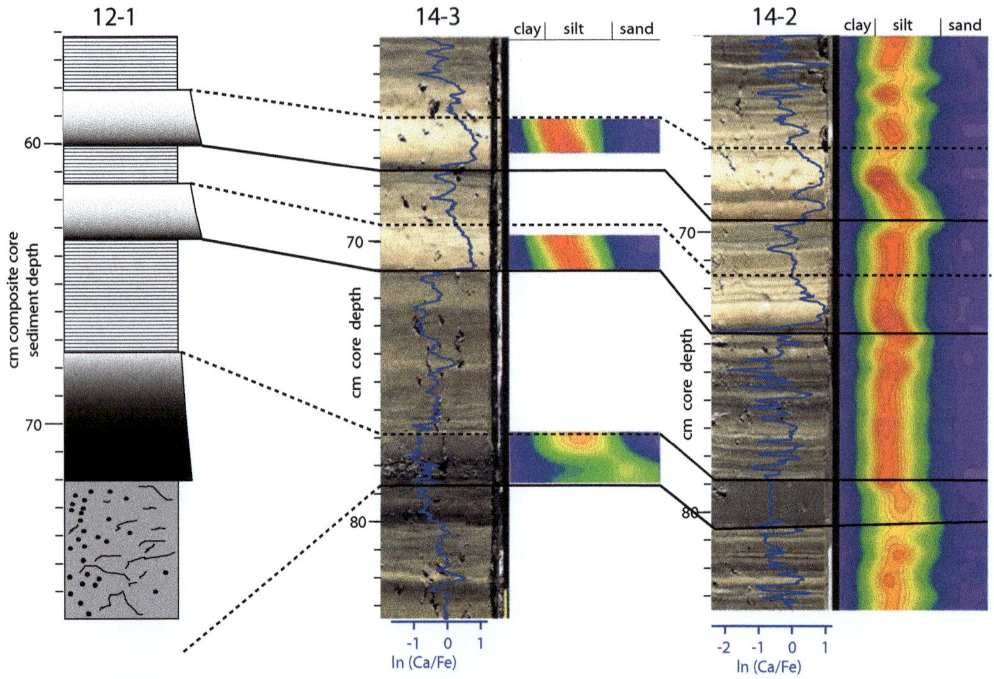

Fig. 6. Characteristics of event layers. Core-to-core correlation of characteristic marker layers (i.e. two prominent yellowish–light grey layers; and an MTD with its correlative turbidite) between cores 12-1, 14-3 and 14-2; based on litholog, XRF-based correlation proxy ln(Ca/Fe) superimposed on core images, and results from laser-diffraction granulometry measurements shown as relative fractions of clay, silt and sand. For legend of symbols used in lithology, see Figure 5.

and a fining-upward trend of only a few millimetres thick towards the top.

Using the distinct correlation proxies of the three event deposits describe above (i.e. MTD-correlative turbidite and homogenite, and the two graded yellowish–light grey detrital carbonate turbidites) as well as the third, younger, graded detrital carbonate layer, a basin-wide core-to-core correlation for the uppermost metre of the sedimentary infill of the southern deep Lake Hallstatt basin can be established (Fig. 7). This succession can be dated using the results from radionuclide dating of core 12-2. The ^{137}Cs profile in this core shows a well-defined peak at 6 cm sediment depth and a more subtle and broader peak around 19–20 cm sediment depth (Fig. 7), which are correlated with the known ^{137}Cs activity peaks related to radioactive fallout from the Chernobyl reactor accident in AD 1986 and the maximum of atmospheric nuclear weapon testing in AD 1963, respectively (e.g. Appleby 2000). Subtracting small turbidites at 9, 12 and 16.5 cm sediment depth, the event-free average sedimentation rate between 6 and 20 cm core depth amounts to 0.43 cm a^{-1}. We estimate uncertainty to $\pm 10\%$ due to uncertainty in event-layer thickness subtraction and core depth reading for the interval. We do not consider the uppermost part of the core when calculating sedimentation rates, as coring disturbance in the uppermost part of core 12-2 is rather strong. Down-core extrapolation using a constant event-free bulk sedimentation rate, subtracting single event layers >0.5 cm in thickness, reveals estimated ages for the three light grey, graded, carbonate layers and the MTD-correlative homogenite in core 12-2 to be AD 1919 $\pm$ 4, AD 1897 $\pm$ 6, AD 1894 $\pm$ 7 and AD 1890 $\pm$ 8. Assuming more or less constant sedimentation rates of 0.43 cm a^{-1} over the last *c.* 130 years is justified by considering the three radiocarbon dates of ‘modern age’ (i.e. younger than 300 cal years BP) obtained from samples down to 278 cm sediment depth (150 cm event-free composite depth) in core 12-1 (Table 1), also suggesting average sedimentation rates in the order of 0.42 cm a^{-1} over the last *c.* 350 years.

Event-stratigraphy and processed-based interpretation of recent mass movements

Estimated ages of the prominent three yellowish–light-grey, graded, event layers correspond to the

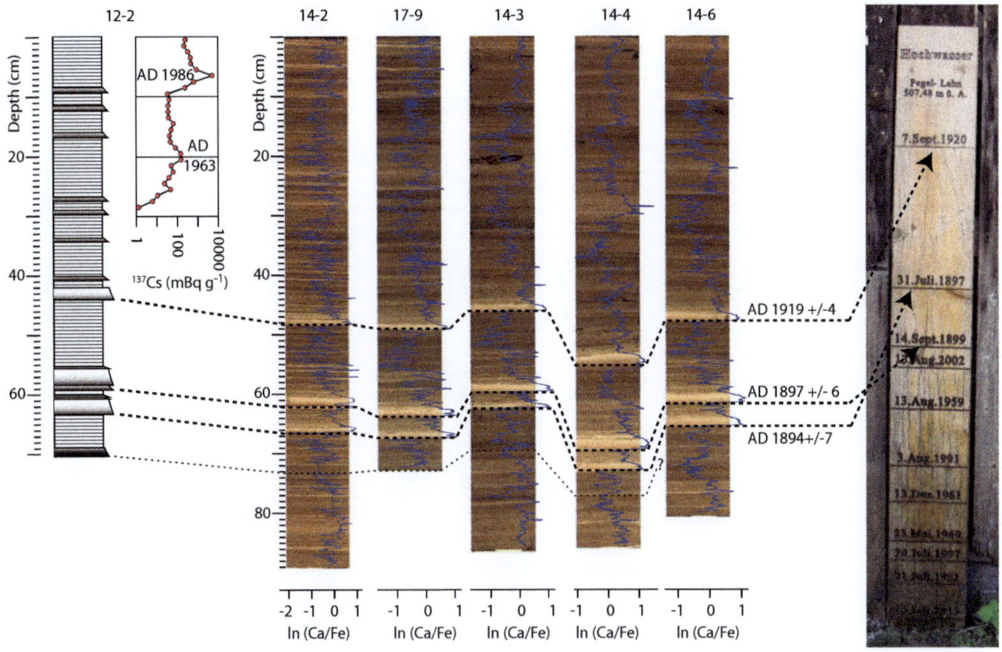

Fig. 7. Lake Hallstatt event-stratigraphy. Core-to-core correlation and dating of the uppermost sedimentary sequence of Lake Hallstatt by gamma spectrometric measurements and historically reported flood event. From left to right: litholog of core HAS_2012-2 (D1o, for legend of symbols used in litholog see Fig. 5); results from gamma spectrometric measurements showing well-defined peaks, which are correlated to the known ^{137}Cs activity peaks related to radioactive fallout from the Chernobyl reactor accident in AD 1986 and maximum of atmospheric nuclear weapon testing in AD 1963, respectively (e.g. Appleby 2000); XRF-based correlation proxy ln(Ca/Fe) superimposed on core images; and photograph of memorial flood marks in Hallstatt, revealing that the three yellowish–light grey, graded, event layers correspond to the dates of highest lake levels of Lake Hallstatt throughout history in AD 1920, 1899 and 1897.

dates of highest lake levels of Lake Hallstatt throughout history in AD 1920, 1897 and 1899, as documented by the memorial flood marks (Fig. 7). Based on the ln(Ca/Fe) proxy and sedimentary structure of these event deposits (i.e. internal lamination in the basal part and grading), they are interpreted to represent major detrital carbonate input by the tributary rivers and deposition from hyperpycnal flows during high-discharge periods (cf. Mulder and Chapron 2011), which is consistent with interpreting deposition during high lake levels. Spatially plotting the grain size D90 values of the basal coarsest sands in one representative event layer further corroborates this interpretation (Fig. 8). The basal sand of this event layers is coarser (D90 ~50 µm) in a more proximal location to the main inflow of the Traun River and finer toward the distal terminal depocentre in the deepest central basin (D90 <30 µm). This suggests that the dominant source of detrital carbonates during such flooding/high lake levels is the Traun River in the SE with a northward particle transport into the deep basin (black arrows in Fig. 8).

In contrast, D90 values of the MTD-correlative turbidite, which has not been positively identified in core 16-6 taken in the basin south of the Hallstatt fan, is more than 100 times coarser than D90 values of the light grey flood layers proximal to the MTD source of the Hallstatt fan and significantly fines within a short transport distance towards the depositional centre of the deep basin (Fig. 8). Furthermore, the ln(Ca/Fe) proxy of this MTD-related event layer suggests remobilization and mixing of lake-internal sediments rather than additional detrital carbonate input.

Given that the yellowish–light grey, graded, event layers can be used as an event-stratigraphic marker of known age, we can improve the estimates of sedimentation rate for the lower part of the cored sedimentary succession. In the main depositional centre of the deep basin (cores 12-2 and 14-2), there is about 1.5–2.0 cm of background sediment between the marker layers of the high lake level stands of summer 1897 and 1899. This suggests that sedimentation rates were slightly higher (*c.* 0.75–1.0 cm a^{-1}) around the end of the

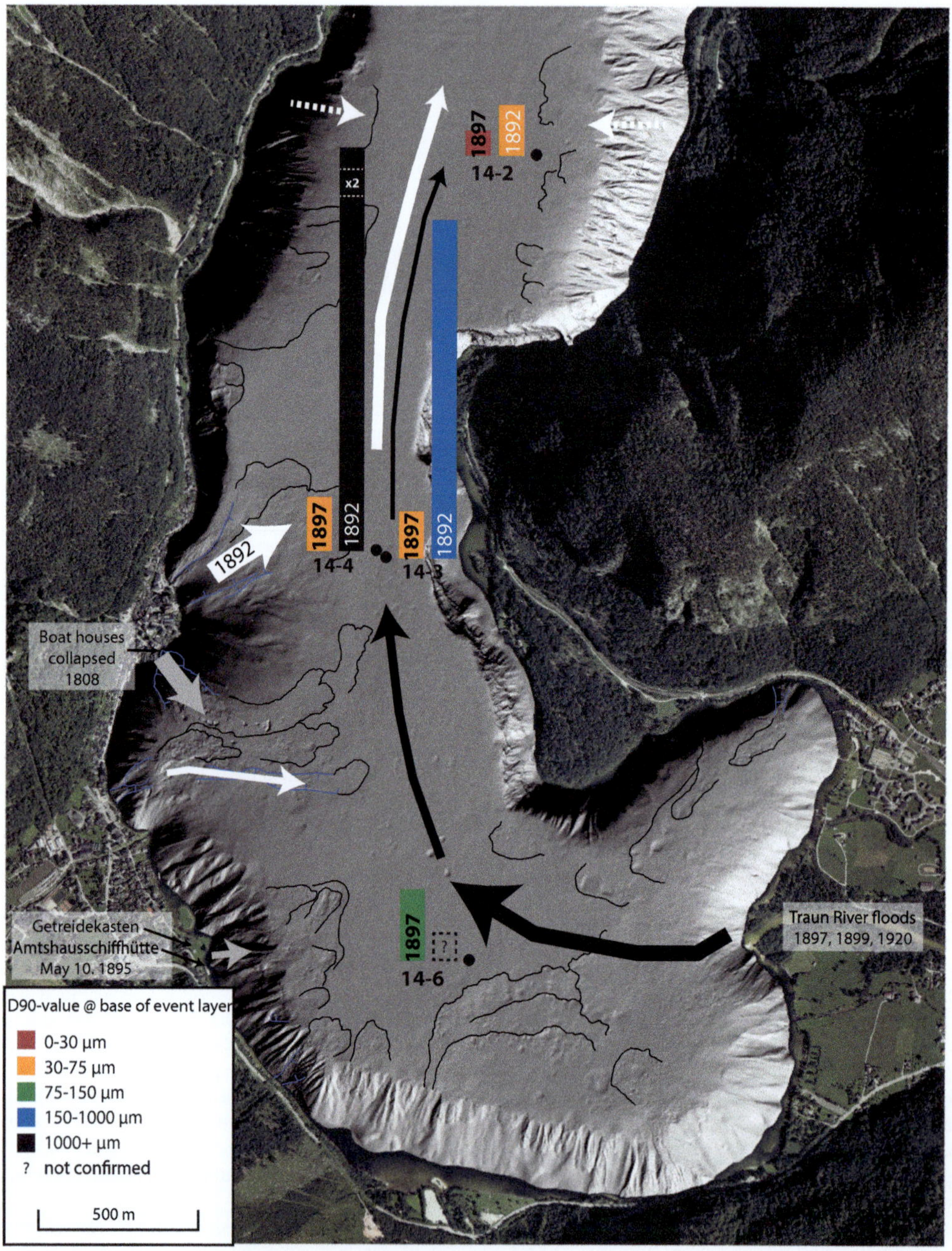

Fig. 8. Summary map of mass-movement processes in Lake Hallstatt over the last *c.* 211 years. Shaded relief from bathymetry with orthophoto of the lake's surroundings. For each core 14-2, 14-3, 14-4 and 14.6, we plot the D90 value from laser-diffraction granulometry measurements of samples taken from the base of the 1897 flood-related (left) and 1892 earthquake-triggered (right) turbidite layers, in order to illustrate the spatial distribution of the sandy base of the respective marker beds. Both the height (relative scaling) as well as the colour code (absolute values, see legend in the lower left corner) indicate D90 values of basal sand samples. Black and white arrows indicate transport direction of sediment transport as interpreted from grain-size data (i.e. from coarser to finer deposits) for the 1897 and 1892 event layer, respectively. Additional dashed-white arrows indicate likely contribution of earthquake-triggered surficial sediment remobilization along the later sediment-covered slopes. The grey arrows locate transport direction of the mass movement linked to historically documented shore collapse events in AD 1895 and 1808 (Urstöger 2000).

nineteenth century compared to the long-term average sedimentation rates estimated from radionuclide dating. This could be explained by higher frequency of mass movements in the catchment in the late nineteenth century, and/or the fact that in the late nineteenth and early twentieth century tributary rivers were heavily engineered (Pürstinger 2017). Using the AD 1897 event layer as a tie point and assuming a sedimentation rate of *c.* 0.75–1.0 cm a^{-1} also for the *c.* 5 cm of background sediment between the AD 1897 event layer and the MTD-correlative turbidite, we can narrow the age estimate for the MTD in core 12-1 to around AD 1891–92.

The Hallstatt chronical (Urstöger 2000) reports that people felt the ground shaking on 27 July 1890, 27 July 1892 and during the night of 9–10 May 1895, suggesting the occurrence of an earthquake series, which, however, lacks verification and analysis in terms of robust historical seismology (e.g. Hammerl 2017). There is also no significant earthquake reported in the database of the European Archive of Historical Earthquake Data (AHEAD; Locati *et al.* 2014) that would match these events. If not misinterpreted in the chronicle, these earthquakes thus must have been local and one or the other could also just link to mining activity or other natural phenomena that might cause local ground motion. Beyond the description that ground motion was felt (which would indicate macroseismic intensities around III to a maximum of V on the Modified Mercalli Intensity (MMI) scale), in just the 27 July 1892 event further descriptions of realistic macroseismic effects (i.e. the bell ringing and furniture moving in the houses) are reported (Urstöger 2000). This substantiates that at least the 27 July 1892 event was an earthquake that caused macroseismic intensities of VI in Hallstatt. This value is within the lowermost range of intensity thresholds for initiating earthquake-triggered subaquatic slope instabilities and lacustrine mass-movements, as established in other lakes in the Alps (Wilhelm *et al.* 2016), Chile (Moernaut *et al.* 2014; Van Daele *et al.* 2015) and New Zealand (Howarth *et al.* 2014). Therefore, and given that the estimated age of the MTD and its time-correlative turbidite also suggests the mass movement to have occurred around 1891–92, we interpret this MTD to be genetically linked to the 27 July 1892 earthquake and infer subaquatic slope instabilities on the Hallstatt alluvial fan delta initiated by earthquake shaking of macro seismic intensity ~VI.

Furthermore, seismic-stratigraphic interpretation reveals that also the youngest flow south of the Hallstatt fan (also assigned to the blue stratigraphic horizon (Fig. 4f) and thus coeval (within seismic resolution) with the AD 1992 earthquake-triggered MTD on the northeastern slope of the Hallstatt fan) probably genetically links to this earthquake. This interpretation is supported by the fact that the typical signature of multiple coeval MTDs in lakes is proven to be a good earthquake indicator (Schnellmann *et al.* 2002; Strasser *et al.* 2013; Praet *et al.* 2017).

An interpretation of earthquake relation is further independently supported by the distinct chemical and textural fingerprint of the correlative turbidite and its spatial distribution across the lake. The thickness of the AD 1892 turbidite in the deeper basin is about the same as immediately overlying the MTD in its proximal location. Although this might partly be explained by the typical ponding of turbidites in enclosed basins (e.g. Vandekerkhove *et al.* 2019), the long distance between the fan and the wide depocentre in the basin further suggests an additional sediment source for the MTD-correlative turbidite in the deep basin. Furthermore, the homogeneous chemical composition of the event layer in the basin corresponds to mixed background sediment and reveals lake-internal sediments as the dominant source. The same 'chemical fingerprint' and lake-internal sediment source from lateral hemipelagic slopes has been documented in other lakes to be a unique characteristic of earthquake-triggered sediment remobilization and allows us to distinguish it from event layers related to floods or resulting from spontaneous delta-slope failure or subaerial mass movements into the lake (e.g. Simonneau *et al.* 2013; Moernaut *et al.* 2014; Van Daele *et al.* 2015). In Lake Hallstatt, such sediment-covered 'hemipelagic' slopes are typical for subaquatic landform SL-2b on the eastern slope of the deep basin (Figs 2, 3c). Yet, the slope-to-basin transition (SL-4c) there does not comprise young MTDs that would correlate to the blue seismic-stratigraphic horizon in reflection-seismic data. We therefore speculate that the earthquake also triggered additional, spatially more extensive, surficial sediment remobilization along the sediment-covered lateral slopes (cf. Moernaut *et al.* 2017; Molenaar *et al.* 2019; dashed white arrows in Fig. 8), which would explain the observed chemical and textural characteristics and thickness distribution of the AD 1892 event bed.

The Hallstatt chronicle further reports that during the night of 9–10 May 1895 parts of the shore collapsed into the lake (locations 'Getreidekasten' and 'Amtshausschiffhütte'; see Fig. 8), followed by another shore collapse on 12 May (Urstöger 2000). However, the sedimentary record of the central deep basin only comprises one obvious event layer that we link to the AD 1892 earthquake (see above). This implies that the shore collapses in AD 1895 did not initiate turbidity currents to form distinct event layers that can be correlated across the basin. We thus interpret that subaquatic mass movements linked to the AD 1895 shore collapses might have been rather slow-moving processes (as also suggested by the historical report, indicating a duration of instability of several days). If the ground motion felt during the night of 9–10 May 1895 was, indeed, caused by an earthquake (and not just related to the slow-moving retrogressive mass movement process itself that may also have caused ground motion strong enough to have been felt by people), then the earthquake was not strong enough to trigger sediment remobilization on the lateral slope of Lake Hallstatt, because no characteristic event deposit is observed in cores at the respective stratigraphic position.

Older mass movements and estimating relative ages from geomorphic indicators

Analyses of the multibeam bathymetry and interpretation of the reflection-seismic data further reveal several striking evidences of older mass-movement processes, which presently cannot be verified and dated based on available core data. Based on seismic-stratigraphic mapping, we hypothesize that the landslide on the southern slope of the Hallstatt fan with several large blocks (Figs 3b, 4g) correlates with the green seismic-stratigraphic horizon at *c.* 1.75 m subsurface depth in the central basin. This horizon occurs roughly twice as deep as the blue horizon dated to AD 1892. The Hallstatt chronicle reports two swells on the lake with amplitudes of 3–4 feet and a period of 8 minutes on 21 July in 1808 (Urstöger 2000). One month later, on 24 August, the same phenomenon occurred accompanied by strong ground motion and the eventual collapse of three boat houses into the lake. The source location of this historically documented shore collapse is consistent with the inferred source area of the MTD based on multibeam data, and the large blocks are of comparable size to the historical boat houses, supporting interpretation that these events are genetically linked (Fig. 8). An age of *c.* 210 years for the green horizon would also be consistent with extrapolating a constant sedimentation rate between the blue and the green horizons. The trigger and kinematics that initiated the subaquatic slope instability remain unknown and require validation by slope stability, landslide initiation and tsunami (swell) modelling. We speculatively link initiation to the first swell reported in the chronicles and hypothesize slow retrogressive instability that eventually led to a shore collapse one month later. Interestingly, we note a report by Alexander van Humboldt describing the possible occurrence of an earthquake in 1808 (Schultes 1809). Yet the occurrence of an earthquake in the Hallstatt area in AD 1808 lacks verification and analysis in terms of robust historical seismology (e.g. Hammerl 2017), and the reported ground motion could also be due to the mass-movement processes itself, but misinterpreted as earthquake-induced shaking. Given that our calibration of the sedimentary event-stratigraphy for the sediments of the last 125 years suggests distinct proxies that allow us to distinguish earthquake-triggered turbidites in the terminal basin (i.e. distinct homogeneous facies both lithologically as well geochemically by means of ln(Ca/Fe)), future analyses of long cores from the deep central basin might allow testing of whether the mass movement was triggered in 1808 by an earthquake and if there are possible signatures for the documented swell events.

Similarly, analyses of longer cores are needed for validating interpretation of the deeper subsurface imaged by 3.5 kHz reflection-seismic data, which shows some evidence of possible older and larger MTDs within the few seismic windows (Fig. 4d). Evidence for older large-scale basin-wide mass-movement processes affecting the entire basin is further provided by the smoothed–hummocky texture of SL-1b (Figs 2, 3b). As indicated in the seismic data, these hummocks are draped by acoustically stratified sediments (Fig. 4e) of *c.* 4 m background sedimentation (or more than *c.* 800 years extrapolating recent sedimentation rates). Due to sedimentation by over- and interflow (draping) and underflow (ponding around the hummocks), the hummocks today have very smooth surfaces, while likely the initial shapes exhibited rougher, probably blocky shapes, typical for blocks mobilized by large landslides and rafted into the basin. Initial size of the blocks is larger than that measured in bathymetric data on the buried and smoothened hummocks. We thus can speculate that large subaquatic landslide blocks occupied large parts of the basin (at least SL-1b), suggesting the occurrence of past landslides of significantly larger magnitude than the documented MTDs of the last 210 years. Whether or not such mega mass movements might link to large-scale collapses of the Traun River or Gosaubach deltas, and/or multiple and potentially earthquake-triggered landslides from several sediment-covered slopes, remains unknown and needs further study. The peculiar concave shape of the southern part of the Gosaubach delta might indeed represent such an ancient large delta-collapse that merits further study.

Conclusion

In summary, our study draws a number of conclusions.

- The new geomorphological map of Lake Hallstatt (Fig. 2) presents the first-ever basin-wide subaquatic geomorphologic characterization of a steeply incised valley-occupying intra-mountainous lake in the eastern Alps.

- Several types of subaquatic mass-movement deposits with clear surface expressions in the bathymetry occur predominantly in the slope-to-basin transition between the flat basin floor and the steep lateral slopes and deltas. The most recent MTDs, as imaged in reflection-seismic data, relate to historically reported events in AD 1992 (moderate to strong earthquake (MMI: V–VI)) and AD 1808 (shore collapse and possible light to moderate earthquake (MMI: III–V)). This suggests that for Lake Hallstatt an earthquake-trigger might be a key process for initiating subaquatic slope instabilities, subsequent landslides and mass transport into the deep basin. This is an interesting finding for subaquatic mass-movement processes in such steeply incised valley-occupying intra-mountainous setting, where different types of mass-movement processes, including subaerial rock falls and debris flows, entering the lake are expected (and documented). These processes, however, result in smaller MTDs compared to those induced by earthquake-triggered subaquatic mass movements.
- The resulting deposits of the earthquake-triggered event layer (in comparison to flood-related turbidites) in the deepest depocentre of the lake exhibit distinctly different characteristics in terms of sedimentary facies, grain size and geochemical fingerprint (ln(Ca/Fe)):
- 'Flood layers': three prominent, light grey turbidite layers show high ln(Ca/Fe) values (documenting enhanced detrital carbonate input) and fining from the main river source in the SE to the distal depocentre in the north. Age estimation based on gamma spectrometric measurements corroborates our event-stratigraphic interpretation, which links these layers to the three highest lake-level stands and major flood events reported in history in AD 1897, 1899 and 1920.
- 'Earthquake layers': the AD 1892 earthquake event layer in the basin core shows very homogeneous sediment, which – based on the XRF-correlation proxy ln(Ca/Fe) – suggests mixing of lake-internal sediments. This implies that this typical earthquake fingerprint of earthquake-triggered lake-internal sediment remobilization, as established for mass movements on sediment-covered lateral slopes triggered by strong earthquakes in peri-alpine piedmont lakes also holds for slope failures in clastic-dominated intra-mountainous lakes. Given that the AD 1892 earthquake is estimated to have reached macroseismic intensities of V–VI in Hallstatt, seismic threshold conditions for initiation of subaqueous mass movements established in other settings are confirmed.
- Smoothed, hummocky surface textures in the flat basin in the northern and southern part of the deep Lake Hallstatt provide evidence of even larger mass-movements processes in the past. Due to very high sedimentation rates (*c.* 0.5 cm a^{-1} for the last 300 years) and limited seismic penetration below 4 m subsurface depth, currently available data do not allow us to investigate the origin and nature of these 'mega' MTDs, which must be older than at least *c.* 800 years. Given the size and spatial distribution of the interpreted landslide blocks, however, these older mass movements must be significantly larger than anything known from the written history of the Lake Hallstatt area.
- The above results and remaining questions suggest the high potential of Lake Hallstatt as a natural laboratory for studying causes and consequences of subaquatic landslides in steeply incised valley-occupying intra-mountainous lakes. In particular the very high sedimentation, suggesting high activity in proximal delta areas, bears potential for future studies focusing on monitoring delta-slope processes. The high sedimentation rate also implies high-resolution sedimentary archives that can be studied further back in time by long-core analyses to test the emerging hypothesis linking the inferred 'mega' MTD from our bathymetry data to prehistoric catastrophic landslides in the salt mining area (Rohn *et al.* 2005). Last but not least, process-understanding gained from this and future studies can also be of value to further submarine mass-movement research of other intra-mountainous lake and fjord basins with comparable settings.

Acknowledgements We acknowledge all helpers during field campaigns on Lake Hallstatt, as well as Georg Schettler for the gamma spectrometric measurements at GFZ Potsdam. IHS Markit Kingdom is acknowledged for their educational grant programs providing seismic interpretation software. We are very grateful to Maarten Van Daele and an anonymous reviewer for very useful comments that helped to improve an earlier manuscript. Many thanks also to volume co-editor Sara Benetti for handling and editing our manuscript.

Funding This research is supported by the Austrian Academy of Sciences (ÖAW Project 'Face Alps' to KK and ÖAW Project 'S^4LIDE-Austria' to MS), the Austrian Science Fund (FWF, Project Nr. M1907-N34 to SL) and the University of Innsbruck (Nachwuchsförderung 2018, W-171508 to J-JSH).

Author contributions **MS**: conceptualization (lead), investigation (equal), writing – original draft (lead); **TB**: investigation (equal), visualization (equal); **SF**: investigation (equal), writing – review & editing (supporting); **MH**: investigation (equal), writing – review & editing (supporting); **J-JSH**: investigation (supporting), methodology (supporting), writing – review & editing (supporting); **SL**: investigation (equal), methodology (equal), writing – review & editing (supporting); **MO**: investigation (supporting), methodology (supporting), visualization (supporting), writing – review & editing (supporting); **HR**: investigation (supporting), writing – review & editing (supporting); **AB**: funding acquisition (supporting), investigation (supporting), resources (supporting); **FA**: conceptualization (supporting), funding acquisition (supporting), investigation (supporting), project administration (supporting), supervision (supporting), writing – review & editing (equal); **KK**: conceptualization (equal), funding acquisition (lead), investigation (supporting), project administration (lead), resources (equal), writing – review & editing (supporting).

References

Appleby, P.G. 2000. Radiometric dating of sediment records in European mountain lakes. *Journal of Limnology*, **59**, 1–14, https://doi.org/10.4081/jlimnol.2000.s1.1

Bellwald, B., Hjelstuen, B.O., Sejrup, H.P. and Haflidason, H. 2016. Postglacial mass movements and depositional environments in a high-latitude fjord system. *Marine Geology*, **379**, 157–175, https://doi.org/10.1016/j.margeo.2016.06.002

Blott, S. and Pye, K. 2001. GRADISTAT: a grain size distribution and statistics package for the analyses of unconsolidated sediments. *Earth Surface Processes and Landforms*, **26**, 1237–1248, https://doi.org/10.1002/esp.261

Bronk Ramsey, C. 2009. Bayesian Analysis of Radiocarbon Dates. *Radiocarbon*, **51**, 337–360, https://doi.org/10.1017/S0033822200033865

Clare, M.A., Hughes Clarke, J.E., Talling, P.J., Cartigny, M.J.B. and Pratomo, D.G. 2016. Preconditioning and triggering of offshore slope failures and turbidity currents revealed by most detailed monitoring yet at a fjord-head delta. *Earth and Planetary Science Letters*, **450**, 208–220, https://doi.org/10.1016/j.epsl.2016.06.021

Croudace, I.W., Rindby, A. and Rothwell, R.G. 2006. ITRAX: description and evaluation of a new multi-function X-ray core scanner. *Geological Society, London, Special Publications*, **267**, 51–63, https://doi.org/10.1144/GSL.SP.2006.267.01.04

Daxer, C., Sammartini, M., Molenaar, A., Piechl, T., Strasser, M. and Moernaut, J. 2020. Morphology and spatio-temporal distribution of lacustrine mass-transport deposits in Wörthersee, eastern Alps, Austria. *Geological Society, London, Special Publications*, **500**, https://doi.org/10.1144/SP500-2019-179

Ficker, H., Gasser, H., Achleitner, D. and Schabetsberger, R. 2011. Ectogenic meromixis of Lake Hallstättersee, Austria induced by waste water intrusions from salt mining. *Water, Air, & Soil Pollution*, **218**, 109–120, https://doi.org/10.1007/s11270-010-0627-2

Gilli, A., Anselmetti, F.S., Glur, L. and Wirth, S.B. 2013. Lake sediments as archives of recurrence rates and intensities of past flood events. *In*: Schneuwly-Bollschweiler, M., Stoffel, M. and Rudolf-Miklau, F. (eds) *Dating Torrential Processes on Fans and Cones*. Advances in Global Change Research, **47**, 225–242. Springer, Dordrecht, https://doi.org/10.1007/978-94-007-4336-6_15

Girardclos, S., Schmidt, O.T., Sturm, M., Ariztegui, D., Pugin, A. and Anselmetti, F.S. 2007. The 1996 AD delta collapse and large turbidite in Lake Brienz. *Marine Geology*, **241**, 137–154, https://doi.org/10.1016/j.margeo.2007.03.011

Hammerl, C. 2017. Historical earthquake research in Austria. *Geoscience Letters*, **4**, 7, https://doi.org/10.1186/s40562-017-0073-8

Hansen, L., Waldmann, N., Storms, J., Eilertsen, R.S., Ariztegui, D., Chapron, E. and Nesje, A. 2016. Morphological signatures of mass wasting and delta processes in a fjord-lake system: insights from Lovatnet, western Norway. *Norwegian Journal of Geology*, **96**, 1–21, https://doi.org/10.17850/njg96-3-02

Harris, P.T., Mcmillan-Lawler, M., Rupp, J. and Baker, E.K. 2014. Geomorphology of the oceans. *Marine Geology*, **352**, 4–24, https://doi.org/10.1016/j.margeo.2014.01.011

Heine, E. 2017. Hydrographische Vermessungen in Seen/Hallstätter See. *In*: Wimmer-Frey, I., Römer, A. and Janda, C. (eds) *Angewandte Geowissenschaften an der GBA*, **2017**, 285–286, https://opac.geologie.ac.at/http://wwwopacx/wwwopac.ashx?command=getcontent&server=images&value=ATA_2017_285.pdf

Hilbe, M. and Anselmetti, F. 2014. Signatures of slope failures and river-delta collapses in a perialpine lake (Lake Lucerne, Switzerland). *Sedimentology*, **61**, 1883–1907, https://doi.org/10.1111/sed.12120

Howarth, J.D., Fitzsimons, S.J., Norris, R.J. and Jacobsen, G.E. 2014. Lake sediments record high intensity shaking that provides insight into the location and rupture length of large earthquakes on the Alpine Fault, New Zealand. *Earth and Planetary Science Letters*, **403**, 340–351, https://doi.org/10.1016/j.epsl.2014.07.008

Huang, J.-J., Löwemark, L., Chang, Q., Lin, T.-Y., Chen, H.-F., Song, S.-R. and Wei, K.-Y. 2016. Choosing optimal exposure times for XRF core-scanning: suggestions based on the analysis of geological reference materials. *Geochemistry, Geophysics, Geosystems*, **17**, 1558–1566, https://doi.org/10.1002/2016GC006256

Kern, A., Kowarik, K., Rausch, A. and Reschreiter, H. (eds) 2009. Kingdom of Salt, 7000 years of Hallstatt. *Veröffentlichungen der Prähistorischen Abteilung*, **3**, Vienna.

Laberg, J.S., Strasser, M., Alves, T.M., Gao, S., Kawamura, K., Kopf, A. and Moore, G.F. 2016. Internal deformation of a muddy gravity flow and its interaction with the seafloor of IODP Expedition 333, Nankai Trough, SE Japan. *Landslides*, **14**, 849–860, https://doi.org/10.1007/s10346-016-0766-7

Land-Oberösterreich 2019. Digitales Geländemodell 50 cm/1 m (XYZ), https://www.land-oberoesterreich.gv.at/211787.htm

Lee, A.-S., Huang, J.-J., Kao, L.C., Wei, K.-Y. and Liou, S.Y.-H. 2019. High resolution record of heavy metals from estuary sediments of Nankan River (Taiwan) assessed by rigorous multivariate statistical analysis. *Quaternary International*, **527**, 44–51, https://doi.org/10.1016/j.quaint.2018.11.018

Locati, M, Rovida, A, Albini, P. and Stucchi, M. 2014. The AHEAD portal: a gateway to European historical earthquake data. *Seismological Research Letters*, **85**, 727–734, https://doi.org/10.1785/0220130113

Löwemark, L., Bloemsma, M. *et al.* 2019. Practical guidelines and recent advances in the Itrax XRF core-scanning procedure. *Quaternary International*, **514**, 16–29, https://doi.org/10.1016/j.quaint.2018.10.044

Mandl, G., Van Husen, D. and Lobitzer, H. 2012. *Erläuterungen zu Blatt 96 Bad Ischl – Geologische Karte der Republik Österreich 1:50 000.* Geologische Bundesanstalt, Wien, https://opac.geologie.ac.at/http://wwwopacx/wwwopac.ashx?command=getcontent&server=images&value=096_Bad_Ischl.pdf

Melzner, S. 2017. Ereignis- und Schadenskataster von Sturzprozessen (Steinschlag/Felssturz) in der Gemeinde Hallstatt. *In*: Wimmer-Frey, I., Römer, A. and Janda, C. (ed.) *Angewandte Geowissenschaften an der GBA*, **2017**, 289–290, https://opac.geologie.ac.at/http://wwwopacx/wwwopac.ashx?command=getcontent&server=images&value=ATA_2017_289.pdf

Micallef, A., Krastel, S. and Savini, A. (eds) 2018. *Submarine Geomorphology.* Springer International, https://doi.org/10.1007/978-3-319-57852-1

Moernaut, J. and De Batist, M. 2011. Frontal emplacement and mobility of sublacustrine landslides: results from morphometric and seismostratigraphic analysis. *Marine Geology*, **285**, 29–45, https://doi.org/10.1016/j.margeo.2011.05.001

Moernaut, J., Van Daele, M. *et al.* 2014. Lacustrine turbidites as a tool for quantitative earthquake reconstruction: new evidence for a variable rupture mode in South-Central Chile. *Journal of Geophysical Research, Solid Earth*, **119,** 1607–1633, https://doi.org/10.1002/2013JB010738

Moernaut, J., Van Daele, M. *et al.* 2017. Lacustrine turbidites produced by surficial slope sediment remobilization: a mechanism for continuous and sensitive turbidite paleoseismic records. *Marine Geology*, **384**, 159–176, https://doi.org/10.1016/j.margeo.2015.10.009

Moernaut, J., Wiemer, G., Kopf, A. and Strasser, M. 2020. Evaluating the sealing potential of young and thin mass-transport deposits: Lake Villarrica, Chile. *Geological Society, London, Special Publications*, **500**, https://doi.org/10.1144/SP500-2019-155

Molenaar, A., Moernaut, J., Wiemer, G., Dubois, N. and Strasser, M. 2019. Earthquake impact on active margins: tracing surficial remobilization and seismic strengthening in a slope sedimentary sequence. *Geophysical Research Letters*, **41**, 1195, https://doi.org/10.1029/2019GL082350

Mulder, T. and Chapron, M. 2011. Flood deposits in continental and marine environments: character and significance, *American Association of Petroleum Geologists (AAPG) Studies in Geology,* **61**, 1–30, https://doi.org/10.1306/13271348St613436

Praet, N., Moernaut, J. *et al.* 2017 Paleoseismic potential of sublacustrine landslide records in a high-seismicity setting (south-central Alaska). *Marine Geology*, **384**, 103–119, https://doi.org/10.1016/j.margeo.2016.05.004

Pürstinger, C. 2017. Verbauungsgeschichte Hallstätter Mühlbach und Revision des Gefahrenzonenplans der Marktgemeinde Hallstatt. *In*: Wimmer-Frey, I., Römer, A. and Janda, C. (eds) *Angewandte Geowissenschaften an der GBA*, **2017**, 132–139, https://opac.geologie.ac.at/wwwopacx/wwwopac.ashx?command=getcontent&server=images&value=ATA_2017_132.pdf

Rapuc, W., Sabatier, P. *et al.* 2018. 6600 years of earthquake record in the Julian Alps (Lake Bohinj, Slovenia). *Sedimentology*, **65**, 1777–1799, https://doi.org/10.1111/sed.12446

Reimer, P.J., Bard, E. *et al.* 2013. IntCal13 and Marine13 Radiocarbon Age Calibration Curves 0–50,000 Years cal BP. *Radiocarbon*, **55**, 869–1887, https://doi.org/10.2458/azu_js_rc.55.16947

Rohn, J., Ehret, D., Moser, M. and Czurda, K. 2005: Prehistoric and recent mass movements of the World Cultural Heritage Site Hallstatt, Austria. *Environmental Geology*, **47**, 702–714, https://doi.org/10.1007/s00254-004-1201-2

Sammartini, M., Moernaut, J., Anselmetti, F., Hilbe, M., Lindhorst, K., Preat, N. and Strasser, M. 2019. An atlas of mass-transport deposits in lakes. *In*: Ogata, K. *et al.* (eds) *Submarine Landslides: Subaqueous Mass Transport Deposits from Outcrops to Seismic Profiles*, 201–226, https://doi.org/10.1002/9781119500513.ch13

Schnellmann, M., Anselmetti, F.S., Giardini, D., McKenzie, J.A. and Ward, S.N. 2002 Prehistoric earthquake history revealed by lacustrine slump deposits. *Geology*, **30**, 1131–1134, https://doi.org/10.1130/0091-7613(2002)030<1131:PEHRBL>2.0.CO;2

Schnurrenberger, D., Russell, J. and Kelts, K. 2003. Classification of lacustrine sediments based on sedimentary components. *Journal of Paleolimnology*, **29**, 141–154, https://doi.org/10.1023/A:1023270324800

Schorn, A. and Neubauer, K. 2014. The structure of the Hallstatt evaporite body (Northern Calcareous Alps, Austria): a compressive diapir superposed by strike-slip shear? *Journal of Structural Geology*, **60**, 70–84, https://doi.org/10.1016/j.jsg.2013.12.008

Schultes, J.A. 1809. *Reisen durch Oberösterreich, in den Jahren 1794, 1795, 1802, 1803, 1804 und 1808.* I. Theil, II. Theil, Tübingen.

Simonneau, A., Chapron, E. *et al.* 2013. Mass-movement and flood-induced deposits in Lake Ledro, southern Alps, Italy: implications for Holocene palaeohydrology and natural hazards. *Climate of the Past*, **9**, 825–840, https://doi.org/10.5194/cp-9-825-2013

Strasser, M., Stegmann, S., Bussmann, F., Anselmetti, F.S., Rick, B. and Kopf, A. 2007. Quantifying subaqueous slope stability during seismic shaking: Lake Lucerne as model for ocean margins. *Marine Geology*, **240**, 77–97, https://doi.org/10.1016/j.margeo.2007.02.016

Strasser, M., Monecke, K., Schnellmann, M. and Anselmetti, F.S. 2013. Lake sediments as natural seismographs: a compiled record of Late Quaternary earthquakes in Central Switzerland and its implication for Alpine deformation. *Sedimentology*, **60**, 319–341, https://doi.org/10.1111/sed.12003

Strupler, M., Anselmetti, F.S., Hilbe, M. and Strasser, M. 2018. Quantitative characterization of subaqueous landslides in Lake Zurich (Switzerland) based on a high-resolution bathymetric dataset. *Geological Society, London, Special Publications*, **477**, https://doi.org/10.1144/SP477.7

Tóth, Z., Spieß, V. and Jensen, J. 2014. Seismo-acoustic signatures of shallow free gas in the Bornholm Basin, Baltic Sea. *Continental Shelf Research*, **88**, 228–239, https://doi.org/10.1016/j.csr.2014.08.007

Urstöger, H.J. 2000. *Hallstatt-Chronik. Von den Anfängen bis zum Jahr 2000. Nach Originalchroniken, Veröffentlichungen über Hallstatt, Unterlagen aus dem Museumsarchiv und Berichten von Zeitzeugen*. Verlag des Musealvereins, Hallstatt.

Van Daele, M., Versteeg, W., Pino, M., Urrutia, R. and De Batist, M. 2013. Widespread deformation of basin-plain sediments in Aysén fjord (Chile) due to impact by earthquake-triggered, onshore-generated mass movements. *Marine Geology*, **337**, 67–79, https://doi.org/10.1016/j.margeo.2013.01.006

Van Daele, M., Moernaut, J. *et al.* 2015. A comparison of the sedimentary records of the 1960 and 2010 great Chilean earthquakes in 17 lakes: implications for quantitative lacustrine palaeoseismology. *Sedimentology*, **62**, 1466–1496, https://doi.org/10.1111/sed.12193

Vandekerkhove, E., Van Daele, M., Praet, N., Cnudde, V., Haeussler, P.J. and De Batist, M. 2019. Flood-triggered v. earthquake-triggered turbidites: a sedimentological study in clastic lake sediments (Eklutna Lake, Alaska). *Sedimentology*, **67**, 364–389, https://doi.org/10.1111/sed.12646

van Husen, D. 2004. Quaternary glaciations in Austria. *Developments in Quaternary Sciences*, **2**, 1–13, https://doi.org/10.1016/S1571-0866(04)80051-4

Weltje, G.J. and Tjallingii, R. 2008. Calibration of XRF core scanners for quantitative geochemical logging of sediment cores: theory and application. *Earth and Planetary Science Letters,* **274**, 423–438, https://doi.org/10.1016/j.epsl.2008.07.054

Wilhelm, B., Nomade, J. *et al.* 2016. Quantified sensitivity of small lake sediments to record historic earthquakes: implications for paleoseismology. *Journal of Geophysical Research: Earth Surface*, **121**, 2–16, https://doi.org/10.1002/2015jf003644

Slope stability hazard in a fjord environment: Douglas Channel, Canada

Cooper D. Stacey[1]*, D. Gwyn Lintern[1], John Shaw[2] and Kim W. Conway[1]

[1]Natural Resources Canada, Geological Survey of Canada Pacific, 9860 W Saanich Rd, Sidney, British Columbia V8L 4B2, Canada

[2]Natural Resources Canada, Geological Survey of Canada Atlantic, 1 Challenger Drive, Dartmouth, Nova Scotia B2Y 4A2, Canada

CDS, 0000-0003-0294-5609; DGL, 0000-0003-1057-2670
*Correspondence: Cooper.stacey@canada.ca

Abstract: Douglas Channel is a 140 km-long fjord system on Canada's west coast where steep topography, high annual precipitation and glacially over-deepened bathymetry have resulted in widespread slope failures. A 5 year project involving numerous marine expeditions to the remote area produced a comprehensive assessment of the magnitude and frequency of slope failures in the region. A classification scheme is presented based on morphology and failure mechanism: (1) debris flows are the most common in all parts of the fjord – they are often small with a subaerial component where fjord wall slope is very high or tend to exceed volumes of 10^6 m^3 where fjord wall slope is lower, allowing for accumulation of marine sediments; (2) large failures of oversteepened glacial sediments occurring at transgressive moraines and glaciomarine plateaus following deglaciation – the largest is at Squally Channel with an estimated volume of 10^9 m^3; (3) fjord wall failures that involve bedrock slump or rock avalanche; (4) translation of marine sediments; (5) composite/other slides; and (6) two scallop-shaped sackungen, or deep-seated gravitational slope deformations of granodiorite with volumes exceeding 60×10^6 m^3. The postglacial marine sedimentary record shows evidence of large-scale slope failures of all styles that were especially active following deglaciation. The Holocene marks a transition to a lower frequency and change to primarily debris flows and smaller rock slides. Slope failures that may be capable of generating tsunamis and may be damaging to coastal infrastructure have occurred in all parts of Douglas Channel through much of the Holocene. Here we present a morphological analysis with volume estimates and age control using multibeam bathymetry, high-resolution sub-bottom data and sediment cores. The study details an extensive analysis of slope failures in a fjord network that can be extended to other fjord environments.

Located on the west coast of British Columbia, Douglas Channel is one of Canada's largest fjords and hosts the towns of Kitimat, Kitimaat Village and Hartley Bay (Fig. 1). In recent years there has been a surge of interest in industrial development, including port facilities, gas pipelines, oil and gas pipeline termini and other shoreline installations. The Geological Survey of Canada has undertaken a 5 year project to complete an analysis of geological hazards in the area. Subaerial terrain stability has been mapped by Maynard *et al.* (2017*a*, *b*, *c*), showing that the terrain is extremely steep and prone to slope failure. Typical subaerial slope gradients are between 30 and 45°, reaching as high as 45–60° in many parts of the fjord. One of the components of the hazard assessment was a regional summary of slope failure activity in the submarine environment that will be used to assist with analysis of possible environmental and geohazard impacts to minimize potential risk to the local population and coastal infrastructure. Here we present the results of a thorough field campaign to reveal an extensive history of submarine slope failures distributed throughout the study area.

Numerous proposals to expand coastal infrastructure in the north coast region of British Columbia require a regional assessment of seafloor and coastal geohazards. Submarine slope failures are known to generate damaging tsunamis that can have a large impact on shipping and coastal infrastructure. For example, the Kitimat area experienced two well-known submarine landslides that occurred in 1974 and 1975, generating destructive 6 and 8 m tsunamis (respectively), which struck the docks of Kitimaat Village (Golder Associates 1975; Murty and Brown 1979; Thomson *et al.* 2012). Assessments undertaken after the slides revealed the presence of a 5 -km-long slide complex with a series of headwalls on the fjord walls on both sides of the active Kitimat Delta (Golder Associates 1975; Murty and Brown 1979; Prior *et al.* 1984; Stacey *et al.* 2018). Further down-fjord, near the town of Hartley Bay, very large bedrock sagging blocks known as sackungen (from the German term for sagging) were identified in two

From: Georgiopoulou, A., Amy, L. A., Benetti, S., Chaytor, J. D., Clare, M. A., Gamboa, D., Haughton, P. D. W., Moernaut, J. and Mountjoy, J. J. (eds) 2020. *Subaqueous Mass Movements and their Consequences: Advances in Process Understanding, Monitoring and Hazard Assessments*. Geological Society, London, Special Publications, **500**, 427–451.
First published online May 20, 2020, https://doi.org/10.1144/SP500-2019-191

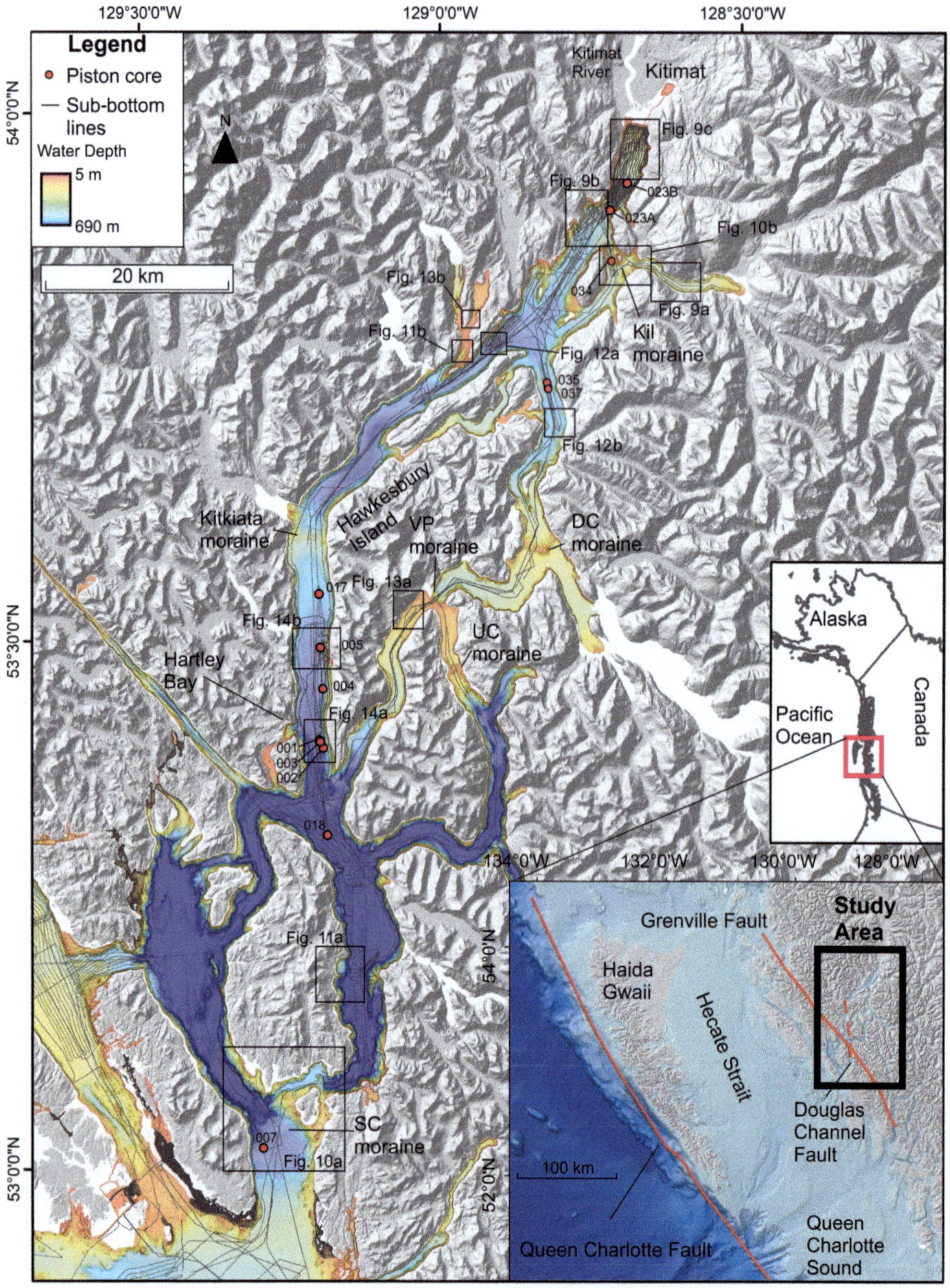

Fig. 1. Map of Douglas Channel showing sub-bottom coverage (grey lines), sediment cores (red dots) and location of Figures 9–14. Locations of local settlements Kitimat and Hartley Bay are shown. The Kitimat River is the primary source of sediment and enters at Kitimat at the north end of the fjord. Locations of recessional moraines are shown at Squally Channel (SC), Ursula Channel (UC), Kitkiata, Verney Passage (VP), Devastation Channel (DC) and Kildala (Kil). Large failures have occurred at all moraines except the Verney Passage moraine. Inset map shows locations of the Queen Charlotte Fault, the Grenville Fault and the Douglas Channel Fault.

locations with volumes of 63×10^6 and 70×10^6 m^3 (Thomson *et al.* 2012; Conway and Barrie 2018).

Setting

The Douglas Channel Fjord system extends 140 km inland from Queen Charlotte Sound and has an extensive associated network of inlets and channels totalling over 320 km (Fig. 1). The fjord is located in the 2000 m-high Coast Range Mountains that are locally characterized by granitic rocks, metasediments and metavolcanic rocks (Roddick 1970; Holland 1976). The active Queen Charlotte Transform Fault is 160 km west of the mouth of the fjord (Fig. 1). Observed earthquakes tend to concentrate along the plate boundary and monitoring indicates low levels of seismicity within the Douglas Channel fjord (Rohr and Tryon 2010; Brillon 2016).

At the Last Glacial Maximum *c.* 20 cal ka BP, the Cordilleran Ice Sheet extended across much of Hecate Strait and Queen Charlotte Sound (Clague and James 2002; Shaw *et al.* 2019). Postglacial rates of ice retreat were variable across the fjords of British Columbia, with stages of rapid retreat interrupted by periods of stability where the ice front remained stationary for periods of time, establishing large recessional moraines (Fig. 1; Clague 1985). High-relief moraines were deposited at Squally Channel, Ursula Channel, Kitkiata, Verney Passage and Devastation Channel (Figs 1–4; Shaw *et al.* 2017). Thick sequences of glaciomarine mud were deposited in basins and on top of moraines (Shaw *et al.* 2017). The relative sea-level fell rapidly at Kitimat, dropping from +100 m at *c.* 10 cal ka BP to below modern sea-level over a short period (Josenhans *et al.* 1997). Sea-level reached its current position by 9–9.5 cal ka BP (Bornhold 1983).

The Kitimat River is the primary source of sediment for Douglas Channel (Fig. 1). It enters at the head of Kitimat Arm with a mean flow rate of 46.3 m^3 s^{-1} in March and 274 m^3 s^{-1} in June during the spring freshet (MacDonald and Shepherd 1983). The Kitimat Delta is an active system characterized by sinuous bed forms and a number of slide scars as recent as 1975 (Stacey *et al.* 2018). Other large deltas are located at the heads of smaller inlet heads, including Kildala Arm and Gardner Canal. Sedimentation in other parts of Douglas Channel is influenced by high regional precipitation resulting in over 109 small fan deltas (Shaw and Lintern 2016) and by strong bottom currents that rework seabed deposits (Shaw *et al.* 2017). Sedimentation is variable throughout the basin; just south of the Kitimat Delta slide complex the ^{210}Pb sediment accumulation rate at Core 23B is 0.2 cm a^{-1} (Fig. 1; Stacey *et al.* 2018). Postglacial sediment accumulation generally decreases with distance from the Kitimat Delta and Kildala Delta; however there are several areas, especially in Squally Channel and Whale Channel, where high bottom current activity reworks sediment into mounds 50–80 m in thickness (Shaw *et al.* 2017).

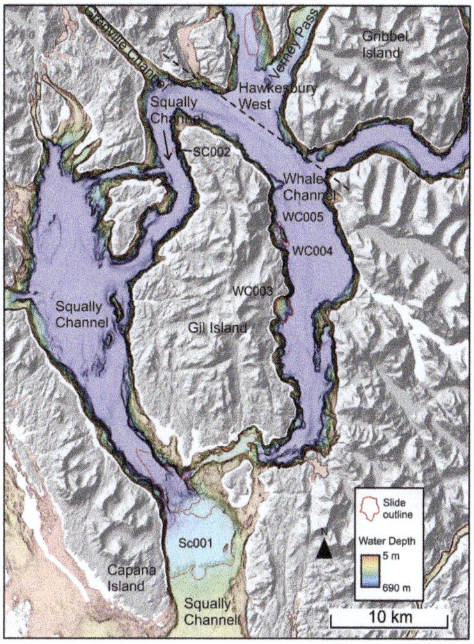

Fig. 2. Slide outlines (red) for Squally Channel (SC) and Whale Channel (WC). SC001 is the largest slide in Douglas Channel and occurred on the Squally Channel moraine. Black dashed line separates Whale Channel from Hawkesbury West.

The Douglas Channel fjord is a complex network of branching channels and inlets. Here we divide the fjord into eight areas to describe slope failure distribution throughout the system starting at the mouth of the fjord (Figs 1–4): (1) Squally Channel (SC) and Whale Channel (WC) from Campania Island in the south to the north end of Gil Island in the north; (2) the channel on the west side of Hawkesbury Island and Maitland Island, or 'Hawkesbury West' (HW); (3) Miskatla Inlet (MI) and Giltoyees Inlet (GI); (4) Verney Passage (VP) and Ursula Channel (UC); (5) Gardner Canal (GC); (6) Devastation Channel (DC) and Sue Channel (Sue); (7) Kitimat Arm (KA) from Hilton Point to Kitimat Delta; and (8) Amos Passage (AP) and Kildala Arm (Kil).

There are five principal sedimentary acoustic units that are recognized from seismic data which have been summarized by Shaw *et al.* (2017) (Table 1). Sediment cores were collected to determine the age of acoustic units and timing of slope failure deposits (Fig. 5). The top of the glaciomarine unit (Table 1) is interpreted as the late stages of

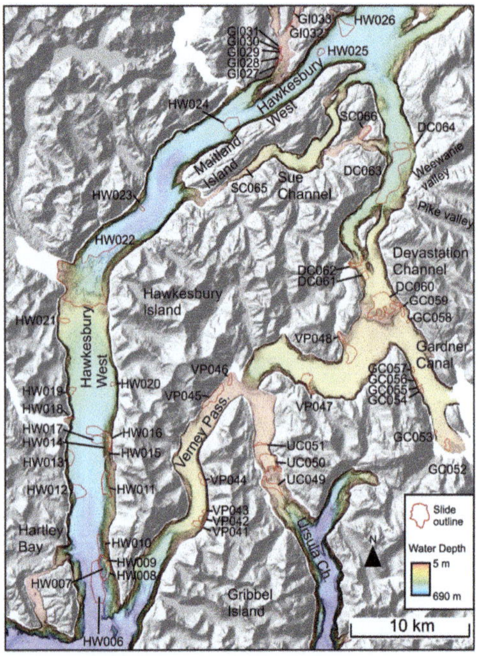

Fig. 3. Slide outlines (red) for Hawkesbury West (HW), Verney Passage (VP), Ursula Channel (UC), Gardner Canal (GC), Devastation Channel (DC) and Sue Channel (SC). HW022 is the second largest slide and occurred at the Kitkiata moraine. Origins of DC063 are not certain but are probably associated with Weewanie Valley or Pike Valley.

deglaciation. Dates from this unit at the outer reaches of the Douglas Channel fjord network (Squally Channel) range between 12.0 and 13.2 cal ka BP

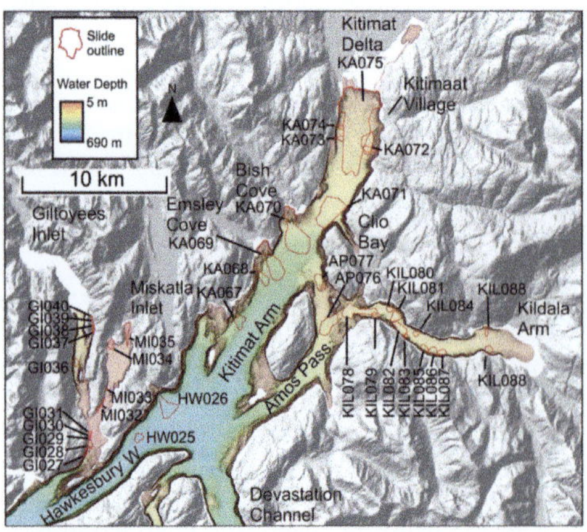

Fig. 4. Slide outlines (red) for Giltoyees Inlet (GI), Miskatla Inlet (MI), Kitimat Arm (KA), Amos Passage (AP) and Kildala Arm (Kil).

(Shaw *et al.* 2017). At Hawkesbury Island just south of the Kitkiata moraine dates range from 11.5 to 13.0 cal ka BP (Fig. 3; Conway and Barrie 2015). In Kitimat Arm dates from the top of this unit range from 9.5 to 11.2 cal ka BP (Fig. 4). Deglaciation at Kitimat occurred by *c.* 11.5 cal ka BP and possibly as early as *c.* 13 cal ka BP (Bornhold 1983). It was not until after these dates that Holocene depositional conditions occurred in Douglas Channel.

The very distinct acoustically transparent interval near the top of the glaciomarine unit is observed in most seismic records north of the Kitkiata and Verney Passage moraines and is interpreted as a glacial outburst flood event that ended at 11.9 cal ka BP (Shaw *et al.* 2017). This deposit can be used as a key isochron for correlation between sub-basins in the Kitimat Arm, Kildala Arm and Devastation Channel area. Radiocarbon dates in the Holocene mud unit range as far back as 12 cal ka BP in cores collected west of Hawkesbury Island (Conway and Barrie 2015). In Kitimat Arm and Devastation Channel the oldest Holocene date is 10 cal ka BP (Shaw *et al.* 2017).

Objectives

In this contribution the authors build on the slope failure assessments of Conway and Barrie (2015, 2018), Shaw *et al.* (2017) and Stacey *et al.* (2018) using additional multibeam bathymetry, sub-bottom profiles and radiocarbon dates from sediment cores to expand the number of known slope failure deposits and establish a classification system, volume estimations and a chronology of events. From this dataset, trends are visible in slope failure distribution throughout the fjord and in their timing. We address the following specific objectives:

- Summarize the findings for the 5 year hazard assessment, which examined the entire fjord system on a regional basis.
- Map all known slope failure deposits geographically and temporally to assess historical and recent slope instability. From this we identify areas of interest and increased risk to infrastructure and public safety.
- Apply a classification system for different types of failures in an effort to understand possible trigger mechanisms.
- Undertake volume assessment of slope failure scars/deposits.

Methods

Seven research expeditions were performed between 2013 and 2017 onboard Canadian Coast Guard Ships to survey the Douglas Channel fjord. Seafloor data

Table 1. *Acoustic units as presented in Shaw* et al. *(2017)*

Unit	Geometry and acoustic character	Backscatter, lithology and chronology	Interpretation
1	Acoustically incoherent; ridges 10–20 m high on inner continental shelf; transverse ridges many hundreds of metres thick in fiords	High backscatter; lithology not sampled – probably diamicton	**Glacial diamicton** formed at ice margins during halts in recession of grounded ice
2A	Draped, parallel, continuous, coherent reflections with acoustically transparent intervals; high backscatter	Laminated – massive grey, gravelly sandy clay; gravel at seafloor; imprinted by iceberg furrows and pits on shelf; dates of 10.1–12.1 cal ka BP near top of unit	**Glaciomarine sediment** deposited by meltwater plumes; interbedded with coeval debris flows
2B	Laterally thinning drape of acoustically transparent sediment just below top of Unit 2A; base is erosional in places	High backscatter; structureless grey clay; similar to unit cored by Bornhold (1983); late glacial age (11.9 cal ka BP)	**Outburst flood/ hyperpycnal flow deposits**
3	Acoustically incoherent, irregular upper surface; infills depressions	High backscatter; muddy gravel at the seafloor; late glacial to present day	**Mass transport deposits**, resulting from submarine slope failure; most frequent in late glacial stage, but includes modern events
4	Sediment drifts; acoustically transparent or weak-moderate internal reflections; pockmarks, gas masking	Low backscatter; mud or sandy mud. Radiocarbon ages ranging from 1.5 to 10.2 cal ka BP	**Holocene mud/sandy mud** derived from fluvial input of suspended sediment in the postglacial period
5	Cone-shaped deposits on the fjord sidewalls; dark-toned, incoherent reflections	Variable backscatter; gravel/ sand/mud	**Fan delta deposits**, late glacial onwards

Observations of sub-bottom geometry and acoustic character and surface backscatter and lithology are summarized. Interpreted depositional conditions based on sub-bottom and surface characteristics.

and samples were collected and analysed at the Pacific Geoscience Centre in Sidney, British Columbia, Canada.

Multibeam bathymetry and high resolution sub-bottom data

Swath multibeam bathymetry surveys were conducted by the Canadian Hydrographic Service with a hull-mounted Kongsberg EM710MKII echosounder with a frequency range of 40–100 kHz as summarized by Conway *et al.* (2013). Data were gridded at 5 m resolution and analysed using ArcGIS. Vertical resolution varies between 10 and 20 cm depending on slope and water depth. High-resolution sub-bottom profiles were acquired using hull-mounted 3.5 kHz Knudsen Chirp sounders and a deep-towed Huntec system consisting of a boomer source and a line-and-cone receiver with vertical resolutions of 20–30 cm. Depths were calculated using an acoustic velocity of 1500 m s^{-1}. Sub-bottom lines were planned on the basis of the available multibeam bathymetry data. As with all sedimentary archives in dynamic environments, the stratigraphic completeness of Douglas Channel is limited in places. This is especially true in channels where bottom currents are high, resulting in a lack of Holocene unit and winnowing of the Pleistocene unit. In places where slope failure deposits consist of coarse material (sand and coarser) the acoustic signal is attenuated and only the more recent deposits are visible in sub-bottom profiles, potentially producing a bias towards increased activity in the late Holocene.

Sediment cores

Sediment cores (piston cores) were collected for ground truthing and age control (Fig. 5). Cores were scanned using the Geological Survey of Canada Pacific whole-core multisensor core logger for gamma density, magnetic susceptibility and P-wave velocity and the Royal Roads University split-core whole-core multisensor core logger high-resolution photographs. X-Radiographs were

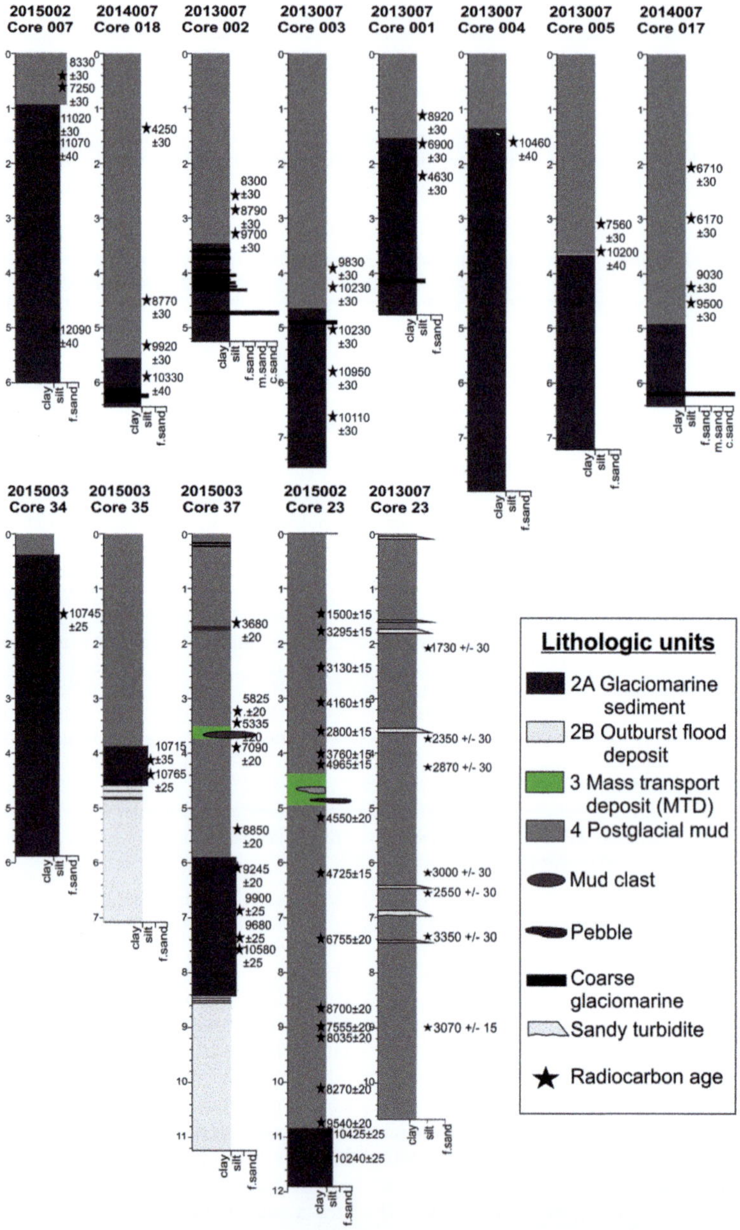

Fig. 5. Graphic logs of sediment cores from Douglas Channel. Cores from Conway and Barrie (2015), Shaw *et al.* (2017) and Stacey *et al.* (2018). Core locations shown on Figure 1. Cores were collected from: 7 , Squally Channel slide headwall (SC001); 18, Whale Channel; 1–3, fjord bottom in front of south sackungen (HW007); 4, 5, fjord bottom in front of north sackungen (HW014); 17, control core from northern Hawkesbury West basin; 34, glaciomarine plateau in Amos Passage; 35, 37, Devastation Channel (only cores to recover clay outburst flood deposit); 23A, Kitimat Arm control core of entire Holocene unit; 23B, south of the Kitimat slide complex (KA075) with turbidite deposits from large delta slide events.

acquired at the Department of Earth Science at Simon Fraser University. Radiocarbon dating was completed on 13 cores using shell and wood material with a 94.5% confidence level. Analysis was performed at the University of California Irvine (Keck Carbon Cycle Accelerator Mass Spectrometry)

Table 2. *Conventional radiocarbon dates with OxCal v. 4.2 calibrated dates at the 95.4% confidence level*

Sample	^{14}C age BP	Cal years BP	Median cal years BP	Material	Laboratory number
2015002PGC core 7, 0.47 m	8330 ± 30	8290–8536	8401	Half valve of *Yoldia beringiana*	Beta-420169
2015002PGC core 7, 0.6 m	7250 ± 30	7257–7480	7373	Half valve of *Yoldia beringiana*	Beta-420170
2015002PGC core 7, 1.19 m	11020 ± 30	11 677–12 246	11 956	Complete *Nuculana pernula*	Beta-420171
2015002PGC core 7, 1.65 m	11070 ± 40	11 795–12 398	12 057	Broken specimen of large bivalve, possibly *Mactra* sp.	Beta-420172
2015002PGC core 7, 5.05 m	12090 ± 40	13 031–13 338	13 189	Complete *N. pernula*	Beta-420173
2015003PGC core 34, 1.44 m	10745 ± 25	11 704–11 206	11 398	Bivalve	UCIAMS-167597
2015003PGC core 35, 4.12 m	10715 ± 35	11 670–11 175	11 347	Shell	UCIAMS-167466
2015003PGC core 35, 4.31 m	10765 ± 25	11 732–11 224	11 440	Shell	UCIAMS-167467
2015003PGC core 37, 1.65 m	3680 ± 20	3294–2959	3123	Shell	UCIAMS-167468
2015003PGC core 37, 3.23 m	5825 ± 20	5918–5663	5805	Bivalve	UCIAMS-167469
2015003PGC core 37, 3.49 m	5335 ± 20	6202–6003	6114	Wood cone	UCIAMS-167578
2015003PGC core 37, 3.98 m	7090 ± 20	7370–7116	7230	Shell	UCIAMS-167470
2015003PGC core 37, 5.42 m	8850 ± 20	9250–8956	9075	Shell	UCIAMS-167471
2015003PGC core 37, 6.15 m	9245 ± 20	9641–9411	9506	Shell	UCIAMS-167472
2015003PGC core 37, 6.94 m	9900 ± 25	10 515–10 220	10361	Shell	UCIAMS-167595
2015003PGC core 37, 7.48 m	9680 ± 25	11 198–10 888	11142	Wood	UCIAMS-167579
2015003PGC core 37, 7.50 m	10580 ± 25	11 311–11 056	11184	Shell	UCIAMS-167596
2015002PGC core 23, 1.50 m	1500 ± 15	1409–1347	1379	Shell fragment	UCIAMS-167569
2015002PGC core 23, 1.86 m	3295 ± 15	2780–2490	2683	Shell fragment	UCIAMS-167458
2015002PGC core 23, 2.48 m	3130 ± 15	3389–3270	3362	Shell fragment	UCIAMS-167570
2015002PGC core 23, 3.10 m	4160 ± 15	3847–3565	3708	Wood	UCIAMS-167459
2015002PGC core 23, 3.58 m	2800 ± 15	2950–2859	2902	Bivalve	UCIAMS-167571
2015002PGC core 23, 4.05 m	3760 ± 15	4224–4013	4119	Wood	UCIAMS-167572
2015002PGC core 23, 4.16 m	4965 ± 15	4920–4610	4786	Wood	UCIAMS-167460
2015002PGC core 23, 5.16 m	4550 ± 20	5315–5066	5157	Shell	UCIAMS-167573
2015002PGC core 23, 6.16 m	4725 ± 15	5579–5329	5475	Wood	UCIAMS-167574
2015002PGC core 23, 7.44 m	6755 ± 20	6975–6690	6833	Wood	UCIAMS-167461
2015002PGC core 23, 8.70 m	8700 ± 20	9696–9557	9632	Bivalve	UCIAMS-167575
2015002PGC core 23, 9.10 m	7555 ± 20	8404–8348	8379	Wood	UCIAMS-167576
2015002PGC core 23, 9.12 m	8035 ± 20	8233–7969	8099	Wood	UCIAMS-167462
2015002PGC core 23, 10.06 m	8270 ± 20	9400–9138	9270	Shell	UCIAMS-167577
2015002PGC core 23, 10.76 m	9540 ± 20	10 128–9715	9928	Wood	UCIAMS-167463
2015002PGC core 23, 10.94 m	10425 ± 25	11 177–10 828	11034	Shell	UCIAMS-167464
2015002pgc core 23, 11.38 m	10240 ± 25	11 001–10 625	10788	Wood	UCIAMS-167465

From Conway and Barrie (2015), Shaw *et al.* (2017) and Stacey *et al.* (2018).

following the procedure outlined by Southon *et al.* (2005) and by Beta Analytic Radiocarbon Dating. OxCal version 4.2 (Ramsey 2009) was used to calibrate the results. Dates are reported in radiocarbon years (^{14}C years BP) (Table 2). A $\Delta^{14}C$ reservoir correction of 390 ± 50 years was used for the shell materials (Shaw *et al.* 2017).

Slope failure chronology

Ages of slope failures are based on the correlation of their deposit to dated sediment cores. Recurrence intervals were calculated by averaging the total number of slope failure deposits within the given window of time. The calculation for the early Holocene begins at the end of the deglacial stage and ends at 6 cal ka BP, and late Holocene is from 6 cal ka BP to present.

Slope failure timing with narrow error bars (<0.5 cal ka BP) is based on cores that targeted the failure or can be directly correlated to a nearby core using sub-bottom data. Slope failure ages with error bar widths of 1000 years or more could not be correlated directly to cores and are estimations based on their depth within the Holocene unit in sub-bottom profiles. This method assumes a fairly constant sedimentation rate throughout the Holocene and was tested with a high level of success on slope failure with known dates.

Volume calculation

An attempt has been made to calculate slope failure volumes based on headwall geometries. This was often not possible because the headwall was not covered by bathymetric datasets, or the headwalls of older slope failures were infilled by Holocene deposition and no accurate calculation could be made based on current bathymetry (if the headwall was visible at all). In several instances, geometries were calculated using trend analysis in ArcGIS to project an inferred pre-failure plane over the headwall (HW025, KA072, KA073 and KA075 – the Kitimat Delta Headwalls A and B). Where headwalls were irregular and a pre-failure plane could not be reproduced, a wedge geometry was calculated where volume = ½ × area × height (McAdoo *et al.* 2000).

In cases where sufficient sub-bottom coverage was available, volumes were calculated using the slope failure deposit. These measurements were presented for older slope failures that are buried and it was not possible to get a reasonable estimate of the headwall volume. Calculations were made by interpreting a base and top surface in sub-bottom profiles of slope failure deposits and then generating a horizon for each surface. Three-dimensional volumes were not available for this study; however, most deposits were covered by at least one seismic line. Volumes were based on runout area and thickness where volume = area × 2/3 maximum thickness (Clare *et al.* 2018). Lobe deposits were used to calibrate headwall volume calculations, and generally speaking, values were comparable or slightly higher.

Results

Volumes of slope failures

Slope failure volumes vary between 0.003×10^6 m^3 and 10^8–10^9 m^3 in Douglas Channel (Table 3). There are 23 slope failure with volumes from 0 to 10^5 m^3, 31 from 10^5 to 10^6 m^3, 30 from 10^6 to 10^7 m^3, six from 10^7 to 10^8 m^3 and three from 10^8 to 10^9 m^3 (Fig. 6). Volumes are further discussed with classification of slope failures (see below).

Timing of slope failures

A chronology of slope failure activity has been established from sediment cores and is shown in Figures 7 & 8. Deposits of slope failures in areas with no Holocene deposition could not be fitted into the relative chronology. Slope failure deposits from Miskatla Inlet, Giltoyees Inlet and Gardner Canal are not presented in Figure 8 because sub-bottom profiles have not been acquired in these areas and ages are unknown. Displacement of HW007 and HW014 is interpreted to have occurred over period of several hundreds to several thousands of years (Conway and Barrie 2018). Timing is further discussed with classification of slope failures (see below).

Classification of slope failures

Classification of slope failures was based on the work of Conway *et al.* (2013) and Shaw *et al.* (2017). Distribution of slope failures is summarized in Table 4.

Debris flow. Debris flows are sediment masses having loosely compacted material that moves downslope as an overall fluid flow (Mulder and Cochonat 1996). Rocks and larger blocks may be entrained and transported. In Douglas Channel debris flows are widespread on the steep fjord wall (Fig. 6). In many instances the submarine fjord wall is very steep (>40°) and narrow with very little accumulated marine sediment (Fig. 9). The submarine fjord-wall slope of Kildala Arm is very high, and consists primarily of exposed bedrock (Shaw and Lintern 2016; Figs 4 & 9). With the limited amount of mud accumulation, there are no submarine headwalls associated with slide deposits. The deposits are interpreted to be linked to subaerial slides as the volume of the lobe deposits could not possibly have come from the submarine fjord wall alone. The subaerial slope in Kildala Arm is extreme and all slide deposits are associated with subaerial slopes in excess of 40°, while many are at the termination point of small creeks. Slide deposits tend to be around the order of 0.1×10^6 m^3 with limited runout distances. Comparable conditions of extreme subaerial slope with minimal sediment accumulation on submarine fjord walls occur in many parts of Douglas Channel, including Gardener Canal, Giltoyees Inlet and Miskatla Inlet (Figs 3 & 4). In total, 52 debris flows are interpreted to have originated in the subaerial margin.

In Kitimat Arm the fjord wall has a lower slope gradient in places, generally within a range of 10–15°. Here, thick sediment accumulations have occurred on the fjord walls, producing available material for submarine debris flows (Fig. 9). Slides KA068-70, 72, 73 and 74 all have headwalls that were incised into fjord wall sediment (Figs 4 & 9). These slides are mostly large with volumes greater than 10^6 m^3 and tend to have long runout distances. KA068-70 are mid-Holocene stacked slides, indicating that multiple stages of failure have occurred, either by retrogressive failure or by headwall infilling and repetitive failure (Figs 4, 7 & 9).

Table 3. *Summary of all large (>10^6 m^3) slide deposits organized by descending volume*

Slide ID	Sub-basin	Classification	Volume (m^3)	Relative age
SC001	Squally Channel	Oversteepened glacial material	804694350	Late glacial
HW022	Hawkesbury West	Oversteepened glacial material	275000000	Late glacial
WC003	Whale Channel	Rock slide	269405200	Late glacial
DC063	Devastation Channel	Debris flow	70994546	Late glacial
KA075	Kitimat Arm	Debris flow	33476720	Modern
HW014	Hawkesbury West	Block creep	32000000	Late glacial
HW007	Hawkesbury West	Block creep	31000000	Late glacial
HW006	Hawkesbury West	Debris flow	14060220	Early Holocene
VP045	Verney Passage	Composite slide	10665900	Unknown
Kil078	Amos Passage	Oversteepened glacial material	8120253	Late glacial
KA071	Kitimat Arm	Debris flow	7280355	Late Holocene
KA070	Kitimat Arm	Debris flow	6292553	Early Holocene
DC060	Devastation Channel	Oversteepened glacial material	4980645	Early Holocene
HW008	Hawkesbury West	Debris flow	3884590	Early Holocene
HW012	Hawkesbury West	Debris flow	3772526	Late Holocene
KA075C	Kitimat Arm	Debris flow	3719008	Modern
AP076	Amos Passage	Oversteepened glacial material	3437943	Late glacial
KA075D	Kitimat Arm	Debris flow	3282650	Modern
KA075A	Kitimat Arm	Debris flow	3167850	Modern
HW016	Hawkesbury West	Debris flow	3095550	Late Holocene
UC050	Ursula Passage	Oversteepened glacial material	3091840	Unknown
WC005	Whale Channel	Debris flow	3010200	Unknown
KA072B	Kitimat Arm	Debris flow	3009858	Late Holocene
KA069	Kitimat Arm	Debris flow	2924311	Early Holocene
HW024	Hawkesbury West	Debris flow	2907619	Early Holocene
VP048	Verney Passage	Debris flow	2514667	Unknown
HW013	Hawkesbury West	Debris flow	2507791	Late Holocene
WC004	Whale Channel	Debris flow	2273820	Unknown
DC064	Devastation Channel	Translational	2250000	Early Holocene
HW025	Hawkesbury West	Translational	1780853	Late Holocene
HW009	Hawkesbury West	Debris flow	1641195	Early Holocene
HW015	Hawkesbury West	Debris flow	1612375	Late Holocene
KA075B	Kitimat Arm	Debris flow	1581696	Late Holocene
AP077	Amos Passage	Oversteepened glacial material	1578403	Late glacial
Kil082	Kildala Arm	Debris flow	1484952	Early Holocene
KA068	Kitimat Arm	Debris flow	1474237	Early Holocene
HW026	Hawkesbury West	Debris flow	1431193	Late Holocene
UC049	Ursula Passage	Oversteepened glacial material	1381600	Unknown
HW019	Hawkesbury West	Debris Flow	1029307	Late Holocene

All slope failures >10^7 m^3 are from the late glacial or early Holocene except for the very large Kitimat slide deposit (KA075). The largest failures are associated with oversteepened glacial material. Volume calculations are based on headwall geometry wherever possible, or deposit geometry when headwalls are infilled or lacking.

Debris flow deposits are also often associated with fan deltas in the Douglas Channel system. Most fan deltas have experienced a number of small failures that can be detected in bathymetry or in sub-bottom data. Slides were only included in the total count if they were associated with a lobe that has a discernable relief in cross-section. Small debris flows tend to have runout distances of several hundreds of metres and more than half of the slides in this class do not cross the centre axis of the fjord. Large debris flows (>$10^6 \times$ m^3) are capable of travelling greater distances from the source, and may have outrunner blocks that are detached from the main body of the slide lobe by as much as 1 km (MI0354, KA075, Fig. 4). Large debris flow deposits are found throughout the study area and occurred during the last deglaciation and throughout the Holocene (Figs 7 & 8).

A number of debris flow deposits are voluminous enough to have potentially generated tsunamis; however, the most noteworthy debris flow deposit is the Kitimat slide complex (KA075; Fig. 9c), which is associated with at least two destructive tsunamis that occurred in 1974 and 1975 (Golder Associates 1975; Murty and Brown 1979; Prior *et al.* 1984; Stacey *et al.* 2018). The slide is characterized by a series

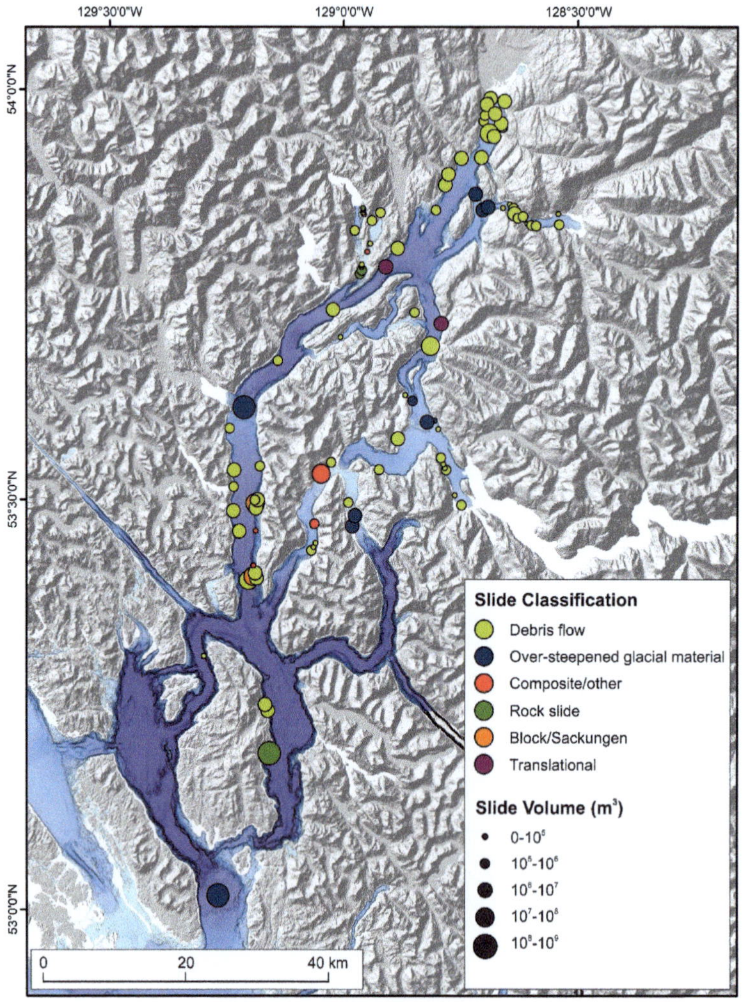

Fig. 6. Submarine slide classes and volumes. Slides of all different classes are distributed throughout Douglas Channel. The largest slide deposits are associated with oversteepened glacial material. Debris flow deposits are most common at the fjord heads.

of clearly defined headwalls; three on the fjord wall west of the Kitimat Delta and one to the east (Stacey *et al.* 2018). Zones of blocky debris escape the headwalls merging as a complex and irregular debris lobe that extends south 5.7 km from the Kitimat Delta. A field of outrunner blocks lie as far as 1 km south of the main slide complex. The blocky material originating from the head scar east of the Kitimat Delta overran the seafloor as a blocky lobe several metres thick. The main slide body on the west side of the delta is blocky and ploughed through the seafloor for a distance of up to 2.3 km, resulting in deformation down to a depth of 6 m. The front edge of the slide has ploughed a large area of the seabed, resulting in a lobe of deformed sediments with a surface morphology characterized by large crescent-shaped pressure ridges. The total slide complex has an estimated volume of 33×10^6 m^3. This slide volume is interpreted as a combination of events that have occurred over several hundred years (Stacey *et al.* 2018). The four headwalls are large: Headwall A off the Alcan terminal is 3.2×10^6 m^3; Headwall B immediately south is 0.6×10^6 m^3; Headwall C at Moon Bay is 3.7×10^6 m^3; and Headwall D between Kitimat Delta and Wathl Fan is 3.3×10^6 m^3. One large deposit is buried beneath the recent slide complex and is similar in depositional style and geometry. A volume estimation was not possible because acoustic energy is attenuated by the sandy Kitimat Delta; however, based on slide thickness of 15 m and

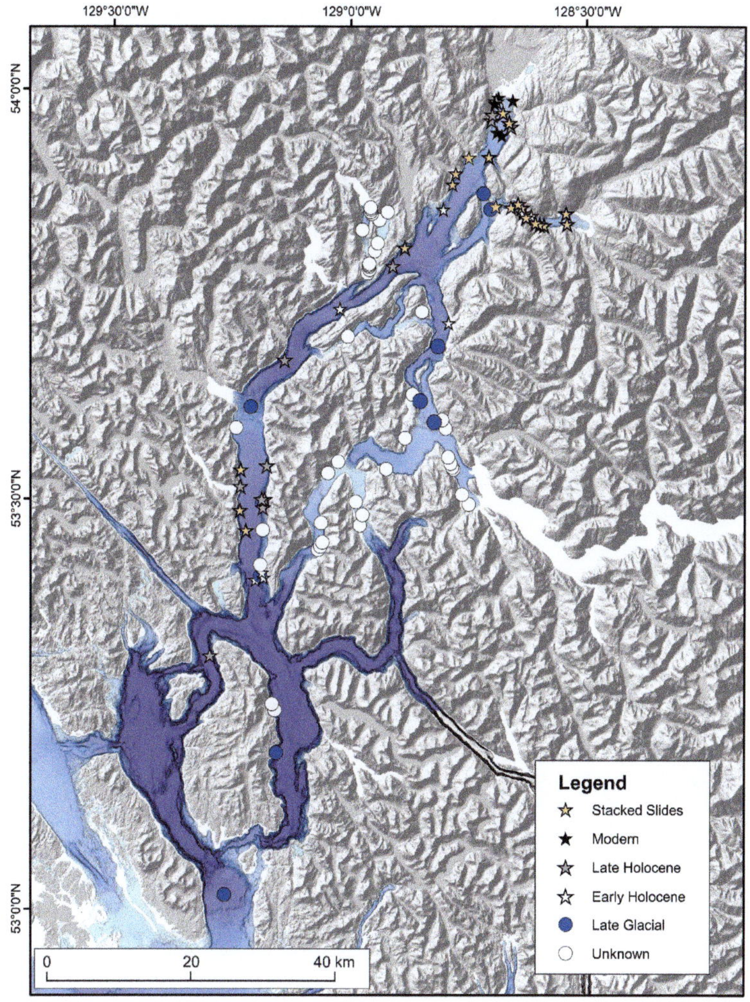

Fig. 7. Timing of submarine slides throughout Douglas Channel (modified from Shaw *et al.* (2017) with permission from Elsevier). Stacked slides indicate multiple slides at the same location through time. Slide ages in Verney Passage and Ursula Channel are unknown because strong bottom currents have not allowed deposition of Holocene sediment and no relative dates could be estimated. Slide ages in Gardner Canal, Giltoyees Inlet, Miskatla Inlet and Sue Channel are unknown because sub-bottom data are not available to produce relative ages.

geometry it is assumed to be comparable in volume with the tsunamigenic 1975 slide (Stacey *et al.* 2018). This deposit has been dated to 2.6 cal ka BP and there is at least one older slide deposit that was destroyed during subsequent flows.

The two fjord heads, Kitimat Arm and Kildala Arm, have the highest density of debris flow deposits and also show the highest number of slides in the late Holocene (Figs 8 & 9a, b). The late Holocene slides are distributed throughout these two tributaries, tend to concentrate in areas of high slope and are often associated with small subaerial creek valleys. The majority of these slide locations are sites of repeat events (Fig. 7).

The largest debris flow deposit is in Devastation Channel with a volume of 71×10^6 m^3 (DC063; Fig. 3). The deposit fills the entire width of the channel, extending 5 km along fjord axis and has a thickness exceeding 20 m. The top surface is irregular and is draped by subsequent Holocene deposition. The deposit pinches out into the top of the glaciomarine unit. The precise origins of the slope failure have not been determined; however, there is likely to be an association with the large and coalescing

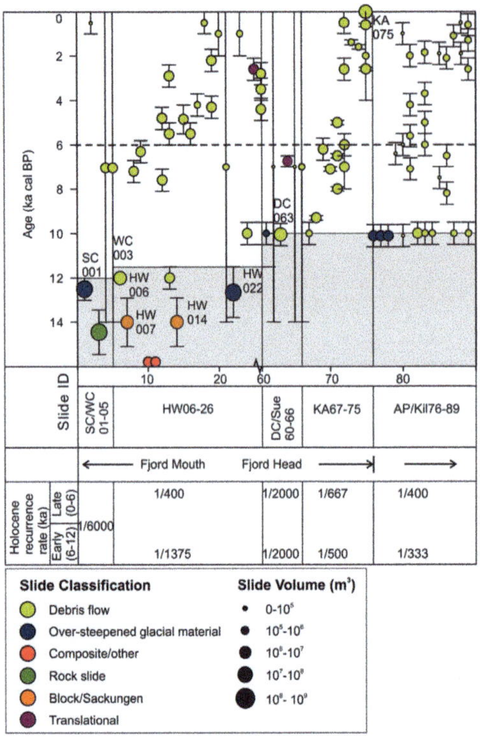

Fig. 8. Slide distribution with approximate ages over the past 16 cal ka BP. Slides are presented in order of slide ID from fjord mouth on the left to fjord head on the right. Slides from Miskatla Inlet, Giltoyees Inlet, Verney Passage and Gardner Canal are not represented here as a Holocene chronology was not established (slide ID sequence jumps from DC026 to Dev060 accordingly as indicated by scale break before 60). Shaded grey fill represents deglacial stage characterized by neotectonics, debutressing of oversteepened slopes and high sedimentation rates from the glacial front (dates from Shaw *et al.* 2017). Slope failures during deglaciation were of all different types. Slope failures in the Holocene (white background) are predominantly debris flows and rock avalanches (Holocene rock avalanches frequent in Giltoyees Inlet, not shown here). Narrow error bars based on dated core, error bars >1000 years are relative ages based on depth in Holocene unit.

Weewanie valley or Pike Valley (Fig. 3). Both of these valleys now terminate with large fan deltas.

Oversteepened glacial material. Slope failures that occur in association with oversteepened glacial material have many different morphologies. Oversteepened slopes occur at positions where grounded ice halted during retreat, building up large moraines or in thick glaciomarine sequences that are truncated by failure surfaces (Figs 1 & 10; Shaw *et al.* 2017). All of the slides occur on glacial features that are high relief with steep surfaces. The resulting headwall slope generally remains between 5 and 15°. The largest slides in Douglas Channel occur at terminal moraines (Fig. 10), and large failures shape every moraine on the side of ice recession except for the Verney Passage moraine (Fig. 1). In some cases failure appears to occur within the moraine material while in others there is a veneer of glaciomarine sediment that has failed.

The largest slope failure in Douglas Channel occurred at a terminal moraine in Squally Channel (SC001; Figs 2 & 10a; Shaw *et al.* 2017). The slide scar forms an escarpment 60 m in height that spans *c.* 5 km across the width of the fjord, cutting into glaciomarine sediment. The slide volume is estimated at 8×10^8 m^3. The slide scar has an area of 30 km^2 and is bound on the north side by a residual plateau of glaciomarine sediment. The residual plateau constricts the evacuation route to the north beyond which a steep drop of nearly 200 m leads to the flat floor of Squally Channel. Bottom currents in the channel are as high as 40 cm s^{-1} inhibiting Holocene deposition over the slide deposit (Shaw *et al.* 2017). The slide deposit has an irregular mounded surface which appears to have been reworked with no clear outline. Sub-bottom profiles of the deposit show a thick and acoustically chaotic unit with no indication of separate slide bodies. An amphitheatre-shaped headwall has incised 1 km upslope of the main headwall and is interpreted as a subsequent slide event. There is a low relief drift of postglacial mud that has accumulated within the slide scar that was cored for age control (Core 007; Figs 2 & 10a), revealing an age predating 13 cal ka BP (Shaw *et al.* 2017).

At the mouth of Kildala Arm a moraine flank has failed to the east leaving a head scar 25 m in height with a slope of 5–8° (Fig. 10b). The headwall has been partly infilled and has a smooth appearance, probably owing to its proximity to the Kitimat and Kildala River. The main slide deposit has a runout of 1.7 km and a volume of 8×10^6 m^3. The deposit erodes down into the glacial outburst deposit and pinches out into the top of the glaciomarine unit, indicating a maximum age of 11.9–10 cal ka BP. At least two subsequent slide lobes extend from the moraine and are stacked on top of the large deposit. A mound of Holocene mud has accumulated on top of these deposits, indicating that the slide occurred before 10 cal ka BP.

In Amos Passage a variation of this style of failure occurs on a submarine plateau of residual stratified glaciomarine mud. The plateau is book-ended by 30–40 m-high escarpments that represent steeply dipping failure surfaces (AP076, 077; Fig. 4; Shaw *et al.* 2017). All of these failures occurred during deglaciation; however, the lack of Holocene

Table 4. *Regional slide distribution by channel (shown in Figs 6 & 8)*

Location	Holocene sedimentation	Slide activity
Squally Channel	Limited deposition – high bottom currents	2 slope failures. SQ001 identified by enormous headwall. Largest slide in Douglas Channel; occurred during the late-glacial stage
Whale Channel	Very thick, gassy Holocene mud interval	3 slope failures. The rocky deposit of WC03 (Fig. 11) was emplaced during the late glacial stage, thus leaving Holocene sediments undeformed. Two headwalls are observed on the west fjord wall
Hawkesbury West	Thick Holocene unit	26 slope failures; large events (HW007, 008 and 022) occurred during deglacial stage. Mostly debris flows that occur throughout the area during Holocene, the majority occurring between 4 and 8 cal ka BP. Many of these are stacked deposits over several thousand years (Fig. 8). Three events in past 2 cal ka BP (HW018, HW020, HW023). Short quiescent period following deglaciation (Fig. 8)
Miskatla Inlet and Giltoyees Inlet	No sub-bottom data or cores	14 slope failures based on bathymetry data. Poor age constraint. Slopes are exceedingly steep in this area and slide deposits are widespread throughout. Observed deposits are small in volume. Most of the slides are debris flows but there is a high number of rock slides relative to other parts of the fjord
Verney Passage and Ursula Channel	Limited deposition – high bottom currents	11 slide deposits based on bathymetry data. Poor age constraint. Two large composite slide deposits are in Verney Channel at the north end of Gribbel Island (VP044, 45). Recessional moraine in Ursula channel with two associated slide deposits (Fig. 3)
Gardener Canal	No sub-bottom data or cores	8 small debris flows deposits have been identified by surface morphology
Devastation Channel	Thick Holocene unit	5 slope failures. All of the very large deposits occurred during deglaciation or transition to Holocene conditions (DC060, 61, 63). The largest deposit is an unusually large debris flow deposit 71×10^6 m^3 (DC063; Figs 3 & 12). A translational slide associated with the Hugh Creek Fan was emplaced *c.* 6.5 cal ka BP
Sue Channel	No sub-bottom data or cores	2 slope failures. One blocky debris flow deposit extends from an embayment on Loretta Island. The deposit has subdued relief, probably indicating that it is an older deposit draped by Holocene deposition
Kitimat Arm	Thick Holocene unit	17 slope failures; all are debris flows with the majority $>10^6$ m^3. The Kitimat Delta slide complex is the largest slide deposit in Kitimat Arm (KA075). There are two populations of slide deposits; a group clustered around the area of Bish Cove and Clio Bay emplaced in the early-mid Holocene; and a second population surrounding the Kitimat Delta. Here, there is a high concentration of smaller fjord wall slides and several larger buried deposits that were all emplaced in the past 2.6 cal ka BP
Amos Passage	Thick Holocene unit except covering glaciomarine plateau	2 slope failures. Submarine plateau composed of glaciomarine sediment that has been truncated at the north and south end by steeply dipping failure surfaces (AP076, AP077). The failure surfaces are onlapped by glaciomarine deposition, implying that the failures occurred in the late glacial stage (Shaw *et al.* 2017).
Kildala Arm	Thick Holocene unit	28 slope failures. Kil078 extends east from a moraine in the direction of ice recession with a maximum age of 11.9 cal ka BP. Two smaller deposits are stacked above at the base of the Holocene unit. All of the observed slide deposits are debris flows and tend to be less than 10^6 m^3; however, Kil081 andKil082 run out across the narrow inlet. Most of the slides are repeat events throughout the Holocene and there are 11 small slides that have occurred in the past 3 cal ka BP

Characteristics of Holocene sediment fill are described. Hawkesbury West, Devastation Channel, Kitimat Arm, Amos Passage and Kildala Arm have thick Holocene units with good acoustic penetration. Where bottom currents are high there is limited Holocene deposition; Squally Channel and Verney Passage are characterized by margins with no Holocene deposition. In such instances relative dating of slope failures is not possible. In Miskatla Inlet and Giltoyees Inlet, Gardner Canal and Sue channel no sub-bottom data are available and as a result relative dating is not possible. Holocene muds in Whale Channel are gassy, resulting in acoustic blanking.

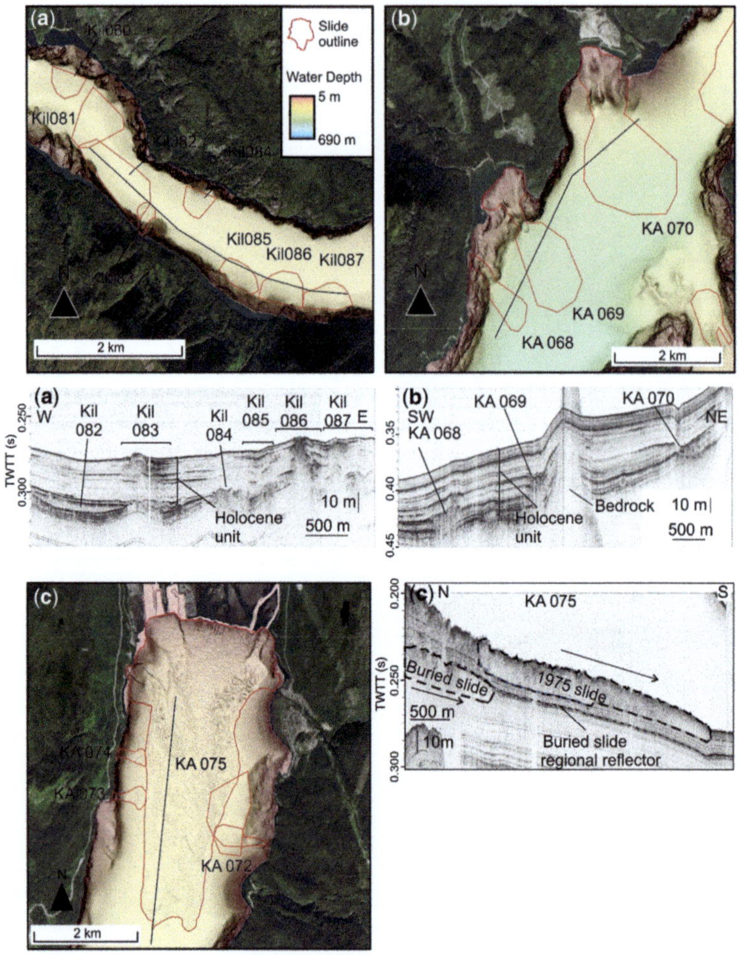

Fig. 9. Debris flow deposits (locations shown on Fig. 1). (**a**) Small deposits that result from very steep fjord walls in Kildala Arm. Slope gradients >40° are common. (**b**) Large slides in Kitimat Arm that have substantial submarine headwalls occur on more gradual slopes (10–15°) where sediment accumulation is thicker. (**c**) Kitimat delta slide complex. Four separate headwalls coalesce into a large slide complex with a total volume of 33×10^6 m^3. Sub-bottom profile shows slide lobes outlined with dashed lines. The 1975 slide lobe is at the surface and a deposit with similar acoustic architecture from 2.6 cal ka BP is buried. Arrows show flow direction.

deposition owing to high bottom currents at several of the moraines results in the inability to use acoustic stratigraphy to determine if any subsequent slides have occurred since deglaciation.

Rock slide. This class encompasses mass failures involving rock slumping or avalanche (Fig. 11). The distinguishing characteristic of rock slides is a rocky lobe deposit that extends from the fjord wall. These deposits tend to have a highly convex profile and do not have a long runout distance, contrasting with blocky debris flows deposits. Only five rock slides have been documented in Douglas Channel (Fig. 6).

One very large slide has occurred in Whale Channel on the east side of Gil Island in metasedimentary rocks of micaceous quartzite, crystalline limestone and schist (Fig. 11a; Roddick 1970). A large deposit interpreted as a bedrock slump extends 1.7 km from the fjord wall with a thickness of 200 m (Shaw *et al.* 2017). The estimated volume is 270×10^6 m^3. Holocene sediment at the toe of the slide are not deformed and thus the slide is inferred to have occurred during deglaciation.

Four small deposits are found in Giltoyees Inlet, which is characterized by quartz diorite, gneiss and migmatite, and gneissic quartz diorite (Fig. 11b; Roddick 1970). Two of these deposits are at the

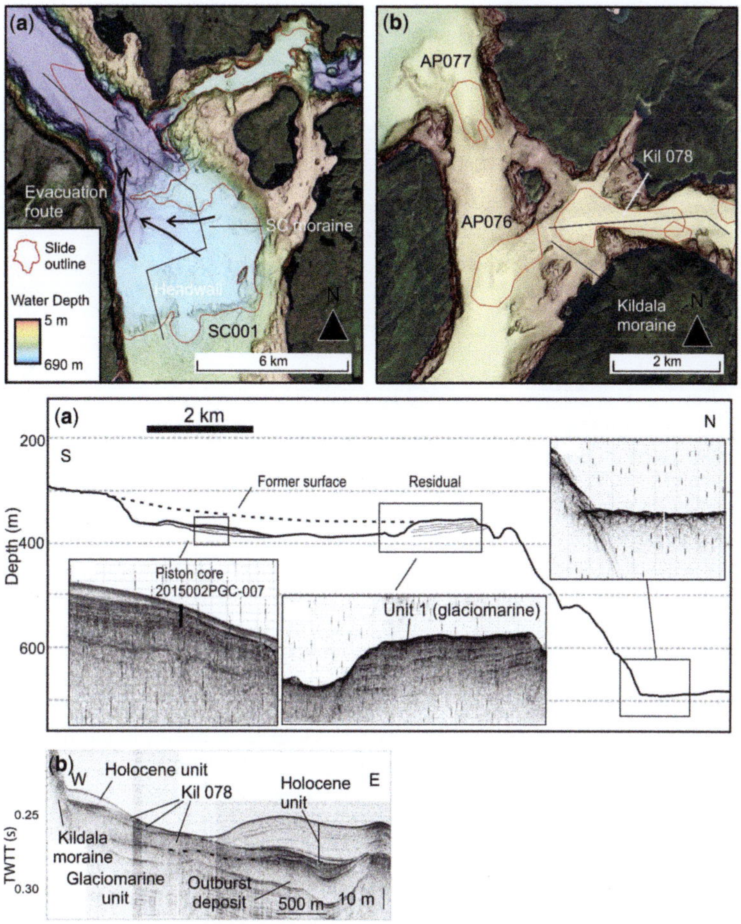

Fig. 10. Oversteepened glacial material. (**a**) The Squally Channel slide is the largest slide in Douglas Channel with a volume exceeding 8×10^8 m^3 (sub-bottom profile from Shaw *et al.* (2017) with permission from Elsevier). Arrows show the evacuation route around a residual plateau. The residual plateau is comprised of glaciomarine material. (**b**) Amos Passage glaciomarine plateau slides (AP077, 078) and Kildala moraine slide (Kil078). At least three wedge-shaped slide deposits originate from the moraine (visible in sub-bottom profile); the lower deposit erodes into the transparent outburst deposit.

mouth of Giltoyees Inlet where the subaerial slope exceeds 50° (GI 27, 28; Figs 4 & 11b). Headwall outlines are visible in satellite images but are obscured by dense forest. Below the water line the fjord wall shows several irregular knobs where the slope decreases; here, a rocky veneer has accumulated on the fjord wall and spilled onto the seafloor (GI 27; Figs 4 & 11b). This deposit has an estimated volume of 0.1×10^6 m^3. A second rock slide immediately to the north has run out onto the seafloor, extending 420 m from the fjord wall with a height of 5 m and an estimated volume of 0.5×10^6 m^3 (GI028; Figs 4 & 11). Neither sub-bottom data nor sediment cores were collected in Giltoyees Inlet, and timing is therefore unknown.

Translational. Translational failures are characterized by the downslope movement of a coherent mass along a plane of weakness. Only two slope failures fit into this category (Figs 6 & 12). HW025 has an arcuate headwall that forms at 340 m bsl (below sea-level), the only observed slide with no association with a fjord wall or steep glacial deposit (Figs 3 & 12a). Here the seafloor is smooth and slopes gradually towards the channel axis with a gradient ranging from 3 to 8°. Much of the deposit remains within the base of the headwall and maximum runout distance is 320 m. The slide deposit is coherent with intact regional reflections (Fig. 12a). Pockmarks are prevalent within the basin, but the closest cluster is 3 km from the NE

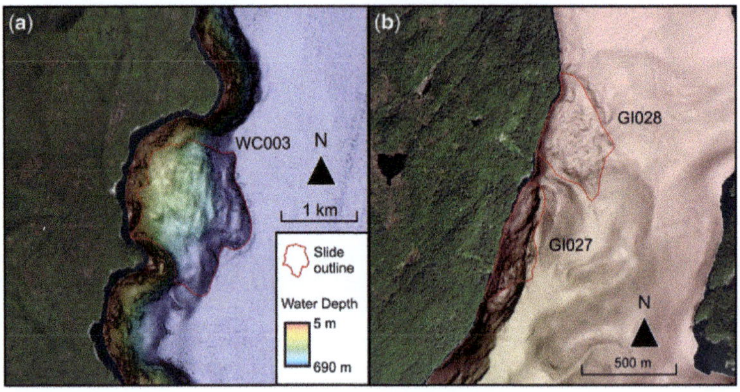

Fig. 11. Rock slides. (**a**) WC003 is in Whale Channel and is interpreted as a late glacial bedrock slump with an estimated volume of 3×10^8 m^3. Holocene sediments at the slide toe are not deformed, suggesting that this slope failure is not active. (**b**) Giltoyees Inlet rock slides are small but have subaerial components. The subaerial slope exceeds 50°. Sub-bottom data are not available in Giltoyees Inlet.

near the debris flow deposit HW026. This may indicate the presence of fluid flow. The slide lobe pinches out just above the regional reflector emplaced by the Kitimat Delta slide of 2.6 cal ka BP.

The second instance is Dev064 in Devastation Channel. The slide originates on the steep fjord wall south of the Hugh Creek fan. A long and narrow headwall is being infilled by sedimentation from Hugh Creek. The lobe has a round morphology in plan view and, although it has chaotic acoustic architecture in places, there are also continuous sections of convoluted primary structure (Fig. 12b). The slide lobe extends nearly the width of the fjord. At least one smaller slide has occurred at this location during the late Holocene, as evidenced by a high-amplitude reflection that is visible in sub-bottom profiles. This deposit has been dated to *c.* 6.5 cal ka BP.

Composite/other. A number of slope failures have unusual morphologies, probably as a result of multiple slides interacting. Four of these share similar characteristics; they are very wide relative to their runout distance (Figs 6 & 13a). These deposits tend to occur in areas where the subaerial slope is very steep (>40°). Two of these deposits are in Verney Pass (VP044, VP045; Fig. 3). The largest is VP045 (Fig. 13a). This deposit is 2500 m wide with a runout of 615 m, a thickness of 15 m and an estimated volume of 11×10^6 m^3. The surface has a shallow dip and is characterized by elongate transverse ridges and elongate blocks. Camera stations and samples have shown that the surface is covered in a sandy gravel lag (Shaw *et al.* 2017).

A second composite slide in Miskatla Inlet (MI032; Figs 4 & 13b) is 1000 m wide and extends 150 m from the fjord wall. The relief is variable between one and several metres. Several small round lobes stand out but the outline of the deposit is irregular with blocks oriented in all directions. Multiple lobes may be indicative of different stages of failure, or different events. The fjord wall is very steep here and with a lower gradient immediately to the north and south of the slide deposit.

Two large breaks in the fjord wall on the west side of Hawkesbury Island resemble headwalls in Douglas Channel (HW010, HW011; Fig. 3). These features are located between the sackungen (HW007, HW014). No deposit is observed in postglacial acoustic units, and as such these features are probably very old. Incision is more than 100 m deep with widths in excess of 1000 m, thus these failures probably involved large volumes of bedrock.

Sackungen (deep-seated gravitational slope displacement). Two very large bedrock blocks (sackungen) have detached from the steep fjord wall on the west side of Hawkesbury Island, resulting in steep, crescent-shaped headwalls (HW007 and HW014; Figs 3 & 14). HW007 has a volume of 63×10^6 m^3 and HW014 has a volume of 70×10^6 m^3 (Fig. 14a; Thomson *et al.* 2012). The southern block (HW007) is developed in gneissic–diorite migmatite and the northern block (HW0014) is developed in quartz diorite (Roddick 1970). LiDAR datasets acquired along the Hawkesbury Island shoreline show well-developed bedrock joint and fault patterns indicating structural control of the form, size and location of the blocks (Fig. 14b; Conway and Barrie 2018). A 50 km lineament interpreted as a fault extends from Hawkesbury Island south to the entrance of Douglas Channel. Evidence for faulting includes aligned stream beds crossing topographic and drainage divides and en echelon fractures at the south end of Hawksebury Island,

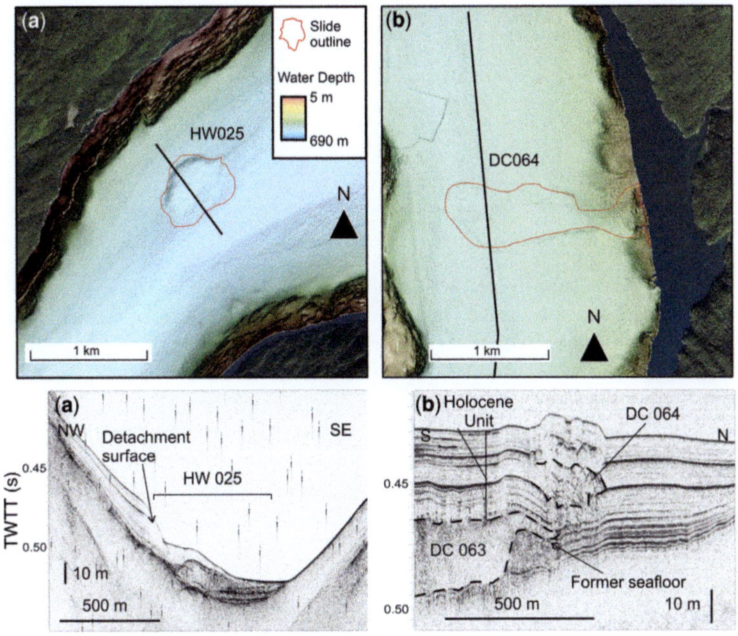

Fig. 12. Translational slides. (**a**) Crescent-shaped headwall west of Maitland Island occurs entirely within Holocene mud on fjord floor, rather than on the fjord wall. Seafloor gradient is 3–8°. Much of the deposit remains within the headwall and the translated block still has coherent acoustic reflectors. (**b**) Slide DC064 with long runout distance. The lobe has a round shape in plan view and has some coherent acoustic structure. Sub-bottom profile shows buried DC063 that was emplaced during the deglacial period.

4 km from the southern block (Conway *et al.* 2012). Soft sediment deformation, including anticlines and faulting, has been observed in seismic profiles in Pleistocene marine sediments (Fig. 14b). Fault traces associated with the block failures align with a fault trace on Gribbel Island (Fig. 14). Locally, faults associated with the sackungen may have been reactivated during glacial loading and unloading or may be part of an existing dextral-shear stress regime. The bedrock blocks were episodically mobile between 15.8 and 13.4 cal ka BP following deglaciation as ice retreat resulted in a debutressing effect on the faulted shoreline (Conway and Barrie 2018).

Based on deformation in the Pleistocene sedimentary record, block movement has occurred incrementally. The sagging block has resulted in a series of anticlines downslope with some evidence of brittle deformation at boundaries of poorly developed folds. Deformation occurs in intervals that are sometimes capped by debris flow deposits and make up a thick sequence which overlies the toe of the sliding block (Conway and Barrie 2015). The deformation sequence is draped by an undeformed interval of glaciomarine and Holocene muds, indicating no recent movement.

Smaller nested slides have occurred in the backfill in the crescent-shaped headwall (Conway and Barrie 2015; HW008, HW009, HW015, HW016; Figs 3 & 14). These slides are classified as debris flow deposits based on depositional characteristics and range in volume from 1.6×10^6 to $3.0 \times 10^6\ m^3$. The figures represent total volume; however, failures probably occurred in different stages. The northernmost of these nested slides (HW016) has a runout of 1.4 km, beyond the central axis of the fjord. The northern block (HW014) also hosts a failure on its leading surface that is interpreted as a surficial mud veneer (HW017).

Discussion

Major regional phase of slope failures immediately following deglaciation

As ice receded inland through Douglas Channel, the region experienced its most extensive phase of failure activity. Slope failures dated from 15.5 to 10.0 cal ka BP are found in every part of the fjord. This phase is also when the largest volume failures were produced. The two largest volume failures occurred at the Squally Channel moraine and the Kitkiata moraine and are directly associated with glacial recession (Figs 2, 3 & 10). Receding ice left an

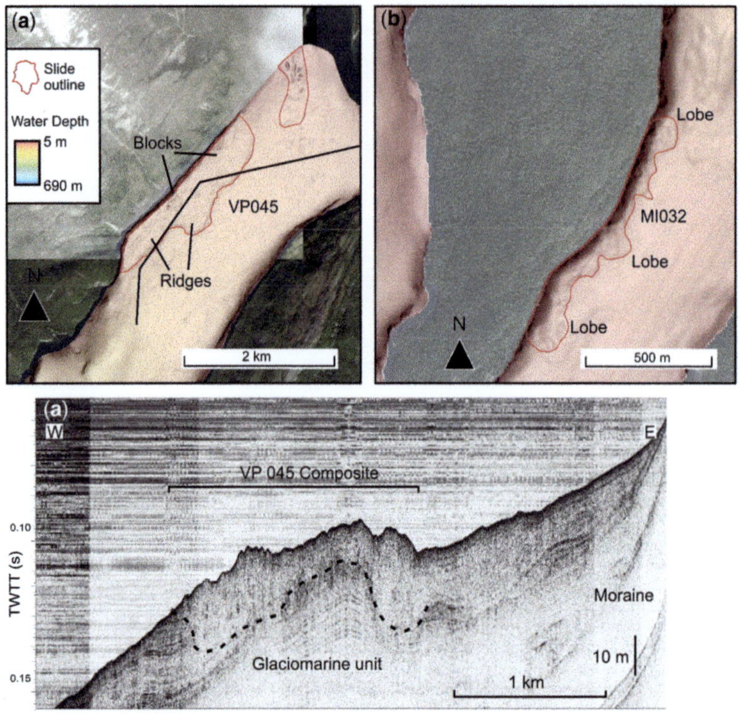

Fig. 13. Composite slides. (**a**) Verney Passage composite slide (VP045) has a width far larger than its runout length and a surface characterized by blocks and ridges. Sub-bottom image shows that the slide eroded down into the glaciomarine unit and is not covered by Holocene deposition. Subaerial slope is >40°. (**b**) Miskatla Inlet composite slide characterized by a number of blocky lobe deposits with short runout. Multiple lobes may indicate different stages of failure.

over-deepened basin and an oversteepened slope of sediment that failed with volumes of 8×10^8 and 3×10^8 m^3, respectively (Shaw *et al.* 2017). Large failures have occurred at all other moraines in the fjord with the exception of the Verney Passage moraine (Fig. 1). Sub-bottom records at the Devastation Channel and Kildala Arm moraines indicate that multiple failures occurred in the upstream direction (Figs 3 & 10). Headwalls on moraines remain very steep, and although there is no indication of Holocene failure activity (in part because bottom currents prevent Holocene deposition), these slopes remain oversteepened and could be subject to further slide activity. As Holocene deposition drapes even very steep fjord wall surfaces, there is potential for smaller-scale failures of this accumulated material well after the deglacial period (Stacey *et al.* 2019).

As ice receded, the effect of glacio-isostatic activity was presumably a factor in further slope instability (Barrie and Conway 2002; Hetherington *et al.* 2004; Hetherington and Barrie 2004). Bedrock block sackung features on the west side of Hawkesbury Island are interpreted to have been active from 15.8 to 13.4 cal ka BP (Conway and Barrie 2018). These blocks each have volumes greater than 60×10^6 m^3 (Thomson *et al.* 2012) and are associated with multiple sets of faults. Glacial loading and unloading may have reactivated these faults contributing to the very large block sackungen. During this time of instability a very large debris flow deposit (HW006) with a volume of 14×10^6 m^3 was emplaced in front of the southern block HW007 (Figs 3 & 14). Other smaller debris flows were also active in Douglas Channel at this time (Fig. 8), probably associated with local faulting and slow movement of the bedrock blocks. In Devastation Channel the largest observed debris flow with a volume of 71×10^6 m^3 occurred during the deglacial stage (DC063; Figs 3 & 12).

Holocene slope failures: shift from period of variable failure styles to debris flows and rock slides

Holocene conditions developed throughout Douglas Channel between *c.* 12 cal ka BP in the outer fjord and 10 cal ka BP in Kitimat Arm. Throughout this phase, slope failure activity generally occurred at a

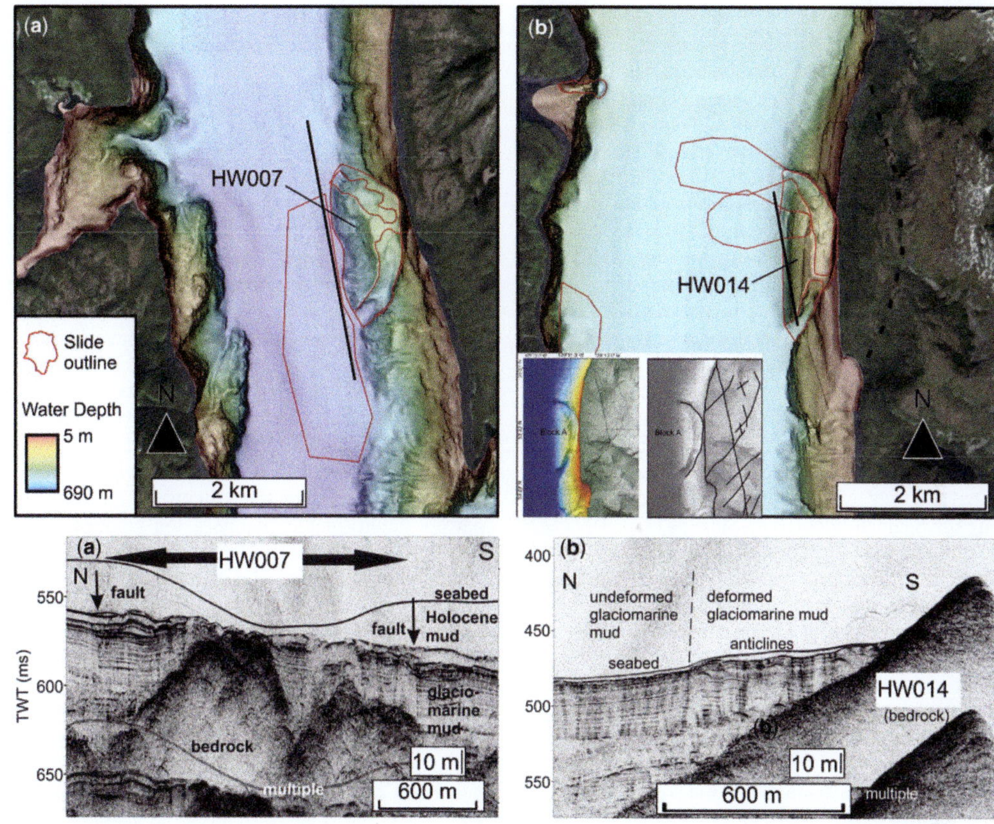

Fig. 14. Block Creep. Two bedrock blocks are detached from the west side of Hawkesbury Island each with volumes in excess of 60×10^6 m^3. Slow creep occurred episodically from 15.8 to 13.4 cal ka BP resulting in deformation of Pleistocene sediment in front of the blocks (Conway and Barrie 2018). (**a**) Several large debris flows have occurred in association with the large bedrock block during the deglacial stage and during the Holocene. Fault offsets are visible in glaciomarine mud unit in sub-bottom profile. (**b**) Fault traces on Hawkesbury Island shown with dashed line on satellite image and with solid lines on Inset map. A series of anticlines have formed owing to compressive stress at the toe of the sackungen. Sub-bottom images from Conway and Barrie (2015).

reduced rate and at lower magnitudes, however with at least marginal amounts of activity in all fjord subsections. A marked transition occurs in the Holocene record of slope failures from a mix of all failure types to primarily debris flows and rock slides (Fig. 8). Debris flows become the most common and most voluminous slope failures.

Debris flow deposits occurred in very high numbers in Kitimat Arm and Kildala Arm in the late Holocene with return intervals of 1/667 and 1/400 years, respectively (Figs 6–8). These regions also have the highest concentration of repeat events (Fig. 7). It is possible that there are buried events that were undetected owing to the attenuation of acoustic signal in recent sandy deposits, so this pattern could extend beyond the late Holocene. A number of preconditioning factors contribute to this high rate of slope failure activity, but most importantly the sedimentation rate and fjord wall slope. The Kitimat River is the primary source of sediment for the fjord and fjord wall sediment accumulations are thick in Kitimat Arm where the submarine fjord-wall slope is low enough to allow deposition (Shaw and Lintern 2016).

The postglacial depositional phase shows no clear regionally distributed coeval events that would indicate seismicity had a substantial effect on slope stability. Coevally triggered MTDs are an indicator for a regional seismic trigger (Moernaut *et al.* 2007; Howarth *et al.* 2014; Bellwald *et al.* 2016). Clusters of slope failures within sub-basins do not correlate well across the entire Douglas Channel fjord (Fig. 8). In the Hawkesbury West basin there are six slides that occurred between 4 and 6 cal ka BP, more than any other 2 cal ka BP window in this sub-basin; however, in Kitimat Arm and Kildala

Arm the larger clusters of slide activity occurred in the early Holocene or over the past *c.* 3 cal ka BP (Fig. 8). There are a number of slides in Kitimat Arm, north of the Kitkiata moraine, that have been dated in close proximity to the 2.6 cal ka BP Kitimat Delta slide, including the translational slide HW025 (Fig. 12) and Kitimat wall slide DC072. It is possible, however, that the regional turbidite produced by the Kitimat Delta 2.6 cal ka BP slide (observed in sub-bottom profiles throughout the basin) acted as a failure plane, resulting in multiple failures at this depth. Kitimat Fjord wall slide KA072 occurred along this failure plane and there is ongoing work investigating the possibility that the tsunamigenic 1975 Kitimat slide occurred along the same failure plane.

Regional slides in other coastal fjords associated with the Cascadia subduction zone have been correlated and used as palaeoseismicity indicators, including the Saanich Inlet (Blais-Stevens *et al.* 1997) and Effingham Inlet (Enkin *et al.* 2013). Two active plate boundaries are worth considering for potential seismic trigger for slope failure activity. The closest active margin is the Queen Charlotte–Fairweather transform fault west of Haida Gwaii (Fig. 1). The fault zone has been known to produce large strike-slip earthquakes including Canada's largest recorded earthquake of 8.1 M_w in 1949 and a 7.8 M_w earthquake in 2012. The only known large landslide activities in Douglas Channel over the past 100 years are the Kitimat Delta slides in 1974 and 1975. These slides do not correspond to the timing of earthquake events and are linked to other failure mechanisms. The risk of a large slope failure in Douglas Channel triggered by seismicity at this fault zone is considered to be low (Brillon 2016). The Douglas Channel fjord system is more than 300 km north of the subducting Juan de Fuca plate. This subduction zone has produced 20 megathrust earthquakes over the past 10 cal ka BP (Goldfinger *et al.* 2012). However, energy from a megathrust earthquake produced here would be greatly dissipated and unlikely to activate slope failure at this distance (Keefer 2002).

Several faults have been observed in the region, including the Grenville Fault and a north–south-trending fault zone within Douglas Channel (Fig. 1). Modelling has indicated that a strike-slip rupture along the 60 km Douglas Channel fault could potentially generate a 7.2 M_w earthquake (Brillon 2016). Such an earthquake could result in moderate to heavy damage within 20 km of the rupture (Brillon 2016), including much of the Douglas Channel fjord. Conway and Barrie (2018) have speculated that some of these faults may have been activated/reactivated during deglaciation; however, deformed sediments associated with faulting have since been buried by Holocene sediment that show no indication of active faulting. The Geological Survey of Canada has monitored seismicity in the Douglas Channel area since 1973 with no record of substantial seismic activity along these fault lines (Brillon 2016).

Geohazard implications

Tsunami risk: is there evidence that modern slope failure activity could be tsunamigenic? Landslide-induced tsunamis in fjord environments are well documented. The largest known tsunami height of 524 m occurred in Lituya Bay, Alaska in 1958 (Weiss *et al.* 2009). In Knight Inlet, British Columbia, the first nations community of Kwalate was devastated by a large tsunami that was generated by a major rockslide, the age of which is unknown (Bornhold *et al.* 2007). Tsunami models in Douglas Channel have shown that, owing to the long and narrow nature of the channels, the greatest risk of tsunami does not occur from offshore earthquake activity but from local landslides (Thomson *et al.* 2012; Kirby *et al.* 2016). The highest risk of damaging tsunamis occurs in close proximity to the landslide source since tsunami height decays quickly with distance along the fjord (Thomson *et al.* 2012; Kirby *et al.* 2016). The main population centres in Douglas Channel are at Kitimat and Kitimaat Village at the head of the fjord, an area of known landslide-generated tsunami activity (KA075), and Hartley Bay across the fjord from the large sagging blocks (HW007 and 014). Landslide-generated tsunamis are a hazard to shipping and other activities elsewhere in the fjord.

The largest slope failures in the Douglas Channel fjord occurred during deglaciation, several having volumes in excess of 10^8 m^3. The primary preconditioning mechanism for the largest failures is oversteepening of glacial material. Although head scarps on the moraines that hosted the largest failures remain steep, there is no evidence of large-scale mass wasting throughout the Holocene. The large Hawkesbury block failures also occurred during deglaciation. Block dislocation was modelled for potential tsunami generation, and based on their rigid structure and volume, a potential surface wave could have reached 40 m in amplitude, but only in a scenario where displacement occurred as one instantaneous event (Thomson *et al.* 2012). This scenario is unlikely, however, as soft sediment deformation and brittle faulting in the late glacial sediment record indicate that displacement occurred episodically over a period of several hundred to thousands of years (Conway and Barrie 2018). Based on the Holocene sediment record these blocks do not appear to have experienced any further displacement during the past *c.* 13.4 cal ka BP (Conway and Barrie 2018).

Many of the larger ($>10^6$ m^3) Holocene events occurred in Kitimat Arm on slopes with lower gradients and thick sediment accumulation. These failures are associated with large headwalls that in many cases are likely to have some subaerial contribution. Their occurrence in shallow water, in combination with a subaerial component, results in increased potential tsunami heights (Fine *et al.* 2003; Thomson *et al.* 2012; Kirby *et al.* 2016). In Kitimat Arm this style of failure occurred in locations from Emsley Cove (KA069) and Bish Cove (KA070) north to the Kitimat Delta (Figs 4 & 9). The 1975 Kitimat slide is an example of this type of failure, occurring on the west fjord wall where thick sediments have accumulated. The headwall area of this slide comprises a sediment volume of 4×10^6 m^3 and involved some subaerial volume. This slide generated an 8 m tsunami that was damaging to the communities of Kitimat and Kitimaat Village (Golder Associates 1975; Murty and Brown 1979; Thomson *et al.* 2012). A factor of safety analysis suggests that the slope angles on the western fjord walls are sufficiently steep that only a little additional loading from recent deposition would be required to produce slope failure (Lintern *et al.* 2019).

Debris flows with large submarine headwall volumes are rare elsewhere in Douglas Channel owing to steep fjord walls and lower sedimentation rates. In locations with fjord wall slopes higher than 40° debris flows are smaller in volume but most of that volume originated in the subaerial environment (Kildala Arm, Giltoyees, Verney Passage). Even though these slides tend to be smaller, the nature of the entire slide volume entering the water at a high velocity is effective for tsunami generation (Fine *et al.* 2003). Rock slides and composite slides represent an equal risk on steep slopes in Verney Passage, Giltoyees Inlet and Miskatla Inlet, although with poor age constraint the level of Holocene activity for these slides is unknown. Several large Holocene debris flow deposits ($>10^6$ m^3) in Hawkesbury West (HW012, 013 and 019; Fig. 3) are linked to subaerial mass wasting on very steep slopes. These slides would pose a greater risk of tsunami generation. All three of these slide locations show stacked deposits throughout the mid-Holocene, probably owing to the steep slope.

Sandy gravel beds interpreted as ancient tsunami deposits have been observed in Minette Bay and Clio Bay, at the head and 11 km south of the head of Kitimat Arm (Huntley *et al.* 2018). These deposits have not been dated; however, there is a possibility that these can be linked to ancient slides at the Kitimat Delta, or buried mid-Holocene slides at Bish Cove or Clio Bay (KA070, 071; Fig. 9). Owing to the steep fjord walls, tsunamite preservation potential is low in other locations of Kitimat Arm and few other sites have been found that would potentially have a sedimentary record of tsunamis generated by landslides. Work is ongoing in Kitimat Arm to link potential palaeo-tsunami deposits to slide events that have been dated in this contribution.

Risk to seabed infrastructure from landslide runout and turbidity current activity. Another known risk associated with submarine slides is damage to seafloor infrastructure from rapid movement of the slide body or by turbidity currents (Piper *et al.* 1999; Pope *et al.* 2017). Large landslides with long runout distances are especially dangerous with an increased ability to destroy submarine infrastructure. Although there is currently no seafloor infrastructure other than marinas and docks, there are a number of industrial projects that are in planning or construction phases at the Head of Kitimat Arm. Any future seafloor construction would have to consider local hazards. There are numerous locations where Holocene slide deposits have crossed the channel axis or in some cases even filled the entire width of the fjord. Of particular concern to developing infrastructure, Kitimat Arm has the greatest concentration of Holocene debris flow deposits that have crossed the fjord axis (KA069, 070, 071, 075; Figs 4 & 9). Careful attention to the impact of landslides on marine infrastructure is required at these sites. Even small slides have the potential to damage infrastructure (Forsberg *et al.* 2016; Clare *et al.* 2017). Small Holocene slides have been observed in all channels of the fjord.

Turbidity currents are sediment-laden flows that move downslope through their excess density and are also hazardous to seafloor infrastructure (Bruschi *et al.* 2006; Parker *et al.* 2009; Clare *et al.* 2015). These destructive flows can accompany debris flows and other modes of slope failure (Sumner *et al.* 2009; Talling 2013) and in fjord environments where the seafloor is flat, turbidity currents can run out for tens of kilometres from sandy delta sources (Prior *et al.* 1986; Stacey *et al.* 2019). Slides on the Kitimat Delta have produced sandy turbidity current activity that has run out as far as 50 km south to Hawkesbury Island (Stacey *et al.* 2018). Any sandy delta in Douglas Channel has the potential to produce a turbidity current with substantial runout distance triggered by landslide activity, therefore posing a risk to future seafloor infrastructure. Kitimat Delta, Kildala Delta and Kitlope Delta at the head of Gardner Canal have the greatest potential to produce long runout turbidity currents owing to their high rates of sandy sedimentation. Small turbidites with limited runout are not documented in this contribution, yet are present at most small fan deltas and distal extents of debris flows and other slides in Douglas Channel, and must be considered a hazard for local infrastructure. River flood events are also capable of generating sediment plumes that can initiate turbidity currents with long runout distances and as such should be considered a risk to seafloor infrastructure (Hizzett *et al.* 2018; Hage *et al.* 2019).

Extension to other fjord environments. Douglas Channel was the subject of intensive survey from 2013 to 2018 owing to large infrastructure projects being proposed there. Prior to this project, the large landslide at Kitimat Delta had been investigated, and only a few others noted. A total of seven expeditions were at least partially dedicated to mapping and dating the remainder of the failures described here. Of these, two expeditions would have captured most of the bathymetry, sub-bottom data and cores required for this analysis, to lead to the discovery of a large number of additional submarine slope failures. The knowledge gained of this level of effort is important for planning similar hazard assessments in other fjords in Canada and globally.

Considering the extensive level of analysis, it might also be possible to translate numbers of magnitude and frequencies of submarine landslides over to lesser known fjords. Most of the fjords in British Columbia are uninhabited. There are very few documented occurrences of tsunamis, but nevertheless there are some oral histories, such as a destructive tsunami in Knight Inlet (Bornhold *et al.* 2007) and a freak wave forming from the centre of Sechelt Inlet (G. Lintern pers. comm.). Landslides are well documented in Howe Sound (Blais-Stevens and Septer 2008; Stacey *et al.* 2019) and Saanich Inlet (Blais-Stevens *et al.* 1997), which are the only other populated and studied fjords. Turbidity currents are also well documented in Howe Sound (Hughes Clarke *et al.* 2012; Stacey *et al.* 2019). Most of the fjords in British Columbia are characterized by steep granitic sidewalls with frequent debris flows. Nearly all have a significant river delta (sometimes two) at the head, which is fed in all cases by glacial or snow melt enhanced by precipitation in a coastal rainforest. With the high sedimentation associated with rivers at fjord heads there is a likelihood of increased slide activity at the heads of other fjords through the late Holocene. Some of the other west coast fjords have intermediate and terminal moraines and similar glacial histories to Douglas Channel (by definition, fjords are formed by glaciers). Given the similar geological structure and history of the Douglas Channel fjord to many of the other fjords, a tsunamigenic hazard of landslides is likely to exist.

Other glaciated regions of the world may expect to share similar conditions. Thorough regional-scale analyses have been completed in Norway and Switzerland, indicating similar results. Bellwald *et al.* (2019) compiled mass transport deposits over 22 Norwegian fjords. Landslides were most frequent following periods of glacial advance and recession. Their comparison of multiple fjords have indicated that seismic activity related to postglacial isostatic rebound is probably a trigger mechanism for coeval slide events. Rapid ice melting also exposed the fjord catchment to increased erosion. Furthermore, the largest mass transport deposits are associated with ice-sheet retreat (Hjelstuen *et al.* 2009; Bellwald *et al.* 2016). Kremer *et al.* (2017) recorded a high number of coeval mass transport deposits associated with seismic activity following deglaciation in lakes of the Swiss Alps.

Summary

- Slope failures in the submarine environment of Douglas Channel were classified with volume estimations and timing determined for all known events. Failures were documented in 89 locations; many of these were repeat events.
- The largest slope failures occurred between 15.5 and 10.0 cal ka BP during deglaciation, including very large slides (up to 8×10^8 m^3) at five of the six recessional moraines and two large bedrock block failures with volumes in excess of 60×10^6 m^3.
- Slope failure activity during the Holocene is characterized by smaller slides that are less frequent; however they have been active in all parts of Douglas Channel, especially Kitimat Arm and Kildala Arm. Fjord heads experience higher sedimentation rates which contribute to increased numbers of submarine slope failures.
- Slides that may be capable of generating tsunamis and damaging seafloor infrastructure have occurred in all parts of Douglas Channel through much of the Holocene.
- Fifty-two slide deposits are linked to the subaerial environment and are associated with fjord walls with very high slope gradient. These slides tend to be smaller than submarine slides that occur on fjord walls where slope is lower, permitting higher sediment accumulation resulting in slides larger than 10^6 m^3.

Acknowledgements The authors would like to thank the following groups for their assistance with this research: Audrey Dallimore and Royal Roads University for the use of their multisensor core logger which collected all core images; and Shahin Dashtgard at Simon Fraser University for the use of their X-ray laboratory. We also thank the captains and crew of CCGS *Vector* and CCGS *John P. Tully*. This work was greatly improved through the thoughtful input of reviewers Michael Clare and Benjamin Bellwald. This work is a product of Natural Resources Canada's North Coast Marine Geohazards project. Funding for this project has been provided by Natural Resources Canada and Transport Canada.

Funding Funding for the North Coast Marine Geohazards project has been provided by Natural Resources Canada and Transport Canada.

Author contributions **CDS**: writing – original draft (lead); **DGL**: writing – original draft (supporting); **JS**: investigation (supporting); **KWC**: writing – original draft (supporting).

References

Barrie, J.V. and Conway, K.W. 2002. Rapid sea-level change and coastal evolution of the Pacific margin of Canada. *Sedimentary Geology*, **150**, 171–183, https://doi.org/10.1016/S0037-0738(01)00274-3

Bellwald, B., Hjelstuen, B.O., Sejrup, H.P. and Haflidason, H. 2016. Postglacial mass movements and depositional environments in a high-latitude fjord system – Hardangerfjorden, Western Norway. *Marine Geology*, **379**, 157–175, https://doi.org/10.1016/j.margeo.2016.06.002

Bellwald, B., Hjelstuen, B.O., Sejrup, H.P., Stokowy, T. and Kuvås, J. 2019. Holocene mass movements in west and mid-Norwegian fjords and lakes. *Marine Geology*, **407**, 192–212, https://doi.org/10.1016/j.margeo.2018.11.007

Blais-Stevens, A. and Septer, D. 2008. *Historical Accounts of Landslides and Flooding Events Along the Sea to Sky Corridor, British Columbia, from 1855–2007*. Geological Survey of Canada, Open File 5741, https://doi.org/10.4095/224755

Blais-Stevens, A., Clague, J.J., Bobrowsky, P.T. and Patterson, R.T. 1997. Late Holocene sedimentation in Saanich Inlet, British Columbia, and its paleoseismic implications. *Canadian Journal of Earth Science*, **34**, 1345–1357, https://doi.org/10.1139/e17-107

Bornhold, B.D. 1983. Sedimentation in Douglas Channel and Kitimat Arm. *In*: Macdonald, R.W. (ed.) *Proceedings of a Workshop on the Kitimat Marine Environment*. Canadian Technical Report of Hydrography and Ocean Sciences, **18**, 89–114.

Bornhold, B.D., Harper, J.R., Mclaren, D. and Thomson, R.E. 2007. Destruction of the first nations village of Kwalate by a rock avalanche-generated tsunami. *Atmosphere–Ocean*, **45**, 123–128, https://doi.org/10.3137/ao.450205

Brillon, C. 2016. *Baseline Assessment of Seismic Hazards in British Columbia's North Coast*. Geological Survey of Canada Open File 7994, https://doi.org/10.4095/297566

Bruschi, R., Bughi, S., Spinazzè, M., Torselletti, E. and Vitali, L. 2006.Impact of debris flows and turbidity currents on seafloor structures *Norsk Geologisk Tidsskrift*, **86**, 317–336.

Clague, J.J. 1985. Deglaciation of the Prince Rupert–Kitimat area, British Columbia. *Canadian Journal of Earth Sciences*, **22**, 256–265.

Clague, J.J. and James, T.S. 2002. History and isostatic effects of the last ice sheet in southern British Columbia. *Quaternary Science Reviews*, **21**, 71–87, https://doi.org/10.1016/S0277-3791(01)00070-1

Clare, M.A., Cartigny, M.J.B. *et al.* 2015. Quantification of near-bed dense layers and implications for seafloor structures: new insights into the most hazardous aspects of turbidity currents. *Proceedings of the Offshore Technology Conference*, Richardson, Texas.

Clare, M.A., Vardy, M.E. *et al.* 2017. Direct monitoring of active geohazards: emerging geophysical tools for deep-water assessments. *Near Surface Geophysics*, **15**, 427–444, https://doi.org/10.3997/1873-0604.2017033

Clare, M., Chaytor, J. *et al.* 2018. A consistent global approach for the morphometric characterization of subaqueous landslides. *Geological Society, London, Special Publications*, **477**, https://doi.org/10.1144/SP477.15

Conway, K.W. and Barrie, J.V. 2015. Large submarine slope failures and associated Quaternary faults in Douglas Channel, British Columbia. *Geological Survey of Canada, Current Research*, **2015-9**, https://doi.org/10.4095/297316

Conway, K.W. and Barrie, J.V. 2018. Large bedrock slope failures in a British Columbia, Canada fjord: first documented submarine sackugen. *Geo-Marine Letters*, **38**, 195–209, https://doi.org/10.1007/s00367-018-0533-y

Conway, K.W., Barrie, J.V. and Thompson, R.E. 2012. Submarine slope failures and tsunami hazard in coastal British Columbia: Douglas channel and Kitimat Arm. *Geological Survey of Canada, Current Research*, **2012-10**, https://doi.org/10.4095/291732

Conway, K.W., Kung, R.B., Barrie, J.V., Hill, P.R. and Lintern, D.G. 2013. *A Preliminary Assessment of the Occurrence of Submarine Slope Failures in Coastal British Columbia by Analysis of Swath Multibeam Bathymetric Data Collected 2001–2011*. Geological Survey of Canada, Open File 7348, https://doi.org/10.4095/292370

Enkin, R.J., Dallimore, A., Baker, J., Southon, J.R. and Ivanochkod, T. 2013. A new high-resolution radiocarbon Bayesian age model of the Holocene and Late Pleistocene from core MD02-2494 and others, Effingham Inlet, British Columbia, Canada; with an application to the paleoseismic event chronology of the Cascadia Subduction Zone. *Canadian Journal of Earth Science*, **50**, 746–760, https://doi.org/10.1139/cjes-2012-0150

Fine, I.V., Rabinovich, A.B., Thomson, R.E. and Kulikov, E.A. 2003. Numerical modeling of tsunami generation by submarine and subaerial landslides. *In*: Yalciner, A.C., Pelinovsky, E.N., Synolakis, C.E. and Okal, E. (eds) *Submarine Landslides and Tsunamis*. Kluwer Academic, Dordrecht, 69–88.

Forsberg, C.F., Heyerdahl, H. and Solheim, A. 2016. Underwater mass movements in lake Mjøsa, Norway. *In*: Lamarche, G., Mountjoy, J. *et al.* (eds) *Submarine Mass Movements and their Consequences*. Advances in Natural and Technological Hazards Research, **41**, 191–199. Springer, Cham. https://doi.org/10.1007/978-3-319-20979-1_19

Golder Associates 1975 *Report to B. C. Water Resources Service in investigation of seawave at Kitimat B.C.* Technical Report **V75077**, Golder, Brawner & Associates.

Goldfinger, C., Nelson, C.H. *et al.* 2012. Turbidite event history – methods and implications for Holocene paleoseismicity of the Cascadia subduction zone. US Geological Survey Professional Papers, 1661-F, https://pubs.usgs.gov/pp/pp1661f/

Hage, S., Cartigny, M.J. *et al.* 2019. Direct monitoring reveals initiation of turbidity currents from extremely dilute river plumes. *Geophysical Research Letters*, **46**, 11310–11320, https://doi.org/10.1029/2019GL084526

Hetherington, R. and Barrie, J.V. 2004. Interaction between local tectonics and glacial unloading on the Pacific margin of Canada. *Quaternary International*, **120**, 65–77, https://doi.org/10.1016/j.quaint.2004.01.007

Hetherington, R., Barrie, J.V., Reid, R.G.B., Macleod, R. and Smith, D.J. 2004. Paleogeography, glacially induced crustal displacement, and Late Quaternary coastlines on the continental shelf of British Columbia, Canada. *Quaternary Science Reviews*, **23**, 295–318, https://doi.org/10.1016/j.quascirev.2003.04.001

Hizzett, J.L., Hughes Clarke, J.E., Sumner, E.J., Cartigny, M.J.B., Talling, P.J. and Clare, M.A. 2018. Which triggers produce the most erosive, frequent, and longest runout turbidity currents on deltas? *Geophysical Research Letters*, **45**, 855–863, https://doi.org/10.1002/2017GL075751

Hjelstuen, B.O., Haflidason, H., Sejrup, H.P. and Lyså, A. 2009. Sedimentary processes and depositional environments in glaciated fjord systems – evidence from Nordfjord, Norway. *Marine Geology*, **258**, 88–99, https://doi.org/10.1016/j.margeo.2008.11.010

Holland, S. 1976. *Landforms of British Columbia – a Physiographic Outline*. British Columbia Department of Mines and Petroleum Resources, Bulletins, **48**.

Howarth, J.D., Fitzsimons, S.J., Norris, R.J. and Jacobsen, G.E. 2014. Lake sediments record high intensity shaking that provides insight into the location and rupture length of large earthquakes on the Alpine Fault, New Zealand. *Earth and Planetary Science Letters*, **403**, 340–351, https://doi.org/10.1016/j.epsl.2014.07.008

Hughes Clarke, J.E., Brucker, S. *et al.* 2012. The Squamish ProDelta: monitoring active landslides and turbidity currents. *Canadian Hydrographic Conference 2012, Proceedings*, Niagara Falls, Canada.

Huntley, D., Bobrowsky, P., Goff, J., Chagué, C., Stead, D., Donati, D. and Mariampillai, D. 2018. Extending the terrestrial depositional record of marine geohazards in coastal NW British Columbia. *Geological Society, London, Special Publications*, **477**, 277–292, https://doi.org/10.1144/SP477.4

Josenhans, H.W., Fedje, D., Pienitz, R. and Southon, J. 1997. Early humans and rapidly changing Holocene sea levels in the Queen Charlotte Islands–Hecate Strait, British Columbia, Canada. *Science*, **277**, 71–74, https://doi.org/10.1126/science.277.5322.71

Keefer, D.K. 2002. Investigating landslides caused by earthquakes – a historical review. *Surveys in Geophysics*, **23**, 473–510, https://doi.org/10.1023/A:1021274710840

Kirby, J.T., Shi, F., Nicolsky, D. and Misra, S. 2016. The 27 April 1975 Kitimat, British Columbia, submarine landslide tsunami: a comparison of modeling approaches. *Landslides*, **13**, 1421–1434, https://doi.org/10.1007/s10346-016-0682-x

Kremer, K., Wirth, S.B. *et al.* 2017. Lake-sediment record as earthquake catalogue in intraplate settings: limitations and Perspectives. *Quaternary Science Reviews*, **168**, 1–18, https://doi.org/10.1016/j.quascirev.2017.04.026

Lintern, G., Blais-Stevens, A. *et al.* 2019. Providing multidisciplinary scientific advice for coastal planning in Kitimat Arm, British Columbia. *Geological Society, London, Special Publications*, **477**, 567–581, https://doi.org/10.1144/SP477.40

MacDonald, D.D. and Shepherd, B.G. 1983. *A review of the Kitimat watershed.* Department of Fisheries and environment Canada Internal report, https://waves-vagues.dfo-mpo.gc.ca/Library/57247.pdf. [last accessed 10 November 2019].

Maynard, D.E., Weiland, I.C., Blais-Stevens, A. and Geertsema, M. 2017*a*. *Surficial geology, North Kitimat Arm, Douglas Channel area.* Canadian Geoscience Map 300 (Preliminary Edition), Geological Survey of Canada 1:25,000 scale map area.

Maynard, D.E., Weiland, I.C., Blais-Stevens, A. and Geertsema, M. 2017*b*. *Surficial geology, south Kitimat Arm, Douglas Channel area.* Canadian Geoscience Map 301 (Preliminary Edition), Geological Survey of Canada, 1:25,000 scale map area.

Maynard, D.E., Weiland, I.C., Blais-Stevens, A. and Geertsema, M. 2017*c*. *Surficial geology, Hartley Bay, Douglas Channel area.* Canadian Geoscience Map 302 (Preliminary Edition), Geological Survey of Canada, 1:25,000 scale map area.

McAdoo, B.G., Pratson, L.F. and Orange, D.L. 2000. Submarine landslide geomorphology, US continental slope. *Marine Geology*, **169**, 103–136, https://doi.org/10.1016/S0025-3227(00)00050-5

Moernaut, J., De Batist, M. *et al.* 2007. Giant earthquakes in South-Central Chile revealed by Holocene mass-wasting events in Lake Puyehue. *Sedimentary Geology*, **213**, 121–135, https://doi.org/10.1016/j.sedgeo.2008.12.002

Mulder, T. and Cochonat, P. 1996. Classification of offshore mass movements. *Journal of Sedimentary Research*, **66**, 43–57.

Murty, T.S. and Brown, R.E. 1979. *The Submarine Slide of 27 April. 1975 in Kitimat Inlet and the Water Waves that Accompanied the Slide*. Pacific Marine Science Report, **79-11**. Institute of Ocean Sciences, Patricia Bay, Sidney.

Parker, E.J., Traverso, C., Del Giudice, T., Evans, T. and Moore, R. 2009. Geohazard risk assessment – vulnerability of subsea structures to geo-hazards-risk implications. *Offshore Technology Conference*, Houston, Texas, https://doi.org/10.4043/OTC-20090-MS

Piper, D.J., Cochonat, P. and Morrison, M.L. 1999. The sequence of events around the epicentre of the 1929 Grand Banks earthquake: initiation of debris flows and turbidity current inferred from sidescan sonar. *Sedimentology*, **46**, 79–97, https://doi.org/10.1046/j.1365-3091.1999.00204.x

Pope, E.L., Talling, P.J., Carter, L., Clare, M.A., Hunt, J.E., 2017. Damaging sediment density flows triggered by tropical cyclones. *Earth and Planetary Science Letters*, **458**, 161–169, https://doi.org/10.1016/j.epsl.2016.10.046

Prior, D.B., Bornhold, B.D. and Johns, M.W. 1984. Depositional characteristics of a submarine debris flow. *Geology*, **92**, 707–727, https://doi.org/10.1086/628907

Prior, D.B., Bornhold, B.D. and Johns, M.W. 1986. Active sand transport along a fjord-bottom channel, Bute Inlet, British Columbia. *Geology*, **14**, 581–584, https://doi.org/10.1130/0091-7613(1986)14<581:ASTAAF>2.0.CO;2

Ramsey, C.B. 2009. Bayesian analysis of radiocarbon dates. *Radiocarbon*, **51**, 337–360, https://doi.org/10.1017/S0033822200033865

Roddick, J.A. 1970. Douglas Channel-Hecate Strait map-area, British Columbia. Geological Survey of Canada Paper no. 70-41, https://doi.org/10.4095/102362

Rohr, K.M.M. and Tryon, A.J. 2010. Pacific–North America plate boundary reorganization in response to a change in relative plate motion: offshore Canada. *Geochemistry, Geophysics, Geosystems*, **11**, https://doi.org/10.1029/2009GC003019

Shaw, J. and Lintern, G. 2016. *Marine geology, geomorphology of the Kitimat fiord system, British Columbia, parts of NTS 103-A, NTS 103-H and NTS 103-I*. Geological Survey of Canada (1 sheet), https://doi.org/10.4095/297482

Shaw, J., Stacey, C.D., Wu, Y. and Lintern, D.G. 2017. Anatomy of the Kitimat fiord system, British Columbia. *Geomorphology*, **293A**, 108–129, https://doi.org/10.1016/j.geomorph.2017.04.043

Shaw, J., Barrie, J.V., Conway, K.W., Lintern, D.G. and Kung, R. 2019. Glaciation of the northern British Columbia continental shelf: the geomorphic evidence derived from multibeam bathymetric data. *Boreas*, **49**, 17–37, https://doi.org/10.1111/bor.12411

Southon, J., Santos, G.M., Druffel, E.R.M., Griffin, S., Trumbore, S.E. and Xu, X. 2005. High throughput, high precision ^{14}C AMS with a small accelerator. *IAEA International Symposium on Utilization of Accelerators*. Dubrovnik, Croatia, IAEA-CN-115/10.

Stacey, C.D., Lintern, D.G. and Enkin, R.J. 2018. Multifaceted re-analysis of the enigmatic Kitimat slide complex, Ceanada. *Sedimentary Geology*, **369**, 46–59, https://doi.org/10.1016/j.sedgeo.2018.01.006

Stacey, C.D., Hill, P.R., Talling, P.J., Enkin, R.J., Hughes Clarke, J. and Lintern, D.G. 2019. How turbidity current frequency and character varies down a fjord-delta system: combining direct monitoring, deposits and seismic data. *Sedimentology*, **66**, 1–31, https://doi.org/10.1111/sed.12488

Sumner, E.J., Talling, P.J. and Amy, L.A. 2009. Deposits of flows transitional between turbidity current and debris flow. *Geology*, **37**, 991–994, https://doi.org/10.1130/G30059A.1

Talling, P.J. 2013. Hybrid submarine flows comprising turbidity current and cohesive debris flow: deposits, theoretical and experimental analyses, and generalized models. *Geosphere*, **9**, 460–488, https://doi.org/10.1130/GES00793.1

Thomson, R., Fine, I., Krassovski, M., Cherniawsky, J., Conway, K. and Wills, P. 2012. *Numerical Simulation of Tsunamis Generated by Submarine Slope Failures in Douglas Channel, British Columbia*. Canadian Science Advisory Secretariat Research Document 1919-50442012/115, Pacific Region.

Weiss, R., Fritz, J.M. and Wünnemann, K. 2009. Hybrid modeling of the megatsunami runup in Lituya Bay after half a century. *Geophysical Research Letters*, **36**, https://doi.org/10.1029/2009GL037814

Submarine canyons, slope failures and mass transport processes in southern Cascadia

Jenna C. Hill*, Janet T. Watt, Daniel S. Brothers and Jared W. Kluesner

US Geological Survey, Pacific Coastal and Marine Science Center, Santa Cruz 95060, CA, USA

JCH, 0000-0002-7475-357X; JTW, 0000-0002-4759-3814;
DSB, 0000-0001-7702-157X
*Correspondence: jhill@usgs.gov

Abstract: Marine turbidite records have been used to infer palaeoseismicity and estimate recurrence intervals for large ($>M_w7$) earthquakes along the Cascadia Subduction Zone. Conventional models propose that upper slope failures are funneled into submarine canyons and develop into turbidity flows that are routed down-canyon to deep-water channel and fan systems. However, the sources and pathways of these turbidity flows are poorly constrained, leading to uncertainties in the connections between ground shaking, slope failure and deep-water turbidites. We examine the spatial distribution of submarine landslides along the southern Cascadia margin to identify source regions for slope failures that may have developed into turbidity flows. Using multibeam bathymetry, sparker multichannel seismic and chirp sub-bottom data, we observe relatively few canyon head slope failures and limited evidence of large landslides on the upper and middle slope. Most of the submarine canyons are draped with sediment infill in the upper reaches and do not appear to be active sediment conduits during the recent sea-level highstand. In contrast, there is evidence of extensive mass wasting of the lower slope and non-channelized downslope flows. Contrary to previous studies, we propose that failures along the lower slope are the primary sources for deep-sea seismoturbidites in southern Cascadia.

Observations suggest that earthquake shaking can lead to relatively synchronous slope failures over broad regions, which subsequently generate diagnostic turbidity flow deposits ('turbidites') that can be dated to infer earthquake recurrence intervals (e.g. Adams 1990; Nakajima and Kanai 2000; Goldfinger *et al.* 2003; Pouderoux *et al.* 2012; Moernaut *et al.* 2014). The interpretation of earthquake-triggered turbidites ('seismoturbidites'), however, hinges on both the correlation of deposits along the abyssal plain and an understanding of the mechanisms and pathways of downslope sediment transport. Recent work has significantly improved our understanding of the processes that shape active submarine canyon systems (e.g. Sweet and Blum 2016; Symons *et al.* 2017; Amblas *et al.* 2018; Paull *et al.* 2018), as well as the role of earthquakes in canyon flushing (e.g. Allin *et al.* 2016; Mountjoy *et al.* 2018) and slope sediment remobilization (e.g. Pope *et al.* 2017; Okutsu *et al.* 2018; Molenaar *et al.* 2019). Concomitantly, advances in the understanding of turbidity flows and the resulting deposits (e.g. Postma and Cartigny 2014; Talling 2014; Hughes-Clarke 2016) allow for more robust interpretations of the stratigraphic record. Together, these studies provide critical insights into the processes governing downslope mass transport and highlight the need to better constrain the parameters that define local and/or regional sedimentary systems.

The fundamental model for seismoturbidite generation suggests that earthquake-generated turbidity flows are funneled into submarine canyons, plunge downslope through canyon/channel networks and deposit turbiditic sediments across the abyssal fan (Adams 1990; Goldfinger *et al.* 2003; Goldfinger 2011). Following this model, the Holocene record of turbidites from the base of Rogue, Smith, Klamath and Trinidad canyons (Fig. 1) has been used to infer palaeoseismicity along the southern Cascadia margin (Goldfinger 2011; Goldfinger *et al.* 2012, 2013). Eel Canyon turbidites are also part of this record; however, this setting has generally been discounted as a key marker site, as turbidity currents also frequently result from flood events in the Eel River watershed that are unrelated to earthquakes (Goldfinger *et al.* 2012; Paull *et al.* 2014). In addition to the conventional, normally graded, sand–silt, Bouma sequence-type deposits (e.g. Lowe 1982), thin layers of sharp-based, fining-upward, silt–clay deposits, referred to as 'mud turbidites', are common in southern Cascadia (Goldfinger *et al.* 2013). These deposits are most prevalent at the base of the Rogue Canyon system, but many of these layers have been correlated with deposits as far south as Trinidad Canyon, where they appear as more sand-rich layers (Goldfinger *et al.* 2012, 2013). Comparison with similar records from northern and central Cascadia, where fewer turbidites are

From: Georgiopoulou, A., Amy, L. A., Benetti, S., Chaytor, J. D., Clare, M. A., Gamboa, D., Haughton, P. D. W., Moernaut, J. and Mountjoy, J. J. (eds) 2020. *Subaqueous Mass Movements and their Consequences: Advances in Process Understanding, Monitoring and Hazard Assessments*. Geological Society, London, Special Publications, **500**, 453–475.
First published online May 20, 2020, https://doi.org/10.1144/SP500-2019-169

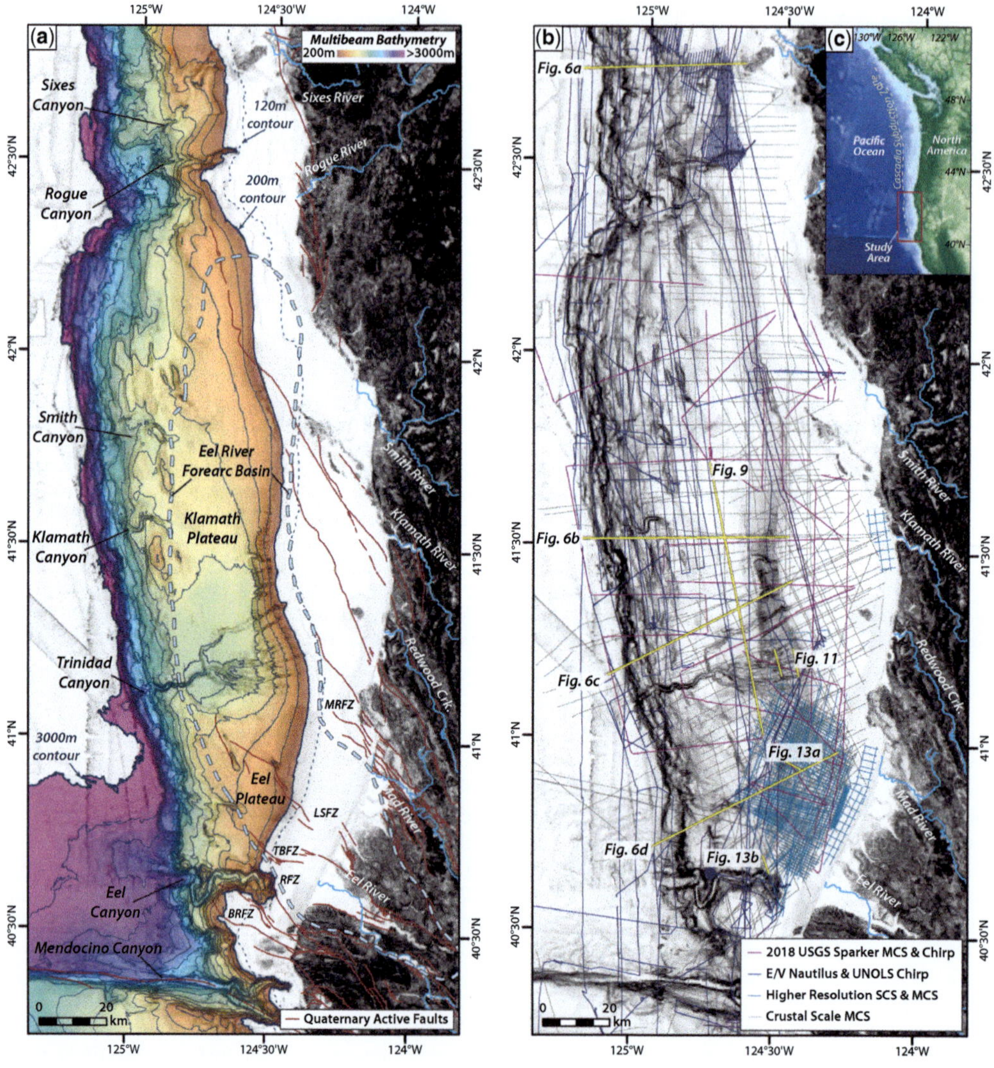

Fig. 1. (**a**) Location map showing the bathymetry, major rivers and structural features of the southern Cascadia region, which is bounded to the north by Rogue and Sixes canyons, and to the south by the MTJ. The Eel River forearc basin (dashed outline) lies primarily offshore. Dark red lines are faults from the USGS Quaternary Fault and Fold Database. Abbreviations: MRFZ, Mad River Fault Zone; LSFZ, Little Salmon Fault Zone; TBFZ, Table Bluff Fault Zone; RFZ, Russ Fault Zone; BRFZ, Bear River River Fault Zone. Bathymetric contours are every 200 m; several key isobaths are noted, including the approximate sea-level lowstand shoreline (*c.* 120 m), the shelf break at *c.* 200 m and the *c.* 3000 m isobath that marks the base of slope along much of the region. (**b**) Trackline coverage of seismic datasets for southern Cascadia, including a range of high-resolution Sparker and Air Gun MCS and Chirp sub-bottom profiles, as well as legacy and crustal-scale MCS from the 1970s to 1990s. Locations of Figures 6, 9, 11 and 13 are also shown. (**c**) The red box indicates the study area in the southern portion of the Cascadia Subduction Zone, located in the Pacific Northwest region of North America.

documented, has led to the interpretation that the frequency of large earthquakes is greater in the south (Goldfinger *et al.* 2012, 2013, 2017); however, uncertainties in the ages and the formative processes linked to the emplacement of seismoturbidite deposits have led to a debate over the reliability of the offshore record (e.g. Atwater and Griggs 2012; Atwater *et al.* 2014). Part of the uncertainty stems from a fundamental lack of understanding on the sources and pathways of submarine slope failure, sediment flows and turbidity currents along this portion of the margin.

Here we analyse the seafloor morphology and shallow stratigraphy of southern Cascadia to examine the spatial distribution of sediment storage, downslope transport pathways, and slope failures. Developing this geological context is a necessary first step in the interpretation of regional turbidite records as seismically generated events. Our results suggest that the canyon-confined, sediment flow model for seismoturbidite generation and routing (Adams 1990; Goldfinger *et al.* 2003) does not work well in southern Cascadia, as a result of tectonically influenced margin physiography that inhibits sediment bypass from the shelf to the deep sea. Instead, we propose that many of the abyssal turbidites in southern Cascadia may be sourced from mass wasting outside of submarine canyon systems, primarily on the lower slope, and routed downslope through relatively unconfined sediment gravity flows.

Geological setting and background

The Cascadia Subduction Zone (CSZ) is characterized by oblique subduction of the Gorda, Juan de Fuca and Explorer plates beneath North America, offshore of northern California, Oregon, Washington and Vancouver Island (Fig. 2). Our study is focused on the southern portion of the CSZ, where subduction of the Gorda Plate beneath North America has created a steep deformation front, often with a stepped morphology on the continental slope (Fig. 1). Large volumes of Plio-Pleistocene and Holocene sediment have filled the trench along the length of the margin, allowing significant accretion to occur as the deformation front progressed seaward over time (Wang and Tréhu 2016; Phrampus *et al.* 2017). The shelf along much of southern Cascadia is relatively narrow (15–20 km wide), with a shelf break occurring around 200 m water depth. Within the southern CSZ, the Eel River forearc basin, a massive catchment for Quaternary sediment, lies primarily offshore, occupying a broad swath of the outer shelf and upper slope (Clarke 1992). The margin is narrowest and the seafloor is steepest where the CSZ interacts with the Blanco and Mendocino Fracture zones, which bound the Gorda plate to the north and south, respectively (Fig. 3). The interaction between the CSZ and the Blanco Fracture Zone appears to have caused shortening of the margin, and led to both interseismic and long-term uplift across the shelf and inland regions (Kelsey 1990; Savage *et al.* 2000; Burgette *et al.* 2009). Similarly, at the southern end of the CSZ, uplift associated with the northward migration of the Mendocino Triple Junction (MTJ) over the past *c.* 500 kyr has generated steep topography both onshore and offshore around Eel and Mendocino canyons (Fig. 1; McCrory 1989; Orange 1999; Burger *et al.* 2002; Gulick and Meltzer 2002). MTJ-related uplift and associated transpression across reactivated, NW-oriented, shoreline-crossing faults has created a marginal plateau that extends *c.* 80 km north of the Mendocino Fracture Zone (Clarke 1992; Gulick and Meltzer 2002).

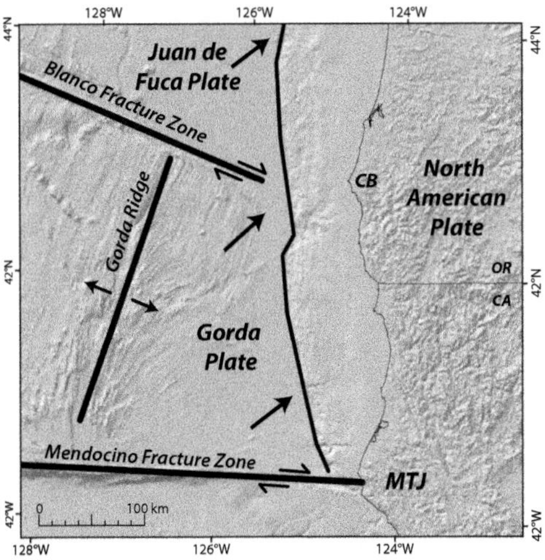

Fig. 2. Simplified structural map showing the oblique convergence of the Gorda plate with North America. The Gorda plate is bounded to the north by the Blanco Fracture Zone and to the south by the Mendocino Fracture Zone, which forms the Mendocino Triple Junction between the Gorda, North American and Pacific plates. To the north, the Juan de Fuca plate is being subducted under central and northern Cascadia.

The steep, mountainous terrain of the southern CSZ is drained by the Eel, Mad, Klamath, Rogue and Smith rivers, which transport large volumes of sediment to the shelf and slope over short time periods, often associated with winter storms (Table 1; Wheatcroft and Sommerfield 2005; Andrews and Antweiler 2012). These northern California rivers are estimated to deliver 30–40 Mt of suspended sediment to the coast, most of which appears to bypass the shelf and accumulate further offshore (Sommerfield *et al.* 2007). The southern Eel margin was the focus of extensive studies as part of the STRATAFORM project in the late 1990s and early 2000s, which combined seafloor sampling, geophysical surveys and instrumented oceanographic monitoring to examine the sedimentary processes across the shelf and upper slope of northern California, primarily between the Mad and Eel River mouths (Nittrouer 1999). The results of these studies showed that sediment transport along the Eel margin is strongly influenced by storm-driven wave resuspension, as well as tides and alongshore currents (Ogston *et al.* 2000; Traykovski *et al.* 2000; Wright *et al.* 2001; Nittrouer *et al.* 2007). During large storm events, the Eel River

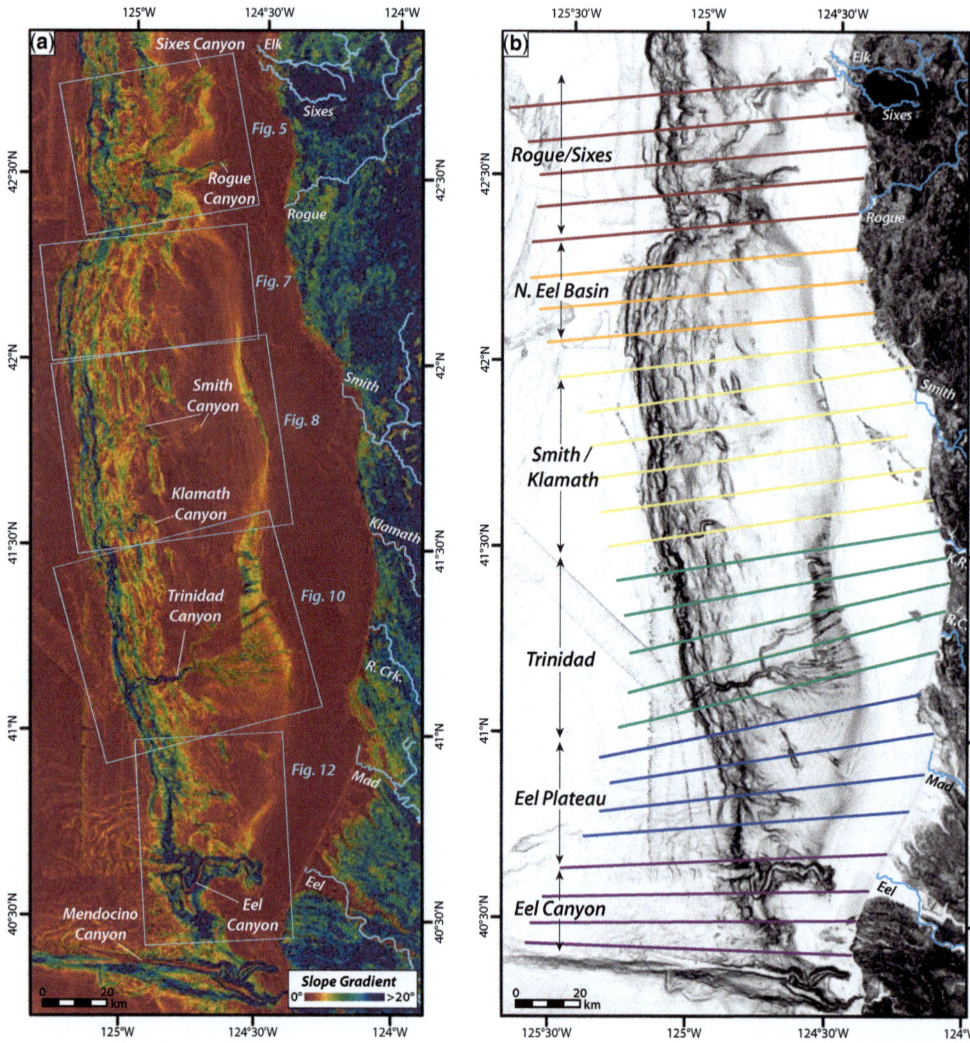

Fig. 3. (**a**) Coloured slope gradient map for southern Cascadia. The seafloor gradient is steepest around Rogue, Sixes and Eel canyons, as well as along the lowermost slope. The Eel and Klamath marginal plateaus separate the shelf from the lower slope across much of the region. The steepest upper slope gradients are found around Trinidad Canyon. The blue boxes indicate the locations of the subregion bathymetry figures. (**b**) Profiles were drawn across the margin roughly perpendicular to the continental slope and grouped into six subregions based on profile morphology: 1, red, Rogue–Sixes; 2, orange, northern Eel Basin; 3, yellow, Smith-Klamath; 4, green, Trinidad; 5, blue, Eel Plateau; 6, purple, Eel Canyon.

discharges massive plumes of suspended sediment that become entrained in a northward-flowing coastal current (Geyer *et al.* 2000). Sediment from these events is often reworked and redistributed the shelf and slope via wave-supported gravity flows and nepheloid layers associated with internal waves (Walsh and Nittrouer 1999; Wright *et al.* 1999; Ogston *et al.* 2004; Nittrouer *et al.* 2007). In addition, tidal forcing in deeper-water initiates internal waves that maintain suspended sediment near and below the shelf break (McPhee-Shaw *et al.* 2004).

STRATAFORM studies also provided some of the initial constraints on upper slope morphology and long-term stratigraphic patterns of the Eel Plateau. Burger *et al.* (2001) mapped a series of channel incisions associated with both glacioisostatic and relative (tectonic uplift induced) lowstands that suggest the Eel River regularly drained into the Eel Canyon over the past 500 kyr. Subsequent work on Eel

Table 1. *Estimates of modern and unaltered sediment loads for the primary coastal drainage systems in southern Cascadia*

River	Modern sediment load (Mt a^{-1})*	Unaltered sediment load (Mt a^{-1})†
Rogue	2.3	
Smith	0.4	1.5
Klamath/Trinity	3.7	20
Redwood Creek	1.3	0.68
Mad	2.6	1.4
Eel	18	14

*Estimates from Wheatcroft and Sommerfield (2005).
†Estimates for river basins prior to alteration by flow storage, regulation and depletion, from Andrews and Antweiler (2012).

Canyon has shown evidence of supercritical flows and frequently recurring turbidity currents (e.g. Lamb *et al.* 2008; Paull *et al.* 2014), which indicate that the Eel Canyon has remained an active sediment transport pathway throughout the Holocene. Export of Eel River sediment to Eel Canyon is estimated to account for 12–19% of the total sediment load; the remainder of this sediment load is presumably dispersed across the shelf and slope north of the canyon (Mullenbach *et al.* 2004; Mullenbach and Nittrouer 2006; Sommerfield *et al.* 2007; Warrick 2014).

Burger *et al.* (2002) developed a sequence stratigraphic model that suggested that most of the sediment preserved on the Eel margin is fluvially derived highstand muds and silts, with lowstand deposits only preserved in fluvial channels. Later work showed that a significant amount of preservation also occurred during transgression, particularly on the inner shelf (Spinelli and Field 2003; Hogarth *et al.* 2012). The thickest post-lowstand sediment deposits (50–60 m) were mapped on the shelf north of the Eel River mouth (Spinelli and Field 2003), upslope of a region described as the 'Humboldt Slide' (HS) owing to the undulating nature of the seafloor between 220 and 650 m water depth (Gardner *et al.* 1999). This undulating stratigraphy was thought to represent evidence of downslope creep (Gardner *et al.* 1999); however, Lee *et al.* (2002) reinterpreted these features as sediment waves resulting from turbidity currents associated with hyperpycnal discharge from the Eel River. Schwehr *et al.* (2007) subsequently developed a magnetic approach that showed no post-depositional deformation in the upper 8 m of sediment from the HS region, reinforcing the interpretation of primary deposition associated with downslope currents. Furthermore, sediment cores collected just above the HS wave field show evidence of graded turbidite-like deposits interpreted to have formed by hyperpycnal discharge in the past 1 kyr (Sommerfield and Wheatcroft 2007).

The initial STRATAFORM bathymetry also revealed a series of shallow gullies and rills between 200 and 700 m water depth between Trinidad Canyon and the Little Salmon Anticline (LSA; Fig. 1; Field *et al.* 1999). Spinelli and Field (2001) proposed an aggradational gully evolution model whereby these gullies were eroded prior to sea-level lowstand, as the Mad River regraded and sediment was sourced from small slope failures and downslope flows. Meanwhile aggradation of both gully thalwegs and interfluves occurred as sediment input from the Eel River outpaced these erosional downslope flows and allowed the channel morphology to persist (Spinelli and Field 2001). Together, these previous studies provided important information on the primary sediment sources and transport pathways along a portion of the study region that can be used as analogues for other areas where geomorphic evidence for recent mass transport is observed.

Methods

During 2018–19, the US Geological Survey (USGS) acquired a suite of high-resolution, sparker multichannel seismic and coincident Chirp sub-bottom data, along with multibeam bathymetry and backscatter across large portions of southern Cascadia. These data were integrated with a large number of existing multibeam bathymetry and sub-bottom datasets acquired from the Ocean Exploration Trust, Monterey Bay Aquarium Research Institute, USGS, National Oceanic and Atmospheric Administration (NOAA) and University-National Oceanographic Laboratory System (UNOLS) vessels and publicly available from government and academic data portals (e.g. https://www.ngdc.noaa.gov; https://walrus.wr.usgs.gov/NAMSS/; http://www-udc.ig.utexas.edu/sdc/). The data integration resulted in nearly complete bathymetric coverage with 10–40 m resolution and dense crossings of seismic reflection profiles throughout the region (Fig. 1).

To examine the first-order margin morphology, topographic profiles were drawn roughly perpendicular to the continental slope, extending 100 km seaward from the coastline (Fig. 3). Seafloor depth and gradient were extracted every 100 m from a composite bathymetry surface gridded at 50 m; profiles were smoothed with a 10 pt moving average window. Where no multibeam data were available, a 50 m Global Multi-Resolution Topography (GRMT v.3.6, Ryan *et al.* 2009) grid was used. The margin profiles were grouped in six distinct subregions, based on a visual assessment of profile morphology (Fig. 4). When combined with the observed substrate architecture, these margin profiles provide a simple, systematic approach to detecting along-margin differences in geomorphic process dominance (e.g.

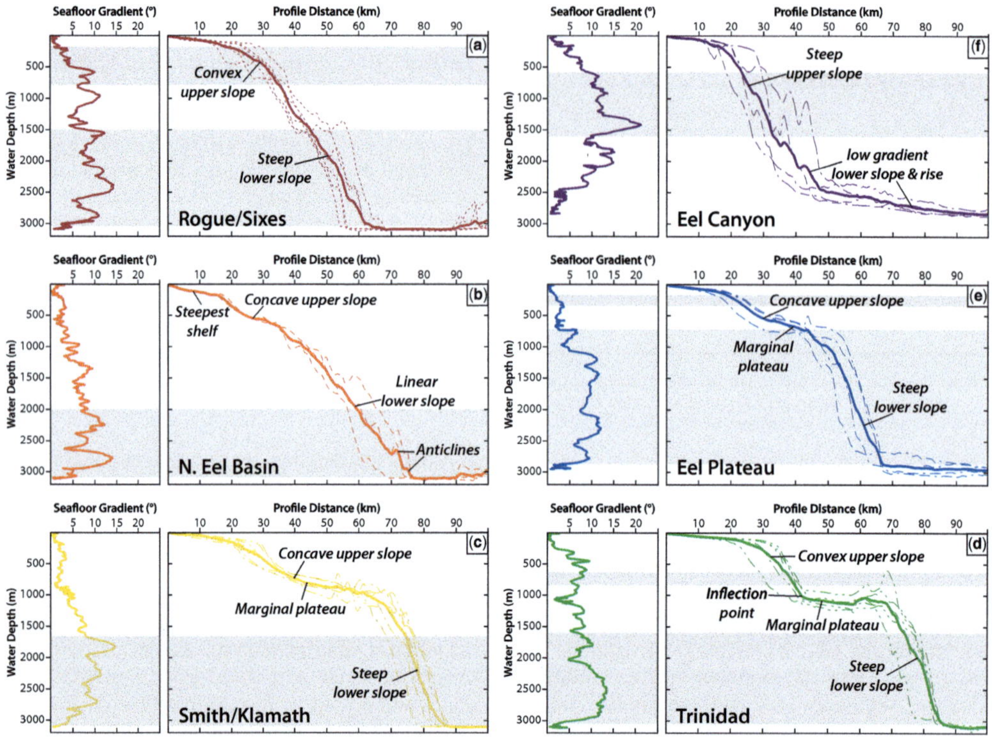

Fig. 4. Plots of seafloor gradient v. water depth and change in bathymetry along profiles drawn extending 100 km from the local coastline. Seafloor depth and gradient were extracted every 100 m from a composite bathymetry surface gridded at 50 m; profiles were smoothed with a 10 point moving average window. See Figure 3b for profile locations. Dashed lines are individual profiles and bold lines are averaged across the subregion. Grey-shaded areas represent mass failure zones where seafloor scarps are most prevalent. (**a**) The Rogue–Sixes subregion has a steep lower slope and stepped morphology; (**b**) northern Eel Basin has a slightly concave upper slope and more linear lower slope; (**c**) the Smith–Klamath subregion has a concave upper slope and steep lower slope separated by a low gradient marginal plateau; (**d**) Trinidad Canyon has a markedly convex upper slope with a sharp inflection point where the basin gradient decreases sharply; (**e**) Eel Plateau also has a marginal plateau that separates a concave upper slope from a steep lower slope; and (**f**) the Eel Canyon subregion has the steepest upper slope gradients and shallowest slope-to-rise transition.

Brothers *et al.* 2013, 2018; Hill *et al.* 2017; Mosher *et al.* 2017).

Results

The cross-margin profiles were used to subdivide the study area into six subregions (Fig. 3), based on variations in slope morphology: (1) Rogue and Sixes canyons; (2) northern Eel Basin, between Rogue and Smith canyons; (3) Smith and Klamath canyons; (4) Trinidad Canyon; (5) Eel Plateau between Trinidad and Eel canyons; and (6) Eel Canyon. Within these groupings (Fig. 4), the Rogue–Sixes and Eel Canyon subregions comprise the narrowest parts of the margin; each has narrow shelves (<20 km) and steep upper slope gradients (i.e. >10° in water depths <1000 m; Fig. 4a, f). In contrast, the northern Eel Basin has the overall lowest slope gradient, with the steeper areas (>10°) only occurring near the base of the slope, where landward-vergent thrust folds have deformed the seafloor (Fig. 4b). In between these end members, the Smith–Klamath, Trinidad and Eel Plateau subregions show variations on a tiered morphology, each with a prominent mid-slope plateau or basin (Fig. 4c–e). Trinidad has the deepest mid-slope basin, which forms a confluence of submarine channels that extend across the slope through the deeply incised Trinidad Canyon (Fig. 1). Across the region, Quaternary sediment thickness is greatest (*c.* 1500 m) in the central Eel River forearc basin, which lies on the mid-slope within the Smith–Klamath and Trinidad subregions. Localized deposition of up to 40 m occurs in small basins, synclines and anticline limbs across the lower slope. The thickest upper slope sediment accumulation (*c.* 800 m) is in the prograded shelf–slope transition of the Trinidad subregion. Below we describe the morphology and stratigraphy of each subregion to assess potential sources of downslope sediment flows that may have contributed to the abyssal turbidite record, including channelized flows in and around canyon systems, open-slope and channel-confined failures, unconfined, open-slope flows and evidence for lower slope mass wasting.

Subregion 1: Rogue and Sixes canyons

The Rogue–Sixes subregion has a convex margin profile with a relatively steep average slope gradient of >5° below 500 m water depth (Figs 3 & 4a). Sections of steeper topography (10–14°) along the slope are punctuated by lower-gradient (*c.* 5°) ledges and ridges that protrude seaward across the margin. Two distinct canyon systems incise the shelf here and converge downslope at a relatively flat (<1°) plateau around 900 m water depth.

The southern canyon head, herein referred to as Rogue Canyon, is incised inshore of the 120 m isobath and is *c.* 22 km from the modern Rogue River outlet (Fig. 1). The canyon is 200–900 m wide and up to 160 m deep (Fig. 5). Several *c.* 3 km-long scarps are observed along the southern wall of the upper slope portion of the canyon (Fig. 5). The uppermost scarp near the canyon head (*c.* 140 m water depth) has 10–40 m of relief. Several stepped, 50 m-high scarps are observed along the canyon wall at *c.* 300 m water depth; the largest of these features is a 200 m high amphitheatre-shaped scarp located *c.* 3 km above the southern canyon wall (Fig. 5). The canyon floor bathymetry near these scarps is shoaled and irregular, with downstream knickpoints (Fig. 5).

The northern canyon system, herein referred to as Sixes Canyon, is closer to the modern Sixes and Elk River outlets (*c.* 16 km; Fig. 1), but is only incised to

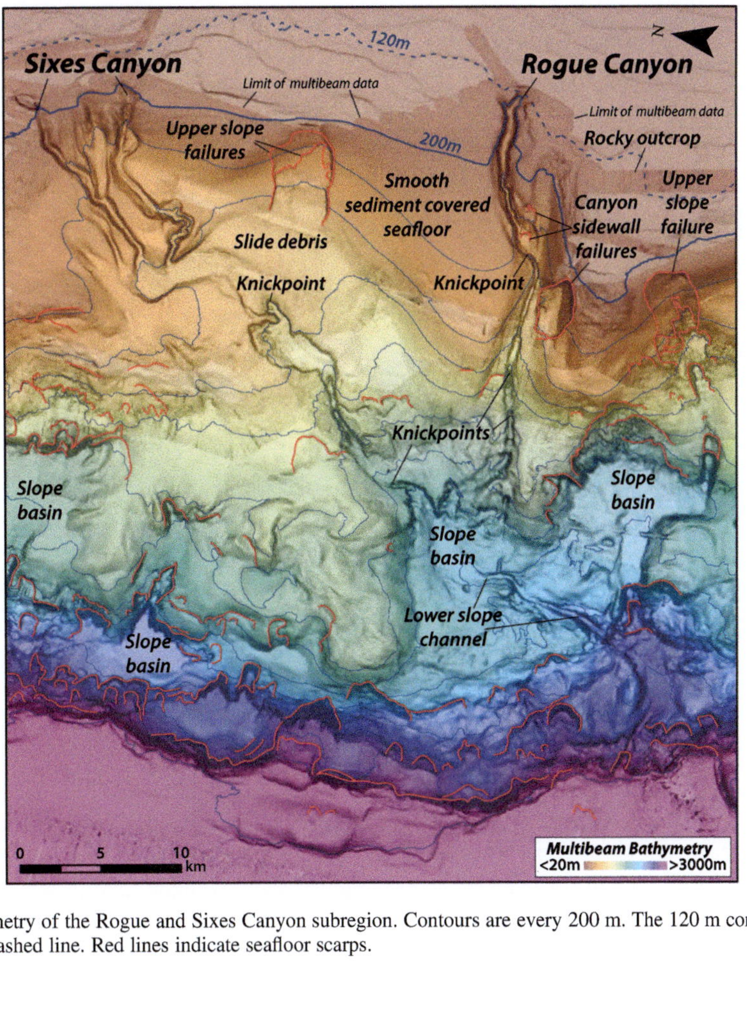

Fig. 5. Bathymetry of the Rogue and Sixes Canyon subregion. Contours are every 200 m. The 120 m contour is denoted by a dashed line. Red lines indicate seafloor scarps.

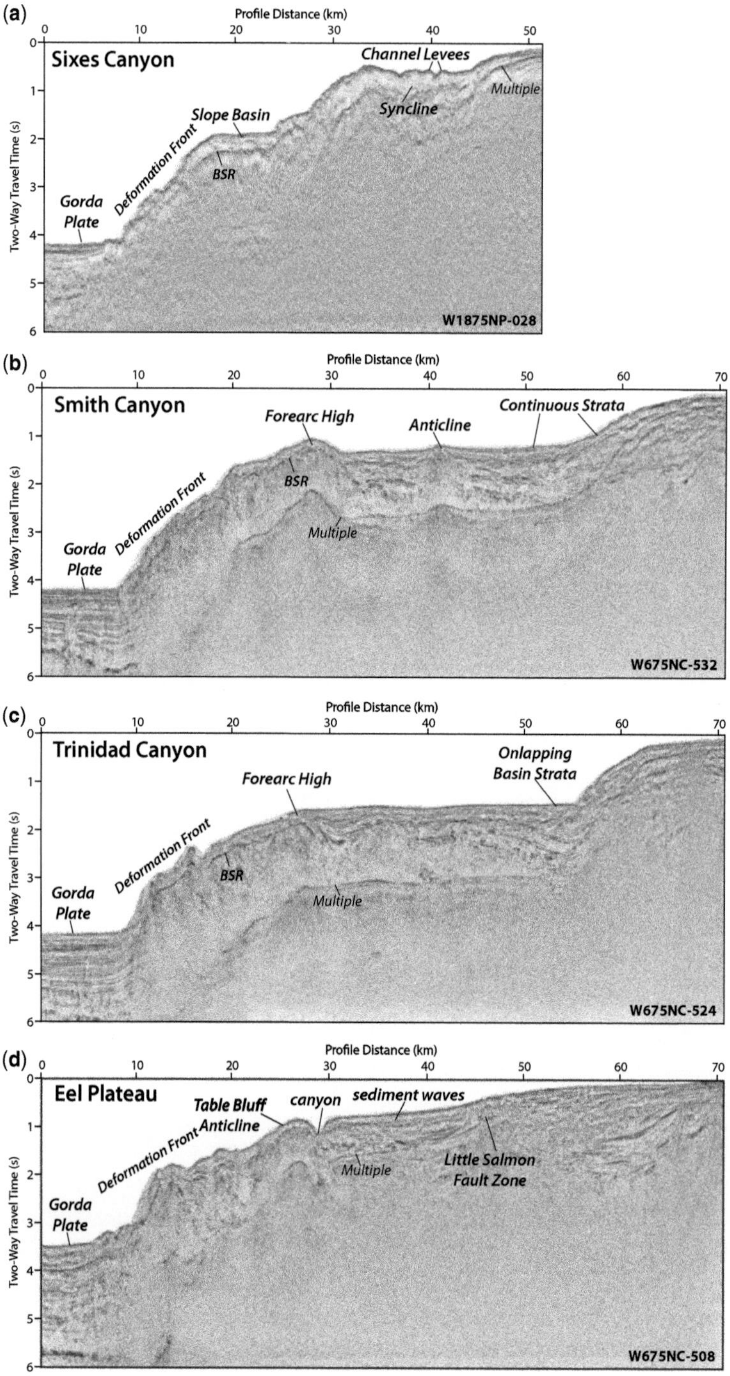

Fig. 6. Legacy MCS profiles showing margin stratigraphy from the outer shelf to the abyssal plain: (**a**) Rogue–Sixes subregion; (**b**) Smith–Klamath subregion; (**c**) Trinidad Canyon subregion; and (**d**) Eel Plateau subregion. See Figure 1b for profile locations.

the 200 m isobath. The Sixes Canyon system has a *c.* 6 km-wide head, made up of multiple small channels that converge downslope to form two larger channels, each *c.* 500 m wide and 50–100 m deep. This canyon system appears to be cut into a small, sediment-filled syncline and shows evidence of depositional levees along the upper channel reaches (Fig. 6). These two channels converge into a larger, meandering, *c.* 1500 m-wide channel that extends across a *c.* 1° slope before stepping down *c.* 50 m into a narrower (*c.* 500 m wide) channel. This narrower channel continues downslope to a basin at *c.* 1700 m water depth where the Sixes and Rogue canyons converge (Fig. 5). A narrow lower slope channel (<400 m wide; *c.* 50 m deep) extends out of this basin across the deformation front (Fig. 5).

The upper slope seafloor between Rogue and Sixes canyons is relatively smooth (Fig. 5), with thick sediment accumulation (hundreds of metres) atop several NW-trending, buried anticlines. A large slide scar (*c.* 20 km^2) observed near the shelf edge has steep (15–20°), 40–70 m-high sidewall scarps and stepped headwall scarps up to *c.* 15 m high (Fig. 5). Seaward of this slide scar, Sixes Canyon shoals and widens, with a >100 m-high knickpoint downstream. Several other 1–3 km-long slide scars are observed south of Rogue Canyon, on the flanks of uplifted anticlines in *c.* 220–250 m water depth (Fig. 5). The largest of these features has a sharp, 50 m-high headwall scarp, gently sloping sidewalls with up to 30 m of relief and multiple smaller scarps inside the slide scar. Gullies within the scar converge downslope into a small, isolated basin at *c.* 1700 m water depth. From this basin, a *c.* 1.2 km-wide, flat-floored channel extends downslope into the Rogue–Sixes channel around 2300 m water depth. Mass wasting morphology is pervasive along the mid to lower slope (>1500 m water depth) in the Rogue–Sixes subregion (Fig. 5), where numerous small failure scarps and gully drainages cut into the steeper (>10°) portions of the slope; the seabed between scarps is rough and irregular.

Subregion 2: northern Eel Basin

The margin width increases by 10–15 km in the northern Eel Basin relative to the Rogue–Sixes subregion (Figs 3 & 4b). The northern Eel Basin subregion has the highest shelf gradient (*c.* 1.5°) and a rounded shelf–slope break that transitions into a low-gradient (<5°), concave upper slope (Fig. 4b). The seafloor gradient begins to increase around 1000 m water depth, at the seaward edge of the forearc basin where north–south-trending anticlines deform the seafloor (Figs 4b & 7). Numerous scarps are observed on the seaward flanks of these anticlines and increase in abundance on the lower slope as the seafloor gradient increases downslope towards the trench (Figs 3 & 7). The slope profile seaward of this point is nearly linear, with a gradient <10° until *c.* 2200 m water depth (Fig. 4b), where steeper flanks of several landward-verging folds define the lower slope along the deformation front (Fig. 7).

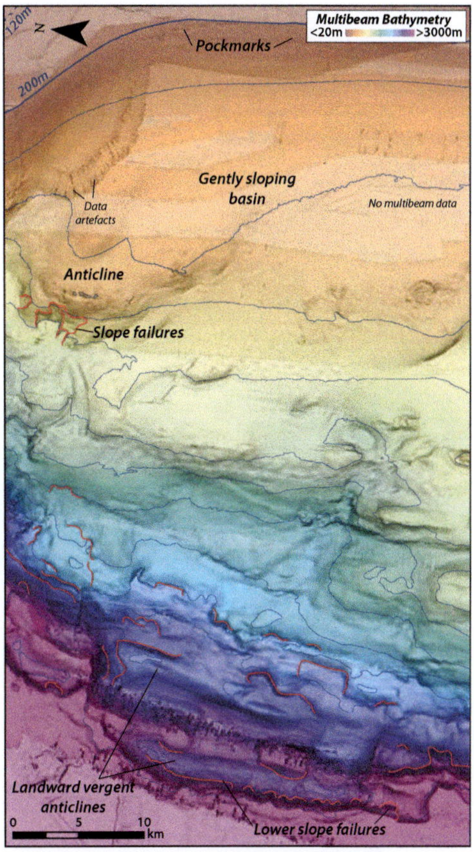

Fig. 7. Bathymetry of the northern Eel Basin subregion. Contours are every 200 m. The 120 m contour is denoted by a dashed line. Red lines indicate seafloor scarps.

Subregion 3: Smith and Klamath canyons

The subregion encompassing Smith and Klamath canyons is the widest part of the margin and displays the lowest-gradient upper slope (Figs 4c & 8). The average slope profile in this subregion is <5° in water depths shallower than 1000 m. The upper slope here is more concave than the northern Eel Basin, with numerous, isolated gullies up to 500 m wide and 40 m deep (Figs 4c & 8). The low-gradient (<2°) Klamath Plateau (Silver 1971) extends 20–30 km west across the central forearc basin (Fig. 1). In between the upper slope gullies, seismic profiles show continuous reflectors from shelf to

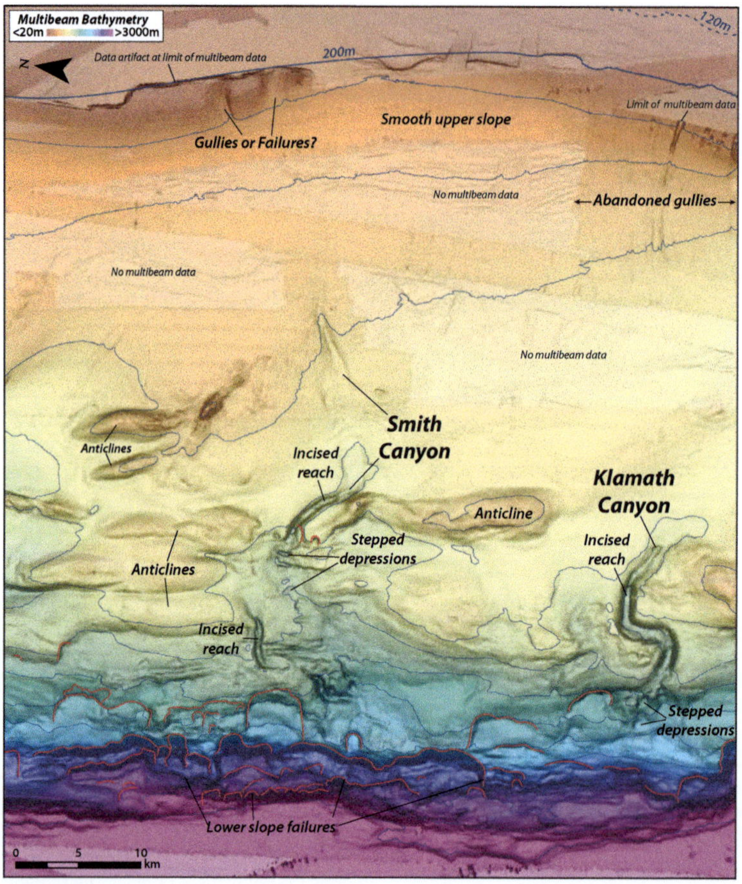

Fig. 8. Bathymetry of the Smith–Klamath Canyon subregion. Contours are every 200 m. The 120 m contour is denoted by a dashed line. Red lines indicate seafloor scarps.

basin (Fig. 6). The seafloor gradient increases >10° around 1500 m water depth, and remains steep across the lower slope and deformation front (Fig. 4c).

The Smith and Klamath canyons are large, lower-slope canyons, incised across the lower slope, with little to no bathymetric expression on the upper to mid slope (Fig. 8). The canyons meander around trench-parallel anticlines on the outer wedge and lose surface expression by *c.* 1800 m water depth, 8–10 km from the trench. North–south seismic transects through the central Eel Basin show no evidence of buried canyons in the upper 500 m of sediment (the maximum penetration depth of our high-resolution seismic data) upslope of these features, indicating that the canyon incisions have been isolated to the lower slope (Fig. 9).

Smith Canyon comprises several sections, or reaches, of varying morphology (Fig. 8). The canyon has two incised reaches on the lower slope where it is confined by uplifted anticlines; the upper incised reach is up to 900 m wide and 50–80 m deep, while the lower incised reach is narrower (750 m wide) and deeper (up to 100 m). These two reaches are separated by trench parallel ridges ranging from 20 to 70 m high. Upslope of the incised reaches, the canyon morphology is only visible in the bathymetry to 800 m water depth. Here the canyon is 1–3 km wide and 25–30 m deep; seismic profiles across this upper reach reveal 70–100 m of sediment infill beneath the thalweg. In contrast, no sediment infill is imaged in the lower slope reaches of the canyon.

Klamath Canyon has an incised valley up to 2 km wide and 180 m deep that extends to *c.* 1600 m water depth (Fig. 8). The head of the canyon, at *c.* 900 m water depth, cuts into *c.* 80 m of Eel Basin sediment. Breaches in the western rim of the canyon, near the base, border two sets of stepped, semicircular depressions that descend the slope (Fig. 8). Little to no sediment fill was imaged in the base of the canyon, but as much as 10 m of sediment appears perched on the flanks of the canyon in some

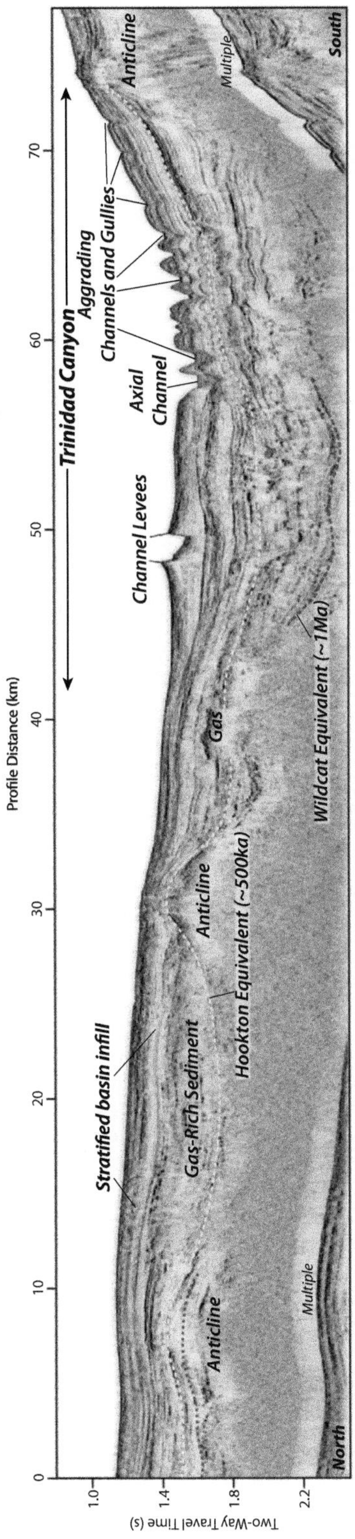

Fig. 9. North–south-trending, high-resolution Sparker multichannel seismic profile (Balster-Gee *et al.* 2020*a*; envelope data shown) highlights the stratified basin fill of the central Eel Basin and shows no evidence of canyon formation in the upper *c.* 500 m of sediment. The Hookton (*c.* 500 ka) and Wildcat Datums (1 Ma) are interpreted from correlation with profiles presented by Gulick and Meltzer (2002). Figure 1b for the profile location.

stretches. Seaward of Smith and Klamath canyons, scalloped head scarp morphology and truncated reflectors on cross-margin seismic profiles suggest that there are numerous, nested, failures on the steep lower slope (Fig. 8).

Subregion 4: Trinidad Canyon

The Trinidad Canyon subregion has the widest (up to 30 km) and flattest shelf of the southern CSZ. The shelf–slope break is rounded and the upper slope is markedly convex, with a sharp inflection point at the transition to a low-gradient (<1.8°) basin around 1000 m water depth (Figs 4d & 10). This basin extends 25 km offshore, where steep anticlines between 1200 and 2000 m water depth form an uplifted shore-parallel ridge (Figs 4d & 10). The seaward edge of this ridge has several large gullies and numerous nested scarps (Fig. 10). Between 2000 and 3000 m water depth, the average seafloor gradient increases to >10°, which is among the steepest portions of the lower slope. Large failure scarps (>5 km long, >1 km high headwalls) originate around 2000 m (Fig. 10). Beneath these scarps, 100 m-high, 1 km^2 blocks are scattered on the abyssal seafloor.

A series of relatively straight, shallow gullies and larger, slightly more sinuous channels are observed on the uppermost slope in and around Trinidad Canyon (Fig. 10). The gullies and channels increase in frequency to the south, becoming wider and more deeply incised, with maximum seafloor expression between 200 and 600 m water depth (Fig. 10). An abrupt change in seafloor gradient occurs at the base of many of these gullies at *c.* 1000 m water depth, where >500 m of Eel Basin sediment onlaps the slope (Figs 3a, 6 & 10). The head of the largest channel north of Trinidad Canyon (100 m deep, up to 1800 m wide) is 35 km offshore of the modern Klamath River mouth (Fig. 1) and *c.* 5 km from the lowstand shoreline (120 m below sea-level; Lambeck and Chappell 2001). Stepped seafloor depressions several 100 m wide and *c.* 5 m deep are found at the mouth of this channel (Fig. 10). To the south, there is an 8 km-long, 40 m-high scarp at 725 m water depth (Fig. 10). Numerous smaller and shorter scarps are cut into the strata below this larger scarp. Some of the larger upper-slope channels exhibit sizeable sidewall failures with scarps up to 40 m high.

The fan-shaped Trinidad Canyon has incised a broad, bowl-shaped depression that forms the deepest part of the Eel River forearc basin (Fig. 1). A nest array of channels and gullies within the canyon system converge downslope into a central, axial channel cut across the outer wedge, which has an apparent plunge pool at the base of the slope (Goldfinger *et al.* 2012; Fig. 10). Our new bathymetry compilation shows that this plunge pool is up to 3.5 km

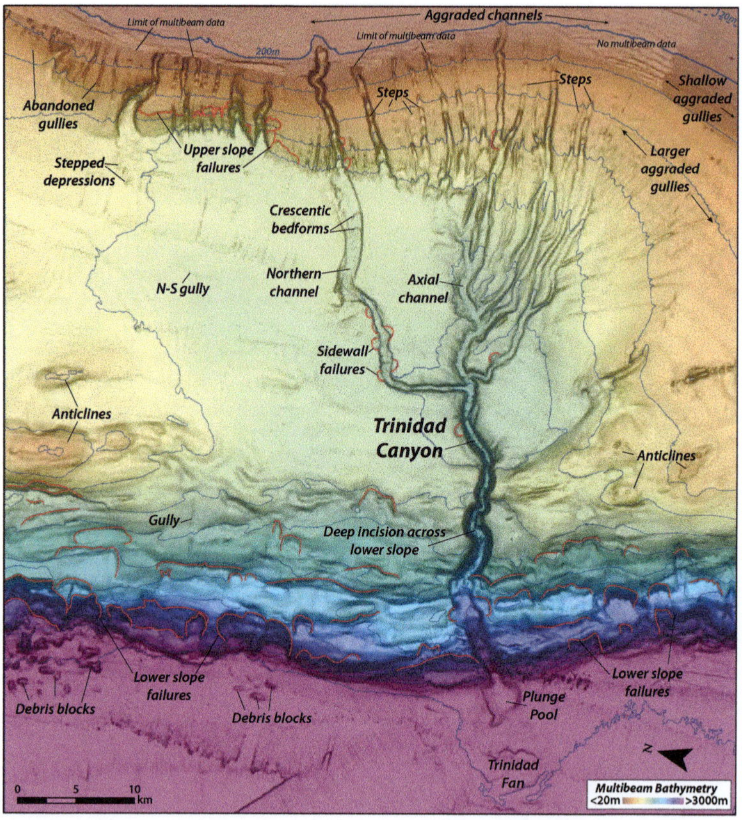

Fig. 10. Bathymetry of the Trinidad Canyon subregion. Contours are every 200 m. The 120 m contour is denoted by a dashed line. Red lines indicate seafloor scarps.

wide, 20–40 m deep, with 40 m-high levees. The axial channel increases in width and incision depth downslope, from 300 to 400 m wide and 50 m deep on the upper slope, to 1.5–2 km wide and 300–400 m deep across the lower slope. Notably only one other large channel extends across Eel Basin into the canyon from the north. This northern channel is *c.* 1500 m wide, with 20 m of sediment in the channel thalweg. Across the Eel Basin, the channel has 15–20 m-high levees that show sediment accumulation on the flanks, numerous scalloped side-wall failures and series of 10 m-high crescent-shaped bedforms that step downslope to circular scour pits (Fig. 10). The northern channel has an abrupt southward bend, where it becomes narrower and more deeply incised (up to a 120 m deep, with an additional 50 m of infill beneath the seafloor), before converging with the axial channel. The seafloor between the northern and axial channels is relatively smooth, underlain by *c.* 700 m of stratified sediment (Fig. 9) that onlaps the base of upper slope channels and gullies. Short (several 100 m-long) gullies drain into the axial channel from the northern edge of this deposit. The axial channel appears to serve as a drainage divide as multiple channels and small gullies feed into channel along the south side as well.

The seafloor within many of the upper slope channels and gullies feeding into the Trinidad Canyon system has a crescentic-step morphology, which is most pronounced on the steeper portions of the slope. Seismic profiles across the outer shelf and upper slope show buried and aggrading channels in the upper 300 m. The stratigraphic patterns of the gullies on the south side of Trinidad Canyon have been described in detail in Field *et al.* (1999), Spinelli and Field (2001) and Burger *et al.* (2003). Many of the larger channels in the central and northern portions of Trinidad Canyon also show a similar aggradational stratigraphy, with >20 m of recent sediment accumulation both in the channels and on the interfluves (Figs 9 & 11). A number of the channels are either partially or completely buried; sub-bottom profiles show sediment infill and aggradation where the gullies decrease in seafloor expression upslope and onto the shelf (Fig. 11).

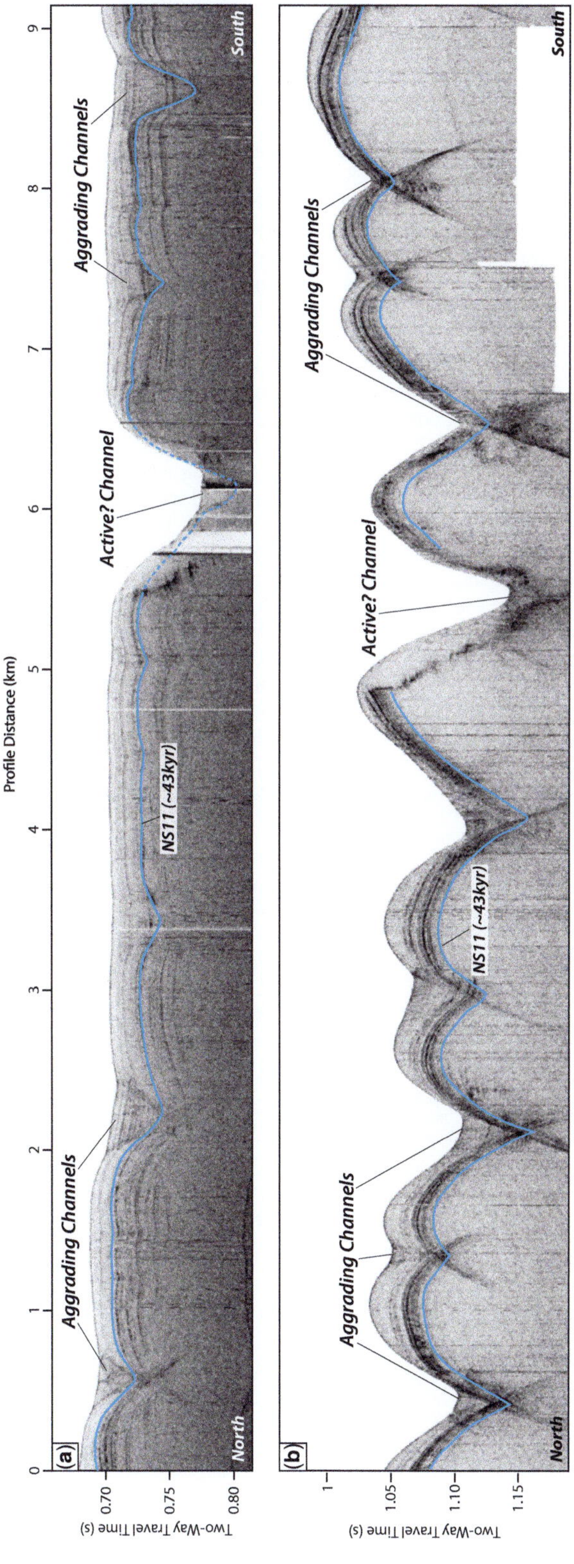

Fig. 11. High-resolution boomer data collected across the upper slope channels of Trinidad Canyon by the USGS in the 1990s (Balster-Gee *et al.* 2020*b*). See Field *et al.* (1999) for acquisition details. The stratigraphy here shows significant aggradation and infilling of channels. The NS11 datum, is interpreted by correlation with Burger *et al.* (2003), and was estimated to be *c.* 43 ka. As much as 20 m of sediment overlie the NS11 horizon in this region. Figure 1b for profile locations.

Subregion 5: Eel Plateau

The Eel Plateau has a relatively low-gradient, concave upper slope that broadens out to a <2° marginal plateau that extends up to 20 km across the margin (Fig. 4). Numerous small, nested scarps (<1 km wide) and open-slope slide scars are observed between 800 and 1200 m water depth, where seafloor gradients are 6–8° (Fig. 12). The average seafloor gradient increases to >10° between 800 and 1200 m water depth at the seaward edge of the plateau. A combination of larger scarps and gullies cut into the steep, lower slope, below *c.* 1400 m water depth (Fig. 12).

A range of sediment wave-type bedforms are observed across the low-gradient portions of Eel Plateau. The largest bedforms (*c.* 1 km wavelength; 5 m amplitude) are found near the shelf–slope break, in water depths of 150–300 m, on the north side of the LSA (Fig. 12). Sub-bottom profiles across this region show a stacked bedform stratigraphy with aggradational bedding on the stoss side and erosional truncation on the lee (seaward) side (Fig. 13). South of the LSA, upslope of the HS, the shelf-edge bedforms are much more muted in seafloor expression <0.5 m amplitude, with limited aggradation on the stoss sides and truncation on the lee sides. The shelf-edge bedforms increase in amplitude (1–2 m) to the south but appear to be overprinted by the downslope gullies, creating a complex morphology and stratigraphy (Fig. 12). Sediment waves from the HS extend downslope into a large canyon (1400 m wide, 200 m deep) that incises up to 600 m water depth and is bounded to the south by the Table Bluff Anticline (Fig. 12). Another large wave field (*c.* 1 km wavelength, 10 m amplitude) is located to the south of the Table Bluff Anticline (Fig. 12).

The upper slope of the Eel Plateau is extensively gullied (Fig. 12). These gullies are much smaller than those found around Trinidad, but have a similar pattern of aggradation in the seismic profiles. There are three distinct gully zones based on seafloor morphology. North of the LSA, where seafloor gradients are <2–4°, the gullies are *c.* 100 m wide and 1–2 m deep (Field *et al.* 1999; Spinelli and Field 2001). Upslope of the HS wave field, the seafloor gradient is similar, but the gullies are up to 10 m deep and most prominent in shallower water depths, between 230 and 380 m (Gardner *et al.* 1999). South of the HS, as seafloor gradient increases to 5–8°, the gullies remain similarly incised, but extend from 175 to 650 m water depth, and have a more stepped morphology, similar to that observed around Trinidad.

Two shelf-edge scarps are observed upslope of the southernmost gullies. A 6 m-high scarp extends roughly parallel to the 175 m isobath for *c.* 4.5 km (Fig. 13). This scarp appears to form the upper headwall for a nested set of slide scars that extend *c.* 8 km downslope (Fig. 13). Seaward of these features, several smaller scarps are observed at the head of a

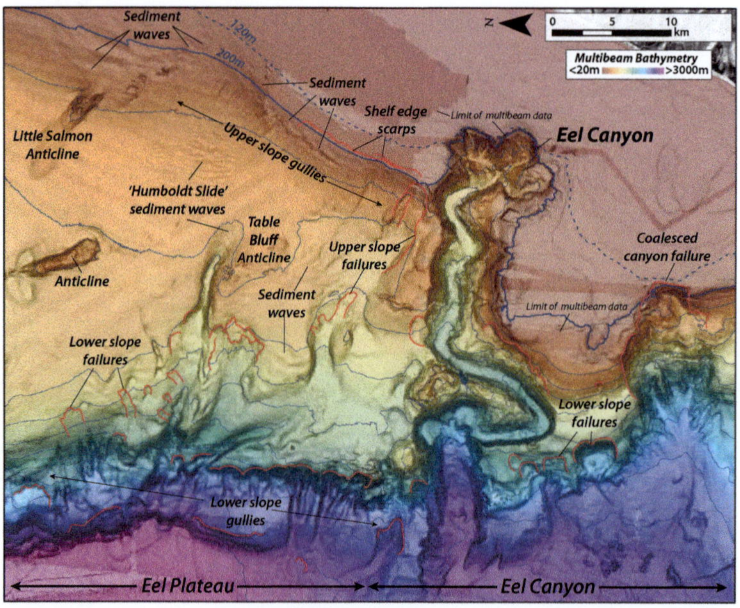

Fig. 12. Bathymetry of the Eel Plateau and Eel Canyon subregions. Contours are every 200 m. The 120 m contour is denoted by a dashed line. Red lines indicate seafloor scarps.

2 km-wide, 200 m-deep lower slope canyon (Fig. 13). Another possible shelf-edge scarp is poorly imaged, but appears to be 1–2 m high, extending up to 10 km along the 185 m isobath (Fig. 13).

Subregion 6: Eel Canyon

The southern Cascadia margin is narrowest around Eel Canyon and has a steep, linear to concave profile, with average slope gradients >10° in water depths between 600 and 2200 m (Fig. 4f). The apparent spike in average slope gradient at *c.* 1500 m occurs where several profiles cross the edges of Eel Canyon and a large coalesced canyon failure to the south. South of Eel Canyon, two large, coalesced canyon failures cut into the slope (Fig. 12). The smaller one, 2.5 km south of Eel Canyon, has a 3.5 km amphitheatre-shaped scar with up to 600 m of relief, originates in *c.* 1200 m water depth and is interpreted as pre-Holocene (Gwiazda *et al.* 2016). The larger failure to the south is up to 7.5 km wide and cuts into the shelf at water depths <200 m. Dendritic channels incise the walls of the failure scar to the abyssal seafloor at >2000 m water depth.

The Eel Canyon is a large, shelf-indenting canyon, cut >15 km landward into the shelf between uplifted anticlines associated with Russ and Bear River Fault zones (Fig. 1). The canyon head is shoreward of the 120 m contour, *c.* 13 km from the modern Eel River mouth. The Eel Canyon has a tripartite, concave head up to 9 km wide with dendritic channels (Fig. 12). The main thalweg is 1.5–2 km wide and has several apparent side wall failures upstream of a sharp bend where the canyon meanders around a bedrock high on the lower slope (Fig. 12). Previous studies have discussed the canyon morphology and sedimentation patterns (e.g. Mullenbach *et al.* 2004; Puig *et al.* 2004; Mullenbach and Nittrouer 2006; Lamb *et al.* 2008; Paull *et al.* 2014).

Discussion

Relationships between slope morphology, sediment delivery and tectonics

To understand the sources and pathways of earthquake-generated turbidity flows along southern Cascadia, it is important to consider the role of margin physiography in controlling patterns of sediment accumulation and bypass (Van Wagoner *et al.* 1988; Ross *et al.* 1994; Galloway 1998), as demonstrated by morphological classifications of siliciclastic margins (e.g. Pratson and Haxby 1996; Adams and Schlager 2000; O'Grady *et al.* 2000; Schlager and Adams 2001; Brothers *et al.* 2013, 2018; Hill *et al.* 2017; Mosher *et al.* 2017). The slope readjustment model of Ross *et al.* (1994) emphasizes the role of changing basin physiography, where oversteepening of the slope above a stable, equilibrium angle can lead to bypass and erosion of slope through downslope gravity flows. This results in deposition and aggradation lower on the slope, where the buildup and onlap of slope fan aprons creates a lower-gradient platform that can support renewed sediment progradation across the upper slope. The equilibrium case in this model is a graded margin that has a balance between depositional and erosional processes, sediment supply, uplift/subsidence and margin physiography (Ross *et al.* 1994). The balance between these conditions is variable among different submarine slopes owing to a range of processes, but several studies have shown that the transition from sediment accumulation to downslope sediment bypass occurs on seafloor gradients of 3–5°, while erosional processes, particularly those associated with downslope turbidity currents, become prevalent on gradients >4–5° (Pratson *et al.* 2007; Gerber *et al.* 2008).

Along active margins, particularly subduction zones such as southern Cascadia, patterns of tectonic uplift and subsidence introduce substantial overprinting, but the fundamental morphodynamic relationship between slope gradient and sediment accumulation remains important. In general, low-gradient (<4–5°) parts of a margin can support sediment accumulation, whereas higher gradients lead to sediment bypass and erosion. In southern Cascadia, there are two end-member morphologies: (1) narrow/steep; and (2) wide/low-gradient. Uplift and shortening associated with subduction at the edges of the Gorda plate has led to the steep, narrow margins characterized by shelf sediment bypass (e.g. Eel Canyon and Rogue–Sixes). These subregions have large, active, shelf-indenting canyon systems that were probably linked to fluvial drainages during sea-level lowstand. The subregions between Rogue–Sixes and Eel Canyon underlain by the Eel River forearc basin comprise the other end member, with wide, low-gradient upper slopes that are the primary depocentres for shelf-derived sediment. These subregions have relict or inactive canyon systems, such that there appears to be no significant transfer of shelf sediment to the deep sea along much of the margin during the Holocene. Complex morphological variability exists across the region though, both in and among these end members, largely dictated by a combination of sediment supply and dispersal patterns, antecedent topography and local neotectonic interactions.

Eel Canyon has a sizeable, well-developed base of slope fan system (Fig. 1), and a record of frequent, canyon-scouring, turbidity flows throughout the Holocene (Lamb *et al.* 2008; Paull *et al.* 2014) that indicate efficient shelf-to-abyss sediment transfer. In contrast, the much smaller apron at the base of Rogue and Sixes canyons (Goldfinger *et al.* 2013) appears to be the result of both a lower sediment

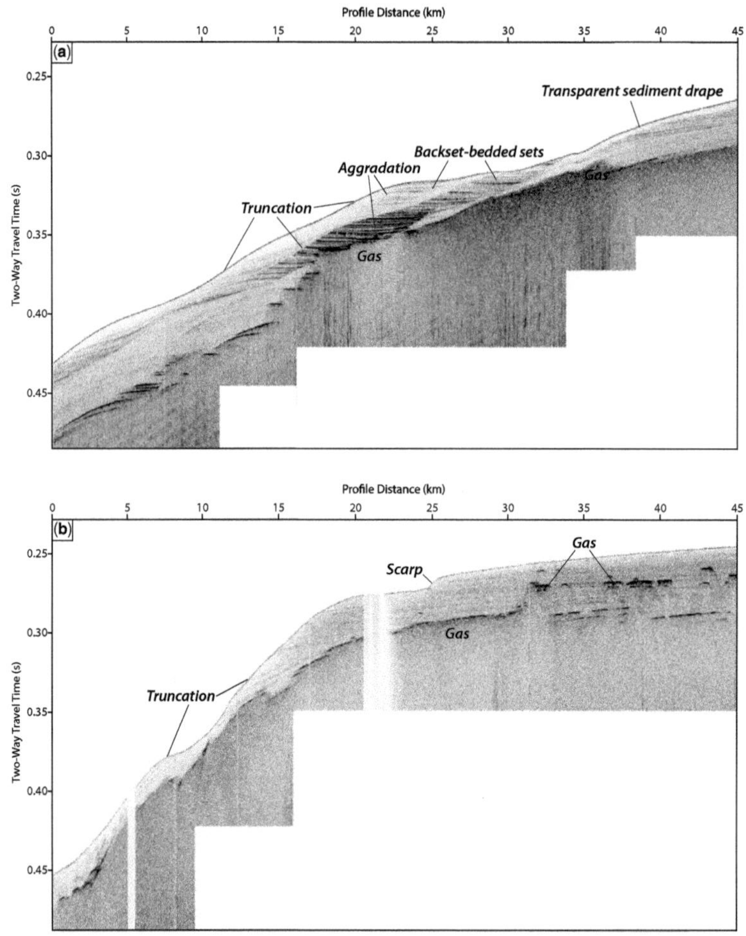

Fig. 13. High-resolution boomer data collected across the outer shelf and upper slope of the Eel Plateau by the USGS in the 1990s, showing (**a**) sediment waves interpreted as partially depositional cyclical step bedforms and (**b**) a *c.* 5 m-high scarp observed on the outer shelf. See Figure 1b for profile locations.

supply and sediment sequestration in these northern canyon systems. Deformation and shortening across the Rogue–Sixes subregion have created a number of stepped synclinal basins along the canyon axes that reduce the downslope gradient across these reaches and serve as local sediment sinks (Figs 5 & 6). In addition, landslide debris from upper slope sediment failures appears to be spread across the main canyon thalwegs in several locations, blocking the channels and restricting downslope transport (Fig. 5). Numerous prominent knickpoints are observed downstream of these deposits, as well as at the seaward edges of the slope basins, suggesting an incomplete regrading of these canyon systems (Fig. 6). Together, this combination of relatively low sediment supply, ongoing tectonic deformation and upper slope failures appears to be restricting the transfer of sediment from the shelf to the deep sea through the canyon systems during sea-level highstand when there is no direct fluvial connection. This implies that turbidites found in the slope apron are probably not sourced from shelf and upper slope material routed through the canyon systems, but rather may be derived from more proximal failures on the lower slope where mass wasting is pervasive (Fig. 5).

The antecedent topography of the Eel River forearc basin dominates the margin morphology between Rogue and Eel canyons (Figs 1 & 4). The offshore portion of this forearc basin forms a series of marginal plateaus underlying northern Eel Basin, Smith–Klamath, Trinidad and Eel Plateau (Figs 1 & 2). The seafloor expression of the forearc basin is most prominent in the Smith–Klamath subregion (Figs 3 & 4) and becomes increasingly overprinted by tectonic uplift and sediment dispersal patterns to the north and south. At the northern end of the basin, a combination of fluvial sediment input and broad regional uplift across the adjacent Cape Blanco area appears to have nearly filled the northern Eel Basin, leaving relatively little bathymetric expression of the forearc basin in this subregion compared with more southern portions of the margin (Fig. 7). As a result, canyon and gully development are largely absent here (Figs 4 & 7). Regionally continuous strata extend from the shelf across the basin in both the northern Eel Basin and Smith–Klamath subregions (Fig. 6). The stratigraphy along the central Eel Basin shows no evidence of canyons in the upper 500 m of sediment, interpreted to represent at least 500 kyr (Fig. 9). This implies that the lower slope Smith and Klamath canyons have had no connection to upper slope gullies or canyons in recent time. With no direct fluvial sources, any sediment inputs to these lower slope canyons must be derived from remobilization of material deposited in the central forearc basin or along the lower slope. These canyons do not extend across the lowermost slope (>1800 m depth) though, thus no channelized turbidity flows reached the abyssal plain. Similar to Rogue–Sixes, mass wasting of the outer wedge appears to be prevalent (Fig. 8) and may have contributed to the seismoturbidite record here as well.

Although Trinidad Canyon is one of the larger submarine canyon systems in southern Cascadia, with a well-developed network of upper slope gullies and a prominent central thalweg incised across the lower slope, the canyon system shows no evidence of direct connections to fluvial systems during sea-level lowstand; during highstand, the canyon heads are more than 30 km from the nearest large river mouths. The uplift associated with the northward migration of the MTJ reduced accommodation space across the Eel Plateau (Orange 1999; Burger *et al.* 2001, 2002; Gulick *et al.* 2002) and directed sediment north into Trinidad Canyon. As result, Trinidad Canyon, which has received distal, fine-grained sediment from the Eel, Mad, Klamath and other small rivers along the coast, appears to be a major sediment sink in the region with substantial sediment storage on the upper slope. Progradation and aggradation of the margin around Trinidad has resulted in a markedly convex and steepened (>5°) upper slope profile where channel and gully erosion, along with small slope failures, dominate the morphology (Figs 4 & 10). Similar to the feeder gullies to the south (Field *et al.* 1999), many of the upper slope channels in and around Trinidad Canyon appear to have been inactive during the present highstand. These channels appear to be infilled, and in some cases buried, from significant accumulation and aggradation across the upper slope since the last sea-level lowstand (Fig. 11). The morphology and stratigraphy of this region are characteristic of the aggradational, abandoned canyons described by Jobe *et al.* (2011), which are dominated by hemipelagic deposition. Although several of the canyon features (e.g. deep incision across the lower slope, the base of slope plunge pool, crescentic bedforms and stepped channel morphology) suggest erosional downslope flows, these may be relict features, given that there is relatively little evidence of energetic erosional flows in the upper reaches of the canyon where large volumes of sediment are stored. Furthermore, Jobe *et al.* (2011) have attributed similar stepped channel morphology to erosional scours associated with fluid escape and not downslope flows. If, as these observations suggest, Trinidad Canyon has been inactive during the Holocene, then the deep sea turbidite record here may be primarily derived from failures of the steep outer wedge, which are common in this subregion (Fig. 10), along with minor contributions from lower-slope canyon sources.

The morphology of the Eel Plateau suggests the potential for significant shelf to deep sea sediment transfer to have occurred outside of large, margin-crossing submarine canyon systems. Sediment accumulation on the Eel Plateau is greatest on low-gradient portions of the outer shelf and upper slope, where expansive sediment wave fields record deposition from the proximal, high-sediment-discharge Eel River source. Many of these bedforms have stratigraphy characteristic of cyclic steps (Figs 12 & 13; Kostic 2011; Covault *et al.* 2017; Slootman and Cartigny 2019), which implies the presence of supercritical flows in relatively unconfined pathways across the outer shelf and upper slope. This is also consistent with previous work that showed evidence for hyperpycnal flows across the plateau (Lee *et al.* 2002; Sommerfield and Wheatcroft 2007). Between the LSA and Eel Canyon, sediment bypass occurs across a steep, narrow upper slope characterized by gully erosion and large slide scars (Fig. 12). Overprinting of the bedforms in the southern portion of plateau by erosional gullies (Fig. 12) suggests that the sediment waves here may be relict deposits (i.e. deposited during sea-level lowstand when shoreface erosion contributed greater volumes of sediment to the shelf-edge; Burger *et al.* 2001). The large shelf-edge failures on the southern end of the plateau (Fig. 12) appear much more recent though. Sub-bottom profiles across the shelf-edge scarps (Fig. 13) at the head of these failures show no evidence of post-failure sediment drape, despite proximity to the high-volume Eel River source. Thus, these failures may present a potential record of recent, local earthquake-triggered mass failure that is independent of Eel Canyon turbidites. Sediment bypass appears to be more channelized further downslope, where mid-slope canyons broaden into more distributed pathways of smaller channels and gullies that extend to the abyssal plain (Fig. 12).

Distribution of mass wasting

Mass wasting morphology (e.g. seafloor scarps, gully erosion, rough and irregular sediment-covered seafloor, debris blocks) representative of slides, slumps, rock falls, debris flows and other slope failure processes is ubiquitous along the lower slope across the region (Figs 5, 7, 8, 10 & 12). Presumably many of these failures are triggered by earthquake shaking (Hampton *et al.* 1996; Locat and Lee 2002) and produce sediment-laden turbidity flows outside of the canyon catchment systems, which may have cascaded downslope in relatively unconfined flows. Such unconfined flows would be consistent with evidence of spatially continuous strata that extend for hundreds of kilometres along the abyssal plain (Goldfinger *et al.* 2013). Given the lack of morphological and stratigraphic evidence for shelf to deep-sea sediment transport in the canyon systems north of Eel Canyon during the Holocene, we propose that much of the abyssal seismoturbidite record is instead derived from non-canyon-confined flows, probably sourced from proximal failures on the steep lower slope and from energetic downslope flows across the southern Eel Plateau.

In southern Cascadia, the majority of mass wasting occurs on the lowermost slope, with several isolated zones of larger upper slope failures in Rogue–Sixes, Trinidad and the Eel Plateau. In contrast, slope failures are far more widespread in central and northern Cascadia (McAdoo *et al.* 2000; McAdoo and Watts 2004; Riedel *et al.* 2018). This leads to the question of why there appear to be so few submarine slope failures in southern Cascadia relative to other portions of the margin. Rapid sediment accumulation and the subsequent development of overpressure, which can reduce the shear strength of unconsolidated sediment (Hampton *et al.* 1996; Locat and Lee 2002), is often cited as a key factor in preconditioning low-angle (<5°) submarine slopes for failure (Urlaub *et al.* 2015). Although triggered primarily by earthquake shaking (Hampton *et al.* 1996; Locat and Lee 2002), large, retrogressive submarine slides are generally thought to occur where thick accumulations of spatially continuous, parallel-bedded strata contribute to the lateral propagation of high-sediment pore pressure and the initiation of failure on the lower slope (Dugan and Flemings 2000; Masson *et al.* 2010; Dugan and Sheahan 2012). Indeed, thick accumulations of regionally continuous, rapidly emplaced sediment blanket much of the upper slope across southern Cascadia, particularly within the Eel River forearc basin (Figs 6 & 9), yet most of these areas appear to have remained relatively intact throughout multiple earthquake cycles.

Most of the open-slope failures in southern Cascadia occur in sediment-covered areas with gradients of 4–6° on surrounding slopes. In the Rogue and Eel Plateau subregions, where these failures are most prevalent, they occur in areas of active uplift, near the shelf-edge or on the flanks of actively deforming anticlines. This suggests that rupture of local, upper plate structures may play an important role in triggering these failures. Elsewhere, the lack of large-scale failures may reflect the influence of seismic strengthening, whereby repeated earthquake shaking along active margins can lead to elevated shear strength and enhanced slope stability through increased consolidation (e.g. Sawyer and Devore 2015; ten Brink *et al.* 2016). Observed seismicity in southern Cascadia is far higher than elsewhere along the CSZ, with frequent small events (≤M6) clustered around the MTJ and BFZ (McCrory *et al.* 2012; Stone *et al.* 2018). This could argue against seismic strengthening as a major factor in slope stability though, since the majority of the failures are observed in the most active regions. Other factors such as fluid migration, gas hydrate dissociation, differences in lithology and/or local variations in the response to earthquake shaking (e.g. Locat and Lee 2002; Bünz *et al.* 2005; Elger *et al.* 2018) could also have a strong influence on the stability across the margin and should be investigated further.

Conclusions

Examination of the along-strike variability in the morphology of the southern Cascadia margin reveals the role of tectonic interactions in influencing patterns of sediment accumulation and dictating areas of sediment bypass and slope failure. Shortening and uplift at the north and south edges of the Gorda plate have led to steep, narrow margins characterized by sediment bypass. The region between is underlain by the Eel River forearc basin, which forms a low-gradient platform for sediment accumulation and prevents shelf-derived sediment from reaching the deep sea. The morphology and stratigraphy across southern Cascadia suggest that most of the canyon systems across southern Cascadia received little to no direct fluvial inputs, even during sea-level lowstand, and are relatively inefficient pathways for the shelf to abyss sediment transport during the Holocene. Rogue and Sixes canyons appear to have retained significant volumes of sediment in slope basins and infilling channels. Smith and Klamath canyons are isolated lower-slope drainages that may serve as conduits for Eel Basin sediment, but show no evidence of channelized flows across the deformation front. The location of Trinidad Canyon on the distal edge of MTJ-related uplift and the proximity to Eel River and other sources has allowed significant upper slope sediment accumulation and storage at this location while Eel Plateau mostly experiences upper slope sediment bypass and shows evidence of open-slope, unconfined hyperpycnal flows. Holocene mass wasting of the steep lower slope appears to be a far more significant contributor to sediment accumulation on the abyssal plain than the local canyon systems, and may instead be the primary source of fine-grained seismoturbidites in southern Cascadia. Detailed source-to-sink and provenance studies are needed to confirm this hypothesis, but if 'mud turbidites' are indeed sourced primarily from disintegration of the steep outer wedge, these deposits could extend further north than previously thought, which would have important implications for the record of earthquake recurrence across the region. Dating of the deposits associated with large open-slope landslides identified in Rogue–Sixes, Trinidad and on the Eel Plateau may offer additional, independent constraints on the earthquake recurrence record in southern Cascadia.

Acknowledgements This manuscript greatly benefited from reviews by David Mosher, Derek Sawyer, Jamie Conrad, Lawrence Amy and Gary Mahon.

Funding This work was funded by the US Geological Survey.

Author contributions **JCH**: conceptualization (lead), investigation (lead), writing – original draft (lead), writing – review & editing (lead); **JTW**: investigation (supporting), writing – review & editing (supporting); **DSB**: investigation (supporting), writing – review & editing (supporting); **JWK**: investigation (supporting), writing – review & editing (supporting).

References

Adams, E.W. and Schlager, W. 2000. Basic types of submarine slope curvature. *Journal of Sedimentary Research*, **70**, 814–828, https://doi.org/10.1306/2DC4093A-0E47-11D7-8643000102C1865D

Adams, J. 1990. Paleoseismicity of the Cascadia Subduction Zone: evidence from turbidites off the Oregon–Washington Margin. *Tectonics*, **9**, 569–583, https://doi.org/10.1029/TC009i004p00569

Allin, J.R., Hunt, J.E., Talling, P.J., Clare, M.A., Pope, E. and Masson, D.G. 2016. Different frequencies and triggers of canyon filling and flushing events in Nazaré Canyon, offshore Portugal. *Marine Geology*, **371**, 89–105, https://doi.org/10.1016/j.margeo.2015.11.005

Amblas, D., Ceramicola, S. *et al.* 2018. Submarine canyons and gullies. *In*: Micallef, A., Krastel, S. and Savini, A. (eds) *Submarine Geomorphology*. Springer, Cham, 251–272, https://doi.org/10.1007/978-3-319-57852-1_14

Andrews, E.D. and Antweiler, R.C. 2012. Sediment fluxes from California coastal rivers: the influences of climate, geology, and topography. *The Journal of Geology*, **120**, 349–366, https://doi.org/10.1086/665733

Atwater, B.F. and Griggs, G.B. 2012. *Deep-sea turbidites as guides to Holocene earthquake history at the Cascadia Subduction Zone – alternative views for a seismic-hazard workshop (USGS Numbered Series No. 2012-1043)*. Open-File Report. US Geological Survey, Reston, VA.

Atwater, B.F., Carson, B., Griggs, G.B., Johnson, H.P. and Salmi, M.S. 2014. Rethinking turbidite paleoseismology along the Cascadia subduction zone. *Geology*, **42**, 827–830, https://doi.org/10.1130/G35902.1

Balster-Gee, A.F., Kluesner, J.W., Watt, J.T., Hill, J.C., Brothers, D.S., Michalak, M.J. and O'Shea, D. 2020*a*. Multichannel minisparker seismic reflection data of USGS field activity 2018-658-FA collected between Cape Blanco and the Mendocino Triple Junction from 2018-10-04 to 2018-10-18. US Geological Survey Data Release, https://doi.org/10.5066/P9R3QM97

Balster-Gee, A.F., Hill, J.C. and O'Brien, T.F. 2020*b*. Archive of boomer sub bottom data collected off shore Eureka, California during USGS field activity W-1-96-NC from 1996-06-29 to 1996-07-07. US Geological Survey Data Release, https://doi.org/10.5066/P950276G

Brothers, D.S., ten Brink, U.S., Andrews, B.D. and Chaytor, J.D. 2013. Geomorphic characterization of the U.S. Atlantic continental margin. *Marine Geology*, **338**, 46–63, https://doi.org/10.1016/j.margeo.2012.12.008

Brothers, D.S., Andrews, B.D. *et al.* 2018. Slope failure and mass transport processes along the Queen Charlotte Fault, southeastern Alaska. *Geological Society, London, Special Publications*, **477**, 69–83, https://doi.org/10.1144/SP477.30

Bünz, S., Mienert, J., Bryn, P. and Berg, K. 2005. Fluid flow impact on slope failure from 3D seismic data: a case study in the Storegga Slide. *Basin Research*, **17**, 109–122, https://doi.org/10.1111/j.1365-2117.2005.00256.x

Burger, R.L., Fulthorpe, C.S. and Austin, J.A., Jr 2001. Late Pleistocene channel incisions in the southern Eel River Basin, northern California: implications for tectonic v. eustatic influences on shelf sedimentation patterns. *Marine Geology*, **177**, 317–330, https://doi.org/10.1016/S0025-3227(01)00166-9

Burger, R.L., Fulthorpe, C.S., Austin, J.A. and Gulick, S.P.S. 2002. Lower Pleistocene to present structural deformation and sequence stratigraphy of the continental shelf, offshore Eel River Basin, northern California. *Marine Geology*, **185**, 249–281, https://doi.org/10.1016/S0025-3227(02)00196-2

Burger, R., Fulthorpe, C.S. and Austin, J.A. 2003. Effects of triple junction migration and glacioeustatic cyclicity on evolution of upper slope morphologies, offshore Eel River Basin, northern California. *Marine Geology*, **1999**, 307–336, https://doi.org/10.1016/S0025-3227(03)00194-4

Burgette, R.J., Weldon, R.J. and Schmidt, D.A. 2009. Interseismic uplift rates for western Oregon and along-strike variation in locking on the Cascadia subduction zone. *Journal of Geophysical Research: Solid Earth*, **114**, https://doi.org/10.1029/2008JB005679

Clarke, S.H. 1992. Geology of the Eel River Basin and Adjacent Region: implications for Late Cenozoic

Tectonics of the Southern Cascadia Subduction Zone and Mendocino Triple Junction (1). *AAPG Bulletin*, **76**, 199–224, https://doi.org/10.1306/BDFF87AE-1718-11D7-8645000102C1865D

Covault, J.A., Kostic, S., Paull, C.K., Sylvester, Z. and Fildani, A. 2017. Cyclic steps and related supercritical bedforms: building blocks of deep-water depositional systems, western North America. *Marine Geology*, **393**, 4–20, https://doi.org/10.1016/j.margeo.2016.12.009

Dugan, B. and Flemings, P.B. 2000. Overpressure and fluid flow in the New Jersey continental slope: implications for slope failure and cold seeps. *Science, New Series*, **289**, 288–291, https://doi.org/10.1126/science.289.5477.288

Dugan, B. and Sheahan, T.C. 2012. Offshore sediment overpressures of passive margins: mechanisms, measurement, and models. *Reviews of Geophysics*, **50**, https://doi.org/10.1029/2011RG000379

Elger, J., Berndt, C., Rüpke, L., Krastel, S., Gross, F. and Geissler, W.H. 2018. Submarine slope failures due to pipe structure formation. *Nature Communications*, **9**, 1–6, https://doi.org/10.1038/s41467-018-03176-1

Field, M.E., Gardner, J.V. and Prior, D.B. 1999. Geometry and significance of stacked gullies on the northern California slope. *Marine Geology*, **154**, 271–286, https://doi.org/10.1016/S0025-3227(98)00118-2

Galloway, W.E. 1998. Siliciclastic slope and base-of-slope depositional systems: component facies, stratigraphic architecture, and classification. *AAPG Bulletin*, **82**, 569–595.

Gardner, J.V., Prior, D.B. and Field, M.E. 1999. Humboldt Slide – a large shear-dominated retrogressive slope failure. *Marine Geology*, **154**, 323–338, https://doi.org/10.1016/S0025-3227(98)00121-2

Gerber, T.P., Pratson, L.F., Wolinsky, M.A., Steel, R., Mohr, J., Swenson, J.B. and Paola, C. 2008. Clinoform progradation by turbidity currents: modeling and experiments. *Journal of Sedimentary Research*, **78**, 220–238, https://doi.org/10.2110/jsr.2008.023

Geyer, W.R., Hill, P., Milligan, T. and Traykovski, P. 2000. The structure of the Eel River plume during floods. *Continental Shelf Research*, **20**, 2067–2093, https://doi.org/10.1016/S0278-4343(00)00063-7

Goldfinger, C. 2011. Submarine paleoseismology based on turbidite records. *Annual Review of Marine Science*, **3**, 35–66, https://doi.org/10.1146/annurev-marine-120709-142852

Goldfinger, C., Nelson, C.H. and Johnson, J.E. 2003. Holocene earthquake records from the Cascadia subduction zone and northern San Andreas fault based on precise dating of offshore turbidites. *Annual Review of Earth and Planetary Sciences*, **31**, 555–577, https://doi.org/10.1146/annurev.earth.31.100901.141246

Goldfinger, C., Nelson, C.H. *et al.* 2012. *Turbidite event history – methods and implications for Holocene paleoseismicity of the Cascadia subduction zone*. USGS Professional Paper No. 1661-F. US Geological Survey, Reston, VA.

Goldfinger, C., Morey, A.E., Black, B., Beeson, J., Nelson, C.H. and Patton, J. 2013. Spatially limited mud turbidites on the Cascadia margin: segmented earthquake ruptures? *Natural Hazards and Earth System Sciences*, **13**, 2109–2146, https://doi.org/10.5194/nhess-13-2109-2013

Goldfinger, C., Galer, S. *et al.* 2017. The importance of site selection, sediment supply, and hydrodynamics: a case study of submarine paleoseismology on the northern Cascadia margin, Washington USA. *Marine Geology*, **384**, 4–46, https://doi.org/10.1016/j.margeo.2016.06.008

Gulick, S.P.S. and Meltzer, A.S. 2002. Effect of the northward-migrating Mendocino triple junction on the Eel River forearc basin, California: structural evolution. *GSA Bulletin*, **114**, 1505–1519, https://doi.org/10.1130/0016-7606(2002)114<1505:EOTNMM>2.0.CO;2

Gulick, S.P.S., Meltzer, A.S. and Clarke, S.H. 2002. Effect of the northward-migrating Mendocino triple junction on the Eel River forearc basin, California: stratigraphic development. *GSA Bulletin*, **114**, 178–191, https://doi.org/10.1130/0016-7606(2002)114<0178:EOTNMM>2.0.CO;2

Gwiazda, R., Paull, C.K. *et al.* 2016. Eel Canyon slump scar and associated fluid venting. *In*: Lamarche, G., Mountjoy, J. *et al.* (eds) *Submarine Mass Movements and Their Consequences: 7th International Symposium*. Advances in Natural and Technological Hazards Research. Springer, Cham, 411–418, https://doi.org/10.1007/978-3-319-20979-1_41

Hampton, M.A., Lee, H.J. and Locat, J. 1996. Submarine landslides. *Reviews in Geophysics*, **34**, 33–59, https://doi.org/10.1029/95RG03287

Hill, J.C., Brothers, D.S., Craig, B.K., ten Brink, U.S., Chaytor, J.D. and Flores, C.H. 2017. Geologic controls on submarine slope failure along the central U.S. Atlantic margin: insights from the Currituck Slide Complex. *Marine Geology*, **385**, 114–130, https://doi.org/10.1016/j.margeo.2016.10.007

Hogarth, L.J., Driscoll, N.W., Babcock, J.M. and Orange, D.L. 2012. Transgressive deposits along the actively deforming Eel River Margin, Northern California. *Marine Geology*, **303–306**, 99–114, https://doi.org/10.1016/j.margeo.2012.02.005

Hughes-Clarke, J.E. 2016. First wide-angle view of channelized turbidity currents links migrating cyclic steps to flow characteristics. *Nature Communications*, **7**, 1–13, https://doi.org/10.1038/ncomms11896

Jobe, Z.R., Lowe, D.R. and Uchytil, S.J. 2011. Two fundamentally different types of submarine canyons along the continental margin of Equatorial Guinea. *Marine and Petroleum Geology*, **28**, 843–860, https://doi.org/10.1016/j.marpetgeo.2010.07.012

Kelsey, H.M. 1990. Late Quaternary deformation of marine terraces on the Cascadia Subduction Zone near Cape Blanco, Oregon. *Tectonics*, **9**, 983–1014, https://doi.org/10.1029/TC009i005p00983

Kostic, S. 2011. Modeling of submarine cyclic steps: controls on their formation, migration, and architecture. *Geosphere*, **7**, 294–304, https://doi.org/10.1130/GES00601.1

Lamb, M.P., Parsons, J.D., Mullenbach, B.L., Finlayson, D.P., Orange, D.L. and Nittrouer, C.A. 2008. Evidence for superelevation, channel incision, and formation of cyclic steps by turbidity currents in Eel Canyon, California. *GSA Bulletin*, **120**, 463–475, https://doi.org/10.1130/B26184.1

Lambeck, K. and Chappell, J. 2001. Sea level change through the last glacial cycle. *Science*, **292**, 679–686, https://doi.org/10.1126/science.1059549

Lee, H.J., Syvitski, J.P.M., Parker, G., Orange, D., Locat, J., Hutton, E.W.H. and Imran, J. 2002. Distinguishing sediment waves from slope failure deposits: field examples, including the 'Humboldt slide', and modelling results. *Marine Geology*, **192**, 79–104, https://doi.org/10.1016/S0025-3227(02)00550-9

Locat, J. and Lee, H.J. 2002. Submarine landslides: advances and challenges. *Canadian Geotechnical Journal*, **39**, 193–212, https://doi.org/10.1139/t01-089

Lowe, D.R. 1982. Sediment gravity flows; II, Depositional models with special reference to the deposits of high-density turbidity currents. *Journal of Sedimentary Research*, **52**, 279–297, https://doi.org/10.1306/212F7F31-2B24-11D7-8648000102C1865D

Masson, D.G., Wynn, R.B. and Talling, P.J. 2010. Large landslides on passive continental margins: processes, hypotheses and outstanding questions. *In*: Mosher, D.C., Shipp, R.C., Moscardelli, L., Chaytor, J.D., Baxter, C.D.P., Lee, H.J. and Urgeles, R. (eds) *Submarine Mass Movements and Their Consequences*. Advances in Natural and Technological Hazards Research. Springer, Dordrecht, 153–165.

McAdoo, B.G. and Watts, P. 2004. Tsunami hazard from submarine landslides on the Oregon continental slope. *Marine Geology*, **203**, 235–245, https://doi.org/10.1016/S0025-3227(03)00307-4

McAdoo, B.G., Pratson, L.F. and Orange, D.L. 2000. Submarine landslide geomorphology, US continental slope. *Marine Geology*, **169**, 103–136, https://doi.org/10.1016/S0025-3227(00)00050-5

McCrory, P.A. 1989. Late Neogene geohistory analysis of the Humboldt Basin and its relationship to convergence of the Juan de Fuca Plate. *Journal of Geophysical Research: Solid Earth*, **94**, 3126–3138, https://doi.org/10.1029/JB094iB03p03126

McCrory, P.A., Blair, J.L., Waldhauser, F. and Oppenheimer, D.H. 2012. Juan de Fuca slab geometry and its relation to Wadati–Benioff zone seismicity. *Journal of Geophysical Research: Solid Earth*, **117**, https://doi.org/10.1029/2012JB009407

McPhee-Shaw, E.E., Sternberg, R.W., Mullenbach, B.L. and Ogston, A.S. 2004. Observations of intermediate nepheloid layers on the northern California continental margin. *Continental Shelf Research*, **24**, 693–720, https://doi.org/10.1016/j.csr.2004.01.004

Moernaut, J., Daele, M.V. *et al.* 2014. Lacustrine turbidites as a tool for quantitative earthquake reconstruction: new evidence for a variable rupture mode in south central Chile. *Journal of Geophysical Research: Solid Earth*, **119**, 1607–1633, https://doi.org/10.1002/2013JB010738

Molenaar, A., Moernaut, J., Wiemer, G., Dubois, N. and Strasser, M. 2019. Earthquake impact on active margins: tracing surficial remobilization and seismic strengthening in a slope sedimentary sequence. *Geophysical Research Letters*, **46**, 6015–6023, https://doi.org/10.1029/2019GL082350

Mosher, D.C., Campbell, D.C., Gardner, J.V., Piper, D.J.W., Chaytor, J.D. and Rebesco, M. 2017. The role of deep-water sedimentary processes in shaping a continental margin: the Northwest Atlantic. *Marine Geology*, **393**, 245–259, https://doi.org/10.1016/j.margeo.2017.08.018

Mountjoy, J.J., Howarth, J.D. *et al.* 2018. Earthquakes drive large-scale submarine canyon development and sediment supply to deep-ocean basins. *Science Advances*, **4**, eaar3748, https://doi.org/10.1126/sciadv.aar3748

Mullenbach, B.L. and Nittrouer, C.A. 2006. Decadal record of sediment export to the deep sea via Eel Canyon. *Continental Shelf Research*, **26**, 2157–2177, https://doi.org/10.1016/j.csr.2006.07.019

Mullenbach, B.L., Nittrouer, C.A., Puig, P. and Orange, D.L. 2004. Sediment deposition in a modern submarine canyon: Eel Canyon, northern California. *Marine Geology*, **211**, 101–119, https://doi.org/10.1016/j.margeo.2004.07.003

Nakajima, T. and Kanai, Y. 2000. Sedimentary features of seismoturbidites triggered by the 1983 and older historical earthquakes in the eastern margin of the Japan Sea. *Sedimentary Geology*, **135**, 1–19, https://doi.org/10.1016/S0037-0738(00)00059-2

Nittrouer, C.A. 1999. STRATAFORM: overview of its design and synthesis of its results. *Marine Geology*, **154**, 3–12, https://doi.org/10.1016/S0025-3227(98)00128-5

Nittrouer, C.A., Austin, J.A., Jr, Field, M.E., Kravitz, J.H., Syvitski, J.P.M. and Wiberg, P.L. 2007. Writing a Rosetta Stone: insights into continental-margin sedimentary processes and strata. *In*: *Continental Margin Sedimentation*, Wiley, 1–48, https://doi.org/10.1002/9781444304398.ch1

O'Grady, D.B., Syvitski, J.P.M., Pratson, L.F. and Sarg, J.F. 2000. Categorizing the morphologic variability of siliciclastic passive continental margins. *Geology*, **28**, 207–210, https://doi.org/10.1130/0091-7613(2000)28<207:CTMVOS>2.0.CO;2

Ogston, A.S., Cacchione, D.A., Sternberg, R.W. and Kineke, G.C. 2000. Observations of storm and river flood-driven sediment transport on the northern California continental shelf. *Continental Shelf Research*, **20**, 2141–2162, https://doi.org/10.1016/S0278-4343(00)00065-0

Ogston, A.S., Guerra, J.V. and Sternberg, R.W. 2004. Interannual variability of nearbed sediment flux on the Eel River shelf, northern California. *Continental Shelf Research*, **24**, 117–136, https://doi.org/10.1016/j.csr.2003.08.007

Okutsu, N., Ashi, J. *et al.* 2018. Evidence for surface sediment remobilization by earthquakes in the Nankai forearc region from sedimentary records. *Geological Society, London, Special Publications*, **477**, 37–45, https://doi.org/10.1144/SP477.22

Orange, D.L. 1999. Tectonics, sedimentation, and erosion in northern California: submarine geomorphology and sediment preservation potential as a result of three competing processes. *Marine Geology*, **154**, 369–382, https://doi.org/10.1016/S0025-3227(98)00124-8

Paull, C.K., McGann, M. *et al.* 2014. Sub-decadal turbidite frequency during the early Holocene: Eel Fan, offshore northern California. *Geology*, **42**, 855–858, https://doi.org/10.1130/G35768.1

Paull, C.K., Talling, P.J. *et al.* 2018. Powerful turbidity currents driven by dense basal layers. *Nature*

Communications, **9**, 1–9, https://doi.org/10.1038/s41467-018-06254-6

Phrampus, B.J., Harris, R.N. and Tréhu, A.M. 2017. Heat flow bounds over the Cascadia margin derived from bottom simulating reflectors and implications for thermal models of subduction. *Geochemistry, Geophysics, Geosystems*, **18**, 3309–3326, https://doi.org/10.1002/2017GC007077

Pope, E., Talling, P.J. and Carter, L. 2017. Which earthquakes trigger damaging submarine mass movements: insights from a global record of submarine cable breaks? *Marine Geology*, **384**, 131–146, https://doi.org/10.1016/j.margeo.2016.01.009

Postma, G. and Cartigny, M.J.B. 2014. Supercritical and subcritical turbidity currents and their deposits – a synthesis. *Geology*, **42**, 987–990, https://doi.org/10.1130/G35957.1

Pouderoux, H., Proust, J.-N., Lamarche, G., Orpin, A.R. and Neil, H. 2012. Postglacial (after 18 ka) deep-sea sedimentation along the Hikurangi subduction margin (New Zealand): characterisation, timing and origin of turbidites. *Marine Geology*, **295–298**, 51–76, https://doi.org/10.1016/j.margeo.2011.11.002

Pratson, L.F. and Haxby, W.F. 1996. What is the slope of the U.S. continental slope? *Geology*, **24**, 3–6, https://doi.org/10.1130/0091-7613(1996)024<0003:WITSOT>2.3.CO;2

Pratson, L.F., Nittrouer, C.A. *et al.* 2007. Seascape Evolution on Clastic Continental Shelves and Slopes. *In*: *Continental Margin Sedimentation*, Wiley, 339–380, https://doi.org/10.1002/97814443 04398.ch7

Puig, P., Ogston, A.S., Mullenbach, B.L., Nittrouer, C.A., Parsons, J.D. and Sternberg, R.W. 2004. Storm-induced sediment gravity flows at the head of the Eel submarine canyon, northern California margin. *Journal of Geophysical Research: Oceans*, **109**, https://doi.org/10.1029/2003JC001918

Riedel, M., Côté, M.M., Urlaub, M., Geersen, J., Scholz, N.A., Naegeli, K. and Spence, G.D. 2018. Slope failures along the deformation front of the Cascadia margin: linking slide morphology to subduction zone parameters. *Geological Society, London, Special Publications*, **477**, 47–67, https://doi.org/10.1144/SP477.33

Ross, W.C., Halliwell, B.A., May, J.A., Watts, D.E. and Syvitski, J.P.M. 1994. Slope readjustment: a new model for the development of submarine fans and aprons. *Geology*, **22**, 511–514, https://doi.org/10.1130/0091-7613(1994)022<0511:SRANMF>2.3.CO;2

Ryan, W.B.F., Carbotte, S.M. *et al.* 2009. Global Multi-Resolution Topography synthesis. *Geochemistry, Geophysics, Geosystems*, **10**, https://doi.org/10.1029/2008GC002332

Savage, J.C., Svarc, J.L., Prescott, W.H. and Murray, M.H. 2000. Deformation across the forearc of the Cascadia subduction zone at Cape Blanco, Oregon. *Journal of Geophysical Research: Solid Earth*, **105**, 3095–3102, https://doi.org/10.1029/1999JB900392

Sawyer, D.E. and DeVore, J.R. 2015. Elevated shear strength of sediments on active margins: evidence for seismic strengthening. *Geophysical Research Letters*, **42**, 10216–10221, https://doi.org/10.1002/2015GL066603

Schlager, W. and Adams, E.W. 2001. Model for the sigmoidal curvature of submarine slopes. *Geology*, **29**, 883–886, https://doi.org/10.1130/0091-7613(2001)029<0883:MFTSCO>2.0.CO;2

Schwehr, K., Tauxe, L. and Driscoll, N.W. 2007. Origin of continental margin morphology: submarine-slide or downslope current-controlled bedforms, a rock magnetic approach. *Marine Geology*, **240**, 19–41, https://doi.org/10.1016/j.margeo.2007.01.012

Silver, E.A. 1971. Transitional tectonics and late cenozoic structure of the continental margin off northernmost California. *GSA Bulletin*, **82**, 1–22, https://doi.org/10.1130/0016-7606(1971)82[1:TTALCS]2.0.CO;2

Slootman, A. and Cartigny, M.J.B. 2019. Cyclic steps: review and aggradation-based classification. *Earth-Science Reviews*, **201**, 102949, https://doi.org/10.1016/j.earscirev.2019.102949

Sommerfield, C.K. and Wheatcroft, R.A. 2007. Late Holocene sediment accumulation on the northern California shelf: oceanic, fluvial, and anthropogenic influences. *GSA Bulletin*, **119**, 1120–1134, https://doi.org/10.1130/B26019.1

Sommerfield, C.K., Ogston, A.S. *et al.* 2007. Oceanic dispersal and accumulation of river sediment. *In*: *Continental Margin Sedimentation*, Wiley, 157–212, https://doi.org/10.1002/9781444304398.ch4

Spinelli, G.A. and Field, M.E. 2001. Evolution of continental slope gullies on the northern California margin. *Journal of Sedimentary Research*, **71**, 237–245, https://doi.org/10.1306/092500710237

Spinelli, G.A. and Field, M.E. 2003. Controls of tectonics and sediment source locations on along-strike variations in transgressive deposits on the northern California margin. *Marine Geology*, **197**, 35–47, https://doi.org/10.1016/S0025-3227(03)00116-6

Stone, I., Vidale, J.E., Han, S. and Roland, E. 2018. Catalog of offshore seismicity in Cascadia: insights into the regional distribution of microseismicity and its relation to subduction processes. *Journal of Geophysical Research: Solid Earth*, **123**, 641–652, https://doi.org/10.1002/2017JB014966

Sweet, M.L. and Blum, M.D. 2016. Connections between fluvial to shallow marine environments and submarine canyons: implications for sediment transfer to deep water. *Journal of Sedimentary Research*, **86**, 1147–1162, https://doi.org/10.2110/jsr.2016.64

Symons, W.O., Sumner, E.J. *et al.* 2017. A new model for turbidity current behavior based on integration of flow monitoring and precision coring in a submarine canyon. *Geology*, **45**, 367–370, https://doi.org/10.1130/G38764.1

Talling, P.J. 2014. On the triggers, resulting flow types and frequencies of subaqueous sediment density flows in different settings. *Marine Geology*, **352**, 155–182, https://doi.org/10.1016/j.margeo.2014.02.006

ten Brink, U.S., Andrews, B.D. and Miller, N.C. 2016. Seismicity and sedimentation rate effects on submarine slope stability. *Geology*, **44**, 563–566, https://doi.org/10.1130/G37866.1

Traykovski, P., Geyer, W.R., Irish, J.D. and Lynch, J.F. 2000. The role of wave-induced density-driven fluid mud flows for cross-shelf transport on the Eel River continental shelf. *Continental Shelf Research*, **20**, 2113–2140, https://doi.org/10.1016/S0278-4343(00)00071-6

Urlaub, M., Talling, P.J., Zervos, A. and Masson, D. 2015. What causes large submarine landslides on low gradient (<2°) continental slopes with slow (~0.15 m/kyr) sediment accumulation? Large submarine landslides on low gradients. *Journal of Geophysical Research: Solid Earth*, **120**, 6722–6739, https://doi.org/10.1002/2015JB012347

Van Wagoner, J.C., Posamentier, H.W., Mitchum, R.M., Vail, P.R., Sarg, J.F., Loutit, T.S. and Hardenbol, J. 1988. An overview of the fundamentals of sequence stratigraphy and key definitions. *SEPM, Special Publications*, **42**, 109–124.

Walsh, J.P. and Nittrouer, C.A. 1999. Observations of sediment flux to the Eel continental slope, northern California. *Marine Geology*, **154**, 55–68, https://doi.org/10.1016/S0025-3227(98)00103-0

Wang, K. and Tréhu, A.M. 2016. Invited review paper: some outstanding issues in the study of great megathrust earthquakes – the Cascadia example. *Journal of Geodynamics*, **98**, 1–18, https://doi.org/10.1016/j.jog.2016.03.010

Warrick, J.A. 2014. Eel River margin source-to-sink sediment budgets: revisited. *Marine Geology*, **351**, 25–37, https://doi.org/10.1016/j.margeo.2014.03.008

Wheatcroft, R.A. and Sommerfield, C.K. 2005. River sediment flux and shelf sediment accumulation rates on the Pacific Northwest margin. *Continental Shelf Research*, **25**, 311–332, https://doi.org/10.1016/j.csr.2004.10.001

Wright, L.D., Kim, S.-C. and Friedrichs, C.T. 1999. Across-shelf variations in bed roughness, bed stress and sediment suspension on the northern California shelf. *Marine Geology*, **154**, 99–115, https://doi.org/10.1016/S0025-3227(98)00106-6

Wright, L.D., Friedrichs, C.T., Kim, S.C. and Scully, M.E. 2001. Effects of ambient currents and waves on gravity-driven sediment transport on continental shelves. *Marine Geology*, **175**, 25–45, https://doi.org/10.1016/S0025-3227(01)00140-2

Tectonic and geomorphic controls on the distribution of submarine landslides across active and passive margins, eastern New Zealand

S. J. Watson[1]*, J. J. Mountjoy[1] and G. J. Crutchley[2]

[1]National Institute of Water and Atmospheric Research (NIWA), PO Box 140901, Wellington, New Zealand

[2]GEOMAR Helmholtz Centre for Ocean Research, Kiel, Germany

SJW, 0000-0002-3666-5034; JJM, 0000-0002-3703-0006; GJC, 0000-0002-6692-0665
*Correspondence: sally.watson@niwa.co.nz

Abstract: Submarine landslides occur on continental margins globally and can have devastating consequences for marine habitats, offshore infrastructure and coastal communities due to potential tsunamigenesis. Therefore, understanding landslide magnitude and distribution is central to marine and coastal hazard planning.

We present the first submarine landslide database for the eastern margin of New Zealand comprising >2200 landslides occurring in water depths from *c.* 300–4000 m. Landslides are more prevalent and, on average, larger on the active margin compared with the passive margin. We attribute higher concentrations of landslides on the active margin to tectonic processes including uplift and oversteepening, faulting and seamount subduction. Submarine landslide scars are concentrated around canyon systems and close to canyon thalwegs. This suggests that not only does mass wasting play a major role in canyon evolution, but also that canyon-forming processes may provide preconditioning factors for slope failure.

Results of this study offer unique insights into the spatial distribution, magnitude and morphology of submarine landslides across different geological settings, providing a better understanding of the causative factors for mass wasting in New Zealand and around the world.

Constraining the influence of potential preconditioning and triggering factors is vital for submarine-landslide hazard assessments. The distribution of submarine landslides can be combined with other spatially varying geological, tectonic and geomorphological features to reveal preconditioning and triggering factors of seafloor slope failure. Few regional spatial datasets have allowed submarine-landslide mapping at the scale of continental margins. Thanks to the wide availability of multibeam echosounders, large regions of the seafloor are now mapped at resolutions that facilitate more detailed and statistical studies of submarine-landslide features.

In this study we provide a margin-wide compilation of landslide features along the eastern seaboard of New Zealand using *c.* 30 kHz multibeam bathymetry (Fig. 1). Analysis of the size and spatial distribution of submarine landslides enables us to investigate spatial patterns related to geological and environmental processes that have previously been implicated in controlling landslide occurrence.

State of knowledge of landslide triggering and preconditioning

Documented examples of landslide triggering are rare but there is good evidence in several cases to isolate a specific process that has been the primary cause of landslide triggering. Direct evidence of landslide triggering mechanisms include dynamic loading during earthquakes (Hampton *et al.* 1993; Piper *et al.* 1999), rapid loading of the slope causing overpressure (Clare *et al.* 2016; Kelner *et al.* 2016), tidal pressure loading (Clare *et al.* 2016), freshwater outflows from a terrestrial source (Dan *et al.* 2007; Kelner *et al.* 2016) and volcanic eruptions (Wright *et al.* 2008). Preconditioning factors are much more difficult to isolate for individual landslides but many contributing factors have been implicated, including weak layers within the stratigraphic sequence (Dan *et al.* 2007; L'Heureux *et al.* 2012; Urlaub *et al.* 2018), canyon erosion removing toe support from slopes (Twichell and Roberts 1982; Baztan *et al.* 2005; Micallef *et al.*

From: Georgiopoulou, A., Amy, L. A., Benetti, S., Chaytor, J. D., Clare, M. A., Gamboa, D., Haughton, P. D. W., Moernaut, J. and Mountjoy, J. J. (eds) 2020. *Subaqueous Mass Movements and their Consequences: Advances in Process Understanding, Monitoring and Hazard Assessments*. Geological Society, London, Special Publications, **500**, 477–494.
First published online March 13, 2020, https://doi.org/10.1144/SP500-2019-165

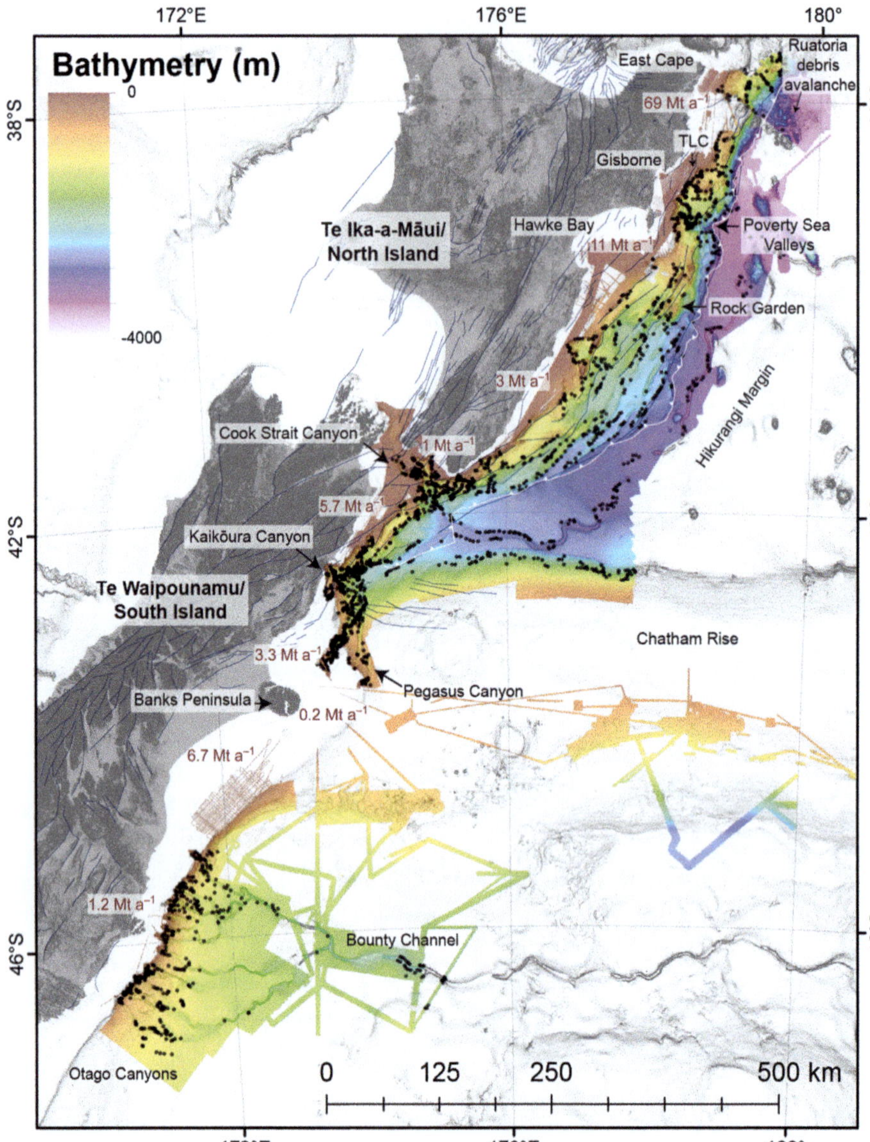

Fig. 1. The eastern margin of New Zealand. Black points are locations of landslides mapped in the current study. Coloured areas represent the region mapped by high-resolution multibeam echo sounders. Solid white line represents the Hikurangi Margin deformation front, with the subduction teeth on the overriding plate (Barnes *et al.* 2010, 2018). Thin blue lines (major active faults) are displayed both onshore and offshore (Litchfield *et al.* 2014). Red italicized numbers represent input of river suspended sediment to the coast (Hicks *et al.* 2011). Features or landmarks mentioned in the text are labelled. TLC, Tuaheni Landslide Complex.

2012), fluid migration creating excess pore pressure (Bünz *et al.* 2005; Berndt *et al.* 2012; Elger *et al.* 2018), sediment overpressure due to rapid sediment loading (Dugan and Flemings 2000; Flemings *et al.* 2008) and gas hydrate dissociation (Paull *et al.* 1996; Lopez *et al.* 2010).

The suggestion in the early 1980s of a causative link between gas hydrates and submarine landslides (e.g. McIver 1982) has led to numerous studies over the last 40 years into the various mechanisms by which gas hydrate systems could promote slope failure (e.g. Popenoe *et al.* 1993; Bünz *et al.* 2003;

Sultan *et al.* 2004; Pecher *et al.* 2005; Crutchley *et al.* 2007; Elger *et al.* 2018). The mechanisms usually relate to excess pore fluid pressure generation from free gas, but other hypotheses relating to a change in sediment rheology due to hydrate occurrence have also been suggested (e.g. Mountjoy *et al.* 2014; Yelisetti *et al.* 2014). To draw a link between marine gas hydrates and slope failure is tempting for two reasons: (1) globally speaking, there is a broad spatial correlation between marine gas hydrate occurrence and submarine landslide distribution – both are common on continental slopes where the seafloor is often inclined and pressure–temperature conditions for hydrate formation are met (Hampton *et al.* 1996; Kvenvolden 1999), and (2) the occurrence of gas hydrate and free gas within sediments can change the bulk rheology and strength of those sediments (e.g. Carpenter 1981; Winters *et al.* 2004). However, despite the decades of research into this supposed correlation, we are not aware of any conclusive evidence anywhere for gas hydrates having been the main contributing factor to a submarine slope failure.

Regional mapping of landslides has been carried out in a number of studies but has been limited by the availability of complete seafloor multibeam data coverage at continental margin scales (Clare *et al.* 2018). Nonetheless, significant insights have been gained from these studies. The relationship between landslides and tectonic processes at the subduction front, based on mapping 50 landslides along the toe of the Cascadia Subduction Wedge, indicates that thrust-ridge geometry is a key parameter controlling the mode of failure as a blocky v. debris flow event (Riedel *et al.* 2018). Mapping of 141 landslides across the USA Atlantic margin shows a log-normal volume distribution centred around 1 km^3, indicating this is a dominant landslide size on this passive margin (Chaytor *et al.* 2009). A compilation of submarine landslides in the Mediterranean Sea that crosses several areas classified as active and passive margins shows that passive margin slopes are more prone to failure than active margin slopes (Camerlenghi *et al.* 2010). In a regional analysis of landslide features around the Queen Charlotte Fault (the Pacific–North America transform plate boundary), Brothers *et al.* (2019) found that most landslides occurred on the part of the margin resembling a passive margin and that the more tectonically-deformed (transpressional) sections of the margin contain fewer landslides. They proposed that this was due to higher sediment bypass in the regions dominated by tectonic geomorphology. Contrary to previous conclusions that landslides are more common during glaciations or following transgressions due to accelerated sedimentation, a global compilation of 68 age-dated large submarine landslides (Urlaub *et al.* 2013) suggested that no robust correlation exists to specific periods in the sea-level curve once age uncertainty is considered.

Local background and submarine landslide occurrence

The 1300 km-long eastern seaboard of New Zealand can be subdivided into three broad tectonic domains. In the North (from East Cape to Cook Strait Canyon) is the Hikurangi Margin, a domain of high tectonic deformation due to oblique subduction of the Pacific Plate beneath the Australian Plate (Fig. 1; Walcott 1978; Pettinga *et al.* 1984). In central New Zealand (from Cook Strait Canyon to Kaikōura; Fig. 1) is a subduction to strike-slip transitional domain (Wallace *et al.* 2012). In the South (from the Chatham Rise southward along eastern South Island to the Otago Canyons; Fig. 1) is a domain of low to no tectonic strain considered to be a passive margin (Litchfield *et al.* 2014). Mapped offshore active faults (Litchfield *et al.* 2014) have been synthesized to major structures and reflect this north to south decrease in active deformation (Fig. 1). Morphotectonic mapping along the Hikurangi Margin has been used to segment the seafloor into primary geomorphic domains (Collot *et al.* 1996).

The region is subject to a vigorous maritime climate and a highly variable oceanographic regime. The sub-Antarctic current from the south has a strong influence on seafloor processes under present conditions (Hillman *et al.* 2018) and an even stronger influence during glacial periods (Barnes 1992). North of the Chatham Rise, ocean currents are less pronounced but can be highly variable over time with periodic eddy development and a general south to north flow (Chiswell 2005). Little is known about the influence of ocean currents on seafloor morphology and sediment transport along the Hikurangi Margin. Tidal currents are exceptionally strong through Cook Strait (Lewis *et al.* 1994) and documented to play a significant role in transporting sediment into canyons (Mountjoy *et al.* 2013).

Seafloor fluid-flow indicators have been documented along both the active Hikurangi Margin (Greinert *et al.* 2010; Watson *et al.* 2020) and along the passive SE margin south of the Chatham Rise (Hillman *et al.* 2015), while subsurface free gas and gas hydrate indicators have been widely documented between Cook Strait and East Cape (Henrys *et al.* 2003; Crutchley *et al.* 2015).

Fluvial sediment input to the coast is highly variable along the eastern seaboard of New Zealand, with some of the highest river-sediment discharge in the world from East Cape to Hawke Bay. Conversely, there is very low sediment discharge along the southeastern North Island (Hicks *et al.* 2011; Fig. 1). At the present day, much of the sediment discharged offshore is captured in continental shelf basins and little makes it to the continental slope. This is reflected in outer shelf sedimentation rates reaching 0.9 cm a^{-1} but being an order of magnitude lower on the slope (Orpin *et al.* 2006). Sedimentation rates are poorly quantified along the shelf and slope along the eastern South Island. Longer-term sedimentary architecture shows a thick sequence of outer-shelf to upper-slope Pleistocene clinoforms from East Cape to Hawke Bay (Paquet *et al.* 2009; Mountjoy and Barnes 2011). Between Hawke Bay and Kaikōura Canyon the shelf is much narrower (Fig. 1) and upper slope clinoform development is very limited. A thick sequence of Pleistocene clinoforms occurs to the south of Banks Peninsula (Browne and Naish 2003).

Preconditioning factors for seafloor slope failure processes in New Zealand

Research on seafloor slope failure off eastern New Zealand has been continuing since the 1970s. Mass failure has been documented along most of the offshore Hikurangi subduction margin (e.g. Collot *et al.* 2001; Mountjoy *et al.* 2009*a*; Micallef *et al.* 2012), but there is limited information south of Banks Peninsula (Fig. 1). Most of the work on submarine landslides in New Zealand waters to date has focused on case studies with the objective of understanding preconditioning and triggering factors or hazard implications of individual slope failures. Extensive multibeam mapping coverage of the eastern seaboard of New Zealand and the range of geological, tectonic, fluid flow and oceanographic conditions provide a sound framework for investigating potential controls on slope instability at a regional scale.

Earthquakes

Pouderoux *et al.* (2014) use marine turbidite palaeoseismology to demonstrate that ground shaking is the most likely trigger mechanism for Holocene slope failures on the northern Hikurangi Margin. Lewis and Barnes (1999) documented potential landslides at the head of the Kaikōura Canyon based on longshore sediment transport and inferred the area to be prone to repeated earthquake-triggered slope failure. Earthquake-triggered slope failures in the head of Kaikōura Canyon were modelled by Walters *et al.* (2006) to generate large tsunami waves; however subsequent geophysical surveys failed to find evidence to support the proposed landslide source area (Lane *et al.* 2018). In 2016 the M_W7.8 Kaikōura Earthquake triggered numerous slope failures around the rim of Kaikōura Canyon as well as canyon axis sediment failure focused in the upper reaches of the canyon (Mountjoy *et al.* 2018), representing one of the few global examples of a documented earthquake triggering submarine landslides. The landslides did not cause any tsunami and were more than an order of magnitude smaller than proposed failure scenarios (Lewis and Barnes 1999). Wang *et al.* (2016) analysed a slope failure up-dip from the Hikurangi Channel near Cook Strait, with an estimated volume of 2.5 km^3. The study focused on developing a coupled slope-failure tsunami model which showed potential for 3–5 m high waves on adjacent coastlines. Earthquake triggering was speculated to be the cause of that landslide.

Subduction zone processes

The geomorphological response of the upper plate to seamounts entering the subduction zone on the lower plate generates significant slope failures. These slope failures, often referred to as wake avalanches, occur as splay fault thrust ridges are oversteepened and collapse (Kukowski *et al.* 2010; Pedley *et al.* 2010). On the northern Hikurangi Margin there are numerous seamounts at various stages of subduction. The most extreme example of such a wake avalanche is the *c.* 3000 km^3 Ruatoria debris avalanche, a margin-scale landslide that extends from the shelf break to the Hikurangi Trough, spanning water depths of *c.* 3000 m (Fig. 1). A very large seamount on the incoming Pacific Plate is inferred to have caused the Ruatoria debris avalanche *c.* 170 ka ago (Collot *et al.* 2001; Lewis *et al.* 2004). No evidence for the subducted seamount remains today.

Gas and gas hydrates

The upper limit of gas hydrate stability in marine sediments on the Hikurangi Margin coincides approximately with the 650 m contour. The occurrence of gas hydrates from this upper limit into sediments at greater water depths (i.e. seafloor depths >2500 m below sea-level (mbsl)) is known from the occurrence of widespread bottom-simulating reflections (BSRs), which mark the base of gas hydrate stability (BGHS) in sediments. The upper limit of hydrate stability, sometimes referred to as the 'feather edge' (Ruppel 2011), is the region where environmental changes (e.g. ocean warming, sea-level change) will have the most pronounced effect on the gas hydrate system. Pecher *et al.* (2005), examining a broad uplifted ridge called 'Rock Garden' in the central Hikurangi Margin (Fig. 1), suggested that repeated formation and dissociation of gas hydrates in this feather-edge region could lead to mechanical weakening, regional erosion and slope failure. Ellis *et al.* (2010) and Crutchley *et al.* (2010) investigated how gas hydrate-induced permeability contrasts and the existence of free gas beneath Rock Garden could also lead to seafloor weakening and slope failure. Elsewhere on the margin, either shallow gas or gas hydrates in the Tuaheni Landslide Complex have been implicated as having a role in ongoing slope failure in this large landslide (Mountjoy *et al.* 2009*b*, 2014; Gross *et al.* 2018). A review by Crutchley *et al.* (2016) considered multiple examples of coincident gas hydrate systems and mass failure off New Zealand and summarized different mechanisms for how the gas hydrate system could affect slope stability.

Submarine canyons

Submarine landslides are extensive in and around the Cook Strait Canyon system (Mountjoy *et al.* 2009*a*, 2013; Micallef *et al.* 2012). The study by Micallef *et al.* (2012) is one of few that has explored the spatial distribution of landslides and its relationship to different preconditioning factors, concluding that canyon incision undercutting canyon walls is a key factor controlling landslide occurrence. The *c.* 30 km^3 Paritu debris avalanche off Gisborne occurs in close proximity to the Poverty Sea Valleys (Fig. 1) and has infilled the Paritu Basin with landslide debris (Mountjoy and Micallef 2012). This large landslide has been inferred to be associated with high rates of sedimentation during sea-level lowstand, with earthquake triggering also potentially playing a role (Pedley *et al.* 2010).

Methods

We compiled bathymetric data from more than 30 marine voyages, covering *c.* 181 000 km^2 (Fig. 1). The wide availability of multibeam echo sounder (MBES) bathymetry data around New Zealand enables us to investigate the size (length, width, area and estimated volume; Fig. 2a–d) and density distribution of submarine landslides across a range of geological settings (e.g. slope, shelf, ridges, canyons, channels); from the convergent tectonic setting in the north to the passive-margin setting in the south. In this study we define the active margin as the seafloor that is subject to tectonic deformation related to subduction zone processes on the Hikurangi Margin, while the passive margin is defined as the area south of the subduction termination where there is no evidence for tectonic deformation of the shallow stratigraphic units in which landslides occur (see Fig. 3a for the mapped division between the active and passive margins and primary geomorphological domains used in this study). Note that we categorize the basin floor region east of North Island's offshore accretionary wedge, that is on the incoming plate east of the broken white line in Figure 3a, together with the passive margin domain. This is because this region on the incoming plate has clearly not (yet) been deformed as part of the active margin. Horizontal MBES bathymetry grid resolution ranges between 10 and 100 m, with an average resolution of 25 m across the study area. Data compilation, landslide mapping and interpretation were performed in ArcGIS 10.6.

Widely recognized landslide features (e.g. landslide scars, head scarps and debris deposits) were identified using MBES bathymetry and bathymetric derivatives (hill shade, slope, rugosity) and manually digitized as polygons (Fig. 2a–d) using a similar approach to other regional-scale landslide mapping surveys (e.g. Queen Charlotte Margin, southeastern Alaska; Brothers *et al.* 2019). Analyses of landslides in this study predominantly focus on landslide scars (evacuated areas) rather than mass-transport deposits (MTDs), due to the relatively small number of MTDs observed. To gain a quantitative estimate of the scale of landslides, we measure a range of morphometric parameters including scar perimeter area, length, width, head-scarp height, deposit area and runout (for an exhaustive list of morphometric parameters measured, see Table 1; Fig. 2a–d; Clare *et al.* 2018). Landslide-volume estimates for all mapped submarine landslides were based on scar dimensions (e.g. McAdoo *et al.* 2000; Table 1).

Due to the range of landslide morphologies observed, landslide features were categorized on a confidence scale from 1 to 4, to convey certainty regarding their origins as a landslide event(s):

(1) *Certain landslide origin.* Clearly defined head scarp and/or lateral ridges observed, with corresponding MTD near base (Fig. 2a).
(2) *Likely landslide origin.* Well-defined landslide head scarp and/or lateral ridges possibly with minor erosional modification, with or without some evidence of MTD downslope (Fig. 2b).
(3) *Reasonable likelihood of landslide origin.* Evidence of scarps/scar, although with some erosional modification. No MTD visible in bathymetric data (Fig. 2c).
(4) *Possible landslide origin.* Some evidence of head scarp visible but heavily eroded and incised since mass wasting event(s) (Fig. 2d).

For those features in the database that may have additional or other processes involved in their formation we have assigned lower confidence values to portray our degree of certainty. For example, an arcuate shape near a canyon may resemble a landslide scar, but it may not have clearly defined head-scarp/lateral ridges (possibly due to subsequent erosion) and it may not have an MTD preserved at the base. In this case it would be assigned a confidence value of 4 to convey that it may have been heavily modified and/or subject to other formative processes.

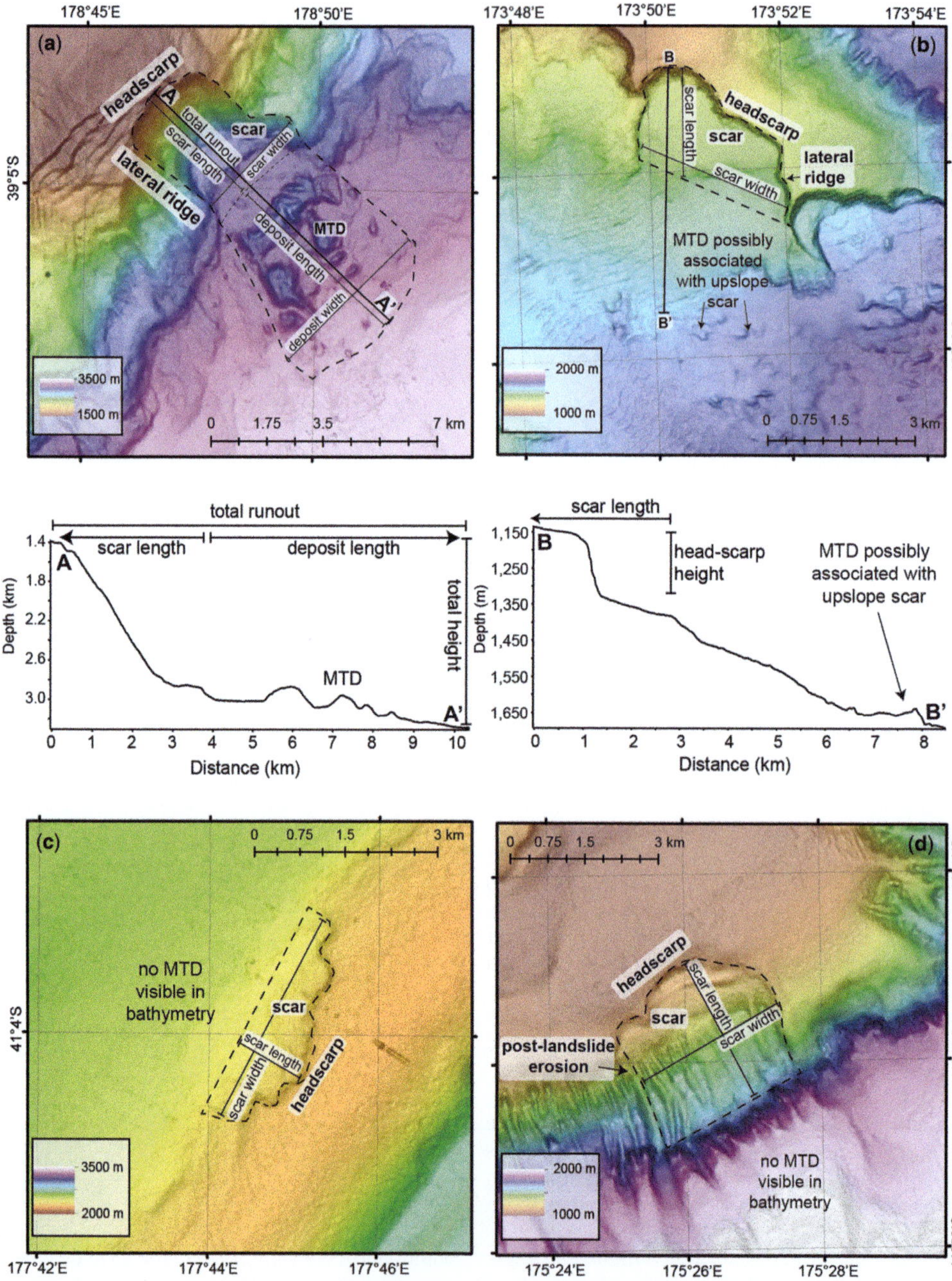

Fig. 2. Methods for parameterizing landslides on the eastern margin of New Zealand. (a–d) Examples of the range of landslide morphologies observed in the study area. Landslide features were assigned confidence values based on features present from: (**a**) clearly defined head scarp and lateral ridges with MTD (confidence rating: 1); (**b**) well-defined head scarp and lateral ridges with possible MTD at base (confidence rating: 2); (**c**) evidence of head scarp, minor erosional modification (confidence rating: 3); (**d**) some evidence of head scarp, highly modified and eroded (confidence rating: 4). Profile A–A' shows cross-section of landslide scar and MTD in (a). Profile B–B' shows cross-section of landslide scar in (b).

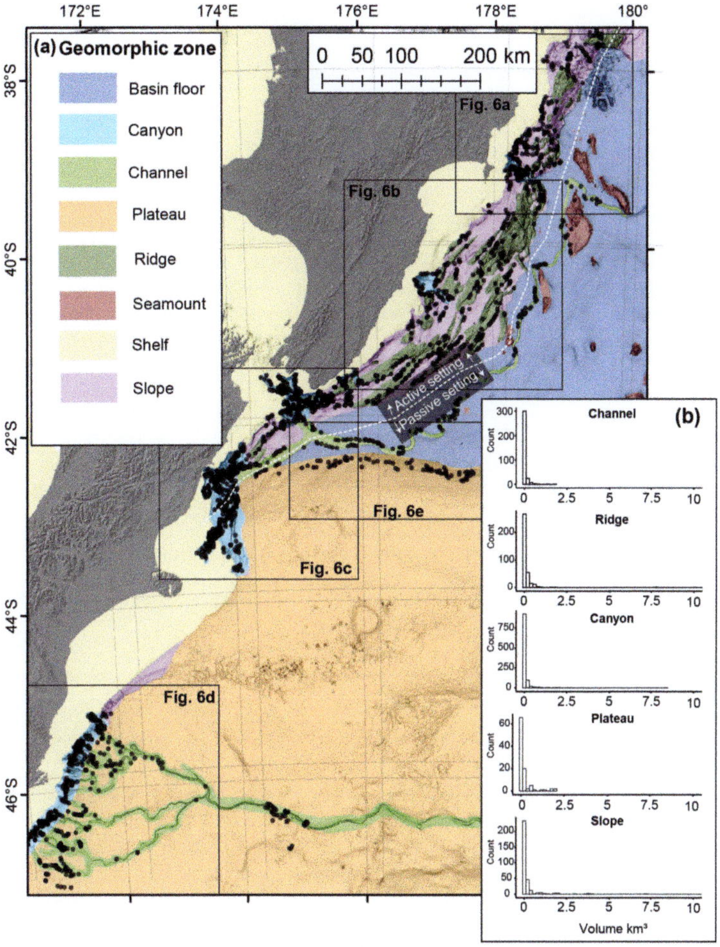

Fig. 3. (**a**) Distribution of landslides (black dots) along the eastern margin of New Zealand according to the broad-scale geomorphology. The boundary between active- and passive-margin settings is marked by the white dotted line. Locations of insets for Figure 6a–e are shown. (**b**) Volume histograms for each geomorphic zone as defined in (a) (shelf, seamount and basin floor zones not included due to small sample sizes). Note different *y*-axis scale for each plot.

Results

Magnitude and spatial distribution of submarine landslides on the eastern margin of New Zealand

This database comprises 2215 individually mapped landslide features spanning 1300 km along the length of the east coast of New Zealand (Figs 1 & 3a). Most (>81%) landslide scars have volumes <0.1 km^3, and only 57 are >1 km^3 (Fig. 4a–c). The frequency–volume distribution of landslides exhibits a power-law distribution (Fig. 4b), with the total estimated volume of failed material for the entire margin exceeding 590 km^3. Landslide scars range from *c.* 140m to 20 km in width and 66 m to 16 km in length, and there is a higher proportion of individual landslide features (70%) compared to landslide complexes (30%). Landslide complexes are more prevalent in the active-margin setting (39%), compared to the passive-margin setting (25%). The largest observed landslide on the margin is the Ruatoria debris avalanche off East Cape; the entire landslide (including both the scar and the debris deposit) has an estimated volume of >3100 km^3 (Collot *et al.* 2001; Fig. 1). The Ruatoria debris avalanche extends over 100 km in length from *c.* 170 to at least 3600 m water depth. More than 50 additional landslide scars are observed within or near the Ruatoria debris avalanche, with estimated individual volumes up to 30 km^3.

Table 1. *Metadata and morphometric parameters gathered for each landslide feature offshore eastern New Zealand*

Parameter	Metadata
Landslide ID	Unique landslide identification
Confidence	Confidence rating assigned to each identified landslide feature [1–4]
Voyage reference	Marine voyage identification number
Bathymetry grid resolution	Horizontal resolution of bathymetry grid used to identify landslide features
Single event or landslide complex	Denoting whether the landslide occurs as a single feature, or within a larger mass-wasting complex
Coordinates of centroid	Latitude and longitude (decimal degrees) of centre of landslide polygon
Presence of scar ± deposit	S = landslide scar, SD = both landslide scar and deposit present
Total feature length	Total length (contour perpendicular) of entire mapped feature (km)
Geomorphic location	Underlying geomorphic region (e.g. slope, shelf, channel, ridge)
Unfailed slope estimate	Estimated unfailed slope based on average slope measured around the centroid coordinate (degrees)
Minimum depth	Minimum depth of feature (m)
Maximum depth	Maximum depth of feature (m)
Total height	Maximum depth – Minimum depth (m)
Landslide scar area	Area of mapped polygon representing scar (km^2)
Landslide scar length	Scar length (contour perpendicular; m)
Landslide scar width	Scar width (widest measurement perpendicular to length; m)
Landslide scar perimeter	Perimeter of scar polygon (km)
Head-scarp height	Height difference between uppermost convex point to lowermost concave point (m)
Volume	(Scar area × head-scarp height)/2 (km^3)

For schematic illustration see Figure 2a–d.

We observe a variety of landslide morphologies across a range of geological settings within the surveyed region, in water depths ranging from *c.* 50 to >3600 m (Figs 1 & 3a). Of the submarine landslides identified in this study, 99% are characterized by a landslide scar, with no bathymetric evidence of an MTD downslope. Only a small proportion of landslides identified in this study are observed with a clearly defined head scarp and lateral ridges, and with a corresponding MTD near the base (i.e. confidence rating 1). All landslides with associated MTD occur along the active portion of the margin, and mostly across the continental slope or on thrust ridges. A further 13% of landslide features mapped in this study have well-defined landslide head scarp and/or lateral ridges possibly with minor erosional modification (i.e. confidence rating 2), 28% of landslide features are characterized by evidence of scarps/scar, although with some erosional modification (i.e. confidence rating 3). More than half (58%) of the mapped landslides in this study are characterized by a heavily eroded and incised head scarp (confidence rating 4). Most (68%) landslides identified on the passive margin are highly modified and eroded by slope failure processes (confidence rating 4), compared to 50% for those in the active-margin setting.

Landslides are more densely distributed on the active margin, with approximately one landslide per 56 km^2, compared to one landslide per 121 km^2 on the passive-margin setting (Fig. 4c; please note that the active portion of the margin makes up 40% of the total mapped region). More than half the mapped scars occur on slopes that are <5°, and 80% occur on slopes <10° (Fig. 4d, e). Landslide scars are, on average, larger by volume, area, length and head-scarp height on the active margin compared to the passive margin. Submarine landslides' scars in this study are predominantly concentrated within canyon systems (49%), followed by ridges (16%), channels (15%) and slopes (14%) (Figs 3a & 4f). The highest density of large landslide (i.e. >1 km^3) scars occur in canyons (namely Pegasus Canyon, Cook Strait Canyon and the Poverty Sea Valleys) and associated channel systems (Figs 3b & 4f). In general, landslide scars on ridges, slopes and on the Chatham Rise (plateau) are smaller than those associated with canyons and channels (Fig. 3b).

Correlating landslide magnitude with potential preconditioning factors and triggers

This margin-wide database enables us to correlate potential preconditioning factors and triggers with landslide magnitude. Interestingly, we do not observe strong spatial correlations between the

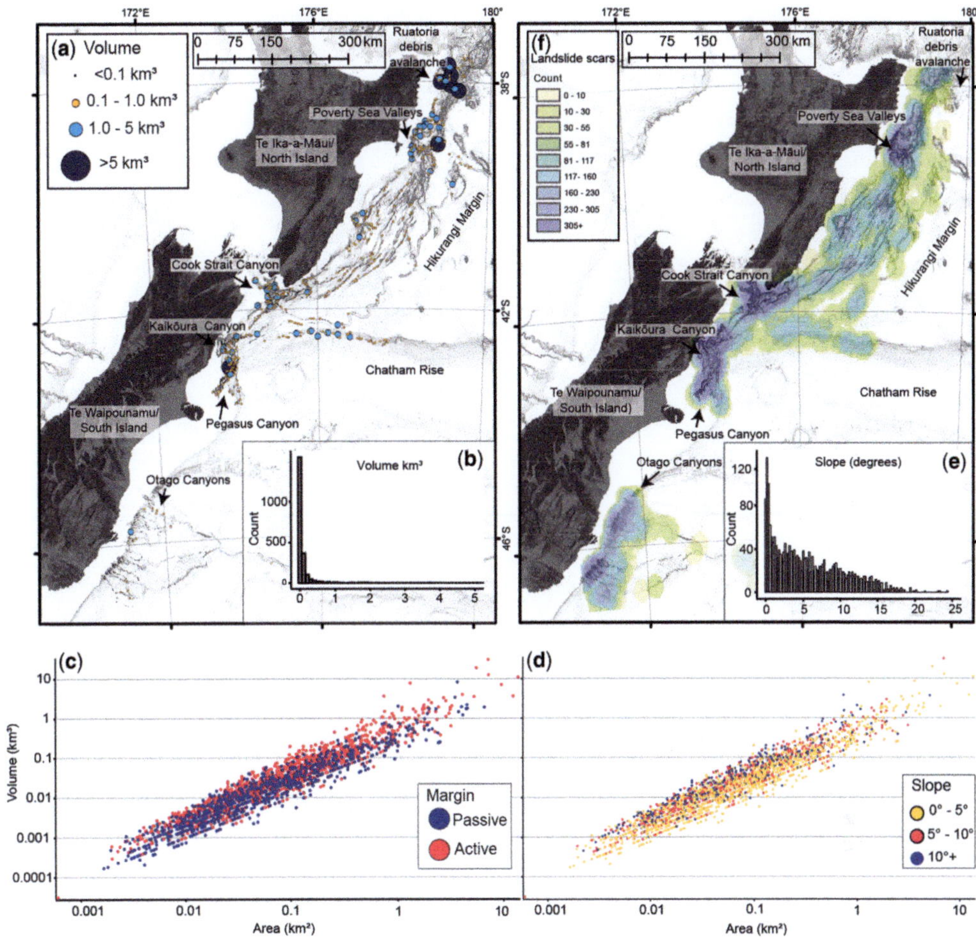

Fig. 4. (**a**) Volume distribution of submarine landslides with graduated symbols to show the locations of moderate, large and very large landslides; (**b**) volume histogram (Ruatoria debris avalanche has been excluded from all graphs for clarity) showing power-law distribution of landslides; (**c**) and (**d**) area v. volume graph (logarithmic scale) with colours denoting active- and passive-margin setting (see Fig. 3a for boundary of active/passive setting; (c) and slope (d)); (**e**) slope histogram showing the distribution of slope values for each landslide feature; (**f**) density analysis of landslide features highlighting the regions (darker blue) where landslides are most prevalent.

magnitude of landslides mapped and potential landslide triggers on the margin (including seismic ground-shaking, depth, slope and proximity to faults, and hydrate stability; Fig. 5a–f). There is no substantial difference between the average slope for active v. passive margin failures, nor the total number of landslides on different slopes. However, 68% of those landslides on slopes >10° are located on the active margin (Fig. 5c). The depths of submarine landslides are relatively consistent across the entire dataset, irrespective of margin setting (Fig. 5d). Approximately 17% of landslide scars overlap the upper limit of hydrate stability (*c.* 650 m water depth; Fig. 5f). Landslide volume does not vary substantially across the active- and passive-margin settings, despite the prominent distinction between the active and passive margins in terms of degree of ground-shaking experienced and the proximity of landslides to major active faults. Landslides within the active-margin setting experience higher average ground-shaking compared to those on the passive margin (Fig. 5a). Landslides on the active margin are also much closer to active faults, with 98% of scars on the active margin occurring within 20 km of a major active fault, compared to 63% for the entire dataset (Fig. 5b). The most striking observation is the apparent association of landslides with canyon systems on the eastern margin of New Zealand, with 68% of all landslides occurring within 5 km of a canyon thalweg (Fig. 5e).

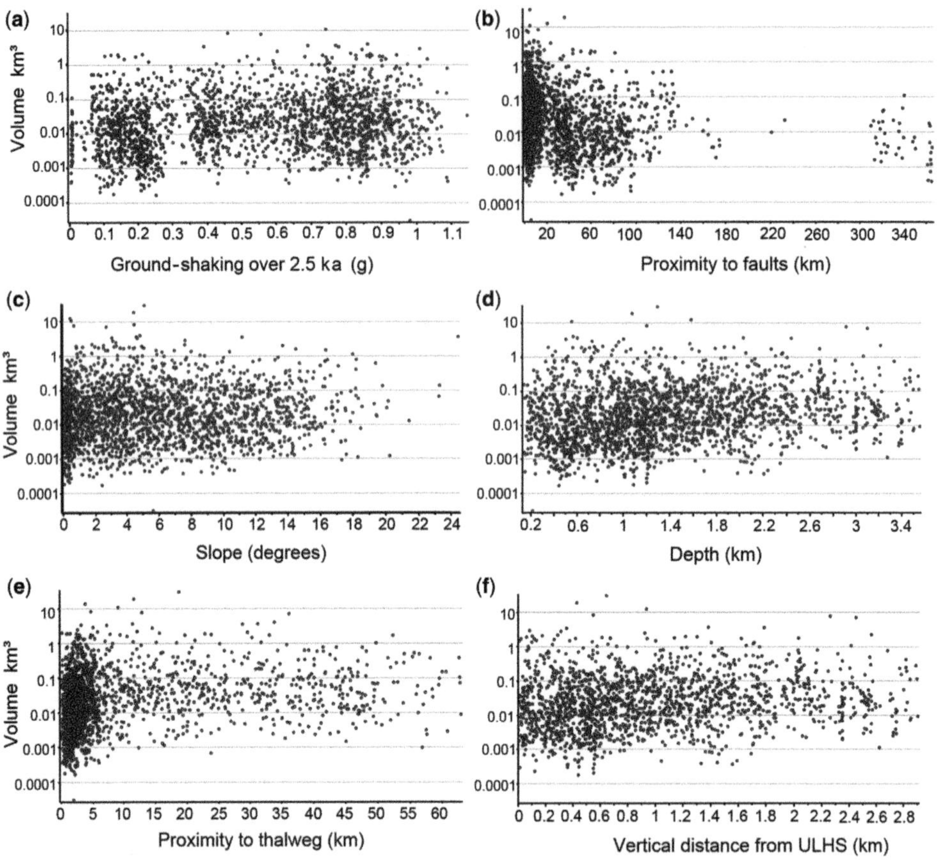

Fig. 5. Graphs of potential preconditioning/trigger mechanisms v. landslide volume on the eastern margin of New Zealand; points are coloured based on active- (red) or passive- (blue) margin setting, see Figure 3a for boundary of active/passive setting. (**a**) Average ground-shaking over 2.5 ka (Stirling *et al.* 2012); (**b**) proximity to major active faults (Litchfield *et al.* 2014); (**c**) estimate of pre-failure slope angle; (**d**) depth at centroid; (**e**) proximity to thalwegs; (**f**) vertical distance to the upper limit of hydrate stability (ULHS; 650 m contour) – in this case, landslides shallower than 650 m are not shown. Note that Ruatoria debris avalanche has been excluded from all graphs for clarity.

Variations in landslide magnitude and morphology with geological setting

Landslides located on the active northernmost Hikurangi Margin are mostly distributed across the continental slope and within the Poverty Sea Valleys (Fig. 6a). Here, landslide head scarps are common at the continental slope break at *c.* 250 m water depth with mostly arcuate and multiple lobe head-scarp morphologies. This portion of the margin has the highest density of very large landslides (seven scars with volumes >5 km^3, including the Ruatoria debris avalanche).

The central Hikurangi Margin (also an active portion of the margin) is characterized by a wide, low-taper accretionary wedge and trench-parallel thrust faults that are common across the continental slope. A high proportion (27%) of landslides' scars on the central Hikurangi Margin are situated along these trench-parallel thrust ridges (Fig. 6b). Landslides along thrust ridges, particularly those situated along the seaward-facing forelimbs, occur across a range of water depths (300–3000 m) and typically have smaller volumes (Fig. 3b), wider, shorter scar morphologies and rectangular-shaped head scarps. Landslides are mostly absent along the slope break on the central Hikurangi Margin, except for scars associated with Madden Canyon (Fig. 6b).

Further south on the eastern margin of New Zealand is the transition from oblique subduction off the North Island to a transpressional tectonic regime in the northeastern reaches of South Island.

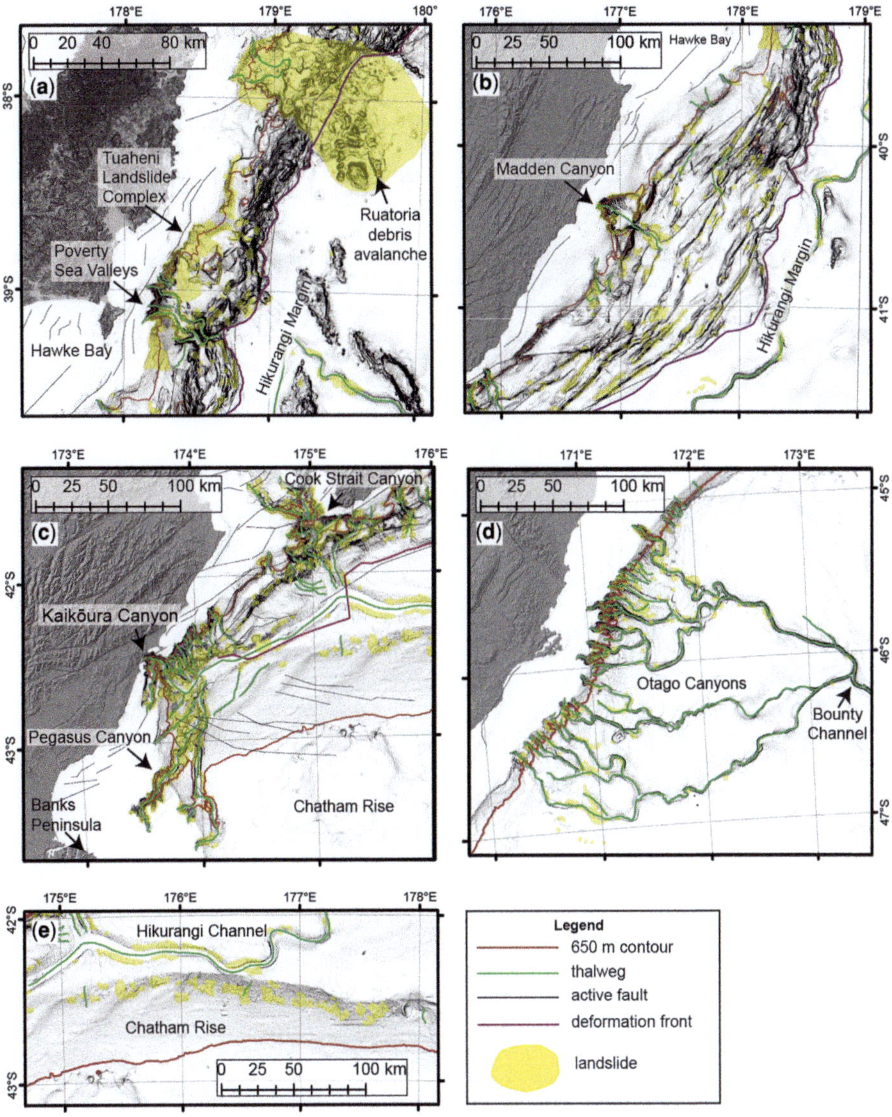

Fig. 6. Enlarged maps showing the similarities of landslide size and location for different geological settings on the eastern margin of New Zealand. See Figure 3a for locations of each inset. (**a**) Northern Hikurangi Margin – active-margin setting with seamount subduction, high sediment yield, high seismic ground-shaking, steep continental slope and gas hydrate presence; (**b**) Central Hikurangi Margin – active-margin setting, trench-parallel ridges prevalent, high seismic ground-shaking and gas hydrate presence; (**c**) Cook Strait – Pegasus Canyon: transition from an active to a passive setting, variable ground-shaking, large canyon systems, and moderate sediment yield (gas hydrates not known as far south as Kaikōura and Pegasus canyons); (**d**) Otago Canyons – passive-margin setting, low ground-shaking, low sediment yield, no known gas hydrate occurrences; (**e**) passive-margin setting, submarine plateau, no proximal onshore sediment yield, low ground-shaking, no known gas hydrate occurrences.

At the nexus of this convergent–transpressional transition lies the Cook Strait Canyon. Landslides are a common feature in the Cook Strait Canyon, covering 19% of the total seafloor within the canyon (Fig. 6c). Landslides in the Cook Strait Canyon are up to 3 km^3 in volume, mostly in water depths between 250 and 1000 m and occur predominantly on the side walls of the canyon channel(s), close to the thalweg. The Pegasus and Kaikōura canyons lie south of the Cook Strait Canyon and south of the actively accreting portion of the margin (Fig. 6c). Landslide scars cover 11% of the total seafloor area within Pegasus and Kaikōura canyons, are up to 8.4 km^3 in size, in water depths between 80 and 2100 m and are mostly situated on the canyon walls. The Cook, Pegasus and Kaikōura canyons have the highest density of landslides scars on the entire margin (Fig. 4a, f) and host 30% of the large landslides (>1 km^3).

The passive-margin setting of the Otago Canyon is characterized by the smallest landslides by volume and area compared to all other canyon systems on the margin (Fig. 4a). Scars are present throughout the canyon system from the canyon head (at *c.* 100 m water depth), to the lower canyon at >2000 m water depth (Fig. 6d). The largest landslide located within the Otago Canyon system is *c.* 2 km^3, 3% of landslides are greater than 0.1 km^3, 29% are between 0.01 and 0.1 km^3 and 67% are <0.01 km^3. The landslide scars in the Otago Canyon are highly eroded and modified by post-failure processes.

Submarine landslide scars are prevalent across the northern mid-slope of the Chatham Rise, up-dip from the Hikurangi Channel (Fig. 6e). These landslides are mostly within a water depth range of 1500 to 2500 m. Head-scarp heights are relatively consistent between 40 and 80 m. There is no evidence of mass wasting within the mapped region of the southern Chatham Rise.

Discussion

Our landslide database includes >2200 landslides that have been mapped from high-resolution bathymetric data, using a consistent margin-wide approach to measuring and quantifying landslides (Clare *et al.* 2018). This database enables us to compare landslide distribution on active v. passive tectonic settings, analyse landslide occurrence in different geomorphic environments, and infer potential preconditioning factors and triggers. Our results indicate there is a higher occurrence of landslides in submarine canyons than any other environment, irrespective of tectonic setting, and that there are more landslides per unit area on the active domain of the margin compared with the passive domain. These are the two key results of the study that we explore in the discussion.

Active v. passive margin landslide distribution

One of the key conclusions drawn from both regional and global studies of submarine landslides is that fewer landslides are observed on active margins than on passive margins (Camerlenghi *et al.* 2010; Sawyer and DeVore 2015). Contrary to this, we observe that landslide scars are more densely distributed on the active margin from Kaikōura Canyon to East Cape (*c.* 1 landslide per 56 km^2), compared to the passive-margin setting from Pegasus to Otago canyons including the Chatham Rise (*c.* 1 landslide per 121 km^2; Fig. 3a). In general, landslide features identified in the passive-margin setting are more eroded and modified by post-failure processes (68% with confidence rating 4) compared to the active margin, which hosts a variety of landslide morphologies with a range of confidence ratings. Landslide morphologies on the active margin range from well-defined scars with associated MTD (confidence rating 1) to highly eroded scars with no MTD present (confidence rating 4).

We identify three geological processes that occur on active margins and not passive margins that may have a role in concentrating landslide occurrence, namely active faulting with associated higher ground-shaking and thrust ridge propagation, gas hydrates and seamount subduction.

Active faults and thrust ridges. We attribute the higher landslide concentration on the active margin to the influence of tectonic-related processes, including the high occurrence of landslides along thrust ridges (*c.* 27% of the active margin dataset; Figs 3a & 6b). The strong spatial correlation between thrust ridges and landslide distribution is particularly clear in the central part of the Hikurangi margin (e.g. Figs 3a & 6b). We identify two factors that may be responsible for the concentration of landslides on ridges. (1) Local slope tectonic uplift and oversteepening – 50% of the landslides located on trench-parallel ridges in the central Hikurangi margin occur on forelimbs which are typically the steepest. (2) All mapped ridges on the active margin are controlled by active faults, so focused high-frequency ground motion during earthquakes may play a role in observed landslide occurrence. High-frequency ground motion is a near-fault phenomenon and is known to be a first-order triggering mechanism both onshore and offshore (Wright and Rathje 2003; Masson *et al.* 2006; Zhu *et al.* 2013; Mountjoy *et al.* 2018).

A high proportion of landslide scars (*c.* 50%) that are located on thrust ridges are primarily focused along the seaward-facing forelimb. The morphology of the landslide scars on the forelimb of thrust ridges is distinct, with smaller volumes (<8 km^3) rectangular head scarps and shorter scar lengths (<5.5 km) that may reflect the short slope length of the ridge, but also a frequency of occurrence factor (i.e. smaller landslides occur more often). Locally uplifted and oversteepened ridge slopes are formed by thrust-fault propagated anticlines with seaward-vergent folds, meaning they commonly have steeper seaward forelimbs and gentler landward back limbs (e.g. Barnes *et al.* 2010). We do not observe a strong correlation between landslides and ground-shaking levels, however (Fig. 5a; discussed further below), leading us to conclude that tectonic uplift and oversteepening related to subduction deformation is most likely to be the principal control on landslide occurrence on ridges. In this context, we are considering oversteepening (induced by uplift and tectonic deformation) as one of many processes that can load a slope toward failure. It does not have to mean that a slope is steepened beyond the repose angle of the slope-forming material.

We do not observe any clear relationship with seismic ground-shaking and landslide-scar distribution in the current study (Fig. 5a). Although seismic strengthening of sediments on active margins may increase slope stability (Sawyer and DeVore 2015), we do not observe any significant difference between active and passive environments, despite significantly different seismic ground-shaking levels at multiple return periods (Fig. 5a). This could be related to the high concentration of landslides within canyon systems along the eastern margin of New Zealand, which may be diminishing any potential correlation between seismic ground-shaking and landslide distribution and magnitude. Alternatively, the magnitude and frequency of ground-shaking may not be exerting a primary control on where slope failure takes place. We conclude that although earthquakes are undoubtedly important landslide triggers in many instances (e.g. Wright and Rathje 2003; Hsu *et al.* 2008; Uri *et al.* 2009; Talling 2014; Mountjoy *et al.* 2018), little can be concluded from results presented in this study in terms of slope failure susceptibility related to ground-shaking alone. We suggest that we can better resolve the influence of other preconditioning factors with the current dataset. We note that the model used to compare ground-shaking across the study area is regional (Stirling *et al.* 2012) and may not be valid at the scale of the thrust ridges.

Seamount subduction. A relatively unique process to subduction systems is repeated seamount subduction in the long term that leads to significant crustal-scale damage zones as large asperities impact upon the frontal wedge and upper plate. We observe failures related to different stages of seamount subduction as noted by previous authors where the largest documented landslides and landscape change have already been linked to this process (Collot *et al.* 1996; Pedley *et al.* 2010). While individual landslides related to the passage of a seamount are not significant in the context of our database, the geomorphic impact of seamounts on the subduction wedge can create an environment where landslides preferentially occur, indicating that this is an important process on active margins. This database shows that the most significant concentrations of landslides occur in areas affected by seamount subduction, the Poverty Sea Valleys and Ruatoria Debris Avalanche (Fig. 4a, f).

Gas hydrates. Our results in this study do not add any weight to the proposed existence of a causative link between gas hydrate systems and slope failure. Despite there being localized examples of slope failure originating near the upper limit of gas hydrate stability (i.e. near 650 mbsl), as summarized by Crutchley *et al.* (2016), we do not observe a significant correlation on a margin-wide scale between slope failure initiation (i.e. head scarps) and the 650 mbsl contour. We acknowledge that gas hydrates, and the distribution of free gas, will play some role in the strength of sediments on the margin. However, we think this contribution to slope destabilization will be dwarfed by other geological factors. The most important 'other geological factors' on this margin seem to be canyon incision (on both the active and passive margins), tectonic uplift and oversteepening by thrust ridge development on the active margin, and rapid sedimentation.

Geomorphic and sedimentary environment control

Submarine canyons. Submarine canyons stand out as dominating the geomorphic environment where landslides occur (Figs 3a & 4f). Our analysis shows that across the entire study area landslides predominantly occur proximal to submarine canyons, with 68% of landslides occurring within 5 km of a canyon thalweg (Figs 4f & 5e). We consider that there are two reasons for this relationship.

(1) Canyons are conduits for sediment and so can form in locations of higher than average sediment loading. Focused sedimentation in these areas may play a role in driving slope instability through abundant material availability and/or rapid sedimentation leading to overpressure.
(2) Canyons are predominantly eroded environments where incision can undercut slopes and expose weaker stratigraphic sequences that may be preferentially prone to failure. Canyon incision is one of the few seafloor processes that can reveal 100 m to kilometre-scale exposures of stratigraphic sequences, and in the process create steep seafloor slopes.

On the Hikurangi margin it has been shown that tectonics rather than high sedimentation play a significant role in submarine canyon development (Mountjoy *et al.* 2009*b*; Pedley *et al.* 2010) and so we surmise that lateral unloading and the exposure of weak stratigraphic horizons may be a critical process in landslide occurrence in canyons. Micallef *et al.* (2012) showed that landslides in the Cook Strait Canyon are more likely to occur on the lower slope, concluding that active undercutting of canyon walls is a first-order control on landslide occurrence.

Sedimentary environment. While submarine canyons are eroded seafloor environments, regions of high sedimentation and long-term deposition are recognized as preferential environments for landslide occurrence (Lee *et al.* 1993). The yield of suspended sediment from river systems north of Hawke's Bay is an order of magnitude higher compared to the margin to the south (Fig. 1). This correlates to sequences of outer shelf-upper slope Quaternary sediments (clinoform packages) that are thick and relatively undeformed. Such sedimentary bodies provide ideal source material for large slope failures. Rapid sedimentation at sea-level lowstand periods may also cause overpressure within the stratigraphic section. The majority and the largest of the landslides on the northern margin (48%) occur between 200 and 1000 m water depth, which is the region of the continental slope underlain by clinoform packages (Figs 4a & 6a; Mountjoy *et al.* 2009*b*). Our results suggest that the concentration of large landslides and landslides with preserved debris deposits (confidence rating 1) on the northern margin is the availability of material, combined with seamount subduction and active margin processes.

It is notable in our dataset that only 1% of the mapped landslides are associated with MTD downslope (confidence rating 1), and all these landslides occur in the active-margin setting. The relatively few MTDs observed in bathymetric data could be related to a high number of smaller, shallower landslides that are composed of weak material that can disaggregate easily and so are not visible in seafloor geomorphology. Also, the rates of landscape change may have eroded or obscured debris deposits in surface geomorphology. Although there are only very few landslides with associated MTDs, we observe that they are all located in the active-margin setting, which may suggest higher frequency failures on this part of the margin, possibly related to preconditioning and triggering factors. A follow-up study mapping landslides in seismic reflection data is likely to shed more light on where debris deposits are being preserved over time within the sedimentary record.

Limitations of this study and future work

Although the extensive survey coverage across the eastern margin of New Zealand represents a broad range of tectonic settings, the grid resolution achievable using hull-mounted, *c.* 30 kHz multibeam echo sounder systems has likely resulted in the under sampling of landslides and MTDs smaller than *c.* 100 m (4 × 4 pixels of *c.* 25 m grid resolution). Given that *c.* 95% of landslides mapped in this study are <1 km^2, we believe that higher resolution mapping would reveal many more small-scale landslides (i.e. smaller than detection limits for acoustic mapping methods used in this study) on the eastern margin of New Zealand.

This study has focused primarily on the spatial distribution of landslide features mapped on the seafloor using multibeam bathymetry data. There is evidence to suggest that there are many more buried MTDs in the subsurface, observable using seismic reflection datasets (e.g. Pedley *et al.* 2010). A detailed analysis of the magnitude, frequency and distribution of buried MTDs using available seismic reflection data would provide valuable insights complementary to this study, into the temporal nature of submarine mass wasting on the eastern margin of New Zealand.

Conclusions

This study represents the first margin-wide submarine landslide database in New Zealand based on the best available multibeam bathymetry data, and one of the largest undertaken globally. We have gained insights into the spatial distribution, magnitude and morphology of submarine landslides across a range of geological settings, allowing us to provide better understanding of the causative factors for landslide preconditioning.

Our results show higher concentration of landslides on the active margin compared to the passive-margin region of our study area. This is at odds with findings from other studies elsewhere around the world. We suggest our observation is related to the prevalence of thrust ridges on the active margin, where slope failures are concentrated due to tectonic processes (including uplift and oversteepening, faulting and seamount subduction).

Areas containing submarine canyons are significantly more prone to slope failure than other geomorphic environments. High proportions of landslides in canyon systems suggest that landslides play a major role in canyon evolution, and that canyon-forming processes, such as the exposure of weak stratigraphic horizons and slope undercutting, may be first-order preconditioning factors for seafloor slope failure.

Higher sediment supply on the northern Hikurangi Margin is likely to be a key factor controlling the concentration of large landslides in this region, while seamount subduction is also an important process.

This regional mapping of landslides and analysis of the relationship to tectonic, geomorphic and sedimentary environments has shown that landslides preferentially concentrate in specific geomorphic environments (namely canyons and ridges) and may not be correlated with other factors (e.g. gas hydrates and ground shaking). This database of seafloor slope failures now offers an exceptional platform to undertake detailed analysis into the more local controls on landslide occurrence, providing valuable information for hazard and risk studies.

Acknowledgements Data used in this study come from numerous marine surveys using several different vessels, but primarily R/V *Tangaroa* voyages funded by the New Zealand Government. We are grateful to the science staff, ship officers and crew who supported the collection of these data. We would like to thank Grace Frontin-Rollet for assisting with figure construction, and Susi Woelz for valuable scientific discussions during the preparation and development of this study. Thank you to Geoffroy Lamarche, Danny Brothers and Derek Sawyer for constructive reviews and feedback that greatly improved this manuscript.

Funding Funding for Watson, Mountjoy and Crutchley is from the New Zealand Ministry of Business, Innovation and Employment (MBIE) Endeavor Fund Programme – Gas Hydrates: Economic opportunities and environmental impact (HYDEE, contract #: CO5X17078). Funding for Watson and Mountjoy is also partly from MBIE Strategic Science Investment fund (SSIF) Marine Geological Processes Programme in NIWA Coasts and Oceans Centre.

Author contributions **SJW**: conceptualization (equal), data curation (lead), formal analysis (lead), methodology (lead), visualization (lead), writing – original draft (equal), writing – review & editing (equal); **JJM**: conceptualization (equal), formal analysis (equal), project administration (lead), visualization (supporting), writing – original draft (equal), writing – review & editing (equal); **GJC**: funding acquisition (lead), project administration (equal), writing – original draft (equal), writing – review & editing (equal).

References

Barnes, P.M. 1992. Mid-bathyal current scours and sediment drifts adjacent to the Hikurangi deep-sea turbidite channel, eastern New Zealand: evidence from character mapping. *Marine Geology*, **106**, 169–187, https://doi.org/10.1016/0025-3227(92)90128-5

Barnes, P.M., Lamarche, G. *et al.* 2010. Tectonic and geological framework for gas hydrates and cold seeps on the Hikurangi subduction margin, New Zealand. *Marine Geology*, **272**, 26–48, https://doi.org/10.1016/j.margeo.2009.03.012

Barnes, P.M., Ghisetti, F.C., Ellis, S. and Morgan, J.K. 2018. The role of protothrusts in frontal accretion and accommodation of plate convergence, Hikurangi subduction margin, New Zealand. *Geosphere*, **14**, 440–468, https://doi.org/10.1130/GES01552.1

Baztan, J., Berné, S. *et al.* 2005. Axial incision: the key to understand submarine canyon evolution (in the western Gulf of Lion). *Marine Petroleum Geology*, **22**, 805–826, https://doi.org/10.1016/j.marpetgeo.2005.03.011

Berndt, C., Costa, S., Canals, M., Camerlenghi, A., de Mol, B. and Saunders, M. 2012. Repeated slope failure linked to fluid migration: the Ana submarine landslide complex, Eivissa Channel, Western Mediterranean Sea. *Earth and Planetary Science Letters*, **319**, 65–74, https://doi.org/10.1016/j.epsl.2011.11.045

Brothers, D.S., Andrews, B.D. *et al.* 2019. Slope failure and mass transport processes along the Queen Charlotte Fault, southeastern Alaska. *Geological Society, London, Special Publications*, **477**, 69–83, https://doi.org/10.1144/SP477.30

Browne, G.H. and Naish, T.R. 2003. Facies development and sequence architecture of a late Quaternary fluvial-marine transition, Canterbury Plains and shelf, New Zealand: implications for forced regressive deposits. *Sedimentary Geology*, **158**, 57–86, https://doi.org/10.1016/S0037-0738(02)00258-0

Bünz, S., Mienert, J. and Berndt, C. 2003. Geological controls on the Storegga gas-hydrate system of the mid-Norwegian continental margin. *Earth and Planetary Science Letters*, **209**, 291–307, https://doi.org/10.1016/S0012-821X(03)00097-9

Bünz, S., Mienert, J., Bryn, P. and Berg, K. 2005. Fluid flow impact on slope failure from 3D seismic data: a case study in the Storegga Slide. *Basin Research*, **17**, 109–122, https://doi.org/10.1111/j.1365-2117.2005.00256.x

Camerlenghi, A., Urgeles, R. and Fantoni, L. 2010. A database on submarine landslides of the Mediterranean Sea. *In*: Mosher, D.C., Shipp, R.C., Moscardelli, L., Chaytor, J.D., Baxter, C.D.P., Lee, H.J. and Urgeles, R. (eds) *Submarine Mass Movements and Their Consequences*. Springer, 503–513.

Carpenter, G. 1981. Coincident sediment slump/clathrate complexes on the US Atlantic continental slope. *Geo-Marine Letters*, **1**, 29–32, https://doi.org/10.1007/BF02463298

Chaytor, J.D., Uri, S., Solow, A.R. and Andrews, B.D. 2009. Size distribution of submarine landslides along the US Atlantic margin. *Marine Geology*, **264**, 16–27, https://doi.org/10.1016/j.margeo.2008.08.007

Chiswell, S.M. 2005. Mean and variability in the Wairarapa and Hikurangi Eddies, New Zealand. *New Zealand Journal of Marine and Freshwater Research*, **39**, 121–134, https://doi.org/10.1080/00288330.2005.9517295

Clare, M.A., Clarke, J.H., Talling, P.J., Cartigny, M.J. and Pratomo, D.G. 2016. Preconditioning and triggering of offshore slope failures and turbidity currents revealed by most detailed monitoring yet at a fjord-head delta. *Earth and Planetary Science Letters*, **450**, 208–220, https://doi.org/10.1016/j.epsl.2016.06.021

Clare, M., Chaytor, J. *et al.* 2018. A consistent global approach for the morphometric characterization of subaqueous landslides. *Geological Society, London, Special Publications*, **477**, 455–477, https://doi.org/10.1144/SP477.15

Collot, J.-Y., Delteil, J. *et al.* 1996. From oblique subduction to intra-continental transpression: structures of the southern Kermadec-Hikurangi margin from multibeam bathymetry, side-scan sonar and seismic reflection. *Marine Geophysical Research*, **18**, 357–381, https://doi.org/10.1007/BF00286085

Collot, J.-Y., Lewis, K., Lamarche, G. and Lallemand, S. 2001. The giant Ruatoria debris avalanche on the northern Hikurangi margin, New Zealand: result of oblique seamount subduction. *Journal of Geophysical Research: Solid Earth*, **106**, 19271–19297, https://doi.org/10.1029/2001JB900004

Crutchley, G.J., Gorman, A.R. and Fohrmann, M. 2007. Investigation of the role of gas hydrates in continental slope stability west of Fiordland, New Zealand. *New Zealand Journal of Geology and Geophysics*, **50**, 357–364, https://doi.org/10.1080/00288300709509842

Crutchley, G.J., Pecher, I.A., Gorman, A.R., Henrys, S.A. and Greinert, J. 2010. Seismic imaging of gas conduits beneath seafloor seep sites in a shallow marine gas hydrate province, Hikurangi Margin, New Zealand. *Marine Geology*, **272**, 114–126, https://doi.org/10.1016/j.margeo.2009.03.007

Crutchley, G.J., Fraser, D.R.A., Pecher, I.A., Gorman, A.R., Maslen, G. and Henrys, S.A. 2015. Gas migration into gas hydrate-bearing sediments on the southern Hikurangi margin of New Zealand. *Journal of Geophysical Research: Solid Earth*, **120**, 725–743, https://doi.org/10.1002/2014JB011503

Crutchley, G., Mountjoy, J.J., Pecher, I., Gorman, A. and Henrys, S.A. 2016. Submarine slope instabilities coincident with shallow gas hydrate systems: insights from New Zealand examples. *In*: Lamarche, G., Mountjoy, J.J. *et al.* (eds) *Submarine Mass Movements and Their Consequences: 7th International Symposium.* 1–4 November 2015, Wellington, New Zealand, Springer, Berlin, 401–409.

Dan, G., Sultan, N. and Savoye, B. 2007. The 1979 Nice harbour catastrophe revisited: trigger mechanism inferred from geotechnical measurements and numerical modelling. *Marine Geology*, **245**, 40–64, https://doi.org/10.1016/j.margeo.2007.06.011

Dugan, B. and Flemings, P.B. 2000. Overpressure and fluid flow in the New Jersey continental slope: implications for slope failure and cold seeps. *Science*, **289**, 288–291, https://doi.org/10.1126/science.289.5477.288

Elger, J., Berndt, C., Rüpke, L., Krastel, S., Gross, F. and Geissler, W.H. 2018. Submarine slope failures due to pipe structure formation. *Nature Communications*, **9**, 715, https://doi.org/10.1038/s41467-018-03176-1

Ellis, S., Pecher, I., Kukowski, N., Xu, W., Henrys, S. and Greinert, J. 2010. Testing proposed mechanisms for seafloor weakening at the top of gas hydrate stability on an uplifted submarine ridge (Rock Garden), New Zealand. *Marine Geology*, **272**, 127–140, https://doi.org/10.1016/j.margeo.2009.10.008

Flemings, P.B., Long, H. *et al.* 2008. Pore pressure penetrometers document high overpressure near the seafloor where multiple submarine landslides have occurred on the continental slope, offshore Louisiana, Gulf of Mexico. *Earth and Planetary Science Letters*, **269**, 309–325, https://doi.org/10.1016/j.epsl.2007.12.005

Greinert, J., Lewis, K.B. *et al.* 2010. Methane seepage along the Hikurangi Margin, New Zealand: overview of studies in 2006 and 2007 and new evidence from visual, bathymetric and hydroacoustic investigations. *Marine Geology*, **272**, 6–25, https://doi.org/10.1016/j.margeo.2010.01.017

Gross, F., Mountjoy, J.J. *et al.* 2018. Free gas distribution and basal shear zone development in a subaqueous landslide: insight from 3D seismic imaging of the Tuaheni Landslide Complex, New Zealand. *Earth and Planetary Science Letters*, **502**, 231–243, https://doi.org/10.1016/j.epsl.2018.09.002

Hampton, M.A., Lemke, R.W. and Coulter, H.W. 1993. Submarine landslides that had a significant impact on man and his activities: Seward and Valdez, Alaska. *USGS Bulletin*, **2002**, 123–134.

Hampton, M.A., Lee, H.J. and Locat, J. 1996. Submarine landslides. *Review of Geophysics*, **34**, 33–59, https://doi.org/10.1029/95RG03287

Henrys, S.A., Ellis, S. and Uruski, C. 2003. Conductive heat flow variations from bottom-simulating reflectors on the Hikurangi margin, New Zealand. *Geophysical Research Letters*, **30**, 37–31, https://doi.org/10.1029/2002GL015772

Hicks, D.M., Shankar, U., McKerchar, A.I., Basher, L.R., Jessen, M., Lynn, I. and Page, M. 2011. Suspended sediment yields from New Zealand rivers. *Journal of Hydrology*, **50**, 81–142.

Hillman, J.I., Gorman, A.R. and Pecher, I.A. 2015. Geostatistical analysis of seafloor depressions on the southeast margin of New Zealand's South Island: investigating the impact of dynamic near seafloor processes on geomorphology. *Marine Geology*, **360**, 70–83, https://doi.org/10.1016/j.margeo.2014.11.016

Hillman, J.I., Klaucke, I., Pecher, I.A., Gorman, A.R., Schneider von Deimling, J. and Bialas, J. 2018. The influence of submarine currents associated with the Subtropical Front upon seafloor depression morphologies on the eastern passive margin of South Island, New Zealand. *New Zealand Journal of Geology and Geophysics*, **61**, 112–125, https://doi.org/10.1080/00288306.2018.1434801

Hsu, S.-K., Kuo, J., Chung-Liang, L., Ching-Hui, T., Doo, W.-B., Ku, C.-Y. and Sibuet, J.-C. 2008. Turbidity currents, submarine landslides and the 2006 Pingtung earthquake off SW Taiwan. *Terrestrial, Atmospheric and Oceanic Sciences*, **19**, 7.

Kelner, M., Migeon, S., Tric, E., Courboulex, F., Dano, A., Lebourg, T. and Taboada, A. 2016. Frequency and triggering of small-scale submarine landslides on decadal timescales: analysis of 4D bathymetric data from the continental slope offshore Nice (France). *Marine Geology*, **379**, 281–297, https://doi.org/10.1016/j.margeo.2016.06.009

Kukowski, N., Greinert, J. and Henrys, S. 2010. Morphometric and critical taper analysis of the Rock Garden region, Hikurangi Margin, New Zealand: implications for slope stability and potential tsunami generation. *Marine Geology*, **272**, 141–153, https://doi.org/10.1016/j.margeo.2009.06.004

Kvenvolden, K.A. 1999. Potential effects of gas hydrate on human welfare. *Proceedings of the National Academy of Sciences*, **96**, 3420–3426, https://doi.org/10.1073/pnas.96.7.3420

Lane, E., Mountjoy, J.J., Orpin, A.R. and Rowden, A. 2018. Submarine landslides and turbidity currents. *In*: Hendtlass, C., Borrerp, J., Neale, D. and Shand, T. (eds) *Shaky Shores: Coastal impacts & responses to the 2016 Kaikoura earthquakes*. New Zealand Coastal Society.

Lee, H.J., Schwab, W.C. and Booth, J.S. 1993. Submarine landslides: an introduction. *USGS Bulletin*, **2002**, 1–13.

Lewis, K.B. and Barnes, P.M. 1999. Kaikoura Canyon, New Zealand: active conduit from near-shore sediment zones to trench-axis channel. *Marine Geology*, **162**, 39–69, https://doi.org/10.1016/S0025-3227(99)00075-4

Lewis, K.B., Carter, L. and Davey, F.J. 1994. The opening of Cook Strait: interglacial tidal scour and aligning basins at a subduction to transform plate edge. *Marine Geology*, **116**, 293–312, https://doi.org/10.1016/0025-3227(94)90047-7

Lewis, K.B., Lallemand, S.E. and Carter, L. 2004. Collapse in a Quaternary shelf basin off East Cape, New Zealand: evidence for passage of a subducted seamount inboard of the Ruatoria giant avalanche. *New Zealand Journal of Geology and Geophysics*, **47**, 415–429, https://doi.org/10.1080/00288306.2004.9515067

L'Heureux, J.-S., Longva, O. *et al.* 2012. Identification of weak layers and their role for the stability of slopes at Finneidfjord, northern Norway. *In*: Yamada, Y., Kawamura, K. *et al.* (eds) *Submarine Mass Movements and Their Consequences: 5th International Symposium*. Springer, 321–330.

Litchfield, N.J., Van Dissen, R. *et al.* 2014. A model of active faulting in New Zealand. *New Zealand Journal of Geology and Geophysics*, **57**, 32–56, https://doi.org/10.1080/00288306.2013.854256

Lopez, C., Spence, G., Hyndman, R. and Kelley, D. 2010. Frontal ridge slope failure at the northern Cascadia margin: margin-normal fault and gas hydrate control. *Geology*, **38**, 967–970, https://doi.org/10.1130/G31136.1

Masson, D.G., Harbitz, C.B., Wynn, R.B., Pedersen, G. and Løvholt, F. 2006. Submarine landslides: processes, triggers and hazard prediction. *Philosophical Transactions of the Royal Society A: Mathematical, Physical and Engineering Sciences*, **364**, 2009–2039, https://doi.org/10.1098/rsta.2006.1810

McAdoo, B.G., Pratson, L.F. and Orange, D.L. 2000. Submarine landslide geomorphology, US continental slope. *Marine Geology*, **169**, 103–136, https://doi.org/10.1016/S0025-3227(00)00050-5

McIver, R.D. 1982. Role of naturally occurring gas hydrates in sediment transport. *AAPG Bulletin*, **66**, 789–792.

Micallef, A., Mountjoy, J.J., Canals, M. and Lastras, G. 2012. Deep-Seated Bedrock Landslides and Submarine Canyon Evolution in an Active Tectonic Margin: Cook Strait, New Zealand. *In*: Yamada, Y., Kawamura, K. *et al.* (eds) *Submarine Mass Movements and Their Consequences: 5th International Symposium*. Springer, 201–212.

Mountjoy, J.J. and Barnes, P.M. 2011. Active upper plate thrust faulting in regions of low plate interface coupling, repeated slow slip events, and coastal uplift: example from the Hikurangi Margin, New Zealand. *Geochemistry, Geophysics, Geosystems*, **12**, https://doi.org/10.1029/2010GC003326

Mountjoy, J.J. and Micallef, A. 2012. Polyphase emplacement of a 30 km^3 blocky debris avalanche and its role in slope-gully development. *In*: Yamada, Y., Kawamura, K. *et al.* (eds) *Submarine Mass Movements and Their Consequences: 5th International Symposium*. Springer, 213–222.

Mountjoy, J.J., Barnes, P.M. and Pettinga, J.R. 2009*a*. Morphostructure and evolution of submarine canyons across an active margin: Cook Strait sector of the Hikurangi Margin, New Zealand. *Marine Geology*, **260**, 45–68, https://doi.org/10.1016/j.margeo.2009.01.006

Mountjoy, J.J., McKean, J., Barnes, P.M. and Pettinga, J.R. 2009*b*. Terrestrial-style slow-moving earthflow kinematics in a submarine landslide complex. *Marine Geology*, **267**, 114–127, https://doi.org/10.1016/j.margeo.2009.09.007

Mountjoy, J.J., Micallef, A., Stevens, C.L. and Stirling, M.W. 2013. Holocene sedimentary activity in a non-terrestrially coupled submarine canyon: Cook Strait Canyon system, New Zealand. *Deep Sea Research Part II: Topical Studies in Oceanography*, **104**, 120–133, https://doi.org/10.1016/j.dsr2.2013.09.001

Mountjoy, J.J., Pecher, I., Henrys, S., Crutchley, G., Barnes, P.M. and Plaza-Faverola, A. 2014. Shallow methane hydrate system controls ongoing, downslope sediment transport in a low-velocity active submarine landslide complex, Hikurangi Margin, New Zealand. *Geochemistry, Geophysics, Geosystems*, **15**, 4137–4156, https://doi.org/10.1002/2014GC005379

Mountjoy, J.J., Howarth, J.D. *et al.* 2018. Earthquakes drive large-scale submarine canyon development and sediment supply to deep-ocean basins. *Science Advances*, **4**, eaar3748, https://doi.org/10.1126/sciadv.aar3748

Orpin, A.R., Alexander, C., Carter, L., Kuehl, S. and Walsh, J.P. 2006. Temporal and spatial complexity in post-glacial sedimentation on the tectonically active, Poverty Bay continental margin of New Zealand. *Continental Shelf Research*, **26**, 2205–2224, https://doi.org/10.1016/j.csr.2006.07.029

Paquet, F., Proust, J.-N., Barnes, P.M. and Pettinga, J.R. 2009. Inner-forearc sequence architecture in response to climatic and tectonic forcing since 150 Ka: Hawke's Bay, New Zealand. *Journal of Sedimentary Research*, **79**, 97–124, https://doi.org/10.2110/jsr.2009.019

Paull, C.K., Buelow, W.J., Ussler Iii, W. and Borowski, W.S. 1996. Increased continental-margin slumping frequency during sea-level lowstands above gas hydrate-bearing sediments. *Geology*, **24**, 143–146, https://doi.org/10.1130/0091-7613(1996)024<0143:ICMSFD>2.3.CO;2

Pecher, I.A., Henrys, S.A., Ellis, S., Chiswell, S.M. and Kukowski, N. 2005. Erosion of the seafloor at the top of the gas hydrate stability zone on the Hikurangi Margin, New Zealand. *Geophysical Research Letters*, **32**, 1–4, https://doi.org/10.1029/2005GL024687

Pedley, K.L., Barnes, P.M., Pettinga, J.R. and Lewis, K.B. 2010. Seafloor structural geomorphic evolution of the accretionary frontal wedge in response to seamount subduction, Poverty Indentation, New Zealand. *Marine Geology*, **270**, 119–138, https://doi.org/10.1016/j.margeo.2009.11.006

Pettinga, J., Lewis, K., Hampton, M.A. and Davey, F.J. 1984. Preliminary interpretations of near-surface multichannel seismic data, across the obliquely convergent Hikurangi margin; offshore eastern North Island. *Geological Society of New Zealand, Miscellaneous Publications*, **31A**.

Piper, D.J., Cochonat, P. and Morrison, M.L. 1999. The sequence of events around the epicentre of the 1929 Grand Banks earthquake: initiation of debris flows and turbidity current inferred from sidescan sonar. *Sedimentology*, **46**, 79–97, https://doi.org/10.1046/j.1365-3091.1999.00204.x

Popenoe, P., Schmuck, E.A., Dillon, W.P., Schwab, W.C., Lee, H.J. and Twichell, D.C. 1993. The Cape Fear landslide: slope failure associated with salt diapirism and gas hydrate decomposition. *USGS Bulletin*, **2002**, 40–53.

Pouderoux, H., Proust, J.-N. and Lamarche, G. 2014. Submarine paleoseismology of the northern Hikurangi subduction margin of New Zealand as deduced from turbidite record since 16 ka. *Quaternary Science Reviews*, **84**, 116–131, https://doi.org/10.1016/j.quascirev.2013.11.015

Riedel, M., Cote, M.M., Urlaub, M., Geersen, J., Scholz, N.A., Naegeli, K. and Spence, G.D. 2018. Slope failures along the deformation front of the Cascadia margin: linking slide morphology to subduction zone parameters. *Geological Society, London, Special Publications*, **477**, 47–67, https://doi.org/10.1144/SP477.33.

Ruppel, C.D. 2011. Methane hydrates and contemporary climate change. *Nature Education Knowledge*, **2**, 12.

Sawyer, D.E. and DeVore, J.R. 2015. Elevated shear strength of sediments on active margins: evidence for seismic strengthening. *Geophysical Research Letters*, **42**, 10216–10221, https://doi.org/10.1002/2015GL066603

Stirling, M., McVerry, G. *et al.* 2012. National seismic hazard model for New Zealand: 2010 update. *Bulletin of the Seismological Society of America*, **102**, 1514–1542, https://doi.org/10.1785/0120110170

Sultan, N., Cochonat, P., Foucher, J.-P. and Mienert, J. 2004. Effect of gas hydrates melting on seafloor slope instability. *Marine Geology*, **213**, 379–401, https://doi.org/10.1016/j.margeo.2004.10.015

Talling, P.J. 2014. On the triggers, resulting flow types and frequencies of subaqueous sediment density flows in different settings. *Marine Geology*, **352**, 155–182, https://doi.org/10.1016/j.margeo.2014.02.006

Twichell, D.C. and Roberts, D.G. 1982. Morphology, distribution, and development of submarine canyons on the United States Atlantic continental slope between Hudson arid Baltimore Canyons. *Geology*, **10**, 408–412, https://doi.org/10.1130/0091-7613(1982)10<408:MDADOS>2.0.CO;2

Uri, S., Lee, H.J., Geist, E.L. and Twichell, D. 2009. Assessment of tsunami hazard to the US East Coast using relationships between submarine landslides and earthquakes. *Marine Geology*, **264**, 65–73, https://doi.org/10.1016/j.margeo.2008.05.011

Urlaub, M., Talling, P.J. and Masson, D.G. 2013. Timing and frequency of large submarine landslides: implications for understanding triggers and future geohazard. *Quaternary Science Reviews*, **72**, 63–82, https://doi.org/10.1016/j.quascirev.2013.04.020

Urlaub, M., Geersen, J., Krastel, S. and Schwenk, T. 2018. Diatom ooze: crucial for the generation of submarine mega-slides? *Geology*, **46**, 331–334, https://doi.org/10.1130/G39892.1

Walcott, R.I. 1978. Present tectonics and late Cenozoic evolution of New Zealand. *Geophysical Journal International*, **52**, 137–164, https://doi.org/10.1111/j.1365-246X.1978.tb04225.x

Wallace, L.M., Barnes, P. *et al.* 2012. The kinematics of a transition from subduction to strike-slip: an example from the central New Zealand plate boundary: Subduction to strike-slip in New Zealand. *Journal of Geophysical Research: Solid Earth*, **117**, B02405, https://doi.org/10.1029/2011JB008640

Walters, R., Barnes, P., Lewis, K. and Goff, J.R. 2006. Locally generated tsunami along the Kaikoura coastal margin: part 2. Submarine landslides. *New Zealand Journal of Marine and Freshwater Research*, **40**, 17–28, https://doi.org/10.1080/00288330.2006.9517400

Wang, X., Mountjoy, J.J., Power, W.L., Lane, E.M. and Mueller, C. 2016. Revealing the failure process of a submarine landslide near Cook Strait and its associated tsunami. *In*: Lamarche, G., Mountjoy, J.J. *et al.* (eds) *Submarine Mass Movements and Their Consequences: 7th International Symposium.* 1–4 November 2015, Wellington, New Zealand, Springer, Berlin, 599–606.

Watson, S.J., Mountjoy, J.J. *et al.* 2020. Focused fluid seepage related to variations in accretionary wedge structure, Hikurangi Margin, New Zealand. *Geology*, **48**, 56–61, https://doi.org/10.1130/G46666.1

Winters, W.J., Pecher, I.A., Waite, W.F. and Mason, D.H. 2004. Physical properties and rock physics models of sediment containing natural and laboratory-formed methane gas hydrate. *American Mineralogist*, **89**, 1221–1227, https://doi.org/10.2138/am-2004-8-909

Wright, S.G. and Rathje, E.M. 2003. Triggering mechanisms of slope instability and their relationship to earthquakes and tsunamis. *Pure Applied Geophysics*, **160**, 1865–1877, https://doi.org/10.1007/s00024-003-2410-4

Wright, I.C., Chadwick, W.W. *et al.* 2008. Collapse and reconstruction of Monowai submarine volcano, Kermadec arc, 1998–2004. *Journal of Geophysical Research: Solid Earth*, **113**, https://doi.org/10.1029/2007JB005138

Yelisetti, S., Spence, G.D. and Riedel, M. 2014. Role of gas hydrates in slope failure on frontal ridge of northern Cascadia margin. *Geophysical Journal International*, **199**, 441–458, https://doi.org/10.1093/gji/ggu254

Zhu, S., Shi, Y., Lu, M. and Xie, F. 2013. Dynamic mechanisms of earthquake-triggered landslides. *Science China Earth Sciences*, **56**, 1769–1779, https://doi.org/10.1007/s11430-013-4582-9

Geological and tectonic controls on morphometrics of submarine landslides of the Spanish margins

Ricardo León[1]*, Roger Urgeles[2], Raul Pérez-López[1], Emilio Payo[1], Amanda Vázquez-Izquierdo[1], Carmen Julia Giménez-Moreno[1] and David Casas[2]

[1]Geological Survey of Spain (IGME), Ríos Rosas, 23. 28003, Madrid, Spain

[2]Institut de Ciències del Mar (CSIC), Passeig Marítim de la Barceloneta 37-49, 08003 Barcelona, Spain

RL, 0000-0001-5598-0710; RU, 0000-0001-9438-1753; RP-L, 0000-0002-9132-4806; EP, 0000-0002-0448-9590
*Correspondence: r.leon@igme.es

Abstract: A geomorphological analysis of the submarine landslides geographical information system catalogue of the Geological Survey of Spain has revealed three main groups of submarine landslides associated with (1) deep-ocean seamount ridges (extinct spreading centres), (2) volcanic islands and (3) continental margins. These three groups have statistically significant morphometric differences, as determined from analysis of variance (ANOVA) and Tukey's HSD Tests, in total length (runout), total area, maximum deposit width and bathymetric depth. Volcanic island-related slope failures affect larger areas of the seafloor and their headwall escarpments often extend above sea-level. Slope failures associated with seamount ridges are the deepest, between 3500 and 5500 m, and display relatively high width-to-length ratios. Finally, landslides on continental margins show two sub-groups. Landslides on tectonically controlled margins have smaller runouts and total area and larger average slope gradients than margins where tectonic controls are limited. These results demonstrate that submarine landslide morphology is strongly controlled by the geological-tectonic setting.

The destructive power of landslides depends on the location, size and velocity of the failure event. Submarine landslides can occur on very gentle slopes (1–5°; Masson *et al.* 2006) where they commonly detach in areas with thick sedimentary sequences. In addition, they are also frequent in steeply sloping areas prone to earthquakes, as well as areas where large storm waves or high excess pore pressures exist (Nisbet and Piper 1998; Canals *et al.* 2004; Clague and Stead 2012).

Submarine landslides are complex phenomena and represent an important process affecting the evolution of basin margins in all oceans and geological settings of the world. Here, landslide deposits are considered *sensu lato* to include the range of products formed by sedimentary slope instability: slides and slumps (rotational and translational), debrites and turbidites (Masson *et al.* 2002; Casas *et al.* 2003*a*, 2015; Droz *et al.* 2006; Urgeles *et al.* 2006; Lastras *et al.* 2007; Camerlenghi *et al.* 2009; Alonso *et al.* 2014).

Submarine landslide databases are an essential tool for hazard assessment. In the North Atlantic (Hühnerbach and Masson 2004; Chaytor *et al.* 2009; Twichell *et al.* 2009; Harders *et al.* 2011), the Mediterranean Sea (Urgeles and Camerlenghi 2013) and lacustrine environments (Moernaut *et al.* 2015) such databases have been used to assess landslide dynamics, trigger mechanisms, control factors and hazards. Elsewhere, the geographical distribution and regional morphometric characteristics of landslide deposits provide valuable information on the evolution, triggers and controls on submarine slope failure in different tectonic environments (Ardizzone *et al.* 2008; Guzzetti *et al.* 2012; Clare *et al.* 2014; Moscardelli and Wood 2015; Casas *et al.* 2016). Landslide morphologies and behaviour have also been analysed in order to establish potential relationships between landslide morphometric parameters and causal mechanisms (Chaytor *et al.* 2009; Urgeles and Camerlenghi 2013; Clare *et al.* 2014; Moscardelli and Wood 2015). These characteristics are useful for determining landslide-generated tsunami hazards that can affect surrounding coastal and offshore infrastructure (Harbitz *et al.* 2014; Macías *et al.* 2015; Janin *et al.* 2019). The present work addresses how different geological–tectonic settings control the morphology of submarine landslides, with a focus on the Spanish continental margins. Using statistical analysis, we aim to test the hypothesis that significant morphometric differences exist between landslides

From: Georgiopoulou, A., Amy, L. A., Benetti, S., Chaytor, J. D., Clare, M. A., Gamboa, D., Haughton, P. D. W., Moernaut, J. and Mountjoy, J. J. (eds) 2020. *Subaqueous Mass Movements and their Consequences: Advances in Process Understanding, Monitoring and Hazard Assessments*. Geological Society, London, Special Publications, **500**, 495–513.
First published online May 20, 2020, https://doi.org/10.1144/SP500-2019-153

that occur on margins in different tectonic settings, and we explore the causes for such differences.

Geological setting

The study area covers different geographical areas and geological settings, including the Canary volcanic province, Cantabrian margin, Galicia margin, Jean Charcot and Coruña seamounts, Alborán Basin, Gulf of Cádiz and the Catalan–Balearic margin (Figs 1 & 2). The Canary volcanic province is a NE–SW-trending 1350 km-long ridge, composed of eight volcanic islands and >100 seamounts located in the Atlantic offshore NW Africa (Fig. 1). Islands are characterized by a very narrow continental shelf (Fig. 2), from 300 to 500 m (e.g. El Hierro Island) to *c.* 13 km (e.g. Gomera and Fuerteventura islands) in width, and steep slopes composed of volcanic substrates typically between 18 and 25° and locally getting up to 30° (Table 1). Its origin is generally accepted as related to a mantle plume (hotspot), and a general age progression is observed from the oldest (68 myr) Lars volcanic seamount to the westernmost La Palma and El Hierro islands, which are younger than 2 myr (van den Bogaard 2013). The flanks of islands are incised by canyons that provide major pathways for the sediment transport from the islands to the volcaniclastic apron and abyssal plains via turbidite channels (Masson *et al.* 2002).

The Cantabrian margin represents the northern boundary of the Iberian plate (Fig. 1) and is a tectonically controlled margin, resulting from the NW–SE oblique Paleocene–Eocene convergence between the Iberian and European plates (Boillot *et al.* 1979). It has a relatively narrow continental shelf, 15–50 km in width and a steep continental slope 10–18° on average (Table 1) (Ercilla *et al.* 2008*b*).

The Galicia margin can be defined as a starved passive margin lacking volcanic activity or significant present-day tectonic activity (Ercilla *et al.* 2011). It has a continental shelf 35–40 km in width

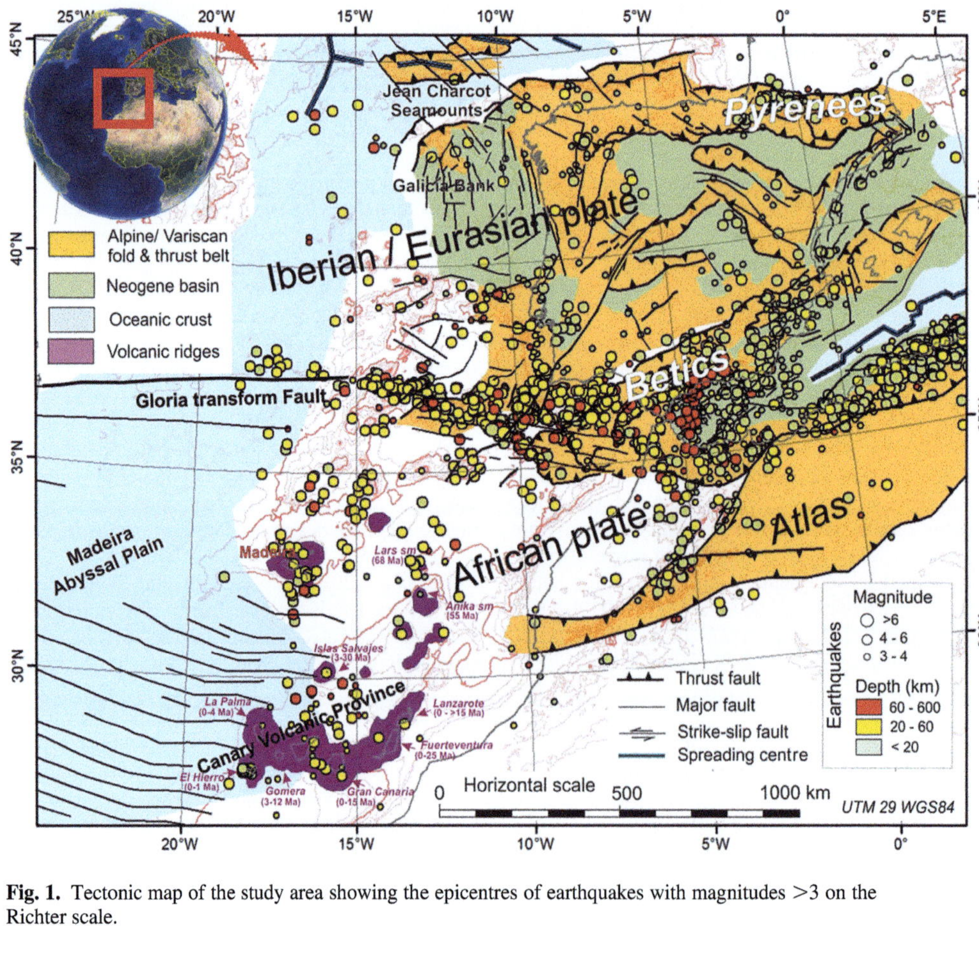

Fig. 1. Tectonic map of the study area showing the epicentres of earthquakes with magnitudes >3 on the Richter scale.

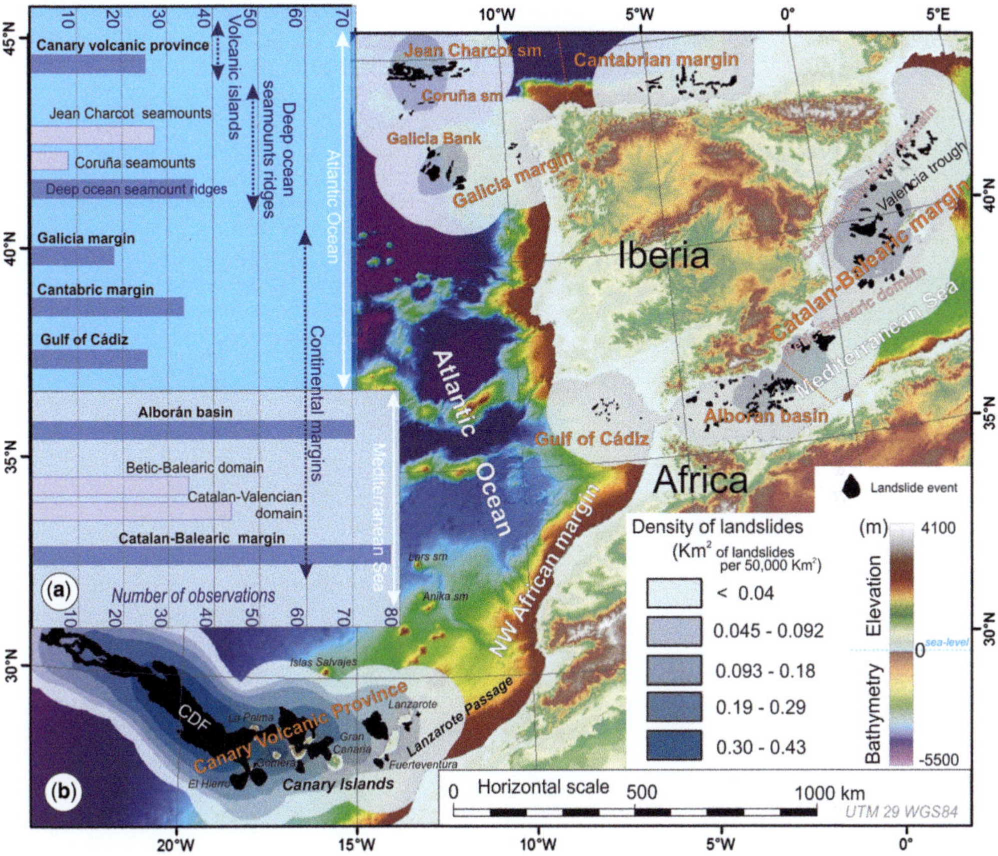

Fig. 2. (**a**) Frequency histogram of submarine landslide events sorted by geological context (text in black and bold). Bars in light purple indicate geological domains and seamounts; text in dark blue denotes the groups of geological contexts with significant morphometric differences in their landslides. (**b**) Density map of landslides (units in km^2 of seafloor affected by submarine landslides per 50 000 km^2 of seafloor study area). Black polygons, submarine landslides bibliographic catalogue of the Geological Survey of Spain (http://mapas.igme.es/gis/rest/services/BasesDatos/IGME_BDMoves_ES/MapServer/).

(Table 1) with a thin (1–15 m) Holocene sedimentary cover due to low sedimentation rates (Rey and Díaz-del-Río 1987). The continental slope has a complex structure, related to its evolution as a hyper-extended margin with structural seamounts and marginal platforms involving half-graben, graben and horsts (Mauffret and Montadert 1987; Vázquez *et al.* 2008). Downslope processes involve mass flows, turbidity flows, mass-transport and slides (Ercilla *et al.* 2011). The Jean Charcot and Coruña seamounts are located to the north of Galicia margin. The Jean Charcot seamounts constitute two east–west oceanic ridges, bounded by reverse faults, considered to be an extinct ocean spreading centre (Williams 1975; Emery and Uchupi 1984). The Coruña seamounts are a deep (between 3970 and 4930 m) NE–SW set of crests of oceanic volcanic rocks generated by seafloor spreading in the North Atlantic Ocean (Emery and Uchupi 1984; Vázquez *et al.* 2008, 2015*b*).

The margins to the south of Iberia, the Gulf of Cádiz and the Alborán Basin comprise both south Iberian and north African margins (Fig. 1). They are located in the context of the Betic–Rifian Orogen both sides of the Gibraltar Strait. They are considered to be moderately active margins (Vázquez *et al.* 2015*b*) with moderate seismicity (Fig. 1). These margins were formed during the north–south convergence between the African and Eurasian tectonic plates that has occurred since the lower Miocene (Platt *et al.* 2006). On the Atlantic side, the Gulf of Cádiz is characterized by the emplacement of an olistostrome during the Miocene, as well as active diapirism and normal, reverse and strike-slip faulting

Table 1. *Physiographic characteristics of the geological contexts described in the paper*

Geological contexts	Continental shelf (km)	Continental slope (°)	Type	Bathymetric range (m)	Slope gradient (°)	Trending
Canary volcanic province	0.3–0.5 13	15–18 max. 30	Volcanic			
Cantabrian margin	16–50	15–18	Sedimentary; tectonically controlled			
Galicia margin	35–40	4–8	Sedimentary; non-tectonically controlled			
Gulf of Cádiz	42–50	200 km 1–6	Sedimentary; tectonically controlled			
Alborán Basin	4–8	4–10	Sedimentary; tectonically controlled			
Catalan–Balearic margin	12–67	5–7 south zone 9–13	Sedimentary; non-tectonically controlled			
Jean Charcot seamount				North zone 2400–−4900 South zone 3300–−5000	Top: 0.5–2 Slope: 17–25, max 30	NNE–SSW
Coruña seamount				−3450–−5300	Average: 2–6 Slope: 20–26 max 30	NE–SW NE zone NNE–SSW

For the continental margins and the Canary volcanic province, the averages of the width and slope gradient are given, as well as the type of margin.

(Medialdea *et al.* 2009). On the Mediterranean side, the Alborán Basin developed through extensional normal faulting (Tortonian) in the back-arc area of the Alborán Domain. A post-Tortonian compressive regime produced the inversion of the basin, leading to the development of reverse and strike-slip faults and folds (Estrada *et al.* 1997; Martínez-García *et al.* 2017). In addition, the Gulf of Cádiz shows a continental shelf of 42–50 km in width and a smooth wide slope of 200 km and 1–6° (Table 1). The Alborán continental shelf is 1–8 km wide with a slope ranging from 4 to 10°.

The Catalan–Balearic margin is considered a relatively seismically quiet passive margin (Urgeles and Camerlenghi 2013) and comprises several basins structurally related to the Alpine Orogeny (Fig. 1). The Valencia Trough opened during the Late Oligocene–Early Miocene (Dañobeitia *et al.* 1990), as a result of the interplay between extensional (in the Catalan–Valencian domain) and compressive (mostly in the Betic–Balearic domain) processes (Roca and Guimerà 1992). The Catalan–Valencian domain has a wide shelf (12–67 km) and a slope (5–7° in average) deeply incised with tectonically controlled, v-shaped canyons (with steep, abrupt and gullied walls), some of them near (<1 km) to the coastline (Maldonado *et al.* 1986, 1987, 1994; Urgeles *et al.* 2006).

Data and methodology

Data

Data used in this work come from the submarine landslide catalogue of the Geological Survey of Spain (Fig. 2). The landslide inventory includes 285 records collected from published secondary sources, including maps, national databases, reports, PhD theses and scientific papers (http://mapas.igme.es/gis/rest/services/BasesDatos/IGME_BDMoves_ES/MapServer/). The geographical information system (GIS) data model structure, INSPIRE compliant, has been standardized to be congruent with its onshore counterpart (León *et al.* 2018).

Morphometric parameters

Seven morphometric variables (Fig. 3 and Table 2) have been measured and analysed in this work: total seafloor area affected by landslides (A_t), total length or runout (L_t), maximum deposit width (W_d), width-to-length ratio (W_d/L_t), maximum elevation (E_{mx}), minimum elevation (E_{mn}) and average slope gradient (S_{av}). These morphometric variables were measured according to Clare *et al.* (2018) and León *et al.* (2018), and the results are summarized in Figure 3. Lengths and areas for each landslide event were calculated using ArcGIS® algorithms over their corresponding UTM Zone (Datum WGS84). A_t was defined on the basis of the area of the seafloor affected by erosion, sediment transfer, landslide deposition and deformation. E_{mx} and E_{mn} are the maximum and minimum values of elevation of each landslide event extracted from the digital elevation model (DEM) of the study area, where positive and negative values indicate elevations above and below sea-level, respectively.

Statistical methods

Morphometric analysis on submarine landslides in the catalogue presented here was carried out using the following statistical analyses. An analysis of variance (ANOVA) (Gelman 2008) was used to test if differences in morphometric variables between different geological contexts are statistically significant. The ANOVA test is performed by comparing two types of variation, along two hypotheses, the variation between the sample means and the variation within each of the samples (or error). The null hypothesis for ANOVA is that the mean is statistically similar for all the groups analysed. The alternative or research hypothesis is that there are statistically significant differences in the mean of the groups. The calculations used in the ANOVA procedure are summarized in Table 3. The F statistic is computed by taking the ratio of the mean square between (MSB)/mean square error (MSE). If there is no effect, F will equal 1. For values of $F > 1$, the evidence is increasingly inconsistent with the null hypothesis, that is there are no significant differences among the groups analysed.

The honestly-significant-difference (HSD) test of Tukey (Abdi and Williams 2010) was also applied to determine which independent groups have significant differences in pairs. The HSD has been applied on the ANOVAs with a probability (p) <0.005 value of the significance, which proves the alternative hypothesis.

The map of density of landslides was generated with the 'kernel density tool' of ESRI's ArcGIS software by inputting the area of each landslide event in km^2 in the 'population field' parameter. A value for landslide density was computed as A_t per surface unit area of seafloor (SU). SU was established as 50 000 km^2 (126 km in searching radius), a round number according to the average size of the sedimentary basins examined (300–400 km long).

The frequency-size or power-law distribution of landslide events was calculated using the total area (km^2), following the methodology of Malamud *et al.* (2004). The equation fitted a reverse power law $N(A_t) = \alpha^* A_t^{-\beta}$, where $N(A_t)$ is the number of submarine landslides with total area greater than A_t, and

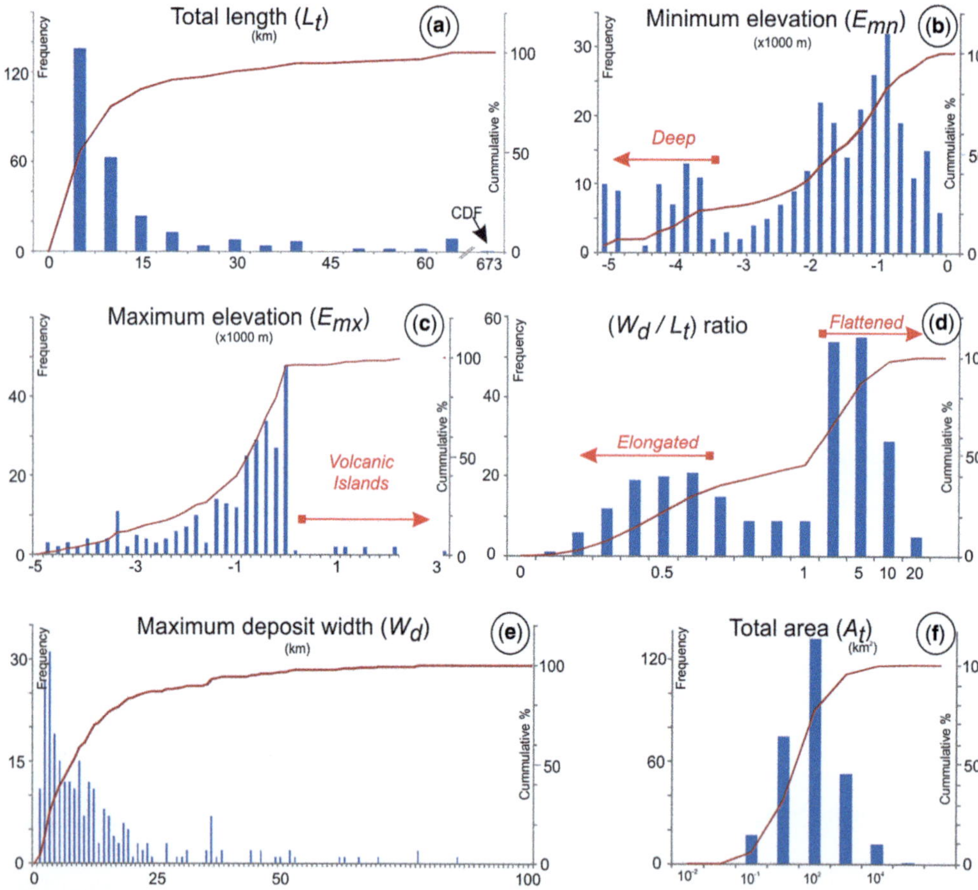

Fig. 3. Frequency histograms and cumulative curves for morphometric parameters for submarine landslides of the Spanish margins, and summary of the descriptive statistics of the main morphometric parameters stored in the GIS catalogue: (**a**) total length or runout; (**b**) minimum and (**c**) maximum elevation (positive values above sea-level and negative values below sea-level); (**d**) W_d/L_t ratio; (**e**) maximum deposit width (mapping distance); (**f**) total area. CDF, Canary debris flow.

Table 2. *Summary of statistical descriptions of the morphometric parameters in the study area*

	L_t (km)	A_t (km^2)	E_{mn}* (m)	E_{mx}* (m)	W_d/L_t	W_d (km)	S_{av} (°)
Average	13.39	277.56	−1997.96	−1162.77	2.12	12.32	5.47
Median	5.01	27.38	−1609.5	−778.5	1.14	7.51	4.49
Mode	5.81	97.93	−1946	−1422	6.14	35.69	3.89
Standard deviation	43.6	2299.46	1397.43	1362.69	2.41	14.84	3.73
Curtosis	194.02	274.93	−0.4	1.03	4.98	6.26	1.48
Asymmetry coefficient	13	16.38	−0.84	−0.66	2.09	2.38	1.18
Range	673.14	38 788.21	5420	8642	14.96	84.68	20.61
Lower limit	0.34	0.34	−5505	−4943	0.1	0.67	0.29
Upper limit	673.47	38 788.54	−85	3699	15.05	85.34	20.9
Count	274	290	290	290	264	264	289

*Positive values above sea-level and negative values below sea-level.

Table 3. *ANOVA procedure*

Source of variation	Sum of squares (SS)	Degrees of freedom (df)	Mean squares (*MS*)	*F*
Between group	$SSB = \sum n_j(\bar{X}_j - \bar{X})^2$	$k - 1$	$MSB = \frac{SSB}{k-1}$	$F = \frac{MSB}{MSE}$
Error (or residual)	$SSE = \sum\sum(X - \bar{X}_j)^2$	$N - k$	$MSE = \frac{SSE}{N-k}$	
Total	$SST = \sum\sum(X - \bar{X})^2$	$N - 1$		

X, individual observation; $\bar{X}_j$, sample mean of the jth group; $\bar{X}$, overall sample mean; *k*, the number of treatments or independent comparison groups; *N*, total number of observations or total sample size; MSB , mean square between; MSE, mean square error; SSB, sum of squares between groups (between group variation); SSE, sum of squared estimate of errors; SST, sum of total squares; *F*, *F* statistic of Fisher (ratio of variances).

α and β are constants. The Pearson coefficient (R^2) gives a value of the linear correlation of this fit.

Results

Morphometric parameters, mean values and ranges

A_t ranges from 0.33 to 38 788 km^2 and the frequency distribution displays a log-normal behaviour (Fig. 3f). Submarine landslides are generally <100 km^2, except in the Canary volcanic province. Some extreme values in total area are located in the Catalan–Balearic margin, with the largest event having an area of *c.* 1600 km^2, and in the Canary volcanic province where the three largest events are 1500, 1600 and 38 800 km^2 in area.

L_t also shows a wide range of values from 0.3 to 673 km (Fig. 3a). The median is 5 km, and the mode is 5.8 km, both of which are somewhat different from the mean of 15 km due to the extreme value of the Canary debris flow (CDF; 673 km). The frequency distribution shows a skewness displaced to the lower values.

W_d (Fig. 3e) shows a similar behaviour to the run-out distance. The range varies widely from 0.6 to 85 km. The frequency distribution shows an asymmetry biased to the lower values, which is clearly shown in the median, average and mode with values of 7, 12 and 35 km, respectively. Larger values are rare, and the cumulative frequency distribution flattens at the upper limit of 85 km (Table 2).

Width-to-length ratios range from 0.09 to 15.05 (Table 2), with low and high ratios being described herein as elongated and flattened shaped shapes, respectively (Fig. 3d). The frequency distribution shows a bimodal pattern, pointing to an elongated population along the direction of transport with an average W_d/L_t ratio of 0.55; and another flattened population, transversally elongated to the direction of transport with a W_d/L_t ratio of 7.5.

E_{mx} (Fig. 3c) frequency distribution shows a wide range interval (from 4943 to 3649 m) with two clear parts: one in the positive side and another in the negative. Events in the positive side are scarce because they are located uniquely in the Canary Islands. Negative values show a frequency distribution with three maxima at 400, 1400 and 3400 m water depth.

E_{mn} (Fig. 3b) shows a frequency distribution with a range interval from 5505 to 85 m water depth. Four main maxima might be recognizable at 800, 1600, 3800 and 5000 m water depth as the lower limit of the range.

ANOVA results

The seven morphometric variables measured in this work were analysed with an ANOVA test, taking into account the geological context of each data point. ANOVA results for L_t, W_d, E_{mx} and E_{mn} elevation (Fig. 4) showed significant morphometric differences between three geological context categories: (1) deep-ocean seamount ridges (Jean Charcot and Coruña seamounts); (2) volcanic islands (Canary volcanic province); and (3) continental margins (Alborán Basin, Gulf of Cádiz, Cantabrian, Galicia and Catalan–Balearic margins).

Results of the ANOVA (Fig. 4) and Tukey's HSD Test (Table 4) from both minimum ($F_{2,253} = 191.34$; $p < 0.005$) and maximum ($F_{2,253} = 247.16$; $p < 0.005$) elevation and W_d/L_t ratio ($F_{2,228} = 11.056$; $p < 0.00003$) show that the morphologies of the landslides of each group differ from one another. This analysis demonstrates that the three groups of geological contexts comprise consistent morphometric characteristics in their landslides.

The ANOVA of A_t ($F_{3,450} = 96.949$; $p < 0.005$) and W_d/L_t ratio ($F_{2,228} = 11.056$; $p < 0.00003$) (Fig. 4c, d) were categorized according to the three groups of geological contexts. Both analyses show

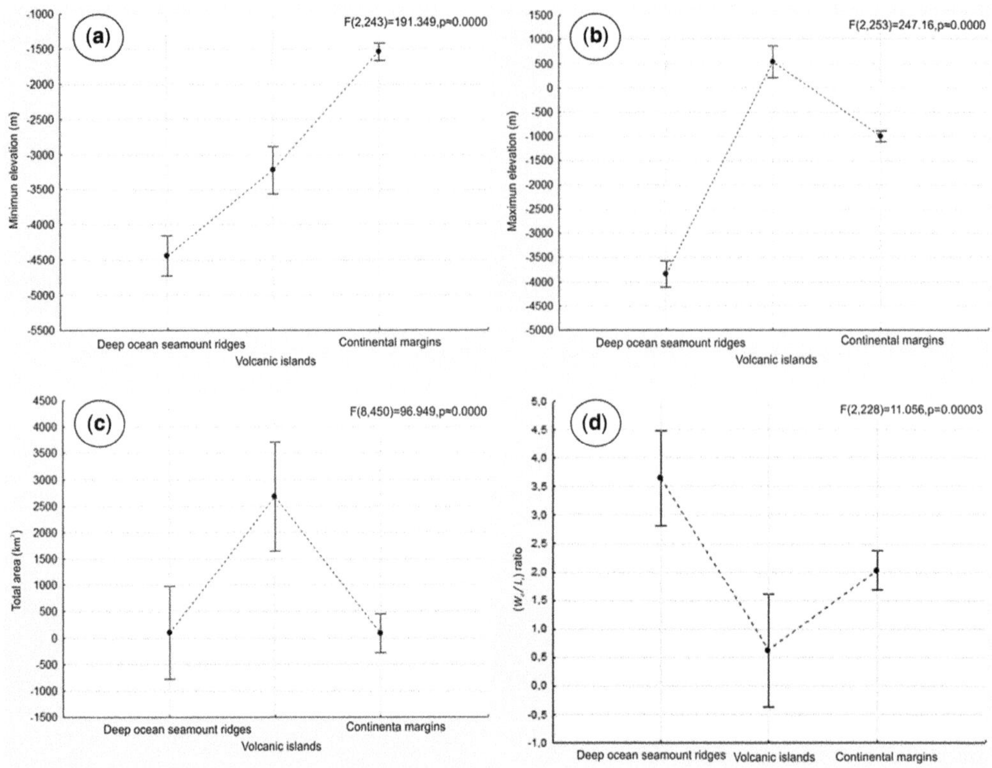

Fig. 4. Analysis of variance (ANOVA) in different submarine landslides sorted by groups of geological context. Black points, mean of values; vertical bars, 90% of observed values. (**a**) Minimum elevation; (**b**) maximum elevation; (**c**) total area; (**d**) W_d/L_t ratio.

significant differences among the cited three groups. These differences are corroborated by the Tukey's HSD test (Table 4). The significant differences in A_t and W_d/L_t ratio of the landslides developed in the volcanic islands reflect significantly larger values in A_t and L_t (elongated bodies of debris flow and debris avalanches) compared to the other geological–tectonic settings. In this way, ANOVA results show that the group comprising volcanic islands is clearly separated from the groups comprising continental margins and deep-ocean seamount ridges.

W_d/L_d *ratio v. maximum elevation*

The scatterplot maximum elevation v. W_d/L_t ratio (Fig. 5) shows how the landslides of deep-ocean seamount ridges, volcanic islands or continental margins are clearly differentiated. The majority of landslides in deep-ocean seamount ridges have the deepest bathymetric range and generally show flattened shapes. They have maximum sediment widths much larger than runout distances (mean of W_d/L_t ratio is 3.67; Fig. 5 and Table 4). Landslides on continental margins cover a relatively wide spectrum of W_d/L_t ratios (0.1–15) and bathymetric range (between 85 and 4075 m water depth). Nevertheless, many landslides have an elongated shape with values of the W_d/L_t ratio between 0.5 and 0.7. Finally, the landslide events in volcanic islands are characterized by a head escarpment above sea-level and an elongated shape (mean of W_d/L_t ratio = 0.6). The CDF, the only volcaniclastic debris flow analysed in this work, plots in a location quite different from the cloud of values of volcanic islands (CDF in Fig. 5).

Frequency–magnitude distributions

The frequency–magnitude distributions of landslides within the three groups of geological contexts can be described with inverse power-law relationships (Fig. 6). The landslide events developed in volcanic islands seem to display two different power-law distributions, one for the sizes smaller than 200 km^2 ($y = 1.2834\ x^{-0.097}$; $R^2 = 0.9634$) and another for those larger than 300 km^2 ($y = 1.253\ x^{-1.202}$;

Table 4. *Tukey HSD test of the different ANOVAs developed over the submarine landslides in the three groups of geological contexts in the study area*

E_{mn}*	1 $-4441^{\|\|}$	2 $-3221^{\|\|}$	3 $-1541^{\|\|}$
1		0 000 022	0 000 022
2	0 000 022		0 000 022
3	0 000 022	0 000 022	
E_{mx}†	1 $-3836^{\|\|}$	2 $535.56^{\|\|}$	3 $-994.0^{\|\|}$
1		0 000 022	0 000 022
2	0 000 022		0 000 022
3	0 000 022	0 000 022	
A_t‡	1 $99.601^{\|\|}$	2 $2681.2^{\|\|}$	3 $86.804^{\|\|}$
1		0 000 535	0 999 610
2	0 000 535		0 000 031
3	0 999 610	0 000 031	
W_d/L_t§	1 $3.6453^{\|\|}$	2 $0.62\,678^{\|\|}$	3 $2.0268^{\|\|}$
1		0 000 033	0 001 218
2	0 000 033		0 023 232
3	0 001 218	0 023 232	

1, Deep-ocean seamounts ridges; 2, volcanic islands; 3, continental margins.
*Approximate probabilities for *post-hoc* (APPH) tests error between mean squared error (MSE) = 7430×10^2, degree of freedom (df) = 25 300.
†APPH tests error between MSE = 6604×10^2, df = 25 300.
‡APPH tests error between MSE = 6103×10^3, df = 228.00.
§APPH tests error between MSE = 1.9175, df = 228.00.
‖Distribution centres of ANOVAs.

$R^2 = 0.8992$). These two different groups are not products of bias in sampling smaller landslides and clearly reflect two distinct geographical areas. The smaller sizes are located on the islands flanking the Lanzarote Passage, and the larger sizes, which display a gentler slope of the power law, are at the base of the Canary Islands. The CDF shows an extreme value and is isolated from the cloud of values of volcanic islands.

In continental margins, the power law for the tectonically controlled ($y = 22.505\ x^{-1.212}$; $R^2 = 0.9006$) show a higher exponent than the non-tectonically controlled ($y = 13.059\ x^{-0.911}$; $R^2 = 0.9228$). This means that in the tectonically controlled continental margins, large landslides are more infrequent than small ones. These smaller sizes in L_t and A_t are also shown in the frequency histograms of tectonically controlled continental margins (Fig. 7). The landslides occurring in tectonically controlled margins are smaller than in the non-tectonically controlled ones. On tectonically controlled margins, the median of L_t is between 1.2 and 4.7 km and of A_t is less than 35 km^2. On non-tectonically controlled margins, these values are between 8 and 35 km for the total length and between 42 and 114 km^2 for the total area (Fig. 7). Regarding S_{av}, the median is between 4 and 8.3° for tectonically controlled margins and *c.* 2.3° for non-tectonically controlled margins (Fig. 7); tectonically controlled margins are therefore steeper.

Discussion

Spatial distribution of landslides v. geological context

Submarine landslides occur in all geological settings of the Spanish continental margins. Their spatial distribution, sorted by geological context, allows the identification of patterns in their frequency and abundance v. seafloor area affected by the landslides.

Some 285 landslide events were identified within the study area. Some of them involve large sediment volumes and affect large areas such as the BIG'95 landslide in the Ebro Delta (*c.* 26 km^3 and *c.* 1600 km^2; Lastras *et al.* 2002) or the CDF in the Canary volcanic province with *c.* 400 km^3 and *c.* 40 000 km^2 (Masson *et al.* 2002). However, in the majority of physiographic domains of the Spanish continental margins, landslide events have sizes <200 km^2 affecting the seafloor or integrated in the sedimentary sequences that build the continental margins and deep basins (Urgeles and Camerlenghi 2013; Casas *et al.* 2015).

The Canary volcanic province shows the highest density of landslides in the study area with 0.43 km^2/SU (Fig. 2). Also, the continental slope of the Canary volcanic province clearly shows one of the largest slope failures of the study area. In this province, the seafloor area affected by landslides is *c.* 61 150 km^2, *c.* 76% of the total area affected (*c.* 80 500 km^2) by landslides in the Spanish continental margins. This area encompasses 49 landslide events (up to *c.* 22 myr old) close to the islands' flanks, which have been deduced from onshore unconformities, volcanoclastic debris flow and seismic reflection and echo-sounder data, and 125 volcaniclastic turbidites levels (up to *c.* 17 myr old) in the Madeira Abyssal Plain (Urgeles *et al.* 1997, 2001; Masson *et al.* 2002; Hunt *et al.* 2011, 2014; Hunt and Jarvis 2017). In the Canary volcanic province, mass transport deposits consist of debris avalanches covering the slopes of volcanic islands and debris flows starting at the foot of the debris avalanche deposits.

The map of density of landslides in the rest of the study area shows five areas with a relatively high

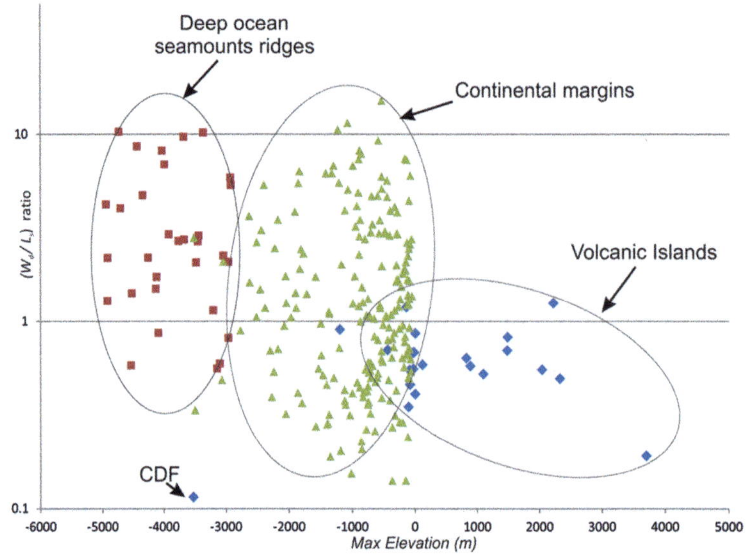

Fig. 5. Scatterplot of maximum elevation v. W_d/L_t ratio of submarine landslides sorted by geological setting: sedimentary margin (green), deep-ocean seamount ridges (red) or volcanic islands (blue). CDF, Canary debris flow.

density of events (0.045–0.092 km^2/SU): Jean Charcot seamounts, Galicia Bank, Catalan–Balearic margin, Alborán Basin and Canary volcanic province. In the Alborán Basin and Betic–Balearic domain, landslides are associated with seamounts in both the Iberian and African continental margins. On the margins of the Alborán Basin, landslides occur on diapiric ridges, canyons and seamounts and are apparently triggered by hydrocarbon seabed fluid flow and earthquakes (Urgeles and Camerlenghi 2013; Casas *et al.* 2015), while in the Catalan–Valencian domain the highest density matches with landslides on the Ebro Delta, Eivissa Channel and Balearic Promontory (Lastras *et al.* 2007; Urgeles and Camerlenghi 2013).

The Galicia Bank is the biggest structural seamount in the Galicia margin where mass-movement features and their sedimentary deposits are some of the most widespread geomorphological elements (Ercilla *et al.* 2011). Here, the higher density of occurrence is associated with the Galicia bank and Jean Charcot seamounts. The Cantabrian margin and Gulf of Cádiz are two representative areas with a lower density of landslides (<0.013 km^2/SU). Here slides are common on the borders of the canyon walls, floors and open slope (Ercilla *et al.* 2008*b*). Transparent seismic reflectors related to mass transport deposits have been identified in the continental slope and the Biscay Abyssal Plain (Llave *et al.* 2018).

The Alborán Basin (71 events) and Catalan–Balearic margin (78 events) are the areas where most landslide events are observed (Fig. 2). In the Mediterranean Sea, there is a conspicuous correlation between well-known slope failure events and available high-resolution seismic and multibeam echo-sounder data available in the catalogue. This correlation points to the fact that the catalogue of submarine slope failures in the Spanish continental margin is still incomplete (especially for the Galicia and Cantabrian margins) and further effort is needed in order to address this situation.

Geological v. physiographical meaning of morphometric parameters

The morphometric analysis of landslide events on the Spanish continental margin shows three groups that can be significantly differentiated using ANOVA (Fig. 4), maximum elevation v. W_d/L_t ratio scatter plots (Fig. 5), and the power-law distributions (Fig. 6). This observation allows us to hypothesize that different geological–tectonic settings control the morphometry of the landslide and subsequently its dynamics after the failure.

Dimensions such as A_t, L_t or runout and W_d reflect geomorphological parameters that can be correlated to the geological processes triggering the landslides as well as failure and transport mechanisms (Ardizzone *et al.* 2008; Chaytor *et al.* 2009; Guzzetti *et al.* 2012; Urgeles and Camerlenghi 2013; Clare *et al.* 2014; Moscardelli and Wood 2015; Casas *et al.* 2016). On the other hand, E_{mx} and E_{mn}, and

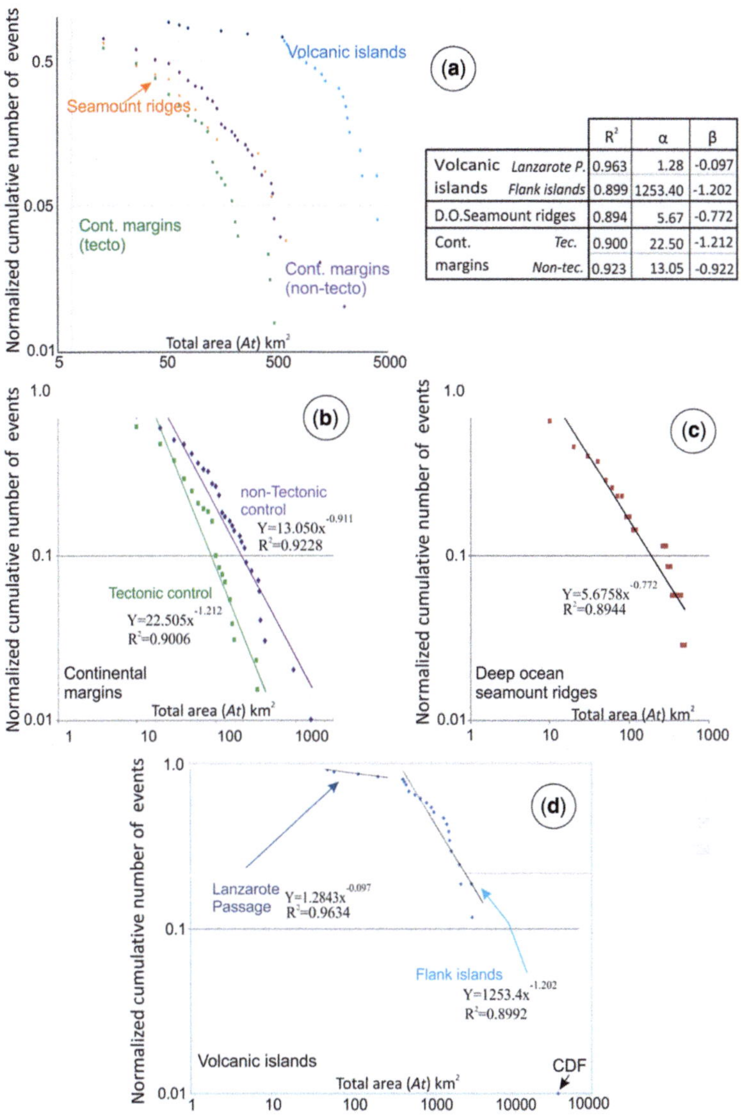

Fig. 6. (**a**) Power-law plots showing the cumulative area distribution of the 290 stored submarine landslides separated by the three groups of landslides with significant morphometric differences. R^2, Pearson coefficient; α and β constants of the power-law distribution. (**b**) Power-law distributions of landslides in continental margins. Tectonically controlled margins in green and non-tectonically controlled margins in purple. (**c**) Power-law distributions of landslides in deep-ocean seamount ridges. (**d**) Power-law distributions of landslides in volcanic islands. Dark blue, landslides in the Lanzarote Passage; light blue in the flank islands; CDF, Canary debris flow.

S_{av} reflect physiographic differences between geographical provinces instead of real landslide differences. In this way, E_{mx}, E_{mn} and S_{av} are indirectly controlled by tectonics. For this reason, these parameters have been taken into account as secondary variables to support the differences found in A_t or L_t. For example, positive values of E_{mn} in the flank collapses in the oceanic islands or the extreme deep values of E_{mn} in the seamounts of ocean ridges.

Size distribution of submarine landslides

Regarding the power-law distributions of the accumulated total area, a bias could be related to our

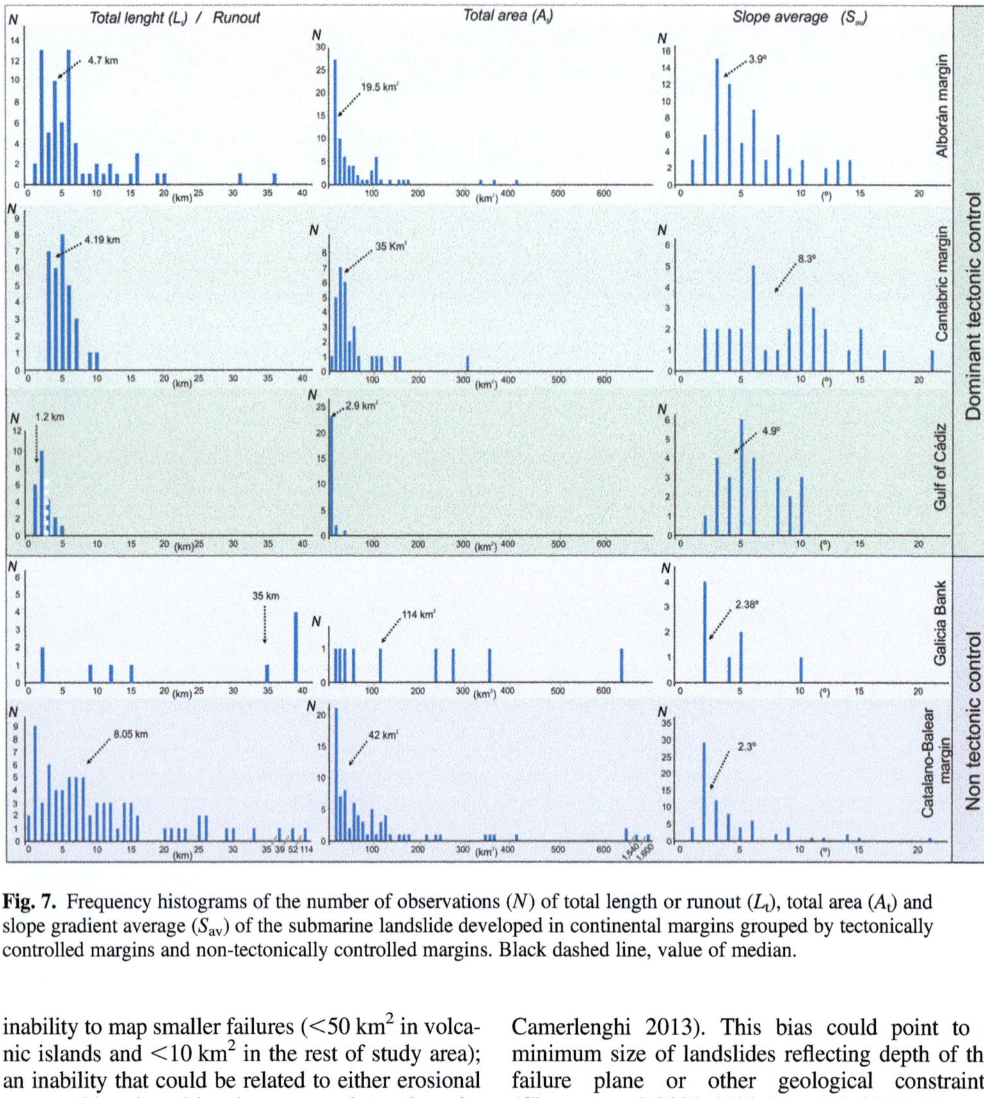

Fig. 7. Frequency histograms of the number of observations (N) of total length or runout (L_t), total area (A_t) and slope gradient average (S_{av}) of the submarine landslide developed in continental margins grouped by tectonically controlled margins and non-tectonically controlled margins. Black dashed line, value of median.

inability to map smaller failures (<50 km^2 in volcanic islands and <10 km^2 in the rest of study area); an inability that could be related to either erosional or reworking depositional processes that reduce the size. Nevertheless, a power-law distribution has therefore been fitted to the size distribution over the range 10–38 000 km^2 (Fig. 6) pointing to a sufficiently complete catalogue (Malamud and Turcotte 2006; Urgeles and Camerlenghi 2013). Landslides represented in a power-law distribution have a non-characteristic scale behaviour (Malamud *et al.* 2004). In the power-law distributions, the process that generates the geological feature has a non-characteristic scale behaviour. In this way, submarine landslides are scale invariant, their morphology does not vary with the scale and the variance is not finite. In the log-normal distribution, the frequency size distribution obeys a biased normal distribution (Malamud *et al.* 2004; Malamud and Turcotte 2006; Urgeles and Camerlenghi 2013). This bias could point to a minimum size of landslides reflecting depth of the failure plane or other geological constraints (Chaytor *et al.* 2009; Milledge *et al.* 2014). Natural processes, such as mechanical failure processes, are more commonly described as scale-invariant obeying a self-similar behaviour, where the tail of the reverse-power law is controlled by the scale of detected failure size (Malamud and Turcotte 2006). Log-normal distributions have, however, also been used to explain submarine landslide size distributions (e.g. Chaytor *et al.* 2009), where submarine landslides mobilize a moderate amount (<1 km^3) of material in the same geological and tectonic framework. In the present paper, different tectonic settings are taken into account where different triggers, controls factors and failure and transport mechanisms take place (Urgeles and Camerlenghi 2013; Casas *et al.* 2015). Our data seem to fit power-law distributions better than log-normal distributions with Pearson coefficients (R^2 in Fig. 6) greater for inverse power law. In the case of oceanic volcanic islands, the roll-over point in the power-law distributions occurs for landslides $<500\,\text{km}^2$. This value probably depends on the bathymetry resolution and/or overlapping of volcanic and tectonic triggers, which could hide a significant number of small submarine landslides.

Finally, morphometrical studies in world landslide databases show that the largest mass transport deposits in the world are found in association with volcanic islands (Hühnerbach and Masson 2004; Moscardelli and Wood 2015). This observation matches with the exponents of the power-law distributions in the present work (Fig. 6) showing two groups of landslides. The first group (tectonic and volcanic landslides) displays a high exponent (*c.* −1.2). This means that the flux of sediment from slope failure is controlled by smaller failures. In the second group (deep-ocean seamount ridges and non-tectonic continental margins), with a power exponent lower than −1 (−0.9), the flux of sediment from slope failure is controlled by the larger (but fewer) failures. In this way, attached mass transport deposits located in passive margins, such as the Atlantic coast of South and North America, have an area and volume bigger than those located in active margins, such as the Pacific coast of North America or Caribbean margin (Chaytor *et al.* 2009; Moscardelli and Wood 2015). The lower exponent of passive margins compared to active margins is in agreement with other works published on the Mediterranean and Black Sea. Landslides located in passive or non-tectonically controlled margins, such as the Nile Delta or Black Sea, have bigger volumes and exponents of their power-law distributions than those located in active areas like the Tyrrhenian Sea or the Calabrian accretionary wedge (Urgeles and Camerlenghi 2013). It has been suggested that sediment has a short residence time in active continental margins and that this sediment is often mobilized in more frequent, but smaller, deep-sea landslides. In contrast, in passive margins (non-tectonically controlled) the sediment has a longer residence time and subsequently slope failures mobilize bigger volumes along more extensive continental slopes.

Submarine landslides in volcanic areas

Landslides in volcanic islands have the most significant morphometric differences (Fig. 4 and Table 4). These differences are confirmed by the ANOVA and the power-law behaviour with an α value of 1253, a thousand times greater than other groups. The geological and tectonic setting of the Canary Islands differs from the rest of the Spanish continental margins. Volcanic debris avalanches are among the largest gravity driven processes on Earth, mobilizing hundreds to thousands of cubic kilometres of rock and sediments (Moore *et al.* 1989; Urgeles *et al.* 1997, 2001; Masson *et al.* 2002). Besides, these types of landslides have not been detected in the continental margins of southern Europe (Urgeles and Camerlenghi 2013; Casas *et al.* 2015). Such large debris avalanches are restricted to volcanic areas related to the evolution of large oceanic volcanic islands characterized by periods of rapid volcanic construction (Hunt and Jarvis 2017; León *et al.* 2019). Oversteepening of the slopes due to volcano growth (Murray and Voight 1996), seismicity (Voight and Elsworth 1997), dyke injection (Famin and Michon 2010; Becerril *et al.* 2015), volcanic spreading (Van Wyk de Vries and Francis 1997; Münn *et al.* 2006) or the presence of hyaloclastites (Schiffman *et al.* 2006; Seisdedos *et al.* 2012) have been proposed as triggers. Besides, flank-collapse processes start the movement in the subaerially exposed area up to 1000 m above sea-level. All the above phenomena are not prevalent on continental margins. The great difference between the volcanic islands and the rest of the geological environment in terms of runout length, A_t and E_{mx} (with positive values) is reflected in all of the statistical analysis shown in our results.

The two functions fitted in the power-law distribution of volcanic islands (Fig. 6d) belong to two different groups of landslides located in two different geographical areas The landslides located in the Lanzarote Passage (Vázquez *et al.* 2015*a*) differ from the typical debris avalanche mapped in the rest of the Canary Islands (Masson *et al.* 2002). The headwall escarpments of the landslides in this area (Acosta *et al.* 2005; Vázquez *et al.* 2015*a*) are always below sea-level and located near the shelf-break. Therefore, the lower slope of the power law could be related to specific geological controls or triggers in the Lanzarote Passage owing to tectonic and volcanic activity in the continental slope, such as fault activity or deep basin volcanism (Vázquez *et al.* 2015*a*) among other reasons.

In the Canary archipelago, submarine landslide events are related to debris avalanches developed from the flanks of the islands and triggered by the volcanic activity (Masson *et al.* 2002; Hunt and Jarvis 2017; León *et al.* 2019). The CDF is displaced from the distribution of other values for volcanic islands (CDF in Fig. 5) and the power-law distribution (CDF in Fig. 6). We argue that this is directly related to the type of failure mechanism. The CDF is the only volcanic debris flow stored in the database. This single event is 18 times bigger than the Orotava debris avalanche, Tenerife Island. The CDF originated in deep-water sediments due to loading of the slope sediments by a debris avalanche (Masson *et al.* 2002).

Deep-ocean seamount ridges (oceanic seafloor and extinct oceanic spreading centres) comprise another setting with a distinct landslide population. In this area, the minimum elevation (5000–3500 m below sea-level) reached by the landslides is similar to that of the volcanic islands' province. Nevertheless, significant differences in the W_d/L_t ratio, area/size and the bathymetric range have been noted. In the study area, the maximum elevation of the deep-ocean seamount ridges is located in between 4900 and 3000 m water depth, while for ocean volcanic islands it is between 3500 m below and 3600 m above sea-level (average 535 m above sea-level). Seafloor morphology of deep-ocean seamounts ridges shows marked structural control. Most of this fabric is related to transform faults and regional lava emission centres (Augustin *et al.* 2016; Watson *et al.* 2016) active during the opening of the Bay of Biscay and its reactivation in compression during the Paleocene and Eocene convergence between the Iberian and European plates (Boillot *et al.* 1979; Emery and Uchupi 1984). In the study area, the summits of deep-ocean seamount ridges are located between 3300 and 3100 m water depth in the Jean Charcot Bank and 4400–4100 m water depth in the Coruña seamounts. Headwall scars are composed of several single, small, slope failures, as observed in the landslides in the Warwick-Cobb seamounts, Juan de Fuca Ridge, Azores Islands or Condor seamount along the Mid-Atlantic Ridge, or Bouvet-Marion islands, along the Southwest Indian Ridge (Mitchell 2003; Chaytor *et al.* 2009; Tempera *et al.* 2013). In these areas, tectonic and geological constraints, such as transform faults and regional linear lava emission centres related to extinct ocean spreading areas, affect the configuration of deep elongated ridges. We therefore argue that these tectonic and geological constraints control the final morphology of the landslide complexes developed in these areas. Linearity, elongation, lateral continuity, low relief and deep-water location in this setting results in landslides characterized by a runout much lower than the maximum width (high W_d/L_t ratio) and deep bathymetric values, which the statistical analysis shows as two distinct morphometric parameters in that population.

Submarine landslides on sedimentary margins

Submarine landslides on continental margins of Spain and adjacent deep-sea areas are spread over the Catalan–Balearic margin (Acosta *et al.* 2002; Lastras *et al.* 2002, 2007; Casas *et al.* 2003*a*; Droz *et al.* 2006; Camerlenghi *et al.* 2009; Lafuerza *et al.* 2012), Alborán Basin (Cattaneo *et al.* 2010; Casas *et al.* 2011; Alonso *et al.* 2014), Gulf of Cádiz (León and Somoza 2011), Galicia (Ercilla *et al.* 2011) and Cantabrian (Ercilla *et al.* 2011) margins. Most of the submarine landslides in the GIS catalogue (León *et al.* 2018) have been generically identified as mass transport deposits, of which debris-flow deposits, shallow- and deep-seated failures and slumps are common failure styles. In most cases, tectonic activity and sedimentary load have played a major role as triggering mechanisms (Camerlenghi *et al.* 2009; Urgeles and Camerlenghi 2013; Casas *et al.* 2015).

The landslides on continental margins occur in areas characterized by different seismicity rates, which suggests a dual morphometric signature to the landslides on these margins. In the Alborán Sea, a large number of mass transport deposits are observed associated with failure of the Plio-Quaternary sequence (Alonso *et al.* 2014). These deposits have been directly related to the post-Messinian cyclical tectonic activity that controls both the physiography and sedimentary architecture of the basin (Vázquez *et al.* 2013; Alonso *et al.* 2014). Tectonic activity also has a dominant control in the Galician and Cantabrian margins (Ercilla *et al.* 2008*a*, *b*; Vázquez *et al.* 2008) as well as in the Moroccan and Portuguese part of the Gulf of Cádiz (Bartolome *et al.* 2012).

In other continental margins, high sedimentation rates, free gas and gas hydrate occurrence, and storm waves have been established as the main landslide triggers. This is the case of the Spanish part of the Gulf of Cádiz (Casas *et al.* 2003*b*; León and Somoza 2011), Alborán Basin and Catalan–Baleric margin (Casas *et al.* 2003*a*; Urgeles *et al.* 2006; Urgeles and Camerlenghi 2013).

Landslides developed in tectonically controlled margins, such as the Alborán Basin and Cantabrian margin, have a smaller runout and area than those occurring in passive margins. In the Alborán Basin and Cantabrian margin, the median is located at *c.* 4.5 km rather than the 8–35 km of non-tectonically controlled margins (Fig. 7). The BIG'95, with an area of 2200 km^2 of affected seafloor (Lastras *et al.* 2002), is the largest event in the Catalan–Balearic margin. The BIG'95 is believed to result from sediment overloading (Urgeles *et al.* 2006) in combination with oversteepening related to differential consolidation around a volcanic dome (Lastras *et al.* 2007).

Finally, the S_{av} in active and passive margins is larger in active margins, such as the Alborán Basin and Cantabrian margin, than in passive margins (Catalan–Balearic margin and Galicia Bank). In the Alborán Basin and Cantabrian margin, the median is located at 3.9 and 8.3° slope, respectively, while in the Catalan–Balearic margin and West Galicia Bank it is at *c.* 2.3° (Fig. 7). We argue that this morphometric character could point to tectonic steepening in active margins. Although the main trigger mechanism in the Cantabrian margin is unclear

(Casas *et al.* 2015), it is probably related to seismicity and tectonic oversteepening (Ercilla *et al.* 2008*b*). In the Alborán Basin, post-Messinian tectonic activity has been established as the main trigger mechanism for most of the landslide deposits observed (Vázquez *et al.* 2013). High-angle faults and uplifting of the margin are responsible for the diverse small-scale, retrogressive landslides from the lower slope to the continental rise (Pérez-Hernández *et al.* 2009).

The Gulf of Cádiz is an area with dominant tectonic control (e.g. Medialdea *et al.* 2009). Here, the mass transport deposits located in the south Portuguese margin, such as the landslides and debris flows related to the Marques de Pombal Fault (Gracia *et al.* 2003; Vizcaino *et al.* 2006), are associated with active faults (Bartolome *et al.* 2012; Lo Iacono *et al.* 2012). However, only landslides located in the Spanish part of the Gulf of Cádiz have been mapped. These events are mainly concentrated in the continental slope and are genetically related to diapirism, the presence of free gas in the sediment (Casas *et al.* 2003*b*; León and Somoza 2011) and the deformation and instability of sediments associated with contouritic sediments (Mulder *et al.* 2009; García *et al.* 2016). We argue that these controlling factors determine the singular morphometric pattern of the Spanish part of the Gulf of Cádiz, with low values in the median runout (1.2 km) and area (2.9 km^2), differing slightly from a tectonically controlled margin, such as the Alborán and Cantabric margins.

According to the features described above, power-law distributions of continental margins (Fig. 6) show a steeper slope for submarine landslides developed in active compared to passive margins, with β values of 1.21 and 0.71, respectively. Although we have used the total area, our results are in accordance with other β values calculated in other margins on both volume and area of deposits (e.g. Micallef *et al.* 2008; Chaytor *et al.* 2009; Urgeles and Camerlenghi 2013; Casas *et al.* 2016) where the β value is larger in active margins.

Uncertainty and error assessment

One of the main factors for determining the uncertainty in the statistical analysis in this work is directly related to the heterogeneity of the information sources. Particularly, heterogeneity of authors, criteria to define the landslide event and differences in the methodology used to characterize each event in the database. This is directly related to the date when the landslide event was published. Well-studied areas, such as the Alborán Basin or Catalan-Balearic margin, have a more complete catalogue than other areas, such as the Galicia margin. The landslides coming from old geological maps of the Spanish continental margin (events mapped only from single-channel seismic profiles), or whose data sources have not been sufficiently documented in the peer-reviewed literature, have been discarded in the morphometric analysis. Landslide parameters considered controversial or incorrect by several authors have been discarded too. These issues are difficult to quantify and return as statistical errors on the submarine landslide density map, for instance.

The lack of landslides in the Portuguese part of the Gulf of Cádiz and in the submarine landslide database of the Geological Survey of Spain has had an effect on the completeness of this work. The Gulf of Cádiz is an active tectonic area where the largest landslides have been defined in Portuguese waters (https://www.emodnet-geology.eu/). Subsequently, we consider the results to be biased in this area.

Other aspects related to cartographic precision also need to be considered, as the majority of inventoried landslides constitute multi-event episodes. In the best-studied areas, inventoried submarine landslides have been usually mapped as multi-event episodes, while in poorly studied areas as single events. This issue could have an effect on the estimation of area, maximum width and W_d/L_t ratio parameters but not on the other parameters used in the morphometric analysis such as runout, maximum and minimum bathymetry or maximum and S_{av}. We postulate that landslide clusters could have been mapped as single events. This could be the case with the largest values of area on the eastern side of the Alborán Basin. Nevertheless, this cluster-effect likely constitutes local and residual effects in the global analysis due to their being few in number compared with the total.

Finally, an additional source of uncertainty is related to inaccuracies in the area calculation. The area (km^2) of each landslide has been automatically calculated by the ArcGIS ERSI© algorithm, which uses the cartographic projected area (on WGS84 UTM-H30) and not the real area over the digital terrain model. The area is calculated based on the Euclidean distance between the vertices of the polygon and not on the orthodromic (great-circle) distance. This implies an error of about 1% in the final area calculated that we consider acceptable in this work.

Conclusions

Different geological–tectonic settings control the morphometry, failure style and trigger mechanisms of submarine landslides. In this work, three main groups of landslides – deep-ocean seamount ridges, volcanic islands and continental margins – have been distinguished based on differences in their minimum or maximum elevation v. W_d/L_t ratio.

Landslides in volcanic islands show the largest morphometric differences compared to the other two groups, based on the ANOVA test and the large power-law constant α value for size distribution, which is three orders of magnitude larger than other groups. These landslides also show relatively large values for their total area, lower values of W_d/L_t ratio and the highest values (positives) in the maximum elevation. Landslides in deep-ocean seamount ridges show relatively large values of W_d/L_t ratio as well as lower values of both maximum and minimum elevation.

Finally, landslides on continental margins can also be separated based on the slope of the power-law scaling relationship. Landslides developed predominantly in tectonically active settings display a steeper slope of the scaling relationship than landslides in passive margin settings. The exponent of the power-law for passive margins (and seamounts ridges) is *c.* −0.7, whereas the exponent for tectonically active margins (and oceanic islands) is *c.* −1.2. These differences can be explained by the underlying process related to specific geological controls or triggers in the Lanzarote Passage owing to the tectonic and volcanic activity in the continental slope that could trigger both groups of landslides. In the rest of the volcanic islands, volcanic eruptions generate larger landslides, although the largest and more destructive are infrequent. A comprehensive morphometric analysis of submarine landslides in different submarine contexts could shed light on triggering mechanisms, especially for passive margins and volcanic islands where those landslides could result in tsunamis.

Acknowledgements We thank Lawrence Amy, John Counts and two anonymous reviewers for corrections and suggestions that helped to improve the article.

Funding This work has been supported by the EU-funded projects GARAH (GeoERA GE-1), EMODnet High Resolution Seabed Mapping (EASME/EMFF/2015/1.3.1.7/SI2.74125) and EMODnet Bathymetry–High Resolution Seabed Mapping (EASME/EMFF/2018/007). Funding was provided by the 'Ministerio de Ciencia, Innovación y Universidades' and the European Regional Development Fund through project INSIGHT (CTM2015-70155-R).

Author contributions **RL**: writing – original draft (lead); **RU**: writing – original draft (equal); **RPL**: writing – original draft (equal); **EP**: writing – original draft (equal); **AVI**: writing – original draft (equal); **CJGM**: writing – original draft (equal); **DC**: writing – original draft (equal).

References

Abdi, H. and Williams, L.J. 2010. *Tukey's Honestly Significant Difference (HSD) Test: Encyclopedia of Research Design.* SAGE, Thousand Oaks, CA.

Acosta, J., Canals, M. *et al.* 2002. The Balearic Promontory geomorphology (western Mediterranean): morphostructure and active processes. *Geomorphology*, **49**, 177–204, https://doi.org/10.1016/S0169-555X(02)00168-X

Acosta, J., Uchupi, E., Muñoz, A., Herranz, P., Palomo, C., Ballesteros, M. and ZEE Working Group 2005. Geologic evolution of the Canarian Islands of Lanzarote, Fuerteventura, Gran Canaria and La Gomera and comparison of landslides at these islands with those at Tenerife, La Palma and El Hierro. *In*: Clift, P. and Acosta, J. (eds) *Geophysics of the Canary Islands; Results of Spain's Exclusive Economic Zone Program.* Springer, Dordrecht, 1–40.

Alonso, B., Ercilla, G. *et al.* 2014. Quaternary North-Eastern Alborán Seamounts (SW Mediterranean Sea). *In*: Krastel, S. *et al.* (eds) *Submarine Mass Movements and Their Consequences.* Advances in Natural and Technological Hazards Research, **37**, 561–570. Springer, Cham, https://doi.org/10.1007/978-3-319-00972-8_50

Ardizzone, F., Cardinali, M., Guzzetti, F. and Reichenbach, P. 2008. Comparing landslide inventory maps. *Geomorphology*, **94**, 268–289, https://doi.org/10.1016/j.geomorph.2006.09.023

Augustin, N., van der Zwan, F.M. *et al.* 2016. Geomorphology of the central red sea rift: determining spreading processes. *Geomorphology*, **274**, 162–179, https://doi.org/10.1016/j.geomorph.2016.08.028

Bartolome, R., Gràcia, E. *et al.* 2012. Evidence for active strike-slip faulting along the Eurasia-Africa convergence zone: implications for seismic hazard in the southwest Iberian margin. *Geology*, **40**, 495–498, https://doi.org/10.1130/G33107.1

Becerril, L., Galindo, I., Martí, J. and Gudmundsson, A. 2015. Three-armed rifts or masked radial pattern of eruptive fissures? The intriguing case of El Hierro volcano (Canary Islands). *Tectonophysics*, **647**, 33–47, https://doi.org/10.1016/j.tecto.2015.02.006

Boillot, G., Dupeuble, P.A. and Malod, J. 1979. Subduction and tectonics on the continental margin off northern Spain. *Marine Geology*, **32**, 53–70, https://doi.org/10.1016/0025-3227(79)90146-4

Camerlenghi, A., Accettella, D., Costa, S., Lastras, G., Acosta, J., Canals, M. and Wardell, N. 2009. Morphogenesis of the SW Balearic continental slope and adjacent abyssal plain, Western Mediterranean Sea. *International Journal of Earth Sciences*, **98**, 735–750, https://doi.org/10.1007/s00531-008-0354-8

Canals, M., Lastras, G. *et al.* 2004. Slope failure dynamics and impacts from seafloor and shallow sub-seafloor geophysical data: case studies from the COSTA project. *Marine Geology*, **213**, 9–72, https://doi.org/10.1016/j.margeo.2004.10.001

Casas, D., Ercilla, G., Baraza, J., Alonso, B. and Maldonado, A. 2003*a*. Recent mass-movement processes on the Ebro continental slope (NW Mediterranean). *Marine and Petroleum Geology*, **20**, 445–457, https://doi.org/10.1016/S0264-8172(03)00078-3

Casas, D., Ercilla, G. and Baraza, J. 2003*b*. Acoustic evidences of gas in the continental slope sediments of the Gulf of Cadiz (E Atlantic). *Geo-Marine Letters*, **23**, 300–310, https://doi.org/10.1007/s00367-003-0160-z

Casas, D., Ercilla, G., Yenes, M., Estrada, F., Alonso, B., García, M. and Somoza, L. 2011. The Baraza Slide: model and dynamics. *Marine Geophysical Research*, **32**, 245–256, https://doi.org/10.1007/s11001-011-9132-2

Casas, D., Casalbore, D., Yenes, M. and Urgeles, R. 2015. Submarine mass movements around the Iberian Peninsula. The building of continental margins through hazardous processes. *Boletín Geológico y Minero*, **126**, 257–278.

Casas, D., Chiocci, F., Casalbore, D., Ercilla, G. and Ortiz De Urbina, J. 2016. Magnitude-frequency distribution of submarine landslides in the Gioia Basin (Southern Tyrrhenian Sea). *Geo-Marine Letters*, **36**, 405–414, https://doi.org/10.1007/s00367-016-0458-2

Cattaneo, A., Babonneau, N. *et al.* 2010. Submarine landslides along the Algerian margin: a review of their occurrence and potential link with tectonic structures. *In*: Mosher, D.C. *et al.* (eds) *Submarine Mass Movements and Their Consequences*. Advances in Natural and Technological Hazards Research, **28**, 515–525. Springer, Dordrecht, https://doi.org/10.1007/978-90-481-3071-9_42

Chaytor, J.D., Uri, S., Solow, A.R. and Andrews, B.D. 2009. Size distribution of submarine landslides along the US Atlantic margin. *Marine Geology*, **264**, 16–27, https://doi.org/10.1016/j.margeo.2008.08.007

Clague, J.J. and Stead, D. (eds) 2012. *Landslides: Types, Mechanisms and Modeling*. Cambridge University Press.

Clare, M.A., Talling, P.J., Challenor, P., Malgesini, G. and Hunt, J. 2014. Distal turbidites reveal a common distribution for large (>0.1 km^3) submarine landslide recurrence. *Geology*, **42**, 263–266, https://doi.org/10.1130/G35160.1

Clare, M., Chaytor, J. *et al.* 2018. A consistent global approach for the morphometric characterization of subaqueous landslides. *Geological Society, London, Special Publications*, **477**, 455–477, https://doi.org/10.1144/SP477.15.

Dañobeitia, J.J., Alonso, B. and Maldonado, A. 1990. Geological framework of the Ebro continental margin and surrounding areas. *Marine Geology*, **95**, 265–287, https://doi.org/10.1016/0025-3227(90)90120-9

Droz, L., dos Reis, A.T. and Rabineau, M. 2006. Quaternary turbidite systems on the northern margins of the Balearic Basin (Western Mediterranean): a synthesis. *Geo-Marine Letters*, **26**, 347–359, https://doi.org/10.1007/s00367-006-0044-0

Emery, K.O. and Uchupi, E. (eds) 1984. *The Geology of the Atlantic Ocean*. Springer Verlag, New York, https://doi.org/10.4319/lo.1986.31.3.0669

Ercilla, G., García-Gil, S. *et al.* 2008*a*. High-resolution seismic stratigraphy of the Galicia Bank Region and neighbouring abyssal plains (NW Iberian continental margin). *Marine Geology*, **249**, 108–127, https://doi.org/10.1016/j.margeo.2007.09.009

Ercilla, G., Casas, D. *et al.* 2008*b*. Morphosedimentary features and recent depositional architectural model of the Cantabrian continental margin. *Marine Geology*, **247**, 61–83, https://doi.org/10.1016/j.margeo.2007.08.007

Ercilla, G., Casas, D. *et al.* ERGAP Project & Cruise Teams. 2011. Imaging the recent sediment dynamics of the Galicia Bank region (Atlantic, NW Iberian Peninsula). *Marine Geophysical Researches*, **32**, 99–126, https://doi.org/10.1007/s11001-011-9129-x

Estrada, F., Ercilla, G. and Alonso, B. 1997. Pliocene-Quaternary tectonic-sedimentary evolution of the NE Alborán Sea (SW Mediterranean Sea). *Tectonophysics*, **282**, 423–442, https://doi.org/10.1016/S0040-1951(97)00227-8

Famin, V. and Michon, L. 2010. Volcano destabilization by magma injections in a detachment. *Geology*, **38**, 219–222, https://doi.org/10.1130/G30717.1

García, M., Hernández-Molina, F.J., Alonso, B., Vázquez, J.T., Ercilla, G., Llave, E. and Casas, D. 2016. Erosive sub-circular depressions on the Guadalquivir Bank (Gulf of Cadiz): interaction between bottom current, mass-wasting and tectonic processes. *Marine Geology*, **378**, 5–19, https://doi.org/10.1016/j.margeo.2015.10.004

Gelman, A. 2008. *'Variance, analysis of'. The new Palgrave dictionary of economics*, 2nd ed. Palgrave Macmillan, Basingstoke.

Guzzetti, F., Mondini, A.C., Cardinali, M., Fiorucci, F., Santangelo, M. and Chang, K. 2012. Landslide inventory maps: new tools for an old problem. *Earth-Science Reviews*, **112**, 42–66, https://doi.org/10.1016/j.earscirev.2012.02.001

Gracia, E., Danobeitia, J., Vergés, J. and Parsifal Team. 2003. Mapping active faults offshore Portugal (36 N–38 N): implications for seismic hazard assessment along the southwest Iberian margin. *Geology*, **31**, 83–86, https://doi.org/10.1130/0091-7613(2003)031<0083:MAFOPN>2.0.CO;2

Harbitz, C.B., Løvholt, F. and Bungum, H. 2014. Submarine landslide tsunamis: how extreme and how likely? *Naural Hazards*, **72**, 1341–1374, https://doi.org/10.1007/s11069-013-0681-3

Harders, R., Ranero, C.R., Weinrebe, W. and Behrmann, J.H. 2011. Submarine slope failures along the convergent continental margin of the Middle America Trench. *Geochemistry, Geophysics, Geosystems*, **12**, Q05S32, https://doi.org/10.1029/2010GC003401

Hühnerbach, V. and Masson, D.G. 2004. Landslides in the North Atlantic and its adjacent seas: an analysis of their morphology, setting and behaviour. *Marine Geology*, **213**, 343–362, https://doi.org/10.1016/j.margeo.2004.10.013

Hunt, J.E. and Jarvis, I. 2017. Prodigious submarine landslides during the inception and early growth of volcanic islands. *Nature Communications*, **8**, 2061, https://doi.org/10.1038/s41467-017-02100-3

Hunt, J.E., Wynn, R.B., Masson, D.G., Talling, P.J. and Teagle, D.A.H. 2011. Sedimentological and geochemical evidence for multistage failure of volcanic island landslides: A case study from Icod landslide on north Tenerife, Canary Islands. *Geochemistry, Geophysics, Geosystems*, **12**, https://doi.org/10.1029/2011GC003740

Hunt, J.E., Talling, P.J., Clare, M.A., Jarvis, I. and Wynn, R.B. 2014. Long-term (17 Ma) turbidite record of the timing and frequency of large flank collapses of the Canary Islands. *Geochemistry, Geophysics, Geosystems*, **15**, 3322–3345, https://doi.org/10.1002/2014GC005232

Janin, A., Rodriguez, M., Sakellariou, D., Lykousis, V. and Gorini, C. 2019. Tsunamigenic potential of a Holocene submarine landslide along the North Anatolian Fault (northern Aegean Sea, off Thasos island): insights from numerical modelling. *Natural Hazards and Earth System Sciences*, **19**, 121–136, https://doi.org/10.5194/nhess-19-121-2019

Lafuerza, S., Sultan, N. *et al.* 2012. Failure mechanisms of Ana Slide from geotechnical evidence, Eivissa Channel, Western Mediterranean Sea. *Marine Geology*, **307**, 1–21, https://doi.org/10.1016/j.margeo.2012.02.010

Lastras, G., Canals, M., Hughes-Clarke, J.E., Moreno, A., De Batist, M., Masson, D.G. and Cochonat, P. 2002. Seafloor imagery from the BIG'95 debris flow, western Mediterranean. *Geology*, **30**, 871–874, https://doi.org/10.1130/0091-7613(2002)030<0871:SIFTBD>2.0.CO;2

Lastras, G., Canals, M., Amblas, D., Frigola, J., Urgeles, R., Calafat, A.M. and Acosta, J. 2007. Slope instability along the northeastern Iberian and Balearic continental margins. *Geologica Acta*, **5**, 35–47.

León, R. and Somoza, L. 2011. GIS-based mapping for marine geohazards in seabed fluid leakage areas (Gulf of Cadiz, Spain). *Marine Geophysical Research*, **32**, 207–223, https://doi.org/10.1007/s11001-011-9135-z

León, R., García-Davalillo, J.C., Casas, D. and Giménez-Moreno, C.J. 2018. Submarine landslide catalogue onshore/offshore harmonization: Spain as a case study. *Geological Society, London, Special Publications*, **477**, 497–510, https://doi.org/10.1144/SP477.38

León, R., Palomino, D., Vázquez, J.T., Medialdea, T. and Somoza, L. 2019. A new scenario for the mass transport deposits west Canary volcanic province. *Earth and Planetary Science Letters*, **509**, 27–37, https://doi.org/10.1016/j.epsl.2018.12.020

Llave, E., Jane, G., Maestro, A., Lopez-Martinez, J., Hernández-Molina, F.J. and Mink, S. 2018. Geomorphological and sedimentary processes of the glacially influenced northwestern Iberian continental margin and abyssal plains. *Geomorphology*, **312**, 60–85, https://doi.org/10.1016/j.geomorph.2018.03.022

Lo Iacono, C., Gràcia, E. *et al.* 2012. Large, deepwater slope failures: implications for landslide-generated tsunamis. *Geology*, **40**, 931–934, https://doi.org/10.1130/G33446.1

Macías, J., Vázquez, J.T. *et al.* 2015. The Al-Borani submarine landslide and associated tsunami. A modelling approach. *Marine Geology*, **361**, 79–95, https://doi.org/10.1016/j.margeo.2014.12.006

Malamud, B.D., Turcotte, D.L., Guzzetti, F. and Reichenbach, R. 2004. Landslide inventories and their statistical properties. *Earth Surface Processes and Landforms*, **29**, 687–711, https://doi.org/10.1002/esp.1064

Malamud, B.D. and Turcotte, D.L. 2006. An inverse cascade explanation for the power-law frequency-area statistics of earthquakes, landslides and wildfires. *Geological Society, London, Special Publications*, **261**, 1–9, https://doi.org/10.1144/GSL.SP.2006.261.01.01

Maldonado, A., Alonso, B., Díaz, J.I., Ferran, M., Giro, S. and Vázquez, A. 1986. *Mapa Geológico de la margen continental y zona terrestre escala 1:200.000, Tortosa – Tarragona 41–42*. Ministerio de Industria y Energía, Madrid.

Maldonado, A., Díaz, J.I., Ferran, M., Giro, S., Escutia, C., Serra, M. and Barnolas, A. 1987. *Mapa Geológico de la margen continental y zona terrestre escala 1:200.000, Barcelona 35–42E*. Instituto Tecnológico y Geominero de España, Madrid.

Maldonado, A., Alonso, B., Díaz, J.I., Ercilla, G. and Ferran, M. 1994. *Mapa Geológico de la margen continental y zona terrestre escala 1:200.000, Figueres 25–25E*. Instituto Tecnológico y Geominero de España, Madrid.

Martínez-García, P., Comas, M., Lonergan, L. and Watts, A.B. 2017. From extension to shortening: tectonic inversion distributed in time and space in the Alborán Sea, Western Mediterranean. *Tectonics*, **36**, 2777–2805, https://doi.org/10.1002/2017TC004489

Masson, D.G., Watts, A.B., Gee, M.J.R., Urgeles, R., Mitchell, N.C., Le Bas, T.P. and Canals, M. 2002. Slope failures on the flanks of the western Canary Islands. *Earth Science Reviews*, **57**, 1–35, https://doi.org/10.1016/S0012-8252(01)00069-1

Masson, D.G., Harbitz, C.B., Wynn, R.B., Pedersen, G. and Lovholt, F. 2006. Submarine Landslides: processes, triggers and hazard prediction. *Philosophical Transactions of the Royal Society A*, **364**, 2009–2039, https://doi.org/10.1098/rsta.2006.1810

Mauffret, A. and Montadert, L. 1987. Rift tectonics on the passive continental margin off Galicia (Spain). *Marine and Petroleum Geology*, **4**, 49–70, https://doi.org/10.1016/0264-8172(87)90021-3

Medialdea, T., Somoza, L. *et al.* 2009. Tectonics and mud volcano development in the Gulf of Cádiz. *Marine Geology*, **261**, 48–63, https://doi.org/10.1016/j.margeo.2008.10.007

Micallef, A., Berndt, C., Masson, D.G. and Stow, D.A.V. 2008. Scale invariant characteristics of the Storegga Slide and implications for large-scale submarine mass movements. *Marine Geology*, **247**, 46–60, https://doi.org/10.1016/j.margeo.2007.08.003

Milledge, D.G., Bellugi, D., McKean, J.A., Densmore, A.L. and Dietrich, W.E. 2014. A multidimensional stability model for predicting shallow landslide size and shape across landscapes. *Journal of Geophysical Research: Earth Surface*, **119**, 2481–2504, https://doi.org/10.1002/2014JF003135

Mitchell, N.C. 2003. Susceptibility of mid-ocean ridge volcanic islands and seamounts to large-scale landsliding. *Journal of Geophysical Research: Solid Earth*, **108**, 2397, https://doi.org/10.1029/2002JB001997

Moernaut, J., Van Daele, M. *et al.* 2015. Lacustrine turbidites produced by surficial slope sediment remobilization: a mechanism for continuous and sensitive turbidite paleoseismic records. *Marine Geology*, **384**, 159–176, https://doi.org/10.1016/j.margeo.2015.10.009

Moore, J.G., Clague, D.A., Holcomb, R.T., Lipman, P.W., Normark, W.R. and Torresan, M.E. 1989. Prodigious submarine landslides on the Hawaiian Ridge. *Journal of Geophysical Research: Solid Earth*, **94**, 17465–17484, https://doi.org/10.1029/JB094iB12p17465

Moscardelli, L. and Wood, L. 2015. Morphometry of mass-transport deposits as a predictive tool. *Geological Society of America Bulletin*, **128**, 47–80, https://doi.org/10.1130/B31221.1

Mulder, T., Gonthier, E., Lecroart, P., Hanquiez, V., Marches, E. and Voisset, M. 2009. Sediment failures and flows in the Gulf of Cadiz (eastern Atlantic). *Marine and Petroleum Geology*, **26**, 660–672, https://doi.org/10.1016/j.marpetgeo.2008.02.009

Münn, S., Walter, T.R. and Klügel, A. 2006. Gravitational spreading controls rift zones and flank instability on El Hierro, Canary Islands. *Geological Magazine*, **143**, 257–268, https://doi.org/10.1017/S0016756806 002019

Murray, J.B. and Voight, B. 1996. Slope stability and eruption prediction on the eastern flank of Mount Etna. *Geological Society, London, Special Publications*, **110**, 111–114, https://doi.org/10.1144/GSL.SP.1996.110. 01.08

Nisbet, E.G. and Piper, D.J. 1998. Giant submarine landslides. *Nature*, **392**, 329–330, https://doi.org/10. 1038/32765

Pérez-Hernández, S., Comas, M., Escutia, C. and Martínez-García, P. 2009. The Águilas Submarine Slides (Palomares Margin, Western Mediterranean). *Geogaceta*, **47**, 93–96.

Platt, J.P., Anczkiewicz, R., Soto, J.I., Kelley, S.P. and Thirlwall, M. 2006. Early Miocene continental subduction and rapid exhumation in the western Mediterranean. *Geology*, **34**, 981–984, https://doi.org/10. 1130/G22801A.1

Rey, J. and Díaz-del-Río, V. 1987. Structure of recent sedimentary units in the Galician continental shelf. *Cuaderno Lab. Xeoloxico de Laxe*, **12**, 35–45.

Roca, E. and Guimerà, J. 1992. The Neogene structure of the eastern Iberian margin: structural constraints on the crustal evolution of the Valencia trough (western Mediterranean). *Tectonophysics*, **203**, 203–218, https://doi.org/10.1016/0040-1951(92)90224-T

Schiffman, P., Watters, R.J., Thompson, N. and Walton, A.W. 2006. Hyaloclastites and the slope stability of Hawaiian volcanoes: insights from the Hawaiian Scientific Drilling Project's 3-km drill core. *Journal of Volcanology and Geothermal Research*, **151**, 217–228, https://doi.org/10.1016/j.jvolgeores.2005.07.030

Seisdedos, J., Ferrer, M. and González de Vallejo, L.I. 2012. Geological and geomechanical models of the pre-landslide volcanic edifice of Güímar and La Orotava mega-landslides (Tenerife). *Journal of Volcanology and Geothermal Research*, **239–240**, 92–110, https://doi.org/10.1016/j.jvolgeores.2012.06.013

Tempera, F., Hipólito, A., Madeira, J., Vieira, S., Campos, A.S. and Mitchell, N.C. 2013. Condor seamount (Azores, NE Atlantic): a morpho-tectonic interpretation. Deep Sea Research Part II. *Topical Studies in Oceanography*, **98**, 7–23, https://doi.org/10.1016/j. dsr2.2013.09.016

Twichell, D.C., Chaytor, J.D., Uri, S. and Buczkowski, B. 2009. Morphology of late Quaternary submarine landslides along the US Atlantic continental margin. *Marine Geology*, **264**, 4–15, https://doi.org/10.1016/j.mar geo.2009.01.009

Urgeles, R. and Camerlenghi, A. 2013. Submarine landslides of the Mediterranean Sea: trigger mechanisms, dynamics, and frequency-magnitude distribution. *Journal of Geophysical Research: Earth Surface*, **118**, 2600–2618, https://doi.org/10.1002/2013JF002720

Urgeles, R., Canals, M., Baraza, J., Alonso, B. and Masson, D. 1997. The most recent megalandslides of the Canary Islands: El Golfo debris avalanche and Canary debris flow, west El Hierro Island. *Journal of Geophysical Research: Solid Earth*, **102**, 20305–20323, https:// doi.org/10.1029/97JB00649

Urgeles, R., Canals, M. and Masson, D.G. 2001. Flank stability and processes off the western Canary Islands: a review from El Hierro and La Palma. *Scientia Marina*, **65**, 21–32, https://doi.org/10.3989/scimar. 2001.65s121

Urgeles, R., Leynaud, D., Lastras, G., Canals, M. and Mienert, J. 2006. Back-analysis and failure mechanisms of a large submarine slide on the Ebro slope, NW Mediterranean. *Marine Geology*, **226**, 185–206, https://doi. org/10.1016/j.margeo.2005.10.004

van den Bogaard, P. 2013. The origin of the Canary Island Seamount Province – new ages of old seamounts. *Scientific Reports*, **3**, 1–7, https://doi.org/10.1038/ srep02107

van Wyk de Vries, B.V.W. and Francis, P.W. 1997. Catastrophic collapse at stratovolcanoes induced by gradual volcano spreading. *Nature*, **387**, 387–390, https://doi. org/10.1038/387387a0

Vázquez, J.T., Medialdea, T. *et al.* 2008. Cenozoic deformational structures on the Galicia Bank Region (NW Iberian continental margin). *Marine Geology*, **249**, 128–149, https://doi.org/10.1016/j.margeo.2007.09.014

Vázquez, J.T., Alonso, B. *et al.* and MONTERA Team 2013. Mass Movement Deposits and tectonics relations as a main factor to control the sratigraphical architecture of the south Alborán Basin (Alborán Sea, western Mediterranean). *In*: Abad de los Santos, M., Izquierdo, T. and Ruiz Muñoz, F. (eds) *Two Decades of Atlantic Neogene Study*. Vth RCANS Congress, 24–26 September, Huelva, Abstracts, **46**, http://hdl.handle.net/10272/ 7006

Vázquez, J.T., Palomino, D. *et al.* 2015*a*. Geomorphological characteristics of the Passage of Lanzarote (East Canary Islands Region). *In*: Díaz-del-Río, V., Bácenas, P. *et al.* (eds) *Volumen de Comunicaciones presentadas en el VII Simposio sobre Margen Ibérico Atlántico*. Ediciones Sia Graf, Málaga, Spain, 589–592.

Vázquez, J.T., Alonso, B. *et al.* 2015*b*. Seamounts along the Iberian continental margins. *Boletín Geológico y Minero*, **121**, 483–514.

Vizcaino, A., Gràcia, E. *et al.* 2006. Sedimentology, physical properties and age of mass transport deposits associated with the Marques de Pombal Fault, Southwest Portuguese Margin. *Norsk Geologisk Tidsskrift*, **86**, 177.

Voight, B. and Elsworth, D. 1997. Failure of volcano slopes. *Geotechnique*, **47**, 1–31, https://doi.org/10. 1680/geot.1997.47.1.1

Williams, C.A. 1975. Sea-floor spreading in the Bay of Biscay and its relationship to the North Atlantic. *Earth and Planetary Science Letters*, **24**, 440–456, https://doi. org/10.1016/0012-821X(75)90151-X

Watson, S.J., Whittaker, J.M., Halpin, J.A., Williams, S.E., Milan, L.A., Daczko, N.R. and Wyman, D.A. 2016. Tectonic drivers and the influence of the Kerguelen plume on seafloor spreading during formation of the early Indian Ocean. *Gondwana Research*, **35**, 97–114, https://doi.org/10.1016/j.gr.2016.03.009

Megaclasts within mass-transport deposits: their origin, characteristics and effect on substrates and succeeding flows

Jefferson Nwoko*, Ian Kane and Mads Huuse

School of Natural Sciences, University of Manchester, Oxford Road, Manchester M13 9PL, UK

JN, 0000-0002-2418-3927; MH, 0000-0002-1766-4343

*Correspondence: Cenjeff@ymail.com

Abstract: Megaclasts transported within submarine landslides can erode the substrate, influence the flow that transports them and, if they form seafloor topography, can influence subsequent flows and their deposits. We document grooves up to 40 km long formed by megaclasts carried in submarine landslides that scoured tens of metres deep into the contemporaneous substrate of the deep-water Taranaki Basin, New Zealand. A 1925 km^2 3D seismic reflection survey records six mass transport deposits (MTDs) interbedded with turbidites. Here, we focus on three MTDs, labelled A (oldest), B and C (youngest). MTD-A features megaclasts that internally have coherent parallel strata, and formed striations 4–15 km long and 2–3 km wide, with protruding megaclasts that are onlapped by younger sediments. The seafloor expression of these megaclasts partially obstructed the submarine landslide that created MTD-B. MTD-B contains megaclasts that incised through the rugose topography of the underlying MTD-A, and formed divergent grooves on the basal surface of MTD-B (8–40 km long and 200–250 m wide), which suggest radial flow expansion where flows exited topographic confinement. MTD-C features grooves 2–6 km long and 100–200 m wide that terminate at megaclasts and which internally are characterized of highly deformed reflectors surrounded by a chaotic matrix. This study directly links megaclasts to the grooves they form, and demonstrates that markedly different styles of scouring and resultant grooves can occur in closely related MTDs.

Mass transport deposits (MTDs) are sedimentary deposits of gravity-driven mass flows (submarine landslides) from topographic or bathymetric highs, over a basal shear surface or zone, to a relative low (Hampton *et al.* 1996; Bull *et al.* 2009; Kneller *et al.* 2016; Sobiesiak *et al.* 2018). MTDs are composed of a range of internal facies that can occur under a range of flow processes, including slumps, slides, debris flows and turbidity currents (Posamentier and Kolla 2003; Joanne *et al.* 2013). Multiple laterally or vertically stacked MTDs form mass transport complexes (MTCs).

The basal surface of a mass transport deposit records its interaction with the substrate over which it moved. The nature of this surface or zone can therefore provide information on flow process operating during the flow event (Butler *et al.* 2016; Ortiz-Karpf *et al.* 2017). The record of erosion during mass transport events and MTD emplacement has been widely documented in both subsurface (e.g. McGilvery and Cook 2003; Moscardelli *et al.* 2006; Bull *et al.* 2009; Posamentier and Martinsen 2011; Omosanya and Alves 2013; Hodgson *et al.* 2018) and outcrop studies (e.g. Lucente and Pini 2003; Dykstra *et al.* 2011; Ogata *et al.* 2012; Dakin *et al.* 2013; Sobiesiak *et al.* 2018). In particular, megaclasts carried within submarine landslides or debris flows may scour the substrate and form a range of erosional features (Nissen *et al.* 1999; McGilvery and Cook 2003; Posamentier and Kolla 2003; Moscardelli *et al.* 2006; Bull *et al.* 2009; Omosanya and Alves 2013; Soutter *et al.* 2018). Our study documents linear and curvilinear erosional features found on the basal surface of MTDs and interpreted herein to have been formed by megaclasts within a laminar flow that gouged into the substrate (Ortiz-Karpf *et al.* 2015). Such features have previously been referred to as 'glide tracks' (Nissen *et al.* 1999), 'grooves' (Posamentier and Kolla 2003; Bull *et al.* 2009), 'basal lineations' (Bull *et al.* 2009; Omeru and Cartwright 2019), 'cat claw marks' (Moscardelli *et al.* 2006; Rusconi 2017), 'monkey fingers' (McGilvery and Cook 2003) and 'striations' (Gee *et al.* 2005, 2007) among others. We use the terms 'grooves' and 'striations'; these differ in that grooves are generally v-shaped in cross-section, similar to those observed in other studies (Bull *et al.* 2009; Posamentier and Martinsen 2011), striations are wide and flat-bottomed (Gee *et al.* 2005, 2007; Bull *et al.* 2009). We use the term 'megaclast' to describe hundreds of metres to kilometre-sized clasts carried by submarine landslides or debris flows and

From: Georgiopoulou, A., Amy, L. A., Benetti, S., Chaytor, J. D., Clare, M. A., Gamboa, D., Haughton, P. D. W., Moernaut, J. and Mountjoy, J. J. (eds) 2020. *Subaqueous Mass Movements and their Consequences: Advances in Process Understanding, Monitoring and Hazard Assessments*. Geological Society, London, Special Publications, **500**, 515–530.
First published online March 31, 2020, https://doi.org/10.1144/SP500-2019-146

preserved within the matrix of a mass transport deposit (McGilvery *et al.* 2004; Moscardelli *et al.* 2006; Hodgson *et al.* 2018). We refer to the seismically chaotic low-amplitude dominated body of the MTDs, as 'debrite' (Posamentier and Kolla 2003; Hodgson *et al.* 2018).

The character and deformation of megaclasts can be used to infer the flow process, and may play a role in transformation of the parent flow as the megaclasts are abraded/spalled (Joanne *et al.* 2013; Ortiz-Karpf *et al.* 2017; Hodgson *et al.* 2018). Using a 1925 km^2 three-dimensional (3D) seismic reflection volume from offshore NW New Zealand, we analyse the grooves and striations on the bases of MTDs to document processes of megaclast transport within the submarine landslides that formed the MTDs, and analyse the interaction of successive submarine landslides with older megaclast-bearing MTDs.

Geological setting

The deep-water Taranaki Basin is located in the NW part of the Taranaki Basin and passes laterally into the New Caledonian Basin (King and Thrasher 1996; Strogen *et al.* 2017). The Taranaki Basin is located within the Australia–Pacific plate boundary zone, with the primary plate boundary faults situated to the east (the Hikurangi subduction thrust) and south (the transpressional Alpine Fault). The Taranaki Basin active volcanic back-arc rift is situated in the continental Australian Plate and is related to subduction of the Pacific Plate along the Hikurangi margin (King and Thrasher 1996; Stagpoole and Nicol 2008; Uruski 2008; Baur *et al.* 2014). Miocene sedimentation comprises about 4 km of deep-marine sedimentary deposits, including submarine channel fill and MTDs (King and Thrasher 1996; Baur *et al.* 2014; Rad 2015). The Pliocene–Pleistocene succession is characterized by MTDs emplaced during repeated submarine landslides associated with failure and remobilization of the prograding slope of the Giant Foresets Formation (Hansen and Kamp 2004*a*, *b*; Anell and Midtkandal 2017). Erosion of the uplifting Southern Alps increased siliciclastic input that progressively filled subsidence-driven accommodation generated between the tectonically active northern graben and tectonically quiescent western stable platform (Fig. 1a) (King and Thrasher 1996; Kamp *et al.* 2004). These terrigenous sediments make up the bulk of the progradational wedge during the Plio-Pleistocene (Tippett and Kamp 1995).

Dataset and methods

The dataset for this study uses post-stack time-migrated (PSTM) seismic reflection data from the offshore deep-water Taranaki Basin, New Zealand. The 3D seismic reflection dataset, the Romney 3D survey, covers an area of approximately 1925 km^2 in water depths ranging 500–1500 m, with slope gradients between 0.5 and 2°, located NW of, and downslope from, the Taranaki Peninsula (Fig. 1a). The upper 3–4 km of the sedimentary succession is dominated by large MTCs. The three MTDs in this study, labelled MTDs A, B and C (oldest to youngest), extend beyond the boundaries of this survey and have been recognized in a different study under an equivalent stratigraphic framework – MTDs 1, 4 and 6 (Omeru 2014; Omeru *et al.* 2016; Omeru and Cartwright 2019). Using a larger regional survey, Omeru (2014) mapped the extents of these three MTDs and found that their MTD 6, the youngest, is a two-lobed deposit, and covers an area of 22 397 km^2; MTD 4 covers an area of 10 128 km^2, while MTD 1 the oldest of the three in this study, covers a minimum area of about 20 887 km^2 (Omeru 2014; Omeru *et al.* 2016; Omeru and Cartwright 2019) (Fig. 1a). For the purpose of the present study and to explicitly define the relationships between the MTDs, we use the alternative nomenclature MTD A, B and C (Fig. 2c).

The MTDs in this study are thought to be Plio-Pleistocene in age based on the Romney-1 well technical report (Rad 2015). Although the well penetrates the interval of interest, no velocity and density information from the Plio-Pleistocene interval are available, necessitating the generation of a synthetic seismogram below the Miocene (Fig. 1e). Checkshot data from the Romney-1 well were used in the resolution calculations to estimate the interval velocity of 1850 m s^{-1} (Rusconi 2017). Spectral analysis of the seismic volume reveals a frequency range of 8–75 Hz with a dominant frequency of 41 Hz, yielding a dominant wavelength of *c.* 45 m and a vertical resolution of *c.* 11 m. Crossline and inline bin spacing of the seismic grid is 12.5 × 25 m and the vertical sample interval is 4 ms. The data are approximately zero phase. SEG-Normal polarity conventions were used, where increased acoustic impedance is represented by positive peak amplitude on seismic sections. Tops and bases of MTDs were mapped based upon their seismic facies, truncations and onlap onto older deposits. The expression of the basal surface of each MTD varies as does the nature of the top surface, but the general internal seismic character of each mass failure is consistently chaotic in nature. Spectral frequency decomposition and attribute analysis of the seismic data allowed clear identification and morphometric characterization of MTD megaclasts and three-dimensional mapping of linear features on the basal surfaces of the MTDs. Spectral decomposition of seismic amplitude data and colour blending of 12, 30 and 68 Hz frequencies provides enhanced imaging of these basal surface features (Figs 3–6). With

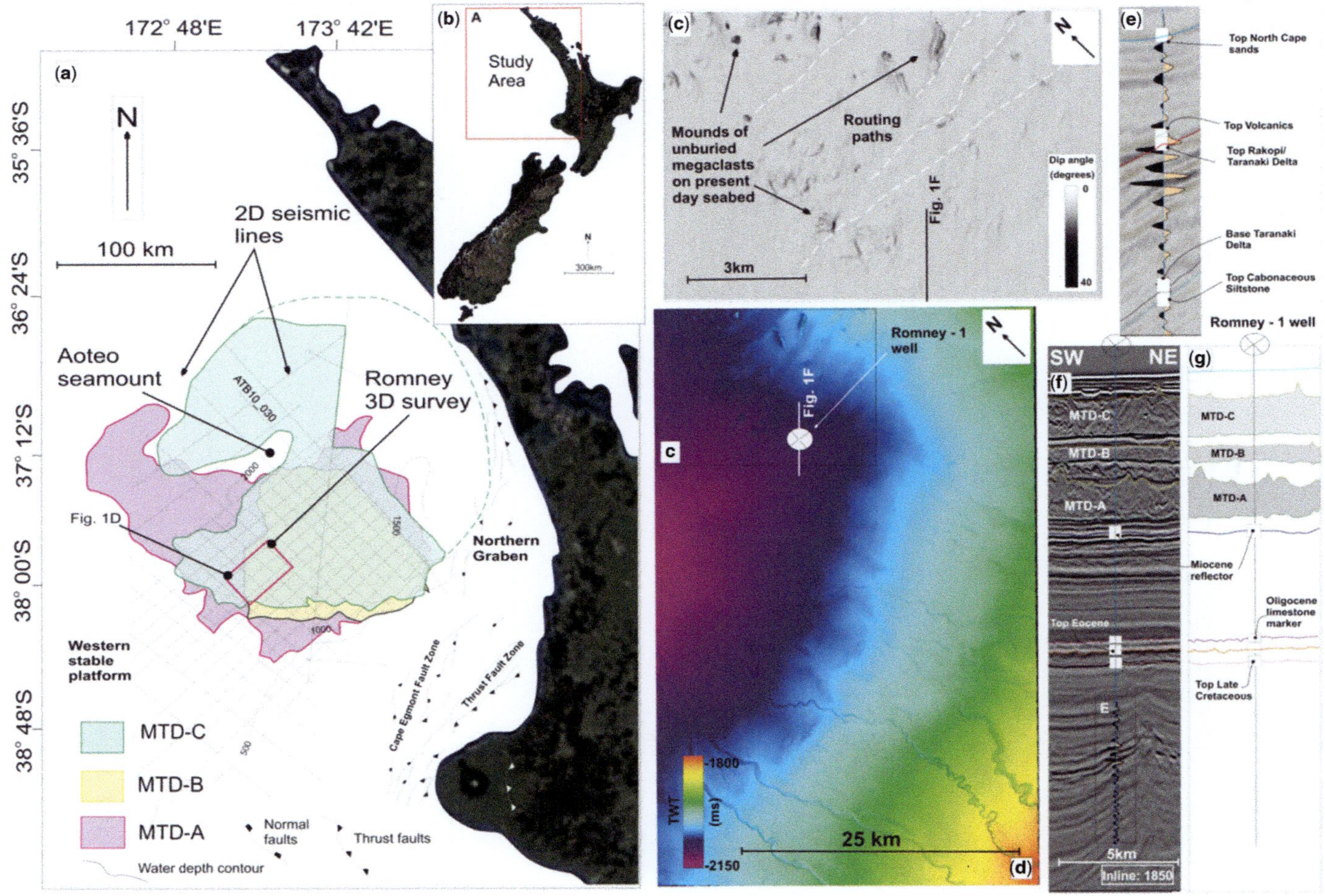

Fig. 1. (**a**) Location map of the study area (Taranaki Basin) and dataset – Romney 3D survey, showing the outline of the identified MTDS (MTD-A, -B and -C) determined by both the 2D and 3D seismic data. The extent of all identified MTC has been inferred (after Omeru 2014). Structural elements of Taranaki Basin are from King and Thrasher (1996). (**b**) Google Earth adaptation of New Zealand map, showing the extent of the study area. (**c**) Dip angle map of the seafloor showing peaks of protruding megaclasts. (**d**) Time structure map of the seafloor showing conspicuous basin floor channels. (**e**), (**f**) Romney well seismogram section with inset of the well section with formation tops and (**g**) geoseismic section of the Romney well with the stratigraphic intervals of the study.

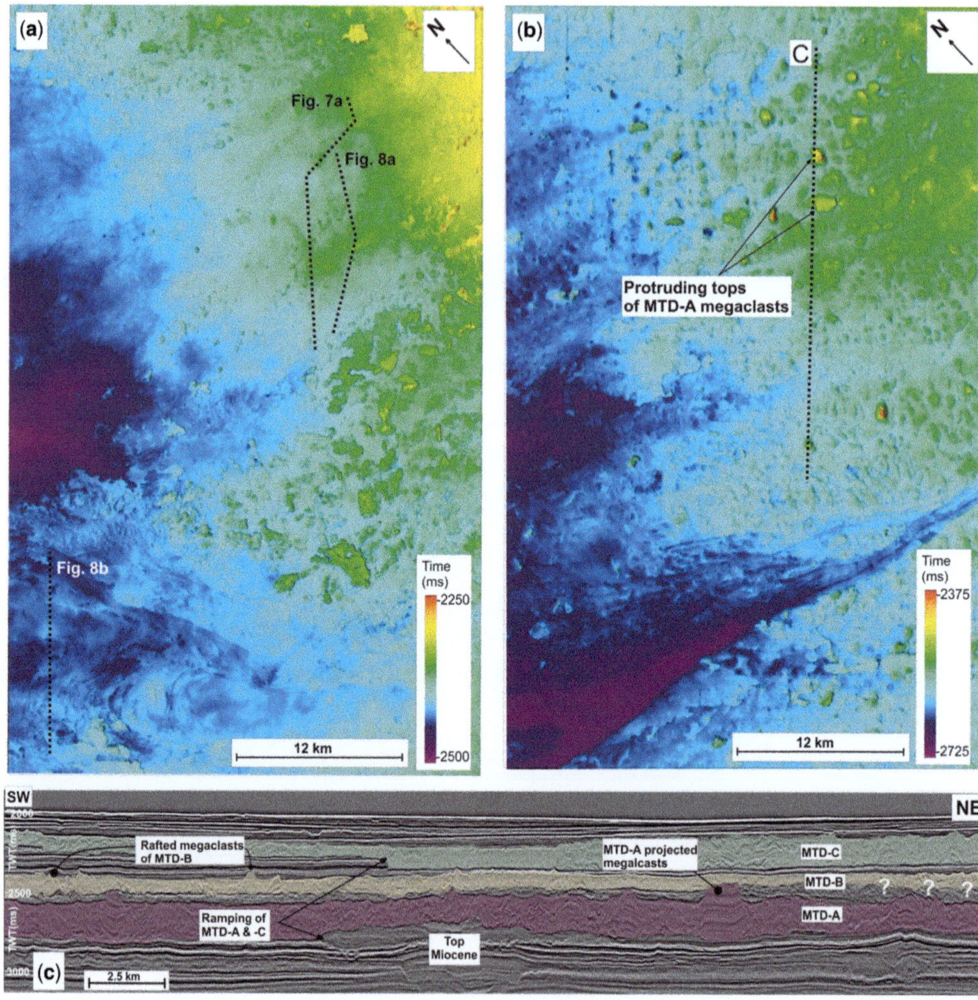

Fig. 2. (**a**) and (**b**) top surface structural maps of MTD-B and -A displayed, with surface map of MTD-A highlighting a planform view of the megaclasts' distribution. (**c**) Generalized stratigraphy showing the studied MTC.

the dominant frequency of 41 Hz for the entire data, the high frequency ranges represented by the blue and green colour blend, are associated with thicker areas of the MTC; whilst the green/red colour blends are restricted to thinner areas.

Results

Megaclasts within MTDs

Megaclasts within MTD-A are clustered towards the NE to SW limits of the Romney 3D seismic reflection dataset (Fig. 2b). Megaclasts within MTD-B are dominantly found in the SW distal area (Fig. 2a); megaclasts of MTD-C are mostly restricted to the NE proximal area (Fig. 3a). In the NE part of the study area, the top surface of MTD-A and the base of MTD–B coalesce, making it challenging to distinguish them (Fig. 3b). The amalgamation of surfaces between MTD-A and MTD-B show a discontinuous and irregular reflection character due to megaclasts in the upper parts of MTD-A, which created a rugose palaeo-topography. In contrast, the basal surface of MTD-C in the NE shows a continuous high negative amplitude reflection character (Fig. 3b). Here we describe the internal character, morphology and occurrence of the megaclasts.

Description of MTD-A megaclasts. MTD-A megaclasts are characterized by continuous parallel and mostly undeformed internal reflections, and are encased within a seismically chaotic facies (Figs 3b

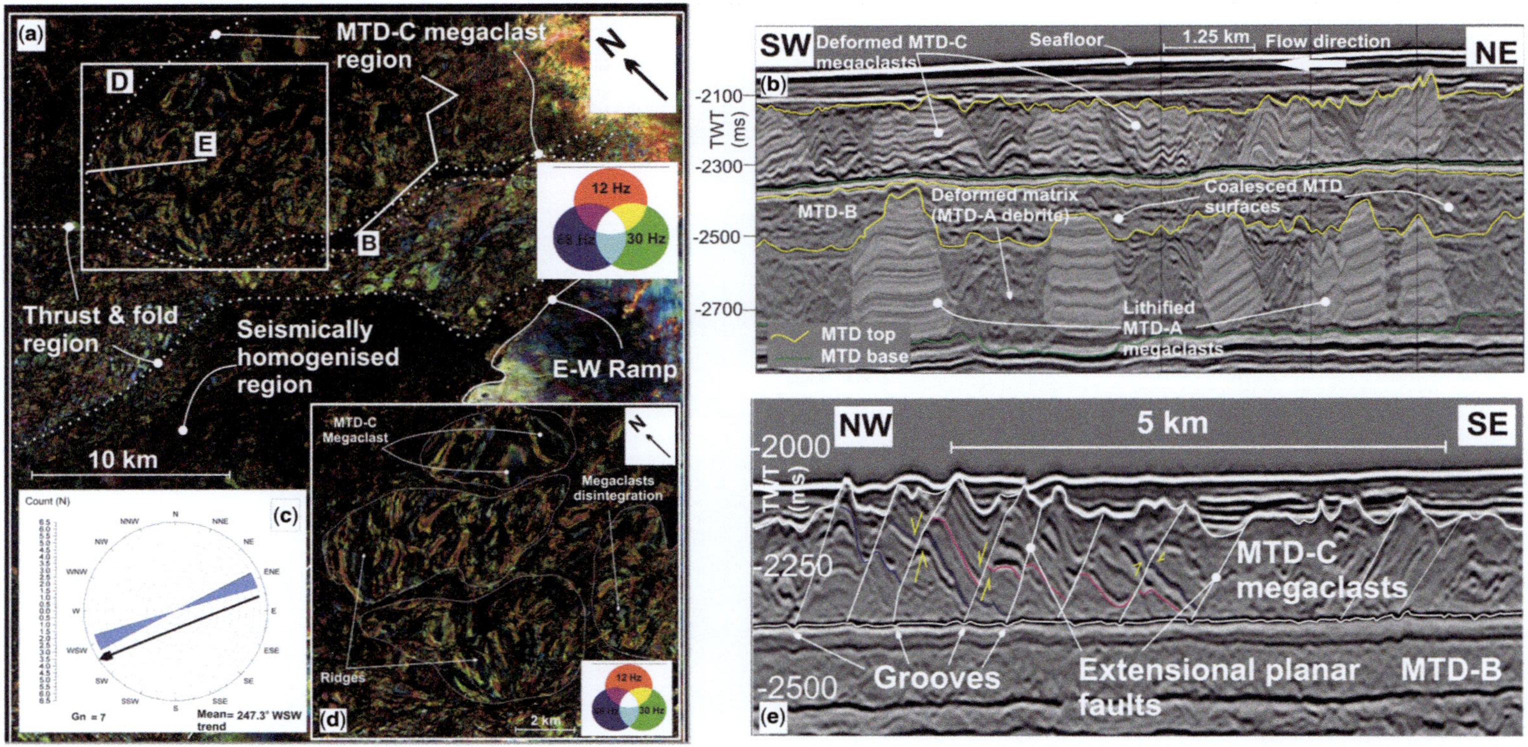

Fig. 3. (**a**) Frequency blend from 30 ms TWT above the basal surface of MTD-C showing the localized distribution of the MTD-C megaclasts adjacent to thrust folds and seismically homogenous zones. (**b**) Arbitrary sections illustrating morphology of megaclasts within the MTDs. MTD-A and -C have similar megaclast dimensions (hundreds of metres to kilometre-scale height and width measurements). Notice the deformed reflection packages and thrust structures in MTD-C megaclasts that are markedly different from the well-preserved parallel to sub-parallel configuration in megaclasts of older MTD-A. (**c**) Rose diagram plot of the identified grooves on MTD-C basal shear surface. (**d**) Spectral decomposition portion above the basal surface of MTD-C highlighting the circular to sub-circular ridges and disintegrated megaclasts. (**e**) Extensional fault-bounded megaclasts within MTD-C, indenting v-shaped marks on its basal surface. The annotated reflectors in colours have been used to illustrate the hanging and footwalls.

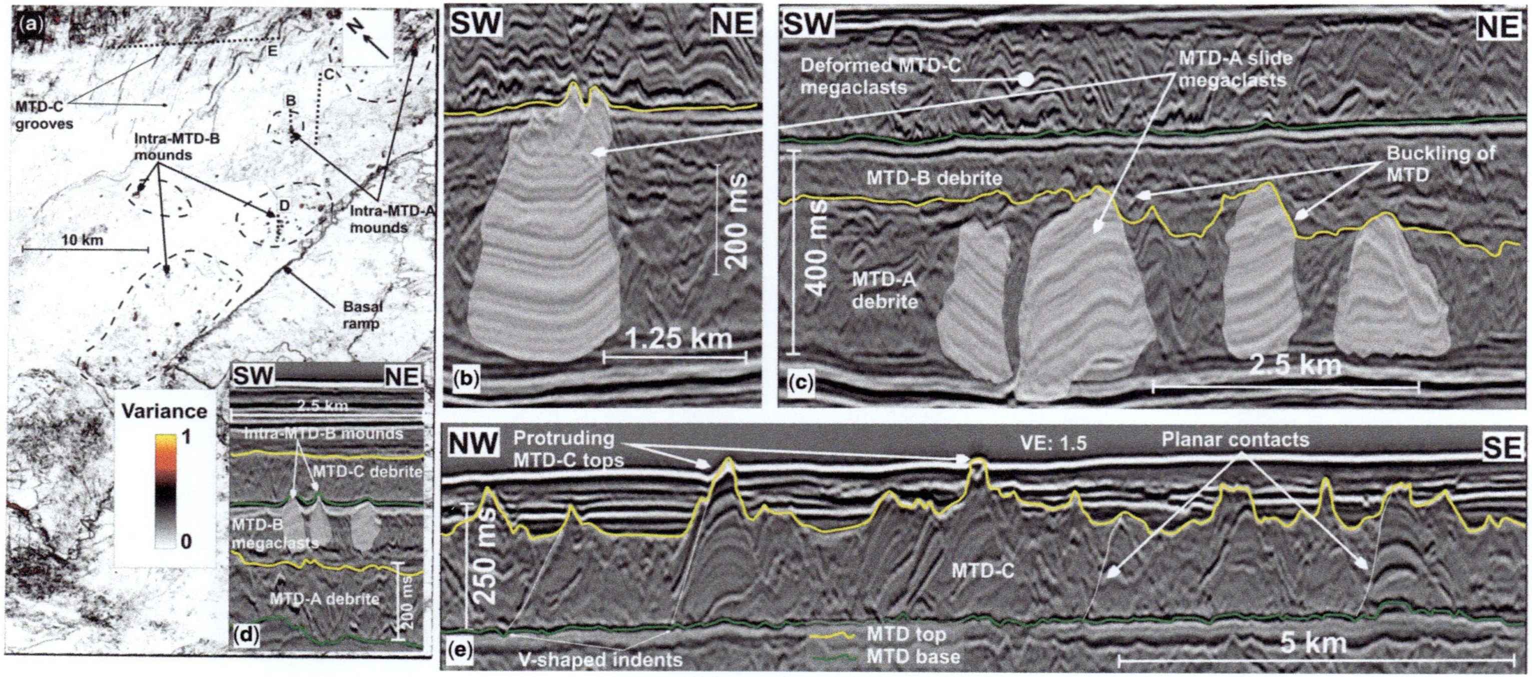

Fig. 4. (**a**) Variance attribute map of the basal surface of MTD-C showing prominent grooves. Intra-MTD mounds of MTD-A and -B megaclasts are seen protruding the base of MTD-C. Notice the prominent east–west-trending basal ramp and the clustering of these subsurface mounds around it. (**b**) Cross-section of intra-MTD mound of MTD-A megaclasts. (**c**) Buckling of undifferentiated MTC against the megaclasts within older MTD-A. (**d**) Cross-section of intra-MTD mound of MTD-B megaclasts. (**e**) Cross-section view of protruding megaclasts of MTD-C, with indented v-shaped marks on its basal surface.

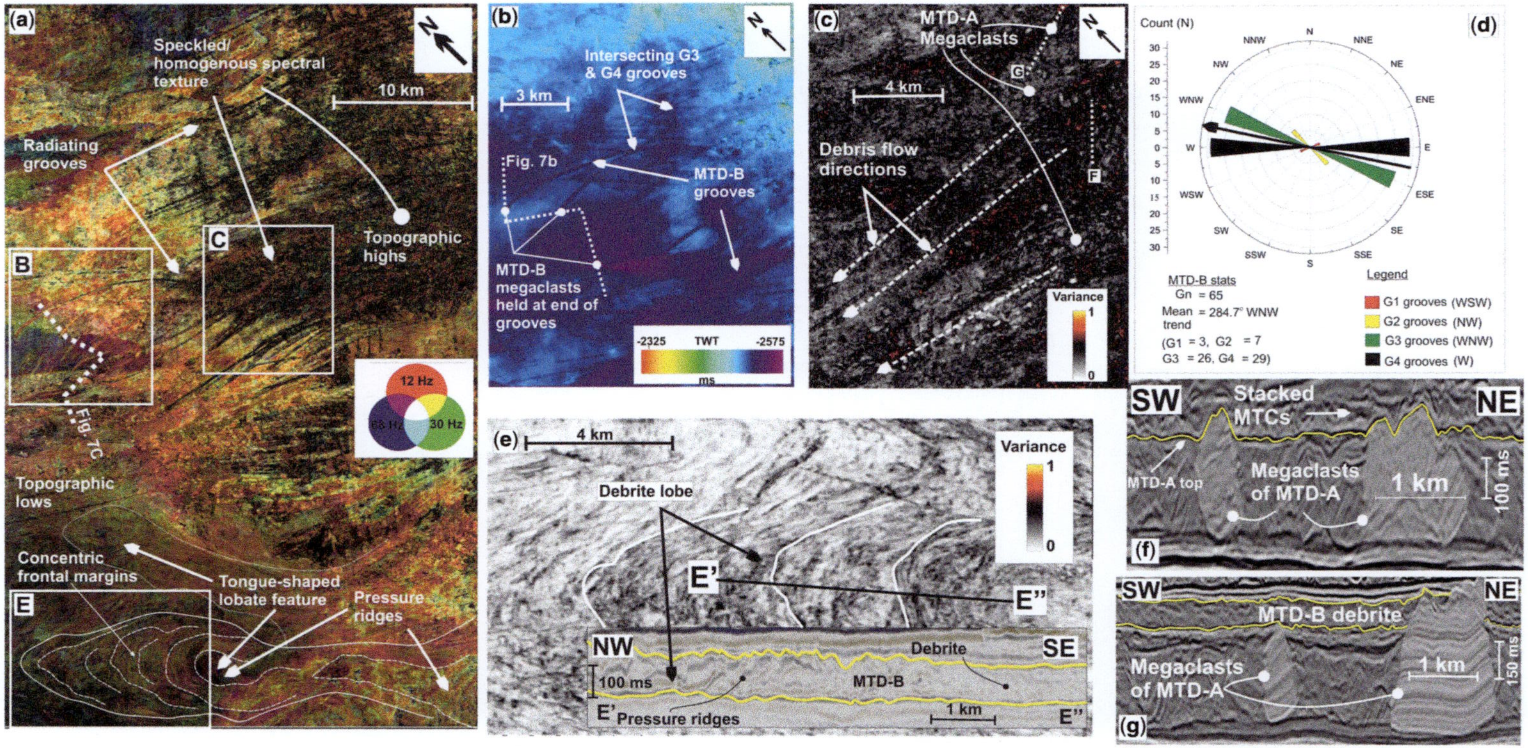

Fig. 5. (**a**) Spectral decomposition amplitude blending on MTD-B base showing kilometre-long (up to 40 km) grooves with associated rafted megaclasts. A prominent tongue-shaped lobate feature is resolved in the distal end. (**b**) Zoom-in of the distinct groove sets intersecting with each other, imaging the groove makers – MTD-B rafted megaclasts, at the end of the grooves. (**c**) Debris flow direction illustrating exit from confinement, represented by the linear features underlying the megaclasts' topography. (**d**) Rose diagram plot of the cross-cutting trends for the different groove sets on the basal surface of MTD-B. (**e**) Inset of the debrite lobe, showing the configuration of the identified pressure ridges in cross-section. (**f**), (**g**) Line sections from (**c**) illustrating the confining topography of MTD-A megaclasts.

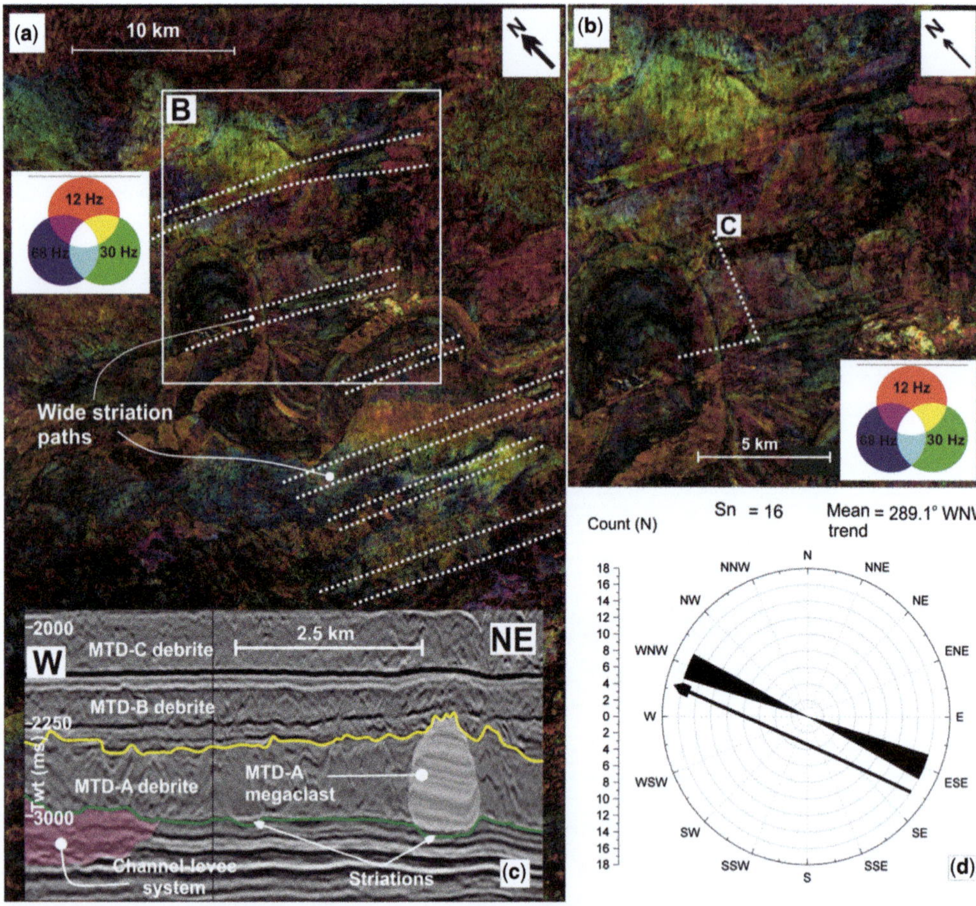

Fig. 6. (**a**) Spectral decomposition of MTD-A base with wide striation paths. (**b**) Inset of the striations showing prominent streaks of green colour blends within the some of the paths. We interpret this to be evidence of substrate erosion as a result of the tooling by MTD-A megaclasts. (**c**) Line section from (**b**) with the large lithic tool in place that created the observed striation; relative to the Miocene channel-levee system. (**d**) Rose diagram of the MTD-A striation trends, dominantly oriented toward the NW.

& 4b, c). Megaclasts within MTD-A are larger than in the other MTDs, measuring between 700 and 1300 m in width and 350–650 m in height, and are angular to sub-angular with sharp contacts between them and the surrounding chaotic seismic facies (Figs 3b, 4b & 7a). Some MTD-A megaclasts are tall enough to remain unburied by MTD-A, protruding through the palaeo-seafloor as evidenced by onlap by younger material including the overlying younger MTD-B (Figs 4b & 7a). In the SW part of the study area, large, internally coherent megaclasts overlie an eroded channel-levee system (Fig. 6b, c).

Interpretation of MTD-A megaclasts. The angular to sub-angular nature and undeformed internal stratigraphy suggests that these megaclasts consisted of sufficiently well-consolidated sediment prior to being entrained and transported (Gee *et al.* 2006; 2007; Hodgson *et al.* 2018). The distribution of megaclasts over and around the channel-levee complex, and within the chaotic seismic facies interpreted to be debrite (i.e. relatively homogenized), may suggest that the megaclasts were entrained into the flow locally (Ortiz-Karpf *et al.* 2017; Hodgson *et al.* 2018); however, given the thickness of the megaclasts, it is more likely that they originated in headwall regions and were deposited over the channel-levee system due to differences in substrate erodibility. The MTD-A megaclasts, which protrude through the top of the encasing MTD, are supported by the encasing matrix and are interpreted to have been carried within a laminar flow during transport (Ortiz-Karpf *et al.* 2015).

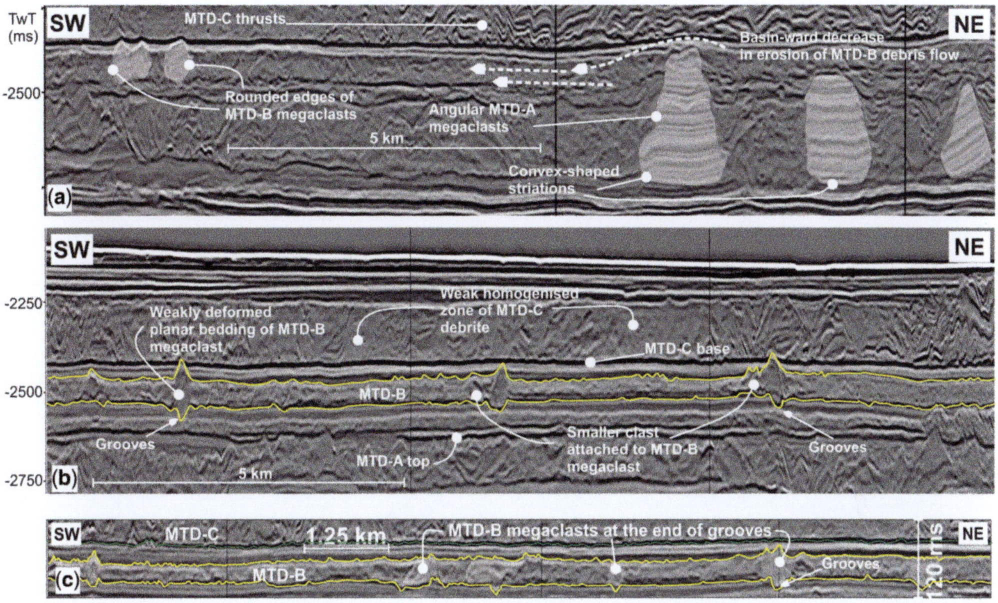

Fig. 7. (**a**) Line section (see Fig. 2a for location) illustrating the basin-ward decrease of the erodibility of MTD-B basal shear surface, showing more rounded edges encased within distinct bounding surfaces of the debris flow as opposed to coalesced surfaces in proximal areas. Notice the angular shapes of MTD-A megaclasts, which, at 5× vertical exaggeration, makes them appear taller than they are wide. (**b**) Line section showing wider v-shaped features, interpreted to be grooves, indented by diamond-shaped moderately deformed megaclasts of MTD-B, held within the debrite but spaced almost evenly apart. See Figure 2a for line location. Notice also the associated smaller clasts adjacent to MTD-B megaclasts; we interpret these to result from abrasive processes. (**c**) Detail of the morphology of MTD-B megaclasts' characteristics at the end of the grooves towards the distal limits.

Description of MTD-B megaclasts. MTD-B megaclasts are between 102 and 178 m wide and 100–180 m in height, and are clustered around the prominent east–west-trending ramp on the basal surface of MTD-C (Fig. 4a). Megaclasts within MTD-B show weakly deformed planar internal reflections (Fig. 7b). Unlike MTD-A megaclasts, there are no parallel or subvertical contacts in the seismically chaotic regions between MTD-B megaclasts (Fig. 7b, c). Some smaller megaclasts are observed attached, or in close proximity, to the largest megaclasts (Fig. 7b). The MTD-B megaclasts mostly contain rounded edges and are distributed in the NW and SW parts of the study area (Figs 5a & 7a). They are observed at the end of grooves in the underlying substrate and are covered by the chaotic seismic facies of the MTD (Fig. 7b, c). Adjacent to these grooves in the distal southwestern limits of the dataset is a prominent lobate feature, which is resolved by spectral-decomposition of seismic frequency data (Fig. 5a). The frequency-blended image shows that the lobate feature is characterized by relatively continuous but mixed blends of red and green; in cross-section a series of steeply dipping features correspond with arcuate ridges on the upper surface (Fig. 5e, cross-section) (Fig. 5e). MTD-B has buckled onlap terminations adjacent to MTD-A megaclast topography (Figs 4c & 8b).

Interpretation of MTD-B megaclasts. The rounded edges of these megaclasts and their association with smaller megaclasts suggest that they were abraded and partly broken up during transport by clast-to-clast interaction and abrasion within the flow matrix (e.g. Mulder and Alexander 2001; McGilvery *et al.* 2004; Jackson 2011; Hodgson *et al.* 2018). The seafloor expression of MTD-A megaclasts (Fig. 5f, g) potentially inhibited basin-ward advancement of MTD-B. These seafloor protrusions may have contributed to the disaggregation of megaclasts in MTD-B as the flow interacted with the topography during flow runout (e.g. Rusconi 2017; Hodgson *et al.* 2018). We interpret the lobate feature to represent the toe-domain of the MTD and the arcuate structures on the upper surface as pressure ridges. These ridges are interpreted to represent frictional freezing and deceleration, which caused compression within the flow and inhibited further downslope movement (Prior *et al.* 1984; Nissen *et al.* 1999; Posamentier and Kolla 2003;

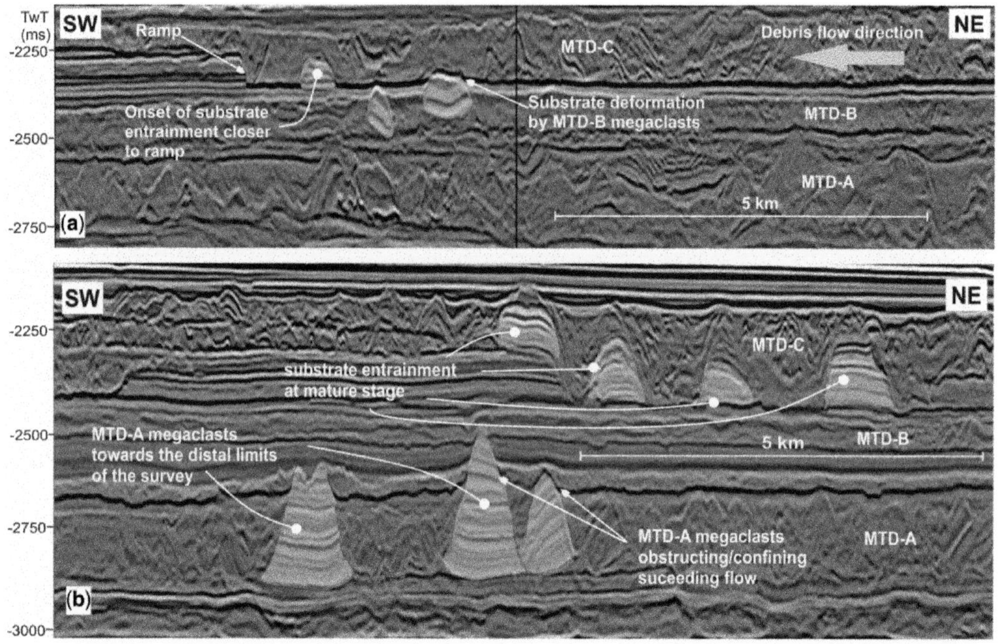

Fig. 8. (**a**) Line section illustrating MTD-C substrate deformation and early onset of substrate entrainment of MTD-B megaclasts closer to the prominent east–west-trending ramp. See Figure 2a for location of seismic cross-section line. (**b**) Another line section from Figure 2a towards the distal limits of the survey, showing MTD-C megaclasts' substrate entrainment now at a mature stage; and megaclasts of MTD-A, obstructing or confining later flows. Notice the onlap termination of the succeeding MTDs.

Moscardelli *et al.* 2006; Bull *et al.* 2009). Frey-Martínez *et al.* (2005, 2006) observed similar pressure ridges in frontally emergent MTDs (Fig. 5a, e). The buckled stratigraphy of MTD-B, adjacent to the protruding megaclasts of MTD-A, is interpreted to be related to localized compression within the flow as it interacted with seafloor protrusions.

Description of MTD-C megaclasts. Megaclasts within MTD-C are between 250–400 m across and 324–370 m high. They are characterized by mostly low- to moderate-amplitude internal reflections. MTD-C megaclasts lie within a matrix characterized by inclined sub-circular planform ridges, separated by parallel or subvertical extensional planar contacts (Figs 3d, e & 4e). Similar to megaclasts within MTD-A, the megaclasts within MTD-C formed a topography that remained unburied by subsequent sedimentation and are thus exposed on the present-day seafloor (Figs 3e & 4e). Internally, these megaclasts feature discontinuous and sometimes folded reflectors in a deformed matrix (Fig. 3b). The MTD-C megaclasts located in the WNW are slightly less deformed than those in the NE (Fig. 3e) where they are thickest (up to 324 m) and more densely concentrated (Fig. 3e). Linear features on the base of the MTD trend ENE–WSW (Fig. 3c).

Interpretation of MTD-C megaclasts. The variability of deformed to weakly deformed megaclasts within MTD-C and the imbricated sequence of folds, ridges and extensional planar contacts within and between the megaclasts all represent evidence of megaclasts associated with retrogressive failure not far from the origin (Frey-Martínez *et al.* 2006; Bull *et al.* 2009; Hodgson *et al.* 2018; Normandeau *et al.* 2019). Additionally, the dense clustering of megaclasts in the NE (Fig. 3a) most likely relates to flow impingement against the adjacent thrust blocks, which inhibited further downslope movement (Omeru and Cartwright 2019). This seafloor relief resulted in the formation of basal surface linear features trending ENE–WSW (Fig. 3c).

Basal surface grooves and striations

Description of MTD-A basal surface features. Linear and curvilinear features are observed on the bases of each of the MTDs (Figs 4–6). On the basal surface of this MTD there are sixteen 4–15 km long, 2–3 km wide features. These elongate features have a mean orientation of 109–289° (ESE–WNW), are mostly sub-parallel and do not diverge downdip (Fig. 6a, b & d). In cross-section, they appear as convex-shaped flat-bottomed features (Figs 6c & 7a). In the frequency blended images some of the more deeply-incised linear features show streaks of green (representing *c.* 30 Hz frequencies) confined in the linear features (Fig. 6b).

Interpretation of MTD-A basal surface features. These elongate and wide features are associated with internally undeformed megaclasts within MTD-A, and are interpreted to be striations given their wider and mostly flat-bottomed impressions in cross-section (Gee *et al.* 2005). Striations recognized in other studies also tend to be sub-parallel to each other and lack downslope divergence (Gee *et al.* 2005, 2006, 2007). The predominantly WNW orientation of the striations on the MTD-A basal surface (Fig. 6d) supports this assumption. The green streaks within the frequency blended images (Fig. 6b, c) represent deeper erosion into the substrate.

Description of MTD-B basal surface features. On the basal surface of MTD-B, in the proximal parts, towards the NE of the study area, the frequency blended data show a discontinuous speckled to homogeneous green response with linear features radiating towards the west (Fig. 5a). The linear features found on the base of MTD-B are between 8 and 40 km long and 200–250 m wide, and v-shaped in cross-section (Fig. 7b, c). Sixty-five linear to curvilinear features comprising four distinct sets of lineations are observed on the base of MTD-B; they may diverge from being parallel and sometimes cross-cut (Fig. 5a, b & d). The mean trend for the observed features is 105–285° (ESE–WNW) (Fig. 5d).

Interpretation of MTD-B basal surface features. The grooves of MTD-B are similar in width and character to those of the debris flow in Kitimat fjord of British Columbia (Prior *et al.* 1984) and on the Nigerian continental slope (Nissen *et al.* 1999). Nissen *et al.* (1999) referred to these tracks as 'glide tracks' and defined them as records of the path of large blocks of sediment *detached* from the leading edge of the mass flow during downslope translation. However, we interpret the observed tracks to have been created by megaclasts within the laminar flow of MTD-B and find that they are v-shaped in cross-section (Bull *et al.* 2009; Ortiz-Karpf *et al.* 2015); we interpret them to be erosional *grooves* incised into the substrate, by tools (megaclasts) carried at the base of a laminar flow (Fig. 7). The surface rugosity of the underlying substrate of MTD-A gives the speckled to homogenous spectral texture in the frequency blend images. Grooves are useful as transport direction indicators (Bull *et al.* 2009). Unlike striations, grooves can record down-system flow divergence (Posamentier and Kolla 2003; Gee *et al.* 2005). Thus, the intersecting orientations of the distinct set of grooves on the MTD-B basal surface (Fig. 5b, d) suggests the radial divergence of megaclasts carried along in the flow. The divergence may be a product of non-uniform frictional forces as the megaclasts gouge into the substrate.

Description of MTD-C basal surface features. Elongate v-shaped cross-sectional marks on the basal surface of MTD-C (Fig. 3e) are shorter in planform than those of MTD-A and -B, measuring about 2–6 km long and 100–200 m wide (Fig. 4a). The mean orientation of seven of the identified linear features is 67–247° (ENE–WSW) (Fig. 3c).

Interpretation of MTD-C basal surface features. Omeru and Cartwright (2019) concluded that the observed lineations on MTD-C basal surface elsewhere are not erosional grooves but instead may represent seismic artefacts; this was based on their orientation in which the grooves were not aligned with the inferred transport direction of the MTD. Based on the v-shaped expressions as observed in cross-section along the length of these features (up to 6 km), we interpret these features not as artefacts but as erosional grooves (Bull *et al.* 2009). The poorly imaged MTD-C grooves may be related to differences in the degree of consolidation of the substrate and/or the size of the well-consolidated blocks held within the flow (Posamentier and Kolla 2003).

Discussion

Association of grooves, groove makers and striations

Although the formative processes and character of grooves differ from striations, they both provide very useful kinematic-boundary layer information (Gee *et al.* 2005; Bull *et al.* 2009; Butler *et al.* 2016). Grooves and striations are associated with megaclasts, evidenced by megaclasts found either *in situ* within grooves or having similar dimensions to the grooves or striations (Posamentier and Kolla 2003; Gee *et al.* 2005; Bull *et al.* 2009). The MTD-B grooves are formed by moderately sized and partially deformed megaclasts (Fig. 7b, c). These grooves are amongst the largest and longest recorded in the subsurface (Table 1). Previous studies have inferred that these lineations are erosional grooves, even though megaclasts were not found within the grooves (Posamentier and Kolla 2003; Bull *et al.* 2009; Ortiz-Karpf *et al.* 2015). These studies attribute the absence of the clasts at the base of the MTD to be due to disaggregation or lifting within the flow matrix away from the flow base (Posamentier and Kolla 2003; McGilvery *et al.* 2004). However, the identification of the megaclasts at the end of the grooves as imaged in this study (Fig. 7b, c)

Table 1. *Comparative track length dimensions of documented subsurface erosional features from published case studies worldwide*

Case study	Survey area (km^2)	Location	Subsurface erosional feature documented	Track length (km)
Nissen *et al.* (1999)	500	Continental margin, offshore Nigeria	Glide tracks	Up to 12
Posamentier and Kolla (2003)	70	Offshore eastern Borneo, Indonesia	Grooves	>20
Gee *et al.* (2005) (Angola landslide)	430	Offshore Angola	Striations	9
Moscardelli *et al.* (2006)	2017	Offshore Trinidad and Venezuela	Megascours	60–70
Gee *et al.* (2007) (Brunei slide)	5300	Offshore Brunei	Striations	30–120
Bull *et al.* (2009) (Storegga slide)	44 000	Norwegian continental margin	Grooves	30
Ortiz-Karpf *et al.* (2015) (Magdalena fan)	*c.* 470	Offshore Columbia	Grooves	Up to 7
Scarselli *et al.* (2016) (Namibian megaslide)	2417	Offshore Namibia	Linear headscarp scours	Up to 12
Soutter *et al.* (2018) (Halibut slide)	*c.* 7000	Central North Sea	Megascours, glide tracks	*c.* 70 up to 10
Omeru and Cartwright (2019) (MTD C)	590	Offshore deep-water Taranaki Basin, New Zealand	Grooves	Up to 6
This study Composite MTC – MTD B)	1925	Offshore deep-water Taranaki Basin, New Zealand	Grooves	8–40
This study Composite MTC – MTD A)	1925	Offshore deep-water Taranaki Basin, New Zealand	Striations	4–15

shows unequivocally that they were carried within the debris flow and were the tools that formed the grooves during the submarine landslide.

Hodgson *et al.* (2018) suggest that different sizes of megaclasts within single deposits may be related to their compositional differences and transport distances. Moreover, the degree of internal deformation within megaclasts can provide an indication of transport distance from the source (Bull *et al.* 2009; Jackson 2011); the results of this study support these assertions. Undeformed MTD-A megaclasts, which were most probably entrained from the headwall, are larger than more deformed MTD-B megaclasts that were transported over longer distances. In addition, local findings from this study suggest a possible link between megaclast size, their degree of deformation and the morphology of their associated basal erosional features (Posamentier and Kolla 2003). The internally coherent megaclasts within MTD-A are associated with wider convex-shaped flat-bottomed striations (Fig. 6c, 7a). Partially deformed and abraded megaclasts within MTD-B were associated with less wide, longer but v-shaped grooves (Fig. 7b, 7c); while structurally deformed megaclasts within MTD-C were associated with narrower and shorter v-shaped grooves (Fig. 3e, 4a). Thus, we present the possibility of generating a continuum of basal surface (drag marks) morphologies as a function of megaclast deformation during runout.

Protruding megaclast influence on succeeding flows

Large megaclasts are thicker than the coeval debrites that they are preserved within and consequently formed seafloor relief. These protruding mounds may have contributed to subsequent megaclast disintegration and distribution (Bull *et al.* 2009; Normandeau *et al.* 2019) (Fig. 4b). The submarine landslide that resulted in MTD-B interacted with the topography created by megaclasts within MTD-A through the process of abrasion, producing rafted megaclasts (Figs 4c, 7b & 8b) (Mulder and Alexander 2001; McGilvery *et al.* 2004; Hodgson *et al.* 2018). The protruding megaclasts resulted in buckling of the MTD-B onlap against MTD-A megaclasts (Fig. 4c). Such protruding megaclasts have been referred to as 'shield blocks' by Rusconi (2017), based on their exposed height on the palaeo-seafloor before and/or after being draped by 'healing phase deposits' (Posamentier and Allen 1993).

Substrate deformation and lateral loading are possible preconditioning mechanisms that have been linked to the entrainment of megaclasts from the substrate (Moscardelli *et al.* 2006; Hodgson *et al.* 2018). Large clusters of protruding megaclasts carried in MTD-B are observed around the east–west-trending ramp on the basal surface of MTD-C (Figs 3a & 4a). The MTD-C submarine landslide interacted with these megaclast clusters resulting in partial entrainment into the MTD-C flow (Fig. 8a, b). A simple model is presented for the interaction of successive but closely related MTDs, as a function of the megaclasts carried within the parent flows (Fig. 9).

Conclusions

We document kilometres-long (up to 40 km) grooves formed by megaclasts, carried in submarine landslides, scoured deep into the substrate of the deep-water Taranaki Basin, New Zealand. We demonstrate a range of linear and curvilinear features found on the basal surface of three MTDs and show that these are related to scour by megaclasts within the submarine landslides that transported and deposited them. We show that the spatial relationship of the grooves and megaclasts indicate that the megaclasts within debris flows can erode the substrate, influence the flow that transports them, and, if they form seafloor topography, can influence subsequent flows and their

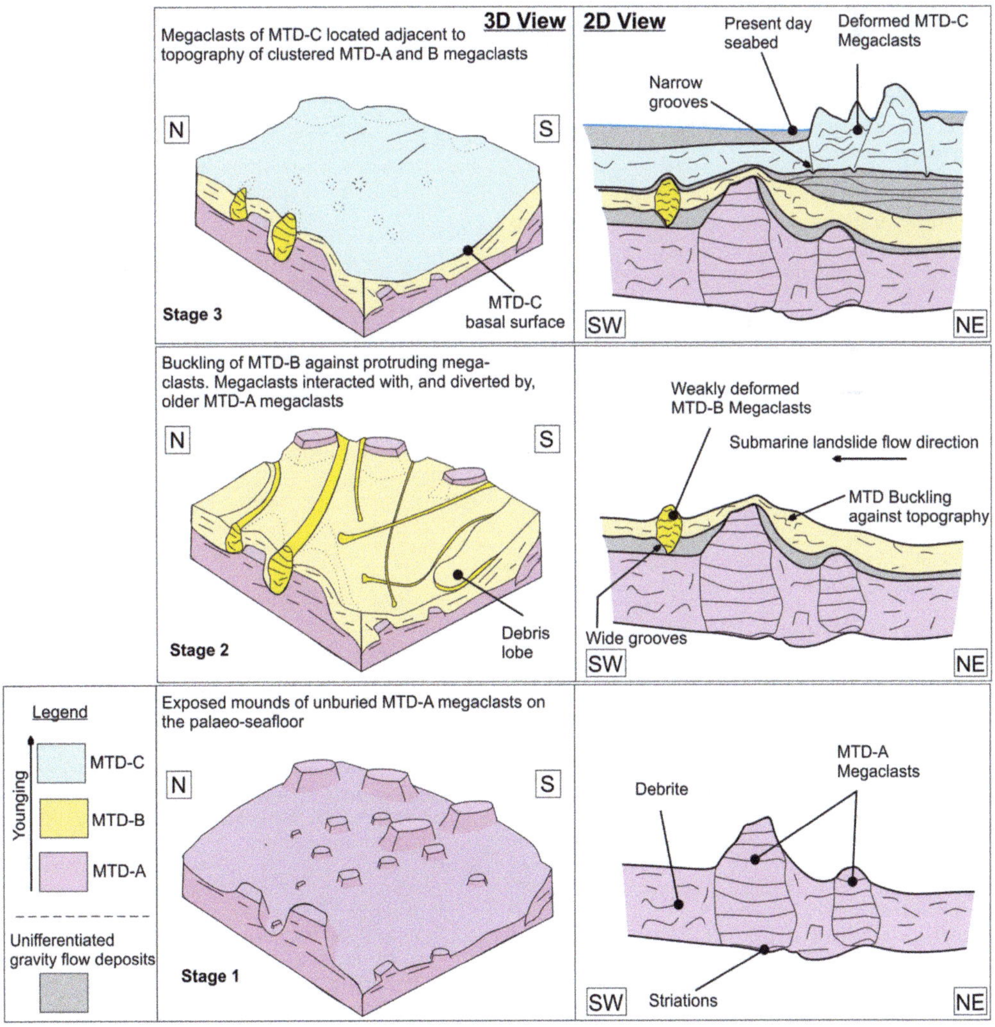

Fig. 9. Schematic model/illustration based on the results of this study, showing the possible successive interactions between megaclasts held within laminar flow.

deposits. MTD-A features megaclasts that have coherent parallel strata that formed sub-parallel striations 4–15 km long and 2–3 km wide, with megaclasts forming seafloor relief, which was onlapped by younger sediments. The seafloor expression of these megaclasts partially obstructed MTD-B, which contains megaclasts with v-shaped grooves of 8–40 km long and 200–250 m wide, extending beyond the main MTD body. These megaclasts incised through the rugose topography of the underlying MTD-A forming lineations on the basal surface of MTD-B, which suggest radial flow expansion where the flow exited topographic confinement. MTD-C features grooves 2–6 km long and 100–200 m wide. These grooves terminate with megaclasts that internally are characterized by contorted reflections surrounded by a chaotic matrix. This study, for the first time, links megaclasts directly to the grooves and striations that they formed, and demonstrates that markedly different styles of scouring and resultant grooves can occur in closely spaced MTDs according to their runout length and induration of the megaclasts.

Acknowledgements This work was conducted in the Sedimentary Basins Research lab at the University of Manchester, UK, using software kindly provided by Schlumberger (Petrel) and ffA (Geoteric). The data were provided by GNS Science. We are very grateful for the insightful suggestions from reviewers David Hodgson, Suzanne Bull and Leslie Wood, to improve this contribution. We also acknowledge the volume editor, Dr Mike Clare, for the opportunity to contribute to this Special Publication.

Funding This work was supported by the Niger Delta Development Commission, with grant NDDC/DEHSS/2016PGFS/ABS/PHD/003 awarded to Jefferson Nwoko.

Author contributions **JN**: writing – original draft (lead); **IK**: supervision (lead), writing – review and editing (lead); **MH**: validation (lead), writing – review and editing (supporting).

References

Anell, I. and Midtkandal, I. 2017. The quantifiable clinothem–types, shapes and geometric relationships in the Plio-Pleistocene giant foresets formation, Taranaki Basin, New Zealand. *Basin Research*, **29**, 277–297, https://doi.org/10.1111/bre.12149

Baur, J., Sutherland, R. and Stern, T. 2014. Anomalous passive subsidence of deep-water sedimentary basins: a prearc basin example, southern New Caledonia Trough and Taranaki Basin, New Zealand. *Basin Research*, **26**, 242–268, https://doi.org/10.1111/bre.12030

Bull, S., Cartwright, J. and Huuse, M. 2009. A review of kinematic indicators from mass transport complexes using 3D seismic data. *Marine and Petroleum Geology*, **26**, 1132–1151, https://doi.org/10.1016/j.marpetgeo.2008.09.011

Butler, R.W., Eggenhuisen, J.T., Haughton, P. and McCaffrey, W.D. 2016. Interpreting syndepositional sediment remobilization and deformation beneath submarine gravity flows; a kinematic boundary layer approach. *Journal of the Geological Society, London*, **173**, 46–58, https://doi.org/10.1144/jgs2014-150

Dakin, N., Pickering, K.T., Mohrig, D. and Bayliss, N.J. 2013. Channel-like features created by erosive submarine debris flows: field evidence from the Middle Eocene Ainsa Basin, Spanish Pyrenees. *Marine Petroleum Geology*, **41**, 62–71, https://doi.org/10.1016/j.marpetgeo.2012.07.007

Dykstra, M., Garyfalou, K. *et al.* 2011. Mass-transport deposits: combining outcrop studies and seismic forward modelling to understand lithofacies distributions, deformation, and their seismic expression. *SEPM Special Publication*, **96**, 293–310.

Frey-Martínez, J., Cartwright, J. and Hall, B. 2005. 3D seismic interpretation of slump complexes: examples from the continental margin of Israel. *Basin Research*, **17**, 83–108, https://doi.org/10.1111/j.1365-2117.2005.00255.x

Frey-Martínez, J., Cartwright, J. and James, D. 2006. Frontally confined v. frontally emergent submarine landslides: a 3D seismic characterisation. *Marine and Petroleum Geology*, **23**, 585–604, https://doi.org/10.1016/j.marpetgeo.2006.04.002

Gee, M.J.R., Gawthorpe, R.L. and Friedmann, J.S. 2005. Giant striations at the base of a submarine landslide. *Marine Geology*, **214**, 287–294, https://doi.org/10.1016/j.margeo.2004.09.003

Gee, M.J.R., Gawthorpe, R.L. and Friedmann, S.J. 2006. Triggering and evolution of a giant submarine landslide, offshore Angola, revealed by 3D seismic stratigraphy and geomorphology. *Journal of Sedimentary Research*, **76**, 9–19, https://doi.org/10.2110/jsr.2006.02

Gee, M.J.R., Uy, H.S., Warren, J., Morley, C.K. and Lambiase, J.J. 2007. The Brunei slide: a giant submarine landslide on the North West Borneo Margin revealed by 3D seismic data. *Marine Geology*, **246**, 9–23, https://doi.org/10.1016/j.margeo.2007.07.009

Hampton, M.A., Lee, H.J. and Locat, J. 1996. Submarine landslides. *Reviews of Geophysics*, **34**, 33–59, https://doi.org/10.1029/95RG03287

Hansen, R.J. and Kamp, P.J. 2004*a*. Late Miocene to early Pliocene stratigraphic record in northern Taranaki Basin: condensed sedimentation ahead of Northern Graben extension and progradation of the modern continental margin. *New Zealand Journal of Geology and Geophysics*, **47**, 645–662, https://doi.org/10.1080/00288306.2004.9515081

Hansen, R.J. and Kamp, P.J. 2004*b*. Rapid progradation of the Pliocene-Pleistocene continental margin, northern Taranaki Basin, New Zealand, and implications. *2004 New Zealand Petroleum Conference Proceedings*, Crown Minerals, Ministry of Economic Development, Wellington, New Zealand, 1–9.

Hodgson, D.M., Brooks, H.L., Ortiz-Karpf, A., Spychala, Y., Lee, D.R. and Jackson, C.L. 2018. Entrainment and abrasion of megaclasts during submarine landsliding and their impact on flow behaviour. *Geological*

Society, London, Special Publications, **477**, 223–240, https://doi.org/10.1144/SP477.26

Jackson, C.A.-L. 2011. Three-dimensional seismic analysis of megaclast deformation within a mass transport deposit; implications for debris flow kinematics. *Geology*, **39**, 203–206, https://doi.org/10.1130/G31767.1

Joanne, C., Lamarche, G. and Collot, J.Y. 2013. Dynamics of giant mass transport in deep submarine environments: the Matakaoa Debris Flow, New Zealand. *Basin Research*, **25**, 471–488, https://doi.org/10.1111/bre.12006

Kamp, P.J., Vonk, A.J. *et al.* 2004. Neogene stratigraphic architecture and tectonic evolution of Wanganui, King Country, and eastern Taranaki Basins, New Zealand. *New Zealand Journal of Geology and Geophysics*, **47**, 625–644, https://doi.org/10.1080/00288306.2004.9515080

King, P.R. and Thrasher, G.P. 1996. *Cretaceous-Cenozoic Geology and Petroleum Systems of the Taranaki Basin, New Zealand. Monograph 13*. Institute of Geological & Nuclear Sciences Ltd., Lower Hutt, NZ, **2**.

Kneller, B., Dykstra, M., Fairweather, L. and Milana, J.P. 2016. Mass-transport and slope accommodation: implications for turbidite sandstone reservoirs. *AAPG Bulletin*, **100**, 213–235, https://doi.org/10.1306/09011514210

Lucente, C.C. and Pini, G.A. 2003. Anatomy and emplacement mechanism of a large submarine slide within a Miocene fore deep in the northern Apennines, Italy: A field perspective. *American Journal of Science*, **303**, 565–602, https://doi.org/10.2475/ajs.303.7.565

McGilvery, T.A. and Cook, D.L. 2003. The influence of local gradients on accommodation space and linked depositional elements across a stepped slope profile, offshore Brunei. *In*: *Shelf Margin Deltas and Linked Down Slope Petroleum Systems: 23rd Annual Research*. Society of Economic Palaeontologists and Mineralogists, 387–419, https://doi.org/10.5724/gcs.03.23.0387

McGilvery, T.A., Haddad, G. and Cook, D.L. 2004. Seafloor and shallow subsurface examples of mass transport complexes, offshore Brunei. *Proceedings, Offshore Technology Conference*, 3–6 May 2004, Houston, Texas, USA, article OTC-16780- MS.

Moscardelli, L., Wood, L. and Mann, P. 2006. Mass-transport complexes and associated processes in the offshore area of Trinidad and Venezuela. *AAPG Bulletin*, **90**, 1059–1088, https://doi.org/10.1306/02210605052

Mulder, T. and Alexander, A. 2001. The physical character of subaqueous sedimentary density flows and their deposits. *Sedimentology*, **48**, 269–299, https://doi.org/10.1046/j.1365-3091.2001.00360.x

Nissen, S.E., Haskell, N.L., Steiner, C.T. and Coterill, K.L. 1999. Debris flow out runner blocks, glide tracks, and pressure ridges identified on the Nigerian continental slope using 3-D seismic coherency. *The Leading Edge*, **18**, 595–599, https://doi.org/10.1190/1.1438343

Normandeau, A., Campbell, D.C., Piper, D.J. and Jenner, K.A. 2019. New evidence for a major late Quaternary submarine landslide on the external western levee of Laurentian Fan. *Geological Society, London, Special Publications*, **477**, 377–387, https://doi.org/10.1144/SP477.14

Ogata, K., Mutti, E., Pini, G.A. and Tinterri, R. 2012. Mass transport-related stratal disruption within sedimentary mélanges: examples from the northern Apennines (Italy) and south-central Pyrenees (Spain). *Tectonophysics*, **568–569**, 185–199, https://doi.org/10.1016/j.tecto.2011.08.021

Omeru, T. 2014. *Mass Transport Deposits: Implications for Reservoir Seals*. Phd thesis, Cardiff University.

Omeru, T. and Cartwright, J.A. 2019. The efficacy of kinematic indicators in a complexly deformed mass transport deposit: insights from the deep-water Taranaki Basin, New Zealand. *Marine and Petroleum Geology*, **106**, 74–87, https://doi.org/10.1016/j.marpetgeo.2019.04.037

Omeru, T., Cartwright, J.A. and Bull, S. 2016. Kinematics of submarine slope failures in the deepwater Taranaki Basin, New Zealand. *In*: Mosher, D.C., Craig Shipp, R. and Moscardelli, L. *et al.* (eds) *Submarine Mass Movements and their Consequences*. Springer, Cham, 61–70.

Omosanya, K.O. and Alves, T.M. 2013. Ramps and flats of mass-transport deposits (MTDs) as markers of seafloor strain on the flanks of rising diapirs (Espírito Santo Basin, SE Brazil). *Marine Geology*, **340**, 82–97, https://doi.org/10.1016/j.margeo.2013.04.013

Ortiz-Karpf, A., Hodgson, D.M. and McCaffrey, W.D. 2015. The role of mass-transport complexes in controlling channel avulsion and the subsequent sediment dispersal patterns on an active margin: the Magdalena Fan, offshore Colombia. *Marine and Petroleum Geology*, **64**, 58–75, https://doi.org/10.1016/j.marpetgeo.2015.01.005

Ortiz-Karpf, A., Hodgson, D.M., Jackson, C.A.L. and McCaffrey, W.D. 2017. Influence of seabed morphology and substrate composition on mass-transport flow processes and pathways: insights from the Magdalena Fan, offshore Colombia. *Journal of Sedimentary Research*, **87**, 189–209, https://doi.org/10.2110/jsr.2017.10

Posamentier, H.W. and Allen, G.P. 1993. Variability of the sequence stratigraphic model: effects of local basin factors. *Sedimentary Geology*, **86**, 91–109, https://doi.org/10.1016/0037-0738(93)90135-R

Posamentier, H.W. and Kolla, V. 2003. Seismic geomorphology and stratigraphy of depositional elements in deep-water settings. *Journal of Sedimentary Research*, **73**, 367–388, https://doi.org/10.1306/111302730367

Posamentier, H.W. and Martinsen, O.J. 2011. The character and genesis of submarine mass transport deposits: insights from outcrop and 3D seismic data. *In*: *Mass-Transport Deposits in Deepwater Settings*. SEPM (Society for Sedimentary Geology), 7–38, https://doi.org/10.2110/sepmsp.096.007

Prior, D.B., Bornhold, B.D. and Johns, M.W. 1984. Depositional characteristics of a submarine debris flow. *The Journal of Geology*, **92**, 707–727, https://doi.org/10.1086/628907

Rad, F. 2015. *PEP 38451 Romney-1 Well Completion Report*. New Zealand Unpublished Petroleum Report, **PR495**.

Rusconi, F.J. 2017. *3D Seismic Interpretation of a Plio-Pleistocene Mass Transport Deposit in the Deepwater Taranaki Basin of New Zealand*. Theses and Dissertations, 2583, University of Arkansas, http://scholarworks.uark.edu/etd/2583

Scarselli, N., McClay, K. and Elders, C. 2016. Seismic geomorphology of Cretaceous megaslides offshore Namibia (Orange Basin): insights into segmentation and degradation of gravity-driven linked systems. *Marine and Petroleum Geology*, **75**, 151–180, https://doi.org/10.1016/j.marpetgeo.2016.03.012

Sobiesiak, M.S., Kneller, B., Alsop, G.I. and Milana, J.P. 2018. Styles of basal interaction beneath mass transport deposits. *Marine and Petroleum Geology*, **98**, 629–639, https://doi.org/10.1016/j.marpetgeo.2018.08.028

Soutter, E.L., Kane, I.A. and Huuse, M. 2018. Giant submarine landslide triggered by Paleocene mantle plume activity in the North Atlantic. *Geology*, **46**, 511–514, https://doi.org/10.1130/G40308.1

Stagpoole, V. and Nicol, A. 2008. Regional structure and kinematic history of a large subduction back thrust: Taranaki Fault, New Zealand. *Journal of Geophysical Research: Solid Earth*, **113**, https://doi.org/10.1029/2007JB005170

Strogen, D.P., Seebeck, H., Nicol, A. and King, P. 2017. Two-phase rifting in the Taranaki Basin region, New Zealand: implications for Gondwana break-up. *Journal of the Geological Society, London*, **174**, 926–946, https://doi.org/10.1144/jgs2016-160

Tippett, J.M. and Kamp, P.J. 1995. Geomorphic evolution of the Southern Alps, New Zealand. *Earth Surface Processes and Landforms*, **20**, 177–192, https://doi.org/10.1002/esp.3290200207

Uruski, C. 2008. Deepwater Taranaki, New Zealand: structural development and petroleum potential. *Exploration Geophysics*, **39**, 94–107, https://doi.org/10.1071/EG08013

Line length balancing to evaluate multi-phase submarine landslide development: an example from the Storegga Slide, Norway

Suzanne Bull[1]* and Joseph A. Cartwright[2]

[1]GNS Science, 1 Fairway Drive, Avalon, Lower Hutt 5010, New Zealand

[2]Department of Earth Sciences, University of Oxford, South Parks Road, Oxford OX1 3AN, UK

SB, 0000-0003-0909-1415; JAC, 0000-0003-3132-6786
*Correspondence: s.bull@gns.cri.nz

Abstract: This study shows how simple structural restoration of a discrete submarine landslide lobe can be applied to large-scale, multi-phase examples to identify different phases of slide-lobe development and evaluate their mode of emplacement. We present the most detailed analysis performed to date on a zone of intense contractional deformation, historically referred to as the compression zone, from the giant, multi-phase Storegga Slide, offshore Norway. 2D and 3D seismic data and bathymetry data show that the zone of large-scale (>650 m thick) contractional deformation can be genetically linked updip with a zone of intense depletion across a distance of 135 km. Quantification of depletion and accumulation along a representative dip-section reveals that significant depletion in the proximal region is not accommodated in the relatively mild amount (*c.* 5%) of downdip shortening. Dip-section restoration indicates a later, separate stage of deformation may have involved removal of a significant volume of material as part of the final stages of the Storegga Slide, as opposed to the minor volumes reported in previous studies.

During emplacement of submarine landslides, compressional structures including thrust faults and folds typically form in the distal margins or 'toe zones', which can yield valuable kinematic information relating to the magnitude and direction of movement (Frey-Martinez *et al.* 2005; Reis *et al.* 2016; Alsop *et al.* 2017; Steventon *et al.* 2018). Such information is critical to the accurate representation of landslide motion in numerical simulations of tsunami generation and propagation, which are applied to hazard analysis, monitoring and mitigation (Lintern *et al.* 2019) . The Storegga Slide, and the tsunami it generated, occurred *c.* 8100 years ago, 120 km off the western coast of Norway (Bugge 1983; Haflidason *et al.* 2004; Bryn *et al.* 2005). It is one of the largest and most intensively studied submarine landslides in the world. The discovery of the Ormen Lange gas field close to its main headscarp led to the acquisition of a large geophysical and geotechnical database, with the overall development summarized in key papers by Haflidason *et al.* (2004, 2005) and Bryn *et al.* (2005). The Storegga region experienced temporally persistent and repeated slope failure linked to the rapid deposition of sediments derived from glacial and interglacial climatic cycles (Solheim *et al.* 2005). An estimated 3200 km^3 of material was involved in the multi-phase and largely retrogressive event that mobilized five main, overlapping slide lobes (lobes 1–5 in stratigraphic order; after Haflidason *et al.* 2004, 2005).

Two end members of submarine landslides have been identified defined by their mode of frontal emplacement (Frey-Martinez *et al.* 2006). Frontally emergent landslides occur where the translated mass ramps up from the original level of detachment and moves freely across the seafloor (Frey-Martinez *et al.* 2006) (Fig. 1a). A relatively thin deposit forms, typically exhibiting pressure ridges (e.g. Prior *et al.* 1984; Frey-Martinez *et al.* 2006) that are interpreted as the surface expression of sub-seismic-scale thrusts (Bull *et al.* 2009*a*). Frontally confined types occur when the translated mass is buttressed downslope against stratigraphically equivalent undisturbed strata, above a deep-seated detachment surface and typically develop a toe zone characterized by large-scale thrust and fold systems (e.g. Huvenne *et al.* 2002; Frey-Martinez *et al.* 2006) (Fig. 1b). Although advancements in seismic reflection data quality have led to several insights regarding the development of fold and thrust systems in toe zones (Frey-Martinez *et al.* 2005; Bull *et al.* 2009*a*; Reis *et al.* 2016; Steventon *et al.* 2018), limitations in the resolution of the data typically preclude the provision of details of local fold and thrust processes (Alsop *et al.* 2017). Consequently, documentation of transport

From: Georgiopoulou, A., Amy, L. A., Benetti, S., Chaytor, J. D., Clare, M. A., Gamboa, D., Haughton, P. D. W., Moernaut, J. and Mountjoy, J. J. (eds) 2020. *Subaqueous Mass Movements and their Consequences: Advances in Process Understanding, Monitoring and Hazard Assessments*. Geological Society, London, Special Publications, **500**, 531–549.
First published online March 23, 2020, https://doi.org/10.1144/SP500-2019-168

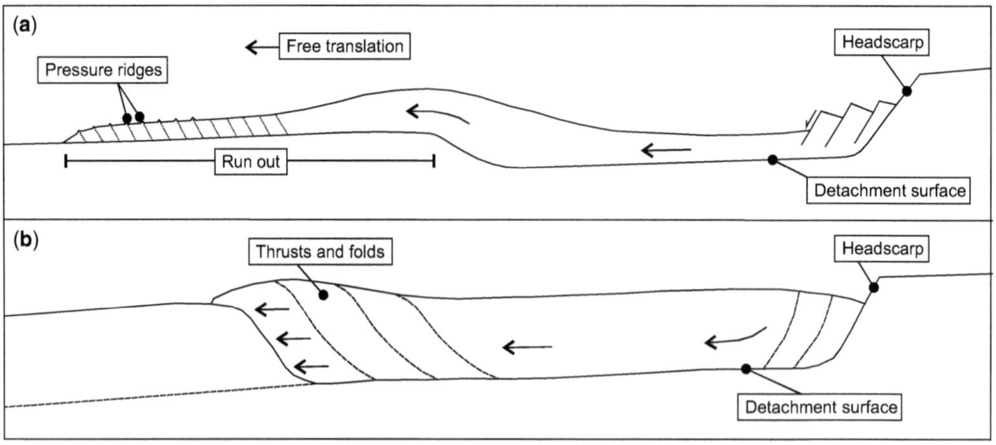

Fig. 1. Schematic representation of the two main types of submarine mass wasting deposits according to their style of frontal emplacement. (**a**) Frontally emergent: the basal shear plane ramps up to the seabed to allow the free translation of failed material to move in an unrestrained fashion across the seafloor. (**b**) Frontally confined: the failed mass is buttressed against the frontal ramp which is formed by the deeply seated basal shear surface ramping up steeply to intersect the seabed. The failed mass does not abandon the original basal shear surface and typically large-scale slump and fold systems develop landward of the distal margin. Modified after Frey-Martinez *et al.* (2006).

distances and volume balances of depleted v. accumulated strata are few (Alves and Cartwright 2009; Steventon *et al.* 2018).

The Storegga Slide exhibits both frontally emergent and frontally confined elements, with a large volume of turbidites and debris flows accumulated in a classic example of long-range run-out typical of a frontally emergent slide (maximum run-out distance of 800 km; Haflidason *et al.* 2004) (Fig. 2). However, during emplacement of Lobe 2, mobilization of sediment from the headscarp region is thought to have been responsible for generation of both the Storegga Slide tsunami and the region of deep-seated, spectacular compressional deformation informally termed the 'compression zone' in previous studies (Haflidason *et al.* 2004; Bryn *et al.* 2005; Gafeira *et al.* 2007; Færseth and Sætersmoen 2008) (Fig. 2). The compression zone is situated along the southern margin of the Storegga Slide scar and in contrast with the characteristics of the remobilized material in the distal toe region, deformation has occurred above a deep-seated detachment surface with a clear frontal ramp, indicative of a frontally confined slide mechanism. Its formation is linked to downslope movement of a large intact slide block (Bryn *et al.* 2005). The variety and complexity of deformational styles and depositional mechanisms observed within the Storegga Slide demonstrates the challenge faced by researchers in determining the mechanisms responsible for subaqueous slope failures.

Previous descriptions of the compression zone are based primarily on 2D reflection seismic data (Haflidason *et al.* 2004; Bryn *et al.* 2005) or limited 3D seismic data (Gafeira *et al.* 2007; Bull *et al.* 2009*a*). Gafeira *et al.* (2007) applied morphological analysis of seabed elements to the broader-scale sequence of sliding in the area. They observed a system of thrust and normal faults and suggested that these were the result of initial compression followed by extension, with extensional reactivation of some thrusts (Gafeira *et al.* 2007). Færseth and Sætersmoen (2008) carried out a detailed analysis of the compression zone and first defined it as an 'intact slump' formed by a single phase of slope failure with concurrent development of contractional and extensional domains. This was attributed to frontal buttressing against the distal margin, and changes in the velocity of dislocation and translation relating to the geometry of the basal detachment surface, respectively (Færseth and Sætersmoen 2008).

This paper aims to build on the intact slump model of Færseth and Sætersmoen (2008) by fully defining a discrete slope failure event pertaining to the compression zone, which we term the Confined Storegga Slide (CSS). We aim to quantify contractional strain and correlate the detachment surface up-dip to a zone of net extension and thinning in the headwall region. We use the results to populate a scaled cross-sectional model of the CSS and will use line length balancing to estimate the expected degree of upslope depletion. We will compare the observed results to further models relating the volume of failed strata to the observed dimensions of headwall, lateral and distal marginal areas which mark the pre-failure position of the seabed.

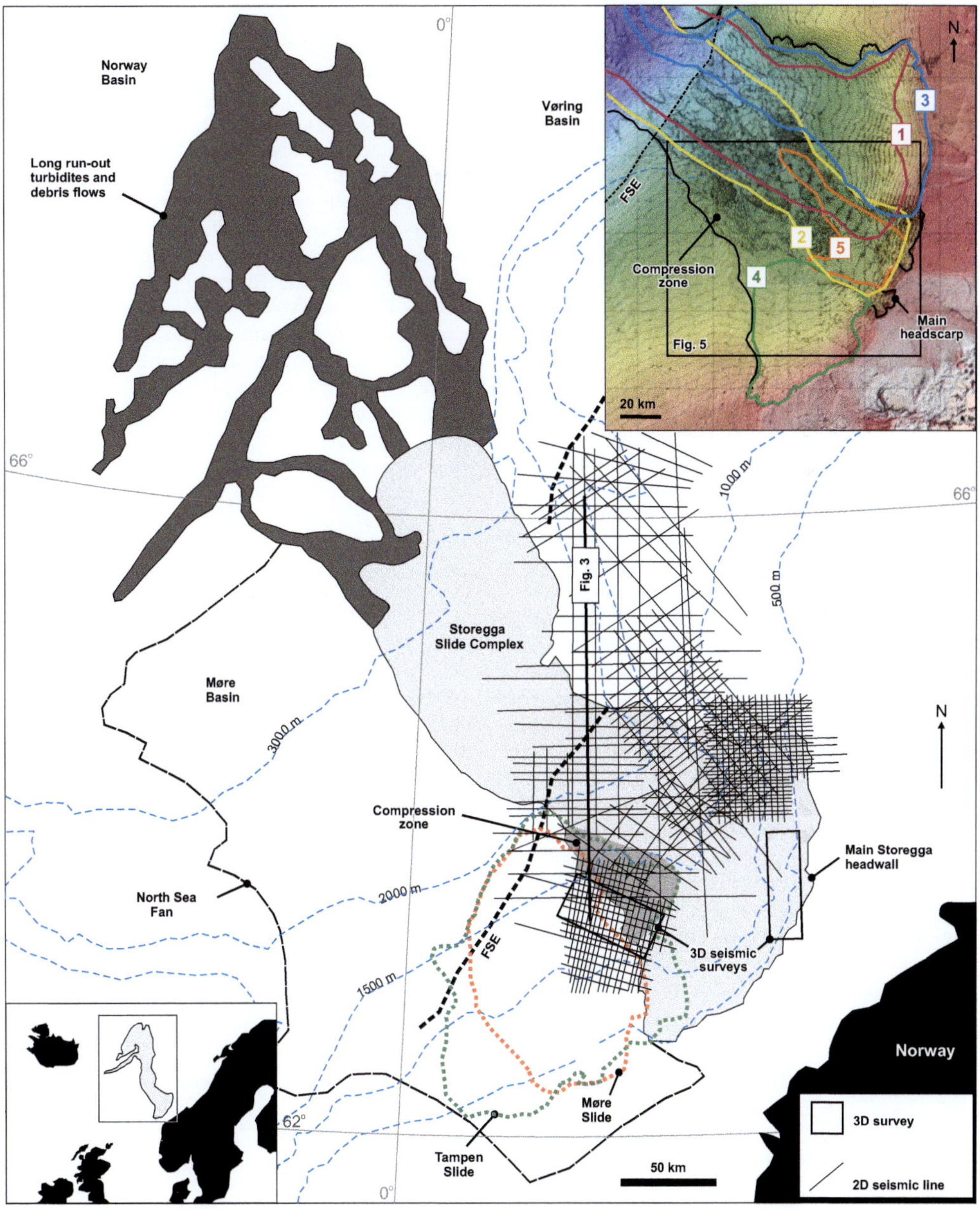

Fig. 2. Map of the study area showing the 'compression zone' (shaded in grey). Also shown the outlines of the Storegga Slide complex, including the distal toe zone where long run-out turbidites and debris flows have been deposited, the Tampen Slide and Møre Slides, which occurred prior to the Storegga Slide and form stratigraphic markers in the southern area of the Storegga Slide (Evans *et al.* 1996, 2005). Bathymetric contours are shown by the blue dashed line in 500 m increments. 2D seismic profiles (black lines) are shown along with 3D seismic surveys (black rectangles). Top right inset: detail of the upper Storegga Slide area showing the bathymetry data available for this study, location of Figure 5 and the sequence of slide lobes numbered in order of their occurrence according to Haflidason *et al.* (2004, 2005). Lobe 2 (yellow line) has been identified as the event associated with the formation of the compression zone (Haflidason *et al.* 2004; Bryn *et al.* 2005; Gafeira *et al.* 2007; Færseth and Sætersmoen 2008).

The results will determine if this simple approach can yield insights into volumetric changes and the sequence of multi-phase slide-lobe development. The large scale of the deformation (affecting sediments up to 650 m thick) and density of high-quality 2D and 3D seismic reflection and bathymetry data offer a rare opportunity to perform a detailed evaluation of the fold and thrust system from an offshore, relatively modern submarine landslide.

Geological setting

Regional setting of the Storegga Slide

The Storegga Slide occured within the Møre and Vøring sedimentary basins (Fig. 2), which formed due to a prolonged period of extension beginning in the Carboniferous, culminating in the Early Eocene with seafloor spreading in the Norway–Greenland Sea and related volcanism (Brekke 2000). Thermal subsidence in the Cretaceous led to the development of over 10 km of sedimentary fill in the Vøring Basin, which remained tectonically active until Tertiary time when a series of compressional dome and arch structures evolved (Brekke 2000). The Møre Basin experienced less tectonism during the Cretaceous and Cenozoic, undergoing continuous thermally driven subsidence (Brekke 2000). Plio-Pleistocene age sediments, on which this study focuses, belong to the Naust Formation (Fig. 3), which is subdivided into five units (T–N; Rise *et al.* 2006) (Fig. 3) to reflect glacial episodes which affected the margin from the Late Pliocene onwards. Large-scale sliding in the region, culminating in the Storegga Slide emplaced at 8.1 ka BP, is thought to be related to climatic variability following onset of regular shelf glaciations at 0.5 Ma (Bryn *et al.* 2005). Glacial debris flows were deposited on the continental slope during peak glacial conditions (Berg *et al.* 2005), while during intervening and comparatively longer periods of reduced ice cover, normal marine to distal glaciomarine sedimentation prevailed, with contourites developing (Berg *et al.* 2005). Laterally continuous layers of increased sensitivity within contourite drifts are thought to have served as the main glide planes for slides in the area, including the Storegga Slide (Bryn *et al.* 2005). The compression zone is located *c.* 100 km downslope of the main headwall of the Storegga Slide, where sediments deformed and dissected by Storegga area events are juxtaposed with the thick progradational wedge of the North Sea Fan (King *et al.* 1996) (Figs 2 & 3) which includes a series of major mass transport deposits including the Tampen Slide (Evans *et al.* 1996, 2005; King *et al.* 1996; Hjelstuen *et al.* 2005), which is disrupted and truncated to the east by Storegga Slide events (Evans *et al.* 1996) (Fig. 2).

Development model for the Storegga Slide

Of the five main lobes of the Storegga Slide identified by Haflidason *et al.* (2004, 2005) Lobe 1 failed first and occupies the northern half of the Storegga Slide scar (Fig. 2, inset), moving sediment from within 20 km of the present-day headscarp into deep water in the form of turbidites and debris flows. Lobe 2 removed material from the central area of the present-day headscarp (Fig. 2, inset) as debris flows along with large blocks of intact material, and is thought to have generated the Storegga Slide tsunami and formed the compression zone

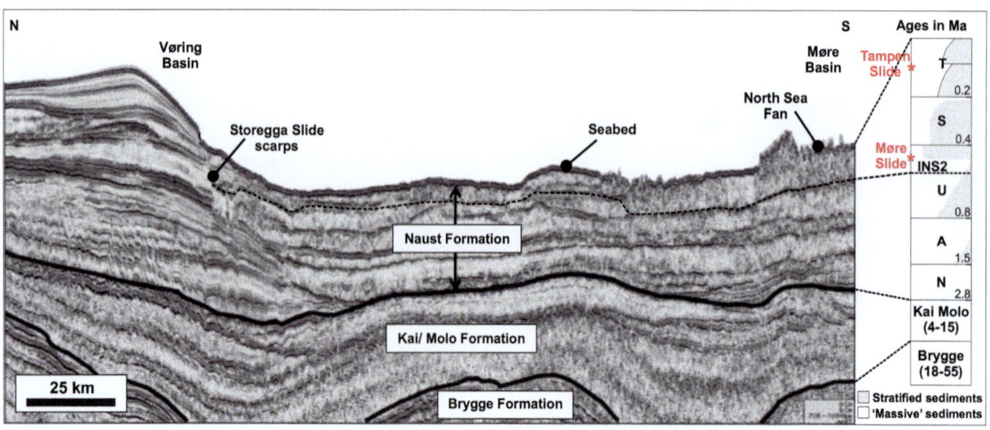

Fig. 3. Regional 2D seismic profile showing the general character and main stratigraphic subdivisions of the Eocene–Recent sediments from the study area (location shown in Fig. 2), along with stratigraphic scheme for the Naust Formation (after Rise *et al.* 2006).

(Haflidason *et al.* 2004, 2005; Bryn *et al.* 2005). Lobe 3 occurred along the same detachment surface as Lobe 1, and in a similar manner, but affected a smaller area (Fig. 2, inset). Lobe 4 occupies the southern corner of the headscarp and is characterized by blocky, partially disintegrated material. Finally, Lobe 5 occurred within the central headscarp area and was also a debris flow (Fig. 2, inset). Along with these main slide lobes >60 smaller slide events have been identified within Lobes 2, 4 and 5 (De Blasio *et al.* 2003; Haflidason *et al.* 2004). Micallef *et al.* (2009) carried out detailed geomorphological analysis of the NE Storegga Slide and further subdivided Lobe 1 into a total of nine individual events representing three major mass movements. In addition to the identification of numerous individual slide lobes, evidence for a range of failure mechanisms has been documented, including sediment evacuation (Bull *et al.* 2009*b*), debris flows (Haflidason *et al.* 2004, 2005), compression (Haflidason *et al.* 2004, 2005; Bryn *et al.* 2005; Gafeira *et al.* 2007; Færseth and Sætersmoen 2008) and spreading (Micallef *et al.* 2016).

Data and methodology

The data used in this study consist of a regional grid of 2D seismic profiles and bathymetry data, and two 3D seismic surveys that image parts of the main headwall and a section of the southern (lateral) margin of the Storegga Slide (Fig. 2). The 3D seismic volumes are time migrated and processed to zero phase, with a bin size of 12.5 × 12.5 m, giving a lateral resolution of *c.* 25 m (Brown 2003) . The dominant frequency of both surveys is *c.* 50 Hz, giving a vertical resolution of 10–15 m for P velocities in the range 1700 to 2000 m s^{-1}. An interval velocity of 1800 m $s^{-1}$was used for depth calculations. The 2D seismic data represent a total line length of *c.* 8000 km, with an average line spacing of 7 km, decreasing to 3 km along the southern margin (Fig. 2). Seismic data were interpreted using Schlumberger's Geoframe digital interpretation software. Interpretation methodology for mapping key features of the various slide units was adopted from Frey-Martinez *et al.* (2006) and Bull *et al.* (2009*a*). We interpreted several key horizons including the seabed, INS2, Top and Base Tampen Slide, and Base Naust Formation across the broader study area to provide regional context as well as the basis for the detailed structural and 3D interpretation. We constructed time structure and isochron (time thickness) maps from the interpreted horizons, and where 3D volumes were available (Fig. 2), also utilized volume-based seismic attribute extractions, time-slice interpretation and 3D visualization.

To estimate shortening, throw and heave values were measured from selected 2D seismic profiles chosen for their orientation approximately parallel to the dominant transport direction of the compression zone (interpreted from its lateral margin and seabed lineations; Gafeira *et al.* 2007; Færseth and Sætersmoen 2008). Shortening resulting from folding was estimated from the same 2D profiles by summing the length of individual folded horizon segments and comparing the total to the sum of the present horizon lengths (illustrated in Fig. 4).

The bathymetry data were loaded into the Geoframe project, allowing us to connect the detailed mapping of seafloor textures to subsurface features visible on the seismic data. The morphology of slide features visible in the bathymetry was interpreted following similar approaches by Masson

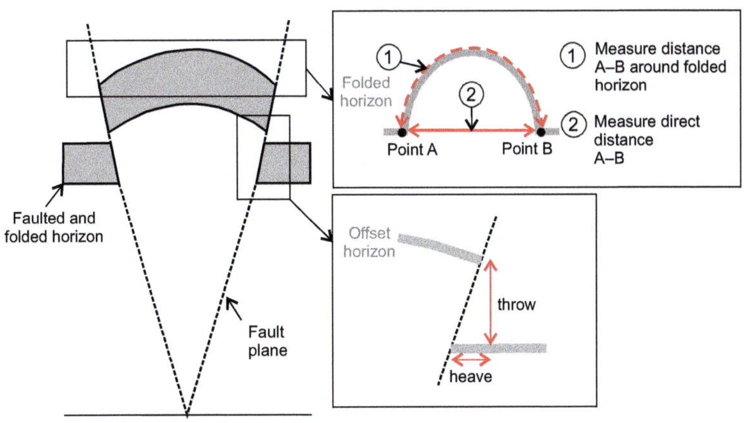

Fig. 4. Diagrammatic representation of measurements taken to quantify shortening from structures imaged within the compression zone.

et al. (1993), Laberg and Vorren (2000) and Lastras *et al.* (2004). The seafloor morphology preserves many diagnostic slide features resulting from the main Storegga Slide events because it formed recently (8.1 ka BP; Bryn *et al.* 2005) and has subsequently experienced low rates of sedimentation (Haflidason *et al.* 2004).

Data interpretation

Structure of the compression zone

We follow previous studies in using the term 'compression zone' to refer to the localized area of compression formed along the southern margin of the Storegga Slide complex (shaded grey on Fig. 2). In the following sections, we describe the compressional features in detail, estimate contractional strain and link the compression zone to further slide-related deformational features in the updip direction. The compression zone is best defined by its morphological expression at the seabed as a series of laterally continuous, subparallel lineations up to 50 km long which are convex in the downslope direction and orientated between NW–SE and north–south (Fig. 5). The distal limit of the compression zone has a relatively narrow frontal ramp that is 10 km wide (Fig. 5). The observed downslope narrowing is highly unusual compared to classical examples of toe regions of most submarine landslides, which typically exhibit a broad, downslope-convex morphology (Prior *et al.* 1984; Frey-Martinez *et al.* 2005; Bull *et al.* 2009*a*). The frontal ramp of the compression zone is associated with fault-bend folding affecting the seabed which has subsequently been partially eroded (Færseth and Sætersmoen 2008).

A representative 2D seismic dip-profile taken through the distal region of the compression zone is shown in Figure 6. The profile shows a deformed series of reflections ranging in thickness from

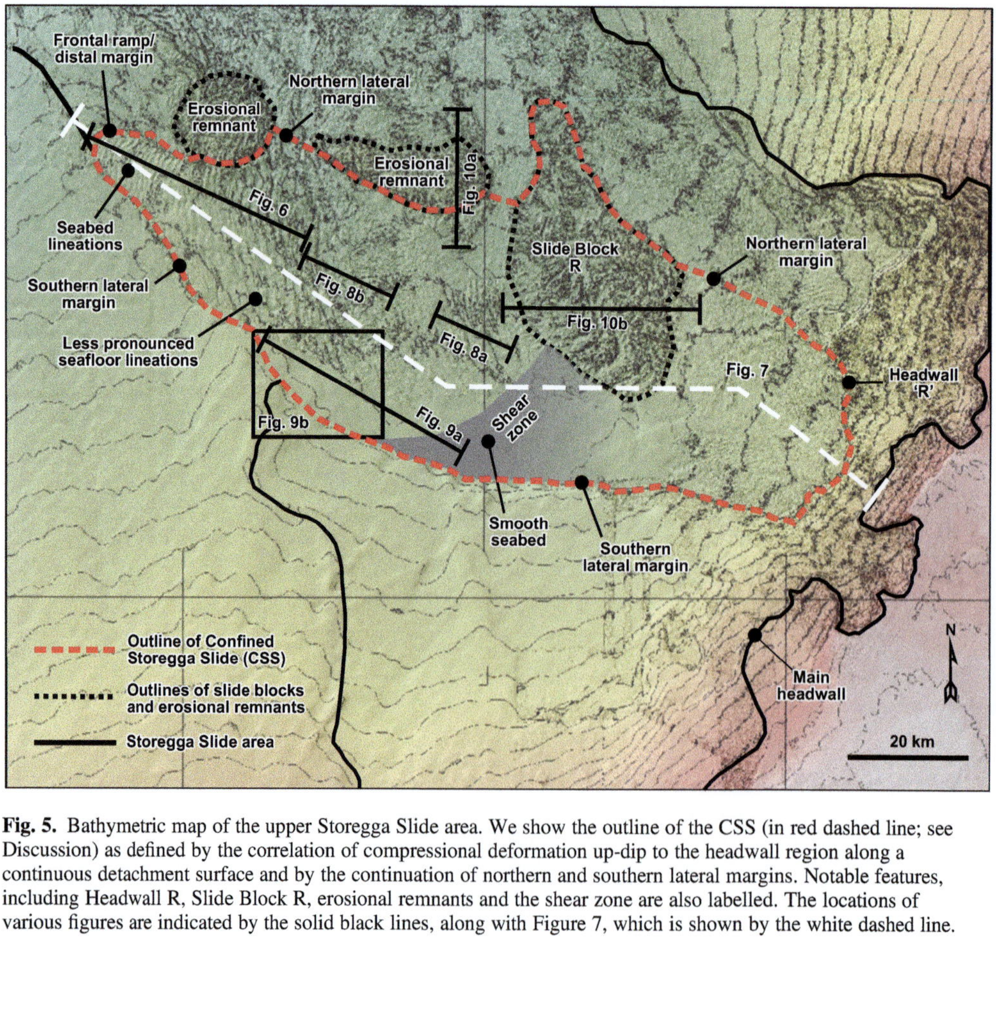

Fig. 5. Bathymetric map of the upper Storegga Slide area. We show the outline of the CSS (in red dashed line; see Discussion) as defined by the correlation of compressional deformation up-dip to the headwall region along a continuous detachment surface and by the continuation of northern and southern lateral margins. Notable features, including Headwall R, Slide Block R, erosional remnants and the shear zone are also labelled. The locations of various figures are indicated by the solid black lines, along with Figure 7, which is shown by the white dashed line.

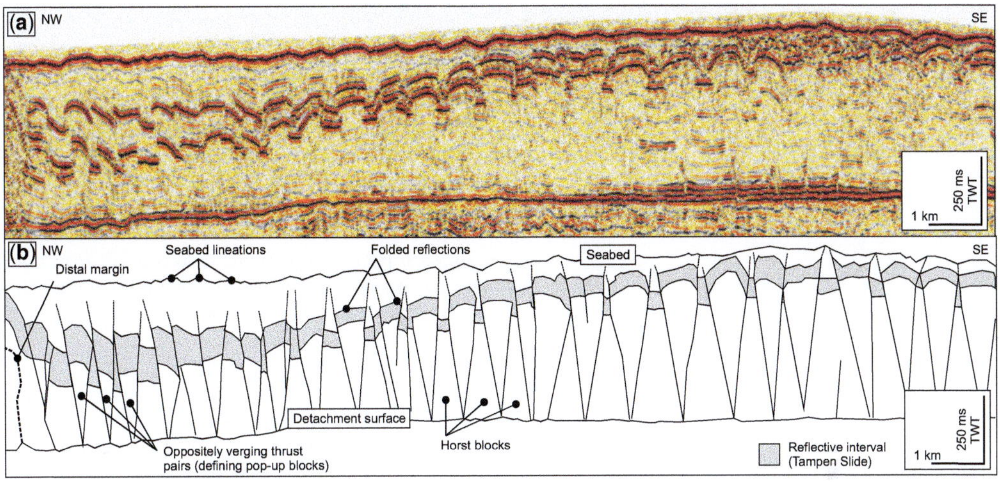

Fig. 6. 2D seismic dip-profile showing a central profile taken through the distal region of the compression zone. (**a**) Uninterpreted profile. (**b**) Interpreted profile. Note: the basal detachment horizon and the thickness of the sediments which overlie it (maximum thickness of 640 m); the numerous thrust faults which occur in pairs of opposing dip and define upwardly displaced 'pop-up blocks', detaching into the detachment horizon; the depositional unit resulting from the Tampen Slide (labelled), which form a chaotic interval bound by continuous, high amplitude top and basal surfaces and provides a sequence of stratigraphic markers. Location shown in Figure 5.

500 to 650 m, which are sharply bounded from underlying undeformed sediments by a distinct high-amplitude reflection. This boundary is interpreted as the basal detachment surface, which from regional correlation is taken as the base of the Storegga Slide (Haflidason *et al.* 2004; Bryn *et al.* 2005; Gafeira *et al.* 2007; Færseth and Sætersmoen 2008).

The seabed lineations visible in the bathymetric dataset (Fig. 5) correspond in seismic profiles to broad ridges with 20–30 m relief (Fig. 6). These ridges are the seabed expression of upwardly displaced, downward-tapering 'pop-up blocks' bound by pairs of oppositely verging thrust faults that sole out onto the detachment surface (Fig. 6). Horizons within the pop-up blocks are offset and often folded (Fig. 6). A strongly reflective package can be seen thinning to the SE within the thrusted interval (Fig. 6). This package is interpreted as the Tampen Slide by regional correlation (Gafeira *et al.* 2007; Færseth and Sætersmoen 2008) and is an excellent marker interval within the compression zone which can be correlated tens of kilometres into the Storegga Slide complex (Fig. 2).

Key fault parameters were measured for a total of 36 individual thrust faults imaged by 2D seismic profiles orientated approximately parallel to the dominant transport direction of the compression zone (Fig. 6). Fault throw and heave measurements were recorded close to the position of maximum observable thrust displacement (close to the Base Tampen Slide horizon; Fig. 6) (Table 1). The thrust faults exhibit dip ranging from 35 to 88°, heights ranging between 470 and 1076 m and maximum offsets (taken at the Base Tampen Slide horizon) of 20–100 m. A total of 4.5 km of shortening is calculated from offset along the thrust faults. Additional shortening resulting from folding was estimated by subtracting the current length of the most intensely folded horizon (the Top Tampen Slide horizon, Fig. 6), from its original length measured from 2D dip-profiles (Fig. 4). From this approach, an estimated 1.85 km of shortening is accommodated through folding.

These measurements assume that slip vectors were in the direction of the 2D profiles used and that true dips have been measured as opposed to apparent dips (i.e. dip measured in a direction other than perpendicular to strike). Error associated with apparent dip has mainly been mitigated by selection of 2D profiles aligned with the dominant transport direction of the most distal portion of the compression zone (see Fig. 5, which shows the location of the 2D profile shown in Fig. 6). However, the seabed ridges show a change in their dominant orientation (i.e. from NW–SE at their most distal point changing to north–south further updip) and the available grid of 2D data in this area is limited to profiles orientated NW–SE and WNW–ESE (Fig. 2). To account for this, we estimate error in our shortening calculation as 20%, representing dip error, interpretational uncertainty in the positions of hanging-wall and footwall cutoffs, and sub-seismic scale deformation

Table 1. *Thrust fault parameters measured from 2D seismic profiles*

Fault	Angle (°)	Height (m)	Maximum throw (m)
1	42.8	599	84
2	73.3	543	59
3	78.8	363	68
4	83.1	523	67
5	39.5	876	59
6	65.4	570	76
7	46.4	701	76
8	86.6	524	48
9	87.5	336	14
10	64.3	560	24
11	50	604	62
12	60.4	505	26
13	34.1	808	69
14	82.7	487	43
15	33.1	833	40
16	78.2	470	48
17	38.5	780	75
18	55.9	594	73
19	45.5	561	34
20	66.8	557	40
21	41.6	729	65
22	71.1	608	50
23	39.2	918	57
24	71.9	551	65
25	37.5	850	100
26	87.9	634	63
27	45	779	62
28	62.3	634	56
29	38.9	910	61
30	61.4	674	64
31	35.4	1076	58
32	67.4	643	52
33	37.2	1043	30
34	77.2	588	24
35	50.3	709	32
36	56.5	661	41

(see Higgins *et al.* 2009 and Steventon *et al.* 2018 for discussion of line balancing errors and underestimation of strain based on seismic data).

Correlation of the basal detachment surface

A distinct basal detachment surface can be identified underlying the compression zone (detachment surface labelled on Fig. 6b) using well-established criteria (i.e. Bull *et al.* 2009*a*). The detachment surface correlates with a continuous high-amplitude reflection that from regional correlation is identified as INS2, a distinctive marker within Naust subunit S, deposited between 420 and 380 ka (Haflidason *et al.* 2004; Berg *et al.* 2005; Bryn *et al.* 2005) (Fig. 3). In the present study we confidently correlate this surface updip over a distance of *c.* 135 km (Fig. 7) directly into the headwall region of the Storegga Slide. In the Storegga Slide headwall region the surface links directly into the major 200 m high headwall scarp referred to as Headwall R (Bryn *et al.* 2005) (Fig. 7). This scarp trends NNE–SSW, comprises two major salients, and is located *c.* 4 km downslope from the main headwall of the Storegga Slide (Fig. 5). The link between Headwall R and the compression zone is further evidenced by direct continuation of the southern extension of Headwall R with a WNW–ESE-trending seabed scarp that defines the southern limit of the compression zone (labelled 'southern lateral margin' on Fig. 5).

Dip measurements for the detachment surface taken across the regional grid of 2D seismic profiles (Fig. 2), range from a maximum of 13° in the vicinity of the Headwall R, to 0.3° in the distal compression zone, although localized variations are observed. Along with the continuity of the detachment surface, Figure 7 also illustrates the variation in deformation and thickness distribution of the overlying sediments. Close to Headwall R, the remnants of the remobilized slope sediments are only 50 m thick and highly disaggregated (Fig. 7). They are expressed as thin, opaque seismic facies, in contrast to the highly reflective thrusted and folded slope sediments in the compression zone where sediments are shortened and thickened compared to its pre-failure state (Figs 6 & 7).

Updip deformation

Having characterized the contractional deformation that defines the compression zone and identified the underlying regional detachment surface, our analysis now moves updip. Approximately 30 km updip from the distal margin of the compression zone, the sediments overlying the basal detachment surface show a distinct change in character marked by a reduction in the amplitude of the seabed lineations (Fig. 5). Along with this change in seabed character the sediments overlying the detachment surface exhibit decreased thickness in the upslope direction, accompanied by increased deformation manifest as decreased continuity and organization of internal reflections (Fig. 7).

Figure 8a shows how the thrust fault pairs and pop-up blocks, and their corresponding positive seabed features, give way updip to pairs of normal faults and intervening graben blocks, with horizons exhibiting normal offsets within a zone which corresponds to a locally depressed, relatively flat area of the seabed (Fig. 8a). Increasingly in the updip direction, further graben blocks are identified, demonstrating increased spacing between the individual detachment points of the faults within each pair (Fig. 8a). The average spacing of the fault pair detachments in this region is 2.3 km. Horst blocks

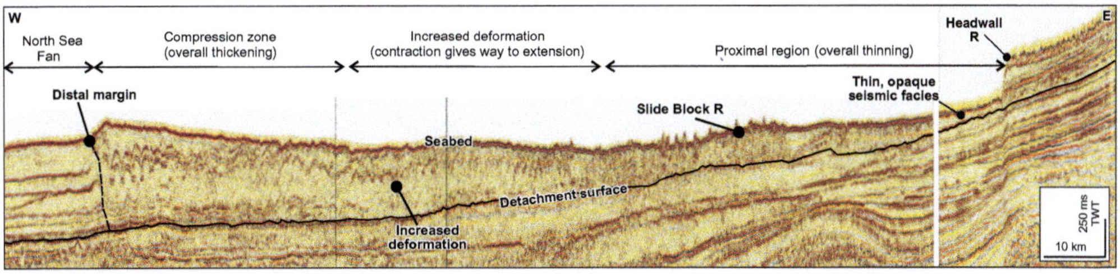

Fig. 7. Composite seismic dip-profile comprising of several individual 2D profiles and a 3D seismic traverse, chosen to provide a continuous image of the CSS. Location shown in Figure 5. The basal detachment horizon is recognized as the regional reflector INS2 (continuous black line; Haflidason *et al.* 2004; Berg *et al.* 2005; Bryn *et al.* 2005).

are interpreted from the presence of intervening zones of relatively undeformed, bed-parallel reflections bound by outward-dipping flanks (Fig. 8a).

Continuing in the landward direction, a further change in the character of deformation is noted (Fig. 8b). A series of seabed ridges and offsets are associated with laterally discontinuous, inclined and deformed reflections in the strata beneath, which define systematic inclinations and offsets of reflections above the detachment horizon (Fig. 8b). These observations are interpreted as the result of extensional faults which affect the full thickness of the succession above the detachment horizon and possess fault angles of between 30 and 40°. Table 2 summarizes the characteristics of the main fault types described in this study.

Given the wide spacing of the 2D grid in this area (Fig. 2) it is difficult to accurately estimate net extension. However, it is possible to ascertain that normal faults are spaced 1–2 km apart, have throws of up to 50 m, and appear to represent a mechanism for thinning the main slide body as is widely observed in other submarine slides (Farrell 1984; Frey-Martinez *et al.* 2005).

Further updip, the degree of deformation increases such that the internal character of sediments overlying the detachment surface appears structureless (Fig. 7). This corresponds to low-relief seabed signature (Fig. 5) and is interpreted as an 'internal shear zone' (cf. Ogata *et al.* 2014; Omeru and Cartwright 2019), within which material is highly deformed and disaggregated to the point of being structureless in seismic section.

Detailed insight into the deformational structures within the compression zone can be gained from the 3D seismic survey that images an area along its southern lateral margin (Fig. 2). A thrust fault defines the boundary between deformed sediments affected by the Storegga Slide and undeformed sediments outside of its bounds (Fig. 9a). Updip of this defining thrust fault, a further series of thrust faults is spectacularly imaged. They occur in pairs of opposite-verging dip and detach at a common point in the basal detachment surface, defining upwardly displaced, downward-tapering 'pop-up' blocks first identified by Gafeira *et al.* (2007). Within the pop-up blocks, reflections exhibit convex-upward geometries interpreted as evidence of folding (Fig. 9a). The presence of intervening 'horst blocks' is interpreted from zones between the pop-up blocks that show much less evidence of deformation and displacement. In the updip direction, the thrust pairs are succeeded by normal faults, followed by a zone of increased deformation in which it is possible to identify several large listric normal faults representing a zone of more disorganized extension (Fig. 9a). This observed sequence of deformation is remarkably similar to that observed in the central region of the compression zone using the 2D data (Figs 6 & 8).

Detailed mapping of horizons within the 3D survey reveals surprising relationships that are not evident on the widely spaced 2D seismic grid. A subtle series of narrow, linear features are mapped trending ESE–WNW, and are observed to cross-cut the zone of thrust faults and pop-up blocks (Fig. 9b). A representative seismic profile taken parallel to the strike of the thrust faults shows the features correspond to a series of normal faults that detach into the same detachment surface as the thrust faults, and affect the entire succession up to and including the seabed (Fig. 9c). Eleven such faults are observed to cross-cut the zone of thrust fault pairs, bisecting both the footwalls and hanging walls of the three pairs of thrust faults that follow the frontal thrust (Fig. 9a). The cross-cutting faults can be traced for a lateral distance of up to 6.5 km. They have an average spacing of 0.53 km, dip between 45 and 56°, exhibit throw values which vary between 20 and 100 m, and fault heights varying from 937 to 1108 m (Fig. 9c; Table 2).

Lateral margins

In order to define a discrete event pertaining to the compression zone, we now attempt to correlate

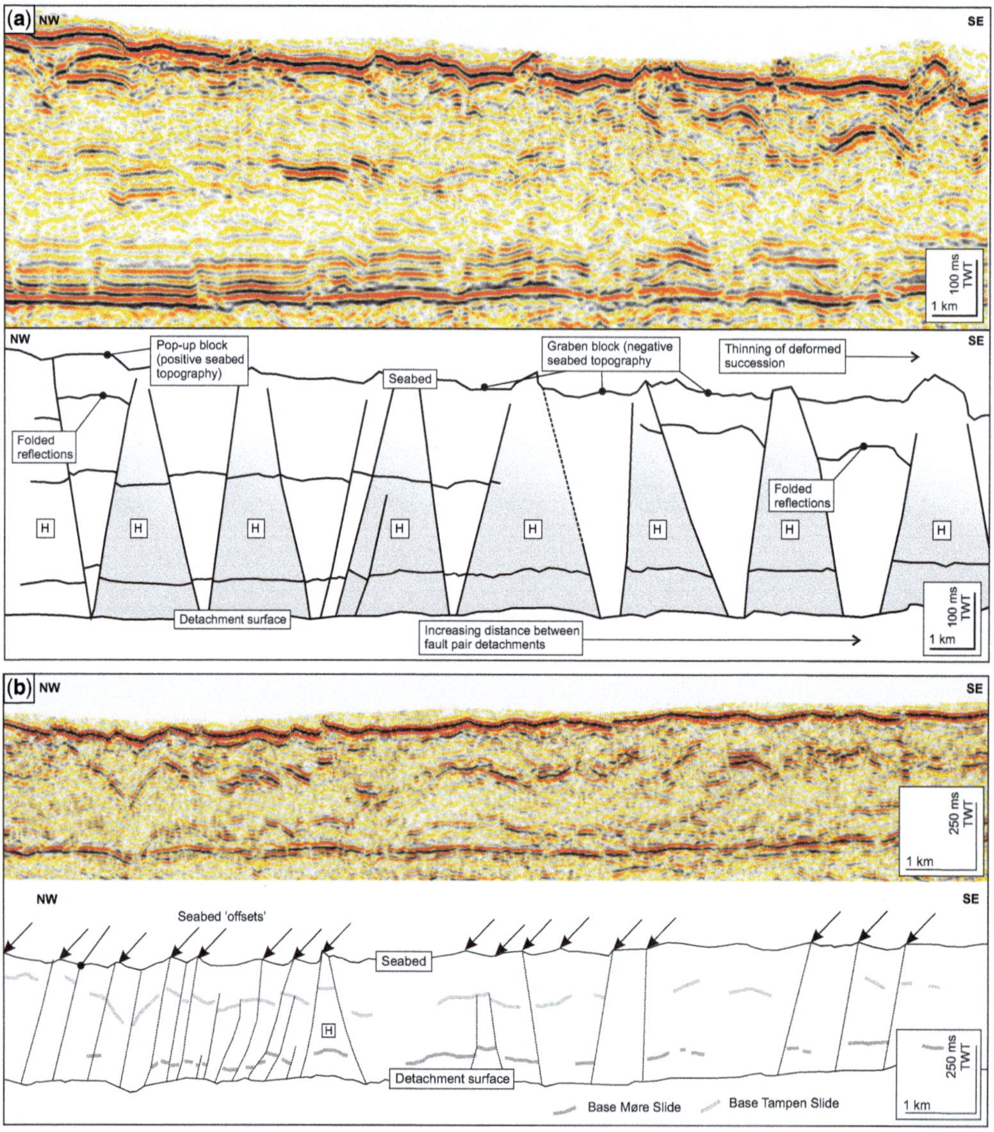

Fig. 8. (**a**) Interpreted and uninterpreted representative 2D seismic dip-profile located *c.* 30 km updip from the distal margin of the compression zone. Location shown in Figure 5. Note: thinning of the sediments overlying the basal detachment horizon in the updip direction (towards the SE); the succession of deformational features which shows upwardly displaced pop-up blocks in the NW of the profile being succeeded by increasingly deformed fault blocks towards the SE. Negative seabed topography and normal reflector offsets together evidence the presence of normal faults (labelled). (**b**) Uninterpreted and interpreted 2D seismic dip-profile *c.* 40 km updip of the distal margin, where a further change in the character of the deformation is observed. Note the systematic offset and inclination of marker horizons (Base Møre and Base Tampen slides), which in conjunction with similarly inclined and offset segments of the seabed define landward-dipping extensional faults that in many cases affect the entire thickness of the sediments above the detachment horizon. Seabed offsets are indicated by arrows. H, horst block. Location shown in Figure 5.

lateral margins that link the compression zone with its updip source area. Lateral margins are a critical component of slide interpretation, since their orientation and kinematic significance help constrain the gross patterns of slide movement (Bull *et al.* 2009*a*). As previously described, the southern lateral margin of the compression zone is readily mapped since it marks a change from the relatively smooth,

Table 2. *Summary of fault parameters for all fault types*

	Thrust fault	Normal faults	Cross-cutting normal faults
Orientation	NNE–SSW changing to NW–SE*	NNW–SSE	WNW–ESE
Height (m)	470–1076	769–956	937–1108
Angle (°)	35–88	30–40	45–56
Throw (m)	20–100	32–50	20–100
Average spacing (km)	1.48†	1–2.3	0.5

Thrust and normal fault parameters measured from 2D seismic profile. Cross-cutting normal fault parameters measured from 3D seismic traverses.
**in most distal region of compression zone.*
†*of thrust fault pair detachment points.*

undeformed seabed of the North Sea Fan to the south, to the area affected by deformation within the compression zone to the north (Fig. 5). The accompanying seabed scarp can be traced directly across the seabed from the compression zone to the scarp of Headwall R (Fig. 5).

Correlation of the northern lateral margin of the compression zone updip into the headwall region is more complex, in part due to the presence of erosional remnants, identified according to established criteria (Frey-Martinez *et al.* 2005; Moscardelli *et al.* 2006; Bull *et al.* 2009*a*) (Fig. 5). Seismic reflection profiles show that erosional remnants consist of partially deformed, but rooted strata which exhibits no basal deformation above the detachment surface and is interpreted not to have been mobilized or translated as part of the Storegga Slide events (i.e. see Moscardelli *et al.* 2006; Bull *et al.* 2009*a*) (Fig. 10a). Therefore, the northern lateral margin is traced to the south of these features (Fig. 5). From

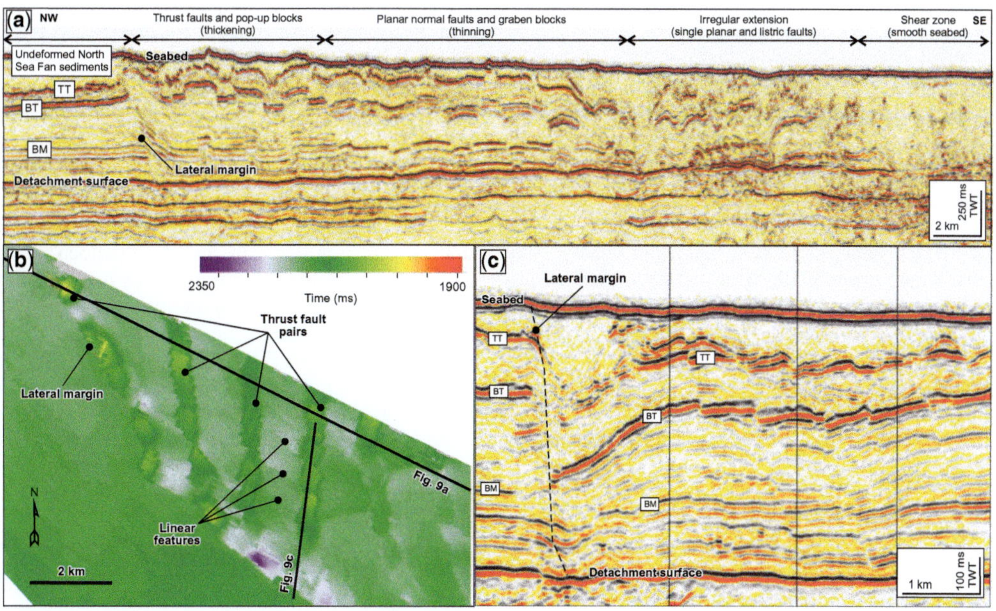

Fig. 9. (**a**) 3D seismic traverse showing the subsurface character of the compression zone through the lateral margin (labelled) which is represented by a thrust fault. Note the undeformed succession NW of the lateral margin, which features the Tampen Slides and a continuation of the basal detachment horizon. A series of further thrust faults, planar normal faults and finally listric normal faults are present within the compression zone (SE of the lateral margin). Note the shear zone in the SE extreme of the section. TT, Top Tampen Slide; BT, Base Tampen Slide; BM, Base Møre Slide. (**b**) Two-way travel time (TWTT) structure map of the Base Møre Slide time structure map showing planform character of the thrust fault pairs and ESE–WNW-trending planar normal faults (indicated by black lines) which cross-cut the thrust faults pairs. Location shown in Figure 5. (**c**) 3D seismic traverse through the zone of cross-cutting planar normal faults. Location shown in (b).

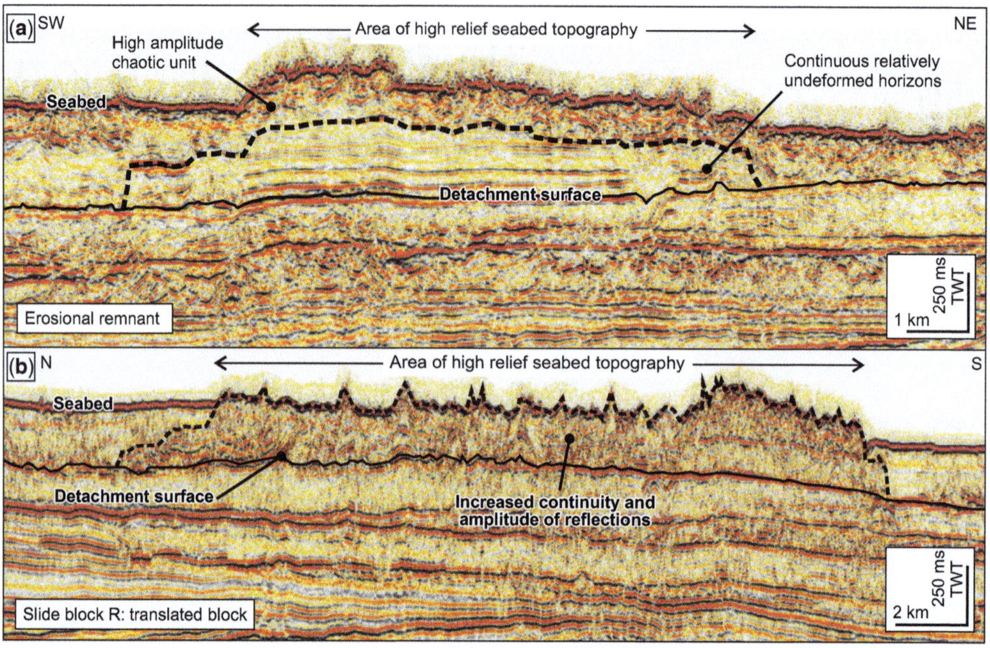

Fig. 10. (**a**) 2D seismic strike profile through one of the erosional remnants. Location shown in Figure 5. (**b**) 2D seismic strike profile through Slide Block R (Bryn *et al.* 2005). The high amplitude, highly deformed seismic reflections within a zone of highly rugose seabed topography is interpreted as an intact block of material that has been translated downslope, and partially disaggregated. Location shown in Figure 5.

this region towards the headscarp, the northern lateral margin is less distinct, but is tentatively correlated across a broad region of low relief seabed to Headwall R (Fig. 5). This encompasses the large translated Slide Block R (Bryn *et al.* 2005) (Figs 5 & 10b), a huge, partially disaggregated intact slide block measuring 50 km wide and 28 km long, comprising an area of *c.* 795 km^2.

Discussion

Identification of a frontally confined submarine landslide: the CSS

It has been shown that the area of the Storegga Slide referred to as the 'compression zone' can be correlated updip to Headwall R over a distance of 135 km via a common basal detachment surface (Fig. 7). Furthermore, Headwall R can be connected to the southern margin of the compression zone by a distinctive surface and subsurface morphological boundary that corresponds to the interface of deformed and undeformed sediments (Fig. 5). As previously suggested by Færseth and Sætersmoen (2008), these correlations and gross morphology are conducive to the interpretation of a discrete slide event, delimited at its toe by the compression zone, and updip by Headwall R (Fig. 5). The analysis presented herein supports the broad conclusions of Færseth and Sætersmoen (2008), although it is noted that the precise boundaries of the lateral limits of this slide body we interpret does differ (see Færseth and Sætersmoen 2008, fig. 2). We interpret this discrete body as a frontally confined submarine landslide, on the basis that the bulk of the shortening in the thickened toe region occurs updip of the frontal ramp (Fig. 7). Beyond the frontal ramp deformed sediments are buttressed *in situ* against its pre-deformational slope template with only a modest downslope translation equivalent to the cumulative shortening seen in the thrust faults and associated folding (Frey-Martinez *et al.* 2006).

The full extent of this frontally confined element of the Storegga Slide is shown in Figure 5, and a summary of interpreted deformational features is presented in Figure 11a. The total area of the CSS is estimated at 3510 km^2, and the volume of deformed sediments calculated as 1230 km^3. The Storegga Slide (including areas of erosion and deposition) has involved a volume of material estimated at 2500–3500 km^3 respectively (Bryn *et al.* 2005). In volumetric terms the CSS is therefore a significant component of the overall Storegga Slide complex, comprising 35% of its volume.

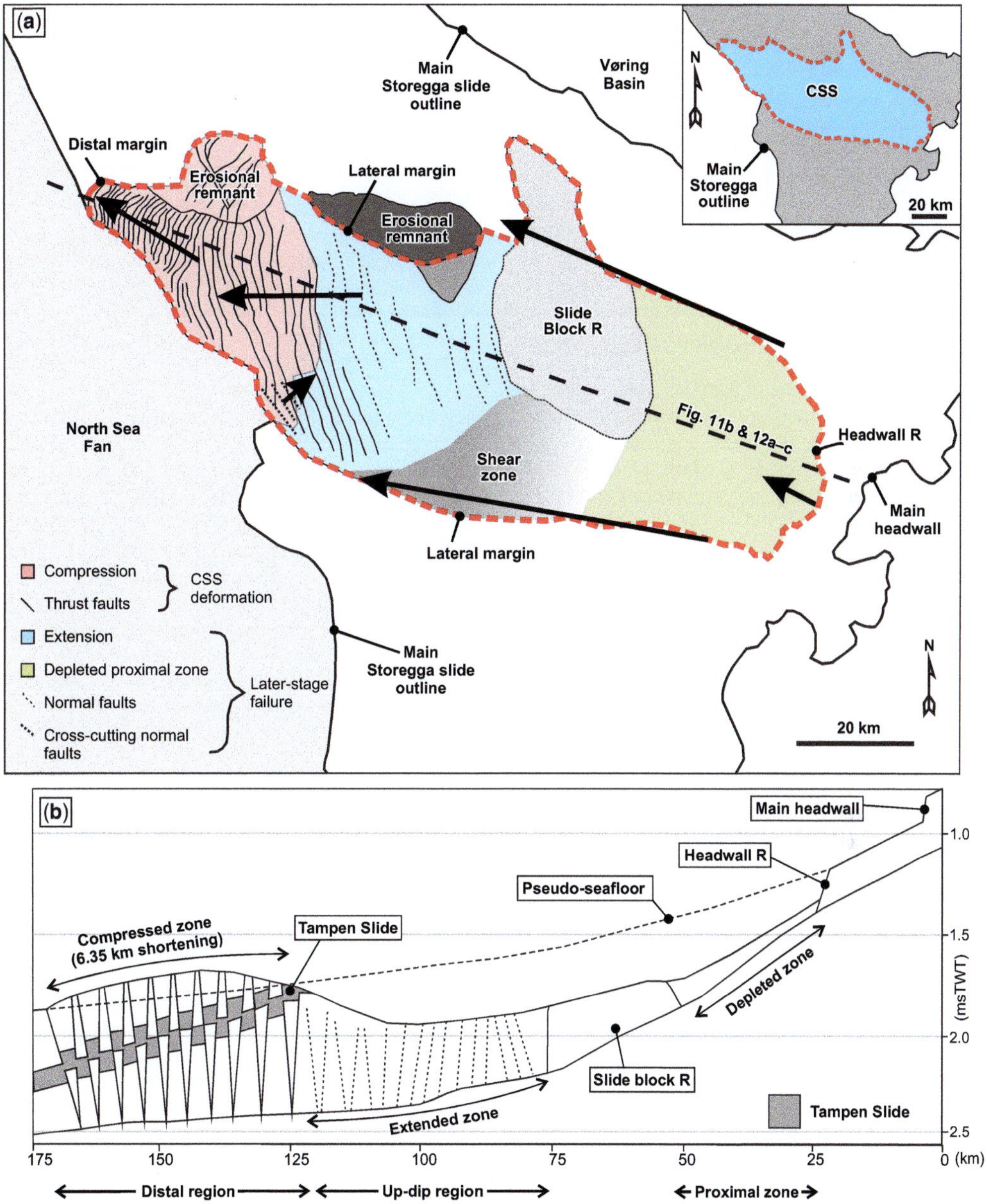

Fig. 11. (**a**) Map to show the interpreted full extent of the CSS, as constrained by the correlation of both surface and subsurface morphological elements including the positions of the distal margin, lateral margins, headwall R, seafloor signature changes and distribution of thrust faults, normal faults and shear zones. These features can be used to infer the main direction of transport, indicated by bold arrows. Inset: Simplified map showing the outline of the CSS in the upper Storegga Slide area. (**b**) Scaled, schematic dip-section through the CSS. Note that the proximal and updip regions are heavily depleted in relation to the distal region. This contradicts classical models for slope failure which predict that accumulation in the distal region is balanced by depletion in the proximal region (*sensu* Varnes 1978). Location shown by dashed black line in (a).

The mapping of well-demarcated zones of contrasting deformational styles observed from the distal and updip regions of the CSS is perhaps the strongest of the gross kinematic indicators, pointing to the dominant transport direction (cf. Varnes 1978; Bull *et al.* 2009*a*). The laterally extensive NNW–SSE strike of thrusts in the distal region, together with NNW–SSE-striking normal faults and graben indicates a dominantly WSW transport direction. This is despite some adjustment in the extreme distal region where thrust strikes alter to a more NE–SW orientation (Figs 5 & 11a). The clearest structural features that conflict with this simple kinematic interpretation are the cross-cutting normal faults (Fig. 9b & c), whose WNW–ESE strike suggests a later phase of extension directed to the NNE (Fig. 11a).

Estimating translation distance from shortening of the compression zone

The translation distance of the CSS can be estimated from the total shortening of the compression zone, where individual thrust faults and associated folds are well preserved (cf. Frey-Martinez *et al.* 2006) (Fig. 6). We note that our estimates of shortening, based on thrust fault heaves taken along dip-profiles orientated perpendicular to the dominant trend of the faults, differs from the estimates of Færseth and Sætersmoen (2008) who concluded that a total shortening of 30–35 km occurred. Their conclusions were based on line length balancing of well-resolved thrusts (3.5 km; Færseth and Sætersmoen 2008), along with comparison of compression zone thicknesses with thicknesses immediately updip (their 'western part of the proximal segment'), and an additional 1.5 km of shortening at the frontal ramp. We consider this value to be too high as estimating shortening from an equivalent thickening is arguable if the pre-failure template is taken from a zone which has itself been deformed. This approach may be used if comparing the thickness of the compression zone with a region to the south of the compression zone, where pre-Storegga sediments are preserved in their original, undeformed state. We recognize that there may be errors in this approach also, in the absence of constraints on the volume of pore fluid that might have been expelled during the shortening. A modest amount of shortening is more consistent with the presence of the erosional remnants (labelled on Figs 5 & 11a) which have remained rooted to the substrate and provide useful pin lines for the assessment of translation distance. The presence of such 'rooted' blocks was interpreted by Frey-Martinez *et al.* (2005) to exclude the possibility of long-range translation of sediments within a submarine landslide. The estimate of Færseth and Sætersmoen (2008) also includes 14 km of shortening due to 'ptygmatic folding' which we do not observe. Folding identified within the compression zone is related to mild and systematic deformation of reflectors (mostly the Top and Base Tampen Slide horizons) within pop-up blocks (Figs 6 & 8) and has been estimated as 1.85 km in the present study based on comparison of original and folded lengths measured

Table 3. *Summary of shortening estimates from the present study and Færseth and Sætersmoen (2008)*

	Færseth and Sætersmoen (2008)	Present study
Thrust fault displacement (km)	3.5	4.5
Frontal ramp (km)	1.5	Not specified
Folding (km)	14	1.85
Total shortening (km)	30–35	6.35
Estimated error	Not specified	20% ($\pm$1.27 km)

perpendicular to the seabed lineations. A summary of shortening estimates from both the present study and those reported by Færseth and Sætersmoen (2008) are presented in Table 3. Figure 11b presents a scaled schematic representative dip-section through the CSS that summarizes and illustrates the kinematic data, distribution of structural styles, sediment thickness and slope angle of the detachment surface and seafloor. The information represented on this cross section is a synthesis of observations taken from both the regional and detailed analysis of all of the available seismic and bathymetric data.

Volume balance

The scaled cross-section presented in Figure 11b clearly demonstrates depletion or thinning in the upslope region of the CSS, and accumulation or shortening in its distal region. The observation of depletion in the upslope region in concert with thickening and shortening in the downslope region is typical of so-called 'depletion' and 'accumulation' zones, respectively, of classical submarine landslide models (Varnes 1978; Frey-Martinez *et al.* 2005, 2006). It is generally expected that upslope depletion will approximately balance with downslope accumulation (e.g. Lewis 1971; Farrell 1984; Steventon *et al.* 2018). Due to the coarse spacing of seismic profiles in the upslope area of the CSS (Fig. 2), the amount of depletion could not be estimated with confidence, but it seems apparent from Figure 11b that the extensive depletion in the upslope region is not balanced by the relatively modest amount of shortening accommodated in the downslope region (i.e. 5%). Various factors may explain why upslope depletion does not precisely balance downslope accumulation, including variations in pre-slide seafloor morphology and sediment thickness, basal erosion during slide emplacement and subsequent sediment compaction (Alves and Cartwright 2009; Omeru 2014; Steventon *et al.* 2018).

To further consider the dynamic evolution and volume transfer involved in the CSS we invoke a simplified model of slide development based on line length balancing whereby downslope accumulation is approximately balanced by upslope depletion. To do this we have used the thickness and dip of intact sediments preserved in the distal and headwall margins of the CSS to provide an estimate of the seafloor position prior to CSS development (cf. Alves and Cartwright 2009). We plotted the resulting 'psuedo-seafloor' surface linking the distal margin and headwall of the CSS on Figure 11b, and used it as the basis for three further scaled cross-sections, to represent time stages before formation of the CSS (Fig. 12a), immediately after its formation (Fig. 12b) and the observed present-day conditions (Fig. 12c):

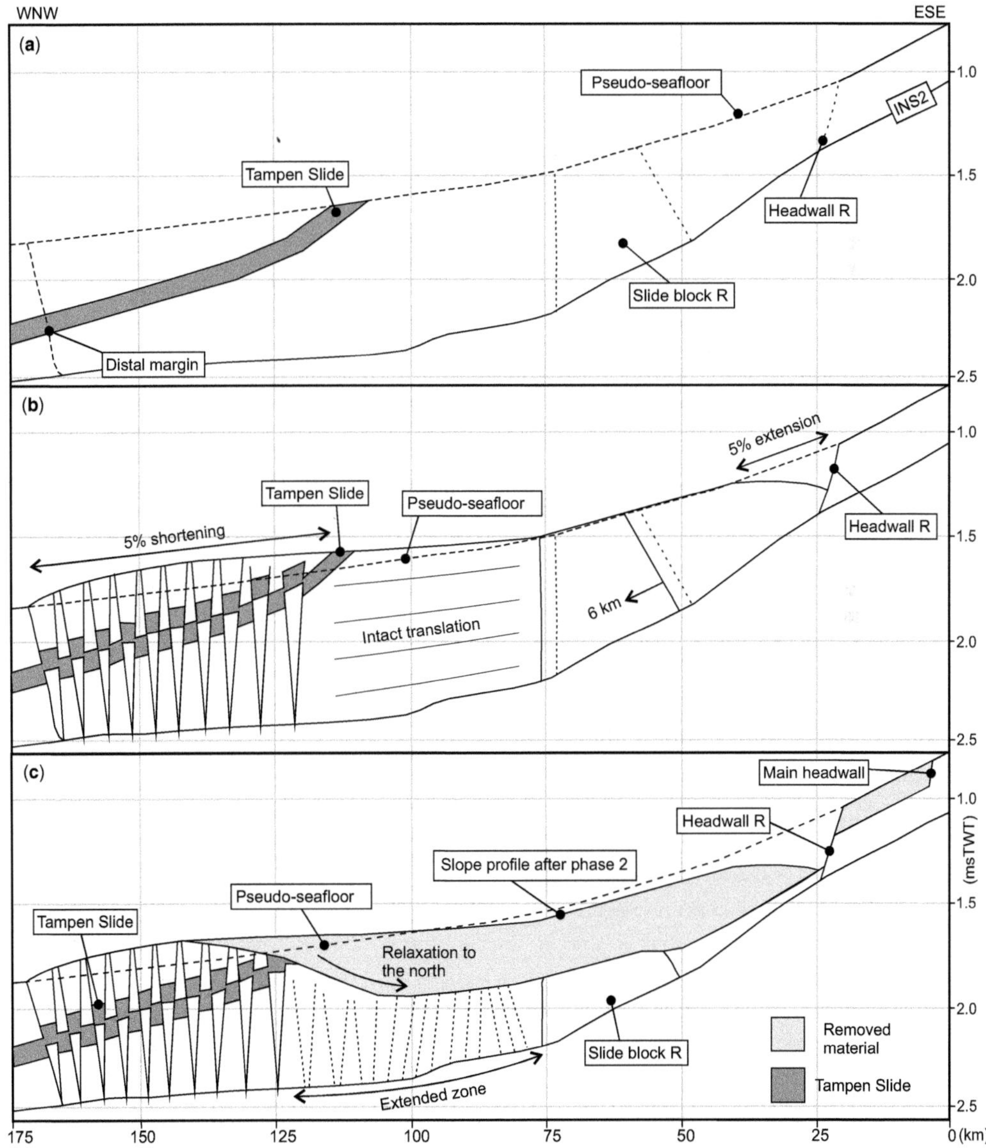

Fig. 12. Scaled, schematic representation of various time-stages to illustrate the development of the CSS. (**a**) Step 1: Pre-failure conditions. (**b**) Step 2: CSS has developed, with downslope translation accommodated in the form of mild depletion (i.e. 5%) in the proximal region, balanced by mild accumulation in the distal region (also 5%). (**c**) Step 3: Later modification of the CSS due to slope failure as part of the multi-stage Storegga Slide. The volume of sediments removed by the inferred later stage failure is shaded grey, to emphasize the contrast between the slope profile expected after development of the CSS, and the actual slope profile observed today. Locations shown in Figure 11a.

(1) Pre-CSS formation (Fig. 12a): the undisturbed seafloor position has been extrapolated based on the thickness and dip of sediments beyond the distal margin and Headwall R (represented on Fig. 12a by dashed lines). Note the Tampen Slide which extends into the CSS but has not yet been overprinted by strain pertaining to the CSS.
(2) Immediately after the CSS (Fig. 12b): contraction resulting from downslope translation and buttressing of the failed CSS sediments results in shortening in the distal region amounting to an estimated 6.35 km, which is an approximation for the maximum translation distance (*sensu* Varnes 1978; Frey-Martinez *et al.* 2006). It is expected that the upslope region undergoes depletion of a similar amount (Varnes 1978; Frey-Martinez *et al.* 2005, 2006). We therefore assume that Slide Block R has been translated over a distance of 6 km, consistent with the value of translation calculated from the contractional deformation. It is assumed for simplicity that a typical tri-partite slide anatomy develops (see Bull *et al.* 2009*a* and references therein), where by the upslope region is characterized by extension, the downslope region by contraction and the intermediate region undergoes relatively intact translation.
(3) Present-day conditions (Fig. 12c): the present-day seafloor profile and distribution of deformation within the CSS is represented in accordance to Figure 11b. A zone dominated by extension (graben blocks and listric faults) has developed updip of the contractional downslope region. In the upslope region close to Headwall R, failed sediments are very thin. Although absolute depletion cannot be quantified from the available data, this model suggests that extensive depletion evidenced by the thin upslope region and normal faults exceeds the relatively modest amount of shortening/translation estimated from the downslope region.

A simple explanation for the observed volume imbalance is that the CSS was further modified following its emplacement, by subsequent phases of slope failure within the wider Storegga Slide complex. This point is underscored by comparison of the present-day seafloor profile with the seafloor profile for each of the previous phases (Fig. 12c). A clear disparity exists between the present-day seafloor profile and the expected profile after phase 2 (shaded grey on Fig. 12c). This suggests that a further intermediate and substantive phase of slope failure or volume transfer (i.e. out with the CSS) has occurred following CSS formation to produce the observed volume imbalance between its upslope and downslope regions.

Evidence for post-event modification of the CCS includes the unique sequence of deformation observed updip of its contractional zone. The sequence of regularly spaced, oppositely verging thrust-fault pairs defining upwardly displaced 'pop-up blocks' gives way to a series of regularly spaced oppositely verging normal fault pairs defining downward-displaced graben blocks (Figs 8a & 9a). The geometry of these sets of contractional v. extensional features is so similar that a logical conclusion is that, originally, all fault pairs formed as thrusts in response to the downslope movement of material and frontal buttressing (*sensu* Frey-Martinez *et al.* 2006). Following post-event modification postulated to involve the further removal of a significant volume of material (Fig. 12c) from the upper CSS area (i.e. to the NE of the CSS), some thrust faults subsequently 'relaxed' towards the evacuated area. This induced localized relief of the contractional regime responsible for forming the thrust faults and caused some faults originally formed as thrusts to reactivate in a normal sense (Fig. 12c). We therefore support the model originally advanced by Gafeira *et al.* (2007), that compression preceded extensional deformation, with the latter attributed to escape of material to the NE and a reduction in lateral confining pressures.

Removal of large volumes of material immediately NE of the CSS would tend to promote extensional spreading in a northerly direction, consistent with the minor extension observed in the form of the normal faults that cross-cut the southern margin of the CSS (Figs 9b, c & 11a). Possible further or later stages of slope failure and removal of material are also evidenced in the increasing intensity of extensional deformation seen as 'degradation' of grabens and intervening horsts updip of the distal region (Figs 7, 8b & 9a).

Wider context of the CSS in the Storegga Slide complex

Establishing the exact timing of the development of the CSS, and propagation of the contractional structures within, relative to the Storegga Slide phases or lobes is made difficult by the provenance attributed to the compression zone in the previous studies. Bugge (1983) was the first to recognize the composite nature of the Storegga Slide and identified three individual events. Haflidason *et al.* (2004) concluded that five major slide lobes and almost 60 smaller slide debris lobe events attest to several key phases of failure, beginning with initiation of the Storegga Slide in the Brygge Formation oozes in the 'lower headwall' area and retrogressive upslope development. Failure then occurred in the central area of the present-day Storegga complex, with deep-seated failure affecting the central main headwall area and leading to the deposition of blocky slide material in an event referred to as 'Lobe II' by Haflidason *et al.* (2004) and Bryn *et al.* (2005) (cf. the second or 'central' failure event of Bugge 1983). The compression zone is historically included in Lobe II (hereafter referred to as Lobe 2), which, critically, is thought to represent the phase of failure which generated the Storegga Slide tsunami (Haflidason *et al.* 2004; Bryn *et al.* 2005). The outline of Lobe 2 (Haflidason *et al.* 2004) clearly indicates its continuous downslope extent throughout the mid-slope region of the Storegga Slide complex and, as such, it is implicitly considered as being characterized as a frontally emergent slide, which would justify its further description as 'erosive', and as the 'strongest single candidate to create the Storegga tsunami' (Haflidason *et al.* 2004, page 222). These attributes are possibly in conflict with our interpretation of the CSS as a frontally confined submarine landslide.

Evidence of post-CSS modification is opposed to the sequence of Storegga Slide development established by previous studies, which broadly conclude that the compression zone (herein considered the downslope region of the CSS) formed during emplacement of Lobe 2 (Haflidason *et al.* 2004). Further lobes are thought to have occurred after development of Lobe 2, notably with Lobe 4 occurring updip and to the south of the CSS and Lobe 5 partially affecting the headwall region of the CSS. However, both the sediment volumes and translation distances involved in Lobes 4 and 5 are thought to be relatively minor (Haflidason *et al.* 2004; Bryn *et al.* 2005). The evidence presented herein and previously by Gafeira *et al.* (2007) suggests that modification of Lobe 2 by subsequent slide lobes may have been more significant. It is beyond the scope of this paper to fully re-examine the interaction of subsequent slide phases involved in the Storegga Slide. However, the application of line length balancing to discrete slide events or lobes where strain can be quantified has been shown to be useful in the evaluation of volumetric transfer and is applicable to both ancient (i.e. Steventon *et al.* 2018) and modern examples sampled by seismic reflection data.

A major open question arising from our study concerns the strain rate of the slide event resulting in the CSS. Given that this volumetric significance of the Storegga Slide has previously been interpreted as a frontally emergent slide, it is entirely reasonable to equate its emplacement with the generation of a tsunami, as has been done in numerous modelling studies (see Bryn *et al.* 2005 for summary). Those studies have indicated velocities in the range of 10–35 m s^{-1} (Harbitz 1992; De Blasio *et al.* 2003; Bondevik *et al.* 2005; Bryn *et al.* 2005). Recognition of the frontally buttressed structure of this 'slump' led Færseth and Sætersmoen (2008) to infer translation of the body downslope at a velocity of 5 m s^{-1} or less. Færseth and Sætersmoen (2008) further postulated that the cumulative total duration of translation and development of the associated deformation is likely to have been 1 h, and they invoked a two-stage model of progression involving a period of initial acceleration at the leading edge, followed by a more gradual slowing down.

Given the divergence of strain rates proposed by previous studies we recommend further modelling be carried out to determine how this volumetrically significant region of the Storegga Slide complex generated tsunami waves, by including the evidence presented here and by Gafeira *et al.* (2007) and Færseth and Sætersmoen (2008). These studies all suggest this major slide element to be reinterpreted as frontally confined rather than frontally emergent. It seems plausible to us that deformation buttressed at a frontal ramp, detaching at a depth of >500 m and is 135 km in length, with a progressive development of >40 thrust faults with displacements of tens of metres and linked shortening along the basal detachment of *c.* 6 km, is unlikely to have propagated with slip rates of >10 m s^{-1}. It seems more likely that this frontally confined component developed early, and slowly, with strain rates commensurate with bulk strength of intact slope sediments, and a later phase of more dramatic deformation resulted in disaggregation of parts of the proximal region of the CSS and the generation of the tsunami.

Conclusions

The area of the Storegga Slide historically referred to as the 'compression zone', has been interpreted as the downslope region of a discrete 'frontally confined' slide (*sensu* Frey-Martinez *et al.* 2006), informally named here the 'Confined Storegga Slide'. This interpretation is based on the deep-seated failure surface, steep frontal ramp, close association with 'pinned' blocks of remnant pre-failure stratigraphy and limited downslope translation.

The combination of extensive subsurface data coverage and the large scale of deformation within the CSS allow a unique opportunity to determine the magnitude and distribution of strain within the toe zone of a discrete submarine landslide event.

Fault-displacement analysis of over 30 pairs of thrust faults evident over a distance of 30 km of the seabed, along with measurement of folded horizons based on 2D dip-orientated seismic profiles, has shown that a total shortening of *c.* 6.35 km occurred.

Analysis of 3D seismic data has revealed the presence of later-stage extensional deformation and translation of sediments in a direction orthogonal to the CSS, interpreted as the result of later-stage failure as part of the multi-phase Storegga Slide.

Volumetric evaluation based on comparison of scaled, schematic models of the likely pre-failure slope configuration, development of the CSS and present-day conditions indicate a significant volume of material has been removed after emplacement of the CSS.

Our results demonstrate the utility of combining line length balancing with reconstruction of the seafloor position prior to slope failure. Such an approach can be used to consider the seafloor deformation, volume transfer and translation magnitude involved in submarine landslides, along with the timing and interaction of discrete elements within complex multi-phase events.

Acknowledgements We are grateful to Equinor (Andreas Helsem) for provision of data and to Schlumberger for providing seismic interpretation software. This work was carried out at Cardiff University's 3D Lab as part of Suzanne Bull's PhD thesis supervised by Joseph Cartwright and Mads Huuse. Greg Browne and Jess Hillman (GNS Science) are thanked for providing constructive comments on the content of the manuscript, and Lorena Moscardelli and Joshu Mountjoy are gratefully acknowledged for providing thorough reviews which improved the final version.

Funding This research received no specific grant from any funding agency in the public, commercial, or not-for-profit sectors.

Author contributions SB: conceptualization (supporting), formal analysis (lead), investigation (lead), methodology (equal), project administration (lead), validation (equal), visualization (lead), writing – original draft (lead), writing – review & editing (lead); **JAC**: conceptualization (lead), formal analysis (supporting), funding acquisition (lead), investigation (supporting), methodology (equal), project administration (supporting), software (lead), supervision (lead), validation (supporting), visualization (supporting), writing – original draft (supporting), writing – review & editing (supporting).

References

Alsop, G.I., Marco, S., Levi, T. and Weinburger, R. 2017. Fold and thrust systems in Mass Transport Deposits. *Journal of Structural Geology*, **94**, 98–115, https://doi.org/10.1016/j.jsg.2016.11.008

Alves, T.M. and Cartwright, J.A. 2009. Volume balance of a submarine landslide in the Espirito Santo Basin, offshore Brazil: Quantifying seafloor erosion, sediment accumulation and depletion. *Earth and Planetary Science Letters*, **288**, 572–580, https://doi.org/10.1016/j.epsl.2009.10.020

Berg, K., Solheim, A. and Bryn, P. 2005. The Pleistocene to recent geological development of the Ormen Lange area. *Marine and Petroleum Geology*, **22**, 45–56, https://doi.org/10.1016/j.marpetgeo.2004.10.009

Bondevik, S., Løvholt, F., Harbitz, C., Mangerud, J., Dawson, A. and Svendsen, J.I. 2005. The Storegga Slide tsunami: comparing field observations with numerical simulations. *Marine and Petroleum Geology*, **22**, 195–208, https://doi.org/10.1016/j.marpetgeo.2004.10.003

Brekke, H. 2000. The tectonic evolution of the Norwegian Sea continental margin with emphasis on the Vøring and Møre basins. *Geological Society, London, Special Publications*, **136**, 327–378, https://doi.org/10.1144/GSL.SP.2000.167.01.13

Brown, A.R. 2003. *Interpretation of Three-Dimensional Seismic Data*. AAPG Memoir. Tulsa, OK, **42**, 541.

Bryn, P., Berg, K., Forsberg, C.F., Solheim, A. and Kvalstad, T.J. 2005. Explaining the Storegga Slide. *Marine and Petroleum Geology*, **22**, 11–19, https://doi.org/10.1016/j.marpetgeo.2004.12.003

Bugge, T. 1983. Submarine slides on the Norwegian continental margin, with special emphasis on the Storegga area. *Institutt for kontinentalsokkelundersøkelser*, **110**, 1–152.

Bull, S., Cartwright, J. and Huuse, M. 2009*a*. A review of kinematic indicators from mass-transport complexes using 3D seismic data. *Marine and Petroleum Geology*, **26**, 1132–1157, https://doi.org/10.1016/j.marpetgeo.2008.09.011

Bull, S., Cartwright, J.A. and Huuse, M. 2009*b*. A subsurface evacuation model for submarine slope failure. *Basin Research*, **21**, 433–443, https://doi.org/10.1111/j.1365-2117.2008.00390.x

De Blasio, F.V., Issler, D. *et al.* 2003. Dynamics, velocity and run-out of the giant Storegga Slide. *In*: Locat, J. and Mienert, J. (eds) *Submarine Mass Movements and their Consequences*. Kluwer Academic Press, 223–230.

Evans, D., King, E.L., Kenyon, N.H., Brett, C. and Wallis, D. 1996. Evidence for long-term instability in the Storegga Slide region off western Norway. *Marine Geology*, **130**, 281–292, https://doi.org/10.1016/0025-3227(95)00135-2

Evans, D., Harrison, Z. *et al.* 2005. Palaeoslides and other mass failures of Pliocene to Pleistocene age along the Atlantic continental margin of NW Europe. *Marine and Petroleum Geology*, **22**, 1131–1148, https://doi.org/10.1016/j.marpetgeo.2005.01.010

Farrell, S.G. 1984. A dislocation model applied to slump structures, Ainsa Basin, South Central Pyrenees. *Journal of Structural Geology*, **6**, 727–736, https://doi.org/10.1016/0191-8141(84)90012-9

Frey-Martinez, J., Cartwright, J. and Hall, B. 2005. 3D seismic interpretation of slump complexes: examples from the continental margin of Israel. *Basin Research*, **17**, 83–108, https://doi.org/10.1111/j.1365-2117.2005.00255.x

Frey-Martinez, J., Cartwright, J. and James, D. 2006. Frontally confined v. frontally emergent submarine landslides: a 3D seismic characterisation. *Marine and Petroleum Geology*, **23**, 585–604, https://doi.org/10.1016/j.marpetgeo.2006.04.002

Færseth, R.B. and Sætersmoen, B.H. 2008. Geometry of a major slump structure in the Storegga slide region offshore western Norway. *Norwegian Journal of Geology*, **88**, 1–11.

Gafeira, J., Bulat, J. and Evans, D. 2007. The southern flank of the Storegga Slide: imaging and geomorphological analyses using 3D seismic. *In*: Lykousis, V., Sakellariou, D. and Locat, J. (eds) *Submarine Mass Movements and Their Consequences*. Springer, Dordrecht, 57–66.

Haflidason, H., Sejrup, H.P. *et al.* 2004. The Storegga Slide: architecture, geometry and slide development. *Marine Geology*, **231**, 201–234, https://doi.org/10.1016/j.margeo.2004.10.007

Haflidason, H., Lien, R., Sejrup, H.P., Forsberg, C.F. and Bryn, P. 2005. The dating and morphometry of the Storegga Slide. *Marine and Petroleum Geology*, **22**, 123–136, https://doi.org/10.1016/j.marpetgeo.2004.10.008

Harbitz, C.B. 1992. Model simulations of tsunamis generated by the Storegga Slides. *Marine Geology*, **105**, 1–21, https://doi.org/10.1016/0025-3227(92)90178-K

Higgins, S., Clarke, B., Davies, R.J. and Cartwright, J.A. 2009. Internal geometry and growth history of a thrust-related anticline in a deep-water fold belt. *Journal of Structural Geology*, **31**, 1597–1611, https://doi.org/10.1016/j.jsg.2009.07.006

Hjelstuen, B.O., Sejrup, H.P., Haflidason, H., Nygård, A., Ceramicola, S. and Bryn, P. 2005. Late Cenozoic glacial history and evolution of the Storegga Slide area and adjacent slide flank regions, Norwegian continental margin. *Marine and Petroleum Geology*, **22**, 57–69, https://doi.org/10.1016/j.marpetgeo.2004.10.002

Huvenne, V.A.I., Croker-Peter, F. and Henriet, J.P. 2002. A refreshing 3D view of an ancient sediment collapse and slope failure. *Terra Nova*, **14**, 33–40, https://doi.org/10.1046/j.1365-3121.2002.00386.x

King, E.L., Sejrup, H.P., Haflidason, H., Elverhøi, A. and Aarseth, I. 1996. Quaternary seismic stratigraphy of the North Sea Fan: glacially-fed gravity flow aprons hemipelagic sediments, and large submarine slides. *Marine Geology*, **130**, 293–315, https://doi.org/10.1016/0025-3227(95)00168-9

Laberg, J.S. and Vorren, T.O. 2000. The Trænadjupet Slide, offshore Norway-morphology, evacuation and triggering mechanisms. *Marine Geology*, **171**, 95–114, https://doi.org/10.1016/S0025-3227(00)00112-2

Lastras, G., Canals, M., Urgeles, R., De Batist, M., Calafat, A.M. and Casamor, J.L. 2004. Characterisation of the recent BIG'95 debris flow on the Ebro margin, Western Mediterranean Sea, after a variety of seismic reflection data. *Marine Geology*, **213**, 235–255, https://doi.org/10.1016/j.margeo.2004.10.008

Lewis, K. 1971. Slumping on a continental slope inclined at 1–4°. *Sedimentology*, **16**, 97–110, https://doi.org/10.1111/j.1365-3091.1971.tb00221.x

Lintern, D.G., Mosher, D.C. and Scherwath, M. 2019. Advancing from subaqueous mass movement case studies to providing advice and mitigation. *Geological Society, London, Special Publications*, **477**, 1–14, https://doi.org/10.1144/SP477-2018-190

Masson, D.G., Huggett, Q.J. and Brunsden, D. 1993. The surface texture of the Saharan debris flow and some speculations on submarine debris flow processes. *Sedimentology*, **40**, 583–598.

Micallef, A., Masson, D.G., Berndt, C. and Stow, D.A.V. 2009. Development and mass movement processes of the north-eastern Storegga Slide. *Quaternary Science Reviews*, **28**, 433–448, https://doi.org/10.1016/j.quascirev.2008.09.026

Micallef, A., Masson, D.G., Berndt, C. and Stow, D.A.V. 2016. Submarine spreading in the Storegga Slide, Norwegian Sea. *Geological Society, London, Memoirs*, **46**, 411–412, https://doi.org/10.1144/M46.88

Moscardelli, L., Wood, L. and Mann, P. 2006. Mass-transport complexes and associated processes in the offshore area of Trinidad and Venezuela. *AAPG Bulletin*, **90**, 1059–1088, https://doi.org/10.1306/02210605052

Ogata, K., Mountjoy, J., Pini, G.A. and Festa, A. 2014. Shear zone liquefaction in mass transport deposit emplacement: A multi-scale integration of seismic reflection and outcrop data. *Marine Geology*, **356**, 50–64, https://doi.org/10.1016/j.margeo.2014.05.001

Omeru, T. 2014. *Mass Transport Complexes: Implications for reservoir seals*. PhD thesis, Cardiff University.

Omeru, T. and Cartwright, J.A. 2019. The efficacy of kinematic indicators in a complexly deformed Mass Transport Deposit: Insights from the deepwater Taranaki Basin, New Zealand. *Marine and Petroleum Geology*, **106**, 74–87, https://doi.org/10.1016/j.marpetgeo.2019.04.037

Prior, D.B., Bornhold, B.D. and Johns, M.W. 1984. Depositional characteristics of a submarine debris flow. *Journal of Geology*, **92**, 707–727, https://doi.org/10.1086/628907

Reis, A.T., Araújob, E. *et al.* 2016. Effects of a regional décollement level for gravity tectonics on late Neogene to recent large-scale slope instabilities in the Foz do Amazonas Basin, Brazil. *Marine and Petroleum Geology*, **75**, 29–52, https://doi.org/10.1016/j.marpetgeo.2016.04.011

Rise, L., Ottesen, D., Longva, O., Solheim, A., Andersen, E.S. and Ayers, S. 2006. The Sklinnadjupet slide and its relation to the Elsterian glaciation on the mid-Norwegian margin. *Marine and Petroleum Geology*, **23**, 569–583, https://doi.org/10.1016/j.marpetgeo.2006.05.005

Solheim, A., Berg, K., Forsberg, C.F. and Bryn, P. 2005. The Storegga Slide: repetitive large scale sliding with similar cause and development. *Marine and Petroleum Geology*, **22**, 97–107, https://doi.org/10.1016/j.marpetgeo.2004.10.013

Steventon, M.J., Jackson, C.A.-L., Hodgson, D.M. and Johnson, H.D. 2018. Strain analysis of a seismically imaged mass-transport complex, offshore Uruguay. *Basin Research*, **31**, 600–620, https://doi.org/10.1111/bre.12337

Varnes, D.J. 1978. Slope movement types and processes. *In*: Schuster, R.L. and Krizek, R.J. (eds) *Landslides, Analysis and Control*. National Academy of Sciences, Washington, Special Report, **176**, 11–33.

A new depositional model for the Tuaheni Landslide Complex, Hikurangi Margin, New Zealand

Benjamin Couvin[1]*, Aggeliki Georgiopoulou[2], Joshu J. Mountjoy[3], Lawrence Amy[1], Gareth J. Crutchley[4], Morgane Brunet[5], Sebastian Cardona[6], Felix Gross[7], Christoph Böttner[4], Sebastian Krastel[7] and Ingo Pecher[8]

[1]Irish Centre for Research in Applied Geosciences, University College Dublin, Ireland

[2]School of Environment and Technology, University of Brighton, UK

[3]National Institute for Water and Atmospheric Research, Wellington, New Zealand

[4]GEOMAR Helmholtz-Zentrum für Ozeanforschung, Kiel, Germany

[5]Géosciences Rennes, Université Rennes 1, France

[6]Colorado School of Mines, Golden, CO, USA

[7]Institute of Geosciences, University of Kiel, Germany

[8]School of Environment, University of Auckland, New Zealand

BC, 0000-0002-0577-0798; AG, 0000-0003-4298-5090; JJM, 0000-0002-3703-0006; LA, 0000-0001-7502-6887; GJC, 0000-0002-6692-0665; MB, 0000-0002-8896-4669; SC, 0000-0001-6800-9608; FG, 0000-0002-0749-829X; CB, 0000-0001-7321-0699; SK, 0000-0002-5899-9748

*Correspondence: benjamin.couvin@ucdconnect.ie

Abstract: The Tuaheni Landslide Complex (TLC) is characterized by areas of compression upslope and extension downslope. It has been thought to consist of a stack of two genetically linked landslide units identified from seismic data. We used 3D seismic reflection, bathymetry data and International Ocean Discovery Program Core U1517C (Expedition 372) to understand the internal structures, deformation mechanisms and depositional processes of the TLC deposits. Units II and III of U1517C correspond to the two chaotic units in 3D seismic data. In the core, Unit II shows deformation, whereas Unit III appears more like an *in situ* sequence. Variance attribute analysis showed that Unit II is split into lobes around a coherent stratified central ridge and is bounded by scarps. By contrast, we found that Unit III is continuous beneath the central ridge and has an upslope geometry, which we interpreted as a channel–levee system. Both units show evidence of lateral spreading due to the presence of the Tuaheni Canyon removing support from the toe. Our results suggest that Units II and III are not genetically linked, are separated substantially in time and had different emplacement mechanisms, but they fail under similar circumstances.

Subaqueous landslides have long been recognized as a geohazard. They can trigger tsunamis, endanger offshore installations and may induce coastal subsidence (Hampton *et al.* 1996; Mulder and Cochonat 1996; Talling *et al.* 2013; Fruergaard *et al.* 2015). Some events like the 1929 Grand Banks earthquake off Newfoundland (Piper *et al.* 1999) or the more recent Nice airport collapse in 1979 (Ioualalen *et al.* 2010; Kelner *et al.* 2016) are examples of cases where human settlements and installations have been directly impacted by subaqueous landslides. Considering the significance of this implication, it is important to better constrain the mechanisms that can lead to the generation and reactivation of subaqueous landslides.

At the Hikurangi Margin, off the eastern coast of New Zealand's North Island, the Tuaheni Landslide Complex (TLC) (Fig. 1) has been interpreted from bathymetry data analysis as a slow-moving landslide (Mountjoy *et al.* 2009). The landslide was recognized by its surface roughness on bathymetry data, while its creeping movement was interpreted on the basis of a reactivated landslide morphology, including a downslope extensional domain, arcuate (concave from downslope) surface scarps and strike-slip faulting along the margins of the landslide body (Mountjoy *et al.* 2009).

Recognition of mass-transport deposits (MTDs) in seismic data is generally based on the identification of a disrupted or chaotic seismic facies underlain

From: Georgiopoulou, A., Amy, L. A., Benetti, S., Chaytor, J. D., Clare, M. A., Gamboa, D., Haughton, P. D. W., Moernaut, J. and Mountjoy, J. J. (eds) 2020. *Subaqueous Mass Movements and their Consequences: Advances in Process Understanding, Monitoring and Hazard Assessments.* Geological Society, London, Special Publications, **500**, 551–566.
First published online May 19, 2020, https://doi.org/10.1144/SP500-2019-180

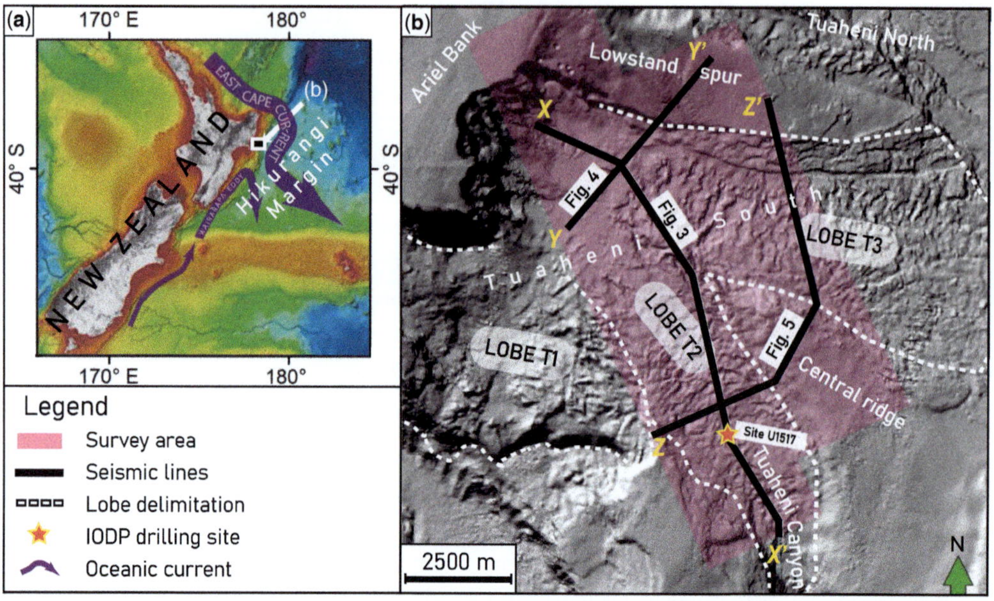

Fig. 1. (**a**) Map of New Zealand and surrounding bathymetry showing the location of the Hikurangi Margin and the study area (white box). Courtesy of National Institute of Water and Atmospheric Research (NIWA). (**b**) Enlarged view of the study area with shaded relief bathymetry, showing the 3D seismic survey area in pink as well as the location of the TLC South lobes (after Mountjoy *et al.* 2009), the International Ocean Discovery Program (IODP) core site (U1517) and the three composite seismic lines shown in this study.

by a basal shear surface (Bull *et al.* 2009). The TLC is thought to be composed of two or more stacked landslide units (Mountjoy *et al.* 2014; Böttner *et al.* 2018; Gross *et al.* 2018; Kuhlmann *et al.* 2018). In 2017, during International Ocean Discovery Program (IODP) Expedition 372 (Exp 372), a 190 m core (U1517C) was retrieved at the TLC (Pecher *et al.* 2018) (Fig. 1b). Physical properties were also acquired during Exp 372, such as density, velocity, magnetic susceptibility and gamma-ray. These data helped correlate the unit boundaries of Core U1517C with reflectors in seismic data (Pecher *et al.* 2018), but they are not used further in this study.

The initial observations of the core highlighted a discrepancy between the observed lithology and the seismic facies; the deposits above what should be the base of the slide at 66 mbsf appear largely undeformed and were therefore interpreted as a remobilized intact block (Pecher *et al.* 2018). This interpretation seems to clash with the previous model of a stack of distinct chaotically mixed landslide events without blocks (Böttner *et al.* 2018; Gross *et al.* 2018). Hence, an integrated approach was taken in this study for a better characterization of the TLC morphology and depositional history.

In this study, we analysed the sedimentary character and depositional environment of the TLC by integrating three-dimensional (3D) seismic data with the newly acquired sedimentological information from Core U1517C. Our aims were to test the hypothesis that the lower landslide unit is actually an *in situ* sedimentary package, despite having the seismic characteristics of a MTD, and to determine from seismic data analysis a most likely depositional history of the TLC.

Geological setting

The Hikurangi Margin is located off the east coast of North Island, New Zealand and is an active subduction zone where the Pacific Plate is subducting beneath the Australian Plate at an average rate of *c.* 4–5 cm a^{-1} relative to the Australian Plate (Anderson and Webb 1994; Wallace *et al.* 2004). The study area is located on the upper continental slope, which is underlain by a Quaternary succession of sea-level-controlled sedimentary units with clinoform geometry (Mountjoy *et al.* 2009; Pedley *et al.* 2010). Rapid tectonic deformation and eustasy have been significant in controlling sediment flux and shifting depocentres on the Hikurangi Margin (Paquet *et al.* 2009). However, higher sediment accumulation rates have been found to be particularly associated with climato-eustatic extremes, correlating with high terrestrial erosion (Paquet *et al.* 2009). The TLC mass consists of failed clinoform material from the upper slope (Mountjoy *et al.* 2009).

In terms of oceanic currents, the zone of study is located within a large zone of mixing where the northward Wairarapa Eddy meets the eastward East Cape Current (Chiswell *et al.* 2015). However, very little is known about local deep-water currents at the Tuaheni Slope. Sea-level fluctuations were of high amplitude and high frequency during the Quaternary, with sea-level variations of 75 to 150 m in New Zealand (Newnham *et al.* 1999). The sea-level has risen 120 m since the Last Glacial Maximum (*c.* 15–25 ka BP) (Wright *et al.* 1995). While South Island was largely covered by ice during the Last Glacial Maximum, North Island had very few glaciers (Newnham *et al.* 1999).

The TLC consists of two mass-transport complexes, Tuaheni North and Tuaheni South. This study focuses on Tuaheni South, which, for means of simplification, will be referred to as 'the TLC'. The TLC consists of three lobes (from south to north): Lobes T1, T2 and T3. Lobes T2 and T3 are separated by a topographic high (central ridge) (Mountjoy *et al.* 2009) (Fig. 1b). The structural regime within the landslide body is unusual in that it displays a compressional zone upslope and an extensional zone downslope (Mountjoy *et al.* 2009, 2014). The extensional part of the TLC downslope overlies a sedimentary sequence deformed by a set of NE–SW-striking, high dip angle, normal faults (Böttner *et al.* 2018).

In previous studies, the TLC was interpreted from seismic data to consist of a stack of at least two MTD units, recognizable from their chaotic facies (Mountjoy *et al.* 2014; Gross *et al.* 2018; Kuhlmann *et al.* 2018). These two units appear, in seismic data, to be separated by a negative impedance reflection with respect to the seafloor reflector, whose origin remains enigmatic (Gross *et al.* 2018). After considering different geological scenarios for this reflection (referred to in Gross *et al.* (2018) as the 'intra-debris reflector'), Gross *et al.* (2018) suggested that it likely represents the basal shear surface of the upper debris unit and possibly a zone of accumulation for excess fluid pressure.

Methods

Seismic and bathymetry data

A high-resolution 3D seismic reflection volume was acquired at the TLC during the RV *Tangaroa* cruise TAN1404 in 2014, using a P-Cable system for high-resolution 3D seismic acquisition. The total area covered by the seismic volume is *c.* 69 km^2 (Böttner *et al.* 2018; Gross *et al.* 2018) (Fig. 1b). The resolution of the data is *c.* 5 m horizontal and 2–3 m vertical, with a penetration of up to 750 m (roughly equivalent to 1 s two-way travel time). The reader is referred to Gross *et al.* (2018) for a full description of the acquisition and processing details of the P-Cable data. The bathymetry data were also acquired during the TAN1404 voyage, using a Kongsberg EM302 SWATH multibeam echo-sounder (Gross *et al.* 2018).

Core logging and lithology

In 2017, a 190 m long sediment core (U1517C) was drilled in the distal part of Lobe T2 of the TLC during Exp 372 (Pecher *et al.* 2018). A recovery of over 90% allowed for a continuous record through the TLC deposit and the base of the slide to the underlying stratigraphy. The core was first bulk described during the expedition, and five lithological units were defined (I to V) based on lithological changes as well as magnetic susceptibility logs (Pecher *et al.* 2018). In order to better understand the internal structure of the landslide deposits, we have described and logged Units I, II and III of Core U1517C in more detail. We identified the lithology and sedimentary structures of the units to help constrain a more precise depth and nature of the boundaries between each of the units. We measured grain size in parts of the core within Units II and III using a Malvern Mastersizer 3000, which helped provide calibration for core-wide grain-size determination. A total of 74 samples were measured for Unit II (Sections 1H-4W, 2H-3W, 5F-2W, 5F-4W and 6F-1W) and a total of 66 for Unit III (Sections 9F-3W, 10F-1W, 10F-2W, 11F-4W and 13F-3W). The samples were chosen so that most of the lithologies found in the core are represented. Considering the consistency of the results, only representative samples are shown in this study.

Seismic attributes and interpretation

We used the Petrel 2018 software to visualize and analyse the data. The seismic data were interpreted in two-way travel time and we focused on four major horizons: the top of Unit II (R1), the base of Unit II (R2), the top of Unit III (R3) and the base of Unit III (R4). The seafloor surface (R0) was also picked, and, due to the limitations in vertical seismic resolution, it was merged with R1 in most of the central and distal domains of the 3D survey. This is due to Unit I being a surface drape, not covering the top of Unit II uniformly throughout the volume, and therefore being only resolvable in the proximal domain where its thickness is greatest. We interpreted the horizons using both manual and automatic picking, and surfaces were created through extrapolation and gridding from the acquired datapoints. Seismic attributes, such as variance, RMS amplitudes, gradient, most negative amplitude and most positive amplitude, were then derived from the 3D seismic volume in order to detect internal deformation, faults, unconformities and deposit geometry. Only the variance and gradient attributes provided results that were useful for this study.

The horizons were interpreted from seismic data according to a number of steps.

(1) The bounding surfaces were defined by correlation of the seismic data with stratigraphic information from Core U1517C at Site U1517.
(2) Units II and III were mapped by picking the bounding surfaces (top and base) throughout the volume (the units themselves were identified as chaotic discontinuous seismic facies or transparent deformed seismic facies).
(3) When the seismic facies changed abruptly laterally into a facies of continuous, parallel reflections, it was considered that a scarp or unit margin was between them.

Results

Unit II

Lithology. From observation in Core U1517C, Unit II comprises many sandbodies, with thicknesses ranging between 2 and 20 cm. Most of the sand units show evidence of deformation, such as folds and injectites, while some of them are abruptly truncated (Fig. 2a). In general, their internal structure is massive and their base is sharp (Fig. 2a). The top of these layers can be either a sharp contact or gradually fines upwards (Fig. 2a). The average observed grain size is very fine sand to fine sand. Grain-size measurements show a well-sorted, very fine sand with a mode at 70 µm (e.g. Fig. 2b). The colour of these sands is olive grey (5Y-5/2 on the Munsell scale) (Fig. 2a). In addition to massive sand layers, discontinuous thin sand lenses (<1 cm thickness/diameter) were observed (Fig. 2a). Similar in colour and grain size to the massive layers, they are often accompanied by shell fragments or reddish-brown specks, likely to be high in iron.

The rest of the sedimentary sequence consists mostly of structureless clayey silt. The colour of these silts is olive grey (5Y-5/2 on the Munsell scale) (Fig. 2a). Grain-size measurements show a poorly sorted sediment ranging from *c.* 0.5 to 100 µm with a mode at *c.* 8 µm (e.g. Fig. 2b). The uppermost 2 m of Unit II are characterized by a colour banding in the silts at 10–20 cm intervals (Fig. 2a).

Seismic character. In seismic reflection data, Unit II is characterized by a chaotic to transparent seismic facies, where any internal reflectors are discontinuous (Fig. 3). In the upslope domain, NW of the central ridge, Unit II features large blocks of subparallel reflectors surrounded by faults (Fig. 3). Unit II is continuous upslope until the head scarp towards the NW (Figs 3 & 4). In the distal part of Lobe T2, Unit II features normal faults (Fig. 5). In the distal part of Lobe T3, Unit II is faulted by normal and strike-slip faults with an east–west orientation (Fig. 5). The fault interpretation is based primarily on the morphology of the seafloor reflection; it is very difficult (or near impossible) to distinguish fault surfaces within the chaotic units themselves, partly due to migration artefacts arising from strong underlying reflections. North of the central ridge, compression occurs, as evidenced by the presence of thrust faults (Fig. 3).

Lateral extent. The lateral margins of the unit are identifiable in seismic data; geographically, they correspond to seafloor scarps that can be seen in bathymetry data (Fig. 1b). The latter have been well described by Mountjoy *et al.* (2009). Upslope, two main head scarps are identified: a 1 km wide amphitheatric one to the west and a narrower 500 m wide one to the east. Their heights are comparable, *c.* 300 m, and their maximum evacuated length is *c.* 2 km (Mountjoy *et al.* 2009) (Fig. 6). At the northern margin of Lobe T3, as well as around the central ridge and at the western margin of Lobe T2 (in the distal domain only), the unit is bounded by side scarps (Figs 4, 5 & 6). At the western margin towards the proximal zone, instead of being bounded by a side scarp, the unit extends to the SW (Lobe T1), ramping down, eroding the stratigraphy directly below, creating a NW–SE-oriented extensional ramp (*sensu* Bull *et al.* 2009) (Fig. 6). Unit II has a toe, identifiable in seismic data by the pinching out of the chaotic reflectors (Fig. 3a, d). At the most downslope extent of the TLC, Lobe T2 is laterally bound by the walls of the Tuaheni Canyon (Fig. 1b).

Boundaries. In core, the top of Unit II corresponds to the base of Unit I and appears as a smooth progressive boundary from featureless clay above to an alternation of deformed clayey silt layers and deformed fine sand layers downcore (Fig. 2a). At the location where the core was drilled, Unit I is a 3 m thick hemipelagic drape and is, therefore, at the limit of the resolution of the seismic data. For seismic interpretation purposes, it was decided to consider the seafloor as the top of Unit II at that location. Only towards the head scarps of Unit II does Unit I increase in thickness to the degree that it could be mapped separately (Figs 3 & 4). In the core, the basal surface of Unit II is distinctly defined by a 2 cm thick siliciclastic dark grey layer (5Y-4/2 on the Munsell scale – this colour has been found nowhere else within the described portion of the core) at 40.74 m depth. It is massive, structureless, apparently undeformed apart from the coring

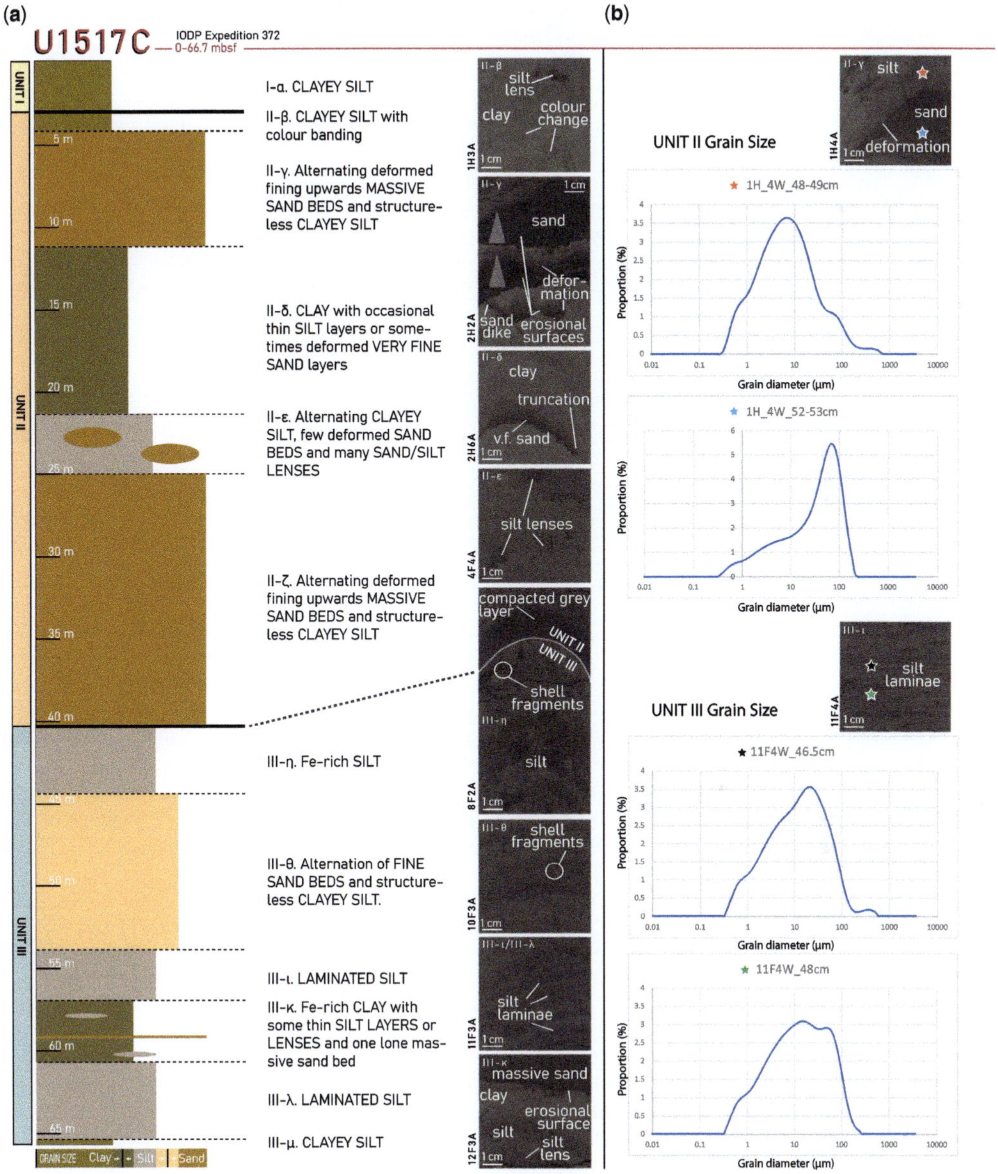

Fig. 2. (**a**) Interpreted sedimentary log of the upper part of IODP Core U1517C, drilled during Expedition 372. Left: Interpretation of sedimentary facies identified within Units I to III. Right: Interpreted photographs of representative sections of the core. Facies number is indicated in the upper left corner of each photograph. Core section number is indicated on the left side of each photograph. (**b**) Grain-size measurements from Sections 1H-4W (Unit II) and 11F-4W (Unit III) of IODP Core U1517C. v.f. sand, very fine sand.

process, has a sharp top and a sharp base and presents a similar grain size to other layers in Unit II (Fig. 2a). However, the layer appears, from direct observation, to be more consolidated than the surrounding layers.

This layer corresponds in seismic data to a reflector of negative acoustic impedance (Reflector R3). Despite covering a large area, this reflector is not completely continuous throughout the volume. It could nonetheless be picked with a good level of confidence in the largest part of the seismic volume and then interpolated.

Seismic attribute analysis. Seismic attributes were calculated from the seismic volume in an attempt

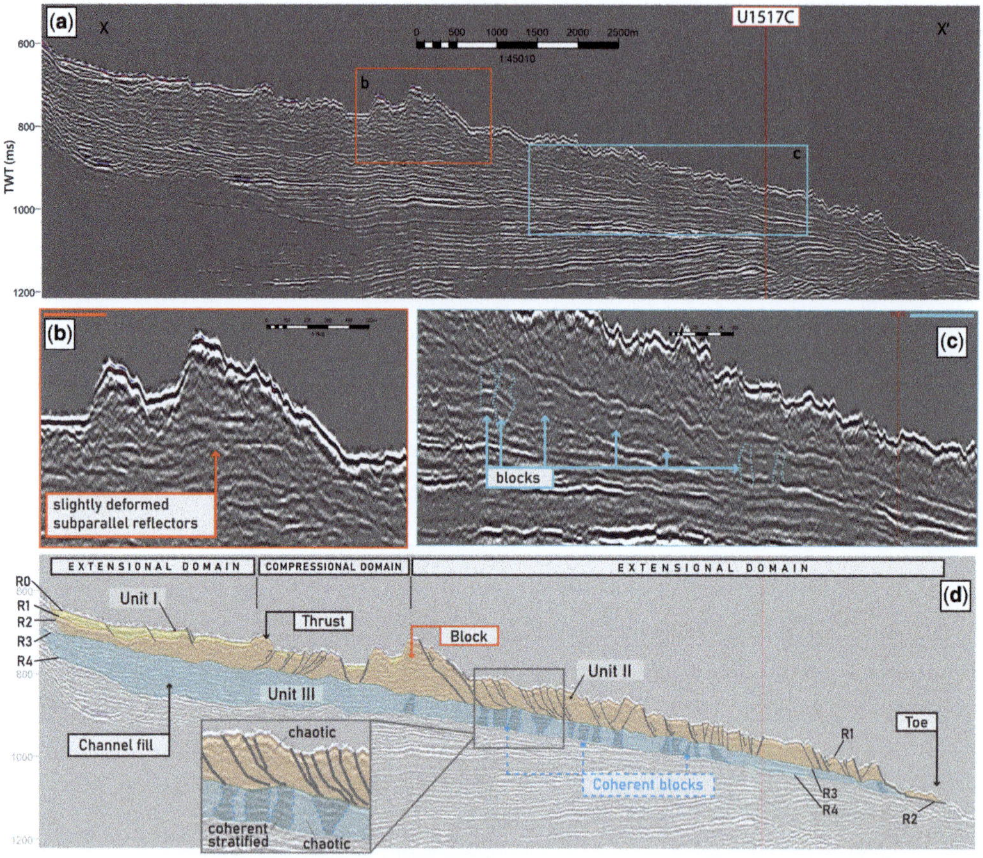

Fig. 3. Seismic line extracted from 3D seismic volume SCHLIP-3D acquired during Expedition TAN1404 (Inline XX′): (**a**) uninterpreted seismic data; (**b**) enlarged view of the block NW of the central ridge (magnification from Inline XX′); (**c**) enlarged view of the small-sized blocks visible on Inline XX′; (**d**) interpretation of Inline XX′. TWT, two-way time; R0, seafloor surface; R1, top of Unit II; R2, base of Unit II; R3, top of Unit III; R4, base of Unit III.

to extract more information about the units' internal character. Variance is a seismic attribute that is generally used to enhance faults, channel and valley margins, prominent changes in dip and lateral discontinuities in general (Randen and Sønneland 2005; Gee and Gawthorpe 2007; Koson *et al.* 2014). A variance map (Fig. 7) of a surface parallel to Reflector R3, 20 ms above it, shows a sharp transition from a zone where variance has a consistently high value (which corresponds to a zone of consistently discontinuous or transparent seismic facies), i.e. the TLC deposits (>0.4), to a zone of low variance value (where the seismic reflectors are continuous), which corresponds to the undisturbed central ridge and lowstand spur sedimentary sequences (0–0.3) (Fig. 7a).

Unit II is consistently thick in the upslope domain (ranging between 50 and 60 ms thickness) and gets progressively thinner downslope to *c.* 20 ms before pinching out (Fig. 8a).

Towards the downslope domain of Lobe T2, a pattern of arcuate ridges appears on thickness data (Fig. 8a). These features are concave in the updip direction and do not extend beyond the margins on Unit II (e.g. the central ridge) (Fig. 8a). On gradient and elevation maps, an elongated ridge is also visible on Horizon R2 within Lobe T2 (Fig. 6c, d).

Unit III

Lithology. In Core U1517C, the sedimentary succession of Unit III (from 40.74 to 66.7 m) consists mainly of fine laminae, with alternations of silty and clay-rich thin beds. Some of the silt beds contain shell fragments (Fig. 2a). The sand fraction is minimal compared to Unit II. Grain size ranges between

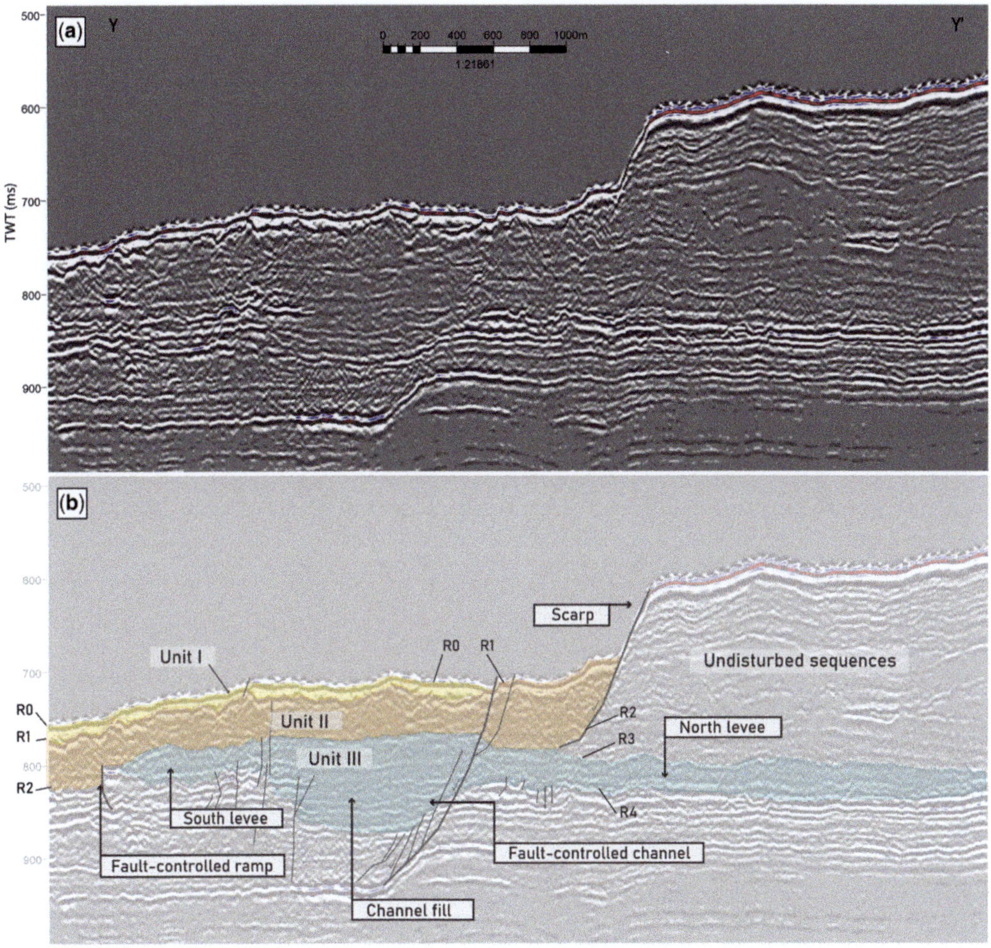

Fig. 4. Seismic line extracted from 3D seismic volume SCHLIP-3D acquired during Expedition TAN1404 (Crossline YY′): (**a**) uninterpreted seismic data; (**b**) interpretation of Crossline YY′.

c. 1 and 120 μm (e.g. Fig. 2b). In terms of macrostructures, little deformation is observed, with scarce, thin, fine sand beds being locally deformed (Fig. 2a).

Seismic character. In seismic data, Unit III is characterized by a chaotic to transparent seismic facies (Figs 3, 4 & 5). At discrete locations of the volume, blocks characterized by a stronger amplitude and a parallel orientation of their reflectors can be observed (Fig. 3). For the same reason as Unit II, it is very difficult to identify faults within the chaotic facies of Unit III. In the upslope domain, we interpreted large faults at each side of the thickest part of the unit (Fig. 4). In the central and distal domains, the seismic facies alternate between transparent, chaotic and parallel. However, no major deformation was observed. The unit seems to be continuous under the central ridge (Fig. 5).

Lateral extent. The lateral margins of Unit III are characterized by the absence of side scarps or clearly defined vertically oriented margins (Figs 4 & 5). We observed a smooth lateral seismic facies change or a progressive pinchout of the chaotic body (Figs 4 & 5). In addition, unlike Unit II, Unit III does not show evidence of a head scarp. As hinted in previous studies (cf. Gross *et al.* 2018, fig. 5), Unit III seems to be confined within a channel in the most upslope domain (Fig. 4). Our reinterpretation of the dataset revealed a more precise location and morphology of this channel. Within the channel, the unit is thicker (50 to 60 ms) than in the rest of the study area (*c.* 20 to 30 ms) (Fig. 8b). At the mouth of the channel in the seaward direction, the unit seems to

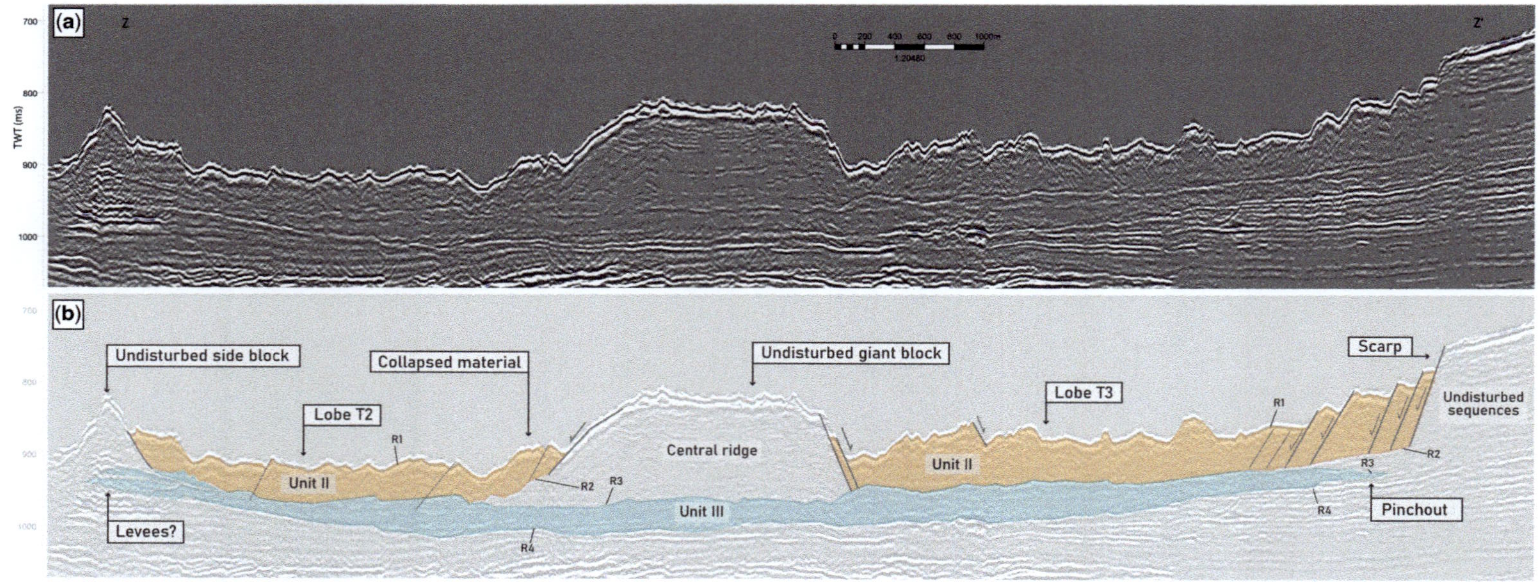

Fig. 5. Seismic line extracted from 3D seismic volume SCHLIP-3D acquired during Expedition TAN1404 (Crossline ZZ′): (**a**) uninterpreted seismic data; (**b**) interpretation of Crossline ZZ′.

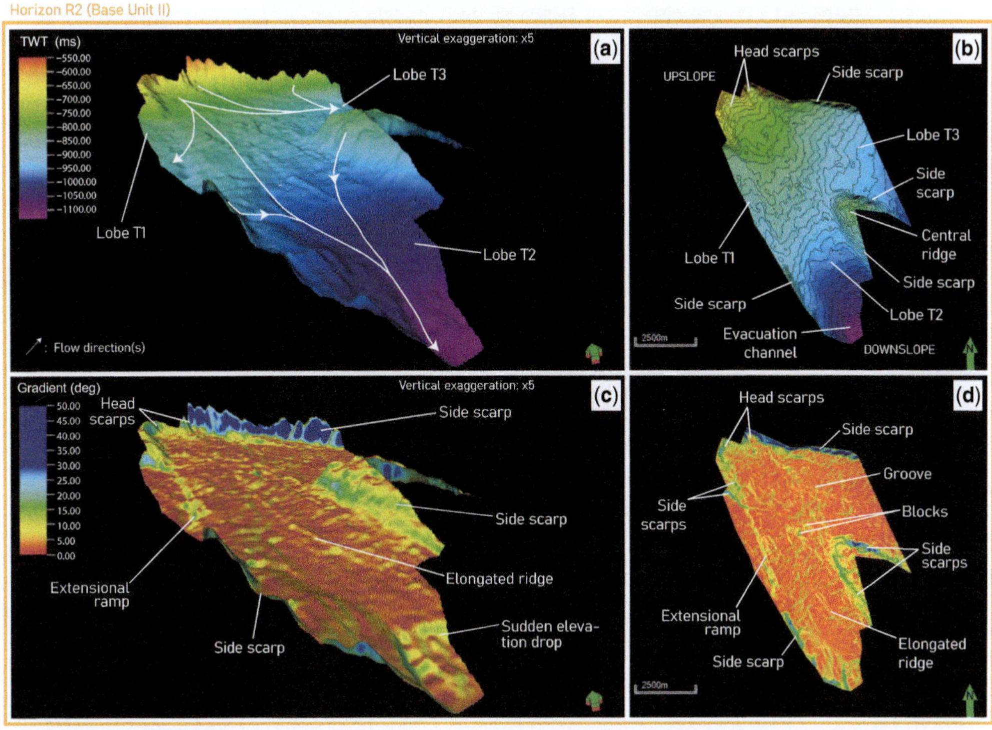

Fig. 6. Horizon R2, corresponding to the base of Unit II. (**a**) 3D view of the base surface of Unit II. The white arrows represent flow directions. (**b**) Two-way time map of Unit II in ms. (**c**) 3D representation of the gradient attribute applied to Reflector R2. The red colour shows the low-angle slopes, whereas blue represents the steepest slopes. (**d**) Gradient map of the base of Unit II.

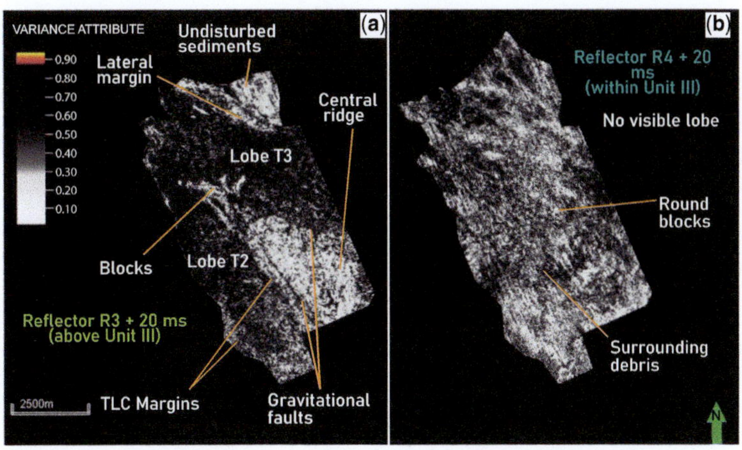

Fig. 7. Variance maps of Unit II and Unit III in time (ms), using the variance attribute. (**a**) Variance map of Reflector R3 + 20 ms (above Unit III). This map shows how Unit II and the central ridge sequence have distinctly different seismic facies. (**b**) Variance map of Reflector R4 + 20 ms (within Unit III). This map shows the dislocated state of the Unit III sedimentary package. White patches correspond to a block of parallel seismic facies around which the seismic facies is chaotic to transparent (in black).

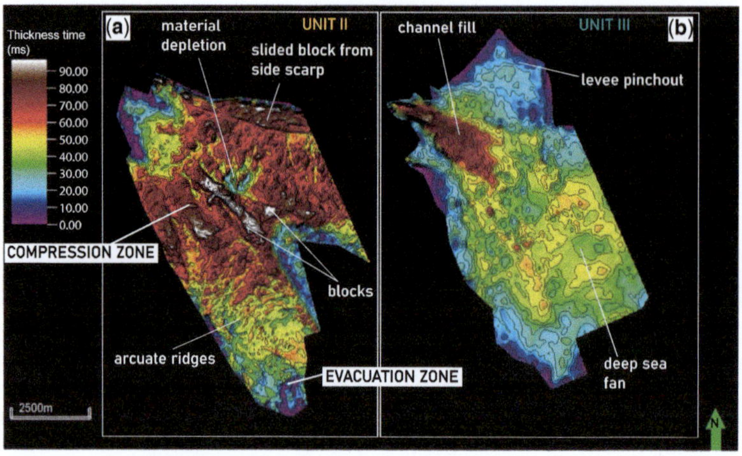

Fig. 8. Thickness maps of Units (**a**) II and (**b**) III in time (ms).

spread out laterally and form two thin levees or channel wings towards the north and south, respectively – the northern one being more extensive than the southern one. A tributary channel was also identified directly NNW of the southern levee (Fig. 9). Unit III has a larger geographical extent than Unit II. It is also continuous under the central ridge and even beyond the side scarps of Unit II (Fig. 9). The extent of Unit

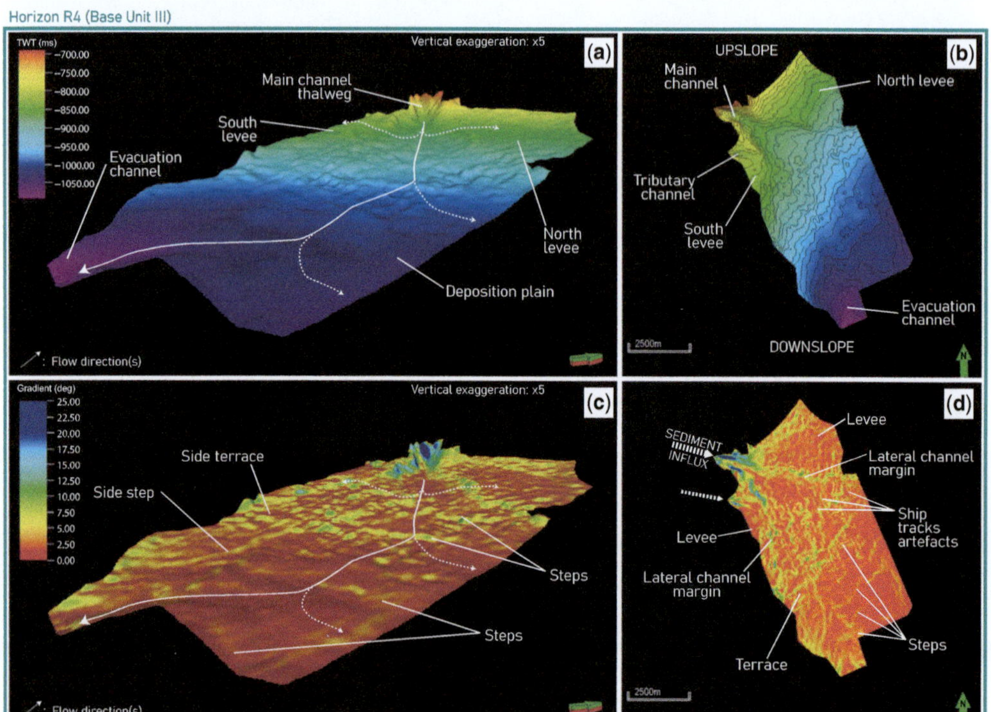

Fig. 9. Horizon R4, corresponding to the base of Unit III. (**a**) 3D view of the base surface of Unit III. The white arrows represent interpreted flow directions and the dashed lines represent interpreted secondary flow directions (levees). (**b**) Two-way time map of Unit III in ms. (**c**) 3D representation of the gradient attribute applied to Reflector R4. (**d**) Gradient map of the base of Unit III using the dip angle attribute. The thick, dashed arrows indicate the source of sediment influx.

III is therefore not bounded by the margins of the TLC as they appear in bathymetry data.

Boundaries. The top surface of Unit III corresponds in Core U1517C to the base of the grey layer identified as the base of Unit II, described above, and is presumed to correspond in seismic data to the R3 reflector.

The base of Unit III was identified in seismic data as a reflector of negative acoustic impedance. In Core U1517C, the base of Unit III was defined at 66.7 m depth based on lithological changes as well as magnetic susceptibility logs (Pecher *et al.* 2018) (Fig. 2a). However, during our visual observations of the core, no obvious basal shear surface could be identified at this depth.

Seismic attribute analysis. The gradient map of the base of Unit III shows a surface with terraces and steps; the surface is rather flat except for discrete locations where the relative gradient is increased locally (Fig. 9c, d). Moreover, variance data show some isolated round patches throughout the volume (Fig. 7b). These patches have a lower variance value (<0.3) (Fig. 7b). The density of these lower variance patches within this slice (Fig. 7b) is *c.* 35%, and their size ranges from 5 to 500 m in diameter. We interpreted from variance attribute analysis that Unit III is not laterally bound by the central ridge (Fig. 7b), as the characteristic low-variance response of the central ridge deposits does not appear on variance results for Unit III.

Central ridge

In seismic data, the central ridge deposits present a parallel continuous to transparent facies (Fig. 5). Overlying Unit III, the reflectors form a wedge that is thickest in its central part (up to 160 ms). The reflectors appear to be truncated by faults. It is surrounded by Unit II debris at its north, south and west sides. Variance data show that the central ridge has a low variance value (<0.3) (Fig. 7a). In its northernmost part, the central ridge is overlain by small blocks of chaotic facies (Fig. 6), which are arguably part of Unit II.

Discussion

The interpretation of the TLC units' lithology and deposition mechanisms will be discussed following the order of emplacement, thus starting with the oldest unit.

Unit III emplacement model

From the interpretation of seismic data and geometric attributes such as variance and gradient, Unit III seems to have significantly different characteristics to Unit II. A key interpretation that we have made in this re-evaluation of the data is that Unit III is continuous beneath the central ridge (Fig. 5). We based this interpretation on a remapping of the reflector that separates Units II and III (the intra-debris reflector, as defined by Gross *et al.* 2018), with the added constraint of sediment character at IODP Site U1517. We are confident that this reflector (here referred to as Reflector R3) can be picked beneath the central ridge (Fig. 5). An important factor to test for our new interpretation is the consistency of the Unit III seismic facies both beneath the central ridge and beneath Unit II.

Unit III upslope seems to present an erosive base, as it creates a channel that cuts through the previously deposited units (Fig. 4). The fill of this channel extends on either side of the channel bed, in a geometry that resembles tapered wedges – a feature usually associated with channel–levee systems and deep-sea fans (Babonneau *et al.* 2002; Migeon *et al.* 2010) (Fig. 4). The identification of a palaeochannel in the upslope domain, combined with core observations, indicates that the Unit III deposits are most likely to have been transported and deposited by turbidity currents through that channel, creating levees and extending broadly into the Tuaheni Basin. The increased thickness of Unit III within the channel thalweg could be explained by a backfilling and subsequent abandonment of the channel (e.g. Cronin *et al.* 2005; Dalla Valle and Gamberi 2011).

Unit III presents various seismic facies within the surveyed area, with the most prevalent one being a chaotic to transparent facies. This type of facies is usually interpreted as deformed material, mixed debris, MTDs etc. (Bull *et al.* 2009). Another type of facies in Unit III is the subparallel stratified facies (Fig. 3), that is likely to correspond to undisturbed strata either *in situ* and undeformed or translated with minimal or no deformation. The round low-variance patches observable on variance slices (Fig. 7b) seem to coincide with the packages of parallel reflectors and could correspond to blocks (Gamboa and Alves 2015). In Core U1517C, the deposits do not appear to have undergone major deformation, and the sedimentary sequence at this precise location could correspond to one of these blocks. Unit III presents a chaotic seismic facies at the location where the core was drilled. However, despite the high resolution of the data, blocks under 5 m in diameter are unlikely to be detected. More blocks have been identified throughout the volume based on observation of sharply contrasted zones in variance slices. We suggest that the presence of blocks within Unit III shows that the unit was remobilized after deposition. Seismic data analysis shows that the Tuaheni Canyon south of the TLC, visible on bathymetry data and thought to be a vector

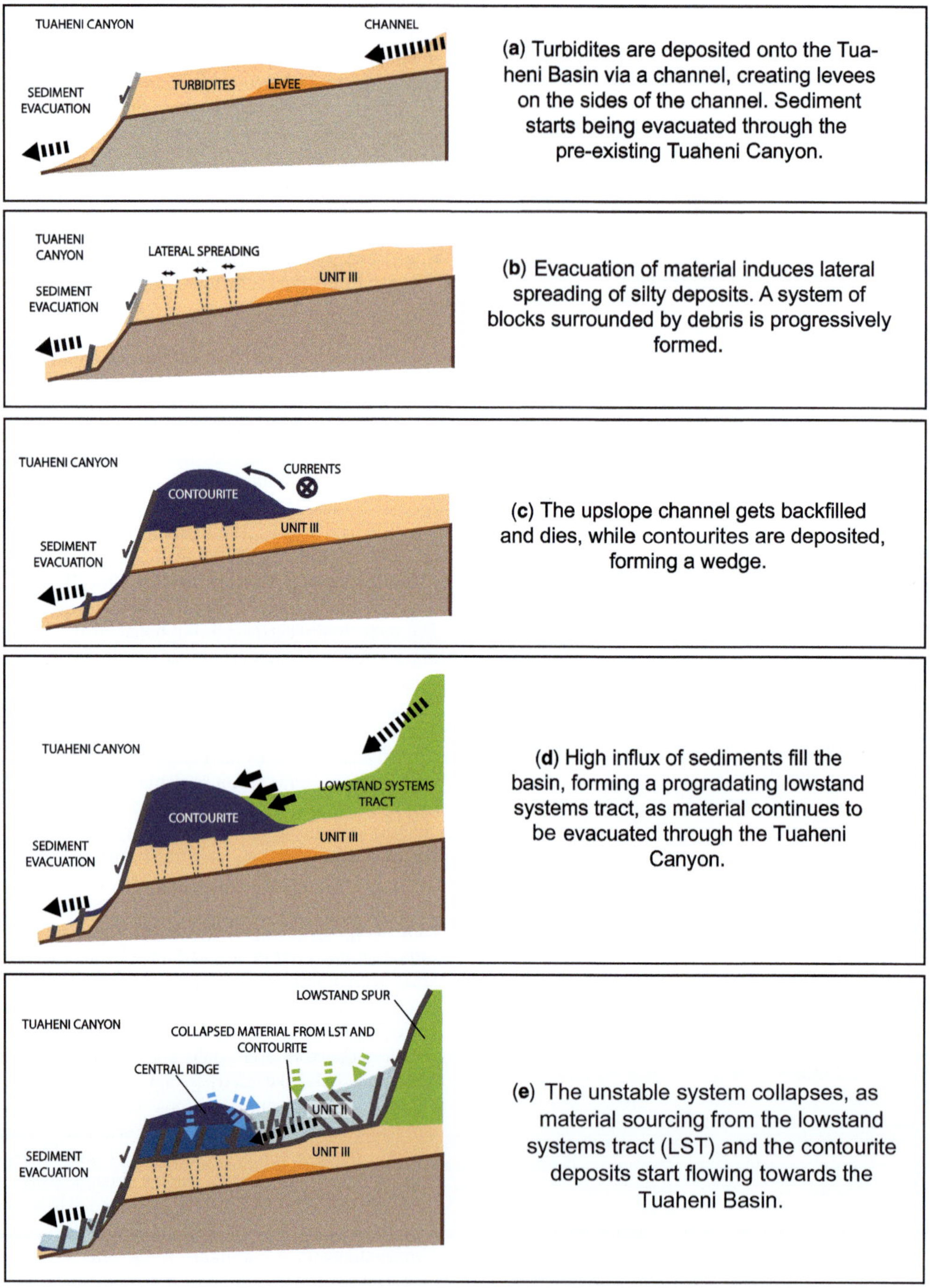

Fig. 10. (**a–e**) A new depositional model for the Tuaheni Landslide Complex, proposed from the interpretation of seismic data and the sediment core. The mid-blue colour in (e) represents landslide material that flows around the central ridge, not beneath it.

of material removal (Mountjoy *et al.* 2009), already existed at the time that Unit III was emplaced. Considering that the conveyor belt model proposed by Mountjoy *et al.* (2009) was already active at that time, we propose that Unit III was subject to a continuous loss of support downslope, which induced the lateral spreading of its sedimentary sequences. This created an extensional system similar to the system observed in the Gulf of Mexico (Sawyer *et al.* 2009), with blocks being pulled away from each other, slowly becoming surrounded by debris depositing *in situ*. On land, an analogue to this mechanism could be peat flows in Irish bogs; peat landslides can form a system of 'rafts' carried by liquefied peat and debris-filled troughs (Dykes and Warburton 2007). Similar onland analogues, in which lateral spread in clay induces displacement of blocks, forming systems of horsts and graben, have been observed in North America (Hungr *et al.* 2014). This emplacement mechanism could explain how Unit III can present a chaotic facies in places and parallel reflections in others and how in Core U1517C little deformation is observable.

Unit II emplacement model

Landslide kinematic indicators (onshore and offshore alike) generally show extension upslope due to sediment depletion and an area of compression downslope due to debris accumulation (Glastonbury and Fell 2008). In the case of the TLC, the mechanics of the landslide differ, as 3D seismic data show that compression occurred in the upslope domain and shows arcuate concave ridges that are evidence of extension in the downslope domain of Lobe T2. These observations concur with previous analysis of bathymetry data at the TLC (Mountjoy *et al.* 2009).

The geometry of the central ridge reflectors on seismic data indicates that the central ridge sedimentary sequence deposited before the emplacement of the Unit II MTD. Using 3D seismic data and attribute analysis, we identified dip-slip gravitational faults at the sides of the central ridge and at the edges of large-scale blocks NW of the central ridge (Figs 3, 5 & 7a), suggesting that the central ridge is the remainder of a larger sedimentary sequence, which partly collapsed. The geometry and internal seismic architecture of the central ridge is indicative of a contouritic deposit (Rebesco *et al.* 2014). Therefore, we propose that Unit II was sourced from a mix of lowstand systems tract sediments that occupied the upslope area (Mountjoy *et al.* 2009) and of contouritic deposits accumulated in the midslope area, which collapsed locally as a result of a loss of support downslope induced by the continuous removal of material in the distal domain due to the presence of the Tuaheni Canyon (conveyor belt model) (Mountjoy *et al.* 2009). Evidence for this is provided by the extensional faults forming concave-upward arcuate ridges in the evacuation channel downslope, in the upper reaches of the Tuaheni Canyon (Figs 3 & 8a). We stress that it remains ambiguous whether the deformation features in the TLC represent active creeping (as proposed by Mountjoy *et al.* (2009)) or whether the landslide complex was reactivated only once in the past and is now immobile.

Proposed sequence of events

We interpreted that Unit III was first deposited on top of previously faulted pre-slide Quaternary clinoforms (Gross *et al.* 2018) by turbidity currents, via at least one channel, forming a deep-sea fan (Fig. 10a). That channel is likely to have backfilled and then been abandoned, possibly due to a relocation of the channel. Removal of material downslope through the Tuaheni Canyon caused these fan deposits to spread laterally, forming rounded blocks between which debris was infilled (Fig. 10b). The sedimentary sequence equivalent to the central ridge was then deposited, possibly due to the activity of bottom currents and/or advance of the Quaternary lowstand systems tract on the slope (Fig. 10c, d). Very little is known about contourite deposits anywhere on the Hikurangi Margin continental slope, making unequivocal identification of the central ridge as a contourite difficult. The remainder of these sequences lie north of Lobe T3, between Tuaheni North and Tuaheni South. The rest failed and became the Tuaheni Landslide Complex, due to the loss of downslope support, as material continued to be evacuated through the Tuaheni Canyon (Fig. 10e). Gravitational faults and blocks developed and left only the contourite deposits of the central ridge unaffected. Unit I, which is a thin drape of hemipelagites at U1517C but possibly more variable at its upslope extent, was then deposited on top of the debris, filling the lows created by the displacement of blocks in Unit II. In our proposed depositional model, Units II and III have different sources and different emplacement mechanisms. We interpreted that the only thing that the units have in common is the cause of failure, i.e. the presence of the Tuaheni Canyon that likely causes constant extension on the deposits on the slope. This new depositional model contrasts with the previous model (Mountjoy *et al.* 2014), which had interpreted the lower unit to be a parent unit to the upper one.

Conclusions

The interpretation of 3D seismic data and analysis of 3D seismic attributes, combined with newly acquired sedimentary information from IODP Expedition 372

(Core U1517C), have allowed a better understanding of the geometry and morphology of the TLC units previously known as the lower and upper landslide units. A key result in understanding the lateral extent of Unit III comes from the calculation of the variance attribute from an arbitrary level (at an offset from Horizon R4) within the interpreted body of Unit III (Fig. 7b). It is clear that the broad character of the unit extends through the subsurface region of the central ridge (Fig. 7b). By contrast, if we extend an arbitrary surface from within Unit II through the central ridge, we observe a marked change in the variance attribute beneath the central ridge. The maps in Figure 7 demonstrate that Unit II is bounded by relatively coherent (i.e. well-stratified, low seismic variance) strata of the central ridge, while Unit III likely continues laterally beneath the central ridge. This means that Unit III was deposited prior to the formation of the central ridge. We find that these two units (II and III, corresponding to upper and lower, respectively), defined in Core U1517C during IODP Expedition 372 and correlated with seismic data thereafter, have distinct emplacement mechanisms and are not genetically linked. The source of the debris and the deformation processes differ. Hence, the status of Unit III being a part of the TLC should be reconsidered as the proposed sequence of events implies that between Units II and III there is a significant time interval, although in this study, it is hard to estimate how long. From these findings, we propose a new depositional model and emplacement history for the TLC. The presence of the Tuaheni Canyon in the downslope extent of both units is significant, as it enabled lateral spreading by the removal of frontal support. Whether its effect on the slope deposits causes a continuous creeping movement of the TLC or whether this happened suddenly remains ambiguous. Considering the results presented in this paper, future research should focus on the geomechanics at the R2 horizon that separates Unit II from Unit III, as it is likely to be the basal shear surface of the TLC. Additionally, if there is creeping movement, the R2 horizon is where it is most likely focused.

Acknowledgements The authors thank the IODP Expedition 372 crew and participants and the expedition staff scientist Dr Leah Levay. We thank the crew and participants of TAN1404. We are grateful to the shipboard party of IODP Exp 372 for all the onboard efforts and valuable discussions regarding the Tuaheni landslides. Professors Ann Cook and Derek Sawyer at Ohio State University are thanked for providing access to their laboratory facilities. Thanks also go to Dr Eoin Dunlevy for his valuable technical assistance regarding the use of the seismic interpretation software. Finally, we thank Drs Matthew J. Owen and Jaume Llopart for their careful reviews, which helped improve the quality of this work.

Funding This publication has emanated from research supported by a grant from Science Foundation Ireland (SFI) under Grant Number 17/RC-PhD/3481 and is co-funded under the European Regional Development Fund. AG received funding support from the Geological Survey Ireland to participate in IODP Expedition 372. JJM and GJC were funded by the Royal Society of New Zealand's Marsden Fund (Grant NIW1603). The collection and processing of the P-Cable 3D volume was jointly funded by the New Zealand Ministry for Business Innovation and Employment (MBIE), the National Institute of Water and Atmospheric Research (NIWA) and the GNS Science Strategic Science Investment Fund (SSIF) and the German Research Foundation (DFG-Grant BI 404/7 I KR 2222/18). The collection and processing of multibeam data were funded by NIWA SSIF.

Author contributions **BC**: conceptualization (supporting), data curation (equal), formal analysis (lead), investigation (lead), methodology (equal), project administration (supporting), validation (equal), visualization (lead), writing – original draft (lead), writing – review & editing (equal); **AG**: conceptualization (lead), data curation (equal), formal analysis (supporting), investigation (supporting), methodology (lead), project administration (lead), resources (equal), supervision (lead), validation (equal), writing – review & editing (lead); **JJM**: conceptualization (lead), data curation (equal), formal analysis (supporting), investigation (supporting), methodology (equal), project administration (supporting), resources (equal), supervision (equal), validation (equal), visualization (supporting), writing – review & editing (equal); **LA**: investigation (supporting), project administration (supporting), supervision (supporting), validation (equal), writing – review & editing (equal); **GJC**: data curation (supporting), formal analysis (supporting), methodology (supporting), resources (equal), writing – review & editing (equal); **MB**: resources (equal), writing – review & editing (equal); **SC**: resources (equal), writing – review & editing (equal); **FG**: writing – review & editing (equal); **CB**: writing – review & editing (equal); **SK**: resources (equal), writing – review & editing (supporting); **IP**: writing – review & editing (supporting).

References

Anderson, H. and Webb, T. 1994. New Zealand seismicity: patterns revealed by the upgraded National Seismograph Network. *New Zealand Journal of Geology and Geophysics*, **37**, 477–493, https://doi.org/10.1080/00288306.1994.9514633

Babonneau, N., Savoye, B., Cremer, M. and Klein, B. 2002. Morphology and architecture of the present canyon and channel system of the Zaire deep-sea fan. *Marine and Petroleum Geology*, **19**, 445–467, https://doi.org/10.1016/S0264-8172(02)00009-0

Böttner, C., Gross, F., Geersen, J., Crutchley, G.J., Mountjoy, J.J. and Krastel, S. 2018. Marine forearc extension in the Hikurangi Margin: new insights from high-resolution 3-D seismic data. *Tectonics*, **37**, 1472–1491, https://doi.org/10.1029/2017TC004906

Bull, S., Cartwright, J. and Huuse, M. 2009. A review of kinematic indicators from mass-transport complexes using 3D seismic data. *Marine and Petroleum Geology*, **26**, 1132–1151, https://doi.org/10.1016/j.marpetgeo.2008.09.011

Chiswell, S.M., Bostock, H.C., Sutton, P.J.H. and Williams, M.J. 2015. Physical oceanography of the deep seas around New Zealand: a review. *New Zealand Journal of Marine and Freshwater Research*, **49**, 286–317, https://doi.org/10.1080/00288330.2014.992918

Cronin, B.T., Akhmetzhanov, A.M., Mazzini, A., Akhmanov, G., Ivanov, M. and Kenyon, N.H. 2005. Morphology, evolution and fill: implications for sand and mud distribution in filling deep-water canyons and slope channel complexes. *Sedimentary Geology*, **179**, 71–97, https://doi.org/10.1016/j.sedgeo.2005.04.013

Dalla Valle, G. and Gamberi, F. 2011. Slope channel formation, evolution and backfilling in a wide shelf, passive continental margin (Northeastern Sardinia slope, Central Tyrrhenian Sea). *Marine Geology*, **286**, 95–105, https://doi.org/10.1016/j.margeo.2011.06.005

Dykes, A.P. and Warburton, J. 2007. Mass movements in peat: a formal classification scheme. *Geomorphology*, **86**, 73–93, https://doi.org/10.1016/j.geomorph.2006.08.009

Fruergaard, M., Piasecki, S., Johannessen, P.N., Noe-Nygaard, N., Andersen, T.J., Pejrup, M. and Nielsen, L.H. 2015. Tsunami propagation over a wide, shallow continental shelf caused by the Storegga slide, southeastern North Sea, Denmark. *Geology*, **43**, 1047–1050, https://doi.org/10.1130/G37151.1

Gamboa, D. and Alves, T.M. 2015. Three-dimensional fault meshes and multi-layer shear in mass-transport blocks: implications for fluid flow on continental margins. *Tectonophysics*, **647**, 21–32, https://doi.org/10.1016/j.tecto.2015.02.007

Gee, M.J.R. and Gawthorpe, R.L. 2007. Early evolution of submarine channels offshore Angola revealed by three-dimensional seismic data. *Geological Society, London, Special Publications*, **277**, 223–235, https://doi.org/10.1144/GSL.SP.2007.277.01.13

Glastonbury, J. and Fell, R. 2008. Geotechnical characteristics of large slow, very slow, and extremely slow landslides. *Canadian Geotechnical Journal*, **45**, 984–1005, https://doi.org/10.1139/T08-021

Gross, F., Mountjoy, J.J. *et al.* 2018. Free gas distribution and basal shear zone development in a subaqueous landslide – insight from 3D seismic imaging of the Tuaheni Landslide Complex, New Zealand. *Earth and Planetary Science Letters*, **502**, 231–243, https://doi.org/10.1016/J.EPSL.2018.09.002

Hampton, M.A., Lee, H.J. and Locat, J. 1996. Submarine landslides. *Geophysics*, **34**, 33–59, https://doi.org/10.1029/95rg03287

Hungr, O., Leroueil, S. and Picarelli, L. 2014. The Varnes classification of landslide types, an update. *Landslides*, **11**, 167–194, https://doi.org/10.1007/s10346-013-0436-y

Ioualalen, M., Migeon, S. and Sardoux, O. 2010. Landslide tsunami vulnerability in the Ligurian Sea: case study of the 1979 October 16 Nice international airport submarine landslide and of identified geological mass failures. *Geophysical Journal International*, **181**, 724–740, https://doi.org/10.1111/j.1365-246X.2010.04572.x

Kelner, M., Migeon, S., Tric, E., Couboulex, F., Dano, A., Lebourg, T. and Taboada, A. 2016. Frequency and triggering of small-scale submarine landslides on decadal timescales: analysis of 4D bathymetric data from the continental slope offshore Nice (France). *Marine Geology*, **379**, 281–297, https://doi.org/10.1016/j.margeo.2016.06.009

Koson, S., Chenrai, P. and Choowong, M. 2014. Seismic attributes and seismic geomorphology. *Bulletin of Earth Sciences of Thailand*, **6**, 1–9.

Kuhlmann, J., Orpin, A.R., Mountjoy, J.J., Crutchley, G.J., Henrys, S., Lunenburg, R. and Huhn, K. 2018. Seismic and lithofacies characterization of a gravity core transect down the submarine Tuaheni Landslide Complex, NE New Zealand. *Geological Society, London, Special Publications*, **477**, https://doi.org/10.1144/sp477.37

Migeon, S., Ducassou, E., Le Gonidec, Y., Rouillard, P., Mascle, J. and Revel-Rolland, M. 2010. Lobe construction and sand/mud segregation by turbidity currents and debris flows on the western Nile deep-sea fan (Eastern Mediterranean). *Sedimentary Geology*, **229**, 124–143, https://doi.org/10.1016/j.sedgeo.2010.02.011

Mountjoy, J.J., McKean, J., Barnes, P.M. and Pettinga, J.R. 2009. Terrestrial-style slow-moving earthflow kinematics in a submarine landslide complex. *Marine Geology*, **267**, 114–127, https://doi.org/10.1016/j.margeo.2009.09.007

Mountjoy, J.J., Pecher, I., Henrys, S., Crutchley, G., Barnes, P.M. and Plaza-Faverola, A. 2014. Shallow methane hydrate system controls ongoing, downslope sediment transport in a low-velocity active submarine landslide complex, Hikurangi Margin, New Zealand. *Geochemistry, Geophysics, Geosystems*, **15**, 4137–4156, https://doi.org/10.1002/2014GC005379

Mulder, T. and Cochonat, P. 1996. Classification of offshore mass movements. *Journal of Sedimentary Research*, **66**, 43–57, https://doi.org/10.1306/D42682AC-2B26-11D7-8648000102C1865D

Newnham, R.M., Lowe, D.J. and Williams, P.W. 1999. Quaternary environmental change in New Zealand: a review. *Progress in Physical Geography: Earth and Environment*, **23**, 567–610, https://doi.org/10.1177/030913339902300406

Paquet, F., Proust, J.-N., Barnes, P.M. and Pettinga, J.R. 2009. Inner-forearc sequence architecture in response to climatic and tectonic forcing since 150 ka: Hawke's Bay, New Zealand. *Journal of Sedimentary Research*, **79**, 97–124, https://doi.org/10.2110/jsr.2009.019

Pecher, I.A., Barnes, P.M., Levay, L.J. and Expedition 372 Scientists 2018. *Expedition 372 Preliminary Report Creeping Gas Hydrate Slides and Hikurangi LWD*, November 2017.

Pedley, K.L., Barnes, P.M., Pettinga, J.R. and Lewis, K.B. 2010. Seafloor structural geomorphic evolution of the accretionary frontal wedge in response to seamount subduction, Poverty Indentation, New Zealand. *Marine Geology*, **270**, 119–138, https://doi.org/10.1016/j.margeo.2009.11.006

Piper, D.J.W., Cochonat, P. and Morrison, M.L. 1999. The sequence of events around the epicentre of the 1929 Grand Banks earthquake: initiation of debris flows and turbidity current inferred from sidescan sonar. *Sedimentology*, **46**, 79–97.

Randen, T. and Sønneland, L. 2005. Atlas of 3D seismic attributes. *In*: Iske, A., Randen, T. (eds) *Mathematical Methods and Modelling in Hydrocarbon Exploration and Production*. Mathematics in Industry, Springer, Berlin, **7**, 23–46.

Rebesco, M., Hernández-Molina, F.J., Van Rooij, D. and Wåhlin, A. 2014. Contourites and associated sediments controlled by deep-water circulation processes: state-of-the-art and future considerations. *Marine Geology*, **352**, 111–154. https://doi.org/10.1016/j.margeo.2014.03.011

Sawyer, D.E., Flemings, P.B., Dugan, B. and Germaine, J.T. 2009. Retrogressive failures recorded in mass transport deposits in the Ursa Basin, Northern Gulf of Mexico. *Journal of Geophysical Research*, **114**, https://doi.org/10.1029/2008JB006159

Talling, P.J., Paull, C.K. and Piper, D.J.W. 2013. How are subaqueous sediment density flows triggered, what is their internal structure and how does it evolve? Direct observations from monitoring of active flows. *Earth-Science Reviews*, **125**, 244–287, https://doi.org/10.1016/j.earscirev.2013.07.005

Wallace, L.M., Beavan, J., McCaffrey, R. and Darby, D. 2004. Subduction zone coupling and tectonic block rotations in the North Island, New Zealand. *Journal of Geophysical Research: Solid Earth*, **109**, 1–21, https://doi.org/10.1029/2004JB003241

Wright, I., McGlone, M., Nelson, C. and Pillans, B. 1995. An integrated latest Quaternary (Stage 3 to present) paleoclimatic and paleoceanographic record from offshore northern New Zealand. *Quaternary Research*, **44**, 283–293, https://doi.org/10.1006/qres. 1995.1073

Mass transport deposits in the Donegal Barra Fan and their association with British–Irish Ice Sheet dynamics

Srikumar Roy[1]*, Aggeliki Georgiopoulou[2], Sara Benetti[3] and Fabio Sacchetti[4]

[1]iCRAG, School of Earth Sciences, University College Dublin, Dublin D04 N2E5, Ireland

[2]School of Environment and Technology, University of Brighton, Cockcroft Building, Lewes Road, Brighton BN2 4GJ, UK

[3]School of Geography and Environmental Science, Ulster University, Cromore Road, Coleraine, BT52 1SA Northern Ireland, UK

[4]Marine Institute, Oranmore, Galway H91 R673, Ireland

AG, 0000-0003-4298-5090; SB, 0000-0001-6286-9842
*Correspondence: Srikumar.roy@icrag-centre.org

Abstract: This study analyses seismic data to investigate the kinematic indicators within the mass transport deposits (MTDs) of the Donegal Barra Fan complex in the Rockall Trough, along the NW European continental margin. Five episodes of mega-scale MTDs (DBF-01, -02, -03, -04 and -05) are identified. DBF-01 is the largest MTD in the NW British continental margin, comprising 1907 km^3 of sediments. Fold-and-thrusts were identified within the MTDs where they attain maximum thickness of *c.* 300–380 ms TWT, but not at the toe region. This indicates that local erosion and deceleration caused bulking up of the MTD volume, but the MTD was not fully arrested due to the high mobility of the mass flow. MTD thickness distribution and thrust fault orientations indicate source areas and flow direction of MTD. The MTDs show a compensational stacking pattern with earlier deposits influencing the position and flow direction of succeeding slides, suggesting that glaciogenic debris flows are sensitive to topographic variability. We propose that increased sediment input associated with at least five expansions of the British–Irish Ice Sheet to the shelf edge led to the development of these MTDs and that the youngest of them, DBF-05, corresponds to the Last Glacial Maximum.

Supplementary material: Seismic cross-section showing buttressing of DBF-02 and -03 against DBF-01 (Fig. 7c); thrusts within DBF-01 (Fig. 7c), DBF-04 overlying DBF-03 and its termination (Fig. 7c), DBF-05 buttressing against DBF-01, -02 and -03 (Fig. 3; refer to Fig. 5e for location) are available at https://doi.org/10.6084/m9.figshare.c.4816911

Submarine mass transport deposits (MTDs) involve upslope depletion and downslope accumulation of sediments along a basal shear surface, with the translated sediments often forming elevated topographic features on the contemporaneous seafloor (Varnes 1978). MTDs are a common occurrence on continental margins and are important in margin evolution. Bull *et al.* (2009) documented various types of kinematic indicators within MTDs including ramps-and-flats, fold-and-thrusts, which are of great importance to the understanding of the initiation, dynamic evolution, flow direction and cessation of MTDs (Frey-Martínez *et al.* 2005; Alsop *et al.* 2017).

MTDs constitute the bulk of sediments in trough mouth fans (TMFs) (Vorren and Laberg 1997). TMFs are stacks of glaciogenic debris flows forming prograding wedges on continental margins in front of palaeo-ice streams, which drained, among others, the former European Ice Sheets, the Greenland and West Antarctic Ice Sheets (Vorren and Laberg 1997; Stow *et al.* 2002; Rebesco *et al.* 2006; Dowdeswell *et al.* 2008). Hemipelagic or contouritic interglacial sediments often separate the glaciogenic debris flow units (Stow *et al.* 2002; Ó Cofaigh *et al.* 2003). TMFs of late Pliocene to Pleistocene age are recognized along the continental margin of NW Europe (Vorren and Laberg 1997; Dahlgren *et al.* 2005), and the southernmost of these is the Donegal Barra Fan (DBF) that was fed by the British–Irish Ice Sheet (BIIS) (Armishaw *et al.* 2000).

Two major MTD complexes have been identified in the Rockall Trough offshore Ireland: the Rockall Bank Slide Complex (RBSC) (Georgiopoulou *et al.* (2019) and references therein), and the DBF complex (Holmes *et al.* 1998; Armishaw *et al.* 2000; Knutz *et al.* 2002; Stow *et al.* 2002) (Fig. 1). The timing, failure mechanisms and number of slumping events of the RBSC have been studied

From: Georgiopoulou, A., Amy, L. A., Benetti, S., Chaytor, J. D., Clare, M. A., Gamboa, D., Haughton, P. D. W., Moernaut, J. and Mountjoy, J. J. (eds) 2020. *Subaqueous Mass Movements and their Consequences: Advances in Process Understanding, Monitoring and Hazard Assessments*. Geological Society, London, Special Publications, **500**, 567–586.
First published online March 18, 2020, https://doi.org/10.1144/SP500-2019-177

using a wide variety of geophysical data and sedimentary cores (Flood *et al.* 1979; Georgiopoulou *et al.* 2013, 2019; Sacchetti *et al.* 2013), as well as simulation modelling studies of the major phases of slumping episodes during and before the Last Glacial Maximum (Salmanidou *et al.* 2018). However, we have a poor understanding of the various slumping events of the DBF complex, and their extent in the Irish sector of the Rockall Trough.

The objective of this study is to improve our understanding of the extent and dominant transport mechanisms of MTDs in the DBF complex by investigating the geometry, distribution, stacking patterns and kinematic indicators of individual MTDs, their interactions with the RBSC and potential links between MTDs and episodes of BIIS advances on to the NW European Irish continental shelf.

Geological setting

The NW European continental margin is characterized by a range of submarine features including TMFs deposited in front of fast-flowing ice streams, and MTDs formed by failure of accumulated glaciogenic and marine sediments and moraines (Dahlgren *et al.* 2005; Sejrup *et al.* 2005; Ó Cofaigh *et al.* 2012). A number of MTDs of various sizes and of Pliocene to Pleistocene age have been identified, including the

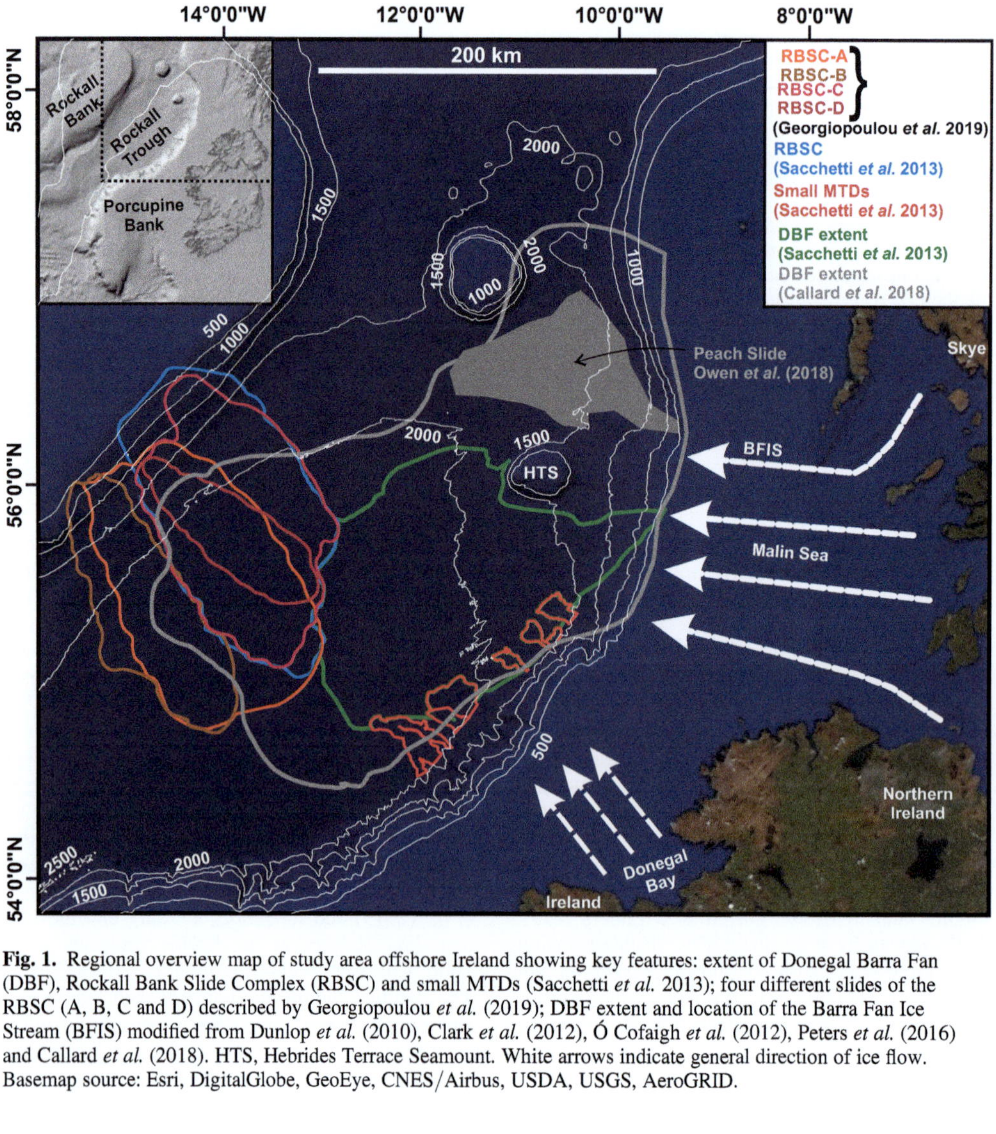

Fig. 1. Regional overview map of study area offshore Ireland showing key features: extent of Donegal Barra Fan (DBF), Rockall Bank Slide Complex (RBSC) and small MTDs (Sacchetti *et al.* 2013); four different slides of the RBSC (A, B, C and D) described by Georgiopoulou *et al.* (2019); DBF extent and location of the Barra Fan Ice Stream (BFIS) modified from Dunlop *et al.* (2010), Clark *et al.* (2012), Ó Cofaigh *et al.* (2012), Peters *et al.* (2016) and Callard *et al.* (2018). HTS, Hebrides Terrace Seamount. White arrows indicate general direction of ice flow. Basemap source: Esri, DigitalGlobe, GeoEye, CNES/Airbus, USDA, USGS, AeroGRID.

Storegga Slide Complex, the North Faroes Slide Complex and the Peach Slide (Haflidason *et al.* 2004; Evans *et al.* 2005; Sejrup *et al.* 2005; Owen *et al.* 2018). Most of the slides and debris flows on the continental slope north of 55° N are associated with glacial depocentres (Evans *et al.* 2005).

The Rockall Trough is the largest of the basins offshore west Ireland, with water depths ranging from 500 m on the eastern and western margins to almost 4000 m in the southern opening of the trough (Fig. 1). The Mid to Late Cenozoic stratigraphic framework for the Rockall Trough comprises three megasequences bounded by unconformities C30, C20 and C10 (Stoker *et al.* 2001). The C10 unconformity is an early Pliocene angular unconformity (*c.* 5 Ma) separating the early Miocene to early Pliocene megasequence from early Pliocene to Holocene sediments (McDonnell and Shannon 2001).

The seafloor and shallow sub-seafloor geomorphology of the deep-water Irish continental margin has been mapped in detail using a wide range of high-resolution geophysical datasets, revealing features such as submarine mass failures of various sizes and extent, escarpments, canyons and channels, sedimentary lobes associated with canyon systems, slab failures and evolved slides, incipient cuspate slides, etc. (Shannon *et al.* 2001; Unnithan *et al.* 2001; Elliott *et al.* 2010; Sacchetti *et al.* 2011, 2012, 2013; Georgiopoulou *et al.* 2013, 2014). Some of these features are the focus of this study and are discussed in more detail below.

The Donegal Barra Fan (DBF) and the British–Irish Ice Sheet (BIIS)

Reconstruction of the advance and retreat of the last BIIS on the continental shelf NW of Ireland is based on relatively recent geophysical and sedimentological datasets (Dunlop *et al.* 2010; Ó Cofaigh *et al.* 2012, 2019; Peters *et al.* 2016; Callard *et al.* 2018). Converging ice flows from NW Ireland and western Scotland merged on the Malin Sea Shelf in the Barra Fan Ice Stream (BFIS) (Fig. 1). The BFIS drained 5–10% of the BIIS (Dove *et al.* 2015) and fed sediments and meltwater towards the largest glacimarine depocentre of the BIIS, known as the DBF (Fig. 1). The BIIS reached the outer shelf and was grounded at the shelf break of the Malin Sea on more than one occasion (Fyfe *et al.* 1993). Bathymetric and sub-bottom data show that the ice was grounded extensively along the shelf edge in the Malin Sea Shelf and offshore of NW Ireland during the last offshore glacial advance (Benetti *et al.* 2010; Peters *et al.* 2015; Callard *et al.* 2018). Some ice streaming was also present in Donegal Bay (Greenwood and Clark 2009; Benetti *et al.* 2010; Ó Cofaigh *et al.* 2012), to the south of the Malin Sea, but the geophysical data on the continental margin do not show the development of a distinct TMF in this region (Sacchetti *et al.* 2012, 2013). The ice sheet reached its maximum extent at the shelf edge sometime before 26.5–26 ka BP (Clark *et al.* 2012; Peters *et al.* 2016; Callard *et al.* 2018). By 25.9 ka BP, the retreat of the BIIS from the shelf edge was already underway. The style of retreat was episodic and characterized by the formation of moraines and grounding zone wedges across the shelf (Dunlop *et al.* 2010; Callard *et al.* 2018; Ó Cofaigh *et al.* 2019). The outer continental shelf in the Malin Sea was free of grounded ice by 23.2 ka BP with the majority of the shelf ice-free by 19.5 ka BP (Callard *et al.* 2018). At this time, while the ice sheet was no longer grounded at the seafloor, it is likely that an ice shelf persisted in this region well into the deglacial period, with meltwater release and ice rafting occurring at least until 15.9 ka BP and resulting in episodes of downslope MTD on the DBF (Tarlati 2018).

The DBF covers an area of about 7000 km^2 and locally approaches 700 m in thickness in the deep-water basin of the Rockall Trough (Fig. 1; Owen and Long 2016). Armishaw *et al.* (2000) recognized a combination of different sedimentary processes dominant during the three-stage glacial to post-glacial evolution of this depositional system, which resulted in the composite nature of the DBF. The first and most significant sediment input was in the NE of the study area, where the DBF overlies the Late Miocene and Early Pliocene sediments and is distal in character. The sedimentary sequence comprises contourites, glaciomarine deposits and hemipelagic sediments (Armishaw *et al.* 2000). Continental uplift later contributed to the growth of the DBF during the mid-Pliocene; however, the majority of the DBF sediments were deposited during the Pleistocene when the fan was a major depocentre for the BIIS. The Peach Slide is the largest known MTD situated in the northern flank of the fan, consisting of four major debrite units (sediment volume *c.* 135–673 km^3), which were deposited during the period between 36.5 and 10.5 ka BP (Holmes *et al.* 1998; Knutz *et al.* 2001; Maslin *et al.* 2004). Shallow subsurface and surface geophysical imaging has allowed recent mapping and interpretation of contourites, hemipelagites and debrites within the Peach Slide (Owen *et al.* 2018).

Slides on the Rockall Trough margins

The RBSC is located along the western margin of the trough and to the east of the Rockall Bank (Fig. 1). It covers *c.* 18 000 km^2 of the base of slope and floor of the Rockall Trough (Elliott *et al.* 2010). Several studies have investigated high-resolution-bathymetry data, GLORIA and TOBI side-scan sonar data, sub-bottom acoustic data, multichannel seismic data and sedimentary core data to study the morphology, multiple episodes of slumping and their ages (Shannon *et al.* 2001; Unnithan *et al.* 2001; Georgiopoulou *et al.* 2012, 2013; Sacchetti *et al.* 2013). The RBSC constitutes at least three voluminous episodes of slope collapse (sediment volume *c.* 125–400 km^3), a fourth less voluminous event, and possibly a fifth more localized event, which occurred during 200 to 22 ka (Georgiopoulou *et al.* 2019). The four larger MTDs are referred to as RBSC slides A, B, C and D in Georgiopoulou *et al.* (2019) and in this paper (Fig. 1).

Mass wasting on the western margin of the trough involved larger sediment volumes as compared to the relatively smaller slope failures of the eastern margin, where sediment was progressively evacuated towards the deeper basin through canyons (Georgiopoulou *et al.* 2014). In the northeastern Rockall Trough, a large mass transport complex was identified as the Erris Wedge in the buried sedimentary sequence directly overlying the C30 unconformity (Elliott *et al.* 2006). The Erris Wedge pinches out at the southern limit of the DBF.

Data and methods

The data used in this study include three surveys of 2D multichannel seismic data acquired over the past decades in offshore Ireland (Fig. 2). The most recent surveys, PAD-13 and PAD-14, were acquired by R/V *BGP Explorer* for ENI Ireland BV during 2013–14. The vessel was equipped with a Sercel G-Gun-II as a source, placed at a depth of 8 m (±1 m), and towed one streamer of length 10 050 m placed at 10 m (±1 m) depth. The survey was acquired with a shot-point interval of 37.5 m, 12 s record length, 2 ms sample rate and 50 ms recording delay. The unpublished PAD-13 and PAD-14 lines are of the best quality, spaced 40–50 km apart, and are the main dataset analysed for this paper. The

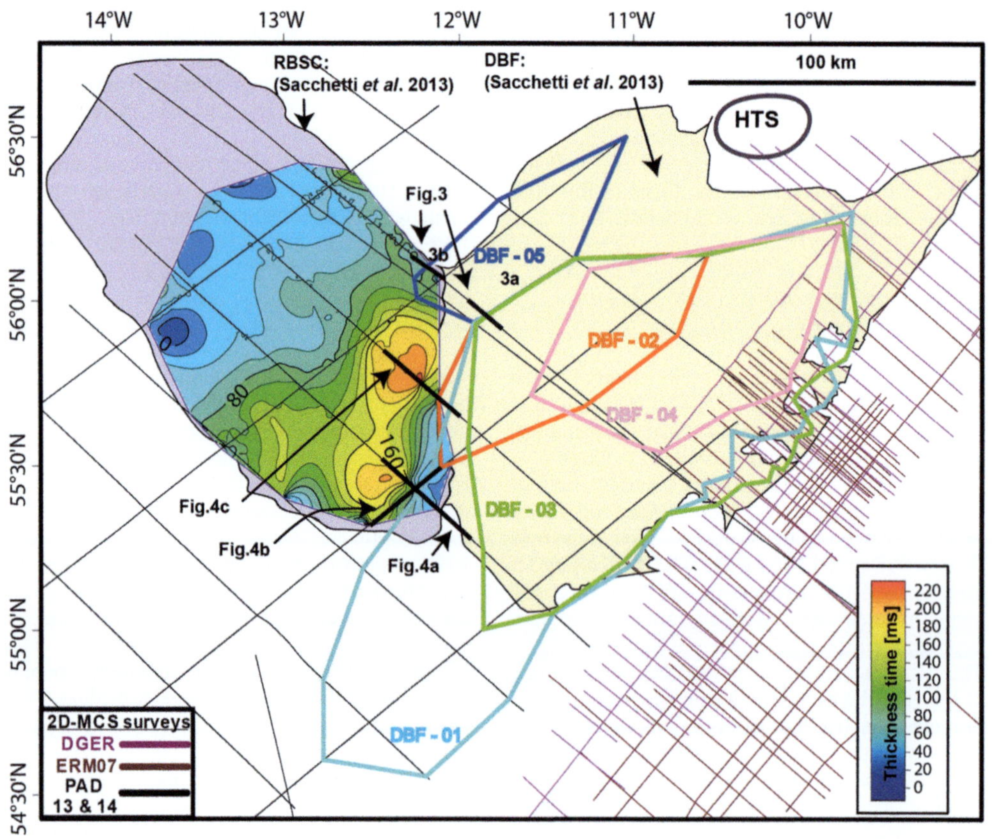

Fig. 2. Extent of previously published Rockall Bank Slide Complex (RBSC) and Donegal Barra Fan (DBF) complex (Sacchetti *et al.* 2013), along with the RBSC time thickness map (TWT, ms) calculated in this study, and the extent of different MTDs comprising the DBF complex in the northeastern part of the Irish Rockall Trough. HTS, Hebrides Terrace Seamount.

ERM07 2D seismic survey was acquired by M/V *Lazarev & Shatskiy* and M/V *Nalivkin* in 2007 for client TGS-Nopec Geophysical. The vessels used an airgun source placed at a depth of 7–10 m, and towed an 8090 m long streamer (648/636 channels) placed at a depth of 12 m. The survey was acquired using a shot-point interval of 25 m, group interval of 12.5 m and 9 s record length. The DGER 2D seismic survey was acquired by the M/V *Polar Princess* in 1996. The vessel used a source gun array (30 × 17.2 m) placed at a depth of 7 m, and towed a streamer 4500 m long placed at a depth of 8 m. The survey was acquired using CDP (common depth points) spacing of 6.25 m, shot-point interval of 25 m and 8 s record length. The above three sets of surveys were processed following standard industry procedure (Yilmaz 2001). The eastern slope of the Rockall Trough is covered by both the ERM07 and DGER surveys with an average spacing of 6–10 km between each seismic line. The ERM07 survey has good to moderate quality of seismic lines, especially in the shallow depths below the seafloor, whereas the DGER survey shows poor quality of seismic imaging in the shallow stratigraphic layers. The 50 km line spacing between the best quality PAD-13 and PAD-14 lines and closely spaced poor to medium quality DGER and ERM07 lines made the correlation of MTDs challenging in this study. All survey data have been made available by the Irish Petroleum Affairs Division (data available on request from https://www.dccae.gov.ie/en-ie/natural-resources/topics/Oil-Gas-Exploration-Production/data/regional-seismic-survey/Pages/Regional-Seismic-Survey.aspx).

The basal shear surface of each MTD, and deformation structures within the MTDs were interpreted in Petrel software, provided by Schlumberger. Thereafter, these surfaces were used to create isochron thickness maps, which were integrated with previously published geophysical interpretation using ArcGIS software for spatial correlation. The sediment volume of each of the MTDs was calculated by converting two-way time (TWT) into metres using an average P-wave velocity of 1700 m s^{-1} for less-consolidated marine sediments (Hamilton and Bachman 1982).

Results and discussion

This section presents and discusses the seismic interpretation of various MTDs that are part of the RBSC and the DBF (Figs 3–9).

RBSC

One of the MTDs within the RBSC has been mapped by Sacchetti *et al.* (2013) using shallow high-resolution geophysical datasets and corresponds with 'Slide C' of Georgiopoulou *et al.* (2019) (Figs 1 & 2). The termination of this MTD was mapped in the northwestern part of the study area. It shows a varied seismic character. The northern part of the MTD is characterized by a chaotic reflection pattern of low to moderate amplitudes (Fig. 3b). In contrast, the toe of the MTD at the southern end is characterized by moderate to high amplitude reflections, showing compressional features, such as reverse faults (Fig. 4a–c). The basal shear surface is typically a continuous, medium-high negative amplitude reflector, easily identified on the seismic data and is parallel to the slope stratigraphy (Figs 3b & 4).

Sediments show two main accumulations within the MTD. The thickness reaches up to 200 ms (TWT) in the southern portion of the MTD (Figs 2 & 4a), and 230 ms (TWT) 40 km north from the former location (Figs 2 & 4c). The volume of sediments calculated for the slide identified in this study is *c.* 660 km^3. The MTD identified as RBSC in this study ramps up a stratigraphic unit which is characterized by chaotic seismic facies (Fig. 4c). This chaotic seismic facies unit could be part of 'Slide A' identified by Georgiopoulou *et al.* (2019). This is observed only on one seismic line, and hence its extent could not be mapped.

DBF complex

After integrated data analysis with previously published results (Knutz *et al.* 2002; Sacchetti *et al.* 2013; Owen and Long 2016), all the MTD events interpreted in this study in the northeastern Rockall Trough were observed to lie above the C10 unconformity and beneath the RBSC, and hence are assumed to be part of the DBF complex. The MTD units have been recognized based on criteria established from previous studies (Evans *et al.* 1996; Frey-Martínez *et al.* 2005; Gamboa *et al.* 2011); each MTD unit was identified as a discrete stratigraphic unit covering a considerably large area, which is characterized internally by chaotic, semi-transparent, highly disrupted seismic facies distinctly different from the adjacent undeformed units and underlying stratigraphy. Detailed interpretation of five basal shear surfaces distinguished five separate MTDs within the DBF complex (Fig. 5). The basal shear surface for each MTD unit was identified in a similar way to unconformities, characterized by abrupt termination of a continuous medium-high amplitude reflection that dips parallel to the underlying chaotic seismic facies of the MTD units (Figs 3, 4, 6–9). Occasionally the basal shear surface is observed to ramp up or down the stratigraphy forming a step-like geometry (Fig. 8b, c). Most of the internal facies of the MTDs are chaotic and/or semi-transparent; however, occasionally, individual semi-continuous reflections within the MTD allow identification of internal facies and therefore better

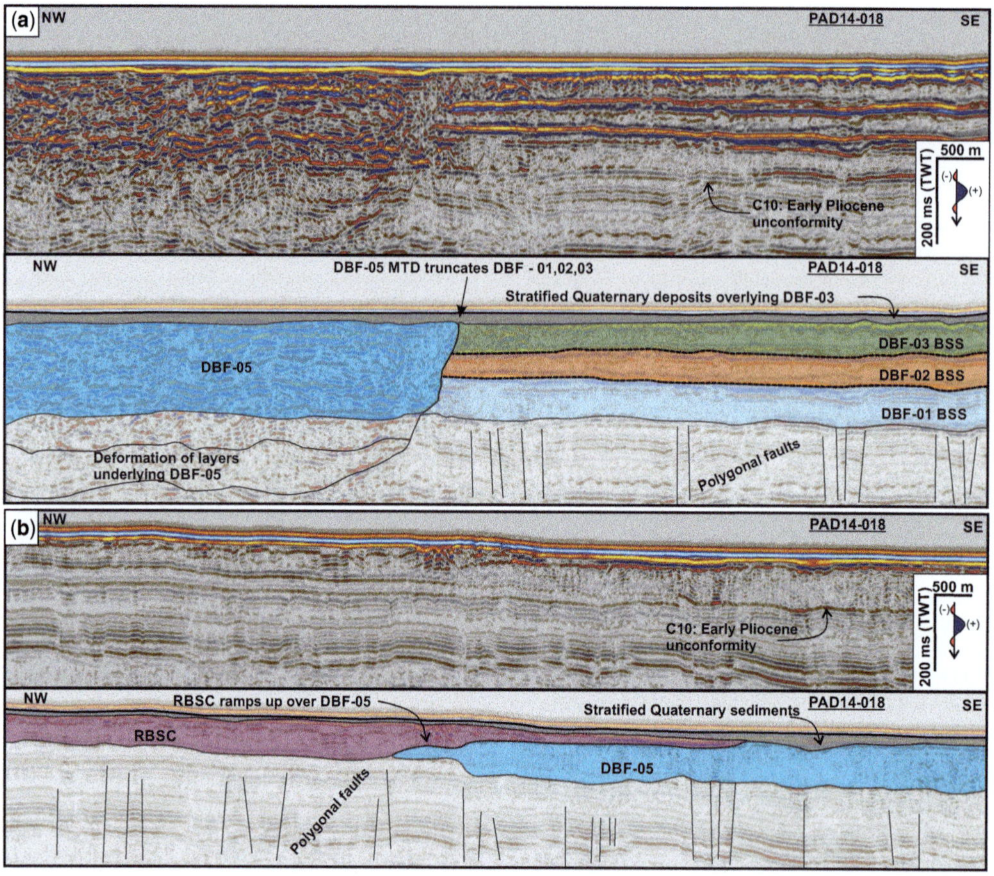

Fig. 3. (**a**) Seismic facies and basal shear zone of DBF-01, DBF-02, DBF-03 and DBF-05. Note the deformation of stratified layers beneath the DBF-05, and the buttressing nature of DBF-05 on DBF-01, DBF-02 and DBF-03. (**b**) RBSC ramps over DBF-05 in the NW Rockall Trough corner. Note thin veneer of stratified sediments (marked with grey colour) above MTDs. Refer to Figure 2 for location of seismic sections.

understanding of the internal architecture. The majority of these semi-continuous reflections are slightly tilted to the basal shear surface and the top of the MTDs. These features are interpreted as fold-and-thrust systems and described separately in the section on 'Flow processes within the DBF'. The areas and sediment volumes of each of the five MTDs interpreted in this study, comprising the DBF complex in the Irish sector of the Rockall Trough, are shown in Figure 10. They are further described below.

DBF-01

DBF-01 extends over the largest surface area out of the five identified MTDs. DBF-01 covers an area of *c.* 16 589 km^2 and consists of *c.* 1907 km^3 of sediments, making it the largest submarine MTD described along the NW British continental margin (Moscardelli and Wood 2016). It stretches over *c.* 258 km in length along a NE–SW direction and is *c.* 100 km wide in its central part. The present-day water depth varies from 1668 m in the NE to 2956 m in the SW over the extent of DBF-01. DBF-01 is not exposed to the seafloor. Almost 60% of DBF-01 is draped by succeeding MTDs (DBF-02, -03 and -04), which are all part of the DBF complex. In the southwestern region, stratified sediments drape *c.* 40% of DBF-01 (4790 km^2); these sediments could be either hemipelagic sediments, contourites or sediment gravity flows related to meltwater pulses. The stratified sediments are highly continuous and easily distinguished from the underlying chaotic seismic facies of the MTD (Fig. 4a, b). Two major depocentres have been identified in the central and southern part of DBF-01 (Fig. 5a). The thickness of these two depocentres varies from 300 to 350 ms (TWT) and are characterized by fold-and-thrust

Fig. 4. (a–c) Compressional features within the RBSC toe section overlying stratified sediments and DBF-01and DBF-02 in the central and northern part of the Rockall Trough. Explanations of DBF-00 and RBSC-A have been given in text. Refer to Figure 2 for location of seismic sections.

systems (Figs 5 & 7). The orientation of individual thrusts is difficult to ascertain in this scenario, as we observe NW-verging thrusts on only one 2D seismic line, which is contradicted by a general NE–SW alignment of thickness map contour lines (Fig. 5a). However, we speculate that the thrusts might have formed due to the accumulation of downslope-translated sediments as part of a slope failure event. Sediments may have been brought in by the BFIS in the Malin Sea (NE–SW palaeoflow direction) and those derived by ice streaming in the Donegal Bay area (SE–NW palaeoflow direction) (Fig. 1; Clark *et al.* 2012).

Ten deformed rafted blocks, as defined by Frey-Martínez *et al.* (2005), have been identified within DBF-01. Nine of them are in the thicker northern part. One is located 63 km NE from the southern toe of the slide. The blocks are usually resting on the basal shear surface of the MTD (Figs 6b & 7a, b), but occasionally they are observed floating in the chaotic matrix of the MTD (Fig. 6c). The top of the rafted blocks are characterized by medium–high amplitude reflections and demonstrate different styles of internal deformation (minor, moderate and major) as classified in Gamboa *et al.* (2012). Irrespective of their location in relation to the main MTD mass, rafted blocks exhibit deformation with increasing translation or duration of sliding and tend to become aligned with their long axis parallel to the direction of flow (Huvenne *et al.* 2002; Bull *et al.* 2009). However, due to the coverage of the seismic data, we are unable to carry out detailed analysis of the spatial distribution of blocks or their internal character. A chaotic seismic unit underlying DBF-01 was identified on only one seismic line (Fig. 4c). This unit could possibly be part of the DBF complex and has been named DBF-00, but no further investigation was possible due to the limited seismic coverage.

DBF-02

This MTD is 115 km long, stretching along a NE–SW axis, and 45 km wide in the central part of the slide, draping the northwestern part of DBF-01 (Fig. 2). DBF-02 covers an area of *c.* 3653 km^2, draping *c.* 30% of DBF-01, and comprising *c.* 287 km^3 of sediments. It has the smallest volume

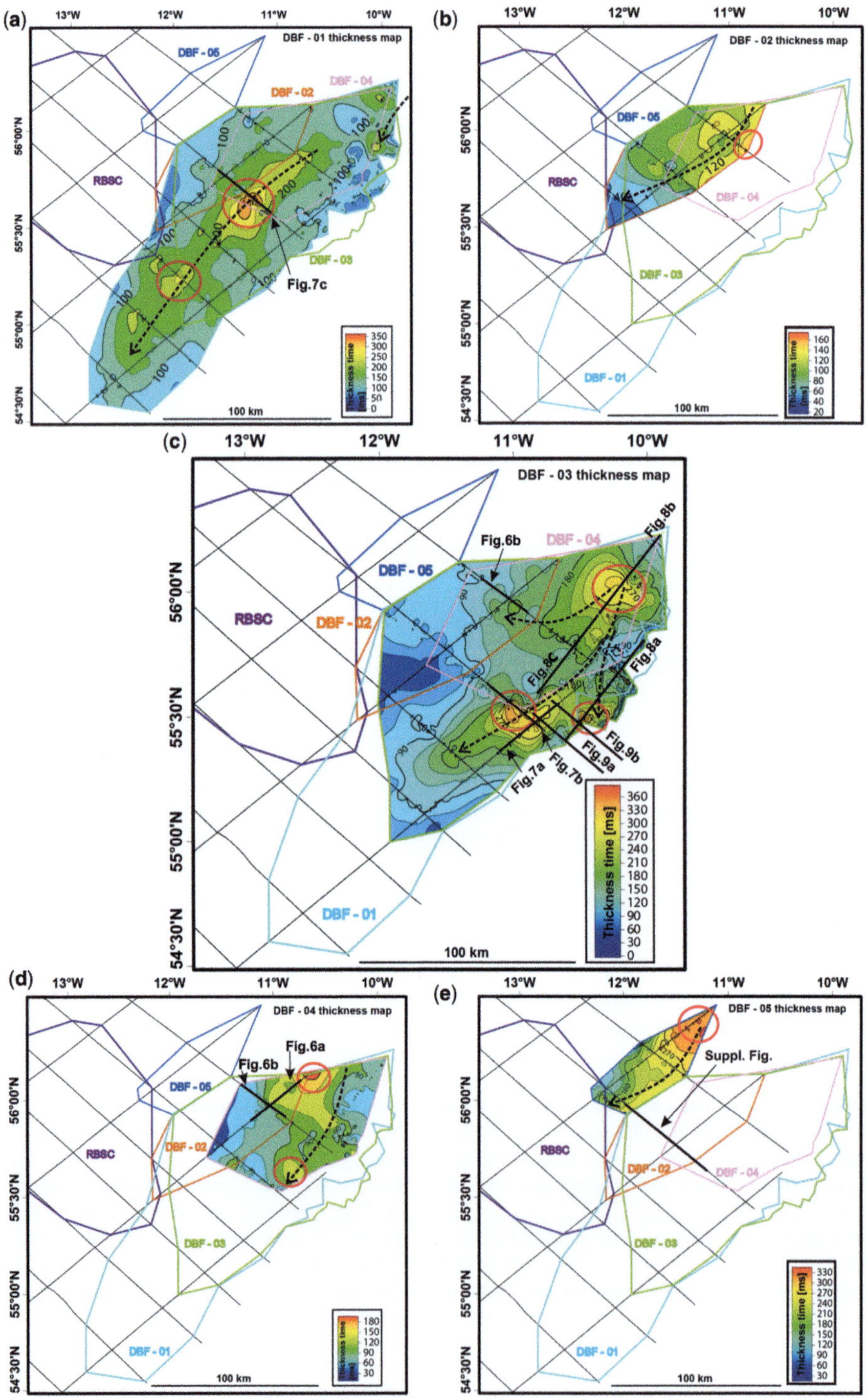

Fig. 5. (**a–e**) Thickness maps (TWT, ms) of different DBF complex MTDs, DBF 01–05. The extent of the RBSC shown in the above maps is the one mapped in this study. Depocentres are shown in red circles, dashed black lines represent flow direction of MTDs inferred from their respective thickness maps. 'Suppl. Fig.' indicates position of figure in Supplementary material.

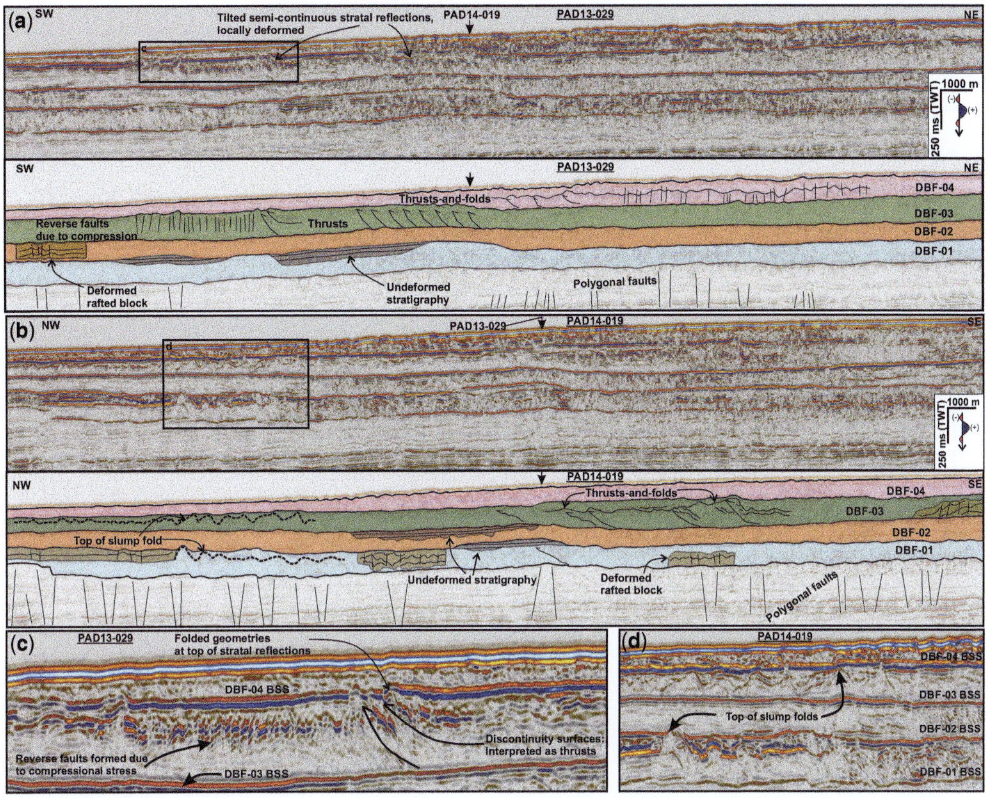

Fig. 6. (**a**, **b**) Multiple episodes of DBF in the NW part of study area. Compressional features such as thrust-and-folds within DBF-03 and DBF-04 are shown along with deformed rafted blocks within DBF-01 and DBF-02. Undeformed stratified sediments are interpreted between DBF-01 and DBF-02, and DBF-02 and DBF-03. (**c**, **d**) Zoom-in of compressional fold-and-thrust features in DBF-03, and top of slump folds in DBF-01 and DBF-03. Refer to Figure 5d for location of seismic sections.

of all the MTDs identified in this study (Fig. 10). The basal shear surface of DBF-02 is a continuous high-amplitude reflector, which separates it from the underlying DBF-01 unit (Fig. 6a, b). The maximum thickness of this slide is *c.* 180 ms (TWT), at a location where it meets the elevated morphology of DBF-01 (Fig. 5b). Most of the slide is draped by DBF-03, leaving *c.* 8–10% of the surface area covered by uniform stratified sediments (Fig. 4c). One minor deformed rafted block with a width of 5.5 km has been identified resting on the basal shear surface, within the thickest region of DBF-02. Undeformed stratigraphic units of stratified sediments have been identified in between DBF-01 and DBF-02 (Fig. 6a, b).

DBF-03

This MTD is 185 km long and 120 km wide in the central part of the slide, covering 65% of DBF-01 and most of DBF-02 (Fig. 2). It covers an area of 11 736 km^2 and comprises *c.* 1378 km^3 of sediments. The basal shear surface is a continuous high-amplitude reflector for most parts of the MTD (Figs 6a, b & 7a), with some exceptions of partially discontinuous, medium–high amplitude reflectors above some of the thrust-and-fold systems identified within DBF-03 (Fig. 7b). The discontinuous basal shear surface might be due to compressional deformation or some sort of seismic data processing artefacts. DBF-03 has the greatest thickness (*c.* 380 ms TWT) of all the MTDs identified in this study. Three depocentres have been identified (Fig. 5c). The southeastern depocentre is characterized by compressional features where it is found to be buttressed in a downslope position against the undisturbed sediments along the northeastern slope of the Rockall Trough (Fig. 9b). The southwestern depocentre is characterized by complex imbricate thrusts and compressional folding within the MTD (Fig. 7a, b). The flow direction of the MTD cannot be easily derived from the thrust

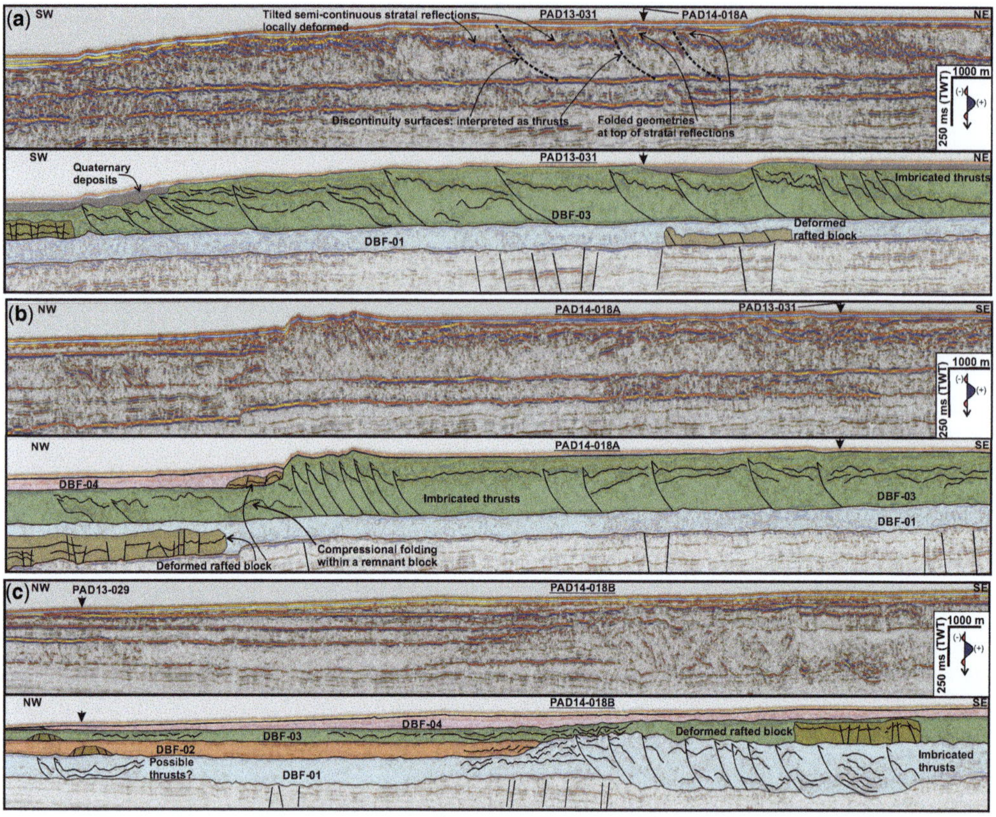

Fig. 7. (**a–c**) Fold-and-thrust and compressional folding structures within DBF-01, and DBF-03. Note the small-scale folded geometries at the tip of the stratal reflections, adjacent to the steeply-dipping discontinuities, which are interpreted as thrusts. Thrusts splay from the basal shear zone and steepen as it propagates upwards to the upper boundary of the MTD. Deformed rafted blocks are shown within DBF-01 and DBF-03. Refer to Figures 5a, c for location of seismic sections.

faults, which are identified using only two seismic lines, but this is discussed further later. The northernmost depocentre is characterized by compressional features formed over a 20 km long remnant block (Fig. 8b, c). Two other remnant blocks and 11 rafted deformed blocks have been identified within the DBF-03. More are expected to be present in between seismic lines. Their overall distribution is therefore not very well constrained; however, they are mostly observed in the central and northern part of the slide. Remnant blocks do not show any basal detachment, and hence represent isolated stationary blocks around which MTD material could have moved slowly. In contrast, rafted deformed blocks might have been subjected to faster-flowing MTD mass and have translated longer distances while being deformed at the same time (Bull *et al.* 2009).

DBF-04

This MTD is 118 km long and 80 km wide in the central part of the slide, covering almost 40% of DBF-03 (Fig. 2). It covers an area of 5036 km^2 and comprises of *c.* 378 km^3 of sediments. The basal shear surface is a continuous high-amplitude reflector for most parts of the MTD (Fig. 6a, b). It is partially discontinuous and characterized as a medium-high amplitude reflector in the southeastern part of the seismic section (Fig. 6b). The discontinuous nature of the basal shear surface is most likely due to the compressional features observed in these regions (Fig. 6a). However, there are regions of the MTD that are characterized by compressional features lying over a continuous high-amplitude basal shear surface. This variability in the imaging of the basal shear surface beneath the compressional

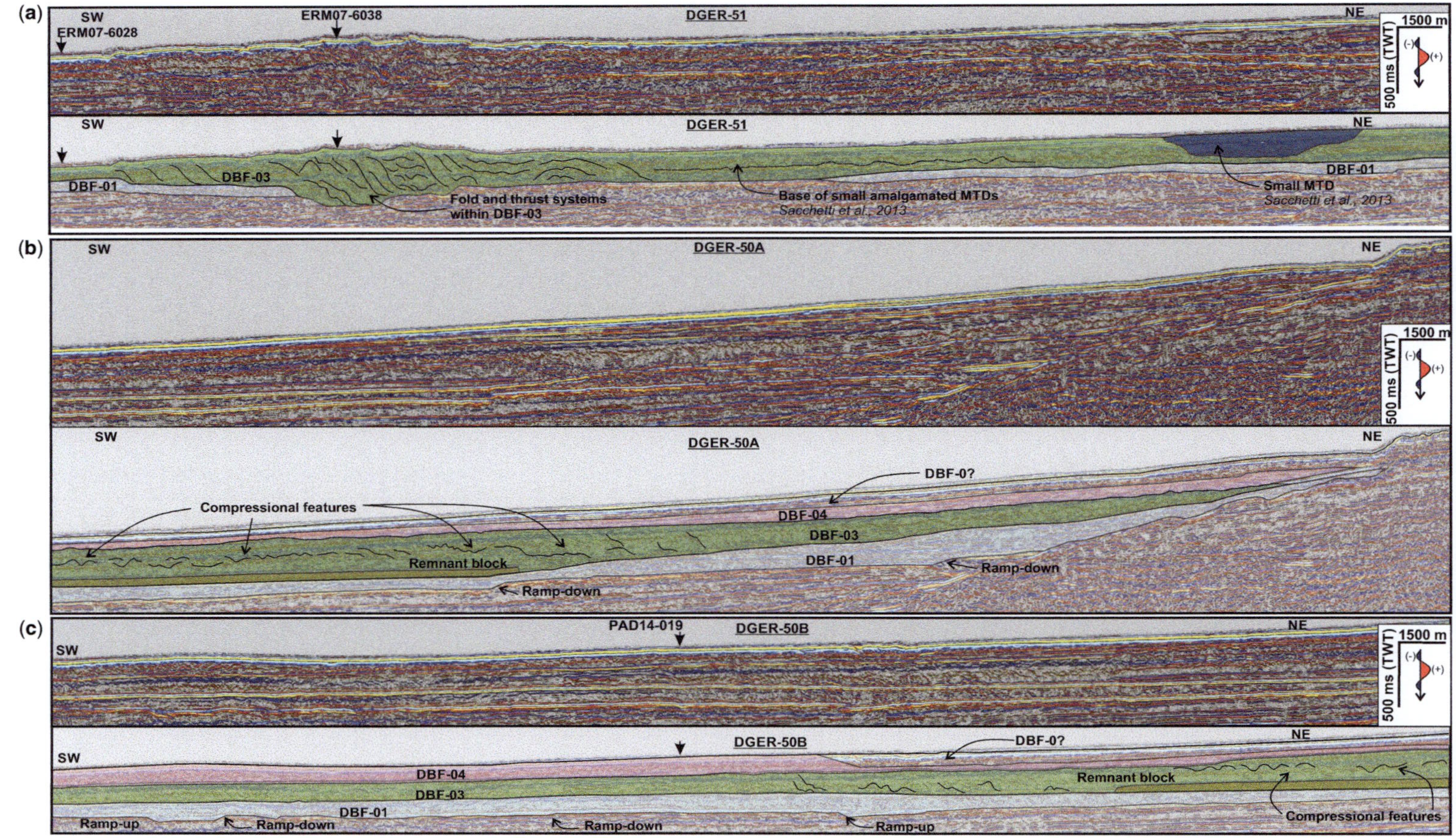

Fig. 8. (**a–c**) Downslope translation of DBF-01, DBF-03 and DBF-04, fold-and-thrusts within DBF-03 in the vicinity of a 20 km long remnant block. The basal shear surface is observed to ramp up and down the stratigraphy to create a series of staircase-like geometries, with clear reflection termination beneath the basal shear surface. Refer to Figure 5c for locations of seismic sections.

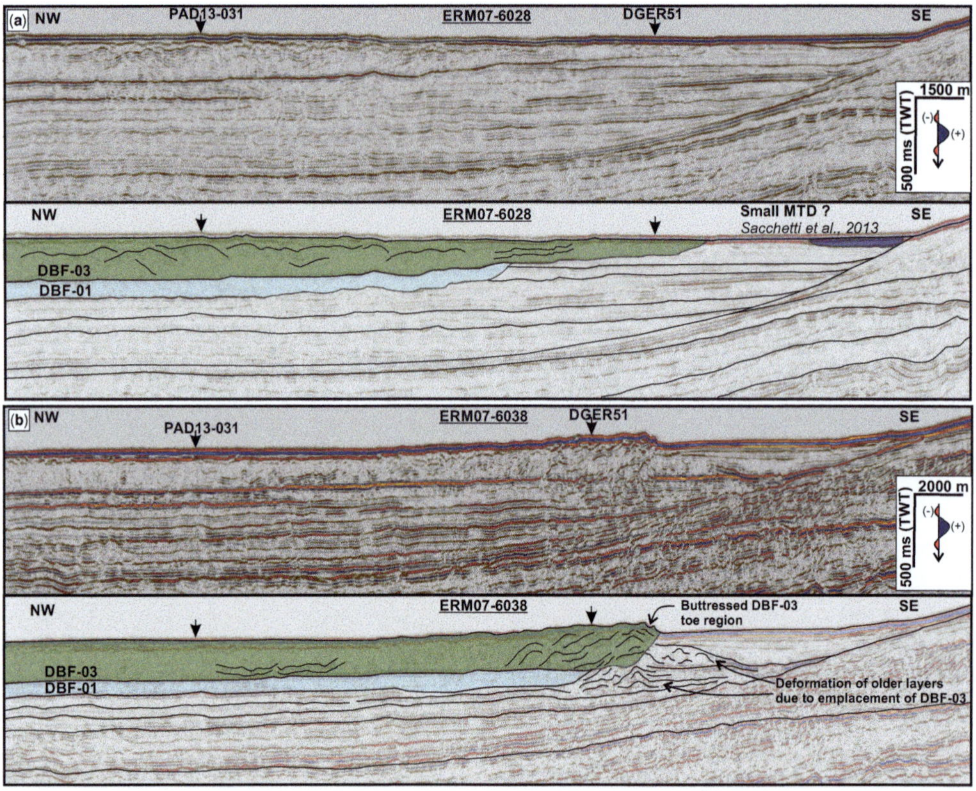

Fig. 9. (**a**) Compressional features within the toe region of the DBF-03, and a possible small MTD identified further SE in the seismic section, which overlaps with MTD interpretation of Sacchetti *et al.* (2013). (**b**) Development of compressional features where the toe region of DBF-03, which is found to be buttressed in a downslope position, is deforming parts of the strata (shown with minor folds). The basal shear surface of DBF-01 changes from high to low amplitude in both the figures; however the seismic facies inside the MTD is chaotic and easily distinguishable from the uniformly stratified underlying strata. Refer to Figure 5c for locations of seismic sections.

features could be related to the plasticity of sediments, degree of compression they have sustained or dewatering. DBF-04 has a maximum thickness of *c.* 200 ms TWT at the northern end of the slide (Fig. 5d). Considering the thickness distribution of the other deposits this implies that DBF-04 is substantially larger, and we are unable to image a large part of it due to lack of data coverage in that direction. However, a second depocentre of *c.* 150 ms TWT has been identified at the southern toe of the slide (Fig. 5d), which marks the region where DBF-04 is buttressed against the thrust faults of DBF-03 (Fig. 7b). A strongly deformed rafted block has also been identified at this meeting point of DBF-03 and DBF-04. Another 10.5 km long rafted block has been identified within this slide in the northern area where it is found to be thickest. This block is slightly-deformed as per the deformation styles illustrated by Gamboa *et al.* (2012). Thrust-reverse faults and compressional folding structures are observed only in the northern part of the slide (Fig. 6a, b).

DBF-05

This MTD is 91 km long and *c.* 34 km wide in the central part of the slide, not overlying or being overlain by any of the other MTDs. It covers an area of 1913 km^2 and comprises *c.* 384 km^3 of sediments. Even though it covers the smallest area out of all the MTDs, it does not have the smallest volume (Fig. 10). It attains a maximum thickness of *c.* 340 ms TWT in the northern part of the slide (Fig. 5e). Similar to DBF-04, the thickness distribution map suggests that a large part of the DBF-05 could not be imaged due to lack of data coverage in the north of the study area. The basal shear surface is a medium-high amplitude semi-continuous

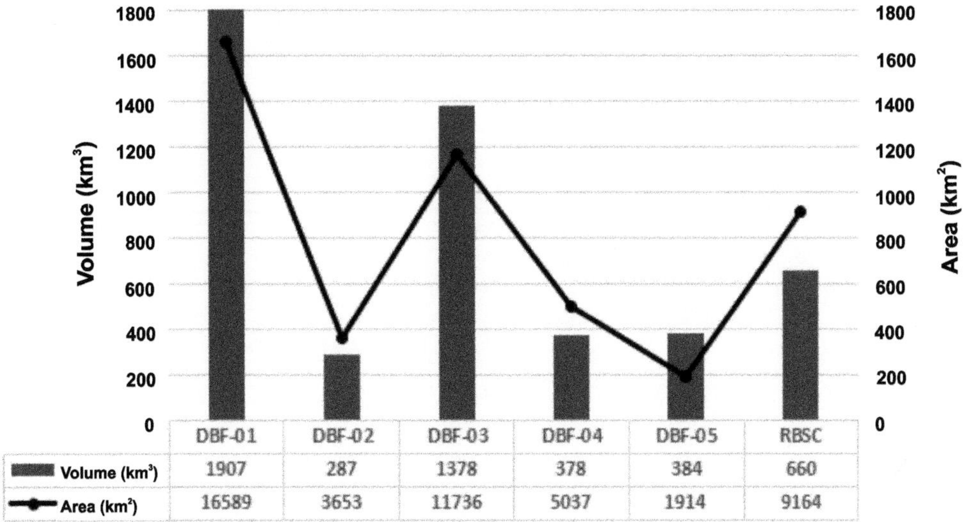

Fig. 10. Area and volume of the MTDs mapped in this study relating to the RBSC and DBF 01–05. Volume of each MTD was calculated by converting TWT into metres using an average P-wave velocity of 1700 m s^{-1} in less-consolidated marine sediments (Hamilton and Bachman 1982).

reflector in the northern part of the slide, and it gradually weakens in amplitude in the southern part. The contrasting difference between the chaotic semi–transparent facies of the MTD and the underlying evenly stratified undeformed layers assist in the interpretation of the basal shear surface (Fig. 3b). It has an impact on the underlying stratified layers and deforms them, as observed from the undulating reflections beneath the slide (Fig. 3a). This might have implications on the composition of sediments, water content and the fact that the DBF-05 toe is constrained by the pre-existing DBF-01, DBF-02 and DBF-03. DBF-05 is observed to truncate parts of DBF-01, DBF-02 and DBF-03 (Fig. 3a), which means that it postdates them. The toe of the RBSC forms a frontal ramp over the DBF-05 in the north-western part of the study area, which suggests RBSC Slide C postdates DBF-05 (Fig. 3b).

Flow processes within DBF

The thickness distribution maps (Fig. 5a–e) allow us to consider how the morphology of each deposit affected the subsequent events. DBF-02 appears to occupy the northern area of DBF-01, which is where DBF-01 is the thinnest. DBF-03 appears to have ridden over the top of both of them, possibly because DBF-02 had smoothed out the topography to some degree. However, the fact that DBF-03's depocentre is off towards the base of slope, suggests that the DBF-02 deposit had probably raised the northern area enough to cause DBF-03 to be constrained to an area closer to the slope. DBF-04 is relatively small, relatively thin and expands over the top of DBF-03, suggesting that it might have been less viscous and more dilute. The depositional area of DBF-05 that differs from all the other four deposits suggests that DBF-01 to DBF-04 created a substantial relief on the seafloor, so that DBF-05 could no longer overcome the topography and had to flow around the Hebrides Terrace Seamount to occupy the lower relief in the basin, while still eroding the margins of the previous MTDs (Fig. 11). This pattern of compensational stacking is derived from the tendency deposits show to preferentially fill topographic lows, smoothing out topographic relief.

A group of semi-continuous stratal reflections are identified extending over *c.* 15–20 km on multiple seismic sections within DBF-01, DBF-03 and DBF-04 (Figs 6a, c & 7a). The stratal reflections are tilted (average: 5–10°) and locally deformed (Figs 6a & 7a). The tilted, deformed stratal reflections are observed to have an offset along surfaces (termed *discontinuities*) that dip more steeply than the dip angle of the stratal reflections (Figs 6c & 7a). Further, we also observe small-scale folded geometries at the tip of the stratal reflections, adjacent to the steeply dipping discontinuities (Figs 6c & 7a). These discontinuities seem to ramp up from the basal shear surface and steepen as they propagate upwards (Figs 6c, 7a & b). We interpret these discontinuity surfaces as thrust faults (Frey-Martínez *et al.* 2005; Bull *et al.* 2009). The maximum reverse throw of the thrusts is of the order 95 ms TWT, and an average of 40 ms TWT. Small thrusts terminate within the top of DBF-03

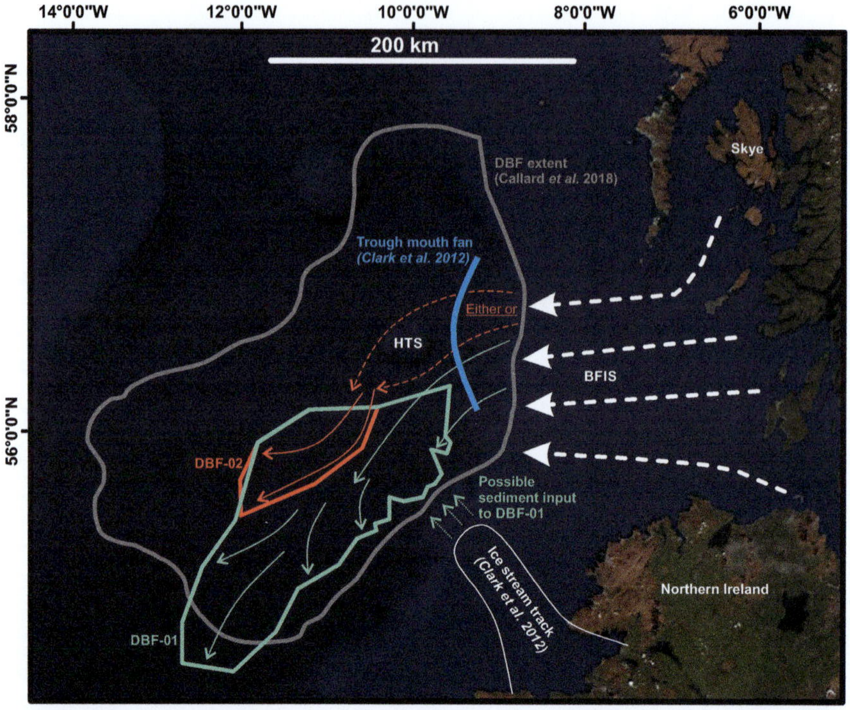

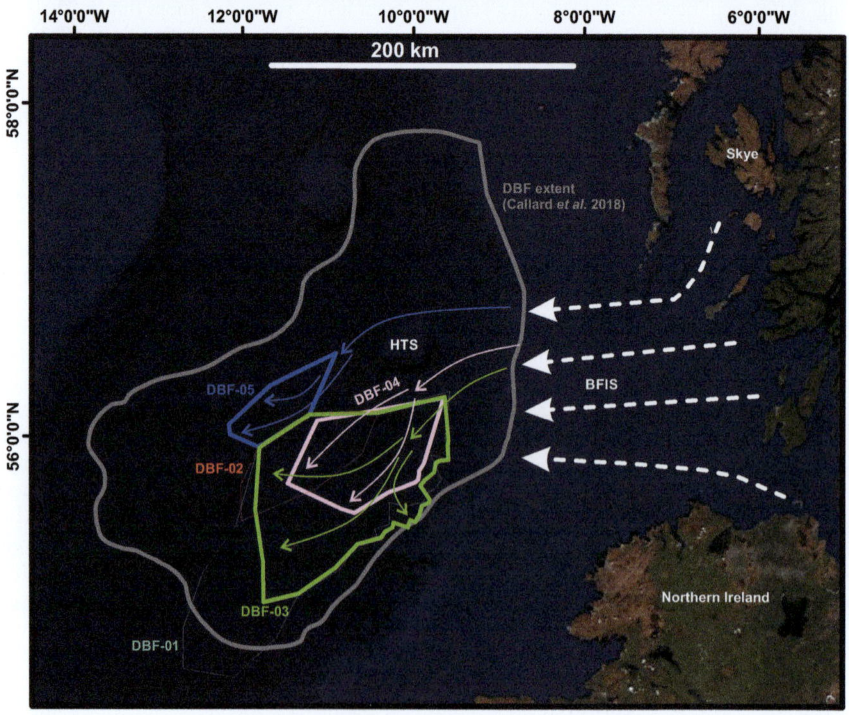

Fig. 11. Conceptual figure showing the Rockall Trough in NW Ireland and Barra Fan Ice Stream (BFIS) directions with estimate ages which have been responsible for the formation of the five MTDs identified within the Donegal Barra Fan (DBF) complex. Note the flow directions of each MTD slide (DBF 01–05), which have been derived from thickness maps and various types of kinematic indicators. HTS, Hebrides Terrace Seamount.

and DBF-04 (Fig. 6a, b). However, the largest thrust faults are observed to terminate at the upper boundary of the DBF-01 and DBF-03 (Fig. 7a, b & c). In general we do not observe any single thrust to emerge above the upper boundary of DBF-03, which is exposed to the seafloor (Fig. 7a, b). However, in the case of DBF-01, the upper tips of the thrusts were eroded in some instances by DBF-03 (Fig. 7c). Compressional folding structures and closely spaced reverse faults are typically observed basinward of the thrust faults in DBF-01 and DBF-03 (Figs 6 & 7b, c). Compressional folding structures are mostly observed above the remnant blocks (Fig. 8a). Usually compressional toe regions in submarine MTDs are dominated by thrusts-and-folds, which are classified as frontally confined MTDs (Frey-Martínez *et al.* 2006; Alsop *et al.* 2017). However, the MTDs of the DBF extend tens of kilometres beyond the regions dominated by thrusts-and-folds within the MTDs, implying that the MTDs still had mobility and were not fully arrested. This may infer the existence of a compressional regime within the MTD unrelated to arrest and freezing. This might also further suggest that the bulking up of the flow with the incorporation of extra material from the substrate changes the flow conditions locally to a more laminar, more plastic flow that causes a local velocity reduction and causes 'pile up' behind it. Similar examples have been described by Haughton *et al.* (2009). This may be similar to the processes that produce hybrid event beds, i.e. debrites encased in turbidites, resulting from rapid entrainment of mud-rich material from the substrate and/or rapid deceleration and vertical top-down flow transformation from turbidity current to debris flow which often lead to thickness increases in the deposits (e.g. Fonnesu *et al.* 2018; Pierce *et al.* 2018).

We observe thrust faults within DBF-01 on only one seismic line (Fig. 7c), within DBF-03 on five seismic lines separated by *c.* 55 km (Figs 6a, b, 7a, b & 8a), and within DBF-04 on only one seismic line (Fig. 6a). Frey-Martínez *et al.* (2006) derived flow directions of MTDs offshore Israel from the strike and dip directions of the thrust splays interpreted on high-resolution 3D seismic data. The actual strike and dip directions of the thrusts cannot be traced in each of the MTDs in this study; however, a general trend can be concluded from the contour lines on the thickness maps (Fig. 5). The NW-verging thrusts within DBF-01 (Fig. 7c) imply flow direction SE–NW. The BIIS ice stream flow lines from onshore Ireland had a similar flow direction (Fig. 11) (Clark *et al.* 2012). However, the contour lines corresponding to 200–350 ms TWT in the central part of the MTD (Fig. 5a) indicate a flow direction of NE–SW, which is more likely the general flow direction of the DBF complex. Thrusts in DBF-04 are observed in one single seismic line, dipping along the NE direction (Fig. 6a), which implies flow direction NE to SW. This aligns with the trend of contour lines corresponding to 90–150 ms TWT (Fig. 5d). Thrusts within the northern part of DBF-03 dip NE (Fig. 8a), which aligns with the general flow direction of the DBF complex and the contour lines observed. The trend of contour lines (120–270 ms TWT) of the thickness map in the northern part of DBF-03 implies two flow directions from the north (sediment source), i.e. north to SE and north to SW. However, in the southern part of DBF-03, two sets of seismic lines provide evidence of two different dip directions of the thrusts, i.e. SE and NE directions (Fig. 7a, b). There might be sediment input from the BIIS in the southern part of DBF-03, which will support the formation of NW-verging thrusts. However, given that we are only looking at one seismic line it might be that the thrusts are dipping in a direction anywhere within an arc between SE and NE, which is supported by the thickness map contour lines corresponding to 90–330 ms TWT in the southern part of DBF-03 (Fig. 5c). Moreover, if we consider that sediments of the MTD are flowing from SE to NW, then we do not expect to observe the buttressed toe of the MTD facing SE direction (Fig. 9b). Possible flow lines derived from a combined study of thrusts and thickness map contour lines are illustrated in Figures 5 and 11.

Susceptibility and triggering factors of MTDs

Various pre-conditioning factors such as sedimentation patterns, climatic history and tectonic events may influence a slope's susceptibility to failure, with slope failure events triggered by short-lived mechanisms such as earthquakes or destabilization of gas hydrates (Locat and Lee 2002; Masson *et al.* 2006).

Owen *et al.* (2007) observed a latitudinal trend in the occurrence of Late Pleistocene MTDs. In high-latitudes (> 38° N), deglaciations increase sediment supply and seismicity related to isostatic uplift, hence increasing the likelihood of continental slope failures. The Pleistocene erosional glacial cycles associated with significant increase in sedimentation, along with the combined effect of continental uplift (Stoker *et al.* 2001), led to the development of MTDs along the northwestern European continental margin, including the DBF (Canals *et al.* 2004; Sultan *et al.* 2004*b*; Evans *et al.* 2005). The sediments were directly supplied from the BFIS and ice streams in Donegal Bay reflecting changes in ice sheet dynamics (Clark *et al.* 2012). In the North Sea Trough Mouth Fan, large-scale voluminous glaciogenic debrites coincide with and reflect periods when the ice sheet was at its maximum extent (King *et al.* 1998). We propose that the BFIS oscillated between shelf edge and mid-shelf at least five times during the Pleistocene glaciations and that in those five times, it remained close enough to the shelf edge to deliver significant volumes of sediment that were then reworked downslope in the form of the five MTDs (DBF 01–05) documented in this study. This would suggest that the youngest of the events, DBF-05 corresponds to the Last Glacial Maximum (LGM), which is probably the last time the BFIS was at its maximum extent. This is corroborated by the fact that Slide C of RBSC ramps up over DBF-05 (Fig. 3b), and that Slide C of RBSC has been relatively accurately dated to be 21 ka (Georgiopoulou *et al.* 2019). It is difficult to assess the ages of the older DBF 01–04 events. They lie above the C10 unconformity, but we have demonstrated that there has been substantial erosion throughout the history of the DBF. Hence any inference about their ages is speculative without additional chronological data but can be attempted by correlation with what (little) is known about the BIIS dynamics in the Pleistocene. Most published records about the BIIS go back only to the LGM and some to Marine Isotope Stage (MIS) 5. Most of these records present counts of ice-rafted debris (IRD) over the last glacial periods. Peaks in IRD represent time of ice wasting rather than times of ice sheet maxima and maximum ice extent can be inferred to be from the time just preceding the peaks to the time of the largest peaks in the records (Scourse *et al.* 2009). Based on these published records, DBF-04 and DBF-05 could represent the latest ice advances on the shelf during the last glacial period, likely at some stage between 70 ka and 26 ka (Peck *et al.* 2007; Scourse *et al.* 2009; Hibbert *et al.* 2010). Another significant IRD peak in MIS6 (175–140 ka) might be related to the deposition of DBF-03 (Hibbert *et al.* 2010). Recent reconstructions suggest extensive glaciation during this time, known as the Penultimate Glacial Maximum (PGM), with a global sea-level lower than during the LGM (Rohling *et al.* 2017). Before the PGM, additional shelf-edge glaciations of Scottish ice are inferred during MIS 8 (300–243 ka) and MIS 10 (374–337 ka) with a first shelf- edge glaciation during MIS 12 (*c.* 0.45 Ma) based on a major seismic unconformity which is traced across the Hebrides shelf (Holmes *et al.* 2003; Sejrup *et al.* 2005). The resolution of the borehole records precludes unambiguous delineation between the age of IRD peaks between MIS 12 and MIS 8 (Sejrup *et al.* 2005), but it seems likely that the deposition of DBF-02 and DBF-01 took place as a result of these early episodes of glacial advances onto the shelf.

Indirect gas hydrate indicators as well as evidence of shallow gas and gas chimneys (fluid escape features) along the eastern slope of the Rockall Trough have been described recently (Minshull *et al.* 2020, fig. 9). Hydrates act as a cement between sediment grains and helps in binding them together (Clennell *et al.* 1999), but natural gas hydrates are very sensitive to changes in sea-bottom temperature and the pressure exerted by the water column above the seafloor (Ruppel and Kessler 2017). Any change of temperature and pressure could lead to the dissociation of gas hydrates which reduces the cohesive strength between sediment grains, and can lead to slope failures (Sultan *et al.* 2004*a*). Sedimentation pulses and increased sedimentation rates along glaciated margins can also trigger widespread gas hydrate dissociation (Karstens *et al.* 2018). The Gas Hydrate Stability Zone (GHSZ) thins down to 0–10 m along the eastern slope of the Rockall Trough (Minshull *et al.* 2020). It is possible that either fluid seepage or dissociation of hydrates along the slope, where the GHSZ is very sensitive to changes in pressure and temperature, could have triggered the MTDs in the Rockall Trough.

Conclusion

Interpretation of new high-resolution 2D multichannel seismic data in the northeastern part of the Rockall Trough of the Irish sector has produced a number of conclusions.

(1) Five episodes of mega-scale mass movements were identified in the Rockall Trough of the Irish sector, which are part of the DBF complex.
(2) The largest of the events, DBF-01, comprises *c.* 1907 km^3 of sediments, is the largest submarine MTD described in the NW British continental margin. A total volume of *c.* 5000 km^3 of sediments was mobilized by these five MTDs.
(3) Each of the MTDs may represent a time of BIIS maximum expansion. This would mean the BIIS reached maximum extent at least five times during the Pleistocene glaciations, and DBF-05 represents the LGM.
(4) The MTDs attained maximum thickness due to local deceleration and formation of fold-and-thrust zones several tens of kilometres before the MTD toe region.
(5) The slides show a compensational stacking pattern as the deposition of each MTD modifies the topography of the contemporaneous seafloor therefore influencing the position and transport direction of succeeding MTDs.

This study is an example of how improved understanding of the transport mechanisms and kinematic indicators of MTDs can be obtained from high-resolution 2D seismic data on continental margins. It also highlights that the investigation of MTDs can improve understanding of long-term glacial histories.

Acknowledgements We are grateful for the valuable suggestions from two reviewers, Suzanne Bull and Uisdean Nicholson, and co-editor Davide Gamboa, who helped us to improve this paper. We acknowledge the Petroleum Affairs Division, Department of Communications, Climate Action & Environment, the Petroleum Infrastructure Programme, the Geological Survey Ireland and the Marine Institute (INFOMAR), for providing the data for this work. We also acknowledge Schlumberger for providing an academic license of Petrel software for this study.

Funding This publication derives from research supported in part by a research grant from Science Foundation Ireland (SFI) under Grant Number 13/RC/2092 and co-funded under the European Regional Development Fund and by iCRAG industry partners. SR also thanks the Irish Research Council Government of Ireland for the Postdoctoral Fellow Award Project No. GOIPD/2018/17).

Author contributions SR: conceptualization (equal), data curation (lead), formal analysis (lead), investigation (equal), methodology (lead), project administration (lead), visualization (lead), writing – original draft (lead), writing – review & editing (equal); **AG**: conceptualization (equal), formal analysis (supporting), investigation (equal), methodology (supporting), supervision (equal), validation (supporting), visualization (supporting), writing – review & editing (equal); **SB**: conceptualization (equal), formal analysis (supporting), supervision (equal), writing – review & editing (equal); **FS**: supervision (supporting), writing – review & editing (equal).

References

Alsop, G.I., Marco, S., Levi, T. and Weinberger, R. 2017. Fold and thrust systems in Mass Transport Deposits. *Journal of Structural Geology*, **94**, 98–115, https://doi.org/10.1016/j.jsg.2016.11.008

Armishaw, J.E., Holmes, R.W. and Stow, D.A.V. 2000. The Barra Fan: a bottom-current reworked, glacially-fed submarine fan system. *Marine and Petroleum Geology*, **17**, 219–238, https://doi.org/10.1016/S0264-8172(99)00049-5

Benetti, S., Dunlop, P. and Ó Cofaigh, C. 2010. Glacial and glacially-related features on the continental margin of northwest Ireland mapped from marine geophysical data. *Journal of Maps*, **6**, 14–29, https://doi.org/10.4113/jom.2010.1092

Bull, S., Cartwright, J. and Huuse, M. 2009. A review of kinematic indicators from mass-transport complexes using 3D seismic data. *Marine and Petroleum Geology*, **26**, 1132–1151, https://doi.org/10.1016/j.marpetgeo.2008.09.011

Callard, S.L., Ó Cofaigh, C. *et al.* 2018. Extent and retreat history of the Barra Fan Ice Stream offshore western Scotland and Northern Ireland during the last glaciation. *Quaternary Science Reviews*, **201**, 280–302, https://doi.org/10.1016/j.quascirev.2018.10.002

Canals, M., Lastras, G. *et al.* 2004. Slope failure dynamics and impacts from seafloor and shallow sub-seafloor geophysical data: case studies from the COSTA project. *Marine Geology*, **213**, 9–72, https://doi.org/10.1016/j.margeo.2004.10.001

Clark, C.D., Hughes, A.L.C., Greenwood, S.L., Jordan, C. and Sejrup, H.P. 2012. Pattern and timing of retreat of the last British–Irish Ice Sheet. *Quaternary Science Reviews*, **44**, 112–146, https://doi.org/10.1016/j.quascirev.2010.07.019

Clennell, M.B., Hovland, M., Booth, J.S., Henry, P. and Winters, W.J. 1999. Formation of natural gas hydrates in marine sediments: 1. Conceptual model of gas hydrate growth conditioned by host sediment properties. *Journal of Geophysical Research: Solid Earth (1978–2012)*, **104**, 22985–23003, https://doi.org/10.1029/1999JB900175

Dahlgren, K.I.T., Vorren, T.O., Stoker, M.S., Nielsen, T., Nygård, A. and Petter Sejrup, H. 2005. Late Cenozoic prograding wedges on the NW European continental margin: their formation and relationship to tectonics and climate. *Marine and Petroleum Geology*, **22**, 1089–1110, https://doi.org/10.1016/j.marpetgeo.2004.12.008

Dove, D., Arosio, R., Finlayson, A., Bradwell, T. and Howe, J.A. 2015. Submarine glacial landforms record Late Pleistocene ice-sheet dynamics, Inner Hebrides, Scotland. *Quaternary Science Reviews*, **123**, 76–90, https://doi.org/10.1016/j.quascirev.2015.06.012

Dowdeswell, J.A., Ó Cofaigh, C. *et al.* 2008. A major trough-mouth fan on the continental margin of the Bellingshausen Sea, West Antarctica: the Belgica Fan. *Marine Geology*, **252**, 129–140, https://doi.org/10.1016/j.margeo.2008.03.017

Dunlop, P., Shannon, R., McCabe, M., Quinn, R. and Doyle, E. 2010. Marine geophysical evidence for ice sheet extension and recession on the Malin Shelf: new evidence for the western limits of the British Irish Ice Sheet. *Marine Geology*, **276**, 86–99, https://doi.org/10.1016/j.margeo.2010.07.010

Elliott, G.M., Shannon, P.M., Haughton, P.D.W., Praeg, D. and O'Reilly, B. 2006. Mid- to Late Cenozoic canyon development on the eastern margin of the Rockall Trough, offshore Ireland. *Marine Geology*, **229**, 113–132, https://doi.org/10.1016/j.margeo.2006. 03.008

Elliott, G.M., Shannon, P.M., Haughton, P.D.W. and Ovrebo, L.K. 2010. The Rockall Bank Mass Flow: collapse of a moated contourite drift onlapping the eastern flank of Rockall Bank, west of Ireland. *Marine and Petroleum Geology*, **27**, 92–107, https://doi.org/10.1016/j.marpetgeo.2009.07.006

Evans, D., King, E.L., Kenyon, N.H., Brett, C. and Wallis, D. 1996. Evidence for long-term instability in the Storegga Slide region off western Norway. *Marine Geology*, **130**, 281–292, https://doi.org/10.1016/0025-3227(95)00135-2

Evans, D., Harrison, Z. *et al.* 2005. Palaeoslides and other mass failures of Pliocene to Pleistocene age along the Atlantic continental margin of NW Europe. *Marine and Petroleum Geology*, **22**, 1131–1148, https://doi.org/10.1016/j.marpetgeo.2005.01.010

Flood, R.D., Hollister, C.D. and Lonsdale, P. 1979. Disruption of the Feni sediment drift by debris flows from Rockall Bank. *Marine Geology*, **32**, 311–334, https://doi.org/10.1016/0025-3227(79)90070-7

Fonnesu, M., Felletti, F., Haughton, P.D.W., Patacci, M. and McCaffrey, W.D. 2018. Hybrid event bed character and distribution linked to turbidite system sub-environments: the North Apennine Gottero Sandstone (north-west Italy). *Sedimentology*, **65**, 151–190, https://doi.org/10.1111/sed.12376

Frey-Martínez, J., Cartwright, J. and Hall, B. 2005. 3D seismic interpretation of slump complexes: examples from the continental margin of Israel. *Basin Research*, **17**, 83–108, https://doi.org/10.1111/j.1365-2117.2005.00255.x

Frey-Martínez, J., Cartwright, J. and James, D. 2006. Frontally confined versus frontally emergent submarine landslides: a 3D seismic characterisation. *Marine and Petroleum Geology*, **23**, 585–604, https://doi.org/10.1016/j.marpetgeo.2006.04.002

Fyfe, J., Long, D., Evans, D. and Abraham, D.A. 1993. *The Geology of the Malin–Hebrides Sea Area*. HMSO for the British Geological Survey, London.

Gamboa, D., Alves, T. and Cartwright, J. 2011. Distribution and characterization of failed (mega)blocks along salt ridges, southeast Brazil: implications for vertical fluid flow on continental margins. *Journal of Geophysical Research: Solid Earth*, **116**, https://doi.org/10.1029/2011jb008357

Gamboa, D., Alves, T. and Cartwright, J. 2012. Seismic-scale rafted and remnant blocks over salt ridges in the Espírito Santo Basin, Brazil. *In*: Yamada, Y., Kawamura, K. *et al.* (eds) *Submarine Mass Movements and Their Consequences*. Advances in Natural and Technological Hazards Research, **31**. Springer, Dordrecht, 629–638.

Georgiopoulou, A., Benetti, S., Shannon, P.M., Haughton, P.D.W. and McCarron, S. 2012. Gravity flow deposits in the deep Rockall Trough, Northeast Atlantic. *In*: Yamada, Y., Kawamura, K. *et al.* (eds) *Submarine Mass Movements and Their Consequences*. Advances in Natural and Technological Hazards Research, **31**. Springer, Dordrecht, 695–707, https://doi.org/10.1007/978-94-007-2162-3_62

Georgiopoulou, A., Shannon, P.M., Sacchetti, F., Haughton, P.D.W. and Benetti, S. 2013. Basement-controlled multiple slope collapses, Rockall Bank Slide Complex, NE Atlantic. *Marine Geology*, **336**, 198–214, https://doi.org/10.1016/j.margeo.2012.12.003

Georgiopoulou, A., Benetti, S., Shannon, P.M., Sacchetti, F., Haughton, P.D.W., Comas-Bru, L. and Krastel, S. 2014. Comparison of mass wasting processes on the slopes of the Rockall Trough, Northeast Atlantic. *In*: Krastel, S., Behrmann, J.-H. *et al.* (eds) *Submarine Mass Movements and Their Consequences: 6th International Symposium*. Advances in Natural and Technological Hazards Research, **37**. Springer, 471–480.

Georgiopoulou, A., Krastel, S. *et al.* 2019. On the timing and nature of the multiple phases of slope instability on Eastern Rockall Bank, Northeast Atlantic. *Geochemistry, Geophysics, Geosystems*, **20**, 594–613, https://doi.org/10.1029/2018gc007674

Greenwood, S.L. and Clark, C.D. 2009. Reconstructing the last Irish Ice Sheet 2: a geomorphologically-driven model of ice sheet growth, retreat and dynamics. *Quaternary Science Reviews*, **28**, 3101–3123, https://doi.org/10.1016/j.quascirev.2009.09.014

Haflidason, H., Sejrup, H.P. *et al.* 2004. The Storegga Slide: architecture, geometry and slide development. *Marine Geology*, **213**, 201–234, https://doi.org/10.1016/j.margeo.2004.10.007

Hamilton, E.L. and Bachman, R.T. 1982. Sound velocity and related properties of marine sediments. *The Journal of the Acoustical Society of America*, **72**, 1891–1904, https://doi.org/10.1121/1.388539

Haughton, P., Davis, C., McCaffrey, W. and Barker, S. 2009. Hybrid sediment gravity flow deposits – Classification, origin and significance. *Marine and Petroleum Geology*, **26**, 1900–1918, https://doi.org/10.1016/j.marpetgeo.2009.02.012

Hibbert, F.D., Austin, W.E.N., Leng, M.J. and Gatliff, R.W. 2010. British Ice Sheet dynamics inferred from North Atlantic ice-rafted debris records spanning the last 175 000 years. *Journal of Quaternary Science*, **25**, 461–482, https://doi.org/10.1002/jqs.1331

Holmes, R., Long, D. and Dodd, L.R. 1998. Large-scale debrites and submarine landslides on the Barra Fan, west of Britain. *Geological Society, London, Special Publications*, **129**, 67–79, https://doi.org/10.1144/gsl.Sp.1998.129.01.05

Holmes, R., Bulat, J., Hamilton, I. and Long, D. 2003. Morphology of an ice-sheet limit and constructional glacially-fed slope front, Faroe–Shetland Channel. *In*: Mienert, J. and Weaver, P. (eds) *European Margin Sediment Dynamics: Side-Scan Sonar and Seismic Images*. Springer, Berlin, 149–152.

Huvenne, V.A.I., Croker, P.F. and Henriet, J.-P. 2002. A refreshing 3D view of an ancient sediment collapse and slope failure. *Terra Nova*, **14**, 33–40, https://doi.org/10.1046/j.1365-3121.2002.00386.x

Karstens, J., Haflidason, H. *et al.* 2018. Glacigenic sedimentation pulses triggered post-glacial gas hydrate dissociation. *Nature Communications*, **9**, 635, https://doi.org/10.1038/s41467-018-03043-z

King, E.L., Haflidason, H., Sejrup, H.P. and Løvlie, R. 1998. Glacigenic debris flows on the North Sea Trough Mouth Fan during ice stream maxima. *Marine Geology*, **152**, 217–246, https://doi.org/10.1016/S0025-3227(98)00072-3

Knutz, P.C., Austin, W.E.N. and Jones, E.J.W. 2001. Millennial-scale depositional cycles related to British Ice Sheet variability and North Atlantic paleocirculation since 45 kyr B.P., Barra Fan, U.K. margin. *Paleoceanography*, **16**, 53–64, https://doi.org/10.1029/1999pa000483

Knutz, P.C., Jones, E.J.W., Howe, J.E., Van Weering, T.J.C. and Stow, D.A.V. 2002. Wave-form sheeted contourite drift on the Barra Fan, NW UK continental margin. *Geological Society, London, Memoirs*, **22**, 85–97, https://doi.org/10.1144/gsl.Mem.2002.022.01.08

Locat, J. and Lee, H.J. 2002. Submarine landslides: advances and challenges. *Canadian Geotechnical Journal*, **39**, 193–212, https://doi.org/10.1139/t01-089

Maslin, M., Owen, M., Day, S. and Long, D. 2004. Linking continental-slope failures and climate change: testing the clathrate gun hypothesis. *Geology*, **32**, 53–56, https://doi.org/10.1130/G20114.1

Masson, D.G., Harbitz, C.B., Wynn, R.B., Pedersen, G. and Løvholt, F. 2006. Submarine landslides: processes, triggers and hazard prediction. *Philosophical Transactions of the Royal Society A: Mathematical, Physical and Engineering Sciences*, **364**, 2009–2039, https://doi.org/10.1098/rsta.2006.1810

McDonnell, A. and Shannon, P.M. 2001. Comparative Tertiary stratigraphic evolution of the Porcupine and Rockall basins. *Geological Society, London, Special Publications*, **188**, 323–344, https://doi.org/10.1144/gsl.sp.2001.188.01.19

Minshull, T.A., Marín-Moreno, H. *et al.* 2020. Hydrate occurrence in Europe: a review of available evidence. *Marine and Petroleum Geology*, **111**, 735–764, https://doi.org/10.1016/j.marpetgeo.2019.08.014

Moscardelli, L. and Wood, L. 2016. Morphometry of mass-transport deposits as a predictive tool. *GSA Bulletin*, **128**, 47–80, https://doi.org/10.1130/b31221.1

Ó Cofaigh, C., Taylor, J., Dowdeswell, J.A. and Pudsey, C.J. 2003. Palaeo-ice streams, trough mouth fans and high-latitude continental slope sedimentation. *Boreas*, **32**, 37–55, https://doi.org/10.1080/0300948031000 1858

Ó Cofaigh, C., Dunlop, P. and Benetti, S. 2012. Marine geophysical evidence for Late Pleistocene ice sheet extent and recession off northwest Ireland. *Quaternary Science Reviews*, **44**, 147–159, https://doi.org/10.1016/j.quascirev.2010.02.005

Ó Cofaigh, C., Weilbach, K. *et al.* 2019. Early deglaciation of the British-Irish Ice Sheet on the Atlantic shelf northwest of Ireland driven by glacioisostatic depression and high relative sea level. *Quaternary Science Reviews*, **208**, 76–96, https://doi.org/10.1016/j.quascirev.2018.12.022

Owen, M.J. and Long, D. 2016. Barra Fan: a major glacial depocentre on the western continental margin of the British Isles. *Geological Society, London, Memoirs*, **46**, 359, https://doi.org/10.1144/M46.76

Owen, M., Day, S. and Maslin, M. 2007. Late Pleistocene submarine mass movements: occurrence and causes. *Quaternary Science Reviews*, **26**, 958–978, https://doi.org/10.1016/j.quascirev.2006.12.011

Owen, M.J., Maslin, M.A., Day, S.J. and Long, D. 2018. Sediment failures within the Peach Slide (Barra Fan, NE Atlantic Ocean) and relation to the history of the British-Irish Ice Sheet. *Quaternary Science Reviews*, **187**, 1–30, https://doi.org/10.1016/j.quascirev.2018.03.018

Peck, V.L., Hall, I.R., Zahn, R. and Scourse, J.D. 2007. Progressive reduction in NE Atlantic intermediate water ventilation prior to Heinrich events: response to NW European ice sheet instabilities? *Geochemistry, Geophysics, Geosystems*, **8**, https://doi.org/10.1029/2006gc001321

Peters, J.L., Benetti, S., Dunlop, P. and Ó Cofaigh, C. 2015. Maximum extent and dynamic behaviour of the last British–Irish Ice Sheet west of Ireland. *Quaternary Science Reviews*, **128**, 48–68, https://doi.org/10.1016/j.quascirev.2015.09.015

Peters, J.L., Benetti, S., Dunlop, P., Ó Cofaigh, C., Moreton, S.G., Wheeler, A.J. and Clark, C.D. 2016. Sedimentology and chronology of the advance and retreat of the last British-Irish Ice Sheet on the continental shelf west of Ireland. *Quaternary Science Reviews*, **140**, 101–124, https://doi.org/10.1016/j.quascirev.2016.03.012

Pierce, C.S., Haughton, P.D.W., Shannon, P.M., Pulham, A.J., Barker, S.P. and Martinsen, O.J. 2018. Variable character and diverse origin of hybrid event beds in a sandy submarine fan system, Pennsylvanian Ross Sandstone Formation, western Ireland. *Sedimentology*, **65**, 952–992, https://doi.org/10.1111/sed.12412

Rebesco, M., Camerlenghi, A., Geletti, R. and Canals, M. 2006. Margin architecture reveals the transition to the modern Antarctic ice sheet ca. 3 Ma. *Geology*, **34**, 301–304, https://doi.org/10.1130/g22000.1

Rohling, E.J., Hibbert, F.D. *et al.* 2017. Differences between the last two glacial maxima and implications for ice-sheet, δ18O, and sea-level reconstructions. *Quaternary Science Reviews*, **176**, 1–28, https://doi.org/10.1016/j.quascirev.2017.09.009

Ruppel, C.D. and Kessler, J.D. 2017. The interaction of climate change and methane hydrates. *Reviews of Geophysics*, **55**, 126–168, https://doi.org/10.1002/2016RG000534

Sacchetti, F., Benetti, S., Georgiopoulou, A., Dunlop, P. and Quinn, R. 2011. Geomorphology of the Irish Rockall Trough, North Atlantic Ocean, mapped from multibeam bathymetric and backscatter data. *Journal of Maps*, **7**, 60–81, https://doi.org/10.4113/jom.2011.1157

Sacchetti, F., Benetti, S. *et al.* 2012. Deep-water geomorphology of the glaciated Irish margin from high-resolution marine geophysical data. *Marine Geology*, **291-294**, 113–131, https://doi.org/10.1016/j.margeo.2011.11.011

Sacchetti, F., Benetti, S., Quinn, R. and Ó Cofaigh, C. 2013. Glacial and post-glacial sedimentary processes in the Irish Rockall Trough from an integrated acoustic analysis of near-seabed sediments. *Geo-Marine Letters*, **33**, 49–66, https://doi.org/10.1007/s00367-012-0310-2

Salmanidou, D., Georgiopoulou, A., Guillas, S. and Dias, F. 2018. Rheological considerations for the modelling of submarine sliding at Rockall Bank, NE Atlantic Ocean. *Physics of Fluids*, **30**, 030705, https://doi.org/10.1063/1.5009552

Scourse, J.D., Haapaniemi, A.I. *et al.* 2009. Growth, dynamics and deglaciation of the last British–Irish ice sheet: the deep-sea ice-rafted detritus record. *Quaternary Science Reviews*, **28**, 3066–3084, https://doi.org/10.1016/j.quascirev.2009.08.009

Sejrup, H.P., Hjelstuen, B.O. *et al.* 2005. Pleistocene glacial history of the NW European continental margin. *Marine and Petroleum Geology*, **22**, 1111–1129, https://doi.org/10.1016/j.marpetgeo.2004.09.007

Shannon, P.M., O'Reilly, B.M., Readman, P.W., Jacob, A.W.B. and Kenyon, N. 2001. Slope failure features on the margins of the Rockall Trough. *Geological Society, London, Special Publications*, **188**, 455–464, https://doi.org/10.1144/gsl.sp.2001.188.01.28

Stoker, M.S., Van Weering, T.C.E. and Svaerdborg, T. 2001. A Mid- to Late Cenozoic tectonostratigraphic framework for the Rockall Trough. *Geological Society, London, Special Publications*, **188**, 411–438, https://doi.org/10.1144/gsl.sp.2001.188.01.26

Stow, D.A.V., Armishaw, J.E. and Holmes, R. 2002. Holocene contourite sand sheet on the Barra Fan slope, NW Hebridean margin. *Geological Society, London, Memoirs*, **22**, 99, https://doi.org/10.1144/GSL.MEM.2002.022.01.09

Sultan, N., Cochonat, P., Foucher, J.P. and Mienert, J. 2004*a*. Effect of gas hydrates melting on seafloor slope instability. *Marine Geology*, **213**, 379–401, https://doi.org/10.1016/j.margeo.2004.10.015

Sultan, N., Cochonat, P. *et al.* 2004*b*. Triggering mechanisms of slope instability processes and sediment failures on continental margins: a geotechnical approach. *Marine Geology*, **213**, 291–321, https://doi.org/10.1016/j.margeo.2004.10.011

Tarlati, S. 2018. *Late Quaternary reconstruction of British-Irish Ice Sheet variability through the analysis of deep-water sediments from the Donegal Barra Fan and the Rockall Trough, North Atlantic*. PhD thesis, Ulster University.

Unnithan, V., Shannon, P.M., McGrane, K., Readman, P.W., Jacob, A.W.B., Keary, R. and Kenyon, N.H. 2001. Slope instability and sediment redistribution in the Rockall Trough: constraints from GLORIA. *Geological Society, London, Special Publications*, **188**, 439–454, https://doi.org/10.1144/gsl.sp.2001.188.01.27

Varnes, D.J. 1978. Slope movement types and processes. *TRB Special Report*, **176**, 11–33.

Vorren, T.O. and Laberg, J.S. 1997. Trough mouth fans – palaeoclimate and ice-sheet monitors. *Quaternary Science Reviews*, **16**, 865–881, https://doi.org/10.1016/S0277-3791(97)00003-6

Yilmaz, Ö. 2001. *Seismic Data Analysis: Processing, Inversion, and Interpretation of Seismic Data*. Society of Exploration Geophysicists, Tulsa.

Short- and long-term movement of mudflows of the Mississippi River Delta Front and their known and potential impacts on oil and gas infrastructure

Jason D. Chaytor[1]*, Wayne E. Baldwin[1], Samuel J. Bentley[2], Melanie Damour[3], Douglas Jones[3], Jillian Maloney[4], Michael D. Miner[5], Jeff Obelcz[6] and Kehui Xu[7,8]

[1]US Geological Survey, Woods Hole, USA

[2]Department of Geology and Geophysics, Louisiana State University, Baton Rouge, USA

[3]Bureau of Ocean Energy Management, New Orleans, USA

[4]Department of Geological Sciences, San Diego State University, San Diego, USA

[5]The Water Institute of the Gulf, New Orleans, USA

[6]NRC Postdoctoral Fellow, US Naval Research Lab, Stennis Space Center, USA

[7]Department of Oceanography and Coastal Sciences, Louisiana State University, Baton Rouge, USA

[8]Coastal Studies Institute, Louisiana State University, Baton Rouge, USA

JDC, 0000-0001-8135-8677; WEB, 0000-0001-5886-0917;
MD, 0000-0002-9911-9591; JO, 0000-0003-2515-2249
*Correspondence: jchaytor@usgs.gov

Abstract: Mudflows on the Mississippi River Delta Front (MRDF) are recognized hazards to oil and gas infrastructure in the shallow (20–300 m water depth) Gulf of Mexico. Preconditioning of the seafloor for failure results from high sedimentation rates coupled with slope over-steepening, under-consolidation and abundant biogenic gas production. Catastrophic failure of production platforms and pipelines due to seafloor displacement during infrequent large hurricanes such as Camille in 1969 and Ivan in 2004, point to cyclical loading of the seafloor by waves as a primary movement trigger. Due to data limitations, the role of smaller storms and background oceanographic processes in driving seafloor movement has remained largely unconstrained but these are thought to contribute to significant seafloor change. With the aid of new high-resolution multibeam mapping and seismic reflection profiling across sections of the MRDF, several moving features within the deforming delta-front environment are investigated and potential hazards to infrastructure installed and adjacent to the region are discussed. Via repeat mapping surveys of selected areas and records of changing shipwreck locations, we highlight significant seafloor displacement across annual to decadal timescales. For example, individual blocks mapped within mudflow gullies adjacent to Southwest Pass show downslope transport of more than 80 m in a single year, while the SS *Virginia*, a 153 m-long oil tanker sunk in 1942, has been relocated and found to have moved downslope more than 400 m in 14 years, without a major hurricane (>Category 2) passing through the region.

Mudflows (channelized and unconfined debris flows) on the Mississippi River Delta Front (MRDF) have long been recognized as an active process (Shepard 1955) and hazard to oil and gas infrastructure in the shallow northern Gulf of Mexico (e.g. Bea 1971). More than 200 platforms, 4000 wells and thousands of kilometres of pipeline segments are located across the MRDF boundary (including those in active and inactive status) (Fig. 1). Increased awareness of the hazards from mudflow activity, illustrated by catastrophic mudflow-related platform failure in 1969 during the passage of Hurricane Camille (SP 70 Platform B; Bea *et al.* 1975; Sterling and Strohbeck 1975; Prior and Coleman 1982) led to focused efforts during the late 1970s and early 1980s to characterize the stability of the MRDF and investigate the conditions driving failure. These efforts resulted in the development of detailed mudflow-system mapping across the entire MRDF (Coleman *et al.* 1980) and new understanding of the influence of hurricanes, changes in sediment delivery and shallow gas in promoting failure of MRDF sediments (e.g. Coleman *et al.* 1974, 1983, 1998; Whelan *et al.* 1976; Coleman and Garrison 1977; Prior and Coleman 1978, 1982).

Mudflows induced by the passage of Hurricane Ivan over the MRDF in 2004 (Fig. 2) caused a failure

From: Georgiopoulou, A., Amy, L. A., Benetti, S., Chaytor, J. D., Clare, M. A., Gamboa, D., Haughton, P. D. W., Moernaut, J. and Mountjoy, J. J. (eds) 2020. *Subaqueous Mass Movements and their Consequences: Advances in Process Understanding, Monitoring and Hazard Assessments*. Geological Society, London, Special Publications, **500**, 587–604.
First published online March 26, 2020, https://doi.org/10.1144/SP500-2019-183

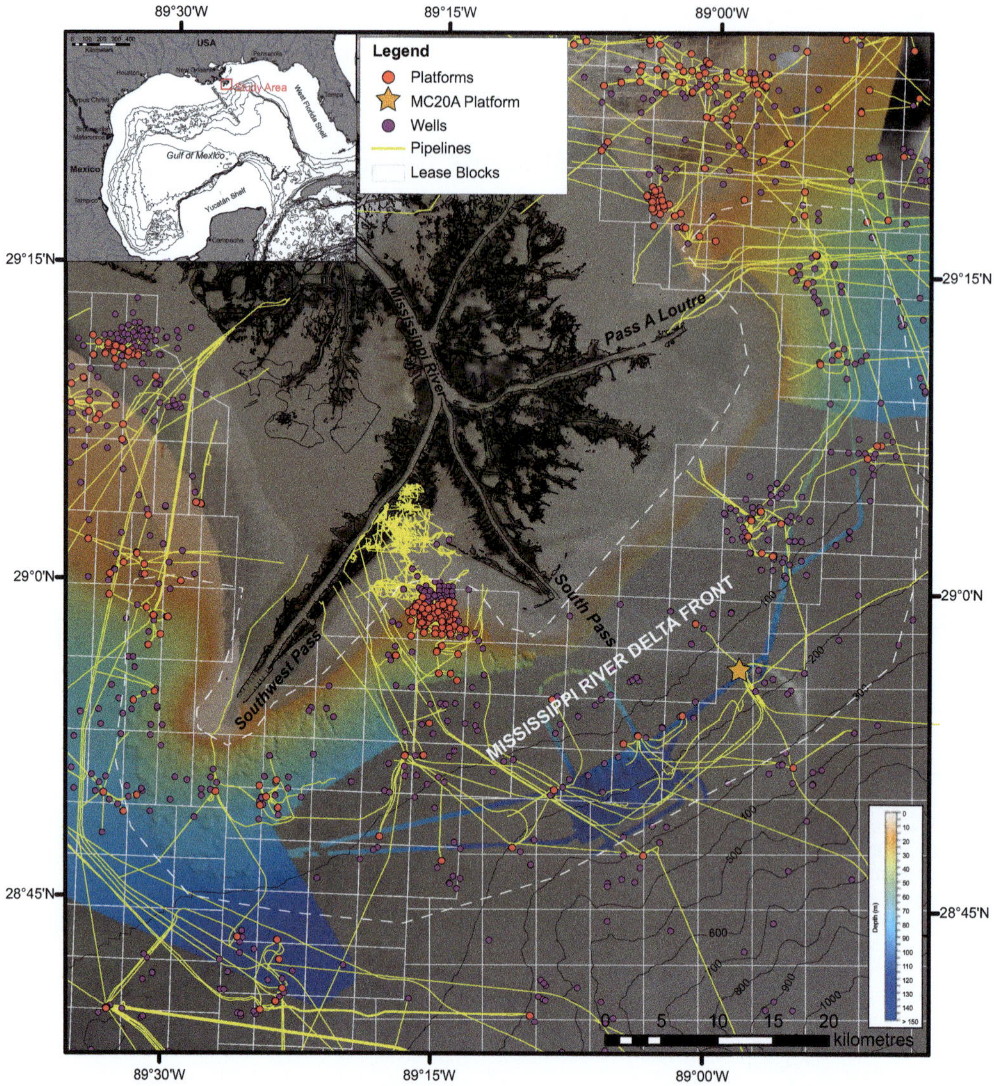

Fig. 1. The Mississippi River Delta Front (MRDF) in relation to the Mississippi River, 'Birdfoot' delta and bottom-installed infrastructure. Colour-shaded bathymetry is a combination of NOAA National Ocean Service (NOS) hydrographic surveys (NOAA 2014–16) and data collected in 2017 by the USGS (Baldwin *et al.* 2018). Approximate boundary of the MRDF is highlighted by the dashed white line. The location of the MC20A platform jacket shown in Figure 3 is highlighted by the orange star off South Pass. Background image: Terra MODIS Corrected Reflectance (25 April 2017). Inset shows the location of the study area.

of the Taylor Energy platform in Mississippi Canyon Lease Block 20 (MC20A platform) (Fig. 3), and also damaged or destroyed numerous pipeline sections and other bottom-mounted infrastructure. These events initiated a renewed interest in further evaluating mudflow hazards and developing mitigation strategies (e.g. Hitchcock *et al.* 2006; Nodine *et al.* 2007, 2009; Guidroz 2009). This need was further underscored by additional damage to infrastructure in the region by subsequent hurricanes Katrina (2005) and Gustav (2008) (Fig. 2). In the last several years, analysis of repeated bathymetric surveys and sediment delivery to the MRDF system have provided new insights into the preconditioning factors for failure, the magnitude of movements and the spatial and temporal scales at which these processes operate (Keller *et al.* 2016; Obelcz *et al.* 2017; Maloney *et al.* 2018).

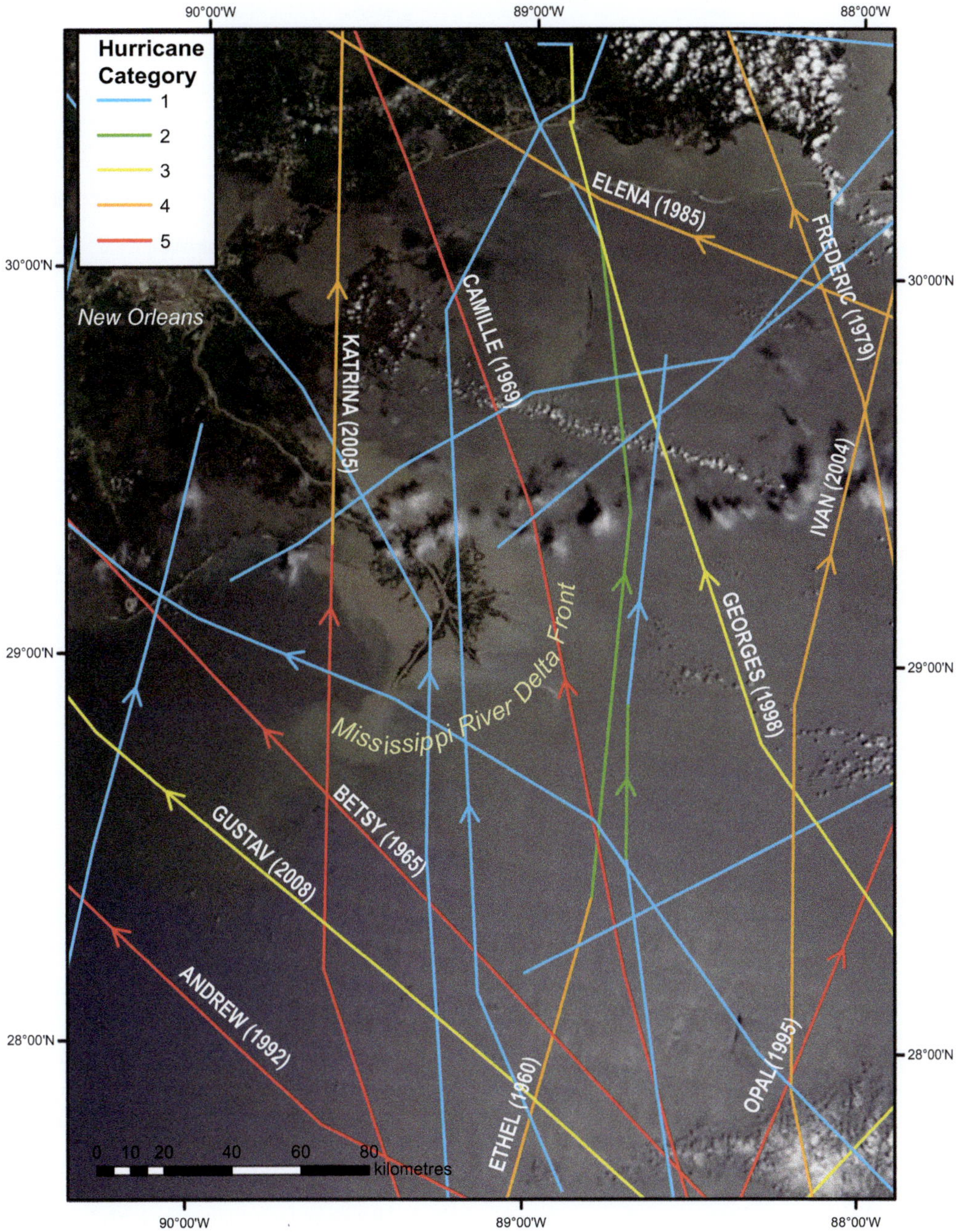

Fig. 2. Category 1 to 5 hurricane tracks that have passed within *c.* 150 km of the MRDF between 1960 and 2018 (Knapp *et al.* 2010 – IBTracs v04). The names of category 3 and above hurricanes and those referred to in the text are highlighted. Background image: Terra MODIS Corrected Reflectance (25 April 2017).

While mudflows triggered by hurricanes that generate localized extreme wave conditions represent a significant instantaneous hazard, progressive or periodic seafloor movement occurring in the absence of readily identifiable external triggers may represent an equal or greater hazard. Utilizing new bathymetric and seismic datasets collected in 2017, we investigate the rates and impact of short- and longer-term

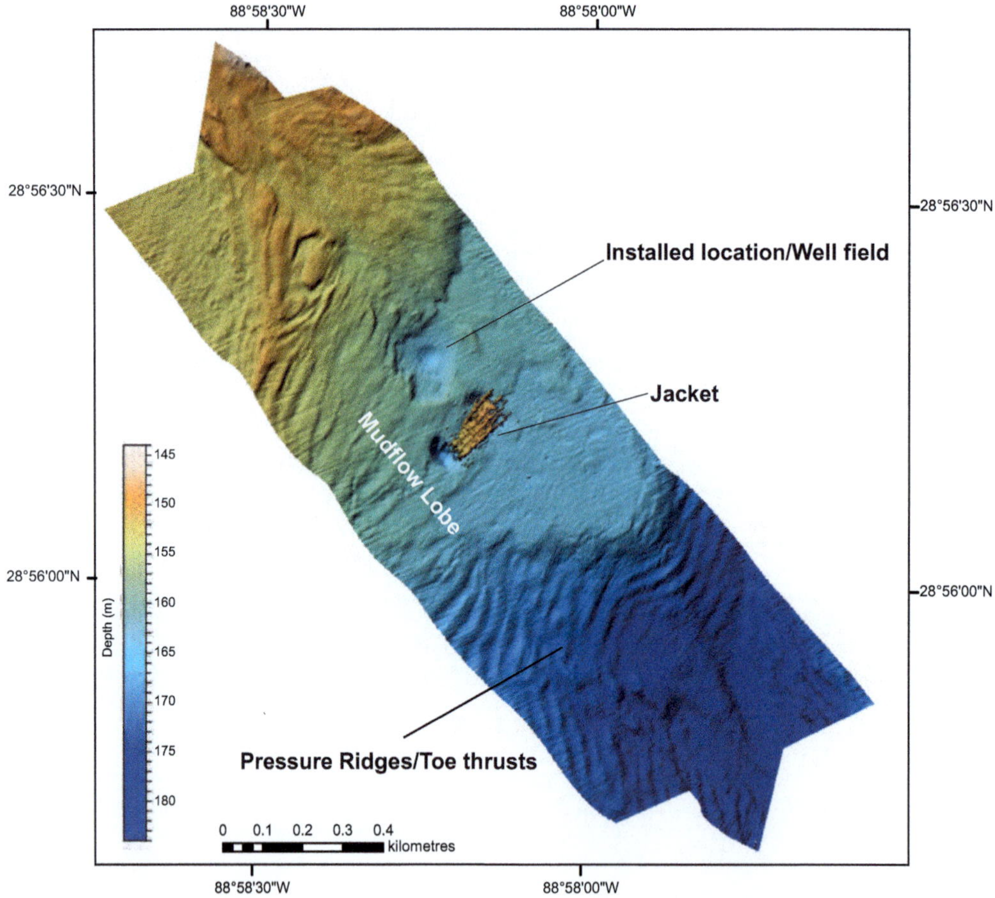

Fig. 3. A portion of the mudflow lobe containing the jacket from the MC20A platform that was destroyed and displaced during Hurricane Ivan in 2004.

seafloor movement of several locations across the MRDF, which have occurred, at least partially, without obvious large external forcing. In addition to using geomorphic features of the seafloor to track short-term (i.e. seasonal to annual timescales) MRDF seafloor movement, as has been done in previous assessments, we evaluate longer-term movement via tracking of a World War II shipwreck, the *c.* 153 m-long oil tanker SS *Virginia* and several major oil and gas pipelines.

Background

The Mississippi River carries an estimated sediment load of 210 × 10^6 tonnes per year, composed predominantly of silt and clay in suspension (Milliman and Meade 1983). Only a small fraction (19% over the 2008–10 period; Allison *et al.* 2012) of this total load reaches the main distributary pathways of the modern Mississippi River delta (the Plaquemines-Balize 'Birdfoot' delta) which has been active since *c.* 0.8 ka (Tornqvist *et al.* 1996; Blum and Roberts 2012; Bentley *et al.* 2016; Maloney *et al.* 2018). Rapid flow decelerations at the mouths of the delta distributary channels, primarily Southwest Pass, South Pass and Pass A Loutre (Fig. 1), result in deposition of bed-load sand and coarser suspended sediment and progradation of channel levees and mouth bars (Fisk 1961; Maloney *et al.* 2018). The fine fraction of the suspended sediment is distributed offshore and has supported construction of a clay-sized fraction dominated delta platform, the MRDF (Fisk and McClelland 1959, Coleman *et al.* 1998; Bentley *et al.* 2016). These fine-grained materials are rapidly deposited with rates varying from >1 m a^{-1} proximal to the distributary channel mouths to 1 cm a^{-1} beyond water depths of 100 m (Coleman *et al.* 1991; Ruttenberg and Goni 1997; Keller *et al.* 2016). The fine grain size, rapid sedimentation, high-water content and large quantities of *in-situ* gases generated by the breakdown of organic material result in low-strength subaqueous deltaic deposits highly susceptible to failure.

The broad-scale morphological regions of the MRDF have been defined based on generalized bathymetry and are as follows (Coleman *et al.* 1998): 0–10 m, delta top and interdistributary bays; 10–70 m, upper delta front; 70–120 m, intermediate delta front; and 120–200 m, lower delta front (Fig. 1). Frequent mass transport processes in this region are largely confined to water depths shallower than 250 m. The slope of this apron is, on average, very gentle, generally less than 0.5°; nevertheless, this low gradient is sufficient, as seen elsewhere even in areas of lower sedimentation rates (e.g. Urlaub *et al.* 2015) to allow downslope flow of sediments under the force of gravity. Primary mass transport-related features of the MRDF (Fig. 4) include collapse depressions, mudflow gullies and mudflow lobes. Seafloor surrounding these transport features is often partially deformed, pre-conditioned for failure or undergoing transformation into gully features. Mudflow gullies form the primary pathway for transporting material from the delta top and upper delta front to deeper water. Upslope (*c.* 20–30 m water depth), mudflow gullies initiate as retrogressive failures that transition into steep-sided depressions funnelling blocky mass transport debris downslope into broad mudflow lobes. Gullies off the mouth of Pass A Loutre differ from those along the southern MRDF, with development of levees that result from remoulding of debris ejected from the gully floor (Coleman *et al.* 1980). Collapse depressions and bottleneck slides are commonly found adjacent to the upslope extensions of mudflow gullies and may act as the point of initial gully development. Mudflow lobes are composed of stacked mudflow deposits sourced from discrete or amalgamated mudflow gullies. Individual mudflow gullies vary in width, converge and diverge with adjacent gullies and are characterized by average relief of 5–10 m, and up to 25 m below the surrounding seafloor (Coleman *et al.* 1980). Estimates of basal shear plane depth below the gullies range from 6 to 30 m below seafloor (Prior and Coleman 1978; Prior and Suhayda 1979). Collection and analysis of new bathymetry and geophysical datasets are providing the opportunity to better define, characterize and temporally compare these mass transport morphologies across the entire MRDF (Fig. 1).

Data and methods

High-resolution bathymetric data collected between 2014 and 2016 on behalf of the National Oceanic and Atmospheric Administration (NOAA) National Ocean Service/Office of Coast Survey are the primary data used here (NOAA 2014–16). These surveys provide full-bottom coverage at resolutions between 0.5 m and 8 m over portions offshore of the Southwest Pass and Pass A Loutre distributaries

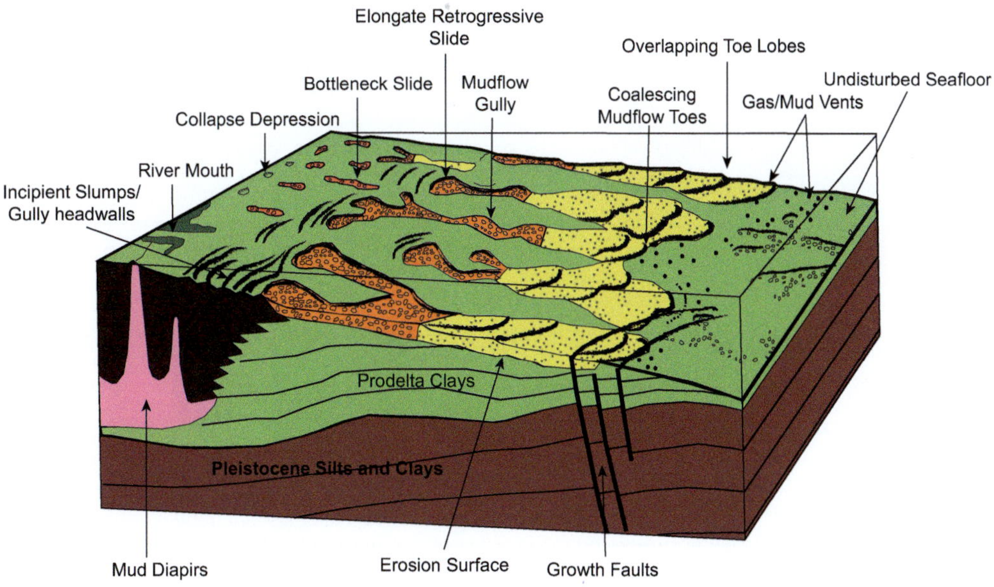

Fig. 4. Components of the Mississippi Delta and MRDF mudflow systems. Modified from Coleman *et al.* (1980).

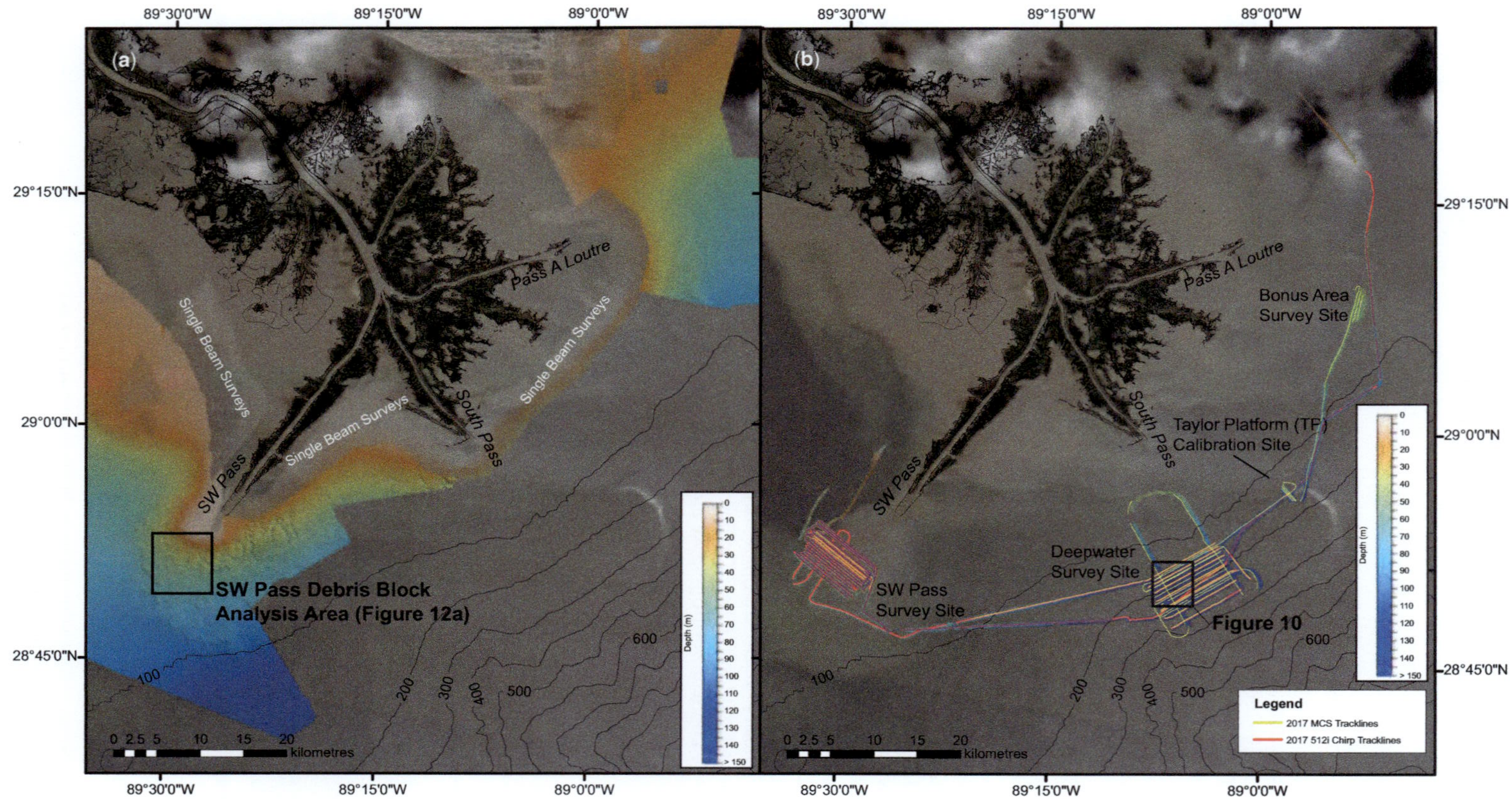

Fig. 5. (**a**) Compilation of NOAA NOS hydrographic surveys covering the West Delta/Southwest Pass, South Pass and Main Pass protraction areas of the MRDF at 8 m resolution (NOAA 2014–16). Note the limited coverage in shallow water around the 'Birdfoot' where dense single-beam soundings were collected. (**b**) Multibeam, Chirp and multichannel seismic data collected by the USGS in 2017 (Baldwin *et al.* 2018). Data were co-acquired, therefore Chirp and MCS data collection generally overlap, except where weather or equipment issues prevented collection of one or both seismic datasets.

of the Mississippi River delta (Fig. 5a). These surveys were combined into a single, 8 m-resolution bathymetry surface for broad-scale morphological interpretation, while the high-resolution surfaces were used for detailed bottom motion analysis and feature discrimination.

In 2017, the US Geological Survey (USGS) collected additional high-resolution bathymetry, Chirp, and multichannel seismic reflection data in several selected areas of the MRDF (Fig. 5b; Baldwin *et al.* 2018). Approximately 600 km of multibeam and acoustic backscatter data with final grid resolutions of 1–2 m were acquired using a dual-head Reson T20P echosounder coupled to an Applanix Pos-MV Wavemaster MRU-navigation system. Water sound speed measurement were regularly acquired using a moving vessel profiler; however, rapid changes in salinity and water structure, such as where freshwater mixes with seawater at the mouth of Southwest Pass, resulted in substantial refraction-related artefacts in parts of the final dataset.

A total of 425 km of Chirp data (Fig. 5b) were acquired with an Edgetech SB-512i, installed on a surface-towed sled with an attached GPS antenna. Component failure on the SB-512i towards the end of the cruise prevented Chirp data collection over part of the Deepwater, TP and Bonus area survey sites (Fig. 5b). Swell filtering was the only processing applied to these Chirp data. Multi-channel seismic data (MCS) were continuously collected using either a 3-plate Applied Acoustic S-Boom (600 J max., 200 J/plate) or SIG ELC 1050 or 1200 mini-sparker (operated at 200 J) source and a solid-state 32-channel Geometrics Geoeel streamer with 1.5625 m channel spacing. MCS coverage using both sources totalled almost 550 km. A constant velocity of 1500 m s^{-1} was used when converting seismic profiles from time to depth.

Feature movement measurement

Seafloor movement was assessed by both repeat measurement of large seafloor targets including the shipwreck *Virginia* and bottom-mounted infrastructure (e.g. pipelines, platform jackets) and tracking of individual blocks within mudflow gullies off Southwest Pass. Displacement of large targets was measured by a combination of historical location records and modern survey imaging to confirm positions or identify movement. Because historic shipwrecks in the region are archaeological sites and, in the case of *Virginia*, a war grave, the shipwreck's specific location is considered sensitive information and is not provided here. Pipeline information, including GIS layers and reports containing as-built locations, repairs and abandonments, are available via the Bureau of Safety and Environmental Enforcement (https://www.data.bsee.gov/Main/Pipeline.aspx). Pipeline displacement distance based on these sources are approximated. The 2017 USGS multibeam data were used to identify and digitize the locations of pipelines visible at the seafloor in South Pass (SP) and Mississippi Canyon (MP) lease blocks SP50, SP51, MC149 and MC150 (Deepwater Survey site, Fig. 5b).

Identifying and tracking individual mudflow debris blocks presents the greatest challenge due to their unstable nature and tendency to change shape during movement, and the variable resolution of bathymetry over the primary area of block movement investigated for this study (Fig. 5a). To carry out this analysis, areas containing blocks of similar dimensions and general spatial arrangement were identified in both a 2016 NOAA hydrographic dataset (2–4 m resolution) and the 2017 USGS dataset (2 m resolution) off Southwest Pass. The basal bounds of corresponding blocks identified in both datasets were then digitized as polygons and assigned matching identifiers in ArcGIS using slope maps derived from the bathymetry datasets. In order to minimize errors associated with measuring displacement by using a particular segment of the digitized bounds, as was done by Galloway *et al.* (2017) for blocks in a nearby mudflow gully, the centroid (or the geometric centre) of each digitized polygon was calculated and used as the reference point for measuring distance and direction for movement of each block (Fig. 6a & b).

Results

SS Virginia shipwreck

Virginia was a bulk oil tanker that sank on 12 May, 1942 after being hit by three torpedoes from the German U-boat, U-507. The shipwreck currently lies intact, listing to its port side in more than 80 m of water offshore Southwest Pass. With its bow pointed towards the NW (Fig. 7), *Virginia* sits entirely within a mudflow lobe, with a small *c.* 2 m deep seafloor depression or area of scour formed around the bow. No significant change in the surface morphology of the mudflow lobe or the location of its toe are noted between the 2016 and 2017 multibeam surveys and there is no obvious evidence in the bathymetry or Chirp data of deformation resulting from localized movement of the ship (e.g. linear troughs or small discrete lobes) (Fig. 7). As such, the wreck appears to be strongly coupled to the larger lobe mass and recent movement of the wreck likely involves movement of the entire lobe or a large portion of it.

The wreck was first imaged in 2001 (Church *et al.* 2007) during oil- and gas-related geophysical

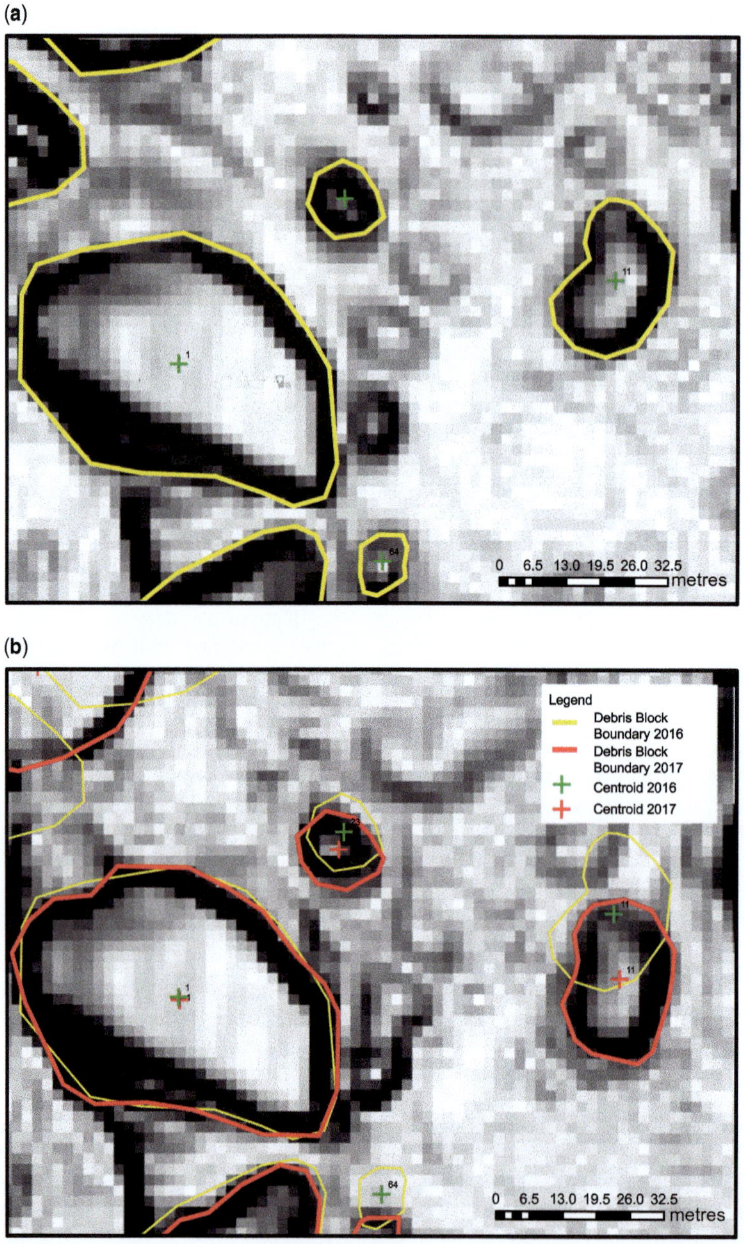

Fig. 6. Method for determining distance and direction of debris-block movement in mudflow gullies off Southwest Pass. (**a**) 2 m resolution black-and-white slope map derived from 2016 NOS data with bounds of debris blocks digitized and respective location of block polygon centroids (NOAA 2014–16). (**b**) 2 m resolution slope map derived from 2017 USGS survey, covering the same area as (a), with bounds of debris blocks from 2016 and 2017 surveys and respective location of block polygon centroids shown. Distance and direction of block motion are measured between corresponding centroids.

surveys off Southwest Pass. In 2003, the wreck was reimaged during another geophysical survey, and confirmed to be the remains of *Virginia* through a Remotely Operated Vehicle (ROV) investigation (Church *et al.* 2007) conducted in 2004. A pipeline survey in 2005–06 after Hurricanes Ivan and Katrina showed that the wreck had moved *c.* 365 m to the SE of its 2004 location (Fig. 8). Unpublished

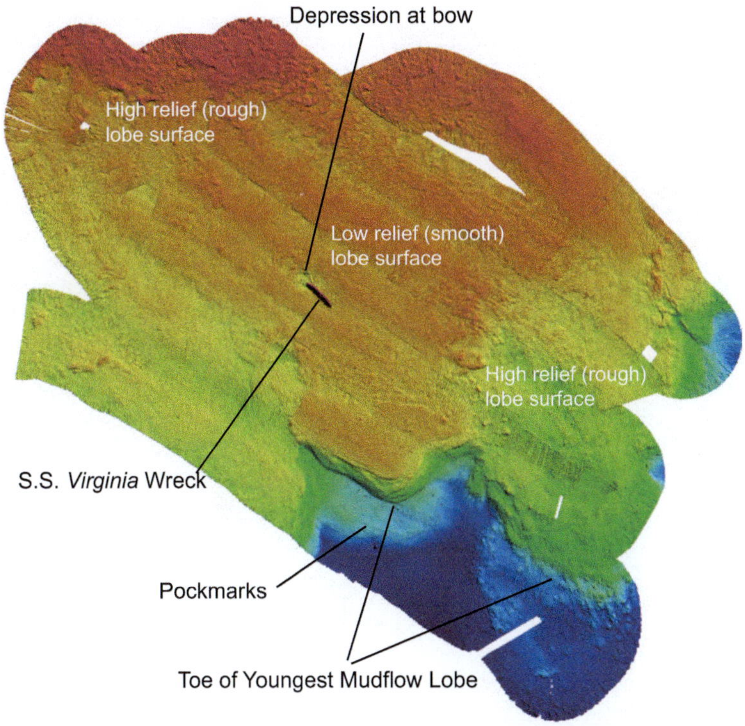

Fig. 7. Mudflow lobe with the *c.* 153 m long SS *Virginia* located just left of centre, with its bow pointed towards the NW. A low-relief (several metres deep) depression or scour pit is visible at the bow of the wreck. The upper surface of the lobe is relatively smooth around the wreck, becoming rougher at the edges. No obvious features are apparent on the surface of the lobe, suggesting that *Virginia* has not been moving independently across the lobe but is rather coupled to the lobe's movement.

geophysical survey reports indicate that in 2007 the wreck had moved from its 2006 location, but had moved SW *c.* 54 m by 2009 following the 2008 hurricane season (e.g. Gustav and Ike) (Fig. 8). More recent hydrographic surveys that imaged the shipwreck were conducted in 2016 but these data were not available for comparative analysis until 2018. The 2017 USGS survey located the wreck *c.* 60 m SW of the 2006 survey location (*c.* 6 m SW of the 2009 location; Fig. 8), which matched the location of the wreck identified during the 2016 hydrographic surveys. In addition to the *c.* 430 m of movement between 2004 and 2017, the relief of the wreck (as measured at the bow) changed by *c.* −4.5 m. No significant change in orientation of the wreck is noted between 2004 and 2017. Based on the 1942 reported sinking location (NOAA 2013), *Virginia* appears to have moved almost 11 km downslope over the 75 years prior to the 2017 survey (Fig. 8). However, given the nature of the sinking event (torpedo attack), the reported sinking location is likely inaccurate, but it does serve as an initial point for estimating maximum possible wreck movement.

Deepwater Survey site pipeline movement

The 2017 USGS Deepwater Survey site (DWS) (Fig. 5b) covers an extensive active mudflow lobe complex of South Pass. An isopach of total mudflow lobe thickness (measured between the seafloor and top prodelta horizons interpreted from Chirp and MCS data; Fig. 9) shows that at least 1.12 km^3 of mudflow sediments have been deposited within the bounds of the DWS alone. Pipeline segments that span DWS (Fig. 10a) have suffered numerous failures from mudflow events over the last 20 years. Two pipelines in particular, the 12″ Tennessee Gas (TGP) and the 20″ Enterprise Products Partners (EPP) pipelines (Fig. 10), suffered multiple displacements and breakages during the 2004–10 period and, as such, are excellent markers of movements within mudflow lobe areas. For ease of reference, the two primary lobe complexes in the survey area are labelled Lobe A (west) and Lobe B (east) (Fig. 10b).

In 2004, Hurricane Ivan-induced mudflows broke the TGP pipeline and displaced a large section of it

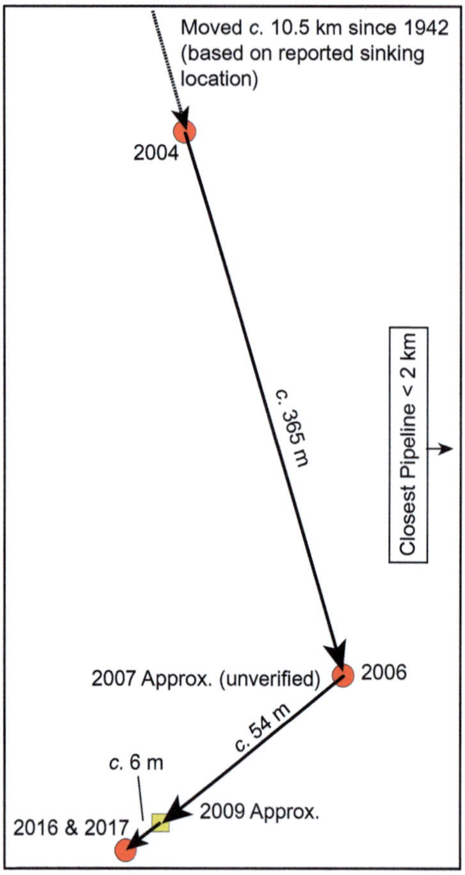

Fig. 8. Schematic illustration of the movement history of the SS *Virginia* since the wreck was surveyed and its identity confirmed. The postulated general direction and distance the wreck has moved from its 1942 sinking location are indicated.

1.85 km or more from its original as-built location (Fig. 10b). Movement appears to have been distributed across the lobe complex spanning the eastern edge of lease blocks SP 51 and all of SP 50, with focused flow in both Lobe A and Lobe B causing noticeable drag-related bends in the pipeline's final shape (Fig. 10b). Some sections of the broken pipeline segments were not relocated. No breakage or record of movement of the EPP pipeline during Hurricane Ivan was noted. The EPP pipeline was displaced and broken the following year by mudflows induced by Hurricane Katrina, again with significant movement across the SP50/SP51 lease block lobe complex. As with the TGP pipeline in Hurricane Ivan, the EPP pipeline was dragged downslope, with 3 km or more of displacement at the western broken end of the pipeline within Lobe B, depositing it in a crenulated shape inside lease block MC150. The eastern end of the pipeline also moved a significant distance downslope but underwent limited deformation during movement due to it being outside the primary deforming mudflow lobes (Fig. 10b).

Both the TGP and EPP pipelines were repaired following the 2004–05 hurricane seasons, with the TGP pipeline restored to its original pre-2004 location and the EPP pipeline restored using a new pipeline connected to sections of the displaced segments near the toe of the lobes. The remainder of the broken and displaced pipeline segments were abandoned in place (Fig. 10c). Maps prepared for partial abandonment and repair of the EPP pipeline in 2007 show further breakage, downslope movement and deformation of TGP pipeline sections displaced during Hurricane Ivan, but the evidence confirming these movements is not provided. As much as 1.1 km of downslope movement of these older TGP sections (between 2005 and 2007) is inferred based on these possible locations with their downslope-elongated shape suggesting movement via discrete small-scale flows (channelized?) within both Lobes A and B. Mudflows induced by Hurricane Gustav in 2008 damaged the repaired TGP pipeline, breaking it into several sections across Lobes A and B, and displacing them as much as 1.6 km downslope (Fig. 10c). Three focal areas of displacement and deformation of the pipeline sections, upslope of those in the older TGP sections, suggest continued activation of flow paths established during earlier mudflow events and multiple active mudflow gully sources. Chirp data across the lobes (Fig. 11a) show the presence of eroded and filled channels at the surface of these lobes, possibly reflecting the development of these discrete flow zones. Survey information reviewed in 2010 prior to abandonment of the EPP pipeline in the South Pass area revealed breakage and displacement of approximately 400 m of the segment at the south end of Lobe A, the driver for which is unknown.

Numerous pipeline sections are visible on the 2017 DWS bathymetry data (Fig. 10d). Several sections of the EPP and TGP pipelines west and downslope of Lobes A and B show no movement since their last recorded locations, although breakage or new burial by lobe movement is possible in two locations along the abandoned EPP pipeline at the toe of the lobes (Figs 10d & 11b). Movement of pipeline sections is difficult to track within the lobes due to the complex surface morphology, but a section of the TGP pipeline displaced by Hurricane Gustav mudflows appears to have moved downslope an additional 800 m, coming to rest just upslope of one of the possible EPP pipeline breakage/cover locations (dashed red line, Fig. 10d).

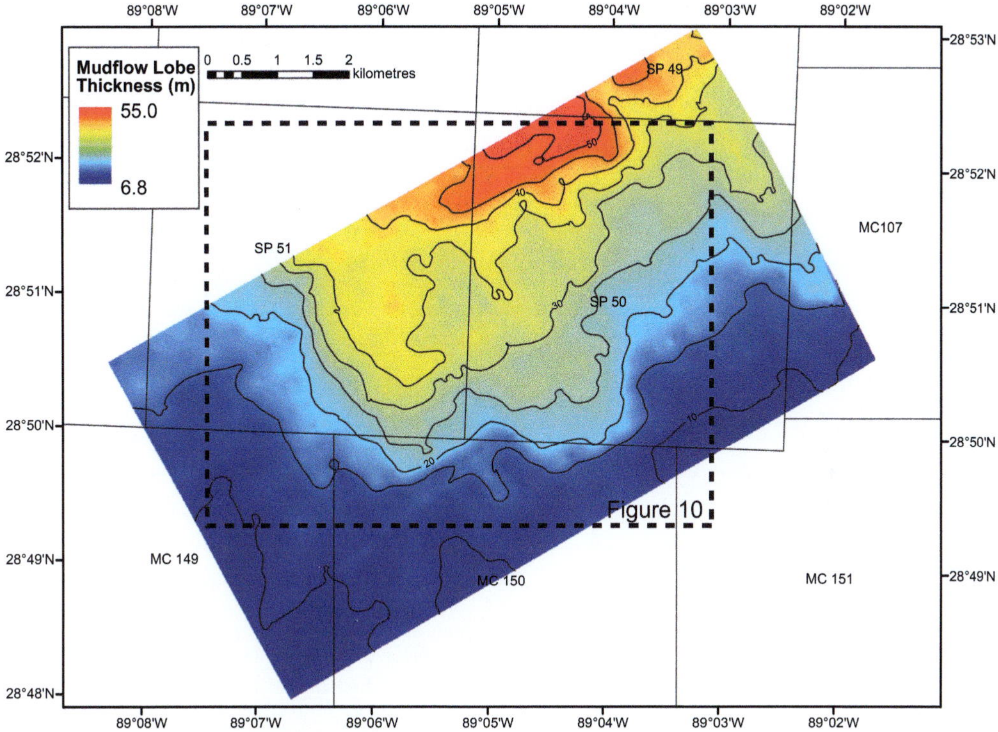

Fig. 9. Mudflow lobe deposit thickness isopach map (seafloor horizon to base mudflow/top prodelta horizon, 1500 m s^{-1} seismic velocity) of the Deepwater Survey site created from Chirp and sparker MCS lines collected in 2017. Lease blocks are shown and the dashed box outlines the area shown in Figure 10.

Southwest Pass debris-block movement

Repeat multibeam survey data spanning the 2016–17 period revealed debris block displacement in just two of the five mudflow gullies off Southwest Pass remapped in 2017 (Fig. 12a). Debris-block movement in these gullies is identified from the northern edge of the 2017 survey to the central section of the mudflow gullies. Data coverage from the 2017 survey was not sufficient to identify block movement at the heads of the mudflow gullies and block movement was not identified in or adjacent to the gully/lobe transition (Fig. 12a).

The movement of 99 individual debris blocks could be identified with certainty (Fig. 12b), although six of these had measured displacement of 2 m or less. The 2 m threshold is a non-statistical boundary below which data resolution and interpolation, georeferencing offsets, data artefacts, changes in debris block shape and digitization of block boundaries could each or all contribute to apparent block movement. Because rigorous calculations have yet to be performed to determine uncertainty thresholds, movements below 2 m are highlighted, but not removed from the analysis (Fig. 12b). Above the 2 m threshold, measured displacement distances ranged from 2.1 m to 84.3 m, with 85% of the blocks moving less than 26 m (Fig. 12b). The blocks with the largest displacement distances were clustered within a *c.* 0.03 km^2 area in the central axis of the westernmost of the two gullies, surrounded on either side by blocks showing significantly less movement (Fig. 12a, inset). In general, the largest displacement distances were observed towards the central axis of the mudflow gullies. Displacement directions (Fig. 12c) track closely with the overall orientation of the gullies, which is primarily towards the SW. Mudflow debris blocks towards the upslope headwalls and close to the gully walls showed the most direction variation due to the less-evolved morphologies of those areas. Some rotation and minor changes in overall shape of the individual blocks were observed, but quantifying these parameters consistently was beyond the resolution of the reviewed datasets.

Discussion

The MRDF is an extremely dynamic system driven by a complex mix of external inputs (e.g. river sediment delivery) and forcing (e.g. hurricanes) that vary

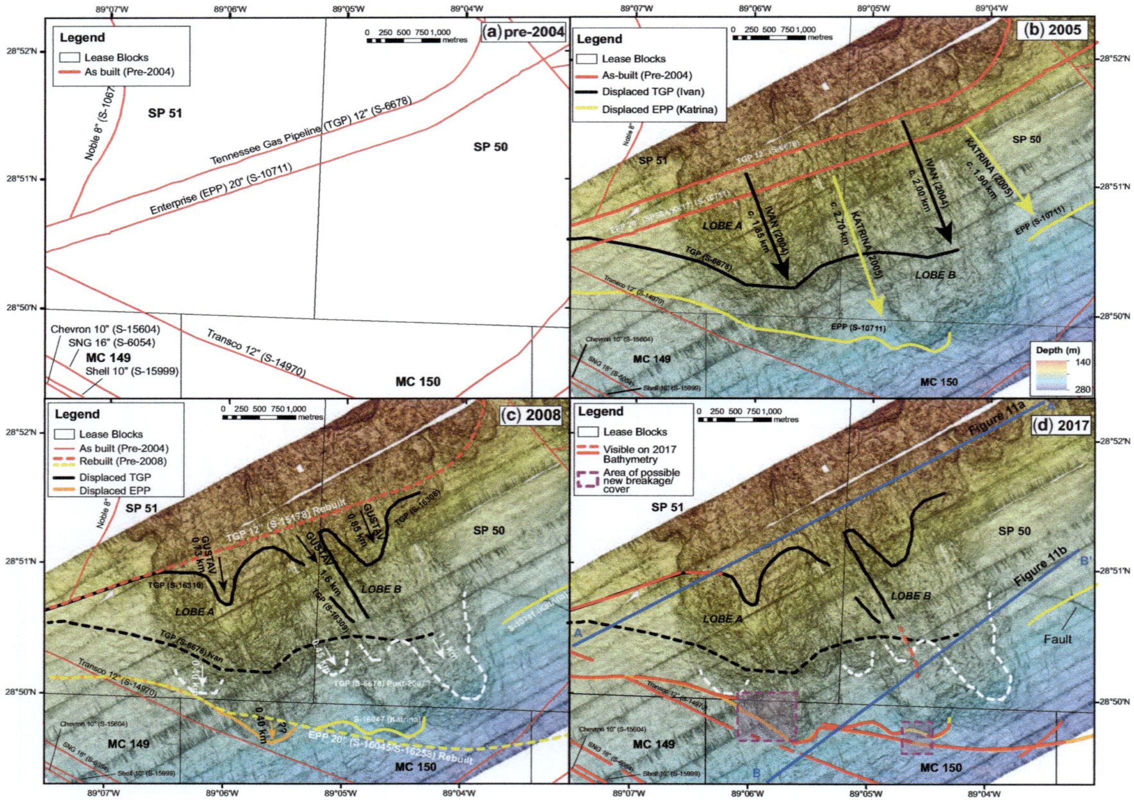

Fig. 10. Time series of gas pipeline locations across a mudflow lobe complex within the Deepwater Survey site based on archived pipeline maps. Some pipelines at the edges of the area have been removed for clarity. ‘S’ numbers in parentheses (e.g. S-6678) refer to assigned pipeline segment numbers, which remain with one section of those segments when abandoned. The 2017 bathymetry is used as a background for pre-2017 panels to show the relationship between the features and the pipeline displacements. (**a**) Pre-2004 ‘as-built’ locations of major pipelines in the area. (**b**) Map showing the displacement and deformation of the TGP (Tennessee Gas Pipeline) and EPP (Enterprise Product Partners) pipelines post hurricanes Ivan (2004) and Katrina (2005). (**c**) Map showing the displacement and deformation of rebuilt TGP and EPP pipeline segments and those moved during the 2004/2005 hurricane season (panel b) post Hurricane Gustav (2008). Note the downslope elongated shapes of the pipelines indicating possible channelized flow within the lobes. (**d**) Pipeline sections visible on the 2017 bathymetry data in relation to 2004–08 pipeline locations. While most sections show no significant movement, minor displacement and burial (highlighted by dashed purple boxes) of the EPP pipeline in lease block MC150 may have occurred. Location of Chirp lines in Figure 11 indicated in blue.

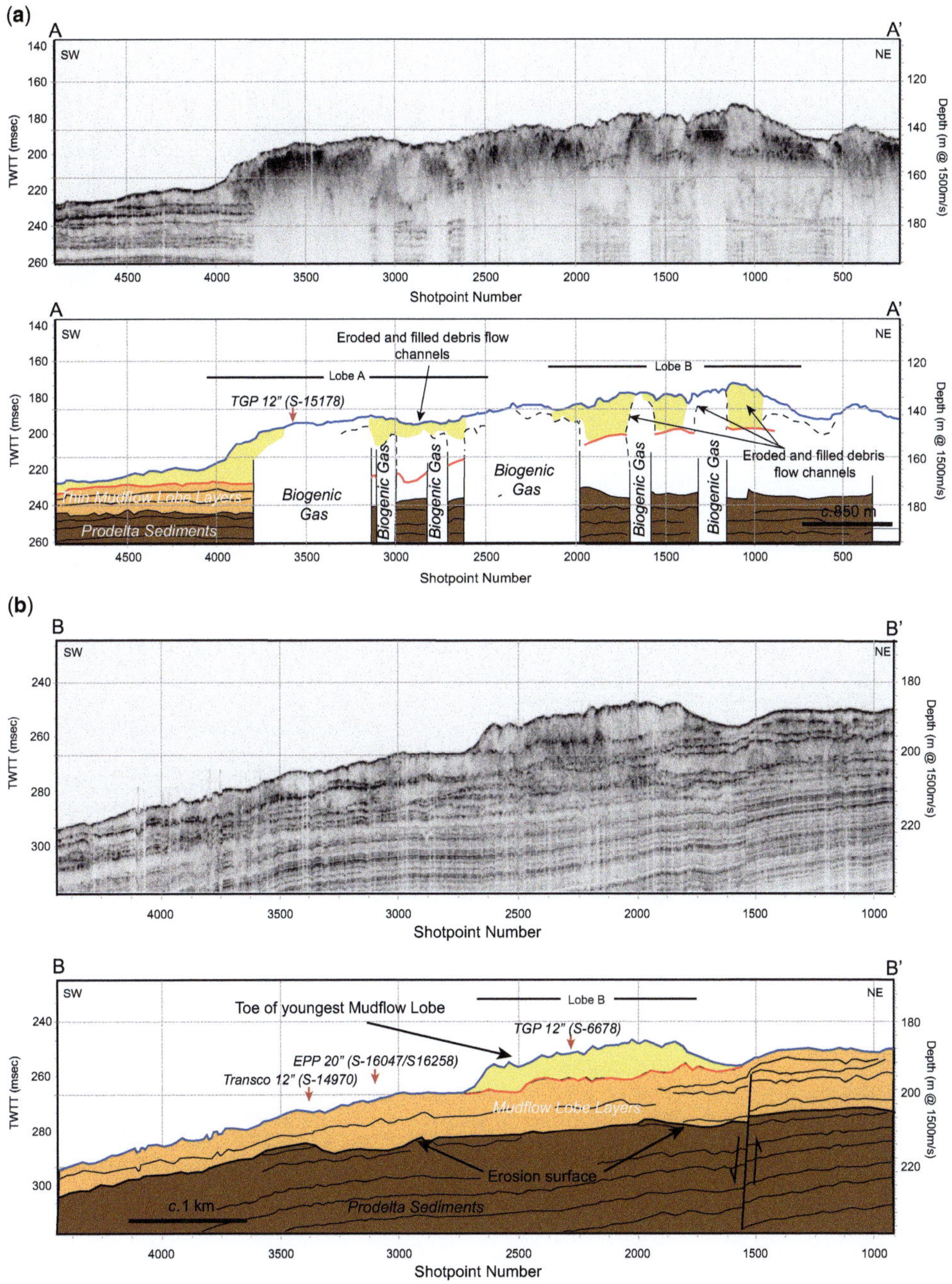

Fig. 11. (**a**) Uninterpreted and interpreted sections of swell-filtered Chirp line 51 across the upper lobe section of the Deepwater Survey area. Stacked mudflow lobes are visible above the laminated and largely undeformed pro-delta sediments. Gas in the mudflow deposits largely obscures the pro-delta sediments below the lobe complex, except where possible channelized flow within the lobes eroded older material and deposited new sediment that has not had time to re-establish the biogenic gas system (highlighted in yellow). Locations of pipeline crossings indicated. (**b**) Uninterpreted and interpreted sections of swell-filtered Chirp line 12 that crosses the toe of the mudflow lobes in the Deepwater Survey site (youngest mudflow lobe deposit highlighted in yellow). Pipeline locations from Figure 10 are indicated. A growth-fault cutting through the entire pro-delta and most of the debris-lobe deposits is highlighted. Location of lines shown in Figure 10d.

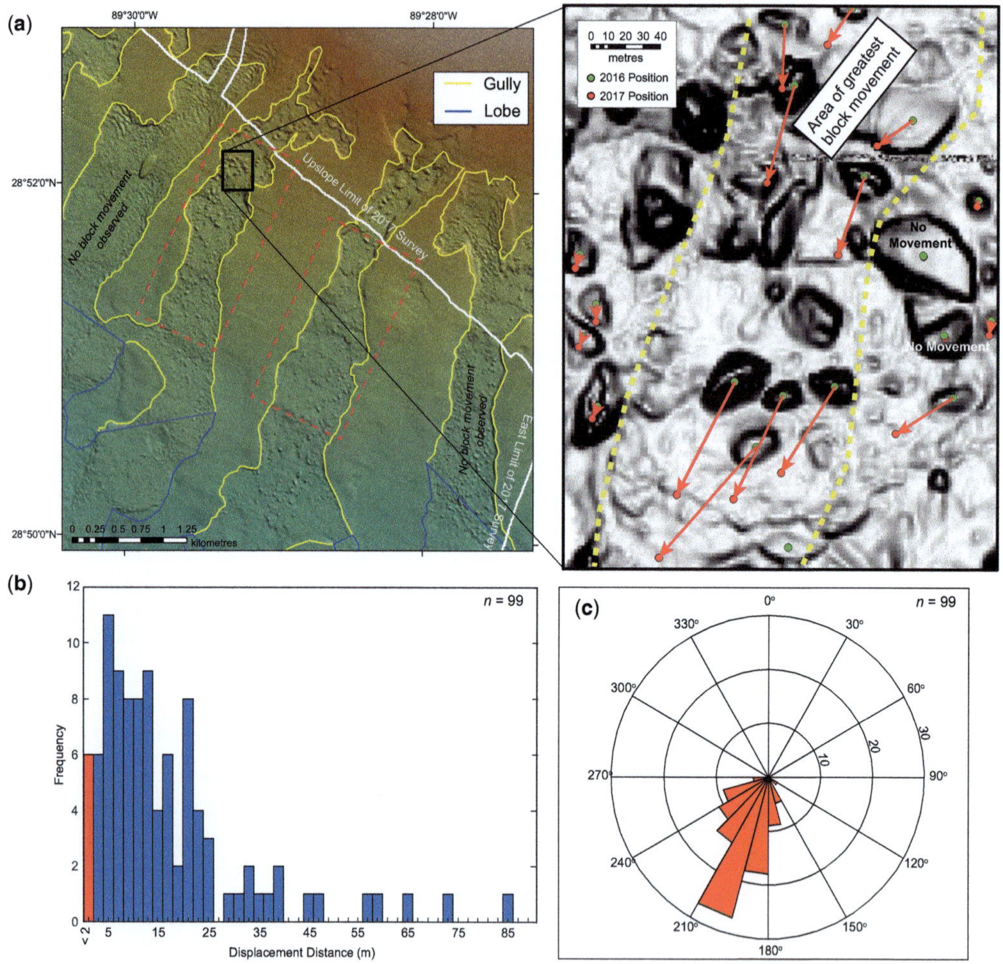

Fig. 12. (**a**) Mudflow gullies offshore Southwest Pass (2016 NOAA bathymetry), with boundaries of 2017 USGS survey shown (location within MRDF study area shown in Fig. 5a). The sections of two mudflow gullies that showed debris blocks movement over the 2016–17 period highlighted by dashed red boxes. Inset: 2016–17 vector displacement map for partial section of a mudflow gully off Southwest Pass, overlain on a black and white slope map generated from the 2016 NOS bathymetry data (NOAA 2014–16). Green dots are 2016 locations of blocks, red dots are 2017 locations of same block. Area of greatest movement during inter-survey period highlighted by yellow dashed lines. (**b**) Histogram showing 2 m binned displacement distance over a one-year period for 99 debris blocks within two mudflow gullies off Southwest Pass. Blocks with less than 2 m of displacement are highlighted in red. (**c**) Rose plot showing the direction of movement of the 99 debris blocks.

seasonally and annually, and relatively steady-state internal processes (e.g. sediment consolidation and biogenic gas production/release). Because of this, the rates and locations of seafloor movement are equally as complex to characterize. While the relationship between hurricane-generated waves, cyclical loading of a pre-conditioned unstable seafloor and generation of instantaneous mudflow failure are now well recognized (Henkel 1970; Bea 1971; Guidroz 2009) and displayed through long-term mapping of seafloor morphology (Coleman *et al.* 1980), and here via tracking of shipwreck and pipeline displacements, the drivers responsible for movement of various sized seafloor areas over shorter timescales without hurricane activity are poorly known but critically important to understand. Although quantifying deformation rates has benefited from the application of modern mapping technology, the intervals between surveys continue to be long enough that it is difficult to definitively identify all the triggers for seafloor movement in the MRDF.

Deformation rates

From the examples presented here and those discussed in more detail in the literature (Coleman *et al.* 1980, 1983; Walsh *et al.* 2006; Nodine *et al.* 2007; Guidroz 2009; Obelcz *et al.* 2017), there is insufficient information available to determine characteristic deformation rates or magnitudes for mudflows in any section of the MRDF seafloor, that is the entire MRDF appears capable of instantaneous or progressive movement over variable timescales or as elements of the mudflow system evolve and mature. Event-driven movement of the seafloor, such as hurricane-induced mudflow generation resulting in repeated movement of the EPP and TGP pipelines and *Virginia* from 2004 to 2008, essentially occurs instantaneously, and is capable of moving tremendous volumes of material (Fig. 9) over distances of several kilometres (e.g. 30–40 m thick mudflow that toppled the MC20A production platform; Chaytor *et al.* 2017). Seafloor movements of these magnitudes likely have the greatest impact on delivering sediment from the shallow delta to deep water (Fig. 9). Yet, these events tend to be sporadic, with recurrence of large-scale seafloor deformation of any particular section of the MRDF occurring at decadal rates.

The dynamic nature of seafloor movement and difficulty in defining deformation rates is well illustrated by *Virginia*, given the movement of the wreck over a 75-year period. Assuming that the 1942-reported sinking location is approximately the point at which the vessel came to rest on the seafloor and it was not subject to hurricane-induced mudflow movement, the wreck would have moved at an annualized rate of *c.* 150 m a^{-1}. This rate significantly exceeds the *c.* 85 m maximum 2016–17 rate of block movement measured in the Southwest Pass mudflow gullies, but is close to the maximum displacement of 141 m measured by Galloway *et al.* (2017) over a two-month period for blocks in a mudflow gully east of Southwest Pass. Looking at just the 13 years between its 2004 reported location and the 2017 USGS survey, the annualized rate would have been *c.* 38 m a^{-1}, becoming 0 m a^{-1} if just the 2016–17 period is evaluated. Given that multibeam surveys following periods of intense hurricane activity show that instantaneous events have contributed to rapid movement of the wreck (e.g. 2004–06, 2007–09), interspersed with periods of low to no movement (2006–07, 2016–17), displacement rates are non-stationary and likely have varied further as the wreck moved from mudflow gully to lobe environments. Movement of the wreck now appears to be entirely coupled to the surrounding mudflow, which will control its movement in the future. Our current limited understanding of the space and time varying rates of movement, even of large and effectively non-deformable features highlights the challenge in determining meaningful deformation rates for the mudflow system and using them for hazard characterization.

The lack of debris-block movement in mudflow gullies adjacent to those active during the 2016–17 period suggests that not all gullies and lobes may be in a 'movement pre-condition state' at any particular time, that triggering processes can be extremely localized even if the external force (e.g. winter storms) is more regional in scale, or that the triggering process is composed of a number of geological, oceanographic and atmospheric factors that remain to be identified. Repeat surveys of the mudflow gullies off Southwest Pass between 2005 and 2014 were used by Obelcz *et al.* (2017) to identify changes in broad-scale bottom elevation across the region evaluated for debris-block movement during the 2016–17 period, finding as much as 4 m of gully deepening over a four-year period. Obelcz *et al.* (2017) concluded that in the absence of major hurricane triggering during the period, smaller storms with one-year return periods (tropical storms, extra-tropical storms) or river flooding would be capable of triggering movement, as previously suggested by Prior and Coleman (1982). Conversely, Galloway *et al.* (2017) found that debris block motion occurred in gullies to the east of Southwest Pass over just a two-month period during which no significant storm activity occurred. Identifying the role of each preconditioning factor and external driver, and the response of the entire MRDF mudflow system to large-scale changes in sediment delivery (e.g. Maloney *et al.* 2018) will remain elusive without active monitoring or, possibly, comprehensive intra-annual repeat surveys of the mudflow system and oceanographic environment similar to that used in Monterey Canyon (Paull *et al.* 2018) or on the Squamish River Delta (Clare *et al.* 2016).

Hazards to infrastructure

The hazards to bottom-mounted infrastructure in the MRDF from hurricane-induced mudflows is well documented, with numerous instances of damage to, or destruction of, pipelines, platforms and wells (Bea 1971; Coyne and Dollar 2005; Thomson *et al.* 2005; Den Norske Veritas 2007; Guidroz 2009). However, the impact of progressive seafloor movement on these same facilities and how infrastructure entrained in the mudflows (e.g. shipwrecks, pipeline segments) behave is not as well studied and their impact on hazards analysis and risk assessment remains largely unconstrained. Guidroz (2009) carried out a detailed assessment of hazard, vulnerability and risk to infrastructure from mudflows in the MRDF, which considered numerous preconditioning and triggering factors including properties of the sediment, seafloor geomorphology, hurricane-induced significant wave heights, spatial change rates of the seafloor, event magnitude and hurricane recurrence. While the risk analysis results from these types of assessments are vital for predicting the effects of future hurricanes on infrastructure, they do not capture or fully account for the impact of progressive or non-hurricane-induced changes in the MRDF mudflow system or consider these processes as hazards that influence vulnerability or modify risk. For example, broken segments of abandoned 12 and 20″ pipelines crossing and imbedded in mudflow lobes in the Deepwater Survey site may now modify the dynamics of the sediments within the lobes and perhaps act to damage, or even entrain into the flow, nearby and downslope infrastructure.

Archaeologically significant sites (e.g. historic shipwrecks) are vulnerable to both hurricane- and non-hurricane-induced mudflow hazards, and act as hazards themselves, yet not all of these sites have been located or relocated (e.g. the World War II shipwreck SS *Rawleigh Warner* in the South Pass protraction area). Continued identification and evaluation of changes to these sites, such as continued movement and rupture of bulk oil or fuel tanks, is critical as part of future multi-component hazard-vulnerability risk assessments. For example, US Federal regulators are required to consider potential effects from their permitted activities on archaeological resources located within the area of potential effect of the proposed action (per the National Historic Preservation Act of 1966 (54 USC § 306108)) and to avoid, minimize, or mitigate adverse effects to those resources. For permitting of oil- and gas-related operations in Federal waters, operators are directed to refrain from conducting any bottom-disturbing activities (e.g. well placement, anchoring, pipeline installation, etc.) within a prescribed radius of a known historic shipwreck, typically a 305 m avoidance zone. However, in the case of the 153 -m-long tanker *Virginia*, the shipwreck is currently located (as of 2017) well outside of its original 305 m avoidance zone (as prescribed in 2004). The shipwreck's movement due to mudflow events, and unknown location at any given time, results in a potential hazard for future installation of infrastructure downslope of the wreck. In an environment of changing baseline information on bottom composition, surficial morphology, subsurface structure, the location of anthropogenic features and the rate at which hazard events occur, hazard definition and mitigation also need to be dynamic.

Conclusions

Instability of the Mississippi River Delta Front is extremely dynamic, with failures occurring over sub-annual to decadal periods, across a large range of water depths, and in areas of enhanced sediment deposition (distributary channel mouths) and limited sediment input. In the absence of regional resurvey of the MRDF, use of a combination of direct seafloor observations and tracking of infrastructure and wreck locations provide a novel way to monitor and quantify seafloor deformation and to mitigate hazards. The large-scale rapid and smaller-magnitude periodic movements of the *Virginia* and pipeline sections off South Pass reflect the complexity in defining the temporal and spatial scales across which mudflows will occur in the future. Although hurricanes generating wave fields capable of triggering destructive mudflow activity remain the primary focus for hazard quantification, the role of seasonal, annual or progressive triggers remains fundamentally important as they are capable of locally moving significant volumes of seafloor sediments. These smaller movements place historically significant wrecks at risk through accelerated destruction or direct impact with bottom infrastructure. Persistent downslope growth of mudflow lobes and the materials they carry must be considered in risk assessments of currently installed and future infrastructure development in the MRDF.

Acknowledgements Reviews by Walter Barnhardt, John Hughes Clarke and Alexandre Normandeau improved this manuscript and are greatly appreciated. We acknowledge the use of imagery provided by services from NASA's Global Imagery Browse Services (GIBS), part of NASA's Earth Observing System Data and Information System (EOSDIS). Any use of trade, product, or firm names is for descriptive purposes only and does not imply endorsement by the US Government.

Funding Funding for the 2017 USGS Survey and data analysis was provided by the USGS Coastal and Marine Hazards and Resources Program. The US Department of the Interior, Bureau of Ocean Energy Management, under Cooperative Agreement number M13AC00013 provided funding to several project participants.

Author contributions **JDC**: conceptualization (lead), formal analysis (lead), funding acquisition (lead), investigation (lead), visualization (lead), writing – original draft (lead); **WEB**: formal analysis (supporting), resources (supporting), visualization (supporting), writing – original draft (supporting); **SJB**: conceptualization (supporting), funding acquisition (supporting), methodology (supporting), writing – review & editing (supporting); **MD**: formal analysis (supporting), methodology (supporting), writing – original draft (supporting); **DJ**: formal analysis (supporting), methodology (supporting), writing – original draft (supporting); **JM**: conceptualization (supporting), formal analysis (supporting), methodology (supporting), writing – original draft (supporting); **MDM**: conceptualization (supporting), funding acquisition (supporting), writing – original draft (supporting); **JO**: formal analysis (supporting), writing – original draft (supporting); **KX**: investigation (supporting), writing – original draft (supporting).

References

Allison, M.A., Demas, C.R. *et al.* 2012. A water and sediment budget for the lower Mississippi–Atchafalaya River in flood years 2008–2010: implications for sediment discharge to the oceans and coastal restoration in Louisiana. *Journal of Hydrology*, **432**, 84–97, https://doi.org/10.1016/j.jhydrol.2012.02.020

Baldwin, W.E., Ackerman, S.D., Worley, C.R., Danforth, W.W. and Chaytor, J.D. 2018. High-resolution geophysical data collected along the Mississippi River Delta front offshore of southeastern Louisiana. US Geological Survey Field Activity 2017-003-FA. US Geological Survey data release, https://doi.org/10.5066/F7X929K6

Bea, R.G. 1971. How sea floor slides affect offshore structures. *Oil and Gas Journal*, 88–92.

Bea, R.G., Bernard, H.A., Arnold, P. and Doyle, E.H. 1975. Soil movements and forces developed by wave-induced slides in the Mississippi Delta. *Journal of Petroleum Technology*, **27**, 500–514, https://doi.org/10.2118/4476-PA

Bentley, S.J., Sr, Blum, M.D., Maloney, J., Pond, L. and Paulsell, R. 2016. The Mississippi River source-to-sink system: Perspectives on tectonic, climatic, and anthropogenic influences, Miocene to Anthropocene. *Earth-Science Reviews*, **153**, 139–174, https://doi.org/10.1016/j.earscirev.2015.11.001

Blum, M.D. and Roberts, H.H. 2012. The Mississippi Delta Region: past, present, and future. *Annual Review of Earth and Planetary Sciences*, **40**, 655–683, https://doi.org/10.1146/annurev-earth-042711-105248

Chaytor, J.D., Baldwin, W.E., Danforth, W.W., Bentley, S.J., Miner, M.D. and Damour, M. 2017. New High-Resolution Multibeam Mapping and Seismic Reflection Imaging of Mudflows on the Mississippi River Delta Front. AGU Fall Meeting Abstracts, EP21B–1837.

Church, R., Warren, D. *et al.* 2007. *Archaeological and Biological Analysis of World War II Shipwrecks in the Gulf of Mexico: Artificial Reef Effect in Deep Water.* U.S. Department of the Interior, Minerals Management Service, Gulf of Mexico OCS Region, New Orleans, LA. OCS Study MMS 2007-015.

Clare, M.A., Hughes Clarke, J.E., Talling, P.J., Cartigny, M.J.B. and Pratomo, D.G. 2016. Preconditioning and triggering of offshore slope failures and turbidity currents revealed by most detailed monitoring yet at a fjord-head delta. *Earth and Planetary Science Letters*, **450**, 208–220, https://doi.org/10.1016/j.epsl.2016.06.021

Coleman, J.M. and Garrison, L.E. 1977. Geological aspects of marine slope stability, northwestern Gulf of Mexico. *Marine Georesources & Geotechnology*, **2**, 9–44, https://doi.org/10.1080/10641197709379769

Coleman, J.M., Suhayda, J.N., Whelan, T. and Wright, L.D. 1974. Mass movement of Mississippi River delta sediments. *Transactions of the Gulf Coast Association of Geological Societies*, **24**, 49–68.

Coleman, J.M., Prior, D.B. and Garrison, L.E. 1980. *Subaqueous sediment instabilities in the offshore Mississippi River delta.* United States Department of Interior, Bureau of Land Management, New Orleans Outer Continental Shelf Office. Open-File Report **80-01**.

Coleman, J.M., Prior, D.B. and Lindsay, J.F. 1983. Deltaic influences on shelf edge instability processes. *SEPM Special Publications*, **33**, 121–137.

Coleman, J.M., Roberts, H.H. and Bryant, W.R. 1991. Late Quaternary sedimentation. *In*: Salvador, A. (ed.) *The Gulf of Mexico Basin.* The Geology of North America. Geological Society of America, Boulder, CO, 325–352.

Coleman, J.M., Roberts, H.H. and Stone, G.W. 1998. Mississippi River delta: an overview. *Journal of Coastal Research*, **14**, 698–716.

Coyne, M.J. and Dollar, J.J. 2005. Shell pipeline's response and repairs after Hurricane Ivan. *Proceedings of the Offshore Technology Conference*, 2–5 May 2005, Houston, TX, 317–326

Den Norske Veritas 2007. Pipeline Damage Assessment from Hurricanes Katrina and Rita in the Gulf of Mexico, MMS Project No. 581, Houston.

Fisk, H.N. 1961. Bar-finger sands of the Mississippi Delta. *AAPG Special Publications*, **22**, 29–52, https://doi.org/10.1306/SV22354C3

Fisk, H.N. and McClelland, B. 1959. Geology of continental shelf off Louisiana: its influence on offshore foundation design. *Geological Society of America Bulletin*, **70**, 1389–1394.

Galloway, N., Levy, T. and George, T. 2017. Mapping active mudflow features off Southwest Pass, Louisiana, Gulf of Mexico. Presented at Hydrographic Society of America 2017 Annual Meeting, 20–23 March, Galveston, TX, http://ushydro2019.thsoa.org/wp-content/uploads/2017/04/Galloway_Hydro2017_MappingMudflowFeatures-1.pdf

Guidroz, W.S. 2009. *Subaqueous, hurricane-initiated shelf failure morphodynamics along the Mississippi River Delta Front, north-central Gulf of Mexico.* PhD thesis, Louisiana State University.

Henkel, D.J. 1970. The role of waves in causing submarine landslides. *Geotechnique*, **20**, 75–80, https://doi.org/10.1680/geot.1970.20.1.75

Hitchcock, C., Angell, M., Givler, R. and Hooper, J. 2006. A pilot study for regionally consistent hazard susceptibility mapping of submarine mudslides, offshore Gulf of Mexico. Minerals Management Service, Engineering Research Branch, Technical Assessment Program Project 550.

Keller, G., Bentley, S.J., Georgiou, I.Y., Maloney, J., Miner, M.D. and Xu, K. 2016. River-plume sedimentation and 210 Pb/7 Be seabed delivery on the Mississippi River delta front. *Geo-Marine Letters*, **37**, 259–272, https://doi.org/10.1007/s00367-016-0476-0

Knapp, K.R., Kruk, M.C., Levinson, D.H., Diamond, H.J. and Neumann, C.J. 2010. The International Best Track Archive for Climate Stewardship (IBTrACS): unifying tropical cyclone best track data. *Bulletin of the American Meteorological Society*, **91**, 363–376, https://doi.org/10.1175/2009BAMS2755.1

Maloney, J.M., Bentley, S.J., Xu, K., Obelcz, J., Georgiou, I.Y. and Miner, M.D. 2018. Mississippi River subaqueous delta is entering a stage of retrogradation. *Marine Geology*, **400**, 23, https://doi.org/10.1016/j.margeo.2018.03.001

Milliman, J.D. and Meade, R.H. 1983. World-wide delivery of river sediment to the oceans. *The Journal of Geology*, **91**, 1–21, https://doi.org/10.1086/628741

National Oceanic and Atmospheric Administration (NOAA) 2013. Screening Level Risk Assessment Package *Virginia*. Office of National Marine Sanctuaries, Office of Response and Restoration, https://nmssanctuaries.blob.core.windows.net/sanctuaries-prod/media/archive/protect/ppw/pdfs/virginia.pdf

National Oceanic and Atmospheric Administration (NOAA) 2014–16. NOS Hydrographic Survey Data. National Centers for Environmental Information, https://www.ngdc.noaa.gov/mgg/bathymetry/hydro.html

Nodine, M.C., Cheon, J.Y., Wright, S.G. and Gilbert, R.B. 2007. *Mudslides during Hurricane Ivan and an assessment of the potential for future mudslides in the Gulf of Mexico: Phase II Project Report Prepared for the Minerals Management Service, MMS/OTRC Cooperative Research Agreement 1435-01-04-CA-35515, Task Order 39239.* MMS TAR Project **Number 552**.

Nodine, M.C., Cheon, J.Y., Wright, S.G. and Gilbert, R.B. 2009. *Addendum to mudslides during Hurricane Ivan and an assessment of the potential for future mudslides in the Gulf of Mexico: Addendum to Phase II Project Report Prepared for the Minerals Management Service, MMS/OTRC Cooperative Research Agreement 1435-01-04-CA-35515, Task Order 39239.* MMS TAR Project **Number 552**, published 2007.

Obelcz, J., Xu, K., Georgiou, I.Y., Maloney, J., Bentley, S.J. and Miner, M.D. 2017. Sub-decadal submarine landslides are important drivers of deltaic sediment flux: Insights from the Mississippi River Delta Front. *Geology*, **45**, 703–706, https://doi.org/10.1130/G38688.1

Paull, C.K., Talling, P.J. *et al.* 2018. Powerful turbidity currents driven by dense basal layers. *Nature Communications*, **9**, 4114, https://doi.org/10.1038/s41467-018-06254-6

Prior, D.B. and Coleman, J.M. 1978. Disintegrating retrogressive landslides on very-low-angle subaqueous slopes, Mississippi delta. *Marine Georesources & Geotechnology*, **3**, 37–60, https://doi.org/10.1080/10641197809379793

Prior, D.B. and Coleman, J.M. 1982. Active slides and flows in underconsolidated marine sediments on the slopes of the Mississippi Delta. *In*: Saxov, S. and Nieuwenhuis, J.K. (eds) *Marine Slides and Other Mass Movements*. Springer, Boston, MA, 21–49.

Prior, D.B. and Suhayda, J.N. 1979. Application of infinite slope analysis to subaqueous sediment instability, Mississippi Delta. *Engineering Geology*, **14**, 1–10, https://doi.org/10.1016/0013-7952(79)90059-0

Ruttenberg, K.C. and Goni, M.A. 1997. Phosphorus distribution, C: N: P ratios, and δ13Coc in arctic, temperate, and tropical coastal sediments: tools for characterizing bulk sedimentary organic matter. *Marine Geology*, **139**, 123–145, https://doi.org/10.1016/S0025-3227(96)00107-7

Shepard, F.P. 1955. Delta-front valleys bordering the Mississippi distributaries. *Geological Society of America Bulletin*, **66**, 1489–1498, https://doi.org/10.1130/0016-7606(1955)66[1489:DVBTMD]2.0.CO;2

Sterling, G.H. and Strohbeck, E.E. 1975. The failure of the South Pass 70 "B" platform in Hurricane Camille. *Proceeding of the 6th Offshore Technology Conference*, 5–8 May, Houston, TX, 719–730, https://doi.org/10.2118/4475-PA

Thomson, J., Garrett, M., Taylor, M., George, T., Melancon, M. and Behrens, K. 2005. Sonar surveys for pipeline inspection show extent of pipeline displacement and seafloor instability following Hurricane Ivan. *Proceedings of the Offshore Technology Conference*, 2–5 May 2005, Houston, TX, 2352–2357.

Tornqvist, T.E., Kidder, T.R., Autin, W.J. and van der Borg, K. 1996. A revised chronology for Mississippi River subdeltas. *Science*, **273**, 1693, https://doi.org/10.1126/science.273.5282.1693

Urlaub, M., Talling, P.J., Zervos, A. and Masson, D. 2015. What causes large submarine landslides on low gradient (<2°) continental slopes with slow (~0.15 m/kyr) sediment accumulation? *Journal of Geophysical Research, Solid Earth*, **120**, 6722–6739, https://doi.org/10.1002/2015JB012347

Walsh, J.P., Corbett, R. *et al.* 2006. Mississippi delta mudflow activity and 2005 Gulf hurricanes. *Eos, Transactions American Geophysical Union*, **87**, 477–478, https://doi.org/10.1029/2006EO440002

Whelan, T., Coleman, J.M., Roberts, H.H. and Suhayda, J.N. 1976. The occurrence of methane in recent deltaic sediments and its effect on soil stability. *Bulletin of the International Association of Engineering Geology*, **13**, 55–64, https://doi.org/10.1007/BF02634759

Lessons learned from the monitoring of turbidity currents and guidance for future platform designs

Michael Clare[1]*, D. Gwyn Lintern[2], Kurt Rosenberger[3], John E. Hughes Clarke[4], Charles Paull[5], Roberto Gwiazda[5], Matthieu J. B. Cartigny[6], Peter J. Talling[6], Daniel Perara[7], Jingping Xu[8], Daniel Parsons[9], Ricardo Silva Jacinto[10] and Ronan Apprioual[10]

[1]National Oceanography Centre, European Way, Southampton SO14 3ZH, UK

[2]Geological Survey of Canada, Institute of Ocean Science, Canada

[3]United States Geologic Survey, Santa Cruz, USA

[4]Center for Coastal and Ocean Mapping/Joint Hydrographic Center, New Hampshire, USA

[5]Monterey Bay Aquarium Research Institute, Moss Landing, USA

[6]Departments of Earth Sciences and Geography, Durham, UK

[7]Canadian Coast Guard, Victoria, BC V8V 4V9, Canada

[8]Southern University of Science and Technology, Shenzhen, China

[9]Energy and Environment Institute, University of Hull, Cottingham Road, Hull HU6 7RX, UK

[10]Marine Geosciences Unit, IFREMER, Centre de Brest, CS10070, 29280 Plouzané, France

DGL, 0000-0003-1057-2670; KR, 0000-0002-5185-5776; JEHC, 0000-0002-3846-9926; RG, 0000-0002-1144-3865; MJBC, 0000-0001-6446-5577
*Correspondence: m.clare@noc.ac.uk

Abstract: Turbidity currents transport globally significant volumes of sediment and organic carbon into the deep-sea and pose a hazard to critical infrastructure. Despite advances in technology, their powerful nature often damages expensive instruments placed in their path. These challenges mean that turbidity currents have only been measured in a few locations worldwide, in relatively shallow water depths (<<2 km). Here, we share lessons from recent field deployments about how to design the platforms on which instruments are deployed. First, we show how monitoring platforms have been affected by turbidity currents including instability, displacement, tumbling and damage. Second, we relate these issues to specifics of the platform design, such as exposure of large surface area instruments within a flow and inadequate anchoring or seafloor support. Third, we provide recommended modifications to improve design by simplifying mooring configurations, minimizing surface area and enhancing seafloor stability. Finally, we highlight novel multi-point moorings that avoid interaction between the instruments and the flow, and flow-resilient seafloor platforms with innovative engineering design features, such as feet and ballast that can be ejected. Our experience will provide guidance for future deployments, so that more detailed insights can be provided into turbidity current behaviour, in a wider range of settings.

Reports of sequential seafloor cable breaks at the start of the last century provided the first direct evidence of subaqueous avalanches of sediment called 'turbidity currents' (Heezen and Ewing 1952, 1955; Shepard 1954; Heezen *et al.* 1964; Ryan and Heezen 1965; Piper *et al.* 1988; Pope *et al.* 2017). These seafloor-hugging flows were shown to be powerful (reaching up to 20 m s^{-1}, sustaining speeds of 3–10 m s^{-1} on slopes of less than 1°; Hsu *et al.* 2008; Carter *et al.* 2014) and capable of transporting large volumes of sand, mud, organic carbon and nutrients across vast distances (tens to hundreds of kilometres) (Krause *et al.* 1970; El-Robrini *et al.* 1985; Piper *et al.* 1988; Mulder *et al.* 1997). More than one million kilometres of seafloor cables now connect the world; transmitting more

From: Georgiopoulou, A., Amy, L. A., Benetti, S., Chaytor, J. D., Clare, M. A., Gamboa, D., Haughton, P. D. W., Moernaut, J. and Mountjoy, J. J. (eds) 2020. *Subaqueous Mass Movements and their Consequences: Advances in Process Understanding, Monitoring and Hazard Assessments*. Geological Society, London, Special Publications, **500**, 605–634.
First published online May 22, 2020, https://doi.org/10.1144/SP500-2019-173

than 98% of all digital data communications, including the internet and financial trading (Burnett and Carter 2017). We are increasingly reliant on this global network and on networks of subsea pipelines that support a growing demand for energy (Yergin 2006; Carter 2010). It is therefore important to understand the hazards posed to this critical seafloor infrastructure by seafloor mass movements, such as turbidity currents, to inform safe routing, geohazard-tolerant design or mitigation measures where necessary (Bruschi *et al.* 2006; Randolph and White 2012; Syahnur and Jaya 2016; Sequeiros *et al.* 2019). In addition to being potential geohazards, turbidity currents are also globally important agents of particulate and pollutant transport (Piper *et al.* 1988; Pohl *et al.* 2020). We want to know information such as: (i) how they are triggered and linked to onshore sedimentary systems; (ii) the frequency at which they recur; (iii) how they interact with the seafloor; (iv) the physical controls on their run-out; and (v) their internal velocity and sediment concentration structure. Inferences can be gleaned from the study of ancient deposits, through analogue modelling of scaled-down flows in the laboratory and from numerical modelling; however, direct field-scale measurements are needed to calibrate and/or validate all of these approaches (Xu 2011; Fildani 2017).

A very brief history of monitoring turbidity currents

Monitoring turbidity currents poses several challenges because deploying instruments on the deep seafloor is logistically challenging, flows may occur infrequently and the powerful nature of flows can damage the instruments intended to measure them (e.g. Inman *et al.* 1976; Talling *et al.* 2013; Puig *et al.* 2014; Clare *et al.* 2017; Lintern *et al.* 2019). Despite these challenges, several studies have prevailed to provide direct measurements of turbidity currents, including seminal field campaigns using point current meters (that measure velocity at one elevation in the water column), in settings ranging from active river-fed fjords (Hay 1987*a*, *b*; Hay *et al.* 1982; Prior *et al.* 1987; Syvitski and Hein 1991; Bornhold *et al.* 1994), lakes (Lambert and Giovanoli 1988) to deep-sea submarine canyons (Gennesseaux *et al.* 1971; Inman *et al.* 1976; Shepard *et al.* 1977; Khripounoff *et al.* 2003, 2009; Vangriesheim *et al.* 2009). These initial pioneering studies demonstrated that some systems can feature tens of turbidity currents in a year and that it is feasible to measure flows of up to 3.5 m s^{-1} (Prior *et al.* 1987). These studies were not without incident, however. Many involved damaged or lost instruments (Table 1). Those early studies were also limited with respect to the temporal resolution of measurements, data storage capabilities, duration of deployments and did not permit depth-resolved flow measurements (Talling *et al.* 2013).

Recent developments in technology, most notably the development of instruments such as Acoustic Doppler Current Profilers (ADCPs) and long-endurance lithium batteries, have enabled depth-resolved measurements of velocity and acoustic backscatter (a proxy measurement for sediment concentration; Thorne and Hanes 2002) (Cacchione *et al.* 2006; Shih 2012). Downward-looking ADCPs avoid the need for numerous individual point measurements made from within flows (Xu 2011; Khripounoff *et al.* 2012). In recent years, a growing number of ADCP-based measurements of turbidity currents have been made in locations including submarine canyons and channels offshore California (Puig *et al.* 2004; Xu *et al.* 2004, 2010; Paull *et al.* 2018), Mississippi (Ross *et al.* 2009), NE Atlantic (de Stigter *et al.* 2007; Martín *et al.* 2011; Mulder *et al.* 2012), Mediterranean (Khripounoff *et al.* 2012; Puig *et al.* 2012; Martín *et al.* 2014; Ribó *et al.* 2015) British Columbia (Hughes Clarke 2016; Lintern *et al.* 2016; Hage *et al.* 2018, 2019), West Africa (Cooper *et al.* 2013, 2016; Azpiroz-Zabala *et al.* 2017*a*, *b*) and Taiwan (Liu *et al.* 2012; Zhang *et al.* 2018).

Modern turbidity current monitoring campaigns typically integrate multiple sensors and tools, such as multibeam sonar (imaging the water column), optical backscatter sensors (to detect suspended particles), acoustic monitoring transponders (to determine seafloor movement) and sediment traps (to collect suspended sediment) (Lintern and Hill 2010; Xu 2011; Khripounoff *et al.* 2012; Hughes Clarke 2016; Lintern *et al.* 2016, 2019; Clare *et al.* 2017; Paull *et al.* 2018; Hage *et al.* 2019; Maier *et al.* 2019*a*, *b*). The tools that can be used to measure turbidity currents are partly covered by a number of reviews (Xu 2011; Talling *et al.* 2013; Puig *et al.* 2014; Clare *et al.* 2017). Here, we focus on the platforms on which these instruments or sensors are mounted, that may include devices such as moorings or frames installed on the seafloor and may be autonomous or connected via a cabled power and communications link. Examples of different types of platforms are illustrated in Figure 1.

Aims

Recent findings enable us to test, refute and refine established hypotheses in turbidity current science; however, direct measurements only exist from a relatively small number of sites worldwide. Many types of system and regions remain completely unrepresented. To date, no detailed measurements of velocity or sediment concentration have been published in water depths of >2 km and none from

Table 1. *Examples of adverse effects to monitoring platforms and instruments caused by turbidity currents from previous studies and sites referenced in this paper*

Location and literature source	Instrument type	Water depth (m)	Maximum measured velocity ($m\ s^{-1}$)	Documented adverse effects, damage, etc.
Scripps Canyon (Inman 1970; Inman *et al.* 1976)	Point current meter 4 m above seafloor connected to shore by a cable	46	1.9	Current meter failed during flow and was subsequently lost. Flows bent a 2.5 cm-thick solid steep rod bolted into canyon bedrock
Lake Geneva, Switzerland (Lambert and Giovanoli 1988)	Point current meters on single-point moorings	Up to 170	Up to 3	Anchor cables broke and moorings floated to surface
La Jolla Canyon, California (Shepard and Marshall 1973)	Point current meters on single-point mooring	200	Up to 0.5 before data recording stopped	Moorings displaced 500 m down-canyon
Open slope, Hawaii (Dengler *et al.* 1984)	Point current meters on single-point mooring	Up to 600	Up to 2	Episodic downslope movement of moorings by 2.4 km
Squamish Delta, British Columbia (Hughes Clarke *et al.* 2012; Hughes Clarke 2016)	Upward-facing ADCP mounted in seabed frame	Up to 150	Up to 1.5	ADCP frame buried by 2 m of sediment.
Fraser Delta, British Columbia (Lintern *et al.* 2016)	Cabled seafloor frame (1 t) fitted with numerous instruments including upward-facing ADCP	40–107	Up to 10	Platform tumbled down delta and severed connection with onshore cable
Bute Inlet, British Columbia (Prior *et al.* 1987)	Point current meters, Anderson-style sediment traps on single point moorings. Seafloor frame vane deflectors	Up to 520	Up to 3.4	Rotors and vanes on current meters broken off or fouled (causing poor data quality), shackles and stainless steel frames bent and sheared, some entire instruments lost. Mooring wires parted, releasing instruments to surface. Moorings displaced along- and down-channel (up to 1 km). Acoustic releases failed to detach due to assumed burial by sand.
Monterey Canyon, California (Paull *et al.* 2002)	Seafloor trapezoidal frame (97×83 cm base and 48 cm tall)	525	N/A	Frame transported 550 m down-canyon and buried in up to 0.7 m of sediment.
Monterey Canyon California (Paull *et al.* 2010)	Trawl-resistant seafloor frames (up to 1360 kg)	289 and 520	N/A	Moved up to 170 m down-canyon and buried in up to 1.5 m sediment.
Monterey Canyon, California (Paull *et al.* 2018)	Array of single-point moorings, a seafloor frame, and an 800 kg frame carrying a transponder	Up to 1850	Up to 7.2	MS1 transported 7.1 km down-canyon before breaking loose and floating to surface, sediment traps torn apart, 800 kg frame transport 4.5 km down canyon and buried in >1 m sand.
Congo Canyon, West Africa (Khripounoff *et al.* 2003; Vangriesheim *et al.* 2009)	Point current meters on single-point mooring	Up to 4790	Up to 3.5	Tilting of mooring prior to parting of mooring anchor line, releasing instruments to surface. Damaged current meter (30 m above seafloor) and sediment trap (40 m above seafloor)
Congo Canyon, West Africa (Cooper *et al.* 2013; Azpiroz-Zabala *et al.* 2017*a*)	Downward-facing ADCP on single-point mooring	2000	Up to 2.5	Rotating ADCP Interference with canyon sidewall

source to deep-water sink (e.g. submarine fan) as the logistics of placing platforms in deep water remains challenging.

Our overarching aim is to share lessons learned from recent campaigns measuring powerful turbidity currents to enable more measurements to be made in a wider variety of locations and settings worldwide. We do this through the following specific objectives. First, we provide an overview of the challenges encountered during the measurement of powerful turbidity currents (up to 10 m s^{-1}), including the tilting, displacement and damage of monitoring platforms. We illustrate these challenges with examples from systems including fjord-head deltas, a major river-fed canyon and an oceanographically fed canyon. Second, we introduce single-point moorings and how a successful design for monitoring turbidity currents may be used for more routine oceanographic purposes. These differences include requirements for extra anchor weighting, positive buoyancy and we discuss the implications of deploying large surface area instruments, such as sediment traps, that can induce excess drag on the mooring string. We outline several methods to reduce drag and enhance mooring stability. Third, we present a method to deploy two- and four-point moorings, anchored either side of a channel; ensuring that neither the instrument, nor the mooring line, interacts with flows. This is important where pronounced erosion or deposition may occur in the channel axis and to reduce mooring drag and tilt. Fourth, we assess the deployment of benthic landers and frame-based platforms, describing methods to enhance stability. Finally, we conclude with a discussion on future advances, in both sensor deployment and platform design, which will enable longer endurance turbidity current monitoring.

Study areas and monitoring data

We now introduce the case study sites discussed in this paper where frequent (sub-annual) turbidity currents have been measured (Fig. 2).

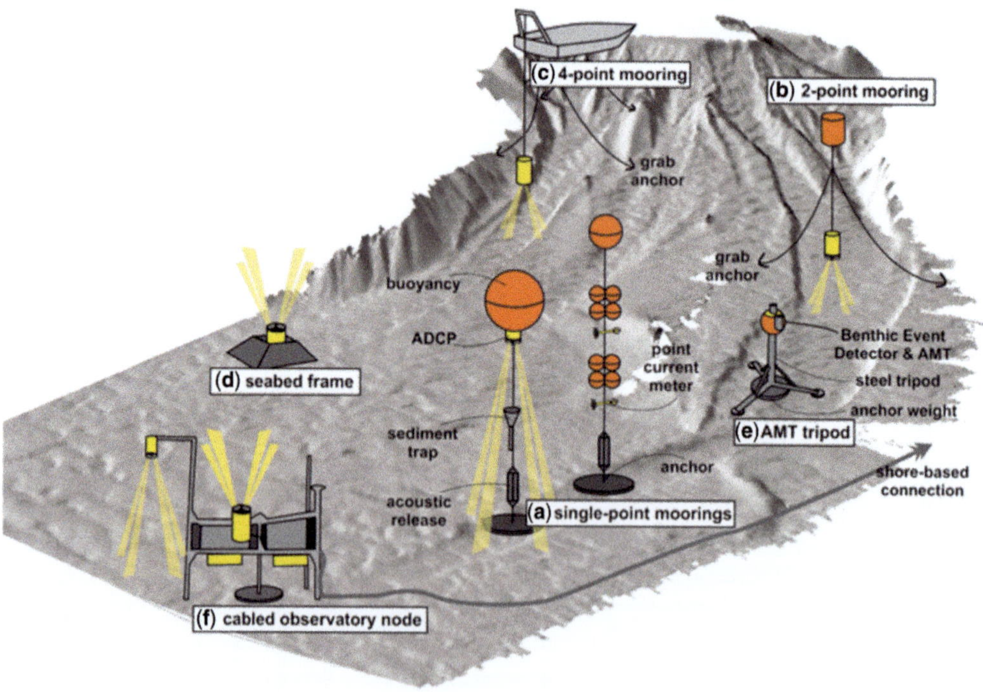

Fig. 1. Illustration depicting examples of some turbidity current monitoring platforms discussed in this paper, including: (**a**) single-point moorings (examples showing older point current meters (right) and more recent ADCP designs (left)) with anchors in the submarine channel axis; (**b**) two-point mooring to suspend a down-looking instrument above an active submarine channel, which avoids placement of the anchor in a channel axis; (**c**) four-point mooring to stabilize the orientation of a vessel and to enable deployment of suspended instruments (Hughes Clarke 2016); (**d**) seabed frame to deploy an upward-facing instrument; (**e**) Acoustic Monitoring Transponder (AMT) tripod with Benthic Event Detector (BED) to track movement (Paull *et al.* 2018); (**f**) platform connected to a seafloor cable network that may host many instruments with real-time communications and power (Lintern *et al.* 2016).

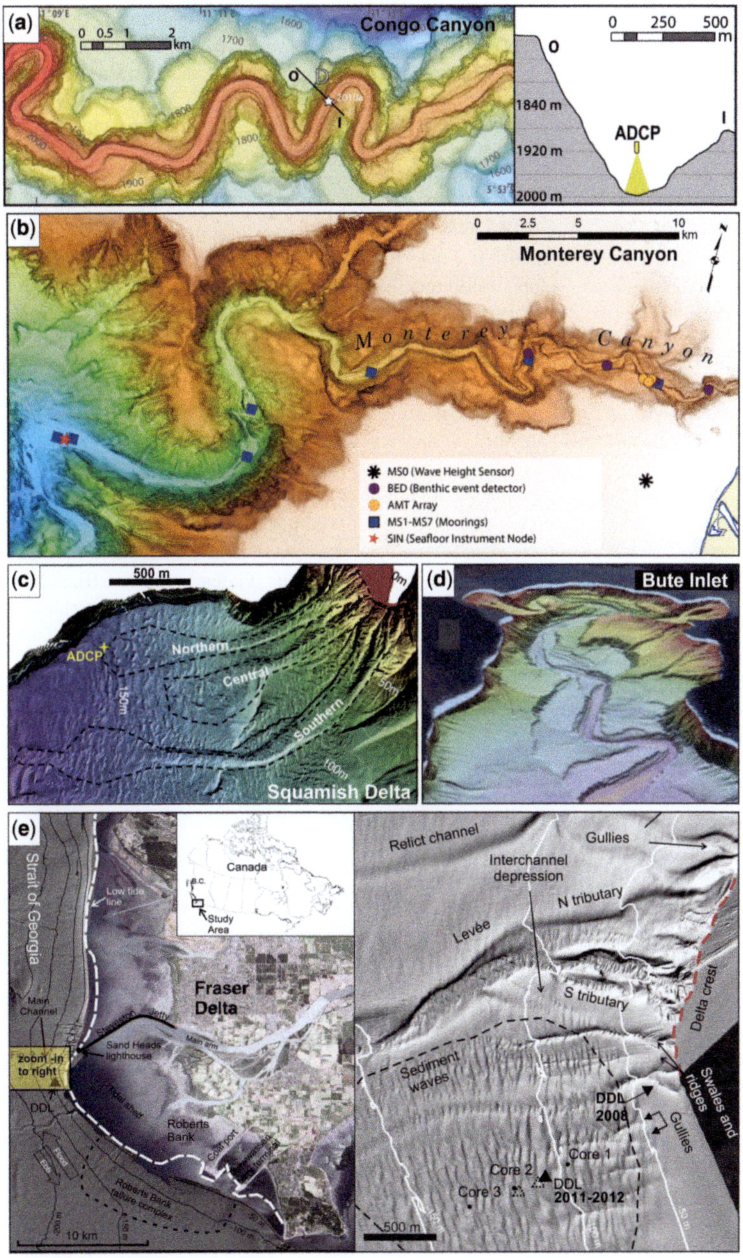

Fig. 2. Location maps and bathymetry for each of the sites discussed in this paper. (**a**) Location of ADCP mooring in Congo Canyon, West Africa at 2000 m water depth (modified from Azpiroz-Zabala *et al.* 2017*a*). (**b**) Configuration of Monterey Canyon CCE instrument deployment, offshore Moss Landing, California, USA. Water depth range of instrument deployment was 30 m to 1840 m (from https://www.mbari.org/cce-instruments-2019/). (**c**) Squamish submarine delta in Howe Sound, British Columbia. Water depth is up to 200 m (modified from Clare *et al.* 2016). (**d**) Bute Inlet, British Columbia, with water depths of up to 700 m. (**e**) Fraser Delta, British Columbia, showing relationship with the Fraser River (left) and detail on offshore delta channel and bedform field (right) where the Delta Dynamics Laboratory (DDL) was deployed in different locations (modified from Lintern *et al.* 2016).

Congo Canyon, West Africa

The Congo Canyon is the proximal part of one of the largest submarine channel systems on the planet and is fed directly by the Congo River (Heezen *et al.* 1964; Babonneau *et al.* 2010; Azpiroz-Zabala *et al.* 2017*b*). Here we focus on previously published ADCP measurements in the upper part of the Congo Canyon (2 km water depth) that revealed a high frequency of turbidity current activity (Fig. 2a; Cooper *et al.* 2013). Eleven turbidity currents were measured using a downward-looking ADCP (measuring every 5 s) deployed from single-point moorings. Flows reached velocities of up to 2.5 m s^{-1} and lasted up to 10 days in duration, accounting for 30% of the four-month monitoring period (Azpiroz-Zabala *et al.* 2017*a*).

Monterey Canyon, Pacific coast, USA

Monterey Canyon extends from its shelf-incising head in Monterey Bay to the deep-sea Monterey Fan, and is one of the largest submarine canyons on the Pacific Coast of North America (Normark *et al.* 2003; Paull *et al.* 2005). Sediment is supplied to the canyon head by long-shore sediment transport cells, rather than directly from a river source (Best and Griggs 1991). Frequent turbidity currents have been recorded by numerous studies in the canyon using downward-looking ADCPs on single-point moorings (e.g. Xu and Noble 2009; Xu *et al.* 2013, 2014). A recent (2015–17) 18-month co-ordinated international experiment installed more than 50 sensors within the canyon to record the passage of 15 turbidity currents; some of which ran out for >50 km in water depths of up to 1840 m and reached velocities of >7.2 m s^{-1} (Paull *et al.* 2018; Heerema *et al.* 2020; Fig. 2b). Here, we focus on four different types of platform: (i) a downward-looking ADCP and sediment trap (at 290 m water depth; Maier *et al.* 2019*a*); (ii) an 800 kg tripod frame (deployed at 300 m water depth) fitted with an Acoustic Monitoring Transponder (AMT) and Benthic Event Detector (BED) to track its movement (Paull *et al.* 2018; Urlaub *et al.* 2018); and (iii) a seafloor frame deployed at the distal end of the monitoring array (1840 m water depth) that hosted numerous instruments including upward-looking ADCPs (Paull *et al.* 2018).

Squamish prodelta, Canadian Pacific Coast

The Squamish prodelta lies offshore from the Squamish River that drains into the Howe Sound fjord, British Columbia. Three submarine channels connect the delta lip to channel lobes in water depths of up to 200 m (Fig. 2c: Hughes Clarke 2016). Repeat seafloor surveys and water column monitoring have revealed extremely frequent (>100 per year) turbidity currents during seasonal peaks in meltwater discharge (Hughes Clarke *et al.* 2012; Clare *et al.* 2016). Here we focus on a seafloor frame containing an upward-looking ADCP (installed on the terminal lobe of one of the channels in 2011; Fig. 2c), and multi-point moorings installed in 2013 and 2015 to measure flows that attained velocities of up to 3 m s^{-1} (Hughes Clarke 2016; Hage *et al.* 2018).

Bute Inlet, Canadian Pacific Coast

Bute Inlet fjord (also in British Columbia) is fed by the Homathko and Southgate rivers, which in turn feed the submarine deltas at the head of a sinuous 50 km-long submarine channel that extends to a terminal lobe at *c.* 700 m water depth (Fig. 2d; Prior *et al.* 1987). Repeated seafloor surveys have shown >metre-scale elevation changes in the channel axis due to erosion and deposition caused by turbidity currents (Gales *et al.* 2019). Some of the earliest direct measurements of turbidity currents were made in Bute Inlet using point current meters on moorings that recorded flows in excess of 3 m s^{-1} (Prior *et al.* 1987; Zeng *et al.* 1991). Here, we focus on more recent ADCP- and 500 kHz multibeam echosounder-based measurements of flows using two- and four-point moorings, deployed in 2016 and 2018.

Fraser Delta, Canadian Pacific Coast

The Fraser submarine delta lies offshore from the Fraser River, British Columbia. The principal offshore distributary channel is located immediately seaward of the river outflow and is flanked to its south by a field of sediment waves on the delta slope (Fig. 2e; Lintern *et al.* 2016). Historical slope failures have been observed from repeat seafloor surveys on the submarine delta slope (e.g. Kostaschuk *et al.* 1992; Hill 2012). Unlike the previous examples, here we focus on an array of monitoring platforms installed outside of a submarine channel – the Delta Dynamics Laboratory (DDL), sited on the open sediment wave field (Fig. 2e). The DDL is part of Ocean Network Canada's Victoria Experimental Network Under the Sea (VENUS) cabled network and has been in operation since 2008 (Lintern and Hill 2010; Lintern *et al.* 2016). The platform can host a wide range of instrumentation due to its cabled power and communications connection, some of which include upward- and downward-looking ADCPs, velocity profilers, turbidity sensors and video cameras (Lintern *et al.* 2016). Other platforms at the site include a Seismic Liquefaction *In situ* Penetrometer (SLIP), which measures pressures and movement within the bed, and a hydrophone array, which listens for landslides and other environmental noises. As with the Bute and Squamish sites, turbidity currents are frequent during the spring and summer when river discharge is elevated (Ayranci *et al.* 2012).

Results from recent direct monitoring of turbidity currents

We now summarize the issues we encountered during recent turbidity current monitoring campaigns, ordered from the smallest to greatest impact.

Temporary instability of single-point moorings: pull down, pitch, roll and rotation

Single-point ADCP moorings in a submarine canyon or channel axis commonly record an abrupt increase in water pressure coincident with the arrival of a turbidity current. In the 2015–17 Monterey Canyon Coordinated Canyon Experiment (CCE), each of the 15 turbidity currents caused an initial increase in water pressure that generally declined over 4–120 minutes (Paull *et al.* 2018). This increase in water pressure is attributed to pull-down of the mooring cable, due to drag imparted by the flow front (which reached velocities of up to 7.2 m s^{-1}) most likely exerted on instruments that were within the flow. A decrease in water pressure occurred when the flows decelerated and the mooring gradually returned to its original vertical position. A similar situation was observed in a previous experiment in Monterey Canyon, where a mooring was severely tilted during the first 15 minutes of a turbidity current, causing a sediment trap (located at 70 m above seafloor) to be pulled down by 37 m into the lower parts of the flow; thus explaining the anomalously coarse material collected by the sediment trap (Symons *et al.* 2017). Mooring tilt and down-canyon transport also occurred during strong internal tidal flows in Monterey Canyon (i.e. tidal frequency flows trapped within the canyon topography, unrelated to turbidity currents). On 30 November 2015, during a particularly strong up-canyon internal tide (*c.* 1 m s^{-1}), the lower current meter was pulled down 2 m and tilted more than 20°.

Such pull-down effects were not observed in the Congo Canyon, where the mooring construction was much simpler and acoustic release links were located much higher (70 m) above the seafloor than in the Monterey Canyon experiments (3 m above seafloor). This is not to say that the Congo Canyon mooring remained unaffected by flows, however. Intervals of increased pitch, roll and tilting (<2°) were recorded by the downward-looking ADCP during turbidity currents; dominantly during the initial passage (<1 h) of the fast frontal cell. These effects (in particular the rotation of the buoy housing the ADCP) resulted in transient interaction of the ADCP beams with the narrow canyon sidewalls, thus limiting the depth range and quality of velocity and backscatter measurements.

Down-canyon transport of single-point moorings and damage to instruments

As well as the reversible pressure changes noted at the start of turbidity currents, several turbidity currents in Monterey Canyon caused permanent pressure and temperature changes, as recorded by ADCPs on single-point moorings. These irreversible changes indicate that, in addition to the buoy-mounted ADCP being temporarily pulled towards the seafloor, single-point moorings were also transported down-canyon. Symons *et al.* (2017) documented the 580 m down-canyon transport of a single-point mooring attached to a 1000 kg anchor at a speed of *c.* 0.5 m s^{-1} from a 2002–03 deployment (Xu *et al.* 2004, 2014). During the CCE (1 December 2015), a single-point mooring (using a 450 kg train wheel for an anchor) was moved down-canyon (as evidenced by an average drop in pressure of 3 m) by a relatively small turbidity current (*c.* 3 m s^{-1}). The most powerful flow event (15 January 2016) caused down-canyon transport of the same mooring by 7.1 km, at an average speed of 4.5 m s^{-1} (Paull *et al.* 2018). This mooring ultimately broke loose from its anchor and was retrieved at the sea surface.

On the final of three deployments in the Monterey Canyon CCE, two train wheels (*c.* 900 kg) were used to anchor the single-point mooring and inline flotation was placed above each sediment trap (as well as additional flotation at the top of the mooring; Fig. 3). Mooring performance was much improved by this revised design. Even in very strong turbidity currents (>5 m s^{-1}) the mooring did not move. Tilt and down-pull during strong internal tides were also considerably reduced (<10° and <1 m, respectively). To make additional measurements within turbidity currents, several instruments were installed on the mooring line beneath the ADCP for the Monterey CCE, including Anderson-style sediment traps, altimeters and a point current meter (Fig. 3a–c). Significant damage was recorded upon retrieval of these instruments, however, including loss of the impellors for the current meter, fouling of instruments with sediment and organic debris, removal of the sediment trap inlet funnel, and sand-blasting, bending and buckling of steel instrument frames (Paull *et al.* 2018; Maier *et al.* 2019*a*; Fig. 3d, e). One particularly important issue also concerned damage to the acoustic release links that are required for remote

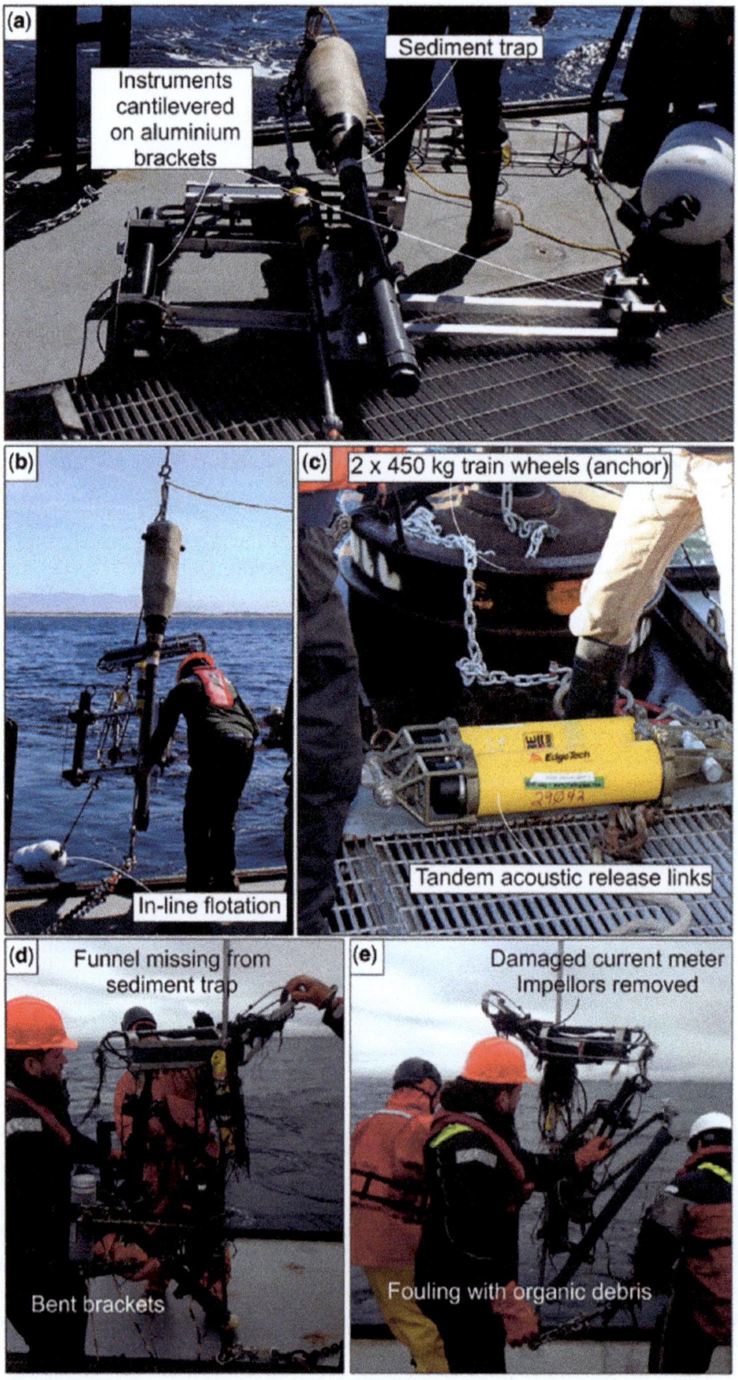

Fig. 3. Photographs of a sediment trap and inline instruments placed within turbidity currents from Monterey Canyon. (**a**) Pre-deployment photograph of a sediment trap and instruments fitted on cantilevered aluminium brackets. (**b**) Deployment of a sediment trap. (**c**) Detail on anchor weight (train wheels) and acoustic release links, which were placed 3 m above seafloor. (**d**, **e**) Sediment trap and instrument brackets following retrieval, showing damage and fouling during interaction with turbidity currents.

release of the mooring and retrieval from the sea surface. Many of the releases (located at 10 m above seafloor) used in the Monterey CCE did not release properly when the command was issued from the support vessel. The extreme case was the final deployment where every mooring required a Remotely Operated Vehicle (ROV) dive to recover the mooring. Some of these required cutting of the mooring string below the release, while others only required tapping the release with the ROV's mechanical arms. These issues were attributed to the presence of sand within the releases and are similar to those encountered by single-point moorings in the submarine channel in Bute Inlet, where Prior *et al.* (1987) recorded: (i) damage, removal and fouling of rotors and vanes on current meters (causing poor data quality); (ii) bent and sheared shackles and stainless steel frames; (iii) up to 1 km down-channel transport of moorings; (iv) failure of acoustic releases to detach due to burial by sand; (v) parting of mooring lines; and (vi) the entire loss of some instruments (also detailed in Zeng *et al.* 1991).

Unlike these examples from Monterey Canyon and Bute Inlet, no irreversible pressure or temperature changes were observed for the single-point mooring in the Congo Canyon. Therefore the Congo Canyon mooring is unlikely to have been moved by any of the eleven turbidity currents that occurred during its deployment (Azpiroz-Zabala *et al.* 2017*a*, *b*). Furthermore, no damage was recorded in this case to either the acoustic release links or the ADCP. No other instruments were placed on the mooring line.

Burial, downslope transport and damage of seabed frames

We now discuss issues that have affected seabed-based platforms. An upward-looking ADCP was mounted on a bottom-mounted tripod in 2011 and deployed at the terminal end of a submarine channel offshore from the Squamish river delta (150 m water depth). This ADCP recorded 22 turbidity currents of up to *c.* 1.5 m s^{-1} over a period of four months (Hughes Clarke *et al.* 2012), with the exception of a 20-day period when the run-out from a delta lip collapse led to the burial of the frame (Clare *et al.* 2016). With a single ensemble averaging interval of 20 s, the ADCP went from recording flow to being completely buried. Thus, no monitoring was possible during this time. Interestingly the ADCP frame was not significantly tilted in this process. Fortunately, a vertically offset surface buoy was attached so that the instrument could be dragged out of the sediment.

In addition to the movement of single-point moorings deployed in the Monterey CCE, down-canyon movement of an 800 kg AMT-tripod-frame (Fig. 4a) was also recorded six times. These episodes of movement corresponded to the timing of turbidity currents. On 15 January 2016, the AMT frame moved 4.2 km down-canyon and was observed from ROV video to be on its side, half-embedded in the seafloor (Paull *et al.* 2018). The heavy-duty steel frame was transported within a dense sediment layer, its feet bent and sheared in places, while much of the pressure-resistant foam coating was abraded from the BED (Fig. 4d–f). Following its redeployment, the mooring was transported 0.9 km on 24 November 2016; also found on its side, but this time buried by at least 2 m of sediment with only one foot protruding at the seafloor (Fig. 4c). Pressure, temperature and accelerometer measurements indicate that once the AMT frame was tilted on to its side it became buried during the initial turbidity current and then remained in that position until it was moved by successive flows. A multi-instrument 'Seafloor Instrument Node' (SIN) was placed in a deeper water location (1840 m), where the Monterey Canyon widens. Flows decelerate from *c.* 4–8 m s^{-1} in the upper part of the canyon where the AMT frame was deployed to *c.* 1–2 m s^{-1} at the SIN location (Fig. 5; Paull *et al.* 2018; Heerema *et al.* 2020). Impacts of turbidity currents were less severe at this more distal location; however, the SIN frame was also transported down-canyon, by 26 m, and was locally buried by up to 34 cm of sediment (Fig. 5c). A high frequency acoustic instrument (Aquadopp) was ripped from the arm that suspended it above seafloor and up to 10 cm of scour was noted from repeated ROV-based bathymetric surveys (Fig. 5b, c).

Even benthic landers sited outside of submarine channels can suffer from adverse impacts that include burial and movement of the platform. The original Delta Dynamics Laboratory platform (DDL), deployed in 2008 on the Fraser Delta (located in a bedform field outside of the main submarine channel; Fig. 2e), was buried by as much as 1 m of sediment. Initially it was thought that this was simply natural sediment deposition from the Fraser River; however, it is now attributed to active turbidity currents (Lintern *et al.* 2016). Recovery using a vessel-deployed crane caused a large ship (the 1800 t CGS *John P. Tully*) to lean uncomfortably and snapped 9000 kg lines. The original platform design at the Fraser Delta had a large surface area, which also made it prone to tumbling during turbidity currents, as recorded by frame-mounted orientation sensors, and was therefore replaced by a lower-profile platform with weighted legs (Fig. 6; Lintern *et al.* 2016). This revised deployment included arms and poles that held instruments

Fig. 4. Photographs of the 800 kg AMT frame deployed at 300 m in Monterey Canyon. (**a**) Prior to deployment of instrument. (**b**) Example of Benthic Event Detectors, one of which was attached to the top of the AMT frame to track the sense of motion of the frame. (**c**) Only the foot of the AMT frame was found protruding from seafloor by an ROV dive video following its burial by a turbidity current. (**d**) AMT frame following retrieval to deck, revealing damage to the frame (**e**) and flotation on the AMT frame (**f**) caused during its down-canyon transport.

away from the platform and above the 2 m powerful flows that were detected (Lintern *et al.* 2019). It also featured feet that snap free on retrieval, as embedment of the original large feet created problems during recovery (Fig. 6b). The second platform design mostly remained upright, but sometimes slid downslope during strong turbidity currents. To make it more resistant to flows, over 900 kg of ballast was suspended below the platform, while the legs penetrate the seafloor by up to 1 m, acting as small piled foundations (Fig. 6c). This enhanced design has so far remained upright for two years, experiencing flows of up to 9 m s^{-1} (Lintern *et al.* 2016, 2019).

Other platforms on the Fraser Delta include a Benthic Boundary Laboratory (BBL) and a SLIP (Fig. 7). The BBL's main design feature was a cantilever to hold instruments away from the main platform to minimize frame turbulence. Despite the increased tipping moment this would appear to cause, it is worth noting that further down the delta slope (140 m v. the DDL 107 m) there has not been a strong enough turbidity current in five years of deployment to topple the deeper BBL.

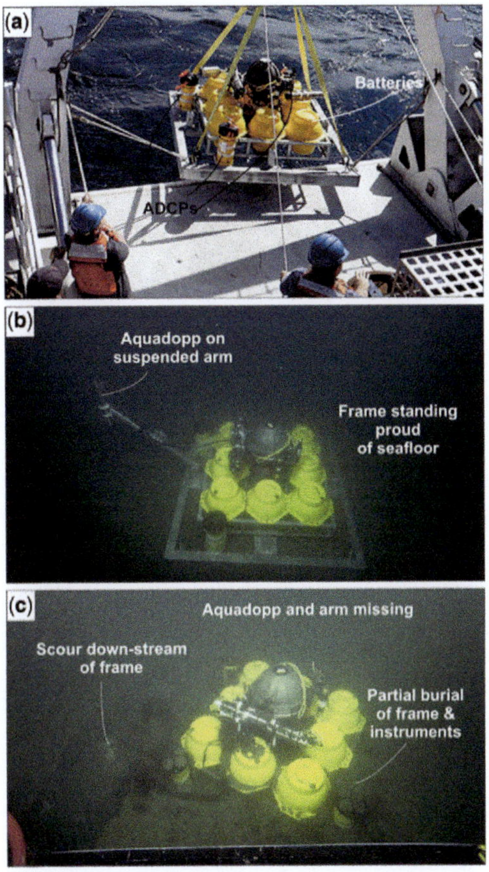

Fig. 5. Photographs of the Seafloor Instrument Node (SIN), deployed at 1840 m water depth in Monterey Canyon. (**a**) SIN prior to deployment. (**b**) ROV video still showing deployed location where the frame sits proud of the seafloor. (**c**) ROV video still at retrieval, following 26 m down-canyon transport, with evidence of local scour and deposition around the frame and removal of the Aquadopp and its mounting arm.

The SLIP is an instrument designed to measure pore pressures that could be associated with subaqueous mass movements (Fig. 7). It is constructed of a fibreglass frame above the seafloor holding a system of valves, data-loggers, instruments and a network plug. The data logging is done on cyclical buffers and has backup battery power in case it is severed from the network. The lower part of the SLIP is a 5 m-long cone tip with multiple pressure and temperature ports. An 800 kg piston core head weight is used to push the SLIP tip into the sediment. The SLIP has been deployed for several years at the site of the DDL and due to its 5 m embedded tip, it has not suffered any translation from the same turbidity currents that have been tumbling the DDL platforms.

Overview of adverse impacts related to turbidity currents

Based on past experiences from recent monitoring campaigns, the following observations can be summarized about the hazards posed by turbidity currents to moorings and seafloor platforms (Fig. 8):

(1) The powerful dense near-bed part of a turbidity current (particularly prone in proximal confined submarine canyons or channels) may be capable of toppling and/or transporting heavy objects (in excess of hundreds of kilograms), including anchors and seabed frames (Fig. 8a). This dense part of the flow can damage platforms, sensors and ancillary mounting equipment through collisional impact or drag, and may even result in short-lived liquefaction of seafloor sediments, causing anchors for single-point moorings or seafloor frames to sink.

(2) Fast flows may pull instruments down towards the seafloor and, in some cases, overcome the tractional forces required to keep the anchor in place, and transport single-point moorings down-channel (Fig. 8a).

(3) Where instruments interact with a turbidity current, this may lead to platform instability and poor-quality data, damage to acoustic releases (jeopardizing successful retrieval of moorings) or, in severe cases, loss of instruments and mooring components (Fig. 8c).

(4) Erosion of the seafloor may change local seafloor elevation and undermine platforms where it occurs as scour around a seafloor structure (Fig. 8d).

(5) Sudden deposition, sometimes involving several metres thickness of sediment, can bury seafloor instruments or low-elevation acoustic releases, limiting instrument performance and causing issues for retrieval (Fig. 8d).

Designing monitoring platforms to successfully measure turbidity currents

In this section we highlight some of the lessons we have learned from previous turbidity current monitoring campaigns, in order to inform future ones.

Finding a 'sweet spot' for the design of single-point moorings

When optimizing mooring design to address one issue, other complications may arise concerning another. We now discuss how mooring designs have been iteratively refined to try and find the ideal configuration for different settings and objectives.

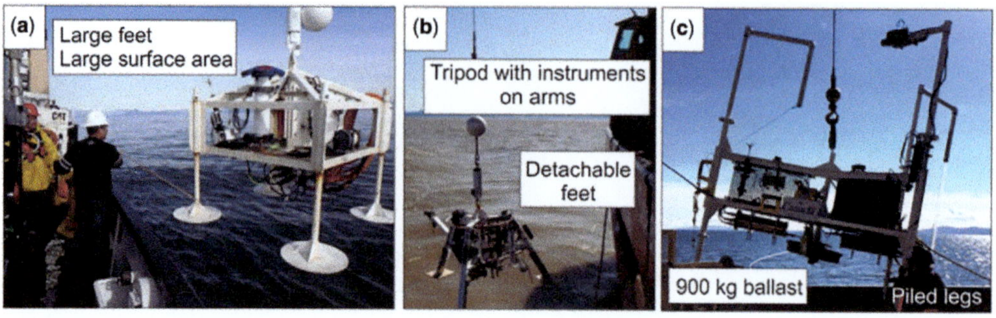

Fig. 6. Development of the Fraser Delta Dynamics Laboratory including (**a**) conventional design with large feet to stop embedment, (**b**) revised tripod design with detachable feet. Both (**a**) and (**b**) tumbled downslope during powerful flows. (**c**) Revised design that has withstood numerous powerful flows to date due to its piled legs and ballast weight that can be ejected. Image modified from Lintern *et al.* (2019).

Reduce the surface area to minimize drag. Single-point moorings are typically the preferred way to monitor turbidity currents as they can be deployed from the back deck of an ocean-going vessel equipped with a suitable winch and A-frame. Successful monitoring of turbidity currents is strongly dependent on the mooring design. Single-point moorings in the Congo Canyon did not show any movement down-canyon during turbidity currents, nor were any of the instruments damaged. We identify three reasons for the stability of this Congo Canyon mooring. First, while the flows in Congo Canyon lasted many hours to days in duration, they were generally muddy and dilute flows (with the exception of a frontal cell of sand-rich, sediment-laden fluid), and relatively slow, reaching maximum velocities of $<3\ \mathrm{m\ s^{-1}}$ with an average of $<1\ \mathrm{m\ s^{-1}}$ (Azpiroz-Zabala *et al.* 2017*a*). Conversely, flows in the Monterey Canyon often reached velocities far in excess of this value, up to $7.2\ \mathrm{m\ s^{-1}}$,

Fig. 7. Prototype benthic monitoring platforms at the Fraser Delta. A large stainless steel container houses data processing and logging instruments and an underwater modem. All components are made from fibreglass or stainless steel in an attempt to minimize corrosion in salt water. Power is provided by the network, and data are transmitted directly to the scientists' offices over the internet in near real time. A battery backup and circular buffer continue to measure data in case of a severed cable, due to a slope failure. (**a**) Overview of the NRCan Cable Caster (note this is a different design to the SLIP in (b)–(d)). The housing above the mud plate contains up to 200 m of cable. When the mud plate impacts the seafloor, the housing, deployment weight and camera are released and retrieved to the surface, unspooling the cable from the bottom for hookup to shore or to a data buoy. (**b**) Deployment of the prototype SLIP using 816 kg weight. (**c**) Detail on the instrumented tip of the SLIP that contains triaxial geophones to measure earthquakes and ground movements, measuring up to 100 times per second. (**d**) The SLIP cable being unspooled 1.5 km to the Victoria Experimental Network Under the Sea (VENUS) node by the manipulator arms of an ROV.

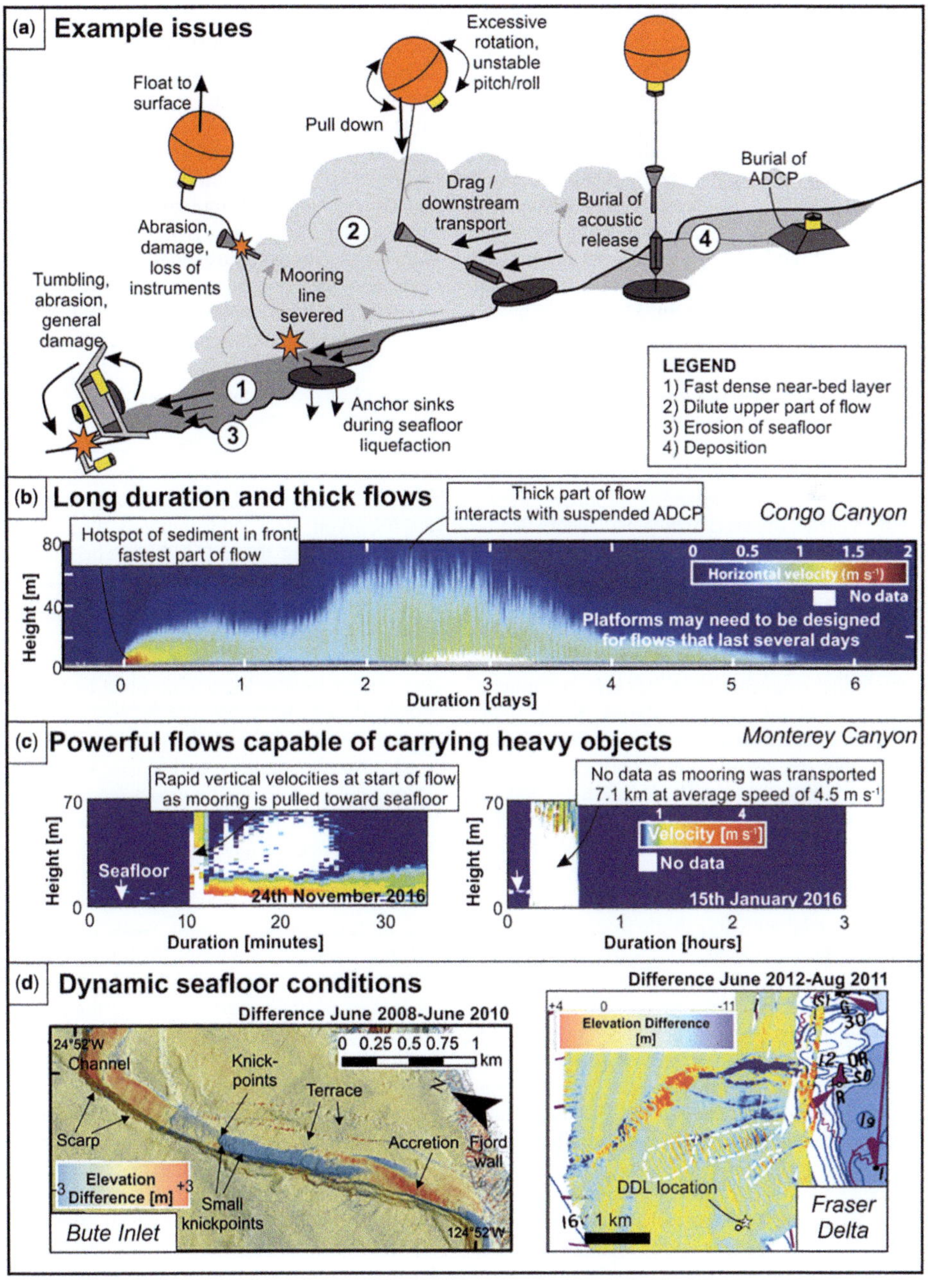

Fig. 8. (**a**) Overview of some of the issues encountered in monitoring active turbidity currents discussed in this paper. (**b**) An example of one of the long-duration turbidity currents measured in the deep-water Congo Canyon that may attain thicknesses of >80 m (modified from Azpiroz-Zabala *et al.* 2017*a*). (**c**) Two turbidity current events measured at the shallowest water mooring in the Monterey coordinated canyon experiment in Monterey Canyon. On the left is a flow that pulled the instruments and buoyancy towards the seafloor at the start of the event due to enhanced drag early on. On the right is the record from an ADCP that was transported by a flow at several m s^{-1}; hence, no reliable data were recorded during the flow. This mooring was transported 7.1 km down-canyon and then broke free from its anchor and was released to the sea surface. (**d**) Repeat multibeam echo-sounder seafloor surveys illustrating how active turbidity currents can both erode and deposit at the seafloor. The location of the Delta Dynamics Laboratory is labelled on the Fraser Delta (right).

and are interpreted to have been denser, with the near-bed part of the flow capable of transporting gravel and cobble-sized material (Paull *et al.* 2018). Second, the mooring design for the Congo Canyon included heavier anchor weighting (*c.* 2000 kg), use of low-drag neutrally buoyant plastic-coated mooring line and a larger syntactic buoy housing the ADCP. This greater buoyancy ensured that the mooring line remained taut during flows (Fig. 9). Third, and perhaps most importantly, the mooring design was much simpler for the Congo Canyon measurements than in Monterey Canyon (Fig. 9). Sediment traps were not deployed, and acoustic release links were placed (*c.* 40–60 m) far above the velocity maximum of the flows, in order to reduce drag on the mooring line imparted by flows. Therefore, one way to maximize the likelihood of successful monitoring is to ensure that any instruments are located above the turbidity currents that you wish to observe, which will decrease the likelihood of drag and also add weight to the mooring line. Previous successful deployments in the Var Canyon (Mediterranean) used lower frequency (75 kHz) downward-looking ADCPs that were placed much higher (300–350 m) above the seafloor than the higher frequency 300–600 kHz instruments in the Monterey and Congo canyons (Khripounoff *et al.* 2012). Coarser vertical resolution was accepted to ensure that the single-point moorings interacted less with turbidity currents. The Var Canyon deployments also featured ADCPs set within gimballed frames that ensure the ADCP can tilt to remain as vertical as possible. Such a situation may be unavoidable, however, if you wish to: (i) measure close to the seafloor using high frequency instruments (e.g. Hughes Clarke *et al.* 2012; Clare *et al.* 2015; Hughes Clarke 2016); (ii) sample sediments within the flow to measure vertical grain size segregation or quantify organic particulate flux (e.g. Maier *et al.* 2019*a*, *b*); (iii) make measurements within the flow to ground-truth other remote sensing style measurements (e.g. Azpiroz-Zabala *et al.* 2017*a*; Hage *et al.* 2019). Sediment traps are typically the largest item on the mooring line; hence the height of the trap above the bed may be critical. The style of sediment trap also makes a difference. McLane-type traps provide a greater cross-sectional area than the narrower Anderson-type traps.

Design anchor weight and flotation appropriately, particularly if multiple instruments are required for single-point moorings. One of the primary goals of the Monterey Canyon CCE was to estimate suspended sediment concentrations during a turbidity current using the acoustic backscatter from the downward-facing ADCP. Given that the acoustic response of the ADCP is both a function of the concentration and the grain size of the material in suspension, it was decided that an inline sediment trap was essential, even if the presence of the trap increased drag on the mooring. While it may seem intuitive that increasing the anchor weight will improve mooring stability, this is not always the case. Moorings deployed in Monterey Canyon in the early 2000s had multiple train wheels for their anchor and long mooring lines with multiple instruments attached (Xu 2011). Some of these moorings were lost due to the drag exerted during turbidity currents and the mooring line parted. Conversely, a mooring has been deployed successfully at 1300 m water depth in the Monterey Canyon almost continuously since 2002 with minimal ballast (scrap steel) (Barry *et al.* 2006; Xu *et al.* 2013). Its light ballast makes this mooring relatively easy to move, but this also ensures that the strain on the mooring line does not reach a critical limit. Thus, one way for a mooring to survive may be to allow it to be dragged down-canyon. This philosophy is also in keeping with minimizing the amount of debris that is left behind following mooring retrieval, as it is difficult to justify leaving iron, cables and potentially fibreglass in the marine environment.

There appears to be a 'sweet spot' for mooring design that involves a compromise between minimizing drag (which may not be possible if several instruments need to be deployed within the flow height), stabilizing the mooring base with an anchor weight and maximizing buoyancy to vertically stabilize the mooring line. The design of the mooring is an iterative process, balancing available anchor weight, surface drag (and weight) from inline mooring elements and both inline and top flotation elements. The mooring design toolbox written in Matlab by Richard Dewey (Mooring Design & Dynamics; Dewey 1999) was used in the Monterey Canyon CCE to evaluate the performance of the single-point taut-wire moorings. The program allows a user to design a surface or subsurface wire mooring and contains a large database of the physical characteristics of standard oceanographic equipment (such as dimensions, submerged weight, surface drag), and will evaluate how a mooring responds to a static flow profile (i.e. does not account for waves). It was thought that turbidity currents in Monterey Canyon did not exceed 2 m s^{-1} (since the most recent data derived from one-hour averages), and this value was used in the initial mooring designs. We now know this was a considerable underestimate. It is best to have contingency and overdesign. Keeping the mooring as upright as possible (increasing the inline tension) requires additional flotation (Fig. 12a), which has the additional negative effect of making the anchor 'lighter' by increasing the upward force on the anchor, thereby making the mooring more likely to move down-canyon during events. Even the type of flotation used was a

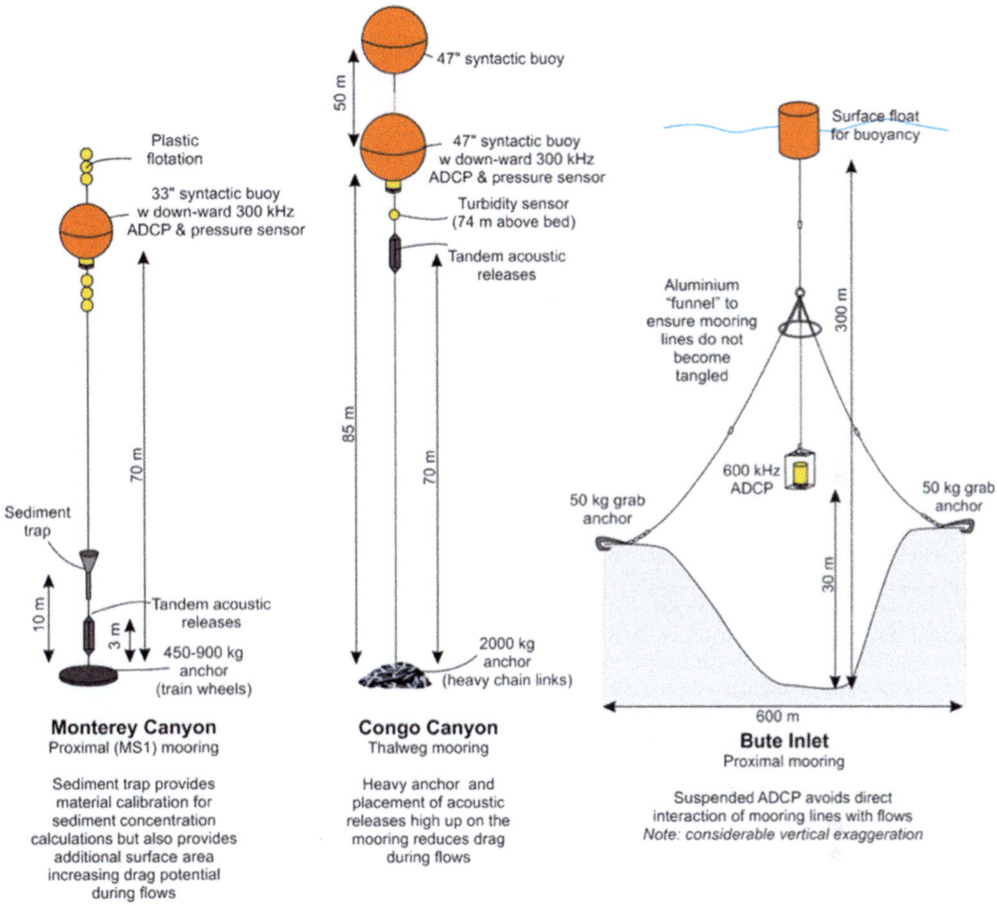

Fig. 9. Comparison of subsurface single-point moorings deployed in Monterey Canyon (Paull *et al.* 2018), Congo Canyon (Cooper *et al.* 2013; Azpiroz-Zabala *et al.* 2017*a*, *b*) and two-point mooring supported by a surface buoy in Bute Inlet. Not drawn fully to scale.

consideration: in shallower water (less than 800 m) plastic flotation was used for inline elements to provide greater flotation per diameter of sphere (and thus surface area, because they weigh less) than comparably-sized glass or syntactic foam elements. Increasing the anchor weight from 450 to 900 kg and increasing the inline flotation above each sediment trap and the top of the mooring dramatically improved mooring performance, demonstrating that it is possible to refine the design successfully. This may require some *a priori* knowledge of the likely flow conditions. Regardless of design, one key lesson learned is to include an iridium beacon on the instrument package such that it can be tracked should it cut loose and float to the surface.

Strengthen the weak points on a monitoring platform: strategic placement of acoustic releases and resilient instrument mounting. Had the Monterey Canyon CCE not been supported by an ROV, then the failure of the acoustic releases (placed close, 3 m, above the seafloor) to return the moorings to the surface would have meant the loss of valuable data and instruments. Many research and industrial expeditions do not have the benefit of a support ROV; hence, we recommend that acoustic release links are placed as high as practicable above the seafloor, where they are away from the damage that may be caused at the sand-rich base of a turbidity current (but low enough such that they do not interfere with the ADCP). A recent study in the Gulf of St Lawrence (eastern Canada) by Normandeau *et al.* (2019*a*) suggested placing the acoustic release a minimum of 1 m above the height of intra-channel bedforms, to avoid interaction with the most vigorous and potentially dense part of the flow. Tandem acoustic release links are routinely deployed for single-point moorings (i.e. to provide redundancy in case one fails) but it may also be sensible to deploy the releases in series, rather than in a parallel twinned deployment so that they are not both subject to impacts at the same elevation within the flow (Xu *et al.* 2004).

Instrument mountings were often found to be weak points in a monitoring platform's design (e.g. Figs 3 & 4). In the case of the Monterey CCE deployments, near-bottom current meters and altimeters (10 m above seafloor) were mounted on protruding brackets (cantilevered) on the single-point moorings 1 m from the sediment trap strong-back, with one-quarter inch aluminium angle stock (instead of stainless steel, to reduce weight; Fig. 3). It was necessary to cantilever them away from the mooring in order to ensure that instruments below the ADCP were not affected by the mooring wire, or other instruments below. This design provided an even larger surface area for drag and also increased the weight on the mooring line, however, and underlines how operational necessities may end up going against the guidance to minimize drag. The aluminium design survived four turbidity currents, but eventually broke. In future, and if resources allow, we suggest that titanium should be used for mounting in similar environments. Heavy metal parts and coated iron wires should be avoided, especially for long-term deployments, as it is impossible to have a visual check on corrosion. Instead, plastic-coated ultra-high molecular weight polyethylene Dyneema rope is preferred as there are no corrosion issues, it is thin and neutrally buoyant and may be used for multiple deployments.

While it may be possible to strengthen brackets and frames, any instruments with moving external parts (e.g. the impellors that were damaged on the current meter deployed in Monterey Canyon) or that protrude away from the platform (e.g. the steel arm that held the near-bed Aquadopp in Monterey Canyon; Fig. 4) are likely to be vulnerable and should be considered to be at high risk during field deployments to measure powerful flows.

Suspended monitoring systems that avoid instrument and mooring-line interaction with the flow

To avoid the damaging effect of a passing turbidity current (e.g. drag, scour, burial), another option is to avoid placing instruments, anchors and mooring lines within the flow at all. Such an approach may also be necessary where the available support vessel for deployment cannot handle the bulky hardware (e.g. *c.* 1 m diameter syntactic buoys and stack of train wheels weighting *c.* 1 t) required for single-point moorings. We now discuss two plausible geometries: (i) hull-mounted systems; and (ii) surface buoy suspended systems with two or more anchors. Both of these are only practical in shallow water (typically <500 m) environments, given the amount of deck space used and the logistics involved with such quantities of mooring line and anchors, and for short-term (months) deployments. Such methods are therefore only generally applicable in fjord or lake environments and not the deep ocean; however, previous deployments in the Var Canyon has demonstrated that subsurface two-point moorings are feasible in water depths as great as 1280 m (Khripounoff *et al.* 2012).

Vessel-mounted monitoring systems

Hull-mounted deployments. Hull-mounted systems include acoustic imaging (downward-looking single or multibeam sonars or ADCP) and rapidly descending underway physical probes (e.g. Moving Vessel Profiler, MVP; Hughes Clarke *et al.* 1996). For any of the sonar systems, the issue becomes resolution – the further away from the seafloor, the poorer the range resolution usually is (longer, narrow-band pulses required); especially angular resolution. For single-beam sonars the width of the projected beam (typically 7–30°) may result in echoes from offset roughness elements (like bedform crests or channel flanks), which can be confused with the real near-seafloor profile. Multibeam systems (with beam widths in the 1–2° range) provide far better definition (see Fig. 10a–c; Hatcher 2017). For ADCPs, just as with the conventional downward-looking single-point moorings, the closest usable data to the seabed are limited by the first echo of the projected side lobes from the beams inclined at 20° (Fig. 12b). This limits the first usable bin to about 10% of the ADCP altitude (using conventional 4-beam systems). Thus, to investigate 5 m thick flows for example, surface-mounted ADCPs would not be of use at elevations much greater than *c.* 50 m, plus the vessel has to be present at the time of the flow.

Therefore, this surface-mounted method is only viable if the flows are known to be frequent and/or of known likely timing. This was the case for the Squamish 2011–13 and 2015 campaigns (Hughes Clarke 2016). Here, a small vessel (CSL *Heron*) deployed an MVP. The MVP consisted of a tow body with a conductivity, temperature and depth (CTD) and an optical backscatter probe that can be released at slow speeds (<6 knots). If the vessel slowed down for the descent duration (typically 2 min), the probe descended to a depth of 100 m. The MVP was deployed daily along the main channel sections to catch evidence of suspended sediment clouds due to a passing turbidity current. On a few occasions, the MVP was able to sample the top of an active turbidity current, which was also observed in the EM710 water column imagery (1 × 2° beam,

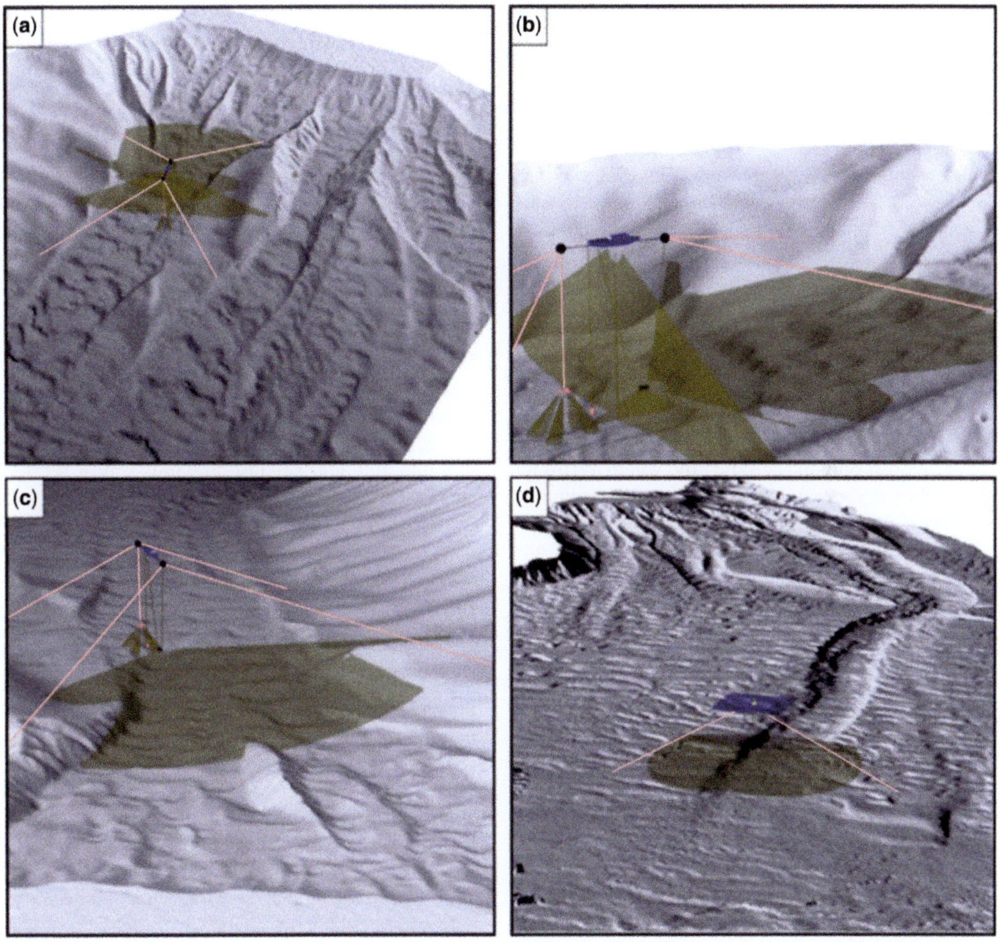

Fig. 10. Schematics illustrating Squamish experimental set-up from 2013 (a–c) and 2017–19 mooring (d). Vessel shown in blue. Acoustic imaging coverage shown in green. (**a**) Location of the four anchors (all located outside the active channel areas) and all acoustic imaging coverage, relative to the delta lip and prodelta channels. (**b**) Details of the offset surface buoys that allowed for azimuth stability of the suspended sonars. (**c**) View from the delta lip that was 300 m away, illustrating the geometry relative to the triggering mechanisms upstream. (**d**) Location of the channel mouth two-point anchor mooring in 2019. The water depths of the mooring were *c.* 160 m on the floor of the channel, which is about 80 m wide and 5 m deep at that point. No vertical exaggeration.

0.2 A–0.5 ms pulses, 70–100 kHz; Hughes Clarke *et al.* 2014; Hage *et al.* 2019).

The MVP has several limitations. The profile is necessarily discrete. The minimum horizontal spacing depends on the time it takes to winch back in the instrument cable, typically 5 min if going to 100 m. The instrument package is deliberately designed to stop free-falling before hitting the seabed. Thus, measurements closer than 5 m from the actual seabed are rare and only the top of an active flow is usually recorded. The use of hull-mounted instrumentation will only be useful in relatively shallow water where the recurrence of active turbidity currents is reasonably predictable. This is not the case for most turbidity current systems, where longer-term campaigns without a crew are required.

AUV-mounted deployments. Autonomous Underwater Vehicles (AUVs) now enable the acquisition of high-resolution seafloor datasets, by flying the AUV close to the seafloor (Wynn *et al.* 2014). These autonomous mobile systems can also hold instruments, such as ADCPs, to monitor the seafloor along transects, in the same manner as river systems are often measured (e.g. Parsons *et al.* 2007). A saline density underflow has been monitored using such an approach, to the north of the Bosphorus Strait in the Black Sea. Along- and across-channel transects of ADCP measurements were acquired using a 1200 kHz ADCP, revealing a range of flow dynamics, which include evidence for secondary circulation cells and the presence of hydraulic jumps. These jumps had previously only been hypothesized from laboratory experiments of submarine channels (Parsons *et al.* 2010; Sumner *et al.* 2013; Wynn *et al.* 2014; Dorrell *et al.* 2016; Azpiroz-Zabala *et al.* 2017*b*). This AUV-based monitoring was also performed in a very narrow and busy shipping lane; hence surface-based monitoring would have been precluded (Wynn *et al.* 2014). Future developments in AUV endurance (e.g. battery performance) may make this type of monitoring more common; however, it is only likely to be used where the timing of the flow is very well constrained or continuous, as in the case of the Bosphorus underflow.

Multi-point anchoring for vessel-based monitoring. If a turbidity current is laterally restricted by canyon or channel flanks, it is possible to use two or more anchors located on either side of the channel to position a surface buoy above the active channel, from which a variety of instruments can then be suspended. In practice, there are depth limitations to this, as the longer the anchor lines, the more the suspended instrument is likely to move. The first test of the two-point anchoring method occurred in 2014 at 200 m water depth in Bute Inlet and then in 2015 in Squamish in 120 m of water. From 2017 to 2019, a two-point mooring was deployed in Squamish at the lobe channel termination in 160 m of water (Fig. 10d). A minimum of two anchors can adequately constrain the buoy across the channel, but any slack in the lines will allow the buoy to move slightly along the channel as a result of wind or tide drag on the surface buoy and the suspended lines. A third anchor helps constrain the along-channel motion.

Four-point moorings were deployed for the 2013 Squamish experiment (Hughes Clarke 2016). This four-anchor approach not only best constrained the surface location, but also allowed the suspended instruments to be held at a fixed azimuth. For any number of anchors, if there is only a single surface buoy, the suspended instrument is free to rotate in azimuth. Thus, the instrument measurement must not be compromised by this rotation. Instruments that have an internal compass can correct for such rotations; however, any system that requires a preferred azimuth (such as the acoustic monitoring of a fixed stretch of channel by forward-looking multibeam (M3) imaging used in the 2013 and 2015 Squamish experiments; Hughes Clarke 2016; Hage *et al.* 2018) would not be usable. To overcome this problem, in the 2013 Squamish experiment, the four anchors were arranged in pairs to come up to two surface buoys located offset along the channel below. The surface buoys were in turn held together by a surface line. The vessel tied up daily between the two buoys and azimuth-sensitive instruments (the M3 sonars described in Hughes Clarke 2016) were suspended on a frame that was attached fore aft so that it could not rotate significantly in azimuth. Such a deployment is only suited to short-term (days to weeks) duration.

Two-point moorings for autonomous deployments. In 2016, and again in 2018, two-point moorings were deployed in water depths of up to 450 m in Bute Inlet (Fig. 9). Such a mooring design was conceived to remove any drag on the instruments or mooring line and because previous repeat seafloor surveys indicated that turbidity currents may be capable of depositing and eroding up to tens of metres of sediment (Conway *et al.* 2012; Gales *et al.* 2019). In this example, the anchor lines no longer connected to the surface float, but instead to a triangular frame that was suspended at about half the water depth from the surface float (Figs 9, 11c, d). This set-up reduces the length of the anchor lines, limiting the movement of the instrument and facilitating the deployment. With the submerged frame at half the water depth, the anchors can be placed one at a time and the anchor can be dropped with slight tension in the line at the final anchor location. Each anchor and chain had a weight of 100 kg, while the surface float carrying the instruments had a buoyancy of 150 kg. The anchor lines were about 300 m to keep the anchors well away from the 200 m wide channel (Fig. 9). Such a deployment is logistically challenging, particularly on smaller vessels, and is therefore unlikely to become a routine mooring configuration. The use of a surface buoy would also be impractical in high latitude settings where seasonal sea ice forms. A completely submerged two-point mooring was deployed in the deep-sea Var Canyon, however, which involved anchoring either side of the canyon in a water depth of 1200 m (Khripounoff *et al.* 2012). Acoustic releases were placed 10 m above the anchor on each of the mooring lines, so that only 10 m lengths of rope and anchors were left on the seafloor following retrieval. Large quantities of mooring lines, buoyancy and anchors meant that this mooring required a spacious back deck of a large ocean-going vessel.

Decision on instrument elevation for multi-point moorings. A two-point mooring with a surface buoy will only fix the elevation of the ADCP with respect to the surface. Any tidal excursions will result in the ADCP moving toward and away from the seabed. Deciding on the optimal height may

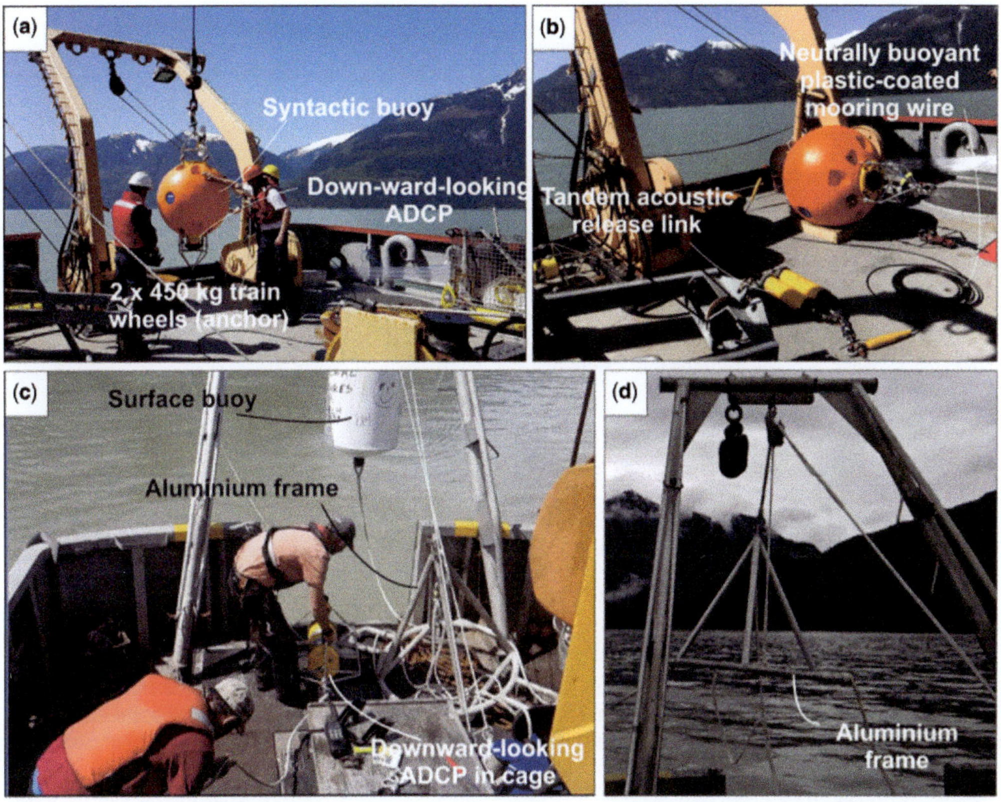

Fig. 11. Photographs comparing the typical hardware for single-point moorings (**a**) and (**b**) with hardware required for two-point moorings (**c**) and (**d**). Not shown are the 50 kg grab anchors used to secure each of the lines for two-point moorings.

require prior information on the likely thickness of the flows. The instrument should be above the active flow, even if partly in the turbid cloud of the wake, but at the same time, as close to the seabed as possible so that side lobe masking is minimized (Fig. 12b). In larger systems and deeper canyons, where the anchors for a two-point mooring are designed to be above the height of the flow (e.g. due to their location on canyon terraces), the ADCP will be at a considerable height above the seafloor; out of the range of high frequency instruments such as 300 or 600 kHz ADCPs (Table 2). A two-point mooring configuration in the Var Canyon enabled the first monitoring of powerful turbidity currents and a debris flow, with thicknesses sometimes in excess of 100 m (Khripounoff *et al.* 2012). At this height a 75 kHz ADCP, placed >300 m above the seafloor, was necessary to have sufficient range to capture this flow (Table 2).

For narrow channels, the greater the height, the higher the likelihood that one of the four ADCP beams will impinge on the channel flanks and thereby obscure details in the lower layers. In Squamish, the ADPC height varied from 10 to 15 m above the seabed at low tide as the high shear part of the flows is significantly thinner than this. This is compounded by the fact that turbidity currents are most likely at low tides (4 m range) at Squamish when elevation is lowest (Hughes Clarke *et al.* 2012; Clare *et al.* 2016; Hage *et al.* 2019). In Bute Inlet, the ADCP height was set at about 20 m above the channel base. The distance to the seafloor strongly varied depending on the tide, and data from a local tide gauge had to be used to extract the tidal signature from the data. These issues should clearly be borne in mind in tide-affected (particularly macro-tidal) settings.

An unexpected phenomenon has been noted twice when ADCPs were suspended above a turbidity current: the instrument package has been 'sucked down' into the flow. As the surface buoy only has an excess of about 50 kg of flotation, it appears that the highest velocity flows have enough turbulence to drag the ADCP frame down-stream, and end up

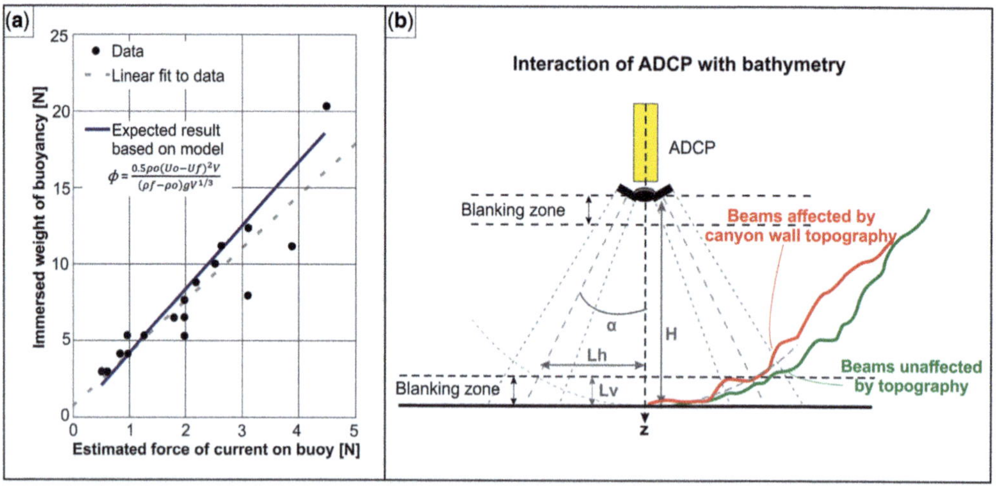

Fig. 12. (**a**) Laboratory test of a theoretical model to determine the force of current exerted on different weights of buoyancy. (**b**) Schematic to show the configuration of ADCP beams, how interaction with a topographically variable seafloor may affect data quality and how the height of the ADCP affects the proximity to the seafloor at which currents can still be monitored (see also Table 2).

pulling the surface buoy underwater because of limited anchoring; thus, entraining the instrument into the flow. In 2017, the package was dragged down on to the seabed, where it sat for 30 min before rising. In 2019, the instrument package was dragged down and buried, this time without release. Fortunately, the flotation was visible in multibeam water column imaging (110 m below the surface) and could be reached with a grapple.

Seafloor platforms and cabled observatories

While there are clear benefits in the deployment of autonomous monitoring platforms, such as moorings, they currently have finite battery power and data memory (which, in turn, limit sampling frequency). To measure power, turbulence and fine structures within flows at high temporal and vertical resolution, high bandwidth data are necessary. This may be possible for moored systems when reliable methods of preconditional sampling can be developed, to record at high bandwidth only during turbidity currents; however, research is still required in this area. Experience at the Fraser Delta has demonstrated that it is possible to design a cabled seafloor observatory that is capable of withstanding powerful turbidity currents and can transmit data in real time (Lintern *et al.* 2019). In many settings, such as the Fraser Delta, turbidity currents occur only at certain times of the year and extreme flows may occur years apart. Capturing these events at high bandwidth and long intervals apart is impossible with battery-

Table 2. *Seafloor footprints of ADCP beams for different vertical heights, assuming a 20° beam angle (typical for the moorings discussed in this paper)*

Height above seafloor, H (m)	Side lobe interference zone – blanked above seafloor, L_v (m)	Radius of ADCP beam footprint at seafloor, L_h (m)	Diameter of ADCP beam footprint at seafloor, $2 \times L_h$ (m)	Example ADCP frequency as discussed in this paper
300	18.1	102.6	205.2	Var Canyon 75 kHz ADCP
85	5.1	29.1	58.1	Congo Canyon 300 kHz ADCP
70	4.2	23.9	47.9	Monterey Canyon 300 kHz ADCP
35	2.1	12.0	23.9	Bute Inlet 600 kHz ADCP
12	0.7	4.1	8.2	Squamish 1200 kHz ADCP

L_v, side lobe interference zone; L_h, radius of ADCP beam footprint (as illustrated in Fig. 12b).

powered instruments. Cabled installations provide both power and the highest bandwidth to a number of instruments. Cabled instruments report live to shore; hence, event detection is possible, which might enable a response to investigate conditions shortly after the event (as was the case of Lintern *et al.* 2016). Due to the array of instruments on the network, the exact environmental conditions under which turbidity currents occur are well understood at the Fraser Delta (strong freshet combined with spring tides) and their onset can be reliably predicted (Lintern *et al.* 2019).

A large cabled observatory requires frequent servicing and with current technologies can only be laid with long-term dedicated resources. An advantage is that once in operation, a scientist can be assured that site visits and platform maintenance and improvement can be done regularly. As mentioned, cabling platforms on the seafloor is a very expensive and intensive undertaking and cannot be readily combined with other systems (unlike more mobile mooring systems). There are only a few organizations worldwide currently able to maintain such a system. Furthermore, the cables that provide power and distribute the data gathered are weak points and are susceptible to rupture by turbidity currents (Carter *et al.* 2014; Clare *et al.* 2017). Therefore, serious consideration should be given to the routing of cable paths and one should also be prepared for the cables to be severed. Currently, ROVs are used to connect cables to platforms. This extends the deployment time from perhaps as little as a few hours on station to a day or two on station, depending on tide and visibility conditions.

Design of seafloor platforms to monitor turbidity currents will necessarily be different from more conventional tripods or other frames that are designed to measure clear-water flows (e.g. Cacchione *et al.* 2006). Lessons learned from the Fraser Delta deployment are similar to those for single-point moorings. The design challenge is to strike a balance between reducing the surface area of the platform to reduce drag and increasing the weight of the structure or type of legs and to ensure that it is stable to withstand toppling. For instance, the final and most successful design to date at the Fraser Delta has the largest surface area and has been stabilized in other ways. Lessons learned for the deployment of seafloor platforms therefore include:

(1) heavy weight (e.g. 900 kg) beneath the platform, which is released when it comes to retrieve the platform;
(2) stable design (e.g. tripod or quadrupedal frame) with legs that can penetrate into the seafloor to act as mini piled foundations. If feet or legs are likely to become embedded or buried, they should be released during retrieval. Where the feet are not removable, the solution to recovering a buried platform is not to winch the platform out of the sediment, but instead to apply tension and let the recovering ship slowly rock the platform free;
(3) where instruments need to be suspended on hanging arms, the frame should be designed such that it can be deployed at seafloor by an ROV, to reduce the amount of deck space needed and to minimize the risk of damage during deployment;
(4) mounting of instruments should be reinforced and use lightweight, durable materials such as titanium. Various mechanisms (hinged arms, telescoping poles) may be used to extend instruments away from the platform-induced vortices, towards the upstream flow to trigger other instruments. It may be appropriate to consider housing instruments such as ADCPs or hydrophones in shrouded cages to minimize environmental noise and vibrations. The ADCPs on the Fraser Delta frame were set in a dual-axis stabilized gimbal, which righted itself and continued to measure flows, even when the platform was completely upside down.

Placement of moorings and seafloor platforms

Given the efforts to ensure that monitoring platforms can successfully withstand and measure turbidity currents, it would be unfortunate if they were not deployed in the correct location. Precise placement also remains a challenge, particularly where support from ROVs (i.e. to verify placement location or assist with re-siting) is unavailable, or is considered too time-consuming or costly. A high-quality base map is essential to ensure that the proposed target is appropriate. As the seafloor elevation and planform can vary considerably in active submarine canyons and channels (e.g. Paull *et al.* 2018; Gales *et al.* 2019; Vendettuoli *et al.* 2019), it is recommended that multibeam bathymetric data be acquired prior to deployment to accurately determine the water depth and seafloor relief to ensure that the proposed location is correct (e.g. the canyon thalweg has not migrated, ADCPs will not be affected by interference with canyon side walls, mooring is not placed on a canyon-wall slump, etc.).

Deployment and siting of moorings. When placing moorings in submarine canyons or channels, the desired seafloor targets are usually very small and may rely on deployment from vessels without dynamic positioning (a computer-controlled system to maintain position and heading using thrusters). Thus, the vessel may drift off location easily during the deployment. Even with dynamic positioning, moorings dropped from the sea surface can drift with the current or during free fall. A triangulated location is typically acquired for moorings by communicating with the acoustic releases; however, this is often inaccurate, difficult in great water depths, and can be complicated by echoes from steep-sided canyon walls or other topographic features.

Another option to determine the location of moorings is to make use of a multibeam echosounder. As long as the mooring array has a series of scattering targets (flotation spheres or instrument housings) that are separated by more than the sonar range resolution, they can usually be discerned from the natural scatterers as you pass over them. This method has been used for detecting location of moorings, as well as to image passing turbidity currents (Hughes Clarke *et al.* 2014) in shallow water fjord settings, and is also feasible in deeper water using the multibeam system constrained to shorter pulse lengths (2 ms) in a narrow swathe. This should therefore enable identification of moorings in up to 2 km of water.

Where moorings are lowered to the seafloor, a position fix can be acquired from an Ultra-Short BaseLine (USBL) system. It is worth including beacons on the moorings that would allow the mooring's actual position (during deployment and monitoring periods) to be determined with the necessary accuracy; however, this technique gets increasingly expensive with greater water depths.

The effects of human interference with the seafloor should be considered when choosing a platform location, as activities such as fishing, trawling, anchor deployment and dredging can snag, displace or damage monitoring platforms. Moorings should be placed in water depths greater than the keel of icebergs in areas affected by seasonal ice cover.

Specifics on deployment of single-point moorings. Two general approaches exist for the deployment of single-point moorings. The first is to deploy the anchor last (i.e. buoyancy and mooring line with instruments attached are offloaded to sea prior to dropping the anchor at the desired location). An anchor-last deployment also allows you to manoeuvre the vessel to above the desired location using USBL and then drop the mooring once on location. This approach has been shown to achieve a precision of $\pm$ 10–20 m horizontal accuracy in water depths of up to 2 km, and 50–60 m at 5 km water depth, and depends firmly upon the vessel's captain, ship handling skills of the mate on watch, maintaining efficient communication between the deck, bridge and science crew, and fair weather conditions and sea state at the time of deployment. The second is to deploy the anchor first, which can be hazardous as the mooring line will be in tension on the back deck of the vessel. For this reason, in particular, an anchor-first strategy is precluded when heavy anchors are required (due to very high line tensions).

Deployment of two-point moorings. In shallow water, where the line suspending the instrument and anchor lines are all connected to a surface buoy, anchors for two-point moorings can first be placed individually. After the anchors are placed with a small surface float, then the anchor lines can be connected to a single point above the channel and the instrument can be lowered from this central surface buoy. In deeper water, the use of submerged frames is more appropriate (given the length of mooring lines required). Two deployment methods have been successful in safely placing these deeper-water two-point moorings. In the first method, the instrument is lowered above the channel, followed by the frame, and roughly kept in place by a small boat. While the small boat holds on to the second anchor line, a larger ship (with winch and A-frame) sets off with the first anchor line. On the larger boat, the anchor line is connected to the chain and anchor before being dropped at the anchor location. Then the larger ship returns to the smaller boat to pick up the second anchor line and drop the second anchor.

A second approach, that has also been successfully applied, involves releasing the central part of the mooring down in one step. For this approach, all the lines, the instrument, the frame and the float need to be carefully laid out on the back deck. The procedure starts by deploying the first anchor and laying out the anchor line, while the ship slowly steams from the first anchor position towards the channel. Just before reaching the channel, the instrument is lowered into the water with a line tied to the submerged frame that is hanging from the A-frame of the ship. As the boat crosses the channel, the first anchor line starts to tighten and the submerged frame is dropped in the water. While the boat keeps steaming slowly towards the second anchor position, the second anchor line and the line connecting the submerged frame to the surface buoy are slowly released. As the line towards the buoy runs out, the buoy is released from the ship. Finally, when the boat reaches the second anchor position the last anchor is dropped. The advantage of the first method is that the deployment is done step by step and is more controlled; however, it requires two vessels and there is a higher chance for the instrument and second anchor line to become tangled during the deployment of the first anchor. The second method requires only one vessel, but needs a larger back deck (and very careful preparations), as the lines are dragged behind the ship and its propeller, the ship will have to continue moving forwards to prevent the lines from tangling in the propeller. So, in the second approach everything needs to be deployed in one go; once the first anchor is dropped there is no way back. Both methods have been used in Bute Inlet four times and all moorings have been placed successfully. Retrieving these two-point moorings is fairly straightforward. After picking up the surface buoy the line is connected to the winch and the whole mooring is pulled out. A 1 t winch has always been successful in retrieving the moorings, although we have had to cut one anchor line, possibly as a result of a buried anchor. Depending on the type of anchor and the angle of the anchor lines, larger forces could be applied to the submerged frame, so it might be advisable to make sure that the link between the anchor lines and the frame are the weakest connection in the mooring, to ensure that the instrument is always recovered. Alternatively, acoustic release links could be incorporated into the mooring design (i.e. one on each mooring line); however, these would add additional weight to the mooring line, which would need to be considered.

Conclusions and final thoughts

The design of monitoring platforms needs to deal with high velocities and sediment concentrations close to the seafloor, capable of tilting, displacing, transporting and even damaging instruments. Our experience shows that, despite the challenges posed, it is possible to make detailed measurements of powerful sediment-laden flows. These challenges can be overcome by simplifying single-point mooring design to reduce drag potential, or deploying two-point moorings (or from vessels), where neither mooring lines nor instruments interact with the flow itself (Fig. 13). Where it is necessary to deploy

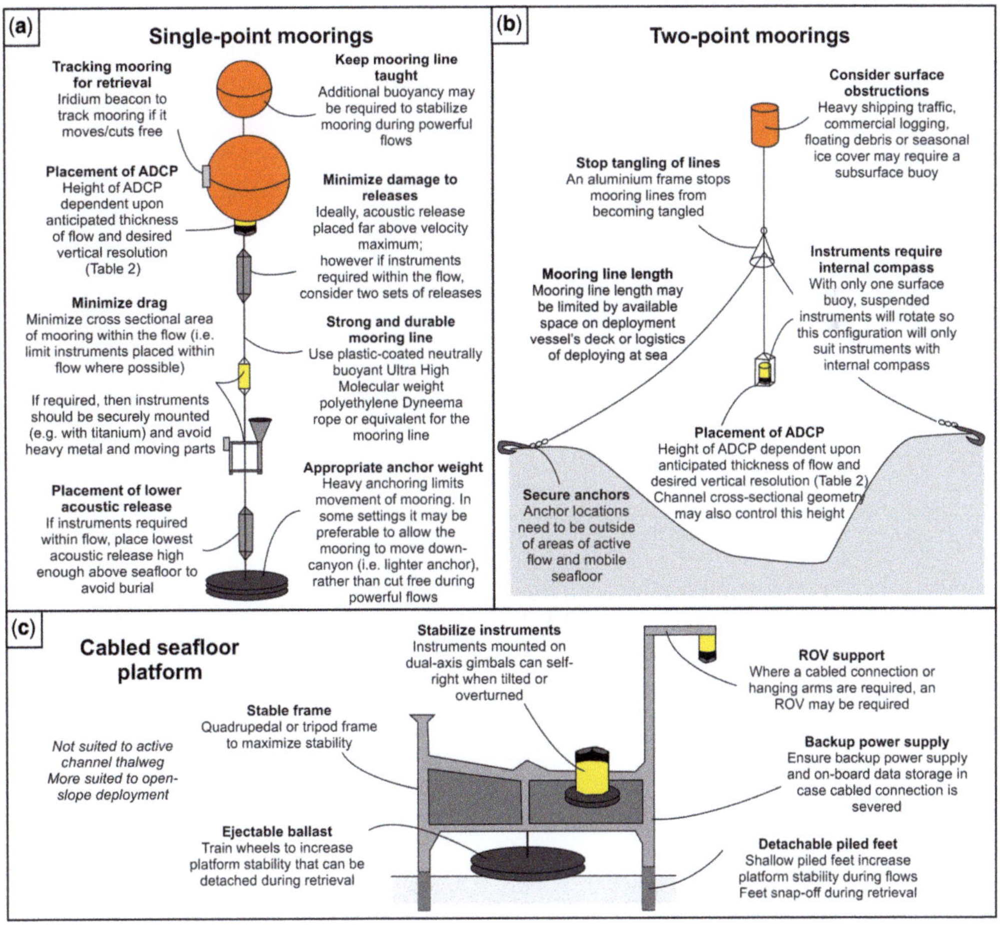

Fig. 13. Summary of lessons learned for designing monitoring platforms, illustrating key considerations when measuring powerful turbidity currents for: (**a**) single-point moorings; (**b**) two-point moorings; and (**c**) cabled seafloor platforms.

Table 3. *Summary of considerations for different types of turbidity current monitoring platforms*

Platform type	Environment	Benefits	Considerations
Single-point mooring	Long-term deployment in canyon/channel thalweg or unconfined slope	• Simple mooring design • Relatively easy to deploy • Simple retrieval using acoustic release link and sacrificial anchor weight	• As it interacts with the flow, mooring may need to be designed to cope with downslope transport or maintain taut line (large anchor weight and high buoyancy) • Ideally, reduce drag by minimizing cross-sectional area (e.g. reducing instruments) on mooring line • Acoustic releases should be placed above velocity maximum of flow; however, if instruments are required within the flow, then releases should be placed below those instruments and also above • Large ocean-going vessel may be required to deploy heavy anchor weights and buoyancy
Two-point mooring	Short-term deployment over channels/ canyons Particularly useful where flows are highly erosive or have dense near-bed layer	• None of mooring interacts with flow • Unaffected if erosion or deposition affect seafloor	• Challenging field deployment requiring considerable lengths of mooring line • Requires larger vessel for retrieval of anchors and mooring lines • Only possible where stable terraces, levees or channel margin permit anchor placement • Surface buoy may pose a problem in areas with seasonal ice cover, busy shipping or logging
Vessel-mounted mooring	Shallow-water settings where timing of turbidity currents is known	• None of mooring interacts with flow • Continuous power to instruments • Possible to adjust instrument settings and acquire calibration samples in real-time	• Only suitable for shallow water settings • Requires vessel with crew; hence, only suitable for relatively short deployments
Benthic lander	Unconfined slope and/ or dilute flows	• Continuous data transmission, enabling near-real time response (e.g. to perform seafloor survey) • Externally powered, allowing for multiple instruments recording at high frequency	• Not suitable for placement in active canyon/channel thalweg • May be buried, or undermined by erosion • To withstand powerful flows, requires ballast that can be ejected and removable feet • Requires support ROV during deployment if cabled links or additional instruments need to be added once platform is on seafloor

instruments within the flow, it may not be possible to reduce drag, hence additional stability is essential, such as extra buoyancy and anchoring for single-point moorings, or piled legs and extra weight for seafloor platforms (Fig. 13). Instrument mounting may be a weak point in such designs; hence brackets and cages should be more robust than standard moorings (see Fig. 13) Table 3 provides a summary of these considerations and scenarios that are most suitable for different monitoring platforms.

There is currently a push to develop next-generation monitoring tools to detect and characterize turbidity currents, relying upon passive detection, rather than direct measurements (e.g. Clare *et al.* 2017; Lintern *et al.* 2019). Such tools include hydrophones and geophones and will enable measurement of turbidity currents and other submarine mass movements, without the need to place moorings or platforms in the path of the flow (particularly where the flow is restricted to channels (Chadwick *et al.* 2012; Caplan-Auerbach *et al.* 2014). This approach requires calibration against ADCPs and other measurements and the initial results are promising. There is clear evidence that acoustic signals can be linked to independently measured turbidity currents (Hatcher 2017; Lintern *et al.* 2019). In addition to measuring transit speeds via arrival times, there is potential to measure some basic features of flow character using hydrophones. For example, the intensity of acoustic signals may be related to internal flow speeds (via intensity of grain collisions), grain size (sand- or mud-dominated flow) or the presence of a dense and coarse near-bed layer. However, further work would be needed to determine what is possible and how flows should be recorded. Other developments in distributed sensing along fibre-optic cables also demonstrate the potential utility of cabled submarine links, such as those that connect the Fraser DDL to the VENUS seafloor cabled network, to measure strain, temperature and to use the optical fibres as distributed acoustic sensors (e.g. Lindsey *et al.* 2017, 2019; Hartog *et al.* 2018).

Finally, there is growing interest in monitoring a wider range of deep-sea sediment transport processes, including the influence of internal tides (Maier *et al.* 2019*b*), thermohaline-driven circulation (Miramontes *et al.* 2019) and the mixed interaction of downslope gravity-driven flows such as turbidity currents with along-slope contour currents (Normandeau *et al.* 2019*b*; Fuhrmann *et al.* 2020). As such flows are typically of lower velocity (generally $<<1$ m s^{-1}; McCave *et al.* 2017) and comprise lower sediment concentrations than turbidity currents, they should be considerably more straightforward to measure. Therefore, many of the issues outlined in this study are unlikely to be a major issue; however, the lessons learned should still be considered – such as minimizing drag and maintaining stability of the platform to ensure that high-quality results are acquired. Burial risk may be greater in areas of high net deposition. To date, limited near-bed measurements of contour currents have been made and none are yet known from mixed turbidity current-contour current systems. Therefore, there is a need for instruments to be placed closer to the seafloor in such systems to fill this knowledge gap. Such systems are also typically much more laterally extensive than 'conventional' turbidity current canyons or channels; hence, it will be necessary to deploy an array of monitoring platforms to characterize the spatial variability in near-bed flow that may be strongly controlled by local variations in seafloor morphology.

We conclude that recent and ongoing advances in technology and mooring design will ensure that key knowledge gaps in turbidity current behaviour can soon be filled, providing valuable information for designing resilient seafloor infrastructure, and understanding of how and when these globally important processes transport sediment, nutrients and organic carbon to the deep sea.

Acknowledgements This study is based on fieldwork and logistical support provided by a large number of people, including: (i) the Monterey Coordinated Canyon Experiment (CCE) Team; (ii) Bute Inlet 2016 and 2018 teams; (iii) Squamish 2015 monitoring team; and most notably, (iv) the patient and diligent crews of the following vessels: *Strickland*, *Heron*, *Vector*, *Rachel Carson* and *Dock Ricketts*. Use of trademark names does not imply endorsement of products by the authors or the USGS.

Funding Clare, Cartigny and Talling acknowledge funding from the Natural Environment Research Council (NERC), including 'Environmental Risks to Infrastructure: Identifying and Filling the Gaps' (NE/P005780/1) and 'New field-scale calibration of turbidity current impact modelling' (NE/P009190/1). The authors also acknowledge discussions with collaborators as part of Talling's NERC International Opportunities Fund grant (NE/M017540/1) 'Coordinating and pump-priming international efforts for direct monitoring of active turbidity currents at global test sites'. Talling was supported by a NERC and Royal Society Industry Fellowship hosted by the International Cable Protection Committee. Funding for the Coordinated Canyon Experiment (CCE) was provided by David and Lucile Packard Foundation, Natural Environment Research Council (grant NE/K011480/1), U.S. Geological Survey (USGS) Coastal and Marine Program, and Ocean University of China. Additional funding for MAC was provided by NERC National Capability Climate Linked Atlantic Sector Science Programme (NE/R015953/1).

Author contributions MC: conceptualization (lead), formal analysis (equal), funding acquisition (equal), investigation (lead), methodology (equal), project administration (equal), visualization (lead), writing – original draft (lead), writing – review & editing (lead); **DGL**: conceptualization (equal), formal analysis (supporting), investigation (equal), methodology (equal), resources (equal), writing – review & editing (supporting); **KR**: formal analysis (equal), investigation (equal), methodology (equal), project administration (supporting), writing – original draft (supporting), writing – review & editing (supporting); **JEHC**: conceptualization (equal), formal analysis (supporting), investigation (equal), methodology (equal), writing – original draft (supporting), writing – review & editing (supporting); **CP**: conceptualization (equal), formal analysis (supporting), funding acquisition (equal), investigation (equal), methodology (equal), project administration (equal), writing – original draft (supporting), writing – review & editing (supporting); **RG**: formal analysis (supporting), investigation (supporting), methodology (supporting), writing – original draft (supporting); **MJBC**: conceptualization (supporting), formal analysis (supporting), investigation (equal), methodology (equal), project administration (equal), writing – original draft (supporting), writing – review & editing (supporting); **PJT**: conceptualization (supporting), funding acquisition (equal), investigation (equal), methodology (supporting), resources (equal), writing – review & editing (supporting); **DPe**: conceptualization (supporting), formal analysis (supporting), investigation (supporting); **JX**: funding acquisition (supporting), methodology (supporting), project administration (supporting), writing - original draft (supporting); **DPa**: conceptualization (supporting), investigation (supporting), methodology (supporting), project administration (supporting); **RSJ**: conceptualization (supporting), methodology (supporting), writing - original draft (supporting), writing – review & editing (supporting); **RA**: methodology (supporting), writing – review & editing (supporting).

References

Ayranci, K., Lintern, D.G., Hill, P.R. and Dashtgard, S.E. 2012. Tide-supported gravity flows on the upper delta front, Fraser River delta, Canada. *Marine Geology*, **326**, 166–170, https://doi.org/10.1016/j.margeo.2012.09.007

Azpiroz-Zabala, M., Cartigny, M.J. *et al.* 2017*a*. Newly recognized turbidity current structure can explain prolonged flushing of submarine canyons. *Science Advances*, **3**, e1700200, https://doi.org/10.1126/sciadv.1700200

Azpiroz-Zabala, M., Cartigny, M.J., Sumner, E.J., Clare, M.A., Talling, P.J., Parsons, D.R. and Cooper, C. 2017*b*. A general model for the helical structure of geophysical flows in channel bends. *Geophysical Research Letters*, **44**, 11–932, https://doi.org/10.1002/2017GL075721

Babonneau, N., Savoye, B., Cremer, M. and Bez, M. 2010. Sedimentary architecture in meanders of a submarine channel: detailed study of the present Congo turbidite channel (Zaiango project). *Journal of Sedimentary Research*, **80**, 852–866, https://doi.org/10.2110/jsr.2010.078

Barry, J.P., Paull, C.K., Xu, J.P., Buck, K.R., Whaling, P., Ussler, W., III and Caress, D. 2006. The tempo and intensity of turbidity flows in Monterey Canyon. AGU Ocean Sciences Meeting, 20–24 February 2006, Honolulu, Hawaii.

Best, T.C. and Griggs, G.B. 1991. The Santa Cruz littoral cell: difficulties in quantifying a coastal sediment budget. *In*: *Coastal Sediments*. American Society of Civil Engineers, 2262–2276.

Bornhold, B.D., Ren, P. and Prior, D.B. 1994. High-frequency turbidity currents in British Columbia fjords. *Geo-Marine Letters*, **14**, 238–243, https://doi.org/10.1007/BF01274059

Bruschi, R., Bughi, S., Spinazzè, M., Torselletti, E. and Vitali, L. 2006. Impact of debris flows and turbidity currents on seafloor structures. *Norwegian Journal of Geology/Norsk Geologisk Forening*, **86**, 317–336.

Burnett, D.R. and Carter, L. 2017. International submarine cables and biodiversity of areas beyond national jurisdiction: the cloud beneath the sea. *Brill Research Perspectives in the Law of the Sea*, **1**, 1–72, https://doi.org/10.1163/24519359-12340002

Cacchione, D.A., Sternberg, R.W. and Ogston, A.S. 2006. Bottom instrumented tripods: history, applications, and impacts. *Continental Shelf Research*, **26**, 2319–2334, https://doi.org/10.1016/j.csr.2006.07.027

Caplan-Auerbach, J., Dziak, R.P., Bohnenstiehl, D.R., Chadwick, W.W. and Lau, T.K. 2014. Hydroacoustic investigation of submarine landslides at West Mata volcano, Lau Basin. *Geophysical Research Letters*, **41**, 5927–5934, https://doi.org/10.1002/2014GL060964

Carter, L. 2010. *Submarine Cables and the Oceans*. Connecting the World, **31**. UNEP/Earthprint.

Carter, L., Gavey, R., Talling, P.J. and Liu, J.T. 2014. Insights into submarine geohazards from breaks in subsea telecommunication cables. *Oceanography*, **27**, 58–67, https://doi.org/10.5670/oceanog.2014.40

Chadwick, W.W., Jr, Dziak, R.P., Haxel, J.H., Embley, R.W. and Matsumoto, H. 2012. Submarine landslide triggered by volcanic eruption recorded by in situ hydrophone. *Geology*, **40**, 51–54, https://doi.org/10.1130/G32495.1

Clare, M.A., Cartigny, M.J.B. *et al.* 2015. Quantification of near-bed dense layers and implications for seafloor structures: new insights into the most hazardous aspects of turbidity currents. Offshore Technology Conference, https://doi.org/10.4043/25705-MS

Clare, M.A., Clarke, J.H., Talling, P.J., Cartigny, M.J. and Pratomo, D.G. 2016. Preconditioning and triggering of offshore slope failures and turbidity currents revealed by most detailed monitoring yet at a fjord-head delta. *Earth and Planetary Science Letters*, **450**, 208–220, https://doi.org/10.1016/j.epsl.2016.06.021

Clare, M.A., Vardy, M.E. *et al.* 2017. Direct monitoring of active geohazards: emerging geophysical tools for deep-water assessments. *Near Surface Geophysics*, **15**, 427–444, https://doi.org/10.3997/1873-0604.2017033

Conway, K.W., Barrie, J.V., Picard, K. and Bornhold, B.D. 2012. Submarine channel evolution: active channels in fjords, British Columbia, Canada. *Geo-Marine Letters*, **32**, 301–312, https://doi.org/10.1007/s00367-012-0280-4

Cooper, C.K., Andrieux, O. and Wood, J. 2013. Turbidity current measurements in the Congo Canyon. Offshore Technology Conference, https://doi.org/10.4043/23992-MS

Cooper, C., Wood, J. *et al.* 2016. Designing for turbidity currents in the Congo Canyon. Offshore Technology Conference, https://doi.org/10.4043/26919-MS

de Stigter, H.C., Boer, W., de Jesus Mendes, P.A., Jesus, C.C., Thomsen, L., van den Bergh, G.D. and van Weering, T.C. 2007. Recent sediment transport and deposition in the Nazaré Canyon, Portuguese continental margin. *Marine Geology*, **246**, 144–164, https://doi.org/10.1016/j.margeo.2007.04.011

Dengler, A.T., Wilde, P., Noda, E.K. and Normark, W.R. 1984. Turbidity currents generated by Hurricane Iwa. *Geo-Marine Letters*, **4**, 5–11, https://doi.org/10.1007/BF02237967

Dewey, R.K. 1999. Mooring Design & Dynamics – a Matlab® package for designing and analyzing oceanographic moorings. *Marine Models*, **1**, 103–157, https://doi.org/10.1016/S1369-9350(00)00002-X

Dorrell, R.M., Peakall, J. *et al.* 2016. Flow dynamics and mixing processes in hydraulic jump arrays: implications for channel-lobe transition zones. *Marine Geology*, **381**, 181–193, https://doi.org/10.1016/j.margeo.2016.09.009

El-Robrini, M., Gennesseaux, M. and Mauffret, A. 1985. Consequences of the El-Asnam earthquakes: turbidity currents and slumps on the Algerian margin (Western Mediterranean). *Geo-Marine Letters*, **5**, 171–176, https://doi.org/10.1007/BF02281635

Fildani, A. 2017. Submarine Canyons: a brief review looking forward. *Geology*, **45**, 383–384, https://doi.org/10.1130/focus042017.1

Fuhrmann, A., Kane, I.A., Clare, M.A., Ferguson, R.A., Schomacker, E., Bonamini, E. and Contreras, F.A. 2020. Hybrid turbidite-drift channel complexes: An integrated multiscale model. *Geology*, https://doi.org/10.1130/G47179.1

Gales, J.A., Talling, P.J., Cartigny, M.J., Hughes Clarke, J., Lintern, G., Stacey, C. and Clare, M.A. 2019. What controls submarine channel development and the morphology of deltas entering deep-water fjords? *Earth Surface Processes and Landforms*, **44**, 535–551, https://doi.org/10.1002/esp.4515

Gennesseaux, M., Guibout, P. and Lacombe, H. 1971. Enregistrement de courants de turbidité dans la vallée sous-marine du Var (Alpes-Maritimes). *Comptes Rendus de l'Academie des Sciences Paris*, **273**, 2456–2459.

Hage, S., Cartigny, M.J. *et al.* 2018. How to recognize crescentic bedforms formed by supercritical turbidity currents in the geologic record: insights from active submarine channels. *Geology*, **46**, 563–566, https://doi.org/10.1130/G40095.1

Hage, S., Cartigny, M.J. *et al.* 2019. Direct monitoring reveals initiation of turbidity currents from extremely dilute river plumes. *Geophysical Research Letters*, **46**, 11310–11320, https://doi.org/10.1029/2019GL084526

Hartog, A.H., Belal, M. and Clare, M.A. 2018. Advances in distributed fiber-optic sensing for monitoring marine infrastructure, measuring the deep ocean, and quantifying the risks posed by seafloor hazards. *Marine Technology Society Journal*, **52**, 58–73, https://doi.org/10.4031/MTSJ.52.5.7

Hatcher, M.G. 2017. *Ambient Noise from Turbidity Currents in Howe Sound*. MSc thesis, Dalhousie University, http://hdl.handle.net/10222/73224

Hay, A.E. 1987*a*. Turbidity currents and submarine channel formation in Rupert Inlet, British Columbia: 1. Surge observations. *Journal of Geophysical Research: Oceans*, **92**, 2875–2881, https://doi.org/10.1029/JC092iC03p02875

Hay, A.E. 1987*b*. Turbidity currents and submarine channel formation in Rupert Inlet, British Columbia: 2. The roles of continuous and surge-type flow. *Journal of Geophysical Research: Oceans*, **92**, 2883–2900, https://doi.org/10.1029/JC092iC03p02883

Hay, A.E., Burling, R.W. and Murray, J.W. 1982. Remote acoustic detection of a turbidity current surge. *Science*, **217**, 833–835, https://doi.org/10.1126/science.217.4562.833

Heerema, C., Talling, P.J. *et al.* 2020. What determines the downstream evolution of turbidity currents? *Earth and Planetary Science Letters*, **532**, 116023, https://doi.org/10.1016/j.epsl.2019.116023

Heezen, B.C. and Ewing, W.M. 1952. Turbidity currents and submarine slumps, and the 1929 Grand Banks [Newfoundland] earthquake. *American Journal of Science*, **250**, 849–873, https://doi.org/10.2475/ajs.250.12.849

Heezen, B.C. and Ewing, M. 1955. Orleansville earthquake and turbidity currents. *AAPG Bulletin*, **39**, 2505–2514.

Heezen, B.C., Menzies, R.J., Schneider, E.D., Ewing, W.M. and Granelli, N.C.L. 1964. Congo submarine canyon. *AAPG Bulletin*, **48**, 1126–1149.

Hill, P.R. 2012. Changes in submarine channel morphology and slope sedimentation patterns from repeat multibeam surveys in the Fraser River delta, western Canada. *International Association of Sedimentologists, Special Publication*, **44**, 47–70.

Hughes Clarke, J.E. 2016. First wide-angle view of channelized turbidity currents links migrating cyclic steps to flow characteristics. *Nature Communications*, **7**, 11896, https://doi.org/10.1038/ncomms11896

Hughes Clarke, J.E., Mayer, L.A. and Wells, D.E. 1996. Shallow-water imaging multibeam sonars: a new tool for investigating seafloor processes in the coastal zone and on the continental shelf. *Marine Geophysical Researches*, **18**, 607–629, https://doi.org/10.1007/BF00313877

Hughes Clarke, J.E., Brucker, S. *et al.* 2012. The Squamish ProDelta: monitoring active landslides and turbidity currents. Canadian Hydrographic Conference 2012, Proceedings, Ontario, Canada, 1–15.

Hughes Clarke, J.E., Marques, C.R.V. and Pratomo, D. 2014. Imaging active mass-wasting and sediment flows on a fjord delta, Squamish, British Columbia. *In*: Krastel, S. *et al.* (eds) *Submarine Mass Movements and their Consequences*. Advances in Natural and Technological Hazards Research, **37**. Springer, Cham, 249–260, https://doi.org/10.1007/978-3-319-00972-8_22

Hsu, S.K., Kuo, J., Chung-Liang, L., Ching-Hui, T., Doo, W.B., Ku, C.Y. and Sibuet, J.C. 2008. Turbidity currents, submarine landslides and the 2006 Pingtung earthquake off SW Taiwan. *TAO: Terrestrial,*

Atmospheric and Oceanic Sciences, **19**, 767–772, https://doi.org/10.3319/TAO.2008.19.6.767(PT)

Inman, D.L. 1970. Strong currents in submarine canyons. *Transactions of the American Geophysical Union*, **51**, 319.

Inman, D.L., Nordstrom, C.E. and Flick, R.E. 1976. Currents in submarine canyons: an air–sea–land interaction. *Annual Review of Fluid Mechanics*, **8**, 275–310, https://doi.org/10.1146/annurev.fl.08.010176.001423

Khripounoff, A., Vangriesheim, A., Babonneau, N., Crassous, P., Dennielou, B. and Savoye, B. 2003. Direct observation of intense turbidity current activity in the Zaire submarine valley at 4000 m water depth. *Marine Geology*, **194**, 151–158, https://doi.org/10.1016/S0025-3227(02)00677-1

Khripounoff, A., Vangriesheim, A., Crassous, P. and Etoubleau, J. 2009. High frequency of sediment gravity flow events in the Var submarine canyon (Mediterranean Sea). *Marine Geology*, **263**, 1–6, https://doi.org/10.1016/j.margeo.2009.03.014

Khripounoff, A., Crassous, P., Bue, N.L., Dennielou, B. and Jacinto, R.S. 2012. Different types of sediment gravity flows detected in the Var submarine canyon (northwestern Mediterranean Sea). *Progress in Oceanography*, **106**, 138–153, https://doi.org/10.1016/j.pocean.2012.09.001

Krause, D.C., White, W.C., Piper, D.J.W. and Heezen, B.C. 1970. Turbidity currents and cable breaks in the western New Britain Trench. *Geological Society of America Bulletin*, **81**, 2153–2160, https://doi.org/10.1130/0016-7606(1970)81[2153:TCACBI]2.0.CO;2

Kostaschuk, R.A., Luternauer, J.L., McKenna, G.T. and Moslow, T.F. 1992. Sediment transport in a submarine channel system; Fraser River delta, Canada. *Journal of Sedimentary Research*, **62**, 273–282.

Lambert, A. and Giovanoli, F. 1988. Records of riverbome turbidity currents and indications of slope failures in the Rhone delta of Lake Geneva. *Limnology and Oceanography*, **33**, 458–468, https://doi.org/10.4319/lo.1988.33.3.0458

Lindsey, N.J., Martin, E.R. *et al.* 2017. Fiber-optic network observations of earthquake wavefields. *Geophysical Research Letters*, **44**, 11–792, https://doi.org/10.1002/2017GL075722

Lindsey, N., Dawe, C. and Ajo-Franklin, J. 2019. Photonic seismology in Monterey Bay: Dark fiber DAS illuminates offshore faults and coastal ocean dynamics [Preprint]. *EarthArXiv*, https://doi.org/10.31223/osf.io/7bf92

Lintern, D.G. and Hill, P.R. 2010. An underwater laboratory at the Fraser River delta. Eos, *Transactions of the American Geophysical Union*, **91**, 333–334, https://doi.org/10.1029/2010EO380001

Lintern, D.G., Hill, P.R. and Stacey, C. 2016. Powerful unconfined turbidity current captured by cabled observatory on the Fraser River delta slope, British Columbia, Canada. *Sedimentology*, **63**, 1041–1064, https://doi.org/10.1111/sed.12262

Lintern, D.G., Mosher, D.C. and Scherwath, M. 2019. Advancing from subaqueous mass movement case studies to providing advice and mitigation. *Geological Society, London, Special Publications*, **477**, 1–14, https://doi.org/10.1144/SP477-2018-190

Liu, J.T., Wang, Y.H., Yang, R.J., Hsu, R.T., Kao, S.J., Lin, H.L. and Kuo, F.H. 2012. Cyclone-induced hyperpycnal turbidity currents in a submarine canyon. *Journal of Geophysical Research: Oceans*, **117**, https://doi.org/10.1029/2011JC007630

Maier, K.L., Gales, J. *et al.* 2019*a*. Linking direct measurements of turbidity currents to submarine canyon-floor deposits. *Frontiers in Earth Science*, **7**, 144, https://doi.org/10.3389/feart.2019.00144

Maier, K.L., Rosenberger, K. *et al.* 2019*b*. Sediment and organic carbon transport and deposition driven by internal tides along Monterey Canyon, offshore California. *Deep Sea Research Part I: Oceanographic Research Papers*, **153**, 103108, https://doi.org/10.1016/j.dsr.2019.103108

Martín, J., Palanques, A., Vitorino, J., Oliveira, A. and De Stigter, H.C. 2011. Near-bottom particulate matter dynamics in the Nazaré submarine canyon under calm and stormy conditions. *Deep Sea Research Part II: Topical Studies in Oceanography*, **58**, 2388–2400, https://doi.org/10.1016/j.dsr2.2011.04.004

Martín, J., Puig, P., Palanques, A. and Ribó, M. 2014. Trawling-induced daily sediment resuspension in the flank of a Mediterranean submarine canyon. *Deep Sea Research Part II: Topical Studies in Oceanography*, **104**, 174–183, https://doi.org/10.1016/j.dsr2.2013.05.036

McCave, I.N., Thornalley, D.J.R. and Hall, I.R. 2017. Relation of sortable silt grain-size to deep-sea current speeds: calibration of the 'Mud Current Meter'. *Deep Sea Research Part I: Oceanographic Research Papers*, **127**, 1–12, https://doi.org/10.1016/j.dsr.2017.07.003

Miramontes, E., Penven, P. *et al.* 2019. The influence of bottom currents on the Zambezi Valley morphology (Mozambique Channel, SW Indian Ocean): in situ current observations and hydrodynamic modelling. *Marine Geology*, **410**, 42–55, https://doi.org/10.1016/j.margeo.2019.01.002

Mulder, T., Savoye, B. and Syvitski, J.P.M. 1997. Numerical modelling of a mid-sized gravity flow: the 1979 Nice turbidity current (dynamics, processes, sediment budget and seafloor impact). *Sedimentology*, **44**, 305–326, https://doi.org/10.1111/j.1365-3091.1997.tb01526.x

Mulder, T., Zaragosi, S. *et al.* 2012. Present deep-submarine canyons activity in the Bay of Biscay (NE Atlantic). *Marine Geology*, **295**, 113–127, https://doi.org/10.1016/j.margeo.2011.12.005

Normandeau, A., Bourgault, D., Neumeier, U., Lajeunesse, P., St-Onge, G., Gostiaux, L. and Chavanne, C. 2019*a*. Storm-induced turbidity currents on a sediment-starved shelf: insight from direct monitoring and repeat seabed mapping of upslope migrating bedforms. *Sedimentology*, **67**, 1045–1058, https://doi.org/10.1111/sed.12673

Normandeau, A., Campbell, D.C. and Cartigny, M.J. 2019*b*. The influence of turbidity currents and contour currents on the distribution of deep-water sediment waves offshore eastern Canada. *Sedimentology*, **66**, 1746–1767, https://doi.org/10.1111/sed.12557

Normark, W.R., Carlson, P.R., Chan, M.A. and Archer, A.W. 2003. Giant submarine canyons: is size any clue to their importance in the rock record? *Geological*

Society of America Special Papers, **370**, 175–190, https://doi.org/10.1130/0-8137-2370-1.175.

Parsons, D.R., Best, J.L., Lane, S.N., Orfeo, O., Hardy, R.J. and Kostaschuk, R. 2007. Form roughness and the absence of secondary flow in a large confluence–diffluence, Rio Paraná, Argentina. *Earth Surface Processes and Landforms*, **32**, 155–162, https://doi.org/10.1002/esp.1457

Parsons, D.R., Peakall, J., Aksu, A.E., Flood, R.D., Hiscott, R.N., Beşiktepe, Ş. and Mouland, D. 2010. Gravity-driven flow in a submarine channel bend: direct field evidence of helical flow reversal. *Geology*, **38**, 1063–1066, https://doi.org/10.1130/G31121.1

Paull, C.K., Ussler, W., Greene, H.G., Keaten, R., Mitts, P. and Barry, J. 2002. Caught in the act: the 20 December 2001 gravity flow event in Monterey Canyon. *Geo-Marine Letters*, **22**, 227–232, https://doi.org/10.1007/s00367-003-0117-2

Paull, C.K., Mitts, P., Ussler, W., III, Keaten, R. and Greene, H.G. 2005. Trail of sand in upper Monterey Canyon: offshore California. *Geological Society of America Bulletin*, **117**, 1134–1145, https://doi.org/10.1130/B25390.1

Paull, C.K., Ussler, W., III *et al.* 2010. Origins of large crescent-shaped bedforms within the axial channel of Monterey Canyon, offshore California. *Geosphere*, **6**, 755–774, https://doi.org/10.1130/GES00527.1

Paull, C.K., Talling, P.J. *et al.* 2018. Powerful turbidity currents driven by dense basal layers. *Nature Communications*, **9**, 4114, https://doi.org/10.1038/s41467-018-06254-6

Piper, D.J., Shor, A.N. and Hughes Clarke, J.E. 1988. The 1929 Grand Banks earthquake, slump and turbidity current. *GSA Special Papers*, **229**, https://doi.org/10.1130/SPE229-p77

Pohl, F., Eggenhuisen, J.T., Kane, I.A. and Clare, M.A. 2020. Transport and burial of microplastics in deep-marine sediments by turbidity currents. *Environmental Science & Technology*, **54**, 4180–4189, https://doi.org/10.1021/acs.est.9b07527

Pope, E.L., Talling, P.J., Carter, L., Clare, M.A. and Hunt, J.E. 2017. Damaging sediment density flows triggered by tropical cyclones. *Earth and Planetary Science Letters*, **458**, 161–169, https://doi.org/10.1016/j.epsl.2016.10.046

Prior, D.B., Bornhold, B.D., Wiseman, W.J. and Lowe, D.R. 1987. Turbidity current activity in a British Columbia fjord. *Science*, **237**, 1330–1333, https://doi.org/10.1126/science.237.4820.1330

Puig, P., Ogston, A.S., Mullenbach, B.L., Nittrouer, C.A., Parsons, J.D. and Sternberg, R.W. 2004. Storm-induced sediment gravity flows at the head of the Eel submarine canyon, northern California margin. *Journal of Geophysical Research: Oceans*, **109**, https://doi.org/10.1029/2003JC001918

Puig, P., Canals, M. *et al.* 2012. Ploughing the deep sea floor. *Nature*, **489**, 286, https://doi.org/10.1038/nature11410

Puig, P., Palanques, A. and Martín, J. 2014. Contemporary sediment-transport processes in submarine canyons. *Annual Review of Marine Science*, **6**, 53–77, https://doi.org/10.1146/annurev-marine-010213-135037

Randolph, M.F. and White, D.J. 2012. Interaction forces between pipelines and submarine slides – A geotechnical viewpoint. *Ocean Engineering*, **48**, 32–37, https://doi.org/10.1016/j.oceaneng.2012.03.014

Ribó, M., Puig, P. and van Haren, H. 2015. Hydrodynamics over the Gulf of Valencia continental slope and their role in sediment transport. *Deep Sea Research Part I: Oceanographic Research Papers*, **95**, 54–66, https://doi.org/10.1016/j.dsr.2014.10.004

Ross, C.B., Gardner, W.D., Richardson, M.J. and Asper, V.L. 2009. Currents and sediment transport in the Mississippi Canyon and effects of Hurricane Georges. *Continental Shelf Research*, **29**, 1384–1396, https://doi.org/10.1016/j.csr.2009.03.002

Ryan, W.B.F. and Heezen, B.C. 1965. Ionian Sea submarine canyons and the 1908 Messina turbidity current. *Geological Society of America Bulletin*, **76**, 915–932, https://doi.org/10.1130/0016-7606(1965)76[915:ISSCAT]2.0.CO;2

Sequeiros, O.E., Pittaluga, M.B. *et al.* 2019. How typhoons trigger turbidity currents in submarine canyons. *Scientific Reports*, **9**, https://doi.org/10.1038/s41598-019-45615-z

Shepard, F.P. 1954. High-velocity turbidity currents, a discussion. *Proceedings of the Royal Society of London, Series A. Mathematical and Physical Sciences*, **222**, 323–326.

Shepard, F.P. and Marshall, N.F. 1973. Storm-generated current in La Jolla Submarine Canyon, California. *Marine Geology*, **15**, M19–M24, https://doi.org/10.1016/0025-3227(73)90018-2

Shepard, F.P., McLoughlin, P.A., Marshall, N.F. and Sullivan, G.G. 1977. Current-meter recordings of low-speed turbidity currents. *Geology*, **5**, 297–301, https://doi.org/10.1130/0091-7613(1977)5<297:CROLTC>2.0.CO;2

Shih, H.H. 2012. Real-time current and wave measurements in ports and harbors using ADCP. *In*: *2012 Oceans-Yeosu*. IEEE, 1–8.

Sumner, E.J., Peakall, J. *et al.* 2013. First direct measurements of hydraulic jumps in an active submarine density current. *Geophysical Research Letters*, **40**, 5904–5908, https://doi.org/10.1002/2013GL05 7862

Syahnur, Y. and Jaya, K.A. 2016. *Geomatics best practices in Saka Indonesia Pangkah Limited (Case Study: Ujung Pangkah Pipeline Integrity)*. 2015 Indonesian Petroleum Association Convention, Jakarta, Indonesia.

Symons, W.O., Sumner, E.J. *et al.* 2017. A new model for turbidity current behavior based on integration of flow monitoring and precision coring in a submarine canyon. *Geology*, **45**, 367–370, https://doi.org/10.1130/G38764.1

Syvitski, J.P. and Hein, F.J. 1991. *Sedimentology of an Arctic Basin: Itirbilung Fiord, Baffin Island, Northwest Territories*. Geological Survey of Canada, Paper 91–11.

Talling, P.J., Paull, C.K. and Piper, D.J. 2013. How are subaqueous sediment density flows triggered, what is their internal structure and how does it evolve? Direct observations from monitoring of active flows. *Earth-Science Reviews*, **125**, 244–287, https://doi.org/10.1016/j.earscirev.2013.07.005

Thorne, P.D. and Hanes, D.M. 2002. A review of acoustic measurement of small-scale sediment processes. *Continental Shelf Research*, **22**, 603–632, https://doi.org/10.1016/S0278-4343(01)00101-7

Urlaub, M., Petersen, F. *et al.* 2018. Gravitational collapse of Mount Etna's southeastern flank. *Science Advances*, **4**, eaat9700, https://doi.org/10.1126/sciadv.aat9700
Vangriesheim, A., Khripounoff, A. and Crassous, P. 2009. Turbidity events observed in situ along the Congo submarine channel. *Deep Sea Research Part II: Topical Studies in Oceanography*, **56**, 2208–2222, https://doi.org/10.1016/j.dsr2.2009.04.004
Vendettuoli, D., Clare, M.A. *et al.* 2019. Daily bathymetric surveys document how stratigraphy is built and its extreme incompleteness in submarine channels. *Earth and Planetary Science Letters*, **515**, 231–247, https://doi.org/10.1016/j.epsl.2019.03.033
Wynn, R.B., Huvenne, V.A. *et al.* 2014. Autonomous Underwater Vehicles (AUVs): their past, present and future contributions to the advancement of marine geoscience. *Marine Geology*, **352**, 451–468, https://doi.org/10.1016/j.margeo.2014.03.012
Xu, J.P. 2011. Measuring currents in submarine canyons: Technological and scientific progress in the past 30 years. *Geosphere*, **7**, 868–876, https://doi.org/10.1130/GES00640.1
Xu, J.P. and Noble, M.A. 2009. Currents in Monterey submarine canyon. *Journal of Geophysical Research: Oceans*, **114**, https://doi.org/10.1029/2008JC004992
Xu, J.P., Noble, M.A. and Rosenfeld, L.K. 2004. In-situ measurements of velocity structure within turbidity currents. *Geophysical Research Letters*, **31**, https://doi.org/10.1029/2004GL019718
Xu, J.P., Swarzenski, P.W., Noble, M. and Li, A.C. 2010. Event-driven sediment flux in Hueneme and Mugu submarine canyons, southern California. *Marine Geology*, **269**, 74–88, https://doi.org/10.1016/j.margeo.2009.12.007
Xu, J.P., Barry, J.P. and Paull, C.K. 2013. Small-scale turbidity currents in a big submarine canyon. *Geology*, **41**, 143–146, https://doi.org/10.1130/G33727.1
Xu, J.P., Sequeiros, O.E. and Noble, M.A. 2014. Sediment concentrations, flow conditions, and downstream evolution of two turbidity currents, Monterey Canyon, USA. *Deep Sea Research Part I: Oceanographic Research Papers*, **89**, 11–34, https://doi.org/10.1016/j.dsr.2014.04.001
Yergin, D. 2006. Ensuring energy security. *Foreign Affairs*, **85**, 69–82, https://doi.org/10.2307/20031912
Zeng, J., Lowe, D.R., Prior, D.B., Wiseman, W.J., Jr and Bornhold, B.D. 1991. Flow properties of turbidity currents in Bute Inlet, British Columbia. *Sedimentology*, **38**, 975–996, https://doi.org/10.1111/j.1365-3091.1991.tb00367.x
Zhang, Y., Liu, Z. *et al.* 2018. Long-term in situ observations on typhoon-triggered turbidity currents in the deep sea. *Geology*, **46**, 675–678, https://doi.org/10.1130/G45178.1

Index

Page numbers in *italics* refer to Figures. Page numbers in **bold** refer to Tables.